AF356647

ABC Transporters in Human Diseases

ABC Transporters in Human Diseases

Editor

Thomas Falguières

MDPI • Basel • Beijing • Wuhan • Barcelona • Belgrade • Manchester • Tokyo • Cluj • Tianjin

Editor
Thomas Falguières
UMR_S 1193 Inserm -
Physiopathogénèse et Traitement
des Maladies du Foie /
Hepatinov
Université Paris-Saclay
Orsay
France

Editorial Office
MDPI
St. Alban-Anlage 66
4052 Basel, Switzerland

This is a reprint of articles from the Special Issue published online in the open access journal *International Journal of Molecular Sciences* (ISSN 1422-0067) (available at: www.mdpi.com/journal/ijms/special_issues/Transporters_Diseases).

For citation purposes, cite each article independently as indicated on the article page online and as indicated below:

LastName, A.A.; LastName, B.B.; LastName, C.C. Article Title. *Journal Name* **Year**, *Volume Number*, Page Range.

ISBN 978-3-0365-3944-7 (Hbk)
ISBN 978-3-0365-3943-0 (PDF)

Contents

About the Editor

Thomas Falguières

Thomas Falguières serves as a research scientist for the French National Institute for Health and Medical Research (Inserm) at the University Paris-Saclay (Orsay, France). With a strong background in cell biology and physiology, he received his Ph.D. degree at the University Paris-Saclay (formerly called University Paris-Sud) for his work on the molecular mechanisms regulating the intracellular traffic of bacterial toxins in cell models. After two post-doc internships in Geneva (Switzerland) and Paris (France), he obtained a tenure-track position from the Inserm. His work is dedicated to the study of the pathobiology and targeted pharmacotherapies of canalicular ABC transporters, which are defective in patients with rare and severe forms of genetic cholestasis.

International Journal of *Molecular Sciences*

Editorial

ABC Transporters in Human Diseases: Future Directions and Therapeutic Perspectives

Thomas Falguières 

Inserm, Université Paris-Saclay, Physiopathogénèse et Traitement des Maladies du Foie, UMR_S 1193, Hepatinov, 91400 Orsay, France; thomas.falguieres@inserm.fr

Citation: Falguières, T. ABC Transporters in Human Diseases: Future Directions and Therapeutic Perspectives. *Int. J. Mol. Sci.* **2022**, *23*, 4250. https://doi.org/10.3390/ijms23084250

Received: 5 April 2022
Accepted: 7 April 2022
Published: 12 April 2022

Publisher's Note: MDPI stays neutral with regard to jurisdictional claims in published maps and institutional affiliations.

The goal of this Special Issue on "ABC Transporters in Human Diseases", for which I was invited as a Guest Editor, was to provide an overview of the state-of-the-art research, understandings, and advances made in recent years on human diseases implicating ATP-binding cassette (ABC) transporters. Mammalian ABC transporters form a protein superfamily composed of 49 members [1], most of which are transmembrane proteins, responsible for the active transport of their substrates (ions, drugs, peptides, and lipids) across biological membranes, owing to their capacity to bind and hydrolyze adenosine triphosphate (ATP) [2]. Importantly, for approximately a third of these transporters, molecular defects have been correlated with human diseases [3,4], the most famous of which being cystic fibrosis, due to mutations of the chloride channel cystic fibrosis transmembrane conductance regulator (CFTR/ABCC7). Concerning CFTR/ABCC7, in this Special Issue, Uliyakina et al. investigated the contribution of the regulatory extension and regulatory regions—two distinct parts of the regulatory domain of the channel—in the rescue efficiency of the corrector VX-809 and the potentiator VX-770 on the F508del variant [5]. To treat patients affected by these ABC-transporter-related diseases, it is important to (1) understand the molecular mechanisms regulating the expression, intracellular traffic, and activity of these ABC transporters in normal and pathological conditions; (2) characterize the molecular and cellular effects of genetic variations identified in patients that relate to the genes encoding these transporters; (3) find new treatments specifically targeting the previously characterized defects (Figure 1). In this Special Issue, authors made significant contributions by presenting original articles and reviews that cover their fields of expertise on their chosen ABC transporters.

In the field of multidrug resistance (MDR) and cancer, findings of Low et al. suggested that ABCC1 (MRP1) AND ABCC4 (MRP4) are expressed in breast cancer cell lines and that these transporters are implicated in cell proliferation and migration, respectively [6]. Hlaváč et al. proposed that two genetic variations in *ABCC11* and *ABCA13* identified in patients can be associated with breast cancer, even if the data were not found to have strong statistical relevance [7]. Using phylogenic and sequence alignment analyses of the 12 members of the A subfamily ABC transporters, the same research group (Dvorak et al.) identified 13 single nucleotide variations in the 5′-untranslated transcribed region (5′-UTR) of *ABCA* genes in a small cohort of 105 patients with breast cancer [8]. Lower expression of D subfamily ABC transporters, localized at peroxisomes and responsible for very-long fatty acid transport, was also claimed to be involved in the progression of cancer, in addition to the well-known implication of ABCD1 in X-linked adrenoleukodystrophy (X-ALD) pathogenesis, reviewed by Tawbeh et al. [9]. The role of ABCC6, an ATP transporter predominantly expressed in the liver and kidneys, in cancer has also been highlighted, an aspect reviewed by Bisaccia et al. [10]. Thus, targeting this transporter would constitute an alternative anticancer strategy to further investigate. However, ABCC6 defects are mostly known for their implication in the development of pseudoxanthoma elasticum (PXE), a rare and severe disease-causing calcification of soft tissues and mostly characterized by skin lesions and cardiovascular complications, as reviewed by Shimada et al. [11]. As regards

ABCC6, Szeri et al. built a new structural model of this transporter based on its homology with bovine Abcc1, which allowed them to highlight the role of specific amino acids in the function of the transporter, using both cell models and 3D analyses [12]. Sasitharan et al. revealed that mutation of the three glutamines into alanines at positions 347, 725, and 990 of ABCB1/MDR1/P-gp, despite their previous implication in taxol binding [13], did not alter taxol transport outside cell models [14].

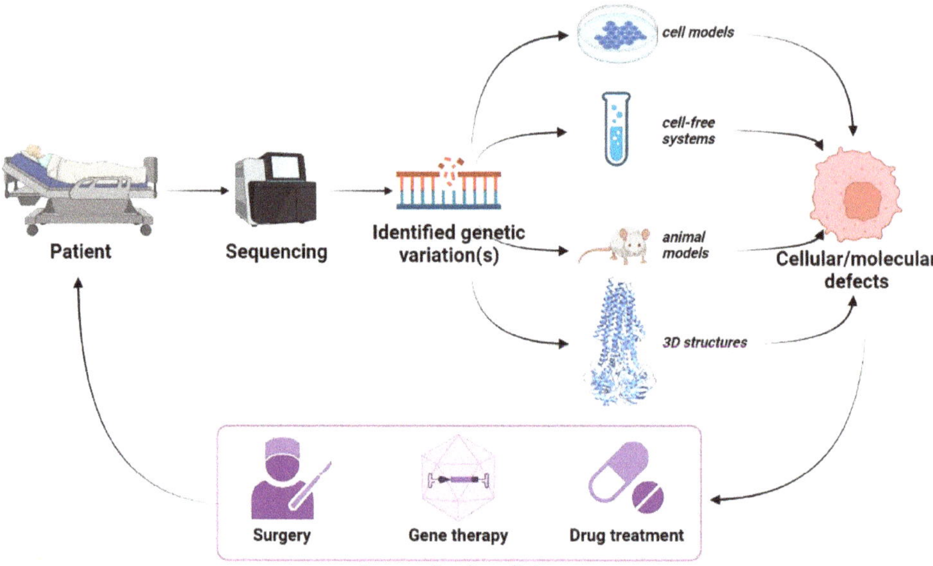

Figure 1. Personalized medicine for ABC-transporter-related diseases. Genes of interest are sequenced from patients to identify genetic variations. Then, the effects of these variations are studied using cell models, cell-free systems, animal models, and 3D modeling, allowing the characterization of the defects induced by the genetic variations. Based on this research, and as a long-term outlook, an adapted strategy for personalized medicine could emerge that would be specific to each patient. Created with BioRender.com.

Addressing the topic of breast cancer resistance protein (BCRP/AGCG2), another ABC transporter also implicated in MDR and gout pathogenesis, Toyoda et al. reported two new genetic variations (M131I and R236X) identified in family members affected by hyperuricemia and gout and responsible for urate transport defects of the transporter [15]. On this topic, Eckenstaler and Benndorf provided a complete review of the literature on the role of ABCG2 in urate homeostasis and its genetic variants involved in the pathogenesis of gout and hyperuricemia [16], while László Homolya proposed a classification of the genetic variants of ABCG2 based on their expression, traffic or function defects, and discussed their implication in human diseases, including cancer [17]. Using phylogenetic, sequence alignment, and structural analyses between members of the G subfamily of human ABC transporters, Mitchell-White et al. identified a conservation pattern that is different in ABCG2, which could confer greater flexibility to this transporter and thus explain its broader range of substrates [18]. Finally, based on already resolved 3D structures of ABCG2 and ABCG5/G8, and in the framework of providing a comprehensive review of the literature and comparative structural analysis, Khunweeraphong and Kuchler proposed a homology model of fungi pleiotropic drug-resistance (PDR) transporters, paving the path to a better understanding of infectious diseases due to pathogenic fungi, thus offering new therapeutic perspectives [19].

Chai et al. described that amyloid-beta peptides accumulating in the brain of patients with Alzheimer's disease (AD) are substrates of ABCB1/MDR1/P-gp [20], setting the groundwork for alternative therapeutic options. On the same topic of AD, Wanek et al. reported that, in line with former studies in the field, ABCB1/ABCG2 dual substrate drugs were normally distributed in the brain of AD mouse models despite less expression of the two transporters [21]. Evidence is also accumulating on the role of ABCA7 genetic variations in the impaired clearance of amyloid-beta peptides in AD, a topic reviewed by Dib et al. in this Special Issue [22]. It is also interesting to note that ABCA1, a ubiquitous exporter of free cholesterol and phospholipids, has also been implicated in AD pathogenesis, among many other pathologies—namely, type 2 diabetes, infectious diseases, cancer, age-related macular disease, and glaucoma, as reviewed by Jacobo-Albavera et al. [23].

Beyond using cell models and sequence alignments, a striking aspect in the recent body of literature on ABC transporters, including most of the contributions gathered in this Special Issue, is the use of 3D structural analyses to unravel the molecular mechanisms regulating the folding and the function of ABC transporters. Indeed, these approaches are very useful for understanding the cell biology of ABC transporters in normal and pathological conditions, as well as to study the potential direct interactions of small molecules with these transporters. Such approaches are now largely facilitated by the exponentially growing numbers of published 3D structures of ABC transporters in several conformational states (PDB structures available at https://www.rcsb.org/, accessed on 4 April 2022), mostly owing to the recent explosion of cryogenic electron microscopy (cryo-EM), which now allows a resolution below 4 Å, almost making X-ray crystallography outdated [24,25]. As stated recently by Shvarev et al., the next steps for these structural analyses now require the combination of *"biochemical and structural cryo-EM studies with molecular dynamics simulations and other biophysical methods"* [25].

Finally, I would like to focus on canalicular ABC transporters specifically expressed at the canalicular membrane of hepatocytes, in which I am particularly interested in the frame of our research projects. These transporters include ABCB4, ABCB11, and ABCG5/G8 implicated in the secretion of phosphatidylcholine, bile acids, and cholesterol into bile, respectively [26]. Regarding the ABCG5/G8 heterodimer, Williams et al. proposed an interesting review of the literature with a focus on sitosterolemia, a very rare disease caused by genetic variations of these half-transporters [27], while Xavier et al. investigated the effect of three ABCG5/G8 disease-causing missense variations by combining ATP hydrolysis measurements in a cell-free system with molecular dynamics approaches [28]. Sohail et al. provided a review article describing the tools (structures, animals, etc.), allowing the study of ABCB11 pathobiology and exploration of new therapeutic strategies [29]. We also proposed a review article describing the molecular regulation of these canalicular ABC transporters [30], as well as an original study highlighting the role of RAB10 as a new ABCB4 interactor regulating the traffic and function of this transporter [31]. As a therapeutic approach for progressive familial intrahepatic cholestasis (PFIC) implicating genetic variations of canalicular ABC transporters, Bosma et al. described recent advances in AAV8-mediated gene therapy aiming at restoring the expression and function of canalicular ABC transporters, mostly ABCB4, for which defects are implicated in PFIC3, and the limitations of such approaches [32]. Interestingly, an alternative mRNA-delivery-based strategy also aiming at restoring ABCB4 expression and function in mouse models was also recently proposed [33]. However, as stated by Bosma et al. [32], gene therapy cannot be the answer to all PFICs; therefore, pharmacotherapy strategies have to be pursued to propose, in the long run, targeted therapies within the framework of personalized medicine (Figure 1).

Funding: T.F. is supported by grants from the Agence Nationale de la Recherche (ANR-21-CE18-0030-01), Association Maladie Foie Enfants (AMFE), and the French Network for Rare Liver Diseases (FILFOIE).

Institutional Review Board Statement: Not applicable.

Informed Consent Statement: Not applicable.

Data Availability Statement: Not applicable.

Conflicts of Interest: The author declares no conflict of interest.

References

1. Vasiliou, V.; Vasiliou, K.; Nebert, D.W. Human ATP-binding cassette (ABC) transporter family. *Hum. Genom.* **2008**, *3*, 281. [CrossRef]
2. Thomas, C.; Tampé, R. Structural and Mechanistic Principles of ABC Transporters. *Annu. Rev. Biochem.* **2020**, *89*, 605–636. [CrossRef]
3. Borst, P.; Elferink, R.O. Mammalian ABC transporters in health and disease. *Annu. Rev. Biochem.* **2002**, *71*, 537–592. [CrossRef]
4. Linton, K.J. Structure and function of ABC transporters. *Physiology* **2007**, *22*, 122–130. [CrossRef]
5. Uliyakina, I.; Botelho, H.M.; Da Paula, A.C.; Afonso, S.; Lobo, M.J.; Felício, V.; Farinha, C.M.; Amaral, M.D. Full Rescue of F508del-CFTR Processing and Function by CFTR Modulators Can Be Achieved by Removal of Two Regulatory Regions. *Int. J. Mol. Sci.* **2020**, *21*, 4524. [CrossRef]
6. Low, F.G.; Shabir, K.; Brown, J.E.; Bill, R.M.; Rothnie, A.J. Roles of ABCC1 and ABCC4 in Proliferation and Migration of Breast Cancer Cell Lines. *Int. J. Mol. Sci.* **2020**, *21*, 7664. [CrossRef]
7. Hlaváč, V.; Václavíková, R.; Brynychová, V.; Koževnikovová, R.; Kopečková, K.; Vrána, D.; Gatěk, J.; Souček, P. Role of Genetic Variation in ABC Transporters in Breast Cancer Prognosis and Therapy Response. *Int. J. Mol. Sci.* **2020**, *21*, 9556. [CrossRef]
8. Dvorak, P.; Hlavac, V.; Soucek, P. 5′ Untranslated Region Elements Show High Abundance and Great Variability in Homologous ABCA Subfamily Genes. *Int. J. Mol. Sci.* **2020**, *21*, 8878. [CrossRef]
9. Tawbeh, A.; Gondcaille, C.; Trompier, D.; Savary, S. Peroxisomal ABC Transporters: An Update. *Int. J. Mol. Sci.* **2021**, *22*, 6093. [CrossRef]
10. Bisaccia, F.; Koshal, P.; Abruzzese, V.; Castiglione Morelli, M.A.; Ostuni, A. Structural and Functional Characterization of the ABCC6 Transporter in Hepatic Cells: Role on PXE, Cancer Therapy and Drug Resistance. *Int. J. Mol. Sci.* **2021**, *22*, 2858. [CrossRef]
11. Shimada, B.K.; Pomozi, V.; Zoll, J.; Kuo, S.; Martin, L.; Le Saux, O. ABCC6, Pyrophosphate and Ectopic Calcification: Therapeutic Solutions. *Int. J. Mol. Sci.* **2021**, *22*, 4555. [CrossRef]
12. Szeri, F.; Corradi, V.; Niaziorimi, F.; Donnelly, S.; Conseil, G.; Cole, S.P.C.; Tieleman, D.P.; van de Wetering, K. Mutagenic Analysis of the Putative ABCC6 Substrate-Binding Cavity Using a New Homology Model. *Int. J. Mol. Sci.* **2021**, *22*, 6910. [CrossRef]
13. Alam, A.; Kowal, J.; Broude, E.; Roninson, I.; Locher, K.P. Structural insight into substrate and inhibitor discrimination by human P-glycoprotein. *Science* **2019**, *363*, 753–756. [CrossRef]
14. Sasitharan, K.; Iqbal, H.A.; Bifsa, F.; Olszewska, A.; Linton, K.J. ABCB1 Does Not Require the Side-Chain Hydrogen-Bond Donors Gln(347), Gln(725), Gln(990) to Confer Cellular Resistance to the Anticancer Drug Taxol. *Int. J. Mol. Sci.* **2021**, *22*, 8561. [CrossRef]
15. Toyoda, Y.; Pavelcová, K.; Bohatá, J.; Ješina, P.; Kubota, Y.; Suzuki, H.; Takada, T.; Stiburkova, B. Identification of Two Dysfunctional Variants in the ABCG2 Urate Transporter Associated with Pediatric-Onset of Familial Hyperuricemia and Early-Onset Gout. *Int. J. Mol. Sci.* **2021**, *22*, 1935. [CrossRef]
16. Eckenstaler, R.; Benndorf, R.A. The Role of ABCG2 in the Pathogenesis of Primary Hyperuricemia and Gout-An Update. *Int. J. Mol. Sci.* **2021**, *22*, 6678. [CrossRef]
17. Homolya, L. Medically Important Alterations in Transport Function and Trafficking of ABCG2. *Int. J. Mol. Sci.* **2021**, *22*, 2786. [CrossRef]
18. Mitchell-White, J.I.; Stockner, T.; Holliday, N.; Briddon, S.J.; Kerr, I.D. Analysis of Sequence Divergence in Mammalian ABCGs Predicts a Structural Network of Residues That Underlies Functional Divergence. *Int. J. Mol. Sci.* **2021**, *22*, 3012. [CrossRef]
19. Khunweeraphong, N.; Kuchler, K. Multidrug Resistance in Mammals and Fungi-From MDR to PDR: A Rocky Road from Atomic Structures to Transport Mechanisms. *Int. J. Mol. Sci.* **2021**, *22*, 4806. [CrossRef]
20. Chai, A.B.; Hartz, A.M.S.; Gao, X.; Yang, A.; Callaghan, R.; Gelissen, I.C. New Evidence for P-gp-Mediated Export of Amyloid-β PEPTIDES in Molecular, Blood-Brain Barrier and Neuronal Models. *Int. J. Mol. Sci.* **2020**, *22*, 246. [CrossRef]
21. Wanek, T.; Zoufal, V.; Brackhan, M.; Krohn, M.; Mairinger, S.; Filip, T.; Sauberer, M.; Stanek, J.; Pekar, T.; Pahnke, J.; et al. Brain Distribution of Dual ABCB1/ABCG2 Substrates Is Unaltered in a Beta-Amyloidosis Mouse Model. *Int. J. Mol. Sci.* **2020**, *21*, 8245. [CrossRef]
22. Dib, S.; Pahnke, J.; Gosselet, F. Role of ABCA7 in Human Health and in Alzheimer's Disease. *Int. J. Mol. Sci.* **2021**, *22*, 4603. [CrossRef]
23. Jacobo-Albavera, L.; Domínguez-Pérez, M.; Medina-Leyte, D.J.; González-Garrido, A.; Villarreal-Molina, T. The Role of the ATP-Binding Cassette A1 (ABCA1) in Human Disease. *Int. J. Mol. Sci.* **2021**, *22*, 1593. [CrossRef]
24. Januliene, D.; Moeller, A. Cryo-EM of ABC transporters: An ice-cold solution to everything? *FEBS Lett.* **2020**, *594*, 3776–3789. [CrossRef]
25. Shvarev, D.; Januliene, D.; Moeller, A. Frozen motion: How cryo-EM changes the way we look at ABC transporters. *Trends Biochem. Sci.* **2022**, *47*, 136–148. [CrossRef]
26. Kroll, T.; Prescher, M.; Smits, S.H.J.; Schmitt, L. Structure and Function of Hepatobiliary ATP Binding Cassette Transporters. *Chem. Rev.* **2020**, *121*, 5240–5288. [CrossRef]

27. Williams, K.; Segard, A.; Graf, G.A. Sitosterolemia: Twenty Years of Discovery of the Function of ABCG5ABCG8. *Int. J. Mol. Sci.* **2021**, *22*, 2641. [CrossRef]

28. Xavier, B.M.; Zein, A.A.; Venes, A.; Wang, J.; Lee, J.Y. Transmembrane Polar Relay Drives the Allosteric Regulation for ABCG5/G8 Sterol Transporter. *Int. J. Mol. Sci.* **2020**, *21*, 8747. [CrossRef]

29. Sohail, M.I.; Dönmez-Cakil, Y.; Szöllősi, D.; Stockner, T.; Chiba, P. The Bile Salt Export Pump: Molecular Structure, Study Models and Small-Molecule Drugs for the Treatment of Inherited BSEP Deficiencies. *Int. J. Mol. Sci.* **2021**, *22*, 784. [CrossRef]

30. Ben Saad, A.; Bruneau, A.; Mareux, E.; Lapalus, M.; Delaunay, J.L.; Gonzales, E.; Jacquemin, E.; Aït-Slimane, T.; Falguières, T. Molecular Regulation of Canalicular ABC Transporters. *Int. J. Mol. Sci.* **2021**, *22*, 2113. [CrossRef]

31. Ben Saad, A.; Vauthier, V.; Lapalus, M.; Mareux, E.; Bennana, E.; Durand-Schneider, A.M.; Bruneau, A.; Delaunay, J.L.; Gonzales, E.; Housset, C.; et al. RAB10 Interacts with ABCB4 and Regulates Its Intracellular Traffic. *Int. J. Mol. Sci.* **2021**, *22*, 7087. [CrossRef] [PubMed]

32. Bosma, P.J.; Wits, M.; Oude-Elferink, R.P. Gene Therapy for Progressive Familial Intrahepatic Cholestasis: Current Progress and Future Prospects. *Int. J. Mol. Sci.* **2020**, *22*, 273. [CrossRef] [PubMed]

33. Wei, G.; Cao, J.; Huang, P.; An, P.; Badlani, D.; Vaid, K.A.; Zhao, S.; Wang, D.Q.; Zhuo, J.; Yin, L.; et al. Synthetic human ABCB4 mRNA therapy rescues severe liver disease phenotype in a BALB/c.Abcb4$^{-/-}$ mouse model of PFIC3. *J. Hepatol.* **2021**, *74*, 1416–1428. [CrossRef] [PubMed]

International Journal of
Molecular Sciences

Article

Full Rescue of F508del-CFTR Processing and Function by CFTR Modulators Can Be Achieved by Removal of Two Regulatory Regions

Inna Uliyakina, Hugo M. Botelho, Ana C. da Paula, Sara Afonso, Miguel J. Lobo, Verónica Felício, Carlos M. Farinha and Margarida D. Amaral *

BioISI—Biosystems & Integrative Sciences Institute, Faculty of Sciences, University of Lisboa, 1749-016 Lisboa, Portugal; iiuliyakina@fc.ul.pt (I.U.); hmbotelho@fc.ul.pt (H.M.B.); acdapaula@fc.ul.pt (A.C.d.P.); scafonso@fc.ul.pt (S.A.); mglobo@fc.ul.pt (M.J.L.); vmfelicio@fc.ul.pt (V.F.); cmfarinha@fc.ul.pt (C.M.F.)
* Correspondence: mdamaral@fc.ul.pt; Tel.: +351-217500861

Received: 22 May 2020; Accepted: 22 June 2020; Published: 25 June 2020

Abstract: Cystic Fibrosis (CF) is caused by mutations in the CF Transmembrane conductance Regulator (CFTR), the only ATP-binding cassette (ABC) transporter functioning as a channel. Unique to CFTR is a regulatory domain which includes a highly conformationally dynamic region—the regulatory extension (RE). The first nucleotide-binding domain of CFTR contains another dynamic region—regulatory insertion (RI). Removal of RI rescues the trafficking defect of CFTR with F508del, the most common CF-causing mutation. Here we aimed to assess the impact of RE removal (with/without RI or genetic revertants) on F508del-CFTR trafficking and how CFTR modulator drugs VX-809/lumacaftor and VX-770/ivacaftor rescue these variants. We generated cell lines expressing ΔRE and ΔRI CFTR (with/without genetic revertants) and assessed CFTR expression, stability, plasma membrane levels, and channel activity. Our data demonstrated that ΔRI significantly enhanced rescue of F508del-CFTR by VX-809. While the presence of the RI seems to be precluding full rescue of F508del-CFTR processing by VX-809, this region appears essential to rescue its function by VX-770, suggesting some contradictory role in rescue of F508del-CFTR by these two modulators. This negative impact of RI removal on VX-770-stimulated currents on F508del-CFTR can be compensated by deletion of the RE which also leads to the stabilization of this mutant. Despite both regions being conformationally dynamic, RI precludes F508del-CFTR processing while RE affects mostly its stability and channel opening.

Keywords: ABC transporters; drug action; regulatory extension; regulatory insertion; mechanism of action

---

1. Introduction

Cystic Fibrosis (CF), a life-threatening recessive disorder affecting ~80,000 individuals worldwide, is caused by mutations in the gene encoding the CF transmembrane conductance regulator (CFTR) protein present at the apical membrane of epithelial cells. This is the only member of the ATP-binding cassette (ABC) transporter family functioning as a channel, more precisely a cyclic Adenosine Monophosphate (cAMP)-dependent chloride (Cl^-)/bicarbonate (HCO_3^-) channel. It consists of two membrane-spanning domains (MSD1/2), two nucleotide binding domains (NBD1/2) and a cytoplasmic regulatory domain (RD), which is unique to CFTR [1]. The MSDs are linked via intra- and extra-cellular loops (ICLs, ECLs, respectively). ATP binding promoting NBD1:NBD2 dimerization preceded by phosphorylation of RD at multiple sites lead to channel gating [2]. The NBD1:NBD2 and ICL4:NBD1 interfaces were shown to represent critical folding conformational sites [3–5] important for the gating and maturation of the CFTR protein [5,6].

Although, >2000 CFTR gene mutations were reported (http://www.genet.sickkids.on.ca), one mutation–F508del–occurs in 85% of CF patients. This NBD1 mutant fails to traffic to the plasma membrane (PM) due to protein misfolding and retention by the endoplasmic reticulum quality control that targets it for premature proteasomal degradation. F508del-CFTR folding is a complex and inefficient process but it can be rescued, at least partially, by several treatments. These include low temperature incubation [7], genetic revertants [4,5,8–13] or pharmacological agents, like "corrector" VX-809 (lumacaftor) [14], one of the first CFTR modulator drugs to receive approval from the United States Food and Drug Administration in combination with potentiator VX-770/ivacaftor. Determining the additive/ synergistic rescue of F508del-CFTR by small molecule correctors together with other rescuing agents/revertants is very valuable to determine the mechanism of action of these CFTR modulators [5].

Comparative studies of CFTR with other ABC transporters are very powerful to understand the uniqueness of some of its regions, their influence on CFTR maturation and function as well as how they affect distinctive binding of CFTR to modulator drugs. Two such unique regions are present in NBD1 of CFTR, which are absent in NBDs of other ABC transporters (Figure S1)—the regulatory extension (RE) and regulatory insertion (RI). Both regions were described as highly conformationally dynamic [3,15] following PKA phosphorylation at certain residues (^{660}Ser, ^{670}Ser, and ^{422}Ser) [16,17]. Importantly, removal of the 32-amino acid RI (ΔRI) was reported to rescue traffic of F508del-CFTR [13].

Our first goal was to assess the impact of removing the 30 amino acid RE (ΔRE) alone or jointly with ΔRI on F508del-CFTR trafficking. Because there is some controversy regarding the RI and RE boundaries [3,13,15,18–20] which have structural implications, we tested the short and long versions of both regions (Figure S1). Secondly, we aimed to evaluate how RE and or RI removal from F508del-CFTR influences the rescue of this mutant by genetic revertants. Our third and final goal was to determine how the traffic and function of these combined variants of F508del-CFTR (ΔRE, ΔRI plus genetic revertants) are rescued by CFTR modulator drugs VX-809 (corrector) and VX-770 (potentiator) to gain further insight into their mechanism of action.

Our data show that although F508del-CFTR without RE did not traffic to the PM, it showed a dramatic stabilization of its immature form (evidencing a turnover rate ~2× lower than that of wt-CFTR. Results also show that, while ΔRI further increased processing of F508del-CFTR with revertants to almost wt-CFTR levels, RE removal completely abolished their processing, thus highlighting the different impact of the two dynamic regions on revertant-rescued F508del-CFTR. Most strikingly, although VX-809 rescued ΔRI-F508del-CFTR and ΔRE-ΔRI-F508del-CFTR processing to wt-CFTR levels, to achieve maximal function of F508del-CFTR, removal of just RI was insufficient, as both RI and RE had to be absent from F508del-CFTR. These data indicate that removal of these two regions has a positive effect on the rescuing efficacy of F508del-CFTR by CFTR modulators.

2. Results

2.1. Removal of Short Regulatory Extension (RE$_S$) Alone or with Regulatory Insertion (RI) has No Impact on 508del-CFTR Processing

Given the controversy in defining both RI and RE, we chose to generate two versions of these regions because of the respective structural implications. Indeed, RI$_L$ and RE$_L$ are the complete regions which are absent in other ABC transporters (Figure S1), whereas, their shorter versions appeared as structurally meaningful to be removed: RI$_S$ is strictly the region described as destructured in the crystal structure and RE$_S$ is the core of RE, i.e., just the β-strand [19]. The impact of deleting RE$_S$—short RE (Δ^{654}Ser-Gly^{673}, Figure S1)—was first assessed, either alone or together with RI in its short and long variants (ΔRI$_S$ and ΔRI$_L$), on the in vivo processing of wt- and F508del-CFTR by Western blot (WB) of Baby Hamster Kidney (BHK) cells stably expressing such variants (Figure 1B,E,F and Table 1). Processing of CFTR was assessed by WB in terms of its fully-glycosylated form (also termed band C) corresponding to post-Golgi forms, assumedly at the PM. Unprocessed CFTR, in turn corresponds to its core-glycosylated, ER-specific form (also termed band B). Removal of RE_S had no impact on processing of either F508del-CFTR or wt-CFTR (Figure 1B, lanes 5,10; Table 1), despite that the immature form of ΔRE_S-F508del-CFTR appeared consistently at increased levels vs. those of F508del-CFTR (Figure 1B, lanes 5,2).

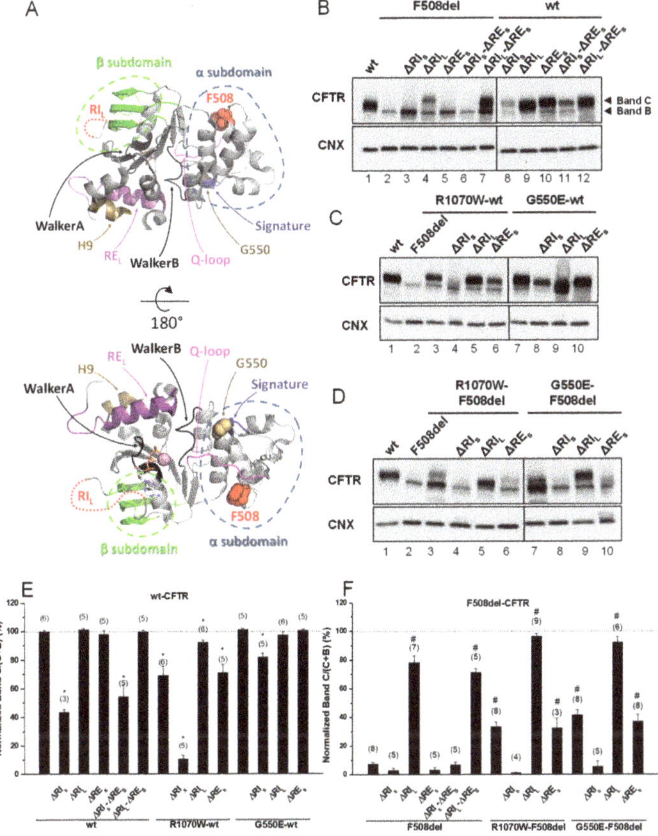

Figure 1. Effect of removal of short regulatory extension (ΔRE_S) on processing of wt- and F508del-Cystic Fibrosis (CF) transmembrane conductance regulator (CFTR). (**A**) Structural features of the CFTR nucleotide binding domain 1 (NBD1) domain. Protein Data Bank identifier (PDB ID): 2BBO. The main structural regions are color coded. F508 is represented as red spheres. G550 is represented as brown spheres. ATP is shown as sticks and Mg^{2+} as a pink sphere. Regulatory insertion (RI) is unstructured in the crystal structure and represented as a dotted red line. The G550E mutation in this structure was reverted in silico. (**B–D**) Samples from Baby Hamster Kidney (BHK) cell lines stably expressing different CFTR variants of ΔRE_S and ΔRI, the latter either in its short "S" or long "L" versions (see Materials and Methods) and alone (**B**) or jointly with (**C,D**) genetic revertants, as indicated, were analyzed for CFTR protein expression by Western blot (WB) with anti-CFTR 596 Ab and also anti-calnexin (CNX) as loading control. All samples were incubated with 3 µM dimethyl sulfoxide (DMSO) for 48 h, as controls for experiments with corrector VX-809 (see Figure 3). Summary of data of CFTR variants on the wt- (**E**) or F508del- (**F**) CFTR backgrounds, expressed as normalized ratios (band C/(band C + band B)) and as a percentage to the respective ratio for wt-CFTR and as mean ± standard error of the mean (SEM). (n) indicates nr. of independent experiments. "*" and "#" indicate significantly different ($p < 0.05$) from wt-CFTR and F508del-CFTR, respectively.

Table 1. Summary of Western blot quantification of original data in Figures 1–3 for different CFTR variants in the presence of VX-809 or DMSO control. Data are expressed as normalized ratios (band C/band (B + C)) for each variant and as a percentage to wt-CFTR ratio.

		wt-Background		F508del-Background	
		Control	VX-809	Control	VX-809
	CFTR Variant	% to wt	% to wt	% to wt	% to wt
	-	100 ± 1	103 ± 1	8 ± 1	17 ± 2
ΔRE_S	-	98 ± 2	100 ± 1	3 ± 2	9 ± 4
	ΔRI_S	54 ± 9	93 ± 3	7 ± 2	20 ± 4
	ΔRI_L	100 ± 1	97 ± 2	71 ± 3	96 ± 2
	R1070W	71 ± 6	90 ± 3	33 ± 7	26 ± 6
	G550E	101 ± 1	102 ± 1	37 ± 5	39 ± 4
ΔRE_L	-	88 ± 5	87 ± 3	9 ± 2	8 ± 2
	$\Delta H9$	55 ± 3	69 ± 3	0 ± 0	0 ± 0
$\Delta H9$		72 ± 4	69 ± 3	1 ± 0	0 ± 0
ΔRI_S	-	44 ± 2	90 ± 3	3 ± 2	4 ± 3
	R1070W	11 ± 3	56 ± 8	2 ± 0	4 ± 2
	G550E	82 ± 3	98 ± 1	6 ± 4	5 ± 3
ΔRI_L	-	101 ± 1	101 ± 0	78 ± 5	92 ± 4
	R1070W	92 ± 2	94 ± 3	96 ± 2	96 ± 3
	G550E	97 ± 2	92 ± 2	92 ± 4	97 ± 2
R1070W		69 ± 7	93 ± 4	34 ± 3	46 ± 4
G550E		101 ± 1	101 ± 1	42 ± 4	73 ± 6

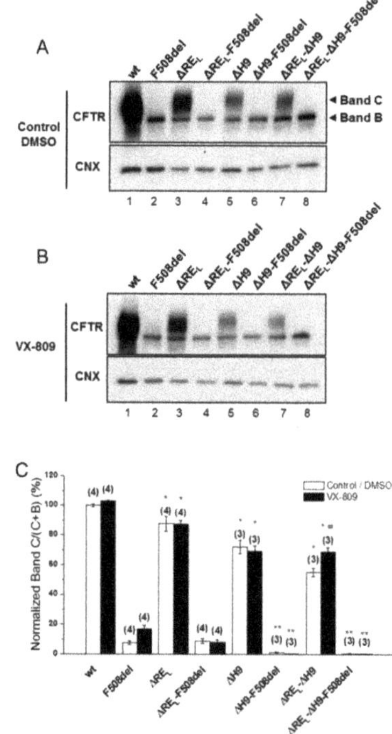

Figure 2. Effect of removal of helix H9 in combination with ΔRE_L and VX-809 on processing of wt- and

F508del-CFTR. (**A**,**B**) WB analysis of samples from BHK cell lines stably expressing wt- or F508del-CFTR variants with ΔRE_L, $\Delta H9$, or ΔRE_L-$\Delta H9$. Cells were incubated with 3 µM VX-809 or DMSO (control) for 48 h at 37 °C. CFTR protein expression was analyzed by WB with the anti-CFTR 596 and anti-CNX Abs. (**C**) Summary of data expressed as normalized ratios (Band C / (Band C + B)) and as a percentage to the respective ratios on the wt- or F508del-CFTR backgrounds and shown as mean ± SEM. (n) indicates number of independent experiments. "#" indicates significantly different from the respective variant treated with DMSO. "*" and "**" indicate significantly different from wt- or F508del-CFTR, respectively.

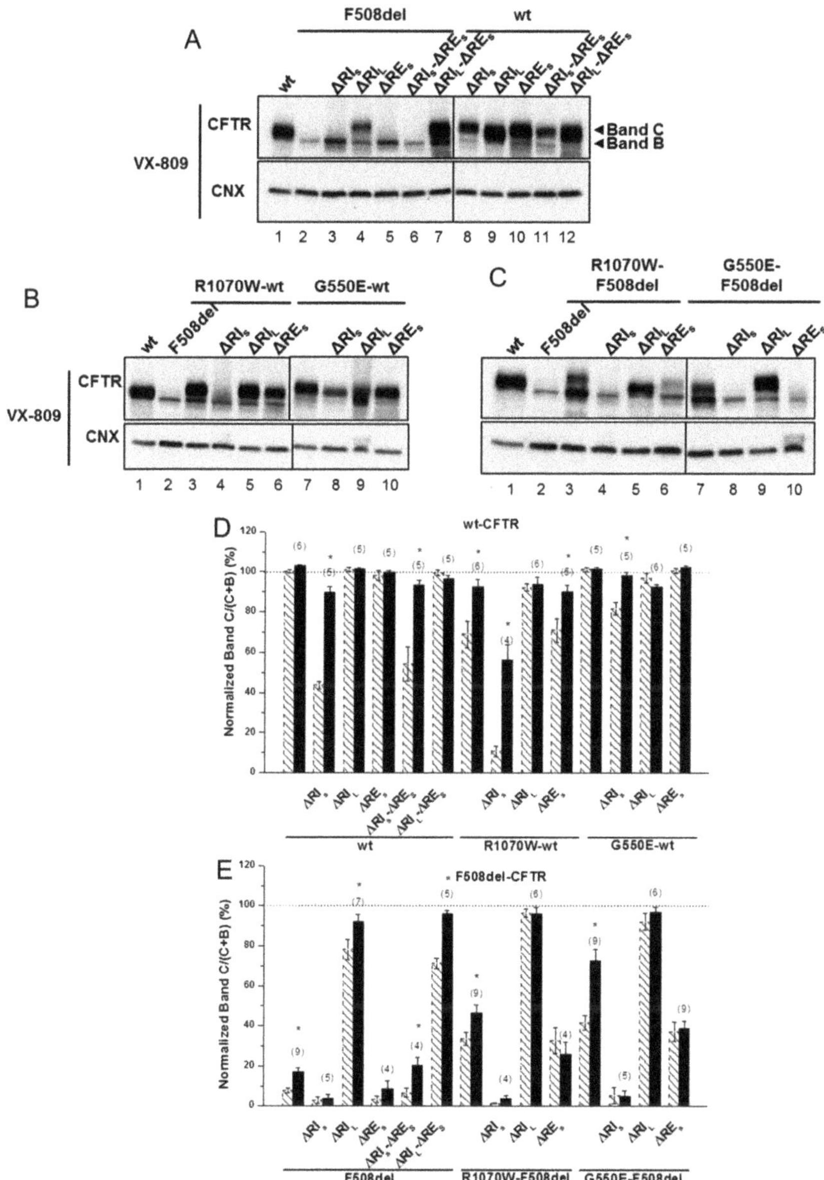

Figure 3. Effect of VX-809 on processing of wt- and F508del-CFTR variants without regulatory extension.

(**A–C**) WB of samples from BHK cell lines stably expressing different CFTR variants of ΔRE and ΔRI, as indicated, alone (**A**) or jointly with (**B,C**) genetic revertants (see Materials and Methods). CFTR protein expression was analyzed by WB with anti-CFTR 596 and anti-CNX Abs as in Figure 1. All samples were incubated with 3 μM VX-809 for 48 h. (**D, E**) Summary of data of variants on the wt- (**D**) or F508del- (**E**) CFTR backgrounds, expressed as normalized ratios (band C/(band C + band B)) and as a percentage to the respective ratio for wt-CFTR and as mean ± SEM. (n) indicates number of independent experiments. "*" indicates significantly different ($p < 0.05$) from respective variant without VX-809 (as shown in Figure 1 and indicated here by dashed bars).

Removal of RE_S together with RI_L—long RI (Δ^{404}Gly-Leu^{435}, Figure S1) led to an increase in processing of F508del-CFTR from $3 \pm 2\%$ to $71 \pm 3\%$ (vs. wt-CFTR levels), similarly to what had been previously reported by Alexandrov et al. for ΔRI_L alone [13]. In contrast, removal of RI_S—short RI (Δ^{412}Ala-Leu^{428}, Figure S1) had no impact on ΔRE_S-F508del-CFTR processing (Figure 1B, lanes 6,7; Figure 1F; Table 1). So in summary, removal of RE_S from the ΔRI_S- or ΔRI_L-F508del-CFTR variants had no further effect on their processing, despite that processing of ΔRE_S-F508del-CFTR is different upon removal of RI_S or RI_L, but this difference remains equivalent to removal RI_S or RI_L alone. Despite this lack of impact on processing, removal of RE_S does affect total protein expression levels of ΔRI_S- or ΔRI_L-F508del-CFTR variants leading to a decrease in RI_S-F508del-CFTR (Figure 1B, lanes 3,6) and an increase in RI_L-F508del-CFTR (Figure 1B, lanes 4,7). Interestingly, when RE_S and RI_S, were jointly removed from wt-CFTR, its processing was significantly reduced to $54 \pm 9\%$, while ΔRE_S jointly with ΔRI_L caused no impact (Figure 1B, lanes11,12, respectively; Figure 1E; Table 1). Again, these data were equivalent to removal of RI_S or RI_L alone on wt-CFTR processing (Figure 1B, lanes 8,9; Figure 1F; Table 1).

The differential effect caused by removal of RI_S vs RI_L on F508del- and wt-CFTR emphasize the importance of those 8 N-term (^{404}Gly-Lys^{411}) and 7 C-term (Phe^{429}-Leu^{435}) amino acid residues that differ between the two RI regions for the folding and processing of CFTR.

Overall, removal of RE_S alone or with RI has no impact on F508del-CFTR processing efficiency.

2.2. Simultaneous Removal of Long Regulatory Extension (RE_L) and Helix H9 Significantly Reduces wt-CFTR Processing but Increases Levels of F508del-CFTR Immature Form

Next, we assessed the impact of removing the long version of RE–RE_L (Δ^{647}Cys-Ser^{678}, Figure 2)— which significantly decreased wt-CFTR processing to $88 \pm 5\%$ (Figure 2A, lane 3; Figure 2C) but had no impact on F508del-CFTR (Figure 2A, lane 4; Figure 2C). As helix H9 (Δ^{637}Gln-Gly^{646}, Figure S1) can be considered to be part of this longer RE region [18] since it interacts with RE (Figure 1A), we also tested the effect of removing helix H9 jointly with ΔRE_L. Deletion of both H9 (ΔH9) and RE_L further decreased the residual processing of ΔRE_L-F508del-CFTR from $9 \pm 2\%$ to $0\pm0\%$ and in fact the same happened for ΔH9 alone (Figure 2A, lanes 8,6, respectively). Curiously, however, levels of immature F508del-CFTR were significantly increased when both RE_L and H9 were removed (Figure 2A, lane 8), similarly to ΔRE_S-F508del-CFTR (Figure 1B) and in contrast to ΔRE_L-F508del-CFTR.

For wt-CFTR, ΔRE_L-ΔH9 also led to a decrease in processing ($55 \pm 3\%$), which was more pronounced than for ΔRE_L alone, but interestingly removal of H9 helix alone only reduced processing to $72 \pm 4\%$ (Figure 2A, lanes 7,5, respectively).

Simultaneous removal of RE_L and helix H9 significantly reduces wt-CFTR processing but increases levels of F508del-CFTR immature form.

2.3. ΔRI_S, but Not ΔRE_S, Abolishes the Plasma Membrane Rescue of F508del-CFTR by Revertants

In order to test whether the stabilized immature form of ΔRE_S-F508del-CFTR could be rescued to mature form, we next tested the effects of removing the RE_S from F508del-CFTR with R1070W and G550E revertants. Indeed, both the R1070W and G550E revertants rescued ΔRE_S-F508del-CFTR to $33 \pm 7\%$ and $37 \pm 5\%$, respectively (Figure 1D, lanes 6,10; Table 1).

However, removal of RI$_S$ from F508del-CFTR with R1070W and G550E revertants, virtually abolished the rescue of F508del-CFTR by both revertants (Figure 1D, lanes 4,8; Table 1). Moreover, processing of ΔRI$_L$-F508del-CFTR was further increased by either R1070W or G550E up to 96 ± 2% and 92 ± 4%, respectively (Figure 1D, lanes 5,9; Table 1). Alone, R1070W and G550E rescued F508del-CFTR from 8 ± 1% to 34 ± 3% and 42 ± 4%, respectively (Figure 1D, lanes 3,7; Table 1), as we previously reported [5].

Interestingly, the impaired processing of ΔRI$_S$-wt-CFTR (44 ± 2%) was significantly rescued by G550E (to 82 ± 3%), but curiously it was further reduced by R1070W (to 11%) i.e., close to levels of F508del-CFTR processing (Figure 1C, lanes 8,4; Table 1). Of note that R1070W alone also reduced wt-CFTR processing to 69 ± 7%, but this reduction could be compensated by the removal of RI$_L$ (92 ± 2%) (Figure 1C, lanes 3,5; Table 1), while G550E alone caused no effect on wt-CFTR processing (Figure 1C, lane 7).

In summary, ΔRI$_S$, but not ΔRE$_S$, abolishes the plasma membrane rescue of F508del-CFTR by genetic revertants R1070W and G550E.

2.4. ΔRI$_L$ Synergises with VX-809, but Not with Revertants to Rescue ΔRE$_S$-F508del-CFTR Processing

To further test how the stabilized immature form of ΔRE$_S$-F508del-CFTR could be pharmacologically rescued, we then assessed the impact of corrector VX-809 (which per se promotes maturation of the F508del-CFTR) on the processing of this and of the other variants, to obtain structural insight on the effects of this novel drug. Although ΔRE$_S$-F508del-CFTR could not be rescued by VX-809, data show that this small molecule was able to rescue ΔRI$_S$-ΔRE$_S$-F508del-CFTR (from 7 ± 2% to 20 ± 4%) and further increased processing of ΔRI$_L$-ΔRE$_S$-F508del-CFTR to wt-CFTR levels (from 71 ± 3% to 96 ± 2%) (Figure 3A, lanes 6,7; Figure 3E; Table 1). These data suggest a strong synergistic effect between VX-809 and ΔRI$_L$ (and with ΔRI$_S$, albeit to a lesser extent) to rescue ΔRE$_S$-F508del-CFTR processing. This is particularly interesting because VX-809 did not rescue processing of the ΔRE$_S$-, nor ΔRI$_S$-F508del-CFTR variants (Figure 3A, lanes 5,3; Table 1) while it recovered the processing of ΔRI$_L$-F508del-CFTR to wt-CFTR levels (Figure 3A, lane4; Table 1). Notably, VX-809 was able to almost completely revert processing impairment of ΔRI$_S$-wt-CFTR and ΔRI$_S$-ΔRE$_S$-CFTR (from 44 ± 2% and 54 ± 9%) to 90 ± 3% and 93 ± 3%, respectively (Figure 3A, lanes 8,11; Figure 3D; Table 1).

Strikingly, VX-809 had no effect on processing of ΔRE$_S$-F508del-CFTR with R1070W or G550E (Figure 3C, lanes 6,10; Figure 3E; Table 1), while it further rescued processing of F508del-CFTR variants with R1070W or G550E alone to 46 ± 4% and 73 ± 6%, respectively (Figure 3C, lanes 3,7; Figure 3E; Table 1), as reported [5]. Similarly, VX-809 caused no significant further rescue on the processing of ΔRI$_L$-F508del-CFTR with those revertants, but these variants already had processing levels close to those of wt-CFTR (Figure 3C, lanes 5,9; Figure 3E; Table 1). Interestingly, the impaired processing of ΔRE$_S$-R1070W-wt-CFTR (71 ± 6%) was also reverted by VX-809 to 90 ± 3% (Figure 3B, lane 6; Table 1).

Corrector VX-809 failed to rescue any of the ΔRE$_L$-, ΔH9-, and ΔRE$_L$-ΔH9-F508del-CFTR variants (Figure 2B,C), while significantly increasing the processing of the ΔRE$_L$-ΔH9-wt-CFTR from 55 ± 3% to 69 ± 3% (Figure 2B, lane 7; Figure 2C).

Altogether, these results suggest that RI$_L$ and the genetic revertants interfere with the rescue of F508del-CFTR by VX-809.

2.5. ΔRI$_L$-ΔRE$_S$-F508del-CFTR Levels at the Plasma Membrane are Equivalent to Those of wt-CFTR

To determine the fraction of the above CFTR variants that localize to the PM, we used quantitative cell surface biotinylation. These data showed that PM levels of ΔRI$_L$-ΔRE$_S$-F508del-CFTR were equivalent to those of wt-CFTR, while those of ΔRI$_L$-F508del-CFTR were significantly lower (Figure 4A, lanes 5,4; Figure 4B). Data also confirmed that ΔRE$_S$ did not induce appearance of F508del-CFTR at the cell surface (data not shown). Corrector VX-809 further increased the PM expression of ΔRI$_L$-ΔRE$_S$-F508del-CFTR to levels that are significantly higher than those of wt-CFTR (Figure 4A, lanes 2,9, Figure 4B). This compound also significantly increased PM levels of ΔRI_L-F508del-CFTR to similar levels of wt-CFTR (Figure 4A, lanes 2,8; Figure 4B).

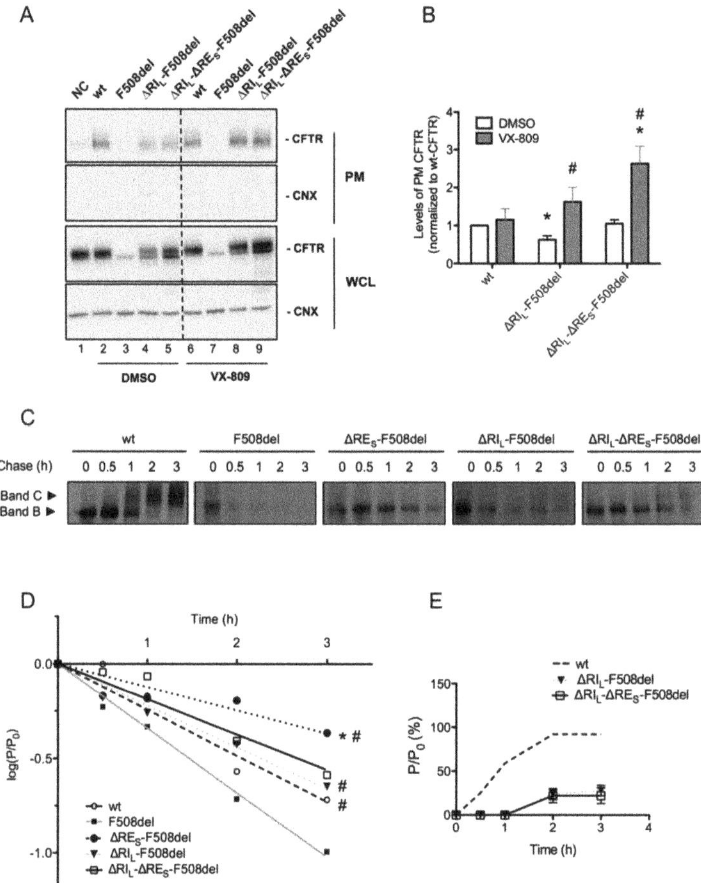

Figure 4. Plasma membrane levels, efficiency of processing and turnover of immature CFTR without Regulatory Extension (RE$_S$). (**A**) BHK cells expressing ΔRI_L-F508del and ΔRI_L-ΔRE_S-F508del-CFTR and treated with 3 μM VX-809 for 48 h (or DMSO control) were subjected to cell surface biotinylation. wt-CFTR samples not treated with biotin were the negative control (NC). After streptavidin pull-down, CFTR was detected by WB. CFTR and CNX are detected in the whole cell lysate (WCL) as controls. (**B**) Quantification of data in (a) for PM CFTR normalized to total protein and shown as fold change relatively to wt-CFTR cells treated with DMSO. "*" and "#" indicate significantly different from wt-CFTR treated with DMSO and from respective variant without VX-809, respectively ($p < 0.05$). (**C**) BHK cells expressing wt-, F508del-CFTR alone or jointly with ΔRI_L, ΔRE_S, and ΔRI_L-ΔRE_S were subjected to pulse-chase (see Materials and Methods) for the indicated times (0, 0.5, 1, 2, and 3h) before lysis and immunoprecipitation (IP) with the anti-CFTR 596 Ab. After electrophoresis and fluorography, images were analyzed by densitometry. (**D**) Turnover of immature (band B) CFTR for different CFTR variants is shown as the percentage of immature protein at a given time point of chase (P) relative to the amount at t = 0 (P$_0$). (**E**) Efficiency of processing of band B into band C is shown as the percentage of band C at a given time of chase relative to the amount of band B at t = 0. "*" and "#" indicate statistical significantly different ($p < 0.05$) from wt-CFTR and F508del-CFTR, respectively. Data represent mean ± SEM ($n = 5$).

Given the very significant stabilization of immature ΔRE_S-F508del-CFTR (Figure 1B), next, we determined how removal of RE_S affected the processing efficiency and the turnover of the F508del- and ΔRI_L-F508del-CFTR variants. To this end, we performed pulse-chase experiments (Figure 4C,D) and indeed our results revealed that ΔRE_S very significantly stabilized immature F508del-CFTR not just relatively to F508del-CFTR but to levels even significantly higher than those of wt-CFTR (Figure 4C,D). Indeed, our data show that although F508del-CFTR without RE did not traffic to the PM, it showed a dramatic stabilization of its immature form (evidencing a turnover rate ~2× lower than that of wt-CFTR. It should be noted that the turnover of F508del-CFTR itself is ~1.4×faster vs. wt-CFTR, thus this stabilization (of ~3× vs. F508del-CFTR) represents indeed a massive stabilization. Interestingly, this stabilizing effect was no longer significant for ΔRI_L-ΔRE_S-F508del-CFTR nor for ΔRI_L-F508del-CFTR, the latter being equivalent to wt-CFTR (Figure 4C,D). Removal of RI_S from F508del-CFTR did not stabilize its turnover (data not shown). As to removal of either RE_S or RI_S from wt-CFTR, it did not affect the processing efficiency or the turnover vs. wt-CFTR (Figure S2).

In summary, levels of ΔRI_L-ΔRE_S-F508del-CFTR at the PM are equivalent to those of wt-CFTR.

2.6. VX-809 Jointly with RE and RI Removal Completely Restored F508del-CFTR Function as Chloride Channel

To investigate the channel function of the ΔRE_S- and ΔRI_L-F508del-CFTR variants, we used the iodide efflux technique. Removal of RE_S alone had no impact on F508del-CFTR function (Figure 5A; black dash line in Figure S3B) as expected from the lack of processed form for this variant (Figure 1B). However, when RE_S and RI_L were removed jointly from F508del-CFTR, functional levels reached 78 ± 7% of wt-CFTR (Figure 5A,C; black dotted line in Figure S3B). As to removal of RI_L alone from F508del-CFTR, it restored function to 92 ± 7% of wt-CFTR (Figure 5A,C; black line in Figure S3B), as described [13]. Also, the delay in peak response observed for F508del-CFTR (min = 4) vs. wt-CFTR (min = 2), was partially corrected (to min = 3) both by removal of RI_L alone from F508del-CFTR or together with RE_S (Figure 5C; Figure S3B, black line and dotted line).

Introduction of G550E and R1070W revertants into the ΔRI_L-F508del-CFTR variant slightly but not significantly decreased its function from 92 ± 7% to 71 ± 8% and 76 ± 8%, respectively (Figure 5A,C; black lines in Figure S3C,D), in parallel to the respective processing levels (Figure 1). In fact, these variants showed higher activity levels than the revertants containing RI_L (29 ± 3% and 56 ± 6%, respectively). Interestingly however, both G550E- and R1070W-ΔRI_L-F508del-CFTR fully corrected (to min = 2) the delay in peak response of F508del-CFTR, which was still partially present in ΔRI_L- and ΔRI_L-RE_S-F508del-CFTR (min = 3).

The effect of VX-809 (3μM, 48 h at 37 °C) was also examined at functional level on the ΔRE_S-, ΔRI_L-ΔRE_S- and ΔRI_L-F508del-CFTR variants. VX-809 shows a tendency to increase in ΔRE_S-ΔRI_L-F508del-CFTR function (84 ± 8% of wt-CFTR) which was parallel to the observed increase in PM expression (Figure 4; Figure 5A,C; black dotted line in Figure S3F). Also, ΔRE_S-ΔRI_L-F508del-CFTR fully corrected (to min = 2) the delay in peak response of F508del-CFTR which was still partially present in ΔRI_L- and ΔRI_L-ΔRE_S-F508del-CFTR (min = 3).

In contrast to the observed increase in processing, VX-809 caused no significant increase in the function of ΔRI_L-F508del-CFTR and actually a slight, but not significant decrease was observed and there was still the same delay in peak response (Figure 5A,C; black line in Figure S3F).

A similar slight decrease in function was observed for VX-809 on the ΔRI_L-R1070W-F508del-CFTR variant (Figure 5A,C; black line in Figure S3H), but the most striking result was the significant decrease caused by VX-809 on function of ΔRI_L-G550E-F508del from 76 ± 8% to 41 ± 3% (Figure 5A; black line in Figure S3G). Nevertheless, both revertant variants of ΔRI_L-F508del-CFTR showed no change in peak response which was still the same as wt-CFTR (min = 2).

Overall, VX-809 jointly with RE and RI removal completely restored F508del-CFTR function as chloride channel.

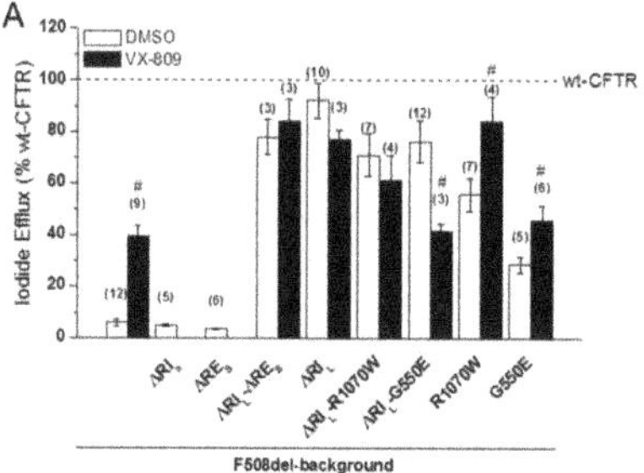

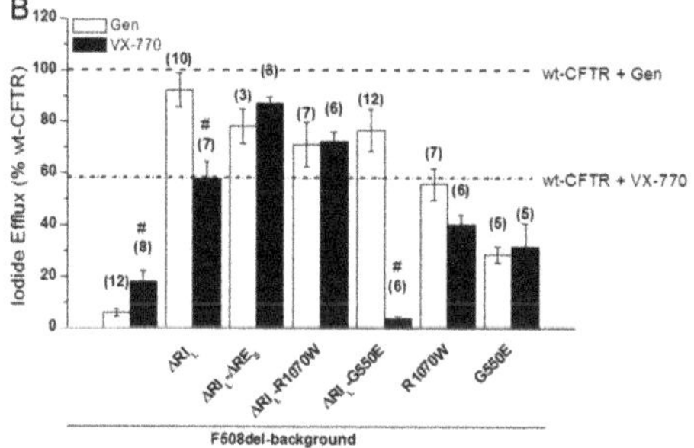

CFTR variant	DMSO			VX-809			VX-770		
	peak (min)	value	% to control wt-CFTR	peak (min)	value	% to control wt-CFTR	peak (min)	value	% to control wt-CFTR
wt	2	51 ± 3	100 ± 5	2	51 ± 4	100 ± 8	2	30 ± 4	58 ± 9
F508del	4	3 ± 1	6 ± 1	4	20 ± 2	39 ± 4	4	9 ± 2	18 ± 4
ΔRI_s-F508del	.	2 ± 0	5 ± 0	.	.	.	.	.	.
ΔRE_s-F508del	.	2 ± 0	3 ± 0	.	.	.	.	.	.
ΔRI_L-ΔRE_s-F508del	3	40 ± 3	78 ± 7	2	43 ± 4	84 ± 8	2	44 ± 1	87 ± 3
ΔRI_L-F508del	3	47 ± 3	92 ± 7	3	39 ± 2	77 ± 4	3	30 ± 3	58 ± 7
ΔRI_L-R1070W-F508del	2	36 ± 4	71 ± 8	2	31 ± 5	61 ± 10	1	37 ± 2	72 ± 4
ΔRI_L-G550E-F508del	2	39 ± 4	76 ± 8	2	21 ± 2	41 ± 3	.	2 ± 0	4 ± 1
R1070W-F508del	2	28 ± 3	56 ± 6	2	43 ± 5	84 ± 10	3	21 ± 2	40 ± 4
G550E-F508del	2	15 ± 2	29 ± 3	2	23 ± 3	46 ± 5	3	16 ± 4	32 ± 9

Figure 5. Functional characterization of the ΔRE_s and ΔRI_l variants of F508del-CFTR with or without VX-809 or VX-770 treatments. (**A,B**) Summary of the data from the iodide efflux peak magnitude generated by BHK cells stably expressing different CFTR variants and treated with 3 μM VX-809 (black bars on panel A) or with DMSO as control (white bars on panel a) for 48 h and stimulated with 10 μM Forskolin (Fsk) and 50 μM Genistein (Gen). In (**B**) cells were not pre-incubated and were stimulated either with 10 μM Fsk and 50 μM Gen (white bars on panel B) or with 10 μM Fsk and 10 μM VX-770 (black bars on panel b), as indicated. Data are shown as a percentage of wt-CFTR activity under Fsk/Gen stimulation and as mean ± SEM. (n) indicates number of independent experiments. "#" indicates significantly different ($p < 0.05$) from the same CFTR variant incubated with DMSO (A) or significantly different from the same CFTR variant under Fsk/Gen stimulation (white bars) (B). (**C**) Summary of iodide efflux data for different F508del-CFTR variants without RE_s, RI_s, and RI_l alone or jointly with revertants R1070W and G550E.

2.7. VX-770-Stimulated Currents of CFTR Variants are Dramatically Decreased by ΔRI_L but not by ΔRI_L-ΔRE_S

Given the interesting and diverse results observed for the ΔRE_S and ΔRI_L under VX-809, next we tested the effects of potentiator VX-770 (Ivacaftor), an approved drug for CF patients with gating mutations (Figure 5B) on these variants and in combination with VX-809, for F508del/F508del patients. Most strikingly, our data show that VX-770/Forskolin (Fsk) significantly decreased the function of ΔRI_L-F508del-CFTRin comparison with potentiation by Gen/Fsk (Figure 5B,C) from 92 ± 7% to 58 ± 7% of wt-CFTR (Figure 5B,C; black line in Figure S3J). As ΔRE_S-F508del-CFTR was not processed, we did not test the effect of VX-770 on this variant, but when the ΔRE_S-ΔRI_L-F508del-CFTR variant was assessed for its function after acute application of VX-770 with Fsk, a tendency for increase in its function (to 87 ± 3% of wt-CFTR) was observed *versus* its function under Gen/Fsk (78 ± 7% of wt-CFTR) (Figure 5B,C; black dotted line in Figure S3J).

The most dramatic result, however, was observed for the ΔRI_L-G550E-F508del-CFTR variant which under VX-770/Fsk only had 4 ± 1% function of wt-CFTR, while under Gen/Fsk it had 76 ± 8% (Figure 5B,C; black line in Figure S3K). For ΔRI_L-R1070W-F508del-CFTR this decrease was not observed (Figure 5B,C; black line in Figure S3L), and in fact this variant had a slightly higher function under VX-770 stimulation (72 ± 4%) than ΔRI_L-F508del-CFTR (58 ± 7%) (Figure 5B,C, black line in Figure S3H). These results suggest that somehow removal of RI_L can increase the chances of VX-770 blocking the channel pore.

The delay in peak response of F508del- comparing with wt-CFTR was not corrected for ΔRI_L-F508del-CFTR (min = 3), but it was corrected for ΔRI_L-ΔRE_S-F508del-CFTR (min = 2) (black and dotted black lines in Figure S3J, respectively). Interestingly, on the other hand ΔRI_L-R1070W-F508del-CFTR showed the quickest peak response (at min = 1), in contrast to either R1070W-F508del-CFTR or ΔRI_L-F508del-CFTR, both at min = 3.

Similarly, VX-770 also significantly decreased the function of ΔRI_L-wt-CFTR from 127 ± 15% under Gen to 57 ± 6% (Figure S4).

These results imply that, VX-770-stimulated currents of CFTR variants are dramatically decreased by ΔRI_L but not by ΔRI_L-ΔRE_S.

3. Discussion

The main goal of this study was to understand how the removal of the regulatory extension (ΔRE) alone or with the regulatory insertion (ΔRI)—two highly conformationally dynamic regions—impact the rescue of F508del-CFTR processing and function by two compounds—VX-809 and VX-770—which in combination were recently approved to clinically treat F508del-homozygous patients. Indeed, CFTR is the sole ABC transporter that functions as a channel and thus, these highly dynamic RI and RE regions which are absent in other ABC transporters, may be of high relevance to understand how CFTR differs from other ABCs, namely in its function as a channel.

These regions RI and RE were originally suggested to be positioned to impede formation of the NBD1-NBD2 dimer required for channel gating [15,19]. Indeed, both RI and RE were shown to be mobile elements in solution that bind transiently to the core of NBD in the β-sheet and α/β subdomains of NBD1 [3]. Similarly to what demonstrated for RI [13], it is plausible to posit that the dynamic flexibility of the RE may also result in exposure of hydrophobic surfaces thus contributing to the dynamic instability of NBD1 and thus contribute to the low folding efficiency of F508del-CFTR. The RE is a ~30-residue segment at NBD1 C-terminus, so-called because it goes beyond canonical ABC NBDs [15]. Although this region was absent from the solved CFTR-NBD1 crystal structure [19], it was described as a helix packing against NBD1 at the NBD1:NBD2 interface and NMR data showed it has significant conformational flexibility [3,21]. Notwithstanding, there is some controversy regarding the RE boundaries. The RE (previously called H9c) was defined as [655]Ala-Ser[670] [15,19], [639]Asp-Ser[670 20], or [654]Ser-Gly[673] [3]. According to the crystal structure (PDB ID 2PZE), NBD1 extends only to [646]Gly, where RD begins, so RE could be proposed to start at [647]Cys [3,18], thus being considered as part of RD, the latter absent in other ABC transporters [1]. As for the RI (previously called S1-S2 loop) its limits are also variable, being firstly described as a ~35-residue segment ([405]Phe-[436]Leu) consisting of α-helices H1b and H1c [15] (Figure S1). Later, however, these limits were proposed to be [404]Gly-Leu[435] [13,20] or [405]Phe-Leu[436] [18].

3.1. Impact of RE and RI on CFTR Processing and Function

Our data shown here on CFTR variants depleted of different versions of the RE dynamic region demonstrate that unlike RI deletion, removal of short RE—ΔRE_S (Δ^{654}Ser-Gly^{673}) did not per se rescue F508del-CFTR processing. Nevertheless, ΔRE_S dramatically stabilized the immature form of F508del-CFTR (see Figure 6). In fact, our pulse-chase experiments show that the immature form of ΔRE_S-F508del-CFTR exhibits a turnover rate which is ~2-fold lower than that of wt-CFTR. Of note, when Aleksandrov et al. removed the dynamic region RI from F508del-CFTR they found a dramatic increase in the channel thermostability, even augmented for higher temperatures [13]. Interestingly however, while RE_S removal from ΔRI_L-F508del-CFTR did not significantly affect processing (71% vs. 78% for ΔRE_S-ΔRI_L-F508del-CFTR and ΔRI_L-F508del-CFTR, respectively, in Table 1, it further reduced its function from 92 to 78% (ΔRI_L-F508del-CFTR vs. ΔRE_S-ΔRI_L-F508del-CFTR, respectively, Figure 5C).

The latter findings on function of ΔRE_S-wt-CFTR and ΔRE_S-ΔRI_L-F508del-CFTR were somewhat surprising since the RE was previously described to impede putative NBD1:NBD2 dimerization that is required for channel gating [20]. Accordingly, an increase in CFTR activity would be expected upon ΔRE_S removal, which is actually the opposite of what we observe for wt-CFTR. Besides the fact that those authors studied a different RE version (^{639}Asp-Ser^{670}), this discrepancy could derive from structural constraints and lack of flexibility of the 'single polypeptide' protein that we used, vs. their 'split' CFTR channels (2 'halves' of 633 and 668 amino acid residues) for which channel function was indistinguishable from wt-CFTR [20]. Moreover, since RD phosphorylation is required for channel gating [24,25], the reduced function of ΔRE_S-wt-CFTR may also result from absence of ^{660}Ser and ^{670}Ser [17,21]. Although the RD contains more than ten PKA phospho-sites and no individual one is essential, phosphorylation of increasing numbers of sites enables progressively greater channel activity [26].

Removal of a longer RE version (ΔRE_L: Δ^{647}Cys-Ser^{678}) was without effect on F508del-CFTR processing and significantly reduced that of wt-CFTR to 88%. Such differential impact on wt- and F508del-CFTR is consistent with the conformational heterogeneity between these two proteins lacking both RI and RE [27].

Removal of RI short version, RI_S (Δ^{412}Ala-Leu^{428}) significantly reduced wt-CFTR processing to less than half of its normal levels, while not rescuing F508del-CFTR processing, thus, being essential for CFTR proper folding. This is in contrast to removal of long RI (ΔRI_L) which led to 78% processing F508del-CFTR, as reported [13] but without impact on wt-CFTR. These data indicate that those 8/7 amino acid residues at the N-term/ C-term of RI_S (and absent in RI_L) impair the folding efficiency and processing of both wt- and F508del-CFTR. Despite the difficulty in speculating how those amino acid residues can specifically impact of F508del-CFTR folding and processing, the fact that structurally RI_S is strictly the region described as destructured in the crystal structure [19] and RI_L includes some flanking residues restricting its mobility may explain the observed difference.

Indeed, the recently published cryo-EM structures of human CFTR [22,28,29] indicate that RI_S is structurally disordered. Consistently, the RI loop in NBD1 is described in the zebrafish CFTR cryo-structure [30,31] to contribute to the amorphous density that is observed between NBD1 and the elbow helix of TMD2 [32]. Our data for the RI are consistent with these findings.

From a structural point of view, models of the full-length CFTR protein suggest that the two regions (RI and RE) behave differently. The distances between the extremities of the deleted parts are indeed different ($\Delta RI_S = 18\text{Å}$, $\Delta RI_L = 10\text{Å}$ versus $\Delta RE_S = 27\text{Å}$, $\Delta RE_L = 30\text{Å}$). The longer distances observed for ΔRE imply a substantial reorganization of the C-terminal parts of NBD1 and NBD2, which precludes a clear understanding of what might happen upon these deletions. In contrast, the lower distances observed for ΔRI (near a possible minimum of 6Å) allows a better simulation of the possible structural behavior of these constructs. Indeed, preliminary molecular dynamics simulations of the ΔRI_L-F508del-CFTR suggested that in this case, part of the RD is reorganized and partially takes space left by the ΔRI deletion, thereby substituting it for tight contacts with NBD2. Meanwhile, following a probable allosteric effect [13,18,32], contact of NBD1-F508 with ICL4 residue L1077 (a likely essential contact for channel opening), is nearly completely restored by I507. Such a feature is not observed in the F508del-CFTR model [33].

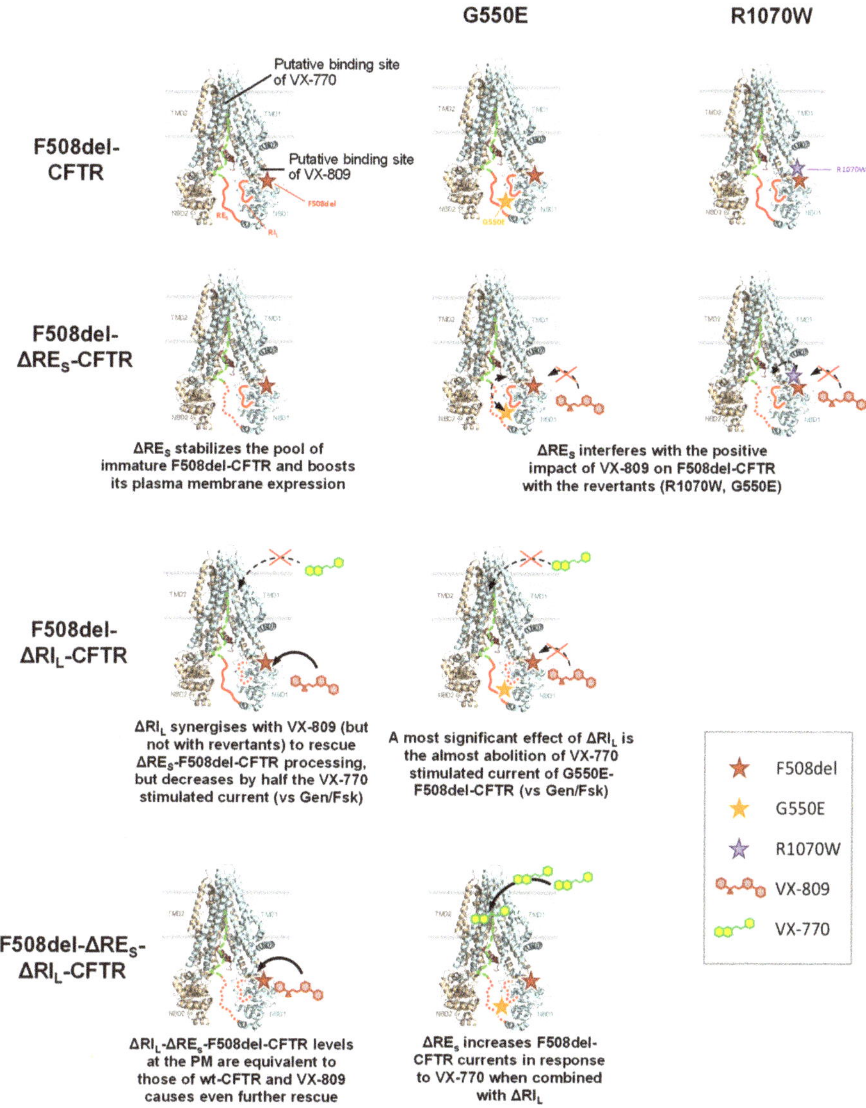

Figure 6. Summary of the most relevant results observed in the present study. CFTR structure used is the one by Liu F et al. [22] and putative bind sites of VX-809 and VX-770 shown are the NBD1:MSD2 (ICL4) interface and to the MSDs (although it is not defined the exact binding site), as described by Farinha et al. [5] and by Jih and Hwang [23], respectively. RI_L and RE_S are shown as red lines, the R region a dashed green line.

3.2. Effect of VX-809 on F508del-CFTR Variants Lacking RE and RI

As for rescue of CFTR variants lacking RE and RI by CFTR modulators, VX-809 restored the processing of both ΔRE_S-ΔRI_L- and ΔRI_L-F508del-CFTR variants equally well and to wt-CFTR levels (from 71-78% to 92-96%. These data suggest a strong synergistic effect between VX-809 and ΔRI_L to rescue ΔRE_S-F508del-CFTR processing and thus some possible interference of the regulatory insertion with VX-809 binding to F508del-CFTR (see Figure 6). Indeed, although previous studies [34] suggested putative binding of VX-809 to NBD1:MSD2 (ICL4) interface (see Figure 6) or to TMD1 [35], they also suggested possibility of further F508del-CFTR correction at distinct conformational sites, so data shown here suggest that VX-809 may also bind to the regulatory insertion.

Interestingly, the processing defect of ΔRI_S-wt-CFTR (but not of ΔRI_S-F508del-CFTR) was rescuable by VX-809 to 90%, indicating that the amino acid stretches of the RI_L that remain present in RI_S do not affect the rescue of ΔRI_S-wt-CFTR but preclude rescue of F508del-CFTR by VX-809.

We also tested the impact of removing helix 9 (H9) which precedes the RE (^{635}Gln-Gly^{646}), just after H8 (^{630}Phe-Leu^{634}), both helices proposed to interact with the NBD1:NBD2 heterodimer interface by folding onto the NBD1 β-subdomain [21,27] as well as to bind ICL4 near Phe508. When H9 is present, ΔRE_L-wt-CFTR processing appears to be favored by VX-809, suggesting some synergy of this small molecule with H9 helix to correct the conformational defect(s) caused by ΔRE_L on wt-CFTR. These data are in contrast to $\Delta H9$-F508del-CFTR which exhibits 0% processing with or without ΔRE_L and no rescue by VX-809. We can assume that H9 also contacts RE, at least in some states, such as when the RE adopts different conformations as previously suggested [19].

Most strikingly, and in contrast to its effect on processing, was the effect of ΔRI_L on the VX-770 stimulated currents which were decreased by almost half vs those stimulated by Gen/Fsk (92% to 58%). These data seem to indicate that the absence of the regulatory insertion could impair (and its presence favor) binding of VX-770 to CFTR (see Figure 6). Surprisingly, the removal of RE_S from ΔRI_L-F508del-CFTR could correct this defect and restore the maximal function by VX-770 (see Figure 6). Indeed, under VX-770, ΔRE_S-ΔRI_L-F508del-CFTR exhibited significantly higher activity (87%) than ΔRI_L-F508del-CFTR (58%). In contrast, removal of RE_S from wt-CFTR significantly reduced its functionin comparison with that of wt-CFTR (down to 70%) but had no effect on processing. Our study does not address the mechanism coupling the RI to F508del-CFTR pharmacological rescue, namely, why is the RI essential for rescue by VX-770 or why is it inhibitory for rescue by VX-809. Nevertheless, insight may be obtained from experimental structural data. In a recent cryo-EM structure of full-length human CFTR [29] VX-770 was found to bind in a transmembrane location coinciding with a hinge involved in gating. The authors suggested that populating this binding pocket may stabilize the intermolecular rotation opening the CFTR channel upon ATP binding at the NBD1-NBD2 interface. Given that the RI is located near the NBD interface and ATP binding site [29] we may speculate that RI deletion hampers NBD dimerization and, therefore, rescue by VX-770. Regarding VX-809, we may speculate that the RI interferes with drug binding.

3.3. Impact of F508del-Revertants on CFTR Variants Lacking RE and RI

Another goal of the present study was to assess how presence of the F508del-CFTR revertants G550E and R1070W [5,10,11] influence variants without RE and RI. Remarkably, our data show that the presence of either of these revertants did not affect ΔRE_S-F508del-CFTR processing, but both of them further increased processing (but not function) of ΔRI_L-F508del-CFTR to almost levels of wt-CFTR: 92–96% (G550E) and 71–76% (R1070W).

In contrast, removal of RI_S from either of these F508del-CFTR revertants completely abolished their processing, emphasizing how important the different residues between RI_S and RI_L are for F508del-CFTR conformers partially rescued by the revertants. We can speculate that RI_S removal from F508del-CFTR has an effect on folding equivalent to that of either G550E or R1070W.

Interestingly, regarding wt-CFTR processing, G550E (at the NBD1:NBD2 dimer interface) partially recovered the negative effect caused by RI_S removal, thus further suggesting that this revertant and the destructured region of RI may be allosterically coupled, since they do not plausibly interact (Figure S5). On the contrary, R1070W (at the NBD1:ICL4 interface), negatively affected processing of wt-CFTR (to 69%) and of ΔRI_S-wt-CFTR (from 44% to 11%), while not affecting ΔRI_L-wt-CFTR. R1070W rescues F508del-CFTR because ^{1070}Trp fills the gap created by deletion of residue ^{508}Phe [36]. It is not surprising that it perturbs CFTR folding due to clashing of ^{508}Phe and ^{1070}Trp residues. It is nevertheless, curious that R1070W affect more the processing ΔRI_S-wt-CFTR than wt-CFTR, to levels of F508del-CFTR, suggesting that both changes affect the same region of the molecule. Rescue of both R1070W-wt-CFTR and R1070W-ΔRI$_S$-wt-CFTR by VX-809 further supports this concept.

The most striking effect of ΔRI$_L$ however, was the almost complete abolition of VX-770-stimulated current of G550E-F508del-CFTR to levels even lower than those observed for ΔRI$_L$-F508del-CFTR (see above). It was suggested that VX-770 binds directly to CFTR to the MSDs (although it is not defined the exact binding site), in phosphorylation-dependent but ATP-independent manner and away from the canonical catalytic site [23,37]. Both RD and RI were suggested by Eckford and colleagues as putative binding regions for VX-770 [37]. Nevertheless, the pharmacological effect of VX-770 remains robust in the absence of the RD [23] and here we demonstrate that it does indeed require RI since VX-770 is unable to stimulate either ΔRI$_L$-F508del-CFTR or ΔRI$_L$-G550E- F508del-CFTR (see Figure 6). We can speculate that removal of RI$_L$ increases the chances of VX-770 blocking the pore. In contrast, this would not occur for genistein, which has been proposed to bind to the NBDs interface [38], thus probably explaining why it does not cause this effect.

4. Materials and Methods

4.1. CFTR Variants, Cells, and Culture Conditions

Several CFTR deletion variants were produced by site-directed mutagenesis corresponding to the removal of residues: ΔRI$_L$–^{404}Gly-Leu^{435}; ΔRI$_S$–^{412}Ala-Leu^{428}; ΔRE$_L$–^{647}Cys-Ser^{678}; ΔRE$_S$–^{654}Ser-Gly^{673}; ΔH9 –^{637}Gln-Gly^{646}; ΔRE$_L$-ΔH9 –^{637}Gln-Ser^{678}; ΔRI$_S$-ΔRE$_S$–both ^{412}Ala-Leu^{428} and ^{654}Ser-Gly^{673}; and ΔRI$_L$-ΔRE$_S$–^{404}Gly-Leu^{435} and ^{654}Ser-Gly^{673}. All the constructs were produced using full length wt-CFTR and F508del-CFTR.

BHK cells lines expressing ΔRI$_L$-, ΔRI$_L$-F508del-, ΔRI$_L$-G550E-, ΔRI$_L$-G550E-F508del, ΔRI$_L$-R1070W-, ΔRI$_L$-R1070W-F508del, ΔRI$_S$-, ΔRI$_S$-F508del, ΔRI$_S$ -G550E-, ΔRI$_S$-G550E-F508del, ΔRI$_S$-R1070W-, ΔRI$_S$-R1070W-F508del, ΔRE$_S$-, ΔRE$_S$-F508del, ΔRE$_S$-G550E, ΔRE$_S$-G550E-F508del, ΔRE$_S$-R1070W-, ΔRE$_S$-R1070W-F508del, ΔRE$_L$-, ΔRE$_L$-F508del, ΔRI$_S$-ΔRE-, ΔRI$_S$-ΔRE$_S$-F508del, ΔRI$_L$-ΔRE-, ΔRI$_L$-ΔRE$_S$-F508del, ΔH9-, ΔH9-F508del, ΔRE$_L$-ΔH9-, and ΔRE$_L$-ΔH9-F508del-CFTR were produced and cultured as previously described [11]. Cells were cultured in DMEM/F-12 medium containing 5% (*v/v*) Fetal bovine serum (FBS) and 500 μM of methotrexate. For some experiments, cells were incubated with 3 μM VX-809 or with the equivalent concentration of Dimethyl sulfoxide (DMSO, Control) for 48 h at 37 °C.

4.2. Western Blot

To study the effect of removal of regulatory extension (RE) and or regulatory insertion (RI) in combination with genetic revertants and VX-809, cells were incubated for 48 h at 37 °C with 3 μM VX-809. After incubation, cells were lysed, and extracts analyzed by Western blot (WB) using the anti-CFTR 596 antibody (Ab) or anti-calnexin Ab as a loading control. Score corresponds to the percentage of band C to total CFTR (bands B + C) as normalized to the same ratio in samples from wt-CFTR expressing cells. Blot images were acquired using BioRad ChemiDoc XRC+ imaging system and band intensities were measured using Image Lab analysis software.

4.3. Pulse-Chase and Immunoprecipitation

BHK cells lines stably expressing CFTR variants were starved for 30 min in methionine-free α-modified Eagle's medium or minimal essential medium and then pulsed for 30 min in the same medium supplemented with 100 μCi/mL [^{35}S] methionine. After chasing for 0, 0.5, 1, 1.5, 2, and 3 h in a-modified Eagle's medium with 8% (*v/v*) fetal bovine serum and 1mM non-radioactive methionine, cells were lysed in 1 mL of Radioimmunoprecipitation assay (RIPA) buffer [1% (*w/v*) deoxycholic acid, 1% (*v/v*) Triton X-100, 0.1% (*w/v*) Sodium dodecyl sulfate (SDS), 50 mM Tris, pH 7.4, and 150 mM NaCl]. The immunoprecipitation (IP) was carried out using the anti-CFTR 596 antibody in independent experiments and Protein G–agarose or Protein A–Sepharose beads. Immunoprecipitated proteins were eluted from the beads with sample buffer for 1h at room temperature and then electrophoretically separated on 7% (*w/v*) polyacrylamide gels. Gels were pre-fixed in methanol/acetic acid (30:10, *v/v*), washed in water and, for fluorography, soaked in 1M sodium salicylate for 60 min. After drying at 80 °C for 2 h, gels were exposed to X-ray films and further analyzed and quantified by densitometry.

4.4. Iodide Efflux

CFTR-mediated iodide effluxes were measured at room temperature using the cAMP agonist forskolin (Fsk 10 μM) and the CFTR potentiator genistein (Gen, 50 μM) or VX-770 (10 μM) and Gen (50 μM).

4.5. Biochemical Determination of the Plasma Membrane Levels of CFTR

To determine plasma membrane levels of CFTR protein, we performed cell surface biotinylation in BHK cells cultured on permeable growth supports or tissue culture plates using cell membrane impermeable EZ-Link™ Sulfo-NHS-SS-Biotin, followed by cell lysis in buffer containing 25 mM HEPES, pH 8.0, 1% (*v/v*) Triton, 10% glycerol (*v/v*), and Complete Protease Inhibitor Mixture, as described previously [39,40]. Biotinylated proteins were isolated by streptavidin-agarose beads, eluted into SDS-sample buffer, and separated by 7.5% (*w/v*) SDS-PAGE.

4.6. Multiple Sequence Alignment

Sequences for NBD domains of ABC transporters were obtained from Uniprot [41] (human and mouse CFTR) or the PDB [42] (1B0U, 1L2T, 1G29, 1G6H, and 1JJ7). Alignments were performed with Jalview [43] using the T-Coffee [44] algorithm with default parameters.

4.7. Data and Statistical Analyses

The data and statistical analyses used in this study comply with the recommendations on experimental design and data analysis in pharmacology [45]. Quantitative results are shown as mean ± SEM of *n* observations. To compare two sets of data, Student's *t*-test was used, and differences considered to be significant for *p*-values ≤ 0.05. In Western blotting (Figures 1–3), pulse chase (Figure 4 and Figure S2); cell surface biotinylation (Figure 4) and in iodide efflux studies (Table 1, Figure 5, Figures S3 and S4), n represents the number of experiments performed with distinct cell cultures on different days, being thus biological replicates.

4.8. Reagents

All reagents used here were of the highest purity grade available. Forskolin and genistein were from Sigma-Aldrich (St. Louis, MI, USA); VX-809 and VX-770 were acquired from Selleck Chemicals (Houston, TX, USA). CFTR was detected with the mouse anti-CFTR monoclonal 596Ab, which recognizes a region of NBD2 (1204–1211) from CFFT—Cystic Fibrosis Foundation Therapeutics CFF Therapeutics [46] and calnexin with rabbit polyclonal anti-calnexin Ab SPA-860, from Stressgen Biotechnologies Corporation (Victoria, BC, Canada). Other specific reagents included: X-ray films (Kodak, supplied by Sigma-Aldrich); Complete Protease Inhibitor Mixture from Roche Applied Science (Penzberg, Germany); EZ-Link™ Sulfo-NHS-SS-Biotin from Pierce Chemical Company (Rockford, IL, USA).

5. Conclusions

Overall, our data show that while the presence of the regulatory insertion (RI) seems to preclude full rescue of F508del-CFTR processing by VX-809, this region appears essential to rescue its function by VX-770, thus suggesting some contradictory role in rescue of F508del-CFTR by these two modulators (Figure 7). Nevertheless, this negative impact of removing RI on VX-770-stimulated currents on F508del-CFTR can be compensated by deletion of the regulatory extension which also leads to the stabilization of this mutant. We thus propose that, despite both these regions being conformationally dynamic, RI precludes F508del-CFTR processing while RE affects mostly its stability and channel opening.

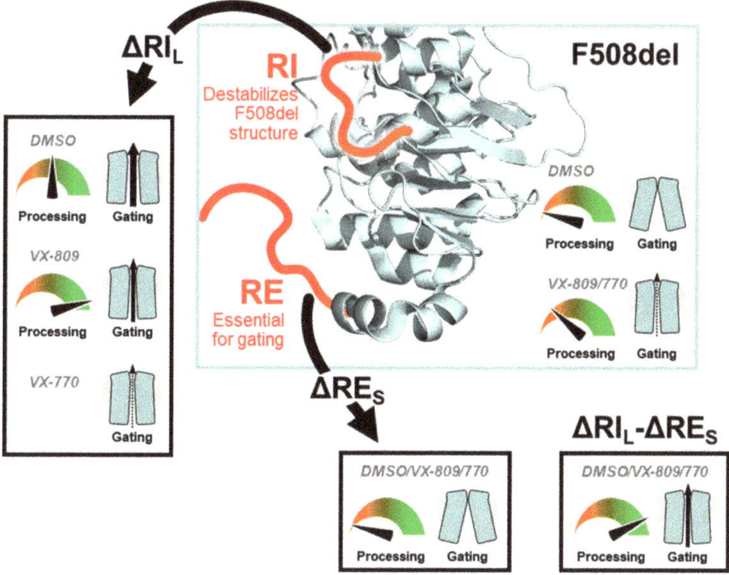

Figure 7. Model for the modulation of F508del processing and gating by RI_L and RE_S. Our data suggest that the RI destabilizes the F508del-CFTR structure. Therefore, deleting the RI (ΔRI_L) rescues F508del-CFTR processing and gating. However, the RI is essential for F508del-CFTR rescuing by VX-770. The RE is essential for gating and removal of its shorter version (ΔRE_S) generates an F508del-CFTR variant which cannot be rescued pharmacologically. Simultaneous removal of both regions (ΔRI_L-ΔRE_S) rescues F508del-CFTR processing and gating. The CFTR structure from Liu F et al. [22] depicts the overall NBD1 structure and the location of the RI and RE.

Supplementary Materials: Supplementary materials can be found at http://www.mdpi.com/1422-0067/21/12/4524/s1, **Figure S1.** ABC transporters alignment. Sequence alignment of NBD sequences from ABC transporters. Sequences correspond to NBD domains with experimentally determined 3D structures: human CFTR (hNBD1) and Tap1, mouse CFTR (mNBD1), HisP from *S. typhimurium*, MJ0796 and MJ1267 from *M. jannaschii* and Malk from *T. litoralis*. Only RI and RE regions are shown. Numbering is the one of human CFTR. The localization of the RI, RE and H9 are shown as arrows and shades. Blue and pink regions respectively highlight structured and unstructured stretches in human NBD1 crystal structures [19], **Figure S2.** Turnover and processing of wt-CFTR, alone or in *cis* with ΔRE_S and ΔRI_S. (A) BHK cells expressing wt, ΔRE_S and ΔRI_S were labelled with ^{35}S-methionine for 30 min and then chased for the indicated times (0, 0.5, 1, 2, 3h) before lysis. IP was performed with the anti-CFTR 596 antibody. After electrophoresis and fluorography, images were analysed by densitometry. (B) Turnover of immature (band B) CFTR for different CFTR variants is shown as the percentage of immature protein at a given time point of chase (P) relative to the amount at t = 0 (P_0). (C) Efficiency of processing of band B into band C is shown as the percentage of band C at a given time of chase relative to the amount of band B at t = 0. Data are mean ± SEM at each point (n = 3), **Figure S3.** Functional characterization of the ΔRE_S and ΔRI_L variants of F508del-CFTR with and without VX-809 or VX-770 treatments. (A-L) Iodide efflux from BHK cells stably expressing ΔRE_S, ΔRI_S- and ΔRI_L-F508del-CFTR variants. Cells were incubated either with 3μM VX-809 (E-H) or equivalent concentration of DMSO (control) for 48h at 37°C (A-D; I-L) and for the assay stimulated with Fsk/Gen (A-H) or Fsk/VX-770 (I-L), as indicated above the bar and for the period indicated by the bar, **Figure S4.** Functional characterization of the ΔRE_S and ΔRI_L variants of wt-CFTR. (A-C) Iodide efflux from BHK cells stably expressing ΔRE_S- or ΔRI-wt-CFTR alone (A) or jointly with revertants R1070W and G550E (B, C). Cells were stimulated during the assay with Fsk/Gen (A, B) or with Fsk/VX-770 (C), as indicated above the bar and for the period indicated by the bar. (D) Graph summarizing data from the iodide efflux peak magnitude generated by different BHK cells stably expressing the various CFTR variants. Data are shown as a percentage of wt-CFTR activity and as mean±SEM. (n) indicates number of independent experiments. "#" indicates significantly different from the wt-CFTR stimulated with Fsk/Gen ($p < 0.05$). (E) Quantification of the iodide efflux data, showing peak response (time and value) and percentage of wt-CFTR activity, **Figure S5.** Relative localization of the RI and G550 in CFTR. The figure shows the structure of full length CFTR bound to ATP and VX-770 [34] viewed from the intracellular side, highlighting the NBDs and rendering secondary structure elements and protein surface. The RI is shown in shades of red: light red highlights the stretch which is resolved in the cryo-EM structure [34] (^{404}Gly-Phe^{409}) and dark red highlights the ^{410}Glu-Leu^{435} stretch, which is not resolved. This stretch is shown as modelled by Serohijos et al. [4], after alignment to NBD1. Other elements are colored as in Figure 1A. (A) Overall structure. (B) Cut through the NBDs. Interaction between the RI and the G550 site is unlikely given their separation and the buried localization of G550.

Author Contributions: Conceptualization, I.U., A.C.d.P., and M.D.A.; data curation, I.U., A.C.d.P., H.M.B., and M.D.A.; formal analysis, I.U., H.M.B., and C.M.F.; funding acquisition, M.D.A.; Investigation, I.U., A.C.d.P., S.A., M.J.L., V.F., H.M.B., and C.M.F.; methodology, I.U., A.C.d.P., H.M.B., and M.D.A.; project administration, I.U.; Supervision, I.U. and M.D.A.; validation, I.U.; writing original draft, I.U., H.M.B., and M.D.A.; review and editing I.U., H.M.B. and M.D.A. All authors have read and agreed to the published version of the manuscript.

Funding: Work supported by grants UIDB/04046/2020 and UIDP/04046/2020 centre grants (to BioISI) and research grant (to M.D.A.): "iDrugCF" (FCT/02/SAICT/2017/28800), both from FCT/MCTES Portugal '. I.U., A.C.P. and V.F. were recipients of SFRH/BD/69180/2010, SFRH/BD/17475/2004, SFRH/BD/87478/2012 PhD fellowships and H.M.B. of SFRH/BPD/93017/2013 post-doctoral fellowship (FCT, Portugal), respectively.

Acknowledgments: Authors acknowledge I Callebaut, B Hoffmann, and J-P Mornon for the revision of the manuscript and for the helpful comments.

Conflicts of Interest: The authors declare no conflict of interest. The funders had no role in the design of the study; in the collection, analyses, or interpretation of data; in the writing of the manuscript, or in the decision to publish the results.

References

1. Riordan, J.R. CFTR Function and Prospects for Therapy. *Annu. Rev. Biochem.* **2008**, *77*, 701–726. [CrossRef] [PubMed]

2. Sheppard, D.N.; Welsh, M.J. Structure and Function of the CFTR Chloride Channel. *Physiol. Rev.* **1999**, *79*, S23–S45. [CrossRef] [PubMed]

3. Kanelis, V.; Hudson, R.P.; Thibodeau, P.H.; Thomas, P.J.; Forman-Kay, J.D. NMR Evidence for Differential Phosphorylation-Dependent Interactions in WT and DeltaF508 CFTR. *EMBO J.* **2010**, *29*, 263–277. [CrossRef] [PubMed]

4. Serohijos, A.W.R.; Hegedus, T.; Aleksandrov, A.A.; He, L.; Cui, L.; Dokholyan, N.V.; Riordan, J.R. Phenylalanine-508 Mediates a Cytoplasmic-Membrane Domain Contact in the CFTR 3D Structure Crucial to Assembly and Channel Function. *Proc. Natl. Acad. Sci. USA* **2008**, *105*, 3256–3261. [CrossRef]

5. Farinha, C.M.; King-Underwood, J.; Sousa, M.; Correia, A.R.; Henriques, B.J.; Roxo-Rosa, M.; Da Paula, A.C.; Williams, J.; Hirst, S.; Gomes, C.M.; et al. Revertants, Low Temperature, and Correctors Reveal the Mechanism of F508del-CFTR Rescue by VX-809 and Suggest Multiple Agents for Full Correction. *Chem. Biol.* **2013**, *20*, 943–955. [CrossRef]

6. He, L.; Aleksandrov, A.A.; Serohijos, A.W.R.; Hegedus, T.; Aleksandrov, L.A.; Cui, L.; Dokholyan, N.V.; Riordan, J.R. Multiple Membrane-Cytoplasmic Domain Contacts in the Cystic Fibrosis Transmembrane Conductance Regulator (CFTR) Mediate Regulation of Channel Gating. *J. Biol. Chem.* **2008**, *283*, 26383–26390. [CrossRef]

7. Denning, G.M.; Anderson, M.P.; Amara, J.F.; Marshall, J.; Smith, A.E.; Welsh, M.J. Processing of Mutant Cystic Fibrosis Transmembrane Conductance Regulator Is Temperature-Sensitive. *Nature* **1992**, *358*, 761–764. [CrossRef]

8. Teem, J.L.; Berger, H.A.; Ostedgaard, L.S.; Rich, D.P.; Tsui, L.C.; Welsh, M.J. Identification of Revertants for the Cystic Fibrosis Delta F508 Mutation Using STE6-CFTR Chimeras in Yeast. *Cell* **1993**, *73*, 335–346. [CrossRef]

9. Teem, J.L.; Carson, M.R.; Welsh, M.J. Mutation of R555 in CFTR-Delta F508 Enhances Function and Partially Corrects Defective Processing. *Recept. Channels* **1996**, *4*, 63–72.

10. DeCarvalho, A.C.V.; Gansheroff, L.J.; Teem, J.L. Mutations in the Nucleotide Binding Domain 1 Signature Motif Region Rescue Processing and Functional Defects of Cystic Fibrosis Transmembrane Conductance Regulator Delta F508. *J. Biol. Chem.* **2002**, *277*, 35896–35905. [CrossRef]

11. Roxo-Rosa, M.; Xu, Z.; Schmidt, A.; Neto, M.; Cai, Z.; Soares, C.M.; Sheppard, D.N.; Amaral, M.D. Revertant Mutants G550E and 4RK Rescue Cystic Fibrosis Mutants in the First Nucleotide-Binding Domain of CFTR by Different Mechanisms. *Proc. Natl. Acad. Sci. USA* **2006**, *103*, 17891–17896. [CrossRef] [PubMed]

12. Loo, T.W.; Bartlett, M.C.; Clarke, D.M. The V510D Suppressor Mutation Stabilizes DeltaF508-CFTR at the Cell Surface. *Biochemistry* **2010**, *49*, 6352–6357. [CrossRef]

13. Aleksandrov, A.A.; Kota, P.; Aleksandrov, L.A.; He, L.; Jensen, T.; Cui, L.; Gentzsch, M.; Dokholyan, N.V.; Riordan, J.R. Regulatory Insertion Removal Restores Maturation, Stability and Function of DeltaF508 CFTR. *J. Mol. Biol.* **2010**, *401*, 194–210. [CrossRef] [PubMed]

14. Van Goor, F.; Hadida, S.; Grootenhuis, P.D.J.; Burton, B.; Stack, J.H.; Straley, K.S.; Decker, C.J.; Miller, M.; McCartney, J.; Olson, E.R.; et al. Correction of the F508del-CFTR Protein Processing Defect in Vitro by the Investigational Drug VX-809. *Proc. Natl. Acad. Sci. USA* **2011**, *108*, 18843–18848. [CrossRef] [PubMed]

15. Lewis, H.A.; Buchanan, S.G.; Burley, S.K.; Conners, K.; Dickey, M.; Dorwart, M.; Fowler, R.; Gao, X.; Guggino, W.B.; Hendrickson, W.A.; et al. Structure of Nucleotide-Binding Domain 1 of the Cystic Fibrosis Transmembrane Conductance Regulator. *EMBO J.* **2004**, *23*, 282–293. [CrossRef] [PubMed]

16. Dahan, D.; Evagelidis, A.; Hanrahan, J.W.; Hinkson, D.A.; Jia, Y.; Luo, J.; Zhu, T. Regulation of the CFTR Channel by Phosphorylation. *Pflüg. Arch. Eur. J. Physiol.* **2001**, *443*, S92–S96. [CrossRef]

17. Pasyk, S.; Molinski, S.; Ahmadi, S.; Ramjeesingh, M.; Huan, L.-J.; Chin, S.; Du, K.; Yeger, H.; Taylor, P.; Moran, M.F.; et al. The Major Cystic Fibrosis Causing Mutation Exhibits Defective Propensity for Phosphorylation. *Proteomics* **2015**, *15*, 447–461. [CrossRef] [PubMed]

18. Dawson, J.E.; Farber, P.J.; Forman-Kay, J.D. Allosteric Coupling between the Intracellular Coupling Helix 4 and Regulatory Sites of the First Nucleotide-Binding Domain of CFTR. *PLoS ONE* **2013**, *8*, e74347. [CrossRef]

19. Lewis, H.A.; Zhao, X.; Wang, C.; Sauder, J.M.; Rooney, I.; Noland, B.W.; Lorimer, D.; Kearins, M.C.; Conners, K.; Condon, B.; et al. Impact of the DeltaF508 Mutation in First Nucleotide-Binding Domain of Human Cystic Fibrosis Transmembrane Conductance Regulator on Domain Folding and Structure. *J. Biol. Chem.* **2005**, *280*, 1346–1353. [CrossRef]

20. Csanády, L.; Chan, K.W.; Nairn, A.C.; Gadsby, D.C. Functional Roles of Nonconserved Structural Segments in CFTR's NH2-Terminal Nucleotide Binding Domain. *J. Gen. Physiol.* **2005**, *125*, 43–55. [CrossRef]

21. Bozoky, Z.; Krzeminski, M.; Muhandiram, R.; Birtley, J.R.; Al-Zahrani, A.; Thomas, P.J.; Frizzell, R.A.; Ford, R.C.; Forman-Kay, J.D. Regulatory R Region of the CFTR Chloride Channel Is a Dynamic Integrator of Phospho-Dependent Intra- and Intermolecular Interactions. *Proc. Natl. Acad. Sci. USA* **2013**, *110*, E4427–E4436. [CrossRef] [PubMed]

22. Liu, F.; Zhang, Z.; Csanády, L.; Gadsby, D.C.; Chen, J. Molecular Structure of the Human CFTR Ion Channel. *Cell* **2017**, *169*, 85–95.e8. [CrossRef] [PubMed]

23. Jih, K.-Y.; Hwang, T.-C. Vx-770 Potentiates CFTR Function by Promoting Decoupling between the Gating Cycle and ATP Hydrolysis Cycle. *Proc. Natl. Acad. Sci. USA* **2013**, *110*, 4404–4409. [CrossRef] [PubMed]

24. Chappe, V.; Irvine, T.; Liao, J.; Evagelidis, A.; Hanrahan, J.W. Phosphorylation of CFTR by PKA Promotes Binding of the Regulatory Domain. *EMBO J.* **2005**, *24*, 2730–2740. [CrossRef]

25. Mense, M.; Vergani, P.; White, D.M.; Altberg, G.; Nairn, A.C.; Gadsby, D.C. In Vivo Phosphorylation of CFTR Promotes Formation of a Nucleotide-Binding Domain Heterodimer. *EMBO J.* **2006**, *25*, 4728–4739. [CrossRef]

26. Hegedus, T.; Serohijos, A.W.R.; Dokholyan, N.V.; He, L.; Riordan, J.R. Computational Studies Reveal Phosphorylation-Dependent Changes in the Unstructured R Domain of CFTR. *J. Mol. Biol.* **2008**, *378*, 1052–1063. [CrossRef]

27. Hudson, R.P.; Chong, P.A.; Protasevich, I.I.; Vernon, R.; Noy, E.; Bihler, H.; An, J.L.; Kalid, O.; Sela-Culang, I.; Mense, M.; et al. Conformational Changes Relevant to Channel Activity and Folding within the First Nucleotide Binding Domain of the Cystic Fibrosis Transmembrane Conductance Regulator. *J. Biol. Chem.* **2012**, *287*, 28480–28494. [CrossRef]

28. Zhang, Z.; Liu, F.; Chen, J. Molecular Structure of the ATP-Bound, Phosphorylated Human CFTR. *Proc. Natl. Acad. Sci. USA* **2018**, *115*, 12757–12762. [CrossRef]

29. Liu, F.; Zhang, Z.; Levit, A.; Levring, J.; Touhara, K.K.; Shoichet, B.K.; Chen, J. Structural Identification of a Hotspot on CFTR for Potentiation. *Science* **2019**, *364*, 1184–1188. [CrossRef]

30. Zhang, Z.; Chen, J. Atomic Structure of the Cystic Fibrosis Transmembrane Conductance Regulator. *Cell* **2016**, *167*, 1586–1597.e9. [CrossRef]

31. Zhang, Z.; Liu, F.; Chen, J. Conformational Changes of CFTR upon Phosphorylation and ATP Binding. *Cell* **2017**, *170*, 483–491.e8. [CrossRef] [PubMed]

32. Aleksandrov, A.A.; Kota, P.; Cui, L.; Jensen, T.; Alekseev, A.E.; Reyes, S.; He, L.; Gentzsch, M.; Aleksandrov, L.A.; Dokholyan, N.V.; et al. Allosteric Modulation Balances Thermodynamic Stability and Restores Function of ΔF508 CFTR. *J. Mol. Biol.* **2012**, *419*, 41–60. [CrossRef]

33. Hoffmann, B.; Callebaut, I.; Mornon, J.-P.; (Sorbonne University, Paris, France). Personal communication, 2016.

34. Hudson, R.P.; Dawson, J.E.; Chong, P.A.; Yang, Z.; Millen, L.; Thomas, P.J.; Brouillette, C.G.; Forman-Kay, J.D. Direct Binding of the Corrector VX-809 to Human CFTR NBD1: Evidence of an Allosteric Coupling between the Binding Site and the NBD1:CL4 Interface. *Mol. Pharmacol.* **2017**, *92*, 124–135. [CrossRef] [PubMed]

35. Ren, H.Y.; Grove, D.E.; De La Rosa, O.; Houck, S.A.; Sopha, P.; Van Goor, F.; Hoffman, B.J.; Cyr, D.M. VX-809 Corrects Folding Defects in Cystic Fibrosis Transmembrane Conductance Regulator Protein through Action on Membrane-Spanning Domain 1. *Mol. Biol. Cell* **2013**, *24*, 3016–3024. [CrossRef] [PubMed]

36. Thibodeau, P.H.; Richardson, J.M.; Wang, W.; Millen, L.; Watson, J.; Mendoza, J.L.; Du, K.; Fischman, S.; Senderowitz, H.; Lukacs, G.L.; et al. The Cystic Fibrosis-Causing Mutation DeltaF508 Affects Multiple Steps in Cystic Fibrosis Transmembrane Conductance Regulator Biogenesis. *J. Biol. Chem.* **2010**, *285*, 35825–35835. [CrossRef]

37. Eckford, P.D.W.; Li, C.; Ramjeesingh, M.; Bear, C.E. Cystic Fibrosis Transmembrane Conductance Regulator (CFTR) Potentiator VX-770 (Ivacaftor) Opens the Defective Channel Gate of Mutant CFTR in a Phosphorylation-Dependent but ATP-Independent Manner. *J. Biol. Chem.* **2012**, *287*, 36639–36649. [CrossRef]

38. Wang, F.; Zeltwanger, S.; Yang, I.C.-H.; Nairn, A.C.; Hwang, T.-C. Actions of Genistein on Cystic Fibrosis Transmembrane Conductance Regulator Channel Gating. *J. Gen. Physiol.* **1998**, *111*, 477–490. [CrossRef]

39. Moyer, B.D.; Loffing, J.; Schwiebert, E.M.; Loffing-Cueni, D.; Halpin, P.A.; Karlson, K.H.; Ismailov, I.I.; Guggino, W.B.; Langford, G.M.; Stanton, B.A. Membrane Trafficking of the Cystic Fibrosis Gene Product, Cystic Fibrosis Transmembrane Conductance Regulator, Tagged with Green Fluorescent Protein in Madin-Darby Canine Kidney Cells. *J. Biol. Chem.* **1998**, *273*, 21759–21768. [CrossRef]

40. Swiatecka-Urban, A.; Duhaime, M.; Coutermarsh, B.; Karlson, K.H.; Collawn, J.; Milewski, M.; Cutting, G.R.; Guggino, W.B.; Langford, G.; Stanton, B.A. PDZ Domain Interaction Controls the Endocytic Recycling of the Cystic Fibrosis Transmembrane Conductance Regulator. *J. Biol. Chem.* **2002**, *277*, 40099–40105. [CrossRef]

41. UniProt Consortium. UniProt: A Hub for Protein Information. *Nucleic Acids Res.* **2015**, *43*, D204–D212. [CrossRef]

42. Berman, H.M.; Westbrook, J.; Feng, Z.; Gilliland, G.; Bhat, T.N.; Weissig, H.; Shindyalov, I.N.; Bourne, P.E. The Protein Data Bank. *Nucleic Acids Res.* **2000**, *28*, 235–242. [CrossRef] [PubMed]

43. Waterhouse, A.M.; Procter, J.B.; Martin, D.M.A.; Clamp, M.; Barton, G.J. Jalview Version 2–a Multiple Sequence Alignment Editor and Analysis Workbench. *Bioinforma. Oxf. Engl.* **2009**, *25*, 1189–1191. [CrossRef] [PubMed]

44. Notredame, C.; Higgins, D.G.; Heringa, J. T-Coffee: A Novel Method for Fast and Accurate Multiple Sequence Alignment. *J. Mol. Biol.* **2000**, *302*, 205–217. [CrossRef]

45. Curtis, M.J.; Bond, R.A.; Spina, D.; Ahluwalia, A.; Alexander, S.P.A.; Giembycz, M.A.; Gilchrist, A.; Hoyer, D.; Insel, P.A.; Izzo, A.A.; et al. Experimental Design and Analysis and Their Reporting: New Guidance for Publication in BJP. *Br. J. Pharmacol.* **2015**, *172*, 3461–3471. [CrossRef] [PubMed]

46. Cui, L.; Aleksandrov, L.; Chang, X.-B.; Hou, Y.-X.; He, L.; Hegedus, T.; Gentzsch, M.; Aleksandrov, A.; Balch, W.E.; Riordan, J.R. Domain Interdependence in the Biosynthetic Assembly of CFTR. *J. Mol. Biol.* **2007**, *365*, 981–994. [CrossRef]

International Journal of
Molecular Sciences

Article

Roles of ABCC1 and ABCC4 in Proliferation and Migration of Breast Cancer Cell Lines

Floren G. Low, Kiran Shabir, James E. Brown, Roslyn M. Bill and Alice J. Rothnie *

College of Health & Life Sciences, Aston University, Aston Triangle, Birmingham B4 7ET, UK;
low_guy@hotmail.com (F.G.L.); shabirk@aston.ac.uk (K.S.); j.e.p.brown@aston.ac.uk (J.E.B.);
R.M.Bill@aston.ac.uk (R.M.B.)
* Correspondence: a.rothnie@aston.ac.uk

Received: 18 September 2020; Accepted: 7 October 2020; Published: 16 October 2020

Abstract: ABCC1 and ABCC4 utilize energy from ATP hydrolysis to transport many different molecules, including drugs, out of the cell and, as such, have been implicated in causing drug resistance. However recently, because of their ability to transport signaling molecules and inflammatory mediators, it has been proposed that ABCC1 and ABCC4 may play a role in the hallmarks of cancer development and progression, independent of their drug efflux capabilities. Breast cancer is the most common cancer affecting women. In this study, the aim was to investigate whether ABCC1 or ABCC4 play a role in the proliferation or migration of breast cancer cell lines MCF-7 (luminal-type, receptor-positive) and MDA-MB-231 (basal-type, triple-negative). The effects of small molecule inhibitors or siRNA-mediated knockdown of ABCC1 or ABCCC4 were measured. Colony formation assays were used to assess the clonogenic capacity, MTT assays to measure the proliferation, and scratch assays and Transwell assays to monitor the cellular migration. The results showed a role for ABCC1 in cellular proliferation, whilst ABCC4 appeared to be more important for cellular migration. ELISA studies implicated cAMP and/or sphingosine-1-phosphate efflux in the mechanism by which these transporters mediate their effects. However, this needs to be investigated further, as it is key to understand the mechanisms before they can be considered as targets for treatment.

Keywords: MRP1; MRP4; breast cancer; proliferation; migration; invasion; cAMP

---

1. Introduction

Breast cancer is the most common cancer in women, affecting more than two million women worldwide per year [1]. The development of new targeted therapeutics and the encouragement of women to carry out early screening programs have significantly improved survival rates in the Western world [2–5]. Breast cancers that express the estrogen receptor and/or the progesterone receptor can be treated with receptor-blocking hormone therapy or with aromatase inhibitors to decrease the levels of estrogen produced [6,7]. Cancers that have high expression of human epidermal growth factor receptor 2 (HER2) can be treated with monoclonal antibodies, which bind to the receptor and block it [7]. However, some breast cancers, termed triple-negative, do not express any of these receptors, and thus, the only treatment available is conventional chemotherapy. Triple-negative breast cancers make up around 10–15% of all breast cancer cases [8], are typically aggressive, and show high levels of metastases and mortality [9,10]. The development of metastasis and/or cancers becoming resistant to therapeutic agents is a growing problem. Women diagnosed at an early stage of breast cancer may have recurrent disease, and at least a quarter of all cases may develop a resistance to therapeutic treatments [2]. Moreover, with the increase in disease progression, the incidence of therapeutic resistance becomes more alarming.

One common cause of therapeutic resistance in breast cancer is the efflux of drugs by membrane proteins of the ATP-binding cassette (ABC) transporter family of proteins [11,12]. ABCC1 (multidrug resistance protein 1/MRP1) and ABCC4 (multidrug resistance protein 4/MRP4) are two members of the C subfamily of ABC transporters that are capable of effluxing several different chemotherapeutic drugs out of cancer cells [13–15]. Previous studies have shown that ABCC1 expression is a negative prognostic marker, associated with a decreased survival rate in breast cancer patients [16–18] and an increased risk of relapse [19]. Following chemotherapeutic treatment, the expression of ABCC1 in breast cancer tumors was found to increase [20], and its expression level was shown to be highest in the most aggressive subtypes of breast cancers [18]. ABCC4 expression is similarly upregulated in chemotherapy-treated breast tumors compared to noncancerous tissue [21], and ABCC4 polymorphisms are linked with the response to aromatase inhibitors for estrogen-receptor-positive breast cancer [22].

However, it has been proposed that ABCC1 and ABCC4 may play a role in the hallmarks of cancer development and progression, independent of their drug efflux capabilities [23]. This is because ABCC1 and ABCC4 are capable of transporting numerous physiological substrates and have roles in metabolism and inflammation [13,24,25]. For example, ABCC1 can transport glutathione and inflammatory mediators such as leukotrienes and prostaglandins, as well as the bioactive lipids sphingosine-1-phosphate (S1P) and lysophosphatidyl inositol (LPI), which are implicated in cell proliferation, migration, and invasion [26–31]. ABCC4 can efflux the cyclic nucleotides cAMP and cGMP involved in cellular signaling, as well as leukotrienes, prostaglandins, and thromboxanes and S1P [32–38]. Studies on neuroblastoma (a rare childhood cancer) have confirmed that both ABCC1 and ABCC4 play an important physiological role in its development, independent of their role in multidrug resistance, affecting cellular proliferation, migration, and differentiation [39]. ABCC4 has also been implicated in cancer cell proliferation in leukemia [40,41], gastric cancer [42], lung cancer [43], renal cancer [44], ovarian cancer [45], and pancreatic cancer [46,47]. However, less is known about whether ABCC1 and ABCC4 have a role in breast cancer development and/or progression.

In this study, the role of ABCC1 and ABCC4 in breast cancer progression was investigated. The breast cancer cell lines MDA-MB-231 and MCF-7 were used. Both are considered to be aggressive, but MCF-7 is a luminal-type breast cancer with the presence of progesterone, estrogen, and human epidermal growth factor 2 (HER2) receptors, whereas MDA-MB-231 is a basal-type triple-negative cell line. The effect of small molecules inhibitors or siRNA knockdown of ABCC1 and ABCC4 on cellular proliferation, clonogenic capacity, cell migration, and invasion were investigated.

2. Results

2.1. Expression of ABCC1 and ABCC4 in Breast Cancer Cell Lines

In order to determine whether the breast cancer cell lines expressed ABCC1 or ABCC4, membrane extracts were prepared and assayed via Western blot, as shown in Figure 1. It can be seen that both MCF-7 and MDA-MB-231 cells express both ABCC1 and ABCC4, with the level of expression of the two transporters being higher in the MDA-MB-231 cells than in the MCF-7 cells. In addition, the cell lines were tested for the well-known multidrug resistance transporters P-glycoprotein/ABCB1 and ABCG2 (Figure S1). Only the MCF-7 cell line showed significant expression of these two transporters.

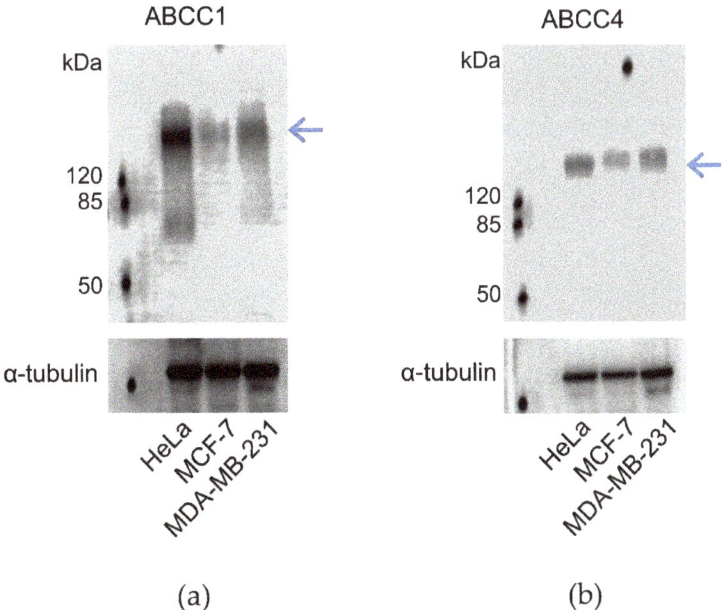

(a) (b)

Figure 1. ABCC1 and ABCC4 are both expressed in MCF-7 and MDA-MB-231 breast cancer cells. Membrane extracts (80 µg protein/well) from MCF-7 and MDA-MB-231 cells were assayed by Western blot for the expression of (**a**) ABCC1 or (**b**) ABCC4, alongside Hela cells as a positive control. ABCC1 was detected using the anti-MRP1 EPR4658 antibody (1:1000). ABCC4 was detected using the anti-ABCC4 M$_4$I-10 antibody (1:100). The blue arrow indicates the band corresponding to ABCC1 (**a**) or ABCC4 (**b**). Comparable sample loading was monitored afterwards using the anti-tubulin primary antibody. Uncropped images can be found in Figure S3.

2.2. The Effect of ABCC Small Molecule Inhibitors on Breast Cancer Cell Proliferation

Having established that both cell lines expressed the ABCC proteins, we investigated the effect of small molecule inhibitors of these proteins on the ability of the cells to proliferate. The small molecule inhibitors used in the study are detailed in Table 1.

Table 1. ABCC inhibitors used in this study.

Inhibitor	Inhibits ABCC1	Inhibits ABCC4	Reference
MK571	√	√	[48–50]
Reversan	√		[51]
Ceefourin 1		√	[52]
Ceefourin 2		√	[52]
Indomethacin		√	[33,53]

The first parameter investigated was the clonogenic capacity of the cells, i.e., the ability to reproduce from a single cell. Examples of the assay are shown in Figure 2a,b, where it can be seen that, following the treatment of MDA-MB-231 cells with MK571 or Reversan, at increasing concentrations, both the number of colonies formed over seven days and the size of those colonies are reduced. The average data shown in Figure 2c–e show that Reversan and MK571 affect the colony formation for both MDA-MB-231 and MCF-7 cells, but the effect is more pronounced for the MDA-MB-231 cells, perhaps correlating with the higher level of ABCC protein expression in these cells. In contrast, Ceefourin 1 and 2 and Indomethacin had no effect at all on the colony formation.

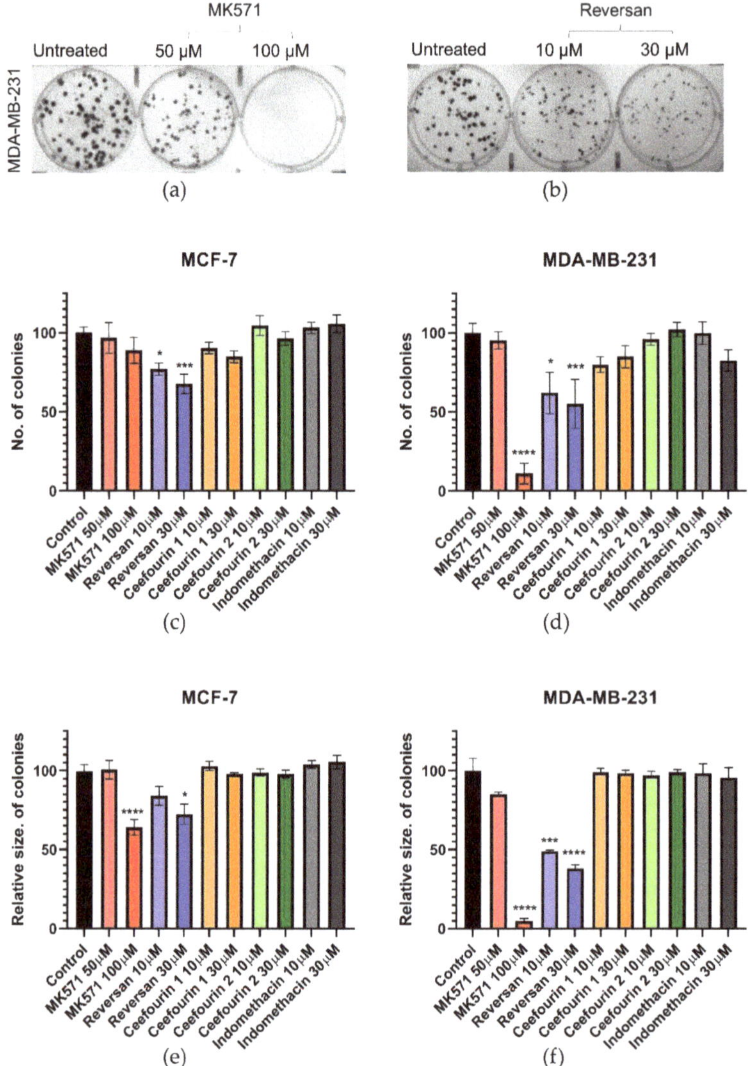

Figure 2. MK571 and Reversan affect the clonogenic capacity of breast cancer cells. 100 cells/well were seeded in 6-well plates and cultured at 37 °C for 24 h, after which cells were treated with the indicated concentrations of inhibitors. Control wells were untreated cells. After 7 days of culture, the colonies formed were fixed with 4% paraformaldehyde and stained with 0.1% crystal violet, the colonies counted and measured. Example results for MDA-MB-231 cells treated with (**a**) MK571 or (**b**) Reversan. Average results for the number (**c,d**) and size (**e,f**) of colonies for MCF-7 cells (**c,e**) and MDA-MB-231 (**d,f**). Data are mean ± sem, n ≥ 6. Data were analyzed using a one-way ANOVA with a Dunnett's post hoc test; * $p < 0.05$, *** $p < 0.001$, and **** $p < 0.0001$ significantly lower than the untreated sample.

The effect of these inhibitors on cellular proliferation was also investigated using an MTT assay. As can be seen in Figure 3, the presence of the inhibitors did not affect the proliferation of either cell line for the first 24 h. However, after this time, MK571 and Reversan had a significant impact on the proliferation of both MCF-7 and MDA-MB-231 cells, whereas Ceefourin 1 and 2 and Indomethacin did not. To confirm the results obtained with the MTT assay, and to make sure it was not due to an indirect effect on the enzyme required to reduce MTT, proliferation was also measured by a trypan blue exclusion and cell counting approach (Figure S2). Although the errors are larger using this approach, the key findings replicate those of the MTT assay.

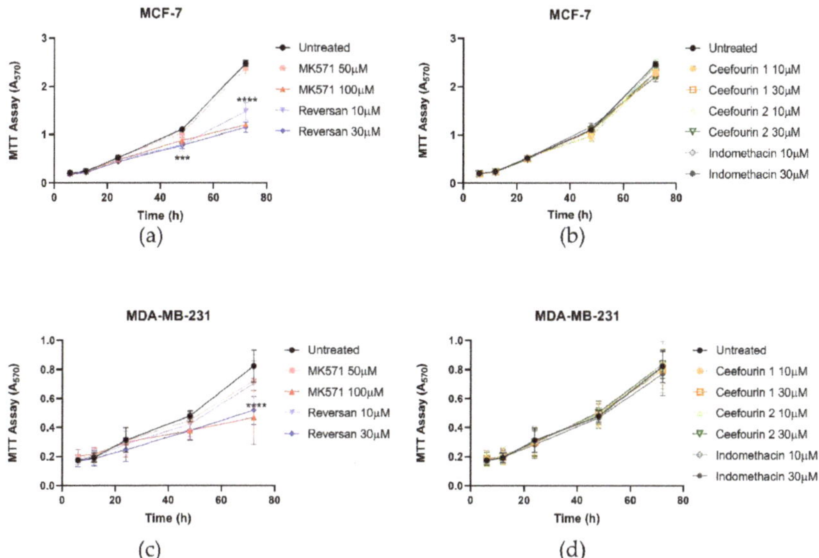

Figure 3. MK571 and Reversan affect the proliferation of breast cancer cells. Fifteen thousand MCF-7 cells (**a,b**) or 6000 MDA-MB-231 cells (**c,d**) were seeded in 24-well plates. After 4 h of culture, cells were treated with inhibitors, as detailed. Cell viability was assessed at 6, 12, 24, 48, and 72 h after treatment using an MTT assay and absorbance measured at 570 nm. Data are mean $\pm$ SD, n $\geq$ 6. Data were analyzed using a two-way ANOVA with a Dunnett's post hoc test. *** $p < 0.001$ and **** $p < 0.0001$ significantly lower than the untreated sample.

2.3. Effect of Inhibitors on Breast Cancer Cell Migration

In addition to rapid proliferation, enhanced migration is a hallmark of aggressive cancers. Therefore, the effects of ABCC inhibitors on breast cancer cell migration was measured using a scratch assay, as shown in Figure 4a. The MDA-MB-231 cells migrated faster than the MCF-7 cells (Figure 4b,c). Most of the inhibitor treatments had no significant effect on the migration. However, the treatment with MK571 did significantly decrease the migration of MDA-MB-231 cells (Figure 4c). This was not due to an effect on proliferation, since after 10–12 h when the migration was most affected, no effect on the proliferation was observed (Figure 3c).

MK571, which inhibits both ABCC1 and ABCC4, and Reversan, which inhibits ABCC1, were the only inhibitors to affect the proliferation of the breast cancer cells. MK571 was the only drug to affect the cell migration. Ceefourin 1 and 2 and Indomethacin, which inhibit ABCC4, had no effects. This might suggest that ABCC1 plays a role in the proliferation of breast cancer cells. Similarly, it might suggest that ABCC1, or maybe both ABCC1 and ABCC4 together, are involved in the migration. However, one of the challenges associated with using inhibitors of multidrug transporters is the lack of specificity. In addition to inhibiting ABCC1 and ABCC4, MK571 can inhibit other ABCC family members, and it is also a leukotriene antagonist, inhibiting the binding of leukotriene D_4 (LTD_4) to cysteinyl leukotriene receptor 1 [50]. Reversan is an inhibitor of ABCC1 but can also inhibit the related protein ABCB1 [54]. Indomethacin inhibits ABCC4 but was developed as an inhibitor of cyclooxygenase. Therefore, the effect of the genetic knockdown of ABCC1 and ABCC4 was investigated.

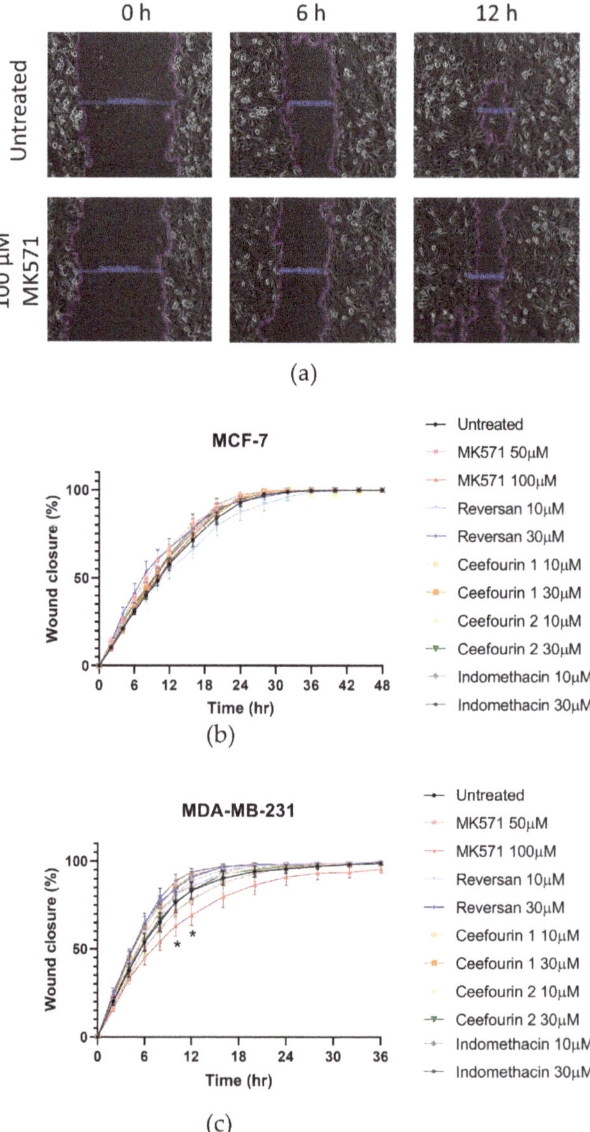

Figure 4. MK571 decreases the rate of migration by MDA-MB-231 cells. Cells were seeded in 24-well plates to reach 100% confluency the day of the assay. A scratch across the monolayer of the cells was carefully made, and the medium was replaced with fresh prewarmed culture medium. Cells were treated with the inhibitors as described above. Three image positions were selected from each well, and images were taken at 1-h intervals using the Cell-IQ. Representative images of MDA-MB-231 scratch assay (**a**). Pink lines represent the scratch edges as defined by the Cell IQ software, and the blue lines are the distance measurement between the edges. Average results for MCF-7 (**b**) and MDA-MB-231 (**c**) cell migration in the presence of inhibitors. Data are mean ± sem, n ≥ 6. Data were analyzed using a two-way ANOVA with a Dunnett's post hoc test. * $p < 0.05$ significantly lower than the untreated sample.

2.4. Knockdown of ABCC1 and ABCC4 Using siRNA

Knockdown of ABCC1 and ABCC4 was carried out using siRNA. Two different siRNA sequences for each protein were tested, alongside a negative siRNA. The effectiveness of the knockdown was measured by both RT-qPCR to monitor the mRNA levels and Western blotting to monitor the protein levels. As can be seen in Figure 5, the knockdown of either ABCC1 or ABCC4 was effective in both cell lines. In addition, since the two proteins have overlapping substrate specificity, the dual knockdown of both proteins was also undertaken (Figure 5e).

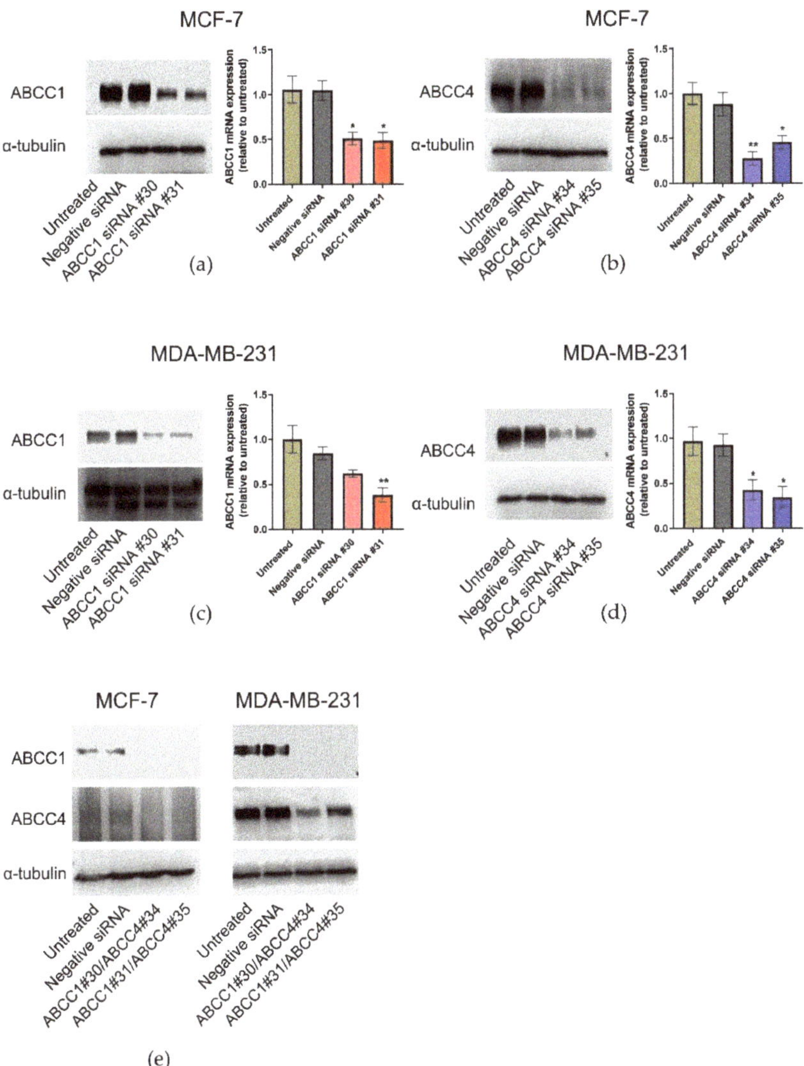

Figure 5. ABCC1 and ABCC4 can be successfully knocked down in breast cancer cells using siRNA. Gene knockdown in breast cancer cells was performed using the INTERFERin-siRNA transfection protocol, with two different ABCC1 siRNAs (#30 or #31), two different ABCC4 siRNAs (#34 or #35),

or in combination (#30/#34 or #31/#35). A negative control siRNA was also used. The effectiveness of the knockdown was measured by both RT-qPCR and Western blotting. (**a**) Knockdown of ABCC1 in MCF-7 cells, (**b**) knockdown of ABCC4 in MCF-7 cells, (**c**) knockdown of ABCC1 in MDA-MB-231 cells, (**d**) knockdown of ABCC4 in MDA-MB231 cells and (**e**) double knockdown of both transporters in each cell line. Uncropped images can be found in Figure S4. RT-qPCR data are mean ± sem, n = 3. Data were analyzed using a one-way ANOVA with a Dunnett's post hoc test. * $p < 0.05$ and ** $p < 0.01$ significantly lower than the negative siRNA sample.

2.5. The Effect of ABCC1 and ABCC4 Knockdown on Breast Cancer Proliferation and Migration

Having established that the ABCC proteins could be knocked down, the effect of this on cell proliferation was investigated. In Figure 6a, it can be seen that one of the knockdowns of ABCC1 in MCF-7 cells caused a decrease in clonogenic capacity. The double knockdown of both ABCC1 and ABCC4 caused an even larger impact. For MDA-MB-231 cells, only one of the double knockdowns had a significant effect (Figure 6b). In contrast, when examining the bulk proliferation, it can be seen in Figure 6c,d that knockdown had no significant effect on the growth of either cell line.

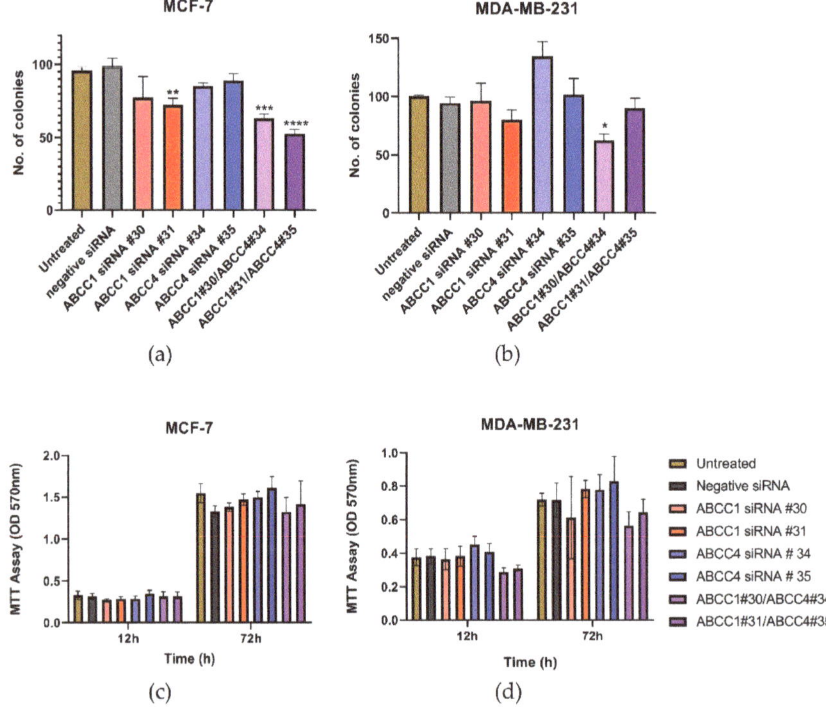

Figure 6. Combined knockdown of ABCC1 and ABCC4 affects the clonogenic capacity of breast cancer cells. (**a**,**b**) Clonogenic capacity of breast cancer cells following the siRNA-mediated knockdown of ABCC1 or ABCC4 was analyzed using a colony formation assay. (**c**,**d**) Proliferation of breast cancer cells following the siRNA-mediated knockdown of ABCC1 or ABCC4 was analyzed using an MTT assay. Data are mean ± sem, n ≥ 6. Data were analyzed by an ANOVA with a Dunnett's post hoc test. * $p < 0.05$, ** $p < 0.01$, *** $p < 0.001$ and **** $p < 0.0001$ significantly lower than with the negative siRNA treatment.

Next, the effect of ABCC knockdown on cell migration was investigated. Figure 7a,b shows the average results from the scratch assays. With the MCF-7 cells (Figure 7a), a knockdown with ABCC4 siRNA #35 caused a significant decrease in migration, and with the MDA-MB-231 cells (Figure 7b), both ABCC4 knockdowns caused a significant decrease in migration. To examine this further, invasion rather than just migration was investigated (Figure 7c,d). With the MCF-7 cells, no significant effects were observed; however, with MDA-MB-231 cells, the knockdown with ABCC4 siRNA #35 and one of the double knockdowns did cause a significant decrease in invasion.

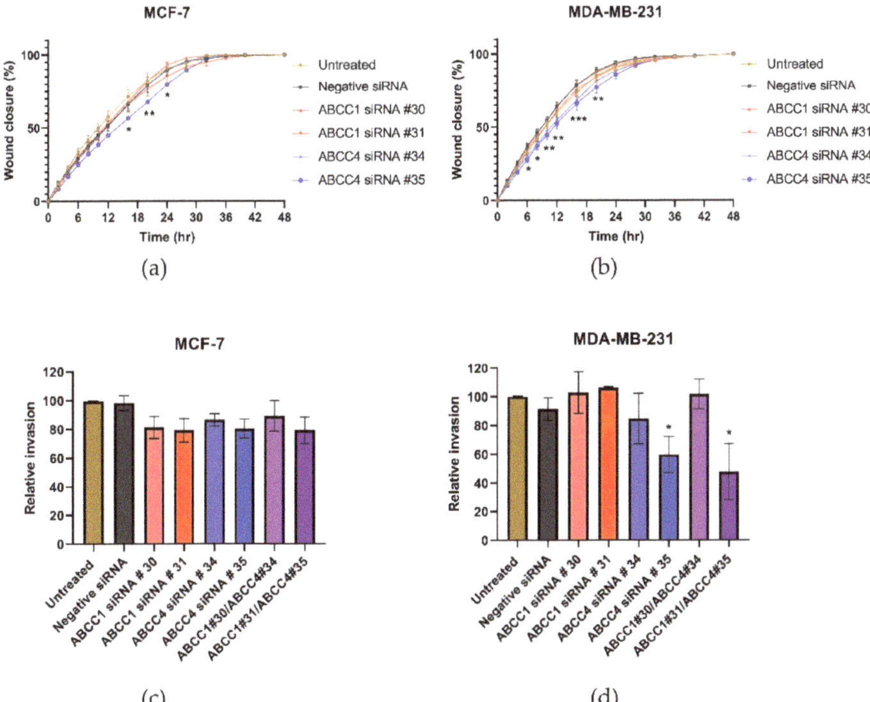

Figure 7. siRNA knockdown of ABCC4 affects the migration of breast cancer cells. (**a**,**b**) Migration of breast cancer cells following the siRNA-mediated knockdown of ABCC1 or ABCC4 was analyzed using a scratch assay. Data are mean ± sem, n ≥ 9. (**c**,**d**) Migration of breast cancer cells following the siRNA-mediated knockdown of ABCC1 or ABCC4 was also investigated using a cellular invasion assay. Data are mean ± sem, n ≥ 3. Data were analyzed by an ANOVA with a Dunnett's post hoc test. * $p < 0.05$, ** $p < 0.01$, and *** $p < 0.001$ significantly lower than with the negative siRNA treatment.

These results with the ABCC knockdowns correlate well with the inhibitor studies, with ABCC1 and the double knockdown having an impact on cellular proliferation, whilst ABCC4 and the double knockdown affect migration.

2.6. Investigation into Potential Mechanisms by Which ABCC Transporters Affect Cellular Proliferation or Migration

ABCC1 and ABCC4 both transport a wide array of different molecules with the potential to impact cellular proliferation and migration, including cyclic nucleotides, eicosanoids, and lipid mediators such as S1P and LPI. It has previously been proposed that ABCC1 is involved in creating a feedback signaling loop with the G protein-coupled receptor, GPR55, whereby ABCC1 exports LPI, which binds to GPR55 and activates it, leading to downstream signaling and increased proliferation [31]. Therefore, we measured the expression of GPR55 in the breast cancer cell lines by Western blot (Figure 8a). GPR55 is indeed expressed in the MDA-MB-231 cells; however, there was little, if any, expression in the MCF-7 cells. Since the effects we observed with ABCC1 and ABCC4 inhibitors and knockdown were with both cell lines, this would argue against it being due to LPI export and GPR55 activation.

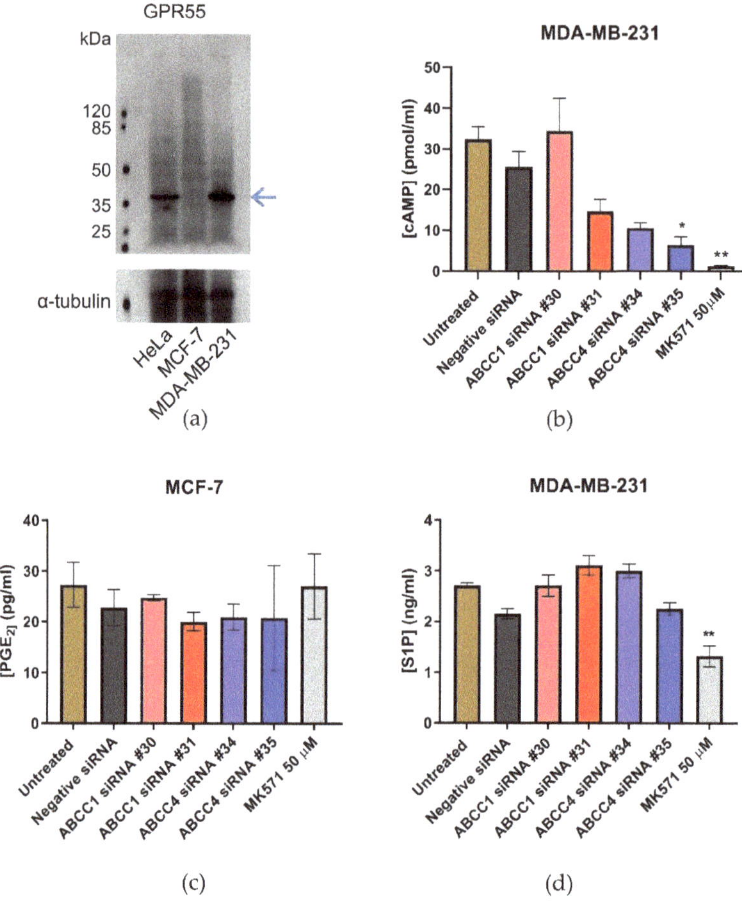

Figure 8. Efflux of cAMP or S1P are possible mechanisms by which ABCC transporters affect breast cancer cells. (**a**) Western blot analysis of the expression of GPR55 in breast cancer cell lines. The blue arrow indicates the band corresponding to GPR55 (**b**) Efflux of cAMP from MDA-MB-231 breast cancer cells was assayed using an ELISA. Data are mean ± sem, n ≥ 2, each in duplicate. (**c**) Efflux of Prostaglandin E_2 (PGE_2) from MCF-7 breast cancer cells was assayed using an ELISA. Data are mean ± sem, n ≥ 2, each in duplicate. (**d**) Efflux of S1P from MDA-MB-231 breast cancer cells was assayed using an ELISA. Data are mean ± sem, n ≥ 3, each in duplicate. Data were analyzed by a one-way ANOVA with a Dunnett's post hoc test. * $p < 0.05$ and ** $p < 0.01$, significantly lower than untreated or negative siRNA treatment.

Next, the efflux of cAMP was investigated as a potential mechanism, as ABCC4 is capable of effluxing cyclic nucleotides [32]. An ELISA was used to assay cAMP in the cell medium (Figure 8b). The concentration of cAMP in the medium from MCF-7 cells was outside the standard range and, therefore, could not be reliably measured. For MDA-MB-231, the knockdown of ABCC1 had no significant effect on the concentration of cAMP in the medium, which is unsurprising, as ABCC1 is not reported to transport cyclic nucleotides. However, the knockdown with ABCC4 siRNA #35 did significantly decrease the concentration of cAMP in the medium, as did the treatment with MK571.

An efflux of eicosanoids such as prostaglandins could be another possible way in which ABCC4 mediates an effect [33]. An ELISA was used to measure Prostaglandin E_2 (PGE$_2$) levels in the cell medium. The concentration of PGE$_2$ in the medium from MDA-MB-231 cells was outside the range of standards used and, thus, could not be reliably measured. The concentration of PGE$_2$ in the medium from MCF-7 cells can be seen in Figure 8c. Neither the knockdown of ABCC proteins nor the treatment with MK571 had any effect on the levels of PGE$_2$ measured.

Finally, S1P was investigated, as there are reports that both ABCC1 and ABCC4 are capable of transporting S1P [29,38]. Figure 8d shows the results of the S1P ELISA. Only the MDA-MB-231 cells effluxed sufficient S1P for reliable measurement. Although none of the siRNA knockdown samples showed any significant difference in S1P concentrations, the treatment with MK571, which inhibits both ABCC1 and ABCC4, did significantly decrease the amount of S1P in the medium.

3. Discussion

ABCC1 and ABCC4 have previously been implicated as negative prognostic indicators for breast cancer [16–19,22]. This may be due to their ability to export chemotherapeutic drugs, leading to multidrug resistance [13–21]. However, ABCC1 and ABCC4 can also transport a wide range of substrates involved in inflammation and cellular signaling and have been shown in several different cancer types to have a role in cancer development and progression that is independent of drug resistance [39–47,55]. Therefore, in this study, we investigated whether ABCC1 or ABCC4 are involved in cellular proliferation or migration in breast cancer cell lines. We found that both MCF-7 and MDA-MB-231 cells expressed both ABCC1 and ABCC4, with MDA-MB-231 cells having a higher expression of both transporters, which is in agreement with previous studies [56,57]. MDA-MB-231 cells are basal-type triple-negative and often reported as more aggressive than MCF-7 cells, so this would correlate with the expression levels observed. It has been reported that ABCC1 is frequently expressed in lymph node metastases of breast cancer patients and that MRP1 expression is more pronounced in lymph node metastases than in corresponding primary tumors [58].

Using both small molecule inhibitors and siRNA knockdown of ABCC1 and ABCC4, we investigated the impact of these transporters on the proliferation, clonogenic capacity, migration, and invasion of the breast cancer cell lines. MK571, which inhibits both ABCC1 and ABCC4, and Reversan, which inhibits ABCC1, both had significant effects on the proliferation and clonogenic capacity of MCF-7 and MDA-MB-231 cells, suggesting a role for ABCC1 or both transporters in cellular proliferation. Similarly, one of the ABCC1 knockdowns and the double ABCC1/ABCC4 knockdowns decreased the colony formation with MCF-7 cells, although only one of the double knockdowns affected colony formation in the MDA-MB-231 cells. It should be noted that MK571 not only inhibits ABCC1 and ABCC4 but, also, other members of the ABCC family, and it is a leukotriene antagonist, inhibiting the binding of LTD$_4$ to the cysteinyl leukotriene receptor 1 [50], which could also impact cell signaling. However, the similar results obtained with the knockdowns would argue that the effects observed are due to ABCC1/ABCC4. Similarly, Reversan is also able to inhibit the related protein ABCB1, but given that Reversan affected both cell lines, and only the MCF-7 cells showed a significant expression of ABCB1 (Figure S1), it is unlikely this is the cause. That knockdown of ABCC1 alone did not affect the colony formation of MDA-MB-231 cells, and only one of the double knockdowns affected it, maybe because of the higher expression level of the transporters in MDA-MB-231 cells. The siRNA method only results in knockdown, not knockout, as seen in Figure 5. Although the protein levels are decreased in MDA-MB-231 cells, they were higher to start with, so it is possible that sufficient protein remains even after knockdown. Furthermore, siRNA knockdown is only transient, and it is not known if all cells were affected or just a proportion of them. We chose to use siRNA, because with complete knockdowns, the overexpression of other ABC transporters often occurs to compensate. In the future, perhaps the use of lentiviral shRNA could be investigated [42,43,46].

When investigating cellular migration, only MK571 had a significant effect, and only with the MDA-MB-231 cells, suggesting perhaps that the dual inhibition of both ABCC1 and ABCC4 was important. However, the siRNA knockdowns showed that the ABCC4 knockdown affected migration in both cell lines. Why the reportedly specific ABCC4 inhibitors, Ceefourin 1 and Ceefourin 2 [52], did not cause any effect on the cellular migration, when ABCC4 knockdown did, is not clear. The cellular invasion assays were more variable but, again, showed an effect of ABCC4 knockdown in MDA-MB-231 cells.

Our results suggest that ABCC1 is important for breast cancer proliferation, whilst ABCC4 has a greater role in cellular migration and invasion (Table 2). This contrasts with a study investigating the role of these transporters in neuroblastoma, where ABCC4 was more associated with proliferation and ABCC1 with migration [39]. Similarly, in pancreatic cancer, ABCC4 was associated with proliferation [47]. However, in epithelial ovarian cancer, ABCC4 was associated with proliferation, migration, and invasion [45]. In agreement with our results, breast cancer cells overexpressing ABCC1 showed an increase in proliferation, which could be inhibited by MK571, and overexpression of ABCC1 enhanced tumor growth in mice [59]. Additionally, mice implanted with breast cancer tumors with varying levels of ABCC4 expression showed no differences in tumor growth, but an increased ABCC4 expression was associated with increased metastases [56]. Therefore, the role(s) played by these transporters is not necessarily the same across different cancers, but within breast cancers, our results appear to agree with other studies in the literature. It should also be noted that, although the current study only investigated ABCC1 and ABCC4, they are not necessarily the only important transporters. In neuroblastoma, ABCC3 was shown to be very important, albeit as a positive prognostic indicator [39]. In contrast, ABCC3 was implicated in breast cancer stem-like features [20]. ABCC11 has also been implicated in aggressive breast cancer and prognosis [18].

Table 2. Summary of findings. ↓ significant decrease. — no significant effect. n/d not determined.

Condition	Clonogenic Capacity	Proliferation	Migration	Invasion	Comparable Findings in the Literature
ABCC1 inhibition	↓	↓	—	n/d	[59]
ABCC4 inhibition	—	—	—	n/d	
Double inhibition	↓	↓	↓	n/d	[59]
ABCC1 knockdown	↓	—	—	—	[59]
ABCC4 knockdown	—	—	↓	↓	[56]
Double knockdown	↓	—	n/d	↓	

Understanding the mechanism by which these transporters mediate their effects is important. Both transporters are able to efflux a wide range of different molecules. It is important to know what molecule is being effluxed that causes the effects observed. It might be that targeting the ABC transporters themselves is not the best approach, as they have historically proven difficult to target, often because they have so many important roles. Instead, understanding the pathway they are involved with and targeting something earlier or later in the signaling pathway might be a more effective approach. Therefore, preliminary investigations into how ABCC1 and/or ABCC4 might elicit their effects on breast cancer cell proliferation and migration were undertaken. ABCC1 has previously been linked to GPR55 in a feedback loop, where ABCC1 exports LPI, which binds to and activates GPR55, leading to downstream signaling, and this was implicated as important for prostate and ovarian cancer cell proliferation [31]. However, we found little/no expression of GPR55 within MCF-7 cells. Given that ABCC1 inhibition or knockdown affected the proliferation and clonogenic capacity of MCF-7 cells, this would argue against LPI export and GPR55 activation being a key feature to explain our findings. The export of cAMP was a second method explored, as ABCC4 is known to be able to efflux cyclic nucleotides [32]. The knockdown of ABCC4 or inhibition with MK571 significantly decreased the amount of cAMP effluxed from MBA-MB231 cells. This agrees well with a previous study [57], and it has been suggested that increasing cellular cAMP levels is a potential method of combatting triple-negative breast cancer [57,60]. The export of PGE_2 has also been suggested as a potential mechanism by which ABCC4 might be important in breast cancer [61]. In our study, the levels of PGE_2 exported by MDA-MB-231 cells were too high to quantify accurately. With MCF-7 cells, no effect of ABCC1/ABCC4 knockdown or inhibition was observed. Similarly, no effect of ABCC4 overexpression on the PGE_2 efflux from MCF-7 cells was observed previously [56]. However, the same report suggests that the knockdown of ABCC4 in MBA-MB-231 cells does decrease the PGE_2 efflux and that this is important for breast cancer metastasis [56]. Given that we observed the effects of ABCC4 knockdown on the migration of both MCF-7 and MDA-MB-231 cells, this might argue that it is not due simply to PGE_2 efflux; however, further investigation with MDA-MB-231 cells is needed. Finally, we investigated the efflux of S1P, as levels of S1P have previously been implicated in breast cancer [62], and both ABCC1 and ABCC4 are capable of transporting S1P [29,38]. Although the knockdown of ABCC1 or ABCC4 did not show a significant effect on extracellular levels of S1P, the inhibition of both ABCC1 and ABCC4 with MK571 did decrease the amount of S1P effluxed. This result agrees with a previous study that also observed a decreased S1P efflux from MCF-7 cells in the presence of MK571 [63]. The same study also showed that siRNA knockdown of ABCC1 decreased the amount of S1P exported from MCF-7 cells, but a large effect was only seen when the cells were stimulated with estradiol, which we did not do, and this might explain why we see no effects of ABCC1 or ABCC4 knockdown.

4. Materials and Methods

4.1. Cell Culture

The MDA-MB-231 and MCF-7 human breast cancer cells were originally obtained from ATCC. They were cultured in Dulbecco's modified Eagle's medium (DMEM) (Lonza, Slough, UK) containing 4.5-g/L glucose and L-glutamine supplemented with 10% *v/v* fetal bovine serum (FBS) (Gibco®, ThermoFisher Scientific, Loughborough, UK) and 100-U/mL penicillin and 0.1-mg/mL streptomycin (sterile-filtered penicillin-streptomycin solution 100×, Biowest, Nuaillé, France). Cells were cultured in an environment of 95% air and 5% CO_2 at 37 °C.

4.2. Colony Formation Assay

Cell growth from a single cell was determined using the colony formation assay, as described previously [39,64]. Briefly, 100 cells/well were seeded in 6-well plates and cultured at 37 °C for 24 h, after which cells were treated with MK571 (Merck Life Science, Gillingham, UK), indomethacin (Merck), Reversan (Tocris Bioscience, Abingdon, UK), Ceefourin 1, or Ceefourin 2 (Abcam, Cambridge, UK). Control wells were untreated cells. After 7 days of culture, the colonies formed were fixed with 4% paraformaldehyde and stained with 0.1% crystal violet. Wells were left to dry in open air at room temperature, and visible colonies of 50 cells or more were manually counted under a light box using a transparent film and a fine marker. The sizes of the colonies were measured using ImageJ 1.52 (NIH, Bethesda, MD, USA).

4.3. Cell Proliferation Assays

Cell viability was analyzed using a standard MTT assay. Fifteen thousand MCF-7 cells or 6000 MDA-MB-231 cells were plated in each well of a 24-well plate in a total volume of 400μL of culture medium. After 4 h of culture, cells were treated with inhibitors, as described above. The cell viability was assessed at 6, 12, 24, 48, and 72 h after treatment. Forty microliters of MTT (5 mg/mL) was added to each well and further incubated for 1 h at 37 °C. Wells were left to dry at 37 °C, and 400 μL of DMSO (dimethylsulfoxide) was added and left at 37 °C for 10 min. One hundred microliters of DMSO from each well was transferred to 96-well plates in triplicates. The absorbance values were read at 570 nm using a Multiskan™ GO microplate spectrophotometer (Thermo Fisher Scientific, Gillingham, UK).

Alternatively, at the given time points after treatment, the cells were harvested, diluted in trypan blue at a ratio of 1:4, viewed using a hemocytometer and microscope, and counted manually.

4.4. Cell Migration Assays

Cell migration was assessed using the Cell IQ (CM Technologies—intelligent cell analysis, Tampere, Finland) scratch assay. Cells were seeded in 24-well plates to reach 100% confluency the day of the assay. A scratch across the monolayer of the cells was carefully made using a 10-µL pipette tip, and the media was replaced with fresh prewarmed culture medium. Cells were treated with the inhibitors, as described above. Three image positions were selected from each well, and images were taken at 1-h intervals. Images were analyzed using the Cell-IQ analysis software (CM Technologies).

4.5. Cell Invasion Assays

The 24-well plate Transwell inserts (pore size 8 µm, translucent) (Greiner bio-one, Stonehouse, UK) were coated with 30 µL of ice-cold Matrigel (Corning, VWR, Lutterworth, UK) mixed with cold serum-free medium at a ratio of 1:3. The coated Transwell inserts were left to set at 37 °C for 2 h. Fifty thousand cells were seeded in each Transwell insert suspended in 300 µL of 0.5% *v/v* FBS culture medium. Eight hundred microliters of chemoattractant (10% *v/v* FBS cell culture medium) was added to the wells, and the Transwell inserts were placed in each of the wells that contain a chemoattractant. The Transwell inserts were incubated for 24 h at 37 °C. After 24 h, the culture medium in the Transwell inserts were discarded, and a cotton swab moistened with Dulbecco's phosphate-buffered saline (DPBS) (Biowest) was used to, gently but putting firm pressure, rub the inside of the Transwell inserts to remove cells, and this process was repeated with a second cotton swab. The cells that moved to the lower surface of the membrane of the Transwell inserts were stained with Differential Quik stain (Modified Giemsa) (Generon, Slough, UK), following the manufacturer's instructions. The Transwell inserts were then rinsed twice with water. A cotton swab moistened with DPBS was used to gently but putting firm pressure wipe again the inside of the Transwell inserts to rid of any cells left. The Transwell inserts were left to dry at room temperature. To determine the percentage cell invasion, the invaded cells were counted from randomly selected five fields of view for each Transwell insert under a light microscope at a magnification of ×40, and the mean number of invaded cells were obtained.

4.6. Gene Knockdown

Gene knockdown was performed following the INTERFERin-siRNA transfection protocol (Polyplus Transfection, Illkirch, France). Briefly, 25,000 cells were seeded in 24-well plates the day before transfection, ensuring 30–50% confluency at the time of transfection. The siRNA transfection complexes were added to cells dropwise and left to incubate for 3 days at 37 °C. The siRNA used were two ABCC1 siRNA (siRNA id: SASI_Hs02_00338730 and SASI_Hs02_00338731, referred to as #30 and #31, respectively) (Gene id: 4363) and 2 ABCC4 siRNA (siRNA id: SASI_Hs02_00324134 and SASI_Hs02_00324135, referred to as #34 and #35, respectively) (Gene id: 10257) (Merck). A negative control siRNA (Merck) was used to distinguish sequence-specific silencing from nonspecific effects.

4.7. Whole Cell Lysates

Harvested cells were placed on ice. Three hundred to four hundred microliters of ice-cold lysis buffer (0.15-M NaCl, 0.05-M Tris, pH 8.0, 1% (*v/v*) Triton-X100, 1-mM EDTA, and 1-mM pepstatin, 1.3-mM benzamidine, and 1.8-mM leupeptin) was added to the cells. The cells were resuspended by vortexing. The cell suspension obtained was kept on ice and vortexed every 10 min for 1 h. Cells suspension were then centrifuged for 15 min at 15,600× *g* at 4 °C, and the supernatant containing the whole cell lysate was quantified for total protein concentrations using a bicinchoninic acid kit (Pierce, ThermoFisher Scientific). The samples were mixed with sample buffer for loading on sodium dodecyl sulphate (SDS) gels.

4.8. Cell Membrane Extraction

Cell pellets harvested and resuspended in 20 mL of homogenization buffer (50-mM Tris, pH 7.4, 250-mM sucrose, and 0.25-mM $CaCl_2$) containing protease inhibitors (1-mM pepstatin, 1.3-mM benzamidine, and 1.8-mM leupeptin). All the subsequent steps were done on ice. The cell suspension was then placed in the cell disruption vessel for nitrogen cavitation at 500 psi (Model 4639 Parr Cell Disruption Vessels (Parr®)). The vessel was incubated on ice for 15 min. Pressure was released slowly, and sample collected dropwise and centrifuged for 10 min at 560× *g* at 4 °C. The supernatant was collected, and the pellet, which contained unbroken cells and debris, was discarded. The supernatant was then ultracentrifuged at 100,000× *g* at 4 °C for 20 min. The supernatant was carefully discarded and the membrane pellet resuspended in 0.5–1-mL Tris sucrose buffer (TSB) (50-mM Tris-HCl, pH 7.4, and 0.25-mM sucrose). A needle syringe (0.45 mm × 13 mm) was used to resuspend the membrane proteins in TSB. Protein concentration was determined using a bicinchoninic acid kit (Pierce, ThermoFisher). Samples were kept on ice during the membrane preparations and were mixed with a sample buffer for loading on sodium dodecyl sulphate (SDS) gels.

4.9. RNA Isolation and Quantitative Real-Time (RT-qPCR)

The total RNA from MDA-MB-231 and MCF-7 cells, respectively, was extracted using a Bioline ISOLATE RNA mini kit according to the manufacturer's protocol (Meridian Biosciences, London, UK). The quantity of the isolated RNA obtained was measured using a NanoDrop™ 1000 spectrophotometer (Thermo Fisher Scientific), and samples were stored at −80 °C. mRNA (600 ng) was reverse-transcribed to cDNA using a Precision NanoScript™ 2 Reverse Transcription kit (Primerdesign, Chandlers Ford, UK) according to the manufacturer's protocol. The samples were then placed in a thermocycler, which was set for 20 min at 42 °C, 10 min at 75 °C, and holding at 4 °C. The cDNA samples obtained were diluted 1 in 10 with RNase/DNase-free water and stored at −20 °C.

The cDNA samples were amplified in qPCR using the Thermo Scientific PiKoReal 96 Real-Time PCR system. For one reaction, a master mix containing 10 μL of PrecisionPlus™ 2 × qPCR Mastermix with Sybr green with inert blue dye (Primerdesign), 1 μL of the reverse primer, 1 μL of the forward primer, and 3 μL of RNase/DNase-free water was prepared. Five microliters of cDNA were first added to each well of a 96-well Piko PCR plate in triplicate, followed by 15 μL of the master mix, ensuring that no bubbles were formed in the wells. The expression levels were normalized to human β-actin and human GAPDH as the house keeping genes. RNase/DNase-free water was used as the negative control. The PCR primers were used as follow: β-actin forward nucleotide: 5′-CTGGAACGGTGAAGGTGACA-3′, reverse nucleotide: 5′-AAGGGACTTCCTGTAACAATGCA-3′; GAPDH forward nucleotide: 5′-TGCACCACCAACTGCTTAGC-3′, reverse nucleotide: 5′-GGCATG GACTGTGGTCATGAG-3′; ABCC1 forward nucleotide 5′-CGACATGACCGAGGCTACATT-3′, reverse nucleotide 5′-AGCAGACGATCCACAGCAAAA-3′; and ABCC4 forward nucleotide 5′-TGTGGCTTTGAACACAGCGTA-3′, reverse nucleotide 5′-CCAGCACACTGAACGTGATAA-3′. The qPCR data were analyzed using the double-delta Ct analysis. Each reaction was performed three times.

4.10. Western Blotting

For Western blot analysis, the total protein samples (80 μg/well) were loaded and separated by SDS-PAGE (sodium dodecyl sulphate polyacrylamide gel electrophoresis), then transferred to a PVDF membrane. The membrane was blocked in blocking buffer (5% *w/v* BSA in Tris-buffered saline (25-mM Tris, pH 7.4, 150-mM NaCl, and 0.05% Tween 20) (TBS-T) for 1 h at room temperature, then incubated in appropriate primary antibodies for 1 h at room temperature or overnight at 4 °C. The primary antibodies used were: anti-ABCC1 derived in rabbit (1:1000, EPR4658(2); Abcam), anti-ABCC4 derived in rat (1:100, M_4I-10; Abcam), anti-ABCB1 derived in mouse (1:200; Thermo Fisher), anti-ABCG2 derived in mouse (1:50, Bxp1; Santa Cruz, Heidelberg, Germany), or anti-GPR55 (rabbit) (1:100;

Abcam). The secondary antibodies used were anti-rabbit HRP (1:3000; Cell Signaling Technology, Leiden, The Netherlands), anti-rat HRP (1:5000; Merck), and anti-mouse HRP (1:4000; Cell Signaling Technology). Blots were visualized using a SuperSignal West Chemiluminescent kit (Pierce, Thermo Fisher) and the Li-Cor C-Digit blot scanner, and the images were analyzed using the Image Studio Lite software imaging system (Li-Cor, Cambridge, UK). Afterwards, blots were re-probed using anti-α-tubulin (mouse) (1:1000; Merck) to check for equal sample loading. Western blot analysis was repeated at least 3 times for each experiment.

4.11. ELISA

Following siRNA knockdown, to determine the levels of molecules exported from cells, ELISAs were carried out on the media. As a control, 3 wells of nontransfected cells were treated with 50-µM MK571. For cAMP, cells were first stimulated with 100-µM forskolin for 2 h at 37 °C. For PGE_2, cells were stimulated with 10-µg/mL lipopolysaccharides (LPS) for 24 h at 37 °C and with 80-nM phorbol 12-myristate 13-acetate (PMA) for 1 h at 37 °C, respectively. For S1P, there was no stimulation. The cell culture medium was then replaced with a fresh cell culture medium, and the cells were incubated for 18 h at 37 °C. The supernatants (conditioned media) were centrifuged at 1000× g to pellet out any floating cells and collected and stored at −80 °C or used immediately for determination of cAMP by using the cAMP ELISA kit (Enzo Life Sciences Inc., Exeter, UK), for the determination of PGE_2, using the PGE_2 ELISA kit (Enzo Life Sciences Inc.), or for S1P, using the human S1P ELISA kit (Abbexa Ltd., Cambridge, UK), according to the manufacturer's instructions. The cells were also harvested for protein quantification.

4.12. Statistical Analysis

Statistical analysis of data was carried out using GraphPad Prism 8.1 (San Diego, CA, USA). For multiple comparisons with one independent variable, a one-way ANOVA was used with Dunnett's post-hoc test. For multiple comparisons with two independent variables, a two-way ANOVA was used with Dunnett's post-hoc test. A value of $p < 0.05$ was considered significant.

5. Conclusions

In summary, our results show that ABCC1 plays a role in breast cancer proliferation, whilst ABCC4 has a greater role in cellular migration and invasion. It may well be that both transporters are important; their overlapping substrate specificity means they can likely compensate for each other. The mechanism by which these transporters (and others) are involved in the development and progression of breast cancer needs to be investigated further. It is key to know exactly how they are involved before they can be considered as targets for treatment.

Supplementary Materials: The supplementary materials can be found at http://www.mdpi.com/1422-0067/21/20/7664/s1. Figure S1: Expression of ABCB1/P-glycoprotein or ABCG2 in breast cancer cell lines. Figure S2: MK571 and Reversan affect the proliferation of breast cancer cells. Figure S3: Original, uncropped Western blots I. Figure S4: Original, uncropped Western blots II.

Author Contributions: Conceptualization, A.J.R. and F.G.L.; methodology, F.G.L., validation, F.G.L.; formal analysis, F.G.L., K.S., and A.J.R.; investigation, F.G.L. and K.S.; resources, A.J.R., J.E.B., and R.M.B., writing—original draft, F.G.L. and A.J.R.; writing—review and editing, K.S., R.M.B., and J.E.B., visualization, F.G.L. and A.J.R.; supervision, A.J.R., R.M.B., and J.E.B.; and funding acquisition, A.J.R., F.G.L., and J.E.B. All authors have read and agreed to the published version of the manuscript.

Funding: F.G.L. was the recipient of an Aston University overseas excellence bursary. K.S. was funded by the Aston Medical School and University Hospitals Coventry and Warwickshire. Thanks to the Aston LHS joint research group fund.

Acknowledgments: Many thanks to Charlie Clarke-Bland for help with the Cell-IQ. The underlying data for this study can be found at the Aston Data Explorer Repository (https://doi.org/10.17036/researchdata.aston.ac.uk.00000483).

Conflicts of Interest: The authors declare no conflict of interest.

Abbreviations

ABC	ATP-Binding Cassette
ATP	Adenosine triphosphate
ABCB1	ABC B subfamily member 1/P-glycoprotein
ABCC1	ABC C subfamily member 1/Multidrug resistance protein 1
ABCC4	ABC C subfamily member 4/Multidrug resistance protein 4
ABCG2	ABC G subfamily member 2/Breast cancer resistance protein
ANOVA	Analysis of variance
BSA	Bovine serum albumin
cAMP	cyclic adenosine monophosphate
cGMP	cyclic guanosine monophosphate
DMSO	dimethylsulfoxide
ELISA	Enzyme-linked immunoadsorbant assay
MCF-7	Michigan cancer foundation 7 cell line
LPI	Lysophosphatidyl inositol
MDA-MB-231	M.D. Anderson metastatic breast cancer cell line
MTT	3-(4,5-dimethylthiazol-2-yl)-2,5-diphenyltetrazolium bromide
mRNA	messenger RNA
PCR	Polymerase chain reaction
PGE_2	prostaglandin E_2
PVDF	Polyvinylidene difluoride
SD	Standard deviation
RT-qPCR	Reverse transcription quantitative PCR
sem	Standard error of the mean
S1P	sphingosine 1 phosphate
siRNA	small interfering RNA
SDS-PAGE	SDS polyacrylamide gel electrophoresis
SYBR green	N′,N′-dimethyl-N-[4-[(E)-(3-methyl-1,3-benzothiazol-2-ylidene)methyl]-1-phenylquinolin-1-ium-2-yl]-N-propylpropane-1,3-diamine

References

1. Bray, F.; Ferlay, J.; Soerjomataram, I.; Siegel, R.L.; Torre, L.A.; Jemal, A. Global cancer statistics 2018: GLOBOCAN estimates of incidence and mortality worldwide for 36 cancers in 185 countries. *CA Cancer J. Clin.* **2018**, *68*, 394–424. [CrossRef] [PubMed]
2. Martin, H.L.; Smith, L.; Tomlinson, D.C. Multidrug-resistant breast cancer: Current perspectives. *Breast Cancer (Dove Med. Press)* **2014**, *6*, 1–13. [CrossRef] [PubMed]
3. Nounou, M.I.; ElAmrawy, F.; Ahmed, N.; Abdelraouf, K.; Goda, S.; Syed-Sha-Qhattal, H. Breast Cancer: Conventional Diagnosis and Treatment Modalities and Recent Patents and Technologies. *Breast Cancer (Auckl.)* **2015**, *9*, 17–34. [CrossRef]
4. Carioli, G.; Malvezzi, M.; Rodriguez, T.; Bertuccio, P.; Negri, E.; La Vecchia, C. Trends and predictions to 2020 in breast cancer mortality in Europe. *Breast* **2017**, *36*, 89–95. [CrossRef] [PubMed]
5. DeSantis, C.E.; Ma, J.; Gaudet, M.M.; Newman, L.A.; Miller, K.D.; Goding Sauer, A.; Jemal, A.; Siegel, R.L. Breast cancer statistics, 2019. *CA Cancer J. Clin.* **2019**, *69*, 438–451. [CrossRef]
6. Scully, O.J.; Bay, B.H.; Yip, G.; Yu, Y. Breast cancer metastasis. *Cancer Genom. Proteom.* **2012**, *9*, 311–320.
7. Maughan, K.L.; Lutterbie, M.A.; Ham, P.S. Treatment of breast cancer. *Am. Fam. Physician* **2010**, *81*, 1339–1346.
8. Mehanna, J.; Haddad, F.G.; Eid, R.; Lambertini, M.; Kourie, H.R. Triple-negative breast cancer: Current perspective on the evolving therapeutic landscape. *Int. J. Womens Health* **2019**, *11*, 431–437. [CrossRef]
9. Cleator, S.; Heller, W.; Coombes, R.C. Triple-negative breast cancer: Therapeutic options. *Lancet Oncol.* **2007**, *8*, 235–244. [CrossRef]
10. Rastelli, F.; Biancanelli, S.; Falzetta, A.; Martignetti, A.; Casi, C.; Bascioni, R.; Giustini, L.; Crispino, S. Triple-negative breast cancer: Current state of the art. *Tumori* **2010**, *96*, 875–888. [CrossRef]
11. Sharom, F.J. ABC multidrug transporters: Structure, function and role in chemoresistance. *Pharmacogenomics* **2008**, *9*, 105–127. [CrossRef]

12. Robey, R.W.; Pluchino, K.M.; Hall, M.D.; Fojo, A.T.; Bates, S.E.; Gottesman, M.M. Revisiting the role of ABC transporters in multidrug-resistant cancer. *Nat. Rev. Cancer* **2018**, *18*, 452–464. [CrossRef] [PubMed]

13. Russel, F.G.; Koenderink, J.B.; Masereeuw, R. Multidrug resistance protein 4 (MRP4/ABCC4): A versatile efflux transporter for drugs and signalling molecules. *Trends Pharmacol. Sci.* **2008**, *29*, 200–207. [CrossRef] [PubMed]

14. Slot, A.J.; Molinski, S.V.; Cole, S.P. Mammalian multidrug-resistance proteins (MRPs). *Essays Biochem.* **2011**, *50*, 179–207. [CrossRef]

15. Deeley, R.G.; Cole, S.P. Substrate recognition and transport by multidrug resistance protein 1 (ABCC1). *FEBS Lett.* **2006**, *580*, 1103–1111. [CrossRef] [PubMed]

16. Nooter, K.; Brutel de la Riviere, G.; Look, M.P.; van Wingerden, K.E.; Henzen-Logmans, S.C.; Scheper, R.J.; Flens, M.J.; Klijn, J.G.; Stoter, G.; Foekens, J.A. The prognostic significance of expression of the multidrug resistance-associated protein (MRP) in primary breast cancer. *Br. J. Cancer* **1997**, *76*, 486–493. [CrossRef] [PubMed]

17. Filipits, M.; Malayeri, R.; Suchomel, R.W.; Pohl, G.; Stranzl, T.; Dekan, G.; Kaider, A.; Stiglbauer, W.; Depisch, D.; Pirker, R. Expression of the multidrug resistance protein (MRP1) in breast cancer. *Anticancer Res.* **1999**, *19*, 5043–5049. [PubMed]

18. Yamada, A.; Ishikawa, T.; Ota, I.; Kimura, M.; Shimizu, D.; Tanabe, M.; Chishima, T.; Sasaki, T.; Ichikawa, Y.; Morita, S.; et al. High expression of ATP-binding cassette transporter ABCC11 in breast tumors is associated with aggressive subtypes and low disease-free survival. *Breast Cancer Res. Treat.* **2013**, *137*, 773–782. [CrossRef]

19. Filipits, M.; Pohl, G.; Rudas, M.; Dietze, O.; Lax, S.; Grill, R.; Pirker, R.; Zielinski, C.C.; Hausmaninger, H.; Kubista, E.; et al. Clinical role of multidrug resistance protein 1 expression in chemotherapy resistance in early-stage breast cancer: The Austrian Breast and Colorectal Cancer Study Group. *J. Clin. Oncol.* **2005**, *23*, 1161–1168. [CrossRef]

20. Balaji, S.A.; Udupa, N.; Chamallamudi, M.R.; Gupta, V.; Rangarajan, A. Role of the Drug Transporter ABCC3 in Breast Cancer Chemoresistance. *PLoS ONE* **2016**, *11*, e0155013. [CrossRef]

21. Hlavac, V.; Brynychova, V.; Vaclavikova, R.; Ehrlichova, M.; Vrana, D.; Pecha, V.; Kozevnikovova, R.; Trnkova, M.; Gatek, J.; Kopperova, D.; et al. The expression profile of ATP-binding cassette transporter genes in breast carcinoma. *Pharmacogenomics* **2013**, *14*, 515–529. [CrossRef] [PubMed]

22. Rumiato, E.; Brunello, A.; Ahcene-Djaballah, S.; Borgato, L.; Gusella, M.; Menon, D.; Pasini, F.; Amadori, A.; Saggioro, D.; Zagonel, V. Predictive markers in elderly patients with estrogen receptor-positive breast cancer treated with aromatase inhibitors: An array-based pharmacogenetic study. *Pharm. J.* **2016**, *16*, 525–529. [CrossRef] [PubMed]

23. Fletcher, J.I.; Haber, M.; Henderson, M.J.; Norris, M.D. ABC transporters in cancer: More than just drug efflux pumps. *Nat. Rev. Cancer* **2010**, *10*, 147–156. [CrossRef] [PubMed]

24. Cole, S.P. Multidrug resistance protein 1 (MRP1, ABCC1), a "multitasking" ATP-binding cassette (ABC) transporter. *J. Biol. Chem.* **2014**, *289*, 30880–30888. [CrossRef]

25. Wijnholds, J.; Evers, R.; van Leusden, M.R.; Mol, C.A.; Zaman, G.J.; Mayer, U.; Beijnen, J.H.; van der Valk, M.; Krimpenfort, P.; Borst, P. Increased sensitivity to anticancer drugs and decreased inflammatory response in mice lacking the multidrug resistance-associated protein. *Nat. Med.* **1997**, *3*, 1275–1279. [CrossRef] [PubMed]

26. Sreekumar, P.G.; Spee, C.; Ryan, S.J.; Cole, S.P.; Kannan, R.; Hinton, D.R. Mechanism of RPE cell death in alpha-crystallin deficient mice: A novel and critical role for MRP1-mediated GSH efflux. *PLoS ONE* **2012**, *7*, e33420. [CrossRef]

27. Robbiani, D.F.; Finch, R.A.; Jager, D.; Muller, W.A.; Sartorelli, A.C.; Randolph, G.J. The leukotriene C(4) transporter MRP1 regulates CCL19 (MIP-3beta, ELC)-dependent mobilization of dendritic cells to lymph nodes. *Cell* **2000**, *103*, 757–768. [CrossRef]

28. Leier, I.; Jedlitschky, G.; Buchholz, U.; Cole, S.P.; Deeley, R.G.; Keppler, D. The MRP gene encodes an ATP-dependent export pump for leukotriene C4 and structurally related conjugates. *J. Biol. Chem.* **1994**, *269*, 27807–27810.

29. Mitra, P.; Oskeritzian, C.A.; Payne, S.G.; Beaven, M.A.; Milstien, S.; Spiegel, S. Role of ABCC1 in export of sphingosine-1-phosphate from mast cells. *Proc. Natl. Acad. Sci. USA* **2006**, *103*, 16394–16399. [CrossRef]

30. Tanfin, Z.; Serrano-Sanchez, M.; Leiber, D. ATP-binding cassette ABCC1 is involved in the release of sphingosine 1-phosphate from rat uterine leiomyoma ELT3 cells and late pregnant rat myometrium. *Cell. Signal.* **2011**, *23*, 1997–2004. [CrossRef]

31. Pineiro, R.; Maffucci, T.; Falasca, M. The putative cannabinoid receptor GPR55 defines a novel autocrine loop in cancer cell proliferation. *Oncogene* **2011**, *30*, 142–152. [CrossRef] [PubMed]

32. Wielinga, P.R.; van der Heijden, I.; Reid, G.; Beijnen, J.H.; Wijnholds, J.; Borst, P. Characterization of the MRP4- and MRP5-mediated transport of cyclic nucleotides from intact cells. *J. Biol. Chem.* **2003**, *278*, 17664–17671. [CrossRef]

33. Reid, G.; Wielinga, P.; Zelcer, N.; van der Heijden, I.; Kuil, A.; de Haas, M.; Wijnholds, J.; Borst, P. The human multidrug resistance protein MRP4 functions as a prostaglandin efflux transporter and is inhibited by nonsteroidal antiinflammatory drugs. *Proc. Natl. Acad. Sci. USA* **2003**, *100*, 9244–9249. [CrossRef] [PubMed]

34. Rius, M.; Hummel-Eisenbeiss, J.; Keppler, D. ATP-dependent transport of leukotrienes B4 and C4 by the multidrug resistance protein ABCC4 (MRP4). *J. Pharmacol. Exp. Ther.* **2008**, *324*, 86–94. [CrossRef] [PubMed]

35. Rius, M.; Thon, W.F.; Keppler, D.; Nies, A.T. Prostanoid transport by multidrug resistance protein 4 (MRP4/ABCC4) localized in tissues of the human urogenital tract. *J. Urol.* **2005**, *174*, 2409–2414. [CrossRef] [PubMed]

36. Chen, Z.S.; Lee, K.; Kruh, G.D. Transport of cyclic nucleotides and estradiol 17-beta-D-glucuronide by multidrug resistance protein 4. Resistance to 6-mercaptopurine and 6-thioguanine. *J. Biol. Chem.* **2001**, *276*, 33747–33754. [CrossRef] [PubMed]

37. Sassi, Y.; Abi-Gerges, A.; Fauconnier, J.; Mougenot, N.; Reiken, S.; Haghighi, K.; Kranias, E.G.; Marks, A.R.; Lacampagne, A.; Engelhardt, S.; et al. Regulation of cAMP homeostasis by the efflux protein MRP4 in cardiac myocytes. *FASEB J.* **2012**, *26*, 1009–1017. [CrossRef]

38. Vogt, K.; Mahajan-Thakur, S.; Wolf, R.; Broderdorf, S.; Vogel, C.; Bohm, A.; Ritter, C.A.; Graler, M.; Oswald, S.; Greinacher, A.; et al. Release of Platelet-Derived Sphingosine-1-Phosphate Involves Multidrug Resistance Protein 4 (MRP4/ABCC4) and Is Inhibited by Statins. *Thromb. Haemost.* **2018**, *118*, 132–142. [CrossRef]

39. Henderson, M.J.; Haber, M.; Porro, A.; Munoz, M.A.; Iraci, N.; Xue, C.; Murray, J.; Flemming, C.L.; Smith, J.; Fletcher, J.I.; et al. ABCC multidrug transporters in childhood neuroblastoma: Clinical and biological effects independent of cytotoxic drug efflux. *J. Natl. Cancer Inst.* **2011**, *103*, 1236–1251. [CrossRef]

40. Copsel, S.; Garcia, C.; Diez, F.; Vermeulem, M.; Baldi, A.; Bianciotti, L.G.; Russel, F.G.; Shayo, C.; Davio, C. Multidrug resistance protein 4 (MRP4/ABCC4) regulates cAMP cellular levels and controls human leukemia cell proliferation and differentiation. *J. Biol. Chem.* **2011**, *286*, 6979–6988. [CrossRef]

41. Copsel, S.; Bruzzone, A.; May, M.; Beyrath, J.; Wargon, V.; Cany, J.; Russel, F.G.; Shayo, C.; Davio, C. Multidrug resistance protein 4/ATP binding cassette transporter 4: A new potential therapeutic target for acute myeloid leukemia. *Oncotarget* **2014**, *5*, 9308–9321. [CrossRef] [PubMed]

42. Chen, L.; Gu, J.; Xu, L.; Qu, C.; Zhang, Y.; Zhang, W. RNAi-mediated silencing of ATP-binding cassette C4 protein inhibits cell growth in MGC80-3 gastric cancer cell lines. *Cell Mol. Biol. (Noisy-le-Grand)* **2014**, *60*, 1–5.

43. Zhao, X.; Guo, Y.; Yue, W.; Zhang, L.; Gu, M.; Wang, Y. ABCC4 is required for cell proliferation and tumorigenesis in non-small cell lung cancer. *Onco Targets Ther.* **2014**, *7*, 343–351. [CrossRef] [PubMed]

44. Colavita, J.P.M.; Todaro, J.S.; de Sousa, M.; May, M.; Gomez, N.; Yaneff, A.; Di Siervi, N.; Aguirre, M.V.; Guijas, C.; Ferrini, L.; et al. Multidrug resistance protein 4 (MRP4/ABCC4) is overexpressed in clear cell renal cell carcinoma (ccRCC) and is essential to regulate cell proliferation. *Int. J. Biol. Macromol.* **2020**, *161*, 836–847. [CrossRef]

45. Jung, M.; Gao, J.; Cheung, L.; Bongers, A.; Somers, K.; Clifton, M.; Ramsay, E.E.; Russell, A.J.; Valli, E.; Gifford, A.J.; et al. ABCC4/MRP4 contributes to the aggressiveness of Myc-associated epithelial ovarian cancer. *Int. J. Cancer* **2020**, *147*, 2225–2238. [CrossRef]

46. Zhang, Z.; Wang, J.; Shen, B.; Peng, C.; Zheng, M. The ABCC4 gene is a promising target for pancreatic cancer therapy. *Gene* **2012**, *491*, 194–199. [CrossRef]

47. Carozzo, A.; Yaneff, A.; Gomez, N.; Di Siervi, N.; Sahores, A.; Diez, F.; Attorresi, A.I.; Rodriguez-Gonzalez, A.; Monczor, F.; Fernandez, N.; et al. Identification of MRP4/ABCC4 as a Target for Reducing the Proliferation of Pancreatic Ductal Adenocarcinoma Cells by Modulating the cAMP Efflux. *Mol. Pharmacol.* **2019**, *96*, 13–25. [CrossRef]

48. Rius, M.; Nies, A.T.; Hummel-Eisenbeiss, J.; Jedlitschky, G.; Keppler, D. Cotransport of reduced glutathione with bile salts by MRP4 (ABCC4) localized to the basolateral hepatocyte membrane. *Hepatology* **2003**, *38*, 374–384. [CrossRef]

49. Reid, G.; Wielinga, P.; Zelcer, N.; De Haas, M.; Van Deemter, L.; Wijnholds, J.; Balzarini, J.; Borst, P. Characterization of the transport of nucleoside analog drugs by the human multidrug resistance proteins MRP4 and MRP5. *Mol. Pharmacol.* **2003**, *63*, 1094–1103. [CrossRef]

50. Gekeler, V.; Ise, W.; Sanders, K.H.; Ulrich, W.R.; Beck, J. The leukotriene LTD4 receptor antagonist MK571 specifically modulates MRP associated multidrug resistance. *Biochem. Biophys. Res. Commun.* **1995**, *208*, 345–352. [CrossRef]

51. Burkhart, C.A.; Watt, F.; Murray, J.; Pajic, M.; Prokvolit, A.; Xue, C.; Flemming, C.; Smith, J.; Purmal, A.; Isachenko, N.; et al. Small-molecule multidrug resistance-associated protein 1 inhibitor reversan increases the therapeutic index of chemotherapy in mouse models of neuroblastoma. *Cancer Res.* **2009**, *69*, 6573–6580. [CrossRef]

52. Cheung, L.; Flemming, C.L.; Watt, F.; Masada, N.; Yu, D.M.; Huynh, T.; Conseil, G.; Tivnan, A.; Polinsky, A.; Gudkov, A.V.; et al. High-throughput screening identifies Ceefourin 1 and Ceefourin 2 as highly selective inhibitors of multidrug resistance protein 4 (MRP4). *Biochem. Pharmacol.* **2014**, *91*, 97–108. [CrossRef]

53. El-Sheikh, A.A.; van den Heuvel, J.J.; Koenderink, J.B.; Russel, F.G. Interaction of nonsteroidal anti-inflammatory drugs with multidrug resistance protein (MRP) 2/ABCC2- and MRP4/ABCC4-mediated methotrexate transport. *J. Pharmacol. Exp. Ther.* **2007**, *320*, 229–235. [CrossRef] [PubMed]

54. Tivnan, A.; Zakaria, Z.; O'Leary, C.; Kogel, D.; Pokorny, J.L.; Sarkaria, J.N.; Prehn, J.H. Inhibition of multidrug resistance protein 1 (MRP1) improves chemotherapy drug response in primary and recurrent glioblastoma multiforme. *Front. Neurosci.* **2015**, *9*, 218. [CrossRef] [PubMed]

55. Munoz, M.; Henderson, M.; Haber, M.; Norris, M. Role of the MRP1/ABCC1 multidrug transporter protein in cancer. *IUBMB Life* **2007**, *59*, 752–757. [CrossRef]

56. Kochel, T.J.; Reader, J.C.; Ma, X.; Kundu, N.; Fulton, A.M. Multiple drug resistance-associated protein 4 (MRP4) exports prostaglandin E2 (PGE2) and contributes to metastasis in basal/triple negative breast cancer. *Oncotarget* **2017**, *8*, 6540–6554. [CrossRef] [PubMed]

57. Wang, W.; Li, Y.; Zhu, J.Y.; Fang, D.; Ding, H.F.; Dong, Z.; Jing, Q.; Su, S.B.; Huang, S. Triple negative breast cancer development can be selectively suppressed by sustaining an elevated level of cellular cyclic AMP through simultaneously blocking its efflux and decomposition. *Oncotarget* **2016**, *7*, 87232–87245. [CrossRef]

58. Zochbauer-Muller, S.; Filipits, M.; Rudas, M.; Brunner, R.; Krajnik, G.; Suchomel, R.; Schmid, K.; Pirker, R. P-glycoprotein and MRP1 expression in axillary lymph node metastases of breast cancer patients. *Anticancer Res.* **2001**, *21*, 119–124.

59. Yamada, A.; Nagahashi, M.; Aoyagi, T.; Huang, W.C.; Lima, S.; Hait, N.C.; Maiti, A.; Kida, K.; Terracina, K.P.; Miyazaki, H.; et al. ABCC1-Exported Sphingosine-1-phosphate, Produced by Sphingosine Kinase 1, Shortens Survival of Mice and Patients with Breast Cancer. *Mol. Cancer Res.* **2018**, *16*, 1059–1070. [CrossRef]

60. Naviglio, S.; Di Gesto, D.; Romano, M.; Sorrentino, A.; Illiano, F.; Sorvillo, L.; Abbruzzese, A.; Marra, M.; Caraglia, M.; Chiosi, E.; et al. Leptin enhances growth inhibition by cAMP elevating agents through apoptosis of MDA-MB-231 breast cancer cells. *Cancer Biol. Ther.* **2009**, *8*, 1183–1190. [CrossRef]

61. Ma, X.; Kundu, N.; Rifat, S.; Walser, T.; Fulton, A.M. Prostaglandin E receptor EP4 antagonism inhibits breast cancer metastasis. *Cancer Res.* **2006**, *66*, 2923–2927. [CrossRef]

62. Tsuchida, J.; Nagahashi, M.; Nakajima, M.; Moro, K.; Tatsuda, K.; Ramanathan, R.; Takabe, K.; Wakai, T. Breast cancer sphingosine-1-phosphate is associated with phospho-sphingosine kinase 1 and lymphatic metastasis. *J. Surg. Res.* **2016**, *205*, 85–94. [CrossRef] [PubMed]

63. Takabe, K.; Kim, R.H.; Allegood, J.C.; Mitra, P.; Ramachandran, S.; Nagahashi, M.; Harikumar, K.B.; Hait, N.C.; Milstien, S.; Spiegel, S. Estradiol induces export of sphingosine 1-phosphate from breast cancer cells via ABCC1 and ABCG2. *J. Biol. Chem.* **2010**, *285*, 10477–10486. [CrossRef]

64. Franken, N.A.; Rodermond, H.M.; Stap, J.; Haveman, J.; van Bree, C. Clonogenic assay of cells in vitro. *Nat. Protoc.* **2006**, *1*, 2315–2319. [CrossRef]

Publisher's Note: MDPI stays neutral with regard to jurisdictional claims in published maps and institutional affiliations.

Article

Role of Genetic Variation in ABC Transporters in Breast Cancer Prognosis and Therapy Response

Viktor Hlaváč [1,2], Radka Václavíková [1,2], Veronika Brynychová [1,2], Renata Koževnikovová [3], Katerina Kopečková [4], David Vrána [5], Jiří Gatěk [6] and Pavel Souček [1,2,*]

[1] Toxicogenomics Unit, National Institute of Public Health, 100 42 Prague, Czech Republic; viktor.hlavac@szu.cz (V.H.); rvaclavikova@seznam.cz (R.V.); veronikabrynychova@seznam.cz (V.B.)

[2] Biomedical Center, Faculty of Medicine in Pilsen, Charles University, 323 00 Pilsen, Czech Republic

[3] Department of Oncosurgery, Medicon Services, 140 00 Prague, Czech Republic; renata.kozevnikovova@onko-centrum.cz

[4] Department of Oncology, Second Faculty of Medicine, Charles University and Motol University Hospital, 150 06 Prague, Czech Republic; katerina.kopeckova@fnmotol.cz

[5] Department of Oncology, Medical School and Teaching Hospital, Palacky University, 779 00 Olomouc, Czech Republic; davvrana@gmail.com

[6] Department of Surgery, EUC Hospital and University of Tomas Bata in Zlin, 760 01 Zlin, Czech Republic; gatekj@gmail.com

* Correspondence: pavel.soucek@szu.cz; Tel.: +420-267-082-711

Received: 19 November 2020; Accepted: 11 December 2020; Published: 15 December 2020

Abstract: Breast cancer is the most common cancer in women in the world. The role of germline genetic variability in ATP-binding cassette (ABC) transporters in cancer chemoresistance and prognosis still needs to be elucidated. We used next-generation sequencing to assess associations of germline variants in coding and regulatory sequences of all human ABC genes with response of the patients to the neoadjuvant cytotoxic chemotherapy and disease-free survival ($n = 105$). A total of 43 prioritized variants associating with response or survival in the above testing phase were then analyzed by allelic discrimination in the large validation set ($n = 802$). Variants in *ABCA4*, *ABCA9*, *ABCA12*, *ABCB5*, *ABCC5*, *ABCC8*, *ABCC11*, and *ABCD4* associated with response and variants in *ABCA7*, *ABCA13*, *ABCC4*, and *ABCG8* with survival of the patients. No association passed a false discovery rate test, however, the rs17822931 (Gly180Arg) in *ABCC11*, associating with response, and the synonymous rs17548783 in *ABCA13* (survival) have a strong support in the literature and are, thus, interesting for further research. Although replicated associations have not reached robust statistical significance, the role of ABC transporters in breast cancer should not be ruled out. Future research and careful validation of findings will be essential for assessment of genetic variation which was not in the focus of this study, e.g., non-coding sequences, copy numbers, and structural variations together with somatic mutations.

Keywords: ABC transporter; therapy response; disease-free survival; breast cancer; next-generation sequencing; competitive allele-specific PCR

---

1. Introduction

Breast cancer (Online Mendelian Inheritance in Man, OMIM no. 114480) is the most common cancer in women in the world [1]. One of the frequently studied reasons for the lack of successful treatment outcomes in a considerable portion of the patients is multidrug resistance [2]. Multidrug resistance can be caused by ATP-binding cassette (ABC) transporters, e.g., by higher efflux of drugs out of tumor cells by P-glycoprotein (Multidrug resistance, MDR coded by *ABCB1* gene) [2,3].

ABC transporters represent a large superfamily of membrane transporter proteins classified into seven families and translocate numerous compounds across intra and extracellular membranes. Their substrates include metabolic products, sterols, lipids, and xenobiotics [3]. However, of the total number of 48 active human ABC transporters, up to only 16 are able to transport anticancer drugs [3]. Most of the drug resistance is ascribed to largely studied multidrug resistance-related transporters *ABCB1* (MDR, OMIM no. 171050), *ABCC1* (MRP, OMIM no. 158343), and *ABCG2* (BCRP, OMIM no. 603756) [4]. Proteins of the family ABCA are mostly lipid sterol transporters and can be associated with several diseases, e.g., Tangier or Alzheimer disease [5]. Their roles in cancer progression and the metastatic potential linked to lipid trafficking recently became the focus of numerous studies [6]. ABCBs participate in antigen processing and immune deficiency. Apart from ABCB1, a promiscuous and ubiquitous efflux pump [2], ABCB members also represent transporters of heme and bile acids [7]. Family C is mostly dedicated to multidrug resistance (MRP1-6) [8,9], but ABCC6, ABCC7, and ABCC8/9 are linked to serious diseases (pseudoxanthoma elasticum, cystic fibrosis, and diabetes mellitus, respectively) [9]. ABCDs are responsible for transport of fatty acids from peroxisomes to the cytoplasm [10] and ABCGs transport various products of metabolism, xenobiotics including anti-cancer drugs, bile acids, and steroids [11]. The rest of the transporters are not involved in transport, but rather act as translational inhibitors or protein synthesis regulators (ABCFs and ABCEs) [12].

Our recent pharmacogenomics study revealed a prognostic and predictive potential in a number of alterations in breast cancer [13]. The studied genes were implicated in the metabolism and transport of drugs administered to breast cancer patients in the clinics, clearly documenting the importance of this tool for the personalized medicine. The study provided a proof-of-the principle for the design and bioinformatics methodology, namely, the assembly and testing of an in-house pipeline for variant prioritization. Given the total number of 509 genes screened by the next generation sequencing (NGS), only a portion of variants could be validated in a subsequent phase. In order to select the most relevant functional alterations from the statistically significant set of variants, we down-sampled the results using information from in silico predictions and according to previously confirmed pharmacogenomic and clinical evidence. Thus, some potentially useful biomarkers of prognosis or prediction of therapy outcome could have been missed.

In the present study, we aimed to use less strict criteria for investigating the importance of germline genetic variability in coding, untranslated regions (UTR), and adjacent regions of all human members of the ABC superfamily for prognosis and response to cytotoxic therapy of breast cancer patients. All variants in ABCs identified in the testing phase were correlated with disease-free survival (DFS) and response of the patients to preoperative cytotoxic therapy, and thoroughly reviewed, including permutation analysis, evaluation of haplotypes, and gene dosage. We have not addressed functional relevance to enable identification of purely correlative biomarkers. Prioritized variants were further validated in a large cohort of breast cancer patients from a single population. Thus, the present study brings a more detailed view of the relevance of genetic variability of ABC transporters for breast cancer prognosis and therapy outcome predictions.

2. Results

2.1. Testing Phase

The clinical characteristics of the patients ($n = 105$) are in Table S1. The subgroup of patients was treated with the neoadjuvant cytotoxic therapy (NACT) ($n = 68$) and the response to this treatment was available. The rest of patients were treated with adjuvant cytotoxic therapy based on monotherapy or combinations of anthracyclines, cyclophosphamide, 5-fluorouracil, and taxanes. DFS was evaluated for all patients and the mean follow-up of the patients was 70 ± 28 months.

The average coverage was 82.3 ± 29.1 and 95% of the captured regions were covered at least ten times. Altogether, we found 2611 variants in exonic and adjacent intronic sequences. Of the total number of 48 genes and one pseudogene (*ABCC13*) subjected to analysis, 46 genes (94%) contained at least one genetic alteration. No alterations were found in *ABCF1*, *TAP1* (alias *ABCB2*), and *TAP2* (*ABCB3*) genes. On the other hand, the most polymorphic genes, with over one hundred alterations, were *ABCA13* (165 alterations), *ABCA4* (114), and *ABCA1* (109). Of the total number of 2611 variants, 636 were in exons, 1544 intronic, and 253 were in 3′UTR or 5′UTR regions according to National Center for Biotechnology Information (NCBI) Reference Sequence Database (RefSeq; https://www.ncbi.nlm.nih.gov/refseq/) in Annovar (Table 1).

Table 1. Observed alterations in ATP-binding cassette (ABC) transporters divided by function according to Annovar.

Type	Total	Percentage
Intronic	1544	59.1
Exonic (coding)	636	24.4
UTR′3	204	7.8
Intergenic	76	2.9
UTR′5	49	1.9
Downstream [1]	28	1.1
Upstream [1]	26	1.0
Splicing [2]	13	0.5
Other [3]	35	1.3

[1] Variants are 1 kb from transcription end/start site; [2] Variants are 2 bp within splice junction. [3] Exonic/intronic non-coding RNA, or variant in overlapping regions (upstream–downstream) of two different genes.

On average, each patient showed 608 ± 33 variants. We found 17 loss-of-function (LOF) truncating variants that were either affecting the stop codon (stop-gain) or frameshift insertions or deletions (indels). There were 355 of the variants that were non-synonymous single nucleotide variants (SNVs) and 263 that were synonymous SNVs (Table 2). In total, 1058 variants (41%) had minor allele frequency (MAF) > 0.05, and the rest of the 1553 variants, had MAF ≤ 0.05.

Table 2. Overview of the observed exonic alterations in ABC transporters by coding consequence.

Classification	Total	Percentage
Non-synonymous SNV	355	55.8
Synonymous SNV	263	41.4
Stop-gain	8	1.3
Frameshift deletion	6	0.9
Frameshift insertion	3	0.5
Non-frameshift deletion	1	0.1

Altogether, 256 (10%) of the variants were novel (i.e., not found in dbSNP Build 150). Out of these, 162 had MAF > 0.01. The distribution of the variants according to their functional classes and frequencies of novel variants in the groups of genes are shown in Figure 1.

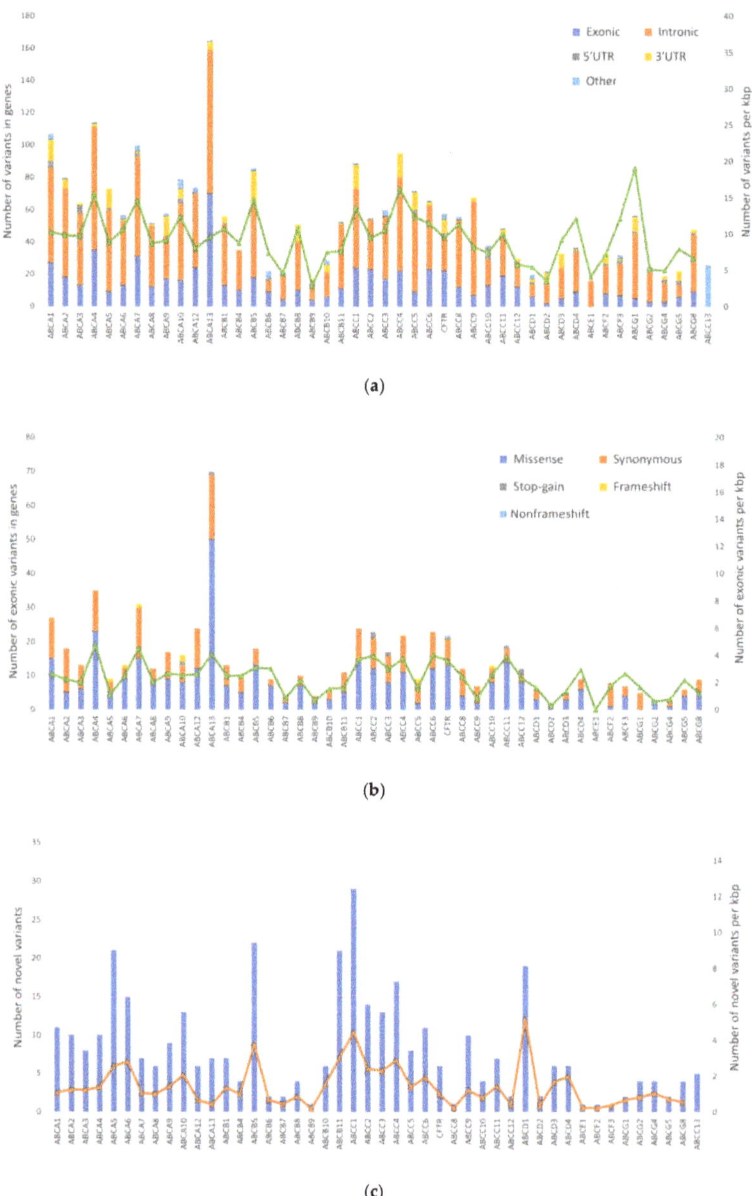

Figure 1. Distribution of alterations in individual ABC transporter genes. The picture shows: (**a**) the frequency of genetic alterations according to their functional classes; (**b**) the frequency of genetic alterations according to their exonic functional classes analyzed by RefSeq: National Center for Biotechnology Information (NCBI) Reference Sequence Database (https://www.ncbi.nlm.nih.gov/refseq/) shown according to individual transporters (excluding *ABCC13* pseudogene); and (**c**) the number of novel variants according to individual transporters. The number of the variants normalized to the transcript length in kilo base pairs (kbp) per each gene are shown for each plot on the right axis and depicted by the overlaid line.

Variants departing from Hardy–Weinberg equilibrium ($p < 0.01$, $n = 101$) were excluded from analyses and further only variants with MAF > 0.05 were considered relevant to achieve adequate statistical power in the validation phase. In addition, variants with the missing data in more than 50% of patients were excluded ($n = 54$). Application of these filtration criteria resulted in a set of 903 variants which were further evaluated for associations with the response of patients to NACT and DFS. We found 56 variants associated with the response to NACT and 47 variants associating with DFS. Six variants reported significant in the previous study [13] were further excluded. Following haplotype evaluations, 38 variants were considered tagged (>80%) with some other variant and were not assessed further. As a result, 22 variants associated with the response and 37 variants associated with DFS were followed. The gene dosage relationship was then evaluated for variants associating with DFS and variants not fulfilling this condition were excluded ($n = 7$). Neither of these variants was significant in the recessive genetic model (variant allele versus wild type). Following these control steps, 52 variants (45 SNVs and 7 indels, Table S2) were prioritized for the validation phase in a larger cohort of breast cancer patients.

2.2. Validation Phase

The clinical characteristics of the patients ($n = 802$) are summarized in Table S3. A subgroup of patients treated with NACT composed of 168 patients. In total, 371 patients received adjuvant cytotoxic therapy. Patients with localized disease and good prognosis were untreated ($n = 83$) and a portion of patients was treated only with hormonal therapy ($n = 311$). The mean follow-up of the patients was 76 ± 30 months.

Out of 52 variants subjected to genotyping, attempts to optimize detection techniques failed in 10 variants (5 indels and 5 SNVs) and could not be further evaluated for clinical associations. One variant was additionally included into the list (rs2893007) based on haplotype tagging ($r^2 = 1$) to replace the variant rs11764054 whose analysis failed. No tagging variants ($r^2 > 0.8$) were found to replace the rest of the failed variants. In the end, 43 variants were successfully genotyped. No variants significantly departed from Hardy–Weinberg equilibrium and less than 1% of theoretical data points were missing due to uncertainty in genotype calling, or an absent signal. MAFs of variants in the validation phase did not substantially differ from those observed in the testing one (Table 3).

Table 3. Distribution of genotypes for variants assessed in the validation phase.

Gene	SNP ID [1]	Genotype Distribution [2]			MAF [3]	
		Common Homozygotes	Heterozygotes	Rare Homozygotes	Validation Set	Testing Set
ABCA1	rs41474449	658	136	3	0.09	0.07
ABCA4	rs537831	377	342	78	0.31	0.31
ABCA4	rs2065711	436	309	55	0.26	0.20
ABCA4	rs2275032	540	230	30	0.18	0.14
ABCA4	rs2275033	270	396	133	0.41	0.40
ABCA4	rs3789398	353	361	84	0.33	0.35
ABCA5	rs1420904	679	113	7	0.08	0.08
ABCA5	rs2067851	681	112	2	0.07	0.07
ABCA7	rs9282562	604	188	8	0.13	0.14
ABCA8	rs4147976	318	358	121	0.38	0.35
ABCA9	rs2302294	368	352	80	0.32	0.34
ABCA9	rs11871944	326	366	103	0.36	0.40
ABCA12	rs71428357	726	70	3	0.05	0.08
ABCA13	rs7780299	597	187	16	0.14	0.12
ABCA13	rs17132289	687	106	6	0.07	0.08
ABCA13	rs17548783	201	400	192	0.49	0.49
ABCA13	rs28637820	628	163	8	0.11	0.13
ABCA13	rs74859514	665	124	10	0.09	0.08
ABCB1	rs9282564	609	168	21	0.13	0.13
ABCB5	rs3210441	283	400	116	0.40	0.44
ABCB5	rs12700230	466	285	49	0.24	0.23
ABCB5	rs2893007	676	120	5	0.08	0.10
ABCB8	rs2303922	336	362	100	0.35	0.34
ABCB11	rs853772	203	403	190	0.49	0.25
ABCC1	rs4148379	456	287	48	0.24	0.20
ABCC2	rs2273697	478	273	39	0.22	0.22
ABCC3	rs8077268	649	147	4	0.10	0.10
ABCC3	rs12604031	271	374	154	0.43	0.44
ABCC4	rs899494	583	198	17	0.15	0.12
ABCC4	rs2274405	339	352	102	0.35	0.37
ABCC5	rs4148579	259	404	137	0.42	0.43
ABCC5	rs12638017	686	111	3	0.07	0.06
ABCC8	rs739689	349	356	91	0.34	0.40
ABCC10	rs75320251	654	135	8	0.09	0.09
ABCC11	rs17822931	592	184	21	0.14	0.14
ABCC13	rs2254297	254	381	160	0.44	0.40
ABCC13	rs2822582	306	369	121	0.38	0.40
ABCD4	rs2301346	394	334	67	0.29	0.32
ABCD4	rs2301347	305	376	120	0.38	0.40
ABCF2	rs79537035	527	242	30	0.19	0.23
ABCG8	rs34198326	685	109	4	0.07	0.06
ABCG8	rs56260466	685	104	9	0.08	0.06
CFTR	rs34855237	538	229	30	0.18	0.08

[1] Reference number in dbSNP (https://www.ncbi.nlm.nih.gov/snp/); [2] Genotypes do not sum up to 802 due to missing data; [3] MAF = minor allele frequency.

We further evaluated associations of variants with the response and DFS of patients in the validation set. For six variants with less frequent homozygous genotypes ($n < 5$), the recessive genetic model was used for the sake of the statistical power of comparisons. The variants that associated with response in both testing and validation phases are listed in Table 4 and thus these variants are considered replicated with putative clinical importance. However, none of these associations passed the false discovery rate (FDR) test for multiple testing ($q = 0.0025$) and, thus, cannot be deemed statistically significant after such correction.

Table 4. Validated variants in ABC transporters significantly associating with the response of patients to the neoadjuvant cytotoxic therapy.

Gene	SNP ID	Genotype	Good Response [1]	Poor Response [1]	χ-Square	*p*-Value
ABCA4	rs2275032	AA	75	33	6.33	0.042
		CA	48	7		
		CC	4	1		
ABCA9	rs11871944	CC	48	24	6.76	0.034
		CT	61	14		
		TT	18	2		
ABCA12	rs71428357	CC	103	41	8.32	0.004
		CT + TT	22	0		
ABCB5	rs3210441	GG	42	11	6.22	0.045
		GA	62	28		
		AA	23	2		
ABCC5	rs4148579	CC	43	6	8.55	0.014
		CT	68	24		
		TT	15	11		
ABCC8	rs739689	AA	55	11	6.81	0.033
		GA	60	21		
		GG	11	9		
ABCC11	rs17822931	CC	89	37	6.42	0.011
		CT + TT	37	4		
ABCD4	rs2301347	CC	53	9	8.59	0.014
		CA	60	21		
		AA	14	11		
ABCD4	rs2301346	TT	73	15	7.28	0.026
		CT	45	20		
		CC	7	6		

[1] Numbers of patients with specified genotypes divided by response to the neoadjuvant treatment.

Subsequently, associations of variants with DFS of all patients and patients stratified according to the received therapy were evaluated. Significant results for all patients with complete follow-up data (n = 744) are displayed in Figure 2a. Results for patients treated with cytotoxic therapy (n = 371) are presented in Figure 2b. No significant association was observed in a subgroup of patients treated only with hormonal therapy (n = 312).

Of these variants, rs74859514 did not pass the gene dosage condition (Figure 2b). None of the associations has passed the FDR test for multiple testing (q = 0.0023) and, thus, cannot be further considered statistically validated.

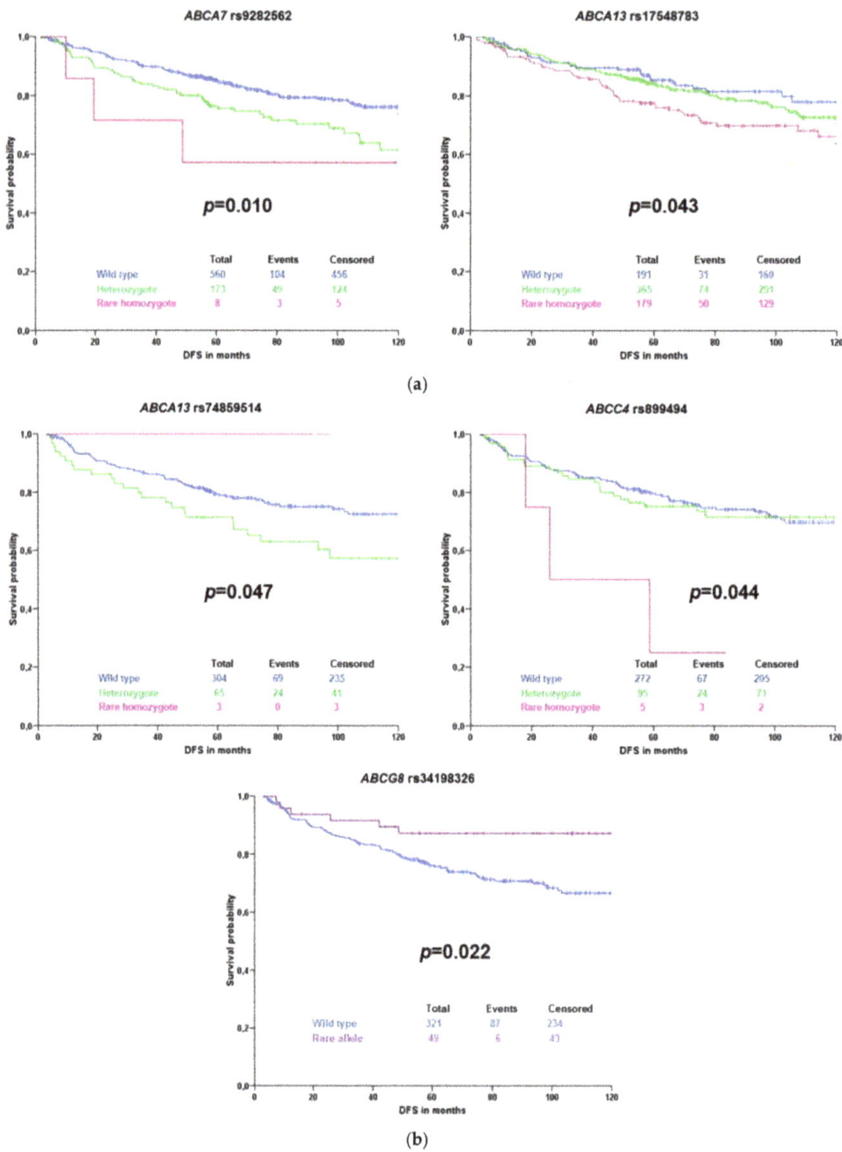

Figure 2. Kaplan–Meier plots with validated associations of variants with disease-free survival (DFS) of all patients unstratified (**a**) or subgroup of patients treated with cytotoxic therapy (**b**). The blue line represents the common homozygous genotype, the green line the heterozygote, and the magenta line the rare homozygote. The violet color represents rare variant carriers. Significance was evaluated by the log-rank test; numbers represent individuals in compared groups.

We further clarified the effect of molecular subtype on prognosis of the patients by their stratification according to the intrinsic molecular subtype. Associations with DFS were then calculated separately for each subtype (Table 5 and Figure S1).

Table 5. Validated associations of variants in ABC transporters associating with DFS of patients treated with cytotoxic therapy according to their molecular subtypes.

Gene	SNP ID	Genotypes	Subtypes [1]			
			Luminal A	Luminal B	HER2	TNBC
All patients (*n* = 744)						
ABCA7	rs9282562	Common homozygous	165	206	39	63
		Rare allele	45	68	17	21
		p-value [2]	0.626	0.316	**0.010**	0.325
ABCA13	rs17548783	Common homozygous	58	67	12	24
		Heterozygous	112	129	31	40
		Rare homozygous	39	74	13	20
		p-value [2]	0.050	0.114	0.492	**0.039**
Patients treated with cytotoxic therapy (*n* = 371)						
ABCA13	rs74859514	Common homozygous	62	125	25	58
		Rare allele	12	27	11	8
		p-value [2]	0.441	0.606	**0.001**	**0.009**
ABCC4	rs899494	Common homozygous	50	114	27	46
		Rare allele	24	38	9	20
		p-value [2]	0.825	0.415	0.050	0.565
ABCG8	rs34198326	Common homozygous	63	135	30	54
		Rare allele	11	17	5	11
		p-value [2]	0.094	**0.040**	0.847	0.091

[1] Numbers of patients for each genotype stratified by molecular subtypes are displayed; HER2 = HER2 enriched, TNBC = triple negative breast cancer. [2] *p*-value departed from log-rank test (significant results in bold).

This analysis showed that prognostic associations differ according to the molecular subtype. In the whole group of patients, rs9282562 in *ABCA7* and rs17548783 in *ABCA13* were prognostic only for HER2 enriched and triple negative (TNBC) subtypes, respectively. In the subgroup treated with cytotoxic therapy, again single nucleotide polymorphisms (SNPs) were prognostic for patients with HER2 enriched and TNBC subtypes (rs74859514 in *ABCA13*). Carriers of the rare variant in *ABCG8* rs34198326 had better DFS than those with the common homozygous genotype—this association was significant only in patients with the luminal B subtype.

In order to clarify associations of variants with gene expression, we used transcriptomic data from previous gene expression profiling [14] and compared it with variants that significantly associated with DFS or response to NACT in the validation set (*n* = 168 patients for whom gene expression was available). We also analyzed expression quantitative trait loci (eQTL) associations of these variants using gene expression in healthy tissues available on the GTEx portal (https://www.gtexportal.org). The SNP rs17548783 was significantly associated with *ABCA13* transcript levels in tumors assessed in the previous study [14] (Table 6), but no eQTL were found in the GTEx database. No eQTL were found also for rs2275032 in *ABCA4*, rs71428357 in *ABCA12*, rs3210441 in *ABCB5*, and rs34198326 in *ABCG8*. Significant results from eQTL analysis are summarized in Table 7 and Figure S2.

Table 6. Association of single nucleotide polymorphism (SNP) rs17548783 in ABCA13 transporter with intratumoral transcript levels.

SNP ID	Genotype	*n*	Expression [1]	S.D. [2]	*p*-Value
rs17548783	Common homozygous	9	−7.29	2.11	0.015
	Heterozygous	18	−9.90	2.67	-
	Rare homozygous	7	−9.45	2.67	-

[1] Transcript levels expressed as Ct (cycle threshold) value of ABCA13 subtracted from mean Ct of reference genes [14]. [2] S.D. = Standard deviation.

Table 7. eQTL analysis of SNPs significantly associating with patients' DFS or response to neoadjuvant cytotoxic therapy (NACT).

SNP ID	Gene	Tissue	Normalized Expression Trend	*p*-Value [1]
rs11871944	*ABCA9*	multiple [2]	CC > TC > TT	3.1×10^{-7}
rs4148579	*ABCC5*	multiple [2]	CC > TC > TT[3]	3.5×10^{-33}
rs739689	*ABCC8*	brain (cerebellum)	AA > AG > GG	8.6×10^{-9}
rs17822931	*ABCC11*	brain	CC > CT > TT	6.9×10^{-5}
	LONP2 [4]	breast	CC > CT > TT	1.4×10^{-4}
rs2301347	*ABCD4*	multiple [2]	AA > CA > CC	5.0×10^{-20}
	lnc-SYNDIG1L-2 [4]	breast	AA > CA > CC	1.8×10^{-16}
rs2301346	*ABCD4*	multiple [2]	CC > TC > TT	1.8×10^{-12}
	lnc-SYNDIG1L-2 [4]	breast	CC > TC > TT	4.2×10^{-15}
rs9282562	*ABCA7*	multiple [2]	ref > het > delTG	1.3×10^{-11}
s74859514	*UPP1* [4]	cerebellum, muscle	GG > GC > CC	2.7×10^{-5}
rs899494	*ABCC4*	thyroid, whole blood	AA > AG > GG	1.9×10^{-16}

[1] *p*-value of the most significant association is shown. [2] Significant results in more than three different tissues. [3] The highest expression is seen for TT genotype in whole blood and esophageal mucosa; the opposite i.e., highest expression of CC genotype is seen for the rest of the tissues. [4] Alternative eQTL.

3. Discussion

The role of germline genetic variability among ABC transporters in prognosis of breast cancer patients as well as in their response to chemotherapy is underexplored. In our previous publication, we dealt with pharmacologically relevant germline genetic polymorphisms in 509 breast cancer-related genes [13]. In the present study, we used the same approach to reveal all associations of genetic variants in human ABC transporters with chemotherapy response and survival of the patients.

A total of 2611 variants were found in a testing set. The majority of variants were found in intronic regions. Lower numbers of variants were found in coding regions and UTRs. Interestingly, no variants were found in *ABCF1*, *TAP1* (alias *ABCB2*), and *TAP2* (*ABCB3*). *TAP1* and *TAP2* are antigen presenting transporters and alterations in these genes associate with autoimmune diseases, susceptibility to infections, or malignancies [15]. Similarly, *ABCF1* plays a role in the regulation of inflammatory processes [16] and alterations in *ABCF1* are linked with autoimmune diseases as well [17]. Therefore, it seems that genetic variants in these genes negatively impacts immunity and inflammatory processes which explain limited variability, in line with our findings. On the other hand, the most variable genes were *ABCA13* (165 alterations), *ABCA4* (114), and *ABCA1* (109). The members of ABCA family are typically large genes (transcript length 7-17 kbp) and thus likely to accumulate variants. When normalized for the length of transcript, *ABCG1*, *ABCC4*, and *ABCA4* have the highest count of variants per kbp, ranging from 16 to 20. Interestingly, *ABCA4*, *ABCA7*, and *ABCA13* had the highest variant counts in exonic regions (4.1-4.8 variants per kbp). We found several LOF variants in ABC transporters—eight stop-gains and nine frameshift indels. These events have high impact on function of the protein. Moreover, all 17 LOF variants were present in genes of the first quartile of the most intolerant genes to LOF events [18]. These facts advocate for further investigation of LOF variants in ABC transporters. Unfortunately, LOF variants are rare. For the sake of maintaining enough statistical power for comparison with clinical data, only common variants (MAF > 0.05) could be used in the present study.

In total, we selected 903 variants and subjected them to a thorough statistical analyses. Of these variants, 43 associated with response or DFS and were capable of validation in a cohort of 802 breast cancer patients. Five associations with DFS and nine with response to NACT were replicated in the validation set. If multiple simultaneous statistical tests are calculated, a type I error (a risk of false positive results) occurs. To prevent this error, correction for multiple testing must be used. There are several methods to do so. Here, we applied the wildly used FDR, a test by Benjamini–Hochberg. After this correction, none of the associations of variants with clinical features remained significant and, thus, cannot be considered validated. Nevertheless, we found some interesting associations which we will discuss further.

ABCA13 is responsible for lipid transport and variants in this gene can cause schizophrenia [5]. Carriage of the rare allele of SNP rs17548783, located in downstream intronic region of *ABCA13,* was associated with shorter DFS of patients in our study. Based on our findings, a rare allele of this variant, significantly associated with lower *ABCA13* intratumoral transcript levels in a validation set (Table 6). Lower transcript levels of *ABCA13* were associated with worse response to NACT in a previous study [14], further underpinning the role of this SNP as a putative poor prognosis biomarker. This consequence is the most interesting link observed at present. Nevertheless, the response to NACT does not significantly associate with DFS in our datasets a fact that clearly calls for further research.

Another variant in *ABCA13*, the missense rs74859514 (Ala2223Pro), associated with DFS in patients treated with chemotherapy, but without gene dosage relationship. Neither of these two SNPs has records in the present literature, although associations of *ABCA13* with patients' outcome have been described in several previous studies. A decreased expression of *ABCA13* was associated with shorter DFS in 51 glioblastoma patients [19] and 51 colorectal cancer patients [20]. The opposite was found for ovarian cancer (n = 77) and higher levels of *ABCA13* predicted worse overall survival in ovarian cancer patients [21]. Amplification of 7p12 (which includes *ABCA13* and *HUS1, EGFR,* and *IKZF1*) predicted worse response to NACT in muscle-invasive bladder cancer [22]. Such contradictory results from different cancers are puzzling. Despite we must bear in mind that none of the associations found in our study passed the FDR test, some may still have clinical potential. Additional studies will be needed to confirm these results.

A synonymous variant rs71428357 in *ABCA12* associated with response to NACT. Patients responding well to chemotherapy were more often carriers of the rare allele. Synonymous variants can affect RNA splicing, folding, and stability [23] and are associated with several diseases, such as Alzheimer disease, pulmonary sarcoidosis, galactosemia, or cancer [24]. The role of this particular *ABCA12* variant in cancer or other diseases is still unknown. However, higher *ABCA12* transcript levels in non-tumor tissues associated with worse response to NACT in breast cancer patients in our previous study [14]. The opposite, i.e., higher levels associating with residual disease, was found by Park et al. [25]. Interestingly, we previously identified this gene among candidate ABCs with predictive or prognostic potential for patients with breast, colorectal, and pancreatic carcinomas [26].

Among other members of the ABCA family, associations with response to NACT were observed for *ABCA4* (variant rs2275032) and *ABCA9* (rs11871944). A deletion in *ABCA7* (rs9282562) associated with shorter DFS of the patients. These variants are not described in the present literature, however, higher transcript levels of *ABCA9* significantly associated with worse survival in high-grade serous ovarian cancer tumors [6]. Silencing of *ABCA7* reduces epithelial to mesenchymal transition in ovarian cancer cell lines and knockdown of *ABCA7* inhibited migration, cell proliferation, and invasion [27]. In addition, lower *ABCA7* levels associated with shorter DFS of colorectal cancer patients [20].

ABCB5 confers 5-fluorouracil resistance and promotes cell invasiveness in colorectal cancer [28]. Variant rs3210441 in *ABCB5* associated with response to NACT in our study, but no eQTL was found and additional supportive data about the role of this SNP or protein in breast cancer is lacking.

Protein coded by *ABCC11* is responsible for transport of bile acids, conjugated steroids, or cyclic nucleotides. Diseases linked with this gene include malfunction of apocrine gland secretion and lateral sinus thrombosis (https://www.genecards.org). *ABCC11* is a transporter of 5-fluorouracil [3]. In our study, a missense *ABCC11* variant rs17822931 (Gly180Arg) associated with response to NACT. Carriers of the wild-type allele had significantly poorer outcomes than patients with an alternative allele. This variant is known for its determination of human earwax type [29]. It is associated with breast cancer risk in the Japanese population [30]. This variant is also linked with axillary osmidrosis, colostrum secretion in the mammary gland, and mastopathies [31]. Wild type allele C also confers to chemotherapy resistance to 5-fluorouracil by exporting active metabolite 5-fluoro-2′-deoxyuridine 5′-monophosphate (FdUMP) [32]. *ABCC11* expression (together with *ABCB1*) is responsible for resistant phenotype of breast cancer cell lines resistant to eribulin and inhibition of *ABCC11* can partially restore the cross-resistance to 5-fluorouracil [33]. Higher *ABCC11* gene expression was also associated with poor response to NACT in breast cancer patients [25]. Interestingly, this SNP is associated with expression of *ABCC11* only in the brain, but with *LONP2*, coding mitochondrial matrix protein, in breast tissue (Table 7). Relations between mastopathy, breast cancer risk, and, after chemotherapy, even drug resistance suggest strong connection of this variant to breast cancer. Association with response to chemotherapy of breast cancer patients has been suggested previously [31], our result corroborates this assertion.

Among other members of the ABCC family, *ABCC5* (rs4148579) and *ABCC8* (rs739689) associated with response to NACT and *ABCC4* (rs899494) with DFS of the patients. *ABCC4* was among amplified genes in resistant cancer cell lines [34]. The *ABCC4* gene was also identified to play a role in cellular migration of breast cancer cell line models MCF-7 and MDA-MB-231 [35]. In our previous study [14], we have seen associations of high *ABCC8* transcript levels with low grade and negative/positive status of estrogen receptor. Additionally, the expression level non-significantly ($p = 0.096$) associated with worse responses of breast cancer patients to NACT [14]. Nevertheless, in the present study we did not find association of rs739689 (intronic A > G transition) with *ABCC8* transcript levels. eQTL associations at the GTEx portal are ambiguous. The wild-type AA genotype has the highest expression of *ABCC8* in cerebellum, but no significant association can be found in breast tissue. This SNP is also highly significantly associated with expression of *NCR3LG1*, *KCNJ11*, and *SNORD14* genes with fragmentary and elusive data on association with breast cancer. From the data discussed above, it can be summarized that the present knowledge is incomplete and, thus, no clear picture can be presented.

Unlike other ABCD transporters, *ABCD4* is not found in peroxisomes, but in lysosomes. It takes part in transport of cobalamin (vitamin B12) and mutations in this transporter cause inherited defects of intracellular cobalamin metabolism [10]. Low transcript levels of this gene were also associated with shorter DFS of colorectal cancer patients [20] and *ABCD4* was among amplified genes in resistant cancer cell lines [34]. In our study, wild-type variants rs2301347 and rs2301346 (both intronic) associated with the good response to NACT. Wild-type genotypes of these two variants show lower transcript levels of long non-coding (lnc) RNA lnc-SYNDIG1L-2 overlapping *ABCD4* in mammary tissue (Table 7) suggesting potential clinical relevance. However, the lack of association with *ABCD4* transcript levels that we found in our dataset precludes any strict conclusions.

ABCG8 is a transporter of sterols from hepatocytes and enterocytes [36]. The rare allele of its SNP rs34198326 was associated with longer DFS of chemotherapy treated patients in our study. Expression of *ABCG8* was downregulated in tumors of breast cancer patients compared to non-neoplastic control tissues [14], but the role of germline polymorphism is unclear.

The role of ABC transporters in cancer has been known for a long time. Multidrug resistance has been studied since 1970, when it was first mentioned [37]. The well-studied *ABCB1* gene (MDR1) was discovered in 1974 by V. Ling, and nearly twenty years later, the discovery of *ABCC1* and *ABCG2* concerning drug resistance was reported [2]. Although associations of *ABCB1* gene expression with breast cancer prognosis were reported repeatedly, evidence for the role of its genetic variability in response to treatment is elusive. A recent review demonstrated that three frequently studied polymorphisms in *ABCB1* (rs1045642, rs1128503, and rs2032582) cannot be considered reliable predictors of response to chemotherapy in breast cancer patients [38]. Similarly, an association of *ABCC1* expression with the survival of breast cancer patients was described [39]. However, only a few studies on genetic polymorphisms can be found. *ABCC1* variants rs4148350, rs45511401, and rs246221 associated with the risk of febrile neutropenia in patients treated with 5-fluorouracil, epirubicin, and cyclophosphamide [40] and it was very recently discovered that *ABCC1* variant burden is a strong predictor of DFS in breast cancer patients rather than the genotype attributed to individual variants [41].

ABCG2 transports several drugs used for breast cancer treatment. In a recent study on the TCGA cohort, *ABCG2* transcript levels associated with a decreased progression-free survival, although, gene variants (either somatic or germline) influenced *ABCG2* expression only moderately [42]. From the above-reviewed information, it can be summarized that despite numerous studies on drug transporters utilization for predicting therapy outcome, strong support is still missing. Other transporters, with rather physiological roles, are much less explored in oncology, and studies were largely dedicated to gene expression in contrast with less studied genetic variability.

In conclusion, genetic variability in ABC transporters might play a role in breast cancer prognosis and help with prediction of therapy outcome of the patients. Although no alterations observed by this study can be considered statistically validated, particularly associations of downstream variant affecting expression, rs17548783 in *ABCA13* with DFS and variant rs17822931 (Gly180Arg) in *ABCC11* with response to NACT attract attention because of their support in the literature. These are interesting candidates for future research. Furthermore, elucidations are needed to explore additional genetic component, e.g., non-coding sequences, copy numbers and structural variations, somatic mutations, etc. of the ABC transporter superfamily.

4. Materials and Methods

4.1. Patients

The evaluation phase of the study included 105 breast cancer patients, diagnosed in the Institute for the Care for Mother and Child and Medicon, both in Prague and in the Hospital Atlas in Zlin (Czech Republic) over the period of 2006–2013. Of these, 68 patients underwent preoperative (neoadjuvant) treatment with regimens containing 5-fluorouracil, anthracyclines, cyclophosphamide (FAC or FEC), and/or taxanes. The rest received adjuvant postoperative treatment with regimens based on the same drugs. Clinical data of these patients are presented in Table S1.

For the validation phase, we used 802 breast cancer patients, recruited over the period of 2001–2013 from the Institute for the Care for Mother and Child, Medicon, the Motol University Hospital, all in Prague and in the Hospital Atlas in Zlin (all in the Czech Republic). Patients received either neoadjuvant or adjuvant chemotherapy or by hormonal therapy. Clinical data of these patients are presented in Table S3.

More details about the patient recruitment can be found elsewhere [13]. DFS was defined as the time between surgery and first disease relapse including local relapses. Response to NACT was evaluated by the Response Evaluation Criteria in Solid Tumors (RECIST [43]) based on imaging data retrieved from medical records.

Procedures performed in the present study were in accordance with the 1964 Helsinki declaration and its later amendments or comparable ethical standards. Study protocol was approved by the Ethical Commission of the National Institute of Public Health in Prague (approvals no. 9799-4, 15-25618A, and 17-28470A). All patients were informed about the study and those who agreed and signed an informed consent further participated in the study.

4.2. Panel Sequencing—Evaluation Phase

Blood samples were collected during the diagnostic procedures using tubes with K3EDTA anticoagulant and genomic DNA was isolated from human peripheral blood lymphocytes by the standard phenol/chloroform extraction and ethanol precipitation.

In the evaluation phase, raw data for 48 ABC transporter genes and one pseudogene were extracted from the previously published study [13]. Briefly, reads were mapped on reference sequence hg19 using Burrows-Wheeler Alignment (BWA) mem [44], base and indel recalibration and short indels and SNVs discovery was done in the Genome Analysis Toolkit (GATK) [45] and annotation of variants was done using Annovar [46] (for details of the library preparation, target enrichment, data processing, and variant calling, see [13]).

4.3. Genotyping—Validation Phase

In total, 42 genetic variants were selected for the validation phase and assessed using commercially provided competitive allele specific PCR (KASP™) genotyping (LGC Genomics, Hoddesdon, UK) in DNA samples from 802 breast cancer patients. Primers and probes were designed by the service provider. 10% of the samples were analyzed in duplicates for the purpose of the quality control. The genotyping concordance between duplicate samples exceeded 99%.

4.4. Statistical Analyses

In the evaluation phase, DFS was calculated with respect to the groups of patients divided by the genotype (common homozygous, heterozygous, and rare homozygous). The log-rank test for each variant was performed and the Kaplan–Meier plot was generated for visual inspection of gene dosage. We set the study follow-up end to 120 months (10 years) and thus, all subjects with DFS exceeding 120 months were censored. The response of patients to NACT was set to "good" in the case of complete or partial pathological remission (CR/PR) and "poor" for stable or progressive disease (SD/PD). We evaluated associations between genotypes (common homozygous, heterozygous, and rare homozygous) and response using the Pearson chi-square test. Adjusted p-value was calculated for each variant and each of these tests. Adjusted p-value for the log-rank test was based on 100 permutations of original data. A p-value of less than 0.05 after adjustment for multiple testing was considered statistically significant. Variants significantly associating with either DFS or response to NACT in the evaluation phase entered the validation phase of the study.

In the validation phase, the Pearson chi-square test and the log-rank tests were used as described above. For the evaluation of allele effect, recessive, dominant, co-dominant, and additive genetic models were used. Association of variants with transcript levels was assessed by the non-parametric Kruskal–Wallis test. Adjusted p-values were calculated using Benjamini–Hochberg false discovery rate (the FDR test) as a correction for multiple testing [47]. Haplotype analysis was conducted in HaploView 4.2 (Broad Institute, Cambridge, MA, USA). Statistical analyses were conducted using R and the statistical program SPSS v16.0 (SPSS, Chicago, IL, USA).

The sequencing data that support the findings of this study are openly available in Sequence Read Archive (SRA, https://www.ncbi.nlm.nih.gov/sra) under accession no. PRJNA510917.

Supplementary Materials: Supplementary Materials can be found at http://www.mdpi.com/1422-0067/21/24/9556/s1.

Author Contributions: Conceptualization, P.S., V.H., and R.V.; methodology, V.H.; software, P.S. and V.H.; validation, P.S., V.B., and V.H.; resources, R.K., K.K., D.V., and J.G.; data curation, R.K., K.K., D.V., and J.G.; writing—original draft preparation, P.S. and V.H.; writing—review and editing, all authors; visualization, P.S., and V.H.; supervision, P.S.; project administration, P.S., V.H., and D.V.; funding acquisition, P.S., V.H., and V.B. All authors have read and agreed to the published version of the manuscript.

Funding: This research was funded by the Czech Medical Council, grant number 17-28470A to P.S., the Czech Ministry of Education, Youth and Sports, grant number CZ.02.1.01/0.0/0.0/16_013/0001634 to V.H., the Grant Agency of Charles University, grant number UNCE/MED/006 to V.B. and the Grant Agency of the Czech Republic, grant number 19-03063S to P.S.

Acknowledgments: Authors would like to thank Pavel Ostašov from Biomedical Center, Faculty of Medicine in Pilsen, Charles University, Czech Republic for help with computing in R software environment and all participating patients for their kind consent to the study and clinical personnel for outstanding support.

Conflicts of Interest: The authors declare no conflict of interest. The funders had no role in the design of the study; in the collection, analyses, or interpretation of data; in the writing of the manuscript, or in the decision to publish the results.

Abbreviations

ABC	ATP-binding cassette
DFS	Disease-free survival
FDR	false discovery rate
LOF	Loss-of-function
MAF	minor allele frequency
NGS	next generation sequencing
SNP	single nucleotide polymorphism
SNV	single nucleotide variant
UTR	untranslated regions

References

1. Bray, F.; Ferlay, J.; Soerjomataram, I.; Siegel, R.L.; Torre, L.A.; Jemal, A. Global cancer statistics 2018: GLOBOCAN estimates of incidence and mortality worldwide for 36 cancers in 185 countries. *CA Cancer J. Clin.* **2018**, *68*, 394. [CrossRef] [PubMed]
2. Robey, R.W.; Pluchino, K.M.; Hall, M.D.; Fojo, A.T.; Bates, S.E.; Gottesman, M.M. Revisiting the role of ABC transporters in multidrug-resistant cancer. *Nat. Rev. Cancer* **2018**, *18*, 452. [CrossRef] [PubMed]
3. Szakacs, G.; Paterson, J.K.; Ludwig, J.A.; Booth-Genthe, C.; Gottesman, M.M. Targeting multidrug resistance in cancer. *Nat. Rev. Drug Discov.* **2006**, *5*, 219–234. [CrossRef] [PubMed]
4. Theodoulou, F.L.; Kerr, I.D. ABC transporter research: Going strong 40 years on. *Biochem. Soc. Trans.* **2015**, *43*, 1033–1040. [CrossRef] [PubMed]
5. Pasello, M.; Giudice, A.M.; Scotlandi, K. The ABC subfamily A transporters: Multifaceted players with incipient potentialities in cancer. *Semin. Cancer Biol.* **2020**, *60*, 57. [CrossRef]
6. Hedditch, E.L.; Gao, B.; Russell, A.J.; Lu, Y.; Emmanuel, C.; Beesley, J.; Johnatty, S.E.; Chen, X.; Harnett, P.; George, J.; et al. ABCA transporter gene expression and poor outcome in epithelial ovarian cancer. *J. Natl. Cancer Inst.* **2014**, 106. [CrossRef]
7. Kenna, J.G.; Taskar, K.S.; Battista, C.; Bourdet, D.L.; Brouwer, K.L.R.; Brouwer, K.R.; Dai, D.; Funk, C.; Hafey, M.J.; Lai, Y.; et al. Can Bile Salt Export Pump Inhibition Testing in Drug Discovery and Development Reduce Liver Injury Risk? An International Transporter Consortium Perspective. *Clin. Pharm.* **2018**, *104*, 916. [CrossRef]
8. Cole, S.P. Multidrug resistance protein 1 (MRP1, ABCC1), a "multitasking" ATP-binding cassette (ABC) transporter. *J. Biol. Chem.* **2014**, *289*, 30880–30888. [CrossRef]
9. Slot, A.J.; Molinski, S.V.; Cole, S.P. Mammalian multidrug-resistance proteins (MRPs). *Essays Biochem.* **2011**, *50*, 179–207. [CrossRef]
10. Szakacs, G.; Abele, R. An inventory of lysosomal ABC transporters. *FEBS Lett.* **2020**. [CrossRef]
11. Chatuphonprasert, W.; Jarukamjorn, K.; Ellinger, I. Physiology and Pathophysiology of Steroid Biosynthesis, Transport and Metabolism in the Human Placenta. *Front. Pharm.* **2018**, *9*, 1027. [CrossRef] [PubMed]
12. Gerovac, M.; Tampé, R. Control of mRNA Translation by Versatile ATP-Driven Machines. *Trends Biochem. Sci.* **2019**, *44*, 167. [CrossRef] [PubMed]
13. Hlavac, V.; Kovacova, M.; Elsnerova, K.; Brynychova, V.; Kozevnikovova, R.; Raus, K.; Kopeckova, K.; Mestakova, S.; Vrana, D.; Gatek, J.; et al. Use of Germline Genetic Variability for Prediction of Chemoresistance and Prognosis of Breast Cancer Patients. *Cancers* **2018**, *10*, 511. [CrossRef] [PubMed]
14. Hlavac, V.; Brynychova, V.; Vaclavikova, R.; Ehrlichova, M.; Vrana, D.; Pecha, V.; Kozevnikovova, R.; Trnkova, M.; Gatek, J.; Kopperova, D.; et al. The expression profile of ATP-binding cassette transporter genes in breast carcinoma. *Pharmacogenomics* **2013**, *14*, 515–529. [CrossRef] [PubMed]
15. Praest, P.; Luteijn, R.D.; Brak-Boer, I.G.J.; Lanfermeijer, J.; Hoelen, H.; Ijgosse, L.; Costa, A.I.; Gorham, R.D.; Lebbink, R.J.; Wiertz, E.J.H.J. The influence of TAP1 and TAP2 gene polymorphisms on TAP function and its inhibition by viral immune evasion proteins. *Mol. Immunol.* **2018**, *101*, 55. [CrossRef] [PubMed]
16. Arora, H.; Wilcox, S.M.; Johnson, L.A.; Munro, L.; Eyford, B.A.; Pfeifer, C.G.; Welch, I.; Jefferies, W.A. The ATP-Binding Cassette Gene ABCF1 Functions as an E2 Ubiquitin-Conjugating Enzyme Controlling Macrophage Polarization to Dampen Lethal Septic Shock. *Immunity* **2019**, *50*, 418. [CrossRef]

17. Zhu, H.; Xia, W.; Mo, X.B.; Lin, X.; Qiu, Y.H.; Yi, N.J.; Zhang, Y.H.; Deng, F.Y.; Lei, S.F. Gene-Based Genome-Wide Association Analysis in European and Asian Populations Identified Novel Genes for Rheumatoid Arthritis. *PLoS ONE* **2016**, *11*, e0167212. [CrossRef]

18. Fadista, J.; Oskolkov, N.; Hansson, O.; Groop, L. LoFtool: A gene intolerance score based on loss-of-function variants in 60 706 individuals. *Bioinformatics* **2017**, *33*, 471. [CrossRef]

19. Drean, A.; Rosenberg, S.; Lejeune, F.X.; Goli, L.; Nadaradjane, A.A.; Guehennec, J.; Schmitt, C.; Verreault, M.; Bielle, F.; Mokhtari, K.; et al. ATP binding cassette (ABC) transporters: Expression and clinical value in glioblastoma. *J. Neuro-Oncol.* **2018**, *138*, 479. [CrossRef]

20. Hlavata, I.; Mohelnikova-Duchonova, B.; Vaclavikova, R.; Liska, V.; Pitule, P.; Novak, P.; Bruha, J.; Vycital, O.; Holubec, L.; Treska, V.; et al. The role of ABC transporters in progression and clinical outcome of colorectal cancer. *Mutagenesis* **2012**, *27*, 187–196. [CrossRef]

21. Nymoen, D.A.; Holth, A.; Hetland Falkenthal, T.E.; Trope, C.G.; Davidson, B. CIAPIN1 and ABCA13 are markers of poor survival in metastatic ovarian serous carcinoma. *Mol. Cancer* **2015**, *14*, 44. [CrossRef] [PubMed]

22. Pichler, R.; Lindner, A.K.; Comperat, E.; Obrist, P.; Schafer, G.; Todenhofer, T.; Horninger, W.; Culig, Z.; Untergasser, G. Amplification of 7p12 Is Associated with Pathologic Nonresponse to Neoadjuvant Chemotherapy in Muscle-Invasive Bladder Cancer. *Am. J. Pathol.* **2020**, *190*, 442. [CrossRef] [PubMed]

23. Sauna, Z.E.; Kimchi-Sarfaty, C.; Ambudkar, S.V.; Gottesman, M.M. Silent polymorphisms speak: How they affect pharmacogenomics and the treatment of cancer. *Cancer Res.* **2007**, *67*, 9609–9612. [CrossRef] [PubMed]

24. Zeng, Z.; Bromberg, Y. Predicting Functional Effects of Synonymous Variants: A Systematic Review and Perspectives. *Front. Genet.* **2019**, *10*, 914. [CrossRef]

25. Park, S.; Shimizu, C.; Shimoyama, T.; Takeda, M.; Ando, M.; Kohno, T.; Katsumata, N.; Kang, Y.K.; Nishio, K.; Fujiwara, Y. Gene expression profiling of ATP-binding cassette (ABC) transporters as a predictor of the pathologic response to neoadjuvant chemotherapy in breast cancer patients. *Breast Cancer Res. Treat.* **2006**, *99*, 9–17. [CrossRef]

26. Dvorak, P.; Pesta, M.; Soucek, P. ABC gene expression profiles have clinical importance and possibly form a new hallmark of cancer. *Tumour Biol.* **2017**, *39*, 1010428317699800. [CrossRef]

27. Xie, W.; Shui, C.; Fang, X.; Peng, Y.; Qin, L. miR-197-3p reduces epithelial-mesenchymal transition by targeting ABCA7 in ovarian cancer cells. *3 Biotech* **2020**, *10*, 375. [CrossRef]

28. Guo, Q.; Grimmig, T.; Gonzalez, G.; Giobbie-Hurder, A.; Berg, G.; Carr, N.; Wilson, B.J.; Banerjee, P.; Ma, J.; Gold, J.S.; et al. ATP-binding cassette member B5 (ABCB5) promotes tumor cell invasiveness in human colorectal cancer. *J. Biol. Chem.* **2018**, *293*, 11166. [CrossRef]

29. Ohashi, J.; Naka, I.; Tsuchiya, N. The impact of natural selection on an ABCC11 SNP determining earwax type. *Mol. Biol. Evol.* **2011**, *28*, 849. [CrossRef]

30. Ishiguro, J.; Ito, H.; Tsukamoto, M.; Iwata, H.; Nakagawa, H.; Matsuo, K. A functional single nucleotide polymorphism in ABCC11, rs17822931, is associated with the risk of breast cancer in Japanese. *Carcinogenesis* **2019**, *40*, 537. [CrossRef]

31. Ishikawa, T.; Toyoda, Y.; Yoshiura, K.; Niikawa, N. Pharmacogenetics of human ABC transporter ABCC11: New insights into apocrine gland growth and metabolite secretion. *Front. Genet.* **2012**, *3*, 306. [CrossRef] [PubMed]

32. Guo, Y.; Kotova, E.; Chen, Z.S.; Lee, K.; Hopper-Borge, E.; Belinsky, M.G.; Kruh, G.D. MRP8, ATP-binding cassette C11 (ABCC11), is a cyclic nucleotide efflux pump and a resistance factor for fluoropyrimidines 2′,3′-dideoxycytidine and 9′-(2′-phosphonylmethoxyethyl)adenine. *J. Biol. Chem.* **2003**, *278*, 29509–29514. [CrossRef] [PubMed]

33. Oba, T.; Izumi, H.; Ito, K.I. ABCB1 and ABCC11 confer resistance to eribulin in breast cancer cell lines. *Oncotarget* **2016**, *7*, 70011. [CrossRef] [PubMed]

34. Yasui, K.; Mihara, S.; Zhao, C.; Okamoto, H.; Saito-Ohara, F.; Tomida, A.; Funato, T.; Yokomizo, A.; Naito, S.; Imoto, I.; et al. Alteration in copy numbers of genes as a mechanism for acquired drug resistance. *Cancer Res.* **2004**, *64*, 1403–1410. [CrossRef]

35. Low, F.G.; Shabir, K.; Brown, J.E.; Bill, R.M.; Rothnie, A.J. Roles of ABCC1 and ABCC4 in Proliferation and Migration of Breast Cancer Cell Lines. *Int. J. Mol. Sci.* **2020**, *21*, 7664. [CrossRef]

36. Xavier, B.M.; Zein, A.A.; Venes, A.; Wang, J.; Lee, J.-Y. Transmembrane Polar Relay Drives the Allosteric Regulation for ABCG5/G8 Sterol Transporter. *Int. J. Mol. Sci.* **2020**, *21*, 8747. [CrossRef]

37. Biedler, J.L.; Riehm, H. Cellular resistance to actinomycin D in Chinese hamster cells in vitro: Cross-resistance, radioautographic, and cytogenetic studies. *Cancer Res.* **1970**, *30*, 1174–1184.

38. Madrid-Paredes, A.; Cañadas-Garre, M.; Sánchez-Pozo, A.; Expósito-Ruiz, M.; Calleja-Hernández, M.Á. ABCB1 gene polymorphisms and response to chemotherapy in breast cancer patients: A meta-analysis. *Surg. Oncol.* **2017**, *26*, 473. [CrossRef]

39. Yamada, A.; Nagahashi, M.; Aoyagi, T.; Huang, W.C.; Lima, S.; Hait, N.C.; Maiti, A.; Kida, K.; Terracina, K.P.; Miyazaki, H.; et al. ABCC1-Exported Sphingosine-1-phosphate, Produced by Sphingosine Kinase 1, Shortens Survival of Mice and Patients with Breast Cancer. *Mol. Cancer Res.* **2018**, *16*, 1059. [CrossRef]

40. Pfeil, A.M.; Vulsteke, C.; Paridaens, R.; Dieudonné, A.S.; Pettengell, R.; Hatse, S.; Neven, P.; Lambrechts, D.; Szucs, T.D.; Schwenkglenks, M.; et al. Multivariable regression analysis of febrile neutropenia occurrence in early breast cancer patients receiving chemotherapy assessing patient-related, chemotherapy-related and genetic risk factors. *BMC Cancer* **2014**, *14*, 201. [CrossRef]

41. Xiao, Q.; Zhou, Y.; Winter, S.; Büttner, F.; Schaeffeler, E.; Schwab, M.; Lauschke, V.M. Germline variant burden in multidrug resistance transporters is a therapy-specific predictor of survival in breast cancer patients. *Int. J. Cancer* **2020**, *146*, 2475. [CrossRef] [PubMed]

42. Reustle, A.; Fisel, P.; Renner, O.; Büttner, F.; Winter, S.; Rausch, S.; Kruck, S.; Nies, A.T.; Hennenlotter, J.; Scharpf, M.; et al. Characterization of the breast cancer resistance protein (BCRP/ABCG2) in clear cell renal cell carcinoma. *Int. J. Cancer* **2018**, *143*, 3181. [CrossRef] [PubMed]

43. Schwartz, L.H.; Litiere, S.; de Vries, E.; Ford, R.; Gwyther, S.; Mandrekar, S.; Shankar, L.; Bogaerts, J.; Chen, A.; Dancey, J.; et al. RECIST 1.1-Update and clarification: From the RECIST committee. *Eur. J. Cancer* **2016**, *62*, 132. [CrossRef] [PubMed]

44. Li, H. Toward better understanding of artifacts in variant calling from high-coverage samples. *Bioinformatics* **2014**, *30*, 2843–2851. [CrossRef] [PubMed]

45. Van der Auwera, G.A.; Carneiro, M.O.; Hartl, C.; Poplin, R.; Del Angel, G.; Levy-Moonshine, A.; Jordan, T.; Shakir, K.; Roazen, D.; Thibault, J.; et al. From FastQ data to high confidence variant calls: The Genome Analysis Toolkit best practices pipeline. *Curr. Protoc. Bioinform.* **2013**, *43*, 11–33. [CrossRef]

46. Wang, K.; Li, M.; Hakonarson, H. ANNOVAR: Functional annotation of genetic variants from high-throughput sequencing data. *Nucleic Acids Res.* **2010**, *38*, e164. [CrossRef]

47. Benjamini, Y.; Hochberg, Y. Controlling the false discovery rate A practical and powerful approach to multiple testing. *J. R. Stat. Soc. Ser. B Methodol.* **1995**, *57*, 289–300. [CrossRef]

Publisher's Note: MDPI stays neutral with regard to jurisdictional claims in published maps and institutional affiliations.

International Journal of
Molecular Sciences

Article

5′ Untranslated Region Elements Show High Abundance and Great Variability in Homologous ABCA Subfamily Genes

Pavel Dvorak [1,2,*], **Viktor Hlavac** [2,3] and **Pavel Soucek** [2,3]

1   Department of Biology, Faculty of Medicine in Pilsen, Charles University, 32300 Pilsen, Czech Republic
2   Biomedical Center, Faculty of Medicine in Pilsen, Charles University, 32300 Pilsen, Czech Republic;
    viktor.hlavac@szu.cz (V.H.); pavel.soucek@szu.cz (P.S.)
3   Toxicogenomics Unit, National Institute of Public Health, 100 42 Prague, Czech Republic
*   Correspondence: Pavel.Dvorak@lfp.cuni.cz; Tel.: +420-377593263

Received: 7 October 2020; Accepted: 20 November 2020; Published: 23 November 2020

Abstract: The 12 members of the ABCA subfamily in humans are known for their ability to transport cholesterol and its derivatives, vitamins, and xenobiotics across biomembranes. Several ABCA genes are causatively linked to inborn diseases, and the role in cancer progression and metastasis is studied intensively. The regulation of translation initiation is implicated as the major mechanism in the processes of post-transcriptional modifications determining final protein levels. In the current bioinformatics study, we mapped the features of the 5′ untranslated regions (5′UTR) known to have the potential to regulate translation, such as the length of 5′UTRs, upstream ATG codons, upstream open-reading frames, introns, RNA G-quadruplex-forming sequences, stem loops, and Kozak consensus motifs, in the DNA sequences of all members of the subfamily. Subsequently, the conservation of the features, correlations among them, ribosome profiling data as well as protein levels in normal human tissues were examined. The 5′UTRs of ABCA genes contain above-average numbers of upstream ATGs, open-reading frames and introns, as well as conserved ones, and these elements probably play important biological roles in this subfamily, unlike RG4s. Although we found significant correlations among the features, we did not find any correlation between the numbers of 5′UTR features and protein tissue distribution and expression scores. We showed the existence of single nucleotide variants in relation to the 5′UTR features experimentally in a cohort of 105 breast cancer patients. 5′UTR features presumably prepare a complex playground, in which the other elements such as RNA binding proteins and non-coding RNAs play the major role in the fine-tuning of protein expression.

Keywords: 5′ untranslated region; cis-acting elements; ABC transporters; ABCA subfamily; bioinformatics

---

1. Introduction

The proteins in the ABC (ATP-binding cassette) family can be found in every group of living organisms, from bacteria to primates, and are generally known for their ability to translocate a wide range of substrates across extracellular as well as intracellular biomembranes [1,2]. Typically, ABC transport proteins contain two nucleotide-binding domains and two transmembrane domains. ABC proteins are organized as full- or half-transporters in eukaryotes. The products of half-transporters have to homodimerize or heterodimerize to create a functional transporter. Forty-eight ABC protein-coding genes, which have been described in the human genome, are divided into seven subfamilies according to the similarity in their amino acid (aa) sequences and organization of protein domains [3]. The ABCA subfamily is represented by 12 full transporters, which belong to the largest molecules among ABC proteins, with a median of 1925 aa. They have been reported to play important roles in the transport of cholesterol and its derivatives, as well as some vitamins and xenobiotics [4,5]. Several members of the ABCA subfamily have been causatively linked to a diverse set of human inborn diseases such as familial high-density lipoprotein (HDL) deficiency (*ABCA1*), neonatal surfactant deficiency (*ABCA3*), degenerative retinopathies (*ABCA4*), and congenital keratinization disorders (*ABCA12*) [6]. Gene expression studies conducted in our laboratories have demonstrated associations of intra-tumoral transcript levels of several ABCA genes with the response of patients to oncological therapy or disease-free survival [7–11]. Their roles in cancer progression and metastasis attributed mainly to lipid trafficking are a matter of intensive research [4,12]. Phylogenetic analyses suggest that current ABCA genes evolved by many duplication and loss events from a common ancestor gene [6,13,14]. *ABCA5*-related genes (*ABCA5/6/8/9/10*), which evolved from the *ABCA5* gene by duplications, form a cluster on the q-arm of the human chromosome 17 (17q24). The remaining ABCA genes are dispersed on six other human chromosomes [5,15].

Gene expression at the protein level does not reflect the mRNA level in normal human tissues perfectly, not only in the case of ABC genes [4,16]. The reason for this difference is believed to lie in post-transcriptional regulation. Regulation of the initiation of translation has been implicated as the major mechanism in this complex process. Several features in the 5′ untranslated regions (5′UTR, also leader sequence) of genes, such as the length of 5′UTRs, upstream ATG start codons (uATG), upstream open-reading frames (uORF), introns, RNA G-quadruplex-forming sequences (RG4), diverse secondary structures like stem loops as well as the Kozak consensus motif in the vicinity of start codons, act as *cis*-acting regulatory factors (Figure 1) [17,18]. RNA structures such as stem loops and RG4s, as well as uORFs and uATGs, mainly inhibit translation. RNA modifications, or RNA-binding proteins (RBP) and long non-coding RNAs (lncRNA) that interact with RNA binding sites, as well as the Kozak motif, can additionally stimulate translation initiation. It is still not clear how the actions of these elements interact, when multiple factors are present together, or if some of them have a superior role. Several highly conserved elements have been revealed in our recent bioinformatics study focusing on the 5′UTRs of the human *ABCA1* gene and its vertebrate orthologs [19]. The 5′UTRs of the other ABCA subfamily genes have not yet been studied in detail. Mapping of the 5′UTR features that are known to have the potential to regulate translation, among the whole subfamily, is addressed in the current work. Those interpreting the significance of new mutations and polymorphisms can take our findings into consideration.

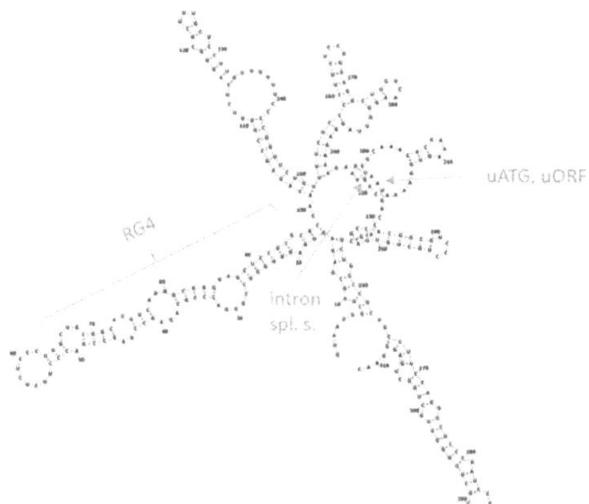

Figure 1. Secondary structure of the 5′UTR of the human *ABCA1* gene. The positions of the features influencing the initiation of translation are depicted; intron spl. s., intron splice-sites; RG4, RNA G-quadruplex-forming sequence; uATG, upstream ATG codon; uORF, upstream open-reading frame.

2. Results

2.1. Alignment and Phylogenetic Tree of 5′ UTRs of Human ABCA Genes

Based on the alignment analyses of the individual human ABCA genes with their vertebrate orthologs, we were able to define the sub-regions of the 5′UTRs showing a very high level of conservation. The names, IDs and basic characteristics of the transcripts analyzed are disclosed in Table S1 and positions of these sub-regions are recorded in Table S2. Notably, we did not find any comparable highly conserved sub-regions in the alignment of all human ABCA genes together. Figures S1 and S2 document the alignment of human *ABCA1* and its orthologs for ClustalO and Mafft algorithms, respectively. Figures S3 and S4 show the alignment of all human ABCA genes in a similar manner. A phylogenetic tree, based on the 5′UTR sequences of all human ABCA genes, was constructed and is shown in Figure 2.

The order in which 5′UTRs of ABCA genes are described in the following sections reflects their relationships calculated in the phylogenetic analysis.

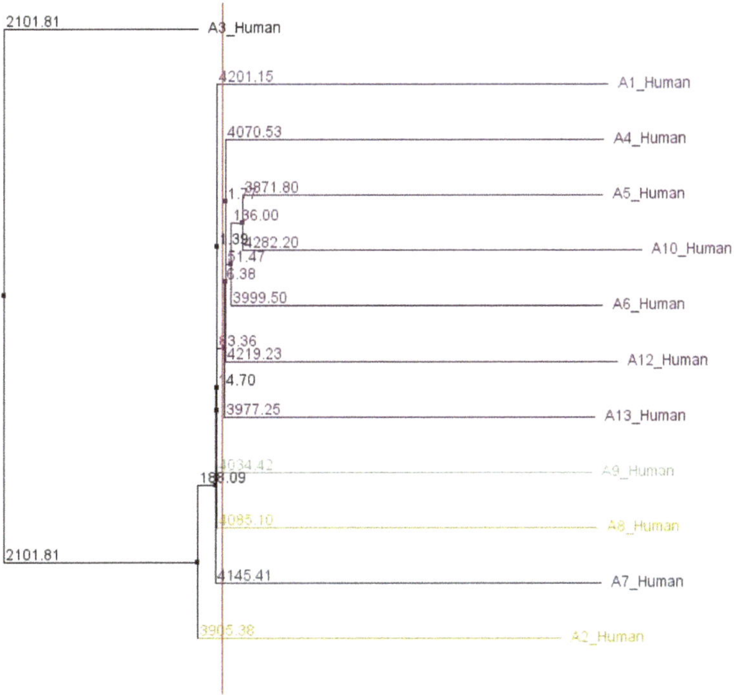

Figure 2. Phylogenetic analysis of 5′UTR sequences of 12 human ABCA genes. The neighbor-joining tree was constructed using DNA model distance measure of Clustal Omega multiple sequence alignment.

2.2. ABCA3

The ABCA3-201 transcript, coding for the main functional isoform, with 33 exons is 6602 bp long. It has the second longest 5′UTR among ABCA genes spanning 694 bps. The 5′UTR is divided into four parts by the three introns—Intron 1–2 (10718 bp), Intron 2–3 (890) and Intron 3–4 (1960). The most conserved region was localized to −519 to −503 from the start ATG codon of the main ORF (sATG). Four uATGs were described at the following positions: 1) −525, 2) −498, 3) −329, and 4) −262. The third and fourth uATGs are conserved in primates (cat. 1), the first uATG in placental mammals (cat. 3) and the second uATG in placental mammals as well as reptiles and birds (cat. 4). The second and fourth showed weak contexts, the first and third adequate contexts. High TIS scores were calculated for the first and second, and middle scores for the third and fourth. Two probable uORFs were predicted: one is 150 nt long (starting at −525) and the second overlapping with the main ORF (starting at −262); and one RG4-forming sequence at −608 to −576. Five stem loops were predicted in this region. The ABCA3 protein was found to be expressed in many human tissues (42 out of 45 tissues tested). The Mode expression score was Medium, enhanced in e. g. brain, glands, lung, testis and spleen. Table 1 discloses an overview of the 5′UTR features for the 12 human ABCA genes. A more detailed overview, including the positions of all features studied, can be viewed in Table S2.

Table 1. Overview of the 5′UTR features in human ABCA genes sorted according to their phylogenetic relationships.

	DNA/RNA							PROTEIN		
Gene	5′UTR Length [nt]	No. of uATGs	No. of uORFs	No. of 5′UTR Introns	No. of RG4-Forming Seq.	No. of 5′UTR Stem Loops	No. of All 5′UTR Elements	Protein Tissue Distrib.	Protein Mode Expres. Score	Protein Expres. Score Range
ABCA3	694	4	2	3	1	5	13	3	3	2–4
ABCA1	313	1	1	1	1	6	9	4	3	2–4
ABCA4	103	1	0	0	0	2	3	1	4	n.a.
ABCA5	97	0	0	1	0	4	5	4	4	3–4
ABCA10	910	14	6	3	0	5	22	3	3	2–3
ABCA6	196	2	0	1	0	3	6	4	2	2–4
ABCA12	418	4	1	0	0	6	10	3	3	2–4
ABCA13	26	0	0	0	0	0	0	n.a.	n.a.	n.a.
ABCA9	75	3	2	1	0	2	6	4	4	2–4
ABCA8	340	6	2	2	0	3	11	2	2	2–4
ABCA7	227	1	1	1	1	3	6	3	2	2–4
ABCA2	97	1	1	0	1	2	4	3	3	2–4

Abbreviations: Distrib., distribution; Expres., expression; n.a., not available; No., number; seq., sequence; uATG, upstream ATG; uORF, upstream ORF. Protein tissue distribution (in normal human tissues): 1-One/2-Some (less than a half)/3-Many/4-All. Protein expression score: 1-Not detected/2-Low/3-Medium/4-High.

2.3. ABCA1

The human ABCA1-202 transcript is 10,408 bp long and has 50 exons. The 5′UTR region covers 313 bps and is divided into two parts by one intron (Intron 1–2), which is the longest among ABCA genes with 24,163 bp. The most conserved sub-region of the 5′UTR was localized to −119 to −59. One uATG was found at −89. This uATG is highly conserved among vertebrates (category 5 = placental mammals + Reptiles and birds + coelacanth/ray-finned fishes); however, it shows a weak flanking sequence context and middle TIS score. One uORF starting at this uATG and overlapping with the main ORF was predicted. The most probable RG4-forming sequence was placed to −251 to −218. Six stem loops were predicted in this region. The ABCA1 protein was found to be expressed in all human tissues (45 tissues). The mode expression score was medium, and the expression was high in lung, stomach, and placenta.

2.4. ABCA4

The transcript ABCA4-201 is 7328 bp long and has 50 exons. The 5′UTR is 103 bp long. The sub-region −65 to −50 was determined to be the most conserved. One uATG (−62) is conserved in placental mammals (cat. 3); however, it has a weak context and low TIS score. No probable uORF or RG4-forming sequence was detected. Two stem loops were predicted to this region. The ABCA4 protein was found to be expressed only in the retina with a high score.

2.5. ABCA5

The transcript ABCA5-201 is 8252 bp long and contains 39 exons. The 5′UTR covers 97 bps and is interrupted by one intron-Intron 1–2 (12,621 bp). Only a very short sequence −7 to −2 was identified as the most conserved. No uATG, uORF, or RG4-forming sequences were detected and four stem loops were predicted in this region. The ABCA5 protein was found to be expressed in all human tissues (45 tissues). The mode expression score was high.

2.6. ABCA10

The human *ABCA10* gene has no known orthologs in ray-finned fishes. The ABCA10-201 transcript is 6362 bp long and has 40 exons. It has the longest 5′UTR among ABCA genes encompassing 910 nts. Three introns (Intron 1–2, 15,668 bp; Intron 2–3, 1295 bp; Intron 3–4, 1615 bp) are annotated within the 5′UTR. The most conserved sub-region was determined to −459 to −368. Fourteen uATGs (1. −755, 2. −722, 3. −709, 4. −645, 5. −625, 6. −574, 7. −571, 8. −482, 9. −464, 10. −279, 11. −265, 12. −197, 13. −169, 14. −56) were discovered. The seventh uATG is present only in humans (cat. 0), the first, second, third, fifth, sixth and eighth are conserved within primates (cat. 1), the ninth to fourteenth (last five) are conserved within placental mammals (cat. 3), and the fourth within placental mammals as well as reptiles and birds (cat. 4). The third, fifth, sixth, tenth, and fourteenth uATGs show weak contexts; the first, second, fourth, seventh, and eighth adequate contexts; the ninth, eleventh, twelfth, and thirteenth strong contexts. The fourteenth uATG shows a low TIS score, the fourth and eleventh high TIS scores, and the others middle TIS scores. Six uORFs (start positions-1. −722, 2. −625, 3. −482, 4. −279, 5. −65, 6. −197) were predicted; 33, 117, 42, 33, 171, and 39 bp long, respectively. Interestingly, the sATG of this transcript shows only an adequate flanking sequence context. No probable RG4-forming sequence and five stem loops were predicted in this region. The ABCA10 protein was found to be expressed in many human tissues (40/45). The mode expression score was medium.

2.7. ABCA6

The human *ABCA6* gene has no known orthologs in Sauropsida (reptiles and birds) and ray-finned fishes. The transcript ABCA6-201 is 5321 bp long and has 39 exons. The 5′UTR is 196 nt long and divided into two parts by Intron 1–2 (996 bp). The most conserved sub-region is located to −156 to −63. Two uATGs were found within the region (−144 and −32). The first uATG is conserved within placental mammals (cat. 3) and shows an adequate context and middle TIS score; the second is conserved only in primates (cat. 1) and has also an adequate context, but low TIS score. Notably, the sATG of this transcript shows only a weak flanking sequence context. No probable uORF or RG4-forming sequence was detected. Three stem loops were predicted in this region. The ABCA6 protein was found to be expressed in all human tissues (31 tissues tested). The Mode expression score was Low, enhanced in e.g., adipose tissue, bone marrow, brain, esophagus, gallbladder, liver, ovary, testis, and urinary bladder.

2.8. ABCA12

The ABCA12-201 transcript is 9298 bp long and has 53 exons. The 5′UTR spans the first 418 nts. Three separate sub-regions showed the highest level of conservation. Four uATGs (1. −398, 2. −333, 3. −330, 4. −115) were described. The first and third uATGs are conserved within primates (cat. 1), the second and fourth within placental mammals as well as reptiles and birds (cat. 4). The second uATG shows a weak context and the others adequate contexts; all show middle TIS scores. One 69 nt long uORF was predicted starting from −398. No probable RG4-forming sequence was found and six stem loops were predicted in this region. The ABCA12 protein was found to be expressed in many human tissues (17/31). The mode expression score was medium, enhanced in duodenum, kidney, ovary, and small intestine.

2.9. ABCA13

The ABCA13-204 transcript is 17,188 bp long and has 62 exons. The 5′UTR is the smallest among ABCA genes and has only 26 nts. No uATG, uORF or probable RG4-forming sequences were located within the 5′UTR. RNA expression was found in many human tissues, enhanced in bone marrow; however, protein expression has not been annotated yet.

2.10. ABCA9

The human *ABCA9* gene has no known orthologs in Sauropsida (reptiles and birds) and ray-finned fishes. The ABCA9-201 transcript is 6377 bp long and has 39 exons. One 9726 bp long intron (Intron 1-2) is annotated within the 75 nt long 5′UTR. Two sub-regions fulfilled the criteria for the most conserved ones. Three uATGs (1. −70, 2. −58, 3. −54) were found; the first and third are conserved within placental mammals (cat. 3) and second only within primates (cat. 1). The first and second show adequate context and third weak context; however, all have low TIS scores. Two uORFs (1. −70, 2. −54) were described, the first 33 and second 39 nt long. No probable RG4-forming sequence was detected and two stem loops were predicted in this region. The ABCA9 protein was found to be expressed in all human tissues (45). The mode expression score was high.

2.11. ABCA8

The human *ABCA8* gene has no known orthologs in Sauropsida (reptiles and birds) and ray-finned fishes. The ABCA8-208 transcript is 6002 bp long and has 40 exons. Intron 1-2 (5746 bp) and intron 2-3 (7272) divide the 340 nt long 5′UTR into three parts. One sub-region (−234 to −180) displayed the highest level of conservation. Six uATGs (1. −265, 2. −243, 3. −152, 4. −149, 5. −135, 6. −79) were recognized; all of them are present only in humans (cat. 0) with the exception of the second one, which can also be found in other primates (cat. 1). All uATGs show adequate contexts and middle TIS scores with the exceptions of the second one, which shows a high TIS score, and the sixth one, which shows a low TIS score. Two uORFs were predicted; the first is 39 nt long and starts from the second uATG and the second 75 nt long from the sixth uATG. No probable RG4-forming sequence was detected and three stem loops were predicted in this region. The ABCA8 protein was found to be expressed in some human tissues (14/45). The mode expression score was low, enhanced in adrenal gland, testis, ovary, liver, and adipose tissue.

2.12. ABCA7

The length of the ABCA7-201 transcript is 6815 bp and it has 47 exons. The 5′UTR spans 227 nt with one annotated intron-intron 1-2 (1028 bp). One sub-region (−137 to −73) shows the highest level of conservation. One uATG (−103), which is conserved in placental mammals (cat. 3) and has an adequate context and middle TIS score, was found. One uORF overlapping with the main ORF was predicted to start from this uATG. One probable RG4-forming sequence was placed to −217 to −158. Three stem loops were predicted in this region. The ABCA7 protein was found to be expressed in many human tissues (29/45). The mode expression score was low, enhanced only in bone marrow and spleen.

2.13. ABCA2

The ABCA2-202 transcript is 8103 bp long and contains 49 exons. The 5′UTR is 97 bp long. The most conserved sub-region was localized to −65 to −12. One uATG (−88) is conserved within primates and rodents (cat. 2); it has an adequate context, however, a low TIS score. Prediction programs found one probable 66 nt long uORF starting at the uATG and one RG4-forming sequence (−81 to −34). Two stem loops were described within this region. The ABCA2 protein was found to be expressed in almost all human tissues lacking only in adipose tissue (44 out of 45). The mode expression score was medium, enhanced only in brain.

2.14. Descriptive Statistics Shows High Abundance and Great Variability of 5′UTR Features

Among the 12 ABCA genes, the minimum 5′UTR length is 26 nts (*ABCA13*), maximum length 910 (*ABCA10*) and mean/average (M) 291 (median (Mdn) 212 and mode (Mo) 97). The minimum number of uATGs is 0 (*ABCA5* and *ABCA13*), maximum 14 (*ABCA10*) and M = 3 (Mdn = 2 and Mo = 1). The minimum number of uORFs is 0 (*ABCA4*, *ABCA5*, *ABCA6*, and *ABCA13*), maximum 6 (*ABCA10*) and M = 1 (Mdn = 1 and Mo = 1). The minimum number of 5′UTR introns is 0 (*ABCA2*, *ABCA4*, *ABCA12*, and *ABCA13*), maximum 3 (*ABCA3* and *ABCA10*) and M = 1 (Mdn = 1 and Mo = 1). The minimum 5′UTR intron length is 890 nts, maximum 24,163 and M = 7208 (Mdn = 5746 and Mo not applicable). Only four genes have one RG4 each (*ABCA1*, *ABCA2*, *ABCA3*, and *ABCA7*). The minimum number of 5′UTR stem loops is 0 (*ABCA13*), maximum 6 (*ABCA1* and *ABCA12*) and M = 3 (Mdn = 3 and Mo = 2). The minimum number of all 5′UTR elements is 0 (*ABCA13*), maximum 22 (*ABCA10*) and M = 8 (Mdn = 6 and Mo = 6). Table 2 presents the summary of the 5′UTRs' descriptive statistics.

Table 2. Descriptive statistics of the 5′UTR features in human ABCA genes.

	No. uATGs	No. uORFs	No. 5′UTR Introns	No. RG4s	No. Stem Loops	No. All 5′UTR Elements
No. genes	12	12	12	12	12	12
Min	0	0	0	0	0	0
Max	14	6	3	1	6	22
Sum	37	16	13	4	41	95
Mean	3	1	1	0	3	8
Variance	15	3	1	0	3	33
Median	2	1	1	0	3	6
Mode	1	1	1	0	2	6

No., number; RG4, RNA G-quadruplex; uATG, upstream ATG; uORF, upstream ORF.

2.15. Positive Correlations among 5′UTR Features

Correlations among the following independent variables: 5′UTR length, no. of uATGs, sATG flanking sequence context, no. of 5′UTR introns, length of all 5′UTR introns in a gene, presence of RG4-forming sequence, no. of stem loops, protein tissue distribution, and protein mode expression score were tested by non-parametric tests (Spearman's rs and Kendall's tau) in the first correlation analysis. Some positive correlations were found to be statistically significant (Table S3 and Figure 3). The variable 5′UTR length correlated with No. of uATGs (p = 0.004 and 0.006, respectively), no. of 5′UTR introns (0.037/0.016), and no. of stem loops (0.002/0.004). Furthermore, no. of uATGs correlated with no. of 5′UTR introns (0.041/0.022), sATG flanking sequence context with the presence of RG4-forming sequence (0.046/0.010), no. of 5′UTR introns with Length of all 5′UTR introns in a gene (<0.001/<0.001), and no. of stem loops with length of all 5′UTR introns in a gene (0.042/0.011). The variables protein tissue distribution and protein mode expression score did not correlate significantly with any other variable.

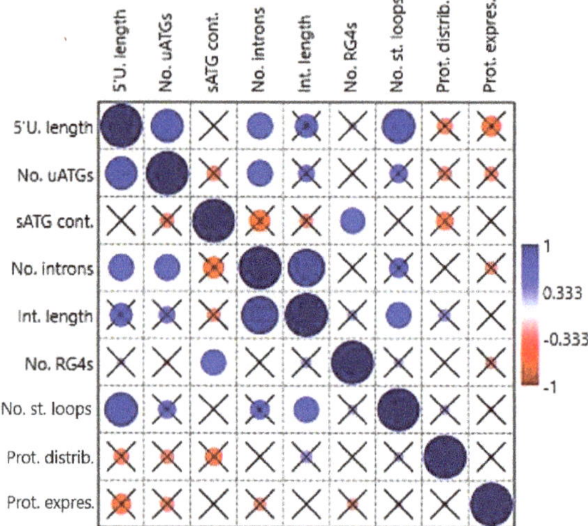

Figure 3. Correlation table plot. The circles represent the Spearman's *rs* correlation coefficients on a given scale; if the relevant *p* value is >0.05 the circle is crossed. 5′U., 5′UTR; Cont., context; Distrib., distribution; Expres., expression; Int., intron; No., number; Prot., protein; RG4, RNA G-quadruplex; sATG, start ATG of the main ORF; St., stem; uATG, upstream ATG.

The special features of uATGs-uATG position (from transcription start site, TSS), uATG conservation, uATG flanking sequence context, and uATG TIS score were tested similarly in the second correlation analysis. The variable uATG TIS score correlated positively with uATG position (0.034/0.031) and uATG flanking sequence context (0.011/0.003) (Table S4 and Figure 4).

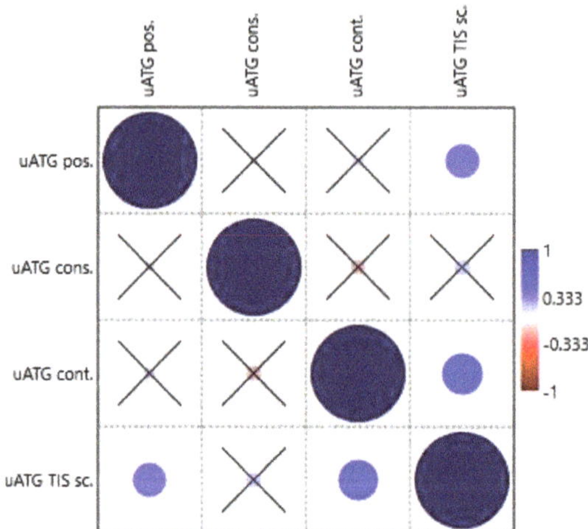

Figure 4. Correlation table plot. The circles represent the Spearman's *rs* correlation coefficients on a given scale; if the relevant *p* value is > 0.05 the circle is crossed. uATG cons., uATG conservation; uATG cont., uATG flanking sequence context; uATG pos., uATG position (from transcription start site); uATG TIS sc., uATG TIS score (from NetStart).

2.16. Ribo-Seq Data Confirmed Translation at Some of the Predicted uORFs

The analysis of the Ribo-seq coverage data confirmed a high concentration of ribosomes within some of the regions where probable uORFs were predicted by the bioinformatics algorithms. The uORFs in the genes *ABCA1*, *ABCA3*, *ABCA7*, and *ABCA8* displayed a high level of ribosome coverage. Notably, a significant ribosome coverage peak was also detected at the site of the first uATG of *ABCA6* (−144), which was not connected to any uORF. An example of the Ribo-seq data analysis in the GWIPS-viz browser is disclosed in Figure 5 (on a case of the *ABCA1* gene); all results regarding uORFs are summarized in Table 3.

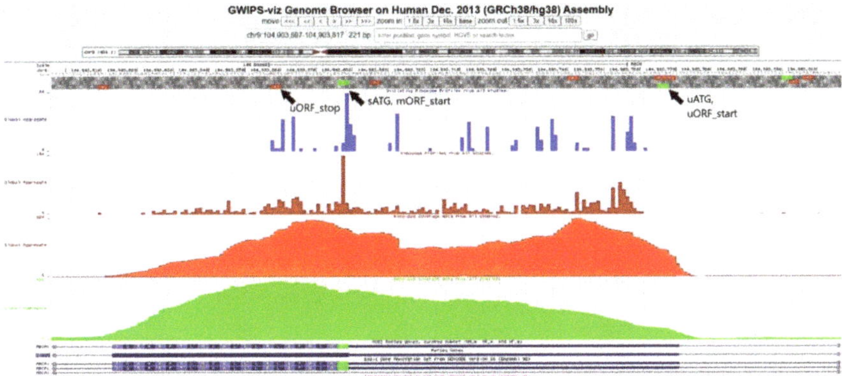

Figure 5. Ribosome profiling analysis in the GWIPS-viz browser, an example of the *ABCA1* gene. High concentrations of ribosomes can be seen at the sites of the predicted upstream ORF and beginning of the main ORF. Blue histograms represent Initiating Ribosome (P-site) Profiles from all studies, dark red—Elongating Ribosome (A-site) Profiles from all studies, red—Ribo-seq coverage data from all studies (Elongating Ribosomes—Footprints), and green—mRNA-seq coverage data from all studies (mRNA-seq Reads). mORF, main open reading frame; sATG, start ATG of the main ORF; uATG, upstream ATG; uORF, upstream open reading frame.

Table 3. Summary of the predicted uORFs in ABCA genes with Ribo-seq coverage data.

Gene	No. of uORFs	uORF Start	uORF End	uORF Length (nt)	Ribo-Seq Cov.
ABCA3	2	−525	−376	150	3
		−262	>+1	>261	2
ABCA1	1	−89	>+1	>87	3
ABCA4	0				
ABCA5	0				
ABCA10	6	−722	−690	33	1
		−625	−509	117	1
		−482	−441	42	1
		−279	−247	33	1
		−265	−95	171	1
		−197	−159	39	1
ABCA6	0				
ABCA12	1	−398	−330	69	1
ABCA13	0				
ABCA9	2	−70	−38	33	1
		−54	−16	39	1
ABCA8	2	−243	−205	39	3
		−79	−5	75	1
ABCA7	1	−103	>+1	>102	3
ABCA2	1	−88	−23	66	1

Abbreviations: Cov., coverage; No., number; uORF, upstream ORF; Ribo-seq coverage: 1 = Low, 2 = Medium, 3 = High; All positions are described in relation to the start ATG of the main ORF.

2.17. Thirteen Single Nucleotide Variants in 5′UTRs of ABCA Genes Were Found in Breast Cancer Patients

Thirteen single nucleotide variants (SNV) in 5′UTRs of main functional isoforms of ABCA genes were found within our cohort of 105 breast cancer patients by targeted sequencing. These SNVs are located within 5′UTRs of *ABCA1*, *ABCA3*, *ABCA7*, and *ABCA10*. An overview of these SNVs is presented in Table 4. One variant in *ABCA10* (rs1238052530) is changing the eighth uATG to GTG and therefore disrupting the start of the third uORF. Eight other variants are located within uORFs (five synonymous changes, two nonsynonymous, and one frameshift) and one within the most conserved region. The location of all 13 SNVs in relation to the 5′UTR features of ABCA genes is visualized in Figure S5.

Table 4. Overview of single nucleotide variants in 5′UTRs of ABCA genes in 105 breast cancer patients.

Gene	SNV	Position within 5′UTR	Genotypes [1]			Variation Type and Localization in Relation to 5′UTR Features
			Common Homo-Zygous	Hetero-Zygous	Rare Homo-Zygous	
ABCA1	rs1800978	−18	80	24	1	SNV (C > G) within uORF, synonymous change
ABCA1	rs1799777	−77−−76	80	25	0	Indel variant (dupC) within uORF, frameshift change
ABCA1	rs111292742	−279	97	8	0	SNV (G > C)
ABCA3	rs45487892	−67	103	2	0	SNV (G > A) within uORF2, nonsynonymous change
ABCA3	rs45518738	−182	104	1	0	SNV (G > A) within uORF2, synonymous change
ABCA3	rs146642275	−397	104	1	0	SNV (G > A) within uORF1, synonymous change
ABCA3	rs1029783163	−409	104	1	0	SNV (G > A) within uORF1, synonymous change
ABCA7	rs182233998	−14	101	4	0	SNV (T > C) within uORF, synonymous change
ABCA7	rs3752229	−9	93	12	0	SNV (A > G) within uORF, nonsynonymous change
ABCA10	rs1024510317	−89	104	1	0	SNV (G > C)
ABCA10	rs9302891	−438	0	14	91	SNV (G > T) within the most conserved region
ABCA10	rs1238052530	−482	104	1	0	SNV (T > C) disrupting the start of uORF3 (uATG8 to GTG)
ABCA10	rs563620435	−762	104	1	0	SNV (C > A)

Footnote: [1] Genotypes do not sum up to 105 due to missing data; SNV, single nucleotide variant.

3. Discussion

Our bioinformatics study focused on the 12 members of the human ABCA gene subfamily. These homologous genes are known for their principal role in lipid trafficking and homeostasis; however, other biological functions such as the involvement in signaling pathways activated by lipids and support of tumor progression are discussed in the recent literature. A comprehensive analysis of their 5′UTR sequences, in view of the features known to be involved in translation regulation, was addressed in the current study. We aimed to answer the question if the incidence of these 5′UTR features correlates clearly with protein expression in this group. Since ABCA genes lie behind several human diseases, both inborn and acquired, this information is important for the interpretation of the newly found mutations and polymorphisms in the clinical as well as research settings. Moreover, because similar studies on the gene family level are still missing, we also aimed at a comparison of our results to current knowledge based on the whole-genome level studies, which is addressed in this section.

3.1. Phylogenetic Tree

Phylogenetic analyses of human ABCA genes were performed by several previous studies. Some of them are based on the alignment of the nucleotide sequences or amino acid sequences of nucleotide-binding domains [3,20], the others on the alignment of the full length amino acid sequences [6,13]. Pairwise comparison of the amino acid sequences of all human ABCA members revealed homologies ranging from 28% (ABCA8/ABCA12) to 72% (ABCA8/ABCA9) [20]. The previous studies suggested that all ABCA genes have evolved from a primordial ancestor gene. Furthermore, they demonstrated that the human ABCA1, 2, 3, 4, 7, and 12 transporters cluster in a subgroup, distinct from ABCA5, 6, 8, 9, and 10. The ABCA5-related transporters share strikingly high overall amino acid sequence homologies but differ significantly from other members [3,6,13,20]. In contrast, our phylogenetic analysis, based on the comparison of nucleotide sequences of ABCA 5′UTRs, shows that the 5′UTRs of ABCA2 and ABCA3 cluster distinctly from the rest of the 5′UTRs. These discrepancies suggest that the evolution of 5′UTRs is shaped by different pressures independently from the other gene regions.

3.2. 5′UTR Length

The average/median length of 5′UTRs is 291/212 nts in the human ABCA genes; however, there is a great variation among the 12 members. The smallest 5′UTR is annotated to *ABCA13* (26 nts) and largest to *ABCA10* (910). In the literature, we find divergent data about the average/median length of human gene 5′UTRs based on the input data and year of study. Pesole and coworkers [21] constructed the first database focusing on the UTR sequences from different eukaryotic taxa and named it UTRdb. The average length of mRNA 5′UTR in humans was calculated to be 210 nts, maximum length 2803 and minimum 18 [22]. Rogozin et al. [23] calculated the average length of human 5′UTRs to be 160 nts (retrieved from EMBL database). In the work of Chen et al. [24], the average/median length of 5′UTRs in humans was calculated to be 254/169 nts from the Ensembl database and 220/160 from the UTRdb database. Recently, Leppek et al. [17] stated that the longest known median length of mRNA 5′ UTRs occurs in humans and is 218 nts (based on RefSeq data). We can conclude that the median length of 5′UTRs in the human ABCA subfamily is close to the median length derived from the whole genome data. Indeed, the genome average of 5′UTR lengths is relatively similar across diverse taxonomic classes of eukaryotes, ranging approximately from 100 to 200 nts, while in sharp contrast, the 5′UTR length varies considerably among the genes in a genome, from a few to several thousands of nts, a fact which has been mentioned in several previous studies [17,22,25,26]. Lynch and colleagues [27] suggested that this discrepancy can be explained by random genetic drift and mutational processes that cause stochastic turnover in transcription-initiation sites and premature start codons. Under the simple null model that they presented, natural selection only indirectly influences the lengths of 5′UTRs through the mutational origin of premature initiation codons within the UTR. We further confirmed in this study that the great length variability of 5′UTRs can be observed even in closely related members of a gene subfamily.

3.3. uATGs—Number and Conservation

Among human ABCA genes, there are two (17%; *ABCA5* and *ABCA13*) having no uATG; the others (83%) have at least 1 uATG, ranging from 1 to 14 uATGs (*ABCA10*). The median number of uATGs in the ABCA subfamily was computed to be 2 (average is 3 and mode 1), again showing a great variability among individual members. In total, 37 uATGs were found in the subfamily. The median value of uATG conservation was 1 (on a scale of 0–5), minimum 0 and maximum 5. 49% of the uATGs had a conservation value more than 1, that is, conserved in at least one other vertebrate subgroup except primates. Generally, uATGs and uORFs decrease mRNA translation efficiency and may be considered strong negative translational regulatory signals [28–30]. One of the approaches to address the issue of the functional significance of uATGs is to examine the evolutionary conservation of these triplets. Churbanov et al. [31] reported that the ATG triplet is conserved to a significantly greater extent than any of the other 63 nucleotide triplets in 5′UTRs of mammalian cDNAs, but not in 3′UTRs or coding sequences, by comparing sequences of human, mouse, and rat orthologous genes. Moreover, they observed that 5′UTRs are significantly depleted in overall ATG content. Approximately 25% of the 5′UTRs analyzed in their work contained at least one conserved uATG. In a similar study performed by Iacono and colleagues [32], uATGs and uORFs were detected in about 44% of 5′UTRs. They also concluded that both uATGs and uORFs are less frequent than expected by chance in 5′UTRs. 24% and 38% of human uATGs and uORFs were evolutionary conserved in all three taxa considered (human, mouse, and rat), respectively. Of the population of human and mouse mRNAs with long 5′UTRs (>60 bases), approximately 55% had at least one uAUG, with about 25% having one or more uORFs (average about 1.9 uORFs) [33]. We found that 5′UTRs of ABCA genes contain above-average numbers of uATGs as well as conserved ones. Approximately half of all uATGs in ABCA are conserved not only within primates, but also in other vertebrate subgroups. Considering this high level of conservation among uATGs, they probably play important biological roles in this subfamily.

3.4. uATGs—Flanking Sequence Context

The median value of uATG flanking sequence context in our study was 2 (on a scale of 1–4), minimum 1 and maximum 3. The median value of sATG flanking sequence context was 3, minimum 1 and maximum 3. Churbanov et al. [31] wrote that there was no significant difference between the nucleotide contexts of the conserved and non-conserved uATGs; in both cases, the information content of the uATG context was lower than that of the sATG context. Similarly, the analysis of the oligonucleotide context of uATGs, uORFs and sATGs has shown that a significant preference bias can be observed only for sATGs which, on average, have a much better context than uATGs and uORFs [32]. Our results are also in agreement with Rogozin et al. [23], who found that the presence of ATG triplets in 5′UTR regions of eukaryotic cDNAs correlates with "weaker" contexts of the sATGs. The median sATG context in our study was "strong" (level 3), not "optimal" (level 4). However, newer results on a larger cohorts of annotated transcripts have indicated some specifications, e.g., the proportion of uATGs in optimal contexts for conserved uORFs was noticeably higher than for non-conserved uORFs, while still half the proportion for main ORFs [33].

3.5. uORFs

There were eight genes (67%) among ABCA genes having at least one uORF predicted by the prediction software, 16 uORFs in total. The median for the whole subgroup is 1 uORF, minimum 0 and maximum 6 (*ABCA10*). Recently, Johnstone and co-workers [34] published that the human and mouse transcriptomes had similar uORF content (49.5 and 46.1%, respectively), consistent with previous computational estimates. The above-average incidence of uORFs in the ABCA subfamily is obvious from our results. The biological significance of a majority of uORFs is probably not limited to the regulation of translation, as the uORF products—small polypeptides—can play many different roles within the metabolism of complex organisms [35].

3.6. 5′UTR Introns

Four ABCA genes (33%) have no 5′UTR intron; the others (67%) have at least one. The median as well as mode number of 5′UTR introns is 1, maximum is 3 (*ABCA3* and *ABCA10*). The median length of 5′UTR introns in ABCA subfamily is 5746 nts, minimum 890 and maximum 24,163 (*ABCA1*). Approximately 35% of human genes have been indicated to harbor introns within 5′UTRs, with median intron size 2643 pb [22,36,37]. A strong barrier against the presence of more than 1 intron in 5′UTRs has also been suggested. In some cases, introns in 5′UTRs were reported to enhance gene expression [38]. Although Cenik et al. [36] found no correlation in 5′UTR intron presence or length with variance in expression across tissues, they observed an uneven distribution of 5′UTR introns amongst genes in specific functional categories. Contrary to the study of Cenik et al. [36], Lim et al. [39] described a strong negative correlation between the number of 5′UTR exon-exon junctions (the junctions between exons after intron removal) and main ORF translation efficiency in the five multicellular eukaryotic species studied (human, mouse, zebrafish, fruit fly and thale cress). We demonstrated an above-average occurrence and length of 5′UTR introns in ABCA genes.

3.7. RG4-Forming Sequences

Probable RG4-forming sequences were only predicted in the 5′UTRs of four ABCA genes (33%; *ABCA1*, *ABCA2*, *ABCA3*, and *ABCA7*); three of the sequences were located within the first third of the region. G-quadruplexes (G4) are secondary structures involving four nucleic acid strands that can be adopted by both DNA (DG4) and RNA (RG4) that contain guanine-rich sequences [40,41]. DG4 sequences have been mainly connected with the regulation of transcription and RG4 sequences regulation of translation; however, many other processes such as telomere maintenance, splicing, or RNA localization have also been discussed. Generally, RG4 sequences are believed to play roles as negative regulators of translation, although a positive effect of these structures has also been reported [17,42]. Thus, the regulatory effect of an RG4 sequence is significantly influenced by its position, the surrounding DNA topology, and other factors [43,44]. Huppert et al. [45] mapped the incidence of G4 in human 5′UTRs and calculated that 6.2% of these regions contained putative G4 sequences, with a density of approximately 0.3 per kb. Their data showed the highest density of putative G4 sequences at the 5′-ends of the 5′UTRs, decreasing approximately linearly along the 5′UTR regions. According to the analyses performed by Maizels and Gray [46], in 10–15% of human genes, a G4 motif occurs in the region specifying the 5′UTR of the encoded mRNA. However, the data analyzed in the latter study showed a higher G4 motif frequency at the 3′-ends of the 5′UTRs than at 5′-ends. With a new algorithm named G4Hunter, Bedrat and coworkers [47] found a significantly higher occurrence of G4 sequences in the human genome, with a density of approximately 2.4 per kb. 53.3% of 5′UTRs contained at least 1 G4-forming sequence in their study. In light of the latest results published in the literature, we can conclude that RG4 sequences are present in below-average numbers within the 5′UTRs of ABCA genes.

3.8. Stem Loops

The median number of 5′UTR stem loops in ABCA genes was determined to be 3; the mode was 2. The minimum was 0 (*ABCA13*) and maximum 6 (*ABCA1* and *ABCA12*). Stem loops (hairpins) are the most frequent secondary structures, which naturally fold along single-stranded RNA molecules [48]. The strength of their effect on translation is dependent on the location within 5′UTRs as well as their stability; however, they mainly inhibit this process in eukaryotic cells [17]. The mechanism of stem loop influence on translation has been examined in detail in many studies [49,50] and the prevalence of stem loops in 5′UTRs is generally thought to be high [51,52]. However, the precise stem loop incidence in specific gene groups has not been calculated yet and we were not able to compare our results to the results of others. Our work, therefore, brings largely new information on this topic.

3.9. Correlations and Influence of 5′UTR Features on Protein Expression

We have demonstrated a great variability in the numbers of the individual 5′UTR features among the genes of the ABCA subfamily. Considering phylogenetic relationships among ABCA genes, there is no clear pattern in these numbers. Some positive correlations among 5′UTR features were found to be statistically significant. Some of these correlations are expectable, such as the correlations among 5′UTR length and No. of uATGs, No. of 5′UTR introns and the no. of stem loops, some are interesting and need further exploration, such as the correlations between the no. of uATGs and the no. of 5′UTR introns or sATG flanking sequence context and the presence of RG4-forming sequence. Notably, the variable uATG TIS score correlated positively, however quite weakly, with uATG position (from transcription start site) and uATG flanking sequence context (from NetStart software). In relation to the canonical cap-dependent translation initiation, a possible influence of the ATG position on the ATG context has been mentioned in several works. Rogozin et al. [23] reported a significant negative correlation between the sATG information content and 5′UTR length for several species. They also concluded that this correlation could be explained by the strong positive correlation between the number of uATGs and the length of the UTR. Lynch and coworkers [27] described a strong distance-dependent gradient of the deficit of uAUGs. Since uATG TIS score and uATG flanking sequence context should be based on the same criteria, we would expect a stronger correlation. However, this correlation was not studied by any other study for comparison. The length of a 5′UTR, which is determined mainly by stochastic events, seems to be the major factor influencing the numbers of the other regulatory features.

Although there is also great variability in the distribution and expression of the ABCA proteins, we did not find any significant correlation that would support a clear connection between the numbers of 5′UTR features and protein expression characteristics. Among the genes which have the smallest numbers of 5′UTR features are *ABCA13*, *ABCA4*, *ABCA2*, and *ABCA5*. *ABCA13* is the least-explored member of the subfamily and the information about its protein distribution and expression in physiological conditions is missing. ABCA4 protein was reported to be expressed only in the retina and the expression level is high. ABCA2 and ABCA5 were found to be expressed in many and all human tissues with mainly medium and high levels, respectively. On the other side, among the genes which show the highest numbers of 5′UTR features are *ABCA10*, *ABCA3*, *ABCA8*, and *ABCA12*. ABCA10, ABCA3, and ABCA12 proteins are all expressed in many human tissues with mainly medium levels. The ABCA8 protein was detected only in several tissues with low expression. In the set of normal human tissues; however, the protein expression levels of the ABCA proteins ranged from low to high for all members, with the exception of ABCA5 (medium to high) and ABCA10 (low to medium). Our results therefore support the view that the great variability in 5′UTR features prepares a complex playground where the other elements such as RNA binding proteins and non-coding RNAs play the major role in the fine-tuning of final protein expressions. Notably, tissue-specific translation repression by miRNAs through binding to uAUGs was demonstrated in Ajay et al. [53].

3.10. Limitations

The current study analyzed data available freely on-line and is based on the in silico approaches and analyses. The main disadvantage is the dependence on the quality of the experimental data from the external sources without own experimental validation. The overwhelming amount of biological data, stored and freely available for the research community worldwide; however, calls for in silico filtering of the content as first step. Experimental verification should be performed for the most relevant findings afterwards. We hope to experimentally verify some of the presented results in the future by ourselves or in cooperation with other research groups. Another limitation of the current work lies in the small number of genes studied. However, we aimed to focus on the clinically important and closely related genes of the ABCA subfamily where the available information is scarce. The other ABC gene subfamilies will be added in the ongoing study.

In the support of our gene family approach, it is important to mention that studies which deal with whole genome data sets require some compromises in the data mining. For example, they considered a uORF (or an uAUG) evolutionarily conserved when occurring in an orthologous transcript independently from its sequence, length, and position in the 5′UTR [32]; or, some analyses were performed only on transcripts containing a single uORF [29]. This, of course, simplifies the view on the genome complexity to some degree.

4. Materials and Methods

4.1. DNA and Protein Sequences from Databases

DNA sequences of the 5′UTRs of the 12 human ABC protein-coding genes grouped together in the ABCA subfamily *ABCA1-10*, *ABCA12*, and *ABCA13* and amino acid sequences of the whole proteins were downloaded from the Ensembl database (EMBL-EBI; https://www.ensembl.org/index.html) in the

FASTA format. The transcripts of the main principal isoforms were chosen for further analyses based on the APPRIS classification system, UniProt annotation score and MANE Select system (description of transcript flags on the Ensembl web pages - https://www.ensembl.org/info/genome/genebuild/transcript_quality_tags.html). A survey of the number of protein-coding isoforms of ABCA genes and genome positions of their 5′UTRs was performed at the beginning. We found that the number of protein-coding isoforms of individual ABCA genes ranges from three to seven transcripts. There are 48 protein-coding transcripts altogether (36 non-principal isoforms). Fifteen out of the 36 non-principal isoforms have their 5′UTRs located within the same genomic regions as the relevant principal isoforms and smaller or equal to the 5′UTRs of principal isoforms in length. Another eight out of the 36 non-principal isoforms do not have 5′UTR sequences annotated. Because of this heterogeneity and inequality of data, we decided to include just one 5′UTR representative for each of the ABCA genes.

5′UTR sequences of orthologous genes from 10 other vertebrate species were selected and downloaded from the same database in relation to each human ABCA gene according to the same criteria. Based on the availability of species in the Ensembl, members of the five subgroups of vertebrates increasingly phylogenetically distant from humans—primates, rodents, other placental mammals, reptiles and birds, and ray-finned fishes—were covered. Table S1 discloses the names, IDs and basic characteristics of all the transcripts downloaded. The numbers, positions and lengths of the 5′UTR introns were also collected.

4.2. Alignment Analyses

Multi-sequence alignment analyses were performed with the help of Jalview software (Consortium Project; www.jalview.org/) [54]. Results of the two alignment algorithms–Clustal Omega (EMBL-EBI; https://www.ebi.ac.uk/Tools/msa/clustalo/) and Mafft (EMBL-EBI; https://www.ebi.ac.uk/Tools/msa/mafft/)—were considered. Alignments of individual ABCA genes and orthologs as well as all ABCA genes together were calculated. Figures S1 and S2 show the results of the alignment analysis for the 5′UTRs of the human ABCA1 gene and its vertebrate orthologs with nucleotide percentage identity colored, consensus logos and occupancy score histograms. Figures S3 and S4 show a similar analysis where the 5′UTRs of all 12 human ABCA genes were aligned together. Based on these analyses and criteria, the most conserved subregions within the 5′ UTRs were described in relation to the sATGs. The cut-off values for the identity and occupancy scores were set to 80%.

4.3. Phylogenetic Tree

Phylogenetic analysis of 5′UTR sequences of 12 human ABCA genes was performed and visualized with the help of Jalview software. A neighbor-joining tree using DNA model distance measure of Clustal Omega multiple sequence alignment was constructed. Clustal Omega algorithm was set to the default settings.

4.4. ATG Analyses

All ATG triplets within the 5′ UTRs of human ABCA genes were found and highlighted in text editor files (MS Word). The positions (relative to the sATGs as well as TSSs) and flanking sequence contexts were recorded. A scale of four categories was applied for the comparison of flanking sequence contexts: (1) weak (NNN(C/U)NNAUG(A/C/U), any sequence lacking both key nucleotides); (2) adequate (NNN(A/G)NNAUG(A/C/U) or NNN(C/U)NNAUGG, only one of these nucleotides is present); (3) strong (NNN(A/G)NNAUGG, only the two important nucleotides are present); and (4) optimal (GCC(A/G)CCAUGG). The scoring system was adopted from Hernandez et al. [55]. A TIS (translation initiation start) score generated by the NetStart prediction server (DTU Health Tech; https://services.healthtech.dtu.dk/service.php?NetStart-1.0) was also recorded for each ATG analyzed. The scores are in the range [0.0, 1.0]; when greater than 0.5 they represent a probable translation start. Based on this definition we further subdivided the TIS scores into three levels: (1) low (less than 0.1), (2) middle (0.1 to 0.5), and (3) high (greater than 0.5). We proposed a scale of six categories for the evaluation of ATG conservation statuses: 0) found only in humans, (1) conserved in primates, (2) in primates and rodents, (3) in primates, rodents and other placental mammals, (4) in placental mammals and reptiles and birds, (5) in placental mammals, reptiles and birds, and coelacanth or ray-finned fishes.

4.5. Upstream ORF Analysis

The ORFfinder (NCBI; https://www.ncbi.nlm.nih.gov/orffinder/) online tool was used to search for possible upstream open reading frames (uORFs) within the human 5′UTRs tested. The positions of all uORFs with the same orientation as the main ORFs, in three possible frames, were collected.

4.6. RG4 Analyses

The DNA sequences downloaded were screened for the presence of RNA G-quadruplex-forming sequences by the three on-line prediction servers working with different prediction algorithms-G4CatchAll (Doluca lab; http://homes.ieu.edu.tr/odoluca/G4Catchall/) [56], G4RNA Screener (Scott Group Bioinformatics; http://scottgroup.med.usherbrooke.ca/G4RNA_screener/) and QRGS Mapper (Ramapo College; http://bioinformatics.ramapo.edu/QGRS/index.php). The default settings of the programs were not changed and we followed the recommended cut-off levels. We adhered to the following result interpretation: the intersection of the results calculated by the G4CatchAll and G4RNA Screener were recorded as the most probable RG4-forming sequences. In the cases where no intersection of the two programs existed, the intersections of the G4CatchAll and QRGS Mapper or G4RNA Screener and QRGS Mapper were considered instead, and eventually recorded.

4.7. RNA Secondary Structure Prediction

RNAfold WebServer (University of Vienna; http://rna.tbi.univie.ac.at/cgi-bin/RNAWebSuite/RNAfold.cgi) was employed to generate the optimal secondary structures for minimum free energy prediction in the form of dot-bracket notation, graphical visualizations colored by base-pairing probabilities and mountain plot representations of the minimum free energy (MFE) structures, the thermodynamic ensembles of RNA structures, and the centroid structures (Figure S5). Equivalent predictions made by RNAstructure Web Servers (Mathews group; http://rna.urmc.rochester.edu/RNAstructureWeb/) were checked for comparison. The number of predicted stem loops within the whole 5′UTRs of the human ABCA genes was compared, counted and recorded.

4.8. Protein Expression Level from Databases

The expression of ABCA genes at the protein level was analyzed in the Human Protein Atlas (Consortium Project; http://www.proteinatlas.org) and Expression Atlas (EMBL-EBI; https://www.ebi.ac.uk/gxa/home). The expression of ABCA genes was tested in 45 and 31 normal human tissues in the Human Protein Atlas and Expression Atlas, respectively. The two variables reflecting the overall distribution and expression in human tissues and their following levels were collected: (A) Protein tissue distribution: 1-One tissue/2-Some tissues (less than a half of the tissues studied in the atlases)/3-Many tissues (equal to or greater than half of the tissues)/4-All tissues, and (B) Mode expression score (the most common level of protein expression based on immunohistochemistry scoring): 1-Not detected/2-Low/3-Medium/4-High. The mode expression score was calculated for the set of tissues where protein expression was detected.

4.9. Ribosome Profiling from Databases

Ribosome profiling (Ribo-seq) data were explored in the GWIPS-viz browser (http://gwips.ucc.ie/), an online genome browser for viewing ribosome profiling data. These four tracks were considered: initiating ribosome profiles from all studies, ribosome profiles from all studies, Ribo-seq coverage data from all studies and mRNA-seq coverage data from all studies. At a particular site, Ribo-seq coverage was qualitatively evaluated as 1-Low (less than 25%), 2-Medium (between 25% and 75%), or 3-High (more than 75%) in relation to the relevant sATG and mORF.

4.10. Statistics

Basic descriptive statistics as well as correlation analyses were computed with the help of the PAST software (University of Oslo; Paleontological Statistics, version 4.03, https://www.nhm.uio.no/english/research/infrastructure/past/). Non-parametric correlation coefficients (Spearman's *rs* and Kendall's tau) were calculated for the correlation analyses. Monte Carlo permutation tests were available for all the correlation coefficients. The significance was computed using a two-tailed *t* test with $n - 2$ degrees of freedom. A *p* value less than 0.05 was considered statistically significant.

4.11. Patient and Sample Characteristics

Our cohort was composed of 105 breast cancer patients of Caucasian origin diagnosed in the Institute for the Care for Mother and Child and Medicon in Prague and Hospital Atlas in Zlin (all in the Czech Republic) during 2006–2013. Patients underwent neoadjuvant cytotoxic therapy with regimens based on 5-fluorouracil/anthracyclines/cyclophosphamide (FAC or FEC) and/or taxanes ($n = 68$) or postoperative adjuvant therapy using the same cytotoxic drugs ($n = 37$). The distribution of molecular subtypes was as follows: Luminal A 16%, Luminal B 24%, and triple negative 60%. These patients constituted the testing set in our previous study conducted by Hlavac et al. [57] and are characterized in detail in the article. Collection of blood samples and DNA Extraction were carried out according to standard procedures and described previously.

4.12. Targeted Sequencing

5′UTR sequences of all human ABCA genes were sequenced within a broader set of all exons of 509 genes representing major drug metabolizing and transporting enzymes, nuclear receptors, cell death, chemotherapy target, and signaling pathway genes. The gene panel selection, libraries preparation, sequencing criteria and data analysis, including selection and annotation of variants, were precisely described in Hlavac et al. [57].

5. Conclusions

We showed that the great variability among 5′UTR features seen on a whole genome level can be observed even in the group of homologous ABCA subfamily genes. Our phylogenetic analysis, based on the comparison of nucleotide sequences of ABCA 5′UTRs, shows that the 5′UTRs of *ABCA2* and *ABCA3* cluster distinctly from the rest of the 5′UTRs. The 5′UTRs of ABCA genes contain above-average numbers of uATGs, uORFs and 5′UTR introns as well as conserved ones and these elements probably play important biological roles in this subfamily, unlike RG4s. Our work brings largely new information on the numbers of stem loops in 5′UTRs. Some of the positive correlations among 5′UTR features are likely, however, some are interesting and need further exploration, such as the correlations between number of uATGs and number of 5′UTR introns or sATG flanking sequence context and presence of RG4-forming sequence. The lengths of the ABCA 5′UTRs seem to be the major factor influencing the numbers of the other known 5′UTR regulatory elements. Although there is also great variability in the distribution and expression of the ABCA proteins, we did not find any significant correlation between the numbers of 5′UTR features and protein expression characteristics. However, we confirmed a high concentrations of ribosomes at some of the predicted uORFs in the analysis of Ribo-seq data. We further verified the existence of SNVs in relation to the 5′UTR features, predicted within this study, experimentally in our cohort of 105 breast cancer patients. Our results support the view that the other elements such as RNA binding proteins and non-coding RNAs play the major role in protein expression fine-tuning within the complex background of the highly variable 5′UTRs. These findings extend our view on human genome variability and raise new questions for further investigations.

Supplementary Materials: Supplementary Materials can be found at http://www.mdpi.com/1422-0067/21/22/8878/s1. **Table S1.** Basic characteristics of the transcripts downloaded and analyzed in the study. **Table S2.** A detailed overview of 5'UTR features in the human ABCA genes. **Table S3.** Correlation tables of the first correlation analysis with Spearman's rs and Kendall's tau coefficients and relevant p values. **Table S4.** Correlation tables of the second correlation analysis with Spearman's rs and Kendall's tau coefficients and relevant p values. **Figure S1.** Multi-sequence alignment (Clustal Omega algorithm) of the 5'UTRs of the human ABCA1 gene and its vertebrate orthologs with nucleotide percentage identity colored, consensus logos and occupancy score histograms. **Figure S2.** Multi-sequence alignment (Mafft algorithm) of the 5'UTRs of the human ABCA1 gene and its vertebrate orthologs with nucleotide percentage identity colored, consensus logos and occupancy score histograms. **Figure S3.** Multi-sequence alignment (Clustal Omega algorithm) of the 5'UTRs of all 12 human ABCA protein-coding genes with nucleotide percentage identity colored, consensus logos and occupancy score histograms. **Figure S4.** Multi-sequence alignment (Mafft algorithm) of the 5'UTRs of all 12 human ABCA protein-coding genes with nucleotide percentage identity colored, consensus logos and occupancy score histograms. **Figure S5.** The location of 13 SNVs, found in breast cancer patients, in relation to the 5'UTR features of ABCA genes.

Author Contributions: Conceptualization: P.D.; methodology: P.D.; software: P.D.; validation: P.D., V.H., and P.S.; formal analysis: P.D.; investigation: P.D., V.H., and P.S.; resources: P.D., V.H., and P.S.; data curation: P.D.; writing—original draft preparation: P.D.; writing—review and editing: P.D., V.H., and P.S.; visualization: P.D.; supervision: P.S.; project administration: P.D., V.H., and P.S.; funding acquisition: P.D., V.H., and P.S. All authors have read and agreed to the published version of the manuscript.

Funding: This work was supported by the Ministry of Education, Youth and Sports of the Czech Republic project INTER-COST no. LTC19015 (to P.S.), project no. CZ.02.1.01/0.0/0.0/16_019/0000787 "Fighting INfectious Diseases", awarded by the MEYS CR, financed from EFRR (to P.D.) and Charles University project "Center of clinical and experimental liver surgery" no. UNCE/MED/006 (to V.H.).

Acknowledgments: We would like to thank Sarah Leupen, PhD. from the University of Maryland Baltimore County for English language review.

Conflicts of Interest: The authors declare no conflict of interest. The funders had no role in the design of the study; in the collection, analyses, or interpretation of data; in the writing of the manuscript, or in the decision to publish the results.

Abbreviations

5'UTR	5' untranslated region/leader sequence
ABC	ATP-binding cassette
cat.	Category
RG4	RNA G-quadruplex
sATG	Start ATG of the main ORF
TSS	Transcription start site
uATG	Upstream ATG start codon
uORF	Upstream open-reading frames

References

1. Davidson, A.; Dassa, E.; Orelle, C.; Chen, J. Structure, Function, and Evolution of Bacterial ATP-Binding Cassette Systems. *Microbiol. Mol. Biol. Rev.* **2008**, *72*, 317–364. [CrossRef] [PubMed]
2. Ford, R.; Beis, K. Learning the Abcs One at a Time: Structure and Mechanism of ABC Transporters. *Biochem. Soc. Trans.* **2019**, *47*, 23–36. [CrossRef] [PubMed]
3. Dean, M.; Rzhetsky, A.; Allikmets, R. The Human ATP-Binding Cassette (ABC) Transporter Superfamily. *Genome Res.* **2001**, *11*, 1156–1166. [CrossRef] [PubMed]
4. Pasello, M.; Giudice, A.; Scotlandi, K. The ABC Subfamily a Transporters: Multifaceted Players with Incipient Potentialities in Cancer. *Semin. Cancer Biol.* **2020**, *60*, 57–71. [CrossRef]
5. Piehler, A.; Özcürümez, M.; Kaminski, W. A-Subclass ATP-Binding Cassette Proteins in Brain Lipid Homeostasis and Neurodegeneration. *Front. Psychiatry* **2012**, *3*, 17. [CrossRef]
6. Kaminski, W.; Piehler, A.; Wenzel, J. ABC A-Subfamily Transporters: Structure, Function and Disease. *Biochim. Biophys. Acta BBA Mol. Basis Dis.* **2006**, *1762*, 510–524. [CrossRef]
7. Elsnerova, K.; Mohelnikova-Duchonova, B.; Cerovska, E.; Ehrlichova, M.; Gut, I.; Rob, L.; Skapa, P.; Hruda, M.; Bartakova, A.; Bouda, J.; et al. Gene Expression of Membrane Transporters: Importance for Prognosis and Progression of Ovarian Carcinoma. *Oncol. Rep.* **2016**, *35*, 2159–2170. [CrossRef]

8. Elsnerova, K.; Bartakova, A.; Tihlarik, J.; Bouda, J.; Rob, L.; Skapa, P.; Hruda, M.; Gut, I.; Mohelnikova-Duchonova, B.; Soucek, P.; et al. Gene Expression Profiling Reveals Novel Candidate Markers of Ovarian Carcinoma Intraperitoneal Metastasis. *J. Cancer* **2017**, *8*, 3598–3606. [CrossRef]

9. Dvorak, P.; Pesta, M.; Soucek, P. ABC Gene Expression Profiles Have Clinical Importance and Possibly Form a New Hallmark of Cancer. *Tumor Biol.* **2017**, *39*, 101042831769980. [CrossRef]

10. Hlaváč, V.; Brynychová, V.; Václavíková, R.; Ehrlichová, M.; Vrána, D.; Pecha, V.; Koževnikovová, R.; Trnková, M.; Gatěk, J.; Kopperová, D.; et al. The Expression Profile of ATP-Binding Cassette Transporter Genes in Breast Carcinoma. *Pharmacogenomics* **2013**, *14*, 515–529. [CrossRef]

11. Hlavata, I.; Mohelnikova-Duchonova, B.; Vaclavikova, R.; Liska, V.; Pitule, P.; Novak, P.; Bruha, J.; Vycital, O.; Holubec, L.; Treska, V.; et al. The Role of ABC Transporters in Progression and Clinical Outcome of Colorectal Cancer. *Mutagenesis* **2012**, *27*, 187–196. [CrossRef] [PubMed]

12. Hedditch, E.; Gao, B.; Russell, A.; Lu, Y.; Emmanuel, C.; Beesley, J.; Johnatty, S.; Chen, X.; Harnett, P.; George, J.; et al. ABCA Transporter Gene Expression and Poor Outcome in Epithelial Ovarian Cancer. *JNCI J. Natl. Cancer Inst.* **2014**, *106*, dju149. [CrossRef] [PubMed]

13. Annilo, T.; Chen, Z.; Shulenin, S.; Costantino, J.; Thomas, L.; Lou, H.; Stefanov, S.; Dean, M. Evolution of the Vertebrate ABC Gene Family: Analysis of Gene Birth and Death. *Genomics* **2006**, *88*, 1–11. [CrossRef] [PubMed]

14. Moitra, K.; Dean, M. Evolution of ABC Transporters by Gene Duplication and Their Role in Human Disease. *Biol. Chem.* **2011**, *392*, 29–37. [CrossRef]

15. Li, G.; Shi, P.; Wang, Y. Evolutionary Dynamics of the ABCA Chromosome 17Q24 Cluster Genes in Vertebrates. *Genomics* **2007**, *89*, 385–391. [CrossRef]

16. Vogel, C.; de Sousa Abreu, R.; Ko, D.; Le, S.; Shapiro, B.; Burns, S.; Sandhu, D.; Boutz, D.; Marcotte, E.; Penalva, L. Sequence Signatures and Mrna Concentration Can Explain Two-Thirds of Protein Abundance Variation in a Human Cell Line. *Mol. Syst. Biol.* **2010**, *6*, 400. [CrossRef]

17. Kozak, M. Regulation of Translation via Mrna Structure in Prokaryotes and Eukaryotes. *Gene* **2005**, *361*, 13–37. [CrossRef]

18. Leppek, K.; Das, R.; Barna, M. Functional 5′ UTR Mrna Structures in Eukaryotic Translation Regulation and How to Find Them. *Nat. Rev. Mol. Cell Biol.* **2017**, *19*, 158–174. [CrossRef]

19. Dvorak, P.; Leupen, S.; Soucek, P. Functionally Significant Features in the 5′ Untranslated Region of the ABCA1 Gene and Their Comparison in Vertebrates. *Cells* **2019**, *8*, 623. [CrossRef]

20. Peelman, F.; Labeur, C.; Vanloo, B.; Roosbeek, S.; Devaud, C.; Duverger, N.; Denèfle, P.; Rosier, M.; Vandekerckhove, J.; Rosseneu, M. Characterization of the ABCA Transporter Subfamily: Identification of Prokaryotic and Eukaryotic Members, Phylogeny and Topology. *J. Mol. Biol.* **2003**, *325*, 259–274. [CrossRef]

21. Pesole, G.; Grillo, G.; Larizza, A.; Liuni, S. The Untranslated Regions of Eukaryotic Mrnas: Structure, Function, Evolution and Bioinformatic Tools for Their Analysis. *Brief. Bioinform.* **2000**, *1*, 236–249. [CrossRef] [PubMed]

22. Pesole, G.; Mignone, F.; Gissi, C.; Grillo, G.; Licciulli, F.; Liuni, S. Structural and Functional Features of Eukaryotic Mrna Untranslated Regions. *Gene* **2001**, *276*, 73–81. [CrossRef]

23. Rogozin, I.; Kochetov, A.; Kondrashov, F.; Koonin, E.; Milanesi, L. Presence of ATG Triplets in 5′ Untranslated Regions of Eukaryotic Cdnas Correlates with a Weak′ Context of the Start Codon. *Bioinformatics* **2001**, *17*, 890–900. [CrossRef] [PubMed]

24. Chen, C.; Lin, H.; Pan, C.; Chen, F. The Plausible Reason Why the Length of 5′ Untranslated Region Is Unrelated to Organismal Complexity. *BMC Res. Notes* **2011**, *4*, 312. [CrossRef] [PubMed]

25. Nagalakshmi, U.; Wang, Z.; Waern, K.; Shou, C.; Raha, D.; Gerstein, M.; Snyder, M. The Transcriptional Landscape of the Yeast Genome Defined by RNA Sequencing. *Science* **2008**, *320*, 1344–1349. [CrossRef] [PubMed]

26. Lin, Z.; Li, W. Evolution of 5′ Untranslated Region Length and Gene Expression Reprogramming in Yeasts. *Mol. Biol. Evol.* **2011**, *29*, 81–89. [CrossRef]

27. Lynch, M.; Scofield, D.; Hong, X. The Evolution of Transcription-Initiation Sites. *Mol. Biol. Evol.* **2005**, *22*, 1137–1146. [CrossRef]

28. Kozak, M. Pushing the Limits of the Scanning Mechanism for Initiation of Translation. *Gene* **2002**, *299*, 1–34. [CrossRef]

29. Calvo, S.; Pagliarini, D.; Mootha, V. Upstream Open Reading Frames Cause Widespread Reduction of Protein Expression and Are Polymorphic Among Humans. *Proc. Natl. Acad. Sci. USA* **2009**, *106*, 7507–7512. [CrossRef]

30. Al-Ali, R.; González-Sarmiento, R. Proximity of AUG Sequences to Initiation Codon in Genomic 5′ UTR Regulates Mammalian Protein Expression. *Gene* **2016**, *594*, 268–271. [CrossRef]

31. Churbanov, A.; Rogozin, I.; Babenko, V.; Ali, H.; Koonin, E. Evolutionary Conservation Suggests a Regulatory Function of AUG Triplets in 5′-Utrs of Eukaryotic Genes. *Nucleic Acids Res.* **2005**, *33*, 5512–5520. [CrossRef] [PubMed]

32. Iacono, M.; Mignone, F.; Pesole, G. Uaug and Uorfs in Human and Rodent 5′ Untranslated Mrnas. *Gene* **2005**, *349*, 97–105. [CrossRef] [PubMed]

33. Crowe, M.; Wang, X.; Rothnagel, J. Evidence for Conservation and Selection of Upstream Open Reading Frames Suggests Probable Encoding of Bioactive Peptides. *BMC Genom.* **2006**, *7*, 16. [CrossRef] [PubMed]

34. Johnstone, T.; Bazzini, A.; Giraldez, A. Upstream ORF S Are Prevalent Translational Repressors in Vertebrates. *EMBO J.* **2016**, *35*, 706–723. [CrossRef]

35. Brunet, M.; Levesque, S.; Hunting, D.; Cohen, A.; Roucou, X. Recognition of the Polycistronic Nature of Human Genes Is Critical to Understanding the Genotype-Phenotype Relationship. *Genome Res.* **2018**, *28*, 609–624. [CrossRef]

36. Cenik, C.; Derti, A.; Mellor, J.; Berriz, G.; Roth, F. Genome-Wide Functional Analysis of Human 5′ Untranslated Region Introns. *Genome Biol.* **2010**, *11*, R29. [CrossRef]

37. Hong, X.; Scofield, D.; Lynch, M. Intron Size, Abundance, and Distribution within Untranslated Regions of Genes. *Mol. Biol. Evol.* **2006**, *23*, 2392–2404. [CrossRef]

38. Chorev, M.; Carmel, L. The Function of Introns. *Front. Genet.* **2012**, *3*, 55. [CrossRef]

39. Lim, C.; Wardell, S.T.; Kleffmann, T.; Brown, C. The Exon-Intron Gene Structure Upstream of the Initiation Codon Predicts Translation Efficiency. *Nucleic Acids Res.* **2018**, *46*, 4575–4591. [CrossRef]

40. Bolduc, F.; Garant, J.; Allard, F.; Perreault, J. Irregular G-Quadruplexes Found in the Untranslated Regions of Human Mrnas Influence Translation. *J. Biol. Chem.* **2016**, *291*, 21751–21760. [CrossRef]

41. Huppert, J.; Balasubramanian, S. Prevalence of Quadruplexes in the Human Genome. *Nucleic Acids Res.* **2005**, *33*, 2908–2916. [CrossRef] [PubMed]

42. Fay, M.; Lyons, S.; Ivanov, P. RNA G-Quadruplexes in Biology: Principles and Molecular Mechanisms. *J. Mol. Biol.* **2017**, *429*, 2127–2147. [CrossRef] [PubMed]

43. Kumari, S.; Bugaut, A.; Balasubramanian, S. Position and Stability Are Determining Factors for Translation Repression by an RNA G-Quadruplex-Forming Sequence within the 5′ UTR of Thenrasproto-Oncogene. *Biochemistry* **2008**, *47*, 12664–12669. [CrossRef] [PubMed]

44. Ravichandran, S.; Ahn, J.; Kim, K. Unraveling the Regulatory G-Quadruplex Puzzle: Lessons from Genome and Transcriptome-Wide Studies. *Front. Genet.* **2019**, *10*, 1002. [CrossRef]

45. Huppert, J.; Bugaut, A.; Kumari, S.; Balasubramanian, S. G-Quadruplexes: The Beginning and End of Utrs. *Nucleic Acids Res.* **2008**, *36*, 6260–6268. [CrossRef]

46. Maizels, N.; Gray, L. The G4 Genome. *PLoS Genet.* **2013**, *9*, e1003468. [CrossRef]

47. Bedrat, A.; Lacroix, L.; Mergny, J. Re-Evaluation of G-Quadruplex Propensity with G4hunter. *Nucleic Acids Res.* **2016**, *44*, 1746–1759. [CrossRef]

48. Svoboda, P.; Cara, A. Hairpin RNA: A Secondary Structure of Primary Importance. *Cell. Mol. Life Sci.* **2006**, *63*, 901–908. [CrossRef]

49. Babendure, J.; Babendure, J.; Ding, J.; Tsien, R. Control of Mammalian Translation by Mrna Structure near Caps. *RNA* **2006**, *12*, 851–861. [CrossRef]

50. Weenink, T.; van der Hilst, J.; McKiernan, R.; Ellis, T. Design of RNA Hairpin Modules That Predictably Tune Translation in Yeast. *Synth. Biol.* **2018**, *3*, ysy019. [CrossRef]

51. Varani, G. Exceptionally Stable Nucleic Acid Hairpins. *Annu. Rev. Biophys. Biomol. Struct.* **1995**, *24*, 379–404. [CrossRef] [PubMed]

52. Wan, Y.; Qu, K.; Zhang, Q.; Flynn, R.; Manor, O.; Ouyang, Z.; Zhang, J.; Spitale, R.; Snyder, M.; Segal, E.; et al. Landscape and Variation of RNA Secondary Structure across the Human Transcriptome. *Nature* **2014**, *505*, 706–709. [CrossRef] [PubMed]

53. Ajay, S.; Athey, B.; Lee, I. Unified Translation Repression Mechanism for Micrornas and Upstream Augs. *BMC Genom.* **2010**, *11*, 155. [CrossRef] [PubMed]

54. Waterhouse, A.; Procter, J.; Martin, D.; Clamp, M.; Barton, G. Jalview Version 2—A Multiple Sequence Alignment Editor and Analysis Workbench. *Bioinformatics* **2009**, *25*, 1189–1191. [CrossRef] [PubMed]
55. Hernández, G.; Osnaya, V.; Pérez-Martínez, X. Conservation and Variability of the AUG Initiation Codon Context in Eukaryotes. *Trends Biochem. Sci.* **2019**, *44*, 1009–1021. [CrossRef] [PubMed]
56. Doluca, O. G4catchall: A G-Quadruplex Prediction Approach Considering Atypical Features. *J. Theor. Biol.* **2019**, *463*, 92–98. [CrossRef]
57. Hlavac, V.; Kovacova, M.; Elsnerova, K.; Brynychova, V.; Kozevnikovova, R.; Raus, K.; Kopeckova, K.; Mestakova, S.; Vrana, D.; Gatek, J.; et al. Use of Germline Genetic Variability for Prediction of Chemoresistance and Prognosis of Breast Cancer Patients. *Cancers* **2018**, *10*, 511. [CrossRef]

Publisher's Note: MDPI stays neutral with regard to jurisdictional claims in published maps and institutional affiliations.

International Journal of
Molecular Sciences

Review

Peroxisomal ABC Transporters: An Update

Ali Tawbeh, Catherine Gondcaille, Doriane Trompier and Stéphane Savary *

Laboratoire Bio-PeroxIL EA7270, University of Bourgogne Franche-Comté, 6 Boulevard Gabriel,
21000 Dijon, France; ali.tawbeh@u-bourgogne.fr (A.T.); Catherine.Gondcaille@u-bourgogne.fr (C.G.);
doriane.trompier@u-bourgogne.fr (D.T.)
* Correspondence: stsavary@u-bourgogne.fr; Tel.: +33-380-396-273

Abstract: ATP-binding cassette (ABC) transporters constitute one of the largest superfamilies of conserved proteins from bacteria to mammals. In humans, three members of this family are expressed in the peroxisomal membrane and belong to the subfamily D: ABCD1 (ALDP), ABCD2 (ALDRP), and ABCD3 (PMP70). These half-transporters must dimerize to form a functional transporter, but they are thought to exist primarily as tetramers. They possess overlapping but specific substrate specificity, allowing the transport of various lipids into the peroxisomal matrix. The defects of ABCD1 and ABCD3 are responsible for two genetic disorders called X-linked adrenoleukodystrophy and congenital bile acid synthesis defect 5, respectively. In addition to their role in peroxisome metabolism, it has recently been proposed that peroxisomal ABC transporters participate in cell signaling and cell control, particularly in cancer. This review presents an overview of the knowledge on the structure, function, and mechanisms involving these proteins and their link to pathologies. We summarize the different in vitro and in vivo models existing across the species to study peroxisomal ABC transporters and the consequences of their defects. Finally, an overview of the known and possible interactome involving these proteins, which reveal putative and unexpected new functions, is shown and discussed.

Keywords: ABC transporters; peroxisome; adrenoleukodystrophy; fatty acids

Citation: Tawbeh, A.; Gondcaille, C.; Trompier, D.; Savary, S. Peroxisomal ABC Transporters: An Update. *Int. J. Mol. Sci.* **2021**, *22*, 6093. https://doi.org/10.3390/ijms22116093

Academic Editor: Thomas Falguières

Received: 28 April 2021
Accepted: 3 June 2021
Published: 5 June 2021

Publisher's Note: MDPI stays neutral with regard to jurisdictional claims in published maps and institutional affiliations.

1. Introduction

ATP-binding cassette (ABC) transporters constitute a superfamily of membrane transporter proteins that actively translocate a wide range of molecules, from simple molecules (fatty acids (FAs), sugars, nucleosides, and amino acids) to complex organic compounds (lipids, oligonucleotides, polysaccharides, and proteins) [1]. Transport of substrates is dependent on the hydrolysis of ATP, which releases energy that can be used to accumulate substances in the cellular compartments or export them to the outside. ABC transporters are distributed not only in the plasma membrane of both prokaryotes and eukaryotes, but also in the membranes of the organelles of eukaryotic cells such as peroxisomes, mitochondria, lysosomes, and endoplasmic reticulum (ER). Based on their amino acid homology and structural configuration, ABC transporters in humans are classified into seven subfamilies, A to G, comprising a total of 48 ABC transporters, many of which are implicated in diseases [2]. ABC transporters of subfamily D include four proteins in mammals: ABCD1 [adrenoleukodystrophy protein (ALDP)], ABCD2 [adrenoleukodystrophy-related protein (ALDRP)], ABCD3 [70 kDa peroxisomal membrane protein (PMP70)], and ABCD4 [peroxisomal membrane protein 69 (PMP69)] [3]. ABCD1, ABCD2, and ABCD3 are located in the peroxisomal membrane. ABCD4 was identified by homology search for ALDP and PMP70 related sequences in the database of expressed sequence tags, and was initially considered peroxisomal despite the absence of a membrane peroxisomal targeting signal [4]. More recently, several studies have demonstrated that ABCD4 resides in the endoplasmic reticulum and lysosomes, and that its function is associated with cobalamin metabolism [3,5,6].

The three human peroxisomal ABC transporters play an important role in the transport of various lipid substrates into the peroxisome for their shortening by β-oxidation (Figure 1). β-oxidation of FAs is a conserved process of peroxisomes by which acyl groups are degraded two carbons at a time after being activated to form the corresponding CoA derivative by a specific acyl-CoA synthetase located at the peroxisomal membrane [7]. The β-oxidation process exists in mitochondria for medium- and long-chain fatty acids (MCFAs and LCFAs) and is necessary to terminate degradation of octanoyl-CoA coming from peroxisomes. However, very long-chain fatty acids (VLCFAs, number of carbon atoms >22) are exclusively β-oxidized into the peroxisome, and this organelle is therefore essential, especially in the brain [8]. Moreover, polyunsaturated fatty acid (PUFA) synthesis may require a peroxisomal cycle of β-oxidation, as in the case of docosahexaenoic acid (DHA, C22:6 n-3) synthesis from its precursor (C24:6 n-3) [9]. It is important to note that DHA is not only of great value by itself as a component of cell membranes, but is also the source of eicosanoids associated with several key signaling functions [10].

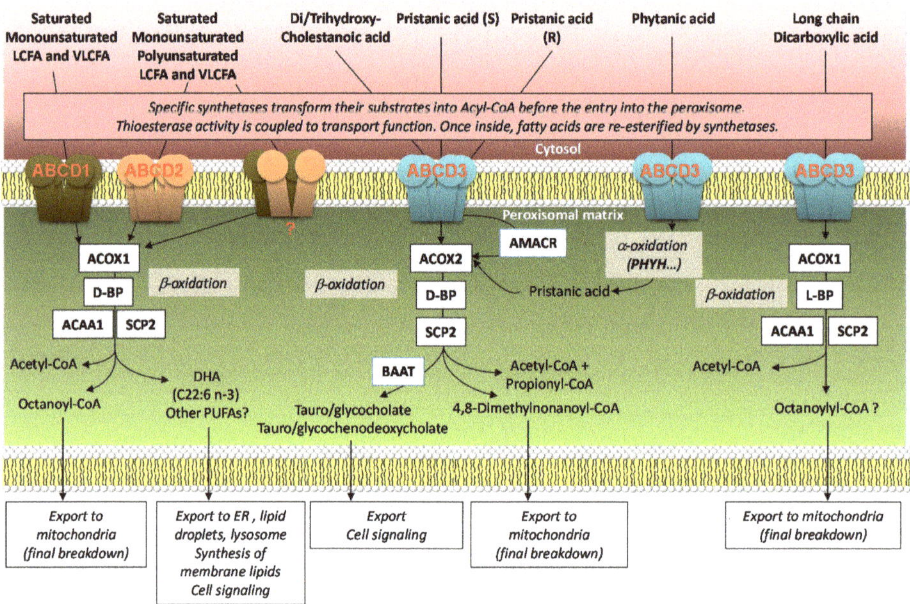

Figure 1. Peroxisomal ABC transporters and their involvement in lipid metabolism. Peroxisomal ABC transporters are represented as homo or heterotetramers with their preferential substrates and their involvement in metabolic routes, including several enzymatic steps, catalyzed by acyl-CoA oxidase 1 and 2 (ACOX1 and ACOX2), D- and L-bifunctional protein (D-BP and L-BP), acetyl-CoA Acyltransferase 1 (3-ketoacyl-CoA thiolase, ACAA1), sterol carrier protein 2 (SCPX thiolase, SCP2), alpha-methylacyl-CoA racemase (AMACR), bile acid-CoA:amino acid N-acyltransferase (BAAT), and phytanoyl-CoA hydroxylase (PHYH).

Thus, peroxisomal β-oxidation may not be considered a simple catabolic process of fatty acids. The role of peroxisomal ABC transporters is therefore not restricted to the catabolic function of peroxisomes, but is fully associated with their various metabolic functions including synthesis and degradation of lipids, cell signaling, inflammation control, and redox homeostasis [11–15].

2. Structure, Function, and Mechanism of Transport

2.1. Structure

The general structure of eukaryotic ABC transporters is a four functional unit organization, comprising two transmembrane domains (TMDs) and two nucleotide binding domains (NBDs). NBDs bind and hydrolyze ATP to trigger conformational changes in the TMDs, resulting in unidirectional transport across the membrane [1]. Human peroxisomal ABC transporters have a half-transporter structure, with only one TMD and one NBD [16]. In 2017, we proposed a structural model of human ABCD1 based on the crystal structure of the mitochondrial ABC transporter ABCB10, which shows not only the putative structure of ABCD1 in a membrane context but also the complex intricacy of α-helices that constitute the whole transmembrane domain (Figure 2) [17]. Therefore, peroxisomal ABC half-transporters need to homo- or heterodimerize in the peroxisomal membrane in order to constitute a full, active transporter [18,19].

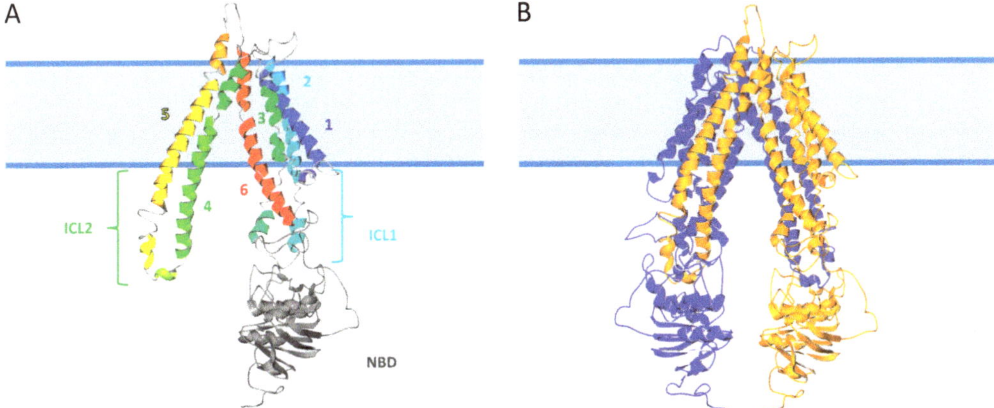

Figure 2. Structural model of human ABCD1 (reprinted from [17]). (**A**) Ribbon representation of the ABCD1 monomer. TMD helices are numbered from 1 to 6 and rainbow colored from dark blue to red. NBD is in light grey, and intracellular loops (ICL) 1 and 2 are indicated. (**B**) Ribbon representation of the ABCD1 homodimer with the two subunits respectively colored in dark blue and yellow.

Data shows that ABCD1, ABCD2, and ABCD3 are able to interact as homodimers or heterodimers [20–22], although both ABCD1 and ABCD3 are mainly found as homodimers in mammalian peroxisomal membranes [23–25]. Moreover, ABCD1 and ABCD2 homodimers are functional [26,27]. However, the fact that nonfunctional ABCD2 has a transdominant negative effect on ABCD1 [20] suggests that heterodimers of ABCD1 and ABCD2 are also functional and can exist within cells and tissues expressing both proteins. Besides, chimeric proteins consisting of homo- and heterodimers of ABCD1 and ABCD2 are functionally active [19]. Concerning ABCD3, although homodimers and heterodimers with ABCD1 and ABCD2 have been described [22,23,25,28,29], no data is available about the functional value of the ABCD3 dimers. Surprisingly, ABCD1 and ABCD3 were found in different detergent-resistant microdomains [29], implying that these proteins have a different environment in the peroxisomal lipid bilayer, questioning the biological relevance of the ABCD1 and ABCD3 heterodimers. Additionally, native PAGE experiments concerning complex oligomerization confirm that ABCD1 and ABCD2 exist predominantly as homo-tetramers, although both homo- and hetero-tetramers are present [28]. Therefore, we cannot rule out the possibility that hetero-interaction between ABCD1 and ABCD2 occurs in hetero-tetramers composed of two distinct homodimers rather than in complexes composed of two heterodimers. Finally, it remains unclear whether the oligomerization of peroxisomal ABC transporters has any influence on substrate specificity.

2.2. Substrate Specificity

Since the cloning of the *ABCD1* gene in 1993 and its association with X-ALD [30], ABCD1 function has been attributed to the transport of saturated and monounsaturated VLCFAs across the peroxisomal membrane for further degradation by β-oxidation. Accumulation of saturated and monounsaturated VLCFAs indeed occurs in the plasma and tissues of X-ALD patients, and is used for diagnosis [31,32]. Due to its importance, studies concerning the structure, function, and defects of ABCD1 have never ceased. Functional complementation experiments in yeast, functional assays in mammalian cells, especially cells coming from X-linked adrenoleukodystrophy (X-ALD) patients, and studies using animal models, mainly knock-out mice, were helpful in clarifying the question of substrate specificity. The *Abcd1* knock-out mice confirmed the human biochemical phenotype, indicating that ABCD1 is indeed involved in the transport of VLCFAs [33–35]. Transfection of X-ALD skin fibroblasts with ABCD1 cDNA corrected the β-oxidation defect and restored normal levels of VLCFAs [36,37]. The preference of ABCD1 for saturated FAs was also confirmed in yeast [26,27].

Cloned by homology using degenerate primers, the *ABCD2* gene was shown to code for ALDRP, the closest homolog of ALDP [38]. Both proteins display overlapping substrate specificities for saturated and monounsaturated LCFAs and VLCFAs. It explains the correction of β-oxidation defect in X-ALD fibroblasts in case of *ABCD2* overexpression after transfection [39]. Using transgenic expression of *Abcd2* in the *Abcd1* knock-out mouse, Pujol et al. demonstrated that VLCFA accumulation and disease phenotype could be corrected in vivo [40]. This set the basis for a new therapeutic strategy for X-ALD patients aiming at inducing *ABCD2* expression with pharmacological, hormonal, or nutritional management [41,42]. Pharmacological induction of *ABCD2* was indeed shown to compensate for ABCD1 defect in vitro and in rare cases, in vivo, opening the way for clinical trials [43–58].

Functional complementation in yeast model and X-ALD fibroblasts confirmed the functional redundancy for saturated VLCFAs, but also demonstrated the specific role of ABCD2 in PUFA transport, especially DHA and its precursor (C24:6 n-3) [26]. Experiments in mammalian cells confirmed such substrate preference [19,20]. Further studies using the *Abcd2* null mice demonstrated a specific role in MUFA transport, especially for erucic acid (C22:1 n-9) in adipose tissue [59,60] and an extended role in FA homeostasis [61].

PMP70, the protein coded by the *ABCD3* gene, was the first identified peroxisomal ABC transporter and is the most abundant peroxisomal membrane protein, at least in hepatocytes [62,63]. Wrongly associated with peroxisome biogenesis [64], ABCD3 is also involved in the transport of various lipids and shows overlapping substrate specificities with ABCD1 when overexpressed [37,65]. Though, ABCD3 clearly has the broadest substrate specificity as it is involved in the transport of LCFAs and VLCFAs but also specifically in the transport of dicarboxylic acids, branched-chain fatty acids, and C27 bile acid intermediates such as di- and tri-hydroxy-cholestanoic acid [65–67]. The *Abcd3* knock-out mice indeed revealed a marked accumulation of bile acid intermediates, and ABCD3 was recently associated with a congenital bile acid defect (CBAS5, see below) [67]. Furthermore, a more recent study performed on manipulated HEK-293 cell models proved that ABCD3 is required for the transport of MCFAs across the peroxisomal membrane [68].

2.3. Mechanism

Conversion of free FAs into CoA esters constitutes an initial activation step before peroxisomal β-oxidation. This reaction is catalyzed by specific acyl-CoA synthetase connected to the cytosolic side of the peroxisomal membrane [69]. It was proved, using protease protection assays, that acyl-CoAs but not free FAs bind to the TMD of the transporter [70]. It is therefore only after activation that the fatty acyl-CoAs are transported to the peroxisomal matrix through peroxisomal ABC transporters. Fatty acyl-CoA are captured on the cytosolic side by the TMD, enhancing the affinity of NBD for ATP. ATP molecules are then hydrolyzed, thus producing the energy needed to switch the conformation of TMD and eventually allowing the translocation of substrates from the cytosol into the peroxisomal matrix [22,71]. However, the exact mechanism of transport remains controversial. Two models are commonly considered. The first implies that esterified FAs are delivered directly to the peroxisomal matrix, whereas in the other model, free FAs are transported into the peroxisomal matrix after the hydrolysis of acyl-CoAs, which are re-esterified by acyl-CoA synthetase when in the peroxisomal lumen.

Although the process of cleavage and reactivation of acyl-CoAs seems to be a waste of energy as two ATP molecules are needed for the activation reaction, such a mechanism is crucial for the specific permeabilization of the substrates of β-oxidation [18]. Several studies have been done in an attempt to figure out the correct model for the transportation mechanism. Early studies on yeast models have demonstrated that fatty acyl-CoAs are hydrolyzed before being transported. This hydrolysis occurs when acyl-CoAs interacts with the heterodimer Pxa1p-Pxa2p at the cytosolic side of the peroxisomal membrane [72]. The peroxisomal ABC transporters would release a free fatty acid that should be re-esterified inside the peroxisome before its catabolic processing. In addition, intrinsic acyl-CoA thioesterase activity has been found in COMATOSE (CTS), a homolog of human ABCD1 in *Arabidopsis thaliana*, proving again that VLCFA-CoA is hydrolyzed prior to transport [73]. Very recently, the work of Kawaguchi et al. provided further proof of the transport mechanism [74]. After expressing human His-tagged ABCD1 in methylotrophic yeast, they directly demonstrated that ABCD1 transports the FA moiety after the hydrolysis of VLCFA-CoA and that acyl-CoA synthetase is required before the β-oxidation of VLCFA-CoA within the peroxisomes. When it comes to the fate of the free CoA, they are released in the peroxisomal lumen, as revealed using isolated peroxisomes from *Saccharomyces cerevisiae* [75].

Finally, after their re-esterification, substrates are directly delivered to specific acyl-CoA oxidases to initiate the β-oxidation process. Peroxisomal acyl-coenzyme A oxidase 1 (ACOX1) catalyzes the first and rate-limiting step of the β-oxidation pathway dedicated to straight-chain fatty acids, which includes LCFAs, VLCFAs, PUFAs, and dicarboxylic acids [76]. Other acyl-CoA oxidases also exist, ACOX2 and ACOX3. ACOX2 is specific to bile acid intermediates [76] whereas the oxidation of branched-chain FAs depends on both ACOX2 and ACOX3 enzymes [77,78]. Of note, mitochondria catalyze the β-oxidation of the majority of short, medium, and long chain FAs but not that of VLCFAs [79]. In yeast and plants, this process of FA β-oxidation occurs exclusively in peroxisomes, whereas in higher eukaryotes, the catabolism of VLCFAs is initiated solely in the peroxisomes [7,80].

3. Human Diseases

3.1. X-Linked Adrenoleukodystrophy

X-linked adrenoleukodystrophy (X-ALD, OMIM # 300100) is the most frequent peroxisomal disorder but is still classified as a rare disease, with an estimated incidence of 1:17,000 [81]. Recent therapeutic successes [82,83], and the feasibility and reliability of a diagnosis method based on VLCFA quantification from blood spot [84,85], have prompted some countries to establish systematic screening of newborns. This complex and fatal neurodegenerative disorder is characterized by a huge clinical variability both in the age of onset and in the symptoms [31]. The two main forms are the childhood cerebral ALD (ccALD), characterized by inflammatory demyelination of the central nervous system and the adult form, called adrenomyeloneuropathy (AMN), consisting of a non-inflammatory, slowly progressive demyelination affecting the spinal cord and peripheral nerves. X-ALD is also the main cause of Addison's disease and adrenal insufficiency may remain the unique symptom of the disease. Since the disease is linked to chromosome X, boys and men are the most severely affected patients. Female carriers usually remain quasi asymptomatic or present only a mild phenotype, but severe forms have also been described [86].

In 1993, using positional cloning, the team of Hugo Moser identified the ABCD1 gene as being responsible for X-ALD [30]. Mutations in the ABCD1 gene have been found in every X-ALD patient and are collected in the X-ALD database (https://adrenoleukodystrophy.info/ accessed on 1 June 2021). In spite of almost 900 non-recurrent mutations, no genotype-phenotype correlation has been described. It is important to note that the majority of missense mutations affect protein stability and result in the absence of the protein. ABCD1 defect results in the accumulation of VLCFAs, mainly C26:0 and C26:1, which accumulate as free FAs or in esterified forms in membrane lipids and cholesteryl esters. This accumulation results from the impossibility of their entry into the peroxisome for their degradation by β-oxidation, but also from an increased endogenous biosynthesis [87]. While the toxicity of VLCFAs has been recognized [88], the sequence of events leading to neurodegeneration and inflammation is still debated. Oxidative stress and cellular components, especially microglial functions, seem to play a major role in the pathogenesis of X-ALD [89–91].

Therapeutic strategies depend on the clinical symptoms of patients. A majority of X-ALD patients present adrenal insufficiency and require a careful patient follow-up, but hormone-replacement therapy successfully manages the adrenal defect and prevents a potentially fatal Addisonian crisis. Hematopoietic stem cell transplantation (HSCT) has proven efficiency to halt neurological involvement in X-ALD. Since 1990 and the first success of this therapy [92], allogeneic graft has been indicated for boys with ccALD at an early stage of the disease when a compatible donor exists. In 2009, autologous HSCT was demonstrated to be successful to halt cerebral demyelination in two boys with no compatible donors who received their own genetically corrected stem cells [82]. Lentiviral correction of bone marrow derived stem cells and autologous transplantation proved effective in 15 patients in 2017 [83]. These promising results suggest that such a therapeutic strategy may be as effective as allogeneic HSCT. In addition, efforts to find pharmacological strategies targeting oxidative stress, inflammation, or compensatory mechanisms (antioxidant cocktail [93], leriglitazone [94], sobetirome [50,54,55]) are still present. It remains to be evaluated whether such treatments would be useful per se or in combination with HSCT strategies, at least to delay the onset of neurological concerns and permit a lengthening of the time window to allow transplantation.

3.2. Congenital Bile Acid Synthesis Defect Type 5

Although mutations were found in the ABCD3 gene of a Zellweger patient [64], further evidence showed that ABCD3 has no link with peroxisomal biogenesis and is definitively not associated with Zellweger Syndrome [95]. ABCD3, which presents partial functional redundancy with ABCD1, has been shown to transport branched-chain FAs, dicarboxylic acids, and bile acid precursors. A few years ago, the accumulation of peroxisomal C27-bile acid intermediates DHCA and THCA, as well as VLCFAs, was described in a young Turkish girl whose parents were consanguineous [67]. The patient presented hepatosplenomegaly and a severe progressive liver disease and she died of complications after liver transplantation. Patient fibroblasts showed reduced numbers of enlarged peroxisomes, as well as reduced β-oxidation of pristanic acid, compared to controls. Immunofluorescence confirmed the absence of ABCD3 in the peroxisomal membrane. A homozygous truncating mutation was identified in the ABCD3 gene of the patient, and the disease was named congenital bile acid synthesis defect (CBAS) type 5 (OMIM # 616278). It should be noted that CBAS type 1, 2, 3, 4, and 6 are associated with mutations in HSD3B7, AKR1D1, CYP7B1, AMACR, and ACOX2 respectively. These genes control key reactions in bile acid synthesis and all the CBAS forms present an autosomal recessive inheritance.

3.3. Peroxisomal ABC Transporters and Cancer

Beyond its recognized role in metabolism and redox homeostasis, the peroxisome is now increasingly regarded as a signaling platform and a key organelle in cellular metabolic reprogramming with major consequences on the immune response, cell cycle, and cell differentiation [12,96]. Elegantly presented in the state of the art of Hlaváč and Souček, several studies have revealed a significant association between the level of expression of peroxisomal ABC transporters and various cancers [97]. This suggests a role of these ABC transporters in cell cycle control, cell differentiation, and tumorigenesis. Downregulation of peroxisomal ABC transporters has been observed in several cases: *ABCD1* in melanoma [98] and renal cell carcinoma [99], *ABCD2* in breast cancer [100], and *ABCD3* in ovarian cancer [101] and colorectal cancer [102]. Moreover, a lower prognostic value has been associated with low expression of *ABCD1* in ovarian cancer [103] and low expression of *ABCD3* in colorectal cancer [102]. On the contrary, *ABCD1* and *ABCD3* were found upregulated in breast carcinoma [104], and a positive correlation was observed between *ABCD3* expression and glioma tumor grades [105]. Recently, VLCFA accumulation was associated with colorectal cancer [106]. Increased endogenous elongation appears to be primarily responsible for this observation, but peroxisomal ABC transporters are also likely involved, and the ability to regulate their expression could potentially represent a therapeutic interest in such cancers. Altogether, further studies are required to understand the link between the transport function and metabolic role of peroxisomal ABC transporters and the control of cell cycle with regard to the complexity of tumor heterogeneity.

4. Cell, Plant, and Animal Models

Phylogenetic analysis of peroxisomal ABC transporters in eukaryotes shows strong conservation, highlighting their fundamental and specific role in the cellular functions. Interestingly, their substrate specificity seems to become more restrictive with the complexification of the biological systems. Although the number of peroxisomal ABC transporters and their specific functions vary between species, each model of study is of scientific interest and has contributed significantly to the knowledge of peroxisomal ABC transporters. Here are described the main eukaryote models.

4.1. Yeast

In addition to its convenience for genomic modification, *Saccharomyces cerevisiae* is a particularly interesting model for studying the peroxisomal metabolism of lipids, since it can use FAs as its only carbon source and β-oxidation of FAs of all length takes place only in peroxisome. The yeast model expresses only two peroxisomal ABC transporters, called Pxa1p and Pxa2p, which function as a strict heterodimer to import fatty acyl-CoAs into the peroxisomal matrix [107–109]. Functional assays and functional complementation experiments of pxa1/pxa2Δ yeast mutants with mammalian peroxisomal ABC transporters were particularly important in studying their transport mechanism and substrate specificity [26,65,75].

4.2. Plant

In *Arabidopsis thaliana*, CTS, the human ABCD1 ortholog, is an integral peroxisomal membrane protein composed of two fused half-size transporters. CTS is involved in the import of FAs and phytohormone precursors into the peroxisome where they are β-oxidized [110,111]. The products of this oxidation are involved in the transition from dormancy to germination, root growth, seedling establishment, and fertility [112]. Expression of human *ABCD1* in *A. thaliana* CTS mutant cannot restore the germination and establishment, whereas human *ABCD2* only restores the germination phenotype [113]. These results are related to the physiological differences between plants and mammals, and highlight the differences in substrate specificity between ABCD1 and ABCD2. The plant model was also very important as it showed for the first time the existence of CTS in high molecular weight complexes and allowed the study of the transport mechanism, especially the role of its thioesterase activity [73].

4.3. Nematode

Caenorhabditis elegans is a well-known worm model in neurobiology studies, but the interest of this model in the field of X-ALD has been shown only very recently. *PMP-4* is one of the five putative peroxisomal ABC transporters identified in *C. elegans* and is the ortholog of human *ABCD1* and *ABCD2*. It is mainly expressed in gut and hypodermis, the main fat storage tissues in the *C. elegans*. Moreover, hypodermal cells have similarities with vertebrate glial cells and participate in neuronal migration [114]. *PMP-4* deficient worms have a normal growth and maturation but show several hallmarks of X-ALD (global VLCFA accumulation, redox imbalance, axonal damage, motility alteration) [115]. Interestingly, the number and the size of lipid droplets (LDs) are increased and can be normalized using a mitochondrial targeted antioxidant. *C. elegans* is therefore a valuable model to study the involvement of FA accumulation and oxidative stress in the pathogenesis of X-ALD but has some limitations since its nervous system is not myelinated.

4.4. Insect

An X-ALD fly model has been generated in *Drosophila melanogaster* using RNA interfering of *dABCD*, the ortholog of *ABCD1*. These flies survive to adulthood but exhibit a specific brain neurodegenerative phenotype with retinal defects including holes and loss of pigment cells associated with death of neurons and glia [116]. Interestingly, cellular targeted disruption of *dABCD* in neurons, but not in glia, triggers the retinal defects. The phenotype is indistinguishable from the one observed in *bgm* (bubblegum) and *dbb* (double-bubble) deficient flies [117]. Both *bgm* and *dbb* genes code for long/very-long-chain acyl-CoA synthetases. The shared neurodegenerative features in *dABCD* and *bgm/dbb* deficient flies show that the lipid metabolic pathway is a key component of the X-ALD-like neurodegenerative disease in *Drosophila*. More specifically, experiments achieved with *bgm* and *dbb* deficient flies indicate that the loss of metabolites is the cause of neurodegenerative disease rather than accumulation of substrates (V/LCFAs), as was commonly thought.

4.5. Fish

Zebrafish (*Danio rerio*) has recently been proved to be a useful model for studying the pathogenesis of X-ALD. Indeed, *Abcd1* (the zebrafish ortholog of *ABCD1*), is expressed during development in spinal cord and in the central nervous system especially in the oligodendrocytes and motor neuron precursors, but also in the interrenal gland (functional equivalent of the adrenal cortex) [118]. Zebrafish *Abcd1* mutant models show key biochemical and nervous system alteration features of X-ALD (increased level of C26:0, accumulation of cholesterol, hypomyelinated spinal cord, modified development of interrenal gland and brain, early alteration of motor behavior, decreased survival, and modified oligodendrocytes pattern associated with apoptosis). Interestingly, the motor alteration and the oligodendrocytes pattern can be corrected by human *ABCD1* expression. Moreover, a recent drug screening study showed that chloroquine can improve motor activity in zebrafish *Abcd1* mutant and reduce saturated VLCFA levels [119].

4.6. Rat and Mouse

Various cellular models have been created in rodent species to study the function of peroxisomal ABC transporters and the consequences of their defect. Considering that the liver is a platform for peroxisomal lipid metabolism in mammals, the hepatic H4IIEC3 cell line was used to create a specific cell model allowing the inducible expression of a normal or mutated rat Abcd2 protein fused to green fluorescent protein [120]. It allowed to precise the substrate specificity of Abcd2 as well as its dimeric status, and even, to demonstrate for the first time its supradimeric structure. [19,20,28]. To better understand the role of peroxisomal ABC transporters in the glial cells, models of ALD astrocytes have been developed. Astrocytes are known to regulate the inflammatory response. In neurodegenerative diseases, reactive astrocytes secrete inflammatory cytokines, which allow the permeability of the blood-brain barrier (BBB) to peripheral infiltrating immune cells. When *Abcd1* and/or *Abcd2* genes are silenced in mouse primary astrocytes, X-ALD biochemical hallmarks are present (decreased C24:0 β-oxidation, increased C26:0 level), but so are redox imbalance and pro-inflammatory features (increased cytokines expression and nitric oxide production) [121]. These characteristics are inverted by treatment with Lorenzo oil and increased by a long-term VLCFA treatment showing the link between VLCFA accumulation and the pro-inflammatory response of these glial cells [122]. These first results obtained in primary astrocytes led to the development of an immortalized astrocyte cell line [123]. This model should be very useful for studying the mechanisms of astrocyte activation and was used to screen therapeutic compounds such as SAHA, an HDAC inhibitor that normalizes ROS production as well as iNOS and TNF expression [53].

Microglia is also considered a major player in the X-ALD pathogenesis, especially in the inflammatory process. To proceed further, *Abcd1* and/or *Abcd2* deficient microglia cell lines have been obtained using CRISPR/Cas9 gene editing in the mouse BV-2 cell line [124]. The *Abcd1*$^{-/-}$*Abcd2*$^{-/-}$ cells, generated to avoid masking effects due to functional redundancy, show classical X-ALD biochemical hallmarks (increased levels of saturated and monounsaturated VLCFAs) but also increased levels of some LCFAs and PUFAs. Like in brain macrophages from X-ALD patients [125], whorled lipid inclusions, probably corresponding to cholesterol esters of VLCFAs, were observed, making these cells particularly interesting for modelling the human disease. Further studies using these cell lines, alone or in co-culture with glial and/or neuronal cells, should bring new insights for understanding the impact of *Abcd1*/*Abcd2* deficiencies in the microglial function, and could be used for the screening of pharmaceutical compounds useful to halt chronic inflammation in the brains of cALD patients.

In order to study the function of peroxisomal ABC transporters and the pathogenesis of X-ALD in integrated mammalian models, *Abcd1-*, *Abcd2-*, and *Abcd3*-deficient mouse models have been generated [33–35,40,67,126]. The *Abcd1* knock-out mice show key biochemical features of X-ALD but develop a late onset progressive neurodegenerative phenotype involving the spinal cord and sciatic nerves without brain damage [127]. In the spinal cord, inflammation is observed in old mice and includes microglia and astrocyte activation [40]. However, microglia activation seems to occur early, probably from eight months of age [91]. VLCFA excess would induce an early oxidative stress leading to mitochondria structural and functional damages as well as an ER stress concomitant with autophagy disruption [128–132]. Although no cerebral phenotype is observed, *Abcd1* knock-out mice can be considered a physiological model of AMN or female myelopathy and can be useful for screening pharmaceutical compounds. Several molecules have thus been tested and have demonstrated their efficacy, including antioxidant compounds that have been proven to reverse oxidative stress in vitro and reduce locomotor impairment [133–135]. These hopeful results led to a prospective phase II pilot study that was carried out for 13 AMN patients treated with a cocktail of antioxidant molecules [93]. The study showed that biomarkers of oxidative damage and inflammation were normalized and that patients' locomotion was improved, paving the way for a hopeful Phase III study.

Even if the mouse model is attractive because of its phylogenic proximity to humans, it doesn't reproduce the human brain phenotype of X-ALD. One possible explanation could be related to species and cell-type differences in the expression levels of *ABCD1–3* and functional redundancy issues. Sustaining this hypothesis, a transcriptomic analysis showed that *ABCD2* is not expressed in human microglia and *ABCD3* is 1.6-fold more expressed than *ABCD1* [136], whereas in mouse BV-2 microglial cells, *Abcd2* is 2.5-fold more expressed than *Abcd1* and *Abcd3* is 1.6-fold more expressed than *Abcd1* [124]. In addition, the biochemical and neurological defects observed in the *Abcd1* knock-out mice can be corrected by ubiquitous transgenic expression of *Abcd2* [40]. On the contrary, *Abcd1/Abcd2* double knock-out mice have an earlier and more severe neurological phenotype associated with inflammatory T lymphocyte infiltration in the spinal cord [40]. The *Abcd2* knock-out mice also develop progressive motor disabilities specifically involving sensitive peripheral neurons and spinal cord dorsal and ventral columns and share subcellular abnormalities with the *Abcd1* knock-out mice (axonal degeneration, C26:0 accumulation, oxidative stress, organelle abnormalities concerning mitochondria, lysosome, endoplasmic reticulum, and Golgi apparatus). This model also revealed the key role of *Abcd2* in adrenals [137] and in adipose tissue and lipid physiology [59–61].

In contrast to the *Abcd1* and *Abcd2* knock-out models, the *Abcd3* knock-out mice do not develop peripheral or central neurodegeneration (like *ABCD3* deficiency in humans), but exhibit hepatomegaly associated with abnormalities in peroxisomal FA metabolism, which seems to represent a suitable model for CBAS5 [67].

4.7. Human

X-ALD patient skin fibroblasts have, for several years, constituted one of the rare in vitro models of the disease. In 1980, Moser et al. demonstrated for the first time that the accumulation of VLCFAs observed in the brain and adrenals of patients is also present in primary fibroblasts, thus validating this model for X-ALD studies, at least at the biochemical level [138]. Since then, this cellular model has become a platform for a broad variety of analyses concerning lipid metabolism, X-ALD diagnosis, functional characterization of peroxisomal ABC transporters, cellular consequences of *ABCD1* deficiency, and screening of therapeutic compounds. Great scientific advances have emerged from this handy model, but its skin origin is a limitation in pathogenesis studies. Indeed, the gene regulation and function in skin fibroblasts are very far from those of neural, glial and microglial cells.

The involvement of peripheral blood mononuclear cells (PBMCs) in the inflammation feature of X-ALD was early suspected, since PBMCs from X-ALD patients produce higher levels of inflammatory cytokines than control ones [139,140]. Used in gene therapy, the CD34+ PBMCs (lymphoid and myeloid progenitors) transduced with normal *ABCD1* can efficiently correct the clinical phenotype of the X-ALD patients [82]. Moreover, AMN monocytes have a pro-inflammatory expression pattern and, after differentiation into macrophages, are not able to switch to an anti-inflammatory regenerative state [141]. *Abcd2*, whose expression level is extremely low in these cells, could be a therapeutic target [142]. Therefore, human monocytes can be used to study the inflammatory process and identify compounds capable of inducing *ABCD2* expression, correcting VLCFA level, β-oxidation, and inflammatory features [44,58].

The development of the iPSC (induced pluripotent stem cell) technology offers the opportunity to study disease-involved cells with a chosen mutation and a phenotype matching physiology. Several iPSC models have successfully been obtained from skin fibroblasts of cALD and AMN patients [143–147]. Gene expression profiling shows that X-ALD iPSCs have differentially expressed genes compared to control iPSCs, among which some are positively correlated to the severity of the disease (cALD versus AMN) [148]. When iPSCs are differentiated into oligodendrocytes or astrocytes, the VLCFA level is increased and is higher in cALD differentiated cells than in AMN cells, whereas no VLCFA accumulation is observed in neurons [144]. iPSC-derived astrocytes show pro-inflammatory features that also correlate with the severity of the phenotype. The differentiation of microglia from iPSC also seems to be a promising model, as differentiated microglia show the main phenotype of primary fetal and adult human microglia including phagocytic and inflammatory capacity [146]. In addition, cALD iPSCs differentiated in brain microvascular endothelial cells show impaired BBB function as well as lipid metabolism modifications and interferon activation [149], and could lead to the study of an important factor of brain pathogenesis in X-ALD. Altogether, these works show that iPSC-derived brain cells should allow the study of the pathogenesis of X-ALD in detail, permit the identification of biomarkers, and screen new therapeutic molecules. Co-culture experiments are expected to provide new insight into intercellular communication in the brain.

In conclusion, for forty years, enormous progress has been made in the knowledge of peroxisomal ABC transporters thanks to the development and the use of cell and animal models. If no model exactly mimics the human X-ALD, there is no doubt that the new technological developments will offer opportunities to progress in the study of the role of peroxisomal ABC transporters in the neuronal, glial, and microglial intercellular communications.

5. Protein Interactions and Unexpected Roles

Physical interaction between peroxisomal ABC transporters and other proteins have been reported in several studies. Most binding partners are involved in lipid metabolism.

Here, we propose to review these binding partners for which strong interaction experiments have been obtained, or for which further investigations are needed to be reliable.

Peroxisomal membrane insertion, substrate binding, transport mechanism, and the potential novel functions of peroxisomal ABC transporters require protein interactions. Since their first identification, many efforts have been developed to understand how peroxisomal ABC transporters are targeted to the peroxisomal membrane. PEX19p, a cytosolic peroxin, was identified as an interactor of ABCD1, ABCD2, and ABCD3 by using the yeast two-hybrid system and in vitro GST pull-down assays [150]. In addition to being involved (in association with PEX3) in the correct peroxisomal targeting of peroxisomal membrane proteins (PMPs), PEX19p may also function as a protein chaperone to prevent aggregation of newly synthesized PMPs [151]. It's worth noting that PEX19p is the only ABCD2 binding partner that has been reported in the literature, probably due to the fact that ABCD2 is much less closely studied than ABCD1 and ABCD3 since it is not responsible for a genetic disease when mutated. To identify potential ABCD2 binding partners, we used the inducible H4IIEC3 cell model, which expresses ABCD2-EGFP depending on the presence of doxycycline [28]. We performed quantitative ABCD2 co-immunoprecipitation assays coupled with tandem mass spectrometry. Differential analysis between cell samples was done to limit detection of false-positive interactions. The list of potential binding partners of ABCD2 is given in Table 1 and includes 13 non-redundant proteins exclusively detected in the positive samples [28]. Only one subunit of the oligosaccharyl transferase (OST) complex that catalyzes the N-glycosylation of newly translated proteins in the endoplasmic reticulum was identified as a potential ABCD2 binding partner: the dolichyl-diphosphooligosaccharide protein glycosyltransferase subunit 2 (RPN2) (Table 1). On the other hand, RPN2 has also been identified by proteomic analyses in a subclass of peroxisome expressing ABCD2 [152]. These data are still quite surprising since peroxisomal ABC transporters, such as most PMPs, are known to be synthesized on free polysomes and to further insert directly from the cytosol into the peroxisomal membrane. It is worth noting that an indirect peroxisomal targeting pathway exists via the ER since several PMPs are found glycosylated [153]. The potential interaction of ABCD2 with the OST complex involved in N-glycosylation is inconsistent with the absence of routing through the ER with respect to peroxisomal ABC transporters. Nevertheless, proteomic data leading to identification is not robust since the protein probability for RPN2 is low (0.7224) (Table 1).

Table 1. List of proteins identified in co-immunoprecipitated ABCD2-EGFP complex by liquid chromatography coupled with tandem mass spectrometry (modified from [28]).

Protein Accession		Protein Name	Protein Probability	Fold Change [a]
Q9QY44	ABCD2	ATP-binding cassette sub-family D member 2	1	12.42
P97612	FAAH1	Fatty-acid amide hydrolase 1	1	5.02
P11507	AT2A2	Sarcoplasmic/endoplasmic reticulum calcium ATPase 2	1	4.71
P07340	AT1B1	Sodium/potassium-transporting ATPase subunit beta	1	2.48
P55159	PON1	Serum paraoxonase/arylesterase 1	1	2.32
D3ZHR2	ABCD1	ATP-binding cassette sub-family D member 1	1	<2
P16970	ABCD3	ATP-binding cassette sub-family D member 3	1	<2
Q7TS56	CBR4	Carbonyl reductase family member 4	1	<2
P11505	AT2B1	Plasma membrane calcium-transporting ATPase 1	1	<2
P16086	SPTN1	Spectrin alpha chain, non-erythrocytic 1	1	<2
Q63151	ACSL3	Long-chain acyl-CoA synthetase 3	0.9997	<2
O88813	ACSL5	Long-chain acyl-CoA synthetase 5	0.9994	<2
P14408	FUMH	Fumarate hydratase, mitochondrial	0.8013	<2
P25235	RPN2	Dolichyl-diphosphooligosaccharide—protein glycosyltransferase subunit 2	0.7224	<2

[a] Statistical significance was obtained for proteins identified with a fold-change >2.

Besides the question of routing and peroxisomal targeting, the main putative ABCD2 partners revealed in this study were associated with lipid metabolism. Unsurprisingly, several binding partners identified have a role in FA activation, which is required on both sides of the peroxisomal membrane. At the cytoplasmic side of the peroxisomal membrane, a complex FA synthesis-transport machinery was evidenced by using a multi-approach method, combining GST pulldown experiments, mass spectrometry (LC/MS), co-immunoprecipitation assays, and bioluminescence resonance energy transfer (BRET) measurements [154]. This machinery consists of the binary interaction of ABCD1/3 with proteins carrying functions associated with FA activation/transport (ACSVL4) and FA synthesis (ACLY, ATP citrate lyase; FASN, FA synthase). On the inner surface of the peroxisomal membrane, studies using a yeast two hybrid system and surface plasmon resonance techniques indicate that the very long-chain acyl-CoA synthetase 1 (ACSVL1) interacts with ABCD1 [155]. In *Saccharomyces cerevisiae*, peroxisomal ABC transporters (Pxa1p and Pxa2p) functionally interact with the acyl-CoA synthetase Faa2p on the inner surface of the peroxisomal membrane for subsequent re-esterification of the VLCFAs [72]. In this model, whether or not a physical interaction with acyl-CoA synthetases exists remains to be investigated. In *Arabidopsis thaliana*, peroxisomal long-chain acyl-CoA synthetases (lacs6 and lacs7) physically and functionally interact with CTS, as assessed by co-immunoprecipitation experiments [73].

In our study aiming at identifying ABCD2 binding partners, the fatty-acid amide hydrolase 1 (FAAH1) exhibited the highest fold change (Table 1, FC = 5.02). This endoplasmic reticulum enzyme is the main enzyme involved in anandamide hydrolysis and plays an important role in endocannabinoid metabolism degrading the FA amides to the corresponding fatty acids, with a PUFA preference over MUFAs and saturated fatty acids [156,157]. Interestingly, FAAH1 catalyzes the conversion of the ethanolamine amide form of DHA (N-docosahexaenoyl ethanolamine) to DHA [158]. The interaction of FAAH1 with ABCD2 could be consistent with the role of FAAH1 as a supplier of ABCD2 substrates (DHA and other PUFAs) for further degradation in the peroxisome by β-oxidation.

Concerning ether lipid biosynthesis, the peroxisomal enzyme alkyl-dihydroxyacetone phosphate synthase (AGPS) is suggested to interact with ABCD1, as assessed by an integrative global proteomic profiling approach based on chromatographic separation [159]. Ether lipid biosynthesis starts in the peroxisome with the transfer of the acyl group of fatty acyl-CoAs to dihydroxyacetonephosphate (DHAP), generating an acyl-DHAP. The second peroxisomal step is catalyzed by AGPS, which exchanges the acyl chain for an alkyl group, yielding an alkyl-DHAP. After a final peroxisomal step, the ether lipid biosynthesis is completed in the ER. This global proteomic analysis showed that AGPS failed to interact with ABCD2, just as our co-immunoprecipitation coupled to proteomic analysis [28].

β-oxidation of MCFAs to LCFAs mainly takes place in the mitochondria, whereas VLCFAs are first metabolized down to octanoyl-CoA in the peroxisome for further degradation in the mitochondria. Surprisingly, proteomic data supported by co-immunoprecipitation experiments evidenced a physical interaction between a long-chain acyl-CoA synthetase 1 (ACSL1) localized in the ER and ABCD3 [159]. This could be in agreement with the role of ABCD3 in the β-oxidation of lauric and palmitic acids [68]. In addition, ACSL1 has been shown to interact with ACBD5 [160], a peroxisomal membrane protein suggested to function as a membrane-bound receptor for VLCFA-CoA in the cytosol to bring them to ABCD1 [161]. Whether ACSL1 transfers other unidentified lipid species to ACBD5, ABCD1, or ABCD3 for peroxisomal degradation needs further investigation.

Other potential binding partners of ABCD2 identified are involved in mitochondrial FA metabolism, such as the carbonyl reductase family member 4 (CBR4), the long-chain acyl-CoA synthetase 3 (ACSL3), and the long-chain acyl-CoA synthetase 5 (ACSL5) (Table 1). These enzymes were identified with less confidence (fold change <2). Although linked to lipid metabolism, CBR4 is a matrix mitochondrial enzyme. ACSL3 and ACSL5 do not activate VLCFAs, nor does ACSL1, which nevertheless has been found to interact with ABCD3 [160] as discussed above.

Peroxisomes contain enzymes involved in the α-oxidation of phytanic acid. Large-scale mapping of protein–protein interactions by mass spectrometry identified a single interaction between peroxisomal proteins i.e., the peroxisome matrix phytanoyl-CoA 2-hydroxylase (PHYH) and ABCD3 [162]. This interaction makes sense since, after activation of phytanic acid, phytanoyl-CoA is imported into the peroxisome by ABCD3 and enters in the peroxisomal α-oxidation pathway of which PHYH is the first enzyme (Figure 1).

The recent demonstration of ABCD1 interaction with M1 spastin, a membrane-bound AAA ATPase found on LDs, suggests the involvement of ABCD1 in inter-organelle FA trafficking [163]. Actually, ABCD1 forms a tethering complex with M1 spastin as assessed by co-immunoprecipitation experiments to connect LDs to peroxisomes. Furthermore, by recruiting IST1 and CHMP1B to LDs, M1 spastin facilitates LD-to-peroxisome FA trafficking. Whether M1 spastin-ABCD1 interaction directly promotes fatty acids channeling into peroxisomes remains unclear. It is worth noting that among proteins detected in our ABCD2 interactome study, the spectrin alpha chain, non-erythrocytic 1 (SPTN1) was identified as a potential ABCD2 binding partner (Table 1). As a cytoskeletal protein, SPTN1 is known to be involved in stabilization of the plasma membrane and to organize intracellular organelles [164]. These data corroborate the existence of peroxisome interconnection with LDs and the cytoskeleton [165,166].

Related to calcium signaling, the sarcoplasmic/endoplasmic reticulum calcium ATPase 2 (AT2A2) was identified with a high fold change (FC = 4.71), which ensures a specific interaction with ABCD2 (Table 1). Besides, its homolog (ATPase1) has been identified as well by proteomic analyses in a subclass of peroxisome expressing ABCD2 [152]. AT2A2 transfers Ca2+ from the cytosol to the ER and is then involved in calcium signaling. Coincidently, disturbed calcium signaling was suggested to be associated with the pathogenesis of X-ALD [122]. Involved in maintaining intracellular calcium homeostasis, the plasma membrane calcium-transporting ATPase 1 (AT2B1) was identified, though with less confidence (fold change <2). Actually, its physical interaction with ABCD2 remains questionable since it is expressed at the plasma membrane. Nevertheless, several high throughput studies using robust affinity purification-mass spectrometry methodologies to elucidate protein interaction networks have revealed the interaction of ABCD1 with AT2B2 [167] and ABCD3 with AT2B2 and AT2A2 [168,169]. Hence, clusters of arguments indicate that peroxisomal ABC transporters could be linked to calcium signaling, but deciphering molecular interaction networks would be required to confirm this hypothesis.

Identification in the putative ABCD2 partners of the serum paraoxonase/arylesterase 1 (PON1), an antioxidant enzyme synthetized and secreted by the liver in the serum [170] where it is closely associated with high density lipoprotein (HDL), could be at first glance intriguing (Table 1). Nevertheless, in the liver, PON1 is primarily localized in microsomal fraction where the enzyme is associated with vesicles derived from the ER [171]. The potential intracellular interaction with ABCD2 remains to be elucidated. Noteworthy, PON1 activity and polymorphisms have been associated with neurodegenerative diseases [172], of which X-ALD is not evoked.

The binding partners of peroxisomal ABC transporters discussed in this review are mainly linked to lipid metabolism (PUFA metabolism, α-oxidation pathway, and ether lipid biosynthesis) and are consequently found in the cytosol, in the peroxisomal membrane, or in the peroxisomal matrix. However, binding partners were identified in other cell compartments. Since peroxisomal lipid metabolism requires cooperation and interaction with mitochondria, ER and LDs, peroxisomal ABC transporters, through their interactome, could therefore actively participate in this intracellular metabolic network. Peroxisome-organelle interactions have physiological relevance [166,173], and peroxisomes are increasingly considered important intracellular signaling platforms that modulate physiological processes such as inflammation, innate immunity and cell fate decision [12,174,175]. Peroxisomal ABC transporters would play an essential part in this emerging role of peroxisomes in signaling pathways such as calcium signaling as highlighted in this review.

6. Conclusions

Transcriptomic, proteomic, and lipidomic studies, which have multiplied in the last few years, have confirmed and/or revealed the involvement of peroxisomal metabolism in various biological processes essential for cellular adaptation, brain homeostasis, or even immune response and inflammation. Peroxisomal ABC transporters constitute a pathway for the entry of various lipid substrates into the peroxisome mainly for their degradation but also for the synthesis of bioactive lipids impacting membranes and signaling pathways. It is therefore quite logical that the role of peroxisomal ABC transporters is now extended to unexpected biological processes. Since their cloning in the 90s, the lack of good antibodies, the rarity of relevant cell models, the fragility of the peroxisomal membrane, and other difficulties have constituted a real handicap towards performing functional assays and in vitro transport reconstitutions, and progressing in the understanding of the role of peroxisomal ABC transporters. The emergence of new cell models and the rise of model organisms, as well as cell reprogramming and CRISPR gene editing technologies, suggest that major new discoveries will be made soon that reveal their role in physiological and pathological situations.

Author Contributions: A.T., C.G., D.T., and S.S. contributed to the bibliography and writing equally. S.S. coordinated and finalized the elaboration of the manuscript. All authors have read and agreed to the published version of the manuscript.

Funding: The authors of this review received no external funding for this bibliographical work.

Acknowledgments: The authors acknowledge the French Ministère de l'Enseignement Supérieur, de la Recherche et de l'Innovation, the University of Bourgogne, as well as the NFRF-Exploration stream (NFRF-E-2019-00007) (Canada) for funding the current scientific projects of the Bio-PeroxIL laboratory.

Conflicts of Interest: The authors declare no conflict of interest.

Abbreviations

ABC	ATP-binding cassette
ACOX1	Acyl-coenzyme A oxidase 1
AMN	Adrenomyeloneuropathy
BBB	Blood-brain barrier
cALD	Cerebral adrenoleukodystrophy
CBAS	Congenital bile acid synthesis defect
CNS	Central nervous system
CoA	Coenzyme A
CTS	Comatose
DHA	Docosahexaenoic acid, C22:6 n-3
ER	Endoplasmic reticulum
FA	Fatty acid
FC	Fold change
HSCT	Hematopoietic stem cell transplantation
iPSC	Induced pluripotent stem cell
LCFA	Long-chain fatty acid
LD	Lipid droplet
MCFA	Medium-chain fatty acid
MUFA	Monounsaturated fatty acid
NBD	Nucleotide binding domain
PBMC	Peripheral blood mononuclear cell
PMP	Peroxisomal membrane protein
PUFA	Polyunsaturated fatty acid
TMD	Transmembrane domain
VLCFA	Very long-chain fatty acid
X-ALD	X-linked adrenoleukodystrophy

References

1. Thomas, C.; Tampé, R. Structural and Mechanistic Principles of ABC Transporters. *Annu. Rev. Biochem.* **2020**, *89*, 605–636. [CrossRef]
2. Dean, M.; Annilo, T. Evolution of the ATP-binding cassette (ABC) transporter superfamily in vertebrates. *Annu. Rev. Genom. Hum. Genet.* **2005**, *6*, 123–142. [CrossRef]
3. Kawaguchi, K.; Morita, M. ABC Transporter Subfamily D: Distinct Differences in Behavior between ABCD1-3 and ABCD4 in Subcellular Localization, Function, and Human Disease. *BioMed Res. Int.* **2016**, *2016*, 6786245. [CrossRef]
4. Shani, N.; Jimenez-Sanchez, G.; Steel, G.; Dean, M.; Valle, D. Identification of a fourth half ABC transporter in the human peroxisomal membrane. *Hum. Mol. Genet.* **1997**, *6*, 1925–1931. [CrossRef]
5. Coelho, D.; Kim, J.C.; Miousse, I.R.; Fung, S.; du Moulin, M.; Buers, I.; Suormala, T.; Burda, P.; Frapolli, M.; Stucki, M.; et al. Mutations in ABCD4 cause a new inborn error of vitamin B12 metabolism. *Nat. Genet.* **2012**, *44*, 1152–1155. [CrossRef] [PubMed]
6. Kashiwayama, Y.; Seki, M.; Yasui, A.; Murasaki, Y.; Morita, M.; Yamashita, Y.; Sakaguchi, M.; Tanaka, Y.; Imanaka, T. 70-kDa peroxisomal membrane protein related protein (P70R/ABCD4) localizes to endoplasmic reticulum not peroxisomes, and NH2-terminal hydrophobic property determines the subcellular localization of ABC subfamily D proteins. *Exp. Cell Res.* **2009**, *315*, 190–205. [CrossRef]
7. Wanders, R.J.; Waterham, H.R. Biochemistry of mammalian peroxisomes revisited. *Annu. Rev. Biochem.* **2006**, *75*, 295–332. [CrossRef] [PubMed]
8. Trompier, D.; Vejux, A.; Zarrouk, A.; Gondcaille, C.; Geillon, F.; Nury, T.; Savary, S.; Lizard, G. Brain peroxisomes. *Biochimie* **2014**, *98*, 102–110. [CrossRef]
9. Ferdinandusse, S.; Denis, S.; Mooijer, P.A.; Zhang, Z.; Reddy, J.K.; Spector, A.A.; Wanders, R.J. Identification of the peroxisomal beta-oxidation enzymes involved in the biosynthesis of docosahexaenoic acid. *J. Lipid Res.* **2001**, *42*, 1987–1995. [CrossRef]
10. Chapkin, R.S.; Kim, W.; Lupton, J.R.; McMurray, D.N. Dietary docosahexaenoic and eicosapentaenoic acid: Emerging mediators of inflammation. *Prostaglandins Leukot. Essent. Fat. Acids* **2009**, *81*, 187–191. [CrossRef]
11. Lodhi, I.J.; Semenkovich, C.F. Peroxisomes: A nexus for lipid metabolism and cellular signaling. *Cell. Metab.* **2014**, *19*, 380–392. [CrossRef]
12. Di Cara, F.; Andreoletti, P.; Trompier, D.; Vejux, A.; Bulow, M.H.; Sellin, J.; Lizard, G.; Cherkaoui-Malki, M.; Savary, S. Peroxisomes in Immune Response and Inflammation. *Int. J. Mol. Sci.* **2019**, *20*, 3877. [CrossRef]
13. Fransen, M.; Nordgren, M.; Wang, B.; Apanasets, O. Role of peroxisomes in ROS/RNS-metabolism: Implications for human disease. *Biochim. Biophys. Acta* **2012**, *1822*, 1363–1373. [CrossRef]
14. Fransen, M.; Lismont, C.; Walton, P. The Peroxisome-Mitochondria Connection: How and Why? *Int. J. Mol. Sci.* **2017**, *18*, 1126. [CrossRef]
15. Lismont, C.; Revenco, I.; Fransen, M. Peroxisomal Hydrogen Peroxide Metabolism and Signaling in Health and Disease. *Int. J. Mol. Sci.* **2019**, *20*, 3673. [CrossRef] [PubMed]
16. Contreras, M.; Sengupta, T.K.; Sheikh, F.; Aubourg, P.; Singh, I. Topology of ATP-binding domain of adrenoleukodystrophy gene product in peroxisomes. *Arch. Biochem. Biophys.* **1996**, *334*, 369–379. [CrossRef] [PubMed]
17. Andreoletti, P.; Raas, Q.; Gondcaille, C.; Cherkaoui-Malki, M.; Trompier, D.; Savary, S. Predictive Structure and Topology of Peroxisomal ATP-Binding Cassette (ABC) Transporters. *Int. J. Mol. Sci.* **2017**, *18*, 1593. [CrossRef] [PubMed]
18. Baker, A.; Carrier, D.J.; Schaedler, T.; Waterham, H.R.; van Roermund, C.W.; Theodoulou, F.L. Peroxisomal ABC transporters: Functions and mechanism. *Biochem. Soc. Trans.* **2015**, *43*, 959–965. [CrossRef]
19. Geillon, F.; Gondcaille, C.; Charbonnier, S.; Van Roermund, C.W.; Lopez, T.E.; Dias, A.M.M.; de Barros, J.-P.P.; Arnould, C.; Wanders, R.J.; Trompier, D.; et al. Structure-function analysis of peroxisomal ATP-binding cassette transporters using chimeric dimers. *J. Biol. Chem.* **2014**, *289*, 24511–24520. [CrossRef] [PubMed]
20. Genin, E.; Geillon, F.; Gondcaille, C.; Athias, A.; Gambert, P.; Trompier, D.; Savary, S. Substrate specificity overlap and interaction between adrenoleukodystrophy protein (ALDP/ABCD1) and adrenoleukodystrophy-related protein (ALDRP/ABCD2). *J. Biol. Chem.* **2011**, *286*, 8075–8084. [CrossRef]
21. Smith, K.D.; Kemp, S.; Braiterman, L.T.; Lu, J.F.; Wei, H.M.; Geraghty, M.; Stetten, G.; Bergin, J.S.; Pevsner, J.; Watkins, P.A. X-linked adrenoleukodystrophy: Genes, mutations, and phenotypes. *Neurochem. Res.* **1999**, *24*, 521–535. [CrossRef]
22. Tanaka, A.R.; Tanabe, K.; Morita, M.; Kurisu, M.; Kasiwayama, Y.; Matsuo, M.; Kioka, N.; Amachi, T.; Imanaka, T.; Ueda, K. ATP binding/hydrolysis by and phosphorylation of peroxisomal ATP-binding cassette proteins PMP70 (ABCD3) and adrenoleukodystrophy protein (ABCD1). *J. Biol. Chem.* **2002**, *277*, 40142–40147. [CrossRef] [PubMed]
23. Hillebrand, M.; Verrier, S.E.; Ohlenbusch, A.; Schafer, A.; Soling, H.D.; Wouters, F.S.; Gartner, J. Live cell FRET microscopy: Homo- and heterodimerization of two human peroxisomal ABC transporters, the adrenoleukodystrophy protein (ALDP, ABCD1) and PMP70 (ABCD3). *J. Biol. Chem.* **2007**, *282*, 26997–27005. [CrossRef] [PubMed]
24. Guimaraes, C.P.; Domingues, P.; Aubourg, P.; Fouquet, F.; Pujol, A.; Jimenez-Sanchez, G.; Sa-Miranda, C.; Azevedo, J.E. Mouse liver PMP70 and ALDP: Homomeric interactions prevail in vivo. *Biochim. Biophys. Acta* **2004**, *1689*, 235–243. [CrossRef]
25. Liu, L.X.; Janvier, K.; Berteaux-Lecellier, V.; Cartier, N.; Benarous, R.; Aubourg, P. Homo- and heterodimerization of peroxisomal ATP-binding cassette half-transporters. *J. Biol. Chem.* **1999**, *274*, 32738–32743. [CrossRef]
26. Van Roermund, C.W.; Visser, W.F.; Ijlst, L.; Waterham, H.R.; Wanders, R.J. Differential substrate specificities of human ABCD1 and ABCD2 in peroxisomal fatty acid beta-oxidation. *Biochim. Biophys. Acta* **2011**, *1811*, 148–152. [CrossRef] [PubMed]

27. Van Roermund, C.W.; Visser, W.F.; Ijlst, L.; van Cruchten, A.; Boek, M.; Kulik, W.; Waterham, H.R.; Wanders, R.J. The human peroxisomal ABC half transporter ALDP functions as a homodimer and accepts acyl-CoA esters. *FASEB J.* **2008**, *22*, 4201–4208. [CrossRef]

28. Geillon, F.; Gondcaille, C.; Raas, Q.; Dias, A.M.M.; Pecqueur, D.; Truntzer, C.; Lucchi, G.; Ducoroy, P.; Falson, P.; Savary, S.; et al. Peroxisomal ATP-binding cassette transporters form mainly tetramers. *J. Biol. Chem.* **2017**, *292*, 6965–6977. [CrossRef]

29. Woudenberg, J.; Rembacz, K.P.; Hoekstra, M.; Pellicoro, A.; van den Heuvel, F.A.; Heegsma, J.; van Ijzendoorn, S.C.; Holzinger, A.; Imanaka, T.; Moshage, H.; et al. Lipid rafts are essential for peroxisome biogenesis in HepG2 cells. *Hepatology* **2010**, *52*, 623–633. [CrossRef] [PubMed]

30. Mosser, J.; Douar, A.M.; Sarde, C.O.; Kioschis, P.; Feil, R.; Moser, H.; Poustka, A.M.; Mandel, J.L.; Aubourg, P. Putative X-linked adrenoleukodystrophy gene shares unexpected homology with ABC transporters. *Nature* **1993**, *361*, 726–730. [CrossRef]

31. Engelen, M.; Kemp, S.; de Visser, M.; van Geel, B.M.; Wanders, R.J.; Aubourg, P.; Poll-The, B.T. X-linked adrenoleukodystrophy (X-ALD): Clinical presentation and guidelines for diagnosis, follow-up and management. *Orphanet J. Rare Dis.* **2012**, *7*, 51. [CrossRef]

32. Rattay, T.W.; Rautenberg, M.; Söhn, A.S.; Hengel, H.; Traschütz, A.; Röben, B.; Hayer, S.N.; Schüle, R.; Wiethoff, S.; Zeltner, L.; et al. Defining diagnostic cutoffs in neurological patients for serum very long chain fatty acids (VLCFA) in genetically confirmed X-Adrenoleukodystrophy. *Sci. Rep.* **2020**, *10*, 15093. [CrossRef]

33. Forss-Petter, S.; Werner, H.; Berger, J.; Lassmann, H.; Molzer, B.; Schwab, M.H.; Bernheimer, H.; Zimmermann, F.; Nave, K.A. Targeted inactivation of the X-linked adrenoleukodystrophy gene in mice. *J. Neurosci. Res.* **1997**, *50*, 829–843. [CrossRef]

34. Kobayashi, T.; Shinnoh, N.; Kondo, A.; Yamada, T. Adrenoleukodystrophy protein-deficient mice represent abnormality of very long chain fatty acid metabolism. *Biochem. Biophys. Res. Commun.* **1997**, *232*, 631–636. [CrossRef] [PubMed]

35. Lu, J.F.; Lawler, A.M.; Watkins, P.A.; Powers, J.M.; Moser, A.B.; Moser, H.W.; Smith, K.D. A mouse model for X-linked adrenoleukodystrophy. *Proc. Natl. Acad. Sci. USA* **1997**, *94*, 9366–9371. [CrossRef] [PubMed]

36. Cartier, N.; Lopez, J.; Moullier, P.; Rocchiccioli, F.; Rolland, M.O.; Jorge, P.; Mosser, J.; Mandel, J.L.; Bougneres, P.F.; Danos, O.; et al. Retroviral-mediated gene transfer corrects very-long-chain fatty acid metabolism in adrenoleukodystrophy fibroblasts. *Proc. Natl. Acad. Sci. USA* **1995**, *92*, 1674–1678. [CrossRef]

37. Braiterman, L.T.; Zheng, S.; Watkins, P.A.; Geraghty, M.T.; Johnson, G.; McGuinness, M.C.; Moser, A.B.; Smith, K.D. Suppression of peroxisomal membrane protein defects by peroxisomal ATP binding cassette (ABC) proteins. *Hum. Mol. Genet.* **1998**, *7*, 239–247. [CrossRef]

38. Lombard-Platet, G.; Savary, S.; Sarde, C.O.; Mandel, J.L.; Chimini, G. A close relative of the adrenoleukodystrophy (ALD) gene codes for a peroxisomal protein with a specific expression pattern. *Proc. Natl. Acad. Sci. USA* **1996**, *93*, 1265–1269. [CrossRef]

39. Netik, A.; Forss-Petter, S.; Holzinger, A.; Molzer, B.; Unterrainer, G.; Berger, J. Adrenoleukodystrophy-related protein can compensate functionally for adrenoleukodystrophy protein deficiency (X-ALD): Implications for therapy. *Hum. Mol. Genet.* **1999**, *8*, 907–913. [CrossRef]

40. Pujol, A.; Ferrer, I.; Camps, C.; Metzger, E.; Hindelang, C.; Callizot, N.; Ruiz, M.; Pampols, T.; Giros, M.; Mandel, J.L. Functional overlap between ABCD1 (ALD) and ABCD2 (ALDR) transporters: A therapeutic target for X-adrenoleukodystrophy. *Hum. Mol. Genet.* **2004**, *13*, 2997–3006. [CrossRef]

41. McGuinness, M.C.; Zhang, H.P.; Smith, K.D. Evaluation of Pharmacological Induction of Fatty Acid beta-Oxidation in X-Linked Adrenoleukodystrophy. *Mol. Genet. Metab.* **2001**, *74*, 256–263. [CrossRef]

42. Bugaut, M.; Fourcade, S.; Gondcaille, C.; Gueugnon, F.; Depreter, M.; Roels, F.; Netik, A.; Berger, J.; Martin, P.; Pineau, T.; et al. Pharmacological induction of redundant genes for a therapy of X-ALD: Phenylbutyrate and other compounds. *Adv. Exp. Med. Biol.* **2003**, *544*, 281–291.

43. Kemp, S.; Wei, H.M.; Lu, J.F.; Braiterman, L.T.; McGuinness, M.C.; Moser, A.B.; Watkins, P.A.; Smith, K.D. Gene redundancy and pharmacological gene therapy: Implications for X-linked adrenoleukodystrophy. *Nat. Med.* **1998**, *4*, 1261–1268. [CrossRef]

44. Weber, F.D.; Weinhofer, I.; Einwich, A.; Forss-Petter, S.; Muneer, Z.; Maier, H.; Weber, W.H.; Berger, J. Evaluation of retinoids for induction of the redundant gene ABCD2 as an alternative treatment option in X-linked adrenoleukodystrophy. *PLoS ONE* **2014**, *9*, e103742. [CrossRef]

45. Rampler, H.; Weinhofer, I.; Netik, A.; Forss-Petter, S.; Brown, P.J.; Oplinger, J.A.; Bugaut, M.; Berger, J. Evaluation of the therapeutic potential of PPARalpha agonists for X-linked adrenoleukodystrophy. *Mol. Genet. Metab.* **2003**, *80*, 398–407. [CrossRef] [PubMed]

46. Fourcade, S.; Savary, S.; Albet, S.; Gauthe, D.; Gondcaille, C.; Pineau, T.; Bellenger, J.; Bentejac, M.; Holzinger, A.; Berger, J.; et al. Fibrate induction of the adrenoleukodystrophy-related gene (ABCD2)—Promoter analysis and role of the peroxisome proliferator-activated receptor PPAR alpha. *Eur. J. Biochem.* **2001**, *268*, 3490–3500. [CrossRef]

47. Fourcade, S.; Savary, S.; Gondcaille, C.; Berger, J.; Netik, A.; Cadepond, F.; El Etr, M.; Molzer, B.; Bugaut, M. Thyroid hormone induction of the adrenoleukodystrophy-related gene (ABCD2). *Mol. Pharmacol.* **2003**, *63*, 1296–1303. [CrossRef]

48. Gondcaille, C.; Depreter, M.; Fourcade, S.; Lecca, M.; Leclercq, S.; Martin, P.; Pineau, T.; Cadepond, F.; El-Etr, M.; Bertrand, N.; et al. Phenylbutyrate up-regulates the adrenoleukodystrophy-related gene as a nonclassical peroxisome proliferator. *J. Cell Biol.* **2005**, *169*, 93–104. [CrossRef] [PubMed]

49. Leclercq, S.; Skrzypski, J.; Courvoisier, A.; Gondcaille, C.; Bonnetain, F.; Andre, A.; Chardigny, J.; Bellenger, S.; Bellenger, J.; Narce, M.; et al. Effect of dietary polyunsaturated fatty acids on the expression of peroxisomal ABC transporters. *Biochimie* **2008**, *90*, 1602–1607. [CrossRef] [PubMed]

50. Genin, E.; Gondcaille, C.; Trompier, D.; Savary, S. Induction of the adrenoleukodystrophy-related gene (ABCD2) by thyromimetics. *J. Steroid Biochem. Mol. Biol.* **2009**, *116*, 37–43. [CrossRef] [PubMed]

51. Gondcaille, C.; Genin, E.C.; Lopez, T.E.; Dias, A.M.M.; Geillon, F.; Andreoletti, P.; Cherkaoui-Malki, M.; Nury, T.; Lizard, G.; Weinhofer, I.; et al. LXR antagonists induce ABCD2 expression. *Biochim. Biophys. Acta* **2014**, *1841*, 259–266. [CrossRef] [PubMed]

52. Trompier, D.; Gondcaille, C.; Lizard, G.; Savary, S. Regulation of the adrenoleukodystrophy-related gene (ABCD2): Focus on oxysterols and LXR antagonists. *Biochem. Biophys. Res. Commun.* **2014**, *446*, 651–655. [CrossRef] [PubMed]

53. Singh, J.; Khan, M.; Singh, I. HDAC inhibitor SAHA normalizes the levels of VLCFAs in human skin fibroblasts from X-ALD patients and downregulates the expression of proinflammatory cytokines in Abcd1/2-silenced mouse astrocytes. *J. Lipid Res.* **2011**, *52*, 2056–2069. [CrossRef]

54. Hartley, M.D.; Kirkemo, L.L.; Banerji, T.; Scanlan, T.S. A Thyroid Hormone-Based Strategy for Correcting the Biochemical Abnormality in X-Linked Adrenoleukodystrophy. *Endocrinology* **2017**, *158*, 1328–1338. [CrossRef]

55. Hartley, M.D.; Shokat, M.D.; DeBell, M.J.; Banerji, T.; Kirkemo, L.L.; Scanlan, T.S. Pharmacological Complementation Remedies an Inborn Error of Lipid Metabolism. *Cell Chem. Biol.* **2020**, *27*, 551–559.e4. [CrossRef]

56. Weinhofer, I.; Kunze, M.; Rampler, H.; Bookout, A.L.; Forss-Petter, S.; Berger, J. Liver X receptor alpha interferes with SREBP1c-mediated Abcd2 expression. Novel cross-talk in gene regulation. *J. Biol. Chem.* **2005**, *280*, 41243–41251. [CrossRef]

57. Weinhofer, I.; Forss-Petter, S.; Zigman, M.; Berger, J. Cholesterol regulates ABCD2 expression: Implications for the therapy of X-linked adrenoleukodystrophy. *Hum. Mol. Genet.* **2002**, *11*, 2701–2708. [CrossRef]

58. Zierfuss, B.; Weinhofer, I.; Kühl, J.S.; Köhler, W.; Bley, A.; Zauner, K.; Binder, J.; Martinović, K.; Seiser, C.; Hertzberg, C.; et al. Vorinostat in the acute neuroinflammatory form of X-linked adrenoleukodystrophy. *Ann. Clin. Transl. Neurol.* **2020**, *7*, 639–652. [CrossRef]

59. Liu, J.; Liang, S.; Liu, X.; Brown, J.A.; Newman, K.E.; Sunkara, M.; Morris, A.J.; Bhatnagar, S.; Li, X.; Pujol, A.; et al. The absence of ABCD2 sensitizes mice to disruptions in lipid metabolism by dietary erucic acid. *J. Lipid Res.* **2012**, *53*, 1071–1079. [CrossRef] [PubMed]

60. Liu, J.; Sabeva, N.S.; Bhatnagar, S.; Li, X.A.; Pujol, A.; Graf, G.A. ABCD2 is abundant in adipose tissue and opposes the accumulation of dietary erucic acid (C22:1) in fat. *J. Lipid Res.* **2010**, *51*, 162–168. [CrossRef] [PubMed]

61. Fourcade, S.; Ruiz, M.; Camps, C.; Schluter, A.; Houten, S.M.; Mooyer, P.A.; Pampols, T.; Dacremont, G.; Wanders, R.J.; Giros, M.; et al. A key role for the peroxisomal ABCD2 transporter in fatty acid homeostasis. *Am. J. Physiol. Endocrinol. Metab.* **2009**, *296*, E211–E221. [CrossRef]

62. Imanaka, T.; Aihara, K.; Takano, T.; Yamashita, A.; Sato, R.; Suzuki, Y.; Yokota, S.; Osumi, T. Characterization of the 70-kDa peroxisomal membrane protein, an ATP binding cassette transporter. *J. Biol. Chem.* **1999**, *274*, 11968–11976. [CrossRef] [PubMed]

63. Kamijo, K.; Taketani, S.; Yokota, S.; Osumi, T.; Hashimoto, T. The 70-kDa peroxisomal membrane protein is a member of the Mdr (P-glycoprotein)-related ATP-binding protein superfamily. *J. Biol. Chem.* **1990**, *265*, 4534–4540. [CrossRef]

64. Gartner, J.; Moser, H.; Valle, D. Mutations in the 70K peroxisomal membrane protein gene in Zellweger syndrome. *Nat. Genet.* **1992**, *1*, 16–23. [CrossRef] [PubMed]

65. Van Roermund, C.W.; Ijlst, L.; Wagemans, T.; Wanders, R.J.; Waterham, H.R. A role for the human peroxisomal half-transporter ABCD3 in the oxidation of dicarboxylic acids. *Biochim. Biophys. Acta* **2014**, *1841*, 563–568. [CrossRef] [PubMed]

66. Jimenez-Sanchez, G.; Silva-Zolezzi, I.; Hebron, K.J.; Mihalik, S.; Watkins, P.; Moser, A.; Thomas, G.; Wood, P.A.; Valle, D. Defective phytanic and pristanic acids metabolism in PMP70 deficient mice results in defective nonshivering thermogenesis and dicarboxylic aciduria. *J. Inherit. Metab. Dis.* **2000**, *23*, 256.

67. Ferdinandusse, S.; Jimenez-Sanchez, G.; Koster, J.; Denis, S.; Van Roermund, C.W.; Silva-Zolezzi, I.; Moser, A.B.; Visser, W.F.; Gulluoglu, M.; Durmaz, O.; et al. A novel bile acid biosynthesis defect due to a deficiency of peroxisomal ABCD3. *Hum. Mol. Genet.* **2015**, *24*, 361–370. [CrossRef]

68. Violante, S.; Achetib, N.; van Roermund, C.W.T.; Hagen, J.; Dodatko, T.; Vaz, F.M.; Waterham, H.R.; Chen, H.; Baes, M.; Yu, C.; et al. Peroxisomes can oxidize medium- and long-chain fatty acids through a pathway involving ABCD3 and HSD17B4. *FASEB J.* **2019**, *33*, 4355–4364. [CrossRef]

69. Watkins, P.A.; Ellis, J.M. Peroxisomal acyl-CoA synthetases. *Biochim. Biophys. Acta Mol. Bas. Dis.* **2012**, *1822*, 1411–1420. [CrossRef]

70. Guimaraes, C.P.; Sa-Miranda, C.; Azevedo, J.E. Probing substrate-induced conformational alterations in adrenoleukodystrophy protein by proteolysis. *J. Hum. Genet.* **2005**, *50*, 99–105. [CrossRef]

71. Roerig, P.; Mayerhofer, P.; Holzinger, A.; Gartner, J. Characterization and functional analysis of the nucleotide binding fold in human peroxisomal ATP binding cassette transporters. *FEBS Lett.* **2001**, *492*, 66–72. [CrossRef]

72. van Roermund, C.W.; Ijlst, L.; Majczak, W.; Waterham, H.R.; Folkerts, H.; Wanders, R.J.; Hellingwerf, K.J. Peroxisomal fatty acid uptake mechanism in Saccharomyces cerevisiae. *J. Biol. Chem.* **2012**, *287*, 20144–20153. [CrossRef] [PubMed]

73. De Marcos Lousa, C.; van Roermund, C.W.; Postis, V.L.; Dietrich, D.; Kerr, I.D.; Wanders, R.J.; Baldwin, S.A.; Baker, A.; Theodoulou, F.L. Intrinsic acyl-CoA thioesterase activity of a peroxisomal ATP binding cassette transporter is required for transport and metabolism of fatty acids. *Proc. Natl. Acad. Sci. USA* **2013**, *110*, 1279–1284. [CrossRef] [PubMed]

74. Kawaguchi, K.; Mukai, E.; Watanabe, S.; Yamashita, A.; Morita, M.; So, T.; Imanaka, T. Acyl-CoA thioesterase activity of peroxisomal ABC protein ABCD1 is required for the transport of very long-chain acyl-CoA into peroxisomes. *Sci. Rep.* **2021**, *11*, 2192. [CrossRef] [PubMed]

75. van Roermund, C.W.; IJlst, L.; Baker, A.; Wanders, R.J.; Theodoulou, F.L.; Waterham, H.R. The Saccharomyces cerevisiae ABC subfamily D transporter Pxa1/Pxa2p co-imports CoASH into the peroxisome. *FEBS Lett.* **2021**, *595*, 763–772. [CrossRef]

76. Baumgart, E.; Vanhooren, J.C.; Fransen, M.; Marynen, P.; Puype, M.; Vandekerckhove, J.; Leunissen, J.A.; Fahimi, H.D.; Mannaerts, G.P.; van Veldhoven, P.P. Molecular characterization of the human peroxisomal branched-chain acyl-CoA oxidase: cDNA cloning, chromosomal assignment, tissue distribution, and evidence for the absence of the protein in Zellweger syndrome. *Proc. Natl. Acad. Sci. USA* **1996**, *93*, 13748–13753. [CrossRef]

77. Ferdinandusse, S.; Denis, S.; van Roermund, C.W.T.; Preece, M.A.; Koster, J.; Ebberink, M.S.; Waterham, H.R.; Wanders, R.J.A. A novel case of ACOX2 deficiency leads to recognition of a third human peroxisomal acyl-CoA oxidase. *Biochim. Biophys. Acta* **2018**, *1864*, 952–958. [CrossRef] [PubMed]

78. Schepers, L.; Van Veldhoven, P.P.; Casteels, M.; Eyssen, H.J.; Mannaerts, G.P. Presence of three acyl-CoA oxidases in rat liver peroxisomes. An inducible fatty acyl-CoA oxidase, a noninducible fatty acyl-CoA oxidase, and a noninducible trihydroxycoprostanoyl-CoA oxidase. *J. Biol. Chem.* **1990**, *265*, 5242–5246. [CrossRef]

79. Reddy, J.K.; Hashimoto, T. Peroxisomal beta-oxidation and peroxisome proliferator-activated receptor alpha: An adaptive metabolic system. *Annu. Rev. Nutr.* **2001**, *21*, 193–230. [CrossRef]

80. Wanders, R.J.; Waterham, H.R.; Ferdinandusse, S. Metabolic Interplay between Peroxisomes and Other Subcellular Organelles Including Mitochondria and the Endoplasmic Reticulum. *Front. Cell Dev. Biol.* **2015**, *3*, 83. [CrossRef]

81. Trompier, D.; Savary, S. *X-Linked Adrenoleukodystrophy*; Morgan and Claypool Life Sciences Publishers: San Rafael, CA, USA, 2013; Volume 2. [CrossRef]

82. Cartier, N.; Hacein-Bey-Abina, S.; Bartholomae, C.C.; Veres, G.; Schmidt, M.; Kutschera, I.; Vidaud, M.; Abel, U.; Dal-Cortivo, L.; Caccavelli, L.; et al. Hematopoietic stem cell gene therapy with a lentiviral vector in X-linked adrenoleukodystrophy. *Science* **2009**, *326*, 818–823. [CrossRef]

83. Eichler, F.; Duncan, C.; Musolino, P.L.; Orchard, P.J.; De Oliveira, S.; Thrasher, A.J.; Armant, M.; Dansereau, C.; Lund, T.C.; Miller, W.P.; et al. Hematopoietic Stem-Cell Gene Therapy for Cerebral Adrenoleukodystrophy. *N. Engl. J. Med.* **2017**, *377*, 1630–1638. [CrossRef] [PubMed]

84. Kemper, A.R.; Brosco, J.; Comeau, A.M.; Green, N.S.; Grosse, S.D.; Jones, E.; Kwon, J.M.; Lam, W.K.; Ojodu, J.; Prosser, L.A.; et al. Newborn screening for X-linked adrenoleukodystrophy: Evidence summary and advisory committee recommendation. *Genet. Med.* **2017**, *19*, 121–126. [CrossRef] [PubMed]

85. Hubbard, W.C.; Moser, A.B.; Liu, A.C.; Jones, R.O.; Steinberg, S.J.; Lorey, F.; Panny, S.R.; Vogt, R.F., Jr.; Macaya, D.; Turgeon, C.T.; et al. Newborn screening for X-linked adrenoleukodystrophy (X-ALD): Validation of a combined liquid chromatography-tandem mass spectrometric (LC-MS/MS) method. *Mol. Genet. Metab.* **2009**, *97*, 212–220. [CrossRef] [PubMed]

86. Jangouk, P.; Zackowski, K.M.; Naidu, S.; Raymond, G.V. Adrenoleukodystrophy in female heterozygotes: Underrecognized and undertreated. *Mol. Genet. Metab.* **2012**, *105*, 180–185. [CrossRef]

87. Ofman, R.; Dijkstra, I.M.; van Roermund, C.W.; Burger, N.; Turkenburg, M.; van Cruchten, A.; van Engen, C.E.; Wanders, R.J.; Kemp, S. The role of ELOVL1 in very long-chain fatty acid homeostasis and X-linked adrenoleukodystrophy. *EMBO Mol. Med.* **2010**, *2*, 90–97. [CrossRef] [PubMed]

88. Savary, S.; Trompier, D.; Andreoletti, P.; Le Borgne, F.; Demarquoy, J.; Lizard, G. Fatty acids—Induced lipotoxity and inflammation. *Curr. Drug Metab.* **2012**, *13*, 1358–1370. [CrossRef]

89. Singh, I.; Pujol, A. Pathomechanisms underlying X-adrenoleukodystrophy: A three-hit hypothesis. *Brain Pathol.* **2010**, *20*, 838–844. [CrossRef]

90. Bergner, C.G.; van der Meer, F.; Winkler, A.; Wrzos, C.; Turkmen, M.; Valizada, E.; Fitzner, D.; Hametner, S.; Hartmann, C.; Pfeifenbring, S.; et al. Microglia damage precedes major myelin breakdown in X-linked adrenoleukodystrophy and metachromatic leukodystrophy. *Glia* **2019**, *67*, 1196–1209. [CrossRef]

91. Gong, Y.; Sasidharan, N.; Laheji, F.; Frosch, M.; Musolino, P.; Tanzi, R.; Kim, D.Y.; Biffi, A.; El Khoury, J.; Eichler, F. Microglial dysfunction as a key pathological change in adrenomyeloneuropathy. *Ann. Neurol.* **2017**, *82*, 813–827. [CrossRef]

92. Aubourg, P.; Blanche, S.; Jambaque, I.; Rocchiccioli, F.; Kalifa, G.; Naud-Saudreau, C.; Rolland, M.O.; Debre, M.; Chaussain, J.L.; Griscelli, C.; et al. Reversal of early neurologic and neuroradiologic manifestations of X-linked adrenoleukodystrophy by bone marrow transplantation. *N. Engl. J. Med.* **1990**, *322*, 1860–1866. [CrossRef] [PubMed]

93. Casasnovas, C.; Ruiz, M.; Schlüter, A.; Naudí, A.; Fourcade, S.; Veciana, M.; Castañer, S.; Albertí, A.; Bargalló, N.; Johnson, M.; et al. Biomarker Identification, Safety, and Efficacy of High-Dose Antioxidants for Adrenomyeloneuropathy: A Phase II Pilot Study. *Neurotherapeutics* **2019**, *16*, 1167–1182. [CrossRef]

94. Rodríguez-Pascau, L.; Britti, E.; Calap-Quintana, P.; Dong, Y.N.; Vergara, C.; Delaspre, F.; Medina-Carbonero, M.; Tamarit, J.; Pallardó, F.V.; Gonzalez-Cabo, P.; et al. PPAR gamma agonist leriglitazone improves frataxin-loss impairments in cellular and animal models of Friedreich Ataxia. *Neurobiol. Dis.* **2021**, *148*, 105162. [CrossRef]

95. Paton, B.C.; Heron, S.E.; Nelson, P.V.; Morris, C.P.; Poulos, A. Absence of mutations raises doubts about the role of the 70-kD peroxisomal membrane protein in Zellweger syndrome. *Am. J. Hum. Genet.* **1997**, *60*, 1535–1539. [CrossRef]

96. Kim, J.A. Peroxisome Metabolism in Cancer. *Cells* **2020**, *9*, 1692. [CrossRef] [PubMed]

97. Hlaváč, V.; Souček, P. Role of family D ATP-binding cassette transporters (ABCD) in cancer. *Biochem. Soc. Trans.* **2015**, *43*, 937–942. [CrossRef] [PubMed]

98. Heimerl, S.; Bosserhoff, A.K.; Langmann, T.; Ecker, J.; Schmitz, G. Mapping ATP-binding cassette transporter gene expression profiles in melanocytes and melanoma cells. *Melanoma Res.* **2007**, *17*, 265–273. [CrossRef]
99. Hour, T.C.; Kuo, Y.Z.; Liu, G.Y.; Kang, W.Y.; Huang, C.Y.; Tsai, Y.C.; Wu, W.J.; Huang, S.P.; Pu, Y.S. Downregulation of ABCD1 in human renal cell carcinoma. *Int. J. Biol. Markers* **2009**, *24*, 171–178. [CrossRef]
100. Soucek, P.; Hlavac, V.; Elsnerova, K.; Vaclavikova, R.; Kozevnikovova, R.; Raus, K. Whole exome sequencing analysis of ABCC8 and ABCD2 genes associating with clinical course of breast carcinoma. *Physiol. Res.* **2015**, *64*, S549–S557. [CrossRef]
101. Elsnerova, K.; Bartakova, A.; Tihlarik, J.; Bouda, J.; Rob, L.; Skapa, P.; Hruda, M.; Gut, I.; Mohelnikova-Duchonova, B.; Soucek, P.; et al. Gene Expression Profiling Reveals Novel Candidate Markers of Ovarian Carcinoma Intraperitoneal Metastasis. *J. Cancer* **2017**, *8*, 3598–3606. [CrossRef]
102. Zhang, Y.; Zhang, Y.; Wang, J.; Yang, J.; Yang, G. Abnormal expression of ABCD3 is an independent prognostic factor for colorectal cancer. *Oncol. Lett.* **2020**, *19*, 3567–3577. [CrossRef]
103. Braicu, E.I.; Darb-Esfahani, S.; Schmitt, W.D.; Koistinen, K.M.; Heiskanen, L.; Poho, P.; Budczies, J.; Kuhberg, M.; Dietel, M.; Frezza, C.; et al. High-grade ovarian serous carcinoma patients exhibit profound alterations in lipid metabolism. *Oncotarget* **2017**, *8*, 102912–102922. [CrossRef] [PubMed]
104. Hlaváč, V.; Brynychová, V.; Václavíková, R.; Ehrlichová, M.; Vrána, D.; Pecha, V.; Koževnikovová, R.; Trnková, M.; Gatěk, J.; Kopperová, D.; et al. The expression profile of ATP-binding cassette transporter genes in breast carcinoma. *Pharmacogenomics* **2013**, *14*, 515–529. [CrossRef] [PubMed]
105. Benedetti, E.; Galzio, R.; Laurenti, G.; D'Angelo, B.; Melchiorre, E.; Cifone, M.G.; Fanelli, F.; Muzi, P.; Coletti, G.; Alecci, M.; et al. Lipid metabolism impairment in human gliomas: Expression of peroxisomal proteins in human gliomas at different grades of malignancy. *Int. J. Immunopathol. Pharmacol.* **2010**, *23*, 235–246. [CrossRef]
106. Hama, K.; Fujiwara, Y.; Hayama, T.; Ozawa, T.; Nozawa, K.; Matsuda, K.; Hashiguchi, Y.; Yokoyama, K. Very long-chain fatty acids are accumulated in triacylglycerol and nonesterified forms in colorectal cancer tissues. *Sci. Rep.* **2021**, *11*, 6163. [CrossRef]
107. Shani, N.; Watkins, P.A.; Valle, D. PXA1, a possible Saccharomyces cerevisiae ortholog of the human adrenoleukodystrophy gene. *Proc. Natl. Acad. Sci. USA* **1995**, *92*, 6012–6016. [CrossRef]
108. Hettema, E.H.; vanRoermund, C.W.T.; Distel, B.; vandenBerg, M.; Vilela, C.; RodriguesPousada, C.; Wanders, R.J.A.; Tabak, H.F. The ABC transporter proteins Pat1 and Pat2 are required for import of long-chain fatty acids into peroxisomes of Saccharomyces cerevisiae. *EMBO J.* **1996**, *15*, 3813–3822. [CrossRef]
109. Verleur, N.; Hettema, E.H.; van Roermund, C.W.; Tabak, H.F.; Wanders, R.J. Transport of activated fatty acids by the peroxisomal ATP-binding-cassette transporter Pxa2 in a semi-intact yeast cell system. *Eur. J. Biochem.* **1997**, *249*, 657–661. [CrossRef]
110. Theodoulou, F.L.; Job, K.; Slocombe, S.P.; Footitt, S.; Holdsworth, M.; Baker, A.; Larson, T.R.; Graham, I.A. Jasmonic acid levels are reduced in COMATOSE ATP-binding cassette transporter mutants. Implications for transport of jasmonate precursors into peroxisomes. *Plant. Physiol.* **2005**, *137*, 835–840. [CrossRef]
111. Kunz, H.H.; Scharnewski, M.; Feussner, K.; Feussner, I.; Flugge, U.I.; Fulda, M.; Gierth, M. The ABC transporter PXA1 and peroxisomal beta-oxidation are vital for metabolism in mature leaves of Arabidopsis during extended darkness. *Plant Cell* **2009**, *21*, 2733–2749. [CrossRef] [PubMed]
112. Footitt, S.; Dietrich, D.; Fait, A.; Fernie, A.R.; Holdsworth, M.J.; Baker, A.; Theodoulou, F.L. The COMATOSE ATP-binding cassette transporter is required for full fertility in Arabidopsis. *Plant. Physiol.* **2007**, *144*, 1467–1480. [CrossRef] [PubMed]
113. Zhang, X.; De Marcos Lousa, C.; Schutte-Lensink, N.; Ofman, R.; Wanders, R.J.; Baldwin, S.A.; Baker, A.; Kemp, S.; Theodoulou, F.L. Conservation of targeting but divergence in function and quality control of peroxisomal ABC transporters: An analysis using cross-kingdom expression. *Biochem. J.* **2011**, *436*, 547–557. [CrossRef]
114. Oikonomou, G.; Shaham, S. The glia of Caenorhabditis elegans. *Glia* **2011**, *59*, 1253–1263. [CrossRef] [PubMed]
115. Coppa, A.; Guha, S.; Fourcade, S.; Parameswaran, J.; Ruiz, M.; Moser, A.B.; Schlüter, A.; Murphy, M.P.; Lizcano, J.M.; Miranda-Vizuete, A.; et al. The peroxisomal fatty acid transporter ABCD1/PMP-4 is required in the C. elegans hypodermis for axonal maintenance: A worm model for adrenoleukodystrophy. *Free Radic. Biol. Med.* **2020**, *152*, 797–809. [CrossRef]
116. Gordon, H.B.; Valdez, L.; Letsou, A. Etiology and treatment of adrenoleukodystrophy: New insights from Drosophila. *Dis. Model Mech.* **2018**, *11*, 11. [CrossRef]
117. Sivachenko, A.; Gordon, H.B.; Kimball, S.S.; Gavin, E.J.; Bonkowsky, J.L.; Letsou, A. Neurodegeneration in a Drosophila model of adrenoleukodystrophy: The roles of the Bubblegum and Double bubble acyl-CoA synthetases. *Dis. Models Mech.* **2016**, *9*, 377–387.
118. Strachan, L.R.; Stevenson, T.J.; Freshner, B.; Keefe, M.D.; Miranda Bowles, D.; Bonkowsky, J.L. A zebrafish model of X-linked adrenoleukodystrophy recapitulates key disease features and demonstrates a developmental requirement for abcd1 in oligodendrocyte patterning and myelination. *Hum. Mol. Genet.* **2017**, *26*, 3600–3614. [CrossRef] [PubMed]
119. Raas, Q.; van de Beek, M.C.; Forss-Petter, S.; Dijkstra, I.M.; DeSchiffart, A.; Freshner, B.C.; Stevenson, T.J.; Jaspers, Y.R.; Nagtzaam, L.M.; Wanders, R.J.; et al. Metabolic rerouting via SCD1 induction impacts X-linked adrenoleukodystrophy. *J. Clin. Invest.* **2021**. [CrossRef]
120. Gueugnon, F.; Volodina, N.; Taouil, J.; Lopez, T.; Gondcaille, C.; Sequeira-Le Grand, A.; Mooijer, P.; Kemp, S.; Wanders, R.; Savary, S. A novel cell model to study the function of the adrenoleukodystrophy-related protein. *Biochem. Biophys. Res. Commun.* **2006**, *341*, 150–157. [CrossRef]
121. Singh, J.; Khan, M.; Singh, I. Silencing of Abcd1 and Abcd2 genes sensitizes astrocytes for inflammation: Implication for X-adrenoleukodystrophy. *J. Lipid Res.* **2009**, *50*, 135–147. [CrossRef]

122. Kruska, N.; Schonfeld, P.; Pujol, A.; Reiser, G. Astrocytes and mitochondria from adrenoleukodystrophy protein (ABCD1)-deficient mice reveal that the adrenoleukodystrophy-associated very long-chain fatty acids target several cellular energy-dependent functions. *Biochim. Biophys. Acta* **2015**, *1852*, 925–936. [CrossRef]
123. Morita, M.; Toida, A.; Horiuchi, Y.; Watanabe, S.; Sasahara, M.; Kawaguchi, K.; So, T.; Imanaka, T. Generation of an immortalized astrocytic cell line from Abcd1-deficient H-2K(b)tsA58 mice to facilitate the study of the role of astrocytes in X-linked adrenoleukodystrophy. *Heliyon* **2021**, *7*, e06228. [CrossRef]
124. Raas, Q.; Gondcaille, C.; Hamon, Y.; Leoni, V.; Caccia, C.; Ménétrier, F.; Lizard, G.; Trompier, D.; Savary, S. CRISPR/Cas9-mediated knockout of Abcd1 and Abcd2 genes in BV-2 cells: Novel microglial models for X-linked Adrenoleukodystrophy. *Biochim. Biophys. Acta Mol. Cell. Biol. Lipids* **2019**, *1864*, 704–714. [CrossRef]
125. Schaumburg, H.H.; Powers, J.M.; Suzuki, K.; Raine, C.S. Adreno-leukodystrophy (sex-linked Schilder disease). Ultrastructural demonstration of specific cytoplasmic inclusions in the central nervous system. *Arch. Neurol.* **1974**, *31*, 210–213. [CrossRef] [PubMed]
126. Ferrer, I.; Kapfhammer, J.P.; Hindelang, C.; Kemp, S.; Troffer-Charlier, N.; Broccoli, V.; Callyzot, N.; Mooyer, P.; Selhorst, J.; Vreken, P.; et al. Inactivation of the peroxisomal ABCD2 transporter in the mouse leads to late-onset ataxia involving mitochondria, Golgi and endoplasmic reticulum damage. *Hum. Mol. Genet.* **2005**, *14*, 3565–3577. [CrossRef]
127. Pujol, A.; Hindelang, C.; Callizot, N.; Bartsch, U.; Schachner, M.; Mandel, J.L. Late onset neurological phenotype of the X-ALD gene inactivation in mice: A mouse model for adrenomyeloneuropathy. *Hum. Mol. Genet.* **2002**, *11*, 499–505. [CrossRef] [PubMed]
128. Fourcade, S.; Lopez-Erauskin, J.; Galino, J.; Duval, C.; Naudi, A.; Jove, M.; Kemp, S.; Villarroya, F.; Ferrer, I.; Pamplona, R.; et al. Early oxidative damage underlying neurodegeneration in X-adrenoleukodystrophy. *Hum. Mol. Genet.* **2008**, *17*, 1762–1773. [CrossRef] [PubMed]
129. McGuinness, M.C.; Lu, J.F.; Zhang, H.P.; Dong, G.X.; Heinzer, A.K.; Watkins, P.A.; Powers, J.; Smith, K.D. Role of ALDP (ABCD1) and mitochondria in X-linked adrenoleukodystrophy. *Mol. Cell. Biol.* **2003**, *23*, 744–753. [CrossRef]
130. Galino, J.; Ruiz, M.; Fourcade, S.; Schluter, A.; Lopez-Erauskin, J.; Guilera, C.; Jove, M.; Naudi, A.; Garcia-Arumi, E.; Andreu, A.L.; et al. Oxidative damage compromises energy metabolism in the axonal degeneration mouse model of X-adrenoleukodystrophy. *Antioxid. Redox Signal.* **2011**, *15*, 2095–2107. [CrossRef]
131. Lopez-Erauskin, J.; Galino, J.; Ruiz, M.; Cuezva, J.M.; Fabregat, I.; Cacabelos, D.; Boada, J.; Martinez, J.; Ferrer, I.; Pamplona, R.; et al. Impaired mitochondrial oxidative phosphorylation in the peroxisomal disease X-linked adrenoleukodystrophy. *Hum. Mol. Genet.* **2013**, *22*, 3296–3305. [CrossRef] [PubMed]
132. Launay, N.; Aguado, C.; Fourcade, S.; Ruiz, M.; Grau, L.; Riera, J.; Guilera, C.; Giros, M.; Ferrer, I.; Knecht, E.; et al. Autophagy induction halts axonal degeneration in a mouse model of X-adrenoleukodystrophy. *Acta Neuropathol.* **2015**, *129*, 399–415. [CrossRef]
133. Morató, L.; Galino, J.; Ruiz, M.; Calingasan, N.Y.; Starkov, A.A.; Dumont, M.; Naudí, A.; Martínez, J.J.; Aubourg, P.; Portero-Otín, M.; et al. Pioglitazone halts axonal degeneration in a mouse model of X-linked adrenoleukodystrophy. *Brain* **2013**, *136*, 2432–2443. [CrossRef]
134. Launay, N.; Ruiz, M.; Grau, L.; Ortega, F.J.; Ilieva, E.V.; Martinez, J.J.; Galea, E.; Ferrer, I.; Knecht, E.; Pujol, A.; et al. Tauroursodeoxycholic bile acid arrests axonal degeneration by inhibiting the unfolded protein response in X-linked adrenoleukodystrophy. *Acta Neuropathol.* **2017**, *133*, 283–301. [CrossRef] [PubMed]
135. Fourcade, S.; Goicoechea, L.; Parameswaran, J.; Schlüter, A.; Launay, N.; Ruiz, M.; Seyer, A.; Colsch, B.; Calingasan, N.Y.; Ferrer, I.; et al. High-dose biotin restores redox balance, energy and lipid homeostasis, and axonal health in a model of adrenoleukodystrophy. *Brain Pathol.* **2020**, *30*, 945–963. [CrossRef] [PubMed]
136. Olah, M.; Patrick, E.; Villani, A.C.; Xu, J.; White, C.C.; Ryan, K.J.; Piehowski, P.; Kapasi, A.; Nejad, P.; Cimpean, M.; et al. A transcriptomic atlas of aged human microglia. *Nat. Commun.* **2018**, *9*, 539. [CrossRef] [PubMed]
137. Lu, J.F.; Barron-Casella, E.; Deering, R.; Heinzer, A.K.; Moser, A.B.; de Mesy Bentley, K.L.; Wand, G.S.; McGuinness, M.C.; Pei, Z.; Watkins, P.A.; et al. The role of peroxisomal ABC transporters in the mouse adrenal gland: The loss of Abcd2 (ALDR), Not Abcd1 (ALD), causes oxidative damage. *Lab. Investig.* **2007**, *87*, 261–272. [CrossRef] [PubMed]
138. Moser, H.W.; Moser, A.B.; Kawamura, N.; Murphy, J.; Suzuki, K.; Schaumburg, H.; Kishimoto, Y. Adrenoleukodystrophy: Elevated C26 fatty acid in cultured skin fibroblasts. *Ann. Neurol.* **1980**, *7*, 542–549. [CrossRef]
139. Griffin, D.E.; Moser, H.W.; Mendoza, Q.; Moench, T.R.; O'Toole, S.; Moser, A.B. Identification of the inflammatory cells in the central nervous system of patients with adrenoleukodystrophy. *Ann. Neurol.* **1985**, *18*, 660–664. [CrossRef]
140. Lannuzel, A.; Aubourg, P.; Tardieu, M. Excessive production of tumour necrosis factor alpha by peripheral blood mononuclear cells in X-linked adrenoleukodystrophy. *Eur. J. Paediatr. Neurol.* **1998**, *2*, 27–32. [CrossRef]
141. Weinhofer, I.; Zierfuss, B.; Hametner, S.; Wagner, M.; Popitsch, N.; Machacek, C.; Bartolini, B.; Zlabinger, G.; Ohradanova-Repic, A.; Stockinger, H.; et al. Impaired plasticity of macrophages in X-linked adrenoleukodystrophy. *Brain* **2018**, *141*, 2329–2342. [CrossRef]
142. Weber, F.D.; Wiesinger, C.; Forss-Petter, S.; Regelsberger, G.; Einwich, A.; Weber, W.H.; Kohler, W.; Stockinger, H.; Berger, J. X-linked adrenoleukodystrophy: Very long-chain fatty acid metabolism is severely impaired in monocytes but not in lymphocytes. *Hum. Mol. Genet.* **2014**, *23*, 2542–2550. [CrossRef]

143. Wang, X.M.; Yik, W.Y.; Zhang, P.; Lu, W.; Dranchak, P.K.; Shibata, D.; Steinberg, S.J.; Hacia, J.G. The gene expression profiles of induced pluripotent stem cells from individuals with childhood cerebral adrenoleukodystrophy are consistent with proposed mechanisms of pathogenesis. *Stem. Cell. Res. Ther.* **2012**, *3*, 39. [CrossRef]
144. Baarine, M.; Khan, M.; Singh, A.; Singh, I. Functional Characterization of IPSC-Derived Brain Cells as a Model for X-Linked Adrenoleukodystrophy. *PLoS ONE* **2015**, *10*, e0143238. [CrossRef] [PubMed]
145. Son, D.; Quan, Z.; Kang, P.J.; Park, G.; Kang, H.C.; You, S. Generation of two induced pluripotent stem cell (iPSC) lines from X-linked adrenoleukodystrophy (X-ALD) patients with adrenomyeloneuropathy (AMN). *Stem Cell Res.* **2017**, *25*, 46–49. [CrossRef] [PubMed]
146. Muffat, J.; Li, Y.; Yuan, B.; Mitalipova, M.; Omer, A.; Corcoran, S.; Bakiasi, G.; Tsai, L.H.; Aubourg, P.; Ransohoff, R.M.; et al. Efficient derivation of microglia-like cells from human pluripotent stem cells. *Nat. Med.* **2016**, *22*, 1358–1367. [CrossRef] [PubMed]
147. Jang, J.; Kang, H.C.; Kim, H.S.; Kim, J.Y.; Huh, Y.J.; Kim, D.S.; Yoo, J.E.; Lee, J.A.; Lim, B.; Lee, J.; et al. Induced pluripotent stem cell models from X-linked adrenoleukodystrophy patients. *Ann. Neurol.* **2011**, *70*, 402–409. [CrossRef]
148. Jang, J.; Park, S.; Jin Hur, H.; Cho, H.J.; Hwang, I.; Pyo Kang, Y.; Im, I.; Lee, H.; Lee, E.; Yang, W.; et al. 25-hydroxycholesterol contributes to cerebral inflammation of X-linked adrenoleukodystrophy through activation of the NLRP3 inflammasome. *Nat. Commun.* **2016**, *7*, 13129. [CrossRef] [PubMed]
149. Lee, C.A.A.; Seo, H.S.; Armien, A.G.; Bates, F.S.; Tolar, J.; Azarin, S.M. Modeling and rescue of defective blood-brain barrier function of induced brain microvascular endothelial cells from childhood cerebral adrenoleukodystrophy patients. *Fluids Barriers CNS* **2018**, *15*, 9. [CrossRef] [PubMed]
150. Gloeckner, C.J.; Mayerhofer, P.U.; Landgraf, P.; Muntau, A.C.; Holzinger, A.; Gerber, J.K.; Kammerer, S.; Adamski, J.; Roscher, A.A. Human adrenoleukodystrophy protein and related peroxisomal ABC transporters interact with the peroxisomal assembly protein PEX19p. *Biochem. Biophys. Res. Commun.* **2000**, *271*, 144–150. [CrossRef]
151. Shibata, H.; Kashiwayama, Y.; Imanaka, T.; Kato, H. Domain architecture and activity of human Pex19p, a chaperone-like protein for intracellular trafficking of peroxisomal membrane proteins. *J. Biol. Chem.* **2004**, *279*, 38486–38494. [CrossRef]
152. Liu, X.; Liu, J.; Lester, J.D.; Pijut, S.S.; Graf, G.A. ABCD2 identifies a subclass of peroxisomes in mouse adipose tissue. *Biochem. Biophys. Res. Commun.* **2015**, *456*, 129–134. [CrossRef]
153. Kim, P.K.; Hettema, E.H. Multiple pathways for protein transport to peroxisomes. *J. Mol. Biol.* **2015**, *427*, 1176–1190. [CrossRef] [PubMed]
154. Hillebrand, M.; Gersting, S.W.; Lotz-Havla, A.S.; Schafer, A.; Rosewich, H.; Valerius, O.; Muntau, A.C.; Gartner, J. Identification of a new fatty acid synthesis-transport machinery at the peroxisomal membrane. *J. Biol. Chem.* **2012**, *287*, 210–221. [CrossRef]
155. Makkar, R.S.; Contreras, M.A.; Paintlia, A.S.; Smith, B.T.; Haq, E.; Singh, I. Molecular organization of peroxisomal enzymes: Protein-protein interactions in the membrane and in the matrix. *Arch. Biochem. Biophys.* **2006**, *451*, 128–140. [CrossRef] [PubMed]
156. Boger, D.L.; Fecik, R.A.; Patterson, J.E.; Miyauchi, H.; Patricelli, M.P.; Cravatt, B.F. Fatty acid amide hydrolase substrate specificity. *Bioorg. Med. Chem. Lett.* **2000**, *10*, 2613–2616. [CrossRef]
157. Wei, B.Q.; Mikkelsen, T.S.; McKinney, M.K.; Lander, E.S.; Cravatt, B.F. A second fatty acid amide hydrolase with variable distribution among placental mammals. *J. Biol. Chem.* **2006**, *281*, 36569–36578. [CrossRef]
158. Brown, I.; Cascio, M.G.; Wahle, K.W.; Smoum, R.; Mechoulam, R.; Ross, R.A.; Pertwee, R.G.; Heys, S.D. Cannabinoid receptor-dependent and -independent anti-proliferative effects of omega-3 ethanolamides in androgen receptor-positive and -negative prostate cancer cell lines. *Carcinogenesis* **2010**, *31*, 1584–1591. [CrossRef]
159. Havugimana, P.C.; Hart, G.T.; Nepusz, T.; Yang, H.; Turinsky, A.L.; Li, Z.; Wang, P.I.; Boutz, D.R.; Fong, V.; Phanse, S.; et al. A census of human soluble protein complexes. *Cell* **2012**, *150*, 1068–1081. [CrossRef]
160. Young, P.A.; Senkal, C.E.; Suchanek, A.L.; Grevengoed, T.J.; Lin, D.D.; Zhao, L.; Crunk, A.E.; Klett, E.L.; Füllekrug, J.; Obeid, L.M.; et al. Long-chain acyl-CoA synthetase 1 interacts with key proteins that activate and direct fatty acids into niche hepatic pathways. *J. Biol. Chem.* **2018**, *293*, 16724–16740. [CrossRef]
161. Ferdinandusse, S.; Falkenberg, K.D.; Koster, J.; Mooyer, P.A.; Jones, R.; van Roermund, C.W.T.; Pizzino, A.; Schrader, M.; Wanders, R.J.A.; Vanderver, A.; et al. ACBD5 deficiency causes a defect in peroxisomal very long-chain fatty acid metabolism. *J. Med. Genet.* **2017**, *54*, 330–337. [CrossRef]
162. Ewing, R.M.; Chu, P.; Elisma, F.; Li, H.; Taylor, P.; Climie, S.; McBroom-Cerajewski, L.; Robinson, M.D.; O'Connor, L.; Li, M.; et al. Large-scale mapping of human protein-protein interactions by mass spectrometry. *Mol. Syst. Biol.* **2007**, *3*, 89. [CrossRef]
163. Chang, C.L.; Weigel, A.V.; Ioannou, M.S.; Pasolli, H.A.; Xu, C.S.; Peale, D.R.; Shtengel, G.; Freeman, M.; Hess, H.F.; Blackstone, C.; et al. Spastin tethers lipid droplets to peroxisomes and directs fatty acid trafficking through ESCRT-III. *J. Cell Biol.* **2019**, *218*, 2583–2599. [CrossRef]
164. Cuevas-Fernández, B.; Fuentes-Almagro, C.; Peragón, J. Proteomics Analysis Reveals the Implications of Cytoskeleton and Mitochondria in the Response of the Rat Brain to Starvation. *Nutrients* **2019**, *11*, 219. [CrossRef] [PubMed]
165. Schrader, M.; Grille, S.; Fahimi, H.D.; Islinger, M. Peroxisome interactions and cross-talk with other subcellular compartments in animal cells. *Subcell. Biochem.* **2013**, *69*, 1–22.
166. Chen, C.; Li, J.; Qin, X.; Wang, W. Peroxisomal Membrane Contact Sites in Mammalian Cells. *Front. Cell Dev. Biol.* **2020**, *8*, 512. [CrossRef] [PubMed]
167. Huttlin, E.L.; Bruckner, R.J.; Paulo, J.A.; Cannon, J.R.; Ting, L.; Baltier, K.; Colby, G.; Gebreab, F.; Gygi, M.P.; Parzen, H.; et al. Architecture of the human interactome defines protein communities and disease networks. *Nature* **2017**, *545*, 505–509. [CrossRef]

168. Hein, M.Y.; Hubner, N.C.; Poser, I.; Cox, J.; Nagaraj, N.; Toyoda, Y.; Gak, I.A.; Weisswange, I.; Mansfeld, J.; Buchholz, F.; et al. A human interactome in three quantitative dimensions organized by stoichiometries and abundances. *Cell* **2015**, *163*, 712–723. [CrossRef]
169. Huttlin, E.L.; Ting, L.; Bruckner, R.J.; Gebreab, F.; Gygi, M.P.; Szpyt, J.; Tam, S.; Zarraga, G.; Colby, G.; Baltier, K.; et al. The BioPlex Network: A Systematic Exploration of the Human Interactome. *Cell* **2015**, *162*, 425–440. [CrossRef] [PubMed]
170. Deakin, S.; Leviev, I.; Gomaraschi, M.; Calabresi, L.; Franceschini, G.; James, R.W. Enzymatically active paraoxonase-1 is located at the external membrane of producing cells and released by a high affinity, saturable, desorption mechanism. *J. Biol. Chem.* **2002**, *277*, 4301–4308. [CrossRef]
171. Gonzalvo, M.C.; Gil, F.; Hernandez, A.F.; Rodrigo, L.; Villanueva, E.; Pla, A. Human liver paraoxonase (PON1): Subcellular distribution and characterization. *J. Biochem. Mol. Toxicol.* **1998**, *12*, 61–69. [CrossRef]
172. Reichert, C.O.; Levy, D.; Bydlowski, S.P. Paraoxonase Role in Human Neurodegenerative Diseases. *Antioxidants* **2020**, *10*, 11. [CrossRef] [PubMed]
173. Silva, B.S.C.; DiGiovanni, L.; Kumar, R.; Carmichael, R.E.; Kim, P.K.; Schrader, M. Maintaining social contacts: The physiological relevance of organelle interactions. *Biochim. Biophys. Acta Mol. Cell Res.* **2020**, *1867*, 118800. [CrossRef] [PubMed]
174. Islinger, M.; Voelkl, A.; Fahimi, H.D.; Schrader, M. The peroxisome: An update on mysteries 2.0. *Histochem. Cell Biol.* **2018**, *150*, 443–471. [CrossRef] [PubMed]
175. He, A.; Dean, J.M.; Lodhi, I.J. Peroxisomes as Cellular Adaptors to Metabolic and Environmental Stress. *Trends Cell Biol.* **2021**. [CrossRef]

International Journal of
Molecular Sciences

MDPI

Review

Structural and Functional Characterization of the ABCC6 Transporter in Hepatic Cells: Role on PXE, Cancer Therapy and Drug Resistance

Faustino Bisaccia, Prashant Koshal, Vittorio Abruzzese, Maria Antonietta Castiglione Morelli and Angela Ostuni *

Department of Sciences, University of Basilicata, 85100 Potenza, Italy; faustino.bisaccia@unibas.it (F.B.); prashantkoshal240@gmail.com (P.K.); v.abruzz@hotmail.it (V.A.); maria.castiglione@unibas.it (M.A.C.M.)
* Correspondence: angela.ostuni@unibas.it

Abstract: Pseudoxanthoma elasticum (PXE) is a complex autosomal recessive disease caused by mutations of ABCC6 transporter and characterized by ectopic mineralization of soft connective tissues. Compared to the other ABC transporters, very few studies are available to explain the structural components and working of a full ABCC6 transporter, which may provide some idea about its physiological role in humans. Some studies suggest that mutations of *ABCC6* in the liver lead to a decrease in some circulating factor and indicate that PXE is a metabolic disease. It has been reported that ABCC6 mediates the efflux of ATP, which is hydrolyzed in PPi and AMP; in the extracellular milieu, PPi gives potent anti-mineralization effect, whereas AMP is hydrolyzed to Pi and adenosine which affects some cellular properties by modulating the purinergic pathway. Structural and functional studies have demonstrated that silencing or inhibition of ABCC6 with probenecid changed the expression of several genes and proteins such as *NT5E* and TNAP, as well as Lamin, and CDK1, which are involved in cell motility and cell cycle. Furthermore, a change in cytoskeleton rearrangement and decreased motility of HepG2 cells makes ABCC6 a potential target for anti-cancer therapy. Collectively, these findings suggested that ABCC6 transporter performs functions that modify both the external and internal compartments of the cells.

Keywords: ABCC6; TNAP; *NT5E*; Pseudoxanthoma elasticum (PXE); cancer

Citation: Bisaccia, F.; Koshal, P.; Abruzzese, V.; Castiglione Morelli, M.A.; Ostuni, A. Structural and Functional Characterization of the ABCC6 Transporter in Hepatic Cells: Role on PXE, Cancer Therapy and Drug Resistance. *Int. J. Mol. Sci.* **2021**, 22, 2858. https://doi.org/10.3390/ijms22062858

Academic Editor: Thomas Falguières

Received: 20 February 2021
Accepted: 9 March 2021
Published: 11 March 2021

Publisher's Note: MDPI stays neutral with regard to jurisdictional claims in published maps and institutional affiliations.

1. Introduction

Pseudoxanthoma elasticum (PXE) is an autosomal recessive disease, which was described in 1881 by a French dermatologist. In 2000, it was first recognized that mutation in *ABCC6* is responsible for PXE [1]. It is affecting approximately 1:50,000 people worldwide, with the prominent characteristic feature of ectopic mineralization of soft tissues like skin, eyes, and arteries (Figure 1), for which no effective curative treatment is available [2,3]. Moreover, PXE shows similar phenotypic characteristics with other common health problems like kidney diseases (chronic kidney disease (CKD) and nephrocalcinosis) and cardiovascular diseases (coronary heart disease, cardiomyopathy, and dyslipidemia), which makes PXE a complex disorder [4,5].

Different hypotheses were proposed for the factors pathologically involved with PXE. The "Metabolic Hypothesis" stated that decrease or loss of ABCC6 functionality especially in the liver may lead to a decrease in some circulating factors in the blood stream, which should be responsible for preventing ectopic mineralization of soft tissues. The "PXE Cell Hypothesis" stated that absence of ABCC6 in PXE tissues leads to an alteration in cell proliferation due to changes in the biosynthetic pathway and alters cells to extracellular matrix interactions. The most recent "ATP Release Hypothesis" stated that ABCC6 mediates the efflux of ATP in extracellular milieu, where it is hydrolyzed into AMP and pyrophosphate and prevents the mineralization of soft tissues [1].

Previous studies of serum analysis either from the *ABCC6* knock-down mouse model or from PXE patients showed an inability to prevent calcium and phosphate deposition and suggested that PXE is a metabolic disease with very slow onset [1,6,7]. It should be noted that the tissues, which mostly express ABCC6, are the liver and, to some lesser extent, the kidney and differ from those in which ectopic mineralization is mostly evident, namely soft tissues of skin, eyes, and the cardiovascular system. The origin of this apparent paradox has not been explained yet (Figure 1) [8].

On the basis of our previous studies of the ABCC6 transporter in hepatic cells, the present review is focused on lightening changes in cellular function associated with ABCC6 transporter activity.

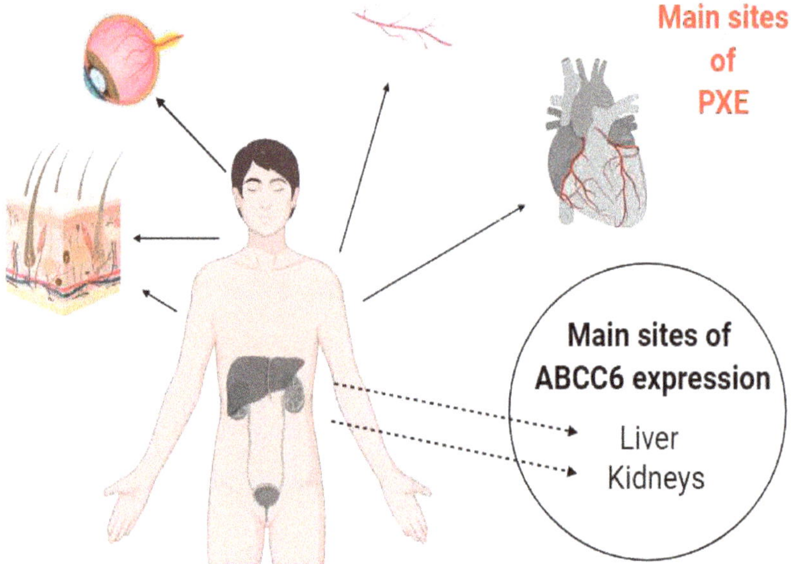

Figure 1. Pictorial presentation of ABCC6 expression and affected tissues in Pseudoxanthoma elasticum (PXE). The protein is expressed mainly in liver and kidney cells but the main areas involved in ectopic calcification are the elastic tissues of heart, blood vessels, skin, and eyes.

2. Structural Properties of ABCC6 Transporter

The ATP-binding cassette (ABC) transporters are a well-known family and ubiquitously found in all living organisms. About 50 different types of ABC transporters have been recognized in humans, divided into seven subfamilies (ABCA to ABCG) based on their structure and genetic sequences. Among them we can find both half transporters, with only one transmembrane domain (TMD) and nucleotide-binding domain (NBD), and full transporters, with two TMDs and NBDs [9,10]. ABCC6 belongs to the ATP-binding cassette (ABC) transporter subfamily C and has been found to be highly expressed in basolateral plasma membrane of hepatocytes and, to some lesser extent, in the proximal tubules of the kidney [11].

The *ABCC6* gene has 31 exons and encodes a protein of 1503 amino acid residues, MRP6. It is made-up of two nucleotide-binding domains (NBD1 and NBD2) and two transmembrane domains (TMD1 and TMD2) with an additional auxiliary NH2-terminal transmembrane domain known as TMD0, which is connected with the canonical components through the cytoplasmic L0 loop [12,13]. The NBDs of ABC transporters consist of different conserved motifs which bind and hydrolyze ATP. Like in other transporters, ABCC6 NBDs domains have Walker A motif (P loop), Walker B motif (Mg^{2+} binding site), histidine loop (Switch region), signature motif (C loop), Q loop (between Walker A and C-loop), and D loop (involved in the intermolecular interactions at the NBD dimmer interface). In the general process of dimerization, the Walker A/B motifs of one NBD are aligned with the C-loop of the other in a yin-yang fashion, forming a sandwich with two molecules of ATP in the middle, which are hydrolyzed in the catalytic cycle. Dimerization induces conformational changes in the TMDs, which are reversed by hydrolysis of ATP, leading to efflux of molecules [1,14,15]. In this situation, the ATP should be sandwiched between two sequence/structural motifs that are both rigorously conserved and form a stable homodimer [16]; this is not always possible with the human transporters, because one of the two NBDs is degenerate and the sequence of motifs which are responsible for hydrolysis of ATP is not identical or canonical [14].

In this context to elucidate the role of NBDs, different experiments were conducted on NBD1 of ABCC1/MRP1, which not only has a high level of homology with ABCC6/MRP6 but also shows a superposition of substrates [17]. Studies revealed that the NBDs of MRP1 are not functionally identical, as both the NBDs bind ATP but only NBD2 is involved in hydrolysis. Moreover, the ATP binding affinity of NBD1 of MRP1 is not completely dependent on NBD2 but the binding of ATP to NBD2 is highly dependent on NBD1 [18,19]. In respect of this, we have gone through by a series of experiments to identify the functional roles of NBDs, TMD0, and L0 loop of ABCC6 (Figure 2).

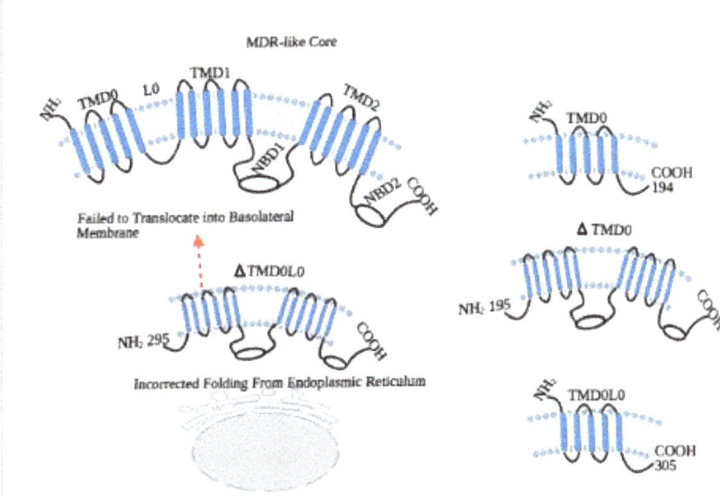

Figure 2. Topology of ABCC6 transporter [20].

In order to evaluate the functioning of NBDs, we have first characterized the NBD1 of ABCC6/MRP6. In this experiment, by using the E748-A785 fragment of MRP6-NBD1, we have compared the helical structure and ATP binding properties of wild type and the R765Q mutated sequence, which is present in PXE patients. The study with circular dichroism analysis revealed that, in both the wild type and mutated (R765Q) NBD1, E748-A785 peptides adopted α-helical conformations, and the helical content was almost the same for both the peptides in aqueous solutions of trifluoroethanol (TFE). No differences in the length of helices have been found between the peptides in NMR spectroscopy. Moreover, Fluorescence Spectroscopy showed no significant difference in ATP binding capacity of both the peptides. These findings suggest that occurrence of PXE symptoms in R765Q mutated patients might be due to different kind of interactions [21].

In subsequent investigations, we have undergone two different experiments with NBD1 and NBD2, because mutational studies of PXE patients showed that domain-domain interaction is important for proper working of ABCC6 transporter and there is a functional difference between both NBDs of MRP6 [22]. In the first study, we constructed the full-length NBD1 (residues from Asp-627 to Leu-851) and short-length NBD1 (residues from Arg-648 to Thr-805) without some key residues; then differences in helical structure, ATP binding, and hydrolysis of both polypeptides were analyzed. Interestingly, both the polypeptides assumed predominantly α-helical conformation. However, only long-length NBD1 showed β-strand conformation, while short-length NBD1 showed higher helix content, which suggested that the sequences D627-H647 and T806-L851, which are only present in the long-length polypeptide, assume a β-strand conformation, similarly to the corresponding regions of MRP1. Although fluorescence quenching experiments revealed that both the polypeptides have the affinity to bind nucleotides (ATP and ADP), the long-length NBD1 showed a higher affinity to bind ATP than the short-length NBD. In addition, no hydrolytic activity was found in short-length NBD1 compared to long-length NBD1 during spectrophotometric analysis of ATPase activity. These findings suggest that short-length NBD1 lacks of some essential residues, which are responsible for ATP hydrolysis. Whereas, long-length NBD1 form homodimer in the presence of ATP, which is the property of half-transporters and physiologically, it is not possible with the full-transporter where NBD1 interacts with NBD2 [12].

In order to explore how the ABCC6 transporter works in-vivo, we have constructed long-length (Thr-1252 to Val-1503) and short-length NBD2 (Val-1295 to Arg-1468), and we created homology models for homodimer of NBD1 and NBD2, and heterodimer of (NBD1-NBD2) [23]. Circular dichroism spectra of long-length NBD2 revealed that it is more structured and contains α-helical conformation. The nucleotide (ATP and ADP) binding affinity of NBD1 and NBD2 were found to be the same during fluorescence quenching analysis. The ATPase activity of both homo and heterodimer were different, as the amount of inorganic phosphate (Pi) produced by NBD2 was lesser then NBD1. Moreover, addition of NBD2 reduced the NBD1 hydrolytic activity. These findings suggest that NBD2 is well structured (it contains α-helix and β-strands) and binds the nucleotides efficiently, but the ATPase activity of NBD2 is lower compared to the NBD1; this reflects that NBD2 is not able to form a functionally active homodimer, as supported by in-silico structural analysis. In addition, decreased ATPase activity of combined NBD1-NBD2 compared to NBD1 alone suggests that NBD1 and NBD2 work together to form a stable heterodimer and function in a regulated manner, whereas NBD1 alone works in an uncontrolled manner. This can be justified by the in-vivo functioning of the transporter [24], where ATP is hydrolyzed after the binding of substrate on TMDs and leads to conformational changes in membrane domains that activate the NBDs.

3. Roles of Additional TMD0 and L0 Domains

The functional role of TMD0 in other proteins like MRP1, MRP2, SUR1 is well defined, which may be involved in stabilization and retention of the transporter in the plasma membrane or in regulation of channel activities. By topological modeling of TMD0 of MRP6, we demonstrated that it contains five transmembrane domains with the N- and C-termini on the external and cytoplasmic side, respectively. These TMs are inserted into the membrane individually on the basis of hydrophobicity and without affecting each other. In addition, we have also found that disease-causing mutations did not affect the membrane insertion of these TMs [25]. In a further study, we have done the structural and functional characterization of L0 loop of ABCC6. We have found that L0 loop of ABCC6 is well structured (Figure 3) as it contains aromatic residues, and three α-helical regions; it resembles the homologous L0 loops of other MRPs and is responsible for plasma membrane localization of TMD0 [13]. However, to understand the exact role of TMD0 and L0 loop (N-terminal Region) of ABCC6, we have constructed two variants of N-terminal lacking TMD0 (ΔTMD0) and a variant lacking both TMD0 and L0 (ΔTMD0L0). Interestingly, we have found that ΔTMD0L0 not only failed to exhibit transport activity, but it was also not able to localize at the basolateral side of the plasma membrane, which reflects that L0 loop is important for both activities. Thus, these findings suggest that L0 not only contains a basolateral sorting signal but that L0 also contributes to folding ABCC6 into a cellular sorting-competent state, which is necessary to pass the endoplasmic reticulum (ER) quality control system and continue through the secretory pathway. In addition, we also found that L0 loop of ABCC6 interacts with the ion channels like other members of ABCC family and might be involved in the modulation of Ca^{2+} channels of plasma membrane [20].

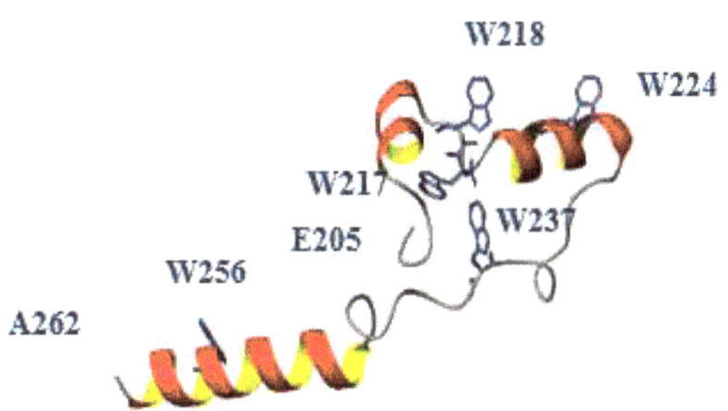

Figure 3. Homology model of L0 loop. The loop includes the region between residues E205 and A262. The PDB structure 5UJ9 corresponding to MRP1 was used as template.

4. The Decrease or Loss of ABCC6 Functionality Changes Extracellular Environment

In early pathological investigations of PXE, Iliás et al. in 2002 proposed that ABCC6 is a transporter of organic anions and actively transports glutathione conjugates, including leukotriene C4 and N-ethylmaleimide S-glutathione (NEM-GS), and their abolishment due to missense mutation in ABCC6 gene is responsible for PXE [26,27]. In the same line of thinking, Borst and co-workers in 2008 proposed that ABCC6 transporter mediates the efflux of vitamin-K as a glutathione or glucuronide conjugate and is involved to prevent calcification of soft tissues [28]. However, in the further investigations, it was found that ABCC6 is not a transporter of vitamin-K, as its administration in PXE mouse model (ABCC6$-/-$) was not able to prevent the mineralization [29,30]. As discussed above, Jansen et al. in 2013 proposed that ABCC6 transporter mediates the efflux of ATP and indirectly produces PPi to prevent ectopic mineralization [2]. However, we have demonstrated that PXE is a complex metabolic disease with the reprogramming of crucial genetic factors in the absence of ABCC6 transporter activity [31–33].

In order to better understand the pathomechanism of PXE (Figure 4), we stably knocked down the *ABCC6* gene in HepG2 cells by using shRNA, and its associated transcriptional/genetic changes were studied. We first examined the production of reactive oxygen species (ROS), which are supposed to increase according to the previous PXE fibroblast studies [34]. On the contrary, in the *ABCC6* knockdown HepG2 cells, the ratio of GSH/GSSG has been found to be increased whereas a significant decrease in ROS level was observed, which means that knockdown cells resembled the reductive stress, which is also required by proliferating cells.

However, we found significant delay in G1 to S transition and slower cell growth in ABCC6 knockdown HepG2 cells (Figure 5). In addition, expression of cyclin-dependent kinase inhibitor (CDKI) p21, which negatively regulates the activity of CDK and is required for the cell entry into the different cycle phases, was found increased in knockdown cells. Moreover, the expression of lamin A/C, which is required to maintain the strength of the nucleus and is pathologically involved in aging process, was decreased in those cells [33].

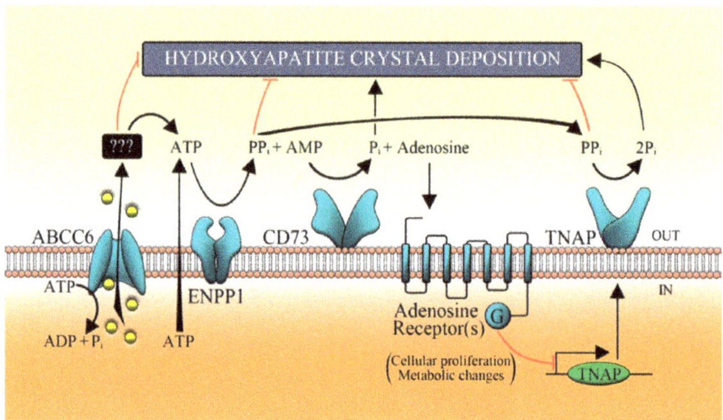

Figure 4. Proposed pathomechanism of PXE. ABCC6 transporter mediates the efflux of ATP, which is metabolized by some ecto-nucleotidases (such as ENPP1) in AMP, which in turn is converted in adenosine and Pi by CD73. In the extracellular milieu, nucleotides regulate the activity of TNAP through the purinergic pathway and prevent the ectopic mineralization [32]. ABCC6, ATP-binding cassette, sub-family C, member 6; ENPP1, ecto-nucleotide pyrophosphatase/phosphodiesterase type I; CD73, cluster of differentiation 73; TNAP, tissue non-specific alkaline phosphatase; Pi, inorganic phosphate; PPi, inorganic pyrophosphate [35].

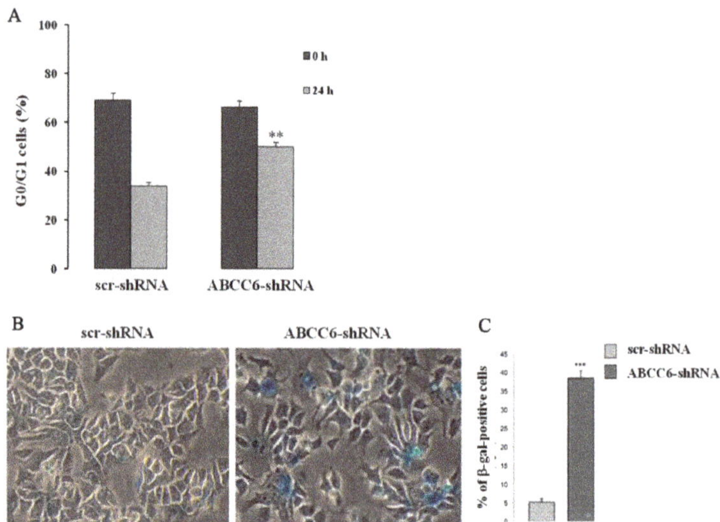

Figure 5. Pictorial presentation of senescent-like phenotype in ABCC6 knockdown HepG2 cell. (**A**)—For the cell cycle analysis, the cells were synchronized at the G1 phase by serum deprivation for 24 h, restimulated with serum for 24 h, and analyzed using flow cytometry after BrDU and PI staining. The percentage of control (scr-shRNA) and ABCC6 knockdown cells (ABCC6-shRNA) in G0/G1 was recorded. (**B**)—Representative images (40 × magnification) of senescence-associated β-galactosidase staining in control and ABCC6 knockdown cells. (**C**)—Quantitative analysis of positive β-galactosidase-stained cells. Data were generated from three independent experiments performed in triplicate and are shown as means ± SD. Statistical analysis was performed using unpaired Student's *t* test: ** $p < 0.01$ and *** $p < 0.001$ [33].

Interestingly, knockdown HepG2 cells were shown to have a decreased expression of the ecto-5′-nucleotidase (*NT5E* or CD73), which regulates the conversion of AMP to adenosine and Pi (inorganic phosphate). Additionally, nonspecific alkaline phosphatase (TNAP), whose activity normally maintains Pi/iPP ratio accelerating the mineralization, was found to be increased in knockdown HepG2 cells. These results suggest that the absence of transporter activity in the hepatic cells decreases the *NT5E* expression and increases the pro-mineralizing TNAP activity, which is also clinically found in patients with arterial calcifications due to deficiency in CD73 (ACDC) [31]. It is clear from the knockdown study that the absence of transporter activity leads to alteration in gene expression, which is required to provide PPi to prevent mineralization. However, in a further study, we also found that ABCC6 plays a crucial role to activate the purinergic pathway, which is important to maintain proper cellular function. In this study, pharmacological inhibition of ABCC6 by probenecid down-regulated the expression of CD73. On the contrary, the expression of CD73 was found increased after the application of adenosine and ATP, which strengthens the idea that ABCC6 not only mediates the efflux of ATP but also regulates the purinergic system [35].

5. Intracellular Consequences Associated with ABCC6 Transporter Activity in HepG2 Cells

It is widely evident that changes in nuclear lamin expression are associated with cellular senescence and age-related diseases [36]. However, cells undergoing aging also adapt a phenomenon to proliferate inappropriately, migrate, and colonize, which are the hallmarks of cancer cells [37]. In both cases, changes in lamin expression traduce in morphological changes in nuclei, which are known as "nuclear atypia". Contrarily, induction of cellular senescence is recognized as an important tumor-suppressive mechanism [38].

In previous studies with *ABCC6* knockdown HepG2 cells, we have found decreased cell growth and lamin A/C expression [33]. Interestingly, in the recent study, pharmacological inhibition of ABCC6 by probenecid or its knockdown in HepG2 cells not only decreased the amount of extracellular ATP content but also decreased the expression of CD73 and lamin A/C proteins. Lamins are the most important structural component of the cytoskeleton and are required to maintain cells in a proper shape. In this context, we examined whether down-regulation of lamin expression affected the actin filaments, required for cytoskeleton rearrangement and cellular movement. In this study, a typical organization of actin filament in filopodia, which is a hallmark of moving cells, was absent in both probenecid-treated and *ABCC6* knockdown HepG2 cells (Figure 6). However, administration of adenosine or ATP restored the normal architecture of filopodia and migration rate (Figure 7). This finding indicates that ABCC6 can be a potential therapeutic target for anti-metastatic treatment and, with coordination of the purinergic system, also regulates the intracellular functions [32].

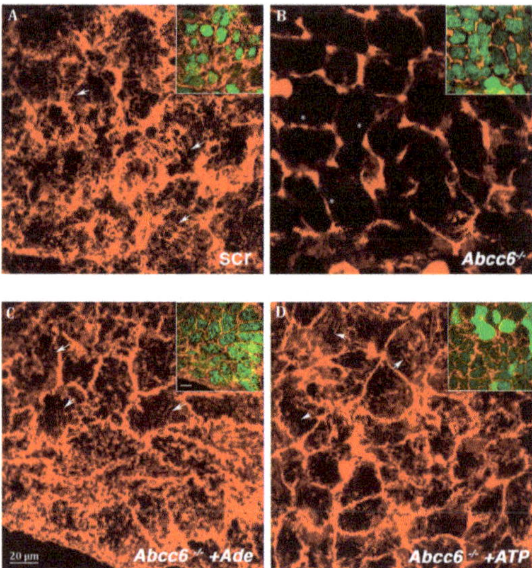

Figure 6. Representative confocal image of (**A**)—scrambled HepG2 cells; (**B**)—Abcc6-shRNA.HepG2 cells; (**C**)—Abcc6-shRNA HepG2 cells treated with 500 µM ATP; (**D**)—Abcc6-shRNA HepG2 cells treated with 100 µM adenosine. F-actin was stained with Texas Red-phalloidin. In the insets superposition of cytoskeleton (red) and EGFP (green) to monitor the infection efficiency. The scale bar in the inserts is 40 µm [32].

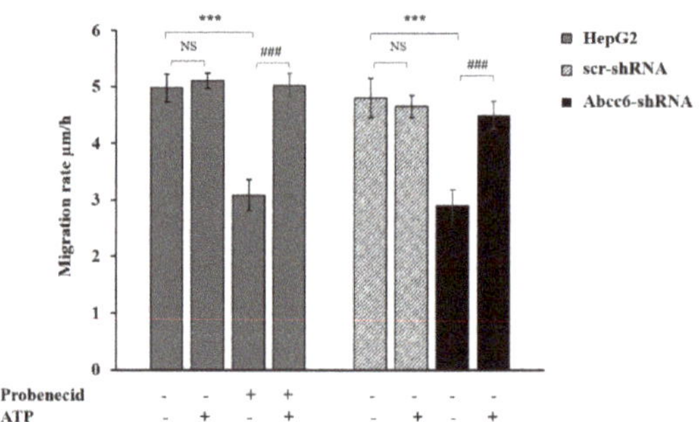

Figure 7. Effect of probenecid and ABCC6 silencing on the migration rate of HepG2 cells. Cells were treated with 250 µM probenecid for 48 h (gray plain bars, Probenecid+). DMSO-treated cells were used as the control (gray plain bars, Probenecid-). A total of 500 µM ATP was added to either the control cells (gray plain bars, Probenecid-, ATP+) or to probenecid-treated cells (gray plain bars, Probenecid+, ATP+). HepG2 cells were transduced with scrambled shRNA (grey-texturized bars, scr-shRNA) or with specific ABCC6-shRNA (black bars, ABCC6-shRNA). A sample of 500 µM ATP was added to either control cells (grey-texturized bars, scr-shRNA, ATP+) or to ABCC6 silenced cells (black bars, ABCC6-shRNA, ATP+). Data are expressed as mean $\pm$ standard error (SE) of three replicates from three independent experiments and were analyzed by one-way ANOVA followed by Dunnett's post hoc: *** $p < 0.001$ probenecid-treated cells vs. control cells in the absence of ATP and ABCC6-shRNA vs. scr-shRNA in the absence of ATP; ### $p < 0.001$ probenecid + ATP-treated cells vs. probenecid-treated cells and ABCC6-shRNA cells + ATP vs. ABCC6-shRNA cells without ATP. NS, not significant [32].

6. Involvement of ABCC6/MRP6 in Drug Resistance

It is well known that ABC transporters play an important role to maintain cellular physiology by transporting different substrates. On the other hand, despite the pathological involvement in certain crucial diseases, these transporters are also involved in drug resistance [39]. Similarly, ABCC family consisted of 13 members, which are involved in the transportation of different substrates. Nine of them are also implicated in the resistance to a variety of chemotherapeutic agents. Martin G. et al., in a study on Chinese hamster ovarian cancer cells (CHO), demonstrated that ABCC6 is not only able to mediate the efflux of glutathione conjugate but is also involved in resistance to some natural anti-cancer drugs, like doxorubicin, etoposide, actinomycin D, daunorubicin, and cisplatin [27]. Tyrosine Kinase Inhibitors (TKIs), like nilotinib and dasatinib, are effective treatments available for Chronic Myeloid Leukaemia (CML) and found to interact with ABCB1 and ABCG2. However, a recent study suggested that nilotinib and dasatinib might be substrates for ABCC6, whose over-expression is responsible for resistance of both TKIs [40].

To investigate the possible involvement of different transporters in drug resistance, we investigated the mRNA expression of ABCC6 and other ABC transporters mainly involved in drug resistance, such as ABCB1, ABCC1, and ABCG2, in the bone marrow samples obtained from acute myeloid leukemia (AML) patients. The samples obtained after diagnosis revealed no difference in the expression level of ABCC6 and ABCB1 between healthy control and AML patients. The expression of ABCC1 from AML patients was higher compared to the healthy individual. Moreover, the expression of ABCG2 was always found down-regulated in AML patients compared to controls. We also observed the age- and gender-associated changes in the expression level of the genes. Interestingly, ABCB1, ABCC1, and ABCC6 were less expressed in older patients compared to younger ones, whereas ABCC6 and ABCC1 were highly expressed in female patients. Additionally, only ABCG2 expression was found higher after chemotherapy, while no variation was found in ABCC1, ABCB1, and ABCC6. However, the expression of ABCC6 and ABCB1 was found to be up-regulated after the treatment with Trichostatin A (an inhibitor of histone deacetylase) and 5-Aza-2′deoxycytidine (an inhibitor of DNA methyltransferase) in AML cell line HL-60. These data reveal that changes in expression of ABCG2 before and after the treatment can be related to disease or as a therapeutic marker. On the other hand, the expression of ABCC6 and ABCB1 is transcriptionally or epigenetically controlled [41].

In addition, a study conducted by Jeon H-Metal [42] suggested that an inhibitor of differentiation 4 (ID4) increases the SOX2-mediated expression of ABCC6 and ABCC3 in glioma stem cells (GSC), which have the potential to initiate a brain tumor, show resistance to chemotherapy, and are responsible for higher recurrence rates of Glioblastoma multiforme (GBM) [43].

Drug resistance is the biggest problem for cancer therapy, for which many molecules have been synthesized and tested, but MDR in cancer therapy still persists. In order to identify the promising molecules to inhibit MRP6 and mitigate marginal drug resistance, we tested 8-(4-chlorophenyl)-5-methyl-8-[(2Z)-pent-2-en-1-yloxy]-8H-[1,2,4] oxadiazolo [3,4-c][1,4] thiazin-3, also known as 2C and structurally similar to diltiazem. Surprisingly, the efflux of doxorubicin was reduced in 2C-treated cells. Moreover, cell esterase activity and H3 histone acetylation were reduced in 2C-treated cells, which suggests that 2C is not only able to mitigate drug resistance but also able to inhibit nucleophilic substitution reactions [44].

These findings confirm that ABCC6 is not only involved in the progression of the genetic disease PXE but is also involved in resistance to many anti-cancer agents [45,46].

7. Conclusions

Different hypotheses were given to elucidate the physiological substrates for the ABCC6 transporter and its pathological involvement in PXE. However, mechanistic details for *ABCC6* mutations leading to ectopic mineralization were yet to be resolved [6,47,48]. In our studies, we not only investigated the structural components of ABCC6 transporter but also lighten the mechanisms, which might be involved in ectopic mineralization. In our experiments, knockdown of *ABCC6* or its inhibition by probenecid decreased the extracellular ATP concentration and altered the expression of *NT5E* and *TNAP*. By these findings, we can propose that the lack of ABCC6 transport activity could lead to a pro-mineralizing state through alteration of extracellular purine metabolites, which ultimately affect TNAP enzymatic activity.

Moreover, in both knockdown and probenecid-treated HepG2 cells, the expression of lamin was found decreased, which suggests that reduced ABCC6 transport activity also leads to cell senescence in PXE and can be beneficial to prevent cancer progression. Interestingly, in both the knockdown and probenecid-treated HepG2 cells, we found a decrease in migration rate, which is restored after ATP administration.

Collectively, these findings suggest that the ABCC6 transporter is not only required to maintain some important circulatory factors like PPi, but is also required to maintain proper functioning of other factors, which are involved in the conversion of extracellular nucleotides, and to feed purine pool to maintain homeostasis between Pi and PPi ratio. Moreover, our studies strengthen the idea that the ABCC6 transporter is not only involved in PXE pathophysiology, but can also be considered a target for anti-cancer therapy.

Author Contributions: Conceptualization, F.B. and A.O.; data curation, A.O.; writing—original draft preparation, V.A. and P.K.; writing—review and editing, P.K. and M.A.C.M.; supervision, F.B.; funding acquisition, F.B. and A.O. All authors have read and agreed to the published version of the manuscript.

Funding: This research was funded by INBIOMED PROJECT (MIUR, ARS01_01081).

Conflicts of Interest: The authors declare no conflict of interest.

References

1. Moitra, K.; Garcia, S.; Jaldin, M.; Etoundi, C.; Cooper, D.; Roland, A.; Dixon, P.; Reyes, S.; Turan, S.; Terry, S.; et al. ABCC6 and pseudoxanthoma elasticum: The face of a rare disease from genetics to advocacy. *Int. J. Mol. Sci.* **2017**, *18*, 1488. [CrossRef] [PubMed]
2. Jansen, R.S.; Küçükosmanoglu, A.; De Haas, M.; Sapthu, S.; Otero, J.A.; Hegman, I.E.M.; Bergen, A.A.B.; Gorgels, T.G.M.F.; Borst, P.; Van De Wetering, K. ABCC6 prevents ectopic mineralization seen in pseudoxanthoma elasticum by inducing cellular nucleotide release. *Proc. Natl. Acad. Sci. USA* **2013**, *110*, 20206–20211. [CrossRef]
3. Luo, H.; Li, Q.; Cao, Y.; Uitto, J. Therapeutics Development for Pseudoxanthoma Elasticum and Related Ectopic Mineralization Disorders: Update 2020. *J. Clin. Med.* **2021**, *10*, 114. [CrossRef] [PubMed]
4. De Vilder, E.Y.; Hosen, M.J.; Vanakker, O.M. The ABCC6 transporter as a paradigm for networking from an orphan disease to complex disorders. *Biomed. Res. Int.* **2015**, *2015*, 1–18. [CrossRef]
5. Lau, W.L.; Liu, S.; Vaziri, N.D. Chronic kidney disease results in deficiency of ABCC6, the novel inhibitor of vascular calcification. *Am. J. Nephrol.* **2014**, *40*, 51–55. [CrossRef]
6. Jiang, Q.; Endo, M.; Dibra, F.; Wang, K.; Uitto, J. Pseudoxanthoma elasticum is a metabolic disease. *J. Investig. Dermatol.* **2009**, *129*, 348–354. [CrossRef]
7. Li, Q.; Jiang, Q.; Pfendner, E.; Váradi, A.; Uitto, J. Pseudoxanthoma elasticum: Clinical phenotypes, molecular genetics and putative pathomechanisms. *Exp. Dermatol.* **2009**, *18*, 1–11. [CrossRef]
8. Klement, J.F.; Matsuzaki, Y.; Jiang, Q.-J.; Terlizzi, J.; Choi, H.Y.; Fujimoto, N.; Li, K.; Pulkkinen, L.; Birk, D.E.; Sundberg, J.P.; et al. Targeted ablation of the abcc6 gene results in ectopic mineralization of connective tissues. *Mol. Cell. Biol.* **2005**, *25*, 8299–8310. [CrossRef]
9. Xiong, J.; Feng, J.; Yuan, D.; Zhou, J.; Miao, W. Tracing the structural evolution of eukaryotic ATP binding cassette transporter superfamily. *Sci. Rep.* **2015**, *5*, 16724. [CrossRef] [PubMed]
10. Liu, X.; Pan, G. *Drug Transporters in Drug Disposition, Effects and Toxicity*; Springer: Berlin/Heidelberg, Germany, 2019. [CrossRef]
11. Favre, G.; Laurain, A.; Aranyi, T.; Szeri, F.; Fulop, K.; Le Saux, O.; Duranton, C.; Kauffenstein, G.; Martin, L.; Lefthériotis, G. The ABCC6 transporter: A new player in biomineralization. *Int. J. Mol. Sci.* **2017**, *18*, 1941. [CrossRef]
12. Ostuni, A.; Miglionico, R.; Castiglione Morelli, M.A.; Bisaccia, F. Study of the nucleotide-binding domain 1 of the human transporter protein MRP6. *Protein Pept. Lett.* **2010**, *17*, 1553–1558. [CrossRef]
13. Ostuni, A.; Castiglione Morelli, M.A.; Cuviello, F.; Bavoso, A.; Bisaccia, F. Structural characterization of the L0 cytoplasmic loop of human multidrug resistance protein 6 (MRP6). *Biochim. Biophys. Acta Biomembr.* **2019**, *1861*, 380–386. [CrossRef]
14. Thomas, C.; Tampé, R. Multifaceted structures and mechanisms of ABC transport systems in health and disease. *Curr. Opin. Struct. Biol.* **2018**, *51*, 116–128. [CrossRef] [PubMed]

15. Ran, Y.; Thibodeau, P.H. Stabilization of nucleotide binding domain dimers rescues ABCC6 mutants associated with pseudoxanthoma elasticum. *J. Biol. Chem.* **2017**, *292*, 1559–1572. [CrossRef] [PubMed]
16. Smith, P.C.; Karpowich, N.; Millen, L.; Moody, J.E.; Rosen, J.; Thomas, P.J.; Hunt, J.F. ATP binding to the motor domain from an ABC transporter drives formation of a nucleotide sandwich dimer. *Mol. Cell* **2002**, *10*, 139–149. [CrossRef]
17. Ran, Y.; Zheng, A.; Thibodeau, P.H. Structural analysis reveals pathomechanisms associated with pseudoxanthoma elasticum–causing mutations in the ABCC6 transporter. *J. Biol. Chem.* **2018**, *293*, 15855–15866. [CrossRef]
18. Dhasmana, D.; Singh, A.; Shukla, R.; Tripathi, T.; Garg, N. Targeting nucleotide binding domain of multidrug resistance-associated protein-1 (MRP1) for the reversal of multi drug resistance in cancer. *Sci. Rep.* **2018**, *8*, 1–11. [CrossRef] [PubMed]
19. Johnson, Z.L.; Chen, J. Structural basis of substrate recognition by the multidrug resistance protein MRP1. *Cell* **2017**, *168*, 1075–1085. [CrossRef] [PubMed]
20. Miglionico, R.; Gerbino, A.; Ostuni, A.; Armentano, M.F.; Monné, M.; Carmosino, M.; Bisaccia, F. New insights into the roles of the N-terminal region of the ABCC6 transporter. *J. Bioenerg. Biomembr.* **2016**, *48*, 259–267. [CrossRef] [PubMed]
21. Ostuni, A.; Miglionico, R.; Bisaccia, F.; Castiglione Morelli, M.A. Biochemical characterization and NMR study of the region E748-A785 of the human protein MRP6/ABCC6. *Protein Pept. Lett.* **2010**, *17*, 861–866. [CrossRef] [PubMed]
22. Le Saux, O.; Beck, K.; Sachsinger, C.; Silvestri, C.; Treiber, C.; Göring, H.H.H.; Johnson, E.W.; De Paepe, A.; Pope, F.M.; Pasquali-Ronchetti, I.; et al. A spectrum of ABCC6 mutations is responsible for pseudoxanthoma elasticum. *Am. J. Hum. Genet.* **2001**, *69*, 749–764. [CrossRef] [PubMed]
23. Ostuni, A.; Miglionico, R.; Monné, M.; Castiglione Morelli, M.A.; Bisaccia, F. The nucleotide-binding domain 2 of the human transporter protein MRP6. *J. Bioenerg. Biomembr.* **2011**, *43*, 465. [CrossRef]
24. Goda, K.; Dönmez-Cakil, Y.; Tarapcsák, S.; Szalóki, G.; Szöllősi, D.; Parveen, Z.; Türk, D.; Szakács, G.; Chiba, P.; Stockner, T. Human ABCB1 with an ABCB11-like degenerate nucleotide binding site maintains transport activity by avoiding nucleotide occlusion. *PLoS Genet.* **2020**, *16*, e1009016. [CrossRef]
25. Cuviello, F.; Tellgren-Roth, Å.; Lara, P.; Selin, F.R.; Monné, M.; Bisaccia, F.; Nilsson, I.; Ostuni, A. Membrane insertion and topology of the amino-terminal domain TMD0 of multidrug-resistance associated protein 6 (MRP6). *FEBS Lett.* **2015**, *589*, 3921–3928. [CrossRef]
26. Iliás, A.; Urbán, Z.; Seidl, T.L.; Le Saux, O.; Sinkó, E.; Boyd, C.D.; Sarkadi, B.; Váradi, A. Loss of ATP-dependent transport activity in pseudoxanthoma elasticum-associated mutants of human ABCC6 (MRP6). *J. Biol. Chem.* **2002**, *277*, 16860–16867. [CrossRef]
27. Belinsky, M.G.; Chen, Z.-S.; Shchaveleva, I.; Zeng, H.; Kruh, G.D. Characterization of the drug resistance and transport properties of multidrug resistance protein 6 (MRP6, ABCC6). *Cancer Res.* **2002**, *62*, 6172–6177.
28. Borst, P.; van de Wetering, K.; Schlingemann, R. Does the absence of ABCC6 (multidrug resistance protein 6) in patients with Pseudoxanthoma elasticum prevent the liver from providing sufficient vitamin K to the periphery? *Cell Cycle* **2008**, *7*, 1575–1579. [CrossRef] [PubMed]
29. Gorgels, T.G.M.F.; Waarsing, J.H.; Herfs, M.; Versteeg, D.; Schoensiegel, F.; Sato, T.; Schlingemann, R.O.; Ivandic, B.; Vermeer, C.; Schurgers, L.J.; et al. Vitamin K supplementation increases vitamin K tissue levels but fails to counteract ectopic calcification in a mouse model for pseudoxanthoma elasticum. *J. Mol. Med.* **2011**, *89*, 1125. [CrossRef] [PubMed]
30. Fülöp, K.; Jiang, Q.; Wetering, K.V.; Pomozi, V.; Szabó, P.T.; Arányi, T.; Sarkadi, B.; Borst, P.; Uitto, J.; Váradi, A. ABCC6 does not transport vitamin K3-glutathione conjugate from the liver: Relevance to pathomechanisms of pseudoxanthoma elasticum. *Biochem. Biophys. Res. Commun.* **2011**, *415*, 468–471. [CrossRef]
31. Miglionico, R.; Armentano, M.F.; Carmosino, M.; Salvia, A.M.; Cuviello, F.; Bisaccia, F.; Ostuni, A. Dysregulation of gene expression in ABCC6 knockdown HepG2 cells. *Cell. Mol. Biol. Lett.* **2014**, *19*, 517–526. [CrossRef]
32. Ostuni, A.; Carmosino, M.; Miglionico, R.; Abruzzese, V.; Martinelli, F.; Russo, D.; Laurenzana, I.; Petillo, A.; Bisaccia, F. Inhibition of ABCC6 Transporter Modifies Cytoskeleton and Reduces Motility of HepG2 Cells via Purinergic Pathway. *Cells* **2020**, *9*, 1410. [CrossRef]
33. Miglionico, R.; Ostuni, A.; Armentano, M.F.; Milella, L.; Crescenzi, E.; Carmosino, M.; Bisaccia, F. ABCC6 knockdown in HepG2 cells induces a senescent-like cell phenotype. *Cell. Mol. Biol. Lett.* **2017**, *22*, 7. [CrossRef] [PubMed]
34. Pasquali-Ronchetti, I.; Garcia-Fernandez, M.I.; Boraldi, F.; Quaglino, D.; Gheduzzi, D.; Paolinelli, C.D.V.; Tiozzo, R.; Bergamini, S.; Ceccarelli, D.; Muscatello, U. Oxidative stress in fibroblasts from patients with pseudoxanthoma elasticum: Possible role in the pathogenesis of clinical manifestations. *J. Pathol. J. Pathol. Soc. Great Br. Irel.* **2006**, *208*, 54–61. [CrossRef]
35. Martinelli, F.; Cuviello, F.; Pace, M.C.; Armentano, M.F.; Miglionico, R.; Ostuni, A.; Bisaccia, F. Extracellular ATP regulates CD73 and ABCC6 expression in HepG2 Cells. *Front. Mol. Biosci.* **2018**, *5*, 75. [CrossRef]
36. Kristiani, L.; Kim, M.; Kim, Y. Role of the Nuclear Lamina in Age-Associated Nuclear Reorganization and Inflammation. *Cells* **2020**, *9*, 718. [CrossRef]
37. Honoki, K.; Fujii, H.; Tsukamoto, S.; Kishi, S.; Tsujiuchi, T.; Tanaka, Y. Crossroads of hallmarks in aging and cancer: Anti-aging and anti-cancer target pathways can be shared. *Tre Can Res* **2016**, *11*, 39–59.
38. Irianto, J.; Pfeifer, C.R.; Ivanovska, I.L.; Swift, J.; Discher, D.E. Nuclear lamins in cancer. *Cell. Mol. Bioeng.* **2016**, *9*, 258–267. [CrossRef]
39. Dean, M.; Allikmets, R. Complete characterization of the human ABC gene family. *J. Bioenerg. Biomembr.* **2001**, *33*, 475–479. [CrossRef]

Ectopic calcification can lead to clinical symptoms when it occurs in cardiovascular tissues and has been the object of intense research focus. Many host, environmental, and genetic factors contributing to this calcification have been identified [3], but there are still gaps in our understanding [4]. In recent years, the identification of mutations in the ATP-binding cassette (ABC) transporter ABCC6 [5–7] and the characterization of its function [8] has provided new molecular insight into the regulation of ectopic calcification inhibition in relation to pyrophosphate (PPi). ABCC6 mediates the cellular efflux of nucleotides, notably ATP, which is rapidly converted into PPi and adenosine at the cellular surface by the ectonucleotidases NPP1 (encoded by *ENPP1*) and CD73 (encoded by *NT5E*) [8–11]. Both PPi and indirectly adenosine are major inhibitors of calcification. ABCC6 deficiencies underlie the calcification disorders PXE (MIM#264800) and a cardiac calcification phenotype (DCC) described in mice [12–14]. GACI (MIM#208000) is primarily linked to NPP1, the key enzyme that generates PPi [15]. However, some cases of GACI (MIM#614473) are caused by *ABCC6* mutations, while some PXE patients carry disease-causing variants only in *ENPP1* [16,17]. CALJA (MIM#211800) is due to mutations in CD73 (*NT5E*) [10]. PXE, together with GACI and CALJA form a spectrum of diseases with overlapping calcification processes but with distinct clinical features. If PXE, GACI, and DCC result from a deficit in PPi production [18,19], CALJA is caused by increased PPi degradation [10,11] (Figure 1). The explanation of these phenotypes places ABCC6 as an upstream modulator of an extracellular purinergic pathway that, among other things, inhibits mineralization by regulating the Pi/PPi ratio in connective tissues (Figure 1). Although it has been recently claimed that ABCC6 acts downstream of NPP1 [20], this hypothesis is highly controversial in the field [21] and contradicts most studies [8,9,11,18]. This pathway (Figure 1) and the molecular cascade leading to PPi generation and TNAP inhibition offers many opportunities for therapeutic interventions in the case of PXE and also GACI.

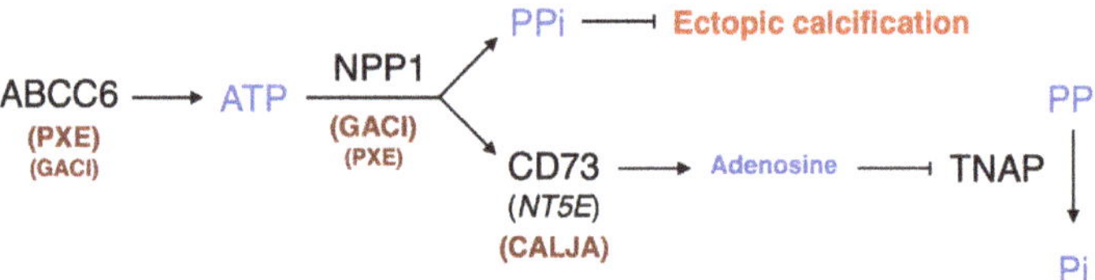

Figure 1. The ABCC6 pathway influences calcification and extracellular purinergic metabolism. ABCC6 facilitates the cellular efflux of ATP from liver and other tissues/cells, which is quickly converted to pyrophosphate (PPi), a potent inhibitor of mineralization. Decreased plasma PPi levels cause calcification in PXE and GACI. CD73 activity leads to adenosine production, which affects many biological activities including the inhibition of TNAP synthesis. TNAP degrades PPi into inorganic phosphate (Pi), an activator of calcification, which leads to vascular calcification in CALJA patients.

At present, only palliative treatments to alleviate some symptoms exist for both PXE and GACI [22–24]. Extensive studies on the mechanism behind calcification has resulted in several novel approaches to treating PXE and GACI. The therapeutic solutions envisioned and tested in animals include strategies focusing on two distinct aspects of the ABCC6 pathway: (1) the correction/replacement/inhibition of dysfunctional genes/proteins involved in the calcification pathway [9,18], and (2) supplementation therapies with exogenous compounds. Drugs targeting ABCC6, NPP1 and TNAP aim at correcting parts of the ABCC6 pathway, whereas exogenous compounds such as magnesium, vitamin K, bisphosphonates (PPi analogs), PPi and recently phytic acid are intended to directly inhibit calcification.

In this article, we provide an overview of ABCC6 as a key regulator of ectopic calcification in PXE (and GACI). We will discuss possible treatment options that have been explored in recent years and speculate as to what future treatments might be. Because ABCC6 and NPP1 are functionally related proteins interchangeably causing PXE and

GACI [17], we address herein some aspects of GACI and NPP1, which can be relevant for therapeutic interventions.

2. The Pathologies Associated with ABCC6 and ENPP1 Deficiencies

2.1. Pseudoxanthoma Elasticum

Pseudoxanthoma elasticum (PXE) is a rare disorder with progressive ocular, vascular and skin abnormalities that result from the accumulation of morphologically abnormal calcified elastic fibers [25]. The skin manifestations are the most prevailing characteristic of PXE, but the ocular and cardiovascular symptoms are responsible for the morbidity of the disease.

The first description of the clinical signs of PXE appeared in the literature more than a century ago [26–28]. Is it Darier who proposed the term "pseudoxanthome élastique", which remains largely used to this day as pseudoxanthoma elasticum [29].

2.1.1. Dermal Manifestations

Skin lesions are generally the first signs of PXE to be observed during childhood or adolescence and often progress slowly and unpredictably. Therefore, a dermatologist frequently makes the initial diagnosis. The accumulation of abnormal calcified elastic fibers in the mid-dermis produces yellow-hued papules and plaques and laxity with loss of elasticity. These lesions can be seen on the neck, axilla, antecubital fossa, popliteal fossa, groin, and periumbilical area [25,30–32].

2.1.2. Ocular Manifestations

Ocular lesions in PXE are due to the accumulation of abnormal elastic fibers in the Bruch's membrane, resulting in angioid streaks [33]. Angioid streaks are completely asymptomatic and can remain undetected until later in life when retinal haemorrhages occur. The majority of PXE patients will develop ocular changes during their second decade of life. Bilateral angioid streaks are normally seen as linear gray or dark red lines with irregular serrated edges lying beneath normal retinal blood vessels and they represent breaks in the Bruch's membrane. The elastic laminae of the Bruch's membrane is located between two layers of collagen and lies in direct contact with the basement membrane of the retinal pigmented epithelium (RPE) and the capillaries of the choroid. Angioid streak formation is likely the direct result of mineralization of elastic fibers in Bruch's membrane. As a consequence of the loss of structural integrity to the Bruch's membrane, PXE patients progressively develop choroidal neovascularization, which can lead to hemorrhagic detachment of the fovea and retinal scarring. Optic nerve drusen may also be associated with angioid streaks and results in visual field deficits.

It is worth noting that a minor yet important palliative treatment exists for some of the ocular manifestations of PXE [24] and more details are provided below (*Cf.* Section 6.2).

2.1.3. Cardiovascular Manifestations

The common cardiovascular complications of PXE are due to the presence of abnormal calcified elastic fibers in the internal elastic lamina of medium-sized arteries. The broad spectrum of phenotypes includes peripheral arterial disease (PAD), resulting from restrictive vascular calcification in the lower limbs, increased susceptibility to atherosclerosis [34–36], intermittent claudication, and possibly myocardial infarction and hypertension. The fibrous thickening of the endocardium and atrioventricular valves can also result in restrictive cardiomyopathy and/or mitral valve prolapse and atrial septal aneurysm [37]. A cardiac evaluation of French PXE patients revealed sporadic cases of left ventricular hypertrophy, aortic stenosis, and valvulopathies. Overall in this cohort, PXE does not appear to be associated with frequent cardiac complications. However, the development of cardiac hypertrophy in old $Abcc6^{-/-}$ mice suggests that aging PXE patients might be susceptible to late cardiopathy [38].

Extra cardiac downstream effects of arterial mineralization include renovascular hypertension, gastrointestinal bleeding and central nervous system complications (such as stroke and dementia) [30,39].

A comprehensive analysis of the arterial consequences (macro- and micro-arterial beds were investigated) of ABCC6 ablation in mice was published in 2014 [40]. This report revealed scattered arterial calcium depositions as a result of osteochondrogenic transdifferentiation of vascular cells as mice age (as seen via increase *Runx2* expression). Lower elasticity and increased myogenic tone without major changes in agonist-dependent contraction was found in older $Abcc6^{-/-}$ mice suggesting reduced control of peripheral blood flow, which in turn may alter vascular homeostasis which is not unlike the PAD observed in PXE patients.

2.1.4. Renal Manifestations

Renal involvement in PXE has received little attention in the literature until recently. Few cases of kidney stones in PXE patients have been described, whereas classic nephrocalcinosis has only been reported in sporadic cases [41–43]. More recent data suggested that nephrolithiasis was an unrecognized and prevalent feature of PXE [44]. The first in depth analysis of renal manifestations in a cohort of 113 French PXE patients showed a history of kidney stones in 40% of patients, which is much higher than the general population. Computed tomography scans revealed evidence of significant papillary calcifications (Randall's plaques) [45,46].

Remarkably, despite the presence of mineralized elastic fibers in pulmonary tissues, there is no lung phenotype associated with PXE [47,48].

2.2. Generalized Arterial Calcification of Infancy (GACI)

Generalized arterial calcification of infancy (GACI [MIM 208000]) is a very rare autosomal recessive disorder characterized by calcification of arterial elastic fibers and associated fibrotic myointimal proliferation of muscular arteries and arterial stenosis [49]. Severe vascular calcification causes hypertension, myocardial ischemia, and congestive heart failure [17]. The majority of patients die within the first six months of life [50,51]. However, patients treated with bisphosphonates can experience a more favorable outcome [52,53]. In addition to vascular mineralization, neonatal patients also present periarticular soft-tissue calcifications. More mildly affected patients may also develop hypophosphatemic rickets [53–55].

2.3. PXE and GACI Are Different Clinical Manifestations of a Phenotypic Continuum

PXE is primarily caused by ABCC6 deficiency, while GACI patients typically present mutations in the *ENPP1* gene. However, some GACI patients only carry *ABCC6* mutations and typical PXE manifestations can be associated with *ENPP1* mutations. The clinical and molecular genetic characterization of PXE and GACI [16,17] has suggested that ABCC6 and NPP1 are functionally related [56] and give rise to overlapping phenotypes with GACI being a severe and acute form of PXE and vice versa [8,16,57].

Calcification of joints and arteries (CALJA, OMIM#211800) is outside of the scope of this review. However, it is worth mentioning briefly because its molecular etiology is intimately related to that of PXE and GACI. CALJA is due to mutated CD73 (encoded by *NT5E*) [10], which is functionally downstream to NPP1 (Figure 1). The disease is characterized by vascular calcification in the lower limbs and periarticular mineralization. CALJA is caused by enhanced PPi degradation [10,11] resulting from reduced adenosine signaling and abnormal activation of tissue non-specific alkaline phosphatase (TNAP/*ALPL*) [11,20] (Figure 1).

2.4. Thalassemia

For the most part, PXE is a monogenic disease caused by mutations in *ABCC6*. However, other than the cases associated with *ENPP1* mutations [17], PXE manifestations can arise from multifactorial inheritance [58,59], environmental exposure [60–63] or secondary to β-thalassemia and sickle cell anemia [64]. β-thalassemia (MIM 141900) is a monogenic disorder caused by mutations in the β-*globin* gene that leads to the underproduction of β-globin chains. The stoichiometric excess of α-chains unbound to β-globin is unstable and precipitates in red blood cell precursors causing the ineffective erythropoiesis. In the past decade, it has become apparent that a large number of Mediterranean patients affected by β-thalassemia or sickle cell anemia also develop manifestations similar to PXE [64]. β-thalassemia and PXE are distinct genetic disorders, yet, the ectopic mineralization phenotype seen in β-thalassemia and sickle cell patients is clinically and structurally identical to inherited PXE [65–68] and arises independently of *ABCC6* mutations [69]. Based on studies with a mouse model of β-thalassemia ($Hbb^{th3/+}$), it was suggested that the β-thalassemia patients could have a suboptimal endowment of ABCC6 expression in liver (and possibly reduced PPi production), thereby increasing susceptibility to ectopic mineralization in a PXE-like manner. If this is indeed the case, then β-thalassemia and sickle cell patients could benefit from several treatment options developed for the inherited forms of PXE.

3. The Structure and Molecular Function of ABCC6

The ATP-binding cassette (ABC) family represents the largest group of transmembrane proteins. These proteins bind ATP and use its energy to drive the efflux and transport of various molecules across cell membranes. ABC transporters are classified based on the sequence and organization of their ATP-binding domains. The human ATP-binding cassette (ABC) gene family consists of 48 members divided into seven subgroups, A through G. Genetic changes in at least 14 of these genes cause heritable diseases [70,71]. About a third of all of these ABC transporter-related diseases are linked to genes from the single sub-group C, which includes ABCC6 and PXE but also the well-known cystic fibrosis associated with *ABCC7* mutations.

3.1. ABCC6 Structure

ABCC6 consists of 1503 amino acids with an approximate molecular weight of 165 kD without glycosylation. This transporter protein possesses three transmembrane segments with 5, 6, and 6 membrane-spanning regions, respectively, and two typical ABC domains involved in the binding and hydrolyzing of ATP used for conformational changes and transport activity. A 3D model of ABCC6 was successfully generated using X-ray structure of the *S. aureus* Sav1866 export pump [72]. Fülöp et al. used this model of ABCC6 and the distribution of PXE-causing mutations to demonstrate the strict relevance of the transmission interface (ICL-ABC contacts) as well as the ABC-ABC domain contacts for the function of the transporter [73].

3.2. ABCC6 Is an Efflux Pump

Transport studies in vesicles isolated from Sf9 cells expressing the rat *Abcc6* identified the anionic cyclopentapeptide and endothelin receptor antagonist BQ-123 as substrates [74]. ATP binding and hydrolysis by rat ABCC6 has been demonstrated in a yeast expression system [75]. Using a similar approach with the human ABCC6, Ilias et al. demonstrated ATP binding and ATP-dependent active transport of the glutathione conjugates leukotriene C_4 and *N*-ethylmaleimide *S*-glutathione and of the cyclopentapeptide BQ-123 [76]. Probenecid, benzbromarone and indomethacin specifically inhibited transport of *N*-ethylmaleimide *S*-glutathione [77]. Another study confirmed ATP-dependent transport of BQ123 and leukotriene C_4 by human ABCC6 and also identified *S*-(2, 4-dinitrophenyl) glutathione as an in vitro substrate. Analysis of the drug sensitivity of ABCC6-transfected cells revealed low levels of resistance to etoposide, teniposide, doxorubicin, and daunorubicin, indicating that ABCC6 is able to confer low levels of resistance to certain anticancer agents [78]. However, this function appears unlikely to play a role in the pathophysiology of PXE and thus has limited bearing on the possible therapeutic solutions. Note that the actual endogenous substrate(s) and the precise molecular mechanism by which ABCC6 achieves the efflux of ATP and other nucleotides/nucleosides are still unknown.

3.3. Cellular Localization of ABCC6

ABCC6 is firmly associated with the basolateral membrane of hepatocytes in mice, rats, and humans [74,79,80], as well as in the proximal kidney tubules [81,82]. Although a report challenged the well-established cellular localization of ABCC6 [83], it is interesting to note that disease-causing missense *ABCC6* mutations can lead to aberrant cellular localization [80].

3.4. NPP1 Structure

NPP1 is a membrane bound enzyme consisting of 840 amino acids (just under 100 kDa) that belongs to the ecto-nucleotide pyrophosphatase/phosphodiesterase (ENPP) family. NPP1 is a type II transmembrane glycoprotein comprised of two identical disulfide-bonded subunits. The polypeptide has a short N-terminal cytoplasmic domain followed by a transmembrane region. The extracellular structure consists of two somatomedin B (SMB)-like domains (SMB1 and SMB2), a phosphodiesterase domain and a nuclease-like domain [84]. The SMB-like domains are disordered and do not interact with the catalytic domain. The mapping of disease-causing mutations highlights the functional significance of the interaction between the catalytic and nuclease-like domains. It is the SMB2 domain of NPP1 that interacts with the insulin receptor and carries out physiological functions other than the regulation of mineralization [85].

4. Mutations in ABCC6 and ENPP1

4.1. ABCC6

Over 400 mutations (https://www.ncbi.nlm.nih.gov/clinvar, accessed on 3 February 2021) have now been identified in *ABCC6*. Most of these are single nucleotide changes leading to a panel of missense, nonsense, splice site, and frameshift mutations, in addition to small and large insertions/deletions. Missense variants are found in areas critical to the stability and/or the function of the protein. Based on a small but representative number of ABCC6 mutants, two possible molecular consequences to *ABCC6* mutations were described: (1) transport deficiency linked to the failure to hydrolyze ATP and (2) abnormal protein folding leading to intracellular retention and/or reduced trafficking. The latter has been the focus of studies designed to rescue trafficking by re-purposing the drug 4-phenylbutyrate (*Cf.* Section 6.3.2). Transport deficiency and abnormal trafficking are likely the reasons behind the loss of physiological function and provide a reasonable explanation for the lack of phenotype–genotype correlation in PXE [86–88].

About 40% of all *ABCC6* gene mutations in patients with PXE consist of premature termination codon mutations and account for 25% of all mutations in PXE patients [88]. However, two mutations are recurrent in the Caucasian population, p.R1141X and g.del23-29, which account for up to ~45% of all mutant alleles [87,88].

These ABCC6 mutations have consequences for the entire calcification pathway, as evidenced by the downregulation of ENPP1 and NT5E gene expression in the absence of ABCC6 [11,18], the reduced plasma levels of PPi [8], and the activation of TNAP as a consequence of lowered adenosine production [11,20].

4.2. ENPP1

On the ClinVar database, there are nearly 100 ENPP1 pathological variants reported. These variants present a similar profile to that found in ABCC6, with a large majority being single nucleotide substitutions leading to missense, nonsense, splice site, and frameshift mutations, plus some deletions and duplications. The main consequence of these disease-causing variants, which affects all critical parts of the protein, is the inactivation of the enzyme resulting in the failure to convert ATP (or ADP) into AMP and PPi. As NPP1 is the only enzyme that generates PPi, plasma levels drop to near zero in its absence [89].

5. Animal Models

5.1. The PXE Mice

Two lines of *Abcc6* knockout ($Abcc6^{-/-}$) mice were generated independently [90,91]. In these $Abcc6^{-/-}$ mice, *Abcc6* exons 15 to 18 were deleted. Both mouse lines lack the ABCC6 protein and develop identical calcification phenotypes that are consistent with the human PXE condition. The mice breed normally and have a life span of about 25+ months. These animals display spontaneous mineralization in vascular, ocular, and renal tissues, as well as in testes and vibrissae in the whiskers. The earliest evidence of mineralization occurs at 5–6 weeks of age in the capsules of vibrissae. The calcification in vibrissae is progressive and quantifiable and thus serves as a reliable marker of disease progression [92]. Heterozygous $Abcc6^{+/-}$ mice do not develop any calcification. Similar to their human counterpart, $Abcc6^{-/-}$ mice have lowered plasma PPi [9,18] altered lipoproteins [93,94], develop Randall's plaques, and have low urinary PPi excretion [45,46]. These animals have been invaluable for understanding the pathobiology of PXE and to test crucial pathophysiological hypotheses [95,96] that in vitro approaches could partially address [97]. These mice were the ultimate tool that allowed the development of therapeutic solutions for PXE patients [18,19,92,98,99].

5.2. The Murine DCC Phenotype and Other PXE Mouse Models

In the past years, two groups of investigators have found that ABCC6 deficiency causes an acute and inducible dystrophic cardiac calcification phenotype (DCC) affecting several strains of inbred mice, including C3H/HeJ, 129S1/SvJ, and DBA/2J [12,14,100]. DCC is an autosomal recessive trait that was described decades ago [101,102]. DCC can either occur spontaneously over the long-term, be initiated by a specific dietary regimen, or be triggered into an acute phenotype by direct injury [103,104] or ischemia [13]. Of note, arteries (most notably the aorta) as well as skeletal muscles, are also susceptible to dystrophic calcification [100,105] (Le Saux et al., unpublished results). DCC is caused by a single *Abcc6* gene mutation in C3H/HeJ, 129S1/SvJ, and DBA/2J mice [12], while it is absent in C57BL/6J mice that are DCC-resistant. Note that the $Abcc6^{-/-}$ mice that have been used thus far were all backcrossed into C57BL/6J. This is important as there are 3 other minor loci affecting the penetrance and the expression of DCC mapping to chromosomes 4, 12, and 14 [104].

Remarkably, DCC-susceptible C3H/HeJ mice develop an attenuated version of the murine PXE phenotype as compared to the $Abcc6^{-/-}$ animals, while the DBA/2J mice present little or no manifestations [106]. It is interesting to note that the murine PXE manifestations reported in KK/H1J mice are remarkably severe [107]. These mice show systemic age-dependent ectopic mineralization, hyperplasia, and fibro-osseous lesions [108]. Calcification mostly affects vibrissae (as for $Abcc6^{-/-}$ animals), but also the heart, the lung and many other organs to a lesser degree. Pancreatic islet hyperplasia was observed as well as fibro-osseous lesions in several bones. More interesting is that these strains of mice carry the exact same *Abcc6* gene mutation and have similar plasma PPi levels (Figure 2), which clearly underlie the possible role of other factors such as the environment [19] and/or modifier genes in the development of ectopic calcification [109,110].

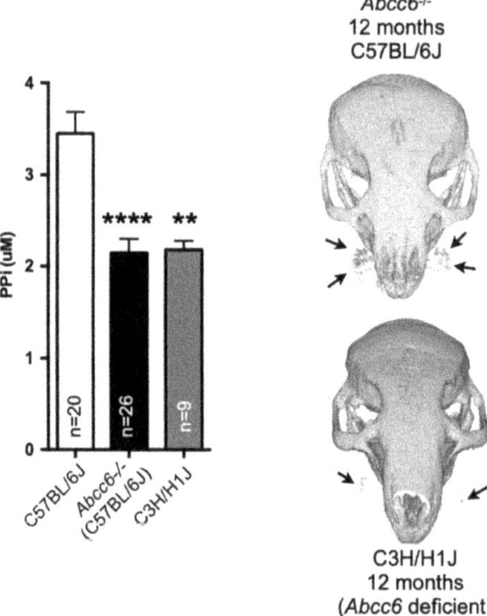

Figure 2. Plasma pyrophosphate levels do not correlate with the calcification phenotype in mice. Plasma pyrophosphate levels in $Abcc6^{-/-}$ mice (C57BL/6J background), and in C3H/H1J mice with a naturally occurring *Abcc6* mutation are virtually identical and significantly lower than wild type C57BL/6J mice. However, at 12 months of age, $Abcc6^{-/-}$ mice present a much more pronounced vibrissae calcification as shown on this µCT scan rendering (right, arrows). Plasma PPi results are shown as means +SEM. *p*-values were determined by Student's *t*-test. ** $p < 0.01$, **** $p < 0.0001$. Derived from [111].

5.3. The PXE Rat

PXE mouse models have been excellent research tools but their small physical size can be an impediment to using certain investigative tools and techniques such as perfusion or organ transplantation. In a recent report, Li et al. described the generation of $Abcc6^{-/-}$ rats using zinc finger nuclease (ZFN) technology [112]. These animals displayed a calcification phenotype similar to that of $Abcc6^{-/-}$ mice, with mineralization in the skin (vibrissae), eyes (Bruch's membrane), kidneys, and arterial beds. Plasma PPi levels were depleted by more than 60%, which is consistent with $Abcc6^{-/-}$ mice and PXE patients. Using in situ perfusion experiments, the authors observed a near abolition of PPi levels in liver (and kidney) perfusates, which has also been described in mice [9,18].

5.4. The "PXE" Zebrafish

Rodent models present remarkable similarity to human diseases and have proven extraordinarily useful. However, like all models, these systems have limitations, i.e., a relatively long life span and the associated cost of development and maintenance. Zebrafish are models that address some of these limitations and provide higher "n" numbers for observation and quantification. Zebrafish carry two orthologs of human *ABCC6* referred to as *abcc6a* and *abcc6b*. These genes encode proteins of 1507 and 1502 amino acids, respectively, which are similar to the human polypeptide at 1503 amino acids. *Abcc6a* and *-b* have ~50% identity and ~70% similarity with the human protein, respectively. The first attempt to generate a zebrafish model used morpholino technology, which is now progressively replaced by genome-targeting methods such as CRISPR/Cas9 because it has often been difficult to discriminate between specific and non-specific effects. Li et al., 2010 reported the expression profiles of both *abcc6a* and *abcc6b* and the lack of phenotype associated with *abcc6b* knockdown [113]. Injection of *abcc6a*-specific morpholinos induced cardiac and developmental malformations followed by death of the animals. This phenotype could be rescued with injection of mouse *Abcc6* cDNA, suggesting some evolutionary conservation in the physiological function of ABCC6. More recently, van Gils and co-workers described a CRISPR/Cas9 *abcc6a* knockout zebrafish [114]. This zebrafish model showed hypermineralization of the spine starting in the embryonic stage and progressing into adulthood with scoliosis, vertebral and rib mineralization and loss of normal bone architecture. These manifestations are not consistent with the human PXE condition and there was no obvious mineralization in the skin or in the eyes. Thus, the *abcc6a* zebrafish model develops an abnormal calcification phenotype but one that does not recapitulate the human PXE. A new zebrafish model was recently created using the transcription activator-like effector nuclease (TALEN) technique [115]. These animals displayed similar skeletal changes as well as calcification in the ocular Bruch's membrane and a range of other previously unreported manifestations, such as cardiac fibrosis. Remarkably, this phenotype could be attenuated by vitamin K treatment, similar to what had been previously reported in 2015 with another zebrafish model [116].

Overall, zebrafish have shown some utility, notably in the characterization of the 4-phenylbutyrate treatment option for PXE [117], but their evolutionary distance and phenotypic divergence limit their usefulness for understanding the pathophysiology of PXE. However, they could be excellent preliminary testing tools for anti-calcification treatments.

5.5. The GACI Models

5.5.1. GACI Mice

The first evidence of a link between defective *Enpp1* expression and abnormal mineralization was first demonstrated in 'tiptoe walking' (*ttw/ttw*) mice [118]. The various *Enpp1* deficient mouse models that exist present similar characteristics. The animals with a naturally occurring mutation (*ttw/ttw*) carry a homozygous nonsense variant in the *Enpp1* coding sequence [118], whereas another mouse model harbors a missense mutation [119]. The phenotype of these mice includes progressive ankylosing intervertebral and peripheral joint hyperostosis, as well as spontaneous arterial and articular cartilage calcification and increased vertebral cortical bone formation. Low plasma PPi and PXE-like calcification are fundamental characteristics of these mice [119,120]. These animals also have abnormal ossification, cardiovascular pathologies and altered glucose homeostasis and diabetes [121,122].

5.5.2. The GACI Zebrafish

Similar to PXE and ABCC6, a zebrafish model of GACI was generated targeting the *Enpp1* gene [123]. Unlike the *abcc6a* KO counterparts, these mutant zebrafish developed ectopic calcifications similar to that seen in GACI and PXE models in a variety of soft tissues that included the skin, cartilage, heart, intracranial space, and the notochord sheet. Treatment with the bisphosphonate etidronate (a PPi analog) rescues some aspects of the phenotype. The report also noted that the expression of *Enpp1* in blood vessels or the floor plate of mutant embryos was able to rescue the notochord mineralization phenotype.

6. Rescue and Therapeutic Solutions

In recent years, basic science has elucidated the molecular etiology underlying PXE [6,8,13,76,87,96,124,125], which has led to rapid advances in the development of many potential therapeutic options for PXE and GACI and the recent completion of three pilot clinical trials [18–20,80,99,117,126–130]. Among the several approaches (summarized in Table 1) that have been considered are enzyme replacement therapy, rescue drugs, and enzyme inhibitors, as well as exogenous compounds such as PPi. All target various steps in the ABCC6 pathway (Figure 3) with the goal of either slowing or reversing the progression of the disease. Figure 3 outlines many of these approaches and indicates the targeted parts of the molecular pathway. We discern two main categories of therapeutic solutions that have undergone development and testing, thus far. Most of these therapeutic solutions have been tested in pre-clinical animal models (mice and zebrafish) and a few phase I/II clinical trials. Beyond what is described in these paragraphs, note that there are larger scale trials in preparation and further drug development that are taking place based on past lessons and data. We cannot describe these ongoing efforts here for reasons of confidentiality or simply for lack of detailed information.

Table 1. Preclinical studies and early clinical trials for PXE (and GACI).

Treatment/Therapy	Rationale/Target	Diagnosis	References
Correction, replacement, or inhibition of dysfunctional genes/proteins			
PTC-124 (Ataluren or Translarna)	Allows read-through of PTC codons, targets nonsense mutations	PXE	[131]
4-Phenylbutyrate (4-PBA)	Corrects missense mutations allowing for correct cellular localization	PXE	[80,99,117]
Rh-NPP1-Fc	Replacement for ENPP1	GACI	[89,132,133]
SBI-425 (TNAP inhibitor)	Inhibits the enzymatic activity of TNAP	PXE	[20,127]
Fetuin-A	Glycoprotein that forms complexes with calcium and phosphate ions, acts as an inhibitor of ectopic calcification	PXE	[134,135]
Adenovirus with ABCC6 cDNA	Transiently express ABCC6 in the liver	PXE	[136]
Bevacizumab (Anti-VEGF) †	Anti-angiogenic therapy; Preserves ocular function in advanced and early disease stages	PXE	[24]
Supplementation therapies for direct inhibition of calcification			
Sevelamer hydrochloride (Renagel)	Phosphate binder	PXE	[130,137]
Magnesium	Inhibits the formation of apatite	PXE	[129,138,139]
Vitamin K (phylloquinone/menaquinone)	Correct for insufficient carboxylation of matrix gla protein (MGP)	PXE	[92,115,116,140–143]
Bisphosphonate (etidronate *, zoledronate)	Non-hydrolyzable analog of PPi, inhibits enzymes that utilize pyrophosphate	PXE and GACI	[18,55,126,144–148]
Pyrophosphate *	Potent inhibitor of calcification, ABCC6 modulates PPi production	PXE	[18,19,120], PROPHECI
INS-3001 (IP6 derivative)	Known inhibitor of calcification	PXE	[149]
Sodium thiosulfate	Approved for calciphlaxis	PXE	[59]

* Denotes treatments undergoing clinical trials, † currently in use in PXE patients. PXE, pseudoxanthoma elasticum; GACI, generalized arterial calcification of infancy; MGP, matrix gla protein; PPi, pyrophosphate.

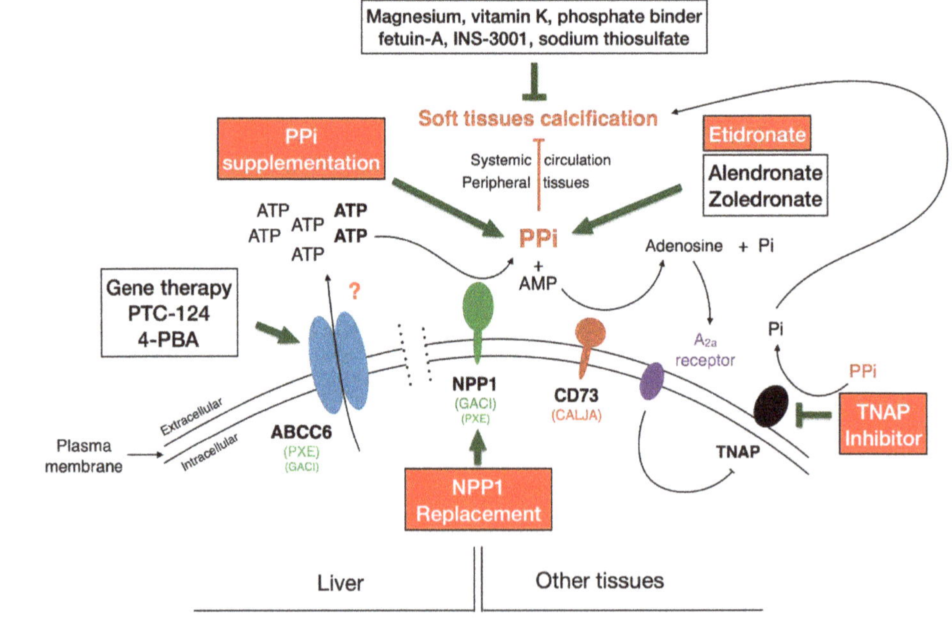

Figure 3. Tested therapeutic Interventions targeting different steps in the ABCC6 pathway to prevent calcification in PXE/GACI. Red boxes point to approaches that were tested in preclinical models and that are currently under human evaluation/trials.

6.1. Treatment Outcome Conundrum

Before we begin describing treatment options, there is one important point to address. With the multiplicity of therapeutic solutions developed and tested, there has been strong consideration given by both scientists and physicians as to what clinical criteria should be used to reliably evaluate and quantify treatment outcomes for PXE patients. This debate, which has been ongoing for several years, has yet to be settled due to the complexity of the PXE phenotype as multiple organ systems (skin, eyes, cardiovascular, renal, etc. ...) are affected with various degree of distribution, severity and considerable inter- and intra-familial variability [150–152]. Indeed, identical mutations can cause the relatively mild PXE or severe GACI [17] in humans but also in mice [111]. The lack of genotype/phenotype correlation only compounds the problem. However, a very recent clinical study performed with 289 PXE patients found that mixed genotype led to less severe arterial calcification and ocular manifestations than patients with two truncating *ABCC6* variants and suggested that environmental or modifier genes could also contribute to the still unexplained phenotypic variability in PXE [153]. The presence of modifier genes modulating calcification in PXE has been investigated in human and mice but their predictive values for clinical applications are not clear [109,110,147,150,154–156]. Moreover, PXE is a slow and chronic disease with stepwise progression of the phenotype occurring over a period of time measured in years, if not decades [157]. The progression of calcification in *Abcc6*$^{-/-}$ mice mirrors that of humans and typically takes many months to develop [18,92]. PPi has been considered as a potential biomarker as it is central to the etiology of PXE [8] but its plasma levels only bring partial information as to the status of the disease [158] (See Figure 2). Furthermore, the daily and long-term natural variations of plasma PPi are not well known. Measuring calcification is an obvious approach but its quantification in a non-invasive or minimally invasive manner is not a trivial task. Invasive biopsies were tested in early trials in semi-quantitative evaluations [129,130]. However, these trials did not yield positive conclusions and it is possible that the quantification methodology combined with the heterogeneity of the manifestations may have contributed to the inconclusive results. The capacity of ^{18}F-sodium fluoride (^{18}F-NaF) positron emission tomography/computed tomography (PET/CT) to identify and evaluate metabolically active osteogenesis in skin and arteries in PXE patients has been demonstrated [159,160]. This method was applied in the recent TEMP clinical trial and was able to discern some phenotypical changes within a year following etidronate administration (*Cf.* Section 6.7.4) [126]. Although the potential of some novel methodology is being investigated, at this time [161], PET/CT scan is probably the best option available to evaluate treatment outcome in PXE patients, but it requires access to well-equipped medical centers. However, it is very likely that a combination of methodology and multiyear trials will be needed to achieve clear and objective results in the future.

6.2. Palliative Treatment: An Ocular Therapy for PXE

As described above, PXE leads to severe visual impairment as a consequence of choroidal neovascularizations resulting from angioid streaks. Visual loss is one of the PXE manifestations that impacts quality of life the most. Photodynamic therapy has been explored but was not found to be effective [162]. Because the monoclonal antibody against the vascular endothelial growth factor (VEGF–Bevacizumab) has been shown to be beneficial for the treatment of choroidal neovascularizations secondary to age-related macular degeneration [163], which present many similarities with PXE, Finger et al. performed a retrospective analysis with a small cohort of PXE patients to determine the effectiveness of repeated intravitreal injections of Bevacizumab on visual acuity and morphologic outcomes [24]. The results were unequivocal in showing that this therapy preserved ocular function in cases of advanced disease and improved function in early stages. As a result, anti-VEGF therapy has become a common symptomatic treatment for PXE patients.

6.3. Targeting ABCC6

ABCC6 deficiency causes a broad spectrum of manifestations beyond calcification that includes vascular malformation (rete mirabile, carotid hypoplasia) [164], dyslipidemia, and atherosclerosis [93,94], as well as ischemic stroke [165], inflammation [159], premature cellular senescence in hepatic cells and dermal fibroblasts [166,167], increased infarct size, and apoptosis [168]. ATP released by ABCC6 and the nucleotides/nucleosides generated downstream by the ectonucleotidases NPP1 and CD73 regulates cellular signaling towards P2 (nucleotides) and P1 (Ado) receptors, which have a wide range of physiological influences that could well explain these other manifestations [124]. The primary advantage and benefit from therapies that directly aim at restoring ABCC6 function would be the potential rescue of the full spectrum of manifestations, not just calcification.

6.3.1. PTC-124

About 400 distinct disease-causing variants have so far been reported in the *ABCC6* gene (https://www.ncbi.nlm.nih.gov/clinvar, accessed on 3 February 2021). A little more than 1/3 of all variants are nonsense that most likely result in nonsense-mediated decay or possibly truncated ABCC6 proteins. The most common and recurrent variant (and the first to be identified) is the nonsense p.R1141X, which accounts for ~30% of all pathogenic alleles in PXE patients of Caucasian descent [169,170].

Correct protein synthesis is at the core of biological processes of all living organisms. Therefore, many compounds with the ability to overcome premature termination codons (PTC) have been studied over the years with the hope of restoring full-length protein synthesis as a therapeutic approach for heritable diseases [171]. Indeed, about 12% of human genetic disorders are caused by nonsense mutations, leading to the generation of PTC. This strategy was tried previously with cystic fibrosis patients. In a phase II clinical trial, these patients were given 1,2,4-oxadiazole (PTC-124), a non-aminoglycoside nonsense mutation suppressor, with some success [172]. However, follow-up phase III trials were relatively unsuccessful [173]. This drug is approved for clinical use by the European Medicines Agency (EMA) to treat Duchenne muscular dystrophy but not by the Federal Drug Administration in the US. It is commercialized under the names of Ataluren or Translarna and can be delivered orally. It has not yet been tried clinically for PXE patients.

One preclinical study examined the efficacy of PTC-124 in a "PXE" zebrafish morpholino model. Zhou et al. utilized HEK293 cells transfected with ABCC6 expression vectors harboring seven different PXE-nonsense mutations, including the most common stop codon mutation, p.R1141X, to evaluate whether PTC-124 could facilitate read-through of these variants [131]. Using immunostaining, they found that the amount of full-length protein synthesis was increased after 72 h of treatment with PTC-124 at a concentration of 5 µg/mL. To test the functionality of these potential full-length read-through ABCC6 proteins, they employed a zebrafish morpholino rescue system. As PTC-124 primarily replaces UAG PTC with tRNA corresponding to Trp, Cys, and Arg (here replaced by Gly), the authors of this study showed that these three possible read-through *ABCC6* variants rescued an *Abcc6a* morpholino-induced phenotype in zebrafish. Of note, in the *ABCC6* mutation database, a single missense variant affecting codon R1141 is reported as p.R1141Q and is classified as benign. It is therefore possible that substitutions at the R1141 site may not have negative impact. Despite these encouraging results, the study did not address calcification and there has been no follow-up study in mouse models of PXE.

One important argument for using PTC-124 would be restoration of the full physiological function of the entire ABCC6 → NPP1 → NT5E→TNAP pathway (Figures 1 and 3) as the *Enpp1, Nt5e* and *Alpl* genes are affected in the absence of *Abcc6* [11,18]. In addition, a large fraction of *ABCC6* disease-causing variants are nonsense, notably the R1141X variant, which is particularly prevalent in Caucasian PXE and GACI patients [169,170]. Another advantage favoring PTC-124 is its oral delivery and that only one allele needs to be targeted, as PXE and GACI are recessive diseases. However, these advantages are also limitations, as treatment would only be restricted to those carrying nonsense alleles.

6.3.2. 4-Phenylbutyrate (4-PBA)

Drug repurposing has gained attention over the past years to cut development cost in part because the safety of the drugs is known and the risk of failure is reduced. This is particularly advantageous for rare diseases like PXE and GACI. 4-phenylbutyrate (4-PBA) is an aromatic fatty acid normally used to treat urea cycle disorders and thalassemia [174–176]. One of its metabolites, phenylacetylglutamine, contains the same nitrogen amount as urea, and is thus an alternative to capturing excess nitrogen before excretion by the kidneys. 4-PBA is classified as an orphan drug commercialized under the names of sodium phenylbutyrate in the U.S. and BUPHENYL or AMMONAPS in other countries. Glycerol phenylbutyrate (RAVICTI) is a related compound. It is a triglyceride pro-drug containing three molecules of 4-PBA linked to a glycerol backbone. RAVICTI is an FDA-approved alternative to sodium 4-PBA, recommended for pediatric use due to its improved palatability [177]. 4-PBA also influences the transcription of endoplasmic reticulum chaperones [178] and this off-target property has been exploited to treat patients with diseases caused by improper translocation of proteins to the plasma membrane, including other ABC transmembrane proteins [179–182].

The repurposing of 4-PBA for use in PXE has been explored in vitro as well as with *Abcc6*$^{-/-}$ mice [80,99,117]. 4-PBA was first tested to determine if it could restore the normal cellular trafficking in vitro and in vivo of 10 frequently occurring disease-causing *ABCC6* missense mutants as well as phenotypic rescue in zebrafish. Seven of the mutants were transport-competent but were retained intracellularly. 4-PBA successfully restored the plasma membrane localization and functionality of four of these ABCC6 mutants thus providing the first evidence that 4-PBA therapy was possible for selected patients with PXE and GACI.

In a follow-up study, the combination of 4-PBA treatment and the transient expression of human *ABCC6* mutants in the liver of *Abcc6*$^{-/-}$ mice was tested against the DCC phenotype, which is a reliable indicator of ABCC6 function [99]. Treatments restored the physiological function of human *ABCC6* mutants and inhibited calcification, but interestingly, failed to restore PPi levels in plasma. However, collectively these studies demonstrated the credibility of 4-PBA for the treatment of both PXE and GACI [99].

The main advantage of 4-PBA is similar to that of PTC-124 and this drug has been in clinical use for several decades. This would certainly facilitate and expedite clinical trials if it were to be tested in eligible PXE/GACI patients. The limitations of 4-PBA would also be similar to those of PTC-124, being an allele-specific drug that can only be applied to *ABCC6* missense mutants with verified sensitivity to the drug. However, the premise of potentially superior approaches focused on restoring PPi levels have supplanted both PTC-124 and 4-PBA and no clinical studies are currently envisioned for either drug.

6.4. Targeting NPP1

As most cases of GACI are caused by mutations in *ENPP1*, enzyme replacement therapy has been proposed as a potential strategy to counteract the clinical manifestations of the disease. NPP1 is a key enzyme in the calcification pathway, converting extracellular ATP into adenosine monophosphate and PPi. There has been success with this strategy in preclinical studies. Albright et al. demonstrated that replacement therapy with soluble recombinant human NPP1 (rhNPP1-Fc) successfully reduced ectopic mineralization, improved plasma PPi levels, and prevented mortality in a rodent GACI model [132]. Another study evaluating cardiac outcomes found recombinant NPP1 reduced not only aortic calcification, but also improved cardiovascular function and blood pressure [133].

GACI presents manifestations beyond calcification that can potentially be attributed to the influence of NPP1 enzyme activity on extracellular purinergic metabolism [89], which is consistent with what has been found for ABCC6 and PXE [124]. In a recent study, Nitschke and co-workers showed that nearly $\frac{3}{4}$ of GACI patients display arterial stenoses due to intimal proliferation of vascular smooth muscle cells (VSMC) [89]. This phenotype can be reversed effectively in culture and in vivo with rhNPP1-Fc or with adenosine. However, neither PPi nor the bisphosphonate etidronate (an off-label treatment for GACI) had an effect on VSMC proliferation. This is an important finding as it pertains directly to PXE. Indeed, NPP1, which is immediately downstream of ABCC6, is not only essential for the generation of extracellular PPi but also for the cleavage of extracellular ATP into adenosine via the action of CD73 (Figures 1 and 3). Remarkably, CALJA patients with *NT5E* mutations (and inactive CD73) develop similar vascular obstructions in the lower extremities in adulthood [10].

rhNPP1-Fc has yet to be tested in rodent models of PXE; however, there are some indications that it could potentially be used in this case as well. It has been observed, even with complete ABCC6 deficiency, that there are fairly high residual plasma levels of PPi (~40%), indicating that other sources of extracellular ATP exist [9,112]. However, it is possible that this strategy would require combination therapy to increase ATP release from cells other than hepatocytes and replace NPP1 in order to restore plasma PPi levels. However, preclinical studies are needed to demonstrate the viability of this therapeutic strategy for PXE.

6.5. Targeting TNAP

Alkaline phosphatases are a group of isoenzymes located at the cell surface catalyzing the hydrolysis of organic phosphate esters in a variety of phosphate-containing physiological compounds. They contribute to a range of physiological functions such as DNA synthesis, in addition to regulating calcification [183–186]. Alkaline phosphatases are classified as tissue-specific and tissue non-specific. The isoenzymes found in the intestine, placenta, and germinal tissue are tissue-specific, whereas the tissue non-specific (TNAP), often referred to as the liver/bone/kidney alkaline phosphatase, is found in these and other tissues. These alkaline phosphatases are encoded by four different genes and each has distinct functions.

The first evidence of a physiological connection between TNAP and a pathology closely related to PXE was reported in 2011 by St Hilaire et al. [187], even though a molecular connection was established a few years later [10,168]. This was followed by a more elaborate series of experiments using cell culture and several mouse models targeting the transporter/enzymes of the ABCC6 pathway (Figures 1 and 3) [20]. Using primary skin fibroblasts from PXE patients with confirmed *ABCC6* mutations, Ziegler et al. found that these cells had elevated *ALPL* gene expression and associated TNAP enzymatic activity as compared to controls. The authors also investigated the effects of a TNAP inhibitor, SBI-425 (30 mg/kg/day), administered in the food of $Abcc6^{-/-}$ mice vs. etidronate (240 mg/kg/day) or normal chow starting at 6 weeks of age for 14 weeks. MicroCT scans at 20 weeks revealed significant attenuation of calcification in both the mice treated with the TNAP inhibitor or etidronate. Furthermore, SBI-425 inhibited TNAP activity, whereas etidronate did not. More promising still as a potential therapeutic, $Abcc6^{-/-}$ mice aged to 20 weeks and treated with the TNAP inhibitor did not show the same progressive calcification as control animals, although the treatment did not reverse existing calcification in this experimental context. Remarkably, Ziegler and colleagues found that plasma PPi levels did not increase after SBI-425 treatment, despite decreased plasma TNAP activity. This is somewhat surprising as PPi concentration in the plasma of $Nt5e^{-/-}$ mice is significantly reduced [188]. Indeed, the lack of CD73 (NT5E) function leads to less adenosine and higher TNAP activity, resulting in higher rates of PPi hydrolysis. This apparent discrepancy is consistent with other circumstantial evidence [111,144], suggesting that plasma levels do not adequately reflect the steady state levels of PPi in connective tissues where it is most relevant.

A follow-up study used a slightly different approach to explore the role of TNAP in the calcification phenotype of PXE and GACI using $Abcc6^{-/-}Alpl^{+/-}$ double-mutant mice and $Enpp1^{-/-}$ animals [127]. Mice heterozygous for *Alpl* in an *Abcc6*-deficient background showed both reduced plasma TNAP activity and mineralization as compared to controls, and the administration of SBI-425 led to similar results. By contrast, SBI-425 treatment of $Enpp1^{-/-}$ mice did not produce any significant change in mineralization.

Overall, these studies demonstrated that inhibition of TNAP is a convincing treatment strategy for PXE (but maybe not for GACI). In fact, it is convincing enough that a pharmaceutical company is investing in this approach.

6.6. Gene Therapy

In 2010, an esoteric approach was explored based on the observation that plasma levels of fetuin-A in patients with PXE as well as in $Abcc6^{-/-}$ mice are reduced by up to 30%. Fetuin-A is a circulating "hepatokine" predominantly synthesized in the liver, which possesses diverse physiological functions, including bone metabolism regulation, vascular calcification, and insulin resistance. Because fetuin-A is able to form complexes with calcium and phosphate ions, it acts as an inhibitor of ectopic calcification. Mice deficient in this glycoprotein show systemic calcification of soft tissues [134]. In this study [135], a fetuin-A cDNA was transiently expressed in the liver of $Abcc6^{-/-}$ mice and resulted in an approximately 70% reduction of whisker calcification. However, the positive effects on calcification were not persistent.

More recently, the transient expression of the human *ABCC6* in mouse liver was shown to have a positive effect on DCC [13], while permanent expression via transgenes produced remarkable effects on ectopic calcification in several tissues (whiskers, kidneys, heart), despite having only modest effects on plasma PPi [18].

A more elaborate methodology was explored with the intravenous administration to $Abcc6^{-/-}$ mice of a recombinant adenovirus carrying a cDNA encoding the normal human ABCC6 [136]. The mice showed high-level expression of the human ABCC6 in liver for several weeks post-delivery. This resulted in the normalization of plasma pyrophosphate levels. For sustained expression up to three repeated adenovirus injections 4 weeks apart were also tested with success. At the time of sacrifice the mice were 4–5 months old and showed very limited signs of mineralization in vibrissae. By contrast, treatments applied to older mice (11 months old) had no effect on existing mineralization. The human species C adenovirus serotype 5 (Ad5) used in this study is one the most frequently used gene delivery systems in animal and clinical studies [189] and presents a high predilection for liver transduction, which is the primary site of expression of ABCC6 [79,81]. This proof-of-concept study suggested that adenovirus-mediated *ABCC6* gene delivery may be possible to treat PXE and ABCC6-related GACI patients, when gene therapy, especially one targeting the liver, has reached sufficient maturity for clinical use.

6.7. Supplementation Therapies for Direct Inhibition of Calcification

Treatments with exogenous compounds such as magnesium, bisphosphonates (PPi analogs), and PPi are focused on reducing or counteracting the excessive susceptibility to mineralization found in PXE and GACI patients and to slow or interrupt the clinical progression of the disease.

6.7.1. Calcium and Phosphate

Quickly after the generation of mouse models for PXE [90,91], several studies were initiated to look at the effect of dietary minerals (calcium, phosphate, magnesium) that could influence ectopic mineralization. Interestingly, a first attempt at altering the level of dietary calcium fed to $Abcc6^{-/-}$ mice produced no significant effect on the calcification phenotype [138], mostly likely because plasma calcium levels are normally kept within a narrow range. However, changes to dietary minerals, notably phosphate, produced significant effects on soft tissue calcification in $Abcc6^{-/-}$ mice [137]. These results prompted the exploration of phosphate binders as therapy. However, experimental testing in mice and human PXE patients showed this approach to be ineffective [130,137].

6.7.2. Magnesium

Magnesium (Mg^{2+}) is another mineral that inhibits the formation of apatite. In healthy individuals, serum magnesium concentrations are carefully balanced in a narrow range [190]. The kidneys are the main organs controlling systemic Mg^{2+} homeostasis, where its transport is highly regulated by hormonal and other intrarenal factors [191]. Bones are the main reservoir of Mg^{2+}, holding ~60% of the total body content. Mg^{2+} in bones is primarily embedded at the surface of the apatite crystals [192]. Because of its inhibitory properties towards calcification and oral bioavailability, Mg^{2+} supplementation appeared to be a good option to decrease soft tissue calcification in pathologies with calcification such as chronic kidney disease (CKD) [193], but also in PXE, as both share some phenotypical characteristics [194]. Preliminary testing in $Abcc6^{-/-}$ mice with increased dietary Mg^{2+} showed results [138,139] encouraging enough that a prospective human clinical trial was initiated (NCT01525875). However, the translation to humans proved to be inconclusive [129], probably because the dosages used in animal models exceeded what could actually be tolerated in humans and/or as a result of the method for phenotype evaluation methodology (*Cf.* Section 6.1). Therefore, Mg^{2+} supplementation is probably not a viable treatment strategy, at least on its own.

6.7.3. Vitamin K

Normal bone mineralization is a highly regulated process, requiring a fine balance between inhibitors and promoters of calcification [195,196]. Among these regulators of mineralization, Matrix Gla Protein (MGP) and osteocalcin (OC) are two proteins reliant on carboxylation for activation and function [196,197]. Gheduzzi et al. described in 2007 high levels of undercarboxylated MGP in the circulation of PXE patients. Uitto et al. also reported the presence of mostly undercarboxylated MGP in the calcified vibrissae of the $Abcc6^{-/-}$ mice [198]. At about the same time, a remarkably PXE-like phenotype was described in patients carrying mutations in the gamma-glutamyl carboxylase gene (*GGCX*) that encodes an enzyme essential for the carboxylation of MGP and OC as well as clotting factors [199]. Because vitamin K is an essential component in the post-translational carboxylation of glutamate residues (Glu) in these proteins, it was logically proposed that a vitamin K precursor or a conjugated form was the potential substrate(s) for ABCC6 and that insufficient carboxylation of MGP was the cause of abnormal calcification in PXE [200]. The fact that PXE patients have low serum levels of vitamin K [201], and that skin fibroblasts from patients present molecular signatures of impaired vitamin K-dependent carboxylation [202] gave further support to this notion. It was the first hypothesis that not only provided a strong and credible explanation to the calcification phenotype of PXE, but was also easily testable and offered the prospect of an affordable treatment. The PXE scientific community lost no time in developing competing studies using $Abcc6^{-/-}$ mice to test various active forms of vitamin K, primarily K1 and K2 (i.e., phylloquinone or menaquinone), through dietary supplementation. In quick succession, three publications reported disappointing results, as none of the vitamin K forms tested at high or low concentrations stopped or reversed PXE mineralization [92,140,141]. Despite these negative results, interest in the relationship between vitamin K metabolism and abnormal calcification in PXE and GACI continues and several follow-up studies have since been published [115,116,142,143]. However, neither phylloquinone or menaquinone are considered to be viable treatment options for the time being.

6.7.4. Bisphosphonate Treatment for PXE and GACI

Bisphosphonates are non-hydrolyzable analogs of PPi with two phosphonate (PO_3) groups covalently linked to a central carbon (instead of an oxygen) and two side chains that determine chemical characteristics. There are two categories of bisphosphonates based on the presence of a non-nitrogenous or nitrogenous side chain. Because bisphosphonates mimic the structure of PPi, they have similar properties and can inhibit enzymes that utilize pyrophosphate. Bisphosphonates have been used clinically for decades in the treatment of osteoporosis and Paget's disease of bone, as well as other applications related to mineralization.

The bisphosphonate etidronate has been used to treat GACI patients and early treatments improved GACI outcome [145]. Retrospective analyses have confirmed that survival beyond infancy is markedly improved with etidronate therapy [55,146]. However, this treatment seems to induce undesirable side-effects as a recent case report highlighted significant inhibition of skeletal calcification with paradoxical joint calcifications in a 7-year-old GACI patient [145]. Furthermore, studies on uremic rats or $Enpp1^{-/-}$ mice suggested that this bisphosphonate is incapable of preventing *de novo* calcification without alteration to bone structures [203,204].

For PXE, a first study reported the anti-calcification properties of zoledronate (nitrogenous) in an in vitro series of experiments with primary fibroblasts [147]. However, more elaborate studies with mouse models showed that only administration of etidronate (non-nitrogenous), but not alendronate (nitrogenous), prevented ectopic mineralization. However, etidronate treatment did not reverse existing mineralization [18,144,148]. The effect of etidronate was accompanied by alterations in the trabecular bone microarchitecture similar to what was described for $Enpp1^{-/-}$ mice. Although the results suggested that biphosphonates could be used as potential treatments for PXE, the high dosages and/or the bone alterations are potentially detrimental in the long term.

However, these animal studies and the fact that etidronate is an approved drug already tested for off-label applications against calcification in GACI and basal ganglia calcifications [205] were sufficiently convincing that a clinical trial termed Treatment of Ectopic Mineralization in PXE or TEMP was initiated in the Netherlands [126]. This investigation primarily tested effectiveness and safety of etidronate administration in PXE. The 12-month long investigation used a cyclical treatment of 20 mg/kg for 2 weeks every 12 weeks in a randomized, placebo-controlled phase I/II with 74 participants. The primary outcome was ectopic calcification as determined by ^{18}fluoride positron emission tomography scans in femoral arterial tissues. Secondary outcomes were computed tomography arterial calcification and ophthalmological changes. Safety outcomes were bone density, serum calcium, and phosphate. The overall results were somewhat mixed, with PXE patients that received etidronate seeing reduced arterial calcification and subretinal neovascularization events. However, they did not experience lower femoral ^{18}fluoride positron emission tomography activity, suggesting that active mineralization was still ongoing. Overall, no adverse effects from the treatment were reported.

Of note, another phase I/II clinical trial with a French cohort is in preparation and is called PyROphosPHate supplementation to fight ECtopIc calcification in PXE (PROPHECI, registration: https://scanr.enseignementsup-recherche.gouv.fr/project/PHRCN-PHRC-19-0402, accessed on 2 March 2021). This trial will use oral disodium PPi, but etidronate will be used as a control instead of a placebo.

6.7.5. Pyrophosphate (PPi)

The idea of using PPi as a therapeutic for PXE and GACI came quickly after the discovery that ABCC6 modulates PPi production [8,9], though the idea of using PPi to prevent ectopic calcification was not new [206,207]. Considering the short half-life of PPi in plasma, several experimental approaches were devised to counteract calcification in PXE and GACI mouse models. Two groups proceeded to test different methodologies, with either a single daily bolus injection [18] or continuous delivery via drinking water [120]. The first study found that daily injections achieved near complete suppression of ectopic mineralization in several tissues despite a relatively poor bioavailability of about 0.5%. Importantly, and similar to other forms of treatments tested thus far, established mineralization could not be reversed [18].

Because daily injections are not necessarily practical for the lifelong treatment needed for PXE or GACI patients, oral administration was also evaluated [120]. Adding sodium PPi to drinking water sustainably raised plasma PPi in both $Abcc6^{-/-}$ mice as well as in human volunteers, and significantly decreased calcification in both PXE and GACI mouse models. However, calcification inhibition was not complete. One of the remarkable discoveries of the latter study was that, contrary to previous beliefs [208], PPi is bioavailable when administered orally, although poorly (~0.1%), but is still effective against mineralization.

Furthermore, there has been a recent and inadvertent discovery that modulating PPi from dietary sources can also be effective at reducing calcification in $Abcc6^{-/-}$ mice [19]. These findings were prompted by a routine change in the supplier of institutional rodent diet. The new chow was enriched in PPi (a fact unknown to the manufacturer and the research laboratory), leading to a doubling of plasma PPi and a halving of calcification in $Abcc6^{-/-}$ mice. Dietary PPi is also readily absorbed in humans with a bioavailability comparable to that of drinking water (~0.1%), which suggests that dietary preference could contribute to the considerable and still unexplained phenotypic heterogeneity in PXE [19].

These recent studies unambiguously showed that PPi supplementation via oral delivery or even by injection are arguably the most promising and credible strategies for treating PXE (and GACI) patients. Indeed, PPi would be a simple and low-cost medication and it has a very safe profile. It is a nontoxic, physiological metabolite as per the World Health Organization (WHO). The US Food and Drug Administration (FDA) lists it as Generally Recognized As Safe and it is designated as food additive E450(a) in Europe. PPi is thus widely used in the food industry as a preservative in canned seafood, in baking soda, in cured meat, etc. . . . and is even present in toothpaste.

For these reasons, a phase II clinical trial using capsulized disodium PPi is currently being conducted (NCT04441671) while the PROPHECI clinical trial (see above) for PXE patients is currently in preparation and also proposes to test the oral delivery of encapsulated disodium PPi.

Of note, several pharmaceutical companies are now investing in the testing and manufacturing of new PPi formulations for the treatment of calcification in PXE, GACI, and other diseases.

6.7.6. Phytic Acid (IP6)

Phytic acid (*myo*-inositol hexakisphosphate; IP6) is a phosphate ester of inositol. Its anti-calcification properties have been known for some time [209,210]. IP6 has poor bioavailability and pharmacokinetics, which has prompted the development of a series of derivatives [211]. These were tested in $Abcc6^{-/-}$ mice [149]. Compound INS-3001 was found to be superior in calcification inhibition both in vitro and in vivo and displayed a better plasma half-life as compared with IP6 [211]. INS-3001 was administered to $Abcc6^{-/-}$ mice before and after the onset of mineralization at various dosages via two methods, either microosmostic pumps for continuous delivery or sub-cutaneous bolus injections. With the highest dosages of 4 to 20 mg/kg/day, calcification inhibition in vibrissae was effective, though calcification reversal was not attained in the longest duration of the treatment (12 weeks). Although phytic acid (IP6) is a safe compound found in food such as plant seed and had the FDA designation Generally Recognized as Safe (GRAS), INS-3001 is a new product with good potential but it remains to be fully evaluated for clinical use.

6.7.7. Sodium Thiosulfate

Sodium thiosulfate ($Na_2S_2O_3$) is an industrial compound with a long clinical history [212] and it is commonly employed as a food preservative. It was originally used as an intravenous medication for metal poisoning [213]. It is now approved for the treatment of certain rare medical conditions notably calciphylaxis [214]. For this reason, intravenous sodium thiosulfate was recently administered to a single PXE patient with polygenic inheritance and severe early-onset manifestations [59]. This treatment achieved a remarkable regression of calcific stenosis in the coeliac and mesenteric arteries. However, significant side-effects resulted in discontinuation of the treatment and in the relapse of the symptoms. It is unclear if this treatment could be generalized to PXE and/or GACI patients but its potential use for the reversal of existing calcification should be explored, perhaps as a temporary measure to reduce existing calcification before a long-term inhibitory treatment can be applied.

7. Conclusions

We have learned much about ABCC6 [21] since its initial discovery 20 years ago [5–7], from its transcriptional regulation [215–217], expression profile [79,81,128], and molecular function and connections to an existing molecular pathway [8,9], to means and ways to rescue this transporter and treatments for PXE [18,19,92,98,99]. In two decades, PXE has gone from gene discovery to early clinical trials [126,218], a remarkable feat for rare disorders with significant morbidity but low mortality. At the time of this writing, PXE alone has been the subject of more than 2000 peer-reviewed publications (PubMed citations) as well as many other book chapters and reviews.

The study of PXE and GACI (and CALJA) has brought to light new fundamental knowledge about abnormal calcification and may have important implications for many other pathological contexts in humans. Indeed, the molecular pathway and mechanisms involved in PXE and the related diseases described above are likely to apply to common age-associated disorders such as chronic kidney insufficiency, diabetes, atherosclerosis, inflammatory skin diseases (sarcoidosis, systemic lupus erythematosus, scleroderma, and dermatomyositis, to name a few). Simple aging is also accompanied by soft tissue mineralization. Factors such as inflammation, infections, metabolic alterations, and genetics greatly influence and can exacerbate the mineralization process. When present, vascular calcification, an age-old condition [219], is predictive of worse clinical outcomes, with individual risk for cardiovascular mortality, or any cardiovascular event, dramatically increasing [220]. We estimate the frequency of heterozygous carriers of *ABCC6* mutations could be as high as 1/80 in the general population [86,221]. Thus, understanding how ABCC6 (and NPP1) influences the homeostasis of connective tissues may one day be used to enhance tissue repair and lessen mineralization-associated morbidity and mortality in the general population. This is particularly important in light of recent reports suggesting that three of the genes in the pathway shown in Figures 1 and 3 that cause ectopic calcification in PXE, GACI, and CALJA [6,8–11,17,18,20] have also been shown to play a role in dyslipidemia and atherosclerosis [93,94,222–225].

Of all the treatment options explored above, none of them have focused and/or achieved the removal or reversal of existing calcification in PXE and GACI, save for one case of sodium thiosulfate usage. If past history has taught us anything, for diseases as complex as PXE and GACI, no single treatment will cover all aspects of these pathologies and there will be room for multiple forms of treatments for either disease, whether applied in combination or in single application.

What is next? The central role of ABCC6 in PXE, GACI, and DCC is now well established in humans [56] and animal models [112,226]. However, many aspects of the pathophysiology of ABCC6 dysfunction are still unexplained. The liver expression of ABCC6 is necessary but not sufficient for calcification inhibition [13,20]. The identification and contribution of peripheral tissues to calcification regulation remains unresolved. Plasma PPi levels fail to explain the wide range of calcification severity in humans [17] and mice [111] (*Cf* Figure 1) and the clinical relevance of potential modifier genes [109,110,147,150,154–156] is still unknown. Finally, what is (if any) the role of inflammation in the progression of PXE [159]? Answering some of these questions will not only inform the scientific community on these intriguing diseases but will ultimately help to refine and expand upon the various treatment options explored today.

Author Contributions: Bibliographical research, drafting and editing manuscript (B.K.S. and O.L.S.), editing, critical revisions and comments (B.K.S., J.Z., V.P., S.K., L.M. and O.L.S.), approval of the final version of manuscript (B.K.S., V.P., J.Z., S.K., L.M. and O.L.S.). All authors have read and agreed to the published version of the manuscript.

Funding: Financial support to Olivier Le Saux came from National Institutes of Health P20GM113134. The Hungarian grant OTKA 128003 provided financial support to V.P. The funding agencies were not involved in the design or execution of this study.

Institutional Review Board Statement: Not applicable.

Informed Consent Statement: Not applicable.

Data Availability Statement: Not applicable.

Conflicts of Interest: The authors have no conflict of interest to report.

References

1. He, K.; Sawczyk, M.; Liu, C.; Yuan, Y.; Song, B.; Deivanayagam, R.; Nie, A.; Hu, X.; Dravid, V.P.; Lu, J.; et al. Revealing nanoscale mineralization pathways of hydroxyapatite using in situ liquid cell transmission electron microscopy. *Sci. Adv.* **2020**, *6*, eaaz7524. [CrossRef] [PubMed]

2. Atzeni, F.; Sarzi-Puttini, P.; Bevilacqua, M. Calcium deposition and associated chronic diseases (atherosclerosis, diffuse idiopathic skeletal hyperostosis, and others). *Rheum. Dis. Clin. N. Am.* **2006**, *32*, 413–426. [CrossRef] [PubMed]

3. Rutsch, F.; Nitschke, Y.; Terkeltaub, R. Genetics in arterial calcification: Pieces of a puzzle and cogs in a wheel. *Circ. Res.* **2011**, *109*, 578–592. [CrossRef]

4. Ruiz, J.L.; Hutcheson, J.D.; Aikawa, E. Cardiovascular calcification: Current controversies and novel concepts. *Cardiovasc. Pathol.* **2015**, *24*, 207–212. [CrossRef] [PubMed]

5. Bergen, A.A.; Plomp, A.S.; Schuurman, E.J.; Terry, S.F.; Breuning, M.H.; Dauwerse, H.G.; Swart, J.; Kool, M.; van Soest, S.; Baas, F.; et al. Mutations in ABCC6 cause pseudoxanthoma elasticum. *Nat. Genet.* **2000**, *25*, 228–231. [CrossRef]

6. Le Saux, O.; Urban, Z.; Tschuch, C.; Csiszar, K.; Bacchelli, B.; Quaglino, D.; Pasquali-Ronchetti, I.; Pope, F.M.; Richards, A.; Terry, S.F.; et al. Mutations in a gene encoding an ABC transporter cause pseudoxanthoma elasticum. *Nat. Genet.* **2000**, *25*, 223–227. [CrossRef]

7. Ringpfeil, F.; Lebwohl, M.G.; Christiano, A.M.; Uitto, J. Pseudoxanthoma elasticum: Mutations in the MRP6 gene encoding a transmembrane ATP-binding cassette (ABC) transporter. *Proc. Natl. Acad. Sci. USA* **2000**, *97*, 6001–6006. [CrossRef]

8. Jansen, R.S.; Küçükosmanoglu, A.; de Haas, M.; Sapthu, S.; Otero, J.A.; Hegman, I.E.M.; Bergen, A.A.B.; Gorgels, T.G.M.F.; Borst, P.; van de Wetering, K. ABCC6 prevents ectopic mineralization seen in pseudoxanthoma elasticum by inducing cellular nucleotide release. *Proc. Natl. Acad. Sci. USA* **2013**, *110*, 20206–20211. [CrossRef] [PubMed]

9. Jansen, R.S.; Duijst, S.; Mahakena, S.; Sommer, D.; Szeri, F.; Váradi, A.; Plomp, A.S.; Bergen, A.A.B.; Elferink, R.P.J.O.; Borst, P.; et al. ABCC6–Mediated ATP Secretion by the Liver Is the Main Source of the Mineralization Inhibitor Inorganic Pyrophosphate in the Systemic Circulation—Brief Report. *Arter. Thromb. Vasc. Biol.* **2014**, *34*, 1985–1989. [CrossRef]

10. Markello, T.C.; Pak, L.K.; St. Hilaire, C.; Dorward, H.; Ziegler, S.G.; Chen, M.Y.; Chaganti, K.; Nussbaum, R.L.; Boehm, M.; Gahl, W.A. Vascular pathology of medial arterial calcifications in NT5E deficiency: Implications for the role of adenosine in pseudoxanthoma elasticum. *Mol. Genet. Metab.* **2011**, *103*, 44–50. [CrossRef]

11. Miglionico, R.; Armentano, M.F.; Carmosino, M.; Salvia, A.M.; Cuviello, F.; Bisaccia, F.; Ostuni, A. Dysregulation of gene expression in ABCC6 knockdown HepG2 cells. *Cell. Mol. Biol. Lett.* **2014**, *19*, 1–10. [CrossRef]

12. Aherrahrou, Z.; Doehring, L.C.; Ehlers, E.-M.; Liptau, H.; Depping, R.; Linsel-Nitschke, P.; Kaczmarek, P.M.; Erdmann, J.; Schunkert, H. An Alternative Splice Variant in Abcc6, the Gene Causing Dystrophic Calcification, Leads to Protein Deficiency in C3H/He Mice. *J. Biol. Chem.* **2008**, *283*, 7608–7615. [CrossRef] [PubMed]

13. Brampton, C.; Aherrahrou, Z.; Chen, L.-H.; Martin, L.; Bergen, A.A.; Gorgels, T.G.; Erdfdi, J.; Schunkert, H.; Szabó, Z.; Váradi, A.; et al. The Level of Hepatic ABCC6 Expression Determines the Severity of Calcification after Cardiac Injury. *Am. J. Pathol.* **2014**, *184*, 159–170. [CrossRef]

14. Meng, H.; Vera, I.; Che, N.; Wang, X.; Wang, S.S.; Ingram-Drake, L.; Schadt, E.E.; Drake, T.A.; Lusis, A.J. Identification of Abcc6 as the major causal gene for dystrophic cardiac calcification in mice through inte-grative genomics. *Proc. Natl. Acad. Sci. USA* **2007**, *104*, 4530–4535. [CrossRef] [PubMed]

15. Kalal, I.G.; Seetha, D.; Panda, A.; Nitschke, Y.; Rutsch, F. Molecular diagnosis of generalized arterial calcification of infancy (GACI). *J. Cardiovasc. Dis. Res.* **2012**, *3*, 150–154. [CrossRef] [PubMed]

16. Le Boulanger, G.; Labreze, C.; Croue, A.; Schurgers, L.J.; Chassaing, N.; Wittkampf, T.; Rutsch, F.; Martin, L. An unusual severe vascular case of pseudoxanthoma elasticum presenting as generalized arterial calcifica-tion of infancy. *Am. J. Med. Genet. A* **2010**, *152*, 118–123. [CrossRef]

17. Nitschke, Y.; Baujat, G.; Botschen, U.; Wittkampf, T.; du Moulin, M.; Stella, J.; le Merrer, M.; Guest, G.; Lambot, K.; Tazarourte-Pinturier, M.-F.; et al. Generalized Arterial Calcification of Infancy and Pseudoxanthoma Elasticum Can Be Caused by Mutations in Either ENPP1 or ABCC6. *Am. J. Hum. Genet.* **2012**, *90*, 25–39. [CrossRef]

18. Pomozi, V.; Brampton, C.; van de Wetering, K.; Zoll, J.; Calio, B.; Pham, K.; Owens, J.B.; Marh, J.; Moisyadi, S.; Váradi, A.; et al. Pyrophosphate Supplementation Prevents Chronic and Acute Calcification in ABCC6-Deficient Mice. *Am. J. Pathol.* **2017**, *187*, 1258–1272. [CrossRef] [PubMed]

19. Pomozi, V.; Julian, C.B.; Zoll, J.; Pham, K.; Kuo, S.; Tőkési, N.; Martin, L.; Váradi, A.; le Saux, O. Dietary Pyrophosphate Modulates Calcification in a Mouse Model of Pseudoxanthoma Elasticum: Implication for Treatment of Patients. *J. Investig. Dermatol.* **2019**, *139*, 1082–1088. [CrossRef]

20. Ziegler, S.G.; Ferreira, C.R.; Macfarlane, E.G.; Riddle, R.C.; Tomlinson, R.E.; Chew, E.Y.; Martin, L.; Ma, C.-T.; Sergienko, E.; Pinkerton, A.B.; et al. Ectopic calcification in pseudoxanthoma elasticum responds to inhibition of tissue-nonspecific alkaline phosphatase. *Sci. Transl. Med.* **2017**, *9*, eaal1669. [CrossRef]

21. Borst, P.; Váradi, A.; van de Wetering, K. PXE, a Mysterious Inborn Error Clarified. *Trends Biochem. Sci.* **2019**, *44*, 125–140. [CrossRef]

22. Ali, S.A.; Ng, C.; Votava-Smith, J.K.; Randolph, L.M.; Pitukcheewanont, P. Bisphosphonate therapy in an infant with generalized arterial calcification with an ABCC6 mutation. *Osteoporos. Int.* **2018**, *29*, 2575–2579. [CrossRef]

23. Edouard, T.; Chabot, G.; Miro, J.; Buhas, D.C.; Nitschke, Y.; Lapierre, C.; Rutsch, F.; Alos, N. Efficacy and safety of 2-year etidronate treatment in a child with generalized arterial calcification of infancy. *Eur. J. Nucl. Med. Mol. Imaging* **2011**, *170*, 1585–1590. [CrossRef] [PubMed]

24. Finger, R.P.; Issa, P.C.; Schmitz-Valckenberg, S.; Holz, F.G.; Scholl, H.N. Long-Term Effectiveness of Intravitreal Bevacizumab for Choroidal Neovascularization Secondary to Angioid Streaks in Pseudoxanthoma Elasticum. *Retina* **2011**, *31*, 1268–1278. [CrossRef]

25. Uitto, J.; Shamban, A. Heritable Skin Diseases with Molecular Defects in Collagen or Elastin. *Dermatol. Clin.* **1987**, *5*, 63–84. [CrossRef]

26. Balzer, F. Recherches sur les caractères anatomiques du xanthelasma. *Arch. Physiol.* **1884**, *4*, 65–80.

27. Chauffard, M.A. Xanthélasma disséminé et symétrique et sans insuffisance hépatique. *Bull. Soc. Med. Paris* **1889**, *6*, 412–419.

28. Rigal, D. Observation pour servir à l'histoire de la chéloide diffuse xanthélasmique. *Ann. Dermatol. Syphilol.* **1881**, *2*, 491–501.

29. Darier, J. *Pseudo-Xanthome Elastique*; III Congrès International de Dermatoligie de Londres: London, UK, 1896; pp. 289–295.

30. Neldner, K.H. Pseudoxanthoma Elasticum. *Int. J. Dermatol.* **1988**, *27*, 98–100. [CrossRef] [PubMed]

31. Truter, S.; Rosenbaum-Fiedler, J.; Sapadin, A.; Lebwohl, M. Calcification of elastic fibers in pseudoxan-thoma elasticum. *Mt. Sinai. J. Med.* **1996**, *63*, 210–215.

32. Uitto, J.; Boyd, C.D.; Lebwohl, M.G.; Moshell, A.N.; Rosenbloom, J.; Terry, S. International Centennial Meeting on Pseudoxan-thoma Elasticum: Progress in PXE Research. *J. Investig. Dermatol.* **1998**, *110*, 840–842. [CrossRef] [PubMed]

33. Weenink, A.C.; Dijkman, G.; de Meijer, P.H. Pseudoxanthoma elasticum and its complications: Two case reports. *Neth. J. Med.* **1996**, *49*, 24–29. [CrossRef]

34. Campens, L.; Vanakker, O.M.; Trachet, B.; Segers, P.; Leroy, B.P.; de Zaeytijd, J.; Voet, D.; de Paepe, A.; de Backer, T.; de Backer, J. Characterization of Cardiovascular Involvement in Pseudoxanthoma Elasticum Families. *Arter. Thromb. Vasc. Biol.* **2013**, *33*, 2646–2652. [CrossRef]

35. Germain, D.P.; Boutouyrie, P.; Laloux, B.; Laurent, S. Arterial Remodeling and Stiffness in Patients with Pseudoxanthoma Elasticum. *Arter. Thromb. Vasc. Biol.* **2003**, *23*, 836–841. [CrossRef]

36. Kornet, L.; Bergen, A.A.; Hoeks, A.P.; Cleutjens, J.P.; Oostra, R.-J.; Daemen, M.J.; van Soest, S.; Reneman, R.S. In patients with pseudoxanthoma elasticum a thicker and more elastic carotid artery is associated with elastin fragmentation and proteoglycans accumulation. *Ultrasound Med. Biol.* **2004**, *30*, 1041–1048. [CrossRef] [PubMed]

37. Bertulezzi, G.; Paris, R.; Moroni, M.; Porta, C.; Nastasi, G.; Amadeo, A. Atrial septal aneurysm in a patient with pseudoxanthoma elasticum. *Acta Cardiol.* **1998**, *53*, 223–225. [PubMed]

38. Prunier, F.; Terrien, G.; le Corre, Y.; Apana, A.L.Y.; Bière, L.; Kauffenstein, G.; Furber, A.; Bergen, A.A.B.; Gorgels, T.G.M.F.; Le Saux, O.; et al. Pseudoxanthoma Elasticum: Cardiac Findings in Patients and Abcc6-Deficient Mouse Model. *PLoS ONE* **2013**, *8*, e68700. [CrossRef]

39. Eddy, D.D.; Farber, E.M. Pseudoxathoam elasticum. Internal manifestatons: A report of cases and a review of the literature. *Arch. Dermatol.* **1962**, *86*, 729–740. [CrossRef]

40. Kauffenstein, G.; Pizard, A.; le Corre, Y.; Vessières, E.; Grimaud, L.; Toutain, B.; Labat, C.; Mauras, Y.; Gorgels, T.G.; Bergen, A.A.; et al. Disseminated arterial calcification and enhanced myogenic response are associated with abcc6 deficiency in a mouse model of pseudoxanthoma elasticum. *Arter. Thromb. Vasc. Biol.* **2014**, *34*, 1045–1056. [CrossRef]

41. Fabre, B.; Bayle, P.; Bazex, J.; Durand, D.; Lamant, L.; Chassaing, N. Pseudoxanthoma elasticum and nephrolithiasis. *J. Eur. Acad. Dermatol. Venereol.* **2005**, *19*, 212–215. [CrossRef]

42. Mallette, L.E.; Mechanick, J.I. Heritable syndrome of pseudoxanthoma elasticum with abnormal phosphorus and vitamin D metabolism. *Am. J. Med.* **1987**, *83*, 1157–1162. [CrossRef]

43. Seeger, H.; Mohebbi, N. Pseudoxanthoma elasticum and nephrocalcinosis. *Kidney Int.* **2016**, *89*, 1407. [CrossRef] [PubMed]

44. Legrand, A.; Cornez, L.; Samkari, W.; Mazzella, J.-M.; Venisse, A.; Boccio, V.; Auribault, K.; Keren, B.; Benistan, K.; Germain, D.P.; et al. Mutation spectrum in the ABCC6 gene and genotype–phenotype correlations in a French cohort with pseudoxanthoma elasticum. *Genet. Med.* **2017**, *19*, 909–917. [CrossRef] [PubMed]

45. Letavernier, E.; Bouderlique, E.; Zaworski, J.; Martin, L.; Daudon, M. Pseudoxanthoma Elasticum, Kidney Stones and Pyrophosphate: From a Rare Disease to Urolithiasis and Vascular Calcifications. *Int. J. Mol. Sci.* **2019**, *20*, 6353. [CrossRef] [PubMed]

46. Letavernier, E.; Kauffenstein, G.; Huguet, L.; Navasiolava, N.; Bouderlique, E.; Tang, E.; Delaitre, L.; Bazin, D.; de Frutos, M.; Gay, C.; et al. ABCC6 Deficiency Promotes Development of Randall Plaque. *J. Am. Soc. Nephrol.* **2018**, *29*, 2337–2347. [CrossRef] [PubMed]

47. Gheduzzi, D.; Sammarco, R.; Quaglino, D.; Bercovitch, L.; Terry, S.; Taylor, W.; Ronchetti, I.P. Extracutaneous ultrastructural alterations in pseudoxanthoma elasticum. *Ultrastruct. Pathol.* **2003**, *27*, 375–384. [CrossRef]

48. Lebwohl, M.; Halperin, J.; Phelps, R.G. Brief report: Occult pseudoxanthoma elasticum in patients with premature cardiovascular disease. *N. Engl. J. Med.* **1993**, *329*, 1237–1239. [CrossRef] [PubMed]

49. Rutsch, F.; Vaingankar, S.; Johnson, K.; Goldfine, I.; Maddux, B.; Schauerte, P.; Kalhoff, H.; Sano, K.; Boisvert, W.A.; Superti-Furga, A.; et al. PC-1 Nucleoside Triphosphate Pyrophosphohydrolase Deficiency in Idiopathic Infantile Arterial Calcification. *Am. J. Pathol.* **2001**, *158*, 543–554. [CrossRef]

50. Moran, J.J. Idiopathic arterial calcification of infancy: A clinicopathologic study. *Pathol. Annu.* **1975**, *10*, 393–417.

51. Morton, R. Idiopathic arterial calcification in infancy. *Histopathology* **1978**, *2*, 423–432. [CrossRef]

52. Ramjan, K.A.; Roscioli, T.; Rutsch, F.; Sillence, D.; Munns, C.F. Generalized arterial calcification of infancy: Treatment with bisphosphonates. *Nat. Clin. Pract. Endocrinol. Metab.* **2009**, *5*, 167–172. [PubMed]

53. Rutsch, F.; Boyer, P.; Nitschke, Y.; Ruf, N.; Lorenz-Depierieux, B.; Wittkampf, T.; Weissen-Plenz, G.; Fischer, R.J.; Mughal, Z.; Gregory, J.W.; et al. Hypophosphatemia, hyperphosphaturia, and bisphosphonate treatment are associated with survival beyond infancy in generalized arterial calcification of infancy. *Circ. Cardiovasc. Genet.* **2008**, *1*, 133–140. [CrossRef]

54. Lorenz-Depierieux, B.; Schnabel, D.; Tiosano, D.; Hausler, G.; Strom, T.M. Loss-of-function ENPP1 mutations cause both generalized arterial calcification of infancy and autosomal-recessive hypophosphatemic rickets. *Am. J. Hum. Genet.* **2010**, *86*, 267–272. [CrossRef] [PubMed]

55. Rutsch, F.; Ruf, N.; Vaingankar, S.; Toliat, M.R.; Suk, A.; Hohne, W.; Schauer, G.; Lehmann, M.; Roscioli, T.; Schnabel, D.; et al. Mutations in ENPP1 are associated with 'idiopathic' infantile arterial calcification. *Nat. Genet.* **2003**, *34*, 379–381. [CrossRef] [PubMed]

56. Le Saux, O.; Martin, L.; Aherrahrou, Z.; Leftheriotis, G.; Varadi, A.; Brampton, C.N. The molecular and physiological roles of ABCC6: More than meets the eye. *Front. Genet.* **2012**, *3*, 289. [CrossRef] [PubMed]

57. Nitschke, Y.; Rutsch, F. Generalized arterial calcification of infancy and pseudoxanthoma elasticum: Two sides of the same coin. *Front. Genet.* **2012**, *3*, 302. [CrossRef] [PubMed]

58. Li, Q.; Grange, D.K.; Armstrong, N.L.; Whelan, A.J.; Hurley, M.Y.; Rishavy, M.A.; Hallgren, K.W.; Berkner, K.L.; Schurgers, L.J.; Jiang, Q.; et al. Mutations in the GGCX and ABCC6 Genes in a Family with Pseudoxanthoma Elasticum-Like Phenotypes. *J. Investig. Dermatol.* **2009**, *129*, 553–563. [CrossRef]

59. Omarjee, L.; Nitschke, Y.; Verschuere, S.; Bourrat, E.; Vignon, M.; Navasiolava, N.; Leftheriotis, G.; Kauffenstein, G.; Rutsch, F.; Vanakker, O.; et al. Severe early-onset manifestations of pseudoxanthoma elasticum resulting from the cumulative effects of several deleterious mutations in ENPP1, ABCC6 and HBB: Transient improvement in ectopic calcification with sodium thiosulfate. *Br. J. Dermatol.* **2019**, *183*, 367–372. [CrossRef] [PubMed]

60. Bentley-Phillips, B. Pseudoxanthoma elasticum-like skin changes induced by penicillamine. *J. R. Soc. Med.* **1985**, *78*, 787.

61. Bolognia, J.L.; Braverman, I. Pseudoxanthoma-elasticum-like Skin Changes Induced by Penicillamine. *Dermatology* **1992**, *184*, 12–18. [CrossRef]

62. Coatesworth, A.P.; Darnton, S.J.; Green, R.M.; Cayton, R.M.; Antonakopoulos, G.N. A case of systemic pseudo-pseudoxanthoma elasticum with diverse symptomatology caused by long-term penicillamine use. *J. Clin. Pathol.* **1998**, *51*, 169–171. [CrossRef]

63. Dhurat, R.; Nayak, C.; Pereira, R.; Kagne, R.; Khatu, S. Penicillamine-induced elastosis perforans serpiginosa with abnormal "lumpy-bumpy" elastic fibers in lesional and non-lesional skin. *Indian J. Dermatol. Venereol. Leprol.* **2011**, *77*, 55–58. [CrossRef]

64. Aessopos, A.; Farmakis, D.; Loukopoulos, D. Elastic tissue abnormalities resembling pseudoxanthoma elasticum in beta thalassemia and the sickling syndromes. *Blood* **2002**, *99*, 30–35. [CrossRef]

65. Baccarani-Contri, M.; Bacchelli, B.; Boraldi, F.; Quaglino, D.; Taparelli, F.; Carnevali, E.; Francomano, M.A.; Seidenari, S.; Bettoli, V.; de Sanctis, V.; et al. Characterization of pseudoxanthoma elasticum-like lesions in the skin of patients with beta-thalassemia. *J. Am. Acad. Dermatol.* **2001**, *44*, 33–39. [CrossRef]

66. Cianciulli, P.; Sorrentino, F.; Maffei, L.; Amadori, S.; Cappabianca, M.P.; Foglietta, E.; Carnevali, E.; Pasquali-Ronchetti, I. Cardiovascular involvement in thalassaemic patients with pseudoxanthoma elasticum-like skin lesions: A long-term follow-up study. *Eur. J. Clin. Investig.* **2002**, *32*, 700–706. [CrossRef]

67. Farmakis, D.; Moyssakis, I.; Perakis, A.; Rombos, Y.; Deftereos, S.; Giakoumis, A.; Polymeropoulos, E.; Aessopos, A. Unstable angina associated with coronary arterial calcification in a thalassemia intermedia patient with a pseudoxanthoma elasticum-like syndrome. *Eur. J. Haematol.* **2003**, *70*, 64–66. [CrossRef]

68. Farmakis, D.; Vesleme, V.; Papadogianni, A.; Tsaftaridis, P.; Kapralos, P.; Aessopos, A. Aneurysmatic dilatation of ascending aorta in a patient with beta-thalassemia and a pseudoxanthoma elasticum-like syndrome. *Ann. Hematol.* **2004**, *83*, 596–599. [CrossRef] [PubMed]

69. Hamlin, N.; Beck, K.; Bacchelli, B.; Cianciulli, P.; Pasquali-Ronchetti, I.; le Saux, O. Acquired Pseudoxanthoma elasticum-like syndrome in beta-thalassaemia patients. *Br. J. Haematol.* **2003**, *122*, 852–854. [CrossRef] [PubMed]

70. Klein, I.; Sarkadi, B.; Váradi, A. An inventory of the human ABC proteins. *Biochim. Biophys. Acta Biomembr.* **1999**, *1461*, 237–262. [CrossRef]

71. Stefková, J.; Poledne, R.; Hubácek, J.A. ATP-binding cassette (ABC) transporters in human metabolism and diseases. *Physiol. Res.* **2004**, *53*, 235–243. [PubMed]
72. Dawson, R.J.; Locher, K.P. Structure of the multidrug ABC transporter Sav1866 from *Staphylococcus aureus* in complex with AMP-PNP. *FEBS Lett.* **2007**, *581*, 935–938. [CrossRef]
73. Fülöp, K.; Barna, L.; Symmons, O.; Závodszky, P.; Váradi, A. Clustering of disease-causing mutations on the domain–domain interfaces of ABCC6. *Biochem. Biophys. Res. Commun.* **2009**, *379*, 706–709. [CrossRef]
74. Madon, J.; Hagenbuch, B.; Landmann, L.; Meier, P.J.; Stieger, B. Transport Function and Hepatocellular Localization of mrp6 in Rat Liver. *Mol. Pharmacol.* **2000**, *57*, 634–641. [CrossRef]
75. Cai, J.; Daoud, R.; Alqawi, O.; Georges, E.; Pelletier, J.; Gros, P. Nucleotide binding and nucleotide hydrolysis properties of the ABC transporter MRP6 (ABCC6). *Biochemistry* **2002**, *41*, 8058–8067. [CrossRef]
76. Ilias, A.; Urban, Z.; Seidl, T.L.; le Saux, O.; Sinko, E.; Boyd, C.D.; Sarkadi, B.; Varadi, A. Loss of ATP-dependent transport activity in pseudoxanthoma elasticum-associated mutants of human ABCC6 (MRP6). *J. Biol. Chem.* **2002**, *277*, 16860–16867. [CrossRef]
77. Hirohashi, T.; Suzuki, H.; Sugiyama, Y. Characterization of the transport properties of cloned rat multi-drug resistance-associated protein 3 (MRP3). *J. Biol. Chem.* **1999**, *274*, 15181–15185. [CrossRef] [PubMed]
78. Belinsky, M.G.; Chen, Z.-S.; Shchaveleva, I.; Zeng, H.; Kruh, G.D. Characterization of the drug resistance and transport properties of multidrug resistance protein 6 (MRP6, ABCC6). *Cancer Res.* **2002**, *62*, 6172–6177. [PubMed]
79. Beck, K.; Hayashi, K.; Nishiguchi, B.; le Saux, O.; Hayashi, M.; Boyd, C.D. The Distribution of Abcc6 in Normal Mouse Tissues Suggests Multiple Functions for this ABC Transporter. *J. Histochem. Cytochem.* **2003**, *51*, 887–902. [CrossRef]
80. Le Saux, O.; Fülöp, K.; Yamaguchi, Y.; Iliás, A.; Szabó, Z.; Brampton, C.N.; Pomozi, V.; Huszár, K.; Arányi, T.; Váradi, A. Expression and In Vivo Rescue of Human ABCC6 Disease-Causing Mutants in Mouse Liver. *PLoS ONE* **2011**, *6*, e24738. [CrossRef] [PubMed]
81. Beck, K.; Hayashi, K.; Dang, K.; Hayashi, M.; Boyd, C.D. Analysis of ABCC6 (MRP6) in normal human tissues. *Histochem. Cell Biol.* **2005**, *123*, 517–528. [CrossRef] [PubMed]
82. Sinkó, E.; Iliás, A.; Ujhelly, O.; Homolya, L.; Scheffer, G.L.; Bergen, A.A.B.; Sarkadi, B.; Váradi, A. Subcellular localization and N-glycosylation of human ABCC6, expressed in MDCKII cells. *Biochem. Biophys. Res. Commun.* **2003**, *308*, 263–269. [CrossRef]
83. Martin, L.J.; Lau, E.; Singh, H.; Vergnes, L.; Tarling, E.J.; Mehrabian, M.; Mungrue, I.; Xiao, S.; Shih, D.; Castellani, L.; et al. ABCC6 localizes to the mitochondria-associated membrane. *Circ. Res.* **2012**, *111*, 516–520. [CrossRef]
84. Kato, K.; Nishimasu, H.; Okudaira, S.; Mihara, E.; Ishitani, R.; Takagi, J.; Aoki, J.; Nureki, O. Crystal structure of Enpp1, an extracellular glycoprotein involved in bone mineralization and insulin signaling. *Proc. Natl. Acad. Sci. USA* **2012**, *109*, 16876–16881. [CrossRef] [PubMed]
85. Roberts, F.; Zhu, D.; Farquharson, C.; Macrae, V.E. ENPP1 in the Regulation of Mineralization and Beyond. *Trends Biochem. Sci.* **2019**, *44*, 616–628. [CrossRef] [PubMed]
86. Chassaing, N.; Martin, L.; Calvas, P.; le Bert, M.; Hovnanian, A. Pseudoxanthoma elasticum: A clinical, pathophysiological and genetic update including 11 novel ABCC6 mutations. *J. Med Genet.* **2005**, *42*, 881–892. [CrossRef] [PubMed]
87. Le Saux, O.; Beck, K.; Sachsinger, C.; Silvestri, C.; Treiber, C.; Goring, H.H.; Johnson, E.W.; de Paepe, A.; Pope, F.M.; Pasquali-Ronchetti, I.; et al. A spectrum of abcc6 mutations is responsible for pseudoxanthoma elasticum. *Am. J. Hum. Genet.* **2001**, *69*, 749–764. [CrossRef]
88. Pfendner, E.G.; Vanakker, O.M.; Terry, P.F.; Vourthis, S.; McAndrew, E.P.; McClain, M.R.; Fratta, S.; Marais, A.-S.; Hariri, S.; Coucke, P.J.; et al. Mutation detection in the ABCC6 gene and genotype phenotype analysis in a large international case series affected by pseudoxanthoma elasticum. *J. Med Genet.* **2007**, *44*, 621–628. [CrossRef]
89. Nitschke, Y.; Yan, Y.; Buers, I.; Kintziger, K.; Askew, K.; Rutsch, F. ENPP1-Fc prevents neointima formation in generalized arterial calcification of infancy through the generation of AMP. *Exp. Mol. Med.* **2018**, *50*, 1–12. [CrossRef]
90. Gorgels, T.G.; Hu, X.; Scheffer, G.L.; van der Wal, A.C.; Toonstra, J.; de Jong, P.T.; van Kuppevelt, T.H.; Levelt, C.N.; de Wolf, A.; Loves, W.J.; et al. Disruption of Abcc6 in the mouse: Novel insight in the pathogenesis of pseudoxanthoma elasticum. *Hum. Mol. Genet.* **2005**, *14*, 1763–1773. [CrossRef]
91. Klement, J.F.; Matsuzaki, Y.; Jiang, Q.-J.; Terlizzi, J.; Choi, H.Y.; Fujimoto, N.; Li, K.; Pulkkinen, L.; Birk, D.E.; Sundberg, J.P.; et al. Targeted Ablation of the Abcc6 Gene Results in Ectopic Mineralization of Connective Tissues. *Mol. Cell. Biol.* **2005**, *25*, 8299–8310. [CrossRef]
92. Brampton, C.; Yamaguchi, Y.; Vanakker, O.; van Laer, L.; Chen, L.-H.; Thakore, M.; de Paepe, A.; Pomozi, V.; Szabó, P.T.; Martin, L.; et al. Vitamin K does not prevent soft tissue mineralization in a mouse model of pseudoxanthoma elasticum. *Cell Cycle* **2011**, *10*, 1810–1820. [CrossRef]
93. Brampton, C.; Pomozi, V.; Chen, L.-H.; Apana, A.; McCurdy, S.; Zoll, J.; Boisvert, W.A.; Lambert, G.; Henrion, D.; Blanchard, S.; et al. ABCC6 deficiency promotes dyslipidemia and atherosclerosis. *Sci. Rep.* **2021**, *11*, 3881. [CrossRef]
94. Ibold, B.; Tiemann, J.; Faust, I.; Ceglarek, U.; Dittrich, J.; Gorgels, T.G.M.F.; Bergen, A.A.B.; Vanakker, O.; van Gils, M.; Knabbe, C.; et al. Genetic deletion of Abcc6 disturbs cholesterol homeostasis in mice. *Sci. Rep.* **2021**, *11*, 2137. [CrossRef]
95. Jiang, Q.; Endo, M.; Dibra, F.; Wang, K.; Uitto, J. Pseudoxanthoma Elasticum Is a Metabolic Disease. *J. Investig. Dermatol.* **2009**, *129*, 348–354. [CrossRef]
96. Jiang, Q.; Oldenburg, R.; Otsuru, S.; Grand-Pierre, A.E.; Horwitz, E.M.; Uitto, J. Parabiotic heterogenetic pairing of Abcc6−/−/Rag1−/− mice and their wild-type counterparts halts ectopic mineralization in a murine model of pseudoxanthoma elasticum. *Am. J. Pathol.* **2010**, *176*, 1855–1862. [CrossRef]

97. Le Saux, O.; Bunda, S.; van Wart, C.M.; Douet, V.; Got, L.; Martin, L.; Hinek, A. Serum Factors from Pseudoxanthoma Elasticum Patients Alter Elastic Fiber Formation In Vitro. *J. Investig. Dermatol.* **2006**, *126*, 1497–1505. [CrossRef]

98. Li, Q.; Sundberg, J.P.; Levine, M.A.; Terry, S.F.; Uitto, J. The effects of bisphosphonates on ectopic soft tissue mineralization caused by mutations in the ABCC6 gene. *Cell Cycle* **2015**, *14*, 1082–1089. [CrossRef] [PubMed]

99. Pomozi, V.; Brampton, C.; Szeri, F.; Dedinszki, D.; Kozak, E.; van de Wetering, K.; Hopkins, H.; Martin, L.; Varadi, A.; le Saux, O. Functional Rescue of ABCC6 Deficiency by 4-Phenylbutyrate Therapy Reduces Dystrophic Calcification in Abcc6(−/−) Mice. *J. Investig. Dermatol.* **2017**, *137*, 595–602. [CrossRef] [PubMed]

100. Doehring, L.C.; Kaczmarek, P.M.; Ehlers, E.-M.; Mayer, B.; Erdmann, J.; Schunkert, H.; Aherrahrou, Z. Arterial calcification in mice after freeze-thaw injury. *Ann. Anat. Anat. Anz.* **2006**, *188*, 235–242. [CrossRef] [PubMed]

101. Eaton, G.J.; Custer, R.P.; Johnson, F.N.; Stabenow, K.T. Dystrophic cardiac calcinosis in mice: Genetic, hormonal, and dietary influences. *Am. J. Pathol.* **1978**, *90*, 173–186.

102. Everitt, J.I.; Olson, L.M.; Mangum, J.B.; Visek, W.J. High mortality with severe dystrophic cardiac calcinosis in C3H/OUJ mice fed high fat purified diets. *Vet. Pathol.* **1988**, *25*, 113–118. [CrossRef] [PubMed]

103. Aherrahrou, Z.; Axtner, S.B.; Kaczmarek, P.M.; Jurat, A.; Korff, S.; Doehring, L.C.; Weichenhan, D.; Katus, H.A.; Ivandic, B.T. A locus on chromosome 7 determines dramatic up-regulation of osteopontin in dystrophic cardiac calcification in mice. *Am. J. Pathol.* **2004**, *164*, 1379–1387. [CrossRef]

104. Ivandic, B.T.; Utz, H.F.; Kaczmarek, P.M.; Aherrahrou, Z.; Axtner, S.B.; Klepsch, C.; Lusis, A.J.; Katus, H.A. New Dyscalc loci for myocardial cell necrosis and calcification (dystrophic cardiac calcinosis) in mice. *Physiol. Genom.* **2001**, *6*, 137–144. [CrossRef] [PubMed]

105. Brunnert, S.R. Morphologic response of myocardium to freeze-thaw injury in mouse strains with dystrophic cardiac calcification. *Lab. Anim. Sci.* **1997**, *47*, 11–18.

106. Smolen, K.K.; Gelinas, L.; Franzen, L.; Dobson, S.; Dawar, M.; Ogilvie, G.; Krajden, M.; Fortuno, E.S., III; Kollmann, T.R. Age of recipient and number of doses differentially impact human B and T cell immune memory responses to HPV vaccination. *Vaccine* **2012**, *30*, 3572–3579. [CrossRef]

107. Li, Q.; Berndt, A.; Guo, H.; Sundberg, J.P.; Uitto, J. A Novel Animal Model for Pseudoxanthoma Elasticum: The KK/HlJ Mouse. *Am. J. Pathol.* **2012**, *181*, 1190–1196. [CrossRef] [PubMed]

108. Berndt, A.; Sundberg, B.A.; Silva, K.A.; Kennedy, V.E.; Richardson, M.A.; Li, Q.; Bronson, R.T.; Uitto, J.; Sundberg, J.P. Phenotypic Characterization of the KK/HlJ Inbred Mouse Strain. *Veter. Pathol.* **2013**, *51*, 846–857. [CrossRef] [PubMed]

109. De Vilder, E.Y.; Hosen, M.J.; Martin, L.; de Zaeytijd, J.; Leroy, B.P.; Ebran, J.; Coucke, P.J.; de Paepe, A.; Vanakker, O.M. VEGFA variants as prognostic markers for the retinopathy in pseudoxanthoma elasticum. *Clin. Genet.* **2020**, *98*, 74–79. [CrossRef]

110. Luo, H.; Faghankhani, M.; Cao, Y.; Uitto, J.; Li, Q. Molecular Genetics and Modifier Genes in Pseudoxanthoma Elasticum, a Heritable Multisystem Ectopic Mineralization Disorder. *J. Investig. Dermatol.* **2020**. [CrossRef]

111. Le Corre, Y.; le Saux, O.; Froeliger, F.; Libouban, H.; Kauffenstein, G.; Willoteaux, S.; Leftheriotis, G.; Martin, L. Quantification of the calcification phenotype of abcc6-deficient mice with microcomputed tomography. *Am. J. Pathol.* **2012**, *180*, 2208–2213. [CrossRef]

112. Li, Q.; Kingman, J.; van de Wetering, K.; Tannouri, S.; Sundberg, J.P.; Uitto, J. Abcc6 Knockout Rat Model Highlights the Role of Liver in PPi Homeostasis in Pseudoxanthoma Elasticum. *J. Investig. Dermatol.* **2017**, *137*, 1025–1032. [CrossRef]

113. Li, Q.; Sadowski, S.; Frank, M.M.; Chai, C.; Váradi, A.; Ho, S.-Y.; Lou, H.; Dean, M.; Thisse, C.; Thisse, B.; et al. The abcc6a Gene Expression Is Required for Normal Zebrafish Development. *J. Investig. Dermatol.* **2010**, *130*, 2561–2568. [CrossRef]

114. Van Gils, M.; Willaert, A.; de Vilder, E.; Coucke, P.; Vanakker, O. Generation and Validation of a Complete Knockout Model of abcc6a in Zebrafish. *J. Investig. Dermatol.* **2018**, *138*, 2333–2342. [CrossRef]

115. Sun, J.; She, P.; Liu, X.; Gao, B.; Jin, D.; Zhong, T.P. Disruption of Abcc6 Transporter in Zebrafish Causes Ocular Calcification and Cardiac Fibrosis. *Int. J. Mol. Sci.* **2020**, *22*, 278. [CrossRef]

116. Mackay, E.W.; Apschner, A.; Schulte-Merker, S. Vitamin K reduces hypermineralisation in zebrafish models of PXE and GACI. *Development* **2015**, *142*, 1095–1101. [CrossRef]

117. Pomozi, V.; Brampton, C.; Fülöp, K.; Chen, L.-H.; Apana, A.; Li, Q.; Uitto, J.; le Saux, O.; Váradi, A. Analysis of Pseudoxanthoma Elasticum–Causing Missense Mutants of ABCC6 In Vivo; Pharmacological Correction of the Mislocalized Proteins. *J. Investig. Dermatol.* **2014**, *134*, 946–953. [CrossRef] [PubMed]

118. Okawa, A.; Nakamura, I.; Goto, S.; Moriya, H.; Nakamura, Y.; Ikegawa, S. Mutation in Npps in a mouse model of ossification of the posterior longitudinal ligament of the spine. *Nat. Genet.* **1998**, *19*, 271–273. [CrossRef]

119. Li, Q.; Guo, H.; Chou, D.W.; Berndt, A.; Sundberg, J.P.; Uitto, J. Mutant Enpp1asj mice as a model for generalized arterial calcification of infancy. *Dis. Model. Mech.* **2013**, *6*, 1227–1235. [CrossRef]

120. Dedinszki, D.; Szeri, F.; Kozak, E.; Pomozi, V.; Tokesi, N.; Mezei, T.R.; Merczel, K.; Letavernier, E.; Tang, E.; Le Saux, O.; et al. Oral administration of pyrophosphate inhibits con-nective tissue calcification. *EMBO Mol. Med.* **2017**, *9*, 1463–1470. [CrossRef] [PubMed]

121. Huesa, C.; Zhu, D.; Glover, J.D.; Ferron, M.; Karsenty, G.; Milne, E.M.; Millan, J.L.; Ahmed, S.F.; Farquharson, C.; Morton, N.M.; et al. Deficiency of the bone mineralization inhibitor NPP1 protects mice against obesity and diabetes. *Dis. Model. Mech.* **2014**, *7*, 1341–1350. [CrossRef] [PubMed]

122. Mackenzie, N.; Huesa, C.; Rutsch, F.; MacRae, V. New insights into NPP1 function: Lessons from clinical and animal studies. *Bone* **2012**, *51*, 961–968. [CrossRef]
123. Apschner, A.; Huitema, L.F.; Ponsioen, B.; Peterson-Maduro, J.; Schulte-Merker, S. Zebrafish enpp1 mutants exhibit pathological mineralization, mimicking features of generalized arterial calcification of in-fancy (GACI) and pseudoxanthoma elasticum (PXE). *Dis. Model Mech.* **2014**, *7*, 811–822. [CrossRef]
124. Kauffenstein, G.; Yegutkin, G.G.; Khiati, S.; Pomozi, V.; le Saux, O.; Leftheriotis, G.; Lenaers, G.; Henrion, D.; Martin, L. Alteration of Extracellular Nucleotide Metabolism in Pseudoxanthoma Elasticum. *J. Investig. Dermatol.* **2018**, *138*, 1862–1870. [CrossRef] [PubMed]
125. Uitto, J.; Li, Q.; van de Wetering, K.; Váradi, A.; Terry, S.F. Insights into Pathomechanisms and Treatment Development in Heritable Ectopic Mineralization Disorders: Summary of the PXE International Biennial Research Symposium-2016. *J. Investig. Dermatol.* **2017**, *137*, 790–795. [CrossRef] [PubMed]
126. Kranenburg, G.; de Jong, P.A.; Bartstra, J.W.; Lagerweij, S.J.; Lam, M.G.; Norel, J.O.-V.; Risseeuw, S.; van Leeuwen, R.; Imhof, S.M.; Verhaar, H.J.; et al. Etidronate for Prevention of Ectopic Mineralization in Patients with Pseudoxanthoma Elasticum. *J. Am. Coll. Cardiol.* **2018**, *71*, 1117–1126. [CrossRef]
127. Li, Q.; Huang, J.; Pinkerton, A.B.; Millan, J.L.; van Zelst, B.D.; Levine, M.A.; Sundberg, J.P.; Uitto, J. Inhibition of Tissue-Nonspecific Alkaline Phosphatase Attenuates Ectopic Mineralization in the Abcc6 (−/−) Mouse Model of PXE but Not in the Enpp1 Mutant Mouse Models of GACI. *J. Investig. Dermatol.* **2019**, *139*, 360–368. [CrossRef] [PubMed]
128. Pomozi, V.; le Saux, O.; Brampton, C.; Apana, A.; Iliás, A.; Szeri, F.; Martin, L.; Monostory, K.; Paku, S.; Sarkadi, B.; et al. ABCC6 Is a Basolateral Plasma Membrane Protein. *Circ. Res.* **2013**, *112*, e148–e151. [CrossRef] [PubMed]
129. Rose, S.; On, S.J.; Fuchs, W.; Chen, C.; Phelps, R.; Kornreich, D.; Haddican, M.; Singer, G.; Wong, V.; Baum, D.; et al. Magnesium Supplementation in the Treatment of Pseudoxanthoma Elasticum (PXE): A randomized trial. *J. Am. Acad. Dermatol.* **2019**, *81*, 263–265. [CrossRef]
130. Yoo, J.Y.; Blum, R.R.; Singer, G.K.; Stern, D.K.; Emanuel, P.O.; Fuchs, W.; Phelps, R.G.; Terry, S.F.; Lebwohl, M.G. A randomized controlled trial of oral phosphate binders in the treatment of pseudoxanthoma elasticum. *J. Am. Acad. Dermatol.* **2011**, *65*, 341–348. [CrossRef]
131. Zhou, Y.; Jiang, Q.; Takahagi, S.; Shao, C.; Uitto, J.; Takahagi, S. Premature termination codon read-through in the ABCC6 gene: Potential treatment for pseudoxanthoma elasticum. *J. Investig. Dermatol.* **2013**, *133*, 2672–2677. [CrossRef]
132. Albright, R.A.; Stabach, P.; Cao, W.; Kavanagh, D.; Mullen, I.; Braddock, A.A.; Covo, M.S.; Tehan, M.; Yang, G.; Cheng, Z.; et al. ENPP1-Fc prevents mortality and vascular calcifications in rodent model of generalized arterial calcification of infancy. *Nat. Commun.* **2015**, *6*, 10006. [CrossRef] [PubMed]
133. Khan, T.; Sinkevicius, K.W.; Vong, S.; Avakian, A.; Leavitt, M.C.; Malanson, H.; Marozsan, A.; Askew, K.L. ENPP1 enzyme replacement therapy improves blood pressure and cardiovascular function in a mouse model of generalized arterial calcification of infancy. *Dis. Model. Mech.* **2018**, *11*, dmm035691. [CrossRef] [PubMed]
134. Schafer, C.; Heiss, A.; Schwarz, A.; Westenfeld, R.; Ketteler, M.; Floege, J.; Muller-Esterl, W.; Schinke, T.; Jahnen-Dechent, W. The serum protein alpha 2-Heremans-Schmid glycoprotein/fetuin-A is a systemically acting inhibitor of ectopic calcification. *J. Clin. Investig.* **2003**, *112*, 357–366. [CrossRef]
135. Jiang, Q.; Dibra, F.; Lee, M.D.; Oldenburg, R.; Uitto, J. Overexpression of fetuin-a counteracts ectopic mineralization in a mouse model of pseudoxanthoma elasticum (abcc6 (−/−)). *J. Investig. Dermatol.* **2010**, *130*, 1288–1296. [CrossRef] [PubMed]
136. Huang, J.; Snook, A.E.; Uitto, J.; Li, Q. Adenovirus-Mediated ABCC6 Gene Therapy for Heritable Ectopic Mineralization Disorders. *J. Investig. Dermatol.* **2019**, *139*, 1254–1263. [CrossRef] [PubMed]
137. La Russo, J.; Jiang, Q.; Li, Q.; Uitto, J. Ectopic mineralization of connective tissue in Abcc6(−/−) mice: Effects of dietary modifications and a phosphate binder—A preliminary study. *Exp. Dermatol.* **2008**, *17*, 203–207. [CrossRef] [PubMed]
138. Gorgels, T.G.M.F.; Waarsing, J.H.; de Wolf, A.; Brink, J.B.T.; Loves, W.J.P.; Bergen, A.A.B. Dietary magnesium, not calcium, prevents vascular calcification in a mouse model for pseudoxanthoma elasticum. *J. Mol. Med.* **2010**, *88*, 467–475. [CrossRef]
139. La Russo, J.; Li, Q.; Jiang, Q.; Uitto, J. Elevated dietary magnesium prevents connective tissue mineralization in a mouse model of pseudoxanthoma elasticum (Abcc6 (−/−)). *J. Investig. Dermatol.* **2009**, *129*, 1388–1394. [CrossRef] [PubMed]
140. Gorgels, T.G.M.F.; Waarsing, J.H.; Herfs, M.; Versteeg, D.; Schoensiegel, F.; Sato, T.; Schlingemann, R.O.; Ivandic, B.; Vermeer, C.; Schurgers, L.J.; et al. Vitamin K supplementation increases vitamin K tissue levels but fails to counteract ectopic calcification in a mouse model for pseudoxanthoma elasticum. *J. Mol. Med.* **2011**, *89*, 1125–1135. [CrossRef] [PubMed]
141. Jiang, Q.; Li, Q.; Grand-Pierre, A.E.; Schurgers, L.J.; Uitto, J. Administration of vitamin K does not counteract the ectopic mineralization of connective tissues in Abcc6 (−/−) mice, a model for pseudoxanthoma elasticum. *Cell Cycle* **2011**, *10*, 701–707. [CrossRef]
142. Carrillo-Linares, J.L.; García-Fernández, M.I.; Morillo, M.J.; Sánchez, P.; Rioja, J.; Barón, F.J.; Ariza, M.J.; Harrington, D.J.; Card, D.; Boraldi, F.; et al. The Effects of Parenteral K1 Administration in Pseudoxanthoma Elasticum Patients Versus Controls. A Pilot Study. *Front. Med.* **2018**, *5*, 86. [CrossRef]
143. Nollet, L.; van Gils, M.; Verschuere, S.; Vanakker, O. The Role of Vitamin K and Its Related Compounds in Mendelian and Acquired Ectopic Mineralization Disorders. *Int. J. Mol. Sci.* **2019**, *20*, 2142. [CrossRef] [PubMed]

144. Bauer, C.; le Saux, O.; Pomozi, V.; Aherrahrou, R.; Kriesen, R.; Stölting, S.; Liebers, A.; Kessler, T.S.H.; Erdmann, J.; Aherrahrou, Z. Etidronate prevents but does not completely reverse dystrophic cardiac calcification by.2 inhibiting macrophage aggregation. *Sci. Rep.* **2018**, *8*, 5812. [CrossRef]

145. Otero, J.E.; Gottesman, G.S.; McAlister, W.H.; Mumm, S.; Madson, K.L.; Kiffer-Moreira, T.; Sheen, C.; Millán, J.L.; Ericson, K.L.; Whyte, M.P. Severe skeletal toxicity from protracted etidronate therapy for generalized arterial calcification of infancy. *J. Bone Miner. Res.* **2013**, *28*, 419–430. [CrossRef]

146. Chong, C.R.; Hutchins, G.M. Idiopathic infantile arterial calcification: The spectrum of clinical presentations. *Pediatr. Dev. Pathol.* **2008**, *11*, 405–415. [CrossRef] [PubMed]

147. Dabisch-Ruthe, M.; Kuzaj, P.; Götting, C.; Knabbe, C.; Hendig, D. Pyrophosphates as a major inhibitor of matrix calcification in *Pseudoxanthoma elasticum*. *J. Dermatol. Sci.* **2014**, *75*, 109–120. [CrossRef] [PubMed]

148. Li, Q.; Kingman, J.; Sundberg, J.P.; Levine, M.A.; Uitto, J. Etidronate prevents, but does not reverse, ectopic mineralization in a mouse model of pseudoxanthoma elasticum (Abcc6−/−). *Oncotarget* **2016**, *9*, 30721. [CrossRef] [PubMed]

149. Jacobs, I.J.; Li, D.; Ivarsson, M.E.; Uitto, J.; Li, Q. A phytic acid analogue INS-3001 prevents ectopic calcification in an Abcc6(−/−) mouse model of pseudoxanthoma elasticum. *Exp. Dermatol.* **2021**. [CrossRef] [PubMed]

150. Hosen, M.J.; van Nieuwerburgh, F.; Steyaert, W.; Deforce, D.; Martin, L.; Leftheriotis, G.; de Paepe, A.; Coucke, P.J.; Vanakker, O.M. Efficiency of Exome Sequencing for the Molecular Diagnosis of Pseudoxanthoma Elasticum. *J. Investig. Dermatol.* **2015**, *135*, 992–998. [CrossRef] [PubMed]

151. Plomp, A.S.; Bergen, A.A.; Florijn, R.J.; Terry, S.F.; Toonstra, J.; van Dijk, M.R.; de Jong, P.T. Pseudoxanthoma elasticum: Wide phenotypic variation in homozygotes and no signs in heterozygotes for the c.3775delT mutation in ABCC6. *Genet. Med.* **2009**, *11*, 852–858. [CrossRef]

152. Uitto, J.; Jiang, Q.; Varadi, A.; Bercovitch, L.G.; Terry, S.F. Pseudoxanthoma Elasticum: Diagnostic Features, Classification, and Treatment Options. *Expert Opin. Orphan Drugs* **2014**, *2*, 567–577. [CrossRef]

153. Bartstra, J.W.; Risseeuw, S.; de Jong, P.A.; van Os, B.; Kalsbeek, L.; Mol, C.; Baas, A.F.; Verschuere, S.; Vanakker, O.; Florijn, R.J.; et al. Genotype-phenotype correlation in pseudoxanthoma elasticum. *Atherosclerosis* **2021**, *324*, 18–26. [CrossRef] [PubMed]

154. Boraldi, F.; Costa, S.; Rabacchi, C.; Ciani, M.; Vanakker, O.; Quaglino, D. Can APOE and MTHFR poly-morphisms have an influence on the severity of cardiovascular manifestations in Italian Pseudoxanthoma elasticum affected patients? *Mol. Genet. Metab. Rep.* **2014**, *1*, 477–482. [CrossRef] [PubMed]

155. Hendig, D.; Knabbe, C.; Götting, C. New insights into the pathogenesis of pseudoxanthoma elasticum and related soft tissue calcification disorders by identifying genetic interactions and modifiers. *Front. Genet.* **2013**, *4*, 114. [CrossRef]

156. Vanakker, O.M.M.; Hosen, M.J.; de Paepe, A. The ABCC6 transporter: What lessons can be learnt from other ATP-binding cassette transporters? *Front. Genet.* **2013**, *4*, 203. [CrossRef] [PubMed]

157. Navasiolava, N.; Gnanou, M.; Douillard, M.; Saulnier, P.; Aranyi, T.; Ebran, J.-M.; Henni, S.; Humeau, H.; Lefthériotis, G.; Martin, L. The extent of pseudoxanthoma elasticum skin changes is related to cardiovascular complications and visual loss: A cross-sectional study. *Br. J. Dermatol.* **2019**, *180*, 207–208. [CrossRef] [PubMed]

158. Zhao, J.; Kingman, J.; Sundberg, J.P.; Uitto, J.; Li, Q. Plasma PPi Deficiency Is the Major, but Not the Exclusive, Cause of Ectopic Mineralization in an Abcc6 Mouse Model of PXE. *J. Investig. Dermatol.* **2017**, *137*, 2336–2343. [CrossRef]

159. Mention, P.J.; Lacoeuille, F.; Leftheriotis, G.; Martin, L.; Omarjee, L. 18F-Flurodeoxyglucose and 18F-Sodium Fluoride Positron Emission Tomography/Computed Tomography Imaging of Arterial and Cu-taneous Alterations in Pseudoxanthoma Elasticum. *Circ. Cardiovasc. Imaging* **2018**, *11*, e007060. [CrossRef]

160. Oudkerk, S.F.; de Jong, P.A.; Blomberg, B.A.; Scholtens, A.M.; Mali, W.P.; Spiering, W. Whole-Body Visualization of Ectopic Bone Formation of Arteries and Skin in Pseudoxanthoma Elasticum. *JACC Cardiovasc. Imaging* **2016**, *9*, 755–756. [CrossRef]

161. Humeau-Heurtier, A.; Colominas, M.A.; Schlotthauer, G.; Etienne, M.; Martin, L.; Abraham, P. Bidimensional unconstrained optimization approach to EMD: An algorithm revealing skin perfusion alterations in pseudoxanthoma elasticum patients. *Comput. Methods Programs Biomed.* **2017**, *140*, 233–239. [CrossRef]

162. Gass, J.D. "Comet" lesion: An ocular sign of pseudoxanthoma elasticum. *Retina* **2003**, *23*, 729–730. [CrossRef]

163. Ciulla, T.A.; Rosenfeld, P.J. Anti-vascular endothelial growth factor therapy for neovascular ocular diseases other than age-related macular degeneration. *Curr. Opin. Ophthalmol.* **2009**, *20*, 166–174. [CrossRef]

164. Vasseur, M.; Carsin-Nicol, B.; Ebran, J.; Willoteaux, S.; Martin, L.; Leftheriotis, G. Carotid Rete Mirabile and Pseudoxanthoma Elasticum: An Accidental Association? *Eur. J. Vasc. Endovasc. Surg.* **2011**, *42*, 292–294. [CrossRef] [PubMed]

165. Pavlovic, A.M.; Zidverc-Trajkovic, J.; Milovic, M.M.; Pavlovic, D.M.; Jovanovic, Z.; Mijajlovic, M.; Petrovic, M.; Kostic, V.S.; Sternic, N. Cerebral Small Vessel Disease in Pseudoxanthoma Elasticum: Three Cases. *Can. J. Neurol. Sci.* **2005**, *32*, 115–118. [CrossRef]

166. Miglionico, R.; Ostuni, A.; Armentano, M.F.; Milella, L.; Crescenzi, E.; Carmosino, M.; Bisaccia, F. ABCC6 knockdown in HepG2 cells induces a senescent-like cell phenotype. *Cell. Mol. Biol. Lett.* **2017**, *22*, 1–10. [CrossRef] [PubMed]

167. Tiemann, J.; Wagner, T.; Lindenkamp, C.; Plumers, R.; Faust, I.; Knabbe, C.; Hendig, D. Linking ABCC6 Deficiency in Primary Human Dermal Fibroblasts of PXE Patients to p21-Mediated Premature Cellular Se-nescence and the Development of a Proinflammatory Secretory Phenotype. *Int. J. Mol. Sci.* **2020**, *21*, 24. [CrossRef] [PubMed]

168. Mungrue, I.N.; Zhao, P.; Yao, Y.; Meng, H.; Rau, C.; Havel, J.V.; Gorgels, T.G.M.F.; Bergen, A.A.; MacLellan, W.R.; Drake, T.A.; et al. Abcc6 Deficiency Causes Increased Infarct Size and Apoptosis in a Mouse Cardiac Ischemia-Reperfusion Model. *Arter. Thromb. Vasc. Biol.* **2011**, *31*, 2806–2812. [CrossRef] [PubMed]
169. Hu, X.; Peek, R.; Plomp, A.; Brink, J.T.; Scheffer, G.; van Soest, S.; Leys, A.; de Jong, P.T.V.M.; Bergen, A.A.B. Analysis of the Frequent R1141X Mutation in theABCC6Gene in Pseudoxanthoma Elasticum. *Investig. Ophthalmol. Vis. Sci.* **2003**, *44*, 1824–1829. [CrossRef]
170. Pfendner, E.G.; Uitto, J.; Gerard, G.F.; Terry, S.F. Pseudoxanthoma elasticum: Genetic diagnostic markers. *Expert Opin. Med. Diagn.* **2008**, *2*, 63–79. [CrossRef]
171. Keeling, K.M.; Bedwell, D.M. Suppression of nonsense mutations as a therapeutic approach to treat genetic diseases. *Wiley Interdiscip. Rev. RNA* **2011**, *2*, 837–852. [CrossRef]
172. Wilschanski, M.; Miller, L.L.; Shoseyov, D.; Blau, H.; Rivlin, J.; Aviram, M.; Cohen, M.; Armoni, S.; Yaakov, Y.; Pugatch, T.; et al. Chronic ataluren (PTC124) treatment of nonsense mutation cystic fibrosis. *Eur. Respir. J.* **2011**, *38*, 59–69. [CrossRef]
173. Kerem, E.E.; Konstan, M.M.; de Boeck, K.; Accurso, F.F.; Sermet-Gaudelus, I.; Wilschanski, M.M.; Elborn, S.J.; Melotti, P.P.; Bronsveld, I.I.; Fajac, I.; et al. Ataluren for the treatment of nonsense-mutation cystic fibrosis: A randomised, double-blind, placebo-controlled phase 3 trial. *Lancet Respir. Med.* **2014**, *2*, 539–547. [CrossRef]
174. Dover, G.J.; Brusilow, S.; Samid, D. Increased Fetal Hemoglobin in Patients Receiving Sodium 4-Phenylbutyrate. *N. Engl. J. Med.* **1992**, *327*, 569–570. [CrossRef]
175. Maestri, N.E.; Brusilow, S.W.; Clissold, D.B.; Bassett, S.S. Long-Term Treatment of Girls with Ornithine Transcarbamylase Deficiency. *N. Engl. J. Med.* **1996**, *335*, 855–860. [CrossRef]
176. Perrine, S.P.; Ginder, G.D.; Faller, D.V.; Dover, G.H.; Ikuta, T.; Witkowska, H.E.; Cai, S.P.; Vichinsky, E.P.; Olivieri, N.F. A short-term trial of butyrate to stimulate fetal-globin-gene expression in the beta-globin disorders. *N. Engl. J. Med.* **1993**, *328*, 81–86. [CrossRef]
177. McGuire, B.M.; Zupanets, I.A.; Lowe, M.E.; Xiao, X.; Syplyviy, V.A.; Monteleone, J.; Gargosky, S.; Dickinson, K.; Martinez, A.; Mokhtarani, M.; et al. Pharmacology and safety of glycerol phenyl-butyrate in healthy adults and adults with cirrhosis. *Hepatology* **2010**, *51*, 2077–2085. [CrossRef] [PubMed]
178. Iannitti, T.; Palmieri, B. Clinical and experimental applications of sodium phenylbutyrate. *Drugs R D* **2011**, *11*, 227–249. [CrossRef] [PubMed]
179. Gonzales, E.; Grosse, B.; Schuller, B.; Davit-Spraul, A.; Conti, F.; Guettier, C.; Cassio, D.; Jacquemin, E. Targeted pharmacotherapy in progressive familial intrahepatic cholestasis type 2: Evidence for improvement of cholestasis with 4-phenylbutyrate. *Hepatology* **2015**, *62*, 558–566. [CrossRef] [PubMed]
180. Hayashi, H.; Naoi, S.; Hirose, Y.; Matsuzaka, Y.; Tanikawa, K.; Igarashi, K.; Nagasaka, H.; Kage, M.; Inui, A.; Kusuhara, H. Successful treatment with 4-phenylbutyrate in a patient with benign recurrent intrahepatic cholestasis type 2 refractory to biliary drainage and bilirubin absorption. *Hepatol. Res.* **2016**, *46*, 192–200. [CrossRef]
181. Rubenstein, R.C.; Zeitlin, P.L. Sodium 4-phenylbutyrate downregulates Hsc70: Implications for intra-cellular trafficking of DeltaF508-CFTR. *Am. J. Physiol. Cell. Physiol.* **2000**, *278*, C259–C267. [CrossRef]
182. Sorrenson, B.; Suetani, R.J.; Williams, M.J.A.; Bickley, V.M.; George, P.M.; Jones, G.T.; McCormick, S.P.A. Functional rescue of mutant ABCA1 proteins by sodium 4-phenylbutyrate. *J. Lipid Res.* **2013**, *54*, 55–62. [CrossRef]
183. Millán, J.L.; Whyte, M.P. Alkaline Phosphatase and Hypophosphatasia. *Calcif. Tissue Int.* **2016**, *98*, 398–416. [CrossRef] [PubMed]
184. Narisawa, S.; Harmey, D.; Yadav, M.C.; O'Neill, W.C.; Hoylaerts, M.F.; Millán, J.L. Novel Inhibitors of Alkaline Phosphatase Suppress Vascular Smooth Muscle Cell Calcification. *J. Bone Miner. Res.* **2007**, *22*, 1700–1710. [CrossRef] [PubMed]
185. Sheen, C.R.; Kuss, P.; Narisawa, S.; Yadav, M.C.; Nigro, J.; Wang, W.; Chhea, T.N.; Sergienko, E.A.; Kapoor, K.; Jackson, M.R.; et al. Pathophysiological Role of Vascular Smooth Muscle Alkaline Phosphatase in Medial Artery Calcification. *J. Bone Miner. Res.* **2015**, *30*, 824–836. [CrossRef]
186. Savinov, A.Y.; Salehi, M.; Yadav, M.C.; Radichev, I.; Millan, J.L.; Savinova, O.V. Transgenic Overexpression of Tissue-Nonspecific Alkaline Phosphatase (TNAP) in Vascular Endothelium Results in Generalized Arterial Calcification. *J. Am. Heart Assoc.* **2015**, *4*, 12. [CrossRef] [PubMed]
187. St Hilaire, C.; Ziegler, S.G.; Markello, T.C.; Brusco, A.; Groden, C.; Gill, F.; Carlson-Donohoe, H.; Lederman, R.J.; Chen, M.Y.; Yang, D.; et al. NT5E mutations and arterial calcifications. *N. Engl. J. Med.* **2011**, *364*, 432–442. [CrossRef] [PubMed]
188. Li, Q.; Price, T.P.; Sundberg, J.P.; Uitto, J. Juxta-articular joint-capsule mineralization in CD73 deficient mice: Similarities to patients with NT5E mutations. *Cell Cycle* **2014**, *13*, 2609–2615. [CrossRef]
189. Lee, C.S.; Bishop, E.S.; Zhang, R.; Yu, X.; Farina, E.M.; Yan, S.; Zhao, C.; Zheng, Z.; Shu, Y.; Wu, X.; et al. Adenovirus-Mediated Gene Delivery: Potential Applications for Gene and Cell-Based Therapies in the New Era of Personalized Medicine. *Genes Dis.* **2017**, *4*, 43–63. [CrossRef]
190. De Baaij, J.H.F.; Hoenderop, J.G.J.; Bindels, R.J.M. Magnesium in Man: Implications for Health and Disease. *Physiol. Rev.* **2015**, *95*, 1–46. [CrossRef]
191. Ter, B.A.D.; Shanahan, C.M.; de Baaij, J.H.F. Magnesium Counteracts Vascular Calcification: Passive Interference or Active Modulation? *Arter. Thromb Vasc. Biol.* **2017**, *37*, 1431–1445.
192. Alfrey, A.C.; Miller, N.L.; Trow, R. Effect of Age and Magnesium Depletion on Bone Magnesium Pools in Rats. *J. Clin. Investig.* **1974**, *54*, 1074–1081. [CrossRef] [PubMed]

193. Sakaguchi, Y.; Hamano, T.; Isaka, Y. Magnesium and Progression of Chronic Kidney Disease: Benefits Beyond Cardiovascular Protection? *Adv. Chronic Kidney Dis.* **2018**, *25*, 274–280. [CrossRef]
194. D'Marco, L.; Lima-Martínez, M.; Karohl, C.; Chacín, M.; Bermúdez, V. Pseudoxanthoma Elasticum: An Interesting Model to Evaluate Chronic Kidney Disease-Like Vascular Damage without Renal Disease. *Kidney Dis.* **2020**, *6*, 92–97. [CrossRef]
195. Schinke, T.; McKee, M.D.; Karsenty, G. Extracellular matrix calcification: Where is the action? *Nat. Genet.* **1999**, *21*, 150–151. [CrossRef] [PubMed]
196. Gheduzzi, D.; Boraldi, F.; Annovi, G.; Devincenzi, C.P.; Schurgers, L.J.; Vermeer, C.; Quaglino, D.; Ronchetti, I.P. Matrix Gla protein is involved in elastic fiber calcification in the dermis of pseudoxanthoma elasticum patients. *Lab. Investig.* **2007**, *87*, 998–1008. [CrossRef]
197. Vermeer, C. Gamma-carboxyglutamate-containing proteins and the vitamin K-dependent carboxylase. *Biochem. J.* **1990**, *266*, 625–636. [CrossRef]
198. Li, Q.; Jiang, Q.; Schurgers, L.J.; Uitto, J. Pseudoxanthoma elasticum: Reduced gamma-glutamyl carboxylation of matrix gla protein in a mouse model (Abcc6−/−). *Biochem. Biophys. Res. Commun.* **2007**, *364*, 208–213. [CrossRef]
199. Vanakker, O.M.; Martin, L.; Gheduzzi, D.; Leroy, B.P.; Loeys, B.L.; Guerci, V.I.; Matthys, D.; Terry, S.F.; Coucke, P.J.; Pasquali-Ronchetti, I.; et al. Pseudoxanthoma Elasticum-Like Phenotype with Cutis Laxa and Multiple Coagulation Factor Deficiency Represents a Separate Genetic Entity. *J. Investig. Dermatol.* **2007**, *127*, 581–587. [CrossRef]
200. Borst, P.; van de Wetering, K.; Schlingemann, R. Does the absence of ABCC6 (multidrug resistance protein 6) in patients with Pseudoxanthoma elasticum prevent the liver from providing sufficient vitamin K to the periphery? *Cell Cycle* **2008**, *7*, 1575–1579. [CrossRef]
201. Vanakker, O.M.; Martin, L.; Schurgers, L.J.; Quaglino, D.; Costrop, L.; Vermeer, C.; Pasquali-Ronchetti, I.; Coucke, P.J.; de Paepe, A. Low serum vitamin K in PXE results in defective carboxylation of mineralization inhibitors similar to the GGCX mutations in the PXE-like syndrome. *Lab. Investig.* **2010**, *90*, 895–905. [CrossRef] [PubMed]
202. Boraldi, F.; Annovi, G.; Guerra, D.; Devincenzi, C.P.; Garcia-Fernandez, M.I.; Panico, F.; de Santis, G.; Tiozzo, R.; Ronchetti, I.; Quaglino, D. Fibroblast protein profile analysis highlights the role of oxidative stress and vitamin K recycling in the pathogenesis of pseudoxanthoma elasticum. *Proteom. Clin. Appl.* **2009**, *3*, 1084–1098. [CrossRef]
203. Huesa, C.; Staines, K.A.; Millan, J.L.; MacRae, V.E. Effects of etidronate on the Enpp1(-)/(-) mouse model of generalized arterial calcification of infancy. *Int. J. Mol. Med.* **2015**, *36*, 159–165. [CrossRef]
204. Lomashvili, K.A.; Monier-Faugere, M.C.; Wang, X.; Malluche, H.H.; O'Neill, W.C. Effect of bisphosphonates on vascular calcification and bone metabolism in experimental renal failure. *Kidney Int.* **2009**, *75*, 617–625. [CrossRef] [PubMed]
205. Oliveira, J.R.M.; Oliveira, M.F. Primary brain calcification in patients undergoing treatment with the biphosphanate alendronate. *Sci. Rep.* **2016**, *6*, 22961. [CrossRef] [PubMed]
206. Fleisch, H.; Bisaz, S. Mechanism of Calcification: Inhibitory Role of Pyrophosphate. *Nat. Cell Biol.* **1962**, *195*, 911. [CrossRef] [PubMed]
207. O'Neill, W.C.; Lomashvili, K.A.; Malluche, H.H.; Faugere, M.C.; Riser, B.L. Treatment with pyrophosphate inhibits uremic vascular calcification. *Kidney Int.* **2011**, *79*, 512–517. [CrossRef]
208. Orriss, I.R.; Arnett, T.R.; Russell, R.G.G. Pyrophosphate: A key inhibitor of mineralisation. *Curr. Opin. Pharmacol.* **2016**, *28*, 57–68. [CrossRef]
209. Grases, F.; Sanchis, P.; Perello, J.; Isern, B.; Prieto, R.M.; Fernandez-Palomeque, C.; Fiol, M.; Bonnin, O.; Torres, J.J. Phytate (Myo-inositol hexakisphosphate) inhibits cardiovascular calcifications in rats. *Front. Biosci.* **2006**, *11*, 136–142. [CrossRef]
210. Van den Berg, C.J.; Hill, L.F.; Stanbury, S.W. Inositol phosphates and phytic acid as inhibitors of biological calcification in the rat. *Clin. Sci.* **1972**, *43*, 377–383. [CrossRef]
211. Schantl, A.E.; Verhulst, A.; Neven, E.; Behets, G.J.; D'Haese, P.C.; Maillard, M.; Mordasini, D.; Phan, O.; Burnier, M.; Spaggiari, D.; et al. Inhibition of vascular calcification by inositol phosphates derivatized with ethylene glycol oligomers. *Nat. Commun.* **2020**, *11*, 721. [CrossRef]
212. Dennie, W.; McBride, C. Treatment of arsphenamine dermatitis and certain other metallic poisonings. *Arch. Dermatol. Syphilol.* **1923**, *11*, 63.
213. Halliday, H.M.; Sutherland, C.E. Arsenical Poisoning Treated by Sodium Thiosulphate. *BMJ* **1925**, *1*, 407. [CrossRef]
214. Hayden, M.R.; Goldsmith, D.J.A. Sodium Thiosulfate: New Hope for the Treatment of Calciphylaxis. *Semin. Dial.* **2010**, *23*, 258–262. [CrossRef]
215. Arányi, T.; Bacquet, C.; de Boussac, H.; Ratajewski, M.; Pomozi, V.; Fülöp, K.; Brampton, C.N.; Pulaski, L.; le Saux, O.; Váradi, A. Transcriptional regulation of the ABCC6 gene and the background of impaired function of missense disease-causing mutations. *Front. Genet.* **2013**, *4*, 27. [CrossRef]
216. Douet, V.; Heller, M.B.; le Saux, O. DNA methylation and Sp1 binding determine the tissue-specific transcriptional activity of the mouse Abcc6 promoter. *Biochem. Biophys. Res. Commun.* **2007**, *354*, 66–71. [CrossRef] [PubMed]
217. Douet, V.; van Wart, C.M.; Heller, M.B.; Reinhard, S.; le Saux, O. HNF4alpha and NF-E2 are key transcriptional regulators of the murine Abcc6 gene expression. *Biochim. Biophys. Acta* **2006**, *1759*, 426–436. [CrossRef]
218. Väärämäki, S.; Pelttari, S.; Uusitalo, H.; Tokesi, N.; Varadi, A.; Nevalainen, P. Pyrophosphate Treatment in Pseudoxanthoma Elasticum (PXE)-Preventing ReOcclusion After Surgery for Critical Limb Ischaemia. *Surg. Case Rep.* **2019**, *2*, 1–3. [CrossRef]
219. Leopold, J.A. Vascular calcification: An age-old problem of old age. *Circulation* **2013**, *127*, 2380–2382. [CrossRef] [PubMed]

220. Rennenberg, R.J.M.W.; Kessels, A.G.H.; Schurgers, L.J.; Van Engelshoven, J.M.A.; de Leeuw, P.; Kroon, A.A. Vascular calcifications as a marker of increased cardiovascular risk: A meta-analysis. *Vasc. Health Risk Manag.* **2009**, *5*, 185–197. [CrossRef]
221. Kranenburg, G.; Baas, A.F.; de Jong, P.A.; Asselbergs, F.W.; Visseren, F.L.J.; Spiering, W. The prevalence of pseudoxanthoma elasticum: Revised estimations based on genotyping in a high vascular risk cohort. *Eur. J. Med. Genet.* **2019**, *62*, 90–92. [CrossRef] [PubMed]
222. Buchheiser, A.; Ebner, A.; Burghoff, S.; Ding, Z.; Romio, M.; Viethen, C.; Lindecke, A.; Köhrer, K.; Fischer, J.W.; Schrader, J. Inactivation of CD73 promotes atherogenesis in apolipoprotein E-deficient mice. *Cardiovasc. Res.* **2011**, *92*, 338–347. [CrossRef] [PubMed]
223. Jalkanen, J.; Hollmén, M.; Jalkanen, S.; Hakovirta, H. Regulation of CD73 in the development of lower limb atherosclerosis. *Purinergic Signal.* **2016**, *13*, 127–134. [CrossRef] [PubMed]
224. Nitschke, Y.; Weissen-Plenz, G.; Terkeltaub, R.; Rutsch, F. Npp1 promotes atherosclerosis in ApoE knockout mice. *J. Cell. Mol. Med.* **2011**, *15*, 2273–2283. [CrossRef]
225. Sutton, N.R.; Bouïs, D.; Mann, K.M.; Rashid, I.M.; McCubbrey, A.L.; Hyman, M.C.; Goldstein, D.R.; Mei, A.; Pinsky, D.J. CD73 Promotes Age-Dependent Accretion of Atherosclerosis. *Arter. Thromb. Vasc. Biol.* **2020**, *40*, 61–71. [CrossRef]
226. Li, Q.; Guo, H.; Chou, D.W.; Berndt, A.; Sundberg, J.P.; Uitto, J. Mouse Models for Pseudoxanthoma Elasticum: Genetic and Dietary Modulation of the Ectopic Mineralization Phenotypes. *PLoS ONE* **2014**, *9*, e89268. [CrossRef] [PubMed]

International Journal of
Molecular Sciences

Article

Mutagenic Analysis of the Putative ABCC6 Substrate-Binding Cavity Using a New Homology Model

Flora Szeri [1,2], Valentina Corradi [3], Fatemeh Niaziorimi [1], Sylvia Donnelly [1], Gwenaëlle Conseil [4], Susan P. C. Cole [4], D. Peter Tieleman [3] and Koen van de Wetering [1,*]

[1]   Department of Dermatology and Cutaneous Biology and PXE Center of Excellence in Research and Clinical Care, Thomas Jefferson University, Philadelphia, PA 19107, USA; szeri.flora@ttk.hu (F.S.); Fatemeh.Niaziorimi@jefferson.edu (F.N.); Sylvia.Donnelly@jefferson.edu (S.D.)

[2]   Research Centre for Natural Sciences, Institute of Enzymology, 1117 Budapest, Hungary

[3]   Department of Biological Sciences and Centre for Molecular Simulation, University of Calgary, Calgary, AB T2N 1N4, Canada; vcorradi@ucalgary.ca (V.C.); tieleman@ucalgary.ca (D.P.T.)

[4]   Department of Pathology and Molecular Medicine, Queen's University, Kingston, ON K7L 3N6, Canada; conseilg@queensu.ca (G.C.); spc.cole@queensu.ca (S.P.C.C.)

*   Correspondence: Koen.vandeWetering@jefferson.edu; Tel.: +1-(215)-503-5701

Abstract: Inactivating mutations in ABCC6 underlie the rare hereditary mineralization disorder pseudoxanthoma elasticum. ABCC6 is an ATP-binding cassette (ABC) integral membrane protein that mediates the release of ATP from hepatocytes into the bloodstream. The released ATP is extracellularly converted into pyrophosphate, a key mineralization inhibitor. Although ABCC6 is firmly linked to cellular ATP release, the molecular details of ABCC6-mediated ATP release remain elusive. Most of the currently available data support the hypothesis that ABCC6 is an ATP-dependent ATP efflux pump, an un-precedented function for an ABC transporter. This hypothesis implies the presence of an ATP-binding site in the substrate-binding cavity of ABCC6. We performed an extensive mutagenesis study using a new homology model based on recently published structures of its close homolog, bovine Abcc1, to characterize the substrate-binding cavity of ABCC6. Leukotriene C4 (LTC_4), is a high-affinity substrate of ABCC1. We mutagenized fourteen amino acid residues in the rat ortholog of ABCC6, rAbcc6, that corresponded to the residues in ABCC1 found in the LTC_4 binding cavity. Our functional characterization revealed that most of the amino acids in rAbcc6 corresponding to those found in the LTC_4 binding pocket in bovine Abcc1 are not critical for ATP efflux. We conclude that the putative ATP binding site in the substrate-binding cavity of ABCC6/rAbcc6 is distinct from the bovine Abcc1 LTC_4-binding site.

Keywords: ABC transporter; pseudoxanthoma elasticum; homology modeling; substrate-binding site; cellular ATP efflux; mutagenesis

Citation: Szeri, F.; Corradi, V.; Niaziorimi, F.; Donnelly, S.; Conseil, G.; Cole, S.P.C.; Tieleman, D.P.; van de Wetering, K. Mutagenic Analysis of the Putative ABCC6 Substrate-Binding Cavity Using a New Homology Model. *Int. J. Mol. Sci.* **2021**, *22*, 6910. https://doi.org/10.3390/ijms22136910

Academic Editor: Satoshi Kametaka

Received: 8 May 2021
Accepted: 23 June 2021
Published: 27 June 2021

Publisher's Note: MDPI stays neutral with regard to jurisdictional claims in published maps and institutional affiliations.

1. Introduction

Inactivating mutations in the gene encoding ATP-binding cassette (ABC) subfamily C member 6 (ABCC6) underlie the autosomal recessive disease pseudoxanthoma elasticum (PXE, OMIM #264800) [1–3], characterized by ectopic mineralization in the skin, eyes, and vascular system [4–6]. PXE is a slowly progressing connective tissue disorder that affects approximately 1 in 50,000 individuals worldwide [7]. There is currently no specific and effective therapy for PXE and the disease slowly progresses after initial diagnosis [8].

ABCC6 is predominantly expressed in the liver [9] where it mediates the release of ATP from hepatocytes into the bloodstream [10,11]. Outside the hepatocytes, yet still in the liver niche, the released ATP is converted into AMP and the mineralization inhibitor pyrophosphate (PPi), by ectonucleotide pyrophosphatase phosphodiesterase 1 (ENPP1) [12]. The absence of ABCC6-mediated ATP release in both PXE patients and Abcc6 null mice results in plasma PPi levels that are < 40% of those found in ABCC6-proficient individuals [11], providing a plausible biochemical explanation for their ectopic mineralization.

Moreover, plasma PPi levels decline during pregnancy, which might explain the increased risk of vascular calcification in multiparous individuals [13]. Recent data indicate that ATP efflux by the progressive ankylosis protein (ANK) is also a major determinant of plasma PPi levels [14]. Intriguingly, an ABC protein other than ABCC6 has been reported to also be involved in cellular ATP release, albeit indirectly, as two ABCG1 variants were found to control volume-regulated anion channel-dependent ATP release by regulating cholesterol levels in the plasma membrane [15]. Neither of these ABCG1 variants, however, have been implicated in the pathology of PXE.

Although low levels of circulating PPi explain why PXE patients suffer from ectopic mineralization, the molecular details of ABCC6-mediated ATP release remain elusive. Most ABC proteins of the C-branch function as ATP-dependent efflux transporters, though there are several exceptions. Thus, ABCC7 is the ATP-gated chloride channel cystic fibrosis transmembrane conductance regulator (CFTR) with inactivating mutations causing cystic fibrosis [16], and ABCC8 and ABCC9 are regulatory subunits of complex potassium channels [17].

Most of the currently available data indicate that ABCC6 is an ATP-dependent ATP efflux transporter: ATP efflux rates from ABCC6-transfected HEK293 cells are very similar to rates at which ABCC1, ABCC2, and ABCC3 transport morphine-3-glucuronide out of cells [18]. Moreover, our recent work indicates ABCC6 does not function as an ATP channel [19] and nor does it induce the exocytosis of ATP-loaded vesicles (our unpublished data). ABCC6 was initially implicated in the transport of glutathione conjugates in in vitro vesicular uptake assays [20,21] but these results proved difficult to reproduce in later studies [6].

In 2017, the structure of LTC_4-bound bovine Abcc1 (bAbcc1) in the ATP-free state, with a bipartite transmembrane cavity open towards the cytosol (inward-facing) was reported using cryogenic electron microscopy (cryoEM) [22]. This report was later followed by the cryoEM structure of the ATP-bound, outward-facing state of bAbcc1, with the transmembrane cavity open to the opposite side of the membrane [23]. Given that (1) ABCC6 shares most sequence similarity with ABCC1 [24], (2) the genes encoding both proteins arose from a recent gene duplication [25], and (3) in vitro studies suggested both proteins might share LTC_4 as a substrate [4,21,26,27] though attempts to connect the transport of LTC_4 to the potential role of ABCC6 failed [6].

We used the ATP-free, LTC_4-bound and ATP-bound, substrate-free, bAbcc1 cryoEM structures as templates to build inward- and outward-facing homology models of hABCC6 and rat Abcc6 (rAbcc6) as a means of identifying amino acids potentially forming the binding cavity for ATP. Amino acids in ABCC6 at the same positions as those in bAbcc1 comprising the proposed bipartite binding cavity of LTC_4 were subsequently mutated in rAbcc6 expression vectors and the mutant rAbcc6 proteins functionally characterized to determine if they play a role in ABCC6-dependent ATP release. Several of the introduced mutations did not markedly alter rAbcc6 activity and thus are not essential for ATP efflux. Strikingly, the generation of a rAbcc6 mutant in which all amino acids of the modeled binding cavity were changed into their ABCC1 counterparts, showed ATP efflux similar to the wild-type protein.

2. Results

2.1. Homology Models of Human and Rat ABCC6/Abcc6

ABC transporters have a common structural core that in ABCC6 and all other ABCC subfamily members consist of transmembrane domain 1 (TMD1, encompassing transmembrane helices (TM) 6 to 11), nucleotide-binding domain 1 (NBD1), TMD2 (TM12 to 17) and NBD2. We built a sequence alignment for TMD1-NBD1 and TMD2-NBD2 (see Supporting Information files TMD1-NBD1_Alignment.pdf and TMD2-NBD2_Alignment.pdf), using several orthologues of ABCC1, ABCC6, and ABCC5, for which negatively charged substrates have been reported. From these alignments, we observed a 46% and 48% sequence identity for rAbcc6-bAbcc1 and hABCC6-bAbcc1 TMD1-NBD1, respectively, and 52% and

53% sequence identities for rAbcc6-bAbcc1 and hABCC6-bAbcc1 TMD2-NBD2, respectively. The % sequence identity calculated for TMD1 (rAbcc6 residues 298–608) and TMD2 (rAbcc6 residues 933–1242) are 39% and 48%, respectively, for rAbcc6 and bAbcc1, and 40% for TMD1 and 48% for TMD2 between hABCC6 and bAbcc1. Considering the high sequence identity with bAbcc1, we modeled the structural core for rAbcc6 and hABCC6 in two distinct conformational states (Figure S1), based on the cryoEM structures reported for the LTC$_4$-bound, ATP-free, inward-facing state and on the ATP-bound outward-facing state of bAbcc1 [22,23].

Consistent with their relatively high degree of sequence identity, both hABCC6/rAbcc6 and bAbcc1 showed a strong positive potential in the cavity along TMs of both TMD1 and TMD2 (Figure 1). Another common feature of hABCC6/rAbcc6 and bAbcc1 is the presence of a more negative potential on the extracellular end of the TMDs (Figure S2A), which following the conformational change to the ATP-bound, outward-facing state, appears less prominent (Figure S2).

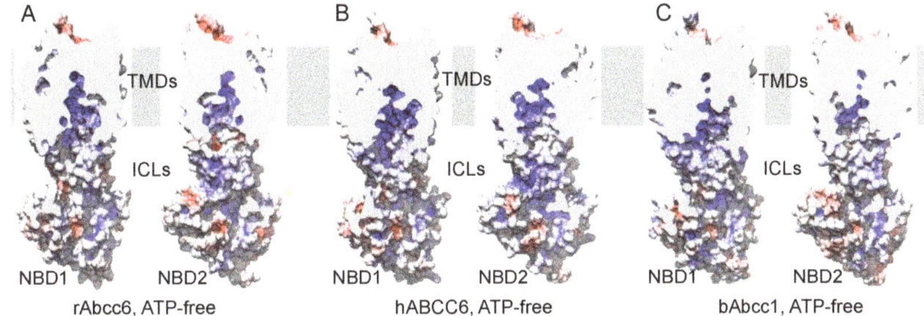

Figure 1. Electrostatic potential of the inward-facing state of hABCC6/rAbcc6 and bAbcc1. Electrostatic potential mapped on the molecular surface of the ATP-free, inward-facing (**A**) rAbcc6 model, (**B**) hABCC6 model, and (**C**) bAbcc1 cryoEM structure. The isovalue was set at -10 k_BT/e for the negative potential (red) and $+10$ k_BT/e for the positive potential (blue). For each transporter, the surface is clipped, and the two halves are shown side by side. The region of the transporters embedded in the membrane is highlighted by the gray slab. TMDs, transmembrane domains; ICLs, intracellular loops, i.e., the intracellular extension of the TMDs; NBD1 and NBD2, nucleotide binding domain 1 and 2.

In the transmembrane cavity of bAbcc1, several residues have been proposed as participating in the recognition of LTC$_4$ through a network of hydrogen bonds, salt bridges, and van der Waals contacts (Figure S3, Figure 2 and Table 1) [22], although not all of the proposed interactions are supported by biochemical studies [28,29].

Table 1. Amino acid residues in rat and human ABCC6, human ABCC1 and human ABCC5 at the same positions proposed to form the LTC$_4$ binding cavity in bAbcc1. In the last column it is indicated to which transmembrane helix (TM) and transmembrane domain (TMD) the residues belong. bAbcc1, bovine Abcc1; hABCC1, human ABCC1; hABCC6, human ABCC6; rAbcc6, rat Abcc6; hABCC5, human ABCC5.

bAbcc1	hABCC1	hABCC6	rAbcc6	hABCC5	TM Helix
K332	K332	L318	L316	L186	TMD1, TM6
H335	H335	S321	S319	T189	TMD1, TM6
L381	L381	E367	E365	L236	TMD1, TM7
F385	F385	M371	M369	W240	TMD1, TM7
Y440	Y440	Y426	H424	V293	TMD1, TM8
T550	T550	L536	L534	V403	TMD1, TM10

Table 1. *Cont.*

bAbcc1	hABCC1	hABCC6	rAbcc6	hABCC5	TM Helix
W553	W553	F539	F537	A406	TMD1, TM10
F594	F594	K579	K578	Q432	TMD1, TM11
M1092	M1093	S1065	T1064	M992	TMD2, TM14
R1196	R1197	R1169	R1168	R1096	TMD2, TM16
Y1242	Y1243	T1215	T1214	L1142	TMD2, TM17
N1244	N1245	Q1217	Q1216	Q1144	TMD2, TM17
W1245	W1246	W1218	W1217	F1145	TMD2, TM17
R1248	R1249	R1221	R1220	R1148	TMD2, TM17

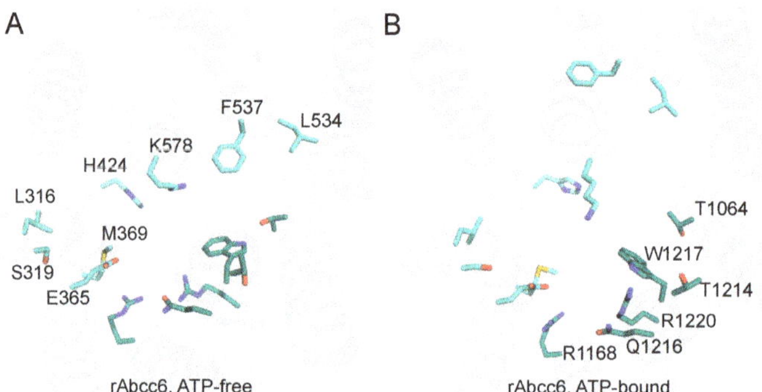

Figure 2. rAbcc6 residues in the transmembrane cavity corresponding to those in the bAbcc1 cavity surrounding LTC$_4$. View from the extracellular side of the transmembrane cavity of the inward-facing, ATP-free (**A**) and outward-facing, ATP-bound (**B**) models of rAbcc6. The residues corresponding to those of the bAbcc1 LTC$_4$ binding cavity are shown as sticks in light cyan for TMD1 and in teal for TMD2.

The residues in the cavity are proposed to form a bipartite binding site [30] with a more prominent positive charge on one side (residues K332, H335, L381, F385, Y440, F594 of TMD1, and R1196, N1244, and R1248 of TMD2, namely the P-pocket) to bind the hydrophilic glutathione moiety of LTC$_4$ and a more hydrophobic pocket (namely, the H-pocket) to accommodate the lipid tail of LTC$_4$ (residues T550, W553 of TMD1, and M1092, Y1242, W1245 of TMD2) [22]. Our goal was to address if ABCC6 (the corresponding residues shown in Figure 2 for rAbcc6 and in Figure S3 for hABCC6), in both the ATP-free and ATP-bound states, might be involved in ATP interaction and efflux. First, we analyzed how the proposed P-pocket and H-pocket residues are conserved across the sequences taken into account to build the models. The analysis of the sequence alignment (Supplemental information files TMD1-NBD1_Alignment.pdf and TMD2-NBD2_Alignment.pdf) showed that bAbcc1 W553 and W1245 in the hydrophobic pocket are conserved among the ABCC6, ABCC1, and ABCC5 sequences, and correspond to rAbcc6 residues F537 (TM10 in TMD1) and W1217 (TM17 in TMD2) (Figure 2). In the P-pocket, the more conserved residues are bAbcc1 R1196, R1248, and N1244, which correspond to rAbcc6 R1168 (TM16 in TMD2), R1220 and Q1216 (TM17 in TMD2) (Figure 2).

The degrees of similarity for the other residues of the bAbcc1 P- and H-pockets vary across the ABCC1, ABCC6, and ABCC5 sequences considered in the alignment. Charged residues that are not conserved are: (1) bAbcc1 K332, which is a leucine in ABCC6 (L316 in rAbcc6) and ABCC5, (2) H335 in hABCC1/bAbcc1, which is a serine in the hABCC6/rAbcc6 sequences, and (3) rAbcc6 E365, which is a glutamate only in ABCC6 sequences and a leucine in hABCC1 (L381 in bAbcc1) and ABCC5. Of note, in hABCC1, K332, and to a lesser extent H335, are indispensable for LTC_4 binding and transport, indicating K332 and H335 are crucial amino acid residues in its LTC_4-binding site [29,31]. An additional alignment performed using sequences of human ABC transporters of the C subfamily confirmed these observations and demonstrate exceptionally high conservation of R1168 and R1220 (numbering of rAbcc6) among ABCC proteins (Figure S5).

2.2. Functional Analysis of Single Amino Acid rAbcc6 Mutants

Our aim was to determine whether the residues in ABCC6 corresponding to those thought to be important in interaction with the physiological ABCC1 substrate LTC_4, play a role in ABCC6-mediated ATP efflux. We used rAbcc6 in these studies, because it has higher activity in HEK293 cells than hABCC6 [10]. ATP and other nucleoside triphosphates (NTPs), the putative physiological substrates of ABCC6, carry multiple negative charges. We hypothesized such negatively charged substrates may be "coordinated" by positively charged residues in the substrate binding cavity of rAbcc6. Therefore, positively charged amino acid moieties (i.e., lysine, arginine, and histidine) at these positions, were replaced with uncharged residues (i.e., glutamine and alanine). Non-charged amino acid residues, according to the canonical/conservative mutagenesis practices, were changed into cavity-creating alanine residues, aimed at retaining the overall structure of the protein. The single amino acid rAbcc6 mutants generated for our study are summarized and positioned in a topology model of rAbcc6 below (Figure 3A).

A

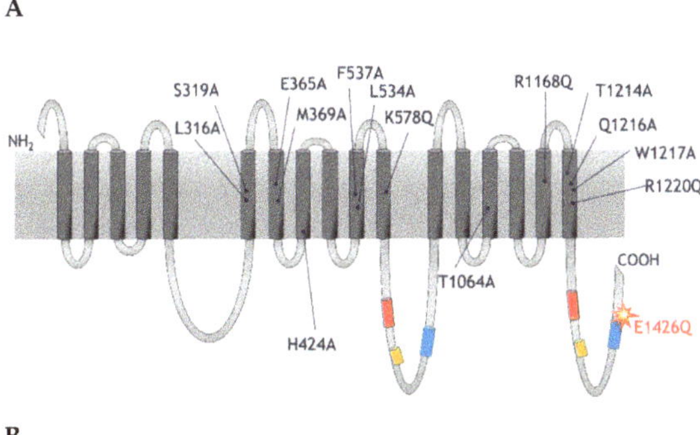

B

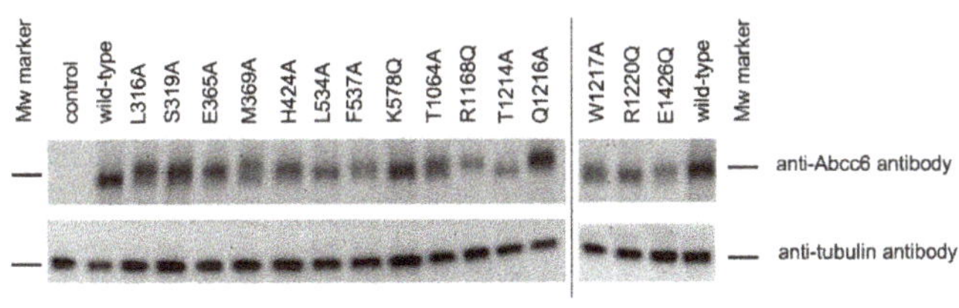

Figure 3. (**A**): Topology of the rAbcc6 amino acids analogous to those that comprise the LTC_4 binding cavity in bAbcc1. The inactivating mutation in the NBD2, E1426Q, is also indicated (**B**): Expression of the rAbcc6 single amino acid mutants in HEK293 cells. Of the total cell protein, 5 μg was fractionated on a 7.5%-polyacrylamide gel and bands corresponding to wild-type and mutant rAbcc6 proteins and the housekeeping protein tubulin were detected by Western-blot analysis using the K14 anti-rat Abcc6 antibody and the anti-tubulin antibody, respectively. The slight differences in electrophoretic mobility of some of the mutants may be attributed to altered glycosylation or other post-translational modifications.

Levels of the mutant rAbcc6 proteins in HEK293 cells varied but were within the same range as those of wild-type Abcc6 (Figure 3B). We then characterized the functionality of the rAbcc6 mutants by following PPi accumulation in the culture medium as an indirect measure of NTP release (Figure 4A,B), as well as by directly determining ATP efflux using a luciferin/luciferase-based assay (Figure 4B). In both assays the untransfected, parental, HEK293 cell line as well as the cell line expressing the catalytically inactive E1426Q mutant, did not release substantial amounts of ATP into the culture medium. In contrast, cells overproducing wild-type rAbcc6 released large amounts of ATP, resulting in robust PPi accumulation in the culture medium (Figure 4). These results demonstrate the suitability of these assays for measuring the consequences of the mutations introduced into rAbcc6.

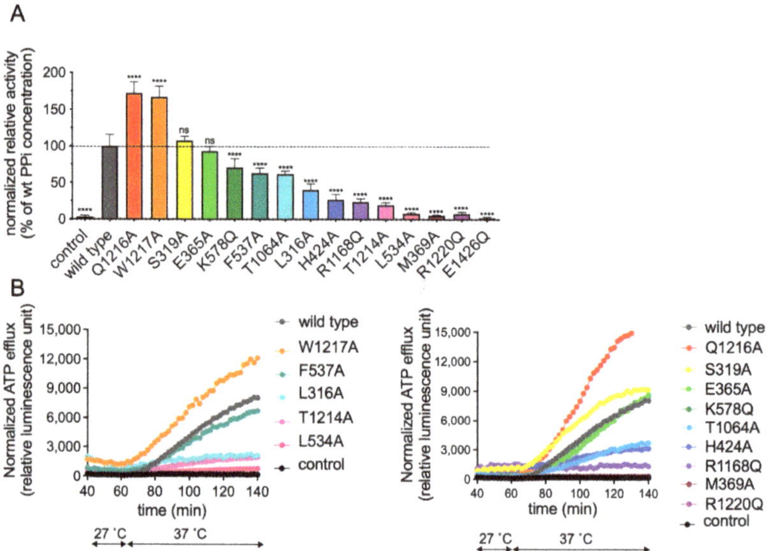

Figure 4. Activity of rAbcc6 mutants in HEK293 cells. (**A**): PPi accumulation in culture medium and (**B**): ATP efflux from cell lines overexpressing rAbcc6 in which amino acid residues corresponding to those forming the bAbcc1 LTC_4 binding cavity were mutated. Data are presented as means ± SD for (**A**). For (**B**), means of representative experiments, each with at least 4 replicates are shown. In (**B**) data are presented in two graphs to better see results of individual mutants. Wild type: wild-type rAbcc6, control: parental HEK293 cells. The dashed line in (**A**) indicates the average amount of PPi in medium of HEK293 cells overproducing wild-type rAbcc6, which was set at 100%. Values have been adjusted to take any differences in protein expression of the mutants relative to wild type rAbcc6 into account. The same color coding was used for each mutant in panels A and B. **** $p < 0.001$ (ANOVA and subsequent Dunnett's multiple comparison test). Changes were considered biologically relevant when reduced by >50% compared to wild-type rAbcc6.

Many of the rAbcc6 single amino acid mutants allowed cellular ATP efflux similar to that seen for wild-type rAbcc6, as determined by both PPi accumulation in the medium (Figure 4A) and the direct ATP efflux assay (Figure 4B). ATP release was substantially reduced (>75%) when M369, L534, R1168, T1214, and R1220 were mutated (Figure 4A,B). Changing L316 and H424 residues into alanine moderately reduced (>50%) efflux activity of rAbcc6. The substitution of the other residues did not reduce, or less substantially reduced, ATP efflux. The two arginine residues critical for function (rAbcc6 R1168 and R1220) belong to TMD2. In the ATP-free, LTC_4-bound, inward-facing conformation, R1220 (TM17) localizes near one of the entrances of the modelled substrate-binding cavity, lined by TM15 and TM17 (Figure 5A). M369 (TM7), L534 (TM10), and R1168 (TM16) approximately lie on the same plane as R1220, and their side chains are exposed to the main cavity, with L534 located on the opposite side (Figure 5A). Among the residues that abolish ATP efflux when mutated, T1214 (TM17) is the one located further up in the transmembrane cavity (Figure 5A). In this conformation, among the other residues considered in this study, the side chains of L316 (TM6) and S319 (TM6) are those further away from the main cavity (Figure 5A). In the outward-facing, ATP-bound state, the cavity opens towards the opposite side of the membrane. In this state many of the residues are more buried and located towards the bottom of the outward-facing cavity (Figure 5B). As mentioned earlier, R1168 and R1220 are conserved among all the sequences considered for model building (see Supporting Information files TMD1-NBD1_Alignment.pdf and TMD2-NBD2_Alignment.pdf), as well as all ABCC family members and their orthologs. The analogous amino acids are indispensable for all the transport activities of hABCC1, suggesting a key function in overall protein structure [32,33], although we cannot completely rule out interaction with its substrates. M369 is present as a phenylalanine and a tryptophan in the ABCC1 and ABCC5 sequences, respectively, while rAbcc6 L534 is a threonine in ABCC1 and a valine in ABCC5 (see Supporting Information file TMD2-NBD1_Alignment.pdf). rAbcc6 T1214 is also not conserved among the sequences here considered, and it is present primarily as a tyrosine in ABCC6 and as a leucine in ABCC5 sequences (see Supporting Information file TMD2-NBD2_Alignment.pdf).

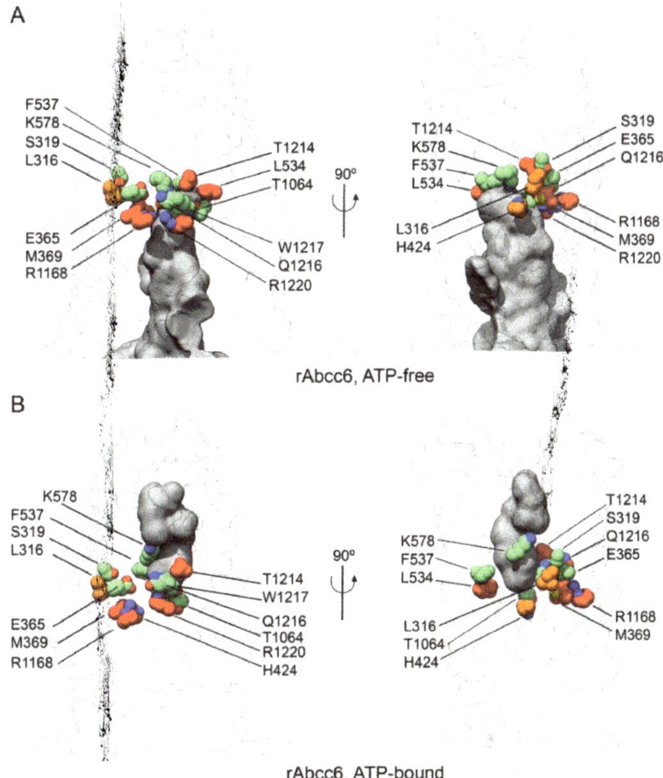

Figure 5. Mutation of the amino acids forming the modelled rAbcc6 substrate-binding cavity affect ATP efflux to different degrees. View of the (**A**) ATP-free inward-facing and (**B**) ATP-bound, outward-facing models of rAbcc6. rAbcc6 residues that when mutated abolished and reduced ATP efflux are shown as red and orange spheres, respectively. Residues that once mutated did not affect ATP efflux are shown as green spheres. The protein is shown as a white transparent surface and the volume of the main cavity in both models is shown as a gray surface. The volume was calculated with the 3V webserver [34].

2.3. Subcellular Localization of the Single Amino Acid rAbcc6 Mutants That Showed Reduced ATP Efflux Activity

To exert its function, ABCC6 needs to reside in the plasma membrane. Confocal microscopy demonstrated that all rAbcc6 mutants with reduced activity routed to the plasma membrane, similar to wild-type rAbcc6 (Figure 6). This indicates that reduced plasma membrane localization was not the underlying cause of the reduced activity of these rAbcc6 mutants. Notably, although a significant proportion of the rAbcc6 mutant proteins also resided in intracellular compartments, this was not different from wild-type rAbcc6 and is consistent with our previous observations [19].

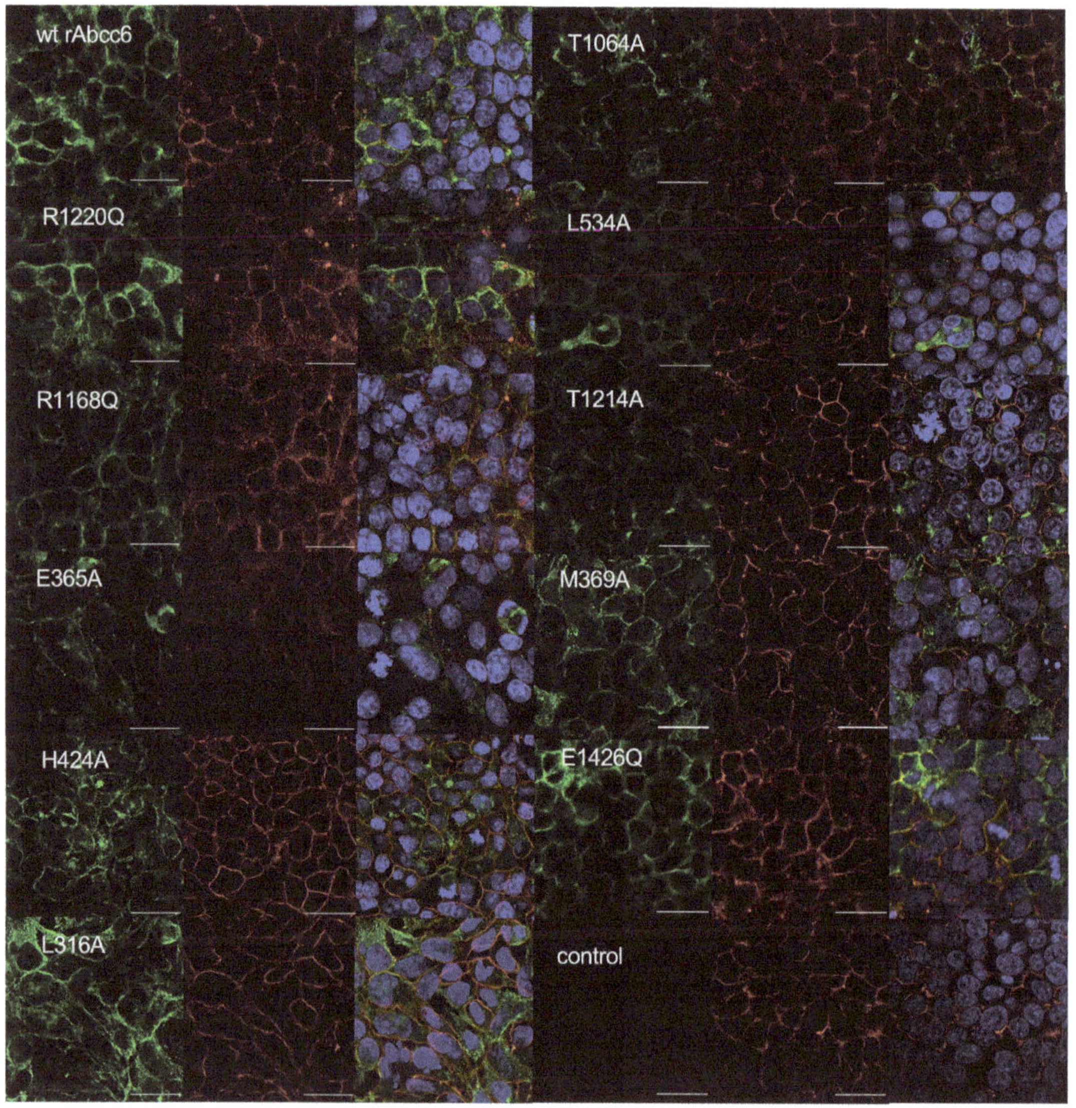

Figure 6. Subcellular localization of rAbcc6 mutants with reduced ATP efflux activity. Representative images of the

subcellular localization of wild-type and single mutant rAbcc6 in HEK293 cells, as determined by confocal microscopy using the K14 anti-rAbcc6 rabbit polyclonal antibody. Red: Na^+/K^+-ATPase, a marker for the plasma membrane; Green: rAbcc6.; Blue: DAPI nuclear staining; wt rAbcc6: wild-type rAbcc6, control: parental HEK293 cells. All scale bars represent 30 μm.

2.4. Functional Consequences of Changing All 11 Amino Acids of the Modeled rAbcc6 Substrate-Binding Site into Those That Comprise the bAbcc1 LTC_4 Binding Site

As most mutants with single amino acid changes in the modeled rAbcc6 substrate binding cavity retained at least 25% activity, we wondered what the consequences would be if the entire modeled substrate binding cavity was altered such that it more closely mimicked that of ABCC1. Of the fourteen amino acids corresponding to those in the bAbcc1 cryoEM structure forming the LTC_4-binding cavity, three are identical in rAbcc6 (rAbcc6 R1168, W1217 and R1220) (Table 1 and Figure 2). Thus, the remaining eleven amino acids of the modeled substrate-binding cavity were changed into residues found in bAbcc1 at the same positions (L316K, S319H, E365L, M369F, H424Y, F537W, L534T, K578F, T1064M, T1214Y, and Q1216N) to generate a rAbcc6 mutant protein we have termed rAbcc6-11aa.

As shown in Figure 7A, the rAbcc6-11aa was expressed at about 6.5-fold lower levels than the wild-type rAbcc6 protein when overexpressed in HEK293 cells and appeared to have a faster electrophoretic mobility (Figure 7A). The reason for the altered mobility of rAbcc6-11a is not known but may be due to changes in glycosylation or other post-translational modification. If underglycosylated, the mutant rAbcc6-11aa protein may have a faster turn-over time thus explaining why it is expressed at lower levels than the wild-type protein. However, even if the mutations caused some misfolding of the transporter during its biosynthesis that impaired its glycosylation, the altered protein structure has remained stable enough to traffic to the plasma membrane where it can still carry out active transport. Thus, rAbcc6-11aa still mediated ATP release, as illustrated by both PPi accumulation in the medium and by real-time ATP efflux assays (Figure 7C,D, respectively). After adjustment for lower protein levels, rAbcc6-11aa seemed to even display higher activity than the wild-type protein. These data indicate that changing these 11 amino acids may have affected the stability of rAbcc6 but had minimal effect on the intrinsic activity of the protein. Consistent with the significant activity of the rAbcc6-11aa mutant protein, its abundance in the plasma membrane was also relatively high (Figure 7B).

We next set out to test if the rAbcc6-11aa mutant with the hABCC1/bAbcc1 LTC_4-binding cavity residues transports LTC_4. Of note, an initial characterization of hABCC6 indicated it could transport LTC_4 [6], although this was not the case for its ortholog rAbcc6 [35]. Nevertheless, we reasoned that introducing the amino acids in bAbcc1 thought to form the bipartite LTC_4-binding cavity at the corresponding positions in rAbcc6 might establish LTC_4 transport in the latter ABC transporter. The cellular ATP efflux capacity of the rAbcc6-11aa indicated the protein retained activity. However, in vitro vesicular transport experiments (widely held to be the gold-standard to confirm that a molecule is a substrate of a specific ABC transporter) failed to demonstrate LTC_4 transport by the rAbcc6-11aa protein (data not shown). Low levels of LTC_4 transport might have been missed, however, because of the low expression levels of rAbcc6-11aa in our system (HEK293 cells).

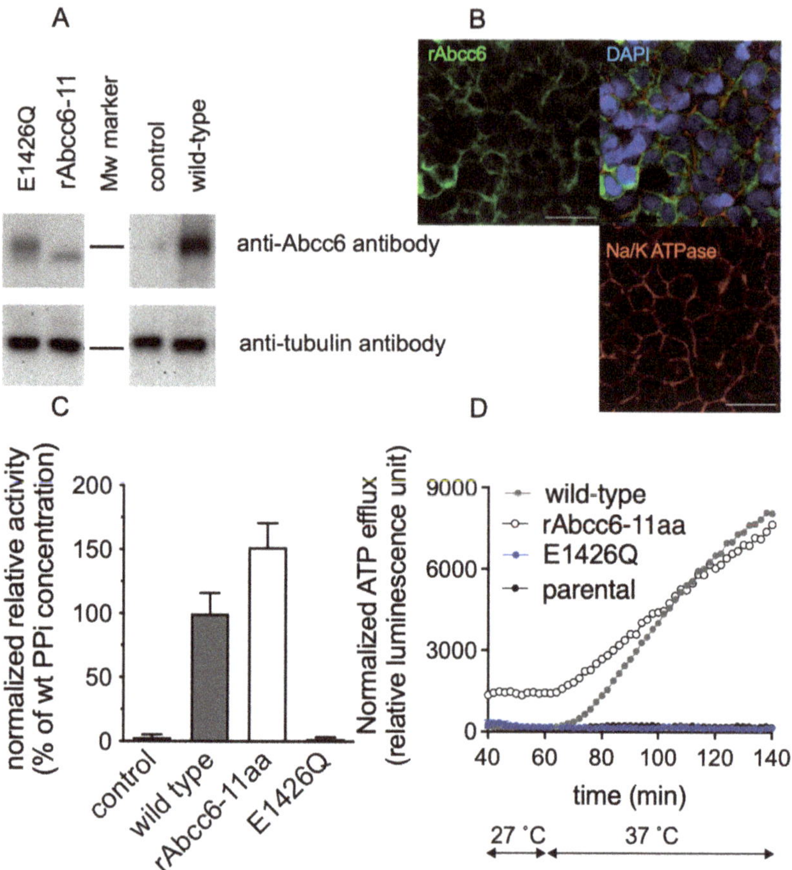

Figure 7. Expression, activity, and subcellular localization of the rAbcc6-11aa mutant. (**A**): Detection of rAbcc6-11aa in HEK293 cells by immunoblot analysis. (**B**): Subcellular localization of rAbcc6-11aa in HEK293 cells. Red: Na^+/K^+-ATPase, a marker for the plasma membrane; Green: rAbcc6; Blue: DAPI. All scale bars represent 30 μm. (**C**): PPi accumulation in the medium of the indicated HEK293 cell lines (**D**): ATP efflux from the indicated HEK293 cell lines. Data are presented as means ± SD for (**C**), while means of representative experiments with at least 4 replicates are shown for (**D**). wild-type: wild-type rAbcc6, control: parental HEK293 cells. The dashed line in (**C**) indicates the average PPi level in medium of HEK293 cells overproducing wild-type rAbcc6, which was set at 100%. The slight differences in electrophoretic mobility of some of the mutants may be attributed to altered glycosylation or other post-translational modification. In panels C and D values have been adjusted to take differences in protein expression of the mutants relative to wild type rAbcc6 into account.

3. Discussion

ABC transporters use the energy derived from ATP hydrolysis to mediate transport of a wide range of substrates across membranes. Several members of the ABCC subfamily have been studied for their role in drug resistance and human diseases. Many of these transporters translocate negatively charged solutes. For example, ABCC1 transports organic anions, including LTC_4, a cysteinyl leukotriene with proinflammatory properties [26,27,36], while ABCC5 transports cyclic nucleotides such as cAMP and cGMP [37,38], important for signal transduction. Most of the available data indicate ABCC6 transports ATP and other NTPs out of cells [10,11,39,40]. However, even now, ABCC6-mediated ATP efflux has not been shown in vesicular uptake experiments, widely used to demonstrate that a given compound is actively transported by an ABC transporter [6,19]. Being involved in the efflux of ATP confers a unique physiological role for ABCC6 among ABCC transporters [19]. We hypothesized that the substrate-binding cavity of ABCC6 is also unique among ABC transporters, involving amino acids that are not necessarily evolutionarily conserved among members of the ABCC subfamily. To address this, we built homology models of hABCC6/rAbcc6 in two different conformations (ATP-free, inward-facing and ATP-bound, outward-facing) (Figure S1). The previous homology modeling studies of ABCC6 were performed using available structures of bacterial ABC transporters or the related mouse P-glycoprotein, a member of the ABCB subfamily [41,42]. Here, we used the cryoEM structures of bAbcc1 [22,23], and compared the residues in the bipartite LTC$_4$ binding pocket of bAbcc1 with other ABCC1 sequences, as well as ABCC6 and ABCC5 sequences (see the Supporting Information alignment files). This choice was dictated by the shared ability of ABCC1, ABCC5 and ABCC6 transporters to translocate negatively charged substrates including cyclic nucleotides, as well as an early study reporting the ability of hABCC6 to transport LTC$_4$ [20], which suggested there may be similarities in substrate recognition by hABCC6, hABCC1, and ABCC5. Of note, however, is that rAbcc6 has never been shown to transport LTC$_4$ [35] and later studies also failed to confirm LTC$_4$ transport by hABCC6 [6]. Nevertheless, the overall electrostatic properties of the transmembrane cavity appear remarkably similar between the hABCC6/rAbcc6 models and the bAbcc1 cryoEM structures, with a strong positive potential that might contribute to the driving force for the uptake of negatively charged substrates from the cytosol (Figure 1 and Figure S2). Interestingly, the negative potential on the extracellular end of the TMDs (Figure S2A) appeared less prominent following the conformational change to the ATP-bound state (Figure S2B), possibly facilitating the negatively charged ATP to leave the substrate binding cavity. Among the bAbcc1 residues found in the proposed LTC$_4$-binding cavity (Table 1) are three residues that are identical to the corresponding R1168, W1217, and R1220 in rAbcc6 (Figure 2 and Figure S4), which are conserved among all human ABCC transporters (Figure S4) and the sequences considered in this study for model building, likely indicating a common function in maintaining the integrity of the substrate binding cavity. In the ATP-free, inward-facing homology model, their side chain faces the main cavity, and the residues belong to TM helices 16 and 17 of TMD2, known to be crucial for substrate binding and transport in ABCC1 [32,43]. Interestingly, in ABCC1 substitutions of W1246 (corresponding to rAbcc6 W1217) adversely affects transport of estradiol-17β-glucuronide and methotrexate but not of LTC$_4$ [29,43], also implicating a role of this amino acid in transporter substrate selectivity. Furthermore, even conservative same charge substitutions of R1197 and R1249 in hABCC1 cause a global loss of transport activity [32].

We proceeded with functional studies to test if the putative transmembrane ATP-binding site of ABCC6 overlaps with the LTC$_4$-binding cavity of bAbcc1 using a rAbcc6 model system [22]. Of the fourteen single amino acid changes introduced into the putative ABCC6 substrate binding cavity, five were found to reduce ABCC6-dependent ATP release by >75%, M369, L534, R1168, T1214, and R1220. Mutating the amino acid residue corresponding to rAbcc6 L534 in ABCC1 (T550) did not affect organic anion transport [44]. In contrast, even conservative mutations of hABCC1 F385 and Y1243 [45] (corresponding to M369 and T1214 in rAbcc6) adversely affected the transport capacity of one or more organic anions by hABCC1 (Unpublished, Conseil and Cole). Regarding the rAbcc6 R1168 position, previous studies (Table S2) have shown that opposite charge but also like-charge substitutions of hABCC1 at R1197, corresponding to rABCC6 R1168 and hABCC6 R1169, respectively, substantially reduced overall organic anion transport activity (all 4 organic anion substrates tested) as well as LTC$_4$ binding [32]. Regarding rAbcc6 R1220, mutations of the corresponding R1221 in hABCC6 have been reported to be disease causing (R1221C) [46–48] and pathogenic (R1221H) [47,49]. The corresponding amino acid in hABCC1, R1249, is crucial for overall organic anion transport activity as well, not just glutathione-dependent binding of substrates and transport of LTC_4 [32,33]. In hABCC2, the analogous amino acid R1257 is also indispensable for activity, as the mutant protein is deficient in glutathione conjugate transport, despite correctly routing to the plasma membrane [50]. In hABCC4, substitution of R998, which is analogous to R1221 in hABCC6 (and R1249 in hABCC1), by alanine completely abolishes the transport of cyclic guanosine monophosphate (cGMP) [51]. Based on the fact that the rAbcc6 R1168 and R1220 are highly conserved in ABCC1-6 and that the mutation of these residues hampers the transport function in the paralogs, we consider it likely that the presence of the charged residues at these positions is indispensable for all ABCC proteins and the requirement for a positive charged residue at this position is not specific to ABCC6. The other residues corresponding to those that form the binding cavity for LTC_4 in bAbcc1 had less impact on the rAbcc6-mediated ATP efflux. Of these, only L316A and H424A had activity that was <50% that of wild-type rAbcc6 (Figure 4). Somewhat surprisingly, despite evidence of misfolding, the mutant rAbcc6-11aa protein containing the same LTC_4 binding cavity amino acids as bAbcc1 was still functional and able to efflux ATP. This suggests that the amino acids corresponding to those proposed to form the bAbcc1 LTC_4 binding cavity, in ABCC6/rAbcc6 are not essential for interaction with or the recognition of its physiological substrate, ATP. Our conclusion, therefore, is that the binding site for ATP in the transmembrane cavity of ABCC6 is clearly distinct from the LTC_4 binding cavity in bAbcc1. The possible exceptions are two highly conserved positively charged residues described above, namely rAbcc6 R1168 and R1220, which are common to the substrate-binding cavity of the ABCC transporters characterized and likely are essential for proper folding and assembly into a transport competent protein. Despite the fact that ABCC1 and ABCC6 arose from a recent gene duplication, simple evolution of a common substrate-binding site most likely does not explain the structurally very distinct substrates effluxed by the two proteins.

The molecular details of ABCC6-mediated cellular ATP release remain unknown. As outlined above, an attractive hypothesis is that ABCC6 functions as an ATP-dependent ATP efflux pump. There are three sets of observations that support the idea that ABCC6 is an ATP efflux pump. First, most members of the C-branch of the ABC superfamily, including ABCC6's closest homolog ABCC1, are bona fide organic anion efflux transporters [24,26,52] and ABCC6 has been shown to transport several organic anions in vitro, albeit sluggishly [6]. Second, the ATP efflux rates found in HEK293-ABCC6 cells [10] are compatible with a direct transport mechanism for ATP as these rates are very similar to the rates by which ABCC1, ABCC2, and ABCC3 pump morphine-3-glucuronide out of transfected HEK293 cells [18]. Third, ATP efflux from ABCC6-containing cells can be blocked by the general ABCC transport inhibitors benzbromarone, indomethacin and MK571 (data not shown).

As mentioned, vesicular transport experiments are often used to establish substrates of ABC transporters [24]. However, such assays have so far failed to directly demonstrate the ABCC6-dependent transport of radiolabeled ATP into inside-out membrane vesicles. We can, therefore, not completely exclude the possibility that ABCC6 mediates cellular ATP release other than by direct transport. Purification of ABCC6 and subsequent reconstitution in proteoliposomes should provide a cleaner experimental system to study ATP transport, for instance by using dual color fluorescence burst analysis (DCFBA) [53], a technique that has a more favorable signal-to-noise ratio than the standard vesicular transport assays that employ radiolabeled ATP. The elucidation of the molecular structure of ABCC6, for instance by cryoEM, might also in the future provide clues about the molecular mechanism by which ABCC6 mediates ATP release. Despite many years of intense work on ABCC6, this ABC protein has not given many of its secrets away.

4. Materials and Methods

4.1. Model Building

We built homology models for the structural ABC transporter core of hABCC6 and rAbcc6, including residues of TMD1, NBD1, TMD2, and NBD2. First, we generated a multiple sequence alignment (MSA) using MAFFT [54] including the sequences of the hABCC6, hABCC1, and hABCC5 proteins from multiple organisms (Table S1), retrieved from UniProtKB [55]. Based on the alignment of hABCC6 and rAbcc6 sequences with bAbcc1, homology models of hABCC6 and rAbcc6 were generated with Modeller 9v15 [56], using the inward- and outward-facing cryoEM structures of bAbcc1 as templates [22,23]. For hABCC6 and rAbcc6, 20 models were generated for both the inward- and the outward-facing states, by applying a slow refinement protocol and 20 cycles of simulated annealing as in our previous work on hABCC7 (CFTR) [57]. For the inward-facing state, the ABCC6 models were generated considering the presence of LTC$_4$ in the template. The final hABCC6 and rAbcc6 models for each conformation were chosen based on the Discrete Optimized Protein Energy (DOPE) value implemented in Modeller. The rAbcc6 and hABCC6 models are provided in the Supporting Information as PDB files. The MSA is provided in the form of two separate files, covering residues of TMD1 and NBD1 (see TMD1-NBD1_Alignment.pdf), and residues of TMD2 and NBD2 (see TMD2-NBD2_Alignment.pdf). These files were generated using Jalview [58] and the residues are colored according to the Clustal X coloring scheme implemented in Jalview. The residues highlighted in bold correspond to the residues investigated in the present study and the rAbcc6 numbering of amino acids is indicated. The residues of the TM helices of TMD1 and TMD2 are also annotated. The percentage amino acid identity among bAbcc1, rAbcc6, and hABCC6 was calculated on the TMD1-NBD1 and TMD2-NBD2 alignment using the id_table command available in Modeller. Sequences of the hABCC1 and hABCC6 proteins shown in Figure S5 were retrieved from UniProtKB [55] and were aligned using Clustal Omega [59].

Electrostatic potential calculations were performed using the PDB2PQR and APBS webservers [60,61], using a pH of 7 and a NaCl concentration of 0.15 M. The electrostatic potential was visually mapped on the molecular surface of the models using UCSF Chimera [62], with a surface offset parameter of 1.4. Figure 5 was also generated using UCSF Chimera, after calculating the cavity volumes with the 3V webserver, using the default parameters for the Channel Finder module [34]. Other figures were generated using PyMOL [63].

4.2. Mutagenesis

Mutagenesis was performed as described previously [19]. Briefly, mutations were introduced into the Gateway entry vector pEntr223-rAbcc6 by uracil-specific excision reagent (USER) cloning with the primers listed in Table 2 using Phusion U PCR master mix (Thermo Scientific, Waltham, MA, USA). PCR fragments were purified using the Nucleospin gel and PCR cleanup kit (Macherey-Nagel, Düren, Germany) and assembled using the USER enzyme mix (New England Biolabs, Ipswich, MA, USA), according to the manufacturer's instructions. Resulting circular constructs were verified by Sanger sequencing and transformed into competent E. coli DH5alpha cells. The cDNAs encoding pEnter223-rAbcc6 mutants were subsequently subcloned into a Gateway compatible pQCXIP expression vector using LR Clonase-II (Thermo Scientific, Waltham, MA, USA).

Table 2. Primers used to generate the various rAbcc6 mutants.

Construct	Mutation	Forward Primer	Reverse Primer
rAbcc6-L316A	L316A	AGC<u>GCC</u>GUCATTAGCGATGCCTTCAGGTTTG	AC<u>GGC</u>GCUGAGGGTCCCCAGCAGGAAA
rAbcc6-S319A	S319A	ATT<u>GCC</u>GAUGCCTTCAGGTTTGCTGTT	ATC<u>GGC</u>AAUGACCAGGCTGAGGGTCCC
rAbcc6-E365A	E365A	ACTGTTU<u>GCC</u>CAGCAGTACATGTACAGA	AAACAGUGTCTGTAGGCAGGCCGACAA
rAbcc6-M369A	M369A	AGTAC<u>GCC</u>UACAGAGTCAAGGTCCTGCAGATG	A<u>GGC</u>GTACUGCTGTTCAAACAGTGTCTG

Table 2. *Cont.*

Construct	Mutation	Forward Primer	Reverse Primer
rAbcc6-H424A	H424A	ATCCTCGCCCUCAACGGGCTGTGGCTGCTCTT	AGGGCGAGGAUGCTCTCGACCAGCCGCTG
rAbcc6-L534A	L534A	AAGTGTCUACATTTCTGGTGGCGCTGGTTGT	AGACACTUGGAAGGACACGGCAGAC-ACAGAGAAGAGGAAG
rAbcc6-F537A	F537A	AAGTGTCUACATTTCTGGTGGCGCTGGTTGT	AGACACTUGGGCGGACACGAGAGACACAGA
rAbcc6-K578Q	K578Q	AGCCAGGCCUTCCTCCCCTTCTCTGTGC	AGGCCTGGGCUTGGTTAAGGATG
rAbcc6-T1064A	T1064A	AGGGCCCUGCTGACCTATGCCTTTGG	AGGGCCCUCATCTTGTCTGGGATGTCCACAT
rAbcc6-R1168Q	R1168Q	ACCAGTGGCUGGCTGCCAACCTGGAGCT	AGCCACTGGUCAGCCACCAGCCTCGGGA
rAbcc6-T1214A	T1214A	AGGCTCUGCAGTGGGTGGTCCGCAGCTG	AGAGCCUGTGTTACCTGGAGGGCAG-CAGAAACCG
rAbcc6-Q1216A	Q1216A	ACTCTGGCCUGGGTGGTCCGCAGCTGGAC	AGGCCAGAGUCTGTGTTACCTGGAGGGC
rAbcc6-W1217A	W1217A	AGGCCGUGGTCCGCAGCTGGACAGATC	ACGGCCUGCAGAGTCTGTGTTACCT
rAbcc6-R1220Q	R1220Q	AGTGGGTGGUCCAATCTGGAGAACAG	ACCACCCACUGCAGTCTGTGTTACCT
rAbcc6-11AA	L316K & S319H	AGGTCATUCACGATGCCTTCAGGTTT-GCTGTTCCCAAGC	AATGACCUTGCTGAGGGTCCCCAGCAGGAAAGT
	E365L & M369F	AGCAGTACUTCTACAGAGTCAAGGTCCTG-CAGATGAGGCTG	AGTACTGCUGCAGAAACAGTGTCTG-TAGGCAGGCCGACAAG
	H424Y	ATCCTCUACCTCAACGGGCTGTGGCTGC	AGAGGAUGCTCTCGACCAGCCGCTG
	L534T	ACCGTGUCCTGGCAAGTGTCTACATTTCTGGTGGC	ACACGGUAGACACAGAGAAGAGG-AAGGCGGAGGTCT
	F537W	AAGTGTCUACATTTCTGGTGGCGCTGGTTG	AGACACTUGCCAGGACACGAGAGACAC-AGAGAAGAGGAAGGC
	K578F	ATCCTTAACUTCGCCCAGGCCTTCCTCCCCTTC	AGTTAAGGAUGCTGAGCACCGTGAGCGT
	T1064M	AGGATGCUGCTGACCTATGCCTTTGGACTCCTGG	AGCATCCUCATCTTGTCTGGGATGTCCACATCCAC
	T1214Y	AGTATCUGAACTGGGTGGTCCGCAGCTGG	AGATACUGTGTTACCTGGAGGGCAGCAGAAACCG
	Q1216N	AACTGGGUGGTCCGCAGCTGGACAGATC	ACCCAGTUCAGAGTCTGTGTTACCTGGAGGGCAGC

4.3. Cell Culture and Generation of Mutant Cell Lines

Cell culture and generation of mutant cell lines were performed as described previously [14,19]. Briefly, HEK293 cells overproducing wild-type rAbcc6 (HEK293-rAbcc6) and control, untransfected cells (HEK293-control) were cultured at 37 °C in a 5% CO_2 atmosphere under humidifying conditions in DMEM (HyClone, GE Healthcare, Chicago, IL, USA) with 100 U pen/strep per mL (Gibco, Waltham, MA, USA) supplemented with 5% FBS (Fisher Scientific, Waltham, MA, USA). rAbcc6 mutants cloned into the pQCXIP expression vector were transfected into HEK293 cells using the calcium phosphate precipitation method. Transfected cells were selected in medium containing 2 μM puromycin. Cell lines were established from clones showing high expression of the respective rAbcc6 mutants. Of note, several clones were generated for each rAbcc6 mutation and these subclones behaved very similarly with respect to PPi accumulation in the culture medium.

4.4. Immunoblot Analysis of Wild-Type and Mutant rAbcc6

The expression of rAbcc6 was confirmed by immunoblot analysis as described previously [19]. Briefly, cell lysates were prepared in lysis buffer (0.1% Triton-x-100, 10 mM KCl, 10 mM Tris-HCl and 1.5 mM MgCl2, pH 7.4) supplemented with protease inhibitors (EDTA-free Protease Inhibitor Cocktail, Sigma Aldrich, St. Louis, MO, USA. Samples containing

5 µg of total protein determined by BCA assay were separated on 7.5% SDS-polyacrylamide gels (Bio-Rad, Hercules, CA, USA and transferred to a PVDF membranes using a semi-dry blotting system (Trans-Blot Turbo, Bio-Rad, Hercules, Ca, USA). rAbcc6 was detected with the polyclonal K14 rabbit anti-rAbcc6 antibody (diluted 1:3000) (kind gift of Dr. Bruno Stieger) and horseradish peroxidase (HRP)-conjugated donkey anti-rabbit secondary antibody (1:5000) (Fisher Scientific, Waltham, MA, USA). Mouse anti-α-tubulin (1:1000) (Sc-23948, Santa-Cruz Biotechnology, Dallas, TX, USA) followed by HRP-conjugated polyclonal rabbit anti-mouse IgG employed as secondary antibody (1:5000) (P0161, Dako, Agilent, Santa Clara, CA, USA), was used as a protein loading control. Antibody binding was visualized by ECL (Pierce Western blotting substrate, Thermo Scientific, Waltham, MA, USA).

4.5. Subcellular Localization of rAbcc6 in HEK293 Cells

The localization of rAbcc6 in intact HEK293 cells was detected as described previously [19]. Briefly, HEK293 cells were seeded and grown for 2 days on 4 well µ-Slides (ibiTreat 1.5 polymer coverslip, 80426, Ibidi) coated with poly-D-lysine. The cells were fixed in 4% PFA and subsequently in -20 °C cold methanol for 5 min each and then samples were blocked with Protein Block (Fisher Scientific) for 60 min. Samples were then incubated for 60 min with the polyclonal rabbit anti-rAbcc6 antibody K14 diluted 1:100 and the mouse monoclonal anti-Na$^+$/K$^+$-ATPase antibody (ab7671, Abcam, Cambridge, UK) diluted 1:250. Then samples were incubated for 60 min with Alexa Fluor 488-conjugated goat anti-rabbit secondary antibody (A11008, Fisher Scientific, Waltham, MA, USA) and Alexa Fluor 568 conjugated goat anti-mouse antibody (A11004, Fisher Scientific, Waltham, MA, USA), both diluted 1:1000. The samples were subsequently incubated with 300 nM DAPI for 5 min to stain nuclei. The subcellular localization of wild-type and mutant forms of rAbcc6 was then analyzed using a Nikon (Tokyo, Japan) Eclipse Ti two point-scanning laser confocal microscope equipped with a Nikon A1R+. A Plan Fluor 40× Oil DIC H N2 objective with a 1× optical zoom was used with 405.5, 490.0 and 561.3 nm excitation and 450/50, 525/50 and 595/50 nm emission filters, respectively. Images were acquired with the pinhole set to 1 airy unit.

4.6. Quantification of PPi Levels in Medium Samples

We have previously found that HEK293 cells endogenously express ENPP1 [10] and that PPi accumulation in the medium can be used as a robust secondary assay to determine the amount of ATP released by cells into the culture medium. Using two independent assays to follow ATP efflux provides robust data on the activity of the studied rAbcc6 mutants. PPi concentrations in cell culture medium samples were determined as described previously [14,19]. Briefly, PPi was quantitatively converted into ATP by incubating samples and standards in an assay containing 50 mM HEPES pH 7.4, 80 µM MgCl2, 32 mU/mL ATP sulfurylase (New England Biolabs, Ipswich, MA, USA), and 8 µM adenosine 5'-phosphosulfate (Santa Cruz Biotechnology, Dallas, TX, USA) at 30 °C for 30 min and the reaction was terminated by incubating the samples at 90 °C for 10 min. ATP levels were then determined in the reaction mix by bioluminescence by adding BacTiterGlo reagent (Promega, Madison, WI, USA) in a 1:1 ratio in a total volume of 40 µL. PPi concentrations in medium samples were then calculated by interpolation from a standard curve. The values were adjusted by subtracting background provided by controls in which ATP sulfurylase was omitted.

4.7. Real-Time ATP Efflux Assay

The ability of the transfected HEK293 cells to release ATP into the culture medium was determined using confluent monolayers as described previously [14,19]. In brief, the medium was removed and replaced with 50 µL efflux buffer, consisting of 11.5 mM HEPES (pH 7.4), 130 mM NaCl, 5 mM MgCl2, 1.5 mM CaCl2, and 11.5 mM glucose. The cells were then incubated for 1 hr at 27 °C. Next, 50 µL BactiterGlo reagent (Promega) dissolved in efflux buffer was added to each well. Bioluminescence was subsequently determined in real time in a Flex Station 3 microplate reader (Molecular Devices, San Jose, CA, USA) as detailed previously [14,19]. The real-time ATP efflux assay was run at 27 °C for the first 1 h and then at 37 °C for 2 hr. The initial low temperature allowed endogenous ecto-nucleotidases to degrade the Abcc6-independent background ATP efflux induced by the medium change.

Supplementary Materials: The following are available online at https://www.mdpi.com/article/10.3390/ijms22136910/s1, Figure S1: Structural models of rABCC6 and hABCC6. Figure S2: Electrostatic potential at the extracellular end of the TMDs for ABCC6 and ABCC1, Figure S3: The LTC$_4$ binding cavity in bABCC1 and hABCC6, Figure S4: Alignment of ABCC1-6 amino acid sequences. Table S1: List of the ABCC1 and ABCC6 sequences and their corresponding UniProtKB codes used to generate the alignment for the homology models. Table S2: Functional consequences of mutating rAbcc6 R1168 and R1220 and corresponding residues in hABCC6, hABCC1 and hABCC2. Supporting information files: TMD1-NBD1_Alignment.pdf, TMD2-NBD2_Alignment.pdf, hABCC6_inward-facing.pdb, hABCC6_outward-facing.pdb, rABCC6_inward-facing.pdb, rABCC6_outward-facing.pdb.

Author Contributions: Conceptualization, K.v.d.W. and V.C.; methodology, K.v.d.W., D.P.T. and V.C.; formal analysis, F.S., V.C. and K.v.d.W.; investigation, F.S., S.D., F.N. and G.C.; writing—original draft preparation, V.C., F.S. and K.v.d.W.; writing—review and editing, F.S., V.C., D.P.T., S.P.C.C. and K.v.d.W.; supervision, K.v.d.W.; project administration, K.v.d.W.; funding acquisition, K.v.d.W., D.P.T. and F.S. All authors have read and agreed to the published version of the manuscript.

Funding: This research was funded by National Institutes of Health, Grant R01AR072695 (K.v.d.W.), U.S. Department of State (Fulbright Visiting Scholar Program), National Research, Development and Innovation Office (OTKA FK131946), Hungarian Academy of Sciences (Bolyai János Fellowship BO/00730/19/8, Mobility grant) and the Ministry for Innovation and Technology from the source of the National Research, Development and Innovation Fund (ÚNKP-2020 New National Excellence Program) to F.S., PXE International to K.v.d.W. and F.S. and the Canadian Institutes of Health Research, Grant MOP 133584, to SPCC. Work in the group of D.P.T. is supported by the Natural Sciences and Engineering and Research Council (Canada) and the Canada Research Chairs Program.

Institutional Review Board Statement: Not applicable.

Informed Consent Statement: Not applicable.

Data Availability Statement: Not applicable.

Conflicts of Interest: The authors declare no conflict of interest.

References

1. Bergen, A.A.; Plomp, A.S.; Schuurman, E.J.; Terry, S.F.; Breuning, M.H.; Dauwerse, H.G.; Swart, J.; Kool, M.; Van Soest, S.; Baas, F.; et al. Mutations in ABCC6 Cause Pseudoxanthoma Elasticum. *Nat. Genet.* **2000**, *25*, 228–231. [CrossRef]
2. Le Saux, O.; Urban, Z.; Tschuch, C.; Csiszar, K.; Bacchelli, B.; Quaglino, D.; Pasquali-Ronchetti, I.; Pope, F.M.; Richards, A.; Terry, S.; et al. Mutations in a Gene Encoding an ABC Transporter Cause Pseudoxanthoma Elasticum. *Nat. Genet* **2000**, *25*, 223–227. [CrossRef]
3. Ringpfeil, F.; Lebwohl, M.G.; Christiano, A.M.; Uitto, J. Pseudoxanthoma Elasticum: Mutations in the MRP6 Gene Encoding a Transmembrane ATP-Binding Cassette (ABC) Transporter. *Proc. Natl. Acad. Sci. USA* **2000**, *97*, 6001–6006. [CrossRef]
4. Verschuere, S.; Van Gils, M.; Nollet, L.; Vanakker, O.M. From membrane to mineralization: The curious case of the ABCC6 transporter. *FEBS Lett.* **2020**, *594*, 4109–4133. [CrossRef]
5. Favre, G.; Laurain, A.; Aranyi, T.; Szeri, F.; Fulop, K.; Le Saux, O.; Duranton, C.; Kauffenstein, G.; Martin, L.; Lefthériotis, G. The ABCC6 Transporter: A New Player in Biomineralization. *Int. J. Mol. Sci.* **2017**, *18*, 1941. [CrossRef]
6. Borst, P.; Váradi, A.; van de Wetering, K. PXE, a Mysterious Inborn Error Clarified. *Trends Biochem. Sci.* **2019**, *44*, 125–140. [CrossRef] [PubMed]
7. Kranenburg, G.; Baas, A.F.; De Jong, P.A.; Asselbergs, F.W.; Visseren, F.L.J.; Spiering, W. The prevalence of pseudoxanthoma elasticum: Revised estimations based on genotyping in a high vascular risk cohort. *Eur. J. Med. Genet.* **2019**, *62*, 90–92. [CrossRef] [PubMed]
8. Luo, H.; Li, Q.; Cao, Y.; Uitto, J. Therapeutics Development for Pseudoxanthoma Elasticum and Related Ectopic Mineralization Disorders: Update 2020. *J. Clin. Med.* **2020**, *10*, 114. [CrossRef] [PubMed]
9. Pomozi, V.; Le Saux, O.; Brampton, C.; Apana, A.; Iliás, A.; Szeri, F.; Martin, L.; Monostory, K.; Paku, S.; Sarkadi, B.; et al. ABCC6 is a basolateral plasma membrane protein. *Circ. Res.* **2013**, *112*, e148–e151. [CrossRef]

10. Jansen, R.S.; Kucukosmanoglu, A.; de Haas, M.; Sapthu, S.; Otero, J.A.; Hegman, I.E.M.; Bergen, A.A.B.; Gorgels, T.G.M.F.; Borst, P.; van de Wetering, K. ABCC6 prevents ectopic mineralization seen in pseudoxanthoma elasticum by inducing cellular nucleotide release. *Proc. Natl. Acad. Sci. USA* **2013**, *110*, 20206–20211. [CrossRef]

11. Jansen, R.S.; Duijst, S.; Mahakena, S.; Sommer, D.; Szeri, F.; Váradi, A.; Plomp, A.S.; Bergen, A.A.B.; Elferink, R.P.J.O.; Borst, P.; et al. ABCC6–mediated ATP secretion by the liver is the main source of the mineralization inhibitor inorganic pyrophosphate in the systemic circulation—Brief report. *Arter. Thromb. Vasc. Biol.* **2014**, *34*, 1985–1989. [CrossRef]

12. Orriss, I.R. Extracellular pyrophosphate: The body's water softener. *Bone* **2020**, *134*, 115243. [CrossRef]

13. Veiga-Lopez, A.; Sethuraman, V.; Navasiolava, N.; Makela, B.; Olomu, I.; Long, R.; Van De Wetering, K.; Martin, L.; Aranyi, T.; Szeri, F. Plasma inorganic pyrophosphate deficiency links multiparity to cardiovascular disease risk. *Front. Cell Dev. Biol.* **2020**, *8*. [CrossRef]

14. Szeri, F.; Lundkvist, S.; Donnelly, S.; Engelke, U.F.H.; Rhee, K.; Williams, C.J.; Sundberg, J.P.; Wevers, R.A.; Tomlinson, R.E.; Jansen, R.S.; et al. The membrane protein ANKH is crucial for bone mechanical performance by mediating cellular export of citrate and ATP. *PLoS Genet.* **2020**, *16*, e1008884. [CrossRef] [PubMed]

15. Dunn, P.J.; Salm, E.; Tomita, S. ABC transporters control ATP release through cholesterol-dependent volume-regulated anion channel activity. *J. Biol. Chem.* **2020**, *295*, 5192–5203. [CrossRef]

16. Gadsby, D.C.; Vergani, P.; Csanády, L. The ABC protein turned chloride channel whose failure causes cystic fibrosis. *Nat. Cell Biol.* **2006**, *440*, 477–483. [CrossRef]

17. Bryan, J.; Muñoz, A.; Zhang, X.; Düfer, M.; Drews, G.; Krippeit-Drews, P.; Aguilar-Bryan, L. ABCC8 and ABCC9: ABC transporters that regulate K+ channels. *Pflügers Arch. Eur. J. Physiol.* **2006**, *453*, 703–718. [CrossRef]

18. van de Wetering, K.; Zelcer, N.; Kuil, A.; Feddema, W.; Hillebrand, M.; Vlaming, M.L.; Schinkel, A.H.; Beijnen, J.H.; Borst, P. Multidrug resistance proteins 2 and 3 provide alternative routes for hepatic excretion of morphine-glucuronides. *Mole. Pharmacol.* **2007**, *72*, 387–394. [CrossRef] [PubMed]

19. Szeri, F.; Niaziorimi, F.; Donnelly, S.; Orndorff, J.; Wetering, K. Generation of fully functional fluorescent fusion proteins to gain insights into ABCC6 biology. *FEBS Lett.* **2021**, *595*, 799–810. [CrossRef] [PubMed]

20. Ilias, A.; Urban, Z.; Seidl, T.L.; Le Saux, O.; Sinko, E.; Boyd, C.D.; Sarkadi, B.; Varadi, A. Loss of ATP-dependent transport activity in pseudoxanthoma elasticum-associated mutants of human ABCC6 (MRP6). *J. Biol. Chem.* **2002**, *277*, 16860–16867. [CrossRef] [PubMed]

21. Belinsky, M.G.; Chen, Z.-S.; Shchaveleva, I.; Zeng, H.; Kruh, G.D. Characterization of the drug resistance and transport properties of multidrug resistance protein 6 (MRP6, ABCC6). *Cancer Res.* **2002**, *62*, 6172–6177.

22. Johnson, Z.L.; Chen, J. Structural Basis of Substrate Recognition by the Multidrug Resistance Protein MRP1. *Cell* **2017**, *168*, 1075–1085. [CrossRef]

23. Johnson, Z.L.; Chen, J. ATP Binding Enables Substrate Release from Multidrug Resistance Protein 1. *Cell* **2018**, *172*, 1–2. [CrossRef] [PubMed]

24. Borst, P.; Evers, R.; Kool, M.; Wijnholds, J. The multidrug resistance protein family. *Biochim. et Biophys. Acta BBA Biomembr.* **1999**, *1461*, 347–357. [CrossRef]

25. Symmons, O.; Váradi, A.; Arányi, T. How Segmental Duplications Shape Our Genome: Recent Evolution of ABCC6 and PKD1 Mendelian Disease Genes. *Mol. Biol. Evol.* **2008**, *25*, 2601–2613. [CrossRef] [PubMed]

26. Cole, S. Multidrug resistance protein 1 (MRP1, ABCC1), a "multitasking" ATP-binding cassette (ABC) transporter. *J. Biol. Chem.* **2014**, *289*, 30880–30888. [CrossRef] [PubMed]

27. Leier, I.; Jedlitschky, G.; Buchholz, U.; Cole, S.P.; Deeley, R.G.; Keppler, D. The MRP gene encodes an ATP-dependent export pump for leukotriene C4 and structurally related conjugates. *J. Biol. Chem.* **1994**, *269*, 27807–27810. [CrossRef]

28. Conseil, G.; Arama-Chayoth, M.; Tsfadia, Y.; Cole, S.P.C. Structure-guided probing of the leukotriene C4binding site in human multidrug resistance protein 1 (MRP1; ABCC1). *FASEB J.* **2019**, *33*, 10692–10704. [CrossRef] [PubMed]

29. Maeno, K.; Nakajima, A.; Conseil, G.; Rothnie, A.; Deeley, R.G.; Cole, S.P.C. Molecular basis for reduced estrone sulfate transport and altered modulator sensitivity of transmembrane helix (TM) 6 and TM17 mutants of multidrug resistance protein 1 (ABCC1). *Drug Metab. Dispos.* **2009**, *37*, 1411–1420. [CrossRef]

30. Loe, D.W.; Almquist, K.C.; Deeley, R.G.; Cole, S.P.C. Multidrug resistance protein (MRP)-Mediated transport of leukotriene C4 and chemotherapeutic agents in membrane vesicles. *J. Biol. Chem.* **1996**, *271*, 9675–9682. [CrossRef]

31. Haimeur, A.; Deeley, R.G.; Cole, S.P.C. Charged amino acids in the sixth transmembrane helix of multidrug resistance protein 1 (MRP1/ABCC1) are critical determinants of transport activity. *J. Biol. Chem.* **2002**, *277*, 41326–41333. [CrossRef] [PubMed]

32. Situ, D.; Haimeur, A.; Conseil, G.; Sparks, K.E.; Zhang, D.; Deeley, R.G.; Cole, S.P.C. Mutational analysis of ionizable residues Proximal to the cytoplasmic interface of membrane spanning domain 3 of the multidrug resistance protein, MRP1 (ABCC1). *J. Biol. Chem.* **2004**, *279*, 38871–38880. [CrossRef] [PubMed]

33. Ren, X.-Q.; Furukawa, T.; Aoki, S.; Sumizawa, T.; Haraguchi, M.; Nakajima, Y.; Ikeda, R.; Kobayashi, M.; Akiyama, S.-I. A Positively charged amino acid proximal to the C-Terminus of TM17 of MRP1 is indispensable for GSH-Dependent binding of substrates and for transport of LTC$_4$†. *Biochemistry* **2002**, *41*, 14132–14140. [CrossRef]

34. Voss, N.R.; Gerstein, M. 3V: Cavity, channel and cleft volume calculator and extractor. *Nucleic Acids Res.* **2010**, *38*, W555–W562. [CrossRef]

35. Madon, J.; Hagenbuch, B.; Landmann, L.; Meier, P.J.; Stieger, B. Transport function and hepatocellular localization of mrp6 in rat liver. *Mol. Pharmacol.* **2000**, *57*, 634–641. [CrossRef]
36. Lone, A.M.; Taskén, K. Proinflammatory and immunoregulatory roles of eicosanoids in T cells. *Front. Immunol.* **2013**, *4*, 130. [CrossRef]
37. Wielinga, P.R.; van der Heijden, I.; Reid, G.; Beijnen, J.H.; Wijnholds, J.; Borst, P. Characterization of the MRP4- and MRP5-mediated transport of cyclic nucleotides from intact cells. *J. Biol. Chem.* **2003**, *278*, 17664–17671. [CrossRef]
38. Borst, P.; de Wolf, C.; van de Wetering, K. Multidrug resistance-associated proteins 3, 4, and 5. *Pflügers Arch. Eur. J. Physiol.* **2007**, *453*, 661–673. [CrossRef]
39. Bäck, M.; Aranyi, T.; Cancela, M.L.; Carracedo, M.; Conceição, N.; Leftheriotis, G.; Macrae, V.; Martin, L.; Nitschke, Y.; Pasch, A.; et al. Endogenous Calcification Inhibitors in the Prevention of Vascular Calcification: A Consensus Statement From the COST Action EuroSoftCalcNet. *Front. Cardiovasc. Med.* **2019**, *5*, 196. [CrossRef] [PubMed]
40. Uitto, J.; Li, Q.; Jiang, Q. Pseudoxanthoma Elasticum: Molecular Genetics and Putative Pathomechanisms. *J. Investig. Dermatol.* **2010**, *130*, 661–670. [CrossRef]
41. Hosen, M.J.; Zubaer, A.; Thapa, S.; Khadka, B.; De Paepe, A.; Vanakker, O.M. Molecular Docking Simulations Provide Insights in the Substrate Binding Sites and Possible Substrates of the ABCC6 Transporter. *PLoS ONE* **2014**, *9*, e102779. [CrossRef] [PubMed]
42. Fülöp, K.; Barna, L.; Symmons, O.; Závodszky, P.; Váradi, A. Clustering of disease-causing mutations on the domain–domain interfaces of ABCC6. *Biochem. Biophys. Res. Commun.* **2009**, *379*, 706–709. [CrossRef]
43. Ito, K.-I.; Olsen, S.L.; Qiu, W.; Deeley, R.G.; Cole, S.P. Mutation of a single conserved tryptophan in multidrug resistance protein 1 (MRP1/ABCC1) results in loss of drug resistance and selective loss of organic anion transport. *J. Biol. Chem.* **2001**, *276*, 15616–15624. [CrossRef]
44. Zhang, D.; Nunoya, K.; Vasa, M.; Gu, H.-M.; Cole, S.P.C.; Deeley, R.G. Mutational analysis of polar amino acid residues within predicted transmembrane helices 10 and 16 of multidrug resistance protein 1 (ABCC1): Effect on substrate specificity. *Drug Metab. Dispos.* **2006**, *34*, 539–546. [CrossRef] [PubMed]
45. Zhang, D.W.; Cole, S.P.; Deeley, R.G. Determinants of the substrate specificity of multidrug resistance protein 1: Role of amino acid residues with hydrogen bonding potential in predicted transmembrane helix 17. *J. Biol. Chem.* **2002**, *277*, 20934–20941. [CrossRef]
46. Miksch, S.; Lumsden, A.; Guenther, U.P.; Foernzler, D.; Christen-Zäch, S.; Daugherty, C.; Ramesar, R.K.S.; Lebwohl, M.; Hohl, D.; Neldner, K.H.; et al. Molecular genetics of pseudoxanthoma elasticum: Type and frequency of mutations inABCC6. *Hum. Mutat.* **2005**, *26*, 235–248. [CrossRef]
47. Pfendner, E.G.; Vanakker, O.M.; Terry, P.F.; Vourthis, S.E.; McAndrew, P.; McClain, M.R.; Fratta, S.; Marais, A.-S.; Hariri, S.; Coucke, P.J.; et al. Mutation Detection in the ABCC6 Gene and Genotype Phenotype Analysis in a Large International Case Series Affected by Pseudoxanthoma Elasticum. *J. Med. Genet.* **2007**, *44*, 621–628. [CrossRef] [PubMed]
48. Noji, Y.; Inazu, A.; Higashikata, T.; Nohara, A.; Kawashiri, M.-A.; Yu, W.; Todo, Y.; Nozue, T.; Uno, Y.; Hifumi, S.; et al. Identification of two novel missense mutations (p.R1221C and p.R1357W) in the ABCC6 (MRP6) gene in a japanese patient with pseudoxanthoma elasticum (PXE). *Intern. Med.* **2004**, *43*, 1171–1176. [CrossRef]
49. Nitschke, Y.; Baujat, G.; Botschen, U.; Wittkampf, T.; Du Moulin, M.; Stella, J.; Le Merrer, M.; Guest, G.; Lambot, K.; Tazarourte-Pinturier, M.-F.; et al. Generalized Arterial Calcification of Infancy and Pseudoxanthoma Elasticum Can Be Caused by Mutations in Either ENPP1 or ABCC6. *Am. J. Hum. Genet.* **2012**, *90*, 25–39. [CrossRef]
50. Ryu, S.; Kawabe, T.; Nada, S.; Yamaguchi, A. Identification of basic residues involved in drug export function of human multidrug resistance-associated protein 2. *J. Biol. Chem.* **2000**, *275*, 39617–39624. [CrossRef]
51. El-Sheikh, A.; Heuvel, J.J.M.W.V.D.; Krieger, E.; Russel, F.; Koenderink, J.B. Functional role of arginine 375 in transmembrane helix 6 of multidrug resistance protein 4 (MRP4/ABCC4). *Mol. Pharmacol.* **2008**, *74*, 964–971. [CrossRef]
52. Slot, A.J.; Molinski, S.; Cole, S.P. Mammalian multidrug-resistance proteins (MRPs). *Essays Biochem.* **2011**, *50*, 179–207. [CrossRef]
53. Zollmann, T.; Moiset, G.; Tumulka, F.; Tampé, R.; Poolman, B.; Abele, R. Single liposome analysis of peptide translocation by the ABC transporter TAPL. *Proc. Natl. Acad. Sci. USA* **2015**, *112*, 2046–2051. [CrossRef]
54. Katoh, K.; Misawa, K.; Kuma, K.; Miyata, T. MAFFT: A novel method for rapid multiple sequence alignment based on fast Fourier transform. *Nucleic. Acids Res.* **2002**, *30*, 3059–3066. [CrossRef]
55. Bateman, A.; Martin, M.J.; O'Donovan, C.; Magrane, M.; Alpi, E.; Antunes, R.; Bely, B.; Bingley, M.; Bonilla, C.; Britto, R.; et al. UniProt: The universal protein knowledgebase. *Nucleic Acids Res.* **2017**, *45*, D158–D169.
56. Sali, A. Comparative protein modeling by satisfaction of spatial restraints. *Mol. Med. Today* **1995**, *1*, 270–277. [CrossRef] [PubMed]
57. Corradi, V.; Vergani, P.; Tieleman, D. Cystic Fibrosis Transmembrane Conductance Regulator (CFTR): Closed and open state channel models. *J. Biol. Chem.* **2015**, *290*, 22891–22906. [CrossRef]
58. Waterhouse, A.M.; Procter, J.; Martin, D.; Clamp, M.; Barton, G.J. Jalview Version 2-A multiple sequence alignment editor and analysis workbench. *Bioinformatics* **2009**, *25*, 1189–1191. [CrossRef] [PubMed]
59. Madeira, F.; Park, Y.M.; Lee, J.; Buso, N.; Gur, T.; Madhusoodanan, N.; Basutkar, P.; Tivey, A.R.N.; Potter, S.C.; Finn, R.D.; et al. The EMBL-EBI search and sequence analysis tools APIs in 2019. *Nucleic Acids Res.* **2019**, *47*, W636–W641. [CrossRef] [PubMed]
60. Baker, N.A.; Sept, D.; Joseph, S.; Holst, M.J.; McCammon, J.A. Electrostatics of nanosystems: Application to microtubules and the ribosome. *Proc. Natl. Acad. Sci. USA* **2001**, *98*, 10037–10041. [CrossRef] [PubMed]

61. Dolinsky, T.J.; Nielsen, J.E.; McCammon, J.A.; Baker, N.A. PDB2PQR: An automated pipeline for the setup of Poisson-Boltzmann electrostatics calculations. *Nucleic Acids Res.* **2004**, *32*, W665–W667. [CrossRef] [PubMed]
62. Pettersen, E.; Goddard, T.; Huang, C.; Couch, G.; Greenblatt, D.; Meng, E.; Ferrin, T. UCSF Chimera-A visualization system for exploratory research and analysis. *J. Comput. Chem.* **2004**, *25*, 1605–1612. [CrossRef] [PubMed]
63. Schrödinger, L. The PyMOL Molecular Graphics System. Version 2.5. Available online: https://www.schrodinger.com/products/pymol (accessed on 7 May 2021).

International Journal of
Molecular Sciences

Article

ABCB1 Does Not Require the Side-Chain Hydrogen-Bond Donors Gln^{347}, Gln^{725}, Gln^{990} to Confer Cellular Resistance to the Anticancer Drug Taxol

Keerthana Sasitharan, Hamzah Asad Iqbal, Foteini Bifsa, Aleksandra Olszewska and Kenneth J. Linton *

Blizard Institute, Barts and the London School of Medicine and Dentistry, Queen Mary University of London, 4 Newark St, London E1 2AT, UK; k.sasitharan@se17.qmul.ac.uk (K.S.); h.a.iqbal@se17.qmul.ac.uk (H.A.I.); f.bifsa@se17.qmul.ac.uk (F.B.); a.olszewska@se18.qmul.ac.uk (A.O.)
* Correspondence: k.j.linton@qmul.ac.uk; Tel.: +44-(0)207-882-8997

Citation: Sasitharan, K.; Iqbal, H.A.; Bifsa, F.; Olszewska, A.; Linton, K.J. ABCB1 Does Not Require the Side-Chain Hydrogen-Bond Donors Gln^{347}, Gln^{725}, Gln^{990} to Confer Cellular Resistance to the Anticancer Drug Taxol. *Int. J. Mol. Sci.* **2021**, *22*, 8561. https://doi.org/10.3390/ijms22168561

Academic Editor: Jose J.G. Marin

Received: 28 May 2021
Accepted: 28 July 2021
Published: 9 August 2021

Publisher's Note: MDPI stays neutral with regard to jurisdictional claims in published maps and institutional affiliations.

Abstract: The multidrug efflux transporter ABCB1 is clinically important for drug absorption and distribution and can be a determinant of chemotherapy failure. Recent structure data shows that three glutamines donate hydrogen bonds to coordinate taxol in the drug binding pocket. This is consistent with earlier drug structure-activity relationships that implicated the importance of hydrogen bonds in drug recognition by ABCB1. By replacing the glutamines with alanines we have tested whether any, or all, of Gln^{347}, Gln^{725}, and Gln^{990} are important for the transport of three different drug classes. Flow cytometric transport assays show that Q347A and Q990A act synergistically to reduce transport of Calcein-AM, BODIPY-verapamil, and OREGON GREEN-taxol bisacetate but the magnitude of the effect was dependent on the test drug and no combination of mutations completely abrogated function. Surprisingly, Q725A mutants generally improved transport of Calcein-AM and BODIPY-verapamil, suggesting that engagement of the wild-type Gln^{725} in a hydrogen bond is inhibitory for the transport mechanism. To test transport of unmodified taxol, stable expression of Q347/725A and the triple mutant was engineered and shown to confer equivalent resistance to the drug as the wild-type transporter, further indicating that none of these potential hydrogen bonds between transporter and transport substrate are critical for the function of ABCB1. The implications of the data for plasticity of the drug binding pocket are discussed.

Keywords: P-glycoprotein; multidrug resistance; ABCB1; taxol; drug transport; ABC transporter

1. Introduction

Multidrug resistance (MDR) remains a problem for the chemotherapy of cancer patients [1,2]. Polyspecific efflux transporters of the plasma membrane that prevent the accumulation of a range of drugs to cytotoxic levels are a common cause of MDR [3,4]. The primary transporter associated with failure of chemotherapy is ABCB1 (previously known as P-glycoprotein and MDR1). Outside of the cancer clinic, ABCB1 is an important determinant of drug absorption, distribution, and excretion of many drugs including antibiotics, anti-epileptics and antiarrhythmics due to its native expression in a range of tissues including the apical membranes of gut epithelia and endothelial cells of the blood-brain barrier, and the canalicular membrane of the liver hepatocytes [5]. The polyspecificity of ABCB1 for many drugs of different structure and chemical class is a key feature of the transporter that needs to be understood to allow for drug designs which avoid recognition by the transporter and for the design of specific inhibitors. In this regard, the recent report by Alam et al. [6] of the structure of ABCB1 in complex with a transport substrate, the anticancer drug taxol, represents a milestone in its field. The complex was solved by single particle imaging using cryoelectron microscopy (cryoEM) with the taxol molecule occluded within the transmembrane domains of the transporter. Twelve amino acids were involved in drug coordination. These were primarily via weak Van der Waals interactions. However, three glutamines, Gln[347], Gln[725], and Gln[990], located respectively in transmembrane helix (TMH) 6, TMH7, and TMH12, were highlighted to form hydrogen bond contacts (Figure 1). The study was not without limitations. The medium resolution of 3.5 Å, the low level of functional activity of the transporter within the nanodisc particles and the trapping of the transporter conformation using an inhibitory antibody may impact the physiological significance of the findings. The affect these have on the veracity of structural detail remain unclear and the implications of the binding pocket were not tested further.

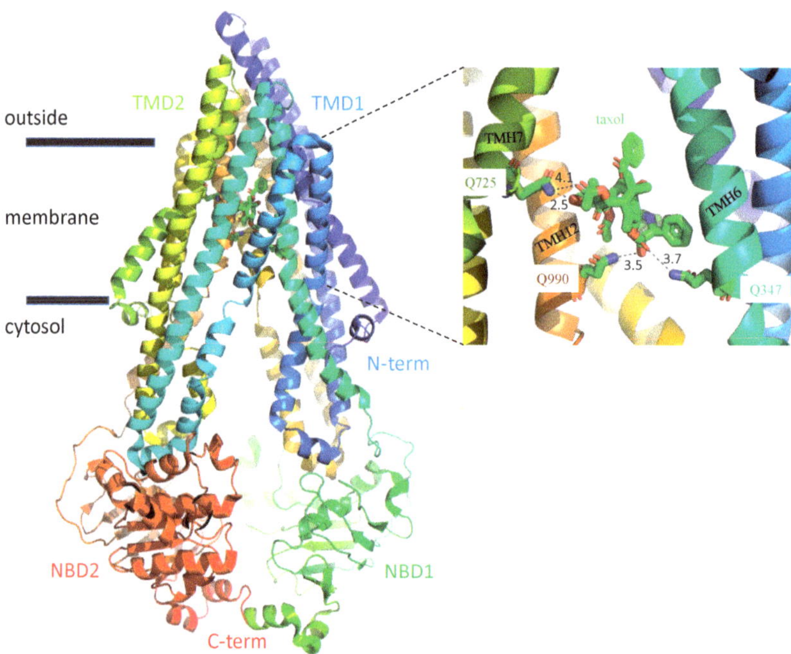

Figure 1. The taxol binding pocket. Ribbon depiction of ABCB1 with taxol occluded by the transmembrane domains (TMD1, blue-turquoise spectrum; TMD2, green-orange spectrum). The nucleotide binding domains, NBD1 and NBD2, are shown in green and red, respectively (pdb: 6QEX). The right-hand panel shows a 12Å slice in the Z plane of the taxol binding site. The three glutamines Gln[347], Gln[725], and Gln[990] highlighted by Alam et al. [6] to hydrogen bond (dashed grey lines) with the taxol are show in single letter code and stick format with the bond lengths (N-O) indicated in black in Ångstroms. The combination of bond angle and length suggests that Gln[725] forms the strongest and only H-bond with the baccatin III tetracyclic ring, while Gln[347] and Gln[990] form weaker H-bonds with the carbonyl and hydroxyl, respectively, which link to the diphenolic tail of the drug.

Prior structure-activity studies, where "activity" refers to whether the one hundred chemicals that were tested are transport substrates of ABCB1 or not, had already suggested the importance of multiple free electron pairs (and their spatial pattern) in defining the transport substrates [7–9]. Taxol is particularly rich in these structural motifs with six pairs (or triplets) of acceptor sites separated by 2.5 or 4.6 Å, respectively, available for electrostatic interaction with the transporter. Two of these motifs were observed to form hydrogen bonds in the structure determined by Alam et al. (Gln[347] and Gln[990] coordinate the different oxygens within the same motif). Taken together, the simplest interpretation of these earlier structure-activity relationship data and the recent empirical structural data is that the three glutamines are likely key to drug recognition.

In the current study, we asked whether any or all of the glutamines are necessary for triggering the transport cycle and whether they have a role in the polyspecificity of the transporter. The glutamines were replaced with alanine residues to create single, paired and triple mutants. We then measured the effect on the transport efficiency of three different classes of drug in transiently-transfected cells and tested whether the mutants are able to confer resistance to taxol after stable expression in Flp-In cells.

2. Results

To test the importance of the hydrogen bond donors implicated in the coordination of taxol, we effectively removed the donor side chains by site directed mutagenesis and measured the function of the mutant protein in its native environment of live human cells. Function analysis used a modified flow cytometric transport assay to correlate ABCB1 expression with the reduced cellular accumulation of a fluorescent transport substrate normalizing to the untransfected cells in the population.

2.1. Endpoint Two-Colour Flow Cytometry Assay Measures the Function of ABCB1 in Live Cells

HEK293T cells were transfected transiently with pABCB1 encoding wild-type ABCB1. The density of ABCB1 on the cell surface was determined using saturating amounts of the anti-ABCB1 monoclonal antibody (4E3) that does not inhibit transporter function. The primary antibody was detected using saturating amounts of anti-mouse secondary antibody conjugated to a red fluorescent fluorophore. Use of saturating levels of the primary and secondary antibodies, which we determined previously [10], is important to confirm that the mutations introduced did not alter the expression level of the transporter in the plasma membrane. The red fluorescent cells are easily detected by flow cytometry and when they express the wild-type transporter these cells accumulate low levels of green-fluorescent transport substrates or drugs (Figure 2). Transient transfection under the conditions used never results in 100% transfection efficiency, so there are always non-expressing cells in the population and these become important internal controls. The drug content (green fluorescence) within the ABCB1-negative (untransfected) cells divided by the drug content of the ABCB1-positive cells (equation in Figure 2b) provides a robust and reproducible measure of the functionality of the transporter. The example given in Figure 2b shows that the wild-type transporter is able to maintain a fold difference of 96 over the untransfected population for Calcein-AM added to a final concentration of 0.5 μM. For comparison, for the Walker B double mutant which is unable to hydrolyse ATP and so cannot function as a primary active transporter, the calculated ratio is 1.3 (Figure 2c).

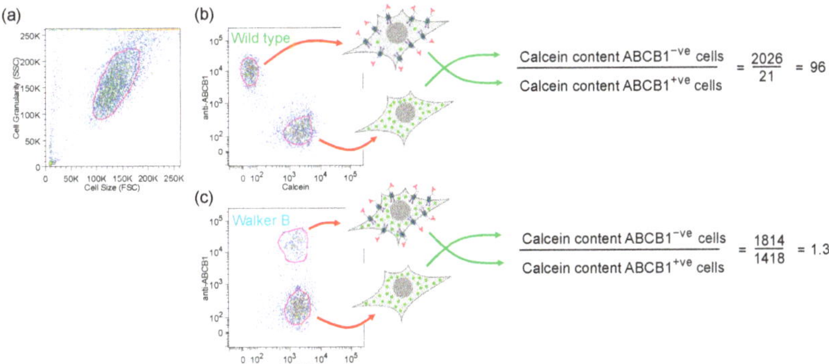

Figure 2. Measurement of functionality of ABCB1 for the transport of Calcein-AM by flow cytometry. (**a**) Dotplot showing the gating (autogated population circled in red) of HEK293T cells of normal size and granularity; (**b**) two-color dotplot of the gated normal cells showing green-fluorescent Calcein on the x-axis and red-fluorescent antibody binding on the y-axis. Cells that express wild-type ABCB1 bind the anti-ABCB1 antibody (red antibody shapes in the cartoon cell) and have a low level of accumulation of Calcein-AM (green circles in the cartoon cell) whilst untransfected cells accumulate high levels. The ratio of drug accumulation between the ABCB1-expressing and non-expressing cells is used to quantify transporter functionality. (**c**) Cells expressing the non-functional Walker B mutant (E556/1201Q) accumulate the same level of Calcein-AM as the untransfected cells in the population.

For these data to reflect the functionality of the expressed transporter in a reproducible manner it is important to control two factors. First, the cells must be exposed to a sufficient concentration of drugs such that all ABCB1 molecules are required to function to limit drug accumulation. This was determined empirically by exposing the cells to increasing levels of drug to determine the concentration at which the ABCB1-positive cells began to accumulate the green fluorophore. The titration of Calcein-AM is shown in Supplementary Figure S1a and shows that 0.5 µM is appropriate. Supplementary Figure S1b,c show the drug accumulation curves for drugs BODIPY-verapamil and OREGON GREEN taxol bisacetate (OG-taxol), respectively. Accumulation of all three drugs is non-saturable at least up to 5 µM for Calcein-AM, 8 µM for BODIPY-verapamil and 4 µM for OG-taxol consistent with a mechanism of entry by passive diffusion. The available data suggests that drug interaction with ABCB1 is directly from the lipid phase of the bilayer and is thus post desolvation of the drug [11]. Second, the experimental set up allows gating on populations of cells that express equivalent levels of the transporter (that have bound equivalent levels of the red fluorescent secondary antibody), as shown for the exemplar Calcein-AM transport experiment in Supplementary Figure S2. This ensures that any differences in drug accumulation between mutants are due to the functionality of the transporter rather than the density of the transporters in the plasma membrane.

2.2. Gln^{347}, Gln^{725} and Gln^{990} Modulate the Transport of the Xanthene Dye Calcein-AM

In order to test the importance of the hydrogen bonds formed between taxol and the side chains of ABCB1 amino acids Gln^{347}, Gln^{725}, and Gln^{990}, the wild-type glutamines were replaced with alanine by site-directed mutagenesis. Alanine with its non-polar side chain is unable to donate an equivalent hydrogen bond to the transport substrate but its smaller side chain should be tolerated in the tertiary structure of the protein. Individual, pairwise mutants and a triple (Qtriple) mutant were generated.

2.2.1. Of the Single Mutants, Only Q347A Shows a Statistically Significant Reduction in Calcein-AM Transport

Single mutants had only a modest effect on the transport of Calcein-AM as shown in the first four columns of Figure 3. Of the single mutants, only Q347A reduced the function of the transporter (to 63% of the wild-type level). Q725A showed a trend towards being more functional than the wild-type but this did not reach statistical significance and no effect was recorded for Q990A. This suggested that none of the hydrogen bonds that can be formed with the side chains of these three glutamines are essential for triggering the transport cycle although it is more efficient at effluxing Calcein-AM with Gln^{347} present.

2.2.2. Q347A and Q990A Act Synergistically to Reduce the Transport of Calcein-AM

All three double mutant combinations and the triple mutant (Qtriple) were generated. Their ability to transport Calcein-AM showed pronounced and unexpected differences. A simple additive effect of Q347A and Q990A would predict a functionality of the Q347/990A double mutant to be 57% (the level of functionality of the Q990A mutant multiplied by the level of functionality of the Q347A mutant; $90/100 \times 63/100 = 57/100$). The observed activity of the Q347/990A mutant was reduced to 8.8% of the wild-type transporter suggesting a synergistic effect of the combined mutations and the importance of these two hydrogen bond donors for the efficient efflux of Calcein-AM. However, it is clearly not essential that these two glutamines are present because the $8.8 \pm 1.87\%$ (mean $\pm$ SEM) level of activity remains statistically higher than the Walker B mutant E556/1201Q ($1.3 \pm 0.59\%$) which is unable to hydrolyze ATP, indicating that the Q347/990A mutant retains measurable function.

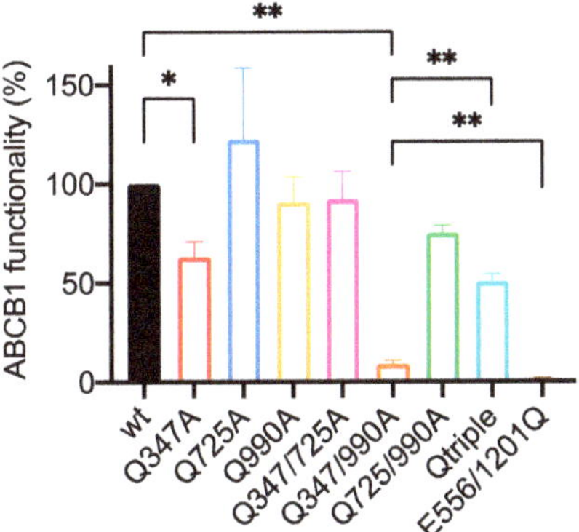

Figure 3. Functionality of glutamine to alanine mutants for the transport of Calcein-AM. Live HEK293T cells transiently expressing equivalent amounts of wild-type (wt) and mutant ABCB1 were challenged with Calcein-AM. Functionality was measured as the ratio of Calcein accumulation (Calcein-AM only becomes fluorescent once it is de-esterified in the cytosol) between the ABCB1-expressing and untransfected cells within the population. This was normalized to 100% for wild-type ABCB1 for the bar graph shown. The mean ± SEM was plotted using GraphPad Prism version 8; sample number was ≥3. Selected statistical analysis (ratio of paired Student's *t*-test, two-tailed) performed on the raw data is shown with *p* values: * <0.05, ** <0.01. The full pairwise comparison of the data is given in Appendix A Table A1.

2.2.3. The Q725A Mutation Improves the Efficiency of Transport of Calcein-AM

The most surprising result is that the mutation of glutamine 725 to alanine improves the functionality of ABCB1 for the transport of Calcein-AM. This is most clearly evident when the Q725A mutation is introduced into the Q347/990A background to generate the Qtriple mutant. The Qtriple mutant has a transport activity of 50.7 ± 4.0% while the Q347/990A double mutant has a transport activity of 8.8 ±1.9%. The *p* value for this comparison by Student's *t*-test is 0.0055, strongly suggesting that the inclusion of the Q725A mutation has made the Qtriple mutant more active than the Q347/990A double mutant and at the same time emphasizing that neither Gln^{347} nor Gln^{990} are absolutely necessary for ABCB1 to transport Calcein-AM. The ostensible increase in the mean activities of the other three constructs that include the Q725A mutation when compared to their respective backbones (wild-type versus Q725A, Q347A versus Q347/725A and Q990A versus Q725/990A) fail to reach statistical significance (Table A1).

2.3. Gln^{347}, Gln^{725} and Gln^{990} Also Modulate the Transport of the Phenylalkylamine BODIPY-Verapamil

To test whether Gln^{347}, Gln^{725}, and Gln^{990} are also important for the transport of a different drug class, the transport assays were repeated with a fluorescent derivative of the phenylalkylamine verapamil, which acts as a calcium channel blocker and is used clinically to treat a variety of heart arrhythmias.

2.3.1. Q725A Improves the Transport of BODIPY-Verapamil in Any Background

The challenge with BODIPY-verapamil gave a clearer indication that the side chain of the native Gln^{725} inhibits transport activity (Figure 4). Comparing the raw transport data of the Q725A single mutant with the wild-type transporter, it is statistically clear that Q725A increased the ability to efflux BODIPY-verapamil. This relationship is maintained for all mutants that include Q725A compared to the backbone into which the mutation was introduced. Thus, Q347/725A is statistically more active for the transport of BODIPY-verapamil compared to Q347A. Likewise, Q725/990A is more active than Q990A, and the Qtriple mutant is more active than Q347/Q990A.

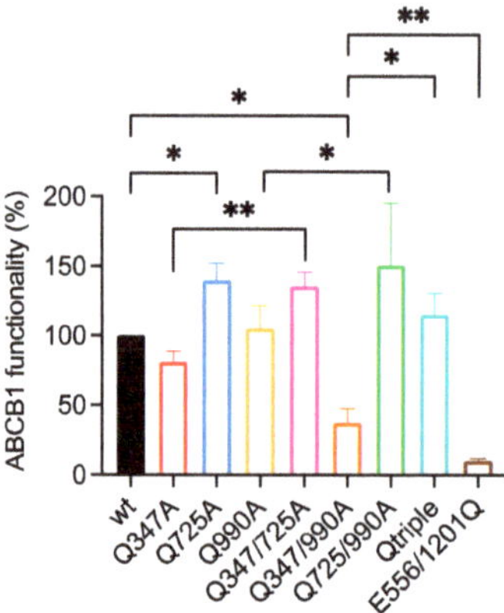

Figure 4. Functionality of glutamine to alanine mutants for the transport of BODIPY-verapamil. Live HEK293T cells transiently expressing equivalent amounts of wild-type (wt) and mutant ABCB1 were challenged with BODIPY-verapamil. Functionality was measured as the ratio of BODIPY-verapamil accumulation between the ABCB1-expressing and untransfected cells within the population. This was normalized to 100% for wild-type ABCB1 for the bar graph shown. The mean $\pm$ SEM was plotted using GraphPad Prism version 8; sample number was $\geq$3. Selected statistical analysis (ratio of paired Student's *t*-test, two-tailed) performed on the raw data is shown with *p* values: * <0.05, ** <0.01. The full pairwise comparison of the data is given in Appendix A Table A2.

2.3.2. Q347A and Q990A Also Act Synergistically to Reduce the Transport of BODIPY-Verapamil

The Q347A and Q990A mutants (normalized transport activity of 81 $\pm$ 8% and 105 $\pm$ 16%, respectively (Appendix A Table A2)) combine in the Q347/990A double mutant to reduce the transport of BODIPY-verapamil to 38 $\pm$ 10% of the wild-type level (Figure 4). This double mutant also retains the ability to efflux BODIPY-verapamil because it is significantly different to the Walker B mutant E556/1201Q.

2.4. Gln^{347}, Gln^{725} and Gln^{990} Have a More Limited Effect on the Transport of the Taxane Diterpenoid Derivative OG-Taxol

Taxol, the transport substrate that was first observed in complex with ABCB1, is not fluorescent but its derivative OREGON-GREEN taxol bisacetate (OG-taxol) fluoresces in the green spectrum and retains an ability to bind to microtubules in live cells. We tested whether, like taxol itself, it is also a transport substrate of ABCB1. A drug titration experiment showed that ABCB1-expressing cells accumulate less OG-taxol than non-expressing control cells and indicated that 0.4 µM OG-taxol was sufficient to require all of the ABCB1 molecules on the surface of our transiently-transfected HEK293T cells to limit accumulation of the drug (Supplementary Figure S1c).

2.4.1. Of the Single Mutants Only Q990A Appears to Reduce the Transport of OG-Taxol

The Q347A and Q725A mutants were not distinguishable from the wild-type transport activity. However, the reduced activity of the Q990A mutant reaches statistical significance only after the raw data are paired (Figure 5). The effect is subtle with the Q990A mutant retaining 66 $\pm$ 8% transport activity for OG-taxol when normalized to wild-type ABCB1.

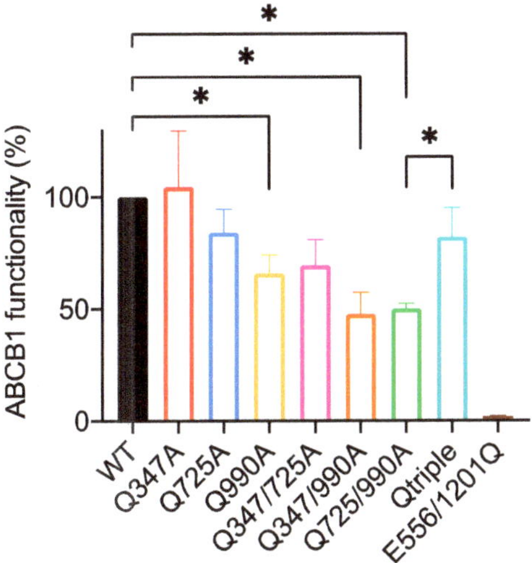

Figure 5. Functionality of glutamine to alanine mutants for the transport of OG-taxol. Live HEK293T cells transiently expressing equivalent amounts of wild-type (wt) and mutant ABCB1 were challenged with OG-taxol. Functionality was measured as the ratio of OG-taxol accumulation between the ABCB1-expressing and untransfected cells within the population. This was normalized to 100% for wild-type ABCB1 for the bar graph shown. The mean ± SEM was plotted using GraphPad Prism version 8; sample number was ≥3. Selected statistical analysis (unpaired Student's *t*-test, two-tailed performed on the raw data except for the comparison of the wild-type with Q990A for which the raw data are paired) is shown with *p* value: * <0.05. The full pairwise comparison of the data is given in Appendix A Table A3.

2.4.2. The Double Mutants Q347/990A and Q725/990A Reduce the Transport Activity Further but the Triple Mutant Restores Wild-Type Levels of OG-Taxol Transport

The Q347/725A double mutant is trending towards reduced transport of OG-taxol but does not reach statistical significance. However, both Q347/990A and Q725/990A have reduced transport activity for OG-taxol, emphasizing the negative effect of the Q990A mutation. Perhaps surprisingly, given that all pairwise mutants seem to have reduced transport of OG-taxol, the triple mutant restores transport activity to wild-type levels.

2.4.3. There Is No Indication That Gln^{725} Is Inhibitory for the Transport of OG-Taxol

In contrast to the transport of Calcein-AM and BODIPY-verapamil, there is no evidence from the data that transport of OG-taxol is improved in any mutant harboring the Q725A change. Although the Q347/990A mutant has a lower mean transport activity (48 ± 9%) than the Qtriple (82 ± 13%) these are not statistically different and none of the other mutants to which Q725A has been introduced come close to a statistically relevant difference to the parent plasmid (e.g., Q990A compared to Q725/990A).

2.5. The Q347/990A and the Qtriple Mutant Are Indistinguishable from the Wild-Type Transporter in Conferring Taxol Resistance to Cells in Culture

The more subtle differences observed for the transport of OG-taxol suggested that either the hydrogen bonds donated by Gln^{347}, Gln^{725}, and Gln^{990} were not particularly important for the transport cycle or that OG-taxol, despite being a transport substrate for ABCB1, does not replicate the geometry of taxol in the binding pocket. To test this, stable cell lines were derived to express the Q347/990A double mutant and the Qtriple mutant. The Q347/990A mutant was chosen because it consistently had the biggest effect on the transport of the three transport substrates tested and the Qtriple was chosen in case the three hydrogen bonds were critical only for the binding of taxol that they had been observed to coordinate. To ensure that like for like comparisons could be made, the cDNAs for Q347/990A and Qtriple were subcloned into pcDNA5/FRT. This allowed site-directed recombination to introduce a single copy of the plasmid into the same site within the genome of Flp-In HEK293. Along with Flp-In-ABCB1wt which was generated previously these stable cell lines ensured uniform levels of wild-type and mutant ABCB1 expression compared with the vector-only (integrated pcDNA5/FRT) negative control (Figure 6a).

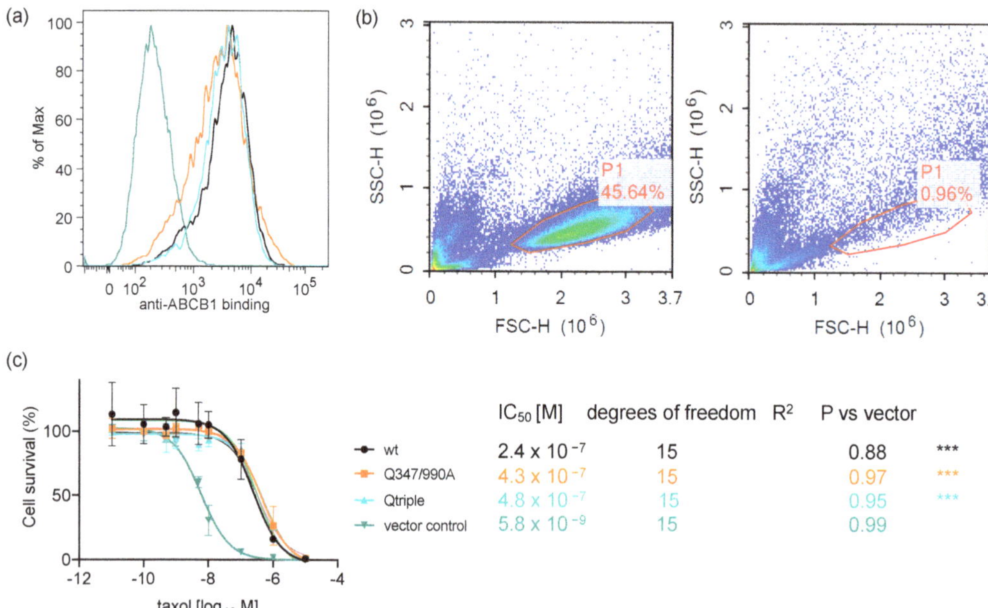

Figure 6. Stable expression of ABCB1-Q347/990A or ABCB1-Qtriple confers the same level of resistance to taxol as wild-type ABCB1. (**a**) Surface expression of ABCB1 (4E3 antibody binding) is similarly elevated for the two mutants (Q347/990A in orange and Qtriple in cyan) and the wild-type transporter (black) compared to the vector-only control (teal); (**b**) Flow cytometric exemplar dotplots in the presence and absence of taxol. Plots for the Flp-In ABCB1-Q347/990A cell line shows the forward scatter (FSC) and side scatter (SSC) heights in the absence (left hand plot) and the presence of 10 µM taxol for 72 h (right hand plot). All cells of normal size and granularity in the well were counted in a NovoCyte flow cytometer and analysed in NovoExpress software. The P1 gate which defines the healthy cell population in the absence of taxol was copied to all other conditions. In the examples shown, there were 71,969 cells in the P1 gate in the absence of taxol and 943 cells in the P1 gate following exposure to 10 µM taxol; (**c**) Non-linear regression analysis of cell survival on challenge with increasing concentration of taxol (colour code as above). Cell number in each well of the taxol dilution series was normalized to 100% for the P1 gate of the zero-taxol condition. The mean ± SEM was plotted with curve fitting by non-linear regression in GraphPad Prism version 8; sample number, n = 2 biological repeats. The biological repeats were averaged from duplicate technical repeats. *** $p < 0.0001$ compared to the vector-only cell line. The cell lines expressing the double and triple mutant ABCB1 are not significantly different to that expressing the wild-type transporter.

The cell lines were challenged with increasing concentrations of taxol for three days, after which the cells with normal size and granularity were counted (Figure 6b; the use of a NovoCyte flow cytometer allowed all cells in the well to be counted in this experiment thus there is no estimation by counting only a small fraction of population). This allowed the IC_50 for taxol to be calculated for the two test cell lines and compared to positive (cells expressing the wild-type transporter) and negative (cells with an integrated empty vector) controls (Figure 6c). It was clear from the survival curves that taxol is a potent cytotoxic, killing the vector-only control cells with an IC_50 = 5.8 nM. Stable expression of wild-type ABCB1 shifts that IC_50 more than 40-fold to an IC_50 = 240 nM. The measured half maximal inhibitory concentrations for the Flp-In-Q347/990A and Flp-In-Qtriple cells are 430 nM and 480 nM, respectively, and are statistically indistinguishable from the effect of taxol on the Flp-In-ABCB1 wild-type cells.

3. Discussion

Structure-activity relationship (SAR) analyses [7] have indicated the importance of free electron pairs in the transport substrates of ABCB1 while structure data of the transporter in complex with taxol [6] has identified three glutamines within the drug binding pocket that donate hydrogen bonds to electron pairs in taxol. Molecular modelling studies [12] replicate these H-bond interactions in silico but their importance for drug recognition and to trigger the transport cycle remains unclear. Alam et al. [6] highlighted hydrogen bonds donated by Gln^{347}, Gln^{725}, and Gln^{990} to coordinate taxol in the binding pocket of ABCB1. We have tested whether this H-bonding pattern was key to triggering the transport cycle with three different pharmacophores, a xanthene, a phenylalkylamine, and two forms of a taxane diterpenoid. Several conclusions can be drawn from this study with the simplest being that none of these hydrogen bonds are absolutely essential for transport. Even the most impaired double mutant, Q347/990A, which retains only 8.8% of the wild-type level of activity for the transport of Calcein-AM, is still able to reduce the accumulation of the dye by cells in comparison to the non-functional Walker B mutant E556/1201Q. This equates to a 25-fold reduction in accumulation of Calcein-AM (the ratio of dye accumulation in the untransfected cells/Q347/990A-expressing cells) which is statistically different to the non-functional Walker B mutant which averages 1.3. The wild-type transporter, for comparison, can reduce accumulation of Calcein-AM by up to 258-fold in these experiments. It is thus clear that these mutant transporters which should lack the ability to donate hydrogen bonds to the transport substrate retain at least some level of transport activity for all three of the different classes of drug tested.

The situation, of course, is more nuanced. There is some consistency in the transport of different drugs. For example, the Q347/990A mutant has significantly reduced activity for the transport of all three fluorescent drugs but the level of impairment is to a different degree (8.8% of the transport activity of the wild-type for Calcein-AM, 37.8% for BODIPY-verapamil and 48% for OG-taxol). Thus, it would appear that the hydrogen bonding capacity of the side chains of Gln^{347} and Gln^{990} are involved in drug transport. There are also drug specific effects. There is a clear indication that introduction of the Q725A mutation improves the transport of Calcein-AM (cf. Q347/990A and Qtriple) and BODIPY-verapamil, but this is not true for OG-taxol. Perhaps the most surprising finding was that the Qtriple mutant, in which all three glutamines are replaced by alanine, retained activity (or regained activity compared to some of the double mutants) for the transport of all three drugs to achieve 51% transport activity for Calcein-AM, 116% activity for BODIPY-verapamil and 82% activity for OG-taxol. This observation also emphasizes that Gln^{347} and Gln^{990} are not critical for efficient transport because they are also absent from the Qtriple mutant which is indistinguishable from the wild-type transporter for the transport of BODIPY-verapamil and OG-taxol.

3.1. A Possible Allosteric Explanation for the Increased Transport Activity of Q725A

The negative effect of the wild-type Gln^{725} on the apparent transport activity is consistent with an earlier study by Loo et al. [13] during which they characterized a Q725C mutant (in an otherwise cysteine-less version of ABCB1). They observed that the basal ATPase activity of Q725C measured in vitro was raised 2.6-fold, offering a possible explanation for the improved transport of BODIPY-verapamil and Calcein-AM by the Q725A mutant if increased ATPase activity leads to increased drug efflux. It is possible that both the observed increase in ATPase activity and the increase in transport activity of fluorescent drugs when Gln^{725} is mutated is not due to the loss of an H-bond to the drug in the binding pocket but to the loss of an intra-molecular H-bond in a distinct conformation of the protein. In 2018, Kim and Chen reported the first medium resolution structure of human ABCB1 at 3.4 Ångstrom resolution [14]. They made use of the same E556/1201Q mutant used in the current study to prevent ATP hydrolysis and so were able to trap the protein with ATP bound in an 'outward-facing' conformation that is considered to show ABCB1 post drug release but prior to ATP hydrolysis. In this conformation, the binding cavity is closed to the membrane but open extracellularly. The side chains of Gln[347] and Gln[990] are not involved in electrostatic interactions with any other residue, but Gln[725] forms a hydrogen bond with Asn[842] of TMH9 (Figure 7). We speculate that the loss of this H-bond in the Q725A mutants may be more likely to increase the ATPase activity as TMH 9 is connected directly to the third 'coupling helix' (located at the base of the intracellular loop formed between TMH8 and TMH9). The coupling helices are thought to allosterically couple the drug binding pocket to the sites of ATP hydrolysis [15,16].

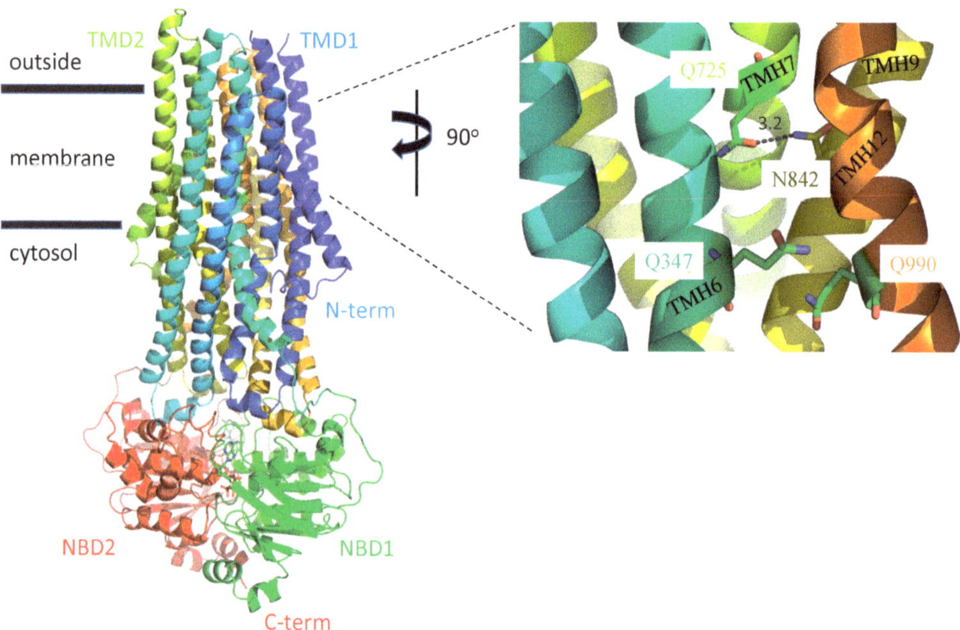

Figure 7. ABCB1 in the ATP-bound conformation focussing on the TMDs showing the positions of Gln[347], Gln[725] and Gln[990] 'post drug release'. Ribbon depiction of ABCB1 mutant E556/1201Q with ATP Mg^{2+} bound (pdb: 6C0V). The transmembrane domains (TMD1, blue-turquoise spectrum; TMD2, green-orange spectrum). The nucleotide binding domains, NBD1 and NBD2, are shown in green and red, respectively with two ATP molecules sandwiched at their shared interface depicted in stick format. The right-hand panel shows a 12 Å slice in the Z plane showing the positions of Gln[347], Gln[725], Gln[990], and Asn[842] in stick format and identified by single letter code. The hydrogen bond between Gln[725] and Asn[842] in this conformation is shown as a dashed grey line with the bond length (N-O) indicated in black in Ångstroms.

This leads us to an important caveat. The measurement of fluorescent drug accumulation by ABCB1-expressing cells compared to non-expressing cells is a robust test of ABCB1 function in live cells in its native environment of the plasma membrane, not just drug binding to the transporter. The rate of accumulation of drugs by cells depends also on the physicochemical properties of the drug to diffuse across the plasma membrane. The available evidence suggests that ABCB1 scans the membrane to identify and efflux hydrophobic compounds intercalated between the fatty acyl chains to preserve the chemical barrier [17]. The transport cycle begins when the drug complexes with the TMDs, triggering conformational change such that the NBDs bind ATP. The binding energy of ATP and the formation of the NBD:NBD interface is sufficient to change the conformation of the TMDs such that the drug binding site is reorientated to open extracellularly and affinity is lowered. Completion of the transport cycle requires ATP hydrolysis (the step that is absent from the E556/1201Q mutant) to drive the transporter back into the inward open conformation. Whilst our study of Gln^{347}, Gln^{725}, and Gln^{990} is predicated on the coordination of taxol it is also possible that the changes introduced could have a role to play as the transporter transitions through the conformational changes required to complete a full cycle.

3.2. Induced Fit of the TMDs around the Transport Substrate Likely Explains the Lack of Importance of Gln^{347}, Gln^{725} and Gln^{990} in Conferring Resistance to Taxol

The flow cytometric assay is also limited by its dependence on fluorescent dyes and drug analogues. In OG-taxol bisacetate, while the oxygen atoms in the taxol pharmacophore that were observed to hydrogen bond with Gln^{347}, Gln^{725}, and Gln^{990} remain available, the molecular weight of the fluorescent drug is 1.5 times greater than that of taxol which must affect the geometry of the drug within the binding pocket. To rule out a possible artefact we generated stable cell lines and, to our surprise, demonstrated that the mutant Q347/990A and the Qtriple transporters conferred resistance to taxol to the same degree as the wild-type transporter. So, in a direct test against unmodified taxol that would be unable to hydrogen bond with the side chains of alanines at positions 347 and 990 (plus or minus position 725) we observed no difference in the survival curves of ABCB1-expressing cells, making it clear that the hydrogen bonds observed in the cryoEM data are not particularly important for the transport of taxol.

A second structure of ABCB1 in complex with a transport substrate, the vinca alkaloid vincristine, has now been reported by Nosol et al. [18]. The binding pocket of vincristine overlaps with that of taxol sharing six amino acids in common, including Gln^{347} and Gln^{990} (the latter considered close enough to hydrogen bond). Six further amino acids are unique to the vincristine pocket and five more for taxol. Some of these differences involve a subtle turn of a helix (for example Tyr^{307} in TMH5 is implicated in taxol binding while the adjacent amino acid Ile^{306} is implicated in vincristine binding). The main contributors to both binding pockets are from TMH5, 6, 11 and 12 while the vincristine pocket also includes Met^{68} and Met^{69} from TMH1 and Glu^{875} from TMH10, and the taxol pocket includes Gln^{725} from TMH7. It is perhaps not coincidental that Seelig had already noted in 1998 that TMH4, 5, 6, 11, and 12 are enriched in amino acids with hydrogen bond donor side chains [7]. It is possible to reconcile the drug SAR data, the empirical structural data, and the lack of effect of Q347/990A or the Qtriple to change the level of taxol resistance if the transmembrane domains are sufficiently flexible to fold around the transport substrate. An induced fit model has long been postulated to explain the unusually broad polyspecificity of ABCB1 and the first evidence in support of induced fit was reported by Clarke's group in 2003 in which they showed a changing cross-linking pattern within the transmembrane domains in response to different drugs [19]. With this in mind, it seems perfectly reasonable to suggest that taxol might hydrogen bond to Gln^{347}, Gln^{725}, and Gln^{990}, but in their absence different hydrogen bonds (or other electrostatic or weaker Van der Waals interactions) may be formed as the transmembrane domains close around the drug in the cavity. Further experiments will be required to test whether the lack of effect of the double or triple glutamine to alanine mutants are due to redundancy among the hydrogen bond donors within ABCB1. However, it is clear that neither Gln^{347}, Gln^{725}, nor Gln^{990} are essential for taxol efflux.

4. Materials and Methods

4.1. Site-Directed Mutagenesis

The cDNA encoding ABCB1 including a 12 histidine carboxy-terminal tag was described previously [20]. This cDNA was subcloned into pCI-neo to generate pCI-neo-ABCB1-12his (henceforth designated pABCB1). The coding sequence was modified by lightning site-directed mutagenesis (Agilent Technologies, Santa Clara, CA, USA) following the manufacturer's recommendations except for generation of the Q347A mutant where a lower annealing temperature of 50 °C was required. The individual mutants for Q347A, Q725A, and Q990A were made first followed by the sequential addition of the second and third mutations. As a negative control for the transport experiments, the catalytic glutamates within NBD1 and NBD2 (Glu556 and Glu1201, respectively) were mutated to glutamines, thus preventing activation of the water for nucleophilic attack on the bound ATP. This mutant was generated previously [21]. Being unable to hydrolyze ATP, this mutant becomes trapped in the ATP-bound state [22]. Each cDNA was fully sequenced to ensure veracity.

Mutagenic oligonucleotides with the new alanine codons emboldened (Table 1).

Table 1. Mutagenic oligonucleotides with the new alanine codons emboldened.

	Q347A
Forward	5′-ttaattggggctttagtgttgga**gcg**gcatctccaagcat-3′
Reverse	5′-atgcttggagatgcc**gc**tccaacactaaaagccccaattaa-3′
	Q725A
Forward	5′-gtgccattataaatggaggcctg**gca**ccagcatttgcaataatatttt-3′
Reverse	5′-aaaatattattgcaaatgctgg**tgc**caggcctccatttataatggcac-3′
	Q990A
Forward	5′-gccatggccgtggggg**gca**gtcagttcatttgc-3′
Reverse	5′-gcaaatgaactgact**gc**ccccacggccatggc-3′

4.2. Transient Expression of ABCB1

HEK293T cells were grown in DMEM high glucose (ThermoFisher Sci, Waltham, MA, USA) supplemented with 10% fetal bovine serum (FBS; ThermoFisher Sci, Waltham, MA, USA) at 37 °C in a humidified incubator with 5% CO_2. For transient transfection 1.2×10^6 cells were seeded onto a T25 culture flask. Twenty-four hours post seeding the cells were transfected with 10 μg plasmid DNA in complex with 15 μg linear polyethylenimine (Sigma-Aldrich, St. Louis, MO, USA). Cells were cultured for a further 48 h before harvesting with TrypLE (ThermoFisher Sci, Waltham, MA, USA) and quenching of the trypsin with culture medium.

4.3. Drug Transport Assay

Transiently-transfected HEK293T cells (5×10^5) were incubated at 4 °C for 20 min with saturating levels (0.5 μg) of anti-ABCB1 antibody (4E3; Abcam, Cambridge, UK) in transport buffer (DMEM, high glucose (4.5 g/L), minus phenol red, supplemented with 1% FBS. The cells were pelleted at 500 G for 2 min and the supernatant discarded. The cells were washed once in transport buffer and resuspended in warm (37 °C) transport buffer containing saturating levels (2.5 μg) of goat anti-mouse secondary antibody, conjugated to recombinant phycoerythrin (RPE; Dako, Santa Clara, CA, USA) and one of three green-fluorescent drugs at a final concentration, unless otherwise stated, of 0.4 μM OREGON-GREEN Taxol bis-acetate (OG-taxol; Invitrogen, Waltham, MA, USA), 0.5 μM Calcein-AM (ThermoFisher Sci, Waltham, MA, USA), or 0.8 μM BODIPY-verapamil (Invitrogen, Waltham, MA, USA). The cells were incubated at 37 °C for 20 min before pelleting and washing as before. The cells were then resuspended in a 400 μL transport buffer and kept on ice until flow cytometry. Single fluorophore samples were also included to control for spectral spillover during flow cytometry. Two-color flow cytometry was performed on a LSRII (Becton Dickinson, Franklin Lakes, NJ, USA). Fluorescence data from 10,000 cells of normal size and granularity were acquired in CellQuest software (Becton Dickinson, Franklin Lakes, NJ, USA) and analyzed in Flowjo (Becton Dickinson, Franklin Lakes, NJ, USA).

4.4. Stable Expression of ABCB1 in HEK293 Flp-in Cells

HEK293 Flp-In cells with stable expression of ABCB1 wild-type were generated previously [10]. The cDNAs encoding ABCB1-Q347/990A and ABCB1-Q347/725/990A were excised from their parent pCI-neo plasmids using BamHI/NotI restriction endonuclease double digests and subcloned into the equivalent sites of pcDNA5/FRT (ThermoFisher Sci, Waltham, MA, USA). The resulting pcDNA5/FRT-ABCB1-Q347/990A and pcDNA5/FRT-ABCB1-Q347/725/990A (Qtriple) were used to co-transfect HEK293 Flp-In cells (ThermoFisher Sci, Waltham, MA, USA) along with the pOG44 (ThermoFisher Sci, Waltham, MA, USA) as a source of Flp recombinase as described by the manufacturer. Stable transfected cells (Flp-In ABCB1-Q347/990A, Flp-In ABCB1-Q347/725/990A (Qtriple) and pcDNA/FRT as a vector-only negative control) were selected with hygromycin (200 µg/mL) and, once uniform expression of the mutant ABCB1 were confirmed, maintained in hygromycin (100 µg/mL).

4.5. Taxol Survival Curve

Flp-In-ABCB1, Flp-In-ABCB1-Q347/990A, Flp-In ABCB1-Qtriple, and Flp-In-vector control cells (1×10^4) were seeded into a 96 well dish in 100 µL of DMEM with high glucose and 10% FBS but without hygromycin and allowed to attach for several hours. Taxol (Cambridge Bioscience, Cambridge, UK) was added to a final concentration ranging from 0 nM to 10 µM and the cells cultured at 37 °C in a humidified incubator with 5% CO_2. After 72 h the media was aspirated, the cells were detached with 30 µL TrypLE Express (ThermoFisher Sci, Waltham, MA, USA) which was quenched with 75 µL transport buffer and transferred to flow cytometry tubes. The entire population of cells of normal size and granularity, gated on the zero-drug condition, were counted in an ACEA NovoCyte flow cytometer (Agilent Technologies, Santa Clara, CA, USA). Cell number data were analyzed in Prism version 8 (GraphPad Software, San Diego, CA, USA). Curve fitting was achieved used non-linear regression.

Supplementary Materials: The following are available online at https://www.mdpi.com/article/10.3390/ijms22168561/s1. Figure S1: Titration of transport substrates for development of flow cytometry assay; Figure S2: Confirmation of equivalent levels of ABCB1 at the plasma membrane ensured that any differences in transport activity were due to ABCB1 functionality rather than the transporter expression level.

Author Contributions: Conceptualization, K.J.L.; formal analysis, K.J.L. and F.B.; investigation, K.S., H.A.I., A.O. and K.J.L.; data curation, K.J.L.; writing—original draft preparation, K.J.L.; writing— review and editing, K.S., H.A.I., A.O., F.B. and K.J.L.; supervision, K.J.L. All authors have read and agreed to the published version of the manuscript.

Funding: This research received no external funding.

Institutional Review Board Statement: Not applicable.

Informed Consent Statement: Not applicable.

Data Availability Statement: The raw data and materials generated in this study are available directly from K.J.L.

Acknowledgments: The authors would like to thank Gary Warnes, manager of the flow cytometry facility in the Blizard Institute for guidance and help with the flow cytometers and flow cytometry software used in this study.

Conflicts of Interest: The authors declare no conflict of interest.

Appendix A

Table A1. Functionality of mutants for the transport of Calcein-AM normalized to the wild-type transporter and pairwise statistical comparison of the raw data.

	Wild Type 100%	E556/1201Q 1.3 ± 0.6	Q347A 63.1 ± 8.0	Q725A 122.3 ± 36.4	Q990A 90.7 ± 12.8	Q347/725A 92.1 ± 14.2	Q347/990A 8.8 ± 1.8	Q725/990A 75.1 ± 4.0
E556/1201Q 1.3 ± 0.6	<0.0001 ****	-						
Q347A 63.1 ± 8.0	0.0363 *	0.0061 **	-					
Q725A 122.3 ± 36.4	ns	0.0379 *	ns	-				
Q990A 90.7 ± 12.8	ns	0.0025 **	0.0485 *	ns	-			
Q347/725A 92.1 ± 14.2	ns	0.0002 ***	ns	ns	ns	-		
Q347/990A 8.8 ± 1.8	0.0097 **	0.0024 **	0.0025 **	0.0358 *	0.0031 **	0.0010 **	-	
Q725/990A 75.1 ± 4.0	0.0301 *	0.0005 ***	ns	ns	ns	ns	0.0106 *	-
Qtriple 50.7 ± 4.0	0.0131 *	0.0031 **	ns	ns	ns	0.0296 *	0.0055 **	ns

The raw data has been used for statistical comparisons. Ratio and paired Student's *t*-test was used for all comparisons to wild-type ABCB1 and the Walker B mutants E556/1201Q activities since these positive and negative controls were included in every experiment. Where this was not possible (because of the large the number of mutants tested and when their construction was completed), unpaired Student's *t*-test was applied to the data normalized to one hundred percent wild-type activity (highlighted in yellow). *p* values: * <0.05, ** <0.01, *** <0.001, **** <0.0001, ns = not significant.

Table A2. Functionality of mutants for the transport of BODIPY-verapamil normalized to the wild-type transporter activity and pairwise statistical comparison of the raw data.

	Wild Type 100%	E556/1201Q 10.5 ± 1.6	Q347A 81.4 ± 7.7	Q725A 140.5 ± 12.3	Q990A 105.6 ± 16.3	Q347/725A 136.3 ± 10.0	Q347/990A 37.9 ± 10.1	Q725/990A 151.3 ± 44.3
E556/1201Q 10.5 ± 1.6	<0.0001 ****	-						
Q347A 81.4 ± 7.7	ns	0.0007 ***	-					
Q725A 140.5 ± 12.3	0.0410 *	<0.0001 ****	0.0061 **	-				
Q990A 105.6 ± 16.3	ns	0.0026 **	ns	ns	-			
Q347/725A 136.3 ± 10.0	ns	0.0040 **	0.0089 **	ns	0.0422 *	-		
Q347/990A 37.9 ± 10.1	0.0258 *	0.0042 **	ns	0.0029 **	ns	ns	-	
Q725/990A 151.3 ± 44.3	ns	0.0032 **	0.0018 **	0.0286 *	0.0182 *	ns	0.0225 *	-
Qtriple 115.7 ± 15.3	ns	0.0008 ***	0.0022 **	ns	ns	ns	0.0300 *	ns

The raw data has been used for statistical comparisons. Ratio or paired Student's *t*-test was used for all comparisons to wild-type ABCB1 and the Walker B mutants E556/1201Q as above. Where this was not possible, unpaired Student's *t*-test was applied to the raw data (highlighted in yellow). *p* values: * <0.05, ** <0.01, *** <0.001, **** <0.0001, ns = not significant.

Table A3. Functionality of mutants for the transport of OREGON-GREEN taxol bisacetate normalized to the wild-type transporter activity and pairwise statistical comparison of the raw data.

	Wild Type 100 %	E556/1201Q 2.0 ± 0.2	Q347A 104 ± 25	Q725A 84 ± 10	Q990A 66 ± 8	Q347/725A 69 ± 11	Q347/990A 48 ± 9	Q725/990A 50 ± 2
E556/1201Q 2.0 ± 0.2	0.0075 **	-						
Q347A 104 ± 25	ns	0.0148 *	-					
Q725A 84 ± 10	ns	0.0032 **	ns	-				
Q990A 66 ± 8	0.0467 *	0.0034 **	ns	ns	-			
Q347/725A 69 ± 11	ns	0.0003 ***	ns	ns	ns	-		
Q347/990A 48 ± 9	0.0184 *	0.0001 ***	ns	ns	ns	0.0292 *	-	
Q725/990A 50 ± 2	0.0290 *	0.028 *	0.0203 *	0.0305 *	ns	0.0234 *	ns	-
Qtriple 82 ± 13	ns	0.0231 *	ns	ns	ns	ns	ns	0.027 *

Unpaired Student's *t*-test on the raw data was used for all comparisons except for Q990A vs. wild-type where a paired Student's *t*-test was applied to the raw data (highlighted in yellow). p values: * <0.05, ** <0.01, *** <0.001, ns = not significant.

References

1. Robey, R.W.; Pluchino, K.M.; Hall, M.D.; Fojo, A.T.; Bates, S.E.; Gottesman, M.M. Revisiting the role of ABC transporters in multidrug-resistant cancer. *Nat. Rev. Cancer* **2018**, *18*, 452–464. [CrossRef]
2. Gottesman, M.M. Mechanisms of cancer drug resistance. *Annu. Rev. Med.* **2002**, *53*, 615–627. [CrossRef]
3. Maloney, S.M.; Hoover, C.A.; Morejon-Lasso, L.V.; Prosperi, J.R. Mechanisms of Taxane Resistance. *Cancers* **2020**, *12*, 3323. [CrossRef]
4. Gottesman, M.M.; Pastan, I.H. The Role of Multidrug Resistance Efflux Pumps in Cancer: Revisiting a JNCI Publication Exploring Expression of the MDR1 (P-glycoprotein) Gene. *J. Natl. Cancer Inst.* **2015**, *107*, djv222. [CrossRef]
5. Nigam, S.K. What do drug transporters really do? *Nat. Rev. Drug. Discov.* **2015**, *14*, 29–44. [CrossRef]
6. Alam, A.; Kowal, J.; Broude, E.; Roninson, I.; Locher, K.P. Structural insight into substrate and inhibitor discrimination by human P-glycoprotein. *Science* **2019**, *363*, 753–756. [CrossRef]
7. Seelig, A. A general pattern for substrate recognition by P-glycoprotein. *Eur. J. Biochem.* **1998**, *251*, 252–261. [CrossRef]
8. Seelig, A.; Blatter, X.L.; Wohnsland, F. Substrate recognition by P-glycoprotein and the multidrug resistance-associated protein MRP1: A comparison. *Int. J. Clin. Pharmacol. Ther.* **2000**, *38*, 111–121. [CrossRef]
9. Seelig, A.; Landwojtowicz, E. Structure-activity relationship of P-glycoprotein substrates and modifiers. *Eur. J. Pharm. Sci.* **2000**, *12*, 31–40. [CrossRef]
10. Zolnerciks, J.K.; Wooding, C.; Linton, K.J. Evidence for a Sav1866-like architecture for the human multidrug transporter P-glycoprotein. *Faseb. J.* **2007**, *21*, 3937–3948. [CrossRef]
11. Shapiro, A.B.; Ling, V. Extraction of Hoechst 33342 from the cytoplasmic leaflet of the plasma membrane by P-glycoprotein. *Eur. J. Biochem.* **1997**, *250*, 122–129. [CrossRef]
12. Mora Lagares, L.; Minovski, N.; Novic, M. Multiclass Classifier for P-Glycoprotein Substrates, Inhibitors, and Non-Active Compounds. *Molecules* **2019**, *24*, 2006. [CrossRef] [PubMed]
13. Loo, T.W.; Bartlett, M.C.; Clarke, D.M. Transmembrane segment 7 of human P-glycoprotein forms part of the drug-binding pocket. *Biochem. J.* **2006**, *399*, 351–359. [CrossRef]
14. Kim, Y.; Chen, J. Molecular structure of human P-glycoprotein in the ATP-bound, outward-facing conformation. *Science* **2018**, *359*, 915–919. [CrossRef]
15. Higgins, C.F.; Linton, K.J. The ATP switch model for ABC transporters. *Nat. Struct. Mol. Biol.* **2004**, *11*, 918–926. [CrossRef]
16. Thomas, C.; Tampe, R. Structural and Mechanistic Principles of ABC Transporters. *Annu. Rev. Biochem.* **2020**, *89*, 605–636. [CrossRef] [PubMed]
17. Raviv, Y.; Pollard, H.B.; Bruggemann, E.P.; Pastan, I.; Gottesman, M.M. Photosensitized labeling of a functional multidrug transporter in living drug-resistant tumor cells. *J. Biol. Chem.* **1990**, *265*, 3975–3980. [CrossRef]
18. Nosol, K.; Romane, K.; Irobalieva, R.N.; Alam, A.; Kowal, J.; Fujita, N.; Locher, K.P. Cryo-EM structures reveal distinct mechanisms of inhibition of the human multidrug transporter ABCB1. *Proc. Natl. Acad. Sci. USA* **2020**, *117*, 26245–26253. [CrossRef]

19. Loo, T.W.; Bartlett, M.C.; Clarke, D.M. Substrate-induced conformational changes in the transmembrane segments of human P-glycoprotein. Direct evidence for the substrate-induced fit mechanism for drug binding. *J. Biol. Chem.* **2003**, *278*, 13603–13606. [CrossRef]
20. Akkaya, B.G.; Zolnerciks, J.K.; Ritchie, T.K.; Bauer, B.; Hartz, A.M.; Sullivan, J.A.; Linton, K.J. The multidrug resistance pump ABCB1 is a substrate for the ubiquitin ligase NEDD4-1. *Mol. Membr. Biol.* **2015**, *32*, 39–45. [CrossRef] [PubMed]
21. Zolnerciks, J.K.; Akkaya, B.G.; Snippe, M.; Chiba, P.; Seelig, A.; Linton, K.J. The Q loops of the human multidrug resistance transporter ABCB1 are necessary to couple drug binding to the ATP catalytic cycle. *Faseb. J.* **2014**, *28*, 4335–4346. [CrossRef] [PubMed]
22. Tombline, G.; Bartholomew, L.A.; Urbatsch, I.L.; Senior, A.E. Combined mutation of catalytic glutamate residues in the two nucleotide binding domains of P-glycoprotein generates a conformation that binds ATP and ADP tightly. *J. Biol. Chem.* **2004**, *279*, 31212–31220. [CrossRef] [PubMed]

International Journal of
Molecular Sciences

Article

Identification of Two Dysfunctional Variants in the ABCG2 Urate Transporter Associated with Pediatric-Onset of Familial Hyperuricemia and Early-Onset Gout

Yu Toyoda [1], Kateřina Pavelcová [2,3], Jana Bohatá [2,3], Pavel Ješina [4], Yu Kubota [1], Hiroshi Suzuki [1], Tappei Takada [1] and Blanka Stiburkova [2,4,*]

[1] Department of Pharmacy, The University of Tokyo Hospital, Tokyo 113-8655, Japan; ytoyoda-tky@umin.ac.jp (Y.T.); yukubota-tky@umin.ac.jp (Y.K.); suzukihi-tky@umin.ac.jp (H.S.); tappei-tky@umin.ac.jp (T.T.)
[2] Institute of Rheumatology, 128 00 Prague, Czech Republic; pavelcova@revma.cz (K.P.); bohata@revma.cz (J.B.)
[3] Department of Rheumatology, First Faculty of Medicine, Charles University, 121 08 Prague, Czech Republic
[4] Department of Pediatrics and Inherited Metabolic Disorders, First Faculty of Medicine, Charles University and General University Hospital, 121 00 Prague, Czech Republic; pavel.jesina@vfn.cz
* Correspondence: stiburkova@revma.cz; Tel.: +420-234-075-319

Citation: Toyoda, Y.; Pavelcová, K.; Bohatá, J.; Ješina, P.; Kubota, Y.; Suzuki, H.; Takada, T.; Stiburkova, B. Identification of Two Dysfunctional Variants in the ABCG2 Urate Transporter Associated with Pediatric-Onset of Familial Hyperuricemia and Early-Onset Gout. *Int. J. Mol. Sci.* **2021**, 22, 1935. https://doi.org/10.3390/ijms22041935

Academic Editor: Thomas Falguières
Received: 30 November 2020
Accepted: 10 February 2021
Published: 16 February 2021

Publisher's Note: MDPI stays neutral with regard to jurisdictional claims in published maps and institutional affiliations.

Abstract: The *ABCG2* gene is a well-established hyperuricemia/gout risk locus encoding a urate transporter that plays a crucial role in renal and intestinal urate excretion. Hitherto, p.Q141K—a common variant of ABCG2 exhibiting approximately one half the cellular function compared to the wild-type—has been reportedly associated with early-onset gout in some populations. However, compared with adult-onset gout, little clinical information is available regarding the association of other uricemia-associated genetic variations with early-onset gout; the latent involvement of ABCG2 in the development of this disease requires further evidence. We describe a representative case of familial pediatric-onset hyperuricemia and early-onset gout associated with a dysfunctional ABCG2, i.e., a clinical history of three generations of one Czech family with biochemical and molecular genetic findings. Hyperuricemia was defined as serum uric acid (SUA) concentrations 420 µmol/L for men or 360 µmol/L for women and children under 15 years on two measurements, performed at least four weeks apart. The proband was a 12-year-old girl of Roma ethnicity, whose SUA concentrations were 397–405 µmol/L. Sequencing analyses focusing on the coding region of *ABCG2* identified two rare mutations—c.393G>T (p.M131I) and c.706C>T (p.R236X). Segregation analysis revealed a plausible link between these mutations and hyperuricemia and the gout phenotype in family relatives. Functional studies revealed that p.M131I and p.R236X were functionally deficient and null, respectively. Our findings illustrate why genetic factors affecting ABCG2 function should be routinely considered in clinical practice as part of a hyperuricemia/gout diagnosis, especially in pediatric-onset patients with a strong family history.

Keywords: *ABCG2* genotype; clinico-genetic analysis; ethnic specificity; genetic variations; precision medicine; rare variant; Roma; serum uric acid; SUA-lowering therapy; urate transporter

1. Introduction

Serum urate concentration is a complex phenotype influenced by both genetic and environmental factors, as well as interactions between them. Hyperuricemia results from an imbalance between endogenous production and excretion of urate. This disorder is a central feature in the pathogenesis of gout [1], which progresses through several degrees, i.e., asymptomatic hyperuricemia, acute gouty arthritis, intercritical gout, and chronic tophaceous gout. While not all individuals with hyperuricemia develop symptomatic gout, the risk of gout increases in proportion to the elevation of urate in circulation. In addition to hyperuricemia, the risk is also associated with gender, weight, age, environmental, and genetic factors [2,3], and interactions between them all. Recent data suggest that the number of gout patients under the age of 40 years (early-onset) is increasing [4]. These early-onset patients may have different clinical signs and co-morbidities from those who present with gout at a later age [5,6]. Given the development of earlier metabolic disorders in the early-onset gout patients compared with common gout patients [5], together with the need for continuous management of health from their younger age, understanding the risks of early-onset gout is clinically important.

More and more evidence suggests that the net amount of excreted uric acid is mainly regulated by physiologically important urate transporters, such as urate transporter 1 (URAT1, known as SLC22A12, a renal urate re-absorber) [7], glucose transporter member 9 (GLUT9, also known as SLC2A9) [8,9], and ATP-binding cassette transporter G2 (ABCG2, a high capacity urate exporter expressed in the kidneys and intestines) [10–13]. Dysfunction of URAT1 and GLUT9 reportedly cause inherited hypouricemia type 1 and type 2, respectively, while dysfunction of ABCG2 is a risk factor for hyperuricemia and gout [1,14]. Additionally, the ABCG2 population-attributable percent risk for hyperuricemia has been reported to be 29.2%, which is much higher than for those with more typical environmental risks such as BMI $\geq$ 25.0 (18.7%), heavy drinking (15.4%), and age $\geq$ 60 years old (5.74%) [15]. Hence, dysfunctional variants of ABCG2 may affect clinical outcomes by influencing the accumulation of uric acid in the body.

The ABCG2 protein, which is an *N*-linked glycoprotein composed of 655-amino acid, is a homodimer membrane transporter found in a variety of tissues [16–18]. ABCG2 is expressed on the brush border membranes of renal and intestinal epithelial cells, where ABCG2 is involved in the ATP-dependent excretion of numerous substrates from the cytosol into the extracellular space. ABCG2 was historically first described as a drug transporter linked to breast cancer resistance [19–21], which led to many studies that focused on its critical role in drug pharmacokinetics. To date, not only xenobiotics but also endogenous substances, including uremic toxin [22] and urate [11,12], have been identified as ABCG2 substrates.

In the context of hyperuricemia/gout, there are about 50 allelic variants, including a number of rare variants with minor allele frequencies (MAF) < 0.01%, which have been found in the *ABCG2* gene. Wide ethnic differences have been found relative to the frequencies of these alleles. There are two well-studied, common ABCG2 allelic variants—p.V12M (c.34G>A, rs2231137) and p.Q141K (c.421C>A, rs2231142) that have highly variable frequencies depending on ethnicity. Both are commonly found in Asians (in a relatively large number of ethnic groups) but are rarely found in Europeans [23]. A minor allele of p.V12M appears to be protective regarding susceptibility to gout [24]; however, this apparent effect is due to linkage disequilibrium between p.V12M and other dysfunctional ABCG2 variants [25]. In other words, the V12M mutation has little impact on the function of ABCG2. On the other hand, the p.Q141K variant decreases ABCG2 levels, which reduces the cellular function of ABCG2, as a urate exporter, by 50% [12]. In addition to p.Q141K, p.Q126X (c.376C>T, rs72552713), which is common in the Japanese population (MAF, 2.8%) [26] but rare in other populations, has been identified as hyperuricemia- and gout-risk allele [12]. Given that p.Q141K and p.Q126X variants are both associated with a significantly increased risk of gout, the effects of dysfunctional variants of ABCG2, relative to gout susceptibility, are genetically strong [27].

In a recent study focusing on 10 single nucleotide polymorphisms in 10 genes (*ABCG2, GLUT9/SLC2A9, SLC17A1, SLC16A9, GCKR, SLC22A11, INHBC, RREB1, PDZK1,* and *NRXN2*) that are strongly associated with serum uric acid (SUA) concentrations, only ABCG2 p.Q141K was associated with early-onset gout (< 40 years of age) in European and Polynesian subjects [28]. Additionally, in a previous study, we found that the MAF of p.Q141K in a cohort of hyperuricemia and gout with pediatric-onset was 38.7%, which was significantly higher than adult-onset (21.2%) as well as normouricemic controls (8.5%) [29]. This information suggests that ABCG2 dysfunction could be strongly associated with pediatric-onset hyperuricemia and early-onset gout; however, compared with adult-onset gout, little clinical information is available, except for the ABCG2 p.Q141K, regarding early-onset gout linked to other SUA-associated mutations. Furthermore, the latent involvement of ABCG2 in the development of this disease requires further evidence.

In this study, we investigated the genetic cause of pediatric-onset hyperuricemia and early-onset gout over three generations of a single family. Based on a positive family history of hyperuricemia or gout, we identified two rare mutations, c.393G>T (p.M131I) and c.706C>T (p.R236X), in the *ABCG2* gene. A series of biochemical assays revealed that ABCG2 p.M131I and p.R236X were functionally deficient and null, respectively. Our results also contributed to a more in-depth understanding of the effects of rare *ABCG2* variants, which is a highly polymorphic gene, on its function as a physiologically important transporter.

2. Results

2.1. Subjects

The clinical and biochemical data from this study are summarized in Table 1. Repeated biochemical analysis of the proband (a 12-year-old girl with chronic asymptomatic hyperuricemia) showed elevated serum urate (397–405 µmol/L; 6.67–6.81 mg/dL) with a decreased fractional excretion of uric acid (FE-UA) (2.2–3.5%). No clinical or laboratory symptoms of renal disease were present in the patient. The girl's mother (38-years-old) had previously presented with asymptomatic hyperuricemia (420–439 µmol/L; 7.06–7.38 mg/dL of serum urate); both the maternal grandfather (66-years-old) and the maternal uncle (40-years-old) had presented with hyperuricemia (500–537 µmol/L; 8.41–9.03 mg/dL) and early-onset gout, with the first gout attack occurring in the uncle when he was 35-years-old. The girl's father (38-years-old) presented with asymptomatic hyperuricemia (487 µmol/L; 8.19 mg/dL of serum urate), which seemed to be associated with metabolic syndrome (i.e., hypertriglyceridemia, hypertension, and central obesity). Metabolic investigation for purine metabolism (i.e., hypoxanthine and xanthine levels in the urine) found normal urinary excretion, suggesting that hyperuricemia in our patients and family relatives was not caused by an excess production of uric acid. Thus, we focused on the physiologically important urate transporters, especially ABCG2, which regulate urate handling in the body as described below.

Table 1. Clinical and biochemical data of a family with early-onset hyperuricemia and gout.

	Proband	Mother	Maternal uncle	Maternal grandfather	Father
Age of onset (HA/gout)	12 years (HA)	35 years (HA)	35 years (gout)	43 years (gout)	30–35 (HA)
Main symptoms	Asymptomatic	Asymptomatic	Gout	Gout	Asymptomatic
Other symptoms or diseases	Astigmatism; Bilateral cataracts; Myopia; Bronchial asthma; Obesity	Thyropathy; Vertebrogenic Algic syndrome	T2D; Hypertension; Obesity; Thyropathy	T2D; Hypertension	Bronchial asthma; Hypertriglyceridemia; Hypertension; Central obesity
SUA before treatment [µmol/L (mg/dL)]	397–405 (6.67–6.81)	420–439 (7.06–7.38)	>500 (>8.41)	537 (9.03)	487 (8.19)
UUA [mmol/mol creatinine]	3.34	1.47	2.10	n.d.	2.45
FE-UA [%]	2.2	4.6	2.9	n.d.	3.9
Therapy	No	No	Allopurinol (100 mg per day)	Colchicine; Allopurinol (300 mg per day)	No
SUA during treatment [µmol/L (mg/dL)]	n.d.	n.d.	446 (7.50) (non-compliance)	422 (7.09) (non-compliance)	n.d.

HA, hyperuricemia; SUA, serum uric acid; UUA, urine uric acid; FE-UA, fractional excretion of uric acid; T2D, type 2 diabetes mellitus; n.d., not determined. Reference ranges are as follows. SUA: 120–360 µmol/L (2.02–6.05 mg/dL) (< 15 years and female), 120–420 µmol/L (2.02–7.06 mg/dL) (male); UUA: 0.1–1.0 mmol/mol creatinine (< 15 years), 0.1–0.8 mmol/mol creatinine (≥15 years); FE-UA: 5–20% (<13 years), 5–12% (male), 5–15% (female).

2.2. Genetic Analysis

Targeted exon sequencing analyses of *ABCG2* revealed that the proband had two rare heterozygous non-synonymous mutations, i.e., (1) c.393G>T (p.M131I, rs759726272) and (2) c.706C>T (p.R236X, rs140207606). No other exonic mutations were found, including previously characterized common genetic risk factors for hyperuricemia/gout, such as *ABCG2* c.376C>T (p.Q126X) and c.421C>A (p.Q141K). Moreover, a segregation analysis of the family confirmed the two mutations as probably being disease-associated mutations. Heterozygous c. 393G>T and c.706C>T were identified through the maternal line in the mother, uncle, and grandfather of the proband. The pedigree is shown in Figure 1 along with representative electropherograms of partial *ABCG2* sequences showing a heterozygous missense mutation (c.393G>T) in exon 5 and a nonsense mutation (c.706C>T) in exon 7. Moreover, sequencing analyses, including analyses of previously known SUA-associating loci, found no disease-associated non-synonymous mutations in *GLUT9* and *URAT1*. Although two mutations, c.73G>A (p.G25R, rs2276961) and c.844G>A (p.V282I, rs16890979), were found in *GLUT9*; however, previous studies have shown that they would not cause hyperuricemia/gout [30,31].

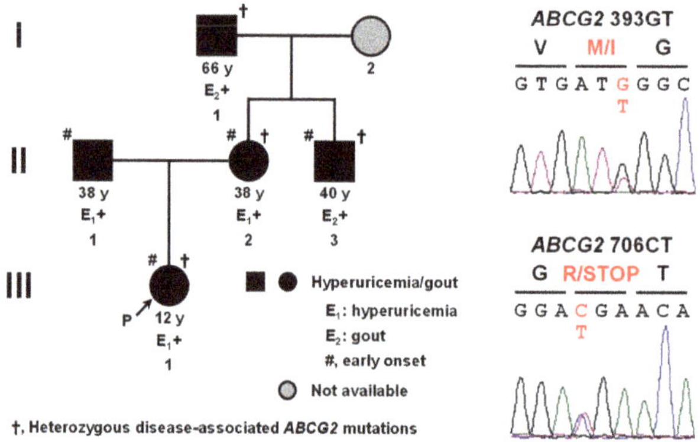

Figure 1. Identification of novel disease-associated mutations c.393G>T (p.M131I) and c.706C>T (p.R236X) in *ABCG2*. Left, the pedigree of a Czech family with early-onset hyperuricemia and gout; Right, representative electropherograms of partial sequences of *ABCG2* showing the heterozygous point mutations (red) discovered in our present study. I–III, generation of the family; P, proband; y, years old.

Further sequencing analyses revealed that neither of these *ABCG2* mutations was detected in a previously enrolled cohort of 360 patients with hyperuricemia/gout and a cohort of 132 normouricemia control subjects of European descendent [32]. However, in another control group of 60 subjects of Roma ethnicity [33], although serum urate and FE-UA values were not available, we found one copy of c.393G>T (p.M131I), but without c.706C>T (p.R236X), in a male hemodialysis patient with severe chronic kidney disease. This finding suggested that c.393G>T (p.M131I) and c.706C>T (p.R236X) can occur independently, which led us to examine the effect of each non-synonymous mutation on the ABCG2 protein.

Additionally, the independent genetic origin of these two *ABCG2* mutations was also strongly supported by supplementary analyses of MAF data that is openly available from the Exome Aggregation Consortium (http://exac.broadinstitute.org/ (accessed on 30 November 2020)). With regard to c.393G>T (p.M131I), only six heterozygotes of the minor allele (393T) were found in a cohort of 64,397 European subjects (MAF, 0.0047%); with even fewer carriers in other populations (total worldwide MAF frequency = 0.0021%). With regard to c.706C>T (p.R236X), 48 heterozygotes of the minor allele (706T) were found in a cohort of 645,050 European subjects (MAF = 0.0372%, which is almost 8 times greater than the MAF of c.393G>T); in other populations, frequencies ranging from 0.0000% in European Finnish to 0.1254% (the highest) in Ashkenazi Jewish were found.

2.3. Functional Analysis

To investigate the effect of the two identified non-synonymous mutations (p.M131I and p.R236X) on the intracellular processing and function of the ABCG2 protein, we conducted several biochemical analyses using transiently ABCG2-expressing mammalian cells (Figure 2). First, immunoblotting for *N*-glycosidase (PNGase F)-treated whole cell lysates (Figure 2A) demonstrated that p.M131I had a minimal effect on protein levels and *N*-linked glycosylation status, while the p.R236X variant was detected as truncated forms of the protein with weaker band intensities compared to ABCG2 wild-type (WT), as would be expected based on its amino acid sequence. Confocal microscopy (Figure 2B) showed that, like ABCG2 WT, the p.M131I variant was mainly located on the plasma membrane, while the p.R236X variant exhibited little plasma membrane localization. Expression of the p.M131I variant as a glycoprotein on plasma membrane-derived vesicles was confirmed using immunoblotting (Figure 2C). Next, using a vesicle transport assay [34] with an optimized experimental procedure for determining the initial rate of urate transport by ABCG2 based on our previous studies [12,35], ABCG2 function was evaluated as having ATP-dependent urate transport activity into plasma membrane vesicles (Figure 2D). The functional assay revealed that contrary to the WT, p.M131 had limited ATP-dependent urate transport activity, with the ABCG2-mediated urate transport activity of this variant calculated to be 14 ± 2% of the WT controls. The p.R236X variant was functionally null (Figure 2D), which was consistent with the results of our biochemical analyses (Figure 2A–C).

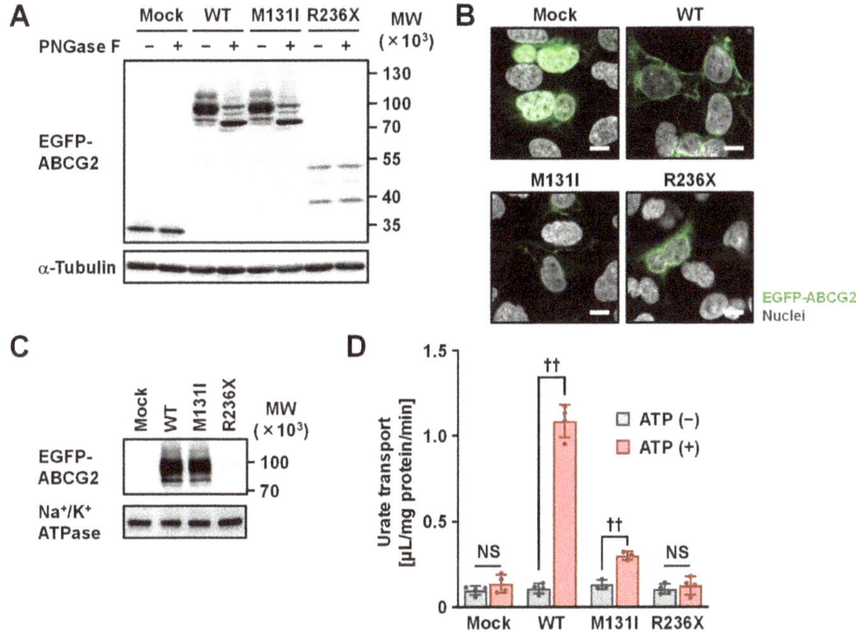

Figure 2. Functional characterization of disease-associated ABCG2 mutations. (**A**) Immunoblot of whole cell lysate samples. α-Tubulin used as the loading control. WT, wild-type. (**B**) Confocal microscopy of intracellular localization. Nuclei were stained with TO-PRO-3 iodide (gray). Bars, 10 µm. (**C**) Immunoblot of plasma membrane vesicles. Na⁺/K⁺ ATPase as the loading control. (**D**) Urate transport activity. The incubation condition was 37 °C for 10 min in the presence of 20 µM radiolabeled urate. Data are expressed as the mean ± SD. *n* = 3–4. ††, *P* < 0.01; NS, not significantly different between groups (two-sided *t*-test).

3. Discussion

In this study, we found and analyzed a representative case of pediatric-onset hyperuricemia and early-onset gout in a Czech family associated with two newly identified, one functionally deficient and the other a null mutation, in *ABCG2*. A positive family history of hyperuricemia/gout in the context of ABCG2 dysfunction was observed in the maternal line (the mother, maternal uncle, and maternal grandfather) of the proband (a young girl, of Roma ethnicity, with chronic asymptomatic hyperuricemia); her father exhibited elevated serum urate that seemed to be associated with metabolic syndrome. Although a decrease in net renal urate excretion was observed in all cases of hyperuricemia in this family, which was characterized by FE-UA < 5%, our findings suggest that the hyperuricemia was linked to heterogeneity. Moreover, the familial hyperuricemia/gout observed in this study was not associated with a common variant, such as p.Q141K, but with a rare ABCG2 variant; this supports the recently proposed genetic concept, i.e., the "Common Disease, Multiple Common and Rare Variant" model [25,36], for the association between hyperuricemia/gout and the *ABCG2* gene.

Our findings of ABCG2 variants will provide deeper insights into amino acid positions that are critical for normal ABCG2 function. Based on cryo-electron microscopy (cryo-EM) of ABCG2 [37], both the M131 and R236 residues are in the cytoplasmic region of the N-terminus of the ABCG2 protein. Regarding p.M131I, the original amino acid (M131) sequence is conserved in several major mammalian species (Figure 3). Unlike the p.Q141K variant, which affects intracellular processing of the ABCG2 protein [38], the p.M131I variant disrupts ABCG2's function as a urate transporter, with little effect on the ABCG2 protein or its cellular localization. A plausible explanation for the molecular mechanism of the p.M131I effect is that this amino acid substitution could affect ABCG2 substrate specificity and/or affect its ATPase activity, including the binding affinity of ATP, which is the driving force for ABC transporters. The latter possibility is supported by a structural feature that the M131 residue is located near the conserved glutamine (Q126) within a Q-loop in the nucleotide-binding domain of ABCG2, a key element for the catalytic cycle of ATP binding and hydrolysis [37,39]. Further studies are needed to address this biochemical issue, which may deepen the mechanistic insight of ABCG2 protein. With regard to p.R236X, the acquired stop codon results in the production of a shortened ABCG2 variant (only about one-third the amino acid length of the native ABCG2 protein) that lacks all the transmembrane domains essential for normal protein function as a membrane transporter. This is consistent with our results demonstrating that the p.R236X variant is functionally null. Thus, the c.[393G>T; c.706C>T] (p.[M131I; R236X]) variant does not function as a urate transporter.

M131I

Homo sapiens	126 Q D D V V **M** G T L T V R E N L Q F S A A 145
Pan troglodytes	Q D D V V **M** G T L T V R E N L Q F S A A
Macaca mulatta	Q D D V V **M** G T L T V R E N L Q F S A A
Sus scrofa	Q D D V V **M** G T L T V R E N L Q F S A A
Bos taurus	Q D D V V **M** G T L T V R E N L Q F S A A
Rattus novergicus	Q D D V V **M** G T L T V R E N L Q F S A A
Mus musculus	Q D D V V **M** G T L T V R E N L Q F S A A

Figure 3. ABCG2 M131 is shown to be evolutionary conserved among seven mammalian species. Regarding the Abcg2 protein in each species, the NCBI Reference Sequence IDs are summarized as *Pan troglodytes* (Chimpanzee), GABE01009237.1; *Macaca mulatta* (Rhesus macaque), NM_001032919.1; *Sus scrofa* (Pig), NM_214010.1; *Bos taurus* (Bovine), NM_001037478.3; *Rattus norvegicus* (Rat), NM_181381.2; *Mus musculus* (Mouse), NM_011920.3. Multiple sequence alignments and homology calculations were carried out using GENETYX software (GENETYX, Tokyo, Japan) and the ClustalW2.1 Windows program, per the protocol used in our previous study [40].

To discuss the expected independence in the genetic origin of two *ABCG2* mutations identified in this study, we will focus on the genetic specificity of each population. The Czech family studied in this study is of Roma ethnicity. The Roma composes a transnational ethnic population of 8–10 million, with the original homeland being India; currently, they are the largest and the most widespread ethnic minority in Europe. The founder effect and subsequent genetic isolation of the Roma have led to a population specificity regarding the genetic background of specific human diseases. In other words, mutations associated with rare diseases found in the Roma population tend to be at extremely low frequencies in other European populations, and vice versa. Indeed, several genetic variants causing rare diseases are unique to the Roma, and many of these mutations have only recently been discovered, e.g., Charcot Marie Tooth disease type 4D and 4G (OMIM 601455 and 605285), Congenital cataract facial dysmorphism neuropathy (OMIM 604166), Gitelman syndrome (OMIM 263800), and Galactokinase deficiency (OMIM 230200) [41,42]. In light of the genetic specificity found in the Roma, we investigated the frequency of two identified *ABCG2* mutants (c.393G>T and c.706C>T) in our control cohort of 60 subjects of Roma origin (see Section 2.2 in Results). MAFs of c.393G>T in the Roma cohort and European population were 0.833% (1 allele/120 subjects) and 0.002% (6 alleles/64,397 subjects), respectively. Although the sample size of our control cohort was very modest, it could be large enough to imply that the origin of c.393G>T might be the Roma population. On the other hand, the prevalence of c.706C>T was higher in Europeans than in the Roma, suggesting that these two *ABCG2* mutations could have different genetic origins.

For clinical practice, our findings suggest a need for further discussion about the potential benefits of urate-lowering therapy after a diagnosis of hyperuricemia in pediatric-onset patients with ABCG2 dysfunction. Interestingly, harboring p.Q141K is reportedly associated with inadequate response to allopurinol (characterized by a smaller reduction in serum urate concentrations compared with WT) [29,43–45]; allopurinol, which inhibits uric acid production, is a well-known and widely used drug for lowering SUA. Although the mechanisms of action for the inadequate response are still unclear, other SUA-lowering drugs may be somewhat better in terms of efficacy for patients with a dysfunctional *ABCG2* allele, which puts them at higher risk of developing hyperuricemia/gout. On the other hand, among the clinically-used inhibitors for the production of uric acid (i.e., allopurinol, febuxostat, and topiroxostat), only febuxostat, to the best of our knowledge, strongly and clinically inhibits ABCG2 function as a urate transporter [35]. This is also supported by a recent clinical study that showed orally-administered febuxostat inhibits intestinal ABCG2 in humans [46]. In this context, the efficacy of febuxostat as an SUA-lowering drug can be partially blocked by ABCG2 inhibition, except in cases of completely null ABCG2 function. Considering this complexity, as well as risks of adverse effects [47], the best SUA-lowering drugs, including uricosuric agents and uric acid production inhibitors, should be carefully determined for each patient.

Additionally, we can emphasize the clinical importance of the documentation regarding the family we addressed in this study, given the infrequency of detailed studies on SUA levels in children. Indeed, pediatric-onset of hyperuricemia is relatively rare in clinical practice; it is often associated with rare conditions (e.g., purine metabolic disorders; kidney disorders including uromodulin-associated disorders; metabolic genetic disorders including glycogen storage disease, hereditary fructose intolerance, and mitochondrial disorders). As we showed previously [48], the levels of SUA and FE-UA are quite dynamic in the first year of life. In brief, the SUA levels are low in infancy (131–149 µmol/L; 2.2–2.5 mg/dL at 2–3 months of age) due to the high FE-UA levels (>10%); the FE-UA levels decrease to approximately 8% at age 1 and then stay through childhood, which is associated with mean SUA levels (208–268 µmol/L; 3.5–4.5 mg/dL) of children. At adolescence (after age 12), the FE-UA levels significantly decrease in boys but not in girls, resulting in a further significant increase in SUA levels in young men but not in young women. These pieces of information support the rarity of our case of which proband is a 12-year-old girl with chronic asymptomatic hyperuricemia characterized by elevated serum urate (397–405 μmol/L; 6.67–6.81 mg/dL) and the decreased FE-UA (2.2–3.5%).

Some limitations warrant mention. It is unclear whether there could have been other genetic factors also affecting the early-onset phenotypes in the family we studied, although harboring dysfunctional *ABCG2* mutations is the most plausible explanation, as the study showed. A previous study showed that in the context of extra-renal underexcretion hyperuricemia, a genetic dysfunction of ABCG2 increased SUA levels and apparent urinary urate excretion, which was coupled with decreased intestinal urate excretion. Thus, we can assume the presence of latent mechanisms causing the decreased (<5%) FE-UA levels observed in our patients. Although there is little information available regarding this, a possible factor could be increased renal urate reabsorption, which can be mediated by up-regulation of renal urate re-absorbers, including *URAT1/SLC22A12, GLUT9/SLC2A9*, and *organic anion transporter 10* (known as *SLC22A13*) [49]. In this context, genetic variations affecting the expression of such genes will be the targets of future studies.

In conclusion, we found a representative case of pediatric hyperuricemia with familial gout that harbored two dysfunctional *ABCG2* mutations. Genetic variations in *ABCG2* should be kept in mind during diagnostic procedures for pediatric-onset hyperuricemia. Considering *ABCG2* genotypes will be beneficial for patients with early-onset and/or familial hyperuricemia and gout. This type of genetic information will also allow a personalized approach regarding the best urate-lowering treatment (i.e., uric acid production inhibitors or uricosuric agents) for patients with dysfunctional ABCG2 variants, as well as the best time to initiate pharmacotherapy for hyperuricemia.

4. Materials and Methods

4.1. Clinical Subjects

The studied proband and her family members were Czechs of Roma ethnicity diagnosed with familial (early-onset) hyperuricemia/gout. Written informed consent was obtained from each subject before enrollment in the study. All tests were performed in accordance with standards set by the institutional ethics committees, which approved 30 June 2015 the project no. 6181/2015. All the procedures were performed in accordance with the Declaration of Helsinki.

Hyperuricemia was defined as serum urate levels greater than 420 μmol/L (7.06 mg/dL) for men or 360 μmol/L (6.05 mg/dL) for women and children under 15 years on two measurements, performed at least four weeks apart. Gouty arthritis was diagnosed according to the American College of Rheumatology criteria, as follows: (1) the presence of sodium urate crystals seen in synovial fluid (using a polarized microscope, Nikon Eclipse E200, Tokyo, Japan) or (2) at least six of 12 clinical criteria being met [50].

The proband was a 12-year-old girl with a complicated perinatal anamnesis. She was born at 31 weeks of gestation with an Apgar score of 4-7-8, a birth weight of 1690 g, and a birth length of 40 cm; additionally, she developed early asphyxia syndrome. She also experienced repeated respiratory infections and was later diagnosed with bronchial asthma. She is also under the care of an ophthalmologist for myopia and astigmatism; she was also investigated for sudden onset mild bilateral cortical cataracts. She was obese (BMI = 27); her psychomotor development corresponded to her age.

4.2. Clinical Investigations and Sequence Analyses

Urate and creatinine levels were measured as described previously [51] using a specific enzymatic method and the Jaffé reaction, which was adapted for an auto-analyzer (Hitachi Automatic Analyzer 902; Roche, Basel, Switzerland). Metabolic investigations of purine metabolism (hypoxanthine and xanthine levels in urine) were also conducted using a method established in a previous study [51]. The proband was screened for metabolic (e.g., glycogen storage disease, hereditary fructose intolerance, and mitochondrial disorders) and kidney disease associated with hyperuricemia (e.g., uromodulin-associated disorders), using our previously published diagnostic algorithm [29].

All coding regions and exon-intron boundaries in *ABCG2* and *GLUT9/SLC2A9*, and exons 7 and 9 in *URAT1/SLC22A12* were analyzed from genomic DNA, as described previously [29,33,52]. The reference sequence for *ABCG2* was defined as version ENST00000237612.7 (location: Chromosome 4: 88,090,269–88,158,912 reverse strand) (www.ensembl.org (accessed on 30 November 2020)). For *GLUT9/SLC2A9* (NM_020041.2; NP_064425.2; SNP source dbSNP 132) and *URAT1/SLC22A12* (NM_144585.3), the reference genomic sequence was defined as version NC_000004.12 (Chromosome 4: 9,771,153–10,054,936) and NC_000011.8 (Chromosome 11: 64,114,688–64,126,396), respectively.

It is worth special mention that the world's highest frequency of the main dysfunctional variants of URAT1, p.T467M (MAF, 5.6%) and p.L415_G417del (MAF, 1.9%), was recently identified in a Roma population (1,016 individuals) from specific regions of the Czech Republic, Slovakia, and Spain [33,53]. According to MAF data from the Exome Aggregation Consortium, p.T467M (rs200104135) showed only one heterozygous allele in a cohort of 15,296 in a South Asian population (MAF, 0.003%) and no occurrence in other ethnic populations; no occurrence of the p.L415_G417del allele was seen in the whole population, which supports the Roma-specific prevalence of these two URAT1 variants. For this reason, we looked for both URAT1 variants and confirmed that neither variant was present in our studied family.

4.3. Materials

ATP, AMP, creatine phosphate disodium salt tetrahydrate, and creatine phosphokinase type I from rabbit muscle were purchased from Sigma-Aldrich (St. Louis, MO, USA), and [8-^{14}C]-uric acid (55 mCi/mmol) was purchased from American Radiolabeled Chemicals (St. Louis, MO, USA). All other chemicals were commercially available and of analytical grade.

4.4. Preparation of ABCG2 Variants Expression Vector

To express human ABCG2 (NM_004827.3) fused with EGFP at its N-terminus (EGFP-ABCG2) and EGFP (control), we used an ABCG2/pEGFP-C1 plasmid (open reading frame of ABCG2 was inserted into the *Hind*III and the *Apa* I sites of a pEGFP-C1 vector plasmid) that was from our previous study [32]. Of note, the functionality and expression of such construct were confirmed by previous studies we and other groups conducted [32,36,54–56]. Using a site-directed mutagenesis technique, the ABCG2 p.M131I (c.393G>T)/pEGFP-C1 plasmid and the ABCG2 R236X (c.706C>T)/pEGFP-C1 plasmid were generated from an ABCG2 WT/pEGFP-C1 plasmid, respectively. The introduction of each mutation was confirmed by full sequencing using BigDye Terminator v3.1 (Applied Biosystems, Foster City, CA, USA) and an Applied Biosystems 3130 Genetic Analyzer (Applied Biosystems), as described previously [40].

4.5. Cell Culture and Transfection

Human embryonic kidney 293 cell-derived 293A cells were purchased from Life Technologies (Carlsbad, CA, USA) and cultured in Dulbecco's Modified Eagle's Medium (DMEM; Nacalai Tesque, Kyoto, Japan) supplemented with 10% fetal bovine serum (Life Technologies), 1% penicillin/streptomycin, 2 mM L-glutamine (Nacalai Tesque), and 1 × Non-Essential Amino Acid (Life Technologies) at 37 °C in an atmosphere of 5% CO_2. Each vector plasmid for ABCG2 WT, p.M131I, or p.R236X was transfected into 293A cells by using polyethyleneimine MAX (PEI-MAX; 1 mg/mL in milli-Q water, pH 7.0; Polysciences, Warrington, PA, USA) as described previously [57]. The amount of plasmid DNA used for transfection was adjusted per sample group.

4.6. Preparation of Whole-Cell Lysates

Forty-eight hours after transfection, whole-cell lysates were prepared in ice-cold lysis buffer A containing 50 mM Tris/HCl (pH 7.4), 1 mM dithiothreitol, 1% (w/v) Triton X-100, and a protease inhibitor cocktail for general use (Nacalai Tesque) as described previously [58]. The protein concentration of the whole cell lysate was quantified using a BCA Protein Assay Kit (Pierce, Rockford, IL, USA) with bovine serum albumin (BSA) as a standard according to the manufacturer's protocol. Before glycosidase treatment, the whole cell lysate samples were incubated with PNGase F (New England Biolabs Japan, Tokyo, Japan) (1.25 U/μg of protein) at 37 °C for 10 min as described previously [59], and then subjected to immunoblotting.

4.7. Preparation of ABCG2-Expressing Plasma Membrane Vesicles

Plasma membrane vesicles were prepared from ABCG2-expressing 293A cells, as described previously [36]. The resulting plasma membrane vesicles were rapidly frozen in liquid N_2 and kept at -80 °C until used. The protein concentration of the plasma membrane vesicles was measured using a BCA Protein Assay Kit, as described above.

4.8. Immunoblotting

Expression of ABCG2 protein in whole-cell lysates and plasma membrane vesicles was assessed using immunoblotting as described previously [36], with minor modifications. In brief, the prepared samples were mixed with a sodium dodecyl sulfate-polyacrylamide gel electrophoresis sample buffer solution containing 10% 2-mercaptoethanol, separated by electrophoresis on polyacrylamide gels, and then transferred to Polyvinylidene Difluoride membranes (Immobilon; Millipore, Billerica, MA, USA) by electroblotting at 15 V for 60 min. For blocking, the membrane was incubated in Tris-buffered saline containing 0.05% Tween 20 and 3% BSA (Nacalai Tesque) (TBST-3% BSA). After overnight incubation at room temperature, blots were probed with rabbit anti-EGFP polyclonal antibodies (A11122; Life Technologies; diluted 1,500 fold in TBST-0.1% BSA), a rabbit anti-α-tubulin antibodies (ab15246; Abcam, Cambridge, MA, USA; diluted 1,000 fold), or a rabbit anti-Na^+/K^+-ATPase α antibodies (sc-28800; Santa Cruz Biotechnology, Santa Cruz, CA, USA; diluted 1,000 fold) followed by incubation with a donkey anti-rabbit immunoglobulin G (IgG)-horseradish peroxidase (HRP)-conjugated antibody (NA934V; diluted 4,000 fold for EGFP-ABCG2 or 3,000 fold for α-tubulin and Na^+/K^+-ATPase). HRP-dependent luminescence was developed using ECL^{TM} Prime Western Blotting Detection Reagent (GE Healthcare UK, Buckinghamshire, UK) and detected using a multi-imaging Fusion Solo 4^{TM} analyzer system (Vilber Lourmat, Eberhardzell, Germany).

4.9. Confocal Laser Scanning Microscopic Observation

For confocal laser scanning microscopy, 48 h after transfection, 293A cells were fixed with ice-cold methanol for 10 min, and then the nuclei were visualized with TO-PRO-3 Iodide (Molecular Probes, Eugene, OR, USA) as described previously [36]. To analyze the localization of EGFP-fused ABCG2 protein, fluorescence was detected using an FV10i Confocal Laser Scanning Microscope (Olympus, Tokyo, Japan).

4.10. Urate Transport Assay

The urate transport assay, with ABCG2-expressing plasma membrane vesicles, was conducted using a rapid filtration technique described in our previous studies [27,36], with some minor modifications. In brief, each plasma membrane vesicle (0.5 mg/mL) was incubated with 20 μM of radiolabeled urate in a reaction mixture (total 20 μL: 10 mM Tris/HCl, 250 mM sucrose, 10 mM $MgCl_2$, 10 mM creatine phosphate, 1 mg/ml creatine phosphokinase, pH 7.4, and 50 mM ATP, or AMP as an ATP substitute) for 10 min at 37 °C. After incubation, the reaction mixture was mixed with 980 μL of ice-cold stop buffer (2 mM EDTA, 0.25 M sucrose, 0.1 M NaCl, 10 mM Tris-HCl, at a of pH 7.4); the resulting solution was rapidly filtered through an MF-Millipore Membrane (HAWP02500; 0.45 μm pore size and 25 mm diameter; Millipore). After washing with 5 ml of ice-cold stop buffer five times, the plasma membrane vesicles on the filter were dissolved in Clear-sol II (Nacalai Tesque). Then, the radioactivity in the plasma membrane vesicles was measured using a liquid scintillator (Tri-Carb 3110TR; PerkinElmer, Waltham, MA, USA).

The urate transport activity was calculated as the incorporated clearance (μL/mg protein/min) defined as the incorporated level of urate [disintegrations per minute (DPM)/mg protein/min]/urate level in the incubation mixture [DPM/μL]. ATP-dependent urate transport was calculated by subtracting the urate transport activity in the absence of ATP from that in the presence of ATP; ABCG2-mediated urate transport activity was calculated by subtracting ATP-dependent urate transport activity of control plasma membrane vesicles from that of ABCG2-expressing plasma membrane vesicles.

4.11. Statistical Analysis

All statistical analyses were performed by using EXCEL 2019 (Microsoft, Redmond, WA, USA) with Statcel4 add-in software (OMS publishing, Saitama, Japan). The number of biological replicates (n) is described in the figure legends. In single pairs of quantitative data, after comparing the variances of a data set (using the F-test), an unpaired Student's t-test was performed. Statistical significance was defined in terms of P values less than 0.05 or 0.01.

Author Contributions: Conceptualization, Y.T., T.T., and B.S.; validation, Y.T., Y.K., and B.S.; formal analysis, Y.T.; investigation, Y.T., K.P., J.B., P.J., and Y.K. (i.e., Y.T. performed the experimental work using mammalian cells and analyzed the data; Y.K. supported the experimental work; K.P. and J.B. conducted sequencing analyses for clinical samples; P.J. was responsible for clinical observations; Y.T. performed statistical analyses); data curation, Y.T., T.T., and B.S.; writing—original draft preparation, Y.T., and B.S.; writing—review and editing, Y.T., T.T., and B.S.; supervision, H.S. and B.S.; project administration, B.S.; funding acquisition, Y.T., T.T., and B.S. All authors have read and agreed to the published version of the manuscript.

Funding: The study was supported by the JSPS KAKENHI Grant, numbers 19K16441 (to Y.T.); 16H01808, 18KK0247, and 20H00568 (to T.T.). The study was also supported by grants from the Czech Republic Ministry of Health RVO 00023728 (Institute of Rheumatology), RVO VFN64165, and BBMRI-CZ LM2018125 (to B.S.); T.T. received research grants from "Gout and uric acid foundation of Japan"; "Takeda Science Foundation"; "Suzuken Memorial Foundation"; "Mochida Memorial Foundation for Medical and Pharmaceutical Research"; "The Pharmacological Research Foundation, Tokyo."

Institutional Review Board Statement: The studied proband and her family members were Czechs of Roma ethnicity diagnosed with familial (early-onset) hyperuricemia/gout. Written informed consent was obtained from each subject before enrollment in the study. All tests were performed in accordance with standards set by the institutional ethics committees, which approved the project (no. 6181/2015). All the procedures were performed in accordance with the Declaration of Helsinki.

Informed Consent Statement: Informed consent was obtained from all subjects involved in the study.

Data Availability Statement: Data supporting the findings of this study are included in this published article or are available from the corresponding author on reasonable request.

Acknowledgments: We would like to express our appreciation to all who took part in this study.

Conflicts of Interest: The authors declare no conflict of interest.

Abbreviations

ABCG2	ATP-binding cassette transporter G2
ATP	Adenosine triphosphate
FE-UA	Fractional excretion of uric acid
GLUT9	Glucose transporter 9
MAF	Minor allele frequency
SUA	Serum uric acid
URAT1	Urate transporter 1
WT	Wild-type

References

1. Dalbeth, N.; Choi, H.K.; Joosten, L.A.B.; Khanna, P.P.; Matsuo, H.; Perez-Ruiz, F.; Stamp, L.K. Gout. *Nat. Rev. Dis. Primers* **2019**, *5*, 69. [CrossRef]
2. Kawamura, Y.; Nakaoka, H.; Nakayama, A.; Okada, Y.; Yamamoto, K.; Higashino, T.; Sakiyama, M.; Shimizu, T.; Ooyama, H.; Ooyama, K.; et al. Genome-wide association study revealed novel loci which aggravate asymptomatic hyperuricaemia into gout. *Ann. Rheum. Dis.* **2019**, *78*, 1430–1437. [CrossRef]
3. Dehlin, M.; Jacobsson, L.; Roddy, E. Global epidemiology of gout: Prevalence, incidence, treatment patterns and risk factors. *Nat. Rev. Rheumatol.* **2020**, *16*, 380–390. [CrossRef] [PubMed]
4. Kuo, C.F.; Grainge, M.J.; See, L.C.; Yu, K.H.; Luo, S.F.; Zhang, W.; Doherty, M. Epidemiology and management of gout in Taiwan: A nationwide population study. *Arthritis Res. Ther.* **2015**, *17*, 13. [CrossRef]
5. Pascart, T.; Norberciak, L.; Ea, H.K.; Guggenbuhl, P.; Liote, F. Patients with early-onset gout and development of earlier severe joint involvement and metabolic comorbid conditions: Results from a cross-sectional epidemiologic survey. *Arthritis Care Res.* **2019**, *71*, 986–992. [CrossRef] [PubMed]
6. Zhang, B.; Fang, W.; Zeng, X.; Zhang, Y.; Ma, Y.; Sheng, F.; Zhang, X. Clinical characteristics of early- and late-onset gout: A cross-sectional observational study from a Chinese gout clinic. *Medicine* **2016**, *95*, e5425. [CrossRef] [PubMed]
7. Enomoto, A.; Kimura, H.; Chairoungdua, A.; Shigeta, Y.; Jutabha, P.; Cha, S.H.; Hosoyamada, M.; Takeda, M.; Sekine, T.; Igarashi, T.; et al. Molecular identification of a renal urate anion exchanger that regulates blood urate levels. *Nature* **2002**, *417*, 447–452. [CrossRef]
8. Matsuo, H.; Chiba, T.; Nagamori, S.; Nakayama, A.; Domoto, H.; Phetdee, K.; Wiriyasermkul, P.; Kikuchi, Y.; Oda, T.; Nishiyama, J.; et al. Mutations in glucose transporter 9 gene SLC2A9 cause renal hypouricemia. *Am. J. Hum. Genet.* **2008**, *83*, 744–751. [CrossRef]
9. Vitart, V.; Rudan, I.; Hayward, C.; Gray, N.K.; Floyd, J.; Palmer, C.N.; Knott, S.A.; Kolcic, I.; Polasek, O.; Graessler, J.; et al. SLC2A9 is a newly identified urate transporter influencing serum urate concentration, urate excretion and gout. *Nat. Genet.* **2008**, *40*, 437–442. [CrossRef] [PubMed]
10. Ichida, K.; Matsuo, H.; Takada, T.; Nakayama, A.; Murakami, K.; Shimizu, T.; Yamanashi, Y.; Kasuga, H.; Nakashima, H.; Nakamura, T.; et al. Decreased extra-renal urate excretion is a common cause of hyperuricemia. *Nat. Commun.* **2012**, *3*, 764. [CrossRef]
11. Woodward, O.M.; Kottgen, A.; Coresh, J.; Boerwinkle, E.; Guggino, W.B.; Kottgen, M. Identification of a urate transporter, ABCG2, with a common functional polymorphism causing gout. *Proc. Natl. Acad. Sci. USA* **2009**, *106*, 10338–10342. [CrossRef] [PubMed]
12. Matsuo, H.; Takada, T.; Ichida, K.; Nakamura, T.; Nakayama, A.; Ikebuchi, Y.; Ito, K.; Kusanagi, Y.; Chiba, T.; Tadokoro, S.; et al. Common defects of ABCG2, a high-capacity urate exporter, cause gout: A function-based genetic analysis in a Japanese population. *Sci. Transl. Med.* **2009**, *1*, 5ra11. [CrossRef]
13. Hoque, K.M.; Dixon, E.E.; Lewis, R.M.; Allan, J.; Gamble, G.D.; Phipps-Green, A.J.; Halperin Kuhns, V.L.; Horne, A.M.; Stamp, L.K.; Merriman, T.R.; et al. The ABCG2 Q141K hyperuricemia and gout associated variant illuminates the physiology of human urate excretion. *Nat. Commun.* **2020**, *11*, 2767. [CrossRef]
14. Major, T.J.; Dalbeth, N.; Stahl, E.A.; Merriman, T.R. An update on the genetics of hyperuricaemia and gout. *Nat. Rev. Rheumatol.* **2018**, *14*, 341–353. [CrossRef]
15. Nakayama, A.; Matsuo, H.; Nakaoka, H.; Nakamura, T.; Nakashima, H.; Takada, Y.; Oikawa, Y.; Takada, T.; Sakiyama, M.; Shimizu, S.; et al. Common dysfunctional variants of ABCG2 have stronger impact on hyperuricemia progression than typical environmental risk factors. *Sci. Rep.* **2014**, *4*, 5227. [CrossRef] [PubMed]
16. Vlaming, M.L.; Lagas, J.S.; Schinkel, A.H. Physiological and pharmacological roles of ABCG2 (BCRP): Recent findings in Abcg2 knockout mice. *Adv. Drug Deliv. Rev.* **2009**, *61*, 14–25. [CrossRef] [PubMed]
17. Robey, R.W.; To, K.K.; Polgar, O.; Dohse, M.; Fetsch, P.; Dean, M.; Bates, S.E. ABCG2: A perspective. *Adv. Drug Deliv. Rev.* **2009**, *61*, 3–13. [CrossRef]
18. Sarkadi, B.; Homolya, L.; Hegedus, T. The ABCG2/BCRP transporter and its variants—From structure to pathology. *FEBS Lett.* **2020**, *594*, 4012–4034. [CrossRef]
19. Allikmets, R.; Schriml, L.M.; Hutchinson, A.; Romano-Spica, V.; Dean, M. A human placenta-specific ATP-binding cassette gene (ABCP) on chromosome 4q22 that is involved in multidrug resistance. *Cancer Res.* **1998**, *58*, 5337–5339.
20. Doyle, L.A.; Yang, W.; Abruzzo, L.V.; Krogmann, T.; Gao, Y.; Rishi, A.K.; Ross, D.D. A multidrug resistance transporter from human MCF-7 breast cancer cells. *Proc. Natl. Acad. Sci. USA* **1998**, *95*, 15665–15670. [CrossRef]
21. Miyake, K.; Mickley, L.; Litman, T.; Zhan, Z.; Robey, R.; Cristensen, B.; Brangi, M.; Greenberger, L.; Dean, M.; Fojo, T.; et al. Molecular cloning of cDNAs which are highly overexpressed in mitoxantrone-resistant cells: Demonstration of homology to ABC transport genes. *Cancer Res.* **1999**, *59*, 8–13. [PubMed]
22. Takada, T.; Yamamoto, T.; Matsuo, H.; Tan, J.K.; Ooyama, K.; Sakiyama, M.; Miyata, H.; Yamanashi, Y.; Toyoda, Y.; Higashino, T.; et al. Identification of ABCG2 as an exporter of uremic toxin indoxyl sulfate in mice and as a crucial factor influencing CKD progression. *Sci. Rep.* **2018**, *8*, 11147. [CrossRef]
23. Heyes, N.; Kapoor, P.; Kerr, I.D. Polymorphisms of the multidrug pump ABCG2: A systematic review of their effect on protein expression, function, and drug pharmacokinetics. *Drug Metab. Dispos. Biol. Fate Chem.* **2018**, *46*, 1886–1899. [CrossRef] [PubMed]

24. Stiburkova, B.; Pavelcova, K.; Zavada, J.; Petru, L.; Simek, P.; Cepek, P.; Pavlikova, M.; Matsuo, H.; Merriman, T.R.; Pavelka, K. Functional non-synonymous variants of ABCG2 and gout risk. *Rheumatology* **2017**, *56*, 1982–1992. [CrossRef] [PubMed]

25. Higashino, T.; Takada, T.; Nakaoka, H.; Toyoda, Y.; Stiburkova, B.; Miyata, H.; Ikebuchi, Y.; Nakashima, H.; Shimizu, S.; Kawaguchi, M.; et al. Multiple common and rare variants of ABCG2 cause gout. *RMD Open* **2017**, *3*, e000464. [CrossRef]

26. Maekawa, K.; Itoda, M.; Sai, K.; Saito, Y.; Kaniwa, N.; Shirao, K.; Hamaguchi, T.; Kunitoh, H.; Yamamoto, N.; Tamura, T.; et al. Genetic variation and haplotype structure of the ABC transporter gene ABCG2 in a Japanese population. *Drug Metab. Pharmacokinet.* **2006**, *21*, 109–121. [CrossRef]

27. Matsuo, H.; Ichida, K.; Takada, T.; Nakayama, A.; Nakashima, H.; Nakamura, T.; Kawamura, Y.; Takada, Y.; Yamamoto, K.; Inoue, H.; et al. Common dysfunctional variants in ABCG2 are a major cause of early-onset gout. *Sci. Rep.* **2013**, *3*, 2014. [CrossRef] [PubMed]

28. Zaidi, F.; Narang, R.K.; Phipps-Green, A.; Gamble, G.G.; Tausche, A.K.; So, A.; Riches, P.; Andres, M.; Perez-Ruiz, F.; Doherty, M.; et al. Systematic genetic analysis of early-onset gout: ABCG2 is the only associated locus. *Rheumatology* **2020**, *59*, 2544–2549. [CrossRef]

29. Stiburkova, B.; Pavelcova, K.; Pavlikova, M.; Jesina, P.; Pavelka, K. The impact of dysfunctional variants of ABCG2 on hyperuricemia and gout in pediatric-onset patients. *Arthritis Res. Ther.* **2019**, *21*, 77. [CrossRef]

30. Hurba, O.; Mancikova, A.; Krylov, V.; Pavlikova, M.; Pavelka, K.; Stiburkova, B. Complex analysis of urate transporters SLC2A9, SLC22A12 and functional characterization of non-synonymous allelic variants of GLUT9 in the Czech population: No evidence of effect on hyperuricemia and gout. *PLoS ONE* **2014**, *9*, e107902. [CrossRef]

31. Pavelcova, K.; Bohata, J.; Pavlikova, M.; Bubenikova, E.; Pavelka, K.; Stiburkova, B. Evaluation of the influence of genetic variants of SLC2A9 (GLUT9) and SLC22A12 (URAT1) on the development of hyperuricemia and gout. *J. Clin. Med.* **2020**, *9*, 2510. [CrossRef]

32. Toyoda, Y.; Pavelcova, K.; Klein, M.; Suzuki, H.; Takada, T.; Stiburkova, B. Familial early-onset hyperuricemia and gout associated with a newly identified dysfunctional variant in urate transporter ABCG2. *Arthritis Res. Ther.* **2019**, *21*, 219. [CrossRef] [PubMed]

33. Stiburkova, B.; Sebesta, I.; Ichida, K.; Nakamura, M.; Hulkova, H.; Krylov, V.; Kryspinova, L.; Jahnova, H. Novel allelic variants and evidence for a prevalent mutation in URAT1 causing renal hypouricemia: Biochemical, genetics and functional analysis. *Eur. J. Hum. Genet.* **2013**, *21*, 1067–1073. [CrossRef] [PubMed]

34. Toyoda, Y.; Takada, T.; Suzuki, H. Inhibitors of human ABCG2: From technical background to recent updates with clinical implications. *Front. Pharmacol.* **2019**, *10*, 208. [CrossRef]

35. Miyata, H.; Takada, T.; Toyoda, Y.; Matsuo, H.; Ichida, K.; Suzuki, H. Identification of febuxostat as a new strong ABCG2 inhibitor: Potential applications and risks in clinical situations. *Front. Pharmacol.* **2016**, *7*, 518. [CrossRef]

36. Toyoda, Y.; Mancikova, A.; Krylov, V.; Morimoto, K.; Pavelcova, K.; Bohata, J.; Pavelka, K.; Pavlikova, M.; Suzuki, H.; Matsuo, H.; et al. Functional characterization of clinically-relevant rare variants in ABCG2 identified in a gout and hyperuricemia cohort. *Cells* **2019**, *8*, 363. [CrossRef]

37. Manolaridis, I.; Jackson, S.M.; Taylor, N.M.I.; Kowal, J.; Stahlberg, H.; Locher, K.P. Cryo-EM structures of a human ABCG2 mutant trapped in ATP-bound and substrate-bound states. *Nature* **2018**, *563*, 426–430. [CrossRef]

38. Nakagawa, H.; Toyoda, Y.; Wakabayashi-Nakao, K.; Tamaki, H.; Osumi, M.; Ishikawa, T. Ubiquitin-mediated proteasomal degradation of ABC transporters: A new aspect of genetic polymorphisms and clinical impacts. *J. Pharm. Sci.* **2011**, *100*, 3602–3619. [CrossRef]

39. Eckenstaler, R.; Benndorf, R.A. 3D structure of the transporter ABCG2-What's new? *Br. J. Pharmacol.* **2020**, *177*, 1485–1496. [CrossRef] [PubMed]

40. Toyoda, Y.; Takada, T.; Miyata, H.; Ishikawa, T.; Suzuki, H. Regulation of the axillary osmidrosis-associated ABCC11 protein stability by N-linked glycosylation: Effect of glucose condition. *PLoS ONE* **2016**, *11*, e0157172. [CrossRef] [PubMed]

41. Morar, B.; Gresham, D.; Angelicheva, D.; Tournev, I.; Gooding, R.; Guergueltcheva, V.; Schmidt, C.; Abicht, A.; Lochmuller, H.; Tordai, A.; et al. Mutation history of the Roma/Gypsies. *Am. J. Hum. Genet.* **2004**, *75*, 596–609. [CrossRef] [PubMed]

42. Kalaydjieva, L.; Morar, B.; Chaix, R.; Tang, H. A newly discovered founder population: The Roma/Gypsies. *Bioessays* **2005**, *27*, 1084–1094. [CrossRef]

43. Roberts, R.L.; Wallace, M.C.; Phipps-Green, A.J.; Topless, R.; Drake, J.M.; Tan, P.; Dalbeth, N.; Merriman, T.R.; Stamp, L.K. ABCG2 loss-of-function polymorphism predicts poor response to allopurinol in patients with gout. *Pharm. J.* **2017**, *17*, 201–203. [CrossRef]

44. Wen, C.C.; Yee, S.W.; Liang, X.; Hoffmann, T.J.; Kvale, M.N.; Banda, Y.; Jorgenson, E.; Schaefer, C.; Risch, N.; Giacomini, K.M. Genome-wide association study identifies ABCG2 (BCRP) as an allopurinol transporter and a determinant of drug response. *Clin. Pharmacol. Ther.* **2015**, *97*, 518–525. [CrossRef]

45. Horvathova, V.; Bohata, J.; Pavlikova, M.; Pavelcova, K.; Pavelka, K.; Senolt, L.; Stiburkova, B. Interaction of the p.Q141K variant of the ABCG2 gene with clinical data and cytokine levels in primary hyperuricemia and gout. *J. Clin. Med.* **2019**, *8*, 1965. [CrossRef]

46. Lehtisalo, M.; Keskitalo, J.E.; Tornio, A.; Lapatto-Reiniluoto, O.; Deng, F.; Jaatinen, T.; Viinamaki, J.; Neuvonen, M.; Backman, J.T.; Niemi, M. Febuxostat, but not allopurinol, markedly raises the plasma concentrations of the breast cancer resistance protein substrate rosuvastatin. *Clin. Transl. Sci.* **2020**, *13*, 1236–1243. [CrossRef] [PubMed]

47. Mackenzie, I.S.; Ford, I.; Nuki, G.; Hallas, J.; Hawkey, C.J.; Webster, J.; Ralston, S.H.; Walters, M.; Robertson, M.; De Caterina, R.; et al. Long-term cardiovascular safety of febuxostat compared with allopurinol in patients with gout (FAST): A multicentre, prospective, randomised, open-label, non-inferiority trial. *Lancet* **2020**, *396*, 1745–1757. [CrossRef]

48. Stiburkova, B.; Bleyer, A.J. Changes in serum urate and urate excretion with age. *Adv. Chronic Kidney Dis.* **2012**, *19*, 372–376. [CrossRef]

49. Higashino, T.; Morimoto, K.; Nakaoka, H.; Toyoda, Y.; Kawamura, Y.; Shimizu, S.; Nakamura, T.; Hosomichi, K.; Nakayama, A.; Ooyama, K.; et al. Dysfunctional missense variant of OAT10/SLC22A13 decreases gout risk and serum uric acid levels. *Ann. Rheum. Dis.* **2020**, *79*, 164–166. [CrossRef]

50. Wallace, S.L.; Robinson, H.; Masi, A.T.; Decker, J.L.; McCarty, D.J.; Yu, T.F. Preliminary criteria for the classification of the acute arthritis of primary gout. *Arthritis Rheum.* **1977**, *20*, 895–900. [CrossRef] [PubMed]

51. Mraz, M.; Hurba, O.; Bartl, J.; Dolezel, Z.; Marinaki, A.; Fairbanks, L.; Stiburkova, B. Modern diagnostic approach to hereditary xanthinuria. *Urolithiasis* **2015**, *43*, 61–67. [CrossRef] [PubMed]

52. Stiburkova, B.; Ichida, K.; Sebesta, I. Novel homozygous insertion in SLC2A9 gene caused renal hypouricemia. *Mol. Genet. Metab.* **2011**, *102*, 430–435. [CrossRef]

53. Stiburkova, B.; Gabrikova, D.; Cepek, P.; Simek, P.; Kristian, P.; Cordoba-Lanus, E.; Claverie-Martin, F. Prevalence of URAT1 allelic variants in the Roma population. *Nucleosides Nucleotides Nucleic Acids* **2016**, *35*, 529–535. [CrossRef] [PubMed]

54. Khunweeraphong, N.; Szollosi, D.; Stockner, T.; Kuchler, K. The ABCG2 multidrug transporter is a pump gated by a valve and an extracellular lid. *Nat. Commun.* **2019**, *10*, 5433. [CrossRef] [PubMed]

55. Orban, T.I.; Seres, L.; Ozvegy-Laczka, C.; Elkind, N.B.; Sarkadi, B.; Homolya, L. Combined localization and real-time functional studies using a GFP-tagged ABCG2 multidrug transporter. *Biochem. Biophys. Res. Commun.* **2008**, *367*, 667–673. [CrossRef]

56. Takada, T.; Suzuki, H.; Sugiyama, Y. Characterization of polarized expression of point- or deletion-mutated human BCRP/ABCG2 in LLC-PK1 cells. *Pharm. Res.* **2005**, *22*, 458–464. [CrossRef]

57. Stiburkova, B.; Miyata, H.; Zavada, J.; Tomcik, M.; Pavelka, K.; Storkanova, G.; Toyoda, Y.; Takada, T.; Suzuki, H. Novel dysfunctional variant in ABCG2 as a cause of severe tophaceous gout: Biochemical, molecular genetics and functional analysis. *Rheumatology* **2016**, *55*, 191–194. [CrossRef]

58. Toyoda, Y.; Sakurai, A.; Mitani, Y.; Nakashima, M.; Yoshiura, K.; Nakagawa, H.; Sakai, Y.; Ota, I.; Lezhava, A.; Hayashizaki, Y.; et al. Earwax, osmidrosis, and breast cancer: Why does one SNP (538G>A) in the human ABC transporter ABCC11 gene determine earwax type? *FASEB J.* **2009**, *23*, 2001–2013. [CrossRef]

59. Nakagawa, H.; Wakabayashi-Nakao, K.; Tamura, A.; Toyoda, Y.; Koshiba, S.; Ishikawa, T. Disruption of N-linked glycosylation enhances ubiquitin-mediated proteasomal degradation of the human ATP-binding cassette transporter ABCG2. *FEBS J.* **2009**, *276*, 7237–7252. [CrossRef] [PubMed]

International Journal of
Molecular Sciences

MDPI

Review

The Role of ABCG2 in the Pathogenesis of Primary Hyperuricemia and Gout—An Update

Robert Eckenstaler and Ralf A. Benndorf *

Institute of Pharmacy, Martin Luther University Halle-Wittenberg, 06120 Halle, Germany;
robert.eckenstaler@pharmazie.uni-halle.de
* Correspondence: ralf.benndorf@pharmazie.uni-halle.de; Tel.: +49-345-55-25150

Abstract: Urate homeostasis in humans is a complex and highly heritable process that involves i.e., metabolic urate biosynthesis, renal urate reabsorption, as well as renal and extrarenal urate excretion. Importantly, disturbances in urate excretion are a common cause of hyperuricemia and gout. The majority of urate is eliminated by glomerular filtration in the kidney followed by an, as yet, not fully elucidated interplay of multiple transporters involved in the reabsorption or excretion of urate in the succeeding segments of the nephron. In this context, genome-wide association studies and subsequent functional analyses have identified the ATP-binding cassette (ABC) transporter ABCG2 as an important urate transporter and have highlighted the role of single nucleotide polymorphisms (SNPs) in the pathogenesis of reduced cellular urate efflux, hyperuricemia, and early-onset gout. Recent publications also suggest that ABCG2 is particularly involved in intestinal urate elimination and thus may represent an interesting new target for pharmacotherapeutic intervention in hyperuricemia and gout. In this review, we specifically address the involvement of ABCG2 in renal and extrarenal urate elimination. In addition, we will shed light on newly identified polymorphisms in ABCG2 associated with early-onset gout.

Keywords: gout; early-onset gout; hyperuricemia; urate; uric acid; ABCG2; BCRP; ABC transporter; single nucleotide polymorphism; SNP

Citation: Eckenstaler, R.; Benndorf, R.A. The Role of ABCG2 in the Pathogenesis of Primary Hyperuricemia and Gout—An Update. *Int. J. Mol. Sci.* **2021**, *22*, 6678. https://doi.org/10.3390/ijms22136678

Academic Editor: Thomas Falguières

Received: 31 May 2021
Accepted: 18 June 2021
Published: 22 June 2021

Publisher's Note: MDPI stays neutral with regard to jurisdictional claims in published maps and institutional affiliations.

1. Introduction

Gout is the clinical manifestation of hyperuricemia which is triggered by urate precipitation (deposition of monosodium urate crystals) in the synovial fluid of joints and other tissues [1,2]. The disease is primarily associated with severe arthropathy, which manifests mainly in the metatarsophalangeal joints (podagra), but also in other joints of the foot, ankles, knee, wrist, fingers, and elbows [3]. In the pathogenesis of the disease, urate deposits promote inflammatory responses in the synovial membrane (synovitis) and thus arthritis characterized by sudden, severe attacks of pain, swelling, redness, and tenderness in the affected joints. Depending on the course of the disease, the symptoms of gout can occur both as acute episodic flares (gout attacks) and persist chronically and, if left untreated, can lead to irreversible deformations and impaired mobility of the affected joints [3]. In addition, gout nephropathy, a form of chronic tubulointerstitial nephritis, induced by the deposition of urate precipitates in the distal collecting ducts and the medullary interstitium may cause progressive chronic kidney disease [4]. Furthermore, gout and hyperuricemia have been associated with a subset of comorbidities including metabolic syndrome, diabetes, hypertension as well as cardiovascular and cerebrovascular disease [5–13]. In most patients, the onset of gout occurs after the age of 60, with the incidence being about three times higher in men than in women [14]. However, a significant proportion of patients develop primary hyperuricemia and gout symptoms before the age of 40, which is defined as the pathotype of early-onset gout [15,16]. In addition to environmental factors, genetic predispositions leading to chronic, yet asymptomatic hyperuricemia in childhood and adolescence are considered to be the main causes for the early onset of the disease. Although not every patient with hyperuricemia necessarily develops gout [17], it is considered to be the major risk factor for the development and progression of the disease. In this review, we will specifically address the pathogenesis and genetic background of early-onset gout and highlight the role of intestinal uric acid transport in this context.

2. Gout and Hyperuricemia

Hyperuricemia represents a prolonged pathophysiological increased serum urate concentration, often defined as >6.0 mg/dL (>360 μmol/L) for females and >7.0 mg/dL (>420 μmol/L) for males [1,18], which is either caused by an increased hepatic biosynthesis or a reduced renal or intestinal excretion of urate [19]. Under physiological conditions, urate is derived from the enzymatic degradation of purine nucleobases/nucleotides, which are involved in a multitude of biochemical processes, such as energy metabolism and the formation of RNA and DNA [20]. In humans, urate is the terminal metabolite of purine catabolism derived from purines that do not enter the salvage pathway for the resynthesis of ATP or GTP [19]. Therefore, secondary hyperuricemia can be induced by an excessive intake of purine-rich food (e.g., red meat, offal, seafood) [21], cellular degradation processes, and high cell turnover in the context of leukemia/lymphoma [22] or anticancer treatment with chemo- or radiation therapy [23], which all increase the availability of free purines. In addition to a diet high in purines, other lifestyle-related behaviors such as excessive intake of fructose [24,25] and alcohol abuse [26,27] can also trigger hyperuricemia, which explains the high prevalence in industrialized countries and the increasing prevalence in developing countries [28]. Aside from the aforementioned environmental factors, also genetic defects in enzymes responsible for the biotransformation of purine bases can favor primary hyperuricemia, as is the case in Lesch–Nyhan or Kelley–Seegmiller syndromes [29]. In line with this notion, the heritability of hyperuricemia is substantial, suggesting important genetic contributions to urate homeostasis [30]. In pharmacotherapy, uricostatic drugs like the xanthinoxidase inhibitor allopurinol can be used to normalize hyperuricemia by preventing the last step in urate biosynthesis. Under this treatment, intermediates of the purine metabolism such as inosine, hypoxanthine, and xanthine accumulate, yet exhibit a better water solubility and a lower tendency to form crystals than urate. Unlike secondary hyperuricemias that are triggered by increased urate biosynthesis, the vast number (>90%) of primary hyperuricemia cases result from a decreased ability of the kidney or intestine to excrete urate [31]. The majority of urate (roughly 70%) is eliminated by the kidney, where it is freely filtered by the glomerulus [32]. Urate homeostasis is primarily influenced by renal proximal tubule cells, which express several transporters that either reabsorb urate (e.g., URAT1 at the apical and GLUT9 at the basolateral membrane) [33–36] or are involved in urate excretion (e.g., NPT1/4 at the apical and OAT1/3 at the basolateral membrane) [20,35,37,38]. Indeed, uricosuric drugs such as the URAT1 inhibitors benzbromarone as well as probenecid and lesinurad are used in pharmacotherapy to treat hyperuricemia by inhibiting renal reabsorption of urate [39]. In addition to transporters of the salute carrier (SLC) and the organic anion transporter (OAT) protein families, ABC transporters such as ABCG2 and ABCC4 are also involved in urate excretion [32,37]. As the previously mentioned other transporters, ABCG2 was shown to be located in the apical brush border membrane of renal proximal tubule cell [40]. In the intestine, the major site for the remaining 30% of urate excretion, the mechanisms of urate excretion are less well defined [38]. Urate transporters GLUT9 [41] and in particular ABCG2 [42] are highly expressed in intestinal epithelial cells and may thus represent interesting new pharmacological targets for the treatment of hyperuricemia [43–47]. Nonetheless, with regard to the sites of urate excretion (kidney & intestine) and the complex interplay of transporter-mediated excretion and reabsorption of urate in the kidney, the mechanisms of urate homeostasis are still not fully understood. However, single nucleotide polymorphisms (SNPs) in different genes involved in urate transport have been associated with hyperuricemia [48], thereby emphasizing the multicausal complexity of gout pathology [49]. In this article, we aim to focus on ABCG2, which has been identified as an important urate transporter in the intestine and the kidney [40,50–52], and discuss its role in renal and extra-renal urate excretion as well as in primary hyperuricemia and early-onset gout.

3. ABCG2 and Its Function in Renal Urate Elimination

ABCG2 (also known as BCRP) is a multi-drug efflux pump that has been described to contribute to transport processes in many different tissues and cell types. It belongs to the ABC (ATP-binding cassette) transporter superfamily [53] and has the ability to transport a variety of substrates across the membrane [54,55]. ABCG2 is highly expressed in the placental syncytiotrophoblasts [56], but can also be found at the entry and exit point of the human body including endothelial cells of the cerebral blood-brain barrier [57] and canalicular membrane of the liver [58] as well as polar epithelial cells of the intestine [42] and the kidney [50,59]. Based on its function and localization, ABCG2 is thought to act as a "gatekeeper", preventing endo- or exotoxins and xenobiotics from crossing biological barriers and entering sensitive tissues [60]. Although these functions of ABCG2 are thought to serve to maintain the healthy state of the organism, they also appear to be responsible for ABCG2-related interference with pharmacotherapeutic interventions to treat certain diseases. In this regard, overexpression of ABCG2 has been associated with multidrug resistance to chemotherapy [61,62], which is associated with poor prognosis in the treatment of certain cancers [63,64]. With regard to its protein structure, ABCG2 consists of an ATP-hydrolyzing nucleotide-binding domain, which is located in the cytoplasm and provides energy for the transport process, and a transmembrane domain, which is responsible for the binding of substrates and their transport across the membrane (Figure 1). Moreover, ABCG2 is a so-called "half-transporter" that needs to homodimerize to form a functional transporter [60]. Recently, several high-resolution 3D structures of the ABCG2 protein bound to different substrates and inhibitors have been solved [65–68], which help to understand the molecular mechanisms of substrate selection, substrate binding, and substrate transport of ABCG2 [69,70]. In addition to its role as an efflux pump with broad specificity, ABCG2 has been proposed to be involved in renal and intestinal urate excretion [40,50–52,71]. The function of ABCG2 as a urate transporter was inferred from genome-wide association analyses and subsequent functional studies, which specifically demonstrated a strong association of a missense SNP in the ABCG2 gene (rs2231142; Q141K) with hyperuricemia [72–74], an SNP that could be causally related to at least 10% of all gout cases [50]. The Q141K polymorphism has been associated with a reduced ABCG2 surface expression and decreases cellular urate efflux to approximately half of wild-type ABCG2 levels [50,52,75–77]. In structural predictions derived from homology models [78] as well as structural cryo-EM data [65], Q141 was shown to be located in the nucleotide-binding domain of the transporter and to form a hydrogen bond to N158 of an α-helix within the nucleotide-binding domain adjacent to transmembrane helix 1 of ABCG2. This connection might be responsible to convey conformational changes induced by ATP binding or ATP hydrolysis from the nucleotide-binding domain to the substrate transporting transmembrane domain, thereby potentially explaining the partial loss of function of the Q141K-mutated transporter. However, also misfolding, reduced protein stability, and reduced membrane expression due to increased proteasomal degradation of Q141K-mutated ABCG2 [75,79] are also discussed as causes of the urate excretion deficit. Recent findings suggest that Q141K- and M71V-related dysfunction is due to aberrant trafficking of ABCG2 to the plasma membrane due to quality control mechanisms in the endoplasmic reticulum rather than reduced ABCG2 transport activity [80]. Moreover, the c.C421 > A mutation that leads to the Q141K polymorphism promotes microRNA-mediated suppression of ABCG2 translation, so that cell type-specific processing of the ABCG2 3′UTR along with cell type-specific microRNA expression profiles may have a profound impact on functional ABCG2 bioavailability in individuals carrying the Q141K polymorphism [76]. In the kidney of humans and mice, ABCG2 was shown to be expressed in the apical membrane of the brush border of proximal tubule epithelial cells [40,50], although in the analyses of The Human Protein Atlas consortium ABCG2 could not be detected at the protein level in human kidney biopsies [81]. Nonetheless, also findings from other groups indicate relevant renal ABCG2 expression [59]. The renal expression pattern of ABCG2 partially resembles the expression pattern of urate reabsorbing transporter URAT1, thereby indicating a functional interplay of both transporters in renal urate handling [40]. However, in measurements of renal urate excretion after a purine challenge (oral administration of inosine, which is rapidly metabolized to urate), human subjects carrying the ABCG2 transport function impairing Q141K polymorphism showed no significant differences in urate excretion and a fraction of filtered urate load (FEUA defined as the ratio between the renal clearance of uric acid to the renal clearance of creatinine), although their serum urate levels were significantly elevated [40]. In the same study, renal urate excretion was also investigated in an orthologous Q140K knock-in mouse model. Here, only the male animals displayed elevated serum urate levels and had, in contrast to humans, at least a significantly reduced fraction of filtered urate load but again no change in urinary urate excretion [40]. Interestingly, these sex-related phenotypes were consistent with the increased prevalence of gout in human males [28]. However, the results of both human and mouse experiments suggest that the hyperuricemia induced by the ABCG2 Q141K polymorphism is not caused by a significant effect on renal urate excretion, but is likely to be triggered by different mechanism [40]. These findings, which are in line with the ABCG2 expression data from The Human Protein Atlas consortium [81], also raise the fundamental question of whether ABCG2 is in fact of any significance for renal urate excretion. In this regard, conflicting results regarding the involvement of ABCG2 in renal urate elimination have been obtained in experiments with ABCG2 knockout mice [51,71]. In both studies, serum urate concentrations of ABCG2 knockout animals were elevated compared to their wildtype littermates. However, while one study did not observe significant differences in renal urate elimination [71], the other study found a significant reduction of about 30% [51]. Nevertheless, these two animal studies, as well as the translational study by Hoque and colleagues, indicate that ABCG2 primarily affects extrarenal regulation of urate homeostasis, which is further discussed below. It should be mentioned that in the kidney, ABCG2 is only one of many renal transporters that are able to excrete urate [37] so that ABCG2 loss of function may be compensated by other transporters. In conclusion, the previously assumed relevance of ABCG2 for renal urate elimination has been questioned by recent studies and therefore further future studies are needed to definitively elucidate this issue.

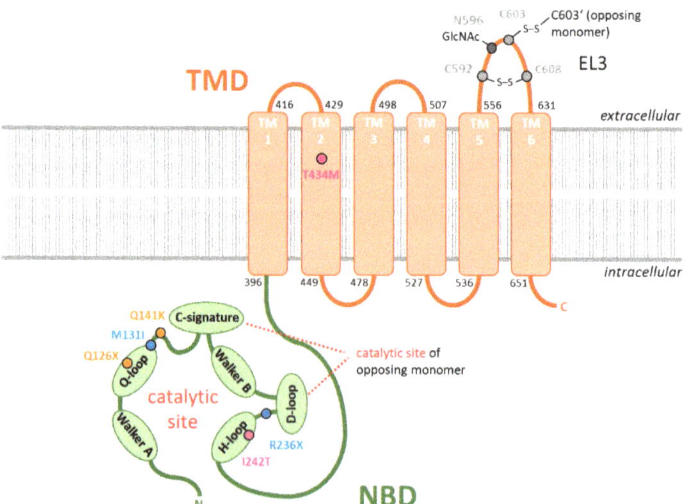

Figure 1. Polymorphisms in ABCG2 protein sequence associated with pediatric-onset hyperuricemia and early-onset gout. Schematic overview of the ABCG2 domain structure consisting of a nucleotide-

binding domain (light green, NBD) and a transmembrane domain (light brown, TMD) modified from [69]. Single membrane-spanning α-helices (TM1–6) were structured according to the information of published protein sequences (NCBI accession number: NP_001335914.1). The catalytic site for ATP hydrolysis is formed by the sequence motifs Walker A, Q-loop, Walker B, and H-loop of one monomer, and the c-signature and D-loop from the other monomer. Cysteine bridge forming residues and N-acetylation sites within extracellular loop 3 (EL3) are marked in grey. SNPs involved in pediatric-onset hyperuricemia and early-onset of gout published in recent seminal publications are highlighted in different colors (yellow, dark blue, and purple).

4. Relevance of ABCG2 in Extrarenal Urate Elimination

As earlier studies indicated that ABCG2 is also expressed in the liver [58] and intestine [42] and that ABCG2 may function as a urate transporter [50,52], it was reasonable to speculate that ABCG2 plays a role in extrarenal urate excretion possibly via the bile or intestine [82]. However, accurate non-invasive measurements of intestinal/biliary urate secretion are not possible in humans because the secreted urate is largely metabolized by the bacterial flora in the intestine. Therefore, the role of ABCG2 in extrarenal urate secretion was revealed in animal experiments using the in situ intestinal "closed-loop" perfusion method [51,71]. Since most vertebrates, including rodents used for animal studies, express the enzyme uricase which converts uric acid to allantoin and which has been lost during human evolution [83], a direct relation of results from animal experiments studying urate metabolism and urate transport to the human organism is inadequate. To account for that, rodents used in both studies were treated with the uricase inhibitor oxonate, a pharmacological intervention that increased the serum urate levels in mice and rats to the magnitude of serum urate levels in humans [51,71]. By administration of radioactive labeled uric acid, Hosomi and colleagues demonstrated that, in addition to the substantial fraction of renal urate elimination [51], there is direct urate excretion via the intestine (mainly in the ileum) and only minor urate excretion via the bile [51,71]. These findings, therefore, suggest that the intestine is the main site of extrarenal urate excretion. In the intestine of mice, ABCG2 expression is mainly located at the villi brush border of epithelial cells of the ileum and the jejunum [40]. Interestingly, in this study, the observed overall expression levels of the protein in the intestine are much higher than in the kidney. In rats, ABCG2 expression in the gut was found to further increase in response to increased blood urate concentrations after oxonate treatment [84]. To study the contribution of ABCG2 in intestinal urate excretion (and renal urate excretion which was already discussed above), oxonate-treated ABCG2 knockout mice were used [51,71] and showed a reduction in intestinal urate elimination by roughly 40–50%. These findings also suggest that other yet unknown transporters besides ABCG2 are involved in intestinal urate secretion [51,71]. A similar severe loss in intestinal urate elimination was also observed in the aforementioned study by Hogue and colleagues in an orthologous Q140K knock-in mouse model but in absence of the uricase inhibitor oxonate [40]. Consistent with these data, Q140K knock-in in mice resulted in a marked reduction in intestinal ABCG2 expression. In contrast, only subtle changes in urate elimination and ABCG2 expression were observed in the kidney in the same knock-in mouse model. These results are further supported by the observation of other authors, which indicate that the ABCG2 Q141K polymorphism and fractional renal clearance both contribute significantly but independently to the risk of hyperuricemia in humans [73]. In addition, impaired intestinal urate excretion induced by the orthologous murine Q140K mutation or complete ABCG2 knockout may explain hyperuricemia despite unaltered renal urate excretion in the respective mouse models [71,85,86] as well as in human individuals carrying the Q141K polymorphism [40]. However, due to the rise in serum urate levels (sUA) caused by the lack of intestinal urate secretion, an indirect increase in the fraction of filtered urate load (FEUA) could be expected in patients with ABCG2 dysfunction although this was not observed in the previously mentioned study [40]. In addition, ABCG2-mediated intestinal urate elimination appears to play an important role in compensating for the loss of renal urate elimination in chronic kidney disease [35]. Taken together, recent publications indicate that the major site of action of the ABCG2 transporter is regulating urate homeostasis in the intestine.

5. ABCG2 Polymorphisms in Pediatric-Onset Hyperuricemia and Early-Onset Gout

Hyperuricemia and gout pathology has often been shown to be related to genetic predisposition [30] and to be affected by SNPs in many of the genes encoding urate transporters [87,88]. Among these, especially SNPs of ABCG2 have been highly associated with pediatric-onset hyperuricemia and early-onset gout [89–93]. These polymorphisms are summarized and organized in Table 1 according to the standard nomenclature rules for molecular diagnostics [94]. Furthermore, the localization of these polymorphisms on the protein sequence of ABCG2 is shown in Figure 1. It should be noted that there is a subset of other function-impairing SNPs in ABCG2 [95,96], but most of them have not yet been associated with pediatric hyperuricemia or early-onset gout. One of the best-studied variations of the ABCG2 amino acid sequence is the previously discussed Q141K polymorphism, which also gives rise to other important clinical phenotypes, such as in the pharmacokinetics and tissue distribution of drugs transported by ABCG2 [97]. The polymorphism is highly associated with early-onset hyperuricemia, gout, and hyperuricemia-associated comorbidities, which cause a high mortality rate in hemodialysis patients [98]. Although the F489L polymorphism has not been as well studied in the context of disease, it shows a similar inhibitory effect on the ABCG2 transport function as the Q141K mutation. As with the Q141K mutation, ABCG2 carriers with the F489L mutation show reduced expression and reduced ABCG2 transport capacity [75]. Inhibition of proteasomal degradation could partially restore the transport function of both ABCG2 variants. In contrast to the Q141K polymorphism, which causes amino acid sequence alterations in the nucleotide-binding domain, the F489L polymorphism is localized in the transmembrane domain. This shows that the impairment of ABCG2 function can be caused by changes in amino acid structure in different domains of the transporter. Polymorphisms in the transmembrane domain have often been associated with decreased surface expression of the ABCG2 transporter and impaired substrate transport abilities [99]. However, there is not much literature to support their clinical impact in both late and early-onset hyperuricemia and gout. The clinical importance of a certain polymorphism on the development of hyperuricemia and gout usually is related to its minor allele frequency in humans and its functional impact on the protein of interest. Due to genetic drift caused by spatial separation of populations, certain polymorphisms have accumulated in different ethnicities. For example, the frequency of V12M polymorphism is high in Mexican Indians but low in Caucasian and Middle Eastern populations [97]. In contrast, the Q141K and Q126X polymorphisms are enriched in Japanese populations, whereas in Caucasians, Q141K is not as common and Q126X is virtually absent [97]. Our understanding of the genetic variations in the ABCG2 sequence associated with hyperuricemia and gout is still incomplete, as evidenced by the recent discovery of less common polymorphisms previously unrecognized or not studied in the context of hyperuricemia and gout [89,90,95,100]. Two of these newly identified rare polymorphisms have been recently described in a case report of a 12-year-old Czech girl of Roma ethnicity with chronic asymptomatic pediatric-onset of hyperuricemia [89]. In this regard, several rare diseases have been found to occur primarily or exclusively in individuals of Roma ethnicity, and many of the mutations underlying these diseases have been recently discovered, such as for Charcot-Marie tooth disease types 4D and 4G [101,102], the congenital cataract facial dysmorphism neuropathy [103], the Gitelman syndrome [104], and the Galactokinase deficiency [105]. In the afore-mentioned case of the 12-year old girl, DNA sequencing analysis of the ABCG2 gene revealed the presence of heterozygously expressed missense (c.393G > T, p.M131I) and nonsense (c.706C > T, p.R236X) mutations (Figure 1, blue residues) causing the pediatric-onset of hyperuricemia observed in the girl's ancestry and the early-onset of gout especially in male individuals of the maternal line of inheritance. In the study, the functional consequences of the mutations were investigated in comparative in vitro experiments. Due to the in-frame stop codon induced by the R236X mutation, the ABCG2 protein sequence was truncated to about 1/3 of the full-length protein, with the mutant protein lacking a functional transmembrane domain. Therefore, no plasma membrane localization and no urate transport activity of the mutant protein could be observed. In contrast, the M131I mutation was translated to a full-length protein with no impairments in N-glycosylation at residue N596 and normal membrane localization. However, the urate transport capabilities of the M131I mutant were reduced to <15% of wildtype levels [89]. M131 itself was found to be a highly conserved residue that is localized close to the Q-loop within the nucleotide-binding domain of ABCG2 (Figure 1). The conserved glutamine Q126 in the center of the Q-loop is responsible for the coordination of the magnesium ion associated with ATP in the catalytic center of the protein [68]. M131I may thus alter the spatial orientation of the Q-loop or sterically hinder the coordination of Mg-ATP, thereby drastically reducing ABCG2′s ATP hydrolysis capabilities necessary for providing the energy for substrate transport across the membrane. Another newly identified polymorphism associated with pediatric hyperuricemia and early-onset gout is I242T, which was found in the lineage of another young European girl and was analyzed in a similar way [93]. Like the aforementioned M131I mutant, the I242T mutant ABCG2 variant showed no impairment in glycosylation and membrane localization, although its urate transport abilities were drastically reduced. This effect could be coincidentally related to the close localization of the mutants at the conserved H243 within the H-loop or also called histidine switch of the catalytic center of ABCG2 (Figure 1). The H-loop is responsible for coordinating the γ-phosphate of ATP, which is responsible for ATP hydrolysis. For further research, I242T and M131I may represent interesting new candidates to study the consequences of ABCG2 loss-of-function without disrupting ABCG2 membrane localization and protein-protein interactions. These representative case reports also show that depending on the severity of the disruption of the urate transportability of ABCG2, homo- or heterozygosity of the dysfunctional polymorphisms and further genetic predispositions in other genes involved in urate homeostasis [106], hyperuricemia can already occur in childhood (pediatric-onset), which increases the risk for the development of early-onset gout. This allows the risk allele of a particular polymorphism to be identified and considered for clinical diagnosis. Interestingly, compared to patients with late-onset gout, patients with early-onset gout also show clinical symptoms that indicate a more severe disease pattern. This includes a prolonged disease duration, a different localization of the first occurring arthritis (with a lower incidence of typical metatarsophalangeal manifestations and a higher incidence of ankle- or mid-foot involvement in early-onset gout), a higher flare frequency (gout attacks), and an increased overall number of involved joints [15,16]. In terms of gout-associated comorbidities, late-onset gout patients are more likely to suffer from chronic kidney disease, metabolic syndrome, and cardiovascular disease, a phenomenon probably related to the age difference between the two patient groups [16]. However, these comorbidities occur at a younger age in patients with early-onset gout. In contrast, a recent study showed that patients diagnosed with gout at age 40 or younger may be at increased risk for cardiovascular disease and recurrent gout compared to those diagnosed later in life [107]. In this study, of 427 adult patients diagnosed with gout at a New England multispecialty group practice, 327 who were aged 40 years or younger at diagnosis were more likely to have cardiovascular risk factors. For example, these younger patients had a significantly higher body mass index than gout patients over 40 years of age, and a substantial proportion of the younger patients also suffered from hypertension or hyperlipidemia. Moreover, early-onset gout patients were less likely to achieve a serum uric acid level below 6.0 mg/dL after therapeutic intervention as compared to late-onset gout patients. Therefore, clinical screening for hyperuricemia in genetically predisposed families and prompt urate-lowering therapy in pediatric, adolescent, or young adult patients with still asymptomatic chronic hyperuricemia could help delay the onset of gout and the development of hyperuricemia-related comorbidities [108–110]. With regard to the treatment of cardiovascular comorbidities in hyperuricemia patients, it should be noted that blood pressure-lowering drugs such as the AT1 receptor blocker telmisartan have been shown to inhibit the transport activity of ABCG2 [75] thereby potentially exacerbating hyperuricemia in patients with a corresponding genetic predisposition. In view of the emerging role of ABCG2 and its importance for intestinal excretion of uric acid, it may in principle represent a novel pharmacotherapeutic target to lower uric acid levels [43–47]. As speculation, this may be accomplished by modifying ABCG2 expression and function in intestinal epithelial cells. For example, in patients expressing mutant forms of ABCG2, this opens up the possibility of developing small molecule drugs with high pre-systemic elimination to target the function, cellular handling, or expression of ABCG2 predominantly in intestinal epithelial cells, thereby locally normalizing the impaired intestinal uric acid excretion in these individuals without interfering with the function of ABCG2 in other tissues (e.g., extrusion of xenobiotics). This area of research, therefore, shows great potential for the development of targeted pharmacotherapies for specific populations of genetically predisposed individuals with early-onset gout and thus warrants innovative research in the near future.

Table 1. Polymorphisms in ABCG2 protein sequence associated with pediatric-onset hyperuricemia and early-onset gout.

rs ID	Coding Sequence	Protein Sequence	Citation
rs72552713	c.376C > T	p.Q126X	[92]
rs759726272	c.393G > T	p.M131I	[89]
rs2231142	c.C421 > A	p.Q141K	[91,92]
rs140207606	c.706C > T	p.R236X	[89]
not annotated	c.725T > C	p.I242T	[93]
rs769734146	c.1301C > T	p.T434M	[90,100]

6. Conclusions

Gout is a major health care burden in developed countries, where it affects about 1% to 2% of the adult population and is the most common cause of inflammatory arthritis in men. In addition to obesity and hyperuricemia, lifestyle changes that have developed in industrialized countries in recent decades, such as a diet rich in red meat and fructose, physical inactivity, and increased alcohol consumption, may play a role in the shift toward a younger age of manifestation of gout in the population and require early intervention. As there is evidence that early onset of hyperuricemia and gout is associated not only with a severe clinical course of gouty arthritis, but also with other comorbidities, such as hypertension, metabolic syndrome, and cardiovascular complications, early detection of hyperuricemia in younger patients with genetic predisposition and early uric acid-lowering therapy should be considered to reduce morbidity and mortality in these patients.

Author Contributions: R.E. and R.A.B. wrote the manuscript. Both authors contributed to the article and approved the submitted version. Both authors have read and agreed to the published version of the manuscript.

Funding: This work was supported by the European Regional Development Fund of the European Commission (W21029490) to R.A.B and the Publication Fund of the Martin-Luther-University Halle/Wittenberg (VAT DE 811353703).

Institutional Review Board Statement: Not applicable.

Informed Consent Statement: Not applicable.

Data Availability Statement: Not applicable.

References

1. Bardin, T.; Richette, P. Definition of hyperuricemia and gouty conditions. *Curr. Opin. Rheumatol.* **2014**, *26*, 186–191. [CrossRef]
2. Masseoud, D.; Rott, K.; Liu-Bryan, R.; Agudelo, C. Overview of Hyperuricaemia and Gout. *Curr. Pharm. Des.* **2005**, *11*, 4117–4124. [CrossRef] [PubMed]
3. Grassi, W.; De Angelis, R. Clinical features of gout. *Reumatismo* **2012**, *63*, 238–245. [CrossRef] [PubMed]
4. Kc, M.; Leslie, S.W. Uric Acid Nephrolithiasis. In *StatPearls*; StatPearls Publishing: Treasure Island, FL, USA, 2021.
5. Lanaspa, M.A.; Andres-Hernando, A.; Kuwabara, M. Uric acid and hypertension. *Hypertens. Res.* **2020**, *43*, 832–834. [CrossRef] [PubMed]
6. Kuwabara, M.; Niwa, K.; Hisatome, I.; Nakagawa, T.; Roncal-Jimenez, C.A.; Andres-Hernando, A.; Bjornstad, P.; Jensen, T.; Sato, Y.; Milagres, T.; et al. Asymptomatic Hyperuricemia Without Comorbidities Predicts Cardiometabolic Diseases: Five-Year Japanese Cohort Study. *Hypertension* **2017**, *69*, 1036–1044. [CrossRef] [PubMed]
7. Sun, Y.; Zhang, H.; Tian, W.; Shi, L.; Chen, L.; Li, J.; Zhao, S.; Qi, G. Association between serum uric acid levels and coronary artery disease in different age and gender: A cross-sectional study. *Aging Clin. Exp. Res.* **2019**, *31*, 1783–1790. [CrossRef]
8. Lin, Y.-H.; Hsu, H.-L.; Huang, Y.-C.; Lee, M.; Huang, W.-Y.; Huang, Y.-C.; Lee, T.-H.; Lee, J.-D. Gouty Arthritis in Acute Cerebrovascular Disease. *Cerebrovasc. Dis.* **2009**, *28*, 391–396. [CrossRef] [PubMed]
9. Thottam, G.E.; Krasnokutsky, S.; Pillinger, M.H. Gout and Metabolic Syndrome: A Tangled Web. *Curr. Rheumatol. Rep.* **2017**, *19*, 60. [CrossRef]
10. Ejaz, A.A.; Nakagawa, T.; Kanbay, M.; Kuwabara, M.; Kumar, A.; Arroyo, F.E.G.; Roncal-Jimenez, C.; Sasai, F.; Kang, D.-H.; Jensen, T.; et al. Hyperuricemia in Kidney Disease: A Major Risk Factor for Cardiovascular Events, Vascular Calcification, and Renal Damage. *Semin. Nephrol.* **2020**, *40*, 574–585. [CrossRef] [PubMed]
11. Sanchez-Lozada, L.G.; Rodriguez-Iturbe, B.; Kelley, E.E.; Nakagawa, T.; Madero, M.; Feig, D.I.; Borghi, C.; Piani, F.; Cara-Fuentes, G.; Bjornstad, P.; et al. Uric Acid and Hypertension: An Update With Recommendations. *Am. J. Hypertens.* **2020**, *33*, 583–594. [CrossRef]
12. Johnson, R.J.; Bakris, G.L.; Borghi, C.; Chonchol, M.B.; Feldman, D.; Lanaspa, M.A.; Merriman, T.R.; Moe, O.W.; Mount, D.B.; Lozada, L.G.S.; et al. Hyperuricemia, Acute and Chronic Kidney Disease, Hypertension, and Cardiovascular Disease: Report of a Scientific Workshop Organized by the National Kidney Foundation. *Am. J. Kidney Dis.* **2018**, *71*, 851–865. [CrossRef] [PubMed]
13. Jiang, Y.; Ge, J.-Y.; Zhang, Y.-Y.; Wang, F.-F.; Ji, Y.; Li, H.-Y. The relationship between elevated serum uric acid and arterial stiffness in a healthy population. *Vascular* **2020**, *28*, 494–501. [CrossRef]
14. Soriano, L.C.; Rothenbacher, D.; Choi, H.K.; Rodríguez, L.A.G. Contemporary epidemiology of gout in the UK general population. *Arthritis Res. Ther.* **2011**, *13*, R39. [CrossRef]
15. Zhang, B.; Fang, W.; Zeng, X.; Zhang, Y.; Ma, Y.; Sheng, F.; Zhang, X. Clinical characteristics of early- and late-onset gout: A cross-sectional observational study from a Chinese gout clinic. *Medicine* **2016**, *95*, e5425. [CrossRef]
16. Pascart, T.; Norberciak, L.; Ea, H.; Guggenbuhl, P.; Lioté, F. Patients With Early-Onset Gout and Development of Earlier Severe Joint Involvement and Metabolic Comorbid Conditions: Results From a Cross-Sectional Epidemiologic Survey. *Arthritis Rheum.* **2019**, *71*, 986–992. [CrossRef] [PubMed]
17. Robinson, P.C. Gout—An update of aetiology, genetics, co-morbidities and management. *Maturitas* **2018**, *118*, 67–73. [CrossRef]
18. Sui, X.; Church, T.S.; Meriwether, R.A.; Lobelo, F.; Blair, S.N. Uric acid and the development of metabolic syndrome in women and men. *Metabolism* **2008**, *57*, 845–852. [CrossRef] [PubMed]
19. Estiverne, C.; Mandal, A.K.; Mount, D.B. Molecular Pathophysiology of Uric Acid Homeostasis. *Semin. Nephrol.* **2020**, *40*, 535–549. [CrossRef] [PubMed]
20. Maiuolo, J.; Oppedisano, F.; Gratteri, S.; Muscoli, C.; Mollace, V. Regulation of uric acid metabolism and excretion. *Int. J. Cardiol.* **2016**, *213*, 8–14. [CrossRef] [PubMed]
21. Choi, H.K.; Atkinson, K.; Karlson, E.W.; Willett, W.; Curhan, G. Purine-Rich Foods, Dairy and Protein Intake, and the Risk of Gout in Men. *N. Engl. J. Med.* **2004**, *350*, 1093–1103. [CrossRef]
22. Pui, C.-H.; Mahmoud, H.H.; Wiley, J.M.; Woods, G.M.; Leverger, G.; Camitta, B.; Hastings, C.; Blaney, S.M.; Relling, M.V.; Reaman, G.H. Recombinant Urate Oxidase for the Prophylaxis or Treatment of Hyperuricemia in Patients with Leukemia or Lymphoma. *J. Clin. Oncol.* **2001**, *19*, 697–704. [CrossRef] [PubMed]
23. Webster, J.S.; Kaplow, R. Tumor Lysis Syndrome: Implications for Oncology Nursing Practice. *Semin. Oncol. Nurs.* **2021**, *37*, 151136. [CrossRef] [PubMed]
24. Zhang, C.; Li, L.; Zhang, Y.; Zeng, C. Recent advances in fructose intake and risk of hyperuricemia. *Biomed. Pharmacother.* **2020**, *131*, 110795. [CrossRef]
25. Sayehmiri, K.; Ahmadi, I.; Anvari, E. Fructose Feeding and Hyperuricemia: A Systematic Review and Meta-Analysis. *Clin. Nutr. Res.* **2020**, *9*, 122–133. [CrossRef]
26. Hernández-Rubio, A.; Sanvisens, A.; Bolao, F.; Pérez-Mañá, C.; García-Marchena, N.; Fernández-Prendes, C.; Muñoz, A.; Muga, R. Association of hyperuricemia and gamma glutamyl transferase as a marker of metabolic risk in alcohol use disorder. *Sci. Rep.* **2020**, *10*, 1–8. [CrossRef]
27. Choi, H.K.; McCormick, N.; Lu, N.; Rai, S.K.; Yokose, C.; Zhang, Y. Population Impact Attributable to Modifiable Risk Factors for Hyperuricemia. *Arthritis Rheumatol.* **2020**, *72*, 157–165. [CrossRef] [PubMed]

28. Dehlin, M.; Jacobsson, L.; Roddy, E. Global epidemiology of gout: Prevalence, incidence, treatment patterns and risk factors. *Nat. Rev. Rheumatol.* **2020**, *16*, 380–390. [CrossRef]

29. Nanagiri, A.; Shabbir, N. Lesch Nyhan Syndrome. In *StatPearls*; StatPearls Publishing: Treasure Island, FL, USA, 2021.

30. Krishnan, E.; Lessov-Schlaggar, C.N.; Krasnow, R.E.; Swan, G.E. Nature Versus Nurture in Gout: A Twin Study. *Am. J. Med.* **2012**, *125*, 499–504. [CrossRef] [PubMed]

31. Stewart, D.J.; Langlois, V.; Noone, D. Hyperuricemia and Hypertension: Links and Risks. *Integr. Blood Press. Control* **2019**, *12*, 43–62. [CrossRef] [PubMed]

32. Woodward, O.M. ABCG2: The molecular mechanisms of urate secretion and gout. *Am. J. Physiol. Physiol.* **2015**, *309*, F485–F488. [CrossRef] [PubMed]

33. Enomoto, A.; Kimura, H.; Chairoungdua, A.; Shigeta, Y.; Jutabha, P.; Cha, S.H.; Hosoyamada, M.; Takeda, M.; Sekine, T.; Igarashi, T.; et al. Molecular identification of a renal urate–anion exchanger that regulates blood urate levels. *Nat. Cell Biol.* **2002**, *417*, 447–452. [CrossRef]

34. Vitart, V.; Rudan, I.; Hayward, C.; Gray, N.K.; Floyd, J.; Palmer, C.N.A.; Knott, S.A.; Kolcic, I.; Polasek, O.; Graessler, J.; et al. SLC2A9 is a newly identified urate transporter influencing serum urate concentration, urate excretion and gout. *Nat. Genet.* **2008**, *40*, 437–442. [CrossRef]

35. Nigam, S.K.; Bhatnagar, V. The systems biology of uric acid transporters: The role of remote sensing and signaling. *Curr. Opin. Nephrol. Hypertens.* **2018**, *27*, 305–313. [CrossRef] [PubMed]

36. Dalbeth, N.; Gosling, A.L.; Gaffo, A.; Abhishek, A. Gout. *Lancet* **2021**, *397*, 1843–1855. [CrossRef]

37. Bobulescu, I.A.; Moe, O.W. Renal Transport of Uric Acid: Evolving Concepts and Uncertainties. *Adv. Chronic Kidney Dis.* **2012**, *19*, 358–371. [CrossRef] [PubMed]

38. So, A.; Thorens, B. Uric acid transport and disease. *J. Clin. Investig.* **2010**, *120*, 1791–1799. [CrossRef]

39. Azevedo, V.F.; Kos, I.A.; Vargas-Santos, A.B.; Pinheiro, G.D.R.C.; Paiva, E.D.S. Benzbromarone in the treatment of gout. *Adv. Rheumatol.* **2019**, *59*, 37. [CrossRef] [PubMed]

40. Hoque, K.M.; Dixon, E.E.; Lewis, R.M.; Allan, J.; Gamble, G.D.; Phipps-Green, A.J.; Kuhns, V.L.H.; Horne, A.M.; Stamp, L.K.; Merriman, T.R.; et al. The ABCG2 Q141K hyperuricemia and gout associated variant illuminates the physiology of human urate excretion. *Nat. Commun.* **2020**, *11*, 2767. [CrossRef]

41. DeBosch, B.J.; Kluth, O.; Fujiwara, H.; Schurmann, A.; Moley, K.H. Early-onset metabolic syndrome in mice lacking the intestinal uric acid transporter SLC2A9. *Nat. Commun.* **2014**, *5*, 4642. [CrossRef] [PubMed]

42. Gutmann, H.; Hruz, P.; Zimmermann, C.; Beglinger, C.; Drewe, J. Distribution of breast cancer resistance protein (BCRP/ABCG2) mRNA expression along the human GI tract. *Biochem. Pharmacol.* **2005**, *70*, 695–699. [CrossRef] [PubMed]

43. Li, Q.; Lin, H.; Niu, Y.; Liu, Y.; Wang, Z.; Song, L.; Gao, L.; Li, L. Mangiferin promotes intestinal elimination of uric acid by modulating intestinal transporters. *Eur. J. Pharmacol.* **2020**, *888*, 173490. [CrossRef]

44. Chen, M.; Ye, C.; Zhu, J.; Zhang, P.; Jiang, Y.; Lu, X.; Wu, H. Bergenin as a Novel Urate-Lowering Therapeutic Strategy for Hyperuricemia. *Front. Cell Dev. Biol.* **2020**, *8*, 703. [CrossRef]

45. Chen, X.; Ge, H.-Z.; Lei, S.-S.; Jiang, Z.-T.; Su, J.; He, X.; Zheng, X.; Wang, H.-Y.; Yu, Q.-X.; Li, B.; et al. Dendrobium officinalis six nostrum ameliorates urate under-excretion and protects renal dysfunction in lipid emulsion-induced hyperuricemic rats. *Biomed. Pharmacother.* **2020**, *132*, 110765. [CrossRef] [PubMed]

46. Ristic, B.; Sikder, M.O.F.; Bhutia, Y.D.; Ganapathy, V. Pharmacologic inducers of the uric acid exporter ABCG2 as potential drugs for treatment of gouty arthritis. *Asian J. Pharm. Sci.* **2020**, *15*, 173–180. [CrossRef] [PubMed]

47. Lu, Y.-H.; Chang, Y.-P.; Li, T.; Han, F.; Li, C.-J.; Li, X.-Y.; Xue, M.; Cheng, Y.; Meng, Z.-Y.; Han, Z.; et al. Empagliflozin Attenuates Hyperuricemia by Upregulation of ABCG2 via AMPK/AKT/CREB Signaling Pathway in Type 2 Diabetic Mice. *Int. J. Biol. Sci.* **2020**, *16*, 529–542. [CrossRef] [PubMed]

48. Dehghan, A.; Köttgen, A.; Yang, Q.; Hwang, S.-J.; Kao, W.L.; Rivadeneira, F.; Boerwinkle, E.; Levy, D.; Hofman, A.; Astor, B.C.; et al. Association of three genetic loci with uric acid concentration and risk of gout: A genome-wide association study. *Lancet* **2008**, *372*, 1953–1961. [CrossRef]

49. Richette, P.; Bardin, T. Gout. *Lancet* **2010**, *375*, 318–328. [CrossRef]

50. Woodward, O.M.; Köttgen, M.; Coresh, J.; Boerwinkle, E.; Guggino, W.B. Identification of a urate transporter, ABCG2, with a common functional polymorphism causing gout. *Proc. Natl. Acad. Sci. USA* **2009**, *106*, 10338–10342. [CrossRef] [PubMed]

51. Hosomi, A.; Nakanishi, T.; Fujita, T.; Tamai, I. Extra-Renal Elimination of Uric Acid via Intestinal Efflux Transporter BCRP/ABCG2. *PLoS ONE* **2012**, *7*, e30456. [CrossRef]

52. Matsuo, H.; Takada, T.; Ichida, K.; Nakamura, T.; Nakayama, A.; Ikebuchi, Y.; Ito, K.; Kusanagi, Y.; Chiba, T.; Tadokoro, S.; et al. Common Defects of ABCG2, a High-Capacity Urate Exporter, Cause Gout: A Function-Based Genetic Analysis in a Japanese Population. *Sci. Transl. Med.* **2009**, *1*, 5ra11. [CrossRef]

53. Dean, M.; Hamon, Y.; Chimini, G. The human ATP-binding cassette (ABC) transporter superfamily. *J. Lipid Res.* **2001**, *42*, 1007–1017. [CrossRef]

54. Mo, W.; Zhang, J.-T. Human ABCG2: Structure, function, and its role in multidrug resistance. *Int. J. Biochem. Mol. Boil.* **2011**, *3*, 1–27.

55. Polgar, O.; Robey, R.W.; Bates, S.E. ABCG2: Structure, function and role in drug response. *Expert Opin. Drug Metab. Toxicol.* **2007**, *4*, 1–15. [CrossRef] [PubMed]

56. Mao, Q. BCRP/ABCG2 in the placenta: Expression, function and regulation. *Pharm. Res.* **2008**, *25*, 1244–1255. [CrossRef]
57. Zhang, W.; Mojsilovic-Petrovic, J.; Andrade, M.F.; Zhang, H.; Ball, M.; Stanimirovic, D.B. Expression and functional characterization of ABCG2 in brain endothelial cells and vessels. *FASEB J.* **2003**, *17*, 1–24. [CrossRef]
58. Maliepaard, M.; Scheffer, G.L.; Faneyte, I.F.; Van Gastelen, M.A.; Pijnenborg, A.C.; Schinkel, A.H.; Van De Vijver, M.J.; Scheper, R.J.; Schellens, J.H. Subcellular localization and distribution of the breast cancer resistance protein transporter in normal human tissues. *Cancer Res.* **2001**, *61*, 3458–3464. [PubMed]
59. Huls, M.; Brown, C.; Windass, A.; Sayer, R.; Heuvel, J.V.D.; Heemskerk, S.; Russel, F.; Masereeuw, R.; Huls, M.; Brown, C.; et al. The breast cancer resistance protein transporter ABCG2 is expressed in the human kidney proximal tubule apical membrane. *Kidney Int.* **2008**, *73*, 220–225. [CrossRef]
60. Sarkadi, B.; Ozvegy-Laczka, C.; Német, K.; Váradi, A. ABCG2—A transporter for all seasons. *FEBS Lett.* **2004**, *567*, 116–120. [CrossRef] [PubMed]
61. Doyle, L.; Ross, D.D. Multidrug resistance mediated by the breast cancer resistance protein BCRP (ABCG2). *Oncogene* **2003**, *22*, 7340–7358. [CrossRef]
62. Robey, R.W.; Polgar, O.; Deeken, J.; To, K.; Bates, S.E. ABCG2: Determining its relevance in clinical drug resistance. *Cancer Metastasis Rev.* **2007**, *26*, 39–57. [CrossRef] [PubMed]
63. Kim, J.E.; Singh, R.R.; Cho-Vega, J.H.; Drakos, E.; Davuluri, Y.; Khokhar, F.A.; Fayad, L.; Medeiros, L.J.; Vega, F. Sonic hedgehog signaling proteins and ATP-binding cassette G2 are aberrantly expressed in diffuse large B-Cell lymphoma. *Mod. Pathol.* **2009**, *22*, 1312–1320. [CrossRef] [PubMed]
64. Damiani, D.; Tiribelli, M.; Calistri, E.; Geromin, A.; Chiarvesio, A.; Michelutti, A.; Cavallin, M.; Fanin, R. The prognostic value of P-glycoprotein (ABCB) and breast cancer resistance protein (ABCG2) in adults with de novo acute myeloid leukemia with normal karyotype. *Haematologica* **2006**, *91*, 825–828. [PubMed]
65. Taylor, N.M.I.; Manolaridis, I.; Jackson, S.M.; Kowal, J.; Stahlberg, H.; Locher, K.P. Structure of the human multidrug transporter ABCG2. *Nat. Cell Biol.* **2017**, *546*, 504–509. [CrossRef]
66. Kowal, J.; Ni, D.; Jackson, S.M.; Manolaridis, I.; Stahlberg, H.; Locher, K.P. Structural Basis of Drug Recognition by the Multidrug Transporter ABCG2. *J. Mol. Biol.* **2021**, *433*, 166980. [CrossRef] [PubMed]
67. Jackson, S.M.; Manolaridis, I.; Kowal, J.; Zechner, M.; Taylor, N.M.I.; Bause, M.; Bauer, S.; Bartholomaeus, R.; Bernhardt, G.; Koenig, B.; et al. Structural basis of small-molecule inhibition of human multidrug transporter ABCG2. *Nat. Struct. Mol. Biol.* **2018**, *25*, 333–340. [CrossRef]
68. Manolaridis, I.; Jackson, S.M.; Taylor, N.M.I.; Kowal, J.; Stahlberg, H.; Locher, K.P. Cryo-EM structures of a human ABCG2 mutant trapped in ATP-bound and substrate-bound states. *Nat. Cell Biol.* **2018**, *563*, 426–430. [CrossRef] [PubMed]
69. Eckenstaler, R.; Benndorf, R.A. 3D structure of the transporter ABCG2-What's new? *Br. J. Pharmacol.* **2020**, *177*, 1485–1496. [CrossRef]
70. Kapoor, P.; Horsey, A.J.; Cox, M.H.; Kerr, I.D. ABCG2: Does resolving its structure elucidate the mechanism? *Biochem. Soc. Trans.* **2018**, *46*, 1485–1494. [CrossRef]
71. Ichida, K.; Matsuo, H.; Takada, T.; Nakayama, A.; Murakami, K.; Shimizu, T.; Yamanashi, Y.; Kasuga, H.; Nakashima, H.; Nakamura, T.; et al. Decreased extra-renal urate excretion is a common cause of hyperuricemia. *Nat. Commun.* **2012**, *3*, 764. [CrossRef]
72. Dong, Z.; Guo, S.; Yang, Y.; Wu, J.; Guan, M.; Zou, H.; Jin, L.; Wang, J. Association between ABCG2 Q141K polymorphism and gout risk affected by ethnicity and gender: A systematic review and meta-analysis. *Int. J. Rheum. Dis.* **2015**, *18*, 382–391. [CrossRef] [PubMed]
73. Kannangara, D.; Phipps-Green, A.J.; Dalbeth, N.; Stamp, L.K.; Williams, K.M.; Graham, G.G.; Day, R.O.; Merriman, T.R. Hyperuricaemia: Contributions of urate transporter ABCG2 and the fractional renal clearance of urate. *Ann. Rheum. Dis.* **2015**, *75*, 1363–1366. [CrossRef]
74. Li, R.; Miao, L.; Qin, L.; Xiang, Y.; Zhang, X.; Peng, H.; Mailamuguli; Sun, Y.; Yao, H. A meta-analysis of the associations between the Q141K and Q126X ABCG2 gene variants and gout risk. *Int. J. Clin. Exp. Pathol.* **2015**, *8*, 9812–9823. [PubMed]
75. Deppe, S.; Ripperger, A.; Weiss, J.; Ergün, S.; Benndorf, R.A. Impact of genetic variability in the ABCG2 gene on ABCG2 expression, function, and interaction with AT1 receptor antagonist telmisartan. *Biochem. Biophys. Res. Commun.* **2014**, *443*, 1211–1217. [CrossRef] [PubMed]
76. Ripperger, A.; Benndorf, R.A. The C421A (Q141K) polymorphism enhances the 3′-untranslated region (3′-UTR)-dependent regulation of ATP-binding cassette transporter ABCG2. *Biochem. Pharmacol.* **2016**, *104*, 139–147. [CrossRef] [PubMed]
77. Woodward, O.M.; Tukaye, D.N.; Cui, J.; Greenwell, P.; Constantoulakis, L.M.; Parker, B.S.; Rao, A.; Köttgen, M.; Maloney, P.C.; Guggino, W.B. Gout-causing Q141K mutation in ABCG2 leads to instability of the nucleotide-binding domain and can be corrected with small molecules. *Proc. Natl. Acad. Sci. USA* **2013**, *110*, 5223–5228. [CrossRef]
78. László, L.; Sarkadi, B.; Hegedűs, T. Jump into a New Fold—A Homology Based Model for the ABCG2/BCRP Multidrug Transporter. *PLoS ONE* **2016**, *11*, e0164426. [CrossRef]
79. Furukawa, T.; Wakabayashi, K.; Tamura, A.; Nakagawa, H.; Morishima, Y.; Osawa, Y.; Ishikawa, T. Major SNP (Q141K) Variant of Human ABC Transporter ABCG2 Undergoes Lysosomal and Proteasomal Degradations. *Pharm. Res.* **2008**, *26*, 469–479. [CrossRef]
80. Bartos, Z.; Homolya, L. Identification of Specific Trafficking Defects of Naturally Occurring Variants of the Human ABCG2 Transporter. *Front. Cell Dev. Biol.* **2021**, *9*, 615729. [CrossRef]

81. Uhlén, M.; Fagerberg, L.; Hallström, B.M.; Lindskog, C.; Oksvold, P.; Mardinoglu, A.; Sivertsson, Å.; Kampf, C.; Sjöstedt, E.; Asplund, A.; et al. Proteomics. Tissue-based map of the human proteome. *Science* **2015**, *347*, 1260419. [CrossRef]

82. Nakayama, A.; Matsuo, H.; Takada, T.; Ichida, K.; Nakamura, T.; Ikebuchi, Y.; Ito, K.; Hosoya, T.; Kanai, Y.; Suzuki, H.; et al. ABCG2 is a High-Capacity Urate Transporter and its Genetic Impairment Increases Serum Uric Acid Levels in Humans. *Nucleosides Nucleotides Nucleic Acids* **2011**, *30*, 1091–1097. [CrossRef]

83. Varela-Echavarría, A.; Oca-Luna, R.M.; Barrera-Saldaña, H.A. Uricase protein sequences: Conserved during vertebrate evolution but absent in humans. *FASEB J.* **1988**, *2*, 3092–3096. [CrossRef]

84. Morimoto, C.; Tamura, Y.; Asakawa, S.; Kuribayashi-Okuma, E.; Nemoto, Y.; Li, J.; Murase, T.; Nakamura, T.; Hosoyamada, M.; Uchida, S.; et al. ABCG2 expression and uric acid metabolism of the intestine in hyperuricemia model rat. *Nucleosides Nucleotides Nucleic Acids* **2020**, *39*, 744–759. [CrossRef] [PubMed]

85. Matsuo, H.; Takada, T.; Nakayama, A.; Shimizu, T.; Sakiyama, M.; Shimizu, S.; Chiba, T.; Nakashima, H.; Nakamura, T.; Takada, Y.; et al. ABCG2 Dysfunction Increases the Risk of Renal Overload Hyperuricemia. *Nucleosides Nucleotides Nucleic Acids* **2014**, *33*, 266–274. [CrossRef]

86. Matsuo, H.; Nakayama, A.; Sakiyama, M.; Chiba, T.; Shimizu, S.; Kawamura, Y.; Nakashima, H.; Nakamura, T.; Takada, Y.; Oikawa, Y.; et al. ABCG2 dysfunction causes hyperuricemia due to both renal urate underexcretion and renal urate overload. *Sci. Rep.* **2014**, *4*, 3755. [CrossRef] [PubMed]

87. Tin, A.; Marten, J.; Halperin Kuhns, V.L.; Li, Y.; Wuttke, M.; Kirsten, H.; Sieber, K.B.; Qiu, C.; Gorski, M.; Yu, Z.; et al. Target genes, variants, tissues and transcriptional pathways influencing human serum urate levels. *Nat. Genet.* **2019**, *51*, 1459–1474. [CrossRef] [PubMed]

88. Lukkunaprasit, T.; Rattanasiri, S.; Turongkaravee, S.; Suvannang, N.; Ingsathit, A.; Attia, J.; Thakkinstian, A. The association between genetic polymorphisms in ABCG2 and SLC2A9 and urate: An updated systematic review and meta-analysis. *BMC Med. Genet.* **2020**, *21*, 1–13. [CrossRef] [PubMed]

89. Toyoda, Y.; Pavelcová, K.; Bohatá, J.; Ješina, P.; Kubota, Y.; Suzuki, H.; Takada, T.; Stiburkova, B. Identification of Two Dysfunctional Variants in the ABCG2 Urate Transporter Associated with Pediatric-Onset of Familial Hyperuricemia and Early-onset Gout. *Int. J. Mol. Sci.* **2021**, *22*, 1935. [CrossRef] [PubMed]

90. Stiburkova, B.; Pavelcova, K.; Pavlikova, M.; Ješina, P.; Pavelka, K. The impact of dysfunctional variants of ABCG2 on hyperuricemia and gout in pediatric-onset patients. *Arthritis Res.* **2019**, *21*, 1–10. [CrossRef]

91. Zaidi, F.; Narang, R.K.; Phipps-Green, A.; Gamble, G.G.; Tausche, A.-K.; So, A.; Riches, P.; Andres, M.; Perez-Ruiz, F.; Doherty, M.; et al. Systematic genetic analysis of early-onset gout: ABCG2 is the only associated locus. *Rheumatology* **2020**, *59*, 2544–2549. [CrossRef] [PubMed]

92. Matsuo, H.; Ichida, K.; Takada, T.; Nakayama, A.; Nakashima, H.; Nakamura, T.; Kawamura, Y.; Takada, Y.; Yamamoto, K.; Inoue, H.; et al. Common dysfunctional variants in ABCG2 are a major cause of early-onset gout. *Sci. Rep.* **2013**, *3*, srep02014. [CrossRef]

93. Toyoda, Y.; Pavelcová, K.; Klein, M.; Suzuki, H.; Takada, T.; Stiburkova, B. Familial early-onset hyperuricemia and gout associated with a newly identified dysfunctional variant in urate transporter ABCG2. *Arthritis Res.* **2019**, *21*, 1–3. [CrossRef]

94. Ogino, S.; Gulley, M.L.; Dunnen, J.T.D.; Wilson, R.B. Standard Mutation Nomenclature in Molecular Diagnostics: Practical and Educational Challenges. *J. Mol. Diagn.* **2007**, *9*, 1–6. [CrossRef]

95. Homolya, L. Medically Important Alterations in Transport Function and Trafficking of ABCG2. *Int. J. Mol. Sci.* **2021**, *22*, 2786. [CrossRef] [PubMed]

96. Zámbó, B.; Mózner, O.; Bartos, Z.; Török, G.; Várady, G.; Telbisz, Á.; Homolya, L.; Orbán, T.I.; Sarkadi, B. Cellular expression and function of naturally occurring variants of the human ABCG2 multidrug transporter. *Cell. Mol. Life Sci.* **2020**, *77*, 365–378. [CrossRef]

97. Heyes, N.; Kapoor, P.; Kerr, I.D. Polymorphisms of the Multidrug Pump ABCG2: A Systematic Review of Their Effect on Protein Expression, Function, and Drug Pharmacokinetics. *Drug Metab. Dispos.* **2018**, *46*, 1886–1899. [CrossRef]

98. Nakashima, A.; Ichida, K.; Ohkido, I.; Yokoyama, K.; Matsuo, H.; Ohashi, Y.; Takada, T.; Nakayama, A.; Suzuki, H.; Shinomiya, N.; et al. Dysfunctional ABCG2 gene polymorphisms are associated with serum uric acid levels and all-cause mortality in hemodialysis patients. *Hum. Cell* **2020**, *33*, 559–568. [CrossRef]

99. Sjöstedt, N.; Heuvel, J.J.; Koenderink, J.B.; Kidron, H. Transmembrane Domain Single-Nucleotide Polymorphisms Impair Expression and Transport Activity of ABC Transporter ABCG2. *Pharm. Res.* **2017**, *34*, 1626–1636. [CrossRef]

100. Toyoda, Y.; Mančíková, A.; Krylov, V.; Morimoto, K.; Pavelcová, K.; Bohatá, J.; Pavelka, K.; Pavlikova, M.; Suzuki, H.; Matsuo, H.; et al. Functional Characterization of Clinically-Relevant Rare Variants in ABCG2 Identified in a Gout and Hyperuricemia Cohort. *Cells* **2019**, *8*, 363. [CrossRef] [PubMed]

101. Li, L.; Liu, G.; Liu, Z.; Gong-Lu, L.; Wu, Z. Identification and functional characterization of two missense mutations in NDRG1 associated with Charcot-Marie-Tooth disease type 4D. *Hum. Mutat.* **2017**, *38*, 1569–1578. [CrossRef] [PubMed]

102. Gabrikova, D.; Mistrik, M.; Bernasovska, J.; Bozikova, A.; Behulova, R.; Tothova, I.; Macekova, S. Founder mutations in NDRG1 and HK1 genes are common causes of inherited neuropathies among Roma/Gypsies in Slovakia. *J. Appl. Genet.* **2013**, *54*, 455–460. [CrossRef]

103. Lassuthova, P.; Sišková, D.; Haberlová, J.; Sakmaryová, I.; Filouš, A.; Seeman, P. Congenital cataract, facial dysmorphism and demyelinating neuropathy (CCFDN) in 10 Czech Gypsy children—Frequent and underestimated cause of disability among Czech Gypsies. *Orphanet J. Rare. Dis.* **2014**, *9*, 46. [CrossRef] [PubMed]

104. Gil-Peña, H.; Coto, E.; Santos, F.; Espino, M.; Cea Crespo, J.M.; Chantzopoulos, G.; Komianou, F.; Gómez, J.; Alonso, B.; Iglesias, S.; et al. A new SLC12A3 founder mutation (p.Val647Met) in Gitelman's syndrome patients of Roma ancestry. *Nefrologia* **2017**, *37*, 423–428. [CrossRef]
105. Schulpis, K.H.; Thodi, G.; Iakovou, K.; Chatzidaki, M.; Dotsikas, Y.; Molou, E.; Triantafylli, O.; Loukas, Y.L. Clinical evaluation and mutational analysis of GALK and GALE genes in patients with galactosemia in Greece: One novel mutation and two rare cases. *J. Pediatr. Endocrinol. Metab.* **2017**, *30*, 775–779. [CrossRef]
106. Butler, F.; Alghubayshi, A.; Roman, Y. The Epidemiology and Genetics of Hyperuricemia and Gout across Major Racial Groups: A Literature Review and Population Genetics Secondary Database Analysis. *J. Pers. Med.* **2021**, *11*, 231. [CrossRef] [PubMed]
107. Li, Y.; Piranavan, P.; Sundaresan, D.; Yood, R. Clinical Characteristics of Early-Onset Gout in Outpatient Setting. *ACR Open Rheumatol.* **2019**, *1*, 397–402. [CrossRef]
108. Singer, R.F.; Walters, G. Uric acid lowering therapies for preventing or delaying the progression of chronic kidney disease. *Cochrane Database Syst. Rev.* **2011**, *10*, 10. [CrossRef]
109. Ying, H.; Yuan, H.; Tang, X.; Guo, W.; Jiang, R.; Jiang, C. Impact of Serum Uric Acid Lowering and Contemporary Uric Acid-Lowering Therapies on Cardiovascular Outcomes: A Systematic Review and Meta-Analysis. *Front. Cardiovasc. Med.* **2021**, *8*, 641062. [CrossRef] [PubMed]
110. Paul, B.J.; Anoopkumar, K.; Krishnan, V. Asymptomatic hyperuricemia: Is it time to intervene? *Clin. Rheumatol.* **2017**, *36*, 2637–2644. [CrossRef]

International Journal of
Molecular Sciences

MDPI

Review

Medically Important Alterations in Transport Function and Trafficking of ABCG2

László Homolya 

Research Centre for Natural Sciences, Institute of Enzymology, H-1117 Budapest, Hungary;
homolya.laszlo@ttk.hu; Tel.: +36-1-382-6608

Abstract: Several polymorphisms and mutations in the human ABCG2 multidrug transporter result in reduced plasma membrane expression and/or diminished transport function. Since ABCG2 plays a pivotal role in uric acid clearance, its malfunction may lead to hyperuricemia and gout. On the other hand, ABCG2 residing in various barrier tissues is involved in the innate defense mechanisms of the body; thus, genetic alterations in *ABCG2* may modify the absorption, distribution, excretion of potentially toxic endo- and exogenous substances. In turn, this can lead either to altered therapy responses or to drug-related toxic reactions. This paper reviews the various types of mutations and polymorphisms in *ABCG2*, as well as the ways how altered cellular processing, trafficking, and transport activity of the protein can contribute to phenotypic manifestations. In addition, the various methods used for the identification of the impairments in ABCG2 variants and the different approaches to correct these defects are overviewed.

Keywords: ABC (ATP-binding cassette) transporters; multidrug resistance; transport; trafficking; urate; mutations; polymorphisms

Citation: Homolya, L. Medically Important Alterations in Transport Function and Trafficking of ABCG2. *Int. J. Mol. Sci.* **2021**, *22*, 2786. https://doi.org/10.3390/ijms 22062786

Academic Editor: Thomas Falguières

Received: 9 February 2021
Accepted: 5 March 2021
Published: 10 March 2021

Publisher's Note: MDPI stays neutral with regard to jurisdictional claims in published maps and institutional affiliations.

1. Introduction

The ABCG2 protein (also named breast cancer resistance protein—BCRP, or mitoxantrone resistance protein—MXR) is a member of the ABC (ATP-binding cassette) protein superfamily. A distinguishing hallmark of ABC proteins is the presence of Walker A, Walker B, and the so-called ABC signature (typically LSGGQ) motifs in their sequences. The members of this large protein family are present in all living organisms, ranging from prokaryotes through fungi, plants, invertebrates to vertebrates. The design of ATP-binding fold and its connection to transport mechanisms seem evolutionarily beneficial, as they have been conserved through evolution [1]. In the human genome, there are 48 genes encoding ABC proteins, which are classified into seven subfamilies (denoted from A to G) primarily on the basis of sequence homology. Since an increasing number of structural and functional data are available for ABC proteins, a new classification based on these parameters has recently been proposed [2]. Most of the human ABC proteins are membrane proteins mediating translocation of substances across biological membranes using the energy of ATP binding and hydrolysis. There are some peculiar members of the family, like the regulatory ABC proteins, exemplified by the sulfonylurea receptors (SUR1/ABCC8 and SUR2/ABCC9), which control the function of other membrane proteins; or the cystic fibrosis transmembrane regulator (CFTR/ABCC7), which is an ion channel facilitating downhill chloride transport across the membrane.

Some human ABC proteins are specialized in the transport of one or a limited number of substrates. For example, MDR3/ABCB4 mediates phosphatidylcholine transport in the canalicular membrane of hepatocytes. In contrast, MDR1 (P-glycoprotein, ABCB1) is rather promiscuous, transporting a large variety of unrelated molecules. Membrane transporter proteins with broad substrate recognition may confer resistance in cells to multiple drugs, i.e., causing cross-resistance in tumor cells. These transporters are called multidrug resistance (MDR) proteins, although they also play a pivotal role at important physiological tissue barriers controlling the uptake and excretion of endo- and xenobiotics. In humans, there are multidrug transporters from the ABCB, the ABCC, and the ABCG subfamilies. These MDR proteins with their broad and partially overlapping substrate recognition, as well as with their tissue- and cell type-specific expression constitute a complex physiological network, called the chemoimmunity system, which is an essential part of the innate defense system against harmful substances [3].

ABCG2 was originally identified as a multidrug transporter in multidrug-resistant cancer cell lines, in which none of the two MDR proteins known at that time (MDR1/ABCB1 and MRP1/ABCC1) was expressed [4,5]. In drug-selected cells and certain tumors, ABCG2 is massively overexpressed and may contribute to the poor clinical outcome of these tumors [6,7]. Its physiological presence in normal tissues has also been demonstrated—first in placenta [8], and subsequently in a large variety of other tissues. The wide-ranging but still specific tissue distribution, combined with the broad substrate recognition, makes ABCG2 an essential element of the chemoimmunity network. This paper will provide an overview of the current knowledge of the structure and function of the ABCG2 protein, as well as on the medical conditions when the transporter improperly operates. A special focus will be put on the naturally occurring mutations and polymorphisms in ABCG2, which cause diminished transport activity and/or impaired cellular routing. Finally, the various methods assessing the function and trafficking of ABCG2, as well as the different efforts to rescue detrimental phenotypes caused by the faulty transporter will be discussed.

2. Architecture of ABCG2

ABCG2 belongs to the ABCG subfamily, the members of which, in addition to sequence homology, exhibit considerable structural similarities. To our recent knowledge, the minimal structure of a functional ABC transporter is composed of two cytoplasmic nucleotide-binding domains (NBDs) and two transmembrane domains (TMDs). The usual arrangement of domains in the core protein is as follows: TMD1-NBD1-TMD2-NBD2. Contrary to the canonical ABC transporters, members of the ABCG subfamily possess only one NBD and one TMD; thus, they are called half-transporters. Moreover, the domain order in ABCG proteins is reversed, i.e., NBD is localized N-terminally to TMD. This reverse domain arrangement could be one of reasons for the sensitivity of ABCG2 to tagging at the C-terminus [9], contrary to many other ABC transporters, which can regularly be tagged C-terminally. To form a functional complex, ABCG proteins, like other ABC half-transporters, assemble into either homo- or heterodimers. While ABCG2 solely forms homodimers, ABCG5 and ABCG8 are obligate heterodimers. In contrast, two other members of the subfamily, ABCG1 and ABCG4 can form both homo- and heterodimers [10].

Membrane topology models of ABCG2 suggest six membrane-spanning helices, a relatively short C-terminal tail, and short loops between the transmembrane helices (TMHs) except for the last extracellular loop (EL3) between TMH5 and TMH6 [11]. N-glycosylation on asparagine 596 located in the EL3 loop has also been demonstrated [12,13]. Although initial reports suggested that glycosylation at N596 was not essential for proper expression, localization, and function, subsequent studies demonstrated N-glycosylation to be an important checkpoint determining the stability and intracellular trafficking of the transporter [14,15]. There are twelve cysteine residues in ABCG2, but only three of them are positioned in an oxidative milieu, and thus capable of forming disulfide bonds. All three of these cysteines are located in the EL3 loop, and while an intramolecular disulfide bond is established between C592 and C608, an intermolecular disulfide bridge is formed between the two halves of the homodimer at C603. Whereas the latter disulfide bond is not required for proper trafficking and function of ABCG2 [16–18], the C592-C608 intramolecular disulfide bond represents another critical checkpoint for protein folding and trafficking [15,18].

Although the X-ray structures of isolated NBDs have been available since the late nineties, the first high-resolution structures of full-length ABC transporters were only published in 2006 and 2007 [19,20]. In the last decade, the spread of cryogenic electron microscopy (cryo-EM) and the substantial progress in crystallography have given a boost to our understanding of ABC protein structures. However, homology modeling was not quite applicable to ABCG proteins, as the members of this family are rather distinct from other 'classical' ABC transporters, such as the P-glycoprotein (MDR1/ABCB1) or the CFTR/ABCC7. The appearance of the first high-resolution structure of an ABCG protein, i.e., that of the heterodimeric sterol transporter ABCG5/ABCG8, was therefore a breakthrough [21], fueling extensive homology modeling of ABCG2 [22–24]. Subsequently, several ABCG2 structures based on cryo-EM analyses have been published [25–28]. The structural characteristics of ACBG2, obtained from these cryo-EM studies and from parallel molecular dynamic stimulations, are recently reviewed in [29,30].

There are some distinguishing structural features in the ACBG2 as compared to the full-length ABC transporters. In general, the structure of ABCG2 is more compact, the NBDs are positioned close to the TMDs. A similar compact arrangement was observed for the ABCG5/ABCG8 crystal structure [21], which resembles the architecture of BtuCD-like bacterial importers, rather than that of MDR1-like transporters. This originates from the relatively short transmembrane helices, possessing no cytosolic extension unlike the helices in the classical ABC proteins, which create a sort of spacer between the TMDs and NBDs. In the MDR1-like proteins, two of the elongated four pairs of helices cross over and bind to the opposite NBD, while the other two pairs interact with the ipsilateral NBD. The interfacings to NBDs are realized by small, so-called coupling helices at the cytosolic tip of the elongated TMHs [19]. In contrast, there is only one coupling helix in each half of the ABCG2 dimer (between TMH2 and TMH3), which does not cross over to the other half. However, an amphipathic helix, called the connecting helix, linked to only TMH1 and reclining against the membrane bilayer, provides an additional TMD-NBD interface in ABCG2.

Another distinct feature of the structure of ABCG2, as well as of ABCG5-ABCG5, is the relatively closed conformation in the absence of ATP. In the classical ABC transporters in this 'apo' form (without ATP), the NBDs are located far from one another, and consequently, the intracellular parts of TMHs also remain apart, forming a large central cavity, the main substrate-binding pocket, which is widely open to the cytoplasm [19]. The presence of a similar cavity at the cytoplasmic side of ABCG2 (cavity 1) has been reported by cryo-EM studies using an anti-ABCG2 antibody to reduce flexibility in the structure [25–27]. Nevertheless, the NBDs, and consequently the intracellular parts of TMHs, are closer to one another in ABCG2 than in MDR-like transporters, resulting in a more compact structure even in the absence of ATP. Residues in this central cavity were shown to be essential not only for transport function but also for biogenesis [24]. Interestingly, a study using no anti-ABCG2 antibodies for structure stabilization reported the lack of cavity 1 [28]. An additional prominent feature in the inward-facing structure (apo form) of ABCG2 is the hydrophobic di-leucine valve (L554 and L555) separating the central substrate-binding pocket from an additional cavity (cavity 2) located toward the extracellular part of ABCG2 [25,26]. Experiments supplemented by molecular dynamic simulations demonstrated an essential role for this di-leucine plug in the transport function [31]. Putting together the different structures in the absence and presence of ATP and/or substrates, MDR1-like proteins seem to alternate between a widely open inward-facing and a fairly open outward-facing conformations, whereas the translocation of substrates through ABCG2 via cavities 1 and 2 rather involves a peristaltic-like movement.

With regard to the NBDs, both sequences and structures are fairly conserved. The two composite ATP-binding pockets are constituted by two separate NBDs in a head-to-tail orientation, i.e., one ATP molecule binds to the Walker A and B motifs of one NBD and to the ABC signature sequence of the other NBD. Unlike in full-length ABC transporters, the cytoplasmic part of the homodimeric ABCG2 is composed of two identical halves, but otherwise the ABC-folds in ABCG2 are structurally similar to that of the classical ABC transporters. It is worth noting that a phenylalanine at position 142 in ABCG2 interacts with the connecting helix, representing a key residue in TMD-NBD interface assembly and a critical checkpoint for protein folding and function [32,33]. Interestingly, this amino acid is analogous to F508 in CFTR/ABCC7, the mutation of which is responsible for diminished trafficking of CFTR, and ultimately the cystic fibrosis (CF) phenotype.

3. The Physiological Functions of ABCG2, and Its Role in Multidrug Resistance

3.1. The Physiological Roles of ABCG2

As mentioned previously, ABCG2 is overexpressed in drug-resistant cell lines and tumors. Habitually, it is expressed at a relatively high level in cell types located at the entry and exit boundaries of the body, as well as in barrier tissues at the borders of sanctuary sites [34,35]. These include the epithelial cells of the gastrointestinal track, especially in small intestine enterocytes [36], the kidney tubular epithelial cells [37], hepatocytes [34], placental syncytiotrophoblasts [38], mammary alveolar epithelial cells (a part of the blood-milk barrier) [39], and brain capillary endothelial cells (a key element of the blood-brain barrier) [40,41]. In these polarized epithelial and endothelial cells, ABCG2 is localized to the apical plasma membrane domain. In addition to these cells constituting tissue barriers, ABCG2 is also expressed in various types of stem cells including hematopoietic stem cells [42], pluripotent stem cells [43,44], and cancer stem cells [45–48]. Interestingly, ABCG2 is also present in the membrane of red blood cells (RBCs) [34,49,50].

As is typical of a multidrug transporter, ABCG2 recognizes a vast variety of compounds as transported substrate molecules. These include uric acid in the first place, but also various endogenous conjugated hormones and metabolites, several hydrophobic and amphipathic drugs, as well as their conjugates [51–53]. This promiscuity and the tissue distribution detailed above delineate the physiological function of this transporter. In general, ABCG2—depending on its location—restricts the uptake or facilitates the excretion of potentially toxic or unwanted substances. Specifically, in the brain capillaries, ABCG2 restricts the passage of substances through the blood-brain barrier, whereas in the placenta, it protects the fetus from maternally derived toxins. For instance, ABCG2 restricts the maternal-fetal transfer of bile acids, which is especially important in expecting mothers with intrahepatic cholestasis of pregnancy, a frequent liver disease leading to augmented serum levels of bile acids [54,55]. In the small intestine, ABCG2 controls the absorption of various molecules and participates in extra-renal clearance of uric acid; in the kidney proximal tubules, it contributes to the elimination of unwanted toxins and metabolites, including uric acid. Impaired ABCG2-mediated urate transport may lead to gout or hyperuricemia, therefore, specific mutations and polymorphisms in ABCG2 are genetic risk factors for these conditions [37,56–58] to be discussed in detail in Section 6. Interestingly, a recent study reported unequal contribution of ABCG2 to renal and extra-renal clearance of uric acid [58]. In mammary alveolar epithelial cells, this transporter influences the milk composition. Endogenous substrates transported by ABCG2 through the blood-milk barrier include riboflavin (vitamin B_2) and bile acids [59,60]. Certainly, vigilance is required for breast-feeding mothers, as various medications can be transported by ABCG2 into the milk [39,61]. The relevance of the Abcg2-mediated drug transport for the dairy industries is also self-evident [62,63].

In these physiological boundaries, ABCG2 accomplishes this 'bouncer duty' in a coordinated fashion together with the other MDR proteins, MDR1/ABCB1 and MRP1/ABCC1, exploiting their partially overlapping substrate recognition and specific subcellular localization. In polarized epithelial cells, ABCG2 is localized to the apical membrane ipsilaterally to MDR1/ABCB1 and contralaterally to MRP1/ABCC1, whereas in cerebral endothelial cells, all three major MDR proteins reside at the same side, i.e., the apical membrane [64]. Accordingly, ABCG2 along with other MDR proteins potentially alters the absorption, distribution, and excretion, as well as, consequently, the metabolism and toxicity (ADME-Tox properties) of pharmaceutical drugs. Especially important is the potential contribution of these transporters to drug–drug interactions, since modification of one (or more) of the MDR proteins by a drug may greatly influence the pharmacokinetics of another one.

Therefore, the examination of drug interactions with MDR proteins, including ABCG2, is a requirement in preclinical drug development [53,65,66]. Interestingly, in mammary epithelial cells, the apically localized ABCG2 and the basolateral MRP4/ABCC4 counteract one another in bile acid transport [60].

The physiological role of ABCG2 in red blood cells and stem cells is enigmatic to some extent. Since phototoxic porphyrins, such as the plant-derived pheophoride A and the heme precursor protoporphyrin IX (PPIX), are noted substrates of ABCG2, its expression in the erythroid precursor cells and in mature RBCs may indicate its involvement in heme metabolism [50,67,68]. It is worth noting, however, that numerous membrane proteins without known function in RBCs are present in their membrane, i.e., the sterol transporter ABCA1 (http://rbcc.hegelab.org/, accessed on 9 February 2021) [69,70]. It is plausible that many of these membrane proteins can be just remnants from previous stages of cell differentiation and maturation. In various stem cell types, a protective role similar to that observed at the border of sanctuary sites has been proposed for ABCG2 [42–44]. Stem cells are poised between self-renewal and differentiation, and are thus exceptionally sensitive to environmental factors. ABCG2 can contribute to the stem cells' self-protective mechanisms. The presence of the transporter may, however, backfire in cancer stem cells, as they can provide tumors with drug-resistant cell populations.

3.2. The Involment of ABCG2 in Multidrug Resistance of Cancer

Beyond its physiological roles, ABCG2 has been implicated in cancer multidrug resistance (recently reviewed in [7]). A large variety of chemotherapeutic agents has been identified as ABCG2 substrates. First, the anti-cancer drug mitoxantrone has been demonstrated to be exported by ABCG2, thus reducing its intracellular accumulation [4,71–73]. Interestingly, a kinetic analysis indicated that mitoxantrone is extruded by ABCG2 not from the cytosol but directly from the plasma membrane, where the drug accumulates [74]. Other anti-cancer drugs identified as ABCG2 substrates include flavopiridol [73,75], methotrexate [76], topotecan, and irinotecan [77,78]. In addition, several prominent tyrosine kinase inhibitors (TKIs) used in chemotherapies, such as gefitinib [79–81], imatinib [81–83], sunitinib [84], and nilotinib [83,85], were proven to be transported by ABCG2. The anti-cancer agents doxorubicin and daunorubicin have also been reported as ABCG2 substrates [4,72,75], but eventually it was revealed that these drugs are transported only by the R482G ABCG2 variant [3,86].

Expression of ABCG2 in tumors often correlates with poor prognosis, especially in hematopoietic malignancies, such as acute myeloid leukemia [87], but also in solid tumors, including diffuse large B-cell lymphoma [88]. However, clinical data are often conflicting like in the case of acute lymphocytic leukemia [89–91], or of breast carcinoma [92–94]. Several other studies demonstrated correlation between ABCG2 expression and response to chemotherapy, even to drugs, which are not ABCG2 substrates. These inconsistencies can originate from the modulatory effect of other drug resistance mechanisms, most evidently the presence of other MDR proteins. In addition, the methods employed to determine ABCG2 expression could be dubious, originating from the potential cross-reactivity of applied antibodies, or from the fact that mRNA levels of membrane proteins often do not correlate with the protein levels. The genetic background of patients could give an extra hue to these clinical studies as mutations and polymorphisms may alter the input of ABCG2 into the clinical outcome or the response to various drugs; therefore, proper stratification of patients is crucial for these analyses. In summary, the role of ABCG2 in tumors has been implicated, but its actual contribution to the clinical multidrug resistance is still unclear [6,7].

4. Mutations and Polymorphisms in ABCG2

4.1. Classifications of Genetic Variants of ABCG2

Normal functioning of ABCG2 can be modulated, attenuated or abolished by mutations or polymorphisms, which in turn may lead to medical conditions. These genetic alterations can affect the transport activity of ABCG2 either by limiting its ATPase activity, or by altering its substrate affinity and/or substrate profile. In addition, when the transporter is not expressed at an adequate level, it cannot fully accomplish its physiological role. Since ABCG2 operates as a pump protein residing in the plasma membrane and expelling various substrates from the cell, not only sufficient expression but also proper cellular localization is a prerequisite for normal function. Diminished cell surface protein level can stem either from reduced overall expression caused by an early stop codon, mRNA instability, protein folding problems, and increased degradation; or from trafficking problems, such as retention in a cellular compartment, halted posttranslational modification, intracellular sequestration, and augmented internalization. The former factors affect general expression of the transporter, whereas trafficking defects alter its steady-state concentration on the cell surface. Based on this, mutations and polymorphisms can be categorized as affecting (i) the general expression level, (ii) the cellular trafficking, or (iii) the transport activity. In many cases, genetic variations alter not only one of these parameters, but various combinations of them.

A classification of different ABCG2 variants has been proposed by Tamuara et al. [95]. This categorization is based on the protein expression level and drug resistance profile of the variants. The four groups are defined as follows: (i) variants with wild type-like drug resistance profile; (ii) mutants with acquired doxorubicin and daurorubicin resistance, as well as with prazosin-stimulated ATPase activity; (iii) non-expressing mutants; (iv) and others possessing normal expression levels but altered drug resistance profile. Recently, a novel and more systematic classification was suggested by Heyes et al. [96]. In this system, the main categories, ranging from one to three, are based on the cell surface expression of the ABCG2 variant: normal (as wild type, wt), reduced, and increased, whereas subcategories denoted by a and *b* indicate whether the transport function is preserved (or even elevated) and reduced, respectively. This classification is simple and reasonable, but does not distinguish between normal variants and gain of function mutants, and does not take into account whether the reduced cell surface appearance is due to general expression problems or trafficking defects despite the fact that rescuing a phenotype caused by one or the other requires distinct interventions.

It is, therefore, worth considering a new classification of ABCG2 mutations similar to that was previously introduced for CFTR variants, which is based on so-called theratypes [97]. In this classification, CFTR variants are categorized according to the nature of their defect and the specific strategies required for phenotype correction. While Class 1 mutations affect protein production, Class 2 mutations impair trafficking. Class 3 and 4 mutations diminish transport function by affecting channel gating and conductance, respectively. Class 5 mutations lead to reduced protein levels, whereas Class 6 mutations reduce plasma membrane half-lives. Finally, Class 7 mutations are the so-called unrescuable genetic variants, e.g., those containing large deletions [97].

With regard to the ABCG2 mutations, we propose here a new classification, which embraces the logic and architecture of the CFTR mutation categorization, but also considers the specific features of ABCG2. Although this classification does not follow the order of sequential cellular event, such as transcription, translation, post-translational modification, trafficking, and degradation, it is rather based on conventions used for CFTR for many years. However, conforming of mutation classifications of various ABC proteins may help to avoid confusion and to adapt interventions to rescue phenotypes caused by the mutations from the same categories (see Section 7).

4.2. Class 0 Mutations

Various databases based on genome and transcriptome sequences list an enormous number (over 1000) of mutations and polymorphisms in the *ABCG2* gene. These also include the synonymous mutations and the genetic variants in the non-coding regions, but from a practical point of view, we focus here primarily on non-synonymous single-nucleotide polymorphisms (SNPs) in the coding region. A comprehensive collection of mutations in ABC proteins, including those of ABCG2, is available at http://abcmutations. hegelab.org/ (accessed on 9 February 2021) [98,99]. This database (named ABCMdb) summarizes not only the naturally occurring but also the artificially generated genetic variants, and annotates them with relevant literature data. Inclusion of artificial mutations in such databases is of great help in gaining insights into structure-function relationships and in designing future experiments.

One group of ABCG mutations can be formed from those that do not affect the function, expression, and trafficking considerably (marked as Class 0). A characteristic representative of this group is the frequent missense polymorphism V12M (rs2231137) [49,100]. The minor allele frequency (MAF) of this SNP falls between 0.19 and 0.33 in the Asian populations and is around 0.06 in Europe [101,102]. Other members of this class are K360del and T434M [103,104]. The specific cellular events affected by the mutations of each group are depicted in Figure 1, whereas detailed data on the major representatives of the various SNP categories are provided in Table 1.

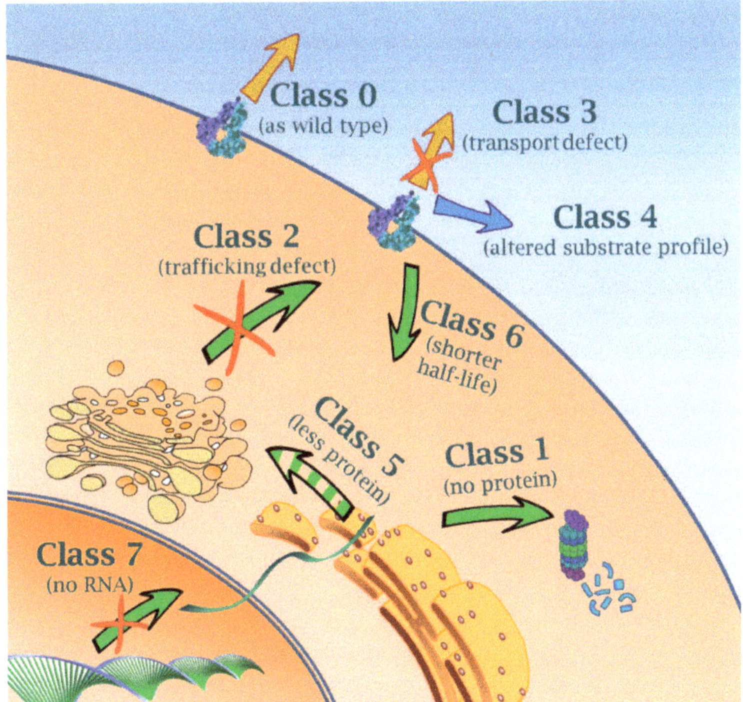

Figure 1. Classification of ABCG2 variants based on the cellular defects caused by the various mutations and polymorphisms. Green arrows—mRNA and protein trafficking, orange arrows—ABCG2-mediated transport, blue arrow—ABCG2-mediated transport with altered substrate preference.

Table 1. Proposed categorization of ABCG2 mutations and polymorphisms. Summary of the main features of the representative mutations of each group.

Class	Description	Variant	SNP Reference	Region	Global MAF	MAF in Asia	References	Assoc. with Gout	References for Gout Assoc.
Class 0	as wt	V12M	rs2231137	N terminal tail	0.158	0.19–0.33	[49,100–102]	dubious	no association in [56,103,105,106] but in [107]
		K360del	rs750972998	Linker	0.004	–	[103,104,107]	–	–
		T434M	rs769734146	TMH2	$<10^{-4}$	–	[104,107]	–	–
Class 1	no protein	Q126X	rs72552713	NBD	0.0012	0.019	[108–111]	yes	[56,103,106]
		R236X	rs140207606	NBD	0.0002	0.0005	[49,104,109,110]	yes **	[112]
		R113X, Q244X, R246X, G262X, E334X, Q531X	miscellaneous	various locations	–	–	[109,110,113]	–	–
		S340del	rs755318857	Linker	$<10^{-4}$	$<10^{-4}$	[109]	–	–
		L264Hfs	rs780593948	NBD	$<10^{-4}$	$<10^{-4}$	[109]	–	–
		R147W	rs372192400	NBD	0.0001	–	[104,107]	yes	[107]
		F208S	rs1061018	NBD (Walker B)	$<10^{-4}$	–	[95,111,114]	–	–
		R383C	–	Linker	–	–	[104]	yes **	[112]
Class 2	trafficking defect	Q141K	rs2231142	NBD	0.119	0.22–0.32	[32,33,95,100,115–118]	yes	[33,37,103,119]
		M71V	rs148475733	NBD	0.001	–	[104,112,120]	yes **	[112]
		F373C	rs752626614	Linker	$<10^{-4}$	–	[104]		–
Class 3	reduced transport activity	S248P	rs3116448	Linker	$<10^{-4}$	–	[95,121]	–	–
		S476P	not annotated	(CL1) TMH3	n.d.	–	[104,107]	–	–
		F489L	rs192169063	TMH3	0.001	0.005	[95,121–123]	–	–
		P269S	rs3116448	NBD:Linker	$<10^{-4}$	–	[100,103,124]	no	[103]
		A528T	rs45605536	TMH4	0.02 *	–	[114]	–	–
		I242T	not annotated	NBD	–	–	[125]	–	–
		K83M	–	NBD, Walker A	–	–	[9,126]	–	–
Class 4	altered substrate recognition	F431L	rs750568956	TMH2	$<10^{-4}$	–	[95,121,127,128]	–	–
		R482G	–	TMH3	–	–	[3,86,129–132]	–	–
		R482T	–	TMH3	–	–	[3,86,129–132]	–	–
Class 5	less protein	T153M	rs199753603	NBD:TM	0.0002	–	[104,107,128,133]	yes	[107]
		D296H	rs41282401	Linker	0.0002	0.02	[114]	–	–
		S441N	rs758900849	TMH2	$<10^{-4}$	–	[95,100,122,127]	–	–
		L525R	rs750568956	TMH4	0.014	–	[114,122]	–	–
		−30477C>G	rs2127861	promoter	–	–	[134]	–	–
		−15622C>T	rs7699188	promoter	–	–	[134]	–	–
		1143G>A	rs2622604	intron 2	–	–	[134]	–	–
Class 6	shorter PM half-life	?	–	–	–	–	–	–	–
Class 7	no RNA	?	–	–	–	–	–	–	–
others	ambiguous	N590Y	rs34264773	EL3	0.0004	–	[101,128,135]	–	–
	ambiguous	D620N	rs34783571	EL3	0.003	–	[107,113,135,136]	yes	[107]
	gain-of-function	I206L	rs12721643	NBD (Walker B)	0.0003	–	[128,135]	–	–

PM, plasma membrane; NBD, nucleotide-binding domain; TMH, transmembrane helix; CL, cytoplasmic loop; EL, extracellular loop; n.d., no data; ?, unknown; * in Caucasians, ** observed in a small cohort of patients with hyperuricemia or gout.

4.3. Class 1 Mutations

The next group is composed of mutations that impair protein production (Class 1). Among them, Q126X (rs72552713) is the most frequent with a MAF of 0.002 in the Asian population [108]. This mutation leads to an early stop codon; thus, no protein is produced. Several other, mainly nonsense mutations, e.g., R113X, R236X, R246X, G262X, E334X, and Q531X, result in a protein with severe structural and folding problems, and consequently in rapid protein degradation. Some other types of mutations, such as the deletion S340del, the frameshift L264Hfs, as well as the missense R147W, F208S, and R383C mutations [104,107], also belong to this group. A link between the lack of ABCG2 expression and the rare blood group Jr(a−) was established in 2012 [109,110]. Interestingly, individuals carrying Class 1 mutations on both alleles have no ABCG2 present in their RBC membranes, but exhibit no apparent signs of diseases. However, they can have transfusion reactions, and can be sensitive to drugs that are ABCG2 substrates, the concentration of which is normally controlled by the transporter.

4.4. Class 2 Mutations

Further embracing the CFTR classification, the next group of mutations (Class 2) consists of genetic alterations, usually mild mutations or polymorphisms, which result in a protein with impaired trafficking to the cell surface, while possessing more or less preserved transport function (recently reviewed in [137]). A typical of Class 2 variants is Q141K (rs2231142), which is the most frequent genetic variant of ABCG2 other than wild type. Its allele frequency is the highest in Southwest Asia (MAF values are between 0.222 and 0.319), whereas it is less common in the Caucasian populations (MAF = 0.107–0.119) and is rather rare in Africa (MAF = 0.009) [115,116]. Because of its frequency and its association with gout (see Section 6.1), ABCG2-Q141K is the most broadly studied variant of ABCG2. It has major folding and trafficking problems, undergoes partial degradation, thus exhibiting reduced overall and cell surface expression [32,33,95,100,115,117,118].

Using a flow cytometry-based screening method for reduced ABCG2 expression in RBCs, we have recently identified a polymorphic variant (M71V) with a character similar to that of Q141K [112,120]. The rare F373C missense mutation also belongs to this group, as the substrate stimulated ATPase activity and the specific transport activity of ABCG2-F373 are preserved, but its cell surface expression is diminished [104].

It should be noted that in most cases, not only the trafficking but also the transport function is affected by these mutations to certain extent. Nevertheless, they are classified into this group, because the defect caused by these mutations can—at least partially—be corrected by pharmacological chaperones (or correctors), whereas application of correctors for other types of mutations is usually unreasonable. The different approaches to rescue the various phenotypes will be discussed later.

4.5. Class 3 Mutations

The following group of ABCG2 mutations (Class 3) contains those that do not affect protein expression considerably, but impair transport function. At this point, the classification proposed here diverges from the categorization of the CFTR mutations, in which genetic alterations causing impaired gating are classified into Class 3, whereas those leading to diminished channel conductance are categorized into Class 4. This classification cannot be applied to ABCG2 for several reasons. Firstly, the transport mechanism of ABCG2 substantially differs from that of CFTR. Secondly, immense information has been accumulated on the mechanism of action of loss-of-function mutations in CFTR, whereas only limited data on ABCG2 mutations are available in this respect.

The genetic alterations in ABCG2 causing impaired transport activity are exemplified by S248P, P269S, F489L, and A528T [95,114,121–123]. The rare S476P also belongs to this group, although the transport function of ABCG2 bearing this mutation is only slightly reduced [104]. A recent addition to this class is I242T, which has normal protein expression and processing, while its transport activity (urate transport) is completely abolished [125]. A peculiar member of this group is K86M, an artificially generated mutation in the Walker A motif, which leads to a catalytic inactive transporter with close to normal cellular localization [9,126]. By means of these beneficial features, K86M is commonly employed in functional assays as a negative control.

It is worth noting that trafficking (Class 2) mutations cause reduced surface expression, and consequently lead to diminished drug efflux and increased drug sensitivity, to an apparent loss of function. Although generally speaking, the functionality of the transporter is affected in these cases, but the underling mechanism is rather different for Class 2 and Class 3 mutations. When a particular ABCG2 variant is evaluated solely on the basis of functionality in cellular systems, e.g., cell-based cytotoxicity or drug efflux assays, this distinction cannot be drawn. The various assessments to explore the particular impairments in the ABCG2 variants caused by mutations and polymorphisms will be discussed in

Section 5. It is noteworthy that Class 3 mutations can potentially be corrected by so-called potentiators, small molecules that are intended to restore the impaired transport function (see Section 7).

4.6. Class 4 Mutations

As mentioned above, Class 4 of ABCG2 mutations differs from that of CFTR variants. According to our proposed classification, this group collects the mutations resulting in a transporter with altered substrate specificity. These mutations are represented by the rare F431L (rs750568956, MAF < 0.0001). ABCG2-F431L is normally expressed in cells, localized to the plasma membrane, mediates porphyrin transports, and confers resistance to mitoxantrone [95,113], but these cells exhibit decreased resistance to methotrexate and various tyrosine kinase inhibitors [95,138]. Conflicting results on altered resistance to camptothecin analog SN-38 (the active metabolite of irinotecan) have been published in connection with this mutant [95,128].

Other members of this class are the R482G and R482T missense mutations, which have been identified from drug-selected cell lines, when ABCG2 was originally cloned [4,5]. In the beginning, the actual sequence of the wild type ABCG2 was ambiguous; causing some controversies in the initial characterization of the transporter, but later it was made clear that there is an arginine at position 482 in the wt ABCG2. Amino acid alterations at R482 strongly influence the substrate profile of the transporter, which is exemplified by methotrexate or uric acid being a substrate of the wt ABCG2 but not of the R482G variant; and inversely, doxorubicin or daunorubicin is exported by ABCG2-R482G, but not by the wild type [3,86,129–132].

4.7. Class 5 Mutations

The next group of mutations consists of those that cause diminished protein expression (Class 5). Class 1 and Class 5 mutations differ from one another in respect of that the former results in no protein expression, whereas the latter only cause reduced protein levels, therefore, different strategies are required for rescuing the phenotype. A representative of this group is T153M (rs199753603), which leads to diminished protein expression, but the smaller amount of protein expressed normally traffics to the cell surface and functions regularly [104,107,128,133].

Although we focus here mostly on the mutations in the coding region, it is worth mentioning that SNPs in the promoter region or in introns (potentially causing alternative splicing) can influence RNA stability and consequently lead to diminished protein expression. These are exemplified by −30477C>G (rs2127861), −15622C>T (rs7699188), and 1143G>A (rs2622604) [134]. Since the transcriptional regulation and RNA splicing is tissue-specific, the manifestation of these SNPs can vary from tissue to tissue. It is noteworthy that correction of aberrant splicing by RNA-based antisense oligonucleotide strategy has been proposed to restore CFTR level caused by splice-altering mutations [139].

4.8. Class 6, Class 7, and Ambigous Mutations

Class 6 and Class 2 mutations are related, as both groups contain genetic alterations leading to trafficking defects, but while Class 2 mutations cause diminished plasma membrane delivery, Class 6 SNPs result in less stable protein on the cell surface. Reduced steady-state plasma membrane level of a membrane protein can be due to either enhanced internalization or diminished recycling back to the cell surface, which are often concomitant with augmented protein degradation. Unfortunately, only limited information is available on the half-life of various ABCG2 mutants on the cell surface; therefore, categorization to this class is rather vague at the moment. The rare missense mutation V441N (rs758900849, MAF < 0.0001) has been reported to cause reduced protein stability [95,100,122,127], thus likely be identified as a Class 6 mutation.

Recently, a new category (Class 7) has been proposed for CFTR mutations, which class collects the so-called unrescuable genetic alterations [97]. This phenotype can be due to a large deletion or other severe mutations generating unstable mRNA. The sparse information available at the moment prevents to classify ABCG2 mutations into this group.

Classification of certain mutations is ambiguous because of the conflicting results published on the particular defect caused by the SNPs. For instance, ABCG2 carrying the relatively frequent D620N (rs34783571 MAF = 0.003) or the rare N590Y (rs34264773, MAF = 0.0004) missense mutation exhibit normal or in some cases even elevated cell surface expression; however, whether these mutations affect the transport function remains to be clarified [107,113,114,135,136,140]. An interesting addition to list of ABCG2 SNPs is the rare I206L gain-of-function mutation (MAF = 0.0003) [135].

5. Assessment of Mutation-Related Defects in ABCG2

Numerous in vitro assay systems and cellular models have been established to evaluate the expression and function of various ABC transporters, including ABCG2. Most of these test systems are intended to identify substrate or inhibitor molecules of a given transporter. These drug-screening methods have been carefully overviewed in comprehensive review papers [141,142]. Here, rather, the approaches capable of identifying the impairments caused by various types of mutations/polymorphisms will be discussed.

As mentioned earlier, several ABCG2 SNPs are frequently miscategorized, as the mutated protein is characterized by one or just few features. For instance, Class 2 (trafficking) variants are often called loss-of-function mutants, although their transport function is in fact preserved. To obtain a comprehensive view on the character of an ABCG2 variant, numerous assessments have to be performed. These include the examination of gene transcription, mRNA level and stability, overall protein expression level, protein stability and degradation, cellular localization and trafficking, half-life in the plasma membrane, ATPase activities, transport function (the specific activity!), and substrate profile. How these various cellular parameters are affected by mutations in each class is summarized in Table 2.

The transcriptional activity of ABCG2 variants bearing mutations in the promoter region can be assessed by nuclear run-on or GRO-Seq (Global Run-On sequencing) assay; the combined outcome of transcription and RNA stability can be detected by RNA-Seq (RNA sequencing) method, whereas splice variants can be determined using RNase protection assay. A detailed analysis demonstrated differential regulation and cell-type specific appearance of four 5′ untranslated exon variants of ABCG2 [143]. Recently, a straightforward test system based on a genome-edited reporter cell was developed to assess transcriptional regulation of ABCG2 [144]. In these cells, a coding sequence for eGFP was targeted to the translational start site of ABCG2. This reporter cell can be adjusted to examine the transcription of ABCG2 variants.

Standard molecular biological and biochemical methods, such as quantitative PCR and Western blotting are used to determine the mRNA and protein levels, respectively. The subcellular localization of the ABCG2 variants is regularly analyzed by immunostaining followed by microscopy, or specifically the plasma membrane expression of ABCG2 can be assessed by cell surface labeling followed by flow cytometry. The commonly used cell surface labeling, however, leaves the question open as to whether the abnormal plasma membrane expression is due to reduced overall expression or trafficking deficiencies. It is also worth noting that no specific markers for subcellular compartments are employed in the majority of studies examining the subcellular localization of an ABCG2 variant, leaving its exact location within the cell unambiguous. In addition, it is frequently disregarded that the trafficking machinery differs from cell type to cell type; thus, application of close-to-physiologic cellular models is highly encouraged for localization studies. The limitation of these widely used immunolabeling approaches is that they provide information only on the steady-state distribution of the protein, and its end-point accumulation in a particular cellular compartment may not reveal the real cause of mislocalization. Recently, we developed a dynamic, synchronization-based method for identifying the specific impairments of Class 2 (trafficking) ABCG2 mutants, Q141K and M71V [120]. Similarly, a dynamic approach is needed to evaluate the protein half-lives on the cell surface, but these kinds of studies on ABCG2 are rather rare. Binding of the ABGC2-specific antibody 5D3 has been demonstrated to induce internalization [145], as well as binding of ABGC2-specific inhibitors has been shown to promote lysosomal degradation of the transporter [146]. However, no data on the internalization rates or the plasma membrane half-lives of different ABCG2 variants have been published thus far.

Table 2. Different cellular parameters to be assessed, and their alterations caused by the mutations of various classes.

Cellular Parameter.	Class 0	Class 1	Class 2	Class 3	Class 4	Class 5	Class 6	Class 7
	as wt	No Protein	Trafficking Defect	Reduced Transport Activity	Altered Substrate Recognition	Less Protein	Shorter PM Half-Life	No RNA
Gene Transcription	+	+	+	+	+	+	+	+/−
mRNA Stability	+	+	+	+	+	+/−	+	+/−
mRNA Level	+	+	+	+	+	+/−	+	no
Protein Stability	+	reduced	+/−	+	+	reduced	+	N/A
Overall Protein Expression	+	no	+/−	+	+	reduced	+	N/A
Localization, Trafficking	normal	N/A	altered, impaired	normal	normal	normal	normal	N/A
Cell Surface Expression	normal	no	reduced	normal	normal	reduced	reduced	N/A
PM Half-Life	normal	N/A	N/A	normal	normal	normal	reduced	N/A
ATPase Activity	+	N/A	+	reduced	+	+	+	N/A
Transport (Specific Activity)	+	N/A	+	reduced	+	+	+	N/A
Substrate Profile	unchanged	N/A	+	+/−	altered	unchanged	unchanged	N/A

wt, wild type; PM, plasma membrane; +/−, normal or altered; N/A, not applicable. Color coding: green—not affected, normal, unchanged; red—altered, impaired; light red—can be altered; white—not applicable.

A wide variety of assay methods has been developed to evaluate the function and substrate recognition of ABCG2 variants (see in [141,142]). Some of these measurements are performed using membrane preparations or membrane vesicles containing the ABCG2 variant to be tested. Substrate molecules typically stimulate the ATPase activity of ABCG2 (like in many other ABC transporters), whereas inhibitors diminish the basal or substrate-stimulated activity. Compounds that modulate the function of the transporter, such as cholesterol, can also be identified using this method [147]. Photoaffinity labeling assay is also a membrane-based method, and indicates the interaction between a test compound and the transporter, while direct transport measurements can be performed employing inside-out membrane vesicles. These assays have the advantage that the specific activities can be determined in this way without being burdened by other factors like transcriptional differences, trafficking defects, or altered membrane half-lives. Moreover, using these approaches, substrate profiling for the different ABCG2 variants can be implemented.

Another group of functional assessments is cell-based methods, which employ cells expressing the ABCG2 variant to be tested. Cytotoxicity assay is the most commonly employed cell-based approach to investigate the transport activity of multidrug ABC transporters. The cell-killing effect of toxic compounds can be examined directly, but substances with no apparent toxic effect can be assayed through their modulatory effect on the cytotoxicity of a toxic drug. Measuring the efflux of a detectable substrate molecule from cells, or inversely, assessing its cellular accumulation (efflux and uptake assays) are other possibilities to determine the functional consequence of a mutation/polymorphism in ABCG2. Here too, the modulatory effect of other compounds on the cellular efflux or uptake of a detected substrate molecule allows mapping of potentially interacting substances. The so-called side-population assay is a representative of these cellular uptake methods [42]. The fluorescent dye Hoechst 33342 is a well-transported substrate of ABCG2, and its fluorescence undergoes a spectral shift upon binding to DNA. These properties make it possible to perform a dye exclusion assay for detecting a subset of cells with functional expression of ABCG2, e.g., stem cells, in a heterogeneous cell population by measuring cellular blue and red fluorescence in parallel [42,46,148,149]. Although cell-based assay systems are straightforward, and also allow substrate profiling, their limitation comes from the fact that the expression, localization, and cell surface stability of the tested ABCG2 variant indirectly affect the net outcome of these examinations.

It is also worth noting that in many heterologous cellular models, the transporter is considerably overexpressed, thus increasing the risk of experimental artefacts. Cell-based systems with a single or controlled number of transgene copies, such as Flp-In-293 system or transposon-based system, are more appropriate for such investigations [14,18,113,117,118,150]. For instance, an initial study using cellular models over-expressing ABCG2 reported that mutations at the glycosylation site (N596) do not affect trafficking of ABCG2 [12], but a subsequent study employing a single-copy Flp-In-293 system demonstrated a harmful effect of N596 mutations and established a stabilizing role for N-glycosylation [14]. Cellular models stably expressing the ABCG2 variant are commonly preferred over the transient systems, even though stable expression of a transgene allows us to investigate solely the steady-state distribution and function; furthermore, it may be accompanied with compensatory mechanisms. Transient and inducible expression systems should also be acknowledged, especially when dynamic cellular events, such as intracellular routing, are to be studied. Moreover, synchronous release methods, such as the RUSH (retention using selective hooks) system, are even more adequate for exploring the trafficking properties of different ABCG2 variants [120].

In addition to these 'wet lab' approaches, the effect of mutations and polymorphisms can be investigated by various in silico methods either by modelling their impact on the 3D structure of the transporter or by predicting their functional consequences, e.g., implicating an effect on protein stability, substrate binding, or intramolecular communication, etc. Although these are intriguing issues, this topic is beyond the scope of this review; thus, it not discussed here. In any case, structural analyses and molecular dynamic simulations in connection with ABCG2 mutations have recently been the subject of a comprehensive overview [30].

6. Medical Conditions Associated with ABCG2 Mutations and Polymorphisms

6.1. The Role of ABCG2 Variants in Hyperuricemia and Gout

One of the important roles of ABCG2 is the clearance of uric acid, the end-product of purine metabolism. Interestingly, uric acid was not identified as a physiologically relevant endogenous substrate of ABCG2 until genome-wide association (GWA) studies revealed the association between the Q141K polymorphism in the *ABCG2* gene and gout [151], a disease characterized by inflammatory arthritis of the joints caused by urate deposition and crystal formation in the synovial fluid. This condition can develop as a result of either hepatic overproduction or diminished excretion of uric acid. The Q141K polymorphism has been established as one of the strongest genetic determinants for high serum urate levels (hyperuricemia) and gout development [37,56,103,105,106,119,152]. Surprisingly, Q141K polymorphism (and its murine ortholog Q140K) diversely affected ABCG2 function in intestine and kidney, i.e., while ABCG2-mediated intestinal urate clearance was substantially reduced (or abolished), the function of ABCG2 was preserved in the kidney, suggesting a tissue-specific regulation/trafficking and differential pathology of this ABCG2 variant [58]. Ironically, another GWA study demonstrated an association between the Q141K polymorphism and a worse response to allopurinol [153], which is the first-line treatment for chronic gout. Allopurinol and its active metabolite, oxypurinol, are intended to lower uric acid levels by inhibiting xanthine oxidase, the major enzyme for the production of uric acid. Allopurinol and oxypurinol have been reported to be transported substrates of ABCG2; thus, they are retained in cells expressing the Q141K ABCG2 as compared to wt-expressing cells. A subsequent study demonstrated that ABCG2 transports oxypurinol, but not allopurinol itself [154]. Further pharmacogenomic analyses established the link between presence of the Q141K allele and increased risk of poor response to allopurinol [155–157], but the underlying mechanism, how Q141K influences allopurinol disposition and the clinical response, is yet to be clarified.

Several other ABCG2 mutations/polymorphisms have been identified as genetically associated risk factors for gout incidence (see Table 1). These include Q126X, R147W, T153M, and D620N [56,103,105–107]. No association between V12M and increased risk of gout have been established [56,103]; indeed, one study reported a protective effect of this SNP on gout susceptibility [105]. Recently, using blood samples from patients with hyperuricemia or gout, in addition to Q141K, we found three other mutations (M71V, R236X, and R383C) associated with reduced protein levels in the RBC membrane [112]. It is worth mentioning that most of these ABCG2 SNPs connected with hyperuricemia or gout belong to Class 1 or Class 2.

6.2. Modulatory Effect of ABCG2 Variants on Drug Pharmacokinetics

As discussed earlier, ABCG2 residing in various physiological barriers plays a pivotal role in tissue and cellular protection by controlling the uptake, distribution, and excretion of potentially toxic endogenous and exogenous substances. Mutations/polymorphisms affecting the functionality of the transporter by any means may also influence the serum level and/or the pharmacokinetic parameters of drugs that are ABCG2 substrates [96,158]. There are two consequences of altered pharmacokinetics: (i) the clinical response to a given drug can be modulated, and (ii) toxic reactions can arise due to shifted drug concentrations or drug–drug interactions. In particular, highly toxic drugs, like chemotherapeutic agents, can evoke adverse drug reaction (ADR) even at the usual dosage, when a polymorphism in transporters alters drug distribution. ABCG2 mutations/polymorphisms in connection with anti-cancer drug-elicited ADRs will be discussed in the following subsection. When drug concentrations (or exposures) are elevated due to the presence of a polymorphic variant, drug dosage should be adjusted accordingly to attain adequate response to treatment but to minimize side effects. However, compiling of dependable personalized treatment protocols demands not only reliable genetic analyses but also well-established knowledge of genotype–phenotype–pharmacokinetics relationships.

The effect of the frequent Q141K ABCG2 variant on drug pharmacokinetics has been extensively investigated, whereas only limited information has been acquired on other ABCG2 SNPs. Interestingly, the alterations in the disposition of several drugs correlate with the allele frequencies of this SNP in various ethnic groups [159]. Based on previous pharmacokinetic and pharmacodynamics data, a recommendation has been made to incorporate examination of ABCG2 Q141K in recent and future drug development [65]. Q141K has been demonstrated to modulate the pharmacokinetics of several types of drugs, including chemotherapeutic agents (discussed later), statins, disease-modifying anti-rheumatic drugs (DMARD), anticoagulants, and anti-viral medications. The pharmacokinetic parameters [the area under the curve (AUC) and/or the maximum concentration (C_{max})] of rosuvastatin, simvastatin, atorvastatin, and fluvastatin was elevated in patients carrying the Q141K allele [160–164]. Therefore, this polymorphism has also been implicated in increased the risk of statin-induced myopathy [165]. Q141K was also associated with higher AUC or C_{max} of DMARDs, such as sulfasalazine and teriflunomide [166–169], as well as of anti-HIV drugs, such as dolutegravir and retegravir [170,171]. It is worth mentioning that in many cases, association was seen only with the homozygous genotype, or with the group involving both hetero- and homozygotes. With regard to the other relatively frequent polymorphisms, V12M has demonstrated not to influence the pharmacokinetics of fluvastatin [162], whereas Q126X was associated with altered disposition of sulfasalazine [169].

Interestingly, a few studies have reported that the polymorphism-related alterations in pharmacokinetic parameters were primarily due to increased intestinal absorption rather than decreased renal excretion [159,171], although reduced hepatic clearance may also contribute to lower C_{max} values [166]. These notions are in concert with the differential pathology of Q141K in the intestine and the kidney observed in gout patients [58]. As discussed earlier, Q141K increases the risk of poor response to allopurinol by means of a not fully understood mechanism. However, further studies focusing on the oxypurinol pharmacokinetics and involving a large cohort of allopurinol-treated gout patients are needed to elucidate causative relationships.

6.3. Significance of ABCG2 Mutations/Polymorphisms in Cancer Therapy

Numerous in vitro studies have demonstrated altered efflux of chemotherapeutic agents and reduced drug resistance in cancer cells expressing ABCG2 carrying various SNPs. While an unambiguous role for Q141K has been established, the results on the effect of the V12M variant were conflicting [100,115,133,172]. It is important to note that the specific activity of both V12M and Q141K, i.e., transport activity normalized by the expression level, remained unaltered [100], confirming that these SNPs do not affect the actual transport function. ABCG2-Q141K has been shown to diminish drug efflux and/or to increase sensitivity to various anti-cancer drugs, including methotrexate, numerous TKIs (gefitinib, erlotinib, lapatinib, imatinib, dasatinib, nilotinib), and the topoisomerase I inhibitor indolocarbazole [100,114,115,133,172,173]. In addition to genetic alterations, epigenetic modifications could also modulate drug responses. It has been demonstrated that promoter hypermethylation, which frequently occurs in tumor cells, leads to repressed transcription of the *ABCG2* gene, lowered protein level, and consequently increased drug sensitivity [174].

An increasing number of clinical studies reported that ABCG2 polymorphisms modulate the pharmakinetics of chemotherapeutic drugs and increase the risk of drug-related adverse reactions. Studies rarely report altered drug response or clinical outcomes, but toxic side effects due to elevated drug concentrations or longer exposures. Higher accumulation of the small molecule TKI gefitinib was observed in patients heterozygous for the Q141K variant [81]. In accordance with this finding, higher incidence of gefitinib-induced diarrhea was found in Q141K heterozygotes [175], although another study found no association between susceptibility to gefitinib-induced ADR and the Q141K (or Q126X) polymorphism [176]. No difference was found in the therapeutic outcomes (response and survival) in diffuse large B-cell lymphoma patients carrying Q141K or V12M polymorphism, when treated with rituximab plus cyclophosphamide/doxorubicin/vincristine/prednisone (R-CHOP) regimen, but chemotherapy-induced diarrhea was associated with the Q141K genotype [177]. Similarly, a higher incidence of sunitinib-induced severe thrombocytopenia was observed in renal cell carcinoma (RCC) patients carrying the Q141K allele [178]. A case report also documented that an RCC patient homologous for Q141K suffered from toxic side effects of sunitinib treatment, such as severe thrombocytopenia, transaminase elevation, severe hypoxia due to pleural effusion and pulmonary edema, in parallel with efficacious treatment of the tumor [179]. Although the expression of ABCG2-Q141K in cancer cells resulted in elevated sensitivity to imatinib [114,173], the in vivo effect of Q141K on imatinib accumulation/response is controversial [180–185].

Similarly, the results on the role of this SNP in the pharmacokinetics of camptothecin analogs, such as diflomotecan, 9-aminocamptothecin, topotecan, and irinotecan, are conflicting [186–192]. An interesting dual role for the Q141K polymorphism was found in prostate cancer patients. On one hand, this ABCG2 variant results in elevated cellular retention of folate, which is a rate limiting factor for prostate cancer cell proliferation, thus leading to higher risk for tumor reoccurrence after prostatectomy in patients, who received no drug treatment. On the other hand, docetaxel-treated prostate cancer patients carrying Q141K SNP have a longer survival time because of the reduced drug efflux [193].

Besides Q141K, some other frequent polymorphisms, such as V12M and Q126X, have been included in a limited number of clinical studies investigating the effect of ABCG2 SNPs on cancer chemotherapy and therapy-related toxicity. V12M did not influence clinical outcomes and has no association with ADR; on the contrary, Q126X promoted toxic side effects of gefitinib [176,177,192]. Only sparse information is available on ABCG2 minor variants in connection with tumor drug therapy. The SNPs in the promoter region (-15622C/T) and in intron 1 (1143C/T) conferring reduced ABCG2 levels (Class 5 mutations) elevated pharmacokinetic parameters (AUC and C_{max}) of erlotinib [194], and increased the incidence of gefitinib-induced diarrhea in non-small-cell lung cancer patients without affecting the clinical outcomes [195]. An SNP in intron 11 (G>A, rs4148157, MAF = 0.096) also altered the pharmacokinetics (absorption rate constant and maximum concentration) of orally administered topotecan in infants and very young children with brain tumors [196].

As mentioned earlier, ABCG2-inhibiting drugs can also modulate the distribution and toxicity of medications, underscoring the importance of ABCG2 in drug–drug or drug–food interactions. On the other hand, mutations and polymorphisms may potentially alter the potency and/or efficacy of these inhibitors, further complicating the evaluation of drug interactions. Nevertheless, this aspect of ABCG2 polymorphic variants remains to be determined.

6.4. Other Disease Conditions in Connection with ABCG2 Variants

Abcg2-deficent mice showed no signs of any noticeable phenotype until they were exposed to light, which in turn induced severe phototoxic lesions on the skin of the animals [67]. This led to the discovery that pheophorbide A, the breakdown product of chlorophyll, is an ABCG2 substrate, and the transporter restricts the intestinal uptake of this potential toxic compound. The level of the heme precursor protoporphyrin IX was also massively increased in the RBCs of Abcg2 knockout mice, implying a role for ABCG2 in heme homeostasis. This was further supported by the observations that ABCG2 is upregulated during erythroid differentiation and lowers the PPIX levels in erythroid cells [50]. A comprehensive in vitro study demonstrated that numerous ABCG2 polymorphisms, such as Q126X, F208S, S248P, E334X, S441N, and F489L, cause abrogated porphyrin transport regarding specific (normalized) activities [113]. It should, however, be noted that among these mutants, the expression level of the Class 1 variants (Q126X, F208S, and E334X) was practically zero, and thus no transport activity is expected anyway. These data and the observations with the Abcg2 knockout mice implicate a role for ABCG2 variants in erythropoietic protoporphyria (EPP), a disease of the heme biosynthesis pathway. EEP is caused by either genetic determinant (mutations in the heme biosynthesis enzymes), or by exposures to toxins or drugs, such as rifampicin and isoniazid. Accumulation of PPIX in EPP patients causes hepatotoxicity and phototoxicity primarily in the skin. A recent study elegantly demonstrated that Abcg2-deficency prevents mice from EPP-associated phototoxicity and hepatotoxicity by altering the disposition of PPIX [68]. In this case, this phototoxin is mostly retained in the RBCs, thus causing a reduced plasma level and preventing PPIX accumulation in the skin and the liver/bile. This surprising result clearly elucidates the multifunctionality of ABCG2: on one hand, it restricts the intestinal absorption of the xenobiotic phototoxin pheophorbide A, preventing light-induced skin damage; on the other hand, it facilitates release of the endogenous phototoxin PPIX from RBCs, contributing to the toxic effects, when PPIX is in excess (e.g., in EPP patients). Seeing pharmacogenomic or human in vivo data in connection with ABCG2 genotypes and ECC would be rather intriguing.

Several ABC transporters, including ABCA1, MDR1/ABCB1, MRP1/ABCC1, ABCG2, and ABCG4, have been implicated in the pathogenesis of Alzheimer's disease (AD) [197,198], a progressive neurodegenerative disorder characterized by the deposition of amyloid-β (Aβ) peptides in the brain. Both in vitro and in vivo data demonstrated an ABCG2-dependent efflux of Aβ1-40, suggesting that ABCG2 at the blood-brain barrier prevents Aβ peptide from entering the brain [198,199]. In addition, an upregulation of ABCG2 has been found in the brains of AD patients with cerebral amyloid angiopathy [199], which was also reflected by increased ABCG2 expression levels observed in the RBCs of late-onset AD patients [69]. The role of ABCG2 in AD can be modulated by the mutations/polymorphisms. A lower prevalence of the Q141K was seen in late-onset AD patients, and interestingly increased susceptibility to AD was associated with the Q141K allele containing genotypes.

Besides AD, another neurodegenerative disorder, Parkinson's disease (PD), has also been linked to ABCG2. Previously, higher levels of the natural antioxidant urate in the serum or the cerebrospinal fluid were associated with the clinical decline of PD patients [200,201]. Accordingly, the Q141K polymorphism was correlated with later disease onset of PD [202]; however, a recent study did not find association of the Q141K allele (or the high plasma level of urate) with the disease [203].

Intrauterine growth restriction (IUGR) is one of the most common forms of pregnancy complications. The idiopathic form of IUGR has also been linked to ABCG2, which is abundantly expressed in the placenta. Markedly decreased ABCG2 expression levels were found in placentas from IUGR pregnancies [204]. Based on this observation and the fact that trophoblasts in IUGR are subjected to excessive oxidative stress and apoptotic signals, a role for ABCG2 has been proposed in the protection of trophoblasts against stress-induced apoptosis, although the actual underlying mechanism is yet to be elucidated.

Recently, it has been reported that the expression level of ABCG2 is reduced in the RBC membranes of patients with type 2 diabetes and carrying the Q141K polymorphism, whereas this difference was not observed in patients homozygous for the wt ABCG2 [205]. The mechanism here too is yet to be clarified, but this observation clearly indicates the differential regulation of the wt and the Q141K polymorphic ABCG2 variants.

The actual role of ABCG2 in various stem cell types is still elusive, but it has become commonly accepted that ABCG2 has a protective role in stem cells, as they are exceptionally sensitive to environmental stresses [42,47,148,206]. The high-level expression of ABCG2 in pluripotent stem cells rapidly declines during differentiation [43], but may also regain in differentiated progeny cells that typically express ABCG2, e.g., in hepatocytes [207,208]. In addition to cell differentiation, various environmental impacts can elicit a drop in ABCG2 expression [43,44]. Not only its overall expression, but also its localization can be altered in response to stresses, e.g., mild oxidative stress evokes a reversible internalization of the transporter in pluripotent stem cells [209]. Nevertheless, little is known about the impact of various mutations/polymorphism in ABCG2 on its role in stem cell defense mechanisms. Exceptional medical relevance has the ABCG2 expression in cancer stem cells or drug-tolerant persisters, which can rapidly adapt to chemotherapy and repopulate the tumor [45–48,210]. The presence of these cell subpopulations in the tumors is an increasingly acknowledged reason for chemotherapy failures.

7. Efforts to Improve Impaired Trafficking or Function of ABCG2 Variants

Advancements in rescuing mutation-derived phenotypes of ABC transporters are largely driven by the research on CFTR, as cystic fibrosis is a frequent hereditary disease with adverse outcomes. The most frequent mutation in CFTR is F508del, which accounts for roughly 90% of all CF cases (~40–45% of the patients are homozygous for this mutation). F508del is considered to be a Class 2 mutation, since the majority of the protein bearing this mutation is retained in the endoplasmic reticulum (ER) by the ER quality control mechanism; nevertheless, the channel opening rate (gating) and the channel selectivity (equivalent to substrate specificity) are also affected by this mutation. Excessive efforts have been made to correct the various defects in CFTR, especially F508del [97]. A wide variety of molecules is employed to rescue not only CFTR mutants but also other deficient ABC transporters (reviewed in [211]).

Numerous small molecules can modulate the trafficking of membrane proteins. The inhibitors of cellular trafficking include fungal antibiotics, such as Brefeldin A, a blocker of the ER to Golgi transfer, Concanamycin A and Destruxin B, inhibitors of V-ATPases, as well as the immunosuppressant mycophenolic acid, a blocker of de novo GMP synthesis. Pharmaceutical drugs that are intended to restore trafficking and cell surface expression of the impaired transporter are named correctors, but are also called chemical or pharmacological chaperones, since most of these variants endure protein folding problems. The application of the correctors is reasonable for the phenotype rescue of Class 2 or conceivably Class 5 mutations. Corrector compounds include non-specific chemicals, such as glycerol or DMSO, but also small molecules with more specific effect. Corr-4a, Lumacaftor (VX-809), VRT-325, and recently Elexacaftor (VX-445) have been specifically developed to promote cell surface delivery of trafficking mutants of CFTR [212–215]. Surprisingly, Lumacaftor was effective in ameliorating cell surface delivery of trafficking-impaired variants of other ABC transporters, such as ABCA4 [216,217]. Moreover, VRT-325 has been shown to rescue the trafficking deficiency of ABCG2-Q141K, and restore urate efflux from cell expressing this ABCG2 variant [33].

The phenotype rescuing capability of the histone deacetylase inhibitor 4-phenylbutirate (4-PBA) has also been demonstrated for numerous ABC transporters, including CFTR [218], MDR1/ABCB1 [219], MDR3/ABCB4 [219,220], BSEP/ABCB11 [221], MRP6/ABCC6 [222], and ABCG2 [32,33,112]. The rescue effect of 4-PBA was attributed to stimulation of the expression of heat shock protein 70, a chaperon protein, which facilities folding of partially folded protein, thus preventing elimination by the ER-associated degradation (ERAD) system. Besides 4-PBA, several other histone deacetylase inhibitors, such as romidepsin, vorinostat, panobinostat, and valproic acid, have been demonstrated to restore cell surface expression of the mislocalized ABCG2-Q141K variant [118]. The underlying mechanism proposed here does not involve facilitated protein processing, but these histone deacetylase inhibitors have been implicated to block the dynein/microtubule retrograde transport, thus preventing the mutated ABCG2 from trafficking from the cell surface to aggresomes, the perinuclear vimentin-coated inclusion bodies [118]. Aggresomes are close to the centrosome and facilitate protein degradation by the autophagy pathway [223], and similar to ABCG2-Q141K, CFTR-F508del also tends to accumulate in this compartment [224]. In concert with the mechanism delineated above, colchicine, a blocker of microtubule polymerization, has been shown to elevate plasma membrane expression of ABCG2-Q141K by inhibiting its traffic to aggresomes [118]. This observation also has a medical relevance, since colchicine is a drug commonly used in gout therapy. It is worth also noting that the anti-cancer agent mitoxantrone also promote cellular processing of various ABCG2 variants [118,225], exemplified by Q141K, the accumulation of which in aggresomes was prevented by mitoxantrone [118].

Several types of ABCG2 defects involve protein degradation. Class 1 mutations cause major protein folding and stability problems; thus, ABCG2 variants carrying this type of mutations (e.g., Q126X) undergo rapid elimination by the ERAD system. Similarly, a fraction of Class 2 variants is subjected to degradative mechanisms; therefore, the overall steady-state protein expression levels of Class 2 variants (e.g., Q141K) are usually lower than that of the wild type. However, pharmacological chaperones can facilitate proper folding and save these variants from protein degradation. If the defect of a Class 5 variant is not due to diminished transcription or mRNA stability, but to reduced protein stability (e.g., S441N), it also undergoes at least partial degradation. Finally, Class 6 variants possessing shorter plasma membrane half-lives are also subjected to protein degradation, although they are primarily eradicated by the lysosomal system, whereas members of the Classes 1, 2, and 5 are rather eliminated by the proteasomal degradation. Most likely autophagosomal degradative mechanisms are also involved in the removal of deficient ABCG2 variants; however, little is known about contribution of this pathway. Blocking the various protein degradation mechanisms is also an option to restore reduced protein levels for these types ABCG2 mutants. According to the considerations above, the expression levels of Class 2 and Class 5 variants are likely to be improved by proteasome inhibition, whereas Class 6 variants can supposedly be corrected by blocking lysosomal degradation. However, this theoretical segregation does not entirely stand for tangible ABCG2 variants. For instance, the expression level of the ABCG-Q141K is increased 2-fold either by the proteasome inhibitor MG132 or by the lysosomal degradation blocker bafilomycin [118], indicating the involvement of both degradative pathways. Moreover, 3-methyladenine, an autophagosome inhibitor also cause a 3-fold increase in the expression level of the Q141K variant [118], implying a marked role for autophagosomal degradation in ABCG2 proteostasis. These findings, along with the observation that colchicine increased cell surface expression level of ABCG-Q141K suggest that a considerable fraction of these variant passes the central quality control mechanisms and reach the plasma membrane. This notion is further supported by the observation that, despite being subjected to proteasomal and autophagosomal degradations, the majority of the Q141K-ABCG2 variant is fully glycosylated, once having traversed the Golgi apparatus. This is in sharp contrast to what is observed with CFTR F508del or MDR1/ABCB1 mutants, which fail to become glycosylated and are entirely retained in the ER or targeted to proteasomal degradation [226–228]. It is worth noting that a large fraction of wt CFTR gets misfolded and consequently degraded [229,230], whereas wt ABCG2 hardly undergoes any proteasomal [14,112,225,231] or autophagosomal [118] degradation. The medical use of these inhibitors of protein degradation is, however, somewhat limited, as a general blocking of these vital cellular processes can be detrimental, despite the fact that proteasome inhibitors, such as MG132 or bortezomibe (Velcade), are used in cancer treatment, mainly as a component of combination therapies.

To restore the impaired transport function caused by Class 3 (and Class 4) mutations, so-called potentiator compounds are used. One of these small molecules is Ivacaftor (VX-770), which effectively restores the chloride channel activity of by loss-of-function CFTR mutants [213,214,232]. Although this drug has also been specifically developed for CFTR and is presumed to be selective for CFTR, surprisingly, it also rescues the phenotype caused by loss-of-function mutations in MDR3/ABCB4 [233], ABCA4 [216,217], and ABCB11 [234]. To our knowledge, no results on the correction of ABCG2 loss-of-function mutations by Ivacaftor have been published thus fur. The underlying mechanism is not quite understood, but Ivacaftor is presumed to somehow stabilize the structure of the impaired transporter. This drug has also been demonstrated to stimulate the ATPase activity of MDR1/ABCB1 [235], but most likely, this is not related to its rescuing effect, but rather indicates that Ivacaftor is a transported substrate. This notion is supported by the further observation that Ivacaftor competitively inhibited the MDR1-mediated transport of the fluorescent dye Hoechst 33,342 [235]. Nevertheless, further studies are needed to understand the potentiator-transporter interaction better and to elucidate the mechanism how Ivacaftor potentiates various ABC proteins.

8. Conclusions

Various mutations and polymorphisms in the *ABCG2* gene diversely affect the structure, stability, trafficking, and function of this multifunctional transporter protein. Gaining insight into the cellular fate of different ABCG2 variants promotes establishing specific interventions to overcome the medical condition caused by either the wild type or the mutated forms of ABCG2. The classification of ABCG2 mutations proposed here is based on the cellular features of various defects; thus, the variants categorized into the same class require similar kinds of efforts to rescue the phenotype. A broad and well-established knowledge base is a prerequisite for prudent therapies; thus, our better understanding of the genetics, biochemistry, and cell biology of ABCG2 variants bearing various mutations could help to establish effective personalized treatments that consider the patients' genetic background.

Funding: This work has been supported by National Research, Development and Innovation Office, Hungary (grant numbers: OTKA_K 128123 and FIEK_16-1-2016-0005).

Institutional Review Board Statement: Not applicable.

Informed Consent Statement: Not applicable.

Data Availability Statement: Not applicable.

Acknowledgments: The author is grateful to Ágnes Enyedi, Balázs Sarkadi, and Tamás Hegedűs for their advices and help in revising the manuscript. The author also thanks Rita Dóra Homolya for her contribution to the graphical design.

Conflicts of Interest: The author declares no conflict of interest.

Abbreviations

4-PBA	4-phenylbutirate
ABC	ATP-binding cassette
AD	Alzheimer's disease
ADME-Tox	absorption, distribution, metabolism, excretion, and toxicity
ADR	adverse drug reaction
AUC	area under the curve
$A\beta$	amyloid-β
BCRP	breast cancer resistance protein
BtuCD	bacterial vitamin B_{12} importer
CF	cystic fibrosis
CFTR	cystic fibrosis transmembrane regulator
CL	cytoplasmic loop
C_{max}	maximum concentration
cryo-EM	cryogenic electron microscopy
DMARD	disease-modifying anti-rheumatic drug
EL	extracellular loop
EPP	erythropoietic protoporphyria
ER	endoplasmic reticulum
ERAD	ER-associated degradation
GRO-Seq	Global Run-On sequencing
GWA	genome-wide association
IUGR	intrauterine growth restriction
MAF	minor allele frequency
MDR	multidrug resistance
MDR1	multidrug resistance protein 1, P-glycoprotein, ABCB1
MRP1	multidrug resistance-related protein 1, ABCC1
MXR	mitoxantrone resistance protein
NBD	nucleotide-binding domain
PD	Parkinson's disease
PM	plasma membrane
PPIX	protoporphyrin IX
RBC	red blood cell
RCC	renal cell carcinoma
R-CHOP	rituximab plus cyclophosphamide/doxorubicin/vincristine/prednisone
RNA-Seq	RNA sequencing
RUSH	retention using selective hooks
SNP	single-nucleotide polymorphism
SUR1	sulfonylurea receptors
TKI	tyrosine kinase inhibitor
TMD	transmembrane domain
TMH	transmembrane helix
wt	wild type

References

1. Ogasawara, F.; Kodan, A.; Ueda, K. ABC proteins in evolution. *FEBS Lett.* **2020**, *594*, 3876–3881. [CrossRef] [PubMed]
2. Thomas, C.; Aller, S.G.; Beis, K.; Carpenter, E.P.; Chang, G.; Chen, L.; Dassa, E.; Dean, M.; Van Hoa, F.D.; Ekiert, D.; et al. Structural and functional diversity calls for a new classification of ABC transporters. *FEBS Lett.* **2020**, *594*, 3767–3775. [CrossRef] [PubMed]
3. Sarkadi, B.; Homolya, L.; Szakács, G.; Váradi, A. Human multidrug resistance ABCB and ABCG transporters: Participation in a chemoimmunity defense system. *Physiol. Rev.* **2006**, *86*, 1179–1236. [CrossRef] [PubMed]
4. Doyle, L.A.; Yang, W.; Abruzzo, L.V.; Krogmann, T.; Gao, Y.; Rishi, A.K.; Ross, D.D. A multidrug resistance transporter from human MCF-7 breast cancer cells. *Proc. Natl. Acad. Sci. USA* **1998**, *95*, 15665–15670. [CrossRef]
5. Litman, T.; Brangi, M.; Hudson, E.; Fetsch, P.; Abati, A.; Ross, D.D.; Miyake, K.; Resau, J.H.; Bates, S.E. The multi-drug-resistant phenotype associated with overexpression of the new ABC half-transporter, MXR (ABCG2). *J. Cell. Sci.* **2000**, *113*, 2011–2021.
6. Robey, R.W.; Polgar, O.; Deeken, J.; To, K.W.; Bates, S.E. ABCG2: Determining its relevance in clinical drug resistance. *Cancer Metastasis Rev.* **2007**, *26*, 39–57. [CrossRef]
7. Robey, R.W.; Pluchino, K.M.; Hall, M.D.; Fojo, A.T.; Bates, S.E.; Gottesman, M.M. Revisiting the role of ABC transporters in multidrug-resistant cancer. *Nat. Rev. Cancer* **2018**, *18*, 452–464. [CrossRef]
8. Allikmets, R.; Schriml, L.M.; Hutchinson, A.; Romano-Spica, V.; Dean, M. A human placenta-specific ATP-binding cassette gene (ABCP) on chromosome 4q22 that is involved in multidrug resistance. *Cancer Res.* **1998**, *58*, 5337–5339.
9. Orbán, T.I.; Seres, L.; Özvegy-Laczka, C.; Elkind, N.B.; Sarkadi, B.; Homolya, L. Combined localization and real-time functional studies using a GFP-tagged ABCG2 multidrug transporter. *Biochem. Biophys. Res. Commun.* **2008**, *367*, 667–673. [CrossRef]
10. Hegyi, Z.; Homolya, L. Functional cooperativity between ABCG4 and ABCG1 isoforms. *PLoS ONE* **2016**, *11*, e0156516. [CrossRef]
11. McDevitt, C.A.; Collins, R.; Kerr, I.D.; Callaghan, R. Purification and structural analyses of ABCG2. *Adv. Drug Deliv. Rev.* **2009**, *61*, 57–65. [CrossRef] [PubMed]
12. Diop, N.K.; Hrycyna, C.A. N-linked glycosylation of the human ABC transporter ABCG2 on asparagine 596 Is not essential for expression, transport activity, or trafficking to the plasma membrane. *Biochemistry* **2005**, *44*, 5420–5429. [CrossRef]
13. Mohrmann, K.; Van Eijndhoven, M.A.J.; Schinkel, A.H.; Schellens, J.H.M. Absence of N-linked glycosylation does not affect plasma membrane localization of breast cancer resistance protein (BCRP/ABCG2). *Cancer Chemother. Pharmacol.* **2005**, *56*, 344–350. [CrossRef] [PubMed]
14. Nakagawa, H.; Wakabayashi-Nakao, K.; Tamura, A.; Toyoda, Y.; Koshiba, S.; Ishikawa, T. Disruption of N-linked glycosylation enhances ubiquitin-mediated proteasomal degradation of the human ATP-binding cassette transporter ABCG2. *FEBS J.* **2009**, *276*, 7237–7252. [CrossRef]
15. Wakabayashi-Nakao, K.; Tamura, A.; Furukawa, T.; Nakagawa, H.; Ishikawa, T. Quality control of human ABCG2 protein in the endoplasmic reticulum: Ubiquitination and proteasomal degradation. *Adv. Drug Deliv. Rev.* **2009**, *61*, 66–72. [CrossRef] [PubMed]
16. Henriksen, U.; Fog, J.U.; Litman, T.; Gether, U. Identification of intra- and intermolecular disulfide bridges in the multidrug resistance transporter ABCG2. *J. Biol. Chem.* **2005**, *280*, 36926–36934. [CrossRef]
17. Kage, K.; Fujita, T.; Sugimoto, Y. Role of Cys-603 in dimer/oligomer formation of the breast cancer resistance protein BCRP/ABCG2. *Cancer Sci.* **2005**, *96*, 866–872. [CrossRef]
18. Wakabayashi, K.; Nakagawa, H.; Adachi, T.; Kii, I.; Kobatake, E.; Kudo, A.; Ishikawa, T. Identification of cysteine residues critically involved in homodimer formation and protein expression of human ATP-binding cassette transporter ABCG2: A new approach using the flp recombinase system. *J. Exp. Ther. Oncol.* **2006**, *5*, 205–222.
19. Dawson, R.J.P.; Locher, K.P. Structure of a bacterial multidrug ABC transporter. *Nat. Cell Biol.* **2006**, *443*, 180–185. [CrossRef] [PubMed]
20. Ward, A.; Reyes, C.L.; Yu, J.; Roth, C.B.; Chang, G. Flexibility in the ABC transporter MsbA: Alternating access with a twist. *Proc. Natl. Acad. Sci. USA* **2007**, *104*, 19005–19010. [CrossRef] [PubMed]
21. Lee, J.-Y.; Kinch, L.N.; Borek, D.M.; Wang, J.; Wang, J.; Urbatsch, I.L.; Xie, X.-S.; Grishin, N.V.; Cohen, J.C.; Otwinowski, Z.; et al. Crystal structure of the human sterol transporter ABCG5/ABCG8. *Nat. Cell Biol.* **2016**, *533*, 561–564. [CrossRef] [PubMed]
22. László, L.; Sarkadi, B.; Hegedűs, T. Jump into a new fold—A homology based model for the ABCG2/BCRP multidrug transporter. *PLoS ONE* **2016**, *11*, e0164426. [CrossRef] [PubMed]
23. Ferreira, R.J.; Bonito, C.A.; Cordeiro, M.N.D.S.; Ferreira, M.-J.U.; Dos Santos, D.J.V.A. Structure-function relationships in ABCG2: Insights from molecular dynamics simulations and molecular docking studies. *Sci. Rep.* **2017**, *7*, 15534. [CrossRef]
24. Khunweeraphong, N.; Stockner, T.; Kuchler, K. The structure of the human ABC transporter ABCG2 reveals a novel mechanism for drug extrusion. *Sci. Rep.* **2017**, *7*, 1–15. [CrossRef]
25. Taylor, N.M.I.; Manolaridis, I.; Jackson, S.M.; Kowal, J.; Stahlberg, H.; Locher, K.P. Structure of the human multidrug transporter ABCG2. *Nat. Cell Biol.* **2017**, *546*, 504–509. [CrossRef] [PubMed]
26. Manolaridis, I.; Jackson, S.M.; Taylor, N.M.I.; Kowal, J.; Stahlberg, H.; Locher, K.P. Cryo-EM structures of a human ABCG2 mutant trapped in ATP-bound and substrate-bound states. *Nat. Cell Biol.* **2018**, *563*, 426–430. [CrossRef]
27. Jackson, S.M.; Manolaridis, I.; Kowal, J.; Zechner, M.; Taylor, N.M.I.; Bause, M.; Bauer, S.; Bartholomaeus, R.; Bernhardt, G.; Koenig, B.; et al. Structural basis of small-molecule inhibition of human multidrug transporter ABCG2. *Nat. Struct. Mol. Biol.* **2018**, *25*, 333–340. [CrossRef]
28. Orlando, B.J.; Liao, M. ABCG2 transports anticancer drugs via a closed-to-open switch. *Nat. Commun.* **2020**, *11*, 1–11. [CrossRef]

29. Eckenstaler, R.; Benndorf, R.A. 3D structure of the transporter ABCG2—What's new? *Br. J. Pharmacol.* **2020**, *177*, 1485–1496. [CrossRef]
30. Sarkadi, B.; Homolya, L.; Hegedűs, T. The ABCG2/BCRP transporter and its variants—From structure to pathology. *FEBS Lett.* **2020**, *594*, 4012–4034. [CrossRef] [PubMed]
31. Khunweeraphong, N.; Szöllősi, D.; Stockner, T.; Kuchler, K. The ABCG2 multidrug transporter is a pump gated by a valve and an extracellular lid. *Nat. Commun.* **2019**, *10*, 1–14. [CrossRef]
32. Sarankó, H.; Tordai, H.; Telbisz, A.; Ozvegy-Laczka, C.; Erdös, G.; Sarkadi, B.; Hegedűs, T. Effects of the gout-causing Q141K polymorphism and a CFTR ΔF508 mimicking mutation on the processing and stability of the ABCG2 protein. *Biochem. Biophys. Res. Commun.* **2013**, *437*, 140–145. [CrossRef] [PubMed]
33. Woodward, O.M.; Tukaye, D.N.; Cui, J.; Greenwell, P.; Constantoulakis, L.M.; Parker, B.S.; Rao, A.; Köttgen, M.; Maloney, P.C.; Guggino, W.B. Gout-causing Q141K mutation in ABCG2 leads to instability of the nucleotide-binding domain and can be corrected with small molecules. *Proc. Natl. Acad. Sci. USA* **2013**, *110*, 5223–5228. [CrossRef]
34. Maliepaard, M.; Scheffer, G.L.; Faneyte, I.F.; Van Gastelen, M.A.; Pijnenborg, A.C.; Schinkel, A.H.; Van De Vijver, M.J.; Scheper, R.J.; Schellens, J.H. Subcellular localization and distribution of the breast cancer resistance protein transporter in normal human tissues. *Cancer Res.* **2001**, *61*, 3458–3464. [PubMed]
35. Leslie, E.M.; Deeley, R.G.; Cole, S.P. Multidrug resistance proteins: Role of P-glycoprotein, MRP1, MRP2, and BCRP (ABCG2) in tissue defense. *Toxicol. Appl. Pharmacol.* **2005**, *204*, 216–237. [CrossRef]
36. Gutmann, H.; Hruz, P.; Zimmermann, C.; Beglinger, C.; Drewe, J. Distribution of breast cancer resistance protein (BCRP/ABCG2) mRNA expression along the human GI tract. *Biochem. Pharmacol.* **2005**, *70*, 695–699. [CrossRef]
37. Woodward, O.M.; Köttgen, M.; Coresh, J.; Boerwinkle, E.; Guggino, W.B. Identification of a urate transporter, ABCG2, with a common functional polymorphism causing gout. *Proc. Natl. Acad. Sci. USA* **2009**, *106*, 10338–10342. [CrossRef]
38. Mao, Q. BCRP/ABCG2 in the placenta: Expression, function and regulation. *Pharm. Res.* **2008**, *25*, 1244–1255. [CrossRef] [PubMed]
39. Jonker, J.W.; Merino, G.; Musters, S.; Van Herwaarden, A.E.; Bolscher, E.; Wagenaar, E.; Mesman, E.; Dale, T.C.; Schinkel, A.H. The breast cancer resistance protein BCRP (ABCG2) concentrates drugs and carcinogenic xenotoxins into milk. *Nat. Med.* **2005**, *11*, 127–129. [CrossRef]
40. Cooray, H.C.; Blackmore, C.G.; Maskell, L.; Barrand, M.A. Localisation of breast cancer resistance protein in microvessel endothelium of human brain. *NeuroReport* **2002**, *13*, 2059–2063. [CrossRef]
41. Zhang, W.; Mojsilovic-Petrovic, J.; Andrade, M.F.; Zhang, H.; Ball, M.; Stanimirovic, D.B. Expression and functional characterization of ABCG2 in brain endothelial cells and vessels. *FASEB J.* **2003**, *17*, 1–24. [CrossRef]
42. Zhou, S.; Schuetz, J.D.; Bunting, K.D.; Colapietro, A.-M.; Sampath, J.; Morris, J.J.; Lagutina, I.; Grosveld, G.C.; Osawa, M.; Nakauchi, H.; et al. The ABC transporter Bcrp1/ABCG2 is expressed in a wide variety of stem cells and is a molecular determinant of the side-population phenotype. *Nat. Med.* **2001**, *7*, 1028–1034. [CrossRef]
43. Apáti, A.; Orbán, T.I.; Varga, N.; Németh, A.; Schamberger, A.; Krizsik, V.; Erdélyi-Belle, B.; Homolya, L.; Várady, G.; Padányi, R.; et al. High level functional expression of the ABCG2 multidrug transporter in undifferentiated human embryonic stem cells. *Biochim. Biophys. Acta BBA Biomembr.* **2008**, *1778*, 2700–2709. [CrossRef]
44. Sarkadi, B.; Orbán, T.I.; Szakacs, G.; Várady, G.; Schamberger, A.; Erdei, Z.; Szebényi, K.; Homolya, L.; Apáti, A. Evaluation of ABCG2 expression in human embryonic stem cells: Crossing the same river twice? *Stem Cells* **2009**, *28*, 174–176. [CrossRef]
45. Dean, M.; Fojo, T.; Bates, S.E. Tumour stem cells and drug resistance. *Nat. Rev. Cancer* **2005**, *5*, 275–284. [CrossRef]
46. Ho, M.M.; Ng, A.V.; Lam, S.; Hung, J.Y. Side Population in human lung cancer cell lines and tumors is enriched with stem-like cancer cells. *Cancer Res.* **2007**, *67*, 4827–4833. [CrossRef]
47. Ding, X.-W.; Wu, J.-H.; Jiang, C.-P. ABCG2: A potential marker of stem cells and novel target in stem cell and cancer therapy. *Life Sci.* **2010**, *86*, 631–637. [CrossRef]
48. Borst, P. Cancer drug pan-resistance: Pumps, cancer stem cells, quiescence, epithelial to mesenchymal transition, blocked cell death pathways, persisters or what? *Open Biol.* **2012**, *2*, 120066. [CrossRef] [PubMed]
49. Kasza, I.; Várady, G.; Andrikovics, H.; Koszarska, M.; Tordai, A.; Scheffer, G.L.; Németh, A.; Szakács, G.; Sarkadi, B. Expression levels of the ABCG2 multidrug transporter in human erythrocytes correspond to pharmacologically relevant genetic variations. *PLoS ONE* **2012**, *7*, e48423. [CrossRef] [PubMed]
50. Zhou, S.; Zong, Y.; Ney, P.A.; Nair, G.; Stewart, C.F.; Sorrentino, B.P. Increased expression of the Abcg2 transporter during erythroid maturation plays a role in decreasing cellular protoporphyrin IX levels. *Blood* **2005**, *105*, 2571–2576. [CrossRef] [PubMed]
51. Borst, P.; Elferink, R.O. Mammalian ABC transporters in health and disease. *Annu. Rev. Biochem.* **2002**, *71*, 537–592. [CrossRef] [PubMed]
52. Van Herwaarden, A.E.; Schinkel, A.H. The function of breast cancer resistance protein in epithelial barriers, stem cells and milk secretion of drugs and xenotoxins. *Trends Pharmacol. Sci.* **2006**, *27*, 10–16. [CrossRef] [PubMed]
53. The International Transporter Consortium. Membrane transporters in drug development. *Nat. Rev. Drug Discov.* **2010**, *9*, 215–236. [CrossRef] [PubMed]
54. Blazquez, A.G.; Briz, O.; Romero, M.R.; Rosales, R.; Monte, M.J.; Vaquero, J.; Macias, R.I.R.; Cassio, R.; Marin, J.J.G. Characterization of the Role of ABCG2 as a bile acid transporter in liver and placenta. *Mol. Pharmacol.* **2011**, *81*, 273–283. [CrossRef]

55. Blazquez, A.G.; Briz, O.; Gonzalez-Sanchez, E.; Perez, M.J.; Ghanem, C.I.; Marin, J.J. The effect of acetaminophen on the expression of BCRP in trophoblast cells impairs the placental barrier to bile acids during maternal cholestasis. *Toxicol. Appl. Pharmacol.* **2014**, *277*, 77–85. [CrossRef]

56. Matsuo, H.; Takada, T.; Ichida, K.; Nakamura, T.; Nakayama, A.; Ikebuchi, Y.; Ito, K.; Kusanagi, Y.; Chiba, T.; Tadokoro, S.; et al. Common defects of ABCG2, a high-capacity urate exporter, cause gout: A function-based genetic analysis in a Japanese population. *Sci. Transl. Med.* **2009**, *1*, 5ra11. [CrossRef]

57. Chen, L.; Manautou, J.E.; Rasmussen, T.P.; Zhong, X.-B. Development of precision medicine approaches based on inter-individual variability of BCRP/ABCG2. *Acta Pharm. Sin. B* **2019**, *9*, 659–674. [CrossRef] [PubMed]

58. Hoque, K.M.; Dixon, E.E.; Lewis, R.M.; Allan, J.; Gamble, G.D.; Phipps-Green, A.J.; Kuhns, V.L.H.; Horne, A.M.; Stamp, L.K.; Merriman, T.R.; et al. The ABCG2 Q141K hyperuricemia and gout associated variant illuminates the physiology of human urate excretion. *Nat. Commun.* **2020**, *11*, 1–15. [CrossRef] [PubMed]

59. Van Herwaarden, A.E.; Wagenaar, E.; Merino, G.; Jonker, J.W.; Rosing, H.; Beijnen, J.H.; Schinkel, A.H. Multidrug transporter ABCG2/breast cancer resistance protein secretes riboflavin (Vitamin B2) into milk. *Mol. Cell. Biol.* **2006**, *27*, 1247–1253. [CrossRef]

60. Blazquez, A.M.G.; Macias, R.I.R.; Cives-Losada, C.; De La Iglesia, A.; Marin, J.J.G.; Monte, M.J. Lactation during cholestasis: Role of ABC proteins in bile acid traffic across the mammary gland. *Sci. Rep.* **2017**, *7*, 1–11. [CrossRef]

61. Vlaming, M.L.; Lagas, J.S.; Schinkel, A.H. Physiological and pharmacological roles of ABCG2 (BCRP): Recent findings in Abcg2 knockout mice. *Adv. Drug Deliv. Rev.* **2009**, *61*, 14–25. [CrossRef]

62. Lindner, S.; Halwachs, S.; Wassermann, L.; Honscha, W. Expression and subcellular localization of efflux transporter ABCG2/BCRP in important tissue barriers of lactating dairy cows, sheep and goats. *J. Vet. Pharmacol. Ther.* **2013**, *36*, 562–570. [CrossRef]

63. Mahnke, H.; Ballent, M.; Baumann, S.; Imperiale, F.; Von Bergen, M.; Lanusse, C.; Lifschitz, A.L.; Honscha, W.; Halwachs, S. The ABCG2 efflux transporter in the mammary gland mediates veterinary drug secretion across the blood-milk barrier into milk of dairy cows. *Drug Metab. Dispos.* **2016**, *44*, 700–708. [CrossRef]

64. Bakos, E.; Homolya, L. Portrait of multifaceted transporter, the multidrug resistance-associated protein 1 (MRP1/ABCC1). *Pflügers Arch. Eur. J. Physiol.* **2006**, *453*, 621–641. [CrossRef]

65. Giacomini, K.M.; Balimane, P.V.; Cho, S.K.; Eadon, M.; Edeki, T.; Hillgren, K.M.; Huang, S.-M.; Sugiyama, Y.; Weitz, D.; Wen, Y.; et al. International transporter consortium commentary on clinically important transporter polymorphisms. *Clin. Pharmacol. Ther.* **2013**, *94*, 23–26. [CrossRef] [PubMed]

66. Prueksaritanont, T.; Chu, X.; Gibson, C.; Cui, D.; Yee, K.L.; Ballard, J.; Cabalu, T.; Hochman, J. Drug–drug interaction studies: Regulatory guidance and an industry perspective. *AAPS J.* **2013**, *15*, 629–645. [CrossRef] [PubMed]

67. Jonker, J.W.; Buitelaar, M.; Wagenaar, E.; Van Der Valk, M.A.; Scheffer, G.L.; Scheper, R.J.; Plösch, T.; Kuipers, F.; Elferink, R.P.J.O.; Rosing, H.; et al. Nonlinear partial differential equations and applications: The breast cancer resistance protein protects against a major chlorophyll-derived dietary phototoxin and protoporphyria. *Proc. Natl. Acad. Sci. USA* **2002**, *99*, 15649–15654. [CrossRef]

68. Wang, P.; Sachar, M.; Lu, J.; Shehu, A.I.; Zhu, J.; Chen, J.; Liu, K.; Anderson, K.E.; Xie, W.; Gonzalez, F.J.; et al. The essential role of the transporter ABCG2 in the pathophysiology of erythropoietic protoporphyria. *Sci. Adv.* **2019**, *5*, eaaw6127. [CrossRef]

69. Várady, G.; Szabó, E.; Fehér, A.; Németh, A.; Zámbó, B.; Pákáski, M.; Janka, Z.; Sarkadi, B. Alterations of membrane protein expression in red blood cells of Alzheimer's disease patients. *Alzheimer's Dement. Diagn. Assess. Dis. Monit.* **2015**, *1*, 334–338. [CrossRef]

70. Hegedus, T. Hegelab. Red Blood Cell Collection. 22 July 2015 Edition. Available online: http://rbcc.hegelab.org/ (accessed on 9 February 2021).

71. Miyake, K.; Mickley, L.; Litman, T.; Zhan, Z.; Robey, R.; Cristensen, B.; Brangi, M.; Greenberger, L.; Dean, M.; Fojo, T.; et al. Molecular cloning of cDNAs which are highly overexpressed in mitoxantrone-resistant cells: Demonstration of homology to ABC transport genes. *Cancer Res.* **1999**, *59*, 8–13. [PubMed]

72. Özvegy, C.; Litman, T.; Szakács, G.; Nagy, Z.; Bates, S.; Váradi, A.; Sarkadi, B. Functional characterization of the human multidrug transporter, ABCG2, expressed in insect cells. *Biochem. Biophys. Res. Commun.* **2001**, *285*, 111–117. [CrossRef]

73. Nakanishi, T.; Doyle, L.A.; Hassel, B.; Wei, Y.; Bauer, K.S.; Wu, S.; Pumplin, D.W.; Fang, H.-B.; Ross, U.D. Functional characterization of human breast cancer resistance protein (BCRP, ABCG2) expressed in the oocytes of Xenopus laevis. *Mol. Pharmacol.* **2003**, *64*, 1452–1462. [CrossRef] [PubMed]

74. Homolya, L.; Orbán, T.I.; Csanády, L.; Sarkadi, B. Mitoxantrone is expelled by the ABCG2 multidrug transporter directly from the plasma membrane. *Biochim. Biophys. Acta BBA Biomembr.* **2011**, *1808*, 154–163. [CrossRef] [PubMed]

75. Robey, R.W.; Medina-Pérez, W.Y.; Nishiyama, K.; Lahusen, T.; Miyake, K.; Litman, T.; Senderowicz, A.M.; Ross, D.D.; Bates, S.E. Overexpression of the ATP-binding cassette half-transporter, ABCG2 (Mxr/BCrp/ABCP1), in flavopiridol-resistant human breast cancer cells. *Clin. Cancer Res.* **2001**, *7*, 145–152. [PubMed]

76. Volk, E.L.; Schneider, E. Wild-type breast cancer resistance protein (BCRP/ABCG2) is a methotrexate polyglutamate transporter. *Cancer Res.* **2003**, *63*, 5538–5543.

77. Burger, H.; Van Tol, H.; Boersma, A.W.M.; Brok, M.; Wiemer, E.A.C.; Stoter, G.; Nooter, K. Imatinib mesylate (STI571) is a substrate for the breast cancer resistance protein (BCRP)/ABCG2 drug pump. *Blood* **2004**, *104*, 2940–2942. [CrossRef]

78. Yang, C.-H.; Schneider, E.; Kuo, M.-L.; Volk, E.L.; Rocchi, E.; Chen, Y.-C. BCRP/MXR/ABCP expression in topotecan-resistant human breast carcinoma cells. *Biochem. Pharmacol.* **2000**, *60*, 831–837. [CrossRef]

79. Elkind, N.B.; Apáti, A.; Várady, G.; Ujhelly, O.; Szabó, K.; Homolya, L.; Buday, L.; Német, K.; Sarkadi, B.; Szentpétery, Z.; et al. Multidrug transporter ABCG2 prevents tumor cell death induced by the epidermal growth factor receptor inhibitor iressa (ZD1839, gefitinib). *Cancer Res.* **2005**, *65*, 1770–1777. [CrossRef]

80. Hegedüs, C.; Truta-Feles, K.; Antalffy, G.; Várady, G.; Német, K.; Özvegy-Laczka, C.; Kéri, G.; Örfi, L.; Szakacs, G.; Settleman, J.; et al. Interaction of the EGFR inhibitors gefitinib, vandetanib, pelitinib and neratinib with the ABCG2 multidrug transporter: Implications for the emergence and reversal of cancer drug resistance. *Biochem. Pharmacol.* **2012**, *84*, 260–267. [CrossRef]

81. Li, J.; Cusatis, G.; Brahmer, J.; Sparreboom, A.; Robey, R.W.; Bates, S.E.; Hidalgo, M.; Baker, S.D. Association of variant ABCG2 and the pharmacokinetics of epidermal growth factor receptor tyrosine kinase inhibitors in cancer patients. *Cancer Biol. Ther.* **2007**, *6*, 432–438. [CrossRef]

82. Ozvegy-Laczka, C.; Hegedűs, T.; Várady, G.; Ujhelly, O.; Schuetz, J.D.; Váradi, A.; Kéri, G.; Örfi, L.; Német, K.; Sarkadi, B.; et al. High-affinity interaction of tyrosine kinase inhibitors with the ABCG2 multidrug transporter. *Mol. Pharmacol.* **2004**, *65*, 1485–1495. [CrossRef]

83. Telbisz, A.; Hegedüs, C.; Özvegy-Laczka, C.; Goda, K.; Várady, G.; Takáts, Z.; Szabó, E.; Sorrentino, B.P.; Váradi, A.; Sarkadi, B. Antibody binding shift assay for rapid screening of drug interactions with the human ABCG2 multidrug transporter. *Eur. J. Pharm. Sci.* **2012**, *45*, 101–109. [CrossRef] [PubMed]

84. Shukla, S.; Robey, R.W.; Bates, S.E.; Ambudkar, S.V. Sunitinib (sutent, SU11248), a small-molecule receptor tyrosine kinase inhibitor, blocks function of the ATP-Binding Cassette (ABC) transporters p-glycoprotein (ABCB1) and ABCG2. *Drug Metab. Dispos.* **2008**, *37*, 359–365. [CrossRef]

85. Ozvegy-Laczka, C.; Hegedus, C.; Szakács, G.; Sarkadi, B. Interaction of ABC multidrug transporters with anticancer protein kinase inhibitors: Substrates and/or inhibitors? *Curr. Cancer Drug Targ.* **2009**, *9*, 252–272. [CrossRef]

86. Robey, R.W.; Honjo, Y.; Morisaki, K.; Nadjem, T.A.; Runge, S.; Risbood, M.; Poruchynsky, M.S.; Bates, S.E. Mutations at amino-acid 482 in the ABCG2 gene affect substrate and antagonist specificity. *Br. J. Cancer* **2003**, *89*, 1971–1978. [CrossRef] [PubMed]

87. Damiani, D.; Tiribelli, M.; Calistri, E.; Geromin, A.; Chiarvesio, A.; Michelutti, A.; Cavallin, M.; Fanin, R. The prognostic value of P-glycoprotein (ABCB) and breast cancer resistance protein (ABCG2) in adults with de novo acute myeloid leukemia with normal karyotype. *Haematology* **2006**, *91*, 825–828.

88. Kim, J.E.; Singh, R.R.; Cho-Vega, J.H.; Drakos, E.; Davuluri, Y.; Khokhar, F.A.; Fayad, L.; Medeiros, L.J.; Vega, F. Sonic hedgehog signaling proteins and ATP-binding cassette G2 are aberrantly expressed in diffuse large B-Cell lymphoma. *Mod. Pathol.* **2009**, *22*, 1312–1320. [CrossRef]

89. Sauerbrey, A.; Sell, W.; Steinbach, D.; Voigt, A.; Zintl, F. Expression of the BCRP gene (ABCG2/MXR/ABCP) in childhood acute lymphoblastic leukaemia. *Br. J. Haematol.* **2002**, *118*, 147–150. [CrossRef]

90. Suvannasankha, A.; Minderman, H.; O'Loughlin, K.L.; Nakanishi, T.; Ford, L.A.; Greco, W.R.; Wetzler, M.; Ross, D.D.; Baer, M.R. Breast cancer resistance protein (BCRP/MXR/ABCG2) in adult acute lymphoblastic leukaemia: Frequent expression and possible correlation with shorter disease-free survival. *Br. J. Haematol.* **2004**, *127*, 392–398. [CrossRef]

91. Plasschaert, S.L.A.; Van Der Kolk, D.M.; De Bont, E.S.J.M.; Vellenga, E.; Kamps, W.A.; De Vries, E.G.E. Breast Cancer Resistance Protein (BCRP) in acute leukaemia. *Leuk. Lymphoma* **2004**, *45*, 649–654. [CrossRef]

92. Burger, H.; Foekens, J.A.; Look, M.P.; Gelder, M.E.M.-V.; Klijn, J.G.M.; Wiemer, E.A.C.; Stoter, G.; Nooter, K. RNA expression of breast cancer resistance protein, lung resistance-related protein, multidrug resistance-associated proteins 1 and 2, and multidrug resistance gene 1 in breast cancer: Correlation with chemotherapeutic response. *Clin. Cancer Res.* **2003**, *9*, 827–836. [PubMed]

93. Faneyte, I.F.; Kristel, P.M.; Maliepaard, M.; Scheffer, G.L.; Scheper, R.J.; Schellens, J.H.; van de Vijver, M.J. Expression of the breast cancer resistance protein in breast cancer. *Clin Cancer Res* **2002**, *8*, 1068–1074.

94. Kanzaki, A.; Toi, M.; Nakayama, K.; Bando, H.; Mutoh, M.; Uchida, T.; Fukumoto, M.; Takebayashi, Y. Expression of multidrug resistance-related transporters in human breast carcinoma. *Jpn. J. Cancer Res.* **2001**, *92*, 452–458. [CrossRef]

95. Tamura, A.; Wakabayashi, K.; Onishi, Y.; Takeda, M.; Ikegami, Y.; Sawada, S.; Tsuji, M.; Matsuda, Y.; Ishikawa, T. Re-evaluation and functional classification of non-synonymous single nucleotide polymorphisms of the human ATP-binding cassette transporter ABCG2. *Cancer Sci.* **2006**, *98*, 231–239. [CrossRef]

96. Heyes, N.; Kapoor, P.; Kerr, I.D. Polymorphisms of the multidrug pump ABCG2: A systematic review of their effect on protein expression, function, and drug pharmacokinetics. *Drug Metab. Dispos.* **2018**, *46*, 1886–1899. [CrossRef] [PubMed]

97. De Boeck, K.; Amaral, M.D. Progress in therapies for cystic fibrosis. *Lancet Respir. Med.* **2016**, *4*, 662–674. [CrossRef]

98. Gyimesi, G.; Borsodi, D.; Sarankó, H.; Tordai, H.; Sarkadi, B.; Hegedűs, T. ABCMdb: A database for the comparative analysis of protein mutations in ABC transporters, and a potential framework for a general application. *Hum. Mutat.* **2012**, *33*, 1547–1556. [CrossRef]

99. Tordai, H.; Jakab, K.; Gyimesi, G.; András, K.; Brózik, A.; Sarkadi, B.; Hegedűs, T. ABCMdb reloaded: Updates on mutations in ATP binding cassette proteins. *Database* **2017**, *2017*. [CrossRef]

100. Kondo, C.; Suzuki, H.; Itoda, M.; Ozawa, S.; Sawada, J.-I.; Kobayashi, D.; Ieiri, I.; Mine, K.; Ohtsubo, K.; Sugiyama, Y. Functional analysis of SNPs variants of BCRP/ABCG2. *Pharm. Res.* **2004**, *21*, 1895–1903. [CrossRef]

101. Zamber, C.P.; Lamba, J.K.; Yasuda, K.; Farnum, J.; Thummel, K.; Schuetz, J.D.; Schuetz, E.G. Natural allelic variants of breast cancer resistance protein (BCRP) and their relationship to BCRP expression in human intestine. *Pharmacogenetics* **2003**, *13*, 19–28. [CrossRef]

102. Kim, K.-A.; Joo, H.-J.; Park, J.-Y. ABCG2 polymorphisms, 34G > A and 421C > A in a Korean population: Analysis and a comprehensive comparison with other populations. *J. Clin. Pharm. Ther.* **2010**, *35*, 705–712. [CrossRef]
103. Higashino, T.; Takada, T.; Nakaoka, H.; Toyoda, Y.; Stiburkova, B.; Miyata, H.; Ikebuchi, Y.; Nakashima, H.; Shimizu, S.; Kawaguchi, M.; et al. Multiple common and rare variants of ABCG2 cause gout. *RMD Open* **2017**, *3*, e000464. [CrossRef] [PubMed]
104. Zámbó, B.; Mózner, O.; Bartos, Z.; Török, G.; Várady, G.; Telbisz, A.; Homolya, L.; Orbán, T.I.; Sarkadi, B. Cellular expression and function of naturally occurring variants of the human ABCG2 multidrug transporter. *Cell. Mol. Life Sci.* **2019**, *77*, 365–378. [CrossRef] [PubMed]
105. Zhou, D.; Liu, Y.; Zhang, X.; Gu, X.; Wang, H.; Luo, X.; Zhang, J.; Zou, H.; Guan, M. Functional polymorphisms of the ABCG2 gene are associated with gout disease in the Chinese Han male population. *Int. J. Mol. Sci.* **2014**, *15*, 9149–9159. [CrossRef] [PubMed]
106. Li, R.; Miao, L.; Qin, L.; Xiang, Y.; Zhang, X.; Peng, H.; Mailamuguli; Sun, Y.; Yao, H. A meta-analysis of the associations between the Q141K and Q126X ABCG2 gene variants and gout risk. *Int. J. Clin. Exp. Pathol.* **2015**, *8*, 9812–9823. [PubMed]
107. Stiburkova, B.; Pavelcova, K.; Zavada, J.; Petru, L.; Simek, P.; Cepek, P.; Pavlikova, M.; Matsuo, H.; Merriman, T.R.; Pavelka, K. Functional non-synonymous variants of ABCG2 and gout risk. *Rheumatology* **2017**, *56*, 1982–1992. [CrossRef]
108. Kim, Y.S.; Kim, Y.; Park, G.; Kim, S.-K.; Choe, J.-Y.; Park, B.L.; Kim, H.S. Genetic analysis of ABCG2 and SLC2A9 gene polymorphisms in gouty arthritis in a Korean population. *Korean J. Intern. Med.* **2015**, *30*, 913–920. [CrossRef] [PubMed]
109. Saison, C.; Helias, V.; Ballif, B.A.; Peyrard, T.; Puy, H.; Miyazaki, T.; Perrot, S.; Vayssier-Taussat, M.; Waldner, M.; Le Pennec, P.-Y.; et al. Null alleles of ABCG2 encoding the breast cancer resistance protein define the new blood group system Junior. *Nat. Genet.* **2012**, *44*, 174–177. [CrossRef]
110. Zelinski, T.; Coghlan, G.; Liu, X.-Q.; Reid, M.E. ABCG2 null alleles define the Jr(a−) blood group phenotype. *Nat. Genet.* **2012**, *44*, 131–132. [CrossRef]
111. Itoda, M.; Saito, Y.; Shirao, K.; Minami, H.; Ohtsu, A.; Yoshida, T.; Saijo, N.; Suzuki, H.; Sugiyama, Y.; Ozawa, S.; et al. Eight novel single nucleotide polymorphisms in ABCG2/BCRP in Japanese cancer patients administered irinotacan. *Drug Metab. Pharmacokinet.* **2003**, *18*, 212–217. [CrossRef] [PubMed]
112. Zámbó, B.; Bartos, Z.; Mózner, O.; Szabó, E.; Várady, G.; Poór, G.; Pálinkás, M.; Andrikovics, H.; Hegedűs, T.; Homolya, L.; et al. Clinically relevant mutations in the ABCG2 transporter uncovered by genetic analysis linked to erythrocyte membrane protein expression. *Sci. Rep.* **2018**, *8*, 1–13. [CrossRef]
113. Tamura, A.; Watanabe, M.; Saito, H.; Nakagawa, H.; Kamachi, T.; Okura, I.; Ishikawa, T. Functional validation of the genetic polymorphisms of human ATP-Binding Cassette (ABC) transporter ABCG2: Identification of alleles that are defective in porphyrin transport. *Mol. Pharmacol.* **2006**, *70*, 287–296. [CrossRef] [PubMed]
114. Skoglund, K.; Moreno, S.B.; Jönsson, J.-I.; Vikingsson, S.; Carlsson, B.; Gréen, H. Single-nucleotide polymorphisms of ABCG2 increase the efficacy of tyrosine kinase inhibitors in the K562 chronic myeloid leukemia cell line. *Pharmacogenet. Genom.* **2014**, *24*, 52–61. [CrossRef] [PubMed]
115. Imai, Y.; Nakane, M.; Kage, K.; Tsukahara, S.; Ishikawa, E.; Tsuruo, T.; Miki, Y.; Sugimoto, Y. C421A polymorphism in the human breast cancer resistance protein gene is associated with low expression of Q141K protein and low-level drug re-sistance. *Mol. Cancer Ther.* **2002**, *1*, 611–616. [PubMed]
116. De Jong, F.A.; Marsh, S.; Mathijssen, R.H.J.; King, C.; Verweij, J.; Sparreboom, A.; McLeod, H.L.; Jones, C.; Ford, E.; Gillett, C.; et al. ABCG2 Pharmacogenetics. *Clin. Cancer Res.* **2004**, *10*, 5889–5894. [CrossRef]
117. Furukawa, T.; Wakabayashi, K.; Tamura, A.; Nakagawa, H.; Morishima, Y.; Osawa, Y.; Ishikawa, T. Major SNP (Q141K) variant of human ABC transporter ABCG2 undergoes lysosomal and proteasomal degradations. *Pharm. Res.* **2008**, *26*, 469–479. [CrossRef]
118. Basseville, A.; Tamaki, A.; Ierano, C.; Trostel, S.; Ward, Y.; Robey, R.W.; Hegde, R.S.; Bates, S.E. Histone deacetylase inhibitors influence chemotherapy transport by modulating expression and trafficking of a common polymorphic variant of the ABCG2 efflux transporter. *Cancer Res.* **2012**, *72*, 3642–3651. [CrossRef]
119. Cleophas, M.; Joosten, L.; Stamp, L.; Dalbeth, N.; Woodward, O.; Merriman, T.R. ABCG2 polymorphisms in gout: Insights into disease susceptibility and treatment approaches. *Pharmacogenomics Pers. Med.* **2017**, *10*, 129–142. [CrossRef]
120. Bartos, Z.; Homolya, L. Identification of specific trafficking defects of naturally occurring variants of the human ABCG2 transporter. *Front. Cell Dev. Biol.* **2021**, *9*. [CrossRef]
121. Deppe, S.; Ripperger, A.; Weiss, J.; Ergün, S.; Benndorf, R.A. Impact of genetic variability in the ABCG2 gene on ABCG2 expression, function, and interaction with AT1 receptor antagonist telmisartan. *Biochem. Biophys. Res. Commun.* **2014**, *443*, 1211–1217. [CrossRef]
122. Sjöstedt, N.; Heuvel, J.J.M.W.V.D.; Koenderink, J.B.; Kidron, H. Transmembrane domain single-nucleotide polymorphisms impair expression and transport activity of ABC transporter ABCG2. *Pharm. Res.* **2017**, *34*, 1626–1636. [CrossRef]
123. Cox, M.H.; Kapoor, P.; Briggs, D.A.; Kerr, I.D. Residues contributing to drug transport by ABCG2 are localised to multiple drug-binding pockets. *Biochem. J.* **2018**, *475*, 1553–1567. [CrossRef] [PubMed]
124. Lee, S.S.; Jeong, H.-E.; Yi, J.-M.; Jung, H.-J.; Jang, J.-E.; Kim, E.-Y.; Lee, S.-J.; Shin, J.-G. Identification and functional assessment of BCRP polymorphisms in a Korean population. *Drug Metab. Dispos.* **2007**, *35*, 623–632. [CrossRef]
125. Toyoda, Y.; Pavelcová, K.; Klein, M.; Suzuki, H.; Takada, T.; Stiburkova, B. Familial early-onset hyperuricemia and gout associated with a newly identified dysfunctional variant in urate transporter ABCG2. *Arthr. Res.* **2019**, *21*, 1–3. [CrossRef] [PubMed]

126. Özvegy-Laczka, C.; Várady, G.; Köblös, G.; Ujhelly, O.; Cervenak, J.; Schuetz, J.D.; Sorrentino, B.P.; Koomen, G.-J.; Váradi, A.; Német, K.; et al. Function-dependent conformational changes of the ABCG2 multidrug transporter modify its interaction with a monoclonal antibody on the cell surface. *J. Biol. Chem.* **2005**, *280*, 4219–4227. [CrossRef] [PubMed]

127. Nakagawa, H.; Tamura, A.; Wakabayashi, K.; Hoshijima, K.; Komada, M.; Yoshida, T.; Kometani, S.; Matsubara, T.; Mikuriya, K.; Ishikawa, T. Ubiquitin-mediated proteasomal degradation of non-synonymous SNP variants of human ABC transporter ABCG2. *Biochem. J.* **2008**, *411*, 623–631. [CrossRef]

128. Yoshioka, S.; Katayama, K.; Okawa, C.; Takahashi, S.; Tsukahara, S.; Mitsuhashi, J.; Sugimoto, Y. The Identification of two germ-line mutations in the human breast cancer resistance protein gene that result in the expression of a low/non-functional protein. *Pharm. Res.* **2007**, *24*, 1108–1117. [CrossRef] [PubMed]

129. Honjo, Y.; Hrycyna, C.A.; Yan, Q.W.; Medina-Pérez, W.Y.; Robey, R.W.; Van De Laar, A.; Litman, T.; Dean, M.; Bates, S.E. Acquired mutations in the MXR/BCRP/ABCP gene alter substrate specificity in MXR/BCRP/ABCP-overexpressing cells. *Cancer Res.* **2001**, *61*, 6635–6639. [PubMed]

130. Volk, E.L.; Farley, K.M.; Wu, Y.; Li, F.; Robey, R.W.; Schneider, E. Overexpression of wild-type breast cancer resistance protein mediates methotrexate resistance. *Cancer Res.* **2002**, *62*, 5035–5040. [PubMed]

131. Mitomo, H.; Kato, R.; Ito, A.; Kasamatsu, S.; Ikegami, Y.; Kii, I.; Kudo, A.; Kobatake, E.; Sumino, Y.; Ishikawa, T. A functional study on polymorphism of the ATP-binding cassette transporter ABCG2: Critical role of arginine-482 in methotrexate transport. *Biochem. J.* **2003**, *373*, 767–774. [CrossRef]

132. Chen, Z.-S.; Robey, R.W.; Belinsky, M.G.; Shchaveleva, I.; Ren, X.-Q.; Sugimoto, Y.; Ross, D.D.; Bates, S.E.; Kruh, G.D. Transport of methotrexate, methotrexate polyglutamates, and 17beta-estradiol 17-(beta-D-glucuronide) by ABCG2: Effects of acquired mutations at R482 on methotrexate transport. *Cancer Res.* **2003**, *63*, 4048–4054.

133. Mizuarai, S.; Aozasa, N.; Kotani, H. Single nucleotide polymorphisms result in impaired membrane localization and reduced atpase activity in multidrug transporter ABCG2. *Int. J. Cancer* **2004**, *109*, 238–246. [CrossRef]

134. Poonkuzhali, B.; Lamba, J.; Strom, S.; Sparreboom, A.; Thummel, K.; Watkins, P.; Schuetz, E. Association of breast cancer resistance protein/ABCG2 phenotypes and novel promoter and intron 1 single nucleotide polymorphisms. *Drug Metab. Dispos.* **2008**, *36*, 780–795. [CrossRef] [PubMed]

135. Vethanayagam, R.R.; Wang, H.; Gupta, A.; Zhang, Y.; Lewis, F.; Unadkat, J.D.; Mao, Q. Functional analysis of the human variants of breast cancer resistance protein: I206L, N590Y, AND D620N. *Drug Metab. Dispos.* **2005**, *33*, 697–705. [CrossRef] [PubMed]

136. Morisaki, K.; Robey, R.W.; Ozvegy-Laczka, C.; Honjo, Y.; Polgar, O.; Steadman, K.; Sarkadi, B.; Bates, S.E. Single nucleotide polymorphisms modify the transporter activity of ABCG2. *Cancer Chemother. Pharmacol.* **2005**, *56*, 161–172. [CrossRef]

137. Mózner, O.; Bartos, Z.; Zámbó, B.; Homolya, L.; Hegedűs, T.; Sarkadi, B. Cellular processing of the ABCG2 transporter—Potential effects on gout and drug metabolism. *Cells* **2019**, *8*, 1215. [CrossRef]

138. Kawahara, H.; Noguchi, K.; Katayama, K.; Mitsuhashi, J.; Sugimoto, Y. Pharmacological interaction with sunitinib is abolished by a germ-line mutation (1291T > C) of BCRP/ABCG2 gene. *Cancer Sci.* **2010**, *101*, 1493–1500. [CrossRef]

139. Igreja, S.; Clarke, L.A.; Botelho, H.M.; Marques, L.; Amaral, M.D. Correction of a Cystic fibrosis splicing mutation by antisense oligonucleotides. *Hum. Mutat.* **2015**, *37*, 209–215. [CrossRef]

140. Toyoda, Y.; Mančíková, A.; Krylov, V.; Morimoto, K.; Pavelcová, K.; Bohatá, J.; Pavelka, K.; Pavlíková, M.; Suzuki, H.; Matsuo, H.; et al. Functional characterization of clinically-relevant rare variants in ABCG2 identified in a gout and hyper-uricemia cohort. *Cells* **2019**, *8*, 363. [CrossRef]

141. Hegedüs, C.; Szakacs, G.; Homolya, L.; Orbán, T.I.; Telbisz, A.; Jani, M.; Sarkadi, B. Ins and outs of the ABCG2 multidrug transporter: An update on in vitro functional assays. *Adv. Drug Deliv. Rev.* **2009**, *61*, 47–56. [CrossRef]

142. Gameiro, M.; Silva, R.; Rocha-Pereira, C.; Carmo, H.; Carvalho, F.; Bastos, M.D.L.; Remião, F. Cellular models and in vitro assays for the screening of modulators of P-gp, MRP1 and BCRP. *Molecules* **2017**, *22*, 600. [CrossRef]

143. Sándor, S.; Jordanidisz, T.; Schamberger, A.; Várady, G.; Erdei, Z.; Apáti, A.; Sarkadi, B.; Orbán, T.I. Functional characterization of the ABCG2 5′ non-coding exon variants: Stem cell specificity, translation efficiency and the influence of drug selection. *Biochim. Biophys. Acta BBA Bioenerg.* **2016**, *1859*, 943–951. [CrossRef]

144. Kovacsics, D.; Brózik, A.; Tihanyi, B.; Matula, Z.; Borsy, A.; Mészáros, N.; Szabó, E.; Németh, E.; Fóthi, A.; Zámbó, B.; et al. Precision-engineered reporter cell lines reveal ABCG2 regulation in live lung cancer cells. *Biochem. Pharmacol.* **2020**, *175*, 113865. [CrossRef]

145. Studzian, M.; Bartosz, G.; Pulaski, L. Endocytosis of ABCG2 drug transporter caused by binding of 5D3 antibody: Trafficking mechanisms and intracellular fate. *Biochim. Biophys. Acta BBA Bioenerg.* **2015**, *1853*, 1759–1771. [CrossRef]

146. Peng, H.; Qi, J.; Dong, Z.; Zhang, J.-T. Dynamic vs. static ABCG2 inhibitors to sensitize drug resistant cancer cells. *PLoS ONE* **2010**, *5*, e15276. [CrossRef] [PubMed]

147. Telbisz, A.; Müller, M.; Özvegy-Laczka, C.; Homolya, L.; Szente, L.; Váradi, A.; Sarkadi, B. Membrane cholesterol selectively modulates the activity of the human ABCG2 multidrug transporter. *Biochim. Biophys. Acta BBA Biomembr.* **2007**, *1768*, 2698–2713. [CrossRef] [PubMed]

148. Hirschmann-Jax, C.; Foster, A.E.; Wulf, G.G.; Nuchtern, J.G.; Jax, T.W.; Gobel, U.; Goodell, M.A.; Brenner, M.K. A distinct "side population" of cells with high drug efflux capacity in human tumor cells. *Proc. Natl. Acad. Sci. USA* **2004**, *101*, 14228–14233. [CrossRef]

149. Moitra, K.; Lou, H.; Dean, M. Multidrug efflux pumps and cancer stem cells: Insights into multidrug resistance and therapeutic development. *Clin. Pharmacol. Ther.* **2011**, *89*, 491–502. [CrossRef]
150. Erdei, Z.; Schamberger, A.; Török, G.; Szebényi, K.; Várady, G.; Orbán, T.I.; Homolya, L.; Sarkadi, B.; Apáti, A. Generation of multidrug resistant human tissues by overexpression of the ABCG2 multidrug transporter in embryonic stem cells. *PLoS ONE* **2018**, *13*, e0194925. [CrossRef]
151. Dehghan, A.; Köttgen, A.; Yang, Q.F.; Hwang, S.-J.; Kao, W.; Rivadeneira, F.; Boerwinkle, E.; Levy, D.; Hofman, A.; Astor, B.; et al. Association of three genetic loci with uric acid concentration and risk of gout: A genome-wide association study. *Lancet* **2008**, *372*, 1953–1961. [CrossRef]
152. Kannangara, D.R.W.; Phipps-Green, A.J.; Dalbeth, N.; Stamp, L.K.; Williams, K.M.; Graham, G.G.; Day, R.O.; Merriman, T.R. Hyperuricaemia: Contributions of urate transporter ABCG2 and the fractional renal clearance of urate. *Ann. Rheum. Dis.* **2015**, *75*, 1363–1366. [CrossRef]
153. Wen, C.C.; Yee, S.W.; Liang, X.; Hoffmann, T.J.; Kvale, M.N.; Banda, Y.; Jorgenson, E.; Schaefer, C.; Risch, N.; Giacomini, K.M. Genome-wide association study identifies ABCG2 (BCRP) as an allopurinol transporter and a determinant of drug response. *Clin. Pharmacol. Ther.* **2015**, *97*, 518–525. [CrossRef] [PubMed]
154. Nakamura, M.; Fujita, K.; Toyoda, Y.; Takada, T.; Hasegawa, H.; Ichida, K. Investigation of the transport of xanthine dehydrogenase inhibitors by the urate transporter ABCG2. *Drug Metab. Pharmacokinet.* **2018**, *33*, 77–81. [CrossRef]
155. Stamp, L.K.; Wallace, M.; Roberts, R.L.; Frampton, C.; Miner, J.N.; Merriman, T.R.; Dalbeth, N. ABCG2 rs2231142 (Q141K) and oxypurinol concentrations in people with gout receiving allopurinol. *Drug Metab. Pharmacokinet.* **2018**, *33*, 241–242. [CrossRef] [PubMed]
156. Roberts, R.L.; Wallace, M.C.; Phipps-Green, A.J.; Topless, R.; Drake, J.M.; Tan, P.; Dalbeth, N.; Merriman, T.R.; Stamp, L.K. ABCG2 loss-of-function polymorphism predicts poor response to allopurinol in patients with gout. *Pharmacogenomics J.* **2016**, *17*, 201–203. [CrossRef]
157. Brackman, D.J.; Yee, S.W.; Enogieru, O.J.; Shaffer, C.; Ranatunga, D.; Denny, J.C.; Wei, W.; Kamatani, Y.; Kubo, M.; Roden, D.M.; et al. Genome-wide association and functional studies reveal novel pharmacological mechanisms for allopurinol. *Clin. Pharmacol. Ther.* **2019**, *106*, 623–631. [CrossRef] [PubMed]
158. Safar, Z.; Kis, E.; Erdo, F.; Zolnerciks, J.K.; Krajcsi, P. ABCG2/BCRP: Variants, transporter interaction profile of substrates and inhibitors. *Expert Opin. Drug Metab. Toxicol.* **2019**, *15*, 313–328. [CrossRef] [PubMed]
159. Li, R.; Barton, H.A. Explaining ethnic variability of transporter substrate pharmacokinetics in healthy asian and caucasian subjects with allele frequencies of OATP1B1 and BCRP: A mechanistic modeling analysis. *Clin. Pharmacokinet.* **2017**, *57*, 491–503. [CrossRef]
160. Zhang, W.; Yu, B.-N.; He, Y.-J.; Fan, L.; Li, Q.; Liu, Z.-Q.; Wang, A.; Liu, Y.-L.; Tan, Z.-R.; Jiang, F.; et al. Role of BCRP 421C>A polymorphism on rosuvastatin pharmacokinetics in healthy Chinese males. *Clin. Chim. Acta* **2006**, *373*, 99–103. [CrossRef] [PubMed]
161. Keskitalo, J.E.; Zolk, O.; Fromm, M.F.; Kurkinen, K.J.; Neuvonen, P.J.; Niemi, M. ABCG2 polymorphism markedly affects the pharmacokinetics of atorvastatin and rosuvastatin. *Clin. Pharmacol. Ther.* **2009**, *86*, 197–203. [CrossRef]
162. Keskitalo, J.E.; Pasanen, M.K.; Neuvonen, P.J.; Niemi, M. Different effects of theABCG2c.421C>A SNP on the pharmacokinetics of fluvastatin, pravastatin and simvastatin. *Pharmacogenomics* **2009**, *10*, 1617–1624. [CrossRef]
163. Zhou, Q.; Ruan, Z.-R.; Yuan, H.; Xu, D.-H.; Zeng, S. ABCB1 gene polymorphisms, ABCB1 haplotypes and ABCG2 c.421c > A are determinants of inter-subject variability in rosuvastatin pharmacokinetics. *Die Pharm.* **2013**, *68*, 129–134.
164. Birmingham, B.K.; Bujac, S.R.; Elsby, R.; Azumaya, C.T.; Wei, C.; Chen, Y.; Mosqueda-Garcia, R.; Ambrose, H.J. Impact of ABCG2 and SLCO1B1 polymorphisms on pharmacokinetics of rosuvastatin, atorvastatin and simvastatin acid in Caucasian and Asian subjects: A class effect? *Eur. J. Clin. Pharmacol.* **2015**, *71*, 341–355. [CrossRef]
165. Feng, Q.; Wilke, R.A.; Baye, T.M. Individualized risk for statin-induced myopathy: Current knowledge, emerging challenges and potential solutions. *Pharmacogenomics* **2012**, *13*, 579–594. [CrossRef] [PubMed]
166. Yamasaki, Y.; Ieiri, I.; Kusuhara, H.; Sasaki, T.; Kimura, M.; Tabuchi, H.; Ando, Y.; Irie, S.; Ware, J.; Nakai, Y.; et al. Pharmacogenetic characterization of sulfasalazine disposition based on NAT2 and ABCG2 (BCRP) gene polymorphisms in humans. *Clin. Pharmacol. Ther.* **2008**, *84*, 95–103. [CrossRef] [PubMed]
167. Kim, K.-A.; Joo, H.-J.; Park, J.-Y. Effect of ABCG2 genotypes on the pharmacokinetics of A771726, an active metabolite of prodrug leflunomide, and association of A771726 exposure with serum uric acid level. *Eur. J. Clin. Pharmacol.* **2010**, *67*, 129–134. [CrossRef]
168. Wiese, M.D.; Schnabl, M.; O'Doherty, C.; Spargo, L.D.; Sorich, M.J.; Cleland, L.G.; Proudman, S.M. Polymorphisms in cytochrome P450 2C19 enzyme and cessation of leflunomide in patients with rheumatoid arthritis. *Arthritis Res. Ther.* **2012**, *14*, R163. [CrossRef]
169. Gotanda, K.; Tokumoto, T.; Hirota, T.; Fukae, M.; Ieiri, I. Sulfasalazine disposition in a subject with 376C>T (nonsense mutation) and 421C>A variants in theABCG2gene. *Br. J. Clin. Pharmacol.* **2015**, *80*, 1236–1237. [CrossRef] [PubMed]
170. Tsuchiya, K.; Hayashida, T.; Hamada, A.; Oki, S.; Oka, S.; Gatanaga, H. High plasma concentrations of dolutegravir in patients with ABCG2 genetic variants. *Pharmacogenetics Genom.* **2017**, *27*, 416–419. [CrossRef]
171. Tsuchiya, K.; Hayashida, T.; Hamada, A.; Oka, S.; Gatanaga, H. Brief report. *JAIDS J. Acquir. Immune Defic. Syndr.* **2016**, *72*, 11–14. [CrossRef]

172. Inoue, Y.; Morita, T.; Onozuka, M.; Saito, K.-I.; Sano, K.; Hanada, K.; Kondo, M.; Nakamura, Y.; Kishino, T.; Nakagawa, H.; et al. Impact of Q141K on the transport of epidermal growth factor receptor tyrosine kinase inhibitors by ABCG2. *Cells* **2019**, *8*, 763. [CrossRef]

173. Gardner, E.; Burger, H.; Vanschaik, R.; Vanoosterom, A.; DeBruijn, E.; Guetens, G.; Prenen, H.; DeJong, F.; Baker, S.; Bates, S. Association of enzyme and transporter genotypes with the pharmacokinetics of imatinib. *Clin. Pharmacol. Ther.* **2006**, *80*, 192–201. [CrossRef]

174. Moon, H.-H.; Kim, S.-H.; Ku, J.-L. Correlation between the promoter methylation status of ATP-binding cassette sub-family G member 2 and drug sensitivity in colorectal cancer cell lines. *Oncol. Rep.* **2015**, *35*, 298–306. [CrossRef] [PubMed]

175. Cusatis, G.; Gregorc, V.; Li, J.; Spreafico, A.; Ingersoll, R.G.; Verweij, J.; Ludovini, V.; Villa, E.; Hidalgo, M.; Sparreboom, A.; et al. Pharmacogenetics of ABCG2 and adverse reactions to gefitinib. *J. Natl. Cancer Inst.* **2006**, *98*, 1739–1742. [CrossRef] [PubMed]

176. Akasaka, K.; Kaburagi, T.; Yasuda, S.; Ohmori, K.; Abe, K.; Sagara, H.; Ueda, Y.; Nagao, K.; Imura, J.; Imai, Y. Impact of functional ABCG2 polymorphisms on the adverse effects of gefitinib in Japanese patients with non–small-cell lung cancer. *Cancer Chemother. Pharmacol.* **2009**, *66*, 691–698. [CrossRef] [PubMed]

177. Kim, I.-S.; Kim, H.-G.; Kim, D.C.; Eom, H.-S.; Kong, S.-Y.; Shin, H.-J.; Hwang, S.-H.; Lee, E.-Y.; Lee, G.-W. ABCG2Q141K polymorphism is associated with chemotherapy-induced diarrhea in patients with diffuse large B-cell lymphoma who received frontline rituximab plus cyclophosphamide/doxorubicin/ vincristine/prednisone chemotherapy. *Cancer Sci.* **2008**, *99*, 2496–2501. [CrossRef]

178. Low, S.-K.; Fukunaga, K.; Takahashi, A.; Matsuda, K.; Hongo, F.; Nakanishi, H.; Kitamura, H.; Inoue, T.; Kato, Y.; Tomita, Y.; et al. Association study of a functional variant on ABCG2 gene with sunitinib-induced severe adverse drug reaction. *PLoS ONE* **2016**, *11*, e0148177. [CrossRef]

179. Miura, Y.; Imamura, C.K.; Fukunaga, K.; Katsuyama, Y.; Suyama, K.; Okaneya, T.; Mushiroda, T.; Ando, Y.; Takano, T.; Tanigawara, Y. Sunitinib-induced severe toxicities in a Japanese patient with the ABCG2 421 AA genotype. *BMC Cancer* **2014**, *14*, 964. [CrossRef] [PubMed]

180. Seong, S.J.; Lim, M.; Sohn, S.K.; Moon, J.H.; Oh, S.-J.; Kim, B.S.; Ryoo, H.M.; Chung, J.S.; Joo, Y.D.; Bang, S.M.; et al. Influence of enzyme and transporter polymorphisms on trough imatinib concentration and clinical response in chronic myeloid leukemia patients. *Ann. Oncol.* **2012**, *24*, 756–760. [CrossRef]

181. Francis, J.; Dubashi, B.; Sundaram, R.; Pradhan, S.C.; Chandrasekaran, A. Influence of Sokal, Hasford, EUTOS scores and pharmacogenetic factors on the complete cytogenetic response at 1 year in chronic myeloid leukemia patients treated with imatinib. *Med. Oncol.* **2015**, *32*. [CrossRef]

182. Petain, A.; Kattygnarath, D.; Azard, J.; Chatelut, E.; Delbaldo, C.; Geoerger, B.; Barrois, M.; Séronie-Vivien, S.; Lecesne, A.; Vassal, G. Population pharmacokinetics and pharmacogenetics of imatinib in children and adults. *Clin. Cancer Res.* **2008**, *14*, 7102–7109. [CrossRef] [PubMed]

183. Takahashi, N.; Miura, M.; Scott, S.A.; Kagaya, H.; Kameoka, Y.; Tagawa, H.; Saitoh, H.; Fujishima, N.; Yoshioka, T.; Hirokawa, M.; et al. Influence of CYP3A5 and drug transporter polymorphisms on imatinib trough concentration and clinical response among patients with chronic phase chronic myeloid leukemia. *J. Hum. Genet.* **2010**, *55*, 731–737. [CrossRef]

184. Au, A.; Baba, A.A.; Goh, A.S.; Fadilah, S.A.W.; Teh, A.; Rosline, H.; Ankathil, R. Association of genotypes and haplotypes of multi-drug transporter genes ABCB1 and ABCG2 with clinical response to imatinib mesylate in chronic myeloid leukemia patients. *Biomed. Pharmacother.* **2014**, *68*, 343–349. [CrossRef] [PubMed]

185. Jiang, Z.-P.; Zhao, X.-L.; Takahashi, N.; Angelini, S.; Dubashi, B.; Sun, L.; Xu, P. Trough concentration andABCG2polymorphism are better to predict imatinib response in chronic myeloid leukemia: A meta-analysis. *Pharmacogenomics* **2017**, *18*, 35–56. [CrossRef]

186. Sparreboom, A. Diflomotecan pharmacokinetics in relation to ABCG2 421C>A genotype*1. *Clin. Pharmacol. Ther.* **2004**, *76*, 38–44. [CrossRef]

187. Sparreboom, A.; Loos, W.J.; Burger, H.; Sissung, T.M.; Verweij, J.; Figg, I.W.; Nooter, K.; Gelderblom, H. Effect of ABCG2 genotype on the oral vioavailability of topotecan. *Cancer Biol. Ther.* **2005**, *4*, 650–653. [CrossRef]

188. Zamboni, W.C.; Ramanathan, R.K.; McLeod, H.L.; Mani, S.; Potter, U.M.; Strychor, S.; Maruca, L.J.; King, C.R.; Jung, L.L.; Parise, R.A.; et al. Disposition of 9-nitrocamptothecin and its 9-aminocamptothecin metabolite in relation to ABC transporter genotypes. *Investig. New Drugs* **2006**, *24*, 393–401. [CrossRef]

189. Han, J.-Y.; Lim, H.-S.; Yoo, Y.-K.; Shin, E.S.; Park, Y.H.; Lee, S.Y.; Lee, J.-E.; Lee, D.H.; Kim, H.T.; Lee, J.S. Associations ofABCB1, ABCC2, andABCG2 polymorphisms with irinotecan-pharmacokinetics and clinical outcome in patients with advanced non-small cell lung cancer. *Cancer* **2007**, *110*, 138–147. [CrossRef]

190. Jada, S.R.; Lim, R.; Wong, C.I.; Shu, X.; Lee, S.C.; Zhou, Q.; Goh, B.C.; Chowbay, B. Role of UGT1A1*6,UGT1A1*28andABCG2c.421C>A polymorphisms in irinotecan-induced neutropenia in Asian cancer patients. *Cancer Sci.* **2007**, *98*, 1461–1467. [CrossRef] [PubMed]

191. Sai, K.; Saito, Y.; Maekawa, K.; Kim, S.-R.; Kaniwa, N.; Nishimaki-Mogami, T.; Sawada, J.-I.; Shirao, K.; Hamaguchi, T.; Yamamoto, N.; et al. Additive effects of drug transporter genetic polymorphisms on irinotecan pharmacokinetics/pharmacodynamics in Japanese cancer patients. *Cancer Chemother. Pharmacol.* **2009**, *66*, 95–105. [CrossRef]

192. Li, N.; Song, Y.; Du, P.; Shen, Y.; Yang, J.; Gui, L.; Wang, S.; Wang, J.; Sun, Y.; Han, X.; et al. Oral topotecan: Bioavailability, pharmacokinetics and impact of ABCG2 genotyping in Chinese patients with advanced cancers. *Biomed. Pharmacother.* **2013**, *67*, 801–806. [CrossRef] [PubMed]

193. Sobek, K.M.; Cummings, J.L.; Bacich, D.J.; O'Keefe, D.S. Contrasting roles of the ABCG2 Q141K variant in prostate cancer. *Exp. Cell Res.* **2017**, *354*, 40–47. [CrossRef] [PubMed]
194. Rudin, C.M.; Liu, W.; Desai, A.; Karrison, T.; Jiang, X.; Janisch, L.; Das, S.; Ramirez, J.; Poonkuzhali, B.; Schuetz, E.; et al. Pharmacogenomic and pharmacokinetic determinants of erlotinib toxicity. *J. Clin. Oncol.* **2008**, *26*, 1119–1127. [CrossRef] [PubMed]
195. Lemos, C.; Giovannetti, E.; Zucali, P.A.; Assaraf, Y.G.; Scheffer, G.L.; Van Der Straaten, T.; D'Incecco, A.; Falcone, A.; Guchelaar, H.-J.; Danesi, R.; et al. Impact of ABCG2 polymorphisms on the clinical outcome and toxicity of gefitinib in non-small-cell lung cancer patients. *Pharmacogenomics* **2011**, *12*, 159–170. [CrossRef] [PubMed]
196. Roberts, J.K.; Birg, A.V.; Lin, T.; Daryani, V.M.; Panetta, J.C.; Broniscer, A.; Robinson, G.W.; Gajjar, A.J.; Stewart, C.F. Population pharmacokinetics of oral topotecan in infants and very young children with brain tumors demonstrates a role of ABCG2 rs4148157 on the absorption rate constant. *Drug Metab. Dispos.* **2016**, *44*, 1116–1122. [CrossRef] [PubMed]
197. Abuznait, A.H.; Kaddoumi, A. Role of ABC Transporters in the pathogenesis of Alzheimer's disease. *ACS Chem. Neurosci.* **2012**, *3*, 820–831. [CrossRef] [PubMed]
198. Do, T.M.; Noel-Hudson, M.-S.; Ribes, S.; Besengez, C.; Smirnova, M.; Cisternino, S.; Buyse, M.; Calon, F.; Chimini, G.; Chacun, H.; et al. ABCG2- and ABCG4- mediated efflux of amyloid-β peptide 1–40 at the mouse blood-brain barrier. *J. Alzheimer's Dis.* **2012**, *30*, 155–166. [CrossRef]
199. Xiong, H.; Callaghan, D.; Jones, A.; Bai, J.; Rasquinha, I.; Smith, C.; Pei, K.; Walker, D.; Lue, L.-F.; Stanimirovic, D.; et al. ABCG2 is upregulated in Alzheimer's brain with cerebral amyloid angiopathy and may act as a gatekeeper at the blood-brain barrier for a 1–40 peptides. *J. Neurosci.* **2009**, *29*, 5463–5475. [CrossRef] [PubMed]
200. Ascherio, A. Urate as a predictor of the rate of clinical decline in Parkinson disease. *Arch. Neurol.* **2009**, *66*, 1460–1468. [CrossRef]
201. Constantinescu, R.; Zetterberg, H. Urate as a marker of development and progression in Parkinson's disease. *Drugs Today* **2011**, *47*, 369. [CrossRef]
202. Matsuo, H.; Tomiyama, H.; Satake, W.; Chiba, T.; Onoue, H.; Kawamura, Y.; Nakayama, A.; Shimizu, S.; Sakiyama, M.; Funayama, M.; et al. ABCG2 variant has opposing effects on onset ages of Parkinson's disease and gout. *Ann. Clin. Transl. Neurol.* **2015**, *2*, 302–306. [CrossRef]
203. Kobylecki, C.J.; Nordestgaard, B.G.; Afzal, S. Plasma urate and risk of Parkinson's disease: A mendelian randomization study. *Ann. Neurol.* **2018**, *84*, 178–190. [CrossRef]
204. Evseenko, D.A.; Murthi, P.; Paxton, J.W.; Reid, G.; Emerald, B.S.; Mohankumar, K.M.; Lobie, P.E.; Brennecke, S.P.; Kalionis, B.; Keelan, J.A. The ABC transporter BCRP/ABCG2 is a placental survival factor, and its expression is reduced in idiopathic human fetal growth restriction. *FASEB J.* **2007**, *21*, 3592–3605. [CrossRef]
205. Szabó, E.; Kulin, A.; Korányi, L.; Literáti-Nagy, B.; Cserepes, J.; Somogyi, A.; Sarkadi, B.; Várady, G. Alterations in erythrocyte membrane transporter expression levels in type 2 diabetic patients. *Sci. Rep.* **2021**, *11*, 1–10. [CrossRef] [PubMed]
206. Krishnamurthy, P.; Ross, D.D.; Nakanishi, T.; Bailey-Dell, K.; Zhou, S.; Mercer, K.E.; Sarkadi, B.; Sorrentino, B.P.; Schuetz, J.D. the stem cell marker Bcrp/ABCG2 enhances hypoxic cell survival through interactions with heme. *J. Biol. Chem.* **2004**, *279*, 24218–24225. [CrossRef]
207. Erdélyi-Belle, B.; Török, G.; Apáti, A.; Sarkadi, B.; Schaff, Z.; Kiss, A.; Homolya, L. Expression of tight junction components in hepatocyte-like cells differentiated from human embryonic stem cells. *Pathol. Oncol. Res.* **2015**, *21*, 1059–1070. [CrossRef]
208. Török, G.; Erdei, Z.; Lilienberg, J.; Apáti, A.; Homolya, L. The importance of transporters and cell polarization for the evaluation of human stem cell-derived hepatic cells. *PLoS ONE* **2020**, *15*, e0227751. [CrossRef]
209. Erdei, Z.; Sarkadi, B.; Brozik, A.; Szebényi, K.; Várady, G.; Makó, V.; Péntek, A.; Orbán, T.I.; Apáti, A. Dynamic ABCG2 expression in human embryonic stem cells provides the basis for stress response. *Eur. Biophys. J.* **2012**, *42*, 169–179. [CrossRef]
210. Robey, R.W.; To, K.K.; Polgar, O.; Dohse, M.; Fetsch, P.; Dean, M.; Bates, S.E. ABCG2: A perspective. *Adv. Drug Deliv. Rev.* **2009**, *61*, 3–13. [CrossRef] [PubMed]
211. Vauthier, V.; Housset, C.; Falguières, T. Targeted pharmacotherapies for defective ABC transporters. *Biochem. Pharmacol.* **2017**, *136*, 1–11. [CrossRef] [PubMed]
212. Loo, T.W.; Bartlett, M.C.; Clarke, D.M. Rescue of ΔF508 and other misprocessed CFTR mutants by a novel quinazoline compound. *Mol. Pharm.* **2005**, *2*, 407–413. [CrossRef] [PubMed]
213. Van Goor, F.; Hadida, S.; Grootenhuis, P.D.J.; Burton, B.; Stack, J.H.; Straley, K.S.; Decker, C.J.; Miller, M.; McCartney, J.; Olson, E.R.; et al. Correction of the F508del-CFTR protein processing defect in vitro by the investigational drug VX-809. *Proc. Natl. Acad. Sci. USA* **2011**, *108*, 18843–18848. [CrossRef] [PubMed]
214. Boyle, M.P.; Bell, S.C.; Konstan, M.W.; McColley, S.A.; Rowe, S.M.; Rietschel, E.; Huang, X.; Waltz, D.; Patel, N.R.; Rodman, D. A CFTR corrector (lumacaftor) and a CFTR potentiator (ivacaftor) for treatment of patients with cystic fibrosis who have a phe508del CFTR mutation: A phase 2 randomised controlled trial. *Lancet Respir. Med.* **2014**, *2*, 527–538. [CrossRef]
215. Keating, D.; Marigowda, G.; Burr, L.; Daines, C.; Mall, M.A.; McKone, E.F.; Ramsey, B.W.; Rowe, S.M.; Sass, L.A.; Tullis, E.; et al. VX-445–tezacaftor–ivacaftor in patients with cystic fibrosis and one or two phe508del alleles. *N. Engl. J. Med.* **2018**, *379*, 1612–1620. [CrossRef] [PubMed]
216. Sabirzhanova, I.; Pacheco, M.L.; Rapino, D.; Grover, R.; Handa, J.T.; Guggino, W.B.; Cebotaru, L. Rescuing trafficking mutants of the ATP-binding cassette protein, ABCA4, with small molecule correctors as a treatment for stargardt eye disease. *J. Biol. Chem.* **2015**, *290*, 19743–19755. [CrossRef]

217. Liu, Q.; Sabirzhanova, I.; Bergbower, E.A.S.; Yanda, M.; Guggino, W.G.; Cebotaru, L. The CFTR corrector, VX-809 (lumacaftor), rescues ABCA4 Trafficking mutants: A potential treatment for stargardt disease. *Cell. Physiol. Biochem.* **2019**, *53*, 400–412. [CrossRef]
218. Rubenstein, R.C.; Egan, M.E.; Zeitlin, P.L. In vitro pharmacologic restoration of CFTR-mediated chloride transport with sodium 4-phenylbutyrate in cystic fibrosis epithelial cells containing delta F508-CFTR. *J. Clin. Investig.* **1997**, *100*, 2457–2465. [CrossRef] [PubMed]
219. Gautherot, J.; Durand-Schneider, A.-M.; Delautier, D.; Delaunay, J.-L.; Rada, A.; Gabillet, J.; Housset, C.; Maurice, M.; Aït-Slimane, T. Effects of cellular, chemical, and pharmacological chaperones on the rescue of a trafficking-defective mutant of the ATP-binding cassette transporter proteins ABCB1/ABCB4. *J. Biol. Chem.* **2012**, *287*, 5070–5078. [CrossRef]
220. Gordo-Gilart, R.; Andueza, S.; Hierro, L.; Jara, P.; Alvarez, L. Functional rescue of trafficking-impaired ABCB4 mutants by chemical chaperones. *PLoS ONE* **2016**, *11*, e0150098. [CrossRef]
221. Hayashi, H.; Sugiyama, Y. 4-phenylbutyrate enhances the cell surface expression and the transport capacity of wild-type and mutated bile salt export pumps. *Hepatology* **2007**, *45*, 1506–1516. [CrossRef]
222. Pomozi, V.; Brampton, C.; Szeri, F.; Dedinszki, D.; Kozák, E.; van de Wetering, K.; Hopkins, H.; Martin, L.; Váradi, A.; Le Saux, O. Functional rescue of ABCC6 deficiency by 4-phenylbutyrate therapy reduces dystrophic calcification in Abcc6−/− Mice. *J. Investig. Dermatol.* **2017**, *137*, 595–602. [CrossRef]
223. Kopito, R.R. Aggresomes, inclusion bodies and protein aggregation. *Trends Cell Biol.* **2000**, *10*, 524–530. [CrossRef]
224. Johnston, J.A.; Ward, C.L.; Kopito, R.R. Aggresomes: A cellular response to misfolded proteins. *J. Cell Biol.* **1998**, *143*, 1883–1898. [CrossRef]
225. Polgar, O.; Ediriwickrema, L.S.; Robey, R.W.; Sharma, A.; Hegde, R.S.; Li, Y.; Xia, D.; Ward, Y.; Dean, M.; Ozvegy-Laczka, C.; et al. Arginine 383 is a crucial residue in ABCG2 biogenesis. *Biochim. Biophys. Acta BBA Bioenerg.* **2009**, *1788*, 1434–1443. [CrossRef]
226. Lukacs, G.; Mohamed, A.; Kartner, N.; Chang, X.; Riordan, J.; Grinstein, S. Conformational maturation of CFTR but not its mutant counterpart (delta F508) occurs in the endoplasmic reticulum and requires ATP. *EMBO J.* **1994**, *13*, 6076–6086. [CrossRef] [PubMed]
227. Ward, C.L.; Kopito, R.R. Intracellular turnover of cystic fibrosis transmembrane conductance regulator. Inefficient pro-cessing and rapid degradation of wild-type and mutant proteins. *J. Biol. Chem.* **1994**, *269*, 25710–25718. [CrossRef]
228. Loo, T.W.; Clarke, D.M. Prolonged association of temperature-sensitive mutants of human P-glycoprotein with calnexin during biogenesis. *J. Biol. Chem.* **1994**, *269*, 28683–28689. [CrossRef]
229. Jensen, T.J.; Loo, M.A.; Pind, S.; Williams, D.B.; Goldberg, A.L.; Riordan, J.R. Multiple proteolytic systems, including the proteasome, contribute to CFTR processing. *Cell* **1995**, *83*, 129–135. [CrossRef]
230. Ward, C.L.; Omura, S.; Kopito, R.R. Degradation of CFTR by the ubiquitin-proteasome pathway. *Cell* **1995**, *83*, 121–127. [CrossRef]
231. Wakabayashi, K.; Nakagawa, H.; Tamura, A.; Koshiba, S.; Hoshijima, K.; Komada, M.; Ishikawa, T. Intramolecular disulfide bond is a critical check point determining degradative fates of ATP-binding cassette (ABC) transporter ABCG2 protein. *J. Biol. Chem.* **2007**, *282*, 27841–27846. [CrossRef]
232. Van Goor, F.; Hadida, S.; Grootenhuis, P.D.J.; Burton, B.; Cao, D.; Neuberger, T.; Turnbull, A.; Singh, A.; Joubran, J.; Hazlewood, A.; et al. Rescue of CF airway epithelial cell function in vitro by a CFTR potentiator, VX-770. *Proc. Natl. Acad. Sci. USA* **2009**, *106*, 18825–18830. [CrossRef] [PubMed]
233. Delaunay, J.; Bruneau, A.; Hoffmann, B.; Durand-Schneider, A.; Barbu, V.; Jacquemin, E.; Maurice, M.; Housset, C.; Callebaut, I.; Aït-Slimane, T. Functional defect of variants in the adenosine triphosphate–binding sites of ABCB4 and their rescue by the cystic fibrosis transmembrane conductance regulator potentiator, ivacaftor (VX-770). *Hepatology* **2016**, *65*, 560–570. [CrossRef] [PubMed]
234. Mareux, E.; Lapalus, M.; Amzal, R.; Almes, M.; Aït-Slimane, T.; Delaunay, J.; Adnot, P.; Collado-Hilly, M.; Davit-Spraul, A.; Falguières, T.; et al. Functional rescue of an ABCB11 mutant by ivacaftor: A new targeted pharmacotherapy approach in bile salt export pump deficiency. *Liver Int.* **2020**, *40*, 1917–1925. [CrossRef] [PubMed]
235. Lingam, S.; Thonghin, N.; Ford, R.C. Investigation of the effects of the CFTR potentiator ivacaftor on human P-glycoprotein (ABCB1). *Sci. Rep.* **2017**, *7*, 1–7. [CrossRef]

International Journal of
Molecular Sciences

Article

Analysis of Sequence Divergence in Mammalian ABCGs Predicts a Structural Network of Residues That Underlies Functional Divergence

James I. Mitchell-White [1,2,*], Thomas Stockner [3], Nicholas Holliday [1], Stephen J. Briddon [1,2] and Ian D. Kerr [1,*]

[1] School of Life Sciences, University of Nottingham, Queen's Medical Centre, Nottingham NG7 2UH, UK; nicholas.holliday@nottingham.ac.uk (N.H.); stephen.briddon@nottingham.ac.uk (S.J.B.)
[2] Centre of Membrane Proteins and Receptors, Universities of Birmingham and Nottingham, The Midlands, Nottingham NG7 2UH, UK
[3] Center for Physiology and Pharmacology, Institute of Pharmacology, Medical University of Vienna, Währingerstrasse 13A, 1090 Vienna, Austria; thomas.stockner@meduniwien.ac.at
* Correspondence: James.Mitchell-White1@nottingham.ac.uk (J.I.M.-W.); Ian.Kerr@nottingham.ac.uk (I.D.K.)

Abstract: The five members of the mammalian G subfamily of ATP-binding cassette transporters differ greatly in their substrate specificity. Four members of the subfamily are important in lipid transport and the wide substrate specificity of one of the members, ABCG2, is of significance due to its role in multidrug resistance. To explore the origin of substrate selectivity in members 1, 2, 4, 5 and 8 of this subfamily, we have analysed the differences in conservation between members in a multiple sequence alignment of ABCG sequences from mammals. Mapping sets of residues with similar patterns of conservation onto the resolved 3D structure of ABCG2 reveals possible explanations for differences in function, via a connected network of residues from the cytoplasmic to transmembrane domains. In ABCG2, this network of residues may confer extra conformational flexibility, enabling it to transport a wider array of substrates.

Keywords: ABC transporter; multidrug resistance; membrane protein; functional divergence; ABCG2

Citation: Mitchell-White, J.I.; Stockner, T.; Holliday, N.; Briddon, S.J.; Kerr, I.D. Analysis of Sequence Divergence in Mammalian ABCGs Predicts a Structural Network of Residues That Underlies Functional Divergence. *Int. J. Mol. Sci.* **2021**, *22*, 3012. https://doi.org/10.3390/ijms22063012

Academic Editor: Thomas Falguières

Received: 1 February 2021
Accepted: 12 March 2021
Published: 16 March 2021

Publisher's Note: MDPI stays neutral with regard to jurisdictional claims in published maps and institutional affiliations.

1. Introduction

ATP-binding cassette (ABC) proteins form a very large family across all domains of life, responsible for the primary active uptake and export of nutrients, toxins, lipids, peptides and other metabolites. ATP hydrolysis is carried out at two cytoplasmic nucleotide-binding domains (NBDs), and the energy released is coupled to conformational changes in two transmembrane domains (TMDs) to power transport of substrate. In mammals, ABC proteins are divided into seven subfamilies, A–G, although the ABCE and F families lack TMDs and are associated with ribosome function [1]. Often, members within a subfamily, though sharing common descent, can have very different functions. The family investigated here is the G subfamily of ABC transporters in mammals (ABCGs). In most mammals, there are five members of this subfamily [2]. All mammalian ABCGs share a common arrangement of domains, all being "half-transporters" with just a single NBD and TMD in the primary amino acid sequence. A unique property of ABCG arrangement is that the NBD is N-terminal to the TMD, so they are referred to as "reverse" half-transporters.

Four of the mammalian ABCGs have a repertoire of substrates limited to lipids. Two of these, ABCG1 and ABCG4, have sequences much more similar to one another than they are to the rest of the ABCGs. They also seem to share much of their function, regulating cholesterol metabolism by transporting cholesterol into high-density lipoprotein [3]. Precise differences in their function are yet to be determined, but they do seem to differ significantly in their tissue expression profiles [3–5].

The two other lipid-transporting ABCGs, ABCG5 and ABCG8, are also more closely related to one another than they are to the other ABCGs, though to a lesser extent than ABCG1 and ABCG4. They have taken on necessarily different roles by forming an obligate heterodimer, neither protomer being trafficked to the membrane if expressed alone [6]. ABCG5/G8 expressed in the liver and intestine limits the uptake of toxic plant and shellfish sterols and is responsible for 35% of the efflux of cholesterol in the intestine. The ABC dimer G5/G8 has only one functional ATP-binding site, indicating that ABCG5 and ABCG8 have diverged in function in this respect.

The final ABCG, ABCG2, has a much broader substrate specificity. It was first isolated in placental tissue and breast cancer cell lines [7,8], and has been since identified as a multidrug resistance (MDR) protein. It can export a wide variety of substrates, including many chemotherapy drugs, making it a target of great therapeutic interest. For this reason, it is the best studied of the ABCG subfamily.

Recently, structures have been solved for ABCG5/G8 and ABCG2. First came the structure of ABCG5/G8 [9], which was used to model a structure of ABCG2 [10]. Docking substrates to this model identified multiple possible binding sites, already suggested by previous biochemical work [11]. With the first structures of ABCG2 [12–14], its broader substrate specificity was explained through a relatively large internal cavity, compared to ABCG5/G8's deep, slit-like cavity, forming part of the transport pathway, though in more recent structures, the cavity is only present in structures with substrates bound [15]. In spite of these structural advances, the molecular basis for differences in function between ABCG family members is largely unknown. As their differences ultimately arise from differences in their sequence, it is possible that comparison of conservation between ABCGs could provide clues to help ascertain this molecular basis.

Families of genes can occur when a gene duplicates and the different copies start to take on different functions, a process known as functional divergence [16,17]. When this happens, the evolutionary pressures on the duplicated genes start to differ, with impact on the sequences of the proteins encoded. Non-synonymous mutations in structural elements with functional importance are less likely to persist [18]. In two functionally divergent proteins, a structural element may be more important to the function of one than the other, which will be reflected in this region being better conserved in the protein for which the element is more important. This has been called type I divergence [16]. A similar phenomenon, type II divergence, occurs if the same element is important for the function of both proteins, but the important properties of the amino acid found there are different. This is reflected in the region being conserved in both proteins, but with different amino acids being conserved. The differences in sequence conservation caused by functional divergence have been used to identify important sites in proteins.

In order to analyse the conservation between the members of the G subfamily of ABC transporters in mammals, we have calculated functional divergence of residues based on Shannon entropy [19] from a large multiple sequence alignment of ABCGs. We have examined residues with particular patterns of type II divergence between ABCGs reflecting some of the functional divergence responsible for their differences. Hypotheses regarding the structural basis of these functional differences were derived by mapping positions in the alignment that share particular types of conservation onto the apo-closed structure of ABCG2. Specifically, we have identified a top-to-bottom signature, passing through the polar relay of ABCG5/G8 [9,20], which may contribute to allosteric differences in the G subfamily responsible for differences in substrate specificity.

2. Results

2.1. Overall Conservation Patterns

A total of 174 ABCG protein sequences (summarised in Supplementary Table S1) were analysed. These were grouped according to the protein they represent, and their conservation calculated as described in methods. A tree constructed from these sequences showing the relationship between the ABCG proteins is shown in Figure 1a. The alignment had a

length of 1269 positions (henceforth "columns"). Of these, 674 columns had gaps in either
>10% of all sequences or >30% of sequences for one of the proteins (see Supplementary
Figure S1a). Of the remaining 595 columns, 594 met the entropy cutoff for conservation in
at least one protein. A total of 61 of these columns were conserved across the ABCG family,
and the remaining 533 had some type of divergence, as summarised in Figure 1b.

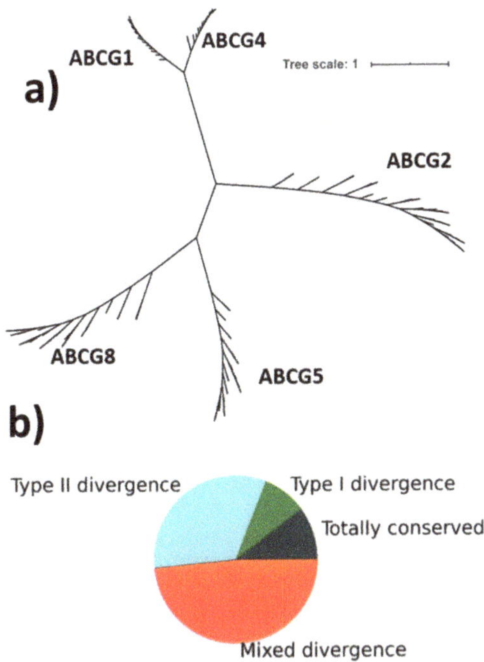

Figure 1. (**a**) Phylogenetic tree of mammalian ABC subfamily G proteins. Tree based on 174 protein
sequences, aligned with multiple alignment fast Fourier transform (MAFFT). Names of taxa have
been removed for clarity. (**b**) Pie chart showing proportions of conservation and divergence. In the
594 columns showing conservation in at least one protein in the G subfamily, 61 are totally conserved
(grey); 52 show simple type I divergence (where one set has conservation, and the others do not)
(green); 193 show type II divergence (where each set is conserved, but with a different residue) (cyan);
and the remaining 288 have some mixture of divergence (e.g., column 891 is a conserved cysteine in
ABCG2, and a conserved leucine in ABCG1 and ABCG4, but is not conserved in other groups. Thus
it has neither purely type I nor type II divergence) (red).

An example of conservation is represented in Figure 2. It shows one part of the interface between TMD and NBD which is vital for transmitting energy from ATP hydrolysis
in ABCGs, often referred to as the "elbow helix" in ABCG literature. Columns in this
region of the alignment display the different types of conservation of relevance; firstly
columns that show total conservation, where not only is the column conserved for each
protein, but it is conserved in the same way (e.g., column 900 in Figure 2 where all ABCG
sequences conserve arginine at this position). Secondly, it shows type I divergence, where
the column is conserved as the same amino acid for at least one protein, with other proteins
not conserving the column. For example, column 895 in Figure 2 is conserved as a cysteine
in ABCG1 and ABCG4 but is not conserved in ABCG2, ABCG5 or ABCG8. Thirdly, type II
divergence, where each protein shows conservation, but different proteins can be conserved
in different ways is evident in columns 893, 894, 897, 901, 904 and 905 in Figure 2. For
example, in position 905, ABCG1 and ABCG4 conserve isoleucine, ABCG2 and ABCG5

conserve leucine and ABCG8 conserves aspartate. Finally, several other columns display a mix of types of divergence; for example, column 890 shows a position conserved only in ABCG1, ABCG2 and ABCG4, and the residue conserved is different in all cases.

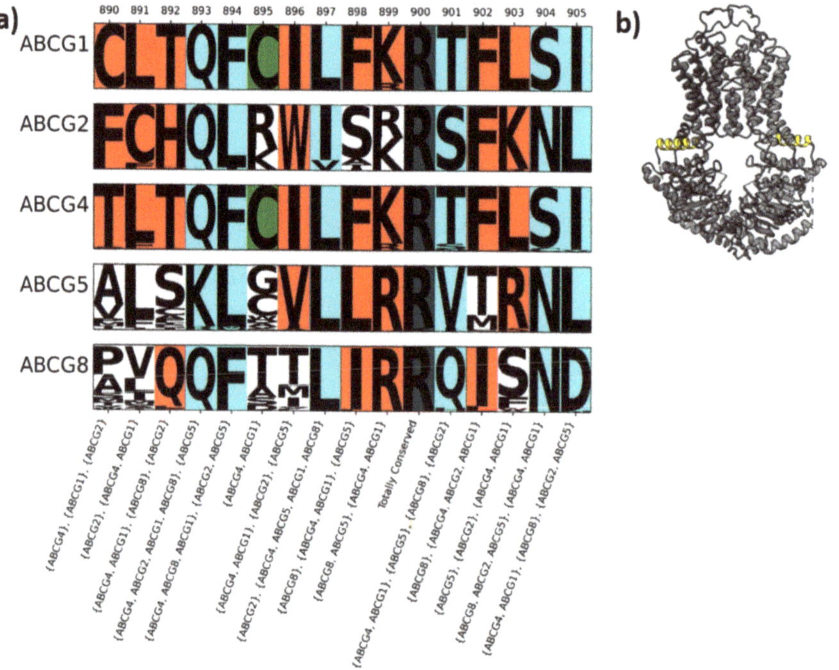

Figure 2. Conservation in the alignment of ABCG protein sequences. (**a**) Sequence logo in which sequences have been divided by the protein they represent. Font size corresponds to the fraction of sequences with that residue in that column. Conserved positions have coloured backgrounds so that totally conserved columns are grey, columns with type I divergence are green, columns with type II divergence are aqua, and columns with mixed divergence are red. Conservation patterns as described in the text are shown at the bottom. (**b**) Structure of ABCG2 (PDBID: 6vxf) highlighting the area represented in the logo. This corresponds to the elbow helix in ABCG2.

Many of the approaches used to investigate functional divergence return a score for each column reflecting how it is conserved across the whole alignment and some are limited to a comparison between two groups. In the case of the ABCGs, one aspect of functional divergence worth exploring might be ABCG2's broader substrate specificity. If comparing two groups, examining both type I and type II divergence is worthwhile. However, the substrates transported by ABCG1 and ABCG4 differ from those transported by ABCG5/G8, so their substrate specificities are achieved in different ways. Considering possible functional divergence within ABCG members highlights some of the difficulties with terminology. Here, we have defined type II divergence to include any column in which each protein is conserved, without conservation across the whole family, and type I divergence to include columns in which one or more proteins have the same amino acid conserved, and all other proteins are not conserved. Rather than calculating scores for the whole alignment, we have classified columns according to the proteins in which they are conserved, allowing inferences to be drawn from differences between multiple groupings.

In this manuscript, the different ways to group proteins to examine their conservation is referred to as a conservation pattern. Columns with a particular conservation pattern are represented by having any family members conserved in that column written in brackets.

If more than one member has the same amino acid conserved at that position, they are held in the same brackets, separated by a comma. To illustrate this nomenclature with respect to Figure 2, among the conservation patterns visible in this section of the alignment are (ABCG1, ABCG4), (ABCG2), (ABCG5), (ABCG8) in column 901 and (ABCG1, ABCG4) (ABCG2, ABCG5, ABCG8) in column 904, both of which represent type II divergence.

There are 202 theoretically possible conservation patterns, of which around half (106) are observed anywhere in the actual alignment. Most of these have very few representatives, with over 60 only having 1–3 representatives. Remarkably, almost half of the divergent positions in the alignment are contributed by just 14 different conservation patterns (Supplementary Table S2). Some of these well-represented patterns have implications for functional divergence when the relationships between the proteins are considered.

2.2. Phylogenetically Significant Type II Divergence in ABCGs

To explore the differences in substrate specificity within the G subfamily, it is necessary to explore columns of the alignment showing functional divergence. Residues essential to maintaining the overall ABCG fold will be either identical across proteins or highly conserved. Other behaviours, such as force transmission and substrate recognition, are likely to be conserved by each protein, but change across the family. The approach adopted here, which classifies columns by their conservation pattern, was deliberately chosen to allow interpretation of differences between multiple groups within the alignment. Though it does not provide a score, classifying columns by conservation pattern allows discrimination of functional divergence at different levels, exploiting existing knowledge of the proteins under investigation. Emphasis here was on the ability to estimate functional divergence in a way that allows interpretation based on what we know of the proteins involved.

The conservation patterns that are most likely to yield insight into differences in substrate binding are those that separate proteins by their substrates. ABCG1 and ABCG4 have a high sequence identity, and are identical in 434 of 595 columns (excluding gapped columns). ABCG1 and ABCG4 also overlap in their function and substrates [3,21], so grouping them together to establish functional divergence is sensible from both an evolutionary and functional perspective. Though ABCG5 and ABCG8 by definition transport the same substrates, their interactions with those substrates may differ, if the shape and chirality of the substrate is reflected in the substrate-binding site. Furthermore, they are less similar in sequence (being conserved in the same way in only 138 columns), and to some extent must carry out different functions due to the asymmetry of their nucleotide-binding sites [6].

2.3. The Conservation Pattern (ABCG1, ABCG4), (ABCG2), (ABCG5), (ABCG8) Defines a Possible Allosteric Pathway in ABCG Proteins

The pattern (ABCG1, ABCG4), (ABCG2), (ABCG5), (ABCG8) separates the proteins by their probable substrate interactions, and is the most populated set of functionally divergent columns, with 33 columns. Residues corresponding to these columns are mapped on to an ABCG2 structure in Figure 3a. In this structure of ABCG2 (and also observed if mapped onto ABCG5/G8, Supplementary Figure S2), these residues form a "corkscrew" pattern from the cytoplasmic face of the NBDs, through the TMDs to the extracellular face of the protein. This distribution implies that some important differences in the function of members the G subfamily are due to differences in allostery, as corresponding residues are ideally placed to form a network of residues coordinating conformational changes throughout the protein. Their distribution can be compared with residues with the conservation pattern (ABCG1, ABCG4), (ABCG2), (ABCG5, ABCG8) (Figure 3b) and the much less common pattern (ABCG1), (ABCG4), (ABCG2), (ABCG5), (ABCG8) (Figure 3c).

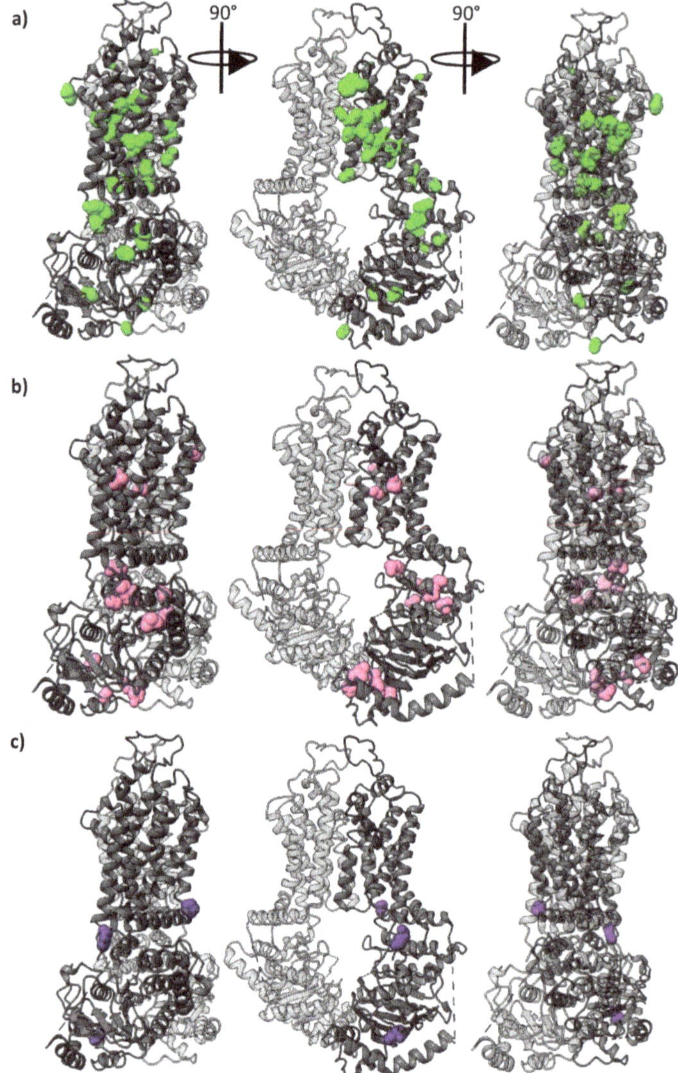

Figure 3. Functionally divergent residues shown on ABCG2 (PDBID: 6vxf) shown as coloured spheres on three views of the structure. (**a**) Residues conserved in the pattern (ABCG1, ABCG4), (ABCG2) (ABCG5), (ABCG8) as green spheres. (**b**) Residues conserved in the pattern (ABCG1, ABCG4), (ABCG2), (ABCG5, ABCG8) as pink spheres. (**c**) Residues conserved differently in each protein, i.e., with the conservation pattern (ABCG1), (ABCG4), (ABCG2), (ABCG5), (ABCG8) as purple spheres.

2.4. Conservation of the Polar Relay

A feature of ABCG5/G8 identified from the ABCG family fold [9] that is also likely to carry out a role in conformational changes is the "polar relay". This comprises 11 residues from ABCG5 and 9 residues from ABCG8. In the multiple sequence alignment, five of these positions overlap, leaving a total of 15 columns in the alignment corresponding to the polar relay (Figure 4). Notably, one of the columns in the polar relay of both ABCG5 and ABCG8 (column 1011) aligns with R482 in transmembrane helix 3 of ABCG2, mutation of which has long been shown to alter substrate specificity [11,22,23].

Figure 4. Conservation patterns in columns corresponding to the polar relay in ABCG5/G8. Columns coloured as in Figure 2. Orange dashed boxes indicate the polar relay for ABCG5 and ABCG8. Columns with the conservation pattern (ABCG1, ABCG4), (ABCG2), (ABCG5), (ABCG8) have that pattern underlined in blue at the bottom.

The conservation patterns of these fifteen residues are shown in Figure 4. Though 12 of these positions have type II divergence for all ABCGs, two columns (915: R389 in ABCG5 and H420 in ABCG8; and 963: N437 in ABCG5 and D466 in ABCG8) are not conserved in ABCG8, and one is not conserved in ABCG5 (1006: V471 in ABCG5 and E500 in ABCG8). One remarkable observation is that 40% of the columns in the polar relay (6/15) have the conservation pattern (ABCG1, ABCG4), (ABCG2), (ABCG5), (ABCG8). This makes this conservation pattern much more common here than in the whole protein, as it is only found in 5.5% (33/595) of the aligned columns (Supplementary Table S3).

The observation that the type II divergence pattern (ABCG1, ABCG4), (ABCG2), (ABCG5), (ABCG8) subsumes much of the polar relay (as shown in Figures 4 and 5a), which has previously been attributed allosteric significance in ABCG5/G8, suggests that the entire corkscrew of residues contributes to the allosteric divergence of the ABCG family.

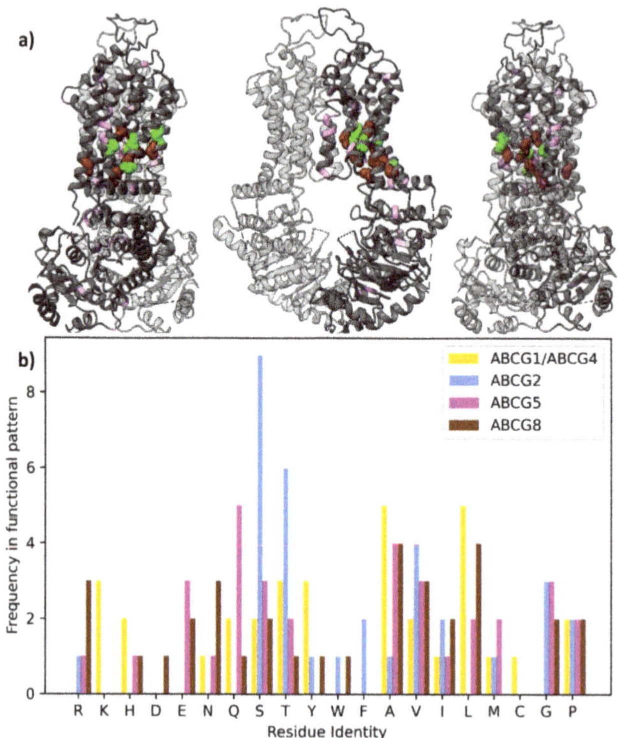

Figure 5. Comparison of the polar relay with functionally divergent residues. (**a**) Distribution on structure of ABCG2. Residues found in the polar relay are shown as spheres. Those with the conservation pattern (ABCG1, ABCG4), (ABCG2), (ABCG5), (ABCG8) are coloured green. Others are coloured red. Residues outside the polar relay with the conservation pattern above are coloured violet within the cartoon representation. (**b**) Identity of residues with conservation pattern (ABCG1, ABCG4), (ABCG2), (ABCG5), (ABCG8). Bars are coloured by protein, and their height represents the number of that residue found in the 33 positions with the above conservation pattern for that group. For each residue, bars are in the order ABCG1 and ABCG4; ABCG2; ABCG5; and ABCG8.

2.5. Sidechain Properties in the Allosteric Corkscrew

In a recent review [24], it was noted that composition of residues in the polar relay of ABCG2 and of ABCG5/G8 differs, with ABCG8 having a relatively high number of charged residues and ABCG2 a relatively low number. In residues with the conservation pattern (ABCG1, ABCG4), (ABCG2), (ABCG5), (ABCG8) this pattern is reiterated, ABCG8 having seven charged residues, ABCG5 five and ABCG2 one (Figure 5b). Perhaps notably, the only charged residue in ABCG2 with this conservation pattern is R482, mutation of which is long associated with altered substrate specificity [11,22,23].

An even greater discriminant between ABCG2 and other ABCG members is that 15 of the 33 residues with this conservation pattern are polar, hydroxylated residues (serine or threonine) in ABCG2 (Supplementary Figure S3). There are relatively few hydroxylated amino acids in these positions for other ABCGs (ABCG1/4: 5, ABCG5: 5, ABCG8: 3). Do these dissimilarities in the corkscrew of type II divergent residues contribute to differences in protein function? Polar and ionisable residues can drive specific helix oligomerisation, but this does not include serine or threonine alone [25–27] and we did not observe the specific motifs predicted to drive helix association [28,29]. Rather, an intriguing possibility is that serine and threonine form intra-helical hydrogen bonds, which can bend the helix in certain conformations, lending ABCG2 unusual flexibility in this region. Other residues significantly contributing to flexibility, such as glycine and proline, are no more common in other proteins compared with ABCG2. This extra flexibility could permit the binding and transport of diverse sizes of substrates, coupled to allosteric motions communicated through this network. Similar influence of hydroxylated amino acids in driving substrate-specific conformational changes is observed in some GPCRs [30–33].

2.6. Conservation of Other Regions

Electron microscopy and X-ray crystallography data on ABCGs have indicated other structural regions that are proposed to be critical for allosteric communication. We analysed the triple helical bundle between the NBD and TMD, which is considered to be a vital region for transmission of force from ATPase activity to the TMD. This region spans 54 columns in the multiple sequence alignment and 28 different conservation patterns are observed here. The hot-spot helix is most highly conserved, with 40% of its residues being conserved across the alignment, but no other patterns are significantly different here from the alignment as a whole (Supplementary Table S4). Though the triple helical bundle is highly conserved, the whole of it is not conserved across the G subfamily. Nor is it a motif that defines the difference between ABCG members well. Part of this comes from its being less well conserved in ABCG5 and ABCG8 (~70% for each), perhaps indicating that heterodimerisation reduces the evolutionary pressure on some of these positions. This may be particularly true for this region, due to its importance for transmitting force from ATP hydrolysis [34], which is altered in ABCG5/G8 due to the degenerate NBS.

Given the differences between the dimerisation behaviour of ABCG members (i.e., that some form homodimers and others are obligate heterodimers), we inspected the dimerisation interfaces (both at the TMD:TMD and the G-family specific NBD:NBD interface [35] to see if this was reflected in the conservation patterns. Residues within 5 Å of the other protomer in the structures 6VXF (ABCG2) and 5DO7 (ABCG5/G8) were found and the conservation patterns of corresponding columns were examined (Supplementary Table S5). A total of 46 different patterns are represented in this set, with completely conserved again being the most frequently observed. However, none of the patterns makes up a statistically significant fraction, meaning that the dimer interface is not a useful discriminant between ABCG members.

Binding pockets for substrates of ABCG2 and ABCG5/G8 have been identified from their structures. These are compared in Supplementary Figure S4 and Table S6. Interestingly, there is little overlap between the residues contributing to these pockets, with two columns contributing to the binding pockets of both ABCG2 and ABCG5, and two contributing to both ABCG5 and ABCG8. Both of the columns contributing to the pockets of both ABCG2 and ABCG5 have the conservation pattern (ABCG1, ABCG4), (ABCG2), (ABCG5), (ABCG8), so are part of the corkscrew. A further five columns with this conservation pattern contribute to the binding pocket of ABCG2, and another to that of ABCG5.

2.7. Other Conservation Patterns

Though the conservation pattern (ABCG1, ABCG4), (ABCG2), (ABCG5), (ABCG8) is the most observed in functionally divergent columns, some other patterns are well represented (the frequencies of well-represented functionally divergent conservation patterns are shown in Supplementary Table S2, and some of these are represented on the structure of ABCG2 in Supplementary Figure S5). Given the evolution of the subfamily, it is instructive to examine the conservation pattern (ABCG1, ABCG4), (ABCG2), (ABCG5, ABCG8), which highlights another 13 residues, shown on ABCG2 in Figure 3b. Notably, ten of these are found either in the NBD:NBD interface or the NBD:TMD interface. The remainder are found in the TMD, and two of these (C438 and I573 in ABCG2, P431/460 and F567/595 in ABCG5/G8) form pairs in the structures of ABCG2 and ABCG5/G8.

An interpretation of these patterns based on their likely evolution is that both of these sets of residues diverged when the ancestors of ABCG1 and ABCG4, ABCG2, and ABCG5 and ABCG8 specialised to transport different substrates. Later, ABCG5 and ABCG8 could take on different parts of the function of a transporter by forming an obligate heterodimer, and the residues corresponding to columns conserved as (ABCG1, ABCG4), (ABCG2), (ABCG5), (ABCG8) represent further functional divergence. Thus, both sets would be responsible for differences in substrate specificity, with 13 residues requiring conservation across ABCG5 and ABCG8. Taken together, these patterns reiterate the likely importance of allostery to differences in the function of the ABCGs.

The three patterns found most frequently other than (ABCG1, ABCG4), (ABCG2), (ABCG5, ABCG8), having 24, 23, and 23 members respectively, are (ABCG1, ABCG4), (ABCG2), (ABCG5); (ABCG1, ABCG4), (ABCG2), (ABCG8); and (ABCG1, ABCG4), (ABCG2). In total, these four patterns make up 103/533 of the functionally divergent columns. In all of these, ABCG1 and ABCG4 conserve the same amino acid, and ABCG2 conserves another, but conservation within ABCG5 and ABCG8 differs. In the conservation patterns not examined more closely in the sections above, ABCG5 or ABCG8 or both do not conserve the column. This indicates positions which have a decreased evolutionary pressure in ABCG5 and ABCG8, perhaps due to their splitting some of the functions which normally both halves of a dimer must maintain due to their forming a heterodimer. That so many of these positions are also sources of functional divergence between ABCG1 and ABCG4 on one hand and ABCG2 on the other is intriguing.

Another set of conservation patterns that is well represented is columns with type II divergence between one member and all the other members. (ABCG1, ABCG4, ABCG2, ABCG8), (ABCG5) has 16 members. Two interesting residues with this pattern are: F439 in ABCG2 (a tyrosine in ABCG5), which serves as a "clamp" for substrates [36], and E451 in ABCG2 (a leucine in ABCG5), which is a key residue in coupling ATPase activity to transport [34]. (ABCG1, ABCG4, ABCG5, ABCG8), (ABCG2) has 12 members. (ABCG1, ABCG4, ABCG2, ABCG5), (ABCG8) has 9 members. Due, probably, to the close relatedness of ABCG1 and ABCG4, there are fewer (0 and 5 respectively) columns with type II divergence between these and the rest of the subfamily. These are tantalising groups, as they show places that each member specialises in a way distinct from the ABCG family on the whole. However, a molecular interpretation is much more difficult.

3. Discussion

In this study, we have identified a corkscrew region in ABCG transporters whose conservation suggests a role in substrate specificity. Though no experimental work has yet been carried out to deliberately explore the functional effects of mutations to the corkscrew, some of its residues have been mutated as part of other studies, or observed as naturally occurring single nucleotide polymorphisms (Supplementary Table S7). Particularly notable in this regard is Cox et al. 2018 [37], which includes mutagenesis of five residues with the conservation pattern (ABCG1, ABCG4), (ABCG2), (ABCG5), (ABCG8), including the very well-studied residue R482A. Mutations of three of these five (T402A, S440A, and I543A) compromise transport of both mitoxantrone and pheophorbide A. T402 mutations have previously been described [38,39] as having decreased transport activity. Several others have observed diminished transport by ABCG2 with mutation of these residues, including mutations to S384, T434, and S441 [38,40–45]. Recently, diminished ATPase activity has been observed in ABCG5 [20] with mutation of A540 to phenylalanine, a residue also sharing this conservation pattern.

Other mutagenesis studies have included residues identified in this analysis as conserved in all aligned proteins. Many of these result in poor expression of mature protein [34,37,46], such as mutation of E138 in ABCG2 [34]. Though some have discernible effects on transport, surprisingly, mutation of P480 to alanine, despite being a mutation to a residue conserved in all sequences used in this analysis, and with dramatic chemical differences, has no effect on transport in ABCG2 [37,47]. This provides a cautionary example that care must be taken when interpreting these results. Other mutations to these positions are found as variants in vivo, some causing sitosterolemia, such as mutations to E146 in ABCG5, analogous to E138 in ABCG2 [48–57]. A summary of this, including disease-causing variants, can be found in the Supplementary Materials.

Comparing the ways mammalian ABCGs are conserved shows functionally and evolutionarily important signatures that are well represented in previous mutagenesis studies. Differences in the substrate specificity of subfamily members correspond to patterns of conservation that, when mapped onto 3D structures, are ideally placed to modulate the communication of conformational change between domains, suggesting that this may be responsible for some of the differences in substrate specificity. Particularly, grouping ABCG1 and ABCG4 together identifies a pattern, which we have named the corkscrew network. This is also important to a previously identified structural feature, the polar relay, and suggests a unifying hypothesis for substrate specificity in the subfamily: that allostery in this network underlies functional divergence. Appropriate experiments to test the importance of the corkscrew network of residues to differences in ABCG function promise to reveal interesting factors in their transport mechanism.

4. Materials and Methods

4.1. Sequence Acquisition

Through the NCBI, the RefSeq database [58] was queried for all nucleotide sequences matching "ABCG AND mammalia [organism]". Analysis was restricted to mammalia to afford greater confidence that function corresponded to the identity of the protein. Initially, 778 sequences from 112 species were identified. Not every species had a full complement of sequences for ABCG1, ABCG4, ABCG2, ABCG5 and ABCG8, so, where possible, these were found in the RefSeq database and added manually. A matching list of protein sequence IDs were used for a submission to Entrez. An in-house Python [59] script was used to check for and remove identical sequences.

Further sequences from some species were removed to prevent sequences from closely related species biasing later analysis. For example, sequences from 29 primates made up a high proportion of the total number of sequences, but presumably a low proportion of the organismal diversity. For this reason, 25 of the sequences were removed, keeping one ape (*Homo sapiens*), one monkey (*Piliocolobus tephrosceles*), one gelada (*Theropithecus gelada*), and one lemur (*Microcebus murinus*). Similar reasoning was used to reduce the number of species to 40. When choosing species to keep, a series of criteria were used. First, any well-studied species (e.g., *Homo sapiens*, *Mus musculus*) were retained. Next, species where one or more ABCG sequences were only tentatively identified (e.g., deposited in the database with the caveat "LOW QUALITY PROTEIN", or that were somewhat shorter than the canonical length of ABCGs (ca. 650 amino acids) were eliminated in preference to species with higher quality sequences. A preliminary alignment of all sequences using multiple alignment fast Fourier transform (MAFFT) was performed. This alignment was processed using MaxAlign, which identifies sequences that align most poorly with the others. If sequences from a species aligned poorly, they were disfavoured in the elimination process. In some cases, a species without an obvious substitute was eliminated—for example, the African elephant has only two ABCG sequences and both are low-quality sequences which aligned poorly. For this reason, the final number of species was reduced to 35. Where species could not be distinguished using these criteria, a random integer between one and the size of the set being reduced was generated, and the sequence matching that number in alphabetical order was kept. A summary of the sequences used can be found in Supplementary Table S1.

4.2. Alignment and Tree Construction

The final 174 protein sequences were aligned with MAFFT using the automatically assigned strategy, and other parameters set automatically by the MAFFT server, except raising the offset value to 0.123, which is the default value for the command line tool. This alignment was used to construct a tree using the Simply Phylogeny tool from ClustalW2, which was then visualised with the interactive Tree of Life [60]. The large number of sequence names reduced the clarity of the figure, so were removed, but the branches were otherwise left intact.

4.3. Calculation of Conservation

First, columns in which at least 10% of the total alignment, or 30% of one protein (e.g., ABCG1 sequences) were gaps, were labelled "Gap" and excluded from further analysis. Next, the conservation of the column across the whole alignment was calculated. Detecting conserved residues was based on information theory. Following Capra and Singh [61], the Shannon entropy of a column (i.e., a position in the multiple protein sequence alignment) was calculated. For amino acids in a column, entropy can take values between zero (all sequences are the same amino acid) and $\log_2(20)$ (each amino acid is equally likely). If entropy was lower than 2/3 of a bit, the column was counted as conserved.

If the Shannon entropy of the column for the whole alignment was <2/3 of a bit, the column was labelled as "All proteins conserved". Columns not labelled as "Gap" or "All proteins conserved" were then analysed by protein, e.g., the Shannon entropy was calculated for the column just in the ABCG1 sequences, or ABCG4 sequences. If it was not conserved (i.e., if the Shannon entropy within any of the proteins was <2/3) the position in the alignment was labelled "Not Conserved". If a column was conserved in one or more proteins, the most common residue found in each protein was recorded. Each of these columns was recorded as a list of pairs of conserved residues and the proteins matching that residue at that column. For example, column 1011 in the alignment corresponds to the well-studied residue 482 in ABCG2. This is conserved in all ABCGs, but differently—in ABCG1 and ABCG4, it is glutamine; in ABCG5, it is serine; and in ABCG8, it is histidine—so the record for that column is:

(1011, [('R', [ABCG2]), ('S', ['ABCG5']), ('Q', ['ABCG1', 'ABCG4']), ('H', ['ABCG8'])]).

To display a summary of sequences, logos were constructed using LogoMaker [62]. The positions in a protein corresponding to columns of interest were displayed on the structure of ABCG2 PDBID: 6VXF [15] using ChimeraX [63].

4.4. Binding Pockets

Residues corresponding to the binding pocket of ABCG2 for imatinib, mitoxantrone and SN38 were identified by taking residues with any part 5 Å from the substrate in structures 6VXH, 6VXI, and 6VXJ, respectively. Residues contributing to potential binding pockets for cholesterol in ABCG5/G8 were taken from Lee et al. 2016 [9]. The columns corresponding to these residues were identified and compared.

4.5. Statistics

To estimate the threshold for significance for the number of columns with a given conservation, the probability of a column being conserved in a particular pattern was modelled as a Poisson distribution with λ of 595/202 (non-gap columns/possible conservation patterns). To find a threshold for significance for the 202 possible conservation patterns, an initial $\alpha = 0.1$ was divided by 202. The cumulative probability of a conservation pattern occurring n times exceeds $1 - (0.1/202)$ at 10 columns, so any conservation pattern with more than 10 columns was treated as significant.

The expected values for the frequency of each conservation pattern were based on the frequency of conserved residues for non-gap positions for each protein. First, assuming conservation between proteins is independent, the probability of any set of proteins being conserved was estimated as the product of probabilities of conservation for the proteins conserved multiplied by the product of the probability of each protein not conserved not being conserved.

For each of these sets, the possible conservation patterns were generated by finding all possible partitions of the set. The relative probabilities of each of these partitions was calculated assuming the residues were conserved for each column independently, so the probability of any two proteins conserving the same residue was 0.05. For each partition, the probability of a column being conserved that way, given that set of proteins is conserved, is then $\frac{19!}{(20-m)!20^{n-1}}$, where n is the size of the set and m is the number of parts. To obtain estimates for the expected value for each conservation pattern, these values were then multiplied by the probability of that set of proteins being conserved, then multiplied by the number of non-gap positions.

Supplementary Materials: The following are available online at https://www.mdpi.com/1422-006 7/22/6/3012/s1, Supplementary Figure S1: Overall alignment properties; Supplementary Figure S2: The conservation pattern (ABCG1, ABCG4), (ABCG2), (ABCG5), (ABCG8) on ABCG5/G8; Supplementary Figure S3: Serines and threonines in the corkscrew; Supplementary Figure S4: Binding pockets of ABCG2, ABCG5, and ABCG8; Supplementary Figure S5: Well-populated type II divergence patterns; Supplementary Table S1: Protein Sequences used to identify functionally divergent positions; Supplementary Table S2: Number of columns with a given conservation pattern across the whole alignment; Supplementary Table S3: Partial contingency table for conservation patterns in the polar relay; Supplementary Table S4: Frequency of conservation patterns in the triple helical bundle; Supplementary Table S5: Frequency of conservation patterns at dimer interfaces; Supplementary Table S6: Residues contributing to the binding pockets of ABCG2 and ABCG5/G8; and Supplementary Table S7: Mutations to positions with conservation patterns of interest, which cites [20,34,37–46,48–51,53–57,64–68].

Author Contributions: J.I.M.-W. designed and carried out sequence analysis. J.I.M.-W., I.D.K., N.H., S.J.B. and T.S. contributed to the final manuscript. All authors have read and agreed to the published version of the manuscript.

Funding: I.D.K., S.J.B. and N.H. were funded by a BBSRC grant (BB/S001611/1).

Informed Consent Statement: Not applicable.

Data Availability Statement: Python code used in this article is available at https://github.com/kuraisle/ABCG_Family_Analysis (accessed on 15 March 2021), which also includes the sequence alignment used and instructions on using the code to explore it.

Conflicts of Interest: The authors declare no conflict of interest. The funders had no role in the design of the study; in the collection, analyses, or interpretation of data; in the writing of the manuscript, or in the decision to publish the results.

References

1. Dean, M.; Hamon, Y.; Chimini, G. The human ATP-binding cassette (ABC) transporter superfamily. *J. Lipid Res.* **2001**, *42*, 1007–1017. [CrossRef]
2. Kerr, I.D.; Haider, A.J.; Gelissen, I.C. The ABCG family of membrane-associated transporters: You don't have to be big to be mighty. *Br. J. Pharmacol.* **2011**, *164*, 1767–1779. [CrossRef] [PubMed]
3. Vaughan, A.M.; Oram, J.F. ABCA1 and ABCG1 or ABCG4 act sequentially to remove cellular cholesterol and generate cholesterol-rich HDL. *J. Lipid Res.* **2006**, *47*, 2433–2443. [CrossRef]
4. Hegyi, Z.; Homolya, L. Functional Cooperativity between ABCG4 and ABCG1 Isoforms. *PLoS ONE* **2016**, *11*, e0156516. [CrossRef]
5. Cserepes, J.; Szentpétery, Z.; Seres, L.; Özvegy-Laczka, C.; Langmann, T.; Schmitz, G.; Glavinas, H.; Klein, I.; Homolya, L.; Váradi, A.; et al. Functional expression and characterization of the human ABCG1 and ABCG4 proteins: Indications for heterodimerization. *Biochem. Biophys. Res. Commun.* **2004**, *320*, 860–867. [CrossRef] [PubMed]
6. Zhang, D.-W.; Graf, G.A.; Gerard, R.D.; Cohen, J.C.; Hobbs, H.H. Functional Asymmetry of Nucleotide-binding Domains in ABCG5 and ABCG8. *J. Biol. Chem.* **2006**, *281*, 4507–4516. [CrossRef]
7. Allikmets, R.; Schriml, L.M.; Hutchinson, A.; Romano-Spica, V.; Dean, M. A human placenta-specific ATP-binding cassette gene (ABCP) on chromosome 4q22 that is involved in multidrug resistance. *Cancer Res.* **1998**, *58*, 5337–5339. [PubMed]
8. Doyle, L.A.; Yang, W.; Abruzzo, L.V.; Krogmann, T.; Gao, Y.; Rishi, A.K.; Ross, D.D. A multidrug resistance transporter from human MCF-7 breast cancer cells. *Proc. Natl. Acad. Sci. USA* **1998**, *95*, 15665–15670. [CrossRef] [PubMed]
9. Lee, J.-Y.; Kinch, L.N.; Borek, D.M.; Wang, J.; Wang, J.; Urbatsch, I.L.; Xie, X.-S.; Grishin, N.V.; Cohen, J.C.; Otwinowski, Z.; et al. Crystal structure of the human sterol transporter ABCG5/ABCG8. *Nat. Cell Biol.* **2016**, *533*, 561–564. [CrossRef]
10. László, L.; Sarkadi, B.; Hegedűs, T. Jump into a New Fold—A Homology Based Model for the ABCG2/BCRP Multidrug Transporter. *PLoS ONE* **2016**, *11*, e0164426. [CrossRef]
11. Clark, R.; Kerr, I.D.; Callaghan, R. Multiple drugbinding sites on the R482G isoform of the ABCG2 transporter. *Br. J. Pharmacol.* **2006**, *149*, 506–515. [CrossRef]

12. Taylor, N.M.I.; Manolaridis, I.; Jackson, S.M.; Kowal, J.; Stahlberg, H.; Locher, K.P. Structure of the human multidrug transporter ABCG2. *Nat. Cell Biol.* **2017**, *546*, 504–509. [CrossRef] [PubMed]

13. Manolaridis, I.; Jackson, S.M.; Taylor, N.M.I.; Kowal, J.; Stahlberg, H.; Locher, K.P. Cryo-EM structures of a human ABCG2 mutant trapped in ATP-bound and substrate-bound states. *Nat. Cell Biol.* **2018**, *563*, 426–430. [CrossRef]

14. Jackson, S.M.; Manolaridis, I.; Kowal, J.; Zechner, M.; Taylor, N.M.I.; Bause, M.; Bauer, S.; Bartholomaeus, R.; Bernhardt, G.; Koenig, B.; et al. Structural basis of small-molecule inhibition of human multidrug transporter ABCG2. *Nat. Struct. Mol. Biol.* **2018**, *25*, 333–340. [CrossRef]

15. Orlando, B.J.; Liao, M. ABCG2 transports anticancer drugs via a closed-to-open switch. *Nat. Commun.* **2020**, *11*, 1–11. [CrossRef]

16. Gu, X. Maximum-Likelihood Approach for Gene Family Evolution Under Functional Divergence. *Mol. Biol. Evol.* **2001**, *18*, 453–464. [CrossRef]

17. Gu, X. Functional Divergence in Protein (Family) Sequence Evolution. *Genetica* **2003**, *118*, 133–141. [CrossRef] [PubMed]

18. Lopez, P.; Casane, D.; Philippe, H. Heterotachy, an Important Process of Protein Evolution. *Mol. Biol. Evol.* **2002**, *19*, 1–7. [CrossRef]

19. Shannon, C.E. A Mathematical Theory of Communication. *Bell Syst. Tech. J.* **1948**, *27*, 379–423. [CrossRef]

20. Xavier, B.M.; Zein, A.A.; Venes, A.; Wang, J.; Lee, J.-Y. Transmembrane Polar Relay Drives the Allosteric Regulation for ABCG5/G8 Sterol Transporter. *bioRxiv* **2020**. [CrossRef]

21. Tarr, P.T.; Edwards, P.A. ABCG1 and ABCG4 are coexpressed in neurons and astrocytes of the CNS and regulate cholesterol homeostasis through SREBP-2. *J. Lipid Res.* **2008**, *49*, 169–182. [CrossRef]

22. Pozza, A.; Pérez-Victoria, J.M.; Sardo, A.; Ahmed-Belkacem, A.; Di Pietro, A. Purification of breast cancer resistance protein ABCG2 and role of arginine-482. *Cell. Mol. Life Sci.* **2006**, *63*, 1912–1922. [CrossRef]

23. Robey, R.W.; Honjo, Y.; Morisaki, K.; Nadjem, T.A.; Runge, S.; Risbood, M.; Poruchynsky, M.S.; Bates, S.E. Mutations at amino-acid 482 in the ABCG2 gene affect substrate and antagonist specificity. *Br. J. Cancer* **2003**, *89*, 1971–1978. [CrossRef] [PubMed]

24. Khunweeraphong, N.; Mitchell-White, J.; Szöllősi, D.; Hussein, T.; Kuchler, K.; Kerr, I.D.; Stockner, T.; Lee, J. Picky ABCG5/G8 and promiscuous ABCG2—a tale of fatty diets and drug toxicity. *FEBS Lett.* **2020**, *594*, 4035–4058. [CrossRef]

25. Zhou, F.; Cocco, M.J.; Russ, W.P.; Brunger, A.T.; Engelman, D.M. Interhelical hydrogen bonding drives strong interactions in membrane proteins. *Nat. Genet.* **2000**, *7*, 154–160. [CrossRef]

26. Zhou, F.X.; Merianos, H.J.; Brunger, A.T.; Engelman, D.M. Polar residues drive association of polyleucine transmembrane helices. *Proc. Natl. Acad. Sci. USA* **2001**, *98*, 2250–2255. [CrossRef] [PubMed]

27. Gratkowski, H.; Lear, J.D.; DeGrado, W.F. Polar side chains drive the association of model transmembrane peptides. *Proc. Natl. Acad. Sci. USA* **2001**, *98*, 880–885. [CrossRef]

28. Dawson, J.P.; Weinger, J.S.; Engelman, D.M. Motifs of serine and threonine can drive association of transmembrane helices. *J. Mol. Biol.* **2002**, *316*, 799–805. [CrossRef]

29. North, B.; Cristian, L.; Stowell, X.F.; Lear, J.D.; Saven, J.G.; DeGrado, W.F. Characterization of a Membrane Protein Folding Motif, the Ser Zipper, Using Designed Peptides. *J. Mol. Biol.* **2006**, *359*, 930–939. [CrossRef]

30. Gray, T.; Matthews, B. Intrahelical hydrogen bonding of serine, threonine and cysteine residues within α-helices and its relevance to membrane-bound proteins. *J. Mol. Biol.* **1984**, *175*, 75–81. [CrossRef]

31. Ballesteros, J.A.; Deupi, X.; Olivella, M.; Haaksma, E.E.; Pardo, L. Serine and Threonine Residues Bend α-Helices in the $\chi 1=g-$ Conformation. *Biophys. J.* **2000**, *79*, 2754–2760. [CrossRef]

32. Deupi, X.; Edwards, P.; Singhal, A.; Nickle, B.; Oprian, D.; Schertler, G.; Standfuss, J. Stabilized G protein binding site in the structure of constitutively active metarhodopsin-II. *Proc. Natl. Acad. Sci. USA* **2011**, *109*, 119–124. [CrossRef]

33. Del Torrent, C.L.; Casajuana-Martin, N.; Pardo, L.; Tresadern, G.; Pérez-Benito, L. Mechanisms Underlying Allosteric Molecular Switches of Metabotropic Glutamate Receptor 5. *J. Chem. Inf. Model.* **2019**, *59*, 2456–2466. [CrossRef]

34. Khunweeraphong, N.; Stockner, T.; Kuchler, K. The structure of the human ABC transporter ABCG2 reveals a novel mechanism for drug extrusion. *Sci. Rep.* **2017**, *7*, 1–15. [CrossRef] [PubMed]

35. Kapoor, P.; Briggs, D.A.; Cox, M.H.; Kerr, I.D. Disruption of the Unique ABCG-Family NBD:NBD Interface Impacts Both Drug Transport and ATP Hydrolysis. *Int. J. Mol. Sci.* **2020**, *21*, 759. [CrossRef] [PubMed]

36. Gose, T.; Shafi, T.; Fukuda, Y.; Das, S.; Wang, Y.; Allcock, A.; McHarg, A.G.; Lynch, J.; Chen, T.; Tamai, I.; et al. ABCG2 requires a single aromatic amino acid to "clamp" substrates and inhibitors into the binding pocket. *FASEB J.* **2020**, *34*, 4890–4903. [CrossRef]

37. Cox, M.H.; Kapoor, P.; Briggs, D.A.; Kerr, I.D. Residues contributing to drug transport by ABCG2 are localised to multiple drug-binding pockets. *Biochem. J.* **2018**, *475*, 1553–1567. [CrossRef]

38. Ni, Z.; Bikádi, Z.; Cai, X.; Rosenberg, M.F.; Mao, Q. Transmembrane helices 1 and 6 of the human breast cancer resistance protein (BCRP/ABCG2): Identification of polar residues important for drug transport. *Am. J. Physiol. Physiol.* **2010**, *299*, C1100–C1109. [CrossRef]

39. Polgar, O.; Ierano', C.; Tamaki, A.; Stanley, B.; Ward, Y.; Xia, D.; Tarasova, N.; Robey, R.W.; Bates, S.E. Mutational Analysis of Threonine 402 Adjacent to the GXXXG Dimerization Motif in Transmembrane Segment 1 of ABCG2. *Biochemistry* **2010**, *49*, 2235–2245. [CrossRef]

40. Tamura, A.; Wakabayashi, K.; Onishi, Y.; Takeda, M.; Ikegami, Y.; Sawada, S.; Tsuji, M.; Matsuda, Y.; Ishikawa, T. Re-evaluation and functional classification of non-synonymous single nucleotide polymorphisms of the human ATP-binding cassette transporter ABCG2. *Cancer Sci.* **2007**, *98*, 231–239. [CrossRef]

41. Tamura, A.; Watanabe, M.; Saito, H.; Nakagawa, H.; Kamachi, T.; Okura, I.; Ishikawa, T. Functional Validation of the Genetic Polymorphisms of Human ATP-Binding Cassette (ABC) Transporter ABCG2: Identification of Alleles That Are Defective in Porphyrin Transport. *Mol. Pharmacol.* **2006**, *70*, 287–296. [CrossRef]

42. Nakagawa, H.; Tamura, A.; Wakabayashi, K.; Hoshijima, K.; Komada, M.; Yoshida, T.; Kometani, S.; Matsubara, T.; Mikuriya, K.; Ishikawa, T. Ubiquitin-mediated proteasomal degradation of non-synonymous SNP variants of human ABC transporter ABCG2. *Biochem. J.* **2008**, *411*, 623–631. [CrossRef]

43. Deppe, S.; Ripperger, A.; Weiss, J.; Ergün, S.; Benndorf, R.A. Impact of genetic variability in the ABCG2 gene on ABCG2 expression, function, and interaction with AT1 receptor antagonist telmisartan. *Biochem. Biophys. Res. Commun.* **2014**, *443*, 1211–1217. [CrossRef]

44. Sjöstedt, N.; Heuvel, J.J.M.W.V.D.; Koenderink, J.B.; Kidron, H. Transmembrane Domain Single-Nucleotide Polymorphisms Impair Expression and Transport Activity of ABC Transporter ABCG2. *Pharm. Res.* **2017**, *34*, 1626–1636. [CrossRef]

45. Toyoda, Y.; Mančíková, A.; Krylov, V.; Morimoto, K.; Pavelcová, K.; Bohatá, J.; Pavelka, K.; Pavlíková, M.; Suzuki, H.; Matsuo, H.; et al. Functional Characterization of Clinically-Relevant Rare Variants in ABCG2 Identified in a Gout and Hyperuricemia Cohort. *Cells* **2019**, *8*, 363. [CrossRef]

46. Polgar, O.; Ediriwickrema, L.S.; Robey, R.W.; Sharma, A.; Hegde, R.S.; Li, Y.; Xia, D.; Ward, Y.; Dean, M.; Ozvegy-Laczka, C.; et al. Arginine 383 is a crucial residue in ABCG2 biogenesis. *Biochim. Biophys. Acta Biomembr.* **2009**, *1788*, 1434–1443. [CrossRef]

47. Ni, Z.; Bikádi, Z.; Shuster, D.L.; Zhao, C.; Rosenberg, M.F.; Mao, Q. Identification of Proline Residues in or near the Transmembrane Helices of the Human Breast Cancer Resistance Protein (BCRP/ABCG2) That Are Important for Transport Activity and Substrate Specificity. *Biochemistry* **2011**, *50*, 8057–8066. [CrossRef]

48. Keller, S.; Prechtl, D.; Aslanidis, C.; Ceglarek, U.; Thiery, J.; Schmitz, G.; Jahreis, G. Increased plasma plant sterol concentrations and a heterozygous amino acid exchange in ATP binding cassette transporter ABCG5: A case report. *Eur. J. Med. Genet.* **2011**, *54*, e458–e460. [CrossRef]

49. Heimer, S.; Langmann, T.; Moehle, C.; Mauerer, R.; Dean, M.; Beil, F.-U.; Von Bergmann, K.; Schmitz, G. Mutations in the human ATP-binding cassette transporters ABCG5 and ABCG8 in sitosterolemia. *Hum. Mutat.* **2002**, *20*, 151. [CrossRef]

50. Niu, D.-M.; Chong, K.-W.; Hsu, J.-H.; Wu, T.J.-T.; Yu, H.-C.; Huang, C.-H.; Lo, M.-Y.; Kwok, C.F.; Kratz, L.E.; Ho, L.-T. Clinical observations, molecular genetic analysis, and treatment of sitosterolemia in infants and children. *J. Inherit. Metab. Dis.* **2010**, *33*, 437–443. [CrossRef] [PubMed]

51. Heyes, N.; Kapoor, P.; Kerr, I.D. Polymorphisms of the Multidrug Pump ABCG2: A Systematic Review of Their Effect on Protein Expression, Function, and Drug Pharmacokinetics. *Drug Metab. Dispos.* **2018**, *46*, 1886–1899. [CrossRef]

52. Abellán, R.; Mansego, M.L.; Martínez-Hervás, S.; Martín-Escudero, J.C.; Carmena, R.; Real, J.T.; Redon, J.; Castrodeza-Sanz, J.J.; Chaves, F.J. Association of selected ABC gene family single nucleotide polymorphisms with postprandial lipoproteins: Results from the population-based Hortega study. *Atherosclerosis* **2010**, *211*, 203–209. [CrossRef] [PubMed]

53. Krawczyk, M.; Lütjohann, D.; Schirin-Sokhan, R.; Villarroel, L.; Nervi, F.; Pimentel, F.; Lammert, F.; Miquel, J.F. Phytosterol and cholesterol precursor levels indicate increased cholesterol excretion and biosynthesis in gallstone disease. *Hepatology* **2012**, *55*, 1507–1517. [CrossRef]

54. Hubacek, J.A.; Berge, K.E.; Cohen, J.C.; Hobbs, H.H. Mutations in ATP-cassette binding proteins G5 (ABCG5) and G8 (ABCG8) causing sitosterolemia. *Hum. Mutat.* **2001**, *18*, 359–360. [CrossRef]

55. Lee, M.-H.; Lu, K.; Patel, S.B. Genetic basis of sitosterolemia. *Curr. Opin. Lipidol.* **2001**, *12*, 141–149. [CrossRef]

56. Berge, K.E.; Tian, H.; Graf, G.A.; Yu, L.; Grishin, N.V.; Schultz, J.; Kwiterovich, P.; Shan, B.; Barnes, R.; Hobbs, H.H. Accumulation of Dietary Cholesterol in Sitosterolemia Caused by Mutations in Adjacent ABC Transporters. *Science* **2000**, *290*, 1771–1775. [CrossRef]

57. Pandit, B.; Ahn, G.-S.; Hazard, S.E.; Gordon, D.; Patel, S.B. A detailed Hapmap of the Sitosterolemia locus spanning 69 kb; differences between Caucasians and African-Americans. *BMC Med. Genet.* **2006**, *7*, 13. [CrossRef]

58. National Center for Biotechnology Information. *The NCBI Handbook*, 2nd ed.; National Center for Biotechnology Information (US): Bethesda, MD, USA, 2013.

59. Rossum, G.V.; Drake, F.L. *Python 3 Reference Manual*; CreateSpace: Scotts Valley, CA, USA, 2009.

60. Letunic, I.; Bork, P. Interactive Tree Of Life (iTOL) v4: Recent updates and new developments. *Nucleic Acids Res.* **2019**, *47*, W256–W259. [CrossRef]

61. Capra, J.A.; Singh, M. Characterization and prediction of residues determining protein functional specificity. *Bioinformatics* **2008**, *24*, 1473–1480. [CrossRef]

62. Tareen, A.; Kinney, J.B. Logomaker: Beautiful sequence logos in Python. *Bioinformatics* **2020**, *36*, 2272–2274. [CrossRef]

63. Goddard, T.D.; Huang, C.C.; Meng, E.C.; Pettersen, E.F.; Couch, G.S.; Morris, J.H.; Ferrin, T.E. UCSF ChimeraX: Meeting modern challenges in visualization and analysis. *Protein Sci.* **2018**, *27*, 14–25. [CrossRef]

64. Haider, A.J.; Cox, M.H.; Jones, N.; Goode, A.J.; Bridge, K.S.; Wong, K.; Briggs, D.; Kerr, I.D. Identification of residues in ABCG2 affecting protein trafficking and drug transport, using co-evolutionary analysis of ABCG sequences. *Biosci. Rep.* **2015**, *35*, e00241. [CrossRef]

65. Miettinen, T.A.; Klett, E.L.; Gylling, H.; Isoniemi, H.; Patel, S.B. Liver Transplantation in a Patient with Sitosterolemia and Cirrhosis. *Gastroenterology* **2006**, *130*, 542–547. [CrossRef]

66. Lu, K.; Lee, M.-H.; Hazard, S.; Brooks-Wilson, A.; Hidaka, H.; Kojima, H.; Ose, L.; Stalenhoef, A.F.; Mietinnen, T.; Bjorkhem, I.; et al. Two Genes That Map to the STSL Locus Cause Sitosterolemia: Genomic Structure and Spectrum of Mutations Involving Sterolin-1 and Sterolin-2, Encoded by ABCG5 and ABCG8, Respectively. *Am. J. Hum. Genet.* **2001**, *69*, 278–290. [CrossRef] [PubMed]
67. Nakanishi, T.; Doyle, L.A.; Hassel, B.; Wei, Y.; Bauer, K.S.; Wu, S.; Pumplin, D.W.; Fang, H.-B.; Ross, U.D. Functional Characterization of Human Breast Cancer Resistance Protein (BCRP, ABCG2) Expressed in the Oocytes of Xenopus laevis. *Mol. Pharmacol.* **2003**, *64*, 1452–1462. [CrossRef] [PubMed]
68. Buch, S.; Schafmayer, C.; Völzke, H.; Becker, C.; Franke, A.; Von Eller-Eberstein, H.; Kluck, C.; Bässmann, I.; Brosch, M.; Lammert, F.; et al. A genome-wide association scan identifies the hepatic cholesterol transporter ABCG8 as a susceptibility factor for human gallstone disease. *Nat. Genet.* **2007**, *39*, 995–999. [CrossRef]

 International Journal of
Molecular Sciences

Review

Multidrug Resistance in Mammals and Fungi—From MDR to PDR: A Rocky Road from Atomic Structures to Transport Mechanisms

Narakorn Khunweeraphong and Karl Kuchler *

Center for Medical Biochemistry, Max Perutz Labs Vienna, Campus Vienna Biocenter, Medical University of Vienna, Dr. Bohr-Gasse 9/2, A-1030 Vienna, Austria; narakorn.khunweeraphong@meduniwien.ac.at
* Correspondence: karl.kuchler@meduniwien.ac.at; Tel.: +43-1-4277-61807; Fax: +43-1-4277-9618

Abstract: Multidrug resistance (MDR) can be a serious complication for the treatment of cancer as well as for microbial and parasitic infections. Dysregulated overexpression of several members of the ATP-binding cassette transporter families have been intimately linked to MDR phenomena. Three paradigm ABC transporter members, ABCB1 (P-gp), ABCC1 (MRP1) and ABCG2 (BCRP) appear to act as brothers in arms in promoting or causing MDR in a variety of therapeutic cancer settings. However, their molecular mechanisms of action, the basis for their broad and overlapping substrate selectivity, remains ill-posed. The rapidly increasing numbers of high-resolution atomic structures from X-ray crystallography or cryo-EM of mammalian ABC multidrug transporters initiated a new era towards a better understanding of structure–function relationships, and for the dynamics and mechanisms driving their transport cycles. In addition, the atomic structures offered new evolutionary perspectives in cases where transport systems have been structurally conserved from bacteria to humans, including the pleiotropic drug resistance (PDR) family in fungal pathogens for which high resolution structures are as yet unavailable. In this review, we will focus the discussion on comparative mechanisms of mammalian ABCG and fungal PDR transporters, owing to their close evolutionary relationships. In fact, the atomic structures of ABCG2 offer excellent models for a better understanding of fungal PDR transporters. Based on comparative structural models of ABCG transporters and fungal PDRs, we propose closely related or even conserved catalytic cycles, thus offering new therapeutic perspectives for preventing MDR in infectious disease settings.

Keywords: yeast; multidrug transporter; anticancer; antifungal resistance; ABC transporters; mechanism

Citation: Khunweeraphong, N.; Kuchler, K. Multidrug Resistance in Mammals and Fungi—From MDR to PDR: A Rocky Road from Atomic Structures to Transport Mechanisms. *Int. J. Mol. Sci.* **2021**, *22*, 4806. https://doi.org/10.3390/ijms22094806

Academic Editor: Thomas Falguières

Received: 25 March 2021
Accepted: 28 April 2021
Published: 30 April 2021

Publisher's Note: MDPI stays neutral with regard to jurisdictional claims in published maps and institutional affiliations.

1. Introduction

1.1. ABC Transporters and Clinical Relevance of MDR

The ATP-binding cassette (ABC) transporter family is one of the largest protein superfamilies present in all living organisms, from prokaryotes to eukaryotes [1–5]. ABC transporters can operate as exporters or importers in an ATP-dependent manner, and mediate the membrane translocation of bewildering substrate spectra against concentration gradients [6–8]. In addition, ABC proteins can function as ion channels, channel regulators, receptors, proteases, protein sensors or are even involved in mRNA translation and ribosome biogenesis [9–11]. Remarkably, conserved architectures offer specific yet broad substrate-binding regions and somehow form a translocation path that operates in a unidirectional way in eukaryotes. The wide substrate range includes cationic anticancer drugs, antifungal drugs, steroids, phospholipids, bile acids, antibiotics, peptides, ions, heavy metals, carbohydrates and glucocorticoids, as well as toxins [12–15]. The hallmark domain organization of ABC transporters entails four core units, two evolutionarily conserved nucleotide-binding domains (NBDs) and two transmembrane-spanning domains (TMDs), typically consisting of twelve hydrophobic transmembrane-spanning helices (TMHs). These four domains are normally arranged as a full-transporter in a single protein as TMD1–NBD1–TMD2–NBD2 or in a reverse configuration NBD1–TMD1–NBD2–TMD2. Alternatively, half-transporters come in NBD–TMD or TMD–NBD arrangements, which require at least homo- or hetero-dimerization for a functional complex [16–19]. In addition, some members contain additional domains or motifs such as the TMD0 domain or the R-domain regulatory motif [8,20,21].

The NBD is a universally conserved domain that consumes ATP and somehow fuels the dynamic switch of the transporter structure from an inward substrate-binding state to an outward substrate-releasing conformation. The catalytic cycle drives the conformational switch at the TMD and enables the substrate translocation through an as yet elusive transport pathway [22,23]. The TMDs are more diverse in sequence and show much less conservation, but clearly are essential for forming putative substrate translocation pores. They must handle a broad spectrum of chemically diverse substrates and inhibitors [24,25]. Notably, the communication between NBD and TMD and the dynamics underlying the entire transport cycle of ABC transporters remain unclear. While certain elements or stages of the transport cycles may be conserved among subfamilies, the expanding number of atomic structures and the resulting mechanistic information for distinct ABC transporters make unifying mechanisms less likely, challenging earlier notions about a unified catalytic cycle [8,16,17,26–45]. No matter what the actual catalytic cycle or mechanism of a given type I or type II exporter may be, a tantalizing possibility is that a basic conserved mechanism operates for PDR and ABCG, but slightly different transport mechanisms could be a consequence of the nature of substrates that would distinctly affect the kinetics and dynamics of the cycle.

The human ABC transporter family of 48 genes served to categorize subfamilies into the ABCA to ABCG nomenclature [46–48], although the surge of recent atomic structures and functional considerations made it clear that a new and improved nomenclature based on structure–function relationships is needed [43]. Remarkably, inborn errors of several human ABC transporters lead to prominent genetic diseases [49], including cystic fibrosis (ABCC7 or CFTR) [50,51], hepatic cholestasis (ABCB11 or BSEP) [52,53], plant sterol sitosterolemia (ABCG5/G8) [54–57], neonatal hyperinsulinemic hypoglycemia or non-insulin-dependent childhood diabetes (ABCC8) [58], gout (ABCG2 or BCRP) [59], Dubin–Johnson syndrome (ABCC2 or MRP2) [60,61] and Stargardt's macular dystrophies and retinophathies (ABCA4) [62], peroxisomal adrenoleukodystrophy or ALD (ABCD1) [63–65], immune deficiency—class I MHC antigen presentation (ABCB2/B3, TAP) [66–68], cholesterol transport and HDL assembly or Tangier's disease (ABCA1, ABCG1) [69–75], pseudoxanthoma elasticum or PXE (ABCC6) [76–82], dilated cardiomyopathy (ABCC9) [83], defective earwax synthesis (ABCC11) [84,85], lung surfactant deficiency (ABCA3) [86–89], lamella and harlequin ichthyosis (ABCA12) [90–92], pregnancy-related cholestasis (ABCB4) [93,94] and sideroblastic anemia (ABCB7) [95–98].

1.2. Mammalian ABC Multidrug Transporters

Most, if not all, eukaryotic ABC transporters function as unidirectional exporters and use ATP consumption to drive transport. Some have been implicated in uptake processes as well, although this remains controversial [39,99]. Importantly, ectopic or dysregulated overexpression of certain ABC transporters often contributes to or promotes MDR phenomena in several but not all human cancer types [2,10,11,100–102]. Based on sequence similarity and domain arrangement, mammalian ABCs fall into two major groups, referred to as type I and type II exporters [12], although recently, a new classification has been proposed [18]. At least three MDR exporters have been linked to MDR in human tumors, including P-glycoprotein (P-gp/MDR1/ABCB1) [103–106], MRP1 (MDR-associated protein 1/ABCC1) [107] and BCRP (Breast Cancer Resistance/ABCG2) [108–110]. All three share rather broad and partially overlapping drug specificity [14,111]. Most substrates of ABCB1 and ABCG2 are cationic hydrophobic compounds [112], which may probably be expelled directly from the lipid phase as originally proposed by the "hydrophobic vacuum cleaner" model [105,113,114] or from the outer membrane leaflet by a floppase-like function [5,14,19,109,111,115,116]. A similar mechanism has been proposed in ABCA1 for the cholesterol loading of apolipoprotein A-I (apoA-I) [5,116,117]. P-gp, MRP1 and ABCG2 are normally residing in the plasma membrane of epithelial organ linings (such as liver, intestine, blood–brain barrier, placenta and mammary epithelium) [118,119]. Their physiological tasks include vital roles in cellular detoxification and in organ protection by excretion of toxic compounds or xenobiotic molecules [120]. Substrates are amphipathic, lipid-soluble compounds of extremely diverse chemical spaces, ranging from small molecules to bulky lipophilic cations and conjugated organic anions [14,112,121].

ABCB1 or P-glycoprotein (P-gp) or MDR1 (encoded by the *MDR1* gene) was identified from a multidrug-resistant KB carcinoma cell line [103,122,123] as the first mammalian type I exporter class. P-gp is expressed on apical membranes of epithelial cells in colon, small intestine, liver, placenta, kidney, gut, pancreatic, bile duct and blood–brain barrier [124,125]. Homozygous P-gp knock-out mice showed a 100-fold increase in drug (ivormectin) permeability at the blood–brain barrier, which led to the discovery of its physiological role in organ protection [126–128]. ABCB1 transports a diverse array of substances, including chemotherapeutic drugs, steroids, several phospholipids, fluorescent dyes, peptides and ionophores [113]. Furthermore, ABCB1 is believed to function as a floppase-like lipid transporter [43,129]. Despite huge therapeutic promises, numerous clinical studies on ABCB1 inhibitors or reversal agents [130–138] showed marginal to no benefits, thus rendering attempts to translate P-gp inhibitors into the clinic so far futile efforts [111,139]. Of note, despite an almost highly conserved primary sequence identity, the closest P-gp homologue, ABCB4/MDR2, has not been implicated in drug transport or cancer MDR [140], as it resides in the canalicular and appears to have a restricted substrate spectrum limited to phosphatidycholine-related phospholipids in the canalicular membrane.

ABCC1 or MRP1 or multidrug resistance-associated protein1 (encoded by the *ABCC1* gene) was discovered as the second member of MDR exporters, cloned from a multidrug-resistant P-gp-negative human lung cancer cell line with doxorubicin tolerance [107]. ABCC1 is mostly on the basolateral surface of polarized epithelial cells, with moderate to high abundance in the gastrointestinal tract, kidney, bladder, testis, ovary, endometrium, adipose tissues, appendix and tonsils. Low-level expression is found in brain, lung, liver, gall bladder, pancreas, bone marrow and skin [141–143] to excrete a variety of endogenous substances, including glutathione, prostaglandins, C4-leukotrienes glucuronide conjugates, sulfate conjugates, heavy metal oxyanions and, most importantly, conjugated metabolites of otherwise hydrophobic compounds [14,112,144,145].

ABCG2 or BCRP or Breast Cancer Resistance Protein (encoded by the *ABCG2* gene) was originally isolated from P-gp-negative multidrug-resistance breast cancer cell lines [109,110]. ABCG2 homes to the apical membranes in many epithelial cells and tissues, including lung, gut, intestine, liver, breast, placenta, hematopoietic stem cells and especially in the blood–brain barrier [109,110,119,146]. ABCG2 is a half-transporter carrying a TMD at the C-terminus, requiring homo-dimerization to form a full functional molecule. ABCG2 is overexpressed in many solid tumors as well as acute myeloid leukemia (AML) and acute lymphocytic leukemia (ALL). Dysregulated ABCG2 overexpression is linked with poor prognosis in several cancer types [139,147], with particularly low survival in AML patients [134,137,148–151]. Like P-gp, ABCG2, as well as PDRs such as Yor1, Pdr5, Cdr1 or Snq2, show extremely broad substrate specificity (Table 1), all in all transporting hundreds of diverse compounds, including dietary xenobiotics, toxins, metabolites, vitamins, lipids, steroids, antibiotics and antifungal as well as anticancer drugs [152–155]. Of note, the many exceptions seen for each transporter make a generalization of substrate preferences for a given ABCG or PDR transporter challenging without experimental evidence. Remarkably, however, despite their pronounced structural conservation, additional ABCG family members such as ABCG1, ABCG4 and the heterodimeric ABCG5/G8 transporter have not been associated with MDR phenotypes in cancer [15,115,156–162].

Table 1. Known ABC transporters in non-pathogenic yeasts.

	Phylogeny	Species	Gene Name	UniProt ID	Length	Substrates	References
Exporter type II	ABCG/PDR	*S. cerevisiae*	PDR5	P33302	1511	Drugs, PC, PE, PS, Steroids, Herbicides	[163–172]
			PDR10	P51533	1564	Regulate PDR12 trafficking, Herbicides, Lipids	[173–175]
			PDR11	P40550	1411	Sterol	[99,176]
			PDR12	Q02785	1511	Weak acids, Fluorescein	[177,178]
			PDR15	Q04182	1529	Herbicide, Detergent	[174,175,179]
			PDR18	P53756	1333	Herbicides, Ethanol, Ergosterol	[180–183]
			AUS1	Q08409	1394	Sterol	[99,158,184]
			SNQ2	P32568	1501	Drugs, Steroids, Mutagens, Chemicals	[169,185,186]
		S. pombe	BFR1	P41820	1530	Brefeldin A, Tributyltin	[187]
Exporter type I	ABCB/MDR	*S. cerevisiae*	ATM1 *	P40416	690	Fe/S proteins	[188–190]
			MDL1 *	P33310	695	Peptides	[191,192]
			STE6	P12866	1290	a-factor	[193,194]
		S. pombe	HMT1 *	Q02592	830	Phytochelatin conjugated Cd2+	[195,196]
			MAM1	P78966	1336	M-factor	[195,196]
	ABCC/MRP	*S. cerevisiae*	YOR1	P53049	1477	Oligomycin, Reveromycin, Beavericin, Metal ions	[197]
			YCF1 *	P39109	1515	GS-conjug. Cd2+, Metals	[170,198–200]
			YBT1 *	P32386	1661	Metal ions, Bile acid, PC	[201,202]
			VMR1 *	P38735	1592	Drugs, Metal ions	[203]
			BPT1 *	P14772	1559	Metal ions, Bile acid, GS-conjugates	[204]
		S. pombe	PMD1	P36619	1362	Drugs	[205]

* All transporters located in the plasma membrane, except for Atm1, Mlt1 and Ybt1, Vmr1, Ycf1 and Hmt1, residing in the mitochondrial and vacuolar membrane. GS: glutathione, PC: phosphatidylcholine, PE: phosphatidylethanolamine, PS: phosphatidylserine.

Collectively, P-gp, MRP1 and ABCG2 act as brothers in arms to ensure the physiological detoxification of endogenous metabolites as well as exogenous xenobiotics across most epithelial barriers, including placenta, testis, mammary epithelium, liver and GI tract as well as the blood–brain barrier [119,139,146,152,206]. However, how, and sometimes even if, they actually cause clinical MDR in cancer has remained a highly controversial issue, often subject to intense discussions in the field [102,207]. As for microbial anti-infective MDR, it has been generally accepted though that bacterial ABC transporters [208–212] and fungal PDR transporters [124,213–216] are key causes for clinical MDR, often setting an unsurmountable roadblock in antimicrobial treatments [10,101,217–219].

1.3. ABC Multidrug Transporters in Fungal Kingdoms

Invasive fungal diseases account for ~1.5 million deaths per year worldwide [220,221]. The increasing numbers of immunosuppressed people, including the elderly, transplant recipients, cancer and HIV/AIDS patients will most likely increase future cases of infections by opportunistic pathogens like *Candida* species (spp.) [222]. Candida spp. are commensal colonizers and part of the microflora present on mucosal and epithelial barriers such as the gastrointestinal and urogenital tracts. *Candida* spp. are a major part of the physiological mycobiome species [223], along with several thousand bacterial species constituting tissue-specific microbiomes [224–229]. Several *Candida* spp. can cause life-threatening invasive systemic disease in severely immunocompromised individuals [230]. The small number of chemical entities in antifungal drugs have been very problematic in the past, especially after prolonged use or after extensive prophylaxis, as was the case for pronounced azole resistance in HIV patients. Hence, the propensity to develop MDR is lower when larger drug arsenals against different targets are available [231]. For example, *C. auris* is a newly emerging pan-resistant fungal pathogen first reported from an ear infection in Japan in 2009 [232]. Within a decade, *C. auris* appeared in more than 40 countries around the globe, causing hospital outbreaks of invasive candidemia [233–240]. Importantly, *C. auris* is also causing severe superinfections with viruses, as seen in a recent co-infection of COVID-19 patients in a Mexican hospital ICU, leading to a dramatic overall mortality of 83%. In fact, intrinsic MDR in other *Candida* spp. such as *C. glabrata* and *C. kruzei* have also been increasing over the last decade [230,241,242]. Importantly, among other mechanisms, such as drug target gene mutations [243,244], efflux-based MDR/PDR has been recognized as a major cause of fungal anti-infective drug resistance [219,235,244–247].

Fungal ABC proteins are also quite diverse and are implicated in many biological functions, contributing to pivotal cellular processes, including cellular detoxification and stress adaptation [214,248,249]. Owing to space constraints, we will limit our discussion to paradigm fungal PDR family members implicated in drug resistance phenomena, but refer the reader to numerous recent reviews discussing fungal ABC proteins at large [249,250]. Fungal ABC proteins of the PDR subfamily are the closest eukaryotic orthologues of human ABCG family exporters [245,251]. For instance, *C. albicans* and *C. glabrata* harbor 27 and 18 ABC proteins in total, respectively [250]. Interestingly, *Cryptococcus neoformans* and *Aspergillus fumigatus* harbor even larger numbers, with 54 and 49 ABC transporters, respectively [252,253]. While we are not discussing transporters which are not linked to MDR phenomena, we still provide a comprehensive list of all known ABC transporters in non-pathogenic yeasts (Table 1) and their phylogenetic relationships (Figure 1A). MRP-like yeast ABC transporters including Yor1, Ycf1, Ybt1, Vmr1 and Bpt1 mediate vacuolar detoxification and heavy metal resistance [214]. Of note, Ycf1, which also has a rudimentary R-domain motif as present in human CFTR [50], was the first yeast homologue of mammalian MRP [254], while pathogenic fungi such as *C. albicans* harbor only Yor1 and Mlt1 (Table 2).

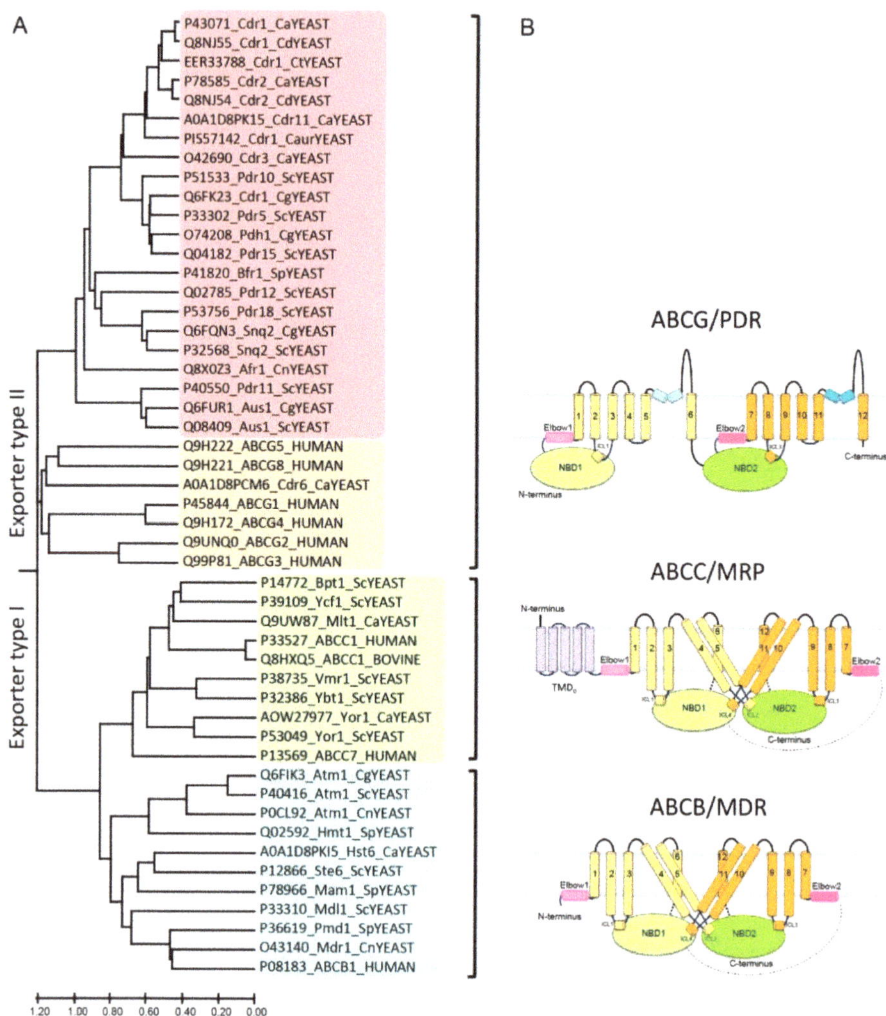

Figure 1. Phylogeny and structural organization of ABC transporters in mammals and yeast. (**A**) The phylogenetic tree shows the evolutionary relationships of ABC transporter subfamilies in yeast and mammals. Some 29 ABC transporters were subjected to amino acid sequence alignments. Branch length was analyzed using MEGA-X, and represents the evolutionary distance in the units of the number of amino acid substitutions per site. Names are given in the UniProt code, protein name and organism, respectively. The analysis reveals two major exporters subfamilies referred to as type I and II. Type I was sub-classified into ABCB/MDR (blue) and ABCC/MRP (green) subgroups. The type II family represents ABCG/PDR subgroups with the fungal (red) and mammalian transporters (orange). Ca: *Candida albicans*, Cg: *Candida glabrata*, Caur: *Candida auris*, Cd: *Candida dubliniensis*, Ct: *Candida tropicalis*, Cn: *Cryptococcus neoformans*, Sc: *Saccharomyces cerevisiae*, Sp: *Schizosaccharomyces pombe*. (**B**) Predicted membrane topologies of three MDR ABC exporter families. The transporters hold several diagnostic hallmark domains, including two NBDs (NBD1: light green; NBD2: green), two TMD regions usually with 6 putative membrane-spanning helices each (TMD1: light yellow; TMD2: bright orange), elbow helix (pink), re-entry helix (blue) and TMD0 (purple), respectively.

The fungal PDR (ABCG) or pleiotropic drug resistance family is the largest subfamily of ABC transporters of the type II exporter class [251]. PDR transporters are the closest structural orthologues of all mammalian ABCG subfamily transporters (Figure 1A), sharing the same topological orientation and domain arrangements with mammalian ABCG5/ABCG8 [255] and ABCG2 [256–259] (Supplementary Figure S1). PDR transporters in fungal pathogens implicated in clinical antifungal resistance in *C. albicans* include Cdr1, Cdr2, Cdr3, Cdr6 and Cdr11, in *C. glabrata* (Cdr1, Pdh1, Snq2 and Aus1) [260], in *C. auris* (Cdr1), in *C. tropicalis* (Cdr1), in *C. dubliniensis* (Cdr1 and Cdr2) and in *Cryptococcus neoformans* (Afr1) (Table 2). All are highly conserved in non-pathogenic yeasts (Table 1). Their overexpression in pathogens causes hallmark MDR phenotypes seen for mammalian or bacterial ABC transporters [111,141,217,218,235,261–264].

Diversity of fungal PDR transporters and evolutionary relationships to mammalian ABC transporters. Most ABC proteins from non-pathogenic baker's yeasts (Table 1) are also found in various numbers and functions in pathogenic fungi (Table 2). A phylogenetic tree analysis suggests two major groups, referred to as exporter type I and type II [4,248,265,266], contain three MDR, MRP and PDR subfamilies (Figure 1A). While type I exporters (MDR/ABCB and MRP/ABCC) hold a TMD1–NBD1–TMD2–NBD2 configuration, all PDR/ABCG type II exporters adopt a "reverse" architecture, such as NBD1–TMD1–NBD2–TMD2 (Figure 1B). By contrast with the mammalian ABCG subfamily, all fungal PDR proteins are full-size transporters (Figure 1B).

Table 2. ABC efflux transporters mediating MDR in pathogenic fungi.

	Phylogeny	Species	Gene	UniProt ID	Length	Substrates	References
Exporter type II	ABCG/PDR	*C. albicans*	*CDR1*	P43071	1501	Drugs, PC, PE, PS, Steroids	[217,260,263, 267–270]
			CDR2	P78595	1499	Drugs, PC, PE, PS, Steroids	[218,260,271]
			CDR3	O42690	1501	Drugs, PC, PE, PS, Steroids	[260,272]
			CDR6/ROA1	A0A1D8PCM6	1274	Azole, Membrane fluidity	[273]
			CDR11	A0A1D8PK15	1512	Drugs, Fosmanogepix	[274]
		C. glabrata	*CDR1*	Q6FK23	1499	Drugs	[275,276]
			PDH1	O74208	1542	Drugs	[277,278]
			SNQ2	Q6FQN3	1507	Drugs	[279]
			AUS1	Q6FUR1	1398	Sterol	[280,281]
		C. auris	*CDR1*	PIS57142	1508	Drugs	[236,237,239]
		C. dubliniensis	*CDR1*	Q8NJ55	1501	Drugs	[282]
			CDR2	Q8NJ54	1500	Drugs	[282]
		C. tropicalis	*CDR1*	EER33788	1498	Drugs	[283–285]
		C. neoformans	*AFR1*	Q8X0Z3	1543	Drugs	[286]
Exporter type I	ABCB/MDR	*C. albicans*	*HST6*	A0A1D8PKI5	1323	A-factor	[287,288]
		C. glabrata	*ATM1 **	Q6FIK3	727	Fe/S	[289]
		C. neoformans	*MDR1*	O43140	1408	Drugs	[290]
			*ATM1 **	P0CL92	734	Fe/S	[290]
	ABCC/MRP	*C. albicans*	*YOR1*	AOW27977	1488	Oligomycin, Beauvericin	[291,292]
			*MLT1 **	Q9UW87	1606	PC, Ni(II)	[180]

* All transporters located in the plasma membrane, except for Atm1 and Mlt1, residing in the mitochondrial and vacuolar membrane, respectively. GS: glutathione, PC: phosphatidylcholine, PE: phosphatidylethanolamine, PS: phosphatidylserine.

2. Atomic Structures of Eukaryotic Multidrug ABC Efflux Exporters

Several high-resolution structures, obtained through X-ray crystallography or cryo-EM approaches in the past five years, provided a better understanding of the structure–function relationships of mammalian type I and type II exporters (Table 3). However, despite ever-increasing structural information, the path from static atomic structures to precise molecular mechanisms has turned out to be a rocky road for scientists and drug discovery. Thus, we are still far from understanding the conformational dynamics as well as the mechanics driving their transport cycles. Each half-molecule of ABCB and ABCC type I exporters harbor at least six transmembrane helices that extend into the intracellular loops linked to their coupling helices, thus connecting to both the proximal NBD and crossing over to the distal NBD [8,31,39,293–296]. In the ATP-free apo state, if it ever exists inside cells, these exporters maintain an inward-facing configuration, whereby both NBDs appear apart to offer access for ATP and possibly substrates (Figure 2A) [297]. Importantly, while the ABCC subfamily shares a highly similar overall fold with ABCB, MRP transporters have an additional N-terminal domain known as TMD0 [20,298,299]. Notably, CFTR also adds the so-called R(egulatory)-domain residing in the core of the channel between the N-terminal and the C-terminal TMDs [21,300,301] (Figure 1B).

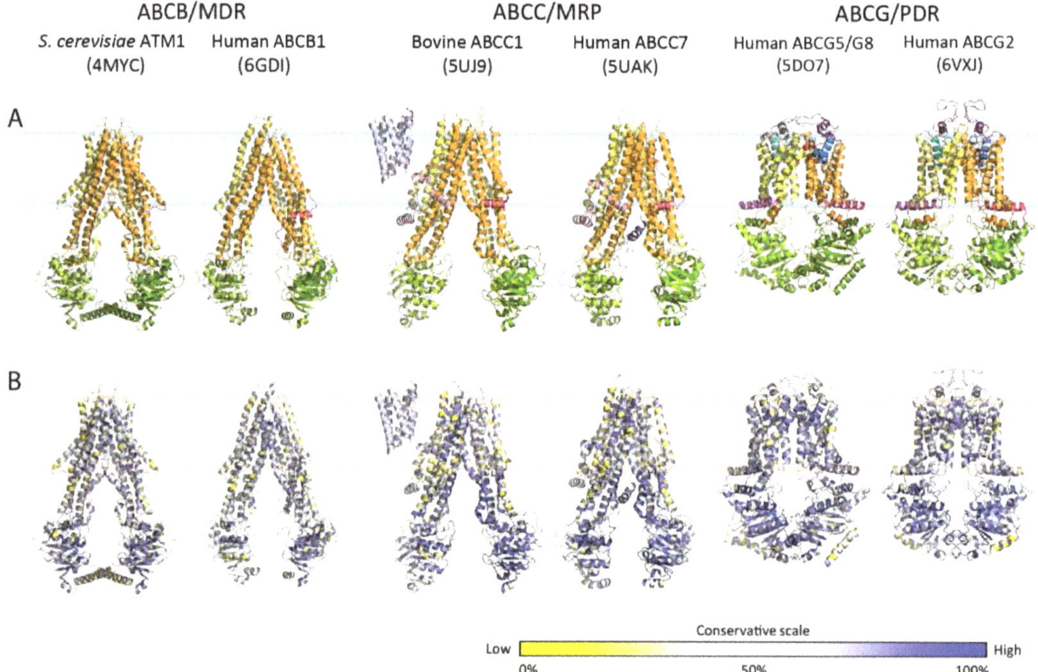

Figure 2. Atomic structures, folds and evolutionary conservation of MDR exporters. (**A**) Crystal or cryo-EM particle structures of multidrug resistance-related ABC exporters from the type I (ABCB and ABCC) and type II families (ABCG subfamily) using the color codes as shown in Figure 1B. (**B**) Conservation analysis of ABC exporters from panel A. Multiple sequence alignments of amino acid sequences from Tables 1 and 2 were generated for all subfamilies. The degree of conservation was calculated and indicated as a color gradient ranging from low conservation (yellow) to high conservation (blue).

At present, the available atomic structures of type I exporters are as listed in Table 3. For the ABCB1 subfamily, many conformations, including inward-facing, outward-facing, occluded state, substrate-bound, inhibitor-bound and antibody-bound, have been solved, all in all yielding 24 structures in the PDB [296,302–310]. For the ABCC subfamily, four cryo-EM structures are available for bovine ABCC1 in the PDB [311,312]. Furthermore, seven structures are available for CFTR [300,301,313–315], and 12 structures for ABCC8 or SUR1 [316–320]. Strikingly, the first atomic structure of the heterodimeric ABCG subfamily member ABCG5/G8 came as a surprise, as the X-ray crystals revealed an unexpected compact fold, in which NBDs were located in close proximity to the TMDs. In addition, the molecule held a lid-like structure at the extracellular roof of the translocation pathway. This fold resembled bacterial importers more than a prototypic export pump [38]. This structure paved the way for solving the human ABCG2 transporter, for which now 11 atomic structures are available in the PDB [256–259] (Table 3).

Table 3. Atomic structures of eukaryotic MDR ABC transporters.

Subfamily	PDB ID	Function	References
ABCB1 (P-gp)	4F4C, 4M1M, 4M2S, 4M2T, 4Q9H, 4Q9J, 4Q9K, 4Q9L, 4XWK, 5KPF, 6KPI, 5KPJ, 5KO2, 5KOY, 6C0V, 6GDI, 6Q81, 6QEX, 6QEE, 6FN4, 6FN1, 3G5U, 3G60, 3G61	Multidrug export, detoxification	[255,296,302–306,308–310]
ABCC1 (MRP1)	5UJA, 5UJ9, 6BHU, 6UY0	Multidrug, leukotriene and sphingolipid export, detoxification	[298,312]
ABCG2 (BCRP)	5NJG, 5NJ3, 6ETI, 6FEQ, 6HIJ, 6HCO, 6HBU, 6HZM, 6VXH, 6VXI, 6VXJ	Multidrug export, detoxification and urate transport	[256–259]

3. Key Residues and Motifs Are Conserved in Multidrug Transporters ABCG/PDR

The mammalian ABCGs are the closest orthologues of yeast PDR transporters [216,245, 250,251,321], especially Pdr5 and Cdr1 [166,322,323] (Supplementary Figure S1). To identify conserved regions of fungal ABC proteins based on mammalian ABC structures, we determined a conservation score for the fungal PDR family, and mapped conserved residues into the atomic structures of mammalian orthologues (Figure 2B). We subjected the fungal ABCG/PDR subfamily to multiple sequence alignments with the mammalian ABCGs (Supplementary Figure S1), showing that NBDs hold several highly conserved motifs related to ATP consumption. Although TMDs are usually more diverse, several regions are also preserved between PDR and ABCG, implying that these domains are pivotal for an evolutionarily related catalytic cycle.

Conservation in the NBD. First, the NBDs in fungal PDRs share highly conserved motifs as well as residues with ABCG required for ATP-binding and hydrolysis [45,153,324–326], including Walker A, Q-loop, Hot spot helix, Signature motif, Pro-loop, Walker B, D-loop and H-loop, respectively [327]. The alignment indicates that fungal NBDs adopt a RecA-like structure, an ATPase-containing fold that was first seen in RecA, which is involved in DNA recombination, and was later found in many ATPases [328] (Supplementary Figure S1 and Table 4). There are only minor differences though, as ABCG2 and ABCG5/G8 are half-transporters that require homo- or hetero-dimerization, while all fungal PDRs are full-size transporters, some of which with asymmetric deviant ATP-binding sites (Table 4). In the first NBD of PDR transporters, the glutamine (Q) in the Q-loop is replaced by glutamate (E), the Pro-loop disappeared and histidine (H) in the H-loop is substituted by tyrosine (Y). Hence, the notion emerged that fungal PDRs would follow an asymmetric catalytic cycle [173,216,306,329,330].

Conservation in the TMD region. Second, the general architecture and configuration of TMDs are maintained, as each TMD in ABCG/PDR transporters contains six putative membrane-spanning helices, and a rather short first intracellular loop contains the coupling helix. The large ECL is part of the lid architecture in the extracellular region. Interestingly, two putative helices residing at both lipid bilayer leaflets, which are important for function, as they restrict dynamic movements during transport cycle, are only found in the ABCG subfamily [40,42]. The amino acid alignments reveal the consensus sites present in conserved motifs and domains (Figure 1B and Supplementary Figure S1).

Table 4. The human ABCG2 and fungal Cdr1 multidrug exporters share conserved motifs.

Location	Conserved Motif	Functional Role	Human ABCG2	*Candida albicans* Cdr1	
				First Half	Second Half
NBD	Walker A	ATP hydrolysis (phosphate binding) *	G79–S88	G187–T195	G895–T903
	Q-loop	TMD–NBD communication	Q126	E238 **	Q942
	Hot spot	Triple helical bundle	L134–A149	L246–P261	S950–S965
	Signature	NBD dimerization and phosphate binding	V186–R193	V303–R310	V996–R1008
	Pro loop	NBD dimerization	P204	?	P1019
	Walker B	ATP hydrolysis	I206–E211	I323–N238	L1021–E1027
	D-loop	NBD dimerization	L216–D217	L333–D334	L1032–D1033
	H-loop	ATP hydrolysis	H243	Y361 **	H1059
Elbow helix	Conserved R	Salt bridge, THB	R383	R503	R1185
ECL1	Conserved R	Salt bridge	R426	R456	?
TMH2	Conserved F	Clamping	F439	F559	F1239
ICL1	Conserved E (1)	Salt bridge and intracellular gating	E451	E570	D1255
	Conserved E (2)	Salt bridge	E458	E576	E1261
	Conserved Y	Salt bridge, THB	Y464	Y584	Y1257
TMH3	Conserved D/E	Intracellular gating	D477	E597	E1280
Valve	Conserved hydrophobic	Valve	G553–L555	G672–V674	G1362–L1364
Re-entry helix	Conserved P	Kinked helix	P574	P692	P1382
	Conserved E	Salt bridge	E585	E704	?
ECL3	Conserved C (1)	Intra/intermolecular disulfide bond	C592	C712	C1418
	conserved C (2)	Intra/intermolecular disulfide bond	C603	?	C1441
	conserved C (3)	Intra/intermolecular disulfide bond	C608	C732	C1444

* Some PDRs may consume GTP as well [186,331]. ** Substitution with another residue.

The elbow helix is a 15-residue amphipathic helix sharing highly conserved residues among all ABCG/PDR members. This elbow helix is an intrinsic part of the triple helical bundle (THB), which also engages the hot spot helix from the NBD and the coupling helix from the first intracellular loop (ICL1) [332]. Importantly, the center of the elbow helix contains a highly conserved arginine (R), which is essential for a salt bridge interaction with a glutamate (E) residue in ICL1 to stabilize the THB [40].

The transmembrane helix 1 (TMH1) is a prototypic 20-residue membrane-spanning helix, following the elbow helix in all ABCG/PDR exporters. The small extracellular loop 1 (ECL1) is a short linker connecting TMH1 and TMH2 at the cell surface. Notably, ECL1 in the first half holds a conserved arginine (R), which is important for salt bridge formation with a conserved glutamate (E) in the re-entry helix of the same molecule [42]. TMH2 is thought to be a part of the substrate/inhibitor binding zone in human ABCG2 [333]. A conserved phenylalanine (F) in the middle of TMH2 may be a recognition site for both substrates and inhibitors [333].

The intracellular loop 1 (ICL1) spans over 30 residues in the ABCG/PDR subfamily and holds a critical U-turn motif. ICL1 operates as the coupling helix for the NBD–TMD communication by participating in a triple helical bundle (THB). The entire THB functions as a molecular spring that controls the catalytic cycle by regulating the conformational switch. Indeed, two conserved negative residues within ICL1 of human ABCG2, E451 and E585, are thought to control the intracellular gating mechanism and engage in salt bridges with the elbow helix, respectively. Remarkably, ICL1 holds an essential tyrosine residue that is conserved in all ABCG/PDR transporters. This Y464 contributes a salt bridge interaction within the THB complex for stabilization of the transmission interface.

TMH3 also has several conserved residues within the ABCG/PDR group such as negative residues (E or D), proline, as well as positive amino acids (K or R). D477 at the start of TMH3 in human ABCG2 is on top of the transmission interface and may contribute to an intracellular gate. ECL2 is again rather short, containing only eight residues with a kink, connecting TMH3 and TMH4 at the cell surface.

The hydrophobic valve is a kinked domain just after TMH5, consisting of the conserved "glycine-Φ-Φ" (Φ is hydrophobic residue) motif that subtends the extracellular bilayer leaflet. The valve is pivotal for controlling substrate transport through a putative translocation pathway whose precise nature and dynamics remains obscure. The re-entry helix is the counterpart of the elbow helix and is unique, establishing a kinked hallmark motif in the ABCG/PDR subfamily with several conserved residues. In addition, ECL3 is the only large extracellular loop, and a main part of the extracellular roof architecture. Interestingly, ECL3 in human ABCG2 has three cysteines, which form intra- and inter-molecular disulfide formations, and perhaps facilitate dimerization and drug release [334,335]. Likewise, conserved cysteines in fungal PDR transporters may engage in an intra-molecular disulfide bond to stabilize the PDR biogenesis. Finally, TMH6 is followed by a very small but highly conserved C-terminus that contains several positive charges at the C-terminus.

4. *C. albicans* PDR/Cdr1 Holds All Conserved Motifs Critical for ABCG Function

The atomic structures from X-ray crystallography and cryo-EM of mammalian ABCGs suggest a rather unique fold resembling an importer rather than an exporter [38,256–259]. Indeed, mammalian ABCGs and fungal PDR exporters share conserved and superimposable topologies with all functional motifs in equivalent places (Figure 1A, Figure 2B, Supplementary Figure S1 and Table 4). Hence, we generated homology models of Cdr1 from *C. albicans*, using human ABCG2 (PDB ID 6VXF) as a template (Figure 3A). Remarkably, the Cdr1 structural model perfectly mirrored the human ABCG2 conformation, since each pivotal motif important for function is also present in Cdr1 (Figure 3).

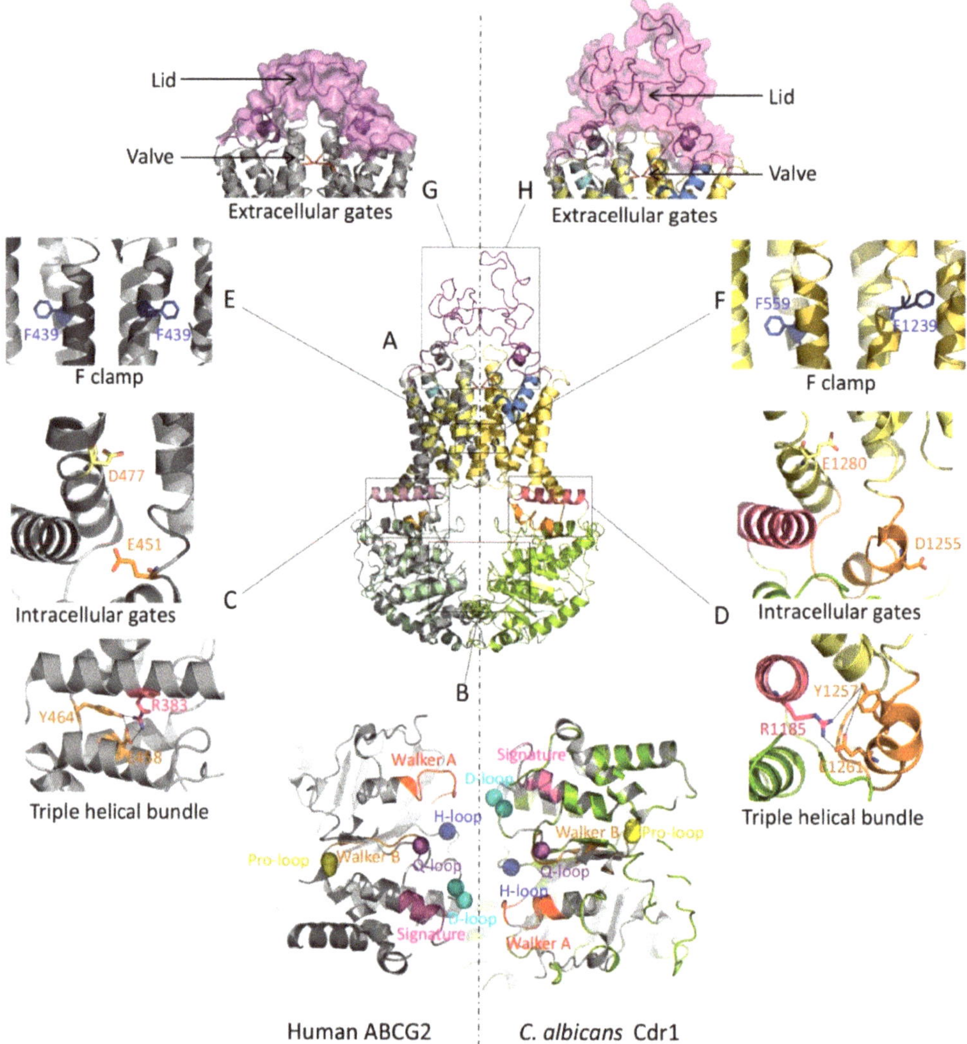

Figure 3. Type II mammalian and fungal ABCG/PDR exporters share functional domains. (**A**) A homology model of the PDR exporter Cdr1 from *C. albicans* was generated using the SWISS-MODEL tool (color ribbon, NBD: green, TMD: yellow, elbow helix: pink, ICL: orange, valve: red, re-entry helix: blue and ECL: purple). The Cdr1 model was superimposed with the cryo-EM structure of human ABCG2 (PDB ID: 6VXF) (gray ribbon). (**B**) Zoom-in top view of the NBD dimer from panel A at the NBD–NBD interface shows essential conserved motifs (Walker A: red, Q-loop: violet, signature: pink, Pro-loop: yellow, Walker B: orange, D-loop: cyan and H-loop: blue). The transmission interfaces of human ABCG2 (**C**) and *C. albicans* Cdr1 (**D**) show a network cluster of triple helical bundles (THB) with conserved tyrosine residues as part of a salt bridge between elbow helix and ICL. Two negative residues are shown as the intracellular gates. Conserved residues are shown as sticks with color-coding as in the topology model. The conserved phenylalanine F clamp F439 in human ABCG2 (**E**) and the putative F clamp in *C. albicans* Cdr1 (**F**) are indicated as sticks. At the extracellular gates, the valve-like structures at the top of central cavity (red) are shown at the corresponding position of human ABCG2 (**G**) and *C. albicans* Cdr1 (**H**). The lid-structure formed by ECLs is shown as a violet ribbon with a surface. The homo-dimeric human ABCG2 transporter has a symmetric lid (**G**), while the full-size *C. albicans* Cdr1 transporter has a larger ECL forming the outer lid and part of the roof architecture (**H**).

The NBDs in Cdr1 are in closer contact and hold all conserved regions required for ABCG2 function (ATPase activity, A-loop, Walker A and B, Q-loop, mutational hot spot helix, signature loop, Pro-loop, D-loop and H-loop). The N-terminal and C-terminal NBDs form a head-to-tail dimer upon ATP-binding [336,337], with a RecA-like and an α-helical subdomain [327] (Figure 3B). Nonetheless, the fungal NBD1 has three minor differences, in that (i) a glutamine residue in the Q-loop is replaced with glutamate, (ii) the histidine residue in the H-loop is substituted with tyrosine and (iii) the Pro-loop is missing but contains a glycine instead [337] (Table 4). The non-identical deviant NBDs in fungal PDRs [41,216,338,339] may support an asymmetric catalytic cycle as proposed for ABCG5/G8 [38,339–341], whereas "symmetric" cycles require the presence of fully conserved "canonical" ATP-binding sites in both NBDs.

Stabilizing salt bridges maintain proper folding and dynamics. At least two salt bridges appear conserved in Cdr1 (Supplementary Figure S1), connecting the elbow helix with the coupling helix (R503 to E576, and R1185 to E1261) (Figure 3D), respectively. However, the salt bridge within ICL1 seems absent in fungal PDR, and a salt bridge at the upper membrane leaflet is found only in the first half of Cdr1 (R456 in ECL1 to E704 in the re-entry helix) (Table 4).

The triple helical bundle (THB) is part of the transmission interface and extremely conserved in Cdr1. Remarkably, the THB is present in diverse ABC transporters, including ABCB1 and LPS extractor, as well as in the antibiotic exporter MacB [332]. This may reflect a universal function in mediating NBD–TMD cross-talk, thus constituting a key element for controlling and driving the conformational switch to drive substrates through the translocation pathway. Moreover, the most highly conserved Y464 residue is essential for ATP consumption, and engages in a salt bridge to stabilize the entire transmission interface in the center of NBD–elbow–ICL1 cluster (Table 4 and Figure 3C,D). Thus, the THB constitutes a cluster of limited conformational flexibility, taking advantage of Y464 and its salt bridge with E458 in ICL1 and/or between elbow helix R383 and E458 [332] (Table 4 and Figure 3C). As for Cdr1, the proposed THB cluster is present in both N- and C-terminal domains. At the N-terminus, Y584 connects to R503 (elbow helix) and E576 (in the coupling helix), whereas at the C-terminal domain, Y1257 bridges to R1185 (second elbow helix) and E1261 in the second coupling helix (Table 4 and Figure 3D). This finding strongly supports the notion for THB as an essential structure acting at the transmission interface to control the entire transport cycle [332].

The mechanism of intracellular gating in PDR/ABCG2 must be crucial for substrate/inhibitor entry into the transport pathway of the exporter, but as yet is little understood. In human ABCG2, two negative residues are conserved in all ABCG/PDR transporters, represented by E451 (between TMH2 and ICL1) and D477 (beginning of TMH3). Interestingly enough, they are pivotal for drug transport, but do not affect ATP hydrolysis [40] (Table 4 and Figure 3C). Since this region around the transmission interface undergoes dynamic movements during the catalytic cycle, E451 and D477 may not be part of a drug-binding zone but rather provide an entry route and gating functions [332]. In fungal PDRs, both negative residues are conserved (Supplementary Figure S1) and Cdr1 indeed contains the corresponding positions with E570 and E597 in the N-terminal part, and D1255 and E1280 in the C-terminal domain (Table 4 and Figure 3D).

The so-called phenylalanine clamp formed by two F residues is located in the substrate-binding zone of ABCG/PDR transporters. Interestingly, the THM2 in the ABCG/PDR subfamily contains at least 4–5 conserved F residues (Supplementary Figure S1). TMH2 occupies space in the middle of the transmembrane core, where a putative binding zone around the central cavity is present. In human ABCG2, F439 (Figure 3E), located in the middle of TMH2, is implicated in binding both substrates and inhibitors [333]. Remarkably, a new cryo-EM structure of ABCG2 illustrates that the aromatic side chains of both phenylalanine residues could contribute to a binding site [259], and thus play a role as a clamp for both substrate and inhibitor recognition [259,333]. Remarkably, the Cdr1 homology model suggests that the conserved residues in both TMDs (Supplementary Figure S1) are F559 in

TMH2 and F1239 in TMH8 (Figure 3F) and equivalent to F439 in ABCG2 (Figure 3E). Hence, conserved phenylalanine in or nearby substrate/inhibitor-binding zones of ABCG/PDR provide a clamping mechanism to trap substrates and/or inhibitors.

Extracellular gating at the membrane interface and subsequent drug release from the outward-facing state is regulated by two conserved motifs, a hydrophobic valve and a flexible lid architecture in the roof [42]. The hydrophobic valve in ABCG/PDR transporters is contributed by two half-molecules, thus generating a physical gate for outward-directed substrate translocation from the central into the upper cavity. The atomic structures of mammalian ABCGs [38,256,259] indeed reveal a unique valve-like motif in the core of the transporter, separating the central cavity from the upper cavity. This conserved "glycine-Φ-Φ" motif plays a critical role as a hydrophobic valve that controls water flow through the transport pathway [42] (Supplementary Figure S1). The glycine adds a flexible kink, whereas two hydrophobic aliphatic leucines build a hydrophobic barrier to prevent water flow or substrate leakage. Human ABCG2 has the "G553-L554-L555" motif, whereas Cdr1 has "G672-F673-V674" and "G1362-V1363-L1364" in the first and second half, respectively (Table 4 and Figure 3G,H). Interestingly, a similar extracellular gating mechanism may also operate in CmABCB1, which possesses a gate at the outer membrane border to regulate substrate translocation [35,295].

A flexible but compact lid is part of the roof architecture formed by a rather large extracellular domain in ABCG/PDR. The roof is another unique motif in the ABCG/PDR subfamily. In human ABCG2, this roof is maintained by an intramolecular C592–C608 disulfide bond that strengthens the compact lid architecture. This lid may establish the second gating mechanism to regulate drug release from the outer cavity [334,335] (Figure 3G). Notably, the covalent C603–C603 inter-molecular link is key for homo-dimer formation in human ABCG2, but not essential for function [334,342]. The last extracellular loop of fungal PDR is slightly larger than the equivalent loop in mammalian ABCG2 (Supplementary Figure S1), where 2–3 conserved cysteine residues are present (Table 4). The overall similarity in the roof architecture also supports the notion of a conserved extracellular gating mechanism in ABCG/PDR transporters (Figure 3G,H).

5. Model for a Conserved Catalytic Transport Cycle of ABCG/PDR Transporters

Mammalian ABCG and fungal PDR share all hallmark domains as well as numerous conserved residues essential for function (Figure 3). Based on these striking similarities, we wish to propose a unified mechanism for the transport cycle of type II ABCG/PDR multidrug transporters (Figure 4). Several studies suggest that more than half of all known ABC transporters including human ABCG5/G8 utilize non-equivalent or deviant NBDs [43,216,259,306,329,343,344]. Interestingly, ABCG2 appears as a perfect homo-dimer molecule, although it may have some asymmetries in the NBD dimers [325,341,345]. Therefore, the "primordial" alternating access model [30,36,324,346,347] forms a rational basis for our model (Figure 4). The ABCG/PDR transporters in the apo drug-free state are in an inward-facing configuration, with the bottom of the NBD dimer connected. We propose that ABCG/PDR subfamilies have asymmetric catalytic cycles, as proposed for human ABCG2 [341]. In this state, only one nucleotide-binding site is occupied by ATP, which could support an "intermediate" NBD dimerization state, with one free site remaining accessible for ATP. The intracellular gate formed by two negative residues at the membrane border of the ICL1 in the transmission interface is open, thus offering a path for substrate/inhibitor entry [40]. The central cavity provides free binding zones to accommodate compounds of variable chemical spaces [157,348]. The aromatic rings in the conserved F clamp establish accessible binding/trapping sites [333]. The hydrophobic valve at the top of the central cavity is almost completely closed and blocks water leakage through the translocation pathway [42]. The lid-forming roof architecture also remains closed in the inward-facing state, with a compact loop structure that limits the space in the upper cavity. Since ABCG/PDR exporters have an uncoupled ATP hydrolysis cycle, the catalytic cycle would still be active even without substrate(s) [33,40,42]. Drug substrates (2a) or inhibitors (2b) can access the central cavity and the translocation pathway through the intracellular gate(s) [40] before getting trapped in the binding zones [157] by the F clamp located in TMH2 in each TMD [333]. This is a critical step which also prevents substrate escape from the central transport pathway. By contrast, the binding of an inhibitor (2b) at the region below the valve, would lock the conformation in the inward–open state and inhibit ATP hydrolysis activity as indicated by cryo-EM particle structures [257]. Whether a compound is an inhibitor or a transport substrate for ABCG/PDR transporters is solely determined by the affinity that sets the on versus off rates, and by the kinetics underlying the interactions in the binding zones [33]. Indeed, biochemistry data also suggest that certain inhibitors inhibit ATPase activity [349–351]. Binding of the second ATP molecule functions as a molecular glue that triggers complete NBD dimerization, thus inducing the conformational switch of the TMDs into a substrate-occluded state (3). The mechanical movement at the transmission interface requires the THB as a rigid structure of limited dynamics [332]. Subsequently, the central cavity is compressed, hence creating peristaltic pressure that drives substrates along the central translocation channel through the concomitantly opening valve. Further, the full NBD dimerization pushes the transporter into a compressed state, imposing a squeezing motion on the central cavity that generates pressure critical for opening the valve. The hydrophobic valve also serves as the first barrier for an extracellular gating to ensure unidirectional transport. A retrograde backflow of substrates is therefore prevented by the outward hydrostatic pressure and by the tight valve that would close when resetting the transporter [42]. Accordingly, the space of the upper cavity is then enlarged to accommodate substrates [42]. The limited dynamics and stability of the extracellular roof is supported by a conserved salt bridge between ECL1 and the re-entry helix [42]. The compact lid, which is mainly formed by ECL, then constitutes the second barrier at the extracellular interface. Once the lid has opened, it allows for substrate release into the extracellular space. Finally, ATP hydrolysis at one (or both sites) releases Pi and ADP, thus initiating the reset of the transporter with two NBDs in the original inward–open-facing state [43]. The resulting accessible ATP site also opens the intracellular gate enabling a new cycle of substrate recognition. This catalytic cycle reflects the current knowledge about the catalytic cycles of ABCG2/PDR transporters, whereby biochemical, structural, genetic and mutational data have been integrated.

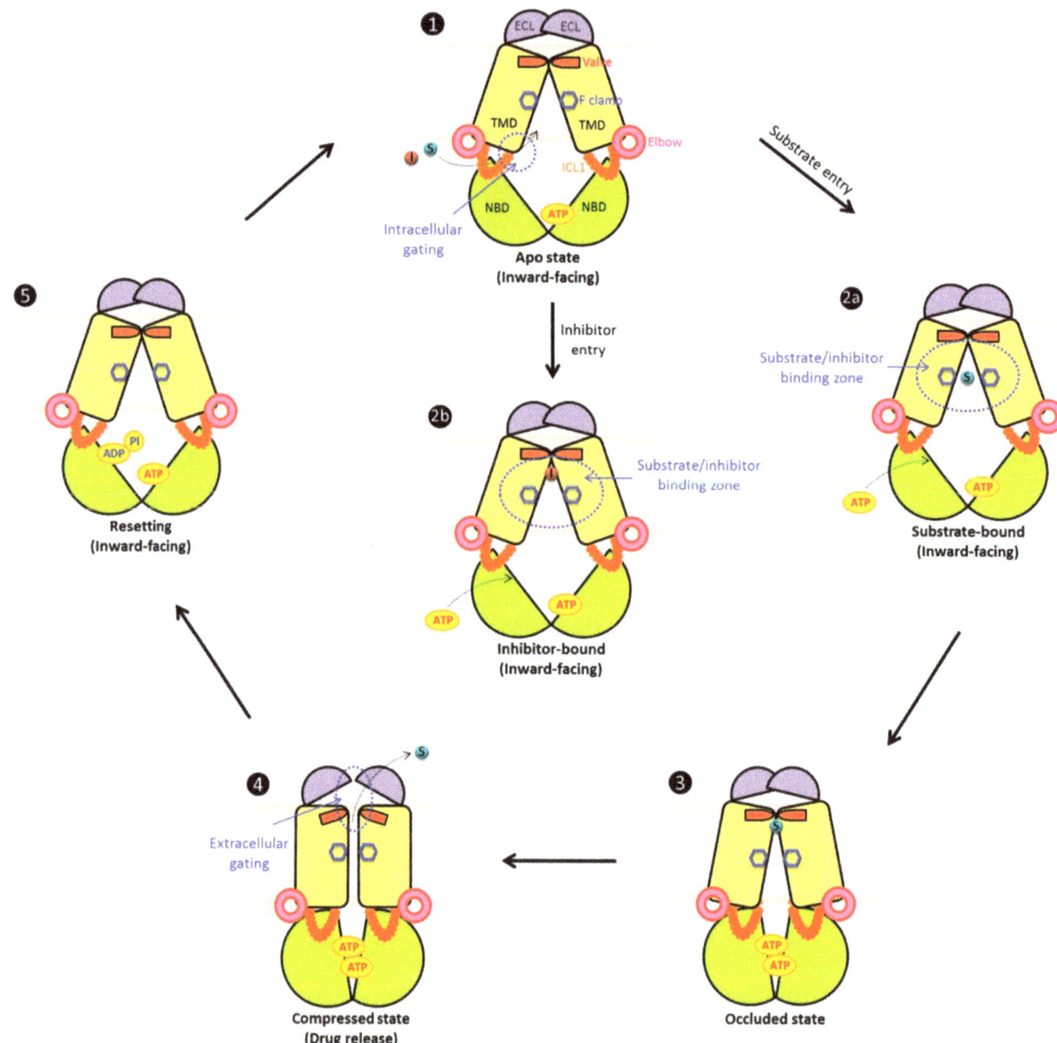

Figure 4. Proposed catalytic cycles of mammalian/fungal type II exporters (ABCG/PDR). In the apo substrate-free state (**1**), the exporter adopts an inward-facing conformation. We propose that NBDs are open with ATP present in at least one NBD or both, which mediates partial NBD dimerization leaving only one accessible ATP-binding region. The intracellular gate(s) at the transmission interface provides access for substrate or inhibitor entry. The aromatic rings at the conserved F clamp form accessible binding sites at the closed transporter valve subtending the closed ECL. Drug substrates (**2a**) or inhibitors (**2b**) can enter through intracellular gate(s), preceding their trapping in distinct binding zones in the central cavity. Binding of ATP at the second binding site or both triggers full NBD dimerization and triggers a first conformational change, setting an occluded state (**3**). The communication between NBD and TMD is regulated via a rigid triple helical bundle (THB) as a key part of the transmission interface. The NBD dimerization compresses the central cavity space to drive substrate movement through the translocation, thus engaging a push and squeeze motion to open the valve. Substrates shift into the upper cavity and are released by the subsequent opening of the ECL lid (**4**). ATP hydrolysis at one NBD site may be enough to reset the catalytic cycle and to convert the transporter molecule into the inward-facing drug-recognizing state (**5**). The structures show NBDs (green), elbow helix (pink), TMDs (yellow), ICL (orange), ECL (purple), phenylalanine clamp (blue hexagon), valve (red), substrates (cyan) and inhibitors (red). For more details and references see the main text.

6. Conclusions and Future Perspectives

The increase in atomic structures of ABC transporters, and data from extensive biochemical and genetic experiments, have been propelling the field, since they have yielded novel insights and a better understanding of ABC transporter mechanisms. At the same time, while atomic structures have been invaluable, they have to be interpreted with caution, especially when biochemistry or genetics are not in line with structural data [265] or when biological relevance appears doubtful. Indeed, the painful history of ABC transporter structures [207,265], shows that even higher resolution structures suffer from their static nature that only reflects snapshots of a catalytic cycle. Thus, we need atomic structures reflecting more than a single conformation and possibly many transition states [352], as well as extensive validation by biochemistry and genetics, to validate their biological relevance and define catalytic cycles. Furthermore, there is an unmet need for more interdisciplinary collaborations that also engage alternative structural approaches like NMR [353–356] as well as complementary biophysical methods [265,357–359] to expand our mechanistic views of ABC transporters in all living kingdoms.

The past few years challenge previous notions that a unified transport mechanism exists for MDR exporters from all three major classes, such as ABCB1/MDR1/P-gp [35,293, 295,302,303,307,310], ABCC1/MRP1 [298,311,312] and ABCG2/BCRP [256–259] (Table 3). While early homology modeling attempts of the ABCG/PDR family yielded incorrect folds due to using type I exporter templates rather than exporter type II [360–363], the use of proper coordinates have now validated the usefulness of modeling for dissecting mechanisms of ABC transport cycles [40,256]. Of note, we have taken a "reverse" approach here, since we exploited human ABCG2 and ABCG5/G8 structures to model evolutionarily conserved fungal PDR transporters. This not only yielded new testable homology models, but also hinted that catalytic cycles may have been conserved at least in orthologous families such as human ABCG and fungal PDR. Of note, four cryo-EM structures of yeast Pdr5 [339] at atomic resolutions from 2.8 to 3.5 Å just emerged [339]. Most remarkably, as we show here for Cdr1, the paradigm yeast Pdr5 efflux pump shows similar transitional movements during the catalytic cycle [339], strongly supporting the proposed catalytic cycle for PDR/ABCG transporters operating as uncoupled peristaltic pumps [339]. *Twist* and *Squeeze* may be used by different types of transporters, and these driving mechanisms may appear related or even similar when looking at it from the mechanics, as either *Twist* or *Squeeze* or a combination of both can result in peristaltic pressure during the switch. While there are many challenges remaining ahead of us, the reversal of clinical MDR phenomena in fungal pathogens in infectious disease settings have regained attention, especially since the catalytic cycle of human ABCG2 likely reflects the mode of action of fungal PDR transporters implicated in anti-infective drug resistance.

Supplementary Materials: The following are available online at https://www.mdpi.com/article/10.3390/ijms22094806/s1. Figure S1. Amino acid sequence alignment of fungal ABC transporters (PDR subfamily) and mammalian ABCG subfamily.

Author Contributions: N.K. performed all analysis and prepared figures. N.K. and K.K. interpreted the data and wrote the manuscript. All authors have read and agreed to the published version of the manuscript.

Funding: This research was funded by the Austrian Science Fund (FWF-SFB-035-20 to K.K.).

Institutional Review Board Statement: Not applicable.

Informed Consent Statement: Not applicable.

Data Availability Statement: The authors confirm that all data are fully available without restrictions. All relevant data are provided in the manuscript and in the Supplementary Materials.

Data and Analysis: Figures and phylogenetic analyses were generated by BioEdit, ClustalX2 and MEGA-X. The homology model was generated using SWISS-MODEL using a Cdr1 sequence from

UniProt (accession no. P43071), using the human ABCG2 as a template (PDB ID 6VXF). The molecular visualizations were performed using PyMOL, v1.8.4.

Acknowledgments: Open Access Funding by the Austrian Science Fund (FWF). This work was supported by grants from the Austrian Science Fund FWF-SFB-035-20 to K.K. We are very much indebted to Ian D. Kerr, Thomas Stockner and Yuan Eric Lee for constructive and helpful discussions. We also appreciate the constructive and helpful comments from reviewer 2.

Conflicts of Interest: The authors have no competing financial interests.

References

1. Higgins, C.F. ABC Transporters: From Microorganisms to Man. *Annu. Rev. Cell Biol.* **1992**, *8*, 67–113. [CrossRef]
2. Kuchler, K. The ABC of ABCs: Multidrug resistance and genetic diseases. *FEBS J.* **2011**, *278*, 3189. [CrossRef] [PubMed]
3. Holland, I.B. Rise and rise of the ABC transporter families. *Res. Microbiol.* **2019**, *170*, 304–320. [CrossRef] [PubMed]
4. Srikant, S. Evolutionary history of ATP-binding cassette proteins. *FEBS Lett.* **2020**, *594*, 3882–3897. [CrossRef]
5. Ogasawara, F.; Kodan, A.; Ueda, K. ABC proteins in evolution. *FEBS Lett.* **2020**, *594*, 3876–3881. [CrossRef] [PubMed]
6. Holland, I.; Blight, M.A. ABC-ATPases, adaptable energy generators fuelling transmembrane movement of a variety of molecules in organisms from bacteria to humans. *J. Mol. Biol.* **1999**, *293*, 381–399. [CrossRef]
7. Rees, D.C.; Johnson, E.; Lewinson, O. ABC transporters: The power to change. *Nat. Rev. Mol. Cell Biol.* **2009**, *10*, 218–227. [CrossRef]
8. Hollenstein, K.; Dawson, R.J.P.; Locher, K.P. Structure and mechanism of ABC transporter proteins. *Curr. Opin. Struct. Biol.* **2007**, *17*, 412–418. [CrossRef] [PubMed]
9. Senior, A.E.; Gadsby, D.C. ATP hydrolysis cycles and mechanism inP-glycoprotein and CFTR. *Semin. Cancer Biol.* **1997**, *8*, 143–150. [CrossRef]
10. Lage, H. ABC-transporters: Implications on drug resistance from microorganisms to human cancers. *Int. J. Antimicrob. Agents* **2003**, *22*, 188–199. [CrossRef]
11. Glavinas, H.; Krajcsi, P.; Cserepes, J.; Sarkadi, B. The Role of ABC Transporters in Drug Resistance, Metabolism and Toxicity. *Curr. Drug Deliv.* **2004**, *1*, 27–42. [CrossRef] [PubMed]
12. Dean, M.; Allikmets, R. Complete Characterization of the Human ABC Gene Family. *J. Bioenerg. Biomembr.* **2001**, *33*, 475–479. [CrossRef] [PubMed]
13. Higgins, C.F. ABC transporters: Physiology, structure and mechanism—An overview. *Res. Microbiol.* **2001**, *152*, 205–210. [CrossRef]
14. Sharom, F.J. ABC multidrug transporters: Structure, function and role in chemoresistance. *Pharmacogenomics* **2008**, *9*, 105–127. [CrossRef]
15. Cuperus, F.J.; Claudel, T.; Gautherot, J.; Halilbasic, E.; Trauner, M. The role of canalicular ABC transporters in cholestasis. *Drug Metab. Dispos.* **2014**, *42*, 546–560. [CrossRef]
16. Loo, T.W.; Clarke, D.M. Mutational analysis of ABC proteins. *Arch. Biochem. Biophys.* **2008**, *476*, 51–64. [CrossRef] [PubMed]
17. Xiong, J.; Feng, J.; Yuan, D.; Zhou, J.; Miao, W. Tracing the structural evolution of eukaryotic ATP binding cassette transporter superfamily. *Sci. Rep.* **2015**, *5*, 16724. [CrossRef] [PubMed]
18. Thomas, C.; Aller, S.G.; Beis, K.; Carpenter, E.P.; Chang, G.; Chen, L.; Dassa, E.; Dean, M.; Van Hoa, F.D.; Ekiert, D.; et al. Structural and functional diversity calls for a new classification of ABC transporters. *FEBS Lett.* **2020**, *594*, 3767–3775. [CrossRef]
19. Higgins, C.F. Multiple molecular mechanisms for multidrug resistance transporters. *Nat. Cell Biol.* **2007**, *446*, 749–757. [CrossRef]
20. Cole, S.P. Targeting Multidrug Resistance Protein 1 (MRP1,ABCC1): Past, Present, and Future. *Annu. Rev. Pharmacol. Toxicol.* **2014**, *54*, 95–117. [CrossRef]
21. Ford, B. CFTR structure: Lassoing cystic fibrosis. *Nat. Struct. Mol. Biol.* **2017**, *24*, 13–14. [CrossRef]
22. Holland, K.A.; Holland, I.B. Adventures with ABC-proteins: Highly Conserved ATP-dependent Transporters. *Acta Microbiol. et Immunol. Hung.* **2005**, *52*, 309–322. [CrossRef]
23. Jones, P.M.; George, A.M. A reciprocating twin-channel model for ABC transporters. *Q. Rev. Biophys.* **2014**, *47*, 189–220. [CrossRef] [PubMed]
24. Davidson, A.L.; Dassa, E.; Orelle, C.; Chen, J. Structure, Function, and Evolution of Bacterial ATP-Binding Cassette Systems. *Microbiol. Mol. Biol. Rev.* **2008**, *72*, 317–364. [CrossRef] [PubMed]
25. Ford, R.C.; Beis, K. Learning the ABCs one at a time: Structure and mechanism of ABC transporters. *Biochem. Soc. Trans.* **2019**, *47*, 23–36. [CrossRef] [PubMed]
26. Locher, K.P.; Lee, A.T.; Rees, D.C. The *E. coli* BtuCD structure: A framework for ABC transporter architecture and mechanism. *Science* **2002**, *296*, 1091–1098. [CrossRef]
27. Schmitt, L.; Tampe, R. Structure and mechanism of ABC transporters. *Curr. Opin. Struct. Biol.* **2002**, *12*, 754–760. [CrossRef]
28. Higgins, C.F.; Linton, K.J. The ATP switch model for ABC transporters. *Nat. Struct. Mol. Biol.* **2004**, *11*, 918–926. [CrossRef]
29. Dawson, R.J.P.; Hollenstein, K.; Locher, K.P. Uptake or extrusion: Crystal structures of full ABC transporters suggest a common mechanism. *Mol. Microbiol.* **2007**, *65*, 250–257. [CrossRef]

30. Oldham, M.L.; Khare, D.; Quiocho, F.A.; Davidson, A.L.; Chen, J. Crystal structure of a catalytic intermediate of the maltose transporter. *Nat. Cell Biol.* **2007**, *450*, 515–521. [CrossRef]
31. Seeger, M.A.; van Veen, H.W. Molecular basis of multidrug transport by ABC transporters. *Biochim. Biophys. Acta* **2009**, *1794*, 725–737. [CrossRef] [PubMed]
32. Procko, E.; O'Mara, M.L.; Bennett, W.F.D.; Tieleman, D.P.; Gaudet, R. The mechanism of ABC transporters: General lessons from structural and functional studies of an antigenic peptide transporter. *FASEB J.* **2009**, *23*, 1287–1302. [CrossRef] [PubMed]
33. Ernst, R.; Kueppers, P.; Stindt, J.; Kuchler, K.; Schmitt, L. Multidrug efflux pumps: Substrate selection in ATP-binding cassette multidrug efflux pumps—First come, first served? *FEBS J.* **2009**, *277*, 540–549. [CrossRef]
34. Xu, K.; Zhang, M.; Zhao, Q.; Yu, F.; Guo, H.; Wang, C.; He, F.; Ding, J.; Zhang, P. Crystal structure of a folate energy-coupling factor transporter from Lactobacillus brevis. *Nat. Cell Biol.* **2013**, *497*, 268–271. [CrossRef] [PubMed]
35. Kodan, A.; Yamaguchi, T.; Nakatsu, T.; Sakiyama, K.; Hipolito, C.J.; Fujioka, A.; Hirokane, R.; Ikeguchi, K.; Watanabe, B.; Hiratake, J.; et al. Structural basis for gating mechanisms of a eukaryotic P-glycoprotein homolog. *Proc. Natl. Acad. Sci. USA* **2014**, *111*, 4049–4054. [CrossRef] [PubMed]
36. Wilkens, S. Structure and mechanism of ABC transporters. *F1000Prime Rep.* **2015**, *7*, 14. [CrossRef]
37. Horsey, A.J.; Cox, M.H.; Sarwat, S.; Kerr, I.D. The multidrug transporter ABCG2: Still more questions than answers. *Biochem. Soc. Trans.* **2016**, *44*, 824–830. [CrossRef]
38. Lee, J.-Y.; Kinch, L.N.; Borek, D.M.; Wang, J.; Wang, J.; Urbatsch, I.L.; Xie, X.-S.; Grishin, N.V.; Cohen, J.C.; Otwinowski, Z.; et al. Crystal structure of the human sterol transporter ABCG5/ABCG8. *Nat. Cell Biol.* **2016**, *533*, 561–564. [CrossRef]
39. Locher, K.P. Mechanistic diversity in ATP-binding cassette (ABC) transporters. *Nat. Struct. Mol. Biol.* **2016**, *23*, 487–493. [CrossRef]
40. Khunweeraphong, N.; Stockner, T.; Kuchler, K. The structure of the human ABC transporter ABCG2 reveals a novel mechanism for drug extrusion. *Sci. Rep.* **2017**, *7*, 1–15. [CrossRef]
41. Wagner, M.; Smits, S.H.J.; Schmitt, L. In vitro NTPase activity of highly purified Pdr5, a major yeast ABC multidrug transporter. *Sci. Rep.* **2019**, *9*, 7761. [CrossRef]
42. Khunweeraphong, N.; Szöllősi, D.; Stockner, T.; Kuchler, K. The ABCG2 multidrug transporter is a pump gated by a valve and an extracellular lid. *Nat. Commun.* **2019**, *10*, 1–14. [CrossRef]
43. Thomas, C.; Tampé, R. Structural and Mechanistic Principles of ABC Transporters. *Annu. Rev. Biochem.* **2020**, *89*, 605–636. [CrossRef] [PubMed]
44. Linton, K.J.; Higgins, C.F. Structure and function of ABC transporters: The ATP switch provides flexible control. *Pflügers Arch. Eur. J. Physiol.* **2006**, *453*, 555–567. [CrossRef]
45. Oswald, C.; Holland, I.B.; Schmitt, L. The motor domains of ABC-transporters. What can structures tell us? *Naunyn Schmiedebergs Arch. Pharmacol.* **2006**, *372*, 385–399. [CrossRef] [PubMed]
46. Dean, M.; Annilo, T. Evolution of the atp-binding cassette (abc) transporter superfamily in vertebrates. *Annu. Rev. Genom. Hum. Genet.* **2005**, *6*, 123–142. [CrossRef] [PubMed]
47. Dassa, E.; Schneider, E. The rise of a protein family: ATP-binding cassette systems. *Res. Microbiol.* **2001**, *152*, 203. [CrossRef]
48. Bouige, P.; Laurent, D.; Piloyan, L.; Dassa, E. Phylogenetic and functional classification of ATP-binding cassette (ABC) systems. *Curr. Protein Pept. Sci.* **2002**, *3*, 541–559. [CrossRef] [PubMed]
49. Theodoulou, F.L.; Kerr, I.D. ABC transporter research: Going strong 40 years on. *Biochem. Soc. Trans.* **2015**, *43*, 1033–1040. [CrossRef]
50. Riordan, J.R.; Rommens, J.M.; Kerem, B.; Alon, N.; Rozmahel, R.; Grzelczak, Z.; Zielenski, J.; Lok, S.; Plavsic, N.; Chou, J.L.; et al. Identification of the cystic fibrosis gene: Cloning and characterization of complementary DNA. *Science* **1989**, *245*, 1066–1073. [CrossRef]
51. Harris, A.; Argent, B.E. The cystic fibrosis gene and its product CFTR. *Semin. Cell Biol.* **1993**, *4*, 37–44. [CrossRef] [PubMed]
52. Stieger, B.; Meier, P.J. Bile acid and xenobiotic transporters in liver. *Curr. Opin. Cell Biol.* **1998**, *10*, 462–467. [CrossRef]
53. Strautnieks, S.S.; Bull, L.N.; Knisely, A.S.; Kocoshis, S.A.; Dahl, N.; Arnell, H.; Sokal, E.; Dahan, K.; Childs, S.; Ling, V.; et al. A gene encoding a liver-specific ABC transporter is mutated in progressive familial intrahepatic cholestasis. *Nat. Genet.* **1998**, *20*, 233–238. [CrossRef]
54. Berge, K.E.; Tian, H.; Graf, G.A.; Yu, L.; Grishin, N.V.; Schultz, J.; Kwiterovich, P.; Shan, B.; Barnes, R.; Hobbs, H.H. Accumulation of Dietary Cholesterol in Sitosterolemia Caused by Mutations in Adjacent ABC Transporters. *Science* **2000**, *290*, 1771–1775. [CrossRef]
55. Patel, S.B.; Salen, G.; Hidaka, H.; Kwiterovich, P.O.; Stalenhoef, A.F.; Miettinen, T.A.; Grundy, S.M.; Lee, M.H.; Rubenstein, J.S.; Polymeropoulos, M.H.; et al. Mapping a gene involved in regulating dietary cholesterol absorption. The sitosterolemia locus is found at chromosome 2p21. *J. Clin. Investig.* **1998**, *102*, 1041–1044. [CrossRef]
56. Hubacek, J.A.; Berge, K.E.; Cohen, J.C.; Hobbs, H.H. Mutations in ATP-cassette binding proteins G5 (ABCG5) and G8 (ABCG8) causing sitosterolemia. *Hum. Mutat.* **2001**, *18*, 359–360. [CrossRef]
57. Hobbs, H.H.; Russell, D.W.; Brown, M.S.; Goldstein, J.L. The LDL receptor locus in familial hypercholesterolemia: Mutational analysis of a membrane protein. *Annu. Rev. Genet.* **1990**, *24*, 133–170. [CrossRef]
58. Aguilar-Bryan, L.; Nichols, C.G.; Wechsler, S.W.; Clement, J.P.; Boyd, A.E.; Gonzalez, G.; Herrera-Sosa, H.; Nguy, K.; Bryan, J.; Nelson, D.A. Cloning of the beta cell high-affinity sulfonylurea receptor: A regulator of insulin secretion. *Science* **1995**, *268*, 423–426. [CrossRef]

59. Woodward, O.M.; Köttgen, M.; Coresh, J.; Boerwinkle, E.; Guggino, W.B. Identification of a urate transporter, ABCG2, with a common functional polymorphism causing gout. *Proc. Natl. Acad. Sci. USA* **2009**, *106*, 10338–10342. [CrossRef]
60. Paulusma, C.C.; Bosma, P.J.; Zaman, G.J.; Bakker, C.T.; Otter, M.; Scheffer, G.L.; Scheper, R.J.; Borst, P.; Elferink, R.P.O. Congenital jaundice in rats with a mutation in a multidrug resistance-associated protein gene. *Science* **1996**, *271*, 1126–1128. [CrossRef] [PubMed]
61. Kartenbeck, J.; Leuschner, U.; Mayer, R.; Keppler, D. Absence of the canalicular isoform of the MRP gene-encoded conjugate export pump from the hepatocytes in Dubin-Johnson syndrome. *Hepatology* **1996**, *23*, 1061–1066. [CrossRef]
62. Allikmets, R. A photoreceptor cell-specific ATP-binding transporter gene (ABCR) is mutated in recessive Stargardt macular dystrophy. *Nat. Genet.* **1997**, *17*, 122. [PubMed]
63. Berger, J.; Forss-Petter, S.; Eichler, F. Pathophysiology of X-linked adrenoleukodystrophy. *Biochimie* **2014**, *98*, 135–142. [CrossRef] [PubMed]
64. Shani, N.; Watkins, P.A.; Valle, D. PXA1, a possible Saccharomyces cerevisiae ortholog of the human adrenoleukodystrophy gene. *Proc. Natl. Acad. Sci. USA* **1995**, *92*, 6012–6016. [CrossRef] [PubMed]
65. Aubourg, P. Adrenoleukodystrophy and other peroxisomal diseases. *Curr. Opin. Genet. Dev.* **1994**, *4*, 407–411. [CrossRef]
66. Seyffer, F.; Tampé, R. ABC transporters in adaptive immunity. *Biochim. Biophys. Acta (BBA) Gen. Subj.* **2015**, *1850*, 449–460. [CrossRef] [PubMed]
67. Suh, W.; Cohen-Doyle, M.; Fruh, K.; Wang, K.; Peterson, P.; Williams, D. Interaction of MHC class I molecules with the transporter associated with antigen processing. *Science* **1994**, *264*, 1322–1326. [CrossRef] [PubMed]
68. De La Salle, H.; Hanau, D.; Fricker, D.; Urlacher, A.; Kelly, A.; Salamero, J.; Powis, S.H.; Donato, L.; Bausinger, H.; Laforet, M.; et al. Homozygous human TAP peptide transporter mutation in HLA class I deficiency. *Science* **1994**, *265*, 237–241. [CrossRef]
69. Gelissen, I.C.; Harris, M.; Rye, K.; Quinn, C.; Brown, A.J.; Kockx, M.; Cartland, S.; Packianathan, M.; Kritharides, L.; Jessup, W. ABCA1 and ABCG1 synergize to mediate cholesterol export to apoA-I. *Arterioscler. Thromb. Vasc. Biol.* **2006**, *26*, 534–540. [CrossRef]
70. Yvan-Charvet, L.; Ranalletta, M.; Wang, N.; Han, S.; Terasaka, N.; Li, R.; Welch, C.; Tall, A.R. Combined deficiency of ABCA1 and ABCG1 promotes foam cell accumulation and accelerates atherosclerosis in mice. *J. Clin. Investig.* **2007**, *117*, 3900–3908. [CrossRef]
71. Brooks-Wilson, A.; Marcil, M.; Clee, S.M.; Zhang, L.-H.; Roomp, K.; Van Dam, M.; Yu, L.; Brewer, C.; Collins, J.A.; Molhuizen, H.O.; et al. Mutations in ABC1 in Tangier disease and familial high-density lipoprotein deficiency. *Nat. Genet.* **1999**, *22*, 336–345. [CrossRef] [PubMed]
72. Bodzioch, M.; Orsó, E.; Klucken, J.; Langmann, T.; Böttcher, A.; Diederich, W.; Drobnik, W.; Barlage, S.; Büchler, C.; Porsch-Özcürümez, M.; et al. The gene encoding ATP-binding cassette transporter 1 is mutated in Tangier disease. *Nat. Genet.* **1999**, *22*, 347–351. [CrossRef]
73. Rust, S.; Rosier, M.; Funke, H.; Real, J.T.; Amoura, Z.; Piette, J.-C.; Deleuze, J.-F.; Brewer, H.B.; Duverger, N.; Denèfle, P.; et al. Tangier disease is caused by mutations in the gene encoding ATP-binding cassette transporter 1. *Nat. Genet.* **1999**, *22*, 352–355. [CrossRef] [PubMed]
74. Tanaka, A.R.; Ikeda, Y.; Abe-Dohmae, S.; Arakawa, R.; Sadanami, K.; Kidera, A.; Nakagawa, S.; Nagase, T.; Aoki, R.; Kioka, N.; et al. Human ABCA1 contains a large amino-terminal extracellular domain homologous to an epitope of Sjogren's Syndrome. *Biochem. Biophys. Res. Commun.* **2001**, *283*, 1019–1025. [CrossRef] [PubMed]
75. Laffitte, B.A.; Repa, J.J.; Joseph, S.B.; Wilpitz, D.C.; Kast, H.R.; Mangelsdorf, D.J.; Tontonoz, P. LXRs control lipid-inducible expression of the apolipoprotein E gene in macrophages and adipocytes. *Proc. Natl. Acad. Sci. USA* **2001**, *98*, 507–512. [CrossRef]
76. Váradi, A.; Szabó, Z.; Pomozi, V.; De Boussac, H.; Fülöp, K.; Arányi, T. ABCC6 as a target in pseudoxanthoma elasticum. *Curr. Drug Targets* **2011**, *12*, 671–682. [CrossRef]
77. Legrand, A.; Cornez, L.; Samkari, W.; Mazzella, J.-M.; Venisse, A.; Boccio, V.; Auribault, K.; Keren, B.; Benistan, K.; Germain, D.P.; et al. Mutation spectrum in the ABCC6 gene and genotype–phenotype correlations in a French cohort with pseudoxanthoma elasticum. *Genet. Med.* **2017**, *19*, 909–917. [CrossRef]
78. Pulkkinen, L.; Nakano, A.; Ringpfeil, F.; Uitto, J. Identification of ABCC6 pseudogenes on human chromosome 16p: Implications for mutation detection in pseudoxanthoma elasticum. *Qual. Life Res.* **2001**, *109*, 356–365. [CrossRef]
79. Le Saux, O.; Beck, K.; Sachsinger, C.; Silvestri, C.; Treiber, C.; Göring, H.H.; Johnson, E.W.; De Paepe, A.; Pope, F.M.; Pasquali-Ronchetti, I.; et al. A Spectrum of ABCC6 Mutations Is Responsible for Pseudoxanthoma Elasticum. *Am. J. Hum. Genet.* **2001**, *69*, 749–764. [CrossRef]
80. Bergen, A.A.; Plomp, A.S.; Schuurman, E.J.; Terry, S.F.; Breuning, M.H.; Dauwerse, H.G.; Swart, J.; Kool, M.; Van Soest, S.; Baas, F.; et al. Mutations in ABCC6 cause pseudoxanthoma elasticum. *Nat. Genet.* **2000**, *25*, 228–231. [CrossRef]
81. Germain, D.P. Pseudoxanthoma elasticum: Evidence for the existence of a pseudogene highly homologous to the ABCC6 gene. *J. Med. Genet.* **2001**, *38*, 457–461. [CrossRef]
82. Le Saux, O.; Urban, Z.; Tschuch, C.; Csiszar, K.; Bacchelli, B.; Quaglino, D.; Pasquali-Ronchetti, I.; Pope, F.M.; Richards, A.; Terry, S.F.; et al. Mutations in a gene encoding an ABC transporter cause pseudoxanthoma elasticum. *Nat. Genet.* **2000**, *25*, 223–227. [CrossRef]
83. Bienengraeber, M.; Olson, T.M.; Selivanov, V.A.; Kathmann, E.C.; O'Cochlain, F.; Gao, F.; Karger, A.B.; Ballew, J.D.; Hodgson, D.M.; Zingman, L.V.; et al. ABCC9 mutations identified in human dilated cardiomyopathy disrupt catalytic KATP channel gating. *Nat. Genet.* **2004**, *36*, 382–387. [CrossRef]

84. Yoshiura, K.-I.; Kinoshita, A.; Ishida, T.; Ninokata, A.; Ishikawa, T.; Kaname, T.; Bannai, M.; Tokunaga, K.; Sonoda, S.; Komaki, R.; et al. A SNP in the ABCC11 gene is the determinant of human earwax type. *Nat. Genet.* **2006**, *38*, 324–330. [CrossRef] [PubMed]

85. Tammur, J.; Prades, C.; Arnould, I.; Rzhetsky, A.; Hutchinson, A.; Adachi, M.; Schuetz, J.D.; Swoboda, K.J.; Ptácek, L.J.; Rosier, M.; et al. Two new genes from the human ATP-binding cassette transporter superfamily, ABCC11 and ABCC12, tandemly duplicated on chromosome 16q12. *Gene* **2001**, *273*, 89–96. [CrossRef]

86. Anandarajan, M.; Paulraj, S.; Tubman, R. ABCA3 Deficiency: An unusual cause of respiratory distress in the newborn. *Ulst. Med. J.* **2009**, *78*, 51–52.

87. Kröner, C.; Wittmann, T.; Reu, S.; Teusch, V.; Klemme, M.; Rauch, D.; Hengst, M.; Kappler, M.; Cobanoglu, N.; Sismanlar, T.; et al. Lung disease caused byABCA3mutations. *Thorax* **2017**, *72*, 213–220. [CrossRef] [PubMed]

88. Shulenin, S.; Nogee, L.M.; Annilo, T.; Wert, S.E.; Whitsett, J.A.; Dean, M. ABCA3Gene Mutations in Newborns with Fatal Surfactant Deficiency. *N. Engl. J. Med.* **2004**, *350*, 1296–1303. [CrossRef] [PubMed]

89. Yoshida, I.; Ban, N.; Inagaki, N. Expression of ABCA3, a causative gene for fatal surfactant deficiency, is up-regulated by glucocorticoids in lung alveolar type II cells. *Biochem. Biophys. Res. Commun.* **2004**, *323*, 547–555. [CrossRef] [PubMed]

90. Scott, C.A.; Rajpopat, S.; Di, W.-L. Harlequin ichthyosis: ABCA12 mutations underlie defective lipid transport, reduced protease regulation and skin-barrier dysfunction. *Cell Tissue Res.* **2012**, *351*, 281–288. [CrossRef]

91. Lefévre, C. Mutations in the transporter ABCA12 are associated with lamellar ichthyosis type 2. *Hum. Mol. Genet.* **2003**, *12*, 2369–2378. [CrossRef]

92. Thomas, A.C.; Cullup, T.; Norgett, E.E.; Hill, T.; Barton, S.; Dale, B.A.; Sprecher, E.; Sheridan, E.; Taylor, A.E.; Wilroy, R.S.; et al. ABCA12 Is the Major Harlequin Ichthyosis Gene. *J. Investig. Dermatol.* **2006**, *126*, 2408–2413. [CrossRef]

93. Johnston, R.C.; Stephenson, M.L.; Nageotte, M.P. Novel heterozygous ABCB4 gene mutation causing recurrent first-trimester intrahepatic cholestasis of pregnancy. *J. Perinatol.* **2014**, *34*, 711–712. [CrossRef] [PubMed]

94. Dixon, P.; Weerasekera, N.; Linton, K.; Donaldson, O.; Chambers, J.; Egginton, E.; Weaver, J.; Nelson-Piercy, C.; De Swiet, M.; Warnes, G.; et al. Heterozygous MDR3 missense mutation associated with intrahepatic cholestasis of pregnancy: Evidence for a defect in protein trafficking. *Hum. Mol. Genet.* **2000**, *9*, 1209–1217. [CrossRef] [PubMed]

95. Pondarre, C.; Campagna, D.R.; Antiochos, B.; Sikorski, L.; Mulhern, H.; Fleming, M.D. Abcb7, the gene responsible for X-linked sideroblastic anemia with ataxia, is essential for hematopoiesis. *Blood* **2006**, *109*, 3567–3569. [CrossRef] [PubMed]

96. Shimada, Y.; Okuno, S.; Kawai, A.; Shinomiya, H.; Saito, A.; Suzuki, M.; Omori, Y.; Nishino, N.; Kanemoto, N.; Fujiwara, T.; et al. Cloning and chromosomal mapping of a novel ABC transporter gene (hABC7), a candidate for X-linked sideroblastic anemia with spinocerebellar ataxia. *J. Hum. Genet.* **1998**, *43*, 115–122. [CrossRef] [PubMed]

97. Allikmets, R.; Raskind, W.H.; Hutchinson, A.; Schueck, N.D.; Dean, M.; Koeller, D.M. Mutation of a Putative Mitochondrial Iron Transporter Gene (ABC7) in X-Linked Sideroblastic Anemia and Ataxia (XLSA/A). *Hum. Mol. Genet.* **1999**, *8*, 743–749. [CrossRef]

98. Bekri, S.; Kispal, G.; Lange, H.; Fitzsimons, E.; Tolmie, J.; Lill, R.; Bishop, D.F. Human ABC7 transporter: Gene structure and mutation causing X-linked sideroblastic anemia with ataxia with disruption of cytosolic iron-sulfur protein maturation. *Blood* **2000**, *96*, 3256–3264. [CrossRef]

99. Kohut, P.; Wüstner, D.; Hronska, L.; Kuchler, K.; Hapala, I.; Valachovic, M. The role of ABC proteins Aus1p and Pdr11p in the uptake of external sterols in yeast: Dehydroergosterol fluorescence study. *Biochem. Biophys. Res. Commun.* **2011**, *404*, 233–238. [CrossRef] [PubMed]

100. Gottesman, M.M. Mechanisms of Cancer Drug Resistance. *Annu. Rev. Med.* **2002**, *53*, 615–627. [CrossRef]

101. Piddock, L.J.V. Multidrug-resistance efflux pumps ? not just for resistance. *Nat. Rev. Genet.* **2006**, *4*, 629–636. [CrossRef] [PubMed]

102. Borst, P. Looking back at multidrug resistance (MDR) research and ten mistakes to be avoided when writing about ABC transporters in MDR. *FEBS Lett.* **2020**, *594*, 4001–4011. [CrossRef]

103. Ueda, K.; Cardarelli, C.; Gottesman, M.M.; Pastan, I. Expression of a full-length cDNA for the human "MDR1" gene confers resistance to colchicine, doxorubicin, and vinblastine. *Proc. Natl. Acad. Sci. USA* **1987**, *84*, 3004–3008. [CrossRef] [PubMed]

104. Kartner, N.; Riordan, J.R.; Ling, V. Cell surface P-glycoprotein associated with multidrug resistance in mammalian cell lines. *Science* **1983**, *221*, 1285–1288. [CrossRef] [PubMed]

105. Gottesman, M.M.; Ling, V. The molecular basis of multidrug resistance in cancer: The early years of P-glycoprotein research. *FEBS Lett.* **2005**, *580*, 998–1009. [CrossRef]

106. Ling, V.; Thompson, L.H. Reduced permeability in CHO cells as a mechanism of resistance to colchicine. *J. Cell. Physiol.* **1974**, *83*, 103–116. [CrossRef]

107. Cole, S.P.; Bhardwaj, G.; Gerlach, J.H.; Mackie, J.E.; Grant, C.E.; Almquist, K.C.; Stewart, A.J.; Kurz, E.U.; Duncan, A.M.; Deeley, R.G. Overexpression of a transporter gene in a multidrug-resistant human lung cancer cell line. *Science* **1992**, *258*, 1650–1654. [CrossRef] [PubMed]

108. Miyake, K.; Mickley, L.; Litman, T.; Zhan, Z.; Robey, R.; Cristensen, B.; Brangi, M.; Greenberger, L.; Dean, M.; Fojo, T.; et al. Molecular cloning of cDNAs which are highly overexpressed in mitoxantrone-resistant cells: Demonstration of homology to ABC transport genes. *Cancer Res.* **1999**, *59*, 8–13.

109. Lee, J.S.; Scala, S.; Matsumoto, Y.; Dickstein, B.; Robey, R.; Zhan, Z.; Altenberg, G.; Bates, S.E. Reduced drug accumulation and multidrug resistance in human breast cancer cells without associated P-glycoprotein or MRP overexpression. *J. Cell. Biochem.* **1997**, *65*, 513–526. [CrossRef]

110. Doyle, L.A.; Yang, W.; Abruzzo, L.V.; Krogmann, T.; Gao, Y.; Rishi, A.K.; Ross, D.D. A multidrug resistance transporter from human MCF-7 breast cancer cells. *Proc. Natl. Acad. Sci. USA* **1998**, *95*, 15665–15670. [CrossRef]

111. Sarkadi, B.; Homolya, L.; Szakács, G.; Váradi, A. Human Multidrug Resistance ABCB and ABCG Transporters: Participation in a Chemoimmunity Defense System. *Physiol. Rev.* **2006**, *86*, 1179–1236. [CrossRef] [PubMed]

112. Leslie, E.M.; Deeley, R.G.; Cole, S.P. Multidrug resistance proteins: Role of P-glycoprotein, MRP1, MRP2, and BCRP (ABCG2) in tissue defense. *Toxicol. Appl. Pharmacol.* **2005**, *204*, 216–237. [CrossRef] [PubMed]

113. Sharom, F.J. The P-glycoprotein multidrug transporter. *Essays Biochem.* **2011**, *50*, 161–178. [CrossRef]

114. Sharom, F.J. Complex Interplay between the P-Glycoprotein Multidrug Efflux Pump and the Membrane: Its Role in Modulating Protein Function. *Front. Oncol.* **2014**, *4*, 41. [CrossRef] [PubMed]

115. Zein, A.A.; Kaur, R.; Hussein, T.O.; Graf, G.A.; Lee, J.-Y. ABCG5/G8: A structural view to pathophysiology of the hepatobiliary cholesterol secretion. *Biochem. Soc. Trans.* **2019**, *47*, 1259–1268. [CrossRef]

116. Okamoto, Y.; Tomioka, M.; Ogasawara, F.; Nagaiwa, K.; Kimura, Y.; Kioka, N.; Ueda, K. C-terminal of ABCA1 separately regulates cholesterol floppase activity and cholesterol efflux activity. *Biosci. Biotechnol. Biochem.* **2020**, *84*, 764–773. [CrossRef]

117. Nagao, K.; Tomioka, M.; Ueda, K. Function and regulation of ABCA1-membrane meso-domain organization and reorganization. *FEBS J.* **2011**, *278*, 3190–3203. [CrossRef]

118. van Herwaarden, A.E.; Schinkel, A.H. The function of breast cancer resistance protein in epithelial barriers, stem cells and milk secretion of drugs and xenotoxins. *Trends Pharmacol. Sci.* **2006**, *27*, 10–16. [CrossRef]

119. Giacomini, K.M.; Huang, S.; Tweedie, D.J.; Benet, L.Z.; Brouwer, K.L.R.; Chu, X.; Dahlin, A.; Evers, R.; Fischer, V.; Hillgren, K.M.; et al. Membrane transporters in drug development. *Nat. Rev. Drug Discov.* **2010**, *9*, 215–236.

120. Martinez, L.; Arnaud, O.; Henin, E.; Tao, H.; Chaptal, V.; Doshi, R.; Andrieu, T.; Dussurgey, S.; Tod, M.; Di Pietro, A.; et al. Understanding polyspecificity within the substrate-binding cavity of the human multidrug resistance P-glycoprotein. *FEBS J.* **2014**, *281*, 673–682. [CrossRef]

121. Kimura, Y.; Morita, S.; Matsuo, M.; Ueda, K. Mechanism of multidrug recognition by MDR1/ABCB1. *Cancer Sci.* **2007**, *98*, 1303–1310. [PubMed]

122. Juliano, R.; Ling, V. A surface glycoprotein modulating drug permeability in Chinese hamster ovary cell mutants. *Biochim. Biophys. Acta (BBA) Biomembr.* **1976**, *455*, 152–162. [CrossRef]

123. Chen, C.-J.; Chin, J.E.; Ueda, K.; Clark, D.P.; Pastan, I.; Gottesman, M.M.; Roninson, I.B. Internal duplication and homology with bacterial transport proteins in the mdr1 (P-glycoprotein) gene from multidrug-resistant human cells. *Cell* **1986**, *47*, 381–389. [CrossRef]

124. Ambudkar, S.V.; Dey, S.; Hrycyna, C.A.; Ramachandra, M.; Pastan, I.; Gottesman, M.M. Biochemical, cellular, and pharmacological aspects of the multidrug transporter. *Annu. Rev. Pharmacol. Toxicol.* **1999**, *39*, 361–398. [CrossRef] [PubMed]

125. Savas, B.; Cole, S.P.; Akoglu, T.F.; Pross, H.F. P-glycoprotein-mediated multidrug resistance and cytotoxic effector cells. *Nat. Immun.* **1992**, *11*, 177–192. [PubMed]

126. Schinkel, A.; Smit, J.; van Tellingen, O.; Beijnen, J.; Wagenaar, E.; van Deemter, L.; Mol, C.; van der Valk, M.; Robanus-Maandag, E.; Riele, H.T.; et al. Disruption of the mouse mdr1a P-glycoprotein gene leads to a deficiency in the blood-brain barrier and to increased sensitivity to drugs. *Cell* **1994**, *77*, 491–502. [CrossRef]

127. Doran, A.; Obach, R.S.; Smith, B.J.; Hosea, N.A.; Becker, S.; Callegari, E.; Chen, C.; Chen, X.; Choo, E.; Cianfrogna, J.; et al. The impact of p-glycoprotein on the disposition of drugs targeted for indications of the central nervous system: Evaluation using the mdr1a/1b knockout mouse modeL. *Drug Metab. Dispos.* **2004**, *33*, 165–174. [CrossRef]

128. Borst, P.; Schinkel, A.H. P-glycoprotein ABCB1: A major player in drug handling by mammals. *J. Clin. Investig.* **2013**, *123*, 4131–4133. [CrossRef]

129. Ruetz, S.; Gros, P. Phosphatidylcholine translocase: A physiological role for the mdr2 gene. *Cell* **1994**, *77*, 1071–1081. [CrossRef]

130. Abraham, J.; Edgerly, M.; Wilson, R.; Chen, C.; Rutt, A.; Bakke, S.; Robey, R.; Dwyer, A.; Goldspiel, B.; Balis, F.; et al. A Phase I Study of the P-Glycoprotein Antagonist Tariquidar in Combination with Vinorelbine. *Clin. Cancer Res.* **2009**, *15*, 3574–3582. [CrossRef]

131. Dantzig, A.H.; Shepard, R.L.; Cao, J.; Law, K.L.; Ehlhardt, W.J.; Baughman, T.M.; Bumol, T.F.; Starling, J.J. Reversal of P-glycoprotein-mediated multidrug resistance by a potent cyclopropyldibenzosuberane modulator, LY335979. *Cancer Res.* **1996**, *56*, 4171–4179.

132. Starling, J.J.; Shepard, R.L.; Cao, J.; Law, K.L.; Norman, B.H.; Kroin, J.S.; Ehlhardt, W.J.; Baughman, T.M.; Winter, M.A.; Bell, M.G.; et al. Pharmacological characterization of LY335979: A potent cyclopropyldibenzosuberane modulator of P-glycoprotein. *Adv. Enzym. Regul.* **1997**, *37*, 335–347. [CrossRef]

133. Benderra, Z.; Faussat, A.-M.; Sayada, L.; Perrot, J.-Y.; Chaoui, D.; Marie, J.-P.; Legrand, O. Breast Cancer Resistance Protein and P-Glycoprotein in 149 Adult Acute Myeloid Leukemias. *Clin. Cancer Res.* **2004**, *10*, 7896–7902. [CrossRef] [PubMed]

134. Benderra, Z.; Faussat, A.M.; Sayada, L.; Perrot, J.-Y.; Tang, R.; Chaoui, D.; Morjani, H.; Marzac, C.; Marie, J.-P.; Legrand, O. MRP3, BCRP, and P-Glycoprotein Activities are Prognostic Factors in Adult Acute Myeloid Leukemia. *Clin. Cancer Res.* **2005**, *11*, 7764–7772. [CrossRef] [PubMed]

135. Damiani, D.; Tiribelli, M.; Calistri, E.; Geromin, A.; Chiarvesio, A.; Michelutti, A.; Cavallin, M.; Fanin, R. The prognostic value of P-glycoprotein (ABCB) and breast cancer resistance protein (ABCG2) in adults with de novo acute myeloid leukemia with normal karyotype. *Haematology* **2006**, *91*, 825–828.

136. Damiani, D.; Tiribelli, M.; Michelutti, A.; Geromin, A.; Cavallin, M.; Fabbro, D.; Pianta, A.; Malagola, M.; Damante, G.; Russo, D.; et al. Fludarabine-based induction therapy does not overcome the negative effect of ABCG2 (BCRP) over-expression in adult acute myeloid leukemia patients. *Leuk. Res.* **2010**, *34*, 942–945. [CrossRef]
137. Heuvel-Eibrink, M.M.V.D.; Wiemer, E.A.C.; Prins, A.; Meijerink, J.P.P.; Vossebeld, P.J.M.; Van Der Holt, B.; Pieters, R.; Sonneveld, P. Increased expression of the breast cancer resistance protein (BCRP) in relapsed or refractory acute myeloid leukemia (AML). *Leukemia* **2002**, *16*, 833–839. [CrossRef] [PubMed]
138. Heuvel-Eibrink, M.M.V.D.; Van Der Holt, B.; Burnett, A.K.; Knauf, W.U.; Fey, M.F.; Verhoef, G.E.G.; Vellenga, E.; Ossenkoppele, G.J.; Löwenberg, B.; Sonneveld, P. CD34-related coexpression of MDR1 and BCRP indicates a clinically resistant phenotype in patients with acute myeloid leukemia (AML) of older age. *Ann. Hematol.* **2007**, *86*, 329–337. [CrossRef]
139. Robey, R.W.; Polgar, O.; Deeken, J.; To, K.W.; Bates, S.E. ABCG2: Determining its relevance in clinical drug resistance. *Cancer Metastasis Rev.* **2007**, *26*, 39–57. [CrossRef] [PubMed]
140. Sticova, E.; Jirsa, M. ABCB4 disease: Many faces of one gene deficiency. *Ann. Hepatol.* **2020**, *19*, 126–133. [CrossRef]
141. Nikaido, H. Multidrug Resistance in Bacteria. *Annu. Rev. Biochem.* **2009**, *78*, 119–146. [CrossRef] [PubMed]
142. Giaccone, G.; Van Ark-Otte, J.; Rubio, G.J.; Gazdar, A.F.; Broxterman, H.J.; Dingemans, A.-M.C.; Flens, M.J.; Scheper, R.J.; Pinedo, H.M. MRP is frequently expressed in human lung-cancer cell lines, in non-small-cell lung cancer and in normal lung. *Int. J. Cancer* **1996**, *66*, 760–767. [CrossRef]
143. Nooter, K.; Bosman, F.T.; Burger, H.; Van Wingerden, K.E.; Flens, M.J.; Scheper, R.J.; Oostrum, R.G.; Boersma, A.W.M.; Van Der Gaast, A.; Stoter, G. Expression of the multidrug resistance-associated protein (MRP) gene in primary non-small-cell lung cancer. *Ann. Oncol.* **1996**, *7*, 75–81. [CrossRef] [PubMed]
144. Keppler, D.; Jedlitschky, G.; Leier, I. Transport function and substrate specificity of multidrug resistance protein. *Methods Enzymol.* **1998**, *292*, 607–616. [CrossRef] [PubMed]
145. Deeley, R.G.; Cole, S.P. Substrate recognition and transport by multidrug resistance protein 1 (ABCC1). *FEBS Lett.* **2005**, *580*, 1103–1111. [CrossRef] [PubMed]
146. Do, T.M.; Noel-Hudson, M.; Ribes, S.; Besengez, C.; Smirnova, M.; Cisternino, S.; Buyse, M.; Calon, F.; Chimini, G.; Chacun, H.; et al. ABCG2- and ABCG4-mediated efflux of amyloid-beta peptide 1-40 at the mouse blood-brain barrier. *J. Alzheimers Dis.* **2012**, *30*, 155–166. [CrossRef]
147. Diestra, J.E.; Scheffer, G.L.; Català, I.; Maliepaard, M.; Schellens, J.H.M.; Scheper, R.J.; Germà-Lluch, J.R.; Izquierdo, M.A. Frequent expression of the multi-drug resistance-associated protein BCRP/MXR/ABCP/ABCG2 in human tumours detected by the BXP-21 monoclonal antibody in paraffin-embedded material. *J. Pathol.* **2002**, *198*, 213–219. [CrossRef]
148. Ross, D.D.; Karp, J.E.; Chen, T.T.; Doyle, L.A. Expression of breast cancer resistance protein in blast cells from patients with acute leukemia. *Blood* **2000**, *96*, 365–368. [CrossRef]
149. Sargent, J.M.; Williamson, C.J.; Maliepaard, M.; Elgie, A.W.; Scheper, R.J.; Taylor, C.G. Breast cancer resistance protein expression and resistance to daunorubicin in blast cells from patients with acute myeloid leukaemia. *Br. J. Haematol.* **2001**, *115*, 257–262. [CrossRef]
150. van der Kolk, D.M.; Vellenga, E.; Scheffer, G.L.; Müller, M.; Bates, S.E.; Scheper, R.J.; de Vries, E.G.E. Expression and activity of breast cancer resistance protein (BCRP) in de novo and relapsed acute myeloid leukemia. *Blood* **2002**, *99*, 3763–3770. [CrossRef]
151. Steinbach, D.; Sell, W.; Voigt, A.; Hermann, J.; Zintl, F.; Sauerbrey, A. BCRP gene expression is associated with a poor response to remission induction therapy in childhood acute myeloid leukemia. *Leukemia* **2002**, *16*, 1443–1447. [CrossRef] [PubMed]
152. van Herwaarden, A.E.; Wagenaar, E.; Merino, G.; Jonker, J.W.; Rosing, H.; Beijnen, J.H.; Schinkel, A.H. Multidrug transporter ABCG2/breast cancer resistance protein secretes riboflavin (vitamin B2) into milk. *Mol. Cell Biol.* **2007**, *27*, 1247–1253. [CrossRef] [PubMed]
153. Robey, R.W.; To, K.K.; Polgar, O.; Dohse, M.; Fetsch, P.; Dean, M.; Bates, S.E. ABCG2: A perspective. *Adv. Drug Deliv. Rev.* **2009**, *61*, 3–13. [CrossRef] [PubMed]
154. Robey, R.W.; Ierano, C.; Zhan, Z.; Bates, S.E. The challenge of exploiting ABCG2 in the clinic. *Curr. Pharm. Biotechnol.* **2011**, *12*, 595–608. [CrossRef]
155. Tarr, P.T.; Tarling, E.J.; Bojanic, D.D.; Edwards, P.A. Baldán, Ángel Emerging new paradigms for ABCG transporters. *Biochim. Biophys. Acta (BBA) Mol. Cell Biol. Lipids* **2009**, *1791*, 584–593. [CrossRef] [PubMed]
156. Patel, S.B.; Graf, G.A.; Temel, R.E. Thematic Review Series: Lipid Transfer Proteins ABCG5 and ABCG8: More than a defense against xenosterols. *J. Lipid Res.* **2018**, *59*, 1103–1113. [CrossRef] [PubMed]
157. Khunweeraphong, N.; Mitchell-White, J.; Szöllősi, D.; Hussein, T.; Kuchler, K.; Kerr, I.D.; Stockner, T.; Lee, J. Picky ABCG5/G8 and promiscuous ABCG2 - a tale of fatty diets and drug toxicity. *FEBS Lett.* **2020**, *594*, 4035–4058. [CrossRef] [PubMed]
158. Jacquier, N.; Schneiter, R. Mechanisms of sterol uptake and transport in yeast. *J. Steroid Biochem. Mol. Biol.* **2012**, *129*, 70–78. [CrossRef]
159. Wang, F.; Li, G.; Gu, H.-M.; Zhang, D.-W. Characterization of the Role of a Highly Conserved Sequence in ATP Binding Cassette Transporter G (ABCG) Family in ABCG1 Stability, Oligomerization, and Trafficking. *Biochemistry* **2013**, *52*, 9497–9509. [CrossRef]
160. Hegyi, Z.; Homolya, L. Functional Cooperativity between ABCG4 and ABCG1 Isoforms. *PLoS ONE* **2016**, *11*, e0156516. [CrossRef]
161. Tarling, E.J.; Edwards, P.A. ATP binding cassette transporter G1 (ABCG1) is an intracellular sterol transporter. *Proc. Natl. Acad. Sci. USA* **2011**, *108*, 19719–19724. [CrossRef] [PubMed]

162. Vaughan, A.M.; Oram, J.F. ABCA1 and ABCG1 or ABCG4 act sequentially to remove cellular cholesterol and generate cholesterol-rich HDL. *J. Lipid Res.* **2006**, *47*, 2433–2443. [CrossRef] [PubMed]

163. Dujon, B.; Albermann, K.; Aldea, M.; Alexandraki, D.; Ansorge, W.; Arino, J.; Benes, V.; Bohn, C.; Bolotin-Fukuhara, M.; Bordonné, R.; et al. The nucleotide sequence of Saccharomyces cerevisiae chromosome XV. *Nat. Cell Biol.* **1997**, *387*, 98–102. [CrossRef]

164. Kolaczkowski, M.; Michel, V.D.R.; Cybularz-Kolaczkowska, A.; Soumillion, J.-P.; Konings, W.N.; André, G. Anticancer Drugs, Ionophoric Peptides, and Steroids as Substrates of the Yeast Multidrug Transporter Pdr5p. *J. Biol. Chem.* **1996**, *271*, 31543–31548. [CrossRef] [PubMed]

165. Balzi, E.; Wang, M.; Leterme, S.; Van Dyck, L.; Goffeau, A. PDR5, a novel yeast multidrug resistance conferring transporter controlled by the transcription regulator PDR1. *J. Biol. Chem.* **1994**, *269*, 2206–2214. [CrossRef]

166. Bissinger, P.H.; Kuchler, K. Molecular cloning and expression of the Saccharomyces cerevisiae STS1 gene product. A yeast ABC transporter conferring mycotoxin resistance. *J. Biol. Chem.* **1994**, *269*, 4180–4186. [CrossRef]

167. Kralli, A.; Bohen, S.P.; Yamamoto, K.R. LEM1, an ATP-binding-cassette transporter, selectively modulates the biological potency of steroid hormones. *Proc. Natl. Acad. Sci. USA* **1995**, *92*, 4701–4705. [CrossRef]

168. Kean, L.S.; Grant, A.M.; Angeletti, C.; Mahé, Y.; Kuchler, K.; Fuller, R.S.; Nichols, J.W. Plasma Membrane Translocation of Fluorescent-labeled Phosphatidylethanolamine Is Controlled by Transcription Regulators, PDR1 and PDR3. *J. Cell Biol.* **1997**, *138*, 255–270. [CrossRef]

169. Mahé, Y.; Lemoine, Y.; Kuchler, K. The ATP Binding Cassette Transporters Pdr5 and Snq2 of Saccharomyces cerevisiae Can Mediate Transport of Steroids in Vivo. *J. Biol. Chem.* **1996**, *271*, 25167–25172. [CrossRef]

170. Watanabe, M.; Mizoguchi, H.; Nishimura, A. Disruption of the ABC transporter genes PDR5, YOR1, and SNQ2, and their participation in improved fermentative activity of a sake yeast mutant showing pleiotropic drug resistance. *J. Biosci. Bioeng.* **2000**, *89*, 569–576. [CrossRef]

171. Khakhina, S.; Johnson, S.S.; Manoharlal, R.; Russo, S.B.; Blugeon, C.; Lemoine, S.; Sunshine, A.B.; Dunham, M.J.; Cowart, L.A.; Devaux, F.; et al. Control of Plasma Membrane Permeability by ABC Transporters. *Eukaryot. Cell* **2015**, *14*, 442–453. [CrossRef] [PubMed]

172. Bauer, B.E.; Wolfger, H.; Kuchler, K. Inventory and function of yeast ABC proteins: About sex, stress, pleiotropic drug and heavy metal resistance. *Biochim. Biophys. Acta (BBA) Biomembr.* **1999**, *1461*, 217–236. [CrossRef]

173. Rockwell, N.C.; Wolfger, H.; Kuchler, K.; Thorner, J. ABC Transporter Pdr10 Regulates the Membrane Microenvironment of Pdr12 in Saccharomyces cerevisiae. *J. Membr. Biol.* **2009**, *229*, 27–52. [CrossRef] [PubMed]

174. Parle-McDermott, A.G.; Hand, N.J.; Goulding, S.E.; Wolfe, K.H. Sequence of 29 kb around the PDR10 locus on the right arm of Saccharomyces cerevisiae chromosome XV: Similarity to part of chromosome I. *Yeast* **1996**, *12*, 999–1004. [CrossRef]

175. Wolfger, H.; Mahé, Y.; Parle-McDermott, A.; Delahodde, A.; Kuchler, K. The yeast ATP binding cassette (ABC) protein genes PDR10 and PDR15 are novel targets for the Pdr1 and Pdr3 transcriptional regulators. *FEBS Lett.* **1997**, *418*, 269–274. [CrossRef]

176. Purnelle, B.; Skala, J.; Goffeau, A. The product of theYCR105 gene located on the chromosome III fromSaccharomyces cerevisiae presents homologies to ATP-dependent permeases. *Yeast* **1991**, *7*, 867–872. [CrossRef]

177. Piper, P.; Mahé, Y.; Thompson, S.; Pandjaitan, R.; Holyoak, C.; Egner, R.; Mühlbauer, M.; Coote, P.; Kuchler, K. The pdr12 ABC transporter is required for the development of weak organic acid resistance in yeast. *EMBO J.* **1998**, *17*, 4257–4265. [CrossRef]

178. Nygård, Y.; Mojzita, D.; Toivari, M.; Penttilä, M.; Wiebe, M.G.; Ruohonen, L. The diverse role of Pdr12 in resistance to weak organic acids. *Yeast* **2014**, *31*, 219–232. [CrossRef]

179. Wolfger, H.; Mamnun, Y.M.; Kuchler, K. The Yeast Pdr15p ATP-binding Cassette (ABC) Protein Is a General Stress Response Factor Implicated in Cellular Detoxification. *J. Biol. Chem.* **2004**, *279*, 11593–11599. [CrossRef]

180. Khandelwal, N.K.; Kaemmer, P.; Förster, T.M.; Singh, A.; Coste, A.T.; Andes, D.R.; Hube, B.; Sanglard, D.; Chauhan, N.; Kaur, R.; et al. Pleiotropic effects of the vacuolar ABC transporter MLT1 of Candida albicans on cell function and virulence. *Biochem. J.* **2016**, *473*, 1537–1552. [CrossRef]

181. Vethanayagam, J.G.; Flower, A.M. Decreased gene expression from T7 promoters may be due to impaired production of active T7 RNA polymerase. *Microb. Cell Factories* **2005**, *4*, 3. [CrossRef] [PubMed]

182. Godinho, C.P.; Prata, C.S.; Pinto, S.; Cardoso, C.; Bandarra, N.M.; Fernandes, F.; Sá-Correia, I. Pdr18 is involved in yeast response to acetic acid stress counteracting the decrease of plasma membrane ergosterol content and order. *Sci. Rep.* **2018**, *8*, 1–13. [CrossRef] [PubMed]

183. Godinho, C.P.; Costa, R.; Sá-Correia, I. The ABC transporter Pdr18 is required for yeast thermotolerance due to its role in ergosterol transport and plasma membrane properties. *Environ. Microbiol.* **2020**, *23*, 69–80. [CrossRef] [PubMed]

184. Henneberry, A.L.; Sturley, S.L. Sterol homeostasis in the budding yeast, Saccharomyces cerevisiae. *Semin. Cell Dev. Biol.* **2005**, *16*, 155–161. [CrossRef] [PubMed]

185. Servos, J.; Haase, E.; Brendel, M. Gene SNQ2 of Saccharomyces cerevislae, which confers resistance to 4-nitroquinoline-N-oxide and other chemicals, encodes a 169 kDa protein homologous to ATP-dependent permeases. *Mol. Genet. Genom.* **1993**, *236–236*, 214–218. [CrossRef] [PubMed]

186. Decottignies, A.; Lambert, L.; Catty, P.; Degand, H.; Epping, E.A.; Moye-Rowley, W.S.; Balzi, E.; Goffeau, A. Identification and Characterization of SNQ2, a New Multidrug ATP Binding Cassette Transporter of the Yeast Plasma Membrane. *J. Biol. Chem.* **1995**, *270*, 18150–18157. [CrossRef] [PubMed]

187. Akiyama, K.; Iwaki, T.; Sugimoto, N.; Chardwiriyapreecha, S.; Kawano, M.; Nishimoto, S.; Sugahara, T.; Sekito, T.; Kakinuma, Y. Bfr1p is responsible for tributyltin resistance in Schizosaccharomyces pombe. *J. Toxicol. Sci.* **2011**, *36*, 117–120. [CrossRef]
188. Srinivasan, V.; Pierik, A.J.; Lill, R. Crystal Structures of Nucleotide-Free and Glutathione-Bound Mitochondrial ABC Transporter Atm1. *Science* **2014**, *343*, 1137–1140. [CrossRef]
189. Leighton, J.; Schatz, G. An ABC transporter in the mitochondrial inner membrane is required for normal growth of yeast. *EMBO J.* **1995**, *14*, 188–195. [CrossRef]
190. Kispal, G.; Csere, P.; Prohl, C.; Lill, R. The mitochondrial proteins Atm1p and Nfs1p are essential for biogenesis of cytosolic Fe/S proteins. *EMBO J.* **1999**, *18*, 3981–3989. [CrossRef]
191. Dean, M.; Allikmets, R.; Gerrard, B.; Stewart, C.; Kistler, A.; Shafer, B.; Michaelis, S.; Strathern, J. XIIXVI. Yeast sequencing reports. Mapping and sequencing of two yeast genes belonging to the ATP-binding cassette superfamily. *Yeast* **1994**, *10*, 377–383. [CrossRef]
192. Young, L.; Leonhard, K.; Tatsuta, T.; Trowsdale, J.; Langer, T. Role of the ABC Transporter Mdl1 in Peptide Export from Mitochondria. *Science* **2001**, *291*, 2135–2138. [CrossRef] [PubMed]
193. Kuchler, K.; Sterne, R.E.; Thorner, J. Saccharomyces cerevisiae STE6 gene product: A novel pathway for protein export in eukaryotic cells. *EMBO J.* **1989**, *8*, 3973–3984. [CrossRef] [PubMed]
194. McGrath, J.P.; Varshavsky, A. The yeast STE6 gene encodes a homologue of the mammalian multidrug resistance P-glycoprotein. *Nat. Cell Biol.* **1989**, *340*, 400–404. [CrossRef] [PubMed]
195. Ortiz, D.F.; Kreppel, L.; Speiser, D.M.; Scheel, G.; McDonald, G.; Ow, D.W. Heavy metal tolerance in the fission yeast requires an ATP-binding cassette-type vacuolar membrane transporter. *EMBO J.* **1992**, *11*, 3491–3499. [CrossRef] [PubMed]
196. Ortiz, D.F.; Ruscitti, T.; McCue, K.F.; Ow, D.W. Transport of Metal-binding Peptides by HMT1, A Fission Yeast ABC-type Vacuolar Membrane Protein. *J. Biol. Chem.* **1995**, *270*, 4721–4728. [CrossRef] [PubMed]
197. Christensen, P.U.; Davey, J.; Nielsen, O. The Schizosaccharomyces pombe mam1 gene encodes an ABC transporter mediating secretion of M-factor. *Mol. Genet. Genom.* **1997**, *255*, 226–236. [CrossRef] [PubMed]
198. Decottignies, A.; Grant, A.M.; Nichols, J.W.; de Wet, H.; McIntosh, D.B.; Goffeau, A. ATPase and Multidrug Transport Activities of the Overexpressed Yeast ABC Protein Yor1p. *J. Biol. Chem.* **1998**, *273*, 12612–12622. [CrossRef]
199. Nagy, Z.; Montigny, C.; Leverrier, P.; Yeh, S.; Goffeau, A.; Garrigos, M.; Falson, P. Role of the yeast ABC transporter Yor1p in cadmium detoxification. *Biochimie* **2006**, *88*, 1665–1671. [CrossRef]
200. Katzmann, D.J.; Hallstrom, T.C.; Voet, M.; Wysock, W.; Golin, J.; Volckaert, G.; Moye-Rowley, W.S. Expression of an ATP-binding cassette transporter-encoding gene (YOR1) is required for oligomycin resistance in Saccharomyces cerevisiae. *Mol. Cell. Biol.* **1995**, *15*, 6875–6883. [CrossRef]
201. Falcón-Pérez, J.M.; Martínez-Burgos, M.; Molano, J.; Mazón, M.J.; Eraso, P. Domain Interactions in the Yeast ATP Binding Cassette Transporter Ycf1p: Intragenic Suppressor Analysis of Mutations in the Nucleotide Binding Domains. *J. Bacteriol.* **2001**, *183*, 4761–4770. [CrossRef] [PubMed]
202. Sasser, T.L.; Lawrence, G.; Karunakaran, S.; Brown, C.; Fratti, R.A. The Yeast ATP-binding Cassette (ABC) Transporter Ycf1p Enhances the Recruitment of the Soluble SNARE Vam7p to Vacuoles for Efficient Membrane Fusion. *J. Biol. Chem.* **2013**, *288*, 18300–18310. [CrossRef] [PubMed]
203. Gulshan, K.; Moye-Rowley, W.S. Vacuolar Import of Phosphatidylcholine Requires the ATP-Binding Cassette Transporter Ybt1. *Traffic* **2011**, *12*, 1257–1268. [CrossRef] [PubMed]
204. Wawrzycka, D.; Sobczak, I.; Bartosz, G.; Bocer, T.; Ułaszewski, S.; Goffeau, A. Vmr 1p is a novel vacuolar multidrug resistance ABC transporter in Saccharomyces cerevisiae. *FEMS Yeast Res.* **2010**, *10*, 828–838. [CrossRef]
205. Klein, M.; Mamnun, Y.M.; Eggmann, T.; Schüller, C.; Wolfger, H.; Martinoia, E.; Kuchler, K. The ATP-binding cassette (ABC) transporter Bpt1p mediates vacuolar sequestration of glutathione conjugates in yeast. *FEBS Lett.* **2002**, *520*, 63–67. [CrossRef]
206. Durmus, S.; Hendrikx, J.J.; Schinkel, A.H. Apical ABC Transporters and Cancer Chemotherapeutic Drug Disposition. *Adv. Cancer Res.* **2015**, *125*, 1–41. [CrossRef]
207. Jones, P.M.; George, A.M. Is the emperor wearing shorts? The published structures of ABC transporters. *FEBS Lett.* **2020**, *594*, 3790–3798. [CrossRef]
208. El-Awady, R.; Saleh, E.; Hashim, A.; Soliman, N.; Dallah, A.; Elrasheed, A.; Elakraa, G. The Role of Eukaryotic and Prokaryotic ABC Transporter Family in Failure of Chemotherapy. *Front. Pharmacol.* **2017**, *7*, 535. [CrossRef]
209. Orelle, C.; Mathieu, K.; Jault, J.-M. Multidrug ABC transporters in bacteria. *Res. Microbiol.* **2019**, *170*, 381–391. [CrossRef]
210. Lubelski, J.; Konings, W.N.; Driessen, A.J.M. Distribution and Physiology of ABC-Type Transporters Contributing to Multidrug Resistance in Bacteria. *Microbiol. Mol. Biol. Rev.* **2007**, *71*, 463–476. [CrossRef]
211. Du, D.; Wang-Kan, X.; Neuberger, A.; Van Veen, H.W.; Pos, K.M.; Piddock, L.J.V.; Luisi, B.F. Multidrug efflux pumps: Structure, function and regulation. *Nat. Rev. Genet.* **2018**, *16*, 523–539. [CrossRef]
212. Piddock, L.J.V. Clinically Relevant Chromosomally Encoded Multidrug Resistance Efflux Pumps in Bacteria. *Clin. Microbiol. Rev.* **2006**, *19*, 382–402. [CrossRef]
213. Sipos, G.; Kuchler, K. Fungal ATP-binding cassette (ABC) transporters in drug resistance & detoxification. *Curr. Drug Targets* **2006**, *7*, 471–481. [PubMed]
214. Jungwirth, H.; Kuchler, K. Yeast ABC transporters—A tale of sex, stress, drugs and aging. *FEBS Lett.* **2005**, *580*, 1131–1138. [CrossRef] [PubMed]

215. Gulshan, K.; Moye-Rowley, W.S. Multidrug Resistance in Fungi. *Eukaryot. Cell* **2007**, *6*, 1933–1942. [CrossRef]
216. Golin, J.; Ambudkar, S.V. The multidrug transporter Pdr5 on the 25th anniversary of its discovery: An important model for the study of asymmetric ABC transporters. *Biochem. J.* **2015**, *467*, 353–363. [CrossRef] [PubMed]
217. Sanglard, D.; Kuchler, K.; Ischer, F.; Pagani, J.L.; Monod, M.; Bille, J. Mechanisms of resistance to azole antifungal agents in Candida albicans isolates from AIDS patients involve specific multidrug transporters. *Antimicrob. Agents Chemother.* **1995**, *39*, 2378–2386. [CrossRef]
218. Sanglard, D.; Ischer, F.; Monod, M.; Bille, J. Susceptibilities of Candida albicans multidrug transporter mutants to various antifungal agents and other metabolic inhibitors. *Antimicrob. Agents Chemother.* **1996**, *40*, 2300–2305. [CrossRef] [PubMed]
219. Prasad, R.; Rawal, M.K. Efflux pump proteins in antifungal resistance. *Front. Pharmacol.* **2014**, *5*, 202. [CrossRef]
220. McDaniel, A.L.; Alger, H.M.; Sawyer, J.K.; Kelley, K.L.; Kock, N.D.; Brown, J.M.; Temel, R.E.; Rudel, L.L. Phytosterol Feeding Causes Toxicity in ABCG5/G8 Knockout Mice. *Am. J. Pathol.* **2013**, *182*, 1131–1138. [CrossRef]
221. Brown, G.D.; Denning, D.W.; Gow, N.A.R.; Levitz, S.M.; Netea, M.G.; White, T.C. Hidden Killers: Human Fungal Infections. *Sci. Transl. Med.* **2012**, *4*, 165. [CrossRef]
222. Kullberg, B.J.; Arendrup, M.C. Invasive Candidiasis. *N. Engl. J. Med.* **2015**, *373*, 1445–1456. [CrossRef]
223. Iliev, I.D.; Underhill, D.M. Striking a balance: Fungal commensalism versus pathogenesis. *Curr. Opin. Microbiol.* **2013**, *16*, 366–373. [CrossRef]
224. Hurabielle, C.; Link, V.M.; Bouladoux, N.; Han, S.-J.; Merrill, E.D.; Lightfoot, Y.L.; Seto, N.; Bleck, C.K.E.; Smelkinson, M.; Harrison, O.J.; et al. Immunity to commensal skin fungi promotes psoriasiform skin inflammation. *Proc. Natl. Acad. Sci. USA* **2020**, *117*, 16465–16474. [CrossRef]
225. Chen, Y.E.; Fischbach, M.A.; Belkaid, Y. Erratum: Skin microbiota–host interactions. *Nat. Cell Biol.* **2018**, *555*, 543. [CrossRef] [PubMed]
226. Chen, Y.E.; Fischbach, M.A.; Belkaid, Y. Skin microbiota-host interactions. *Nature* **2018**, *553*, 427–436. [CrossRef] [PubMed]
227. Mao, K.; Baptista, A.P.; Tamoutounour, S.; Zhuang, L.; Bouladoux, N.; Martins, A.J.; Huang, Y.; Gerner, M.Y.; Belkaid, Y.; Germain, R.N. Innate and adaptive lymphocytes sequentially shape the gut microbiota and lipid metabolism. *Nat. Cell Biol.* **2018**, *554*, 255–259. [CrossRef] [PubMed]
228. Byrd, A.L.; Belkaid, Y.; Segre, J.A. The human skin microbiome. *Nat. Rev. Microbiol.* **2018**, *16*, 143–155. [CrossRef] [PubMed]
229. Linehan, J.L.; Harrison, O.J.; Han, S.-J.; Byrd, A.L.; Vujkovic-Cvijin, I.; Villarino, A.V.; Sen, S.K.; Shaik, J.; Smelkinson, M.; Tamoutounour, S.; et al. Non-classical Immunity Controls Microbiota Impact on Skin Immunity and Tissue Repair. *Cell* **2018**, *172*, 784–796.e18. [CrossRef] [PubMed]
230. Pfaller, M.A.; Huband, M.D.; Flamm, R.K.; Bien, P.A.; Castanheira, M. In Vitro Activity of APX001A (Manogepix) and Comparator Agents against 1706 Fungal Isolates Collected during an International Surveillance Program in 2017. *Antimicrob. Agents Chemother.* **2019**, *63*. [CrossRef]
231. Perfect, J.R. The antifungal pipeline: A reality check. *Nat. Rev. Drug Discov.* **2017**, *16*, 603–616. [CrossRef] [PubMed]
232. Satoh, K.; Makimura, K.; Hasumi, Y.; Nishiyama, Y.; Uchida, K.; Yamaguchi, H. Candida aurissp. nov., a novel ascomycetous yeast isolated from the external ear canal of an inpatient in a Japanese hospital. *Microbiol. Immunol.* **2009**, *53*, 41–44. [CrossRef] [PubMed]
233. Chowdhary, A.; Sharma, C.; Meis, J.F. Candida auris: A rapidly emerging cause of hospital-acquired multidrug-resistant fungal infections globally. *PLoS Pathog.* **2017**, *13*, e1006290. [CrossRef] [PubMed]
234. Lockhart, S.R.; Etienne, K.A.; Vallabhaneni, S.; Farooqi, J.; Chowdhary, A.; Govender, N.P.; Colombo, A.L.; Calvo, B.; Cuomo, C.A.; Desjardins, C.A.; et al. Simultaneous Emergence of Multidrug-Resistant Candida auris on 3 Continents Confirmed by Whole-Genome Sequencing and Epidemiological Analyses. *Clin. Infect. Dis.* **2017**, *64*, 134–140. [CrossRef] [PubMed]
235. Arendrup, M.C.; Patterson, T.F. Multidrug-Resistant Candida: Epidemiology, Molecular Mechanisms, and Treatment. *J. Infect. Dis.* **2017**, *216* (Suppl. 3), S445–S451. [CrossRef]
236. Kwon, Y.J.; Shin, J.H.; Byun, S.A.; Choi, M.J.; Won, E.J.; Lee, D.; Lee, S.Y.; Chun, S.; Lee, J.H.; Choi, H.J.; et al. Candida auris Clinical Isolates from South Korea: Identification, Antifungal Susceptibility, and Genotyping. *J. Clin. Microbiol.* **2019**, *57*. [CrossRef] [PubMed]
237. Rybak, J.M.; Doorley, L.A.; Nishimoto, A.T.; Barker, K.S.; Palmer, G.E.; Rogers, P.D. Abrogation of Triazole Resistance upon Deletion of CDR1 in a Clinical Isolate of Candida auris. *Antimicrob. Agents Chemother.* **2019**, *63*. [CrossRef] [PubMed]
238. Torres, S.R.; Kim, H.C.; Leach, L.; Chaturvedi, S.; Bennett, C.J.; Hill, D.J.; De Jesus, M. Assessment of environmental and occupational exposure while working with multidrug resistant (MDR) fungus Candida auris in an animal facility. *J. Occup. Environ. Hyg.* **2019**, *16*, 507–518. [CrossRef]
239. Kim, S.H.; Iyer, K.R.; Pardeshi, L.; Muñoz, J.F.; Robbins, N.; Cuomo, C.A.; Wong, K.H.; Cowen, L.E. Genetic Analysis of Candida auris Implicates Hsp90 in Morphogenesis and Azole Tolerance and Cdr1 in Azole Resistance. *mBio* **2019**, *10*, e02529-18. [CrossRef]
240. Meis, J.F.; Chowdhary, A. Candida auris: A global fungal public health threat. *Lancet Infect. Dis.* **2018**, *18*, 1298–1299. [CrossRef]
241. Colombo, A.L.; Júnior, J.N.D.A.; Guinea, J. Emerging multidrug-resistant Candida species. *Curr. Opin. Infect. Dis.* **2017**, *30*, 528–538. [CrossRef]
242. Villanueva-Lozano, H.; Treviño-Rangel, R.D.J.; González, G.M.; Ramírez-Elizondo, M.T.; Lara-Medrano, R.; Aleman-Bocanegra, M.C.; Guajardo-Lara, C.E.; Gaona-Chávez, N.; Castilleja-Leal, F.; Torre-Amione, G.; et al. Outbreak of Candida auris infection in a COVID-19 hospital in Mexico. *Clin. Microbiol. Infect.* **2021**. [CrossRef] [PubMed]

243. Hameed, H.M.A.; Islam, M.M.; Chhotaray, C.; Wang, C.; Liu, Y.; Tan, Y.; Li, X.; Tan, S.; Delorme, V.; Yew, W.W.; et al. Molecular Targets Related Drug Resistance Mechanisms in MDR-, XDR-, and TDR-Mycobacterium tuberculosis Strains. *Front. Cell. Infect. Microbiol.* **2018**, *8*, 114. [CrossRef] [PubMed]

244. Healey, K.R.; Perlin, D.S. Fungal Resistance to Echinocandins and the MDR Phenomenon in Candida glabrata. *J. Fungi* **2018**, *4*, 105.

245. Klein, C.; Kuchler, K.; Valachovic, M. ABC proteins in yeast and fungal pathogens. *Essays Biochem.* **2011**, *50*, 101–119. [CrossRef] [PubMed]

246. Sanglard, D. Emerging Threats in Antifungal-Resistant Fungal Pathogens. *Front. Med.* **2016**, *3*, 11. [CrossRef] [PubMed]

247. Jenull, S.T.; Kashko, M.N.; Shivarathri, R.; Stoiber, A.; Chauhan, M.; Petryshyn, A.; Chauhan, N.; Kuchler, K. Transcriptiome Signatures Predict Phenotypic Variations of Candida auris. *Front. Microbiol. Infect. Dis.* **2021**. [CrossRef]

248. Kovalchuk, A.; Driessen, A.J.M. Phylogenetic analysis of fungal ABC transporters. *BMC Genom.* **2010**, *11*, 177. [CrossRef]

249. Buechel, E.R.; Pinkett, H.W. Transcription factors and ABC transporters: From pleiotropic drug resistance to cellular signaling in yeast. *FEBS Lett.* **2020**, *594*, 3943–3964. [CrossRef]

250. Kumari, S.; Kumar, M.; Gaur, N.A.; Prasad, R. Multiple roles of ABC transporters in yeast. *Fungal Genet. Biol.* **2021**, *150*, 103550. [CrossRef]

251. Moreno, A.; Banerjee, A.; Prasad, R.; Falson, P. PDR-like ABC systems in pathogenic fungi. *Res. Microbiol.* **2019**, *170*, 417–425. [CrossRef]

252. Loftus, B.J.; Fung, E.; Roncaglia, P.; Rowley, D.; Amedeo, P.; Bruno, D.; Vamathevan, J.; Miranda, M.; Anderson, I.J.; Fraser, J.A.; et al. The Genome of the Basidiomycetous Yeast and Human Pathogen Cryptococcus neoformans. *Science* **2005**, *307*, 1321–1324. [CrossRef] [PubMed]

253. Nierman, W.C.; Pain, A.; Anderson, M.J.; Wortman, J.R.; Kim, H.S.; Arroyo, J.; Berriman, M.; Abe, K.; Archer, D.B.; Bermejo, C.; et al. Genomic sequence of the pathogenic and allergenic filamentous fungus Aspergillus fumigatus. *Nat. Cell Biol.* **2005**, *438*, 1151–1156. [CrossRef] [PubMed]

254. Szczypka, M.S.; Wemmie, J.A.; Moye-Rowley, W.S.; Thiele, D.J. A yeast metal resistance protein similar to human cystic fibrosis transmembrane conductance regulator (CFTR) and multidrug resistance-associated protein. *J. Biol. Chem.* **1994**, *269*, 22853–22857. [CrossRef]

255. Lee, Y.H.; Kang, H.-M.; Kim, M.-S.; Lee, J.-S.; Jeong, C.-B.; Lee, J.-S. The protective role of multixenobiotic resistance (MXR)-mediated ATP-binding cassette (ABC) transporters in biocides-exposed rotifer Brachionus koreanus. *Aquat. Toxicol.* **2018**, *195*, 129–136. [CrossRef]

256. Taylor, N.M.I.; Manolaridis, I.; Jackson, S.M.; Kowal, J.; Stahlberg, H.; Locher, K.P. Structure of the human multidrug transporter ABCG2. *Nat. Cell Biol.* **2017**, *546*, 504–509. [CrossRef] [PubMed]

257. Jackson, S.M.; Manolaridis, I.; Kowal, J.; Zechner, M.; Taylor, N.M.I.; Bause, M.; Bauer, S.; Bartholomaeus, R.; Bernhardt, G.; Koenig, B.; et al. Structural basis of small-molecule inhibition of human multidrug transporter ABCG2. *Nat. Struct. Mol. Biol.* **2018**, *25*, 333–340. [CrossRef]

258. Manolaridis, I.; Jackson, S.M.; Taylor, N.M.I.; Kowal, J.; Stahlberg, H.; Locher, K.P. Cryo-EM structures of a human ABCG2 mutant trapped in ATP-bound and substrate-bound states. *Nat. Cell Biol.* **2018**, *563*, 426–430. [CrossRef] [PubMed]

259. Orlando, B.J.; Liao, M. ABCG2 transports anticancer drugs via a closed-to-open switch. *Nat. Commun.* **2020**, *11*, 1–11. [CrossRef]

260. Smriti; Krishnamurthy, S.; Dixit, B.L.; Gupta, C.M.; Milewski, S.; Prasad, R. ABC transporters Cdr1p, Cdr2p and Cdr3p of a human pathogen *Candida albicans* are general phospholipid translocators. *Yeast* **2002**, *19*, 303–318. [CrossRef]

261. Wu, C.-P.; Hsieh, C.-H.; Wu, Y.-S. The Emergence of Drug Transporter-Mediated Multidrug Resistance to Cancer Chemotherapy. *Mol. Pharm.* **2011**, *8*, 1996–2011. [CrossRef]

262. Sanglard, D.; Ischer, F.; Monod, M.; Bille, J. Cloning of Candida albicans genes conferring resistance to azole antifungal agents: Characterization of CDR2, a new multidrug ABC transporter gene. *Microbiology* **1997**, *143*, 405–416. [CrossRef] [PubMed]

263. Prasad, R.; De Wergifosse, P.; Goffeau, A.; Balzi, E. Molecular cloning and characterization of a novel gene of Candida albicans, CDR1, conferring multiple resistance to drugs and antifungals. *Curr. Genet.* **1995**, *27*, 320–329. [CrossRef] [PubMed]

264. Smits, S.H.; Schmitt, L.; Beis, K. Self-immunity to antibacterial peptides by ABC transporters. *FEBS Lett.* **2020**, *594*, 3920–3942. [CrossRef] [PubMed]

265. Lewinson, O.; Orelle, C.; Seeger, M.A. Structures of ABC transporters: Handle with care. *FEBS Lett.* **2020**, *594*, 3799–3814. [CrossRef]

266. Lewinson, O.; Livnat-Levanon, N. Mechanism of Action of ABC Importers: Conservation, Divergence, and Physiological Adaptations. *J. Mol. Biol.* **2017**, *429*, 606–619. [CrossRef]

267. Hernaez, M.L.; Gil, C.; Pla, J.; Nombela, C. Induced expression of the Candida albicans multidrug resistance gene CDR1 in response to fluconazole and other antifungals. *Yeast* **1998**, *14*, 517–526. [CrossRef]

268. Krishnamurthy, S.; Gupta, V.; Prasad, R.; Panwar, S.L.; Prasad, R. Expression of CDR1, a multidrug resistance gene of Candida albicans: Transcriptional activation by heat shock, drugs and human steroid hormones. *FEMS Microbiol. Lett.* **1998**, *160*, 191–197. [CrossRef]

269. Shukla, S.; Saini, P.; Smriti; Jha, S.; Ambudkar, S.V.; Prasad, R. Functional Characterization of Candida albicans ABC Transporter Cdr1p. *Eukaryot. Cell* **2003**, *2*, 1361–1375. [CrossRef]

270. Larsen, B.; Anderson, S.; Brockman, A.; Essmann, M.; Schmidt, M. Key physiological differences inCandida albicans CDR1 induction by steroid hormones and antifungal drugs. *Yeast* **2006**, *23*, 795–802. [CrossRef]
271. Lamping, E.; Monk, B.C.; Niimi, K.; Holmes, A.R.; Tsao, S.; Tanabe, K.; Niimi, M.; Uehara, Y.; Cannon, R.D. Characterization of Three Classes of Membrane Proteins Involved in Fungal Azole Resistance by Functional Hyperexpression in Saccharomyces cerevisiae. *Eukaryot. Cell* **2007**, *6*, 1150–1165. [CrossRef]
272. Balan, I.; Alarco, A.M.; Raymond, M. The Candida albicans CDR3 gene codes for an opaque-phase ABC transporter. *J. Bacteriol.* **1997**, *179*, 7210–7218. [CrossRef]
273. Khandelwal, N.K.; Chauhan, N.; Sarkar, P.; Esquivel, B.D.; Coccetti, P.; Singh, A.; Coste, A.T.; Gupta, M.; Sanglard, D.; White, T.C.; et al. Azole resistance in a Candida albicans mutant lacking the ABC transporter CDR6/ROA1 depends on TOR signaling. *J. Biol. Chem.* **2018**, *293*, 412–432. [CrossRef] [PubMed]
274. Liston, S.D.; Whitesell, L.; Kapoor, M.; Shaw, K.J.; Cowen, L.E. Enhanced Efflux Pump Expression in Candida Mutants Results in Decreased Manogepix Susceptibility. *Antimicrob. Agents Chemother.* **2020**, *64*. [CrossRef] [PubMed]
275. Sanglard, D.; Ischer, F.; Calabrese, D.; Majcherczyk, P.A.; Bille, J. The ATP binding cassette transporter gene CgCDR1 from Candida glabrata is involved in the resistance of clinical isolates to azole antifungal agents. *Antimicrob. Agents Chemother.* **1999**, *43*, 2753–2765. [CrossRef] [PubMed]
276. Song, J.W.; Shin, J.H.; Kee, S.J.; Kim, S.H.; Shin, M.G.; Suh, S.P.; Ryang, D.W. Expression of CgCDR1, CgCDR2, and CgERG11 in Candida glabrata biofilms formed by bloodstream isolates. *Med. Mycol.* **2009**, *47*, 545–548. [CrossRef]
277. Miyazaki, H.; Miyazaki, Y.; Geber, A.; Parkinson, T.; Hitchcock, C.; Falconer, D.J.; Ward, D.J.; Marsden, K.; Bennett, J.E. Fluconazole Resistance Associated with Drug Efflux and Increased Transcription of a Drug Transporter Gene, PDH1, inCandida glabrata. *Antimicrob. Agents Chemother.* **1998**, *42*, 1695–1701. [CrossRef] [PubMed]
278. Izumikawa, K.; Kakeya, H.; Tsai, H.-F.; Grimberg, B.; Bennett, J.E. Function ofCandida glabrata ABC transporter gene, PDH1. *Yeast* **2003**, *20*, 249–261. [CrossRef] [PubMed]
279. Torelli, R.; Posteraro, B.; Ferrari, S.; La Sorda, M.; Fadda, G.; Sanglard, D.; Sanguinetti, M. The ATP-binding cassette transporter-encoding gene CgSNQ2 is contributing to the CgPDR1-dependent azole resistance of Candida glabrata. *Mol. Microbiol.* **2008**, *68*, 186–201. [CrossRef]
280. Zavrel, M.; Hoot, S.J.; White, T.C. Comparison of Sterol Import under Aerobic and Anaerobic Conditions in Three Fungal Species, Candida albicans, Candida glabrata, and Saccharomyces cerevisiae. *Eukaryot. Cell* **2013**, *12*, 725–738. [CrossRef] [PubMed]
281. Nakayama, H.; Tanabe, K.; Bard, M.; Hodgson, W.; Wu, S.; Takemori, D.; Aoyama, T.; Kumaraswami, N.S.; Metzler, L.; Takano, Y.; et al. The Candida glabrata putative sterol transporter gene CgAUS1 protects cells against azoles in the presence of serum. *J. Antimicrob. Chemother.* **2007**, *60*, 1264–1272. [CrossRef]
282. Moran, G.; Sullivan, D.; Morschhauser, J.; Coleman, D. The Candida dubliniensis CdCDR1 Gene Is Not Essential for Fluconazole Resistance. *Antimicrob. Agents Chemother.* **2002**, *46*, 2829–2841. [CrossRef] [PubMed]
283. Silva, S.; Hooper, S.; Henriques, M.; Oliveira, R.; Azeredo, J.; Williams, D. The role of secreted aspartyl proteinases in Candida tropicalis invasion and damage of oral mucosa. *Clin. Microbiol. Infect.* **2011**, *17*, 264–272. [CrossRef] [PubMed]
284. Silva, S.; Negri, M.; Henriques, M.; Oliveira, R.; Williams, D.W.; Azeredo, J. Candida glabrata, Candida parapsilosis and Candida tropicalis: Biology, epidemiology, pathogenicity and antifungal resistance. *FEMS Microbiol. Rev.* **2012**, *36*, 288–305. [CrossRef] [PubMed]
285. Zuza-Alves, D.L.; Silva-Rocha, W.P.; Chaves, G.M. An Update on Candida tropicalis Based on Basic and Clinical Approaches. *Front. Microbiol.* **2017**, *8*, 1927. [CrossRef] [PubMed]
286. Sanguinetti, M.; Posteraro, B.; La Sorda, M.; Torelli, R.; Fiori, B.; Santangelo, R.; Delogu, G.; Fadda, G. Role of AFR1, an ABC Transporter-Encoding Gene, in the In Vivo Response to Fluconazole and Virulence of Cryptococcus neoformans. *Infect. Immun.* **2006**, *74*, 1352–1359. [CrossRef]
287. Raymond, M.; Gros, P.; Whiteway, M.; Thomas, D.Y. Functional complementation of yeast ste6 by a mammalian multidrug resistance mdr gene. *Science* **1992**, *256*, 232–234. [CrossRef]
288. Bennett, R.J.; Uhl, M.A.; Miller, M.G.; Johnson, A.D. Identification and Characterization of a Candida albicans Mating Pheromone. *Mol. Cell. Biol.* **2003**, *23*, 8189–8201. [CrossRef]
289. Schwarzmuller, T.; Ma, B.; Hiller, E.; Istel, F.; Tscherner, M.; Brunke, S.; Ames, L.; Firon, A.; Green, B.; Cabral, V.; et al. Systematic phenotyping of a large-scale Candida glabrata deletion collection reveals novel antifungal tolerance genes. *PLoS Pathog.* **2014**, *10*, e1004211. [CrossRef]
290. Chang, M.; Sionov, E.; Lamichhane, A.K.; Kwon-Chung, K.J.; Chang, Y.C. Roles of Three Cryptococcus neoformans and Cryptococcus gattii Efflux Pump-Coding Genes in Response to Drug Treatment. *Antimicrob. Agents Chemother.* **2018**, *62*, e01751-17. [CrossRef]
291. Ramírez-Zavala, B.; Manz, H.; Englert, F.; Rogers, P.D.; Morschhäuser, J. A Hyperactive Form of the Zinc Cluster Transcription Factor Stb5 CausesYOR1Overexpression and Beauvericin Resistance inCandida albicans. *Antimicrob. Agents Chemother.* **2018**, *62*. [CrossRef] [PubMed]
292. Shekhar-Guturja, T.; Tebung, W.A.; Mount, H.; Liu, N.; Köhler, J.R.; Whiteway, M.; Cowen, L.E. Beauvericin Potentiates Azole Activity via Inhibition of Multidrug Efflux, BlocksC. albicansMorphogenesis, and is Effluxed via Yor1 and Circuitry Controlled by Zcf29. *Antimicrob. Agents Chemother.* **2016**, *60*, AAC.01959-16. [CrossRef]

293. Dawson, R.J.P.; Locher, K.P. Structure of a bacterial multidrug ABC transporter. *Nat. Cell Biol.* **2006**, *443*, 180–185. [CrossRef] [PubMed]
294. Locher, K.P. Structure and mechanism of ATP-binding cassette transporters. *Philos. Trans. R. Soc. B Biol. Sci.* **2008**, *364*, 239–245. [CrossRef]
295. Kodan, A.; Yamaguchi, T.; Nakatsu, T.; Matsuoka, K.; Kimura, Y.; Ueda, K.; Kato, H. Inward- and outward-facing X-ray crystal structures of homodimeric P-glycoprotein CmABCB1. *Nat. Commun.* **2019**, *10*, 1–12. [CrossRef]
296. Thonghin, N.; Collins, R.F.; Barbieri, A.; Shafi, T.; Siebert, A.; Ford, R.C. Novel features in the structure of P-glycoprotein (ABCB1) in the post-hydrolytic state as determined at 7.9 Å resolution. *BMC Struct. Biol.* **2018**, *18*, 1–11. [CrossRef] [PubMed]
297. Beis, K. Structural basis for the mechanism of ABC transporters. *Biochem. Soc. Trans.* **2015**, *43*, 889–893. [CrossRef] [PubMed]
298. Johnson, Z.L.; Chen, J. Structural Basis of Substrate Recognition by the Multidrug Resistance Protein MRP1. *Cell* **2017**, *168*, 1075–1085.e9. [CrossRef] [PubMed]
299. Szakács, G.; Jakab, K.; Antal, F.; Sarkadi, B. Diagnostics of multidrug resistance in cancer. *Pathol. Oncol. Res.* **1998**, *4*, 251–257. [PubMed]
300. Liu, F.; Zhang, Z.; Csanády, L.; Gadsby, D.C.; Chen, J. Molecular Structure of the Human CFTR Ion Channel. *Cell* **2017**, *169*, 85–95.e8. [CrossRef]
301. Zhang, Z.; Chen, J. Atomic Structure of the Cystic Fibrosis Transmembrane Conductance Regulator. *Cell* **2016**, *167*, 1586–1597.e9. [CrossRef] [PubMed]
302. Jin, M.S.; Oldham, M.L.; Zhang, Q.; Chen, J. Crystal structure of the multidrug transporter P-glycoprotein from Caenorhabditis elegans. *Nat. Cell Biol.* **2012**, *490*, 566–569. [CrossRef] [PubMed]
303. Li, J.; Jaimes, K.F.; Aller, S.G. Refined structures of mouse P-glycoprotein. *Protein Sci.* **2014**, *23*, 34–46. [CrossRef] [PubMed]
304. Szewczyk, P.; Tao, H.; McGrath, A.P.; Villaluz, M.; Rees, S.D.; Lee, S.C.; Doshi, R.; Urbatsch, I.L.; Zhang, Q.; Chang, G. Snapshots of ligand entry, malleable binding and induced helical movement in P-glycoprotein. *Acta Crystallogr. Sect. D Biol. Crystallogr.* **2015**, *71*, 732–741. [CrossRef] [PubMed]
305. Nicklisch, S.C.T.; Rees, S.D.; McGrath, A.P.; Gökirmak, T.; Bonito, L.T.; Vermeer, L.M.; Cregger, C.; Loewen, G.; Sandin, S.; Chang, G.; et al. Global marine pollutants inhibit P-glycoprotein: Environmental levels, inhibitory effects, and cocrystal structure. *Sci. Adv.* **2016**, *2*, e1600001. [CrossRef] [PubMed]
306. Esser, L.; Zhou, F.; Pluchino, K.M.; Shiloach, J.; Ma, J.; Tang, W.-K.; Gutierrez, C.; Zhang, A.; Shukla, S.; Madigan, J.P.; et al. Structures of the Multidrug Transporter P-glycoprotein Reveal Asymmetric ATP Binding and the Mechanism of Polyspecificity. *J. Biol. Chem.* **2017**, *292*, 446–461. [CrossRef]
307. Kim, Y.; Chen, J. Molecular structure of human P-glycoprotein in the ATP-bound, outward-facing conformation. *Science* **2018**, *359*, 915–919. [CrossRef]
308. Alam, A.; Kowal, J.; Broude, E.; Roninson, I.; Locher, K.P. Structural insight into substrate and inhibitor discrimination by human P-glycoprotein. *Science* **2019**, *363*, 753–756. [CrossRef]
309. Alam, A.; Kung, R.; Kowal, J.; McLeod, R.A.; Tremp, N.; Broude, E.V.; Roninson, I.B.; Stahlberg, H.; Locher, K.P. Structure of a zosuquidar and UIC2-bound human-mouse chimeric ABCB1. *Proc. Natl. Acad. Sci. USA* **2018**, *115*, E1973–E1982. [CrossRef]
310. Aller, S.G.; Yu, J.; Ward, A.; Weng, Y.; Chittaboina, S.; Zhuo, R.; Harrell, P.M.; Trinh, Y.T.; Zhang, Q.; Urbatsch, I.L.; et al. Structure of P-Glycoprotein Reveals a Molecular Basis for Poly-Specific Drug Binding. *Science* **2009**, *323*, 1718–1722. [CrossRef]
311. Johnson, Z.L.; Chen, J. ATP Binding Enables Substrate Release from Multidrug Resistance Protein 1. *Cell* **2018**, *172*, 81–89.e10. [CrossRef] [PubMed]
312. Wang, L.; Johnson, Z.L.; Wasserman, M.R.; Levring, J.; Chen, J.; Liu, S. Characterization of the kinetic cycle of an ABC transporter by single-molecule and cryo-EM analyses. *eLife* **2020**, *9*, 9. [CrossRef]
313. Zhang, Z.; Liu, F.; Chen, J. Conformational Changes of CFTR upon Phosphorylation and ATP Binding. *Cell* **2017**, *170*, 483–491.e8. [CrossRef] [PubMed]
314. Fay, J.F.; Aleksandrov, L.A.; Jensen, T.J.; Cui, L.L.; Kousouros, J.N.; He, L.; Aleksandrov, A.A.; Gingerich, D.S.; Riordan, J.R.; Chen, J.Z. Cryo-EM Visualization of an Active High Open Probability CFTR Anion Channel. *Biochemistry* **2018**, *57*, 6234–6246. [CrossRef] [PubMed]
315. Liu, F.; Zhang, Z.; Levit, A.; Levring, J.; Touhara, K.K.; Shoichet, B.K.; Chen, J. Structural identification of a hotspot on CFTR for potentiation. *Science* **2019**, *364*, 1184–1188. [CrossRef]
316. Martin, G.M.; Yoshioka, C.; Rex, E.A.; Fay, J.F.; Xie, Q.; Whorton, M.R.; Chen, J.Z.; Shyng, S.-L. Cryo-EM structure of the ATP-sensitive potassium channel illuminates mechanisms of assembly and gating. *eLife* **2017**, *6*, 6. [CrossRef]
317. Lee, K.P.K.; Chen, J.; MacKinnon, R. Molecular structure of human KATP in complex with ATP and ADP. *eLife* **2017**, *6*, e32481. [CrossRef]
318. Wu, J.-X.; Ding, D.; Wang, M.; Kang, Y.; Zeng, X.; Chen, L. Ligand binding and conformational changes of SUR1 subunit in pancreatic ATP-sensitive potassium channels. *Protein Cell* **2018**, *9*, 553–567. [CrossRef]
319. Ding, D.; Wang, M.; Wu, J.-X.; Kang, Y.; Chen, L. The Structural Basis for the Binding of Repaglinide to the Pancreatic KATP Channel. *Cell Rep.* **2019**, *27*, 1848–1857. [CrossRef]
320. Martin, G.M.; Sung, M.W.; Yang, Z.; Innes, L.M.; Kandasamy, B.; David, L.L.; Yoshioka, C.; Shyng, S.-L. Mechanism of pharmacochaperoning in a mammalian KATP channel revealed by cryo-EM. *eLife* **2019**, *8*, 8. [CrossRef]

321. Muthu, V.; Gandra, R.R.; Dhooria, S.; Sehgal, I.S.; Prasad, K.T.; Kaur, H.; Gupta, N.; Bal, A.; Ram, B.; Aggarwal, A.N.; et al. Role of flexible bronchoscopy in the diagnosis of invasive fungal infections. *Mycoses* **2021**. [CrossRef]

322. Egner, R.; Rosenthal, F.E.; Kralli, A.; Sanglard, D.; Kuchler, K. Genetic Separation of FK506 Susceptibility and Drug Transport in the Yeast Pdr5 ATP-binding Cassette Multidrug Resistance Transporter. *Mol. Biol. Cell* **1998**, *9*, 523–543. [CrossRef] [PubMed]

323. Cannon, R.D.; Lamping, E.; Holmes, A.R.; Niimi, K.; Tanabe, K.; Niimi, M.; Monk, B.C. Candida albicans drug resistance—Another way to cope with stress. *Microbiology* **2007**, *153 Pt 10*, 3211–3217. [CrossRef]

324. Senior, A.E.; Al-Shawi, M.K.; Urbatsch, I.L. The catalytic cycle of P-glycoprotein. *FEBS Lett.* **1995**, *377*, 285–289. [CrossRef] [PubMed]

325. Kerr, I.D.; Haider, A.J.; Gelissen, I.C. The ABCG family of membrane-associated transporters: You don't have to be big to be mighty. *Br. J. Pharmacol.* **2011**, *164*, 1767–1779. [CrossRef] [PubMed]

326. Polgar, O.; Robey, R.W.; Bates, S.E. ABCG2: Structure, function and role in drug response. *Expert Opin. Drug Metab. Toxicol.* **2007**, *4*, 1–15. [CrossRef]

327. Smith, P.C.; Karpowich, N.; Millen, L.; Moody, J.E.; Rosen, J.; Thomas, P.J.; Hunt, J.F. ATP Binding to the Motor Domain from an ABC Transporter Drives Formation of a Nucleotide Sandwich Dimer. *Mol. Cell* **2002**, *10*, 139–149. [CrossRef]

328. Ye, J.; Osborne, A.R.; Groll, M.; Rapoport, T.A. RecA-like motor ATPases–lessons from structures. *Biochim. Biophys. Acta* **2004**, *1659*, 1–18. [CrossRef] [PubMed]

329. Hohl, M.; Hürlimann, L.M.; Böhm, S.; Schöppe, J.; Grütter, M.G.; Bordignon, E.; Seeger, M.A. Structural basis for allosteric cross-talk between the asymmetric nucleotide binding sites of a heterodimeric ABC exporter. *Proc. Natl. Acad. Sci. USA* **2014**, *111*, 11025–11030. [CrossRef]

330. Stockner, T.; Gradisch, R.; Schmitt, L. The role of the degenerate nucleotide binding site in type I ABC exporters. *FEBS Lett.* **2020**, *594*, 3815–3838. [CrossRef]

331. Golin, J.; Kon, Z.N.; Wu, C.P.; Martello, J.; Hanson, L.; Supernavage, S.; Ambudkar, S.V.; Sauna, Z.E. Complete inhibition of the Pdr5p multidrug efflux pump ATPase activity by its transport substrate clotrimazole suggests that GTP as well as ATP may be used as an energy source. *Biochemistry* **2007**, *46*, 13109–13119. [CrossRef]

332. Khunweeraphong, N.; Kuchler, K. The first intracellular loop is essential for the catalytic cycle of the human ABCG2 multidrug resistance transporter. *FEBS Lett.* **2020**, *594*, 4059–4075. [CrossRef] [PubMed]

333. Gose, T.; Shafi, T.; Fukuda, Y.; Das, S.; Wang, Y.; Allcock, A.; McHarg, A.G.; Lynch, J.; Chen, T.; Tamai, I.; et al. ABCG2 requires a single aromatic amino acid to "clamp" substrates and inhibitors into the binding pocket. *FASEB J.* **2020**, *34*, 4890–4903. [CrossRef] [PubMed]

334. Henriksen, U.; Fog, J.U.; Litman, T.; Gether, U. Identification of Intra- and Intermolecular Disulfide Bridges in the Multidrug Resistance Transporter ABCG2. *J. Biol. Chem.* **2005**, *280*, 36926–36934. [CrossRef] [PubMed]

335. Wakabayashi, K.; Nakagawa, H.; Tamura, A.; Koshiba, S.; Hoshijima, K.; Komada, M.; Ishikawa, T. Intramolecular Disulfide Bond Is a Critical Check Point Determining Degradative Fates of ATP-binding Cassette (ABC) Transporter ABCG2 Protein. *J. Biol. Chem.* **2007**, *282*, 27841–27846. [CrossRef]

336. Jones, P.M.; George, A.M. Opening of the ADP-bound active site in the ABC transporter ATPase dimer: Evidence for a constant contact, alternating sites model for the catalytic cycle. *Proteins: Struct. Funct. Bioinform.* **2009**, *75*, 387–396. [CrossRef]

337. Prasad, R.; Banerjee, A.; Khandelwal, N.K.; Dhamgaye, S. The ABCs of Candida albicans Multidrug Transporter Cdr1. *Eukaryot. Cell* **2015**, *14*, 1154–1164. [CrossRef]

338. Ernst, R.; Kueppers, P.; Klein, C.M.; Schwarzmueller, T.; Kuchler, K.; Schmitt, L. A mutation of the H-loop selectively affects rhodamine transport by the yeast multidrug ABC transporter Pdr5. *Proc. Natl. Acad. Sci. USA* **2008**, *105*, 5069–5074. [CrossRef]

339. Harris, A.; Wagner, M.; Du, D.; Raschka, S.; Gohlke, H.; Smits, S.H.J.; Luisi, B.F.; Schmitt, L. Structure and efflux mechanism of the yeast pleiotropic drug resistance transporter Pdr5. *bioRxiv* **2021**. [CrossRef]

340. Al-Shawi, M.K. Catalytic and transport cycles of ABC exporters. *Essays Biochem.* **2011**, *50*, 63–83. [CrossRef]

341. Kapoor, P.; Horsey, A.J.; Cox, M.H.; Kerr, I.D. ABCG2: Does resolving its structure elucidate the mechanism? *Biochem. Soc. Trans.* **2018**, *46*, 1485–1494. [CrossRef]

342. Kage, K.; Fujita, T.; Sugimoto, Y. Role of Cys-603 in dimer/oligomer formation of the breast cancer resistance protein BCRP/ABCG2. *Cancer Sci.* **2005**, *96*, 866–872. [CrossRef] [PubMed]

343. Szöllősi, D.; Szakács, G.; Chiba, P.; Stockner, T. Dissecting the Forces that Dominate Dimerization of the Nucleotide Binding Domains of ABCB1. *Biophys. J.* **2018**, *114*, 331–342. [CrossRef] [PubMed]

344. Jones, P.M.; George, A.M. Mechanism of the ABC transporter ATPase domains: Catalytic models and the biochemical and biophysical record. *Crit. Rev. Biochem. Mol. Biol.* **2012**, *48*, 39–50. [CrossRef] [PubMed]

345. Ford, R.C.; Hellmich, U.A. What monomeric nucleotide binding domains can teach us about dimeric ABC proteins. *FEBS Lett.* **2020**, *594*, 3857–3875. [CrossRef]

346. Jardetzky, O. Simple Allosteric Model for Membrane Pumps. *Nature* **1966**, *211*, 969–970. [CrossRef]

347. Verhalen, B.; Dastvan, R.; Thangapandian, S.; Peskova, Y.; Koteiche, H.A.; Nakamoto, R.K.; Tajkhorshid, E.; Mchaourab, H.S. Energy transduction and alternating access of the mammalian ABC transporter P-glycoprotein. *Nat. Cell Biol.* **2017**, *543*, 738–741. [CrossRef]

348. Sarkadi, B.; Homolya, L.; Hegedűs, T. The ABCG2/BCRP transporter and its variants—From structure to pathology. *FEBS Lett.* **2020**, *594*, 4012–4034. [CrossRef]

349. Özvegy, C.; Váradi, A.; Sarkadi, B. Characterization of Drug Transport, ATP Hydrolysis, and Nucleotide Trapping by the Human ABCG2 Multidrug Transporter. *J. Biol. Chem.* **2002**, *277*, 47980–47990. [CrossRef]

350. Glavinas, H.; Kis, E.; Pál, Á.; Kovács, R.; Jani, M.; Vági, E.; Molnár, É.; Bánsághi, S.; Kele, Z.; Janaky, T.; et al. ABCG2 (Breast Cancer Resistance Protein/Mitoxantrone Resistance-Associated Protein) ATPase Assay: A Useful Tool to Detect Drug-Transporter Interactions. *Drug Metab. Dispos.* **2007**, *35*, 1533–1542. [CrossRef]

351. Sarkadi, B.; Price, E.M.; Boucher, R.C.; Germann, U.A.; Scarborough, G.A. Expression of the human multidrug resistance cDNA in insect cells generates a high activity drug-stimulated membrane ATPase. *J. Biol. Chem.* **1992**, *267*, 4854–4858. [CrossRef]

352. Januliene, D.; Moeller, A. Cryo-EM of ABC transporters: An ice-cold solution to everything? *FEBS Lett.* **2020**, *594*, 3776–3789. [CrossRef]

353. Bock, C.; Löhr, F.; Tumulka, F.; Reichel, K.; Würz, J.; Hummer, G.; Schäfer, L.; Tampé, R.; Joseph, B.; Bernhard, F.; et al. Structural and functional insights into the interaction and targeting hub TMD0 of the polypeptide transporter TAPL. *Sci. Rep.* **2018**, *8*, 15662. [CrossRef] [PubMed]

354. Kaur, H.; Lakatos-Karoly, A.; Vogel, R.; Nöll, A.; Tampé, R.; Glaubitz, C. Coupled ATPase-adenylate kinase activity in ABC transporters. *Nat. Commun.* **2016**, *7*, 13864. [CrossRef] [PubMed]

355. Rose-Sperling, D.; Tran, M.A.; Lauth, L.M.; Goretzki, B.; Hellmich, U.A. 19F NMR as a versatile tool to study membrane protein structure and dynamics. *Biol. Chem.* **2019**, *400*, 1277–1288. [CrossRef] [PubMed]

356. de Boer, M.; Gouridis, G.; Vietrov, R.; Begg, S.L.; Schuurman-Wolters, G.K.; Husada, F.; Eleftheriadis, N.; Poolman, B.; McDevitt, C.A.; Cordes, T. Conformational and dynamic plasticity in substrate-binding proteins underlies selective transport in ABC importers. *Elife* **2019**, *8*, e44652. [CrossRef] [PubMed]

357. Bordignon, E.; Seeger, M.A.; Galazzo, L.; Meier, G. From in vitro towards in situ: Structure-based investigation of ABC exporters by electron paramagnetic resonance spectroscopy. *FEBS Lett.* **2020**, *594*, 3839–3856. [CrossRef] [PubMed]

358. Egloff, P.; Zimmermann, I.; Arnold, F.M.; Hutter, C.A.J.; Morger, D.; Opitz, L.; Poveda, L.; Keserue, H.-A.; Panse, C.; Roschitzki, B.; et al. Engineered peptide barcodes for in-depth analyses of binding protein libraries. *Nat. Methods* **2019**, *16*, 421–428. [CrossRef]

359. Slotboom, D.J.; Ettema, T.W.; Nijland, M.; Thangaratnarajah, C. Bacterial multi-solute transporters. *FEBS Lett.* **2020**, *594*, 3898–3907. [CrossRef]

360. Rutledge, R.M.; Esser, L.; Ma, J.; Xia, D. Toward understanding the mechanism of action of the yeast multidrug resistance transporter Pdr5p: A molecular modeling study. *J. Struct. Biol.* **2011**, *173*, 333–344. [CrossRef]

361. Rosenberg, M.F.; Bikadi, Z.; Chan, J.; Liu, X.; Ni, Z.; Cai, X.; Ford, R.C.; Mao, Q. The human breast cancer resistance protein (BCRP/ABCG2) shows conformational changes with mitoxantrone. *Structure* **2010**, *18*, 482–493. [CrossRef]

362. Rosenberg, M.F.; Bikádi, Z.; Hazai, E.; Starborg, T.; Kelley, L.; Chayen, N.E.; Ford, R.C.; Mao, Q. Three-dimensional structure of the human breast cancer resistance protein (BCRP/ABCG2) in an inward-facing conformation. *Acta Crystallogr. Sect. D Biol. Crystallogr.* **2015**, *71 Pt 8*, 1725–1735. [CrossRef]

363. Wang, H.; Lee, E.-W.; Cai, X.; Ni, Z.; Zhou, L.; Mao, Q. Membrane Topology of the Human Breast Cancer Resistance Protein (BCRP/ABCG2) Determined by Epitope Insertion and Immunofluorescence†. *Biochemistry* **2008**, *47*, 13778–13787. [CrossRef] [PubMed]

International Journal of
Molecular Sciences

Article

New Evidence for P-gp-Mediated Export of Amyloid-β Peptides in Molecular, Blood-Brain Barrier and Neuronal Models

Amanda B. Chai [1], Anika M. S. Hartz [2,3], Xuexin Gao [4], Alryel Yang [1], Richard Callaghan [4,*]
and Ingrid C. Gelissen [1,*]

[1] School of Pharmacy, Faculty of Medicine and Health, University of Sydney, Sydney, NSW 2006, Australia;
     acha3237@uni.sydney.edu.au (A.B.C.); zyan8789@sydney.edu.au (A.Y.)
[2] Sanders-Brown Center on Aging, University of Kentucky, Lexington, KY 40504, USA; anika.hartz@uky.edu
[3] Department of Pharmacology and Nutritional Sciences, University of Kentucky, Lexington, KY 40504, USA
[4] Research School of Biology and Medical School, Australian National University,
     Canberra, ACT 2601, Australia; xuexin.gao@anu.edu.au
*   Correspondence: richard.callaghan@anu.edu.au (R.C.); ingrid.gelissen@sydney.edu.au (I.C.G.);
     Tel.: +61-2-8627-0357 (I.C.G.)

Abstract: Defective clearance mechanisms lead to the accumulation of amyloid-beta (Aβ) peptides in the Alzheimer's brain. Though predominantly generated in neurons, little is known about how these hydrophobic, aggregation-prone, and tightly membrane-associated peptides exit into the extracellular space where they deposit and propagate neurotoxicity. The ability for P-glycoprotein (P-gp), an ATP-binding cassette (ABC) transporter, to export Aβ across the blood-brain barrier (BBB) has previously been reported. However, controversies surrounding the P-gp–Aβ interaction persist. Here, molecular data affirm that both $A\beta_{40}$ and $A\beta_{42}$ peptide isoforms directly interact with and are substrates of P-gp. This was reinforced ex vivo by the inhibition of $A\beta_{42}$ transport in brain capillaries from P-gp-knockout mice. Moreover, we explored whether P-gp could exert the same role in neurons. Comparison between non-neuronal CHO-APP and human neuroblastoma SK-N-SH cells revealed that P-gp is expressed and active in both cell types. Inhibiting P-gp activity using verapamil and nicardipine impaired $A\beta_{40}$ and $A\beta_{42}$ secretion from both cell types, as determined by ELISA. Collectively, these findings implicate P-gp in Aβ export from neurons, as well as across the BBB endothelium, and suggest that restoring or enhancing P-gp function could be a viable therapeutic approach for removing excess Aβ out of the brain in Alzheimer's disease.

Keywords: P-glycoprotein; ABCB1; amyloid-beta; neuron; SK-N-SH; Alzheimer's disease

Citation: Chai, A.B.; Hartz, A.M.S.; Gao, X.; Yang, A.; Callaghan, R.; Gelissen, I.C. New Evidence for P-gp-Mediated Export of Amyloid-β Peptides in Molecular, Blood-Brain Barrier and Neuronal Models. *Int. J. Mol. Sci.* **2021**, *22*, 246. https://doi.org/10.3390/ijms22010246

Received: 2 December 2020
Accepted: 25 December 2020
Published: 29 December 2020

Publisher's Note: MDPI stays neutral with regard to jurisdictional claims in published maps and institutional affiliations.

1. Introduction

The accumulation of amyloid-beta (Aβ) peptides in the brain is a key pathological hallmark of Alzheimer's disease (AD). These peptides vary between 37–43 amino acids in length and exist in a range of conformations and assembly states, from monomers to oligomers, protofibrils, and finally insoluble fibrils and plaques [1]. $A\beta_{40}$ and the more hydrophobic and aggregation-prone $A\beta_{42}$ are the most common isoforms, with a higher ratio of $A\beta_{42}$: $A\beta_{40}$ being associated with increased neurotoxicity, and accelerated disease pathology and cognitive decline [2]. In the brain, these peptides are constitutively produced in neurons and astrocytes following enzymatic cleavage of the transmembrane amyloid precursor protein (APP), and subsequently rapidly cleared [3–5]. Studies in late-onset AD patients have shown that the production rate of Aβ remains unaltered; rather, impaired cellular clearance mechanisms are responsible for their accumulation in the brain [6]. Excess levels of soluble oligomeric Aβ have been demonstrated to impair long-term potentiation, drive synaptic and receptor dysfunction, propagate tau pathology, neuroinflammation, and oxidative stress, and correlate with disease severity [7–9].

Both intra- and extracellular Aβ accumulation have been implicated in neurotoxicity, with the former reported to precede the latter [9–12]. This raises the question of how intraneuronally-generated peptides are able to exit neurons and enter the extracellular space. The hydrophobic, aggregation-prone and highly membrane-associated nature of the Aβ peptide suggests that its constitutive release from cells relies on active transport [13,14]. P-glycoprotein (P-gp), also known as ATP-binding cassette (ABC) transporter B-family subtype 1 (ABCB1) or multi-drug resistance protein 1 (MDR1), is an ATP-dependent exporter protein with broad substrate specificity, that is ubiquitously expressed on cells with barrier or excretory functions [15]. Several studies have provided evidence that P-gp at the blood-brain barrier (BBB) is responsible for Aβ export out of the brain [16]. Consequently, we hypothesised that such a mechanism could also occur in the neuron. However, there remains overall skepticism about the capacity of P-gp to carry these peptides due to their considerably larger size than most known P-gp substrates [14,17].

The aim of the present study was two-fold: firstly, we sought to provide unequivocal evidence that P-gp is able to transport Aβ peptides using in vitro and ex vivo model systems that have been validated previously [18]. Secondly, we investigated the role of neuronal P-gp, in a cell culture system utilised extensively in AD research, to ascertain whether P-gp plays a role at the site of peptide generation and peptide-mediated damage.

2. Results

2.1. Biochemical Characterisation of the Interaction between Aβ$_{40}$ and Aβ$_{42}$ with P-gp

The baculovirus system provides high capacity expression of membrane proteins, including human P-gp with full retention of function including substrate binding [19–21]. Figure 1a demonstrates that P-gp was purified to near homogeneity from High-Five membranes and eluted specifically in the presence of 400 mM imidazole. The efficiency of reconstitution was routinely assessed as described previously [21] and the concentration of P-gp obtained in the present investigation was 0.092 ± 0.036 mg/mL ($n = 4$). The total yield was 234 ± 86 μg of purified P-gp per 100 mg of crude High-Five membrane, indicating that the expression was >0.2% of total membrane protein. These parameters are similar to previously published values [21].

The tryptophan quenching assay has been widely used in the field to describe the selectivity and apparent potency of ligand/substrate interaction with P-gp. The affinity is classified as apparent since it measures the binding step and subsequent interaction with a proximal tryptophan residue. Consequently, this assay was used to provide further evidence to support a putative interaction between Aβ peptides and P-gp. A recent publication from our group and collaborators using molecular docking [14] indicated that this interaction is complex, and the kinetics are likely to be slow. Consequently, the assay conditions were configured with an elevated temperature (37 °C) and a longer incubation time than typically adopted for ligand binding studies [22]. As shown in Figure 1b,c, both peptides were able to produce a dose-dependent quenching of the endogenous tryptophan fluorescence intensity, with no shift in the maximal wavelength. The extent of reduction in signal was 20–40% of that produced by the P-gp modulator nicardipine. Numerous researchers [19,23–25] have demonstrated considerable variability in the extent of tryptophan quenching by drugs and short peptide substrates of P-gp. These observations suggest that the Aβ peptides have a direct interaction with P-gp that is similar to established drug substrates and inhibitors. This is supported by an earlier study demonstrating that the Aβ peptides quench the fluorescence of an extrinsic probe (MIANS) covalently attached to P-gp [26]. MIANS was attached at the nucleotide-binding domains of P-gp and this indicates a long-distance allosteric interaction, whereas the intrinsic tryptophan quenching is thought to involve residues proximal to the drug binding site and central cavity [23].

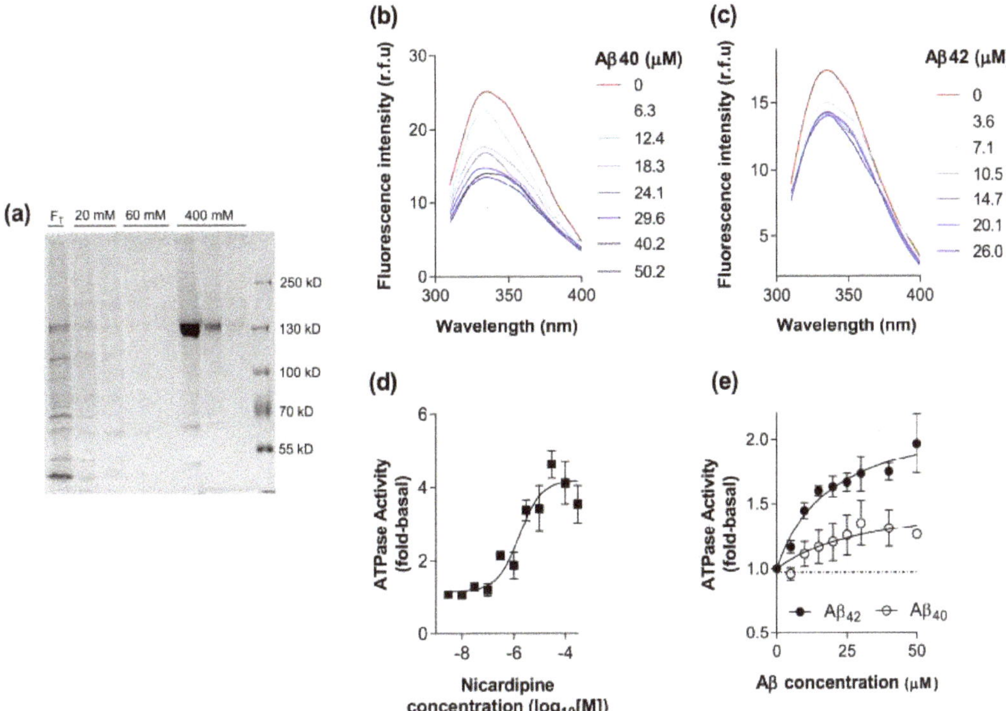

Figure 1. The interaction between $A\beta_{40}$ and $A\beta_{42}$ peptides with purified, reconstituted P-gp. (**a**) Fractions obtained from the metal affinity purification of P-gp were separated on 8% SDS-PAGE and detected with Instant Blue^{TM} protein stain. Fractions (40 µL) from each stage of the purification were loaded onto the gel. F_T corresponds to the unbound detergent extracted material, 20 mM and 60 mM represent washing stages, and 400 mM is the elution phase. (**b**) Representative fluorescence emission spectra were obtained for the endogenous tryptophan residues of purified, reconstituted P-gp (15 µg) with excitation at 295 ± 10 nm at 37 °C. Spectra were recorded in the absence or presence of a series of $A\beta_{40}$ concentrations. (**c**) Fluorescence emission spectra obtained as in (**b**), but with varying concentrations of $A\beta_{42}$ peptide. (**d**) ATPase activity of purified, reconstituted P-gp (0.2–0.5 µg) in the presence of varying concentrations of nicardipine. The activity was measured over 40 min at 37 °C and normalised to the value obtained in the absence of drug (basal). Data were fitted by the general dose-response relationship using non-linear least squares regression and values represent mean ± SEM obtained from four independent preparations. (**e**) ATPase activity of purified, reconstituted P-gp (0.2–0.5 µg) in the presence of varying concentrations of $A\beta_{40}$ and $A\beta_{42}$ peptide. The activity was measured over 40 min at 37 °C and normalised to the value obtained in the absence of drug (basal). Data were fitted by a hyperbolic relationship using non-linear least squares regression and values represent mean ± SEM obtained from three independent preparations.

Binding to the transporter represents the first step in the process of translocation across the membrane and therefore an activity assay was chosen to further explore the interaction between Aβ peptides and P-gp. Transported substrates of P-gp increase the rate of ATP hydrolysis [27–30], whereas pure inhibitors reduce the activity [31,32]. Consequently, measuring substrate effects on ATPase activity provides a rigorous assessment of the interaction with P-gp at the level of initial binding and the subsequent coupling event. Figure 1d shows the dose-dependent stimulation of ATPase activity by nicardipine. The overall activity of purified, reconstituted P-gp was characterised with a basal level of 419 ± 30 nmol/min/mg and a maximal activity of 1815 ± 139 nmol/min/mg for the preparations generated ($n = 4$) in this investigation. These values are in agreement with previously published values [21] and represent a 4.3-fold stimulation by nicardipine. Both Aβ peptide isoforms were also able to elicit a stimulation in the ATPase activity of purified, reconstituted P-gp as shown in

Figure 1e. The estimated maximal degree of stimulation by $A\beta_{40}$ was 1.5-fold, potentially suggesting that it is a relatively weak substrate of P-gp. $A\beta_{42}$ produced a comparatively greater stimulation of approximately 2.2-fold, which is equivalent, or greater, than established substrates vinblastine, paclitaxel, and rhodamine 123 [19]. It was not possible to generate a more extensive range of concentrations for the peptides (see Methods Section 5.5) and consequently, the affinity cannot be reliably estimated.

Overall, the two assays provide further evidence that $A\beta_{40}$ and $A\beta_{42}$ interact with P-gp and the use of purified, reconstituted protein demonstrates a direct mechanism. In addition, the nature of the effect on ATP hydrolysis may indicate their interaction is akin to a transported substrate, for which there is considerable evidence using cellular systems (for review see [16]).

2.2. P-gp Mediates $A\beta_{42}$ Transport from Brain to Capillary Lumen Ex Vivo

Freshly isolated brain capillaries provide a unique ex vivo model of the BBB, which can be used to study endogenous transport processes across the endothelium. P-gp is expressed on the luminal (blood-facing) membrane of the BBB, where it exports substrates from the endothelial cells into the blood [33].

To provide a physiological perspective to the binding and transport assays studied in Section 2.1, we compared the accumulation of fluorescent labelled human $A\beta_{42}$ (HiLyte^{TM}-hAβ_{42}; 5 µM) at steady state in brain capillaries isolated from wild-type (WT) versus P-gp-knockout (KO) mice, using confocal microscopy combined with quantitative image analysis. The $A\beta_{42}$ isoform was selected as it displayed a greater capacity for stimulating ATP-hydrolysis and therefore was deemed a higher affinity substrate of P-gp than $A\beta_{40}$ (Figure 1e). Accumulation of HiLyte^{TM}-hAβ_{42} was lower in the lumen of capillaries isolated from P-gp KO mice compared to capillaries isolated from WT mice (Figure 2a,b). Image analysis revealed that luminal HiLyte^{TM}-hAβ_{42} fluorescence was significantly ($p < 0.001$) reduced in capillaries from P-gp KO (60.1 ± 4.4 (r.f.u.)) versus WT mice (127.8 ± 5.9 (r.f.u.)), indicating that P-gp is necessary for active $A\beta$ transport from the bath to the vascular space. In concordance, luminal HiLyte^{TM}-hAβ_{42} fluorescence was significantly ($p < 0.001$) reduced in WT capillaries treated with the P-gp-specific inhibitor PSC833 (51.9 ± 2.4 (r.f.u.)) versus untreated capillaries (127.8 ± 5.9 (r.f.u.)), whereas fluorescence levels remained comparable between treated (60.1 ± 4.4 (r.f.u.)) versus untreated KO capillaries (55.5 ± 3.9 (r.f.u.)) (Figure 2c). The residual fluorescence present in PSC833-treated capillaries is due to non-specific binding of the HiLyte^{TM}-hAβ42 primarily to the endothelial cell surface. Figure 2c shows specific luminal NBD-CSA fluorescence that was taken as the difference between total luminal fluorescence and fluorescence in the presence of PSC833, which represents the P-gp-specific component of HiLyte^{TM}-hAβ42 transport. Together, these data indicate that the observed differences in fluorescence accumulation are specific to P-gp-mediated transport. Western immunoblot analysis confirmed high P-gp protein expression in capillary membranes from WT mice and lack of P-gp expression in capillary membranes from P-gp KO mice. In contrast, low-density lipoprotein receptor-related protein 1 (LRP-1), the receptor at the abluminal (brain-facing) membrane of the capillaries responsible for $A\beta$ uptake into the endothelial cells, was detected in capillaries from both WT and P-gp KO mice (Figure 2d).

These results confirm our previous findings showing $A\beta$ transport at the BBB is an active and ATP-dependent two-step process, involving LRP-1-mediated $A\beta$ uptake from the brain into capillary endothelial cells, followed by P-gp-mediated $A\beta$ transport from the endothelium into the capillary lumen [18,34].

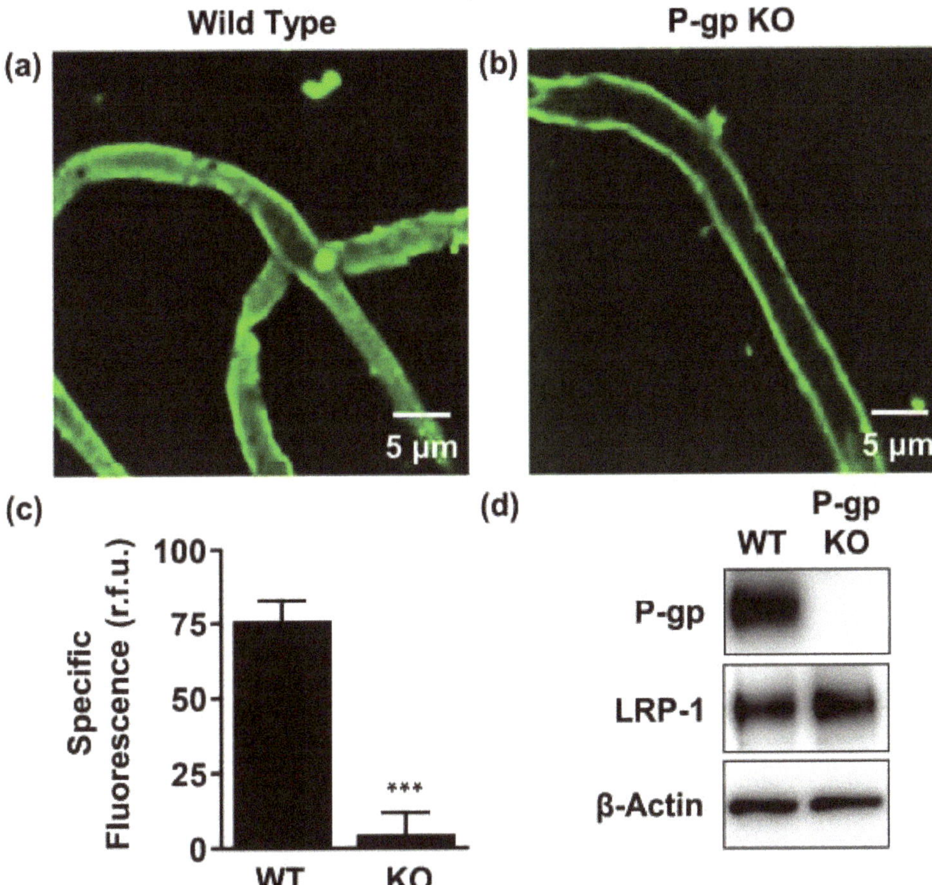

Figure 2. P-gp-mediated human Aβ₄₂ (hAβ₄₂) transport in isolated brain capillaries. (**a**), (**b**) Representative confocal images showing accumulation of HiLyte™-hAβ₄₂ in capillary lumens isolated from wild-type (WT) mice, but not in capillaries from P-gp knockout (KO) mice after a 1-h incubation (steady state; 5 µM HiLyte™-hAβ₄₂). (**c**) Data after digital image analysis using ImageJ. Specific fluorescence refers to the difference between total luminal HiLyte™-hAβ₄₂ fluorescence and HiLyte™-hAβ₄₂ fluorescence in the presence of the P-gp-specific inhibitor PSC833 (5 µM). (**d**) Western blot showing P-gp protein expression in isolated capillaries from WT mice, but not in capillaries isolated from P-gp KO mice. In contrast, LRP-1 is expressed in isolated capillaries from both WT mice and P-gp KO mice. β-actin was used as the loading control. Statistics: data per group are given as mean ± SEM for 10 capillaries from one preparation (pooled tissue: WT ($n = 10$ mice), P-gp KO ($n = 10$ mice)). Shown are relative fluorescence units ((r.f.u.) scale 0–255). *** Significantly lower than control, $p < 0.001$.

2.3. P-gp Protein Is Expressed in Human Neuroblastoma Cells

Expression of P-gp protein in neurons has been a point of contention. For example, several groups have demonstrated P-gp expression in neuronally-derived cell lines [35–38] and on peripheral nerve tissue at the blood-nerve-barrier (BNB) [39–41], whereas others have failed to do so or have demonstrated that it is only expressed in the context of brain injury or pathology [42–47].

We analysed P-gp protein expression by Western blot in cell lysates obtained from three separate human neuroblastoma cell lines that are regularly used in brain and AD-related research.

Figure 3 (left panel) shows that P-gp could be detected in Be(2)C, SH-SY-5Y and SK-N-SH cells, at levels comparable to those found in human brain endothelial hCMEC/D3 cells, which are commonly used as an in vitro BBB endothelial cell model.

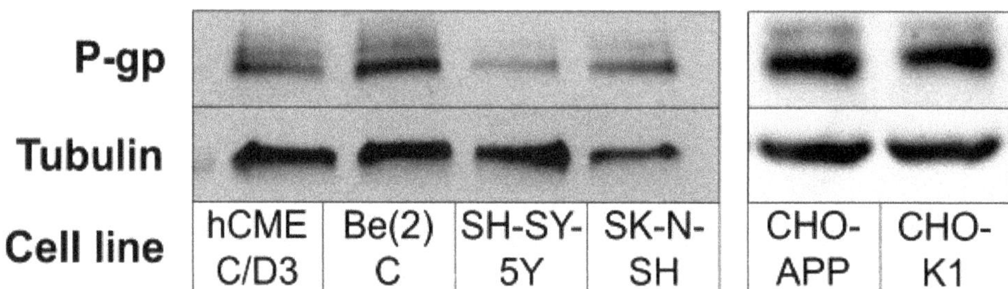

Figure 3. P-gp protein expression in cells. (**Left panel**) Cell lysates obtained from hCMEC/D3 and human neuroblastoma lines Be(2)C, SH-SY-5Y and SK-N-SH were analysed for P-gp protein expression (~170 kDa) via Western blot. (**Right panel**) P-gp expression in the non-neuronal CHO-APP cell line and its parental cell line CHO-K1. Tubulin (~50 kDa) was used as the positive loading control.

2.4. P-gp Is Active and Can Be Chemically Inhibited in CHO-APP and SK-N-SH Cells

Calcein-AM is a soluble hydrophobic non-fluorescent dye that rapidly crosses the plasma membrane and is actively exported by P-gp. When P-gp is active, calcein-AM is efficiently removed from the cell before it can undergo hydrolysis. If P-gp is inactive, intracellular esterases cleave calcein-AM to produce the free acid calcein, which cannot be transported by P-gp and thus remains trapped inside the cell where it produces an intense fluorescence [48]. Therefore, measuring the accumulation of fluorescent calcein is a rapid and sensitive method for studying P-gp activity.

Experiments were firstly performed in the CHO-APP cells, which secrete relatively large quantities of Aβ peptides. P-gp protein was confirmed to be expressed in CHO-APP cells at a level consistent with that of the parental CHO-K1 cell line (Figure 3; right panel). Verapamil and nicardipine, both anti-hypertensive calcium channel blockers, were selected due to their well established and strong P-gp inhibitory activity [49,50]. Figure 4a,b show that addition of these P-gp inhibitors to CHO-APP cells increased fluorescence, indicating increased intracellular calcein accumulation, in a concentration-dependent manner. Experiments were subsequently replicated in SK-N-SH neuroblastoma cells, which have previously been established to secrete $A\beta_{40}$ and $A\beta_{42}$ peptides [51,52]. Data show a similar concentration-dependent effect of P-gp inhibition on intracellular fluorescence (Figure 4c,d). Together, these data indicate that P-gp is active and can be chemically inhibited by these two drugs in both cell lines. Furthermore, nicardipine was approximately three-fold more effective than verapamil at inhibiting P-gp-mediated export of calcein-AM (Figure 4), which corresponds with previously published findings [49,50].

Although calcein-AM is a substrate of both P-gp and the related multi-drug resistance transporter ABCC1/MRP1, verapamil and nicardipine do not directly affect ABCC1/MRP1 activity [53,54]. Furthermore, verapamil does not appear to affect the activity of other transporters including ABCG2/BCRP, ABCG4, LRP-1, or RAGE that have been implicated in Aβ transport [55–58]. Although nicardipine is an effective inhibitor of ABCG2/BCRP [59], expression of this transport protein has not been reported in SK-N-SH and is absent or minimal CHO cells [60–62]. Therefore, the observations described in Figure 4 are deemed to be specific to P-gp and not confounded by activity of other ABC transporters.

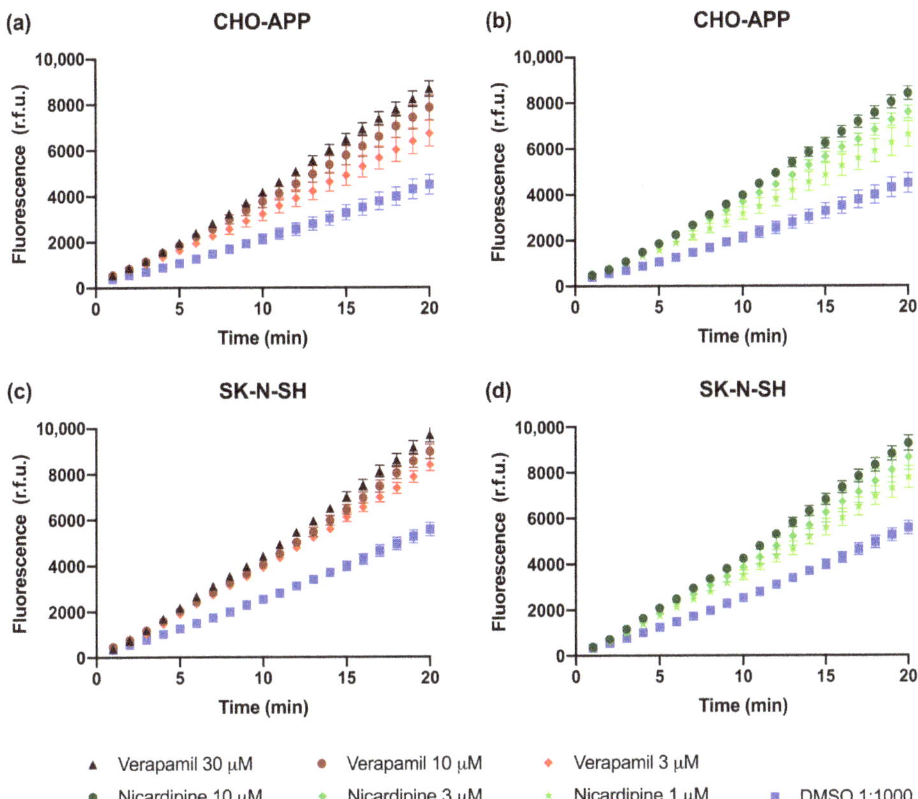

Figure 4. P-gp activity in CHO-APP and SK-N-SH cells. P-gp efflux activity was measured using the fluorogenic P-gp substrate calcein-AM. CHO-APP cells were treated with (**a**) verapamil at 3, 10 or 30 µM or DMSO control, or (**b**) nicardipine at 1, 3, or 10 µM or DMSO control, immediately prior to the addition of 0.1 µM calcein-AM. Similarly, SK-N-SH cells were treated with varying concentrations of either (**c**) verapamil or (**d**) nicardipine or DMSO control. Fluorescence measurements were obtained at 485/535 nm every minute over twenty minutes. Data are presented as the mean ± SEM of three independent experiments, with each condition conducted with six replicates.

2.5. Cell viability Assays

Reduced cell viability compromises the production and secretion of cellular products. Therefore, MTT assays were performed to verify whether incubation with the P-gp inhibitors verapamil and nicardipine at the final experimental concentrations (1–30 µM), would affect CHO-APP or SK-N-SH cell viability. Figure 5a shows that CHO-APP cell viability was not affected by the inhibitors. There was a marginal effect of high verapamil concentrations on SK-N-SH cell viability. However, this was not statistically significant ($p > 0.05$ compared to control) (Figure 5b).

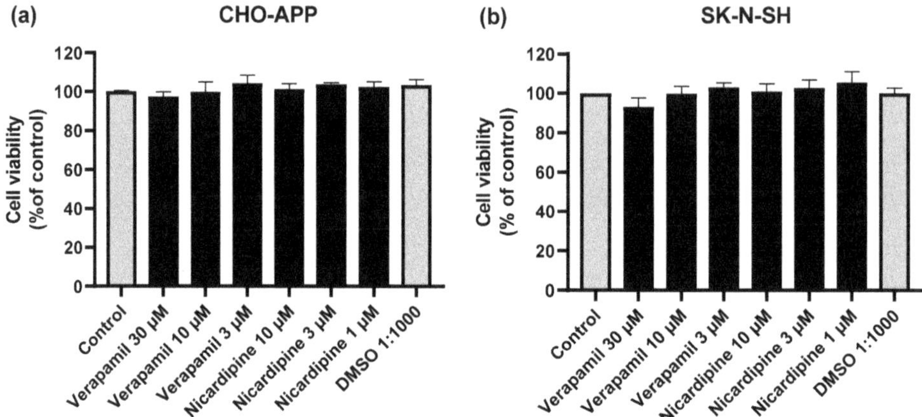

Figure 5. P-gp inhibitors do not affect cell viability. (**a**) CHO-APP and (**b**) SK-N-SH cells were incubated with verapamil or nicardipine at varying concentrations, DMSO, or untreated full culture medium (control) for 24 h to simulate the full experiment duration. The following day, cells were washed and treated with 0.5 mg/mL MTT in serum-free culture medium for two hours at 37 °C. After washing with PBS, cells were lysed with DMSO. Absorbance was measured at 550 nm. Data are presented as mean $\pm$ SEM of three independent experiments, each performed with six replicates.

2.6. Inhibition of P-gp Reduces Aβ Secretion from CHO-APP and SK-N-SH Cells

To investigate whether P-gp is involved in the export of Aβ, we assessed the effect of chemical inhibition of P-gp activity on the secretion of $A\beta_{40}$ and $A\beta_{42}$ peptides into cell media. Initial experiments were conducted using CHO-APP cells, which overexpress human APP and exhibit ample P-gp expression. Considering these cells produce large quantities of Aβ, we were able to use in-house ELISAs to quantify these peptides in the supernatant. Control (untreated) cells in our experiments secreted on average 3.3 and 1.1 ng/mL $A\beta_{40}$ and $A\beta_{42}$, respectively. Treatment of CHO-APP with verapamil for 24 h significantly reduced secretion of $A\beta_{40}$ (Figure 6a) and $A\beta_{42}$ (Figure 6b) into the media compared to control in a concentration-dependent manner. Results were most pronounced with verapamil 30 µM, which reduced $A\beta_{40}$ and $A\beta_{42}$ secretion by approximately half, compared to control. Treatment with nicardipine similarly yielded a dose-dependent effect on $A\beta_{40}$ secretion, with 10 µM reducing $A\beta_{40}$ levels to 48 $\pm$ 7.6% of that of control (Figure 6a); reductions in $A\beta_{42}$ levels were also significant. However, a dose-dependent relationship could not be confirmed due to variability in response to the highest nicardipine concentration (Figure 6b).

Our observations in CHO-APP are in line with findings from other groups that also report the involvement of P-gp in Aβ export in vitro, utilising human embryonic HEK293 cells transfected with APP695 [26], P-gp-transfected Lewis lung carcinoma cells [63] and LS-180 human colon adenocarcinoma cells [64]. However, the relationship between P-gp and Aβ has, until now, not been investigated in neurons where these peptides are predominantly generated. Therefore, we applied the same experimental conditions to SK-N-SH human neuroblastoma cells.

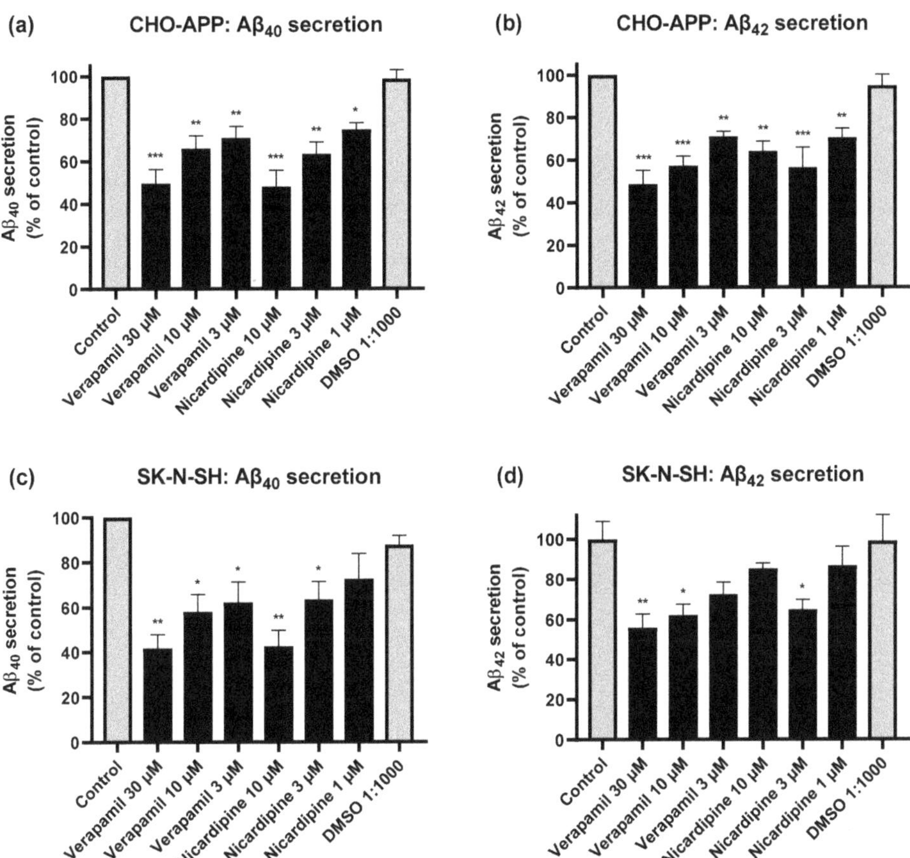

Figure 6. Inhibition of P-gp reduces Aβ efflux from CHO-APP and SK-N-SH cells. CHO-APP cells were treated with verapamil, nicardipine, or DMSO, or left untreated, for 24 h before supernatant was collected and analysed for (**a**) $A\beta_{40}$ and (**b**) $A\beta_{42}$ content by in-house ELISA. Data are displayed as mean ± SEM of three independent experiments for each peptide, and were adjusted for protein concentration. Each experiment was conducted with three biological replicates, each with additional three technical replicates. SK-N-SH cells were similarly treated and analysed for (**c**) $A\beta_{40}$ (mean ± SEM, n = 2) and (**d**) $A\beta_{42}$ (mean ± SEM of one experiment run in triplicate cultures) secretion into media but using commercial ELISA kits. * $p \leq 0.05$, ** $p \leq 0.01$, *** $p \leq 0.001$.

Since SK-N-SH cells secrete considerably smaller quantities of peptides (approximately 100-fold less than CHO-APP cells), the cell supernatants were analysed using commercial ELISA kits. Control SK-N-SH cells were found to secrete on average 29.0 pg/mL $A\beta_{40}$ and 3.6 pg/mL $A\beta_{42}$. As seen in Figure 6c, pharmacological inhibition of P-gp significantly and dose-dependently reduced $A\beta_{40}$ secretion from SK-N-SH cells. Results were most pronounced with verapamil 30 µM and nicardipine 10 µM, which reduced levels to 42 ± 6.1% and 43 ± 4.0% of that of control, respectively. Verapamil similarly yielded a dose-dependent reduction in $A\beta_{42}$ secretion (Figure 6d). Interestingly, the same observation as seen in CHO-APP (Figure 6b) was also observed in SK-N-SH, with the higher nicardipine concentration producing an unexpected increase in $A\beta_{42}$. Verapamil has been shown not to affect cellular $A\beta_{40}$ or $A\beta_{42}$ production [65], suggesting that the observed reductions in Aβ secretion can be attributable to reduced P-gp-mediated export. As anticipated, SK-N-SH cells consistently secreted higher proportions of $A\beta_{40}$ compared to $A\beta_{42}$ (approximately 8:1 ratio), which corresponds with previously reported in vitro data as well as physiological ratios observed in human AD brains [52,66]. Overall, chemical inhibition using P-gp-specific inhibitors was demonstrated to suppress Aβ peptide secretion from both CHO-APP and SK-N-SH cells.

3. Discussion

Positive correlations between P-gp and Aβ transport have previously been described in human and animal studies [16]. However, data from in vitro studies have been more conflicting. The P-gp/Aβ interaction was first proposed by Lam et al., who used a combination of pharmacological inhibition, binding studies, and vesicular transport assays to establish P-gp as an Aβ exporter [26]. Since then, whilst this relationship has been reinforced in several cell models, other groups have reported contradicting data suggesting that P-gp modulation does not affect Aβ transport [16]. In the present study, three lines of evidence collectively point to the involvement of P-gp in the export of $A\beta_{40}$ and $A\beta_{42}$ peptides. Firstly, binding assays utilising purified, reconstituted P-gp demonstrate a direct interaction. Whilst this finding is in line with that reported by Lam et al. [26], the use of intrinsic tryptophan quenching as described here, over quenching of a covalently attached fluorophore probe, has the advantage of stipulating a more direct transporter-peptide binding interaction. Concertedly, Aβ was able to stimulate ATP hydrolysis in a manner comparable with other established P-gp substrates. Secondly, vessels from the P-gp knockout mice model provided firm physiological evidence for P-gp-mediated transport of the $A\beta_{42}$ peptide at the BBB endothelium. Thirdly, in distinction from previously published in vitro studies that have utilised exogenously applied Aβ peptides [17,26,63,67], our findings from two distinct cell lines (non-neuronal CHO-APP and human neuron-like SK-N-SH) demonstrate that endogenously generated Aβ peptides are transported by endogenously expressed P-gp.

Interestingly, not only do our data indicate that P-gp is expressed and active in neuronal cells, they also show that the extent of inhibition of Aβ efflux by P-gp in SK-N-SH neuroblastoma cells was comparably significant to that in CHO-APP cells. This highlights a previously unappreciated role of P-gp in neurons that not only reshapes our understanding of Aβ pathology, but also has potentially significant implications for drug development and drug-drug interactions. Further studies utilising primary neurons and in vivo models are warranted to confirm the clinical significance of these effects. Notably, although Aβ secretion from both CHO-APP and SK-N-SH cells were significantly reduced in the presence of P-gp inhibitors (Figure 6), secretion was not completely eliminated. This is consistent with previously reported in vitro data [26,63]. This may be attributed to several factors, including those pertaining to the drugs themselves, such as concentration, half-life, and efficacy of inhibition, as well as the involvement of auxiliary peptide export mechanisms such as exosomes and other active transport proteins [55,68,69]. In fact, incomplete elimination of cellular Aβ may be favourable in the context of therapeutic applications [70]. Several reports regard APP and Aβ peptides as serving important physiological roles, including maintaining neuronal function, facilitating brain development, and conferring protection against pathogens [70,71]. Rather, P-gp activity could be a potential target for the development of novel therapeutics in AD to limit the neurotoxic effects of excess Aβ in the brain [64,72,73]. One approach would be to upregulate P-gp function to enhance Aβ export; at the neuron, this could remove intracellularly accumulated peptides, and at the BBB, extracellularly deposited peptides could be cleared. It has been established that intraneuronal Aβ accumulation may be just as toxic, and precedes, extracellular accumulation [10,11]. Hence, alleviating the Aβ load within neurons by facilitating the clearance process out of these cells would be potentially beneficial. However, there is growing evidence that the cell-to-cell spread of misfolded and aggregated proteins, including Aβ peptides, tau proteins and α-synuclein contributes to disease progression in AD as well as other neurodegenerative conditions [74–77]. In particular, Aβ peptides have been reported to behave as "seeds" that spread in the brain in a prion-like manner [78]. Therefore, further research is necessary to ensure that any transient intermediary extracellular accumulation resulting from increased neuronal secretion of Aβ peptides does not lead to an increase in "seeding" events. In addition, further clarification is required to determine what happens to these peptides once they are exported from their cells of origin. Although not fully elucidated, association with the lipid carrier apolipoprotein E is known to be an important step in the peptide clearance process [79,80]. Critically, any therapeutics aimed at upregulating P-gp function have the potential for drug-drug interactions or off-target effects in patients with comorbidities (such as multi-drug resistant cancers) that must be considered. An alternate recommendation would be to reconsider the prescribing of medications with P-gp-inhibitory effects in patients who are at risk of developing, or have been diagnosed with AD.

As previously mentioned, P-gp expression has been reported in neurons in the periphery at the BNB. It has been suggested that increased permeability and breakdown of the BNB is a contributor to immune- and inflammatory-related neuropathic and neurodegenerative disorders [39,81]. Although expression of P-gp at the BNB has been indicated by several studies to be significantly lower than expression at the BBB [40,41], further studies are needed to determine whether P-gp serves a protective function in these neurons and whether modulation of activity may be beneficial in the prevention of other neurodegenerative disorders.

Lastly, it has been reported that P-gp function declines with ageing, and moreover, Aβ peptides themselves may directly compromise P-gp expression and activity [82–85]. These factors potentially propagate a vicious cycle that drives AD progression. Therefore, it is imperative to unravel the underlying mechanisms that lead to the decay of P-gp function in the ageing process. So far, post-translational mechanisms such as protein ubiquitination have been described [72,86], which could explain why P-gp expression at the gene level has not yet been identified as a strong genetic risk factor for AD development [87,88]. Further studies are warranted to establish whether curtailing age- and/or disease-related P-gp decline would effectuate any symptomatic improvements or disease-modifying effects.

4. Conclusions

The hydrophobic and membrane-anchored nature of the intraneuronally-generated Aβ peptide suggests that simple diffusion, as it has until now been assumed, does not adequately explain the mechanism of its expulsion into the extracellular space. Data presented here provide compelling evidence to substantiate the ability for P-gp to export Aβ. This not only occurs at the BBB endothelium, but for the first time, we have shown that the clearance of Aβ out of neurons is also an active process mediated by P-gp. Further studies are still needed to examine whether modulating P-gp function affects markers of neurodegeneration in vivo, and to confirm if this is a viable avenue to pursue in the search for effective AD therapies. Nonetheless, clarifying the molecular mechanisms involved in the pathway of the Aβ peptide, from its synthesis in the cell to clearance from the brain, is critical for our understanding of the pathophysiology of AD.

5. Materials and Methods

5.1. Materials

All cell culture materials including media and additives were purchased from Thermo Fisher Scientific (Scoresby, VIC, Australia), except for Hanks' Balanced Salt Solution (HBSS) which was purchased from Sigma-Aldrich (Castle Hill, NSW, Australia).

For P-gp purification and reconstitution steps (Section 5.3), dodecyl-β-D-maltoside was obtained from Anatrace (Ohio, USA), Ni-NTA His-Bind resin from Merck (Bayswater, VIC, Australia), and SM2 BioBeads from BioRad (Gladesville, NSW, Australia).

Reagents for casting SDS-PAGE gels including Tris-HCl, sodium dodecyl sulfate (SDS) and tetramethylethylenediamine (TEMED) were purchased from VWR Life Science (Tingalpa, QLD, Australia). 40% acrylamide/bis-acrylamide solution was from BioRad (Gladesville, NSW, Australia). Ammonium persulfate was from Sigma-Aldrich. Nitrocellulose membranes and enhanced chemiluminescent reagents were purchased from Merck, GE Healthcare and Pierce (Rockford, IL, USA). The Pierce Bicinchoninic acid (BCA) protein assay kit was purchased from Thermo Fisher Scientific.

($\pm$)-Verapamil hydrochloride and nicardipine hydrochloride were obtained from Sigma-Aldrich. Stock solutions were prepared at 25 μM by dissolving the powders in dimethylsulfoxide (DMSO) and stored at -20 °C. Calcein acetoxymethyl ester (calcein-AM), also obtained from Sigma-Aldrich, was diluted in DMSO to produce a 4 μM stock solution and stored at -20 °C.

Lypholised human Aβ 1-40 (Aβ_{40}) and Aβ 1-42 (Aβ_{42}) peptides, and HiLyte^{TM}-hAβ_{42} were purchased from AnaSpec (Fremont, CA, USA).

Chemicals for brain capillary isolation (Section 5.7) were purchased from Sigma-Aldrich (St Louis, MO, USA). All other reagents, unless otherwise specified, including bovine serum albumin (BSA), IGEPAL, protease inhibitor cocktail, copper (II) sulfate pentahydrate, thiazolyl blue tetrazoliumbromide (MTT) powder, DMSO, glycine, Instant Blue^{TM} protein stain, phosphate buffered saline (PBS), sodium carbonate, sodium bicarbonate, sodium hydroxide, sulfuric acid, and Tetramethylbenzidine (TMB) Liquid Substrate System were obtained from Sigma-Aldrich (Castle Hill, NSW, Australia).

5.2. Antibodies

Anti-ABCB1 monoclonal C219 antibodies were obtained from Novus Biologicals and Abcam (Cambridge, MA, USA). Anti-LRP-1 antibody was obtained from Calbiochem-Novabiochem (La Jolla, CA, USA). Anti-β-actin antibody was purchased from Abcam. Anti-α-tubulin monoclonal antibody, and secondary HRP-conjugated anti-mouse and anti-rabbit antibodies were purchased from Sigma-Aldrich. For the in-house Aβ_{40} ELISA, capture and detection antibodies were anti-Aβ_{1-40} (polyclonal rabbit; catalog no. ABN240 from Merck Millipore) and anti-Aβ_{1-16} (monoclonal mouse; clone AB10 from Sigma-Aldrich), respectively. For the in-house Aβ_{42} ELISA, capture and detection antibodies were anti-Aβ_{1-16} and anti-Aβ_{37-42} (polyclonal rabbit; catalog no. Ab34376 from Abcam), respectively.

5.3. Purification and Reconstitution of P-gp

A C-terminal dodecyl-histidine version of human P-gp was expressed in *Trichoplusia ni* (High-Five) insect cells using recombinant baculovirus as previously described [19]. Following expression, crude membranes were prepared using differential ultra-centrifugation and stored at -80 °C. P-gp was extracted from the High-Five crude membranes using the detergent dodecyl-β-D-maltoside (DDM) and purified by metal affinity chromatography on Ni-NTA His-Bind resin. Chromatography buffers contained 0.1% (w/v) DDM and were supplemented with 0.1% (w/v) of a lipid mixture comprising a 4:1 ratio of *E. coli* total lipid extract and cholesterol. This enabled rapid reconstitution into vesicles by detergent adsorption using SM2 BioBeads.

5.4. Tryptophan Fluorescence Quenching Assay for Ligand Binding to Purified, Reconstituted P-gp

Binding of Aβ peptides to purified, reconstituted P-gp was measured by quenching of the intrinsic fluorescence of endogenous tryptophan residues as described [23] and modified [19]. However, imidazole was removed by ultra-centrifugation (70,000 g, 20 min 4 °C) of reconstituted P-gp and subsequent resuspension in binding buffer (20 mM MOPS, pH 8.0, 200 mM NaCl). P-gp (10–15 μg) was added to quartz silica cuvettes in a total volume of 300 μL. A tryptophan fluorescence emission spectrum was measured from 300–400 nm (emission slit width 5 nm) using excitation at 295 $\pm$ 10 nm at a scan speed of 120 nm/min. Lyophilised Aβ_{40} and Aβ_{42} peptides were resuspended in 2 mM NaOH at pH~10.5 to concentrations of 1.6 mg/mL (385 mM) and 1.0 mg/mL (221 mM) as described [89]. The peptides were added to sample cuvettes at concentrations in the range of 1–25 μM (Aβ_{42}) or 1–50 μM (Aβ_{40}). Sample cuvettes were held at a temperature of 37 °C and incubated for 20 min prior to measuring the fluorescence emission spectrum. Nicardipine (5 μM) was added following the final Aβ peptide addition to determine the maximal possible quenching. The spectra obtained in the presence of nicardipine were subtracted from those containing Aβ peptide to remove background signal.

5.5. ATP Hydrolysis by Purified, Reconstituted P-gp

The rate of ATP hydrolysis by purified, reconstituted P-gp was determined spectrophotometrically by the liberation of inorganic phosphate as described [19,90]. Samples of P-gp (0.2–0.5 µg) were incubated in 96-well microplates with disodium-ATP (2 mM) and either Aβ peptide (1–50 µM) or nicardipine ($10^{-9} - 3 \times 10^{-4}$ M) in a total volume of 50 µL at 37 °C for 40 min. The absorbance ($\lambda = 750$ nm) was measured using an iMark plate reader. The activity values were normalised to the basal (i.e., substrate-free) level and plotted as a function of ligand concentration. Data in the presence of nicardipine were analysed using the general dose-response curve:

$$v = v_{initial} + \left(v_{final} - v_{initial}\right) / \left(1 + 10^{\log\left(EC50 - [L]\right)}\right)$$

as described in [19], where: v is the activity, L is the compound added and EC50 is the potency of effect. Complete dose-response curves were not possible for the Aβ peptides due to poor solubility and their expense from commercial suppliers.

5.6. Animals

Animal experiments were approved by the Institutional Animal Care and Use Committee of the University of Kentucky (protocol no.: 2014–1233, PI: Hartz; approved April, 2014) and were carried out in accordance with AAALAC regulations, the US Department of Agriculture Animal Welfare Act, and the Guide for the Care and Use of Laboratory Animals of the NIH.

Male P-gp knockout (KO) mice (CF-1 strain; CF1-Abcb1amds—PGP) and corresponding male CF-1 wild-type (WT) mice were purchased from Charles River Laboratories (Wilmington, MA, USA). Mice were 9 weeks old with an average body weight of 33.7 g (31–35 g) for WT mice and 36.3 g (33–42 g) for P-gp KO mice. All mice were single-housed and kept under controlled environmental conditions (21 °C; 51–62% relative humidity; 12-h light/dark cycle) using an Ecoflo Allentown ventilation system (Allentown Inc., Allentown, NJ, USA). Animals were monitored at least once a day and had free access to tap water and Harlan Teklad Chow 2918 rodent feed (Harlan Laboratories Inc., Indianapolis, IN, USA). After shipping, animals were allowed to acclimate to their new environment for at least 7 days prior to experiments.

5.7. Brain Capillary Isolation

Brain capillaries were isolated as previously described [18,72,86]. Mice were euthanised by CO_2 inhalation and decapitated, brains were removed, dissected, and homogenised in cold PBS buffer (2.7 mM KCl, 1.46 mM KH_2PO_4, 136.9 mM NaCl, 8.1 mM Na_2HPO_4, 5 mM D-glucose, 1 mM sodium pyruvate, pH 7.4). Ficoll® was added to the brain homogenate to a final concentration of 15% and the Ficoll®/brain mixture was centrifuged at 5800× *g* for 15 min at 4 °C. After resuspending the pellet in 1% BSA/PBS, the capillary suspension was passed over a glass bead column to purify capillaries from debris and red blood cells. Capillaries adhering to the glass beads were collected by gentle agitation in 1% BSA/PBS, washed with BSA-free PBS and used for experiments.

5.8. Aβ Transport Assay in Isolated Brain Capillaries

To determine P-gp-mediated transport of Aβ, freshly isolated brain capillaries from WT and P-gp KO mice were incubated for 1 h at room temperature with HiLyte™-hAβ$_{42}$ (5 µM; [18,72,86]). For each group, images of 10 capillaries were acquired by confocal microscopy (Zeiss LSM 710 inverted confocal microscope, 40× 1.2 NA water immersion objective, 488 nm line of argon laser, Carl Zeiss Inc., Thornwood, NY, USA). Images were analysed by quantitating luminal HiLyte™-hAβ$_{42}$ fluorescence using ImageJ software v1.48. Specific, luminal HiLyte™-hAβ$_{42}$ fluorescence was taken as the difference between total luminal fluorescence and fluorescence in the presence of the P-gp-specific inhibitor PSC833 (5 μM; [18,72,86]).

5.9. Brain Capillary Harvest and Western Blot Analysis

Protein expression levels in brain capillaries were analysed by Western blotting as previously described [18,72,86]. Brain capillaries were homogenised in lysis buffer (Sigma-Aldrich, St Louis, MO, USA) containing Complete® protease inhibitor (Roche, Mannheim, Germany). Homogenised samples were centrifuged at 100,000× g for 90 min. Brain capillary membranes were resuspended in buffer containing protease inhibitor and stored at −80 °C.

Western blots were performed using the Invitrogen NuPage™ Bis-Tris electrophoresis and blotting system (Invitrogen, Carlsbad, CA, USA). After electrophoresis and protein transfer, membranes were blocked and incubated overnight with the primary antibody as indicated (P-gp (Abcam): 1:100 (1 μg/mL); β-actin: 1:1000 (1 μg/mL); LRP-1: 1:750 (1 μg/mL)). Membranes were washed and incubated for 1 h with horseradish peroxidase-conjugated ImmunoPure® secondary IgG (1:10,000; Pierce, Rockford, IL, USA). Proteins bands were detected via enhanced chemiluminescence and recorded using a BioRad Gel Doc 2000™ gel documentation system (BioRad, Hercules, CA, USA).

5.10. Cell Culture

Chinese hamster ovary cells stably overexpressing human APP (CHO-APP) were a generous gift from Dr. Woojin Kim (Faculty of Medicine and Health, University of Sydney, Sydney, Australia). The cells were generated by transfecting CHO cells with recombinant vectors expressing human 695-amino acid APP cDNA and a puromycin resistance gene, as described [91]. Cells were routinely maintained in F-12 growth medium supplemented with heat-inactivated foetal bovine serum (FBS; 10% v/v), L-glutamine (2 mM), penicillin (100 units/mL) and streptomycin (100 μg/mL), with the addition of puromycin (7.5 μg/mL), at 37 °C in humidified air containing 5% CO_2.

SK-N-SH human neuroblastoma cells were obtained from Sigma-Aldrich (Castle Hill, NSW, Australia) and maintained in MEM growth medium supplemented with heat-inactivated FBS (10% v/v), L-glutamine (2 mM), penicillin (100 units/mL) and streptomycin (100 μg/mL) at 37 °C in 5% CO_2.

5.11. Cell Harvest and Western Blot Analysis

Cells were washed twice with ice-cold PBS and lysed with cell lysis buffer containing 5 μL/mL protease inhibitor cocktail in IGEPAL. Lysates were syringed with 23-gauge needles to shear cellular DNA and centrifuged at 12,000× g rpm at 4 °C for 5 min. Total cell protein concentrations of the resulting supernatants were determined using BCA protein assays. Equal amounts of cell protein were loaded onto 10% (v/v) acrylamide gels for separation by SDS-PAGE. Proteins were transferred onto nitrocellulose membranes, blocked for 1 h using 5% (w/v) skim milk or 0.1% (w/v) BSA in PBS-Tween (0.05% v/v), then incubated with anti-ABCB1 (1:2000 (Novus Biologicals), overnight 4 °C) or anti-tubulin (1:3000, overnight 4 °C) antibodies. Membranes were rinsed with PBS-Tween, then incubated with HRP-conjugated anti-mouse secondary antibodies (1:10,000) for 1 h at room temperature, and then rinsed again with PBS-Tween. Protein bands were visualised via chemiluminescence and the Bio-Rad ChemiDoc imaging system.

5.12. Calcein-AM Assay

P-glycoprotein activity was measured using the calcein-AM assay. CHO-APP and SK-N-SH cells were seeded into 96-well clear-bottom black-walled plates at 4 × 10^4 cells/well in full culture medium and incubated overnight. The next day when cells reached ~80–90% confluency, media was discarded, and the cells were washed with phenol red-free HBSS or MEM. To assess the inhibitory effect of verapamil and nicardipine on P-gp activity, the inhibitors and DMSO control were prepared at 2× concentrations in HBSS buffer and added to the cells at 100 µL/well. Calcein-AM substrate was also prepared at $2\times$ concentration, and 100 µL was added to each well using a multi-channel pipette to achieve the final working concentrations. Fluorescence measurements were obtained every minute for 20 min, starting immediately after addition of calcein-AM, using the Perkin Elmer Victor X plate reader. The excitation and emission wavelengths were set at 485 and 535 nm, respectively. Temperature was maintained at 37 °C. Measurements were recorded in relative fluorescence units (RFU) and computed using Graphpad Prism software.

5.13. MTT Assay

Cells were seeded into a 96-well plate in full culture medium and allowed to adhere overnight. The next day, media was discarded, and cells were incubated with 200 µL/well of verapamil (1–30 µM), nicardipine (1–10 µM), DMSO (1:1000), or no treatment in full culture medium for 24 h. On the day of the experiment, the cells were washed with serum-free medium, to minimise background effect caused by presence of serum. 10X MTT stock (5 mg/mL, prepared in PBS and filter sterilised using a 0.22 µm syringe filter unit) was diluted to 0.5 mg/mL using serum-free medium to yield the working concentration. Using a multi-channel pipette, 100 µL of $1\times$ MTT reagent was added to each well. Cells were incubated for two hours at 37 °C, washed with PBS, then lysed with 100 µL/well of DMSO. The plate was wrapped in foil and placed on a shaker for approximately 10 min to allow even distribution of colour across the wells. Absorbance was measured at 550 nm using a Bio-Rad Microplate Reader.

5.14. Preparation of Aβ Peptides

$A\beta_{40}$ and $A\beta_{42}$ solutions were prepared by dissolving lyophilised peptides in 2 mM NaOH (pH~10.5). By avoiding the isoelectric point of Aβ (5.5) in the initial solvation step, aggregation and oligomerisation of the peptide is minimised [89]. For fluorescence quenching assays, $A\beta_{40}$ and $A\beta_{42}$ peptides were prepared fresh at 385 mM and 221 mM, respectively. For ELISA standards, $A\beta_{40}$ and $A\beta_{42}$ $2\times$ solutions were prepared and stored at -80 °C in single-use aliquots. These stock solutions were diluted in PBS (1:1) immediately prior to use to readjust the pH from 10.5 to physiological 7.4.

5.15. Enzyme-Linked Immunosorbent Assay (ELISA)

CHO-APP and SK-N-SH cells were seeded onto 24-well plates. The following day, cells were treated with 400 µL/well verapamil, nicardipine, DMSO or plain cell culture media. Because SK-N-SH cells have been reported to secrete low concentrations of Aβ peptides (in the picogram/mL range) [51,52,92,93], an experiment period of 24 h was selected to allow time to accumulate sufficient peptide in the cell supernatant for detection using ELISA. After 24 h, media from each well was collected, treated with 1 µL protease inhibitor cocktail (equivalent to 0.25 mM AEBSF), centrifuged at $3000\times g$ RPM at 4 °C for 5 min to remove debris, and used immediately for ELISA (to avoid peptide degradation due to storage and freeze/thawing). Cellular secretion of $A\beta_{40}$ and $A\beta_{42}$ into media was detected and quantified by sandwich ELISA.

For CHO-APP supernatant, a pair of in-house isoform-specific ELISAs were developed: analyses were performed in 96-well Nunc-Immuno^{TM} MaxiSorp plates (Thermo Fisher Scientific) coated with 100 µL per well of capture antibody at 2.5 µg/mL in 0.1 M carbonate buffer (pH 9.5) at 4 °C overnight. Wells were washed four times with PBS-Tween (0.05% v/v) to remove any unbound antibody, before blocking with 200 µL of blocking buffer (2% BSA, 7.5 g/L glycine in PBS) for 1 h at room temperature. After another four washes, 100 µL peptide standards and samples were loaded into the wells and incubated for 1 h at room temperature on a slow-speed shaker. Washing was repeated, then wells were incubated with 100 µL of the appropriate primary detection antibody (1:1000 in 1% BSA, 3.75 g/L glycine in PBS) for 1 h at room temperature. After washing, 100 µL of secondary horseradish peroxidase-conjugated antibody (1:10,000 in 1% BSA, 3.75 g/L glycine in PBS) was added, and the plate was incubated for 1 h at room temperature. Following another four washes, 100 µL of TMB was added to each well. Plates were incubated in the dark for 30 min at room temperature, then the reaction was terminated by the addition of 50 µL of H_2SO_4 (20% v/v). Absorbance values at 450 nm were measured using a Bio-Rad Microplate Reader. Assay sensitivity was <300 pg/mL for both $A\beta_{40}$ and $A\beta_{42}$.

For SK-N-SH supernatant, the commercial human $A\beta_{40}$ and $A\beta_{42}$ ELISA kits (Invitrogen, catalogue no. KHB3481 and KHB3544) were used in accordance with the manufacturer's protocols.

Data were computed using Graphpad Prism software; a non-linear 4-parameter regression was used to generate a standard curve, from which the unknown concentrations were determined. ELISA data were adjusted for total protein content of each sample (as determined by BCA protein assays) to account for any potential slight variations in cell viability, by multiplying the ELISA concentration by the ratio of control protein concentration to sample protein concentration.

5.16. Statistical Analysis

Data were analysed using Graphpad Prism (v6.01, La Jolla, CA, USA) and Microsoft Excel 2019. Values are expressed as mean $\pm$ SEM. MTT and ELISA data were statistically analysed using one-way ANOVA followed by Dunnett's test. Two-tailed unpaired Student's t test was used to evaluate differences between WT and P-gp KO mice. A p value of 0.05 was considered significant (* $p \leq 0.05$, ** $p \leq 0.01$, *** $p \leq 0.001$).

Author Contributions: Conceptualization, A.B.C., R.C. and I.C.G.; methodology, all authors; software, A.B.C.; validation, all authors.; formal analysis, A.B.C., R.C., A.M.S.H. and I.C.G.; investigation, all authors; writing—original draft preparation, A.B.C.; writing—review and editing, all authors.; supervision, I.C.G. and R.C.; project administration, I.C.G. and R.C.; funding acquisition, I.C.G., A.M.S.H. and R.C. All authors have read and agreed to the published version of the manuscript.

Funding: This research was funded by a seed grant from the University of Sydney, Australia. A.B.C. and A.Y. were recipients of Australian Government scholarships. A.M.S.H. was supported by grant number 2R01AG039621 from the National Institute on Aging. The content is solely the responsibility of the authors and does not necessarily represent the official views of the National Institute on Aging or the National Institutes of Health.

Institutional Review Board Statement: Animal experiments reported in this article were approved by the Institutional Animal Care and Use Committee of the University of Kentucky and carried out in accordance with AAALAC regulations, the US Department of Agriculture Animal Welfare Act, and the Guide for the Care and Use of Laboratory Animals of the NIH.

Data Availability Statement: The data presented in this study are available on request from the corresponding authors.

Conflicts of Interest: The authors declare no conflict of interest.

Abbreviations

$A\beta$	Amyloid-beta
ABC	ATP-binding cassette
AD	Alzheimer's disease
APP	Amyloid precursor protein
ATP	Adenosine triphosphate
BBB	Blood-brain barrier
BNB	Blood-nerve-barrier
Calcein-AM	Calcein-acetoxymethyl ester
CHO	Chinese hamster ovary
ELISA	Enzyme-linked immunosorbent assay
$hA\beta_{42}$	Human $A\beta_{42}$
KO	Knockout
LRP-1	Low density lipoprotein receptor-related protein 1
P-gp	P-glycoprotein
WT	Wild-type

References

1. Steiner, H.; Fukumori, A.; Tagami, S.; Okochi, M. Making the final cut: Pathogenic amyloid-β peptide generation by γ-secretase. *Cell Stress* **2018**, *2*, 292–310. [CrossRef]
2. Murphy, M.P.; LeVine, H., 3rd. Alzheimer's disease and the amyloid-beta peptide. *J. Alzheimers Dis.* **2010**, *19*, 311–323. [CrossRef]
3. Zhao, J.; O'Connor, T.; Vassar, R. The contribution of activated astrocytes to Aβ production: Implications for Alzheimer's disease pathogenesis. *J. Neuroinflamm.* **2011**, *8*, 150. [CrossRef]
4. Busciglio, J.; Gabuzda, D.H.; Matsudaira, P.; Yankner, B.A. Generation of beta-amyloid in the secretory pathway in neuronal and nonneuronal cells. *Proc. Natl. Acad. Sci. USA* **1993**, *90*, 2092–2096. [CrossRef]
5. Bateman, R.J.; Munsell, L.Y.; Morris, J.C.; Swarm, R.; Yarasheski, K.E.; Holtzman, D.M. Human amyloid-β synthesis and clearance rates as measured in cerebrospinal fluid in vivo. *Nat. Med.* **2006**, *12*, 856–861. [CrossRef]
6. Mawuenyega, K.G.; Sigurdson, W.; Ovod, V.; Munsell, L.; Kasten, T.; Morris, J.C.; Yarasheski, K.E.; Bateman, R.J. Decreased clearance of CNS beta-amyloid in Alzheimer's disease. *Science* **2010**, *330*, 1774. [CrossRef]
7. Viola, K.L.; Klein, W.L. Amyloid β oligomers in Alzheimer's disease pathogenesis, treatment, and diagnosis. *Acta Neuropathol.* **2015**, *129*, 183–206. [CrossRef]
8. Kayed, R.; Lasagna-Reeves, C.A. Molecular mechanisms of amyloid oligomers toxicity. *J. Alzheimers Dis.* **2013**, *33* (Suppl. 1), S67–S78. [CrossRef]
9. Cline, E.N.; Bicca, M.A.; Viola, K.L.; Klein, W.L. The Amyloid-β Oligomer Hypothesis: Beginning of the Third Decade. *J. Alzheimers Dis.* **2018**, *64*, S567–S610. [CrossRef]
10. Oddo, S.; Caccamo, A.; Smith, I.F.; Green, K.N.; LaFerla, F.M. A dynamic relationship between intracellular and extracellular pools of Abeta. *Am. J. Pathol.* **2006**, *168*, 184–194. [CrossRef]
11. Cuello, A.C. Intracellular and Extracellular Aβ, a Tale of Two Neuropathologies. *Brain Pathol.* **2005**, *15*, 66–71. [CrossRef]
12. Bayer, T.; Wirths, O. Intracellular accumulation of amyloid-beta—A predictor for synaptic dysfunction and neuron loss in Alzheimer's disease. *Front. Aging Neurosci.* **2010**, *2*. [CrossRef]
13. Mucke, L.; Selkoe, D.J. Neurotoxicity of amyloid β-protein: Synaptic and network dysfunction. *Cold Spring Harb. Perspect. Med.* **2012**, *2*. [CrossRef]
14. Callaghan, R.; Gelissen, I.C.; George, A.M.; Hartz, A.M.S. Mamma Mia, P-glycoprotein binds again. *FEBS Lett.* **2020**. [CrossRef]
15. Ambudkar, S.V.; Kimchi-Sarfaty, C.; Sauna, Z.E.; Gottesman, M.M. P-glycoprotein: From genomics to mechanism. *Oncogene* **2003**, *22*, 7468–7485. [CrossRef]
16. Chai, A.B.; Leung, G.K.F.; Callaghan, R.; Gelissen, I.C. P-glycoprotein: A role in the export of amyloid-β in Alzheimer's disease? *FEBS J.* **2020**, *287*, 612–625. [CrossRef]
17. Bello, I.; Salerno, M. Evidence against a role of P-glycoprotein in the clearance of the Alzheimer's disease Aβ1-42 peptides. *Cell Stress Chaperones* **2015**, *20*, 421–430. [CrossRef]
18. Hartz, A.M.; Miller, D.S.; Bauer, B. Restoring blood-brain barrier P-glycoprotein reduces brain amyloid-beta in a mouse model of Alzheimer's disease. *Mol. Pharmacol.* **2010**, *77*, 715–723. [CrossRef]
19. Mittra, R.; Pavy, M.; Subramanian, N.; George, A.M.; O'Mara, M.L.; Kerr, I.D.; Callaghan, R. Location of contact residues in pharmacologically distinct drug binding sites on P-glycoprotein. *Biochem. Pharmacol.* **2017**, *123*, 19–28. [CrossRef]
20. Crowley, E.; O'Mara, M.L.; Kerr, I.D.; Callaghan, R. Transmembrane helix 12 plays a pivotal role in coupling energy provision and drug binding in ABCB1. *FEBS J.* **2010**, *277*, 3974–3985. [CrossRef]
21. Taylor, A.M.; Storm, J.; Soceneantu, L.; Linton, K.J.; Gabriel, M.; Martin, C.; Woodhouse, J.; Blott, E.; Higgins, C.F.; Callaghan, R. Detailed characterization of cysteine-less P-glycoprotein reveals subtle pharmacological differences in function from wild-type protein. *Br. J. Pharmacol.* **2001**, *134*, 1609–1618. [CrossRef] [PubMed]
22. Kenakin, T. Principles: Receptor theory in pharmacology. *Trends Pharmacol. Sci.* **2004**, *25*, 186–192. [CrossRef] [PubMed]
23. Liu, R.; Siemiarczuk, A.; Sharom, F.J. Intrinsic fluorescence of the P-glycoprotein multidrug transporter: Sensitivity of tryptophan residues to binding of drugs and nucleotides. *Biochemistry* **2000**, *39*, 14927–14938. [CrossRef] [PubMed]
24. Dayan, G.; Jault, J.M.; Baubichon-Cortay, H.; Baggetto, L.G.; Renoir, J.M.; Baulieu, E.E.; Gros, P.; Di Pietro, A. Binding of steroid modulators to recombinant cytosolic domain from mouse P-glycoprotein in close proximity to the ATP site. *Biochemistry* **1997**, *36*, 15208–15215. [CrossRef]
25. Qu, Q.; Chu, J.W.; Sharom, F.J. Transition state P-glycoprotein binds drugs and modulators with unchanged affinity, suggesting a concerted transport mechanism. *Biochemistry* **2003**, *42*, 1345–1353. [CrossRef]
26. Lam, F.C.; Liu, R.; Lu, P.; Shapiro, A.B.; Renoir, J.-M.; Sharom, F.J.; Reiner, P.B. β-Amyloid efflux mediated by p-glycoprotein. *J. Neurochem.* **2001**, *76*, 1121–1128. [CrossRef]
27. Loo, T.W.; Bartlett, M.C.; Clarke, D.M. Substrate-induced conformational changes in the transmembrane segments of human P-glycoprotein. Direct evidence for the substrate-induced fit mechanism for drug binding. *J. Biol. Chem.* **2003**, *278*, 13603–13606. [CrossRef]
28. Loo, T.W.; Clarke, D.M. Drug-stimulated ATPase activity of human P-glycoprotein requires movement between transmembrane segments 6 and 12. *J. Biol. Chem.* **1997**, *272*, 20986–20989. [CrossRef]
29. Müller, M.; Bakos, E.; Welker, E.; Váradi, A.; Germann, U.A.; Gottesman, M.M.; Morse, B.S.; Roninson, I.B.; Sarkadi, B. Altered drug-stimulated ATPase activity in mutants of the human multidrug resistance protein. *J. Biol. Chem.* **1996**, *271*, 1877–1883. [CrossRef]

30. Sharom, F.J.; Yu, X.; Lu, P.; Liu, R.; Chu, J.W.; Szabó, K.; Müller, M.; Hose, C.D.; Monks, A.; Váradi, A.; et al. Interaction of the P-glycoprotein multidrug transporter (MDR1) with high affinity peptide chemosensitizers in isolated membranes, reconstituted systems, and intact cells. *Biochem. Pharmacol.* **1999**, *58*, 571–586. [CrossRef]
31. Orlowski, S.; Garrigos, M. Multiple recognition of various amphiphilic molecules by the multidrug resistance P-glycoprotein: Molecular mechanisms and pharmacological consequences coming from functional interactions between various drugs. *Anticancer Res.* **1999**, *19*, 3109–3123. [PubMed]
32. Martin, C.; Berridge, G.; Mistry, P.; Higgins, C.; Charlton, P.; Callaghan, R. The molecular interaction of the high affinity reversal agent XR9576 with P-glycoprotein. *Br. J. Pharmacol.* **1999**, *128*, 403–411. [CrossRef]
33. Schinkel, A.H. P-Glycoprotein, a gatekeeper in the blood–brain barrier. *Adv. Drug Deliv. Rev.* **1999**, *36*, 179–194. [CrossRef]
34. Storck, S.E.; Hartz, A.M.S.; Bernard, J.; Wolf, A.; Kachlmeier, A.; Mahringer, A.; Weggen, S.; Pahnke, J.; Pietrzik, C.U. The concerted amyloid-beta clearance of LRP1 and ABCB1/P-gp across the blood-brain barrier is linked by PICALM. *Brain Behav. Immun.* **2018**, *73*, 21–33. [CrossRef] [PubMed]
35. Bates, S.E.; Mickley, L.A.; Chen, Y.N.; Richert, N.; Rudick, J.; Biedler, J.L.; Fojo, A.T. Expression of a drug resistance gene in human neuroblastoma cell lines: Modulation by retinoic acid-induced differentiation. *Mol. Cell Biol.* **1989**, *9*, 4337–4344. [CrossRef] [PubMed]
36. Xu, Z.; Sun, Y.; Wang, D.; Sun, H.; Liu, X. SNHG16 promotes tumorigenesis and cisplatin resistance by regulating miR-338-3p/PLK4 pathway in neuroblastoma cells. *Cancer Cell Int.* **2020**, *20*, 236. [CrossRef] [PubMed]
37. Stage, T.B.; Mortensen, C.; Khalaf, S.; Steffensen, V.; Hammer, H.S.; Xiong, C.; Nielsen, F.; Poetz, O.; Svenningsen, Å.F.; Rodriguez-Antona, C.; et al. P-Glycoprotein Inhibition Exacerbates Paclitaxel Neurotoxicity in Neurons and Patients with Cancer. *Clin. Pharmacol. Ther.* **2020**, *108*, 671–680. [CrossRef]
38. Bernstein, H.-G.; Hölzl, G.; Dobrowolny, H.; Hildebrandt, J.; Trübner, K.; Krohn, M.; Bogerts, B.; Pahnke, J. Vascular and extravascular distribution of the ATP-binding cassette transporters ABCB1 and ABCC1 in aged human brain and pituitary. *Mech. Ageing Dev.* **2014**, *141–142*, 12–21. [CrossRef]
39. Palladino, S.P.; Helton, E.S.; Jain, P.; Dong, C.; Crowley, M.R.; Crossman, D.K.; Ubogu, E.E. The Human Blood-Nerve Barrier Transcriptome. *Sci. Rep.* **2017**, *7*, 17477. [CrossRef]
40. Balayssac, D.; Cayre, A.; Authier, N.; Bourdu, S.; Penault-Llorca, F.; Gillet, J.P.; Maublant, J.; Eschalier, A.; Coudore, F. Patterns of P-glycoprotein activity in the nervous system during vincristine-induced neuropathy in rats. *J. Peripher. Nerv. Syst.* **2005**, *10*, 301–310. [CrossRef]
41. Liu, H.; Chen, Y.; Huang, L.; Sun, X.; Fu, T.; Wu, S.; Zhu, X.; Zhen, W.; Liu, J.; Lu, G.; et al. Drug Distribution into Peripheral Nerve. *J. Pharmacol. Exp. Ther.* **2018**, *365*, 336–345. [CrossRef] [PubMed]
42. Langford, D.; Grigorian, A.; Hurford, R.; Adame, A.; Ellis, R.J.; Hansen, L.; Masliah, E. Altered P-Glycoprotein Expression in AIDS Patients with HIV Encephalitis. *J. Neuropathol. Exp. Neurol.* **2004**, *63*, 1038–1047. [CrossRef] [PubMed]
43. Ak, H.; Ay, B.; Tanriverdi, T.; Sanus, G.Z.; Is, M.; Sar, M.; Oz, B.; Ozkara, C.; Ozyurt, E.; Uzan, M. Expression and cellular distribution of multidrug resistance-related proteins in patients with focal cortical dysplasia. *Seizure* **2007**, *16*, 493–503. [CrossRef] [PubMed]
44. Lautz, T.B.; Jie, C.; Clark, S.; Naiditch, J.A.; Jafari, N.; Qiu, Y.-Y.; Zheng, X.; Chu, F.; Madonna, M.B. The Effect of Vorinostat on the Development of Resistance to Doxorubicin in Neuroblastoma. *PLoS ONE* **2012**, *7*, e40816. [CrossRef] [PubMed]
45. Chu, F.; Naiditch, J.A.; Clark, S.; Qiu, Y.-Y.; Zheng, X.; Lautz, T.B.; Holub, J.L.; Chou, P.M.; Czurylo, M.; Madonna, M.B. Midkine Mediates Intercellular Crosstalk between Drug-Resistant and Drug-Sensitive Neuroblastoma Cells In Vitro and In Vivo. *ISRN Oncol.* **2013**, *2013*, 518637. [CrossRef]
46. Blanc, E.; Goldschneider, D.; Ferrandis, E.; Barrois, M.; Le Roux, G.; Leonce, S.; Douc-Rasy, S.; Bénard, J.; Raguénez, G. MYCN enhances P-gp/MDR1 gene expression in the human metastatic neuroblastoma IGR-N-91 model. *Am. J. Pathol.* **2003**, *163*, 321–331. [CrossRef]
47. Sita, G.; Hrelia, P.; Tarozzi, A.; Morroni, F. P-glycoprotein (ABCB1) and Oxidative Stress: Focus on Alzheimer's Disease. *Oxid. Med. Cell. Longev.* **2017**, *2017*, 7905486. [CrossRef]
48. Holló, Z.; Homolya, L.; Davis, C.W.; Sarkadi, B. Calcein accumulation as a fluorometric functional assay of the multidrug transporter. *Biochim. Biophys. Acta Biomembr.* **1994**, *1191*, 384–388. [CrossRef]
49. Wessler, J.D.; Grip, L.T.; Mendell, J.; Giugliano, R.P. The P-Glycoprotein Transport System and Cardiovascular Drugs. *J. Am. Coll. Cardiol.* **2013**, *61*, 2495–2502. [CrossRef]
50. Takara, K.; Sakaeda, T.; Tanigawara, Y.; Nishiguchi, K.; Ohmoto, N.; Horinouchi, M.; Komada, F.; Ohnishi, N.; Yokoyama, T.; Okumura, K. Effects of 12 Ca2+ antagonists on multidrug resistance, MDR1-mediated transport and MDR1 mRNA expression. *Eur. J. Pharm. Sci.* **2002**, *16*, 159–165. [CrossRef]
51. Alley, G.M.; Bailey, J.A.; Chen, D.; Ray, B.; Puli, L.K.; Tanila, H.; Banerjee, P.K.; Lahiri, D.K. Memantine lowers amyloid-beta peptide levels in neuronal cultures and in APP/PS1 transgenic mice. *J. Neurosci. Res.* **2010**, *88*, 143–154. [CrossRef] [PubMed]
52. Haugabook, S.J.; Yager, D.M.; Eckman, E.A.; Golde, T.E.; Younkin, S.G.; Eckman, C.B. High throughput screens for the identification of compounds that alter the accumulation of the Alzheimer's amyloid β peptide (Aβ). *J. Neurosci. Methods* **2001**, *108*, 171–179. [CrossRef]
53. Loe, D.W.; Deeley, R.G.; Cole, S.P. Verapamil stimulates glutathione transport by the 190-kDa multidrug resistance protein 1 (MRP1). *J. Pharmacol. Exp. Ther.* **2000**, *293*, 530–538. [PubMed]

54. Ivnitski-Steele, I.; Larson, R.S.; Lovato, D.M.; Khawaja, H.M.; Winter, S.S.; Oprea, T.I.; Sklar, L.A.; Edwards, B.S. High-throughput flow cytometry to detect selective inhibitors of ABCB1, ABCC1, and ABCG2 transporters. *ASSAY Drug Dev. Technol.* **2008**, *6*, 263–276. [CrossRef] [PubMed]

55. Do, T.M.; Noel-Hudson, M.S.; Ribes, S.; Besengez, C.; Smirnova, M.; Cisternino, S.; Buyse, M.; Calon, F.; Chimini, G.; Chacun, H.; et al. ABCG2- and ABCG4-mediated efflux of amyloid-beta peptide 1-40 at the mouse blood-brain barrier. *J. Alzheimers Dis.* **2012**, *30*, 155–166. [CrossRef] [PubMed]

56. Deane, R.; Wu, Z.; Zlokovic, B.V. RAGE (yin) versus LRP (yang) balance regulates alzheimer amyloid beta-peptide clearance through transport across the blood-brain barrier. *Stroke* **2004**, *35*, 2628–2631. [CrossRef]

57. Wolf, A.; Bauer, B.; Hartz, A.M. ABC Transporters and the Alzheimer's Disease Enigma. *Front. Psychiatry* **2012**, *3*, 54. [CrossRef]

58. Krohn, M.; Lange, C.; Hofrichter, J.; Scheffler, K.; Stenzel, J.; Steffen, J.; Schumacher, T.; Brüning, T.; Plath, A.-S.; Alfen, F.; et al. Cerebral amyloid-β proteostasis is regulated by the membrane transport protein ABCC1 in mice. *J. Clin. Investig.* **2011**, *121*, 3924–3931. [CrossRef]

59. Zhang, Y.; Gupta, A.; Wang, H.; Zhou, L.; Vethanayagam, R.R.; Unadkat, J.D.; Mao, Q. BCRP transports dipyridamole and is inhibited by calcium channel blockers. *Pharm. Res.* **2005**, *22*, 2023–2034. [CrossRef]

60. Yin, J.Y.; Huang, Q.; Yang, Y.; Zhang, J.T.; Zhong, M.Z.; Zhou, H.H.; Liu, Z.Q. Characterization and analyses of multidrug resistance-associated protein 1 (MRP1/ABCC1) polymorphisms in Chinese population. *Pharm. Genom.* **2009**, *19*, 206–216. [CrossRef]

61. Blazquez, A.G.; Briz, O.; Romero, M.R.; Rosales, R.; Monte, M.J.; Vaquero, J.; Macias, R.I.; Cassio, D.; Marin, J.J. Characterization of the role of ABCG2 as a bile acid transporter in liver and placenta. *Mol. Pharmacol.* **2012**, *81*, 273–283. [CrossRef] [PubMed]

62. Zhao, S.; Chen, C.; Liu, S.; Zeng, W.; Su, J.; Wu, L.; Luo, Z.; Zhou, S.; Li, Q.; Zhang, J.; et al. CD147 promotes MTX resistance by immune cells through up-regulating ABCG2 expression and function. *J. Dermatol. Sci.* **2013**, *70*, 182–189. [CrossRef] [PubMed]

63. Kuhnke, D.; Jedlitschky, G.; Grube, M.; Krohn, M.; Jucker, M.; Mosyagin, I.; Cascorbi, I.; Walker, L.C.; Kroemer, H.K.; Warzok, R.W.; et al. MDR1-P-Glycoprotein (ABCB1) Mediates Transport of Alzheimer's amyloid-beta peptides–implications for the mechanisms of Abeta clearance at the blood-brain barrier. *Brain Pathol.* **2007**, *17*, 347–353. [CrossRef] [PubMed]

64. Abuznait, A.H.; Cain, C.; Ingram, D.; Burk, D.; Kaddoumi, A. Up-regulation of P-glycoprotein reduces intracellular accumulation of beta amyloid: Investigation of P-glycoprotein as a novel therapeutic target for Alzheimer's disease. *J. Pharm. Pharmacol.* **2011**, *63*, 1111–1118. [CrossRef] [PubMed]

65. Paris, D.; Bachmeier, C.; Patel, N.; Quadros, A.; Volmar, C.H.; Laporte, V.; Ganey, J.; Beaulieu-Abdelahad, D.; Ait-Ghezala, G.; Crawford, F.; et al. Selective antihypertensive dihydropyridines lower Aβ accumulation by targeting both the production and the clearance of Aβ across the blood-brain barrier. *Mol. Med.* **2011**, *17*, 149–162. [CrossRef]

66. Murphy, M.P.; Hickman, L.J.; Eckman, C.B.; Uljon, S.N.; Wang, R.; Golde, T.E. γ-Secretase, Evidence for Multiple Proteolytic Activities and Influence of Membrane Positioning of Substrate on Generation of Amyloid β Peptides of Varying Length. *J. Biol. Chem.* **1999**, *274*, 11914–11923. [CrossRef]

67. Nazer, B.; Hong, S.; Selkoe, D.J. LRP promotes endocytosis and degradation, but not transcytosis, of the amyloid-beta peptide in a blood-brain barrier in vitro model. *Neurobiol. Dis.* **2008**, *30*, 94–102. [CrossRef]

68. Rajendran, L.; Honsho, M.; Zahn, T.R.; Keller, P.; Geiger, K.D.; Verkade, P.; Simons, K. Alzheimer's disease β-amyloid peptides are released in association with exosomes. *Proc. Natl. Acad. Sci. USA* **2006**, *103*, 11172–11177. [CrossRef]

69. Yuyama, K.; Igarashi, Y. Exosomes as Carriers of Alzheimer's Amyloid-ß. *Front. Neurosci.* **2017**, *11*. [CrossRef]

70. Brothers, H.M.; Gosztyla, M.L.; Robinson, S.R. The Physiological Roles of Amyloid-β Peptide Hint at New Ways to Treat Alzheimer's Disease. *Front. Aging Neurosci.* **2018**, *10*, 118. [CrossRef]

71. Nalivaeva, N.N.; Turner, A.J. The amyloid precursor protein: A biochemical enigma in brain development, function and disease. *FEBS Lett.* **2013**, *587*, 2046–2054. [CrossRef] [PubMed]

72. Hartz, A.M.S.; Zhong, Y.; Shen, A.N.; Abner, E.L.; Bauer, B. Preventing P-gp Ubiquitination Lowers Aβ Brain Levels in an Alzheimer's Disease Mouse Model. *Front. Aging Neurosci.* **2018**, *10*. [CrossRef] [PubMed]

73. Brenn, A.; Grube, M.; Jedlitschky, G.; Fischer, A.; Strohmeier, B.; Eiden, M.; Keller, M.; Groschup, M.H.; St. Vogelgesang, S. John's Wort reduces beta-amyloid accumulation in a double transgenic Alzheimer's disease mouse model-role of P-glycoprotein. *Brain Pathol.* **2014**, *24*, 18–24. [CrossRef] [PubMed]

74. Colin, M.; Dujardin, S.; Schraen-Maschke, S.; Meno-Tetang, G.; Duyckaerts, C.; Courade, J.-P.; Buée, L. From the prion-like propagation hypothesis to therapeutic strategies of anti-tau immunotherapy. *Acta Neuropathol.* **2020**, *139*, 3–25. [CrossRef]

75. Stancu, I.C.; Vasconcelos, B.; Ris, L.; Wang, P.; Villers, A.; Peeraer, E.; Buist, A.; Terwel, D.; Baatsen, P.; Oyelami, T.; et al. Templated misfolding of Tau by prion-like seeding along neuronal connections impairs neuronal network function and associated behavioral outcomes in Tau transgenic mice. *Acta Neuropathol.* **2015**, *129*, 875–894. [CrossRef]

76. Van Den Berge, N.; Ferreira, N.; Gram, H.; Mikkelsen, T.W.; Alstrup, A.K.O.; Casadei, N.; Tsung-Pin, P.; Riess, O.; Nyengaard, J.R.; Tamgüney, G.; et al. Evidence for bidirectional and trans-synaptic parasympathetic and sympathetic propagation of alpha-synuclein in rats. *Acta Neuropathol.* **2019**, *138*, 535–550. [CrossRef]

77. Elfarrash, S.; Jensen, N.M.; Ferreira, N.; Betzer, C.; Thevathasan, J.V.; Diekmann, R.; Adel, M.; Omar, N.M.; Boraie, M.Z.; Gad, S.; et al. Organotypic slice culture model demonstrates inter-neuronal spreading of alpha-synuclein aggregates. *Acta Neuropathol. Commun.* **2019**, *7*, 213. [CrossRef]

78. Walker, L.C.; Schelle, J.; Jucker, M. The Prion-Like Properties of Amyloid-β Assemblies: Implications for Alzheimer's Disease. *Cold Spring Harb. Perspect. Med.* **2016**, *6*. [CrossRef]

79. Kanekiyo, T.; Xu, H.; Bu, G. ApoE and Abeta in Alzheimer's disease: Accidental encounters or partners? *Neuron* **2014**, *81*, 740–754. [CrossRef]

80. Yamazaki, Y.; Zhao, N.; Caulfield, T.R.; Liu, C.-C.; Bu, G. Apolipoprotein E and Alzheimer disease: Pathobiology and targeting strategies. *Nat. Rev. Neurol.* **2019**, *15*, 501–518. [CrossRef]

81. Richner, M.; Ferreira, N.; Dudele, A.; Jensen, T.S.; Vaegter, C.B.; Gonçalves, N.P. Functional and Structural Changes of the Blood-Nerve-Barrier in Diabetic Neuropathy. *Front. Neurosci.* **2018**, *12*, 1038. [CrossRef] [PubMed]

82. Chiu, C.; Miller, M.C.; Monahan, R.; Osgood, D.P.; Stopa, E.G.; Silverberg, G.D. P-glycoprotein expression and amyloid accumulation in human aging and Alzheimer's disease: Preliminary observations. *Neurobiol. Aging* **2015**, *36*, 2475–2482. [CrossRef] [PubMed]

83. Bartels, A.L.; Kortekaas, R.; Bart, J.; Willemsen, A.T.; de Klerk, O.L.; de Vries, J.J.; van Oostrom, J.C.; Leenders, K.L. Blood-brain barrier P-glycoprotein function decreases in specific brain regions with aging: A possible role in progressive neurodegeneration. *Neurobiol. Aging* **2009**, *30*, 1818–1824. [CrossRef] [PubMed]

84. Kania, K.D.; Wijesuriya, H.C.; Hladky, S.B.; Barrand, M.A. Beta amyloid effects on expression of multidrug efflux transporters in brain endothelial cells. *Brain Res.* **2011**, *1418*, 1–11. [CrossRef]

85. Park, R.; Kook, S.Y.; Park, J.C.; Mook-Jung, I. Abeta1-42 reduces P-glycoprotein in the blood-brain barrier through RAGE-NF-kappaB signaling. *Cell Death Dis.* **2014**, *5*, e1299. [CrossRef]

86. Hartz, A.M.S.; Zhong, Y.; Wolf, A.; LeVine, H.; Miller, D.S.; Bauer, B. Aβ40 Reduces P-Glycoprotein at the Blood–Brain Barrier through the Ubiquitin–Proteasome Pathway. *J. Neurosci.* **2016**, *36*, 1930. [CrossRef]

87. Jansen, I.E.; Savage, J.E.; Watanabe, K.; Bryois, J.; Williams, D.M.; Steinberg, S.; Sealock, J.; Karlsson, I.K.; Hägg, S.; Athanasiu, L.; et al. Genome-wide meta-analysis identifies new loci and functional pathways influencing Alzheimer's disease risk. *Nat. Genet.* **2019**, *51*, 404–413. [CrossRef]

88. van Assema, D.M.; Lubberink, M.; Rizzu, P.; van Swieten, J.C.; Schuit, R.C.; Eriksson, J.; Scheltens, P.; Koepp, M.; Lammertsma, A.A.; van Berckel, B.N. Blood-brain barrier P-glycoprotein function in healthy subjects and Alzheimer's disease patients: Effect of polymorphisms in the ABCB1 gene. *EJNMMI Res.* **2012**, *2*, 57. [CrossRef]

89. Teplow, D.B. Preparation of Amyloid β-Protein for Structural and Functional Studies. *Methods Enzymol.* **2006**, *413*, 20–33. [CrossRef]

90. Chifflet, S.; Torriglia, A.; Chiesa, R.; Tolosa, S. A method for the determination of inorganic phosphate in the presence of labile organic phosphate and high concentrations of protein: Application to lens ATPases. *Anal. Biochem.* **1988**, *168*, 1–4. [CrossRef]

91. Li, H.; Kim, W.S.; Guillemin, G.J.; Hill, A.F.; Evin, G.; Garner, B. Modulation of amyloid precursor protein processing by synthetic ceramide analogues. *Biochim. Biophys. Acta Mol. Cell Biol. Lipids* **2010**, *1801*, 887–895. [CrossRef] [PubMed]

92. Ray, B.; Banerjee, P.K.; Greig, N.H.; Lahiri, D.K. Memantine treatment decreases levels of secreted Alzheimer's amyloid precursor protein (APP) and amyloid beta (A beta) peptide in the human neuroblastoma cells. *Neurosci. Lett.* **2010**, *470*, 1–5. [CrossRef] [PubMed]

93. Qi, H.-S.; Liu, P.; Gao, S.-Q.; Diao, Z.-Y.; Yang, L.-L.; Xu, J.; Qu, X.; Han, E.-J. Inhibitory effect of piperlongumi-nine/dihydropiperlonguminine on the production of amyloid beta and APP in SK-N-SH cells. *Chin. J. Physiol.* **2009**, *52*, 160–168. [CrossRef] [PubMed]

International Journal of
Molecular Sciences

Article

Brain Distribution of Dual ABCB1/ABCG2 Substrates Is Unaltered in a Beta-Amyloidosis Mouse Model

Thomas Wanek [1,*], Viktoria Zoufal [1], Mirjam Brackhan [2], Markus Krohn [2], Severin Mairinger [1], Thomas Filip [1], Michael Sauberer [1], Johann Stanek [1], Thomas Pekar [3], Jens Pahnke [2,4,5] and Oliver Langer [1,6,7]

[1] Preclinical Molecular Imaging, AIT Austrian Institute of Technology GmbH, 2444 Seibersdorf, Austria; viktoria.zoufal@chello.at (V.Z.); Severin.Mairinger@chuv.ch (S.M.); thomas.filip@ait.ac.at (T.F.); michael.sauberer@ait.ac.at (M.S.); johann.stanek@ait.ac.at (J.S.); oliver.langer@meduniwien.ac.at (O.L.)

[2] Department of Neuro-/Pathology, University of Oslo (UiO) and Oslo University Hospital (OUS), 0424 Oslo, Norway; mirjam.brackhan@medisin.uio.no (M.B.); markus.krohn@uni-luebeck.de (M.K.); jens.pahnke@medisin.uio.no (J.P.)

[3] Biomedical Analytics, University of Applied Sciences Wiener Neustadt, 2700 Wiener Neustadt, Austria; thomas.pekar@fhwn.ac.at

[4] LIED, University of Lübeck, 23562 Lübeck, Germany

[5] Department of Pharmacology, Faculty of Medicine, University of Latvia, 1586 Rīga, Latvia

[6] Department of Clinical Pharmacology, Medical University of Vienna, 1090 Vienna, Austria

[7] Department of Biomedical Imaging und Image-guided Therapy, Division of Nuclear Medicine, Medical University of Vienna, 1090 Vienna, Austria

[*] Correspondence: thomas.wanek@ait.ac.at; Tel.: +43 (0)-505 50-3496

Received: 12 October 2020; Accepted: 28 October 2020; Published: 3 November 2020

Abstract: Background: ABCB1 (P-glycoprotein) and ABCG2 (breast cancer resistance protein) are co-localized at the blood-brain barrier (BBB), where they restrict the brain distribution of many different drugs. Moreover, ABCB1 and possibly ABCG2 play a role in Alzheimer's disease (AD) by mediating the brain clearance of beta-amyloid (Aβ) across the BBB. This study aimed to compare the abundance and activity of ABCG2 in a commonly used β-amyloidosis mouse model (APP/PS1-21) with age-matched wild-type mice. Methods: The abundance of ABCG2 was assessed by semi-quantitative immunohistochemical analysis of brain slices of APP/PS1-21 and wild-type mice aged 6 months. Moreover, the brain distribution of two dual ABCB1/ABCG2 substrate radiotracers ([11C]tariquidar and [11C]erlotinib) was assessed in APP/PS1-21 and wild-type mice with positron emission tomography (PET). [11C]Tariquidar PET scans were performed without and with partial inhibition of ABCG2 with Ko143, while [11C]erlotinib PET scans were only performed under baseline conditions. Results: Immunohistochemical analysis revealed a significant reduction (by 29–37%) in the number of ABCG2-stained microvessels in the brains of APP/PS1-21 mice. Partial ABCG2 inhibition significantly increased the brain distribution of [11C]tariquidar in APP/PS1-21 and wild-type mice, but the brain distribution of [11C]tariquidar did not differ under both conditions between the two mouse strains. Similar results were obtained with [11C]erlotinib. Conclusions: Despite a reduction in the abundance of cerebral ABCG2 and ABCB1 in APP/PS1-21 mice, the brain distribution of two dual ABCB1/ABCG2 substrates was unaltered. Our results suggest that the brain distribution of clinically used ABCB1/ABCG2 substrate drugs may not differ between AD patients and healthy people.

Keywords: ABCG2; ABCB1; blood-brain barrier; PET; Alzheimer's disease; beta-amyloid; tariquidar; erlotinib

1. Introduction

The major pathohistological hallmarks of Alzheimer's disease (AD) are the accumulation of beta-amyloid (Aβ) plaques and neurofibrillary tangles consisting of hyperphosphorylated tau protein in the brain. It is believed that one of the underlying causes of this cerebral Aβ accumulation is the impaired clearance of Aβ peptides from the brain [1–3]. There are several different mechanisms for the removal of Aβ peptides from the brain [4]; one important mechanism is its transport across the blood-brain barrier (BBB) into the blood. The adenosine triphosphate-binding cassette (ABC) transporter ABCB1 (also known as P-glycoprotein), which is expressed in the luminal (blood-facing) membrane of brain capillary endothelial cells, has been shown to work together with the low-density lipoprotein receptor-related protein 1 (LRP1) in the abluminal membrane of endothelial cells in translocating Aβ peptides across the BBB [5–9]. There is evidence that the abundance and activity of ABCB1 are reduced in the brains of AD patients relative to age-matched healthy control subjects [10–14]. Studies in β-amyloidosis mouse models indicated that the activity of cerebral ABCB1 can be pharmacologically induced (e.g., by treatment with pregnane X receptor activators), leading to enhanced Aβ clearance from the brain, which may constitute a potential therapeutic target in AD [8,15–18].

At the BBB, ABCB1 is co-localized with ABCG2 (also known as breast cancer resistance protein). ABCB1 and ABCG2 have a largely overlapping substrate spectrum and act as highly efficient gate keepers in preventing the brain distribution of a range of different drugs, such as most currently known molecularly targeted anticancer drugs [19,20]. While the role of ABCB1 in mediating Aβ clearance across the BBB has been thoroughly investigated [5–9,21], considerably less is known with respect to ABCG2. It has been shown that ABCG2 can also transport Aβ peptides [22–24]. There are conflicting data regarding the abundance of ABCG2 in the brains of AD patients versus age-matched healthy controls. One study found an increase [22], another study a decrease [12] and four other studies found no changes in the abundance of ABCG2 in AD patients [10,11,25,26]. As the abundance of ABCG2 may not always correlate with its activity, it would be preferable to directly measure ABCG2 activity at the BBB of AD patients to further investigate the possible role of ABCG2 in the brain clearance of Aβ.

Positron emission tomography (PET) imaging with radiolabeled transporter substrates has been proposed as a powerful method to measure the activity of ABCB1 at the BBB [13,14]. While several effective PET tracers for ABCB1 have been described [27], ABCG2-selective PET tracers are currently not available. We have recently developed a PET protocol to measure ABCG2 activity at the mouse and human BBB [28–30]. This protocol is based on PET scans with the dual ABCB1/ABCG2 substrate [^{11}C]tariquidar [31] under conditions of complete ABCB1 inhibition achieved by co-administration of unlabeled tariquidar [28–30]. In mice, the contribution of ABCG2 to the brain efflux of [^{11}C]tariquidar can be revealed by administration of the ABCG2 inhibitor Ko143 [28,29].

In the present study, we first used immunohistochemistry to stain ABCG2 in the brains of a β-amyloidosis mouse model (APP/PS1-21) [32] and control mice (both aged 6 months), which revealed a significant reduction in ABCG2-stained microvessels in APP/PS1-21 mice. We then applied PET imaging to measure the consequences of the decreased abundance of cerebral ABCG2 on the brain distribution of two dual ABCB1/ABCG2 substrate radiotracers ([^{11}C]tariquidar and [^{11}C]erlotinib) in APP/PS1-21 mice. The brain distribution of both radiotracers did not significantly differ between APP/PS1-21 mice and wild-type mice, suggesting that the observed reduction in cerebral ABCG2 abundance may not be sufficient to alter the brain distribution of ABCB1/ABCG2 substrate drugs.

2. Results

2.1. Immunohistochemistry

ABCG2 was immunohistochemically stained in brain slices of APP/PS1-21 mice and wild-type littermates aged 6 months ($n = 3$ per group) (Figure 1). For immunohistochemical analysis, we selected two brain regions in which Aβ load was shown to be high in APP/PS1-21 mice at the investigated age (hippocampus and cortex) and one region with negligible Aβ load (cerebellum) [33]. A semi-quantitative analysis of the stained microvessels indicated a significant decrease (by 29–37%) in the number of ABCG2-stained microvessels in the hippocampus and cerebellum of APP/PS1-21 versus wild-type mice (Figure 2a,c), while no significant difference was found in the cortex (Figure 2b). This suggested that the reduction in the abundance of ABCG2 in APP/PS1-21 mice was independent of Aβ deposition.

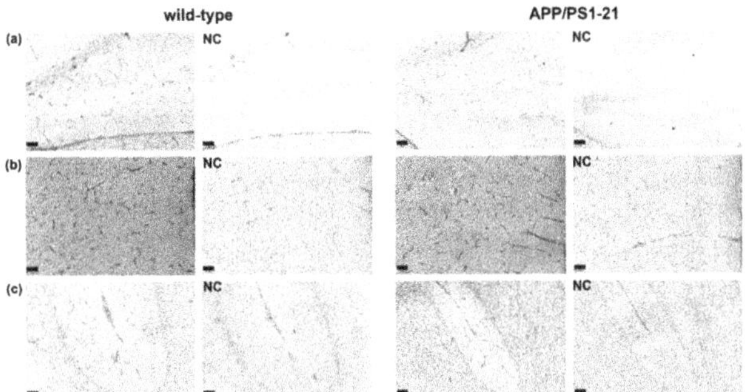

Figure 1. Examples of immunohistochemical staining of mouse ABCG2 in microvessels of 6-months-old wild type and APP/PS1-21 mice in (**a**) hippocampus (dentate gyrus), (**b**) cortex (cingulate cortex) and (**c**) cerebellum (4th and 5th cerebellar lobules with primary fissure) region. Enlarged areas are shown at 20× magnification. Scale bar in lower left indicates 50 μm. (NC, negative control: staining protocol without primary antibody).

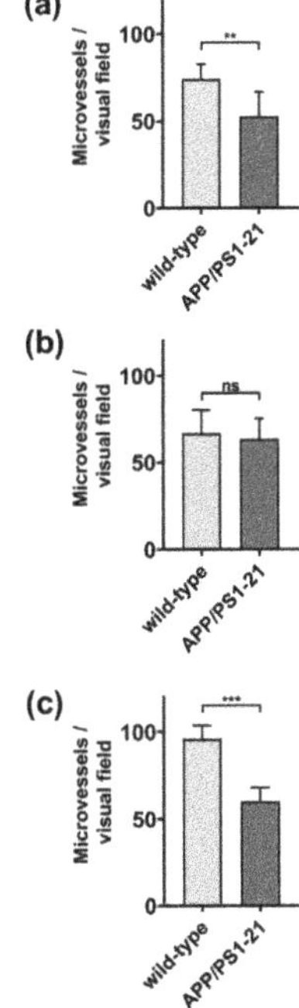

Figure 2. Semi-quantitative analysis of ABCG2-stained microvessels in (**a**) hippocampus, (**b**) cortex and (**c**) cerebellum of 6-months-old APP/PS1-21 mice and age-matched wild-type mice. For each region, the mean of four visual fields (at 20× digital magnification) per animal (n = 3 animals per strain) was used for statistical testing. Error bars indicate SD. (ns, not significant; ** $p < 0.01$; *** $p < 0.001$; 2-sided t-test).

2.2. [^{11}C]Tariquidar PET

We used a previously developed PET protocol [28] which involved PET scans with the dual ABCB1/ABCG2 substrate [^{11}C]tariquidar with co-administration of unlabeled tariquidar (12 mg/kg) to saturate cerebral ABCB1 activity and thereby selectively measure ABCG2 activity, without and with partial ABCG2 inhibition with the ABCG2 inhibitor Ko143 (5 mg/kg) [34]. PET summation images of [^{11}C]tariquidar in APP/PS1-21 and wild-type mice are shown in Figure 3. Without Ko143 pretreatment, brain radioactivity concentrations in both mouse strains were markedly lower than in most of the surrounding head region, while in Ko143-treated animals, brain radioactivity concentrations approached the concentrations in the surrounding head region. The corresponding time-activity-curves (TACs) in whole brains are shown in Figure 4. In both mouse strains, mean TACs were higher under conditions of partial ABCG2 inhibition than under conditions without ABCG2 inhibition. Brain-to-plasma radioactivity concentration ratios ($K_{p,brain}$) were determined as a parameter for the brain distribution of [^{11}C]tariquidar (i.e., the ratio of PET-derived radioactivity concentration at the last time point and the radioactivity concentration in plasma measured at the end of the PET scan) [28]. In Figure 5, $K_{p,brain}$ values are shown for the two mouse strains for the three examined brain regions (hippocampus, cortex and cerebellum) without and with ABCG2 inhibition. No significant differences in $K_{p,brain}$ values were found between APP/PS1-21 and wild-type mice in any of the investigated brain regions, either for scans without or for scans with partial ABCG2 inhibition. For both mouse strains, $K_{p,brain}$ values in the hippocampus and cortex were significantly higher after partial ABCG2 inhibition (Figure 5a,b). In the cerebellum, $K_{p,brain}$ was only significantly increased after partial ABCG2 inhibition in APP/PS1-21 mice but not in wild-type mice (Figure 5c). In all three brain regions, the percentage increase in $K_{p,brain}$ of [^{11}C]tariquidar following ABCG2 inhibition was not significantly different between APP/PS1-21 and wild-type mice (APP/PS1-21: 36–52%, wild-type: 26–41%).

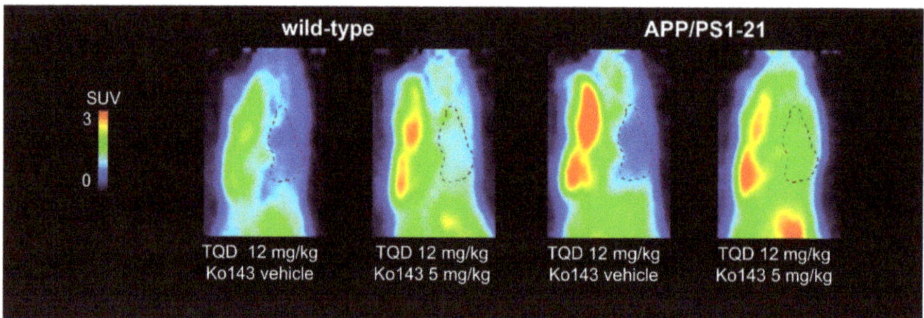

Figure 3. Sagittal (median axis) PET summation images (0–60 min) of 6-months-old APP/PS1-21 mice and age-matched wild-type mice pretreated i.v. with tariquidar (TQD, 12 mg/kg) at 2 h and Ko143 vehicle solution or Ko143 (5 mg/kg) at 1 h prior to [^{11}C]tariquidar PET. Whole brain region is outlined with a red broken line. All images are set to the same intensity scale (0–3 standardized uptake value, SUV).

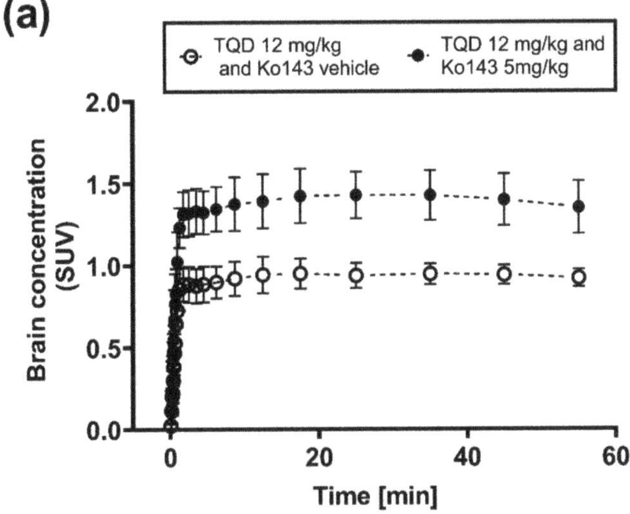

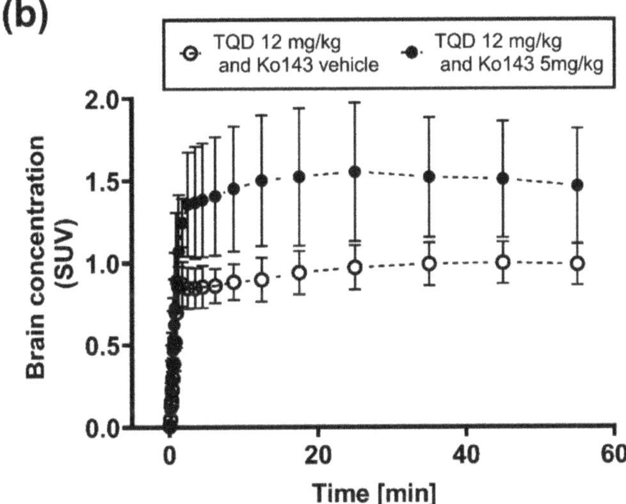

Figure 4. Time-activity curves (mean ± SD) of [^{11}C]tariquidar in whole brains of (**a**) wild-type mice and (**b**) APP/PS1-21 mice pretreated with unlabeled tariquidar (TQD, 12 mg/kg) at 2 h and Ko143 vehicle solution (open circles, wild-type: $n = 6$, APP/PS1-21: $n = 5$) or Ko143 (5 mg/kg, closed circles, wild-type: $n = 7$, APP/PS1-21: $n = 7$) at 1 h prior to PET acquisition.

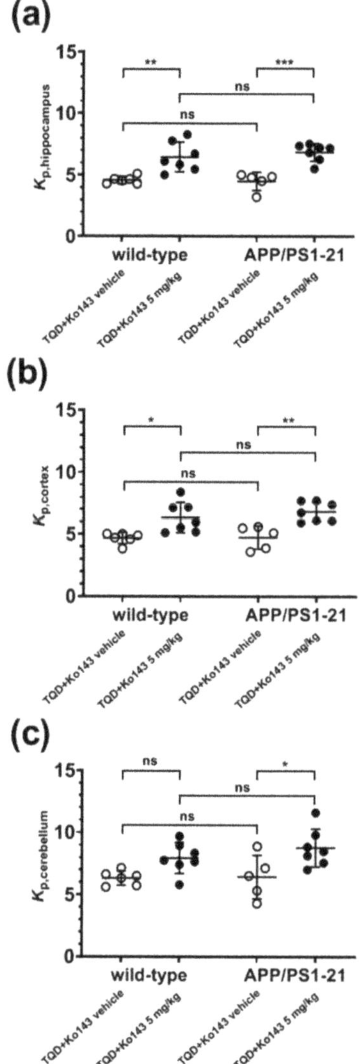

Figure 5. Regional brain-to-plasma radioactivity concentration ratios ($K_{\text{p,brain}}$) in (**a**) hippocampus, (**b**) cortex and (**c**) cerebellum at the end of the [^{11}C]tariquidar PET scan in 6-months-old APP/PS1-21 mice and age-matched wild-type mice pretreated with unlabeled tariquidar (TQD, 12 mg/kg) at 2 h and Ko143 vehicle solution (wild-type: $n = 6$, APP/PS1-21: $n = 5$) or Ko143 (5 mg/kg, wild-type: $n = 7$, APP/PS1-21: $n = 7$) at 1 h prior to the start of the PET scan. (ns, not significant; * $p < 0.05$; ** $p < 0.01$; *** $p < 0.001$; compared to Ko143 vehicle treated animals; one-way ANOVA followed by Tukey's multiple comparison test).

2.3. [^{11}C]Erlotinib PET

To confirm the lack of an effect of the decreased abundance of ABCG2 on the brain distribution of [^{11}C]tariquidar, we performed PET scans with a second ABCB1/ABCG2 substrate radiotracer ([^{11}C]erlotinib) [35]. In contrast to [^{11}C]tariquidar, PET scans were only performed under conditions of full ABCB1/ABCG2 activity, i.e., no ABCB1 or ABCG2 inhibitors were administered. To mimic the therapeutic usage of erlotinib, we co-injected [^{11}C]erlotinib with a pharmacological dose of unlabeled erlotinib (2 mg/kg) [36]. Whole brain TACs of [^{11}C]erlotinib were very similar in the two mouse strains (Figure 6a). Moreover, $K_{p,brain}$ values were not significantly different between APP/PS1-21 and wild-type mice in the three examined brain regions (hippocampus, cortex and cerebellum) (Figure 6b–d).

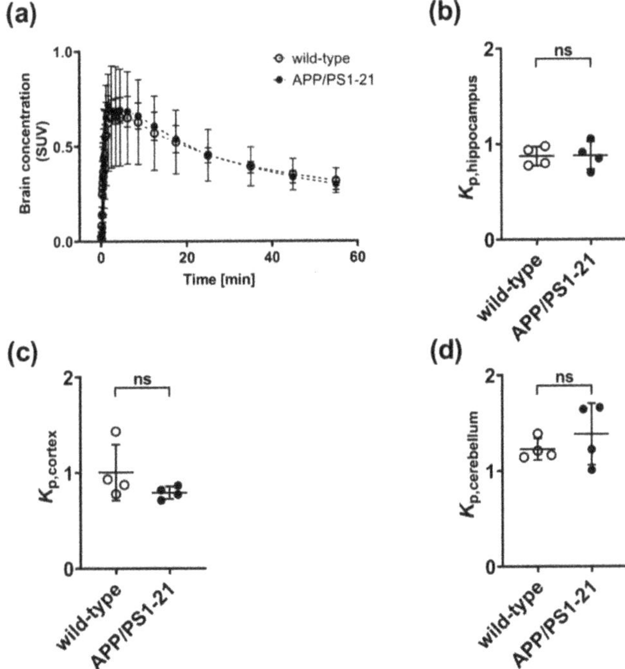

Figure 6. Time-activity curves (mean ± SD) of [^{11}C]erlotinib in whole brains of 6-months-old APP/PS1-21 mice (closed circles, $n = 4$) and age-matched wild-type mice (open circles, $n = 4$) (**a**). Regional brain-to-plasma radioactivity concentration ratios ($K_{p,brain}$) in (**b**) hippocampus, (**c**) cortex and (**d**) cerebellum at the end of the [^{11}C]erlotinib PET scan in APP/PS1-21 mice and wild-type mice (ns, not significant; 2-sided *t*-test).

3. Discussion

ABCG2 was shown to be co-localized with ABCB1 at the BBB, where both transporters limit the brain distribution of many therapeutic drugs [19,37]. Studies in transgenic mice have provided evidence for functional redundancy between ABCB1 and ABCG2 at the BBB [19,37]. In the absence of ABCB1, the transport capacity of ABCG2 usually suffices to restrict the brain distribution of dual ABCB1/ABCG2 substrates and vice versa. Only when both transporters are genetically knocked out or pharmacologically inhibited, do dual ABCB1/ABCG2 substrates show unrestricted brain distribution. Both ABCB1 and ABCG2 have been identified as Aβ transporters [5–9,22–24] and the abundance and activity of cerebral ABCB1 were found to be decreased in AD [10–14], which may be caused by an Aβ-induced ubiquitination, internalization and proteasomal degradation of ABCB1 [38]. It is tempting to speculate that a similar functional redundancy between ABCB1 and ABCG2 exists with respect to Aβ export as described for restricting the brain distribution of small-molecule drugs [19,37]. Xiong et al. have shown an upregulation of ABCG2 in the brains of AD patients and AD mouse models (3XTg and Tg-SwDI) by means of immunohistochemistry and Western blot [22]. Moreover, these authors used optical imaging to show that the brain concentration of fluorescent-labeled Aβ after i.v. injection was higher in *Abcg2*$^{(-/-)}$ mice as compared with wild-type mice, which confirmed Aβ transport by mouse ABCG2 [22]. Other studies, however, were not able to confirm an ABCG2 upregulation in AD brains by using either immunohistochemistry [10–12] or quantitative targeted proteomics [25,26].

To further address these questions, in the present study we examined both the abundance and activity of cerebral ABCG2 in a commonly employed β-amyloidosis mouse model (APP/PS1-21) [32]. This mouse model rapidly develops extensive cerebral Aβ deposits from an age of 2 months onwards [32,33] and has been used in a series of previous studies conducted in our laboratory to assess the activity of ABC transporters implicated in brain Aβ clearance with PET [15,33,39]. In a previous study we used immunohistochemical staining to show that the abundance of ABCB1 is decreased in the brains of APP/PS1-21 mice (i.e., in the hippocampus and the cortex) relative to wild-type mice of the same age range [33]. In the present study, using a comparable methodology we found a reduction in the number of ABCG2-stained microvessels in APP/PS1-21 mice (Figures 1 and 2). This reduction was not only found in the hippocampus, a brain region with high Aβ load, but also in the cerebellum, in which Aβ load is negligible [33]. Our analysis was not able to differentiate whether the reduced abundance of ABCG2 in the brains of APP/PS1-21 mice was caused by a decrease in vascular density [40] or by a decrease in the density of the transporter [11]. Therefore, additional experiments with a different methodology (e.g., Western blot analysis of isolated brain microvessels) will be needed to further examine the regional differences in ABCG2 observed in the present study.

To assess the consequences of these parallel reductions in the abundance of both ABCB1 and ABCG2, utilizing PET imaging, we studied the brain distribution of two radiotracers that are dual ABCB1/ABCG2 substrates ([^{11}C]tariquidar and [^{11}C]erlotinib). As a first approach, we used a previously developed PET protocol dedicated to measuring cerebral ABCG2 activity [28–30]. This protocol uses the dual ABCB1/ABCG2 substrate [^{11}C]tariquidar [31] co-administered with a pharmacological dose of unlabeled tariquidar (12 mg/kg), which leads to complete saturation of ABCB1 activity while ABCG2 remains fully active, thereby ABCG2 selectivity is achieved. The attainment of ABCG2 selectivity is enabled by the great difference in half-maximum inhibitory concentrations (IC_{50}) of tariquidar for in vitro inhibition of its own transport by ABCB1 (IC_{50} = 17.1 nM) and ABCG2 (IC_{50} = 310.4 nM) [30]. To reveal the activity of ABCG2, [^{11}C]tariquidar PET scans were performed without and with pretreatment with the ABCG2 inhibitor Ko143 [34] at a dose that only partially inhibits ABCG2. The employed dose of Ko143 (5 mg/kg) was selected based on a previous dose-response curve generated in *Abcb1a/b*$^{(-/-)}$ mice, which provided a half-maximum effect dose of Ko143 of 4.98 mg to enhance brain uptake of [^{11}C]tariquidar [28]. We determined $K_{p,brain}$ as the outcome parameter of the brain distribution of [^{11}C]tariquidar, which was in a similar range in both APP/PS1-21 and wild-type mice after 5 mg/kg Ko143 ($K_{p,brain}$ range: 6–8, see Figure 5) as in *Abcb1a/b*$^{(-/-)}$ mice pretreated with 5 mg/kg Ko143 [28]. Maximum brain uptake of [^{11}C]tariquidar amounted to a $K_{p,brain}$ of approximately 15 in *Abcb1a/b*$^{(-/-)}$ mice pretreated with 15 mg/kg Ko143, which was comparable to the brain uptake of [^{11}C]tariquidar in *Abcb1a/b*$^{(-/-)}$*Abcg2*$^{(-/-)}$ mice [28]. Taken together, the present data as well as the previous Ko143 dose-response data strongly suggest that the dose of Ko143 employed in our study (5 mg/kg) led to only partial ABCG2 inhibition at the mouse BBB. The experimental paradigm of studying transporter activity with a radiolabeled substrate by employing an inhibitor administered at a dose that only partially inhibits the transporter has been successfully employed by our group and by others to measure the activity of ABCB1 in the rodent and human brain with the ABCB1 substrate (*R*)-[^{11}C]verapamil [33,41–44]. For instance, only PET scans after partial inhibition of ABCB1 revealed significant differences in the brain distribution of (*R*)-[^{11}C]verapamil in APP/PS1-21 versus wild-type mice, while no differences were observed in baseline scans without ABCB1 inhibition [33]. In contrast to this previous study, we failed to detect differences in [^{11}C]tariquidar brain distribution between APP/PS1-21 and wild-type mice, both under conditions of full ABCG2 activity and partial ABCG2 inhibition (Figure 5).

To confirm this apparent lack of difference in ABCG2 activity between APP/PS1-21 and wild-type mice, we also performed PET imaging in APP/PS1-21 and wild-type mice with a second dual

ABCB1/ABCG2 substrate radiotracer. For this we used [11C]erlotinib, which is structurally identical to the epidermal growth factor receptor (EGFR)-targeting tyrosine kinase inhibitor erlotinib, a clinically used drug for the treatment of non-small cell lung cancer. Patients with this type of cancer often develop brain metastases, which are difficult to treat with erlotinib, most likely due to its low brain distribution caused by ABCB1/ABCG2-efflux transport at the BBB. We have previously shown that [11C]erlotinib is transported by mouse ABCB1 and ABCG2 and can be used to assess the functional redundancy between ABCB1 and ABCG2 at the mouse BBB [35]. To mimic the clinical use of this drug we co-injected [11C]erlotinib with a pharmacological dose of unlabeled erlotinib (2 mg/kg) [36]. Similar to the results obtained with [11C]tariquidar, we found no differences in the brain distribution of [11C]erlotinib between APP/PS1-21 and wild-type mice (Figure 6).

Our results are in line with and extend previous studies that failed to detect differences in the brain distribution of different drugs (including several ABCB1 substrates) between different β-amyloidosis mouse models and wild-type mice [15,45–47]. For instance, Gustafsson et al. found no differences in the unbound brain-plasma concentration ratios ($K_{p,uu,brain}$) of the ABCB1 substrates paliperidone and digoxin between tg-APP$_{ArcSwe}$ and age-matched control mice [46]. Similarly, Mehta et al. found no changes in the brain distribution of the ABCB1 substrates loperamide, verapamil and digoxin in triple transgenic AD mice harboring three mutant genes (APP_{swe}, $PS\text{-}1_{M146V}$ and tau_{P301L}), despite a reduction in the abundance of ABCB1 in this mouse model as shown by Western blot analysis of isolated cerebral microvessels [45]. This was explained by Mehta et al. by a thickening of the basement membrane of the BBB in AD mice, which may have impeded transcellular diffusion and thereby counteracted the reduction in the abundance of ABCB1. An alternative explanation for these previous findings may be that the changes in ABCB1 abundance in the AD mouse models were too low to cause significant changes in the brain distribution of ABCB1 substrates. ABCB1 is a high-capacity transporter, whose abundance needs to be reduced by >50% to see >2-fold changes in the brain distribution of its substrates [48]. In line with this hypothesis, we have shown that the brain distribution of the ABCB1 substrate radiotracers [11C]*N*-desmethyl-loperamide and (*R*)-[11C]verapamil was increased by only 1.1- and 1.5-fold, respectively, in heterozygous *Abcb1a/b* knockout mice ($Abcb1a/b^{(+/-)}$), which have a 50% reduction in the abundance of ABCB1 at the BBB, as compared with wild-type mice [49]. In contrast, in homozygous *Abcb1a/b* knockout mice ($Abcb1a/b^{(-/-)}$), which completely lack ABCB1, the brain distribution of [11C]*N*-desmethyl-loperamide and (*R*)-[11C]verapamil was increased by 2.8- and 3.9- fold, respectively, relative to wild-type mice [49]. In our present study, we extended previous findings related to ABCB1 substrates [15,45–47] to dual ABCB1/ABCG2 substrates and showed that despite a concomitant reduction in the abundance of cerebral ABCB1 and ABCG2 in APP/PS1-21 mice as revealed by immunohistochemistry, the brain distribution of the dual ABCB1/ABCG2 substrates [11C]tariquidar and [11C]erlotinib was unaltered. While caution is warranted in extrapolating these results to humans, our results suggest that the brain distribution of clinically used ABCB1/ABCG2 substrate drugs may be unaffected by possible disease-induced alterations in transporter abundances at the BBB of AD patients.

Limitations of our study include the low number of animals used for immunohistochemical analysis, our inability to differentiate between a reduction in vascular density and a reduction in the density of ABCG2, and the lack of further experimental data to confirm the reduction in cerebral ABCG2 (e.g., Western blot analysis of ABCG2 in isolated brain microvessels).

4. Materials and Methods

4.1. General

Unless otherwise stated, all chemicals were purchased from Sigma-Aldrich Chemie (Schnelldorf, Germany) or Merck (Darmstadt, Germany) and were of analytical grade and used without further purification. The ABCB1 inhibitor tariquidar dimesylate [50] was obtained from Haoyuan Chemexpress Co. Ltd. (Shanghai, China). The ABCG2 inhibitor Ko143 [34] was purchased from Enzo Life Sciences

AG (Lausen, Switzerland) or MedChemExpress LLC (Monmouth Junction, NJ, USA). Isoflurane was obtained from VIRBAC S.A. (Carros, France). Directly prior to administration, tariquidar dimesylate was freshly dissolved in 2.5% (*w/v*) aqueous (aq.) dextrose solution and injected intravenously (i.v.) into mice at a volume of 4 mL/kg body weight over a period of 2 min. Ko143 was freshly dissolved in ethanol and formulated for i.v. administration in a solution containing 10% (*v/v*) Kolliphor HS 15 and 80% (*v/v*) sterile aqueous saline to a final ethanol concentration of 10% (*v/v*). Formulated Ko143 solution was injected into mice at a volume of 2 mL/kg body weight over a period of 3–5 min.

4.2. Radiotracer Synthesis and Formulation

[^{11}C]Tariquidar was synthesized as described previously [51]. For i.v. injection, [^{11}C]tariquidar was formulated in sterile aqueous saline containing 0.01 % (*w/v*) Tween 80 to an approximate concentration of 370 MBq/mL. Radiochemical purity, as determined by radio-high performance liquid chromatography, was greater than 98%, and molar activity at the end of synthesis was >100 GBq/μmol.

[^{11}C]Erlotinib was synthesized following established procedures [52] with a radiochemical purity of >98% and a molar activity of >100 GBq/μmol at end of synthesis. For administration into mice, [^{11}C]erlotinb was formulated in 0.1 mM hydrochloric acid in sterile aqueous saline solution to yield a concentration of approximately 370 MBq/mL.

4.3. Animals

Female transgenic mice, which express mutated human amyloid precursor protein (APP) and presenilin 1 (PS1) under control of the Thy1-promoter (APP$_{KM670/671NL}$, PS$_{L166P}$) (referred to as APP/PS1-21 mice) [32,33] and age-matched wild-type littermates in a C57BL/6J genetic background were maintained at the University of Oslo and transferred to the imaging site at least three weeks prior to the PET examinations. In total, 50 mice were used in the experiments. All animals were housed in groups of 3–5 animals in individually ventilated (IVC) type III cages under controlled environmental conditions (22 ± 3 °C, 40% to 70% humidity, 12-h light/dark cycle) and had free access to standard laboratory animal diet (ssniff R/M-H, ssniff Spezialdiäten GmbH, Soest, Germany) and water ad libitum. The study was reviewed by the responsible national authorities (Amt der Niederösterreichischen Landesregierung) and approved under study numbers LF1-TVG-48/003-2014 approval date: 08 January 2015 and LF1-TVG-48/044-2019 approval date: 07 May 2019. All study procedures were in accordance with the European Community's Council Directive of September 22, 2010 (2010/63/EU). The animal experimental data reported in this study are in compliance with the ARRIVE (Animal Research: Reporting In Vivo Experiments) guidelines.

4.4. Experimental Design and Pretreatment

Female APP/PS1-21 and wild-type animals were assigned to the respective groups as shown in Table 1. Two PET imaging approaches for assessing the activity of ABCG2 at the mouse BBB were performed using the two ABCB1/ABCG2 substrate radiotracers [^{11}C]tariquidar [28,31] and [^{11}C]erlotinib [35]:

(1) For PET imaging using [^{11}C]tariquidar, groups of APP/PS1-21 and wild-type mice aged 185 ± 5 days and weighing 27.2 ± 3.0 g were pretreated two hours prior to PET with unlabeled tariquidar administered i.v. at a dose of 12 mg/kg in awake condition. Subsequently, animals were anesthetized using isoflurane/air, a catheter (Instech Lab. Inc., Plymouth Meeting, PA, USA) was introduced into a lateral tail vein and Ko143 at a dose of 5 mg/kg or Ko143 vehicle solution was additionally administered to the animals i.v. one hour prior to the start of PET acquisition. Subsequently, mice underwent a 60-min dynamic [^{11}C]tariquidar PET scan. The doses of tariquidar and Ko143 were selected based on previous work to achieve full inhibition of ABCB1 [49] and partial inhibition of ABCG2 [28] at the mouse BBB.

(2) For PET imaging using [^{11}C]erlotinib, no pretreatment was applied. Groups of APP/PS1-21 and wild-type mice aged 177 ± 2 days and weighing 25.1 ± 2.9 g were prepared for imaging following

the procedures described below and underwent a 60-min dynamic [11C]erlotinib PET scan, in which a pharmacological dose of unlabeled erlotinib (2 mg/kg) was co-injected with [11C]erlotinib [36].

Table 1. Overview of examined animal groups and numbers.

[11C]Tariquidar Study		Animal Group	
Pretreatment		Wild-Type	APP/PS1-21
Tariquidar (12 mg/kg) and Ko143 vehicle	*n*	10 (4) [1]	7 (2)
	Age [days] [2]	189 ± 4	188 ± 1
	Body weight [g] [2]	26.7 ± 2.1	27.9 ± 3.2
	Injected activity [MBq] [3]	28 ± 6	25 ± 3
Tariquidar (12 mg/kg) and Ko143 (5 mg/kg)	*n*	11 (4)	12 (5)
	Age [days]	181 ± 5	183 ± 6
	Body weight [g]	28.9 ± 3.3	25.3 ± 2.4
	Injected activity [MBq]	24 ± 7	23 ± 10
[11C]Erlotinib Study		Animal Group	
		Wild-Type	APP/PS1-21
	n	5 (1)	5 (1)
	Age [days]	176 ± 1	179 ± 2
	Body weight [g]	25.1 ± 1.2	25.2 ± 4.3
	Injected activity [MBq]	19 ± 3	26 ± 4

[1] Number in parentheses indicates drop-outs due to death/intolerability of the protocol; [2] At date of PET examination; [3] Injected amounts of radioactivity.

4.5. PET Imaging Procedure

For PET imaging, mice were pre-anesthetized in an induction chamber using isoflurane and then positioned in a prone position on a double imaging chamber (m2m Imaging Corp, Cleveland, OH, USA). Two mice were imaged simultaneously during one PET acquisition. Animals were warmed throughout the experiment and body temperature and respiratory rate were constantly monitored (SA Instruments Inc, Stony Brook, NY, USA). The level of isoflurane concentration was adjusted (range 1.5–3% in air) during the imaging procedure to achieve a constant level of anesthesia. The imaging chamber was positioned in the gantry of a microPET scanner (Focus 220, Siemens Medical Solutions, Knoxville, TN, USA) and a 10 min transmission scan using a rotating ^{57}Co point source was recorded. Subsequently, 60-min dynamic emission scans (energy window 250–750 keV; timing window = 6 ns) were started with the injection of either [11C]tariquidar (25 ± 7 MBq) or [11C]erlotinib (22 ± 5 MBq) injected i.v. at a volume of 0.1 mL over a period of 120 s.

After completion of the PET scan, a blood sample (20–30 µL) was collected from the retro-orbital venous plexus, the mice were sacrificed by cervical dislocation and transcardially perfused using 10 mL phosphate-buffered saline. Blood was centrifuged (13,000× *g*, 4 °C, 4 min) to obtain plasma and whole brains were removed and processed for immunohistochemistry as described below. Aliquots of blood and plasma were transferred into pre-weighted test tubes and measured in a gamma-counter (HIDEX AMG Automatic Gamma Counter, Turku, Finland). Filled tubes were weighed to obtain tissue weight. The gamma-counter was calibrated using a series of tubes with decreasing activity of a ^{11}C-solution. The measured radioactivity data were decay-corrected to the time of radiotracer injection and expressed as standardized uptake value (SUV), which is calculated as follows: (radioactivity per g (kBq/g)/injected radioactivity (kBq)) × body weight (g).

4.6. PET Data Analysis

The dynamic PET data were binned into 23 frames, which increased incrementally in time length. PET images were reconstructed using Fourier re-binning of the 3-dimensional sinograms followed by a

2-dimensional filtered back-projection with a ramp filter giving a voxel size of $(0.4 \times 0.4 \times 0.796)$ mm^3. Using Amide (version 1.0.4.) [53] or PMOD software (version 3.6, PMOD Technologies Ltd., Zurich, Switzerland) volumes of interest (VOI) covering the whole brain as well as the cortex, hippocampus and cerebellum region were outlined on the PET images with the aid of the Mirrione Mouse Atlas and guided by representative magnetic resonance (MR) images obtained in comparable animals on a 1 Tesla benchtop MR scanner (ICON, Bruker BioSpin GmbH, Ettlingen, Germany). From the VOIs, time-activity curves (TACs) were derived and expressed in SUV units. Whole brain and regional brain uptake was expressed as the brain-to-plasma radioactivity concentration ratio in the last PET frame ($K_{p,brain}$) [28]. For calculation of $K_{p,brain}$, the radioactivity concentration derived from the last PET time frame (from 50–60 min after injection) was divided by the radioactivity concentration measured with the gamma-counter in the plasma sample obtained after the PET scan.

4.7. Immunohistochemistry

Freshly harvested brains from animals that had undergone [^{11}C]erlotinib PET scans were immediately washed with 30% aq. sucrose solution and embedded in freezing medium (Tissue-Tek, Sakura Finetek, Staufen, Germany). Samples were snap frozen in liquid nitrogen and stored at −80 °C until further processed.

After defrosting from −80 °C to −20 °C brains were cut in transversal orientation in 10 µm thick slices using a cryostat (Microm HM 550, Walldorf, Germany). Frozen sections of three brain regions including the hippocampus, cortex, and cerebellum were mounted on coated slides (VWR Superfrost Plus) and stored at −80 °C until the staining procedure was initiated.

For immunohistochemical staining of mouse ABCG2, the thawed brain slices were fixed with methanol/acetone (1:1) for 10 min at 4 °C. After washing with 0.1 M tris-buffered saline solution (TBS), endogenous peroxidase activity was quenched by incubation with 0.5% (v/v) hydrogen peroxide in TBS for 30 min. In order to obtain standardized staining results, the slides were washed and inserted into cover plates (Thermo Scientific™ Shandon™ Glass Coverplates, Fisher Scientific, Vienna, Austria). Blocking solution was added for 1 h at room temperature to suppress non-specific reactions. Anti-BCRP/ABCG2 antibody (1:400, [BXP-53, ab24115], Abcam) or antibody carrier solution (negative control) was used to incubate the brain slices overnight at 4 °C. After washing the slides with TBS, slides were incubated with the secondary antibody (1:500, Biotin-SP (long spacer) AffiniPure Donkey Anti-Rat IgG (H + L), Jackson Immuno Research, West Grove, PA, USA) for 60 min at room temperature. Following three further washing steps, the antibody signals were amplified for 60 min at room temperature employing the VectaStain ABC-Kit (Vector Laboratories Inc, Burlingame, CA, USA). After rinsing, the slides were incubated in nickel/diaminobenzidine solution for 10 min to visualize ABCG2. The slides were then washed, dehydrated and mounted using Entellan$^®$ (Merck Darmstadt, Germany).

The mounted slides were scanned at (0.11×0.11) µm/pixel resolution using a digital slide scanner (Pannoramic Desk, 3dHistech Ltd., Budapest, Hungary) and digital images of comparable regions in the hippocampus, cortex and cerebellum of APP/PS1-21 and wild-type mice were extracted. For the semi-quantitative evaluation of stained microvessels, four visual fields (20× digital magnification) in the same brain region per mouse ($n = 3$ animals per group) were counted manually on the digital images by the same operator. The operator was not blinded to the study groups. However, the digital images were selected randomly for manual counting to avoid bias. Due to high background staining automatic analysis was not possible.

4.8. Statistical Analysis

To analyze differences between two groups a 2-sided *t*-test and between multiple groups a one-way ANOVA followed by a Tukey's multiple comparison test were employed using Prism 8 software (GraphPad Software, La Jolla, CA, USA). The level of statistical significance was set to $p < 0.05$. All values are given as mean ± standard deviation (SD).

5. Conclusions

Despite significant reductions in the abundance of the Aβ transporters ABCB1 and ABCG2 in the brains of APP/PS1-21 mice, the brain distribution of the dual ABCB1/ABCG2 substrates [^{11}C]tariquidar and [^{11}C]erlotinib was unaltered relative to wild-type mice. This may be related to the high transport capacities of ABCB1 and ABCG2 and is in line and extends previous studies, which failed to detect differences in the brain distribution of diverse ABCB1 substrates between different AD mouse models and wild-type mice. While caution is warranted in extrapolating our results to the human BBB, our findings suggest that while disease-induced alterations in the abundance of ABCB1 and ABCG2 may be sufficient to decrease the brain clearance of Aβ peptides, they may not be sufficient to cause large changes in the brain distribution of clinically used ABCB1/ABCG2 substrate drugs.

Author Contributions: Conceptualization, T.W. and O.L.; methodology, T.W., S.M., M.K., M.B., T.P. and T.F.; PET investigation, T.F., M.S., V.Z., S.M. and J.S.; data analysis, V.Z. and T.W.; writing—original draft preparation, T.W. and O.L.; writing—review and editing, V.Z., T.F., T.P. and J.P.; visualization, V.Z. and T.W.; supervision, T.W. and O.L.; project administration, T.W.; funding acquisition, T.W., J.P. and O.L. All authors have read and agreed to the published version of the manuscript.

Funding: This research was funded by the Lower Austria Corporation for Research and Education (NFB) [grant number LS14-008, to T.W.], the Austrian Science Fund (FWF) [grant number I 1609-B24, to O. L.] and the Deutsche Forschungsgemeinschaft (DFG) [grant number DFG PA930/9, to J.P.]. The work of J.P. was also supported by the following grants: Deutsche Forschungsgemeinschaft/Germany (DFG PA930/12); Ministerium für Wirtschaft und Wissenschaft Sachsen-Anhalt/Germany (ZS/2016/05/78617); Leibniz Gemeinschaft/Germany (SAW-2015-IPB-2); Latvian Council of Science/Latvia (lzp-2018/1-0275); Nasjonalforeningen (16154), HelseSØ/Norway (2016062, 2019054, 2019055); Barnekreftforeningen (19008); Norges forskningsrådet/Norway (251290, 260786 (*PROP-AD*), 295910). *PROP-AD* is an EU Joint Programme-Neurodegenerative Disease Research (JPND) project and has received funding from the European Union's Horizon 2020 research and innovation programme under grant agreement #643417 (JPco-fuND).

Acknowledgments: The authors wish to thank Mathilde Löbsch for help in conducting the PET experiments and Alina Zwickl, Jaqueline Koller, Marlies Nemeth and Monika Kühteubl (University of Applied Sciences, Wiener Neustadt, Austria) for support with immunohistochemistry. Open Access Funding by the Austrian Science Fund (FWF).

Conflicts of Interest: The authors declare no conflict of interest.

Abbreviations

ABC	Adenosine triphopshate-binding cassette
ABCB1	ABC subfamily B member 1, also known as P-glycoprotein (P-gp)
ABCG2	ABC subfamily G member 2, also known as breast cancer resistance protein (BCRP)
AD	Alzheimer's disease
Aβ	Beta-amyloid
BBB	Blood-brain barrier
$K_{p,brain}$	Brain-to-plasma radioactivity concentration ratio
PET	Positron Emission Tomography
SUV	Standardized uptake value
TAC	Time-activity curve

References

1. Mawuenyega, K.G.; Sigurdson, W.; Ovod, V.; Munsell, L.; Kasten, T.; Morris, J.C.; Yarasheski, K.E.; Bateman, R.J. Decreased clearance of CNS beta-amyloid in Alzheimer's disease. *Science* **2010**, *330*, 1774. [CrossRef]

2. Hardy, J.; Allsop, D. Amyloid deposition as the central event in the aetiology of Alzheimer's disease. *Trends Pharmacol. Sci.* **1991**, *12*, 383–388.

3. Pahnke, J.; Wolkenhauer, O.; Krohn, M.; Walker, L.C. Clinico-pathologic function of cerebral ABC transporters—implications for the pathogenesis of Alzheimer's disease. *Curr. Alzheimer Res.* **2008**, *5*, 396–405.

4. Boland, B.; Yu, W.H.; Corti, O.; Mollereau, B.; Henriques, A.; Bezard, E.; Pastores, G.M.; Rubinsztein, D.C.; Nixon, R.A.; Duchen, M.R.; et al. Promoting the clearance of neurotoxic proteins in neurodegenerative disorders of ageing. *Nat. Rev. Drug Discov.* **2018**, *17*, 660–688.

5. Storck, S.E.; Hartz, A.M.S.; Bernard, J.; Wolf, A.; Kachlmeier, A.; Mahringer, A.; Weggen, S.; Pahnke, J.; Pietrzik, C.U. The concerted amyloid-beta clearance of LRP1 and ABCB1/P-gp across the blood-brain barrier is linked by PICALM. *Brain. Behav. Immun.* **2018**, *73*, 21–33.

6. Lam, F.C.; Liu, R.; Lu, P.; Shapiro, A.B.; Renoir, J.M.; Sharom, F.J.; Reiner, P.B. beta-Amyloid efflux mediated by p-glycoprotein. *J. Neurochem.* **2001**, *76*, 1121–1128.

7. Kuhnke, D.; Jedlitschky, G.; Grube, M.; Krohn, M.; Jucker, M.; Mosyagin, I.; Cascorbi, I.; Walker, L.C.; Kroemer, H.K.; Warzok, R.W.; et al. MDR1-P-glycoprotein (ABCB1) mediates transport of Alzheimer's amyloid-beta peptides–implications for the mechanisms of Abeta clearance at the blood-brain barrier. *Brain Pathol.* **2007**, *17*, 347–353.

8. Hartz, A.M.; Miller, D.S.; Bauer, B. Restoring blood-brain barrier P-glycoprotein reduces brain amyloid-beta in a mouse model of Alzheimer's disease. *Mol. Pharmacol.* **2010**, *77*, 715–723.

9. Cirrito, J.R.; Deane, R.; Fagan, A.M.; Spinner, M.L.; Parsadanian, M.; Finn, M.B.; Jiang, H.; Prior, J.L.; Sagare, A.; Bales, K.R.; et al. P-glycoprotein deficiency at the blood-brain barrier increases amyloid-beta deposition in an Alzheimer disease mouse model. *J. Clin. Investig.* **2005**, *115*, 3285–3290.

10. Wijesuriya, H.C.; Bullock, J.Y.; Faull, R.L.; Hladky, S.B.; Barrand, M.A. ABC efflux transporters in brain vasculature of Alzheimer's subjects. *Brain Res.* **2010**, *1358*, 228–238.

11. Kannan, P.; Schain, M.; Kretzschmar, W.W.; Weidner, L.; Mitsios, N.; Gulyas, B.; Blom, H.; Gottesman, M.M.; Innis, R.B.; Hall, M.D.; et al. An automated method measures variability in P-glycoprotein and ABCG2 densities across brain regions and brain matter. *J. Cereb. Blood Flow Metab.* **2017**, *37*, 2062–2075.

12. Carrano, A.; Snkhchyan, H.; Kooij, G.; van der Pol, S.; van Horssen, J.; Veerhuis, R.; Hoozemans, J.; Rozemuller, A.; de Vries, H.E. ATP-binding cassette transporters P-glycoprotein and breast cancer related protein are reduced in capillary cerebral amyloid angiopathy. *Neurobiol. Aging* **2014**, *35*, 565–575.

13. van Assema, D.M.; Lubberink, M.; Bauer, M.; van der Flier, W.M.; Schuit, R.C.; Windhorst, A.D.; Comans, E.F.; Hoetjes, N.J.; Tolboom, N.; Langer, O.; et al. Blood-brain barrier P-glycoprotein function in Alzheimer's disease. *Brain* **2012**, *135 (Pt 1)*, 181–189.

14. Deo, A.K.; Borson, S.; Link, J.M.; Domino, K.; Eary, J.F.; Ke, B.; Richards, T.L.; Mankoff, D.A.; Minoshima, S.; O'Sullivan, F.; et al. Activity of P-glycoprotein, a beta-amyloid transporter at the blood-brain barrier, is compromised in patients with mild Alzheimer disease. *J. Nucl. Med.* **2014**, *55*, 1106–1111.

15. Zoufal, V.; Mairinger, S.; Brackhan, M.; Krohn, M.; Filip, T.; Sauberer, M.; Stanek, J.; Wanek, T.; Tournier, N.; Bauer, M.; et al. Imaging P-glycoprotein induction at the blood-brain barrier of a beta-amyloidosis mouse model with ^{11}C-metoclopramide PET. *J. Nucl. Med.* **2020**, *61*, 1050–1057.

16. Qosa, H.; Abuznait, A.H.; Hill, R.A.; Kaddoumi, A. Enhanced brain amyloid-beta clearance by rifampicin and caffeine as a possible protective mechanism against Alzheimer's disease. *J. Alzheimers Dis.* **2012**, *31*, 151–165.

17. Durk, M.R.; Han, K.; Chow, E.C.; Ahrens, R.; Henderson, J.T.; Fraser, P.E.; Pang, K.S. 1alpha,25-Dihydroxyvitamin D3 reduces cerebral amyloid-beta accumulation and improves cognition in mouse models of Alzheimer's disease. *J. Neurosci.* **2014**, *34*, 7091–7101.

18. Brenn, A.; Grube, M.; Jedlitschky, G.; Fischer, A.; Strohmeier, B.; Eiden, M.; Keller, M.; Groschup, M.H.; Vogelgesang, S. St. John's Wort reduces beta-amyloid accumulation in a double transgenic Alzheimer's disease mouse model-role of P-glycoprotein. *Brain Pathol.* **2014**, *24*, 18–24.

19. Agarwal, S.; Hartz, A.M.; Elmquist, W.F.; Bauer, B. Breast cancer resistance protein and P-glycoprotein in brain cancer: Two gatekeepers team up. *Curr. Pharm. Des.* **2011**, *17*, 2793–2802.

20. Durmus, S.; Hendrikx, J.J.; Schinkel, A.H. Apical ABC transporters and cancer chemotherapeutic drug disposition. *Adv. Cancer Res.* **2015**, *125*, 1–41.

21. Vogelgesang, S.; Warzok, R.W.; Cascorbi, I.; Kunert-Keil, C.; Schroeder, E.; Kroemer, H.K.; Siegmund, W.; Walker, L.C.; Pahnke, J. The role of P-glycoprotein in cerebral amyloid angiopathy; implications for the early pathogenesis of Alzheimer's disease. *Curr. Alzheimer Res.* **2004**, *1*, 121–125.

22. Xiong, H.; Callaghan, D.; Jones, A.; Bai, J.; Rasquinha, I.; Smith, C.; Pei, K.; Walker, D.; Lue, L.F.; Stanimirovic, D.; et al. ABCG2 is upregulated in Alzheimer's brain with cerebral amyloid angiopathy and may act as a gatekeeper at the blood-brain barrier for Abeta(1-40) peptides. *J. Neurosci.* **2009**, *29*, 5463–5475.

23. Tai, L.M.; Loughlin, A.J.; Male, D.K.; Romero, I.A. P-glycoprotein and breast cancer resistance protein restrict apical-to-basolateral permeability of human brain endothelium to amyloid-beta. *J. Cereb. Blood Flow Metab.* **2009**, *29*, 1079–1083.

24. Do, T.M.; Noel-Hudson, M.S.; Ribes, S.; Besengez, C.; Smirnova, M.; Cisternino, S.; Buyse, M.; Calon, F.; Chimini, G.; Chacun, H.; et al. ABCG2- and ABCG4-mediated efflux of amyloid-beta peptide 1-40 at the mouse blood-brain barrier. *J. Alzheimers Dis.* **2012**, *30*, 155–166.

25. Storelli, F.; Billington, S.; Kumar, A.R.; Unadkat, J.D. Abundance of P-glycoprotein and other drug transporters at the human blood-brain barrier in Alzheimer's Disease: A quantitative targeted proteomic study. *Clin. Pharmacol. Ther.* **2020**. [CrossRef]

26. Al-Majdoub, Z.M.; Al Feteisi, H.; Achour, B.; Warwood, S.; Neuhoff, S.; Rostami-Hodjegan, A.; Barber, J. Proteomic quantification of human blood-brain barrier SLC and ABC transporters in healthy individuals and dementia patients. *Mol. Pharm.* **2019**, *16*, 1220–1233.

27. Wanek, T.; Mairinger, S.; Langer, O. Radioligands targeting P-glycoprotein and other drug efflux proteins at the blood–brain barrier. *J. Labelled Comp. Radiopharm.* **2013**, *56*, 68–77.

28. Wanek, T.; Kuntner, C.; Bankstahl, J.P.; Mairinger, S.; Bankstahl, M.; Stanek, J.; Sauberer, M.; Filip, T.; Erker, T.; Müller, M.; et al. A novel PET protocol for visualization of breast cancer resistance protein function at the blood-brain barrier. *J. Cereb. Blood Flow Metab.* **2012**, *32*, 2002–2011.

29. Dallas, S.; Salphati, L.; Gomez-Zepeda, D.; Wanek, T.; Chen, L.; Chu, X.; Kunta, J.; Mezler, M.; Menet, M.C.; Chasseigneaux, S.; et al. Generation and characterization of a breast cancer resistance protein humanized mouse model. *Mol. Pharmacol.* **2016**, *89*, 492–504.

30. Bauer, M.; Römermann, K.; Karch, R.; Wulkersdorfer, B.; Stanek, J.; Philippe, C.; Maier-Salamon, A.; Haslacher, H.; Jungbauer, C.; Wadsak, W.; et al. Pilot PET study to assess the functional interplay between ABCB1 and ABCG2 at the human blood-brain barrier. *Clin. Pharmacol. Ther.* **2016**, *100*, 131–141.

31. Bankstahl, J.P.; Bankstahl, M.; Römermann, K.; Wanek, T.; Stanek, J.; Windhorst, A.D.; Fedrowitz, M.; Erker, T.; Müller, M.; Löscher, W.; et al. Tariquidar and elacridar are dose-dependently transported by p-glycoprotein and bcrp at the blood-brain barrier: A small-animal positron emission tomography and in vitro study . *Drug Metab. Dispos.* **2013**, *41*, 754–762.

32. Radde, R.; Bolmont, T.; Kaeser, S.A.; Coomaraswamy, J.; Lindau, D.; Stoltze, L.; Calhoun, M.E.; Jaggi, F.; Wolburg, H.; Gengler, S.; et al. Abeta42-driven cerebral amyloidosis in transgenic mice reveals early and robust pathology. *EMBO Rep* **2006**, *7*, 940–946.

33. Zoufal, V.; Wanek, T.; Krohn, M.; Mairinger, S.; Filip, T.; Sauberer, M.; Stanek, J.; Pekar, T.; Bauer, M.; Pahnke, J.; et al. Age dependency of cerebral P-glycoprotein function in wild-type and APPPS1 mice measured with PET. *J. Cereb. Blood Flow Metab.* **2020**, *40*, 150–162.

34. Allen, J.D.; van Loevezijn, A.; Lakhai, J.M.; van der Valk, M.; van Tellingen, O.; Reid, G.; Schellens, J.H.; Koomen, G.J.; Schinkel, A.H. Potent and specific inhibition of the breast cancer resistance protein multidrug transporter in vitro and in mouse intestine by a novel analogue of fumitremorgin C. *Mol. Cancer Ther.* **2002**, *1*, 417–425.

35. Traxl, A.; Wanek, T.; Mairinger, S.; Stanek, J.; Filip, T.; Sauberer, M.; Müller, M.; Kuntner, C.; Langer, O. Breast cancer resistance protein and P-glycoprotein influence in vivo disposition of ^{11}C-erlotinib. *J. Nucl. Med.* **2015**, *56*, 1930–1936.

36. Linder, M.; Glitzner, E.; Srivatsa, S.; Bakiri, L.; Matsuoka, K.; Shahrouzi, P.; Dumanic, M.; Novoszel, P.; Mohr, T.; Langer, O.; et al. EGFR is required for FOS-dependent bone tumor development via RSK2/CREB signaling. *EMBO Mol. Med.* **2018**, *10*, e9408. [CrossRef]

37. Kodaira, H.; Kusuhara, H.; Ushiki, J.; Fuse, E.; Sugiyama, Y. Kinetic analysis of the cooperation of P-glycoprotein (P-gp/Abcb1) and breast cancer resistance protein (Bcrp/Abcg2) in limiting the brain and testis penetration of erlotinib, flavopiridol, and mitoxantrone. *J. Pharmacol. Exp. Ther.* **2010**, *333*, 788–796.

38. Hartz, A.M.; Zhong, Y.; Wolf, A.; LeVine, H., 3rd; Miller, D.S.; Bauer, B. Abeta40 reduces P-glycoprotein at the blood-brain barrier through the ubiquitin-proteasome pathway. *J. Neurosci.* **2016**, *36*, 1930–1941.

39. Zoufal, V.; Mairinger, S.; Krohn, M.; Wanek, T.; Filip, T.; Sauberer, M.; Stanek, J.; Kuntner, C.; Pahnke, J.; Langer, O. Measurement of cerebral ABCC1 transport activity in wild-type and APP/PS1-21 mice with positron emission tomography. *J. Cereb. Blood Flow Metab.* **2019**, *40*, 954–965.

40. Li, H.C.; Chen, P.Y.; Cheng, H.F.; Kuo, Y.M.; Huang, C.C. In vivo visualization of brain vasculature in Alzheimer's Disease mice by high-frequency micro-Doppler imaging. *IEEE Trans. Biomed. Eng.* **2019**, *66*, 3393–3401.

41. Bauer, M.; Wulkersdorfer, B.; Karch, R.; Philippe, C.; Jager, W.; Stanek, J.; Wadsak, W.; Hacker, M.; Zeitlinger, M.; Langer, O. Effect of P-glycoprotein inhibition at the blood-brain barrier on brain distribution of (*R*)-[^{11}C]verapamil in elderly vs. young subjects. *Br. J. Clin. Pharmacol.* **2017**, *83*, 1991–1999.

42. Bankstahl, J.P.; Bankstahl, M.; Kuntner, C.; Stanek, J.; Wanek, T.; Meier, M.; Ding, X.; Müller, M.; Langer, O.; Löscher, W. A novel PET imaging protocol identifies seizure-induced regional overactivity of P-glycoprotein at the blood-brain barrier. *J. Neurosci.* **2011**, *31*, 8803–8811.

43. Feldmann, M.; Asselin, M.-C.; Liu, J.; Wang, S.; McMahon, A.; Anton-Rodriguez, J.; Walker, M.; Symms, M.; Brown, G.; Hinz, R.; et al. P-glycoprotein expression and function in patients with temporal lobe epilepsy: A case-control study. *Lancet Neurol* **2013**, *12*, 777–785.

44. Shin, J.W.; Chu, K.; Shin, S.A.; Jung, K.H.; Lee, S.T.; Lee, Y.S.; Moon, J.; Lee, D.Y.; Lee, J.S.; Lee, D.S.; et al. Clinical applications of simultaneous PET/MR imaging using (*R*)-[^{11}C]-verapamil with cyclosporin A: Preliminary results on a surrogate marker of drug-resistant epilepsy. *Am. J. Neuroradiol.* **2016**, *37*, 600–606.

45. Mehta, D.C.; Short, J.L.; Nicolazzo, J.A. Altered brain uptake of therapeutics in a triple transgenic mouse model of Alzheimer's disease. *Pharm. Res.* **2013**, *30*, 2868–2879.

46. Gustafsson, S.; Lindström, V.; Ingelsson, M.; Hammarlund-Udenaes, M.; Syvänen, S. Intact blood-brain barrier transport of small molecular drugs in animal models of amyloid beta and alpha-synuclein pathology. *Neuropharmacology* **2018**, *128*, 482–491.

47. Cheng, Z.; Zhang, J.; Liu, H.; Li, Y.; Zhao, Y.; Yang, E. Central nervous system penetration for small molecule therapeutic agents does not increase in multiple sclerosis- and Alzheimer's disease-related animal models despite reported blood-brain barrier disruption. *Drug Metab. Dispos.* **2010**, *38*, 1355–1361.

48. Kalvass, J.C.; Polli, J.W.; Bourdet, D.L.; Feng, B.; Huang, S.M.; Liu, X.; Smith, Q.R.; Zhang, L.K.; Zamek-Gliszczynski, M.J. Why clinical modulation of efflux transport at the human blood-brain barrier is unlikely: The ITC evidence-based position. *Clin. Pharmacol. Ther.* **2013**, *94*, 80–94.

49. Wanek, T.; Römermann, K.; Mairinger, S.; Stanek, J.; Sauberer, M.; Filip, T.; Traxl, A.; Kuntner, C.; Pahnke, J.; Bauer, F.; et al. Factors governing p-glycoprotein-mediated drug-drug interactions at the blood-brain barrier measured with positron emission tomography. *Mol. Pharm.* **2015**, *12*, 3214–3225.

50. Fox, E.; Bates, S.E. Tariquidar (XR9576): A P-glycoprotein drug efflux pump inhibitor. *Expert Rev. Anticancer Ther.* **2007**, *7*, 447–459.

51. Bauer, F.; Kuntner, C.; Bankstahl, J.P.; Wanek, T.; Bankstahl, M.; Stanek, J.; Mairinger, S.; Dörner, B.; Löscher, W.; Müller, M.; et al. Synthesis and in vivo evaluation of [^{11}C]tariquidar, a positron emission tomography radiotracer based on a third-generation P-glycoprotein inhibitor. *Bioorg. Med. Chem.* **2010**, *18*, 5489–5497.

52. Philippe, C.; Mairinger, S.; Pichler, V.; Stanek, J.; Nics, L.; Mitterhauser, M.; Hacker, M.; Wanek, T.; Langer, O.; Wadsak, W. Comparison of fully-automated radiosyntheses of [^{11}C]erlotinib for preclinical and clinical use starting from in target produced [^{11}C]CO$_2$ or [^{11}C]CH$_4$. *EJNMMI Radiopharm. Chem.* **2018**, *3*, 8. [CrossRef]

53. Loening, A.M.; Gambhir, S.S. AMIDE: A free software tool for multimodality medical image analysis. *Mol. Imaging* **2003**, *2*, 131–137.

Publisher's Note: MDPI stays neutral with regard to jurisdictional claims in published maps and institutional affiliations.

International Journal of
Molecular Sciences

Review

Role of ABCA7 in Human Health and in Alzheimer's Disease

Shiraz Dib [1], Jens Pahnke [2,3,4,5] and Fabien Gosselet [1,*]

[1] UR2465, LBHE–Blood–Brain Barrier Laboratory, University Artois, 62300 Lens, France; shiraz_deeb@hotmail.com
[2] Department of Neuro-/Pathology, University of Oslo and Oslo University Hospital, Sognsvannsveien 20, 0372 Oslo, Norway; jens.pahnke@medisin.uio.no
[3] LIED, University of Lübeck, Ratzenburger Allee 160, 23538 Lübeck, Germany
[4] Department of Pharmacology, Faculty of Medicine, University of Latvia, Jelgavas iela 3, 1004 Riga, Latvia
[5] Department of Bioorganic Chemistry, Leibniz-Institute of Plant Biochemistry, Weinberg 3, 06120 Halle, Germany
* Correspondence: fabien.gosselet@univ-artois.fr; Tel.: +33-(0)3-21791733

Abstract: Several studies, including genome wide association studies (GWAS), have strongly suggested a central role for the ATP-binding cassette transporter subfamily A member 7 (ABCA7) in Alzheimer's disease (AD). This ABC transporter is now considered as an important genetic determinant for late onset Alzheimer disease (LOAD) by regulating several molecular processes such as cholesterol metabolism and amyloid processing and clearance. In this review we shed light on these new functions and their cross-talk, explaining its implication in brain functioning, and therefore in AD onset and development.

Keywords: ABCA7; Alzheimer's disease; phagocytosis; cholesterol; Aβ peptides

Citation: Dib, S.; Pahnke, J.; Gosselet, F. Role of ABCA7 in Human Health and in Alzheimer's Disease. *Int. J. Mol. Sci.* **2021**, *22*, 4603. https://doi.org/10.3390/ijms22094603

Academic Editor: Thomas Falguières

Received: 29 March 2021
Accepted: 26 April 2021
Published: 27 April 2021

Publisher's Note: MDPI stays neutral with regard to jurisdictional claims in published maps and institutional affiliations.

1. Introduction

ATP-binding cassette (ABC) transporters are present in all living organisms. In humans, this family comprises 49 members that are classified based on their sequence homologies and their conserved nucleotide binding domains (NBDs) into seven subfamilies, designated from A to G. These transporters play a significant role by transporting ions, small organic or inorganic molecules, peptides, proteins and lipids from a cellular compartment to another one or across the cell membranes [1,2]. Importantly, their physiological functions strongly depend on their cellular localization (membrane, endoplasmic reticulum, Golgi, etc.) as well as the organ and cell type in which they are expressed. For example, members of the ABCB and ABCC subfamilies are expressed in physiological barriers such as intestine and blood–brain interfaces to restrict the entry or mediate the exit of harmful endogenous molecules or xenobiotics [3,4]. Overexpression of these ABC transporters has been reported in several cancers and is responsible for the multidrug resistance (MDR) phenotype of cancer cells [5]. Members of the ABCG and A subfamilies mediate cholesterol and lipid efflux from cells to low- or high-density lipoproteins and are highly expressed by cells involved in lipid homeostasis such as macrophages or hepatocytes.

It is therefore not surprising that ABC transporter deficiencies lead to major diseases including but not limited to atherosclerosis, cystic fibrosis or neurodegenerative diseases such as Alzheimer's disease (AD) [2,6,7]. AD is the most common form of dementia in the world. This disease is mainly characterized by brain atrophy as a consequence of the deposition of amyloid-β (Aβ) peptides and tau protein hyperphosphorylation and aggregation [8]. Since the early 2000s, several ABC transporters such as ABCB1, ABCA1, ABCG2, ABCC1 and ABCG4 are under scrutiny in AD because of their in vitro and in vivo involvement in Aβ peptide synthesis and/or its deposition. These different contributions in the AD field are already summarized elsewhere [4,7,9,10].

The ATP-Binding cassette transporter subfamily A member 7 (ABCA7) was first identified in the early 2000s [11]. As shown in Figure 1, ABCA7 was poorly studied until

2011. These studies mainly investigated ABCA7 involvement in peripheral cholesterol metabolism and phagocytosis processes. In 2011, Hollingworth et al. highlighted a very strong association between ABCA7 polymorphisms and AD patients in a genome-wide association study (GWAS) [12].

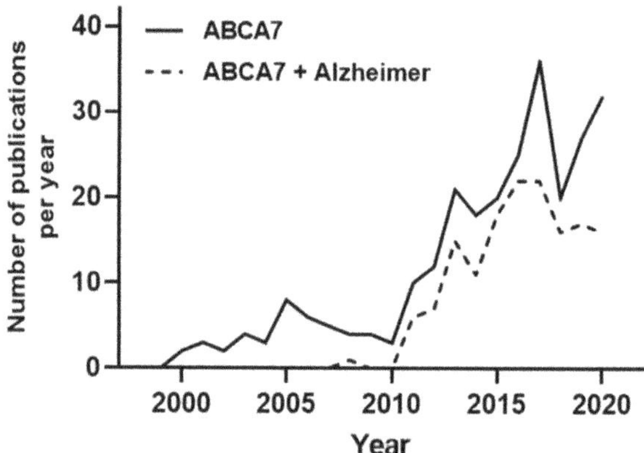

Figure 1. Number of searchable publications in Pubmed per year by the terms specified in the legend. For the "ABCA7 + Alzheimer" search, a second analysis of the obtained publications was performed to exclusively select studies and review investigating the role of ABCA7 in AD. Data source: Pubmed. Search date 31/12/2020. Y-axis shows number of publications, X-axis shows years

Since then, the number of projects studying ABCA7 have been considerably increased, and almost all of them have been realized with the objective to decipher the molecular role of ABCA7 in AD (Figure 1).

This review aims to summarize the first roles of ABCA7 identified in the periphery, but also to describe the most recent findings demonstrating its significant contribution in brain functioning and in AD onset and development.

2. ABCA7 Expression Pattern and Structure

2.1. Tissue Localization of ABCA7

ABCA7 was first cloned in 2000 by Kaminski and colleagues from human macrophages [11]. Distribution/expression of the ABCA7 mRNA in human tissues demonstrated that it is highly expressed in myelo-lymphatic tissues (peripheral leucocytes, thymus, spleen, bone marrow, fetal tissues) [11]. Preferential and high expression of *Abca7* mRNA in lymphomyeloid tissues was immediately confirmed in mice and rats [13,14], strongly suggesting a key role of this transporter in hematopoietic cell lineages. More recently, analysis based upon ImmGen Deep RNA-seq data showed that *Abca7* is one of the most highly expressed ABC transporter in a purified population of mouse immune cells like follicular B cells, NK cells, peritoneal macrophages, thus highlighting a role for ABCA7 in immunity [15]. In addition, it appears that ABCA7 expression is dependent of the differentiation state of the cells since higher expressions were measured in differentiated macrophages comparing to monocytes [11].

Interestingly, no ABCA7 mRNA signal was initially detected in total human brain samples but this result was later refuted by several groups demonstrating mRNA and protein expression of ABCA7 in human and murine samples of the different cell types present in the brain [16–23]. Therefore, it is now largely demonstrated that ABCA7 is expressed in neurons, astrocytes, microglia, endothelial cells of the blood–brain barrier (BBB), brain pericytes, not only in mice but also in humans ([16–24]—http://www.brainrnaseq.org/, http://www.celltypes.org/ and https://www.proteinatlas.org/ENSG00000064687-ABCA7/brain, accessed on 31 December 2020).

2.2. ABCA7 Alternative Splicing

There are two isoforms of ABCA7, arising as a result of alternative splicing, both of which are detected in the human brain [21,25]. The splicing variant is named Type II ABCA7 and gives a protein with 28 amino acids in the N terminal tail instead of the 166 amino acids present in the full-length ABCA7. These variants show a tissue-dependent expression pattern and differential cellular sub-localization. While full length ABCA7 has been detected on the cell surface and intracellularly, the so-called Type II ABCA7 splice variant is only detected in the endoplasmic reticulum [21]. Northern blot analysis performed with human tissue samples showed that Type II ABCA7 is abundant in lymph node, spleen, thymus and trachea whereas full ABCA7 is strongly expressed in brain and bone marrow [21]. Thus, it is possible that these variant proteins have different biological functions.

2.3. ABCA7 Structure

Functional ABC transporters are commonly composed of four domains—two transmembrane domains (TMDs) connected with two cytoplasmic nucleotide-binding domains (NBDs). TMDs are variable in sequence and are responsible for substrate interaction and translocation. NBDs are highly conserved and have characteristic motifs such as Walker A and Walker B, allowing to bind and hydrolyze ATP. As shown in Figure 2, full ABC transporters like transporters of subfamilies A and C have 2 TMDs. ABC transporters of the D and G subfamilies such as ABCG1 or ABCG2 display only one TMD and one NBD. They are therefore considered as half-transporters and have to homo- or hetero-dimerize to be functional. ABC transporters are classified based on their NBD domain but it was recently suggested that it would be more pertinent to use a TMD-based classification system [1].

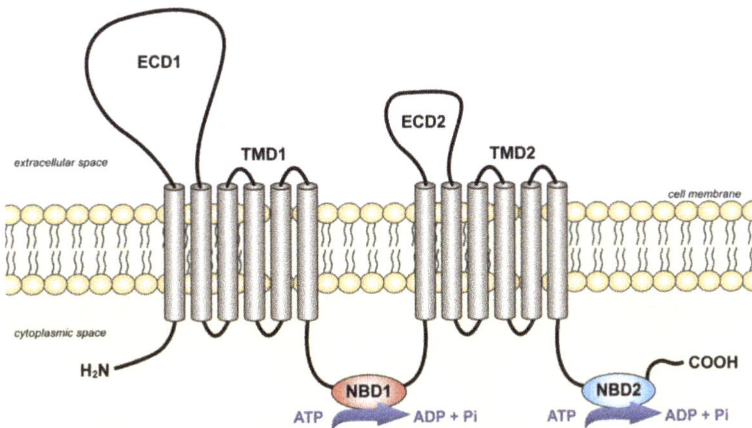

Figure 2. Schematic representation of the full ABC transporters of the subfamily A. Two transmembrane domains (TMDs) are connected with two cytoplasmic nucleotide-binding domains (NBDs). In addition, members of subfamily A are characterized by two large extracellular domains (ECD1 and ECD2). Reprinted from [26] with permission from Elsevier.

ABCA7 is predicted to be a full ABC transporter, as presented in Figure 2. The transport mechanism of these transporters is controlled by alternating the conformation of TMD, through which the transporter switches between inward- and outward-facing states. Recently, alternative mechanisms have been proposed or identified, suggesting that this process remains under debate [27]. In all cases, the exact transport mechanism of ABCA7 is uncharacterized, and, despite the fact that single-particle cryoelectron microscopy (cryo-EM) structure of ABCA1 has been recently obtained [28], the structure of ABCA7 is still unknown.

Sequence comparisons made with other ABC transporters showed high ABCA7 homology with ABCA1 (54%), ABCA2 (45%), ABCA3 (41%) and ABCA4 (49%) [11]. Because these latter are involved in lipid transport, it was rapidly assumed that ABCA7 was also able to transfer lipid from cell membranes to lipoproteins.

3. ABCA7 and the Lipid Metabolism

3.1. Role of ABCA7 in Lipid Release and Trafficking

As indicated above, ABCA7 shares high amino sequence homology with other ABC transporters mediating lipid release from cell membranes, and in particular with ABCA1. This latter is the most studied ABC transporter in this cellular process and its role in cholesterol efflux to high-density lipoproteins (HDL) is very well characterized [29–31]. Mutations in the *ABCA1* gene cause an inherited disease named Tangier disease in which the cholesterol transfer from cells to HDL is decreased, leading to lower ApoA-I synthesis as well as low levels of plasmatic HDL [32]. Consequently, this lipid homeostasis imbalance promotes lipid accumulation inside vascular cells and macrophages, thus provoking coronary artery disease and atherosclerosis in patients [27,33,34].

On this basis, $Abca7^{-/-}$ mice were generated to further characterize the role of this ABC transporter in cholesterol metabolism. Interestingly, only transgenic female mice showed slight abnormalities in their cholesterol metabolism when compared with the wild-type littermates [16]. Indeed, female $Abca7^{-/-}$ mice showed weight gain identical to female controls but less white adipose tissue, plasmatic HDL and cholesterol [16]. No significant differences have been measured in males. In the same study, no alteration of the cholesterol or phospholipid efflux was measured in macrophages purified from the $Abca7^{-/-}$ mice. However, further in vitro studies using different transfected cell types reported that ABCA1, ABCA7 binds ApoA-I, and ApoE, to transfer cholesterol and to generate HDL [18–20]. Of note, this efflux is marginal when compared with cholesterol efflux mediated by ABCA1. On the contrary, ABCA7 is more prone to transfer phospholipids to HDL, in particular sphingomyelin, and (lyso)phosphatidylcholine [18,19,35]. Importantly, almost all of these studies were done by artificially upregulating or downregulating *Abca7* expression that, in turn, provokes a modification of ABCA1 expression, probably by a compensatory mechanism [20,36]. This effect is also reciprocal because the decrease of *Abca1* expression increases *Abca7* level [37]. Abe-Dohmae et al. demonstrated that this compensatory mechanism allows the transfer of cholesterol to lipoproteins when the *Abca1* gene is deficient or absent [35]. However, it remains still unknown why ABCA7 cannot compensate this deficiency in a Tangier disease patient. In light of all these data, it is now widely accepted that ABCA7 is specialized in phospholipid transfer to HDL whereas ABCA1 releases cholesterol.

It is also clear that *Abca7* transcriptional expression is closely regulated by cholesterol metabolism. BALB/3T3 cells loaded with cholesterol showed a significant downregulation of *Abca7* whereas an upregulation of *Abca1* was observed [37]. On the contrary, cholesterol-depleted cells displayed increased *Abca7* expression. In fact, ABCA7 expression is regulated by the sterol-responsive/regulatory element binding protein (SREBP), able to bind specific sterol regulatory element DNA sequences when mammalian cells lack cholesterol, thus triggering the expression of the mRNA of genes involved in cholesterol synthesis or lipoprotein uptake such as HMG-CoA reductase or LDLR, respectively [37]. Blocking HMG-CoA reductase activity with statins, thus blocking cholesterol synthesis, activates SREBP nucleus translocation and increases ABCA7 expression in murine macrophages [38]. Interestingly, *ABCA1* expression is controlled by the Liver X receptor (LXR) nuclear receptors [39], also acting as cholesterol sensors in mammalian cells, thus probably explaining why ABCA1 and ABCA7 expressions are regulated differently to modulate the intracellular pool of cholesterol.

ABCA7 is also implicated in lipid trafficking during keratinocyte differentiation and is reported as a ceramide homeostasis regulator. ABCA7 upregulation was detected in normal human epidermal keratinocytes and HaCaT cells undergoing in vitro differentiation in parallel with an increase in intracellular ceramide levels. In accordance with these results, ABCA7 overexpression in Hela cells showed a ceramide de novo synthesis activation with an increase in intracellular and cell surface ceramide expression [40].

Altogether, these data reinforced the hypothesis that ABCA1 and ABCA7 are closely linked to regulate the cellular lipid metabolism but the role of ABCA7 seems to be rather indirect or negligible when directly compared to ABCA1.

3.2. Role of ABCA7 in Phagocytosis and Immune Response

In their initial investigation about ABCA7 sequence genomic organization, Kaminski et al. also observed that ABCA7 is arranged in a head-to-tail array with the human minor histocompatibility antigen HA-1 in a common locus on the 19p13.3 chromosome [11]. Since HA-1 is implicated in host defense, and based on its physical proximity to ABCA7, the question about possible common regulatory and functional mechanisms between these two genes was quickly raised [11]. Another clue suggesting a role of ABCA7 in immunity is the high sequence homology of the ABCA7 gene with cell death proteins (*Ced*) genes. In *Caenorhabditis elegans*, *ced* genes are key players of engulfment of dying cells. Among all these proteins, CED-7 is required for CED-1 functioning. Interestingly, mammalian orthologues have been proposed for all *Ced* genes, based on their sequence homologies. ABCA1 and ABCA7 have been proposed as being orthologues for CED7 [41], and low-density lipoprotein receptor–related protein 1 (LRP1) has been suggested as a protein with similar function to CED-1 [42]. In macrophages, ABCA7 and LRP1 were relocalized together to the plasma membrane in the presence of apoptotic cells, thus promoting their engulfment. Macrophages blocked with an antibody against LRP1 or $Abca7^{+/-}$ macrophages show both lower phagocytosis of these dying cells, as well as a decrease of the ERK phosphorylation, highlighting links between this ABC transporter, LRP1 and this signaling pathway [41]. ABCA7 expression and mediated phagocytosis are also increased when cholesterol synthesis is inhibited by statins that block HMG-CoA activity [38], demonstrating again strong relationships between cellular cholesterol pool and ABCA7 activity and expression.

As indicated above, ABCA7 is also able to interact with ApoA-I and HDL. Therefore, a potential contribution of (apo)lipoproteins in phagocytosis was investigated by Tanaka et al. They reported that extracellular HDL increases ABCA7-associated phagocytosis by stabilizing ABCA7, thus suggesting an involvement of the HDL components in the host defense system that deserves further investigation [38].

Therefore, a possible role of ABCA7 in host defense was clearly established and investigated by several studies reporting the ABCA7 implication in phagocytosis-mediated processes by macrophages instead of cholesterol regulation.

However, until 2011, no study clearly demonstrated a key role of ABCA7 in a human disease. When Hollingworth et al. published that single nucleotide polymorphisms (SNPs) in ABCA7 sequences were strongly associated with Alzheimer's disease (AD) onset and development [12], a new field of investigation has emerged with the objectives to identify the exact function of ABCA7 in the brain as well as in the apparition and evolution of AD.

4. Roles of ABCA7 in Brain Functioning and in Alzheimer's Disease (AD)

4.1. AD Pathology

AD is a neurodegenerative disease and the number one leading cause of dementia in the elderly [43]. Most AD patients show a sporadic form of the disease (late-onset AD, LOAD) whereas almost 1% of them show a familial early-onset form (FAD), with Mendelian inheritance (<60 years). Nevertheless, these two forms of AD are both characterized by accumulation of amyloid-β (Aβ) peptides and hyperphosphorylated tau protein, both provoking neuronal loss mainly in the hippocampus and cortex. Major AD symptoms are thus memory loss and behavioral complications [8].

Links between Aβ peptides and tau are still elusive despite a large number of studies, but it is now widely accepted that Aβ peptide deposition precedes the tau hyperphosphorylation by several decades [44]. In addition, FAD patients possess mutations in genes involved in Aβ peptide production, as, for example, in amyloid beta precursor protein (APP) sequence and in the enzyme responsible for APP cleavage. For this reason, it is compulsory to better understand the early molecular and cellular events controlling Aβ peptide synthesis and clearance, thus leading to the early step of the cerebral Aβ peptide accumulation.

All cells in the body are able to synthesize these peptides from APP but neurons remain the major sources of Aβ peptides [7]. APP is cleaved by at least three types of proteases (α-, β- and γ-secretases). To date, two processing pathways have been described—the amyloidogenic pathway generates Aβ peptides via β-secretase, whereas the anti-amyloidogenic pathway is initiated by α-secretase and prevents Aβ peptide generation. Interestingly, cellular cholesterol content can strongly influence these pathways and modulate Aβ synthesis, as discussed below and reviewed elsewhere [45].

Cerebral pools of Aβ peptides are cleared by several mechanisms, including phagocytosis processes occurring at the microglial cell level, and elimination across the blood–brain barrier (BBB). BBB physiology and functioning are summarized elsewhere and exhaustive reviews can be consulted for more details [7,10,46]. In brief, this physical and metabolic barrier is composed of the endothelial cells lining the brain microvessels. These cells are sealed together by tight junctions (TJs) in relation with adherent (AJs) and gap (GJ) junctions. Importantly, BBB endothelial cells also express a large panel of ABC transporters involved in the cerebral efflux of Aβ peptides but also restricting their entry to the CNS. Thus, a large body of evidence has confirmed in humans and in several animal models the involvement of P-gp (ABCB1), BCRP (ABCG2), MRP1 (ABCC1) and more recently ABCG4 (reviewed in [4,7]).

Noteworthy, some studies provide strong evidence that a decrease of amyloid clearance across the BBB or a decrease of phagocytosis by microglial cells might be involved in AD rather than an Aβ peptide overproduction by cleavage of the APP [47]. When analyzed in mouse models of AD and in patients, expression of ABCB1, ABCC1 and ABCG2 at the BBB are decreased. Restoration of these expressions increases amyloid clearance across the BBB, but also decreases the entry of peripheral Aβ peptides into the CNS, and thus reduces the brain Aβ burden and alleviates cognitive dysfunction [7,48–51]. However, despite these promising results, there is currently no cure for AD.

Because AD is also closely linked to the cholesterol metabolism and because the brain is one of the most cholesterol-rich organs of the body, researchers also focused their attention on ABCA1. In transgenic mice models, altering the cerebral cholesterol homeostasis by overexpressing or downregulating *Abca1* expression decreases or increases Aβ peptide synthesis and deposition, respectively [52–54].

It has also been largely demonstrated that environmental factors such as diet or lack of activity can promote AD onset. For example, animals fed with cholesterol-rich diets show increased production of Aβ peptides, whereas the use of statins decrease, in vivo and in vitro, this synthesis and deposition processes [45,55]. In addition, genetic factors are also largely involved as suggested since the early 2010s with the discovery of several gene polymorphisms in AD by GWAS. These studies confirmed that *ApoE4*, identified in the early 1990s, and is an important component of the low-density lipoproteins (LDL), remains the most strongly associated allele to AD. However, interestingly, almost 20 other genes have been identified, among them ABCA7, for which very little information in relation to its physiological and cellular functions has been identified so far.

Since 2011, genetic polymorphisms of *Abca7* have been reported in several dozens of genetic studies using samples coming from AD patients worldwide (reviewed in [56,57]). Further investigations of several of these variants demonstrated that they are responsible for an alteration of ABCA7 expression in AD brains. While a protective ABCA7 allele was identified, deleterious alleles are responsible for protein loss-of-function or ABCA7 downregulation or upregulation [56–58]. These ABCA7 loss-of-function mutations increase risk of AD by 80% in African ancestry populations and, risk of early AD onset by 100% to 400% in European ancestry populations [59–61]. Table 1 summarizes the AD-associated ABCA7 epigenetic and genetic variations, but also the available data obtained in patients regarding the amyloid and tau pathology as well brain morphology and clinical symptoms.

Table 1. AD-associated ABCA7 (epi-)genetic variation and their effects in individuals.

Variation	Interpretation	Reported Significant Effect of the Risk Allele	
	Common Risk-Increasing Variants	**Amyloid and Tau pathology**	**Brain Morphology and Clinical Symptoms**
rs3764650	Intronic GWAS SNP, low predicted functional effect	- Increased neuritic plaque burden [62] - Decreased CSF Aβ_{1-42} levels [63].	- Cortical and hippocampal atrophy [64] - Later age at onset and shorter disease duration [65] - Interaction on memory [66] - Association with posterior cortical atrophy variant of AD [67,68] - Memory decline for subjects who eventually developed MCI/LOAD [69] - Cognitive declines in females [70] - Lower immediate recall test and reduced rate of decline in symbol digit modalities test [71]
rs4147929	Intronic GWAS SNP, low predicted functional effect		- Voxel-based morphometry in the left postcentral gyrus [72] - Brain asymmetry in the hippocampus [73] - Increase of symptomatic AD compared to asymptomatic [74]
rs3752246	Missense GWAS SNP, predicted benign	- Increased amyloid deposition [75] - Increased brain amyloidosis [76]	- Decreased mean medial temporal lobe gray-patter density in dementia patients [77] - Interaction on memory [66]
rs115550680	Intronic GWAS SNP, low predicted functional effect		- Dissociation in entorhinal cortex resting state functional connectivity [78] - Behavioral generalization [78]
rs78117248	Intronic GWAS SNP, low predicted functional effect		
rs142076058 ABCA7 VNTR expansions	Loss-of-function Reduced ABCA7 expression, loss of exon 19 encoding an ATP-binding domain		
	Common Protective Variants	**Amyloid and Tau pathology**	**Brain Morphology and Clinical Symptoms**
rs72973581	Missense variant		
	CpG Methylation	**Amyloid and Tau pathology**	**Brain Morphology and Clinical Symptoms**
cg02308560 cg24402332	Hypermethylation in AD, effect on ABCA7 unknown		
cg04587220		- Increased brain amyloidosis and higher tau tangle density [79]	
	Rare Variants	**Amyloid and Tau pathology**	**Brain Morphology and Clinical Symptoms**
Missense and PTC variants	Loss-of-function for PTC variants. Unclear for missense variants.		

Variants and CpG methylation sites in ABCA7 with their interpreted functional effect, and their association with AD and brain morphology and clinical symptoms when investigated. Adapted from [57]. AD: Alzheimer's disease; GWAS: Genome-wide association studies; PTC: premature termination codon; SNP: single nucleotide polymorphism; VNTR: variable number tandem repeat.

4.2. ABCA7 in Brain Functions

Abca7 is mainly expressed by neurons and microglia in human and mouse brains [23,80–82]. Both isoforms described in Section 2.2 are observed by western blots in brain samples of healthy donors and AD patients [25]. Consequences of *Abca7* depletion on brain cholesterol homeostasis were first investigated in mice. Lipidomic analysis of forebrain samples from five male $Abca7^{-/-}$ mice showed alterations in lipid content; of the 275 studied lipids, only 24 were significantly affected by the absence of expression of $Abca7$ [83]. Among them, 12 subspecies in ethanolamine, three in phosphoglycerol, one in lysophosphatidylcholine, and two in shingomyelin were lower in brains of $Abca7^{-/-}$ mice. Three subspecies in phosphatidylcholine, one in ceramide, three in sulfatide, and one in cerebroside were increased, thus suggesting that $Abca7$ deficiency may significantly affect cerebral lipid metabolism.

The same study reported deficiencies in spatial memory in $Abca7^{-/-}$ 20- to 22-month-old male and female mice [83] but these deficiencies were only observed in 19- to 20-week-old females in another study performed by Logge et al. [84]. In the same study, males and females displayed an altered novel object recognition but this was more pronounced in females [84]. No difference between genders or with wild type mice was measured when a battery of tests for behavior was realized. Indeed, $Abca7^{-/-}$ mice showed the same sensory abilities, neurological reflexes, motor functions, anxiety, spatial learning and short-term memory as the wild-type mice [84]. No significant role in neurogenesis or neuron proliferation has been observed in another study in which only 8.5-month-old males were studied [85]. All these data suggest that ABCA7 might play a minor role in behavioral domains, but again further studies considering mouse strains, ages, sex or methodology are necessary to elucidate the role of ABCA7 in brain development and behaviors.

In humans, it was demonstrated more recently in brain samples that levels of ABCA7 modestly but significantly decrease with normal aging [23]. A significant association between ABCA7 SNP (rs3764650) and cognitive decline was only observed in females in a longitudinal study including 3267 females and 3026 males [70].

4.3. Impact of ABCA7 Depletion in Aβ Burden in Animals and Cells

When KO animals are crossbred with transgenic mice overproducing Aβ peptides (J20 mice at $\pm$ 17 months of age), Aβ burden was worsened, reinforcing the link between ABCA7 and AD. Interestingly, this effect is rather the consequence of a higher Aβ peptide accumulation than an overproduction [86]. Indeed, a significant decrease of Aβ peptide efflux across the BBB was reported in an $Abca7$-deficient in vitro model of the BBB, in relation to ApoA-I lipidation status [20], as well as a decrease in the phagocytosis process mediated by microglial cells [81,86]. These cells express high levels of ABCA7 when compared with neurons [82]. In addition, it was reported that $Abca7$ haplodeficiency provokes a microglial abnormal morphology and an altered response to inflammation, thus leading to cerebral amyloid accumulation in mice [87].

On the contrary to the aforementioned study demonstrating that $Abca7$ deficiency does not impact Aβ peptide synthesis [86], other studies reported that the absence of $Abca7$ in AD transgenic mouse models (APP/PS1 and TgCRND8) promotes the Aβ peptide production [22,83] or that ABCA7 upregulation in vitro diminishes this synthesis [82]. Transgenic models used to reproduce AD in mice should be taken into account and probably explain why such discrepancies are observed. It is important to note that in AD mouse models, ABCA1 and ABCA7 expressions seem to act in the same direction in order to modulate Aβ peptide production and deposition.

4.4. Evidence in AD Patients

Numerous works have studied ABCA7 as a risk gene in patients with mild cognitive impairment (MCI), or in AD individuals. All these studies are summarized elsewhere [57], and Table 1 gives an overview of AD-associated ABCA7 (epi-) genetic variation. Recently, an analysis of cerebrospinal fluid (CSF) from AD patients bearing the protective rs3764650 allele reported a decrease of CSF $A\beta_{1\text{-}42}$ although total tau and phosphorylated tau levels remained unchanged [63]. Another recent study investigating genetic associations correlating with Aβ deposition found a strong association with ABCA7 among 1600 investigated genes [88]. These observations confirmed the link between ABCA7 and Aβ peptides in humans but do not explain if this is mainly due to a decrease in Aβ production, or an increase of the clearance mediated by microglia across the BBB. A recent and elegant study from Lyssenko et al. investigated ABCA7 protein level in different Braak stages in AD patients and controls [23]. They observed that patients with low levels of ABCA7 are more prone to develop early AD than patients with the highest ABCA7 levels. Therefore, authors suggested that ABCA7 acts like a blocker of AD in the early stage of disease, particularly by removing toxic lipids from cellular membranes. As largely demonstrated now, lipid composition of cellular membranes can influence secretase activities and then APP processing and Aβ synthesis [89], therefore it might be hypothesized that ABCA7 levels and functioning can directly modulate APP processing and then, Aβ production. Another recent study performed in the frame of the Alzheimer's Disease Neuroimaging Initiative (ADNI) suggests also that ABCA7 acts very early in amyloid deposition [76]. Authors analyzed 18F-florbetapir positron emission tomographic data from 322 cognitively normal control individuals, 496 patients with mild cognitive impairment (MCI), and 159 individuals with AD. These results were compared with genetic data obtained in GWAS. They clearly observed that several AD risk variants of ABCA7 are highly associated with increased amyloid deposition, in particular in the cognitively healthy and MCI populations, but not in AD patients [76].

However, the impact of these variants on ABCA7 expression is still unclear since some studies report that deleterious alleles such as rs4147929 downregulates ABCA7 expression in AD patients [57], whereas other studies report contradictory results [58]. Studies investigating the association between rs3764650 and ABCA7 expression also reported inconsistent findings [65,90]. Therefore, there is a need of further studies that should address how ABCA7 is regulated and how this transporter is involved in brain functioning as well in Aβ peptide clearance or deposition in AD brains. Use of iPSCs cells from patients bearing SNPs of ABCA7 should open new avenues for investigating this transporter in AD. These cells can be differentiated into neurons to investigate not only neurogenesis but also Aβ production. It would be also possible to generate in vitro models of the BBB [91] or microglial cells [92] to further study Aβ clearance from CNS, and therefore to gain insight into the molecular role of ABCA7 in AD.

5. ABCA7 and Cancers

As mentioned previously, due to the GWAS studies, ABCA7 functions were mainly studied in AD mouse models and in AD samples. This is noteworthy because ABC transporters have been closely linked to the MDR phenotype observed in cancer cells in the most recent studies of ABCA7 in cancer. Finally, no relation with the MDR phenotype has been observed, but a possible involvement in the epithelial to mesenchymal transition (EMT process) was reported, in particular in ovarian cancers (OC) [93]. In EMT, cells lose their epithelial markers and express more mesenchymal markers in order to acquire capacities like migration, invasion and proliferation, characteristics of malignancies and metastatic cancers. In this study, authors demonstrated that ABCA7 was upregulated in ovarian cancer (OC) cells from patients when compared to adjacent non-cancer tissues [93]. This upregulation was associated with poor survival rates in OC patients. When ABCA7 expression is suppressed, OC cell lines showed a decrease in migration and an increase in epithelial marker expression such as E-cadherin, correlated with a decrease of the mesenchymal marker, N-cadherin [93,94]. Interestingly, ABCA7 knockdown decreased the TGF beta transcription factor SMAD-4, a key regulator for EMT [93].

Additionally, regulation of ABCA7 in cancer cells was reported to be modulated by the micro-RNA tumor suppressor called Mir-197-3p. This is found downregulated in OVACAR-3 cells as well as in other types of cancer cells such as hepatocarcinoma cells [94], suggesting a potential implication of ABCA7 in other types of cancers. Further investigations for the ABCA7 role in cancer are needed for a better understanding of this transporter functions and for a better diagnosis and prognosis. They will also determine if targeting ABCA7 expression might be a promising approach to prevent or cure certain types of cancers.

6. Conclusions and Future Perspectives

As summarized in Figure 3, recent evidence supports a central role of ABCA7 in AD. Increase of ABCA7 expression in the first step of the disease would help to maintain the cerebral lipid homeostasis and the balance between the amyloid synthesis and clearance. This latter process is controlled by microglial cell degradation and elimination across the blood–brain barrier. Aging, genetic polymorphisms and probably external factors such as diet or physical activity, influence ABCA7 expression, thereby promoting or slowing down amyloid deposition and then AD onset and evolution (Figure 3).

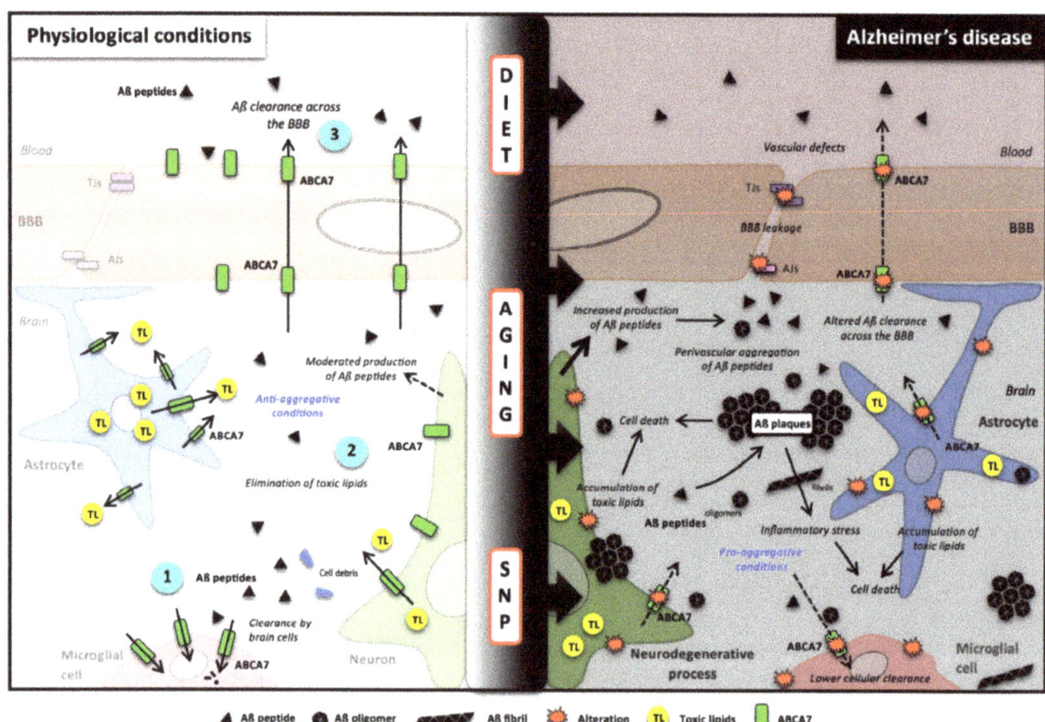

Figure 3. Potential roles of ABCA7 in brain cholesterol homeostasis and Aβ clearance and synthesis. (**Left**), in healthy brains, ABCA7 is normally expressed by neurons, glial cells and blood–brain barrier endothelial cells where it mediates Aβ clearance (1 and 3) as well cellular and toxic lipid (TL) efflux (2). Because Aβ peptide production is also linked with cellular lipid levels, it is likely that ABCA7 can also regulate Aβ synthesis. (**Right**) Single nucleotide polymorphisms (SNPs), aging and diet can affect ABCA7 function and expression, thus downregulating Aβ peptide clearance, decreasing toxic lipid recycling and then promoting Aβ burden. Aβ peptides: amyloid-β peptides; ABCA7: ATP-binding cassette subfamily A member 7; AJs: Adherent junctions of the BBB; BBB: blood–brain barrier; SNP: single nucleotide polymorphism; TJs: tight junctions of the BBB; TL: toxic lipids.

To exploit ABCA7 for therapeutic purposes, more attention needs to be given to its role in phagocytosis processes, in particular at the level of microglial cells that remove amyloid and play a central role in inflammatory processes. Then, it might be essential to better characterize ABCA7 regulatory sequences and signaling pathways in order to develop new molecules or approaches able to promote its cerebral expression, in healthy and AD patients. It might be also suggested that development of new tracers for measuring

ABCA7 activity at the brain level should be useful to detect AD or other neurodegenerative diseases at early stages.

At last, gender differences reported in several studies might also be taken into account to understand if ABCA7 has a different role in females rather in males, as highlighted by the role of ABCA7 in ovarian cancer [93]. This hypothesis is also strongly supported by the fact that AD deficiencies in mouse models affect more females than males [95,96] and that cognitive decline has been observed in females bearing ABCA7 SNPs when compared with males [70].

Author Contributions: S.D., J.P. and F.G. reviewed the literature, drafted and edited the manuscript and all figures. All authors have read and agreed to the published version of the manuscript.

Funding: S.D. has received fellowships from the Region Hauts-de-France and the University of Artois. F.G. has been supported by internal grants from University of Artois (BQR-ABCA7), from foundation France Alzheimer et Maladies apparentées (AAP 2013, #22) and under the ERANET JPcofuND 2-NET-PETABC collaborative project by grants from the French national agency (ANR, grant number ANR-20-JPW2-0002-04). The work of J.P. was supported by the following grants: Deutsche Forschungsgemeinschaft/Germany (DFG 263024513); Ministerium für Wirtschaft und Wissenschaft Sachsen-Anhalt/Germany (ZS/2016/05/78617); Latvian Council of Science/Latvia (lzp-2018/1-0275); Nasjonalforeningen (16154 to Luisa Möhle), HelseSØ/Norway (2019054, 2019055); Barnekreftforeningen (19008); EEA grant/Norges grants (TAČR TARIMAD TO100078); Norges forskningsrådet/Norway (260786 PROP-AD, 295910 NAPI, 327571 PETABC). PROP-AD and PETABC are EU Joint Programme-Neurodegenerative Disease Research (JPND) projects. PROP-AD is supported through the following funding organizations under the aegis of JPND—www.jpnd.eu (accessed on 31 December 2020) (AKA #301228—Finland, BMBF #01ED1605—Germany, CSO-MOH #30000-12631—Israel, NFR #260786—Norway, SRC #2015-06795—Sweden). PETABC is supported through the following funding organisations under the aegis of JPND—www.jpnd.eu (NFR #32757—Norway, FFG #882717—Austria, BMBF #XXXXX—Germany, MSMT #8F21002—Czech Republic, VIAA #ES RTD/2020/26—Latvia, ANR #20-JPW2-0002-04—France, SRC #2020-02905—Sweden). The projects receive funding from the European Union's Horizon 2020 research and innovation programme under grant agreement #643417 (JPco-fuND).

Acknowledgments: Authors would like to acknowledge Julien Saint-Pol for designing the final version of the figure n°3.

Conflicts of Interest: The authors declare no conflict of interest.

References

1. Thomas, C.; Aller, S.G.; Beis, K.; Carpenter, E.P.; Chang, G.; Chen, L.; Dassa, E.; Dean, M.; Duong Van Hoa, F.; Ekiert, D.; et al. Structural and functional diversity calls for a new classification of ABC transporters. *FEBS Lett.* **2020**, *594*, 3767–3775. [CrossRef] [PubMed]
2. Liu, X. ABC Family Transporters. *Adv. Exp. Med. Biol.* **2019**, *1141*, 13–100. [CrossRef] [PubMed]
3. Domínguez, C.J.; Tocchetti, G.N.; Rigalli, J.P.; Mottino, A.D. Acute regulation of apical ABC transporters in the gut. Potential influence on drug bioavailability. *Pharmacol. Res.* **2021**, *163*, 105251. [CrossRef] [PubMed]
4. Gil-Martins, E.; Barbosa, D.J.; Silva, V.; Remião, F.; Silva, R. Dysfunction of ABC transporters at the blood-brain barrier: Role in neurological disorders. *Pharmacol. Ther.* **2020**, *213*, 107554. [CrossRef] [PubMed]
5. Muriithi, W.; Macharia, L.W.; Heming, C.P.; Echevarria, J.L.; Nyachieo, A.; Filho, P.N.; Neto, V.M. ABC transporters and the hallmarks of cancer: Roles in cancer aggressiveness beyond multidrug resistance. *Cancer Biol. Med.* **2020**, *17*, 253–269. [CrossRef]
6. Schmitz, G.; Liebisch, G.; Langmann, T. Lipidomic strategies to study structural and functional defects of ABC-transporters in cellular lipid trafficking. *FEBS Lett.* **2006**, *580*, 5597–5610. [CrossRef]
7. Gosselet, F.; Saint-Pol, J.; Candela, P.; Fenart, L. Amyloid-beta Peptides, Alzheimer's Disease and the Blood-brain Barrier. *Curr. Alzheimer Res.* **2013**, *10*, 1015–1033. [CrossRef]
8. Selkoe, D.J. Alzheimer's disease. *Cold Spring Harb. Perspect. Biol.* **2011**, *3*, 895–897. [CrossRef]
9. Pereira, C.D.; Martins, F.; Wiltfang, J.; da Cruz, E.S.O.A.B.; Rebelo, S. ABC Transporters Are Key Players in Alzheimer's Disease. *J. Alzheimers Dis.* **2018**, *61*, 463–485. [CrossRef]
10. Pahnke, J.; Walker, L.C.; Scheffler, K.; Krohn, M. Alzheimer's disease and blood-brain barrier function-Why have anti-beta-amyloid therapies failed to prevent dementia progression? *Neurosci. Biobehav. Rev.* **2009**, *33*, 1099–1108. [CrossRef]
11. Kaminski, W.E.; Orso, E.; Diederich, W.; Klucken, J.; Drobnik, W.; Schmitz, G. Identification of a novel human sterol-sensitive ATP-binding cassette transporter (ABCA7). *Biochem. Biophys. Res. Commun.* **2000**, *273*, 532–538. [CrossRef]

12. Hollingworth, P.; Harold, D.; Sims, R.; Gerrish, A.; Lambert, J.C.; Carrasquillo, M.M.; Abraham, R.; Hamshere, M.L.; Pahwa, J.S.; Moskvina, V.; et al. Common variants at ABCA7, MS4A6A/MS4A4E, EPHA1, CD33 and CD2AP are associated with Alzheimer's disease. *Nat. Genet.* **2011**, *43*, 429–435. [CrossRef]

13. Sasaki, M.; Shoji, A.; Kubo, Y.; Nada, S.; Yamaguchi, A. Cloning of rat ABCA7 and its preferential expression in platelets. *Biochem. Biophys. Res. Commun.* **2003**, *304*, 777–782. [CrossRef]

14. Broccardo, C.; Osorio, J.; Luciani, M.F.; Schriml, L.M.; Prades, C.; Shulenin, S.; Arnould, I.; Naudin, L.; Lafargue, C.; Rosier, M.; et al. Comparative analysis of the promoter structure and genomic organization of the human and mouse ABCA7 gene encoding a novel ABCA transporter. *Cytogenet. Cell Genet.* **2001**, *92*, 264–270. [CrossRef]

15. Heng, T.S.; Painter, M.W. The Immunological Genome Project: Networks of gene expression in immune cells. *Nat. Immunol.* **2008**, *9*, 1091–1094. [CrossRef]

16. Kim, W.S.; Fitzgerald, M.L.; Kang, K.; Okuhira, K.; Bell, S.A.; Manning, J.J.; Koehn, S.L.; Lu, N.; Moore, K.J.; Freeman, M.W. Abca7 null mice retain normal macrophage phosphatidylcholine and cholesterol efflux activity despite alterations in adipose mass and serum cholesterol levels. *J. Biol. Chem.* **2005**, *280*, 3989–3995. [CrossRef]

17. Gosselet, F.; Candela, P.; Sevin, E.; Berezowski, V.; Cecchelli, R.; Fenart, L. Transcriptional profiles of receptors and transporters involved in brain cholesterol homeostasis at the blood-brain barrier: Use of an in vitro model. *Brain Res.* **2009**, *1249*, 34–42. [CrossRef]

18. Tomioka, M.; Toda, Y.; Mañucat, N.B.; Akatsu, H.; Fukumoto, M.; Kono, N.; Arai, H.; Kioka, N.; Ueda, K. Lysophosphatidylcholine export by human ABCA7. *Biochim. Biophys. Acta Mol. Cell Biol. Lipids* **2017**, *1862*, 658–665. [CrossRef]

19. Wang, N.; Lan, D.; Gerbod-Giannone, M.; Linsel-Nitschke, P.; Jehle, A.W.; Chen, W.; Martinez, L.O.; Tall, A.R. ATP-binding cassette transporter A7 (ABCA7) binds apolipoprotein A-I and mediates cellular phospholipid but not cholesterol efflux. *J. Biol. Chem.* **2003**, *278*, 42906–42912. [CrossRef]

20. Lamartiniere, Y.; Boucau, M.C.; Dehouck, L.; Krohn, M.; Pahnke, J.; Candela, P.; Gosselet, F.; Fenart, L. ABCA7 Downregulation Modifies Cellular Cholesterol Homeostasis and Decreases Amyloid-beta Peptide Efflux in an in vitro Model of the Blood-Brain Barrier. *J. Alzheimers Dis.* **2018**, *64*, 1195–1211. [CrossRef]

21. Ikeda, Y.; Abe-Dohmae, S.; Munehira, Y.; Aoki, R.; Kawamoto, S.; Furuya, A.; Shitara, K.; Amachi, T.; Kioka, N.; Matsuo, M.; et al. Posttranscriptional regulation of human ABCA7 and its function for the apoA-I-dependent lipid release. *Biochem. Biophys. Res. Commun.* **2003**, *311*, 313–318. [CrossRef]

22. Satoh, K.; Abe-Dohmae, S.; Yokoyama, S.; St George-Hyslop, P.; Fraser, P.E. ATP-binding Cassette Transporter A7 (ABCA7) Loss of Function Alters Alzheimer Amyloid Processing. *J. Biol. Chem.* **2015**, *290*, 24152–24165. [CrossRef]

23. Lyssenko, N.N.; Praticò, D. ABCA7 and the altered lipidostasis hypothesis of Alzheimer's disease. *Alzheimer's Dement. J. Alzheimer's Assoc.* **2021**, *17*, 164–174. [CrossRef]

24. Zhang, Y.; Sloan, S.A.; Clarke, L.E.; Caneda, C.; Plaza, C.A.; Blumenthal, P.D.; Vogel, H.; Steinberg, G.K.; Edwards, M.S.; Li, G.; et al. Purification and Characterization of Progenitor and Mature Human Astrocytes Reveals Transcriptional and Functional Differences with Mouse. *Neuron* **2016**, *89*, 37–53. [CrossRef]

25. Allen, M.; Lincoln, S.J.; Corda, M.; Watzlawik, J.O.; Carrasquillo, M.M.; Reddy, J.S.; Burgess, J.D.; Nguyen, T.; Malphrus, K.; Petersen, R.C.; et al. ABCA7 loss-of-function variants, expression, and neurologic disease risk. *Neurol. Genet.* **2017**, *3*, e126. [CrossRef]

26. Pasello, M.; Giudice, A.M.; Scotlandi, K. The ABC subfamily A transporters: Multifaceted players with incipient potentialities in cancer. *Semin. Cancer Biol.* **2020**, *60*, 57–71. [CrossRef]

27. Ye, Z.; Lu, Y.; Wu, T. The impact of ATP-binding cassette transporters on metabolic diseases. *Nutr. Metab.* **2020**, *17*, 61. [CrossRef]

28. Qian, H.; Zhao, X.; Cao, P.; Lei, J.; Yan, N.; Gong, X. Structure of the Human Lipid Exporter ABCA1. *Cell* **2017**, *169*, 1228–1239.e1210. [CrossRef]

29. Attie, A.D.; Kastelein, J.P.; Hayden, M.R. Pivotal role of ABCA1 in reverse cholesterol transport influencing HDL levels and susceptibility to atherosclerosis. *J. Lipid Res.* **2001**, *42*, 1717–1726. [CrossRef]

30. Schmitz, G.; Kaminski, W.E.; Porsch-Ozcürümez, M.; Klucken, J.; Orsó, E.; Bodzioch, M.; Büchler, C.; Drobnik, W. ATP-binding cassette transporter A1 (ABCA1) in macrophages: A dual function in inflammation and lipid metabolism? *Pathobiol. J. Immunopathol. Mol. Cell. Biol.* **1999**, *67*, 236–240. [CrossRef]

31. Jacobo-Albavera, L.; Domínguez-Pérez, M.; Medina-Leyte, D.J.; González-Garrido, A.; Villarreal-Molina, T. The Role of the ATP-Binding Cassette A1 (ABCA1) in Human Disease. *Int. J. Mol. Sci.* **2021**, *22*, 1593. [CrossRef] [PubMed]

32. O'Connell, B.J.; Denis, M.; Genest, J. Cellular physiology of cholesterol efflux in vascular endothelial cells. *Circulation* **2004**, *110*, 2881–2888. [CrossRef] [PubMed]

33. Singaraja, R.R.; Brunham, L.R.; Visscher, H.; Kastelein, J.J.; Hayden, M.R. Efflux and atherosclerosis: The clinical and biochemical impact of variations in the ABCA1 gene. *Arter. Thromb. Vasc. Biol.* **2003**, *23*, 1322–1332. [CrossRef] [PubMed]

34. Iatan, I.; Alrasadi, K.; Ruel, I.; Alwaili, K.; Genest, J. Effect of ABCA1 mutations on risk for myocardial infarction. *Curr. Atheroscler. Rep.* **2008**, *10*, 413–426. [CrossRef]

35. Abe-Dohmae, S.; Ikeda, Y.; Matsuo, M.; Hayashi, M.; Okuhira, K.; Ueda, K.; Yokoyama, S. Human ABCA7 supports apolipoprotein-mediated release of cellular cholesterol and phospholipid to generate high density lipoprotein. *J. Biol. Chem.* **2004**, *279*, 604–611. [CrossRef]

36. Meurs, I.; Calpe-Berdiel, L.; Habets, K.L.; Zhao, Y.; Korporaal, S.J.; Mommaas, A.M.; Josselin, E.; Hildebrand, R.B.; Ye, D.; Out, R.; et al. Effects of deletion of macrophage ABCA7 on lipid metabolism and the development of atherosclerosis in the presence and absence of ABCA1. *PLoS ONE* **2012**, *7*, e30984. [CrossRef]
37. Iwamoto, N.; Abe-Dohmae, S.; Sato, R.; Yokoyama, S. ABCA7 expression is regulated by cellular cholesterol through the SREBP2 pathway and associated with phagocytosis. *J. Lipid Res.* **2006**, *47*, 1915–1927. [CrossRef]
38. Tanaka, N.; Abe-Dohmae, S.; Iwamoto, N.; Fitzgerald, M.L.; Yokoyama, S. Helical apolipoproteins of high-density lipoprotein enhance phagocytosis by stabilizing ATP-binding cassette transporter A7. *J. Lipid Res.* **2010**, *51*, 2591–2599. [CrossRef]
39. Saint-Pol, J.; Vandenhaute, E.; Boucau, M.C.; Candela, P.; Dehouck, L.; Cecchelli, R.; Dehouck, M.P.; Fenart, L.; Gosselet, F. Brain Pericytes ABCA1 Expression Mediates Cholesterol Efflux but not Cellular Amyloid-beta Peptide Accumulation. *J. Alzheimers Dis.* **2012**, *30*, 489–503. [CrossRef]
40. Kielar, D.; Kaminski, W.E.; Liebisch, G.; Piehler, A.; Wenzel, J.J.; Möhle, C.; Heimerl, S.; Langmann, T.; Friedrich, S.O.; Böttcher, A.; et al. Adenosine triphosphate binding cassette (ABC) transporters are expressed and regulated during terminal keratinocyte differentiation: A potential role for ABCA7 in epidermal lipid reorganization. *J. Investig. Dermatol.* **2003**, *121*, 465–474. [CrossRef]
41. Jehle, A.W.; Gardai, S.J.; Li, S.; Linsel-Nitschke, P.; Morimoto, K.; Janssen, W.J.; Vandivier, R.W.; Wang, N.; Greenberg, S.; Dale, B.M.; et al. ATP-binding cassette transporter A7 enhances phagocytosis of apoptotic cells and associated ERK signaling in macrophages. *J. Cell Biol.* **2006**, *174*, 547–556. [CrossRef]
42. Su, H.P.; Nakada-Tsukui, K.; Tosello-Trampont, A.C.; Li, Y.; Bu, G.; Henson, P.M.; Ravichandran, K.S. Interaction of CED-6/GULP, an adapter protein involved in engulfment of apoptotic cells with CED-1 and CD91/low density lipoprotein receptor-related protein (LRP). *J. Biol. Chem.* **2002**, *277*, 11772–11779. [CrossRef]
43. Association, A.s. Alzheimer's Disease Facts and Figures Alzheimer's Dementia 2019. *Annu. Rep.* **2019**, *15*, 321–387.
44. Jack, C.R., Jr.; Knopman, D.S.; Jagust, W.J.; Petersen, R.C.; Weiner, M.W.; Aisen, P.S.; Shaw, L.M.; Vemuri, P.; Wiste, H.J.; Weigand, S.D.; et al. Tracking pathophysiological processes in Alzheimer's disease: An updated hypothetical model of dynamic biomarkers. *Lancet Neurol.* **2013**, *12*, 207–216. [CrossRef]
45. Martins, I.J.; Berger, T.; Sharman, M.J.; Verdile, G.; Fuller, S.J.; Martins, R.N. Cholesterol metabolism and transport in the pathogenesis of Alzheimer's disease. *J. Neurochem.* **2009**, *111*, 1275–1308. [CrossRef]
46. Zlokovic, B.V. Neurovascular pathways to neurodegeneration in Alzheimer's disease and other disorders. *Nat. Rev. Neurosci.* **2011**, *12*, 723–738. [CrossRef]
47. Mawuenyega, K.G.; Sigurdson, W.; Ovod, V.; Munsell, L.; Kasten, T.; Morris, J.C.; Yarasheski, K.E.; Bateman, R.J. Decreased clearance of CNS beta-amyloid in Alzheimer's disease. *Science* **2010**, *330*, 1774. [CrossRef]
48. Saint-Pol, J.; Candela, P.; Boucau, M.C.; Fenart, L.; Gosselet, F. Oxysterols decrease apical-to-basolateral transport of Abeta peptides via an ABCB1-mediated process in an in vitro Blood-brain barrier model constituted of bovine brain capillary endothelial cells. *Brain Res.* **2013**, *1517*, 1–15. [CrossRef]
49. Hofrichter, J.; Krohn, M.; Schumacher, T.; Lange, C.; Feistel, B.; Walbroel, B.; Heinze, H.J.; Crockett, S.; Sharbel, T.F.; Pahnke, J. Reduced Alzheimer's disease pathology by St. John's Wort treatment is independent of hyperforin and facilitated by ABCC1 and microglia activation in mice. *Curr. Alzheimer Res.* **2013**, *10*, 1057–1069. [CrossRef]
50. Qosa, H.; Abuznait, A.H.; Hill, R.A.; Kaddoumi, A. Enhanced Brain Amyloid-beta Clearance by Rifampicin and Caffeine as a Possible Protective Mechanism Against Alzheimer's Disease. *J. Alzheimers Dis.* **2012**, *31*, 151–165. [CrossRef]
51. Krohn, M.; Lange, C.; Hofrichter, J.; Scheffler, K.; Stenzel, J.; Steffen, J.; Schumacher, T.; Bruning, T.; Plath, A.S.; Alfen, F.; et al. Cerebral amyloid-beta proteostasis is regulated by the membrane transport protein ABCC1 in mice. *J. Clin. Investig.* **2011**, *121*, 3924–3931. [CrossRef]
52. Wahrle, S.E.; Jiang, H.; Parsadanian, M.; Hartman, R.E.; Bales, K.R.; Paul, S.M.; Holtzman, D.M. Deletion of Abca1 increases Abeta deposition in the PDAPP transgenic mouse model of Alzheimer disease. *J. Biol. Chem.* **2005**, *280*, 43236–43242. [CrossRef]
53. Wahrle, S.E.; Jiang, H.; Parsadanian, M.; Kim, J.; Li, A.; Knoten, A.; Jain, S.; Hirsch-Reinshagen, V.; Wellington, C.L.; Bales, K.R.; et al. Overexpression of ABCA1 reduces amyloid deposition in the PDAPP mouse model of Alzheimer disease. *J. Clin. Investig.* **2008**, *118*, 671–682. [CrossRef]
54. Hirsch-Reinshagen, V.; Maia, L.F.; Burgess, B.L.; Blain, J.F.; Naus, K.E.; McIsaac, S.A.; Parkinson, P.F.; Chan, J.Y.; Tansley, G.H.; Hayden, M.R.; et al. The absence of ABCA1 decreases soluble ApoE levels but does not diminish amyloid deposition in two murine models of Alzheimer disease. *J. Biol. Chem.* **2005**, *280*, 43243–43256. [CrossRef]
55. Gosselet, F. The Mysterious Link between Cholesterol and Alzheimer's Disease: Is the Blood-Brain Barrier a Suspect? *J. Alzheimer's Dis. Parkinsonism* **2011**, *1*, 2161-0460. [CrossRef]
56. Aikawa, T.; Holm, M.L.; Kanekiyo, T. ABCA7 and Pathogenic Pathways of Alzheimer's Disease. *Brain Sci.* **2018**, *8*, 27. [CrossRef]
57. De Roeck, A.; Van Broeckhoven, C.; Sleegers, K. The role of ABCA7 in Alzheimer's disease: Evidence from genomics, transcriptomics and methylomics. *Acta Neuropathol.* **2019**, *138*, 201–220. [CrossRef]
58. Liu, G.; Zhang, H.; Liu, B.; Wang, T.; Han, Z.; Ji, X. rs4147929 variant minor allele increases ABCA7 gene expression and ABCA7 shows increased gene expression in Alzheimer's disease patients compared with controls. *Acta Neuropathol.* **2020**, *139*, 937–940. [CrossRef]
59. Bellenguez, C.; Charbonnier, C.; Grenier-Boley, B.; Quenez, O.; Le Guennec, K.; Nicolas, G.; Chauhan, G.; Wallon, D.; Rousseau, S.; Richard, A.C.; et al. Contribution to Alzheimer's disease risk of rare variants in TREM2, SORL1, and ABCA7 in 1779 cases and 1273 controls. *Neurobiol. Aging* **2017**, *59*, 220.e221–220.e229. [CrossRef]

60. Cukier, H.N.; Kunkle, B.W.; Vardarajan, B.N.; Rolati, S.; Hamilton-Nelson, K.L.; Kohli, M.A.; Whitehead, P.L.; Dombroski, B.A.; Van Booven, D.; Lang, R.; et al. ABCA7 frameshift deletion associated with Alzheimer disease in African Americans. *Neurology. Genet.* **2016**, *2*, e79. [CrossRef]

61. De Roeck, A.; Van den Bossche, T.; van der Zee, J.; Verheijen, J.; De Coster, W.; Van Dongen, J.; Dillen, L.; Baradaran-Heravi, Y.; Heeman, B.; Sanchez-Valle, R.; et al. Deleterious ABCA7 mutations and transcript rescue mechanisms in early onset Alzheimer's disease. *Acta Neuropathol.* **2017**, *134*, 475–487. [CrossRef] [PubMed]

62. Shulman, J.M.; Chen, K.; Keenan, B.T.; Chibnik, L.B.; Fleisher, A.; Thiyyagura, P.; Roontiva, A.; McCabe, C.; Patsopoulos, N.A.; Corneveaux, J.J.; et al. Genetic susceptibility for Alzheimer disease neuritic plaque pathology. *JAMA Neurol.* **2013**, *70*, 1150–1157. [CrossRef] [PubMed]

63. Ma, F.C.; Zong, Y.; Wang, H.F.; Li, J.Q.; Cao, X.P.; Tan, L. ABCA7 genotype altered Aβ levels in cerebrospinal fluid in Alzheimer's disease without dementia. *Ann. Transl. Med.* **2018**, *6*, 437. [CrossRef] [PubMed]

64. Ramirez, L.M.; Goukasian, N.; Porat, S.; Hwang, K.S.; Eastman, J.A.; Hurtz, S.; Wang, B.; Vang, N.; Sears, R.; Klein, E.; et al. Common variants in ABCA7 and MS4A6A are associated with cortical and hippocampal atrophy. *Neurobiol. Aging* **2016**, *39*, 82–89. [CrossRef]

65. Karch, C.M.; Jeng, A.T.; Nowotny, P.; Cady, J.; Cruchaga, C.; Goate, A.M. Expression of novel Alzheimer's disease risk genes in control and Alzheimer's disease brains. *PLoS ONE* **2012**, *7*, e50976. [CrossRef]

66. Engelman, C.D.; Koscik, R.L.; Jonaitis, E.M.; Okonkwo, O.C.; Hermann, B.P.; La Rue, A.; Sager, M.A. Interaction between two cholesterol metabolism genes influences memory: Findings from the Wisconsin Registry for Alzheimer's Prevention. *J. Alzheimers Dis.* **2013**, *36*, 749–757. [CrossRef]

67. Carrasquillo, M.M.; Khan, Q.; Murray, M.E.; Krishnan, S.; Aakre, J.; Pankratz, V.S.; Nguyen, T.; Ma, L.; Bisceglio, G.; Petersen, R.C.; et al. Late-onset Alzheimer disease genetic variants in posterior cortical atrophy and posterior AD. *Neurology* **2014**, *82*, 1455–1462. [CrossRef]

68. Schott, J.M.; Crutch, S.J.; Carrasquillo, M.M.; Uphill, J.; Shakespeare, T.J.; Ryan, N.S.; Yong, K.X.; Lehmann, M.; Ertekin-Taner, N.; Graff-Radford, N.R.; et al. Genetic risk factors for the posterior cortical atrophy variant of Alzheimer's disease. *Alzheimer's Dement. J. Alzheimer's Assoc.* **2016**, *12*, 862–871. [CrossRef]

69. Carrasquillo, M.M.; Crook, J.E.; Pedraza, O.; Thomas, C.S.; Pankratz, V.S.; Allen, M.; Nguyen, T.; Malphrus, K.G.; Ma, L.; Bisceglio, G.D.; et al. Late-onset Alzheimer's risk variants in memory decline, incident mild cognitive impairment, and Alzheimer's disease. *Neurobiol. Aging* **2015**, *36*, 60–67. [CrossRef]

70. Nettiksimmons, J.; Tranah, G.; Evans, D.S.; Yokoyama, J.S.; Yaffe, K. Gene-based aggregate SNP associations between candidate AD genes and cognitive decline. *Age* **2016**, *38*, 41. [CrossRef]

71. Andrews, S.J.; Das, D.; Anstey, K.J.; Easteal, S. Late Onset Alzheimer's Disease Risk Variants in Cognitive Decline: The PATH Through Life Study. *J. Alzheimers Dis.* **2017**, *57*, 423–436. [CrossRef]

72. Roshchupkin, G.V.; Adams, H.H.; van der Lee, S.J.; Vernooij, M.W.; van Duijn, C.M.; Uitterlinden, A.G.; van der Lugt, A.; Hofman, A.; Niessen, W.J.; Ikram, M.A. Fine-mapping the effects of Alzheimer's disease risk loci on brain morphology. *Neurobiol. Aging* **2016**, *48*, 204–211. [CrossRef]

73. Wachinger, C.; Nho, K.; Saykin, A.J.; Reuter, M.; Rieckmann, A. A Longitudinal Imaging Genetics Study of Neuroanatomical Asymmetry in Alzheimer's Disease. *Biol. Psychiatry* **2018**, *84*, 522–530. [CrossRef]

74. Monsell, S.E.; Mock, C.; Fardo, D.W.; Bertelsen, S.; Cairns, N.J.; Roe, C.M.; Ellingson, S.R.; Morris, J.C.; Goate, A.M.; Kukull, W.A. Genetic Comparison of Symptomatic and Asymptomatic Persons With Alzheimer Disease Neuropathology. *Alzheimer Dis. Assoc. Disord.* **2017**, *31*, 232–238. [CrossRef]

75. Hughes, T.M.; Lopez, O.L.; Evans, R.W.; Kamboh, M.I.; Williamson, J.D.; Klunk, W.E.; Mathis, C.A.; Price, J.C.; Cohen, A.D.; Snitz, B.E.; et al. Markers of cholesterol transport are associated with amyloid deposition in the brain. *Neurobiol. Aging* **2014**, *35*, 802–807. [CrossRef]

76. Apostolova, L.G.; Risacher, S.L.; Duran, T.; Stage, E.C.; Goukasian, N.; West, J.D.; Do, T.M.; Grotts, J.; Wilhalme, H.; Nho, K.; et al. Associations of the Top 20 Alzheimer Disease Risk Variants With Brain Amyloidosis. *JAMA Neurol.* **2018**, *75*, 328–341. [CrossRef]

77. Stage, E.; Duran, T.; Risacher, S.L.; Goukasian, N.; Do, T.M.; West, J.D.; Wilhalme, H.; Nho, K.; Phillips, M.; Elashoff, D.; et al. The effect of the top 20 Alzheimer disease risk genes on gray-matter density and FDG PET brain metabolism. *Alzheimer's Dement.* **2016**, *5*, 53–66. [CrossRef]

78. Sinha, N.; Reagh, Z.M.; Tustison, N.J.; Berg, C.N.; Shaw, A.; Myers, C.E.; Hill, D.; Yassa, M.A.; Gluck, M.A. ABCA7 risk variant in healthy older African Americans is associated with a functionally isolated entorhinal cortex mediating deficient generalization of prior discrimination training. *Hippocampus* **2019**, *29*, 527–538. [CrossRef]

79. Yu, L.; Chibnik, L.B.; Srivastava, G.P.; Pochet, N.; Yang, J.; Xu, J.; Kozubek, J.; Obholzer, N.; Leurgans, S.E.; Schneider, J.A.; et al. Association of Brain DNA methylation in SORL1, ABCA7, HLA-DRB5, SLC24A4, and BIN1 with pathological diagnosis of Alzheimer disease. *JAMA Neurol.* **2015**, *72*, 15–24. [CrossRef]

80. Kim, W.S.; Guillemin, G.J.; Glaros, E.N.; Lim, C.K.; Garner, B. Quantitation of ATP-binding cassette subfamily-A transporter gene expression in primary human brain cells. *Neuroreport* **2006**, *17*, 891–896. [CrossRef]

81. Fu, Y.; Hsiao, J.H.; Paxinos, G.; Halliday, G.M.; Kim, W.S. ABCA7 Mediates Phagocytic Clearance of Amyloid-beta in the Brain. *J. Alzheimers Dis.* **2016**, *54*, 569–584. [CrossRef]

82. Chan, S.L.; Kim, W.S.; Kwok, J.B.; Hill, A.F.; Cappai, R.; Rye, K.A.; Garner, B. ATP-binding cassette transporter A7 regulates processing of amyloid precursor protein in vitro. *J. Neurochem.* **2008**, *106*, 793–804. [CrossRef]
83. Sakae, N.; Liu, C.C.; Shinohara, M.; Frisch-Daiello, J.; Ma, L.; Yamazaki, Y.; Tachibana, M.; Younkin, L.; Kurti, A.; Carrasquillo, M.M.; et al. ABCA7 Deficiency Accelerates Amyloid-beta Generation and Alzheimer's Neuronal Pathology. *J. Neurosci.* **2016**, *36*, 3848–3859. [CrossRef]
84. Logge, W.; Cheng, D.; Chesworth, R.; Bhatia, S.; Garner, B.; Kim, W.S.; Karl, T. Role of Abca7 in mouse behaviours relevant to neurodegenerative diseases. *PLoS ONE* **2012**, *7*, e45959. [CrossRef]
85. Li, H.; Karl, T.; Garner, B. Abca7 deletion does not affect adult neurogenesis in the mouse. *Biosci. Rep.* **2016**, *36*, 1–6. [CrossRef]
86. Kim, W.S.; Li, H.; Ruberu, K.; Chan, S.; Elliott, D.A.; Low, J.K.; Cheng, D.; Karl, T.; Garner, B. Deletion of Abca7 increases cerebral amyloid-beta accumulation in the J20 mouse model of Alzheimer's disease. *J. Neurosci.* **2013**, *33*, 4387–4394. [CrossRef]
87. Aikawa, T.; Ren, Y.; Yamazaki, Y.; Tachibana, M.; Johnson, M.R.; Anderson, C.T.; Martens, Y.A.; Holm, M.L.; Asmann, Y.W.; Saito, T.; et al. ABCA7 haplodeficiency disturbs microglial immune responses in the mouse brain. *Proc. Natl. Acad. Sci. USA* **2019**, *116*, 23790–23796. [CrossRef]
88. Vacher, M.; Porter, T.; Villemagne, V.L.; Milicic, L.; Peretti, M.; Fowler, C.; Martins, R.; Rainey-Smith, S.; Ames, D.; Masters, C.L.; et al. Validation of a priori candidate Alzheimer's disease SNPs with brain amyloid-beta deposition. *Sci. Rep.* **2019**, *9*, 17069. [CrossRef]
89. Chew, H.; Solomon, V.A.; Fonteh, A.N. Involvement of Lipids in Alzheimer's Disease Pathology and Potential Therapies. *Front. Physiol.* **2020**, *11*, 598. [CrossRef] [PubMed]
90. Allen, M.; Zou, F.; Chai, H.S.; Younkin, C.S.; Crook, J.; Pankratz, V.S.; Carrasquillo, M.M.; Rowley, C.N.; Nair, A.A.; Middha, S.; et al. Novel late-onset Alzheimer disease loci variants associate with brain gene expression. *Neurology* **2012**, *79*, 221–228. [CrossRef] [PubMed]
91. Wellens, S.; Dehouck, L.; Chandrasekaran, V.; Singh, P.; Loiola, R.A.; Sevin, E.; Exner, T.; Jennings, P.; Gosselet, F.; Culot, M. Evaluation of a human iPSC-derived BBB model for repeated dose toxicity testing with cyclosporine A as model compound. *Toxicol. In Vitro* **2021**, *73*, 105112. [CrossRef] [PubMed]
92. Wurm, J.; Konttinen, H.; Andressen, C.; Malm, T.; Spittau, B. Microglia Development and Maturation and Its Implications for Induction of Microglia-Like Cells from Human iPSCs. *Int. J. Mol. Sci.* **2021**, *22*, 3088. [CrossRef] [PubMed]
93. Liu, X.; Li, Q.; Zhou, J.; Zhang, S. ATP-binding cassette transporter A7 accelerates epithelial-to-mesenchymal transition in ovarian cancer cells by upregulating the transforming growth factor-β signaling pathway. *Oncol. Lett.* **2018**, *16*, 5868–5874. [CrossRef] [PubMed]
94. Xie, W.; Shui, C.; Fang, X.; Peng, Y.; Qin, L. miR-197-3p reduces epithelial-mesenchymal transition by targeting ABCA7 in ovarian cancer cells. *3 Biotech* **2020**, *10*, 375. [CrossRef] [PubMed]
95. Carroll, J.C.; Rosario, E.R.; Kreimer, S.; Villamagna, A.; Gentzschein, E.; Stanczyk, F.Z.; Pike, C.J. Sex differences in β-amyloid accumulation in 3xTg-AD mice: Role of neonatal sex steroid hormone exposure. *Brain Res.* **2010**, *1366*, 233–245. [CrossRef]
96. Schäfer, S.; Wirths, O.; Multhaup, G.; Bayer, T.A. Gender dependent APP processing in a transgenic mouse model of Alzheimer's disease. *J. Neural Transm.* **2007**, *114*, 387–394. [CrossRef] [PubMed]

International Journal of
Molecular Sciences

Review

The Role of the ATP-Binding Cassette A1 (ABCA1) in Human Disease

Leonor Jacobo-Albavera [1,†], Mayra Domínguez-Pérez [1,†], Diana Jhoseline Medina-Leyte [1,2],
Antonia González-Garrido [1] and Teresa Villarreal-Molina [1,*]

[1] Laboratorio de Genómica de Enfermedades Cardiovasculares, Dirección de Investigación, Instituto Nacional de Medicina Genómica (INMEGEN), Mexico City CP14610, Mexico; ljacobo@inmegen.gob.mx (L.J.-A.); mdominguez@inmegen.gob.mx (M.D.-P.); dianajhos18@gmail.com (D.J.M.-L.); agonzalezg@uchicago.edu (A.G.-G.)

[2] Posgrado en Ciencias Biológicas, Universidad Nacional Autónoma de México (UNAM), Coyoacán, Mexico City CP04510, Mexico

* Correspondence: mvillareal@inmegen.gob.mx
† These authors contributed equally to this work.

Abstract: Cholesterol homeostasis is essential in normal physiology of all cells. One of several proteins involved in cholesterol homeostasis is the ATP-binding cassette transporter A1 (ABCA1), a transmembrane protein widely expressed in many tissues. One of its main functions is the efflux of intracellular free cholesterol and phospholipids across the plasma membrane to combine with apolipoproteins, mainly apolipoprotein A-I (Apo A-I), forming nascent high-density lipoprotein-cholesterol (HDL-C) particles, the first step of reverse cholesterol transport (RCT). In addition, ABCA1 regulates cholesterol and phospholipid content in the plasma membrane affecting lipid rafts, microparticle (MP) formation and cell signaling. Thus, it is not surprising that impaired ABCA1 function and altered cholesterol homeostasis may affect many different organs and is involved in the pathophysiology of a broad array of diseases. This review describes evidence obtained from animal models, human studies and genetic variation explaining how ABCA1 is involved in dyslipidemia, coronary heart disease (CHD), type 2 diabetes (T2D), thrombosis, neurological disorders, age-related macular degeneration (AMD), glaucoma, viral infections and in cancer progression.

Keywords: ATP-binding cassette transporter A1 (ABCA1); cholesterol homeostasis; reverse cholesterol transport; HDL-C; dyslipidemia; type 2 diabetes; microparticles

Citation: Jacobo-Albavera, L.; Domínguez-Pérez, M.; Medina-Leyte, D.J.; González-Garrido, A.; Villarreal-Molina, T. The Role of the ATP-Binding Cassette A1 (ABCA1) in Human Disease. *Int. J. Mol. Sci.* **2021**, *22*, 1593. https://doi.org/10.3390/ijms22041593

Academic Editor: Thomas Falguières
Received: 31 December 2020
Accepted: 27 January 2021
Published: 5 February 2021

Publisher's Note: MDPI stays neutral with regard to jurisdictional claims in published maps and institutional affiliations.

1. Introduction

Cholesterol is an essential biomolecule, involved in a wide array of physiological and pathological processes. In the plasma membrane, changes in free cholesterol content and phospholipid species modulate signaling of multiple receptors [1]. A physiological free cholesterol/phospholipid ratio in cellular membranes is necessary to maintain membrane fluidity [2], and altered membrane fluidity adversely affects the conformation and function of certain integral membrane proteins that can be inhibited by a high free cholesterol/phospholipid ratio [3]. Excess plasma membrane cholesterol also disrupts the function of certain signaling molecules that normally reside in non-raft domains. In addition, excess intracellular cholesterol levels can also cause toxicity by mechanisms including intracellular cholesterol crystallization, oxidation of cholesterol to oxysterols and triggering of apoptotic signaling pathways [4].

One of several proteins involved in cholesterol homeostasis is the ATP-binding cassette transporter A1 (ABCA1), a transmembrane protein widely expressed in many tissues where it may have many different functions. Its most studied function is the efflux of intracellular free cholesterol and phospholipids across the plasma membrane to combine with apolipoproteins, mainly apolipoprotein A-I (ApoA-I), forming nascent high-density lipoprotein particles (HDLs), the first step of reverse cholesterol transport (RCT) [5]. RCT is the process by which the body removes excess cholesterol from peripheral tissues and delivers this cholesterol to the liver, where it is redistributed to other tissues or removed from the body by the gallbladder. HDL-cholesterol (HDL-C) particles are the main lipoproteins involved in this process [6]. In addition to HDL-C formation, ABCA1 regulates cholesterol and phospholipid content in the plasma membrane and is involved in microparticle formation and thus in cell signaling. For all these reasons, it is not surprising that altered cholesterol homeostasis may affect many different organs and is involved in the pathophysiology of a broad array of diseases (Figure 1). The present review focuses on the role of the ABCA1 cholesterol transporter in human disease.

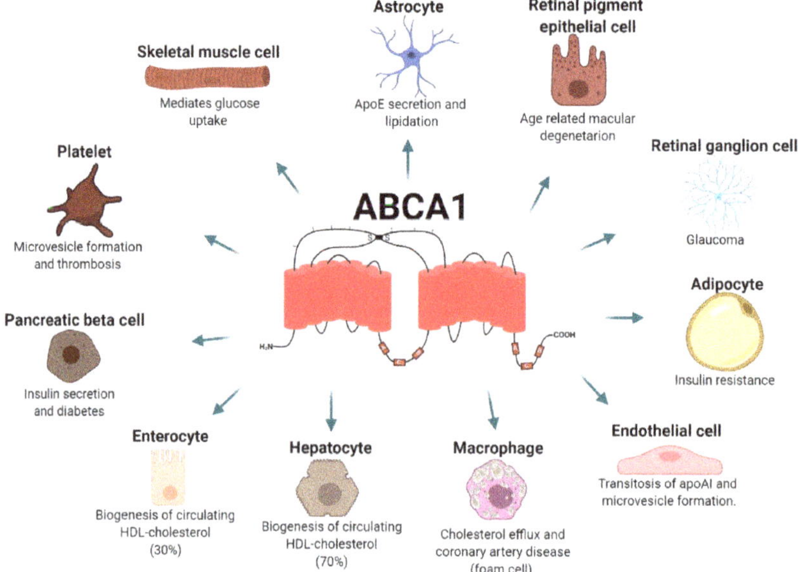

Figure 1. ATP-binding cassette transporter A1 (ABCA1) functions in different cell types and associated diseases. ABCA1 is widely expressed and participates in a broad array of physiological and pathological processes. ApoE: apolipoprotein E; HDL: high-density lipoproteins; apoAI: apolipoprotein I.

2. Global ABCA1 Deficiency: Tangier Disease

Tangier disease (TD) is a rare autosomal recessive disease caused by homozygous or compound heterozygous loss of function variants in both alleles of the *ABCA1* gene (OMIM #205400). TD is characterized by severe deficiency or absence of circulating HDL-C particles and accumulation of cholesteryl-esters in cells throughout the body, particularly in the reticuloendothelial system [7,8]. The major clinical signs of TD are very low HDL-C levels (<5 mg/dL), hyperplastic yellow orange tonsils and hepatosplenomegaly; while peripheral neuropathy occurs in approximately 50%, and premature coronary heart disease (CHD), occurs in 30 to 50% of TD patients [9–11]. Carriers of a single *ABCA1* mutation (heterozygotes) have variable reductions in plasma HDL-C levels and a variable increased risk for CHD [12]. Other less frequent symptoms include corneal opacity and hematologic manifestations, such as thrombocytopenia, altered platelet morphology and function, mild bleeding tendency, reticulocytosis, stomatocytosis and hemolytic anemia [13].

Macrophages and other cells from TD patients are overloaded with cholesterol (foam cells) because the ABCA1-mediated efflux of cellular free (unesterified) cholesterol and phospholipids to ApoA-I is defective [14]. These foam cells play a crucial role in the pathogenesis of atherosclerosis and CHD. However, it is not clear why not all TD patients develop premature CHD. A review of 185 TD cases reported that 51% of patients aged 40 to 65 years had premature CHD and suggested that reduced low-density lipoprotein-cholesterol (LDL-C) levels in TD patients provide cardiovascular protection, while TD patients with normal LDL-C levels are likely to develop premature CHD [10]. A more recent review reported angina in 24.8%, and other vascular diseases in 21.8% of TD cases. Patients with CHD had a higher mean age, and while total cholesterol and LDL-C levels were higher in CHD than in non-CHD TD patients, the differences were only statistically significant in women [15]. The presence of small-dense LDL-C particles in some TD patients are also thought contribute to the development of CHD [10,15].

3. ABCA1 and Plasma Lipid Levels

3.1. ABCA1, Reverse Cholesterol Transport and Plasma HDL-C Levels

The key role of ABCA1 in RCT and HDL metabolism is evident as *ABCA1* gene mutations causing Tangier disease are associated with extremely low plasma HDL-C levels, a characteristic feature of the disease [16,17]. Murine tissue-specific knockout (KO) models have shown that cholesterol efflux via hepatic *Abca1* is responsible for 70% [18], whereas intestinal *Abca1* is responsible for 30%, of the biogenesis of HDL-cholesterol [19], thus leaving only a minute fraction of total cholesterol efflux to arterial wall macrophages.

In the first step of RCT, ABCA1 exports excess cellular cholesterol and phosphatidyl-choline (PC) to circulating lipid-free ApoA-I [20]. This generates nascent HDL, a bilayer fragment formed by 200 to 700 lipids wrapped by two to four ApoA-I molecules [21,22]. Two different models have been proposed to explain how nascent HDL-C particles are formed. According to the direct loading model, ABCA1 transfers lipids to ApoA-1 directly while it is bound to the transporter. In the indirect model, the phospholipid translocation activity of the ABCA1 protein forms specific membrane domains, and ApoA-I acquires lipids through these domains. The existence of two types of ApoA-I binding sites on the plasma membranes of cells expressing ABCA1 (a high-affinity/low-capacity binding site and a low-affinity/high-capacity binding site) supports the indirect model [23,24]. This model is also supported by the observation that ApoA-I alone can bind to high curvature liposomes and spontaneously form discoidal HDL particles in vitro [25]. Recently, in baby hamster kidney/ABCA1 cells, Ishigami et al. reported that trypsin treatment causes rapid release of PC and cholesterol, suggesting that these lipids are temporarily sequestered at trypsin-sensitive sites on the surface of cells in an ATP-dependent manner. Thus, these sites may be the large extracellular domains (ECDs) of ABCA1, and the lipids may be temporarily sequestered within these ECDs during nascent HDL formation [26]. Although further studies are required to establish the molecular details of the mechanistic links between the ECDs of ABCA1 and the known functions of the transporter, it is clear that ABCA1 function is the first and a crucial step for HDL-C formation.

3.2. ABCA1 Gene Variation Is Associated with HDL-C Levels

ABCA1 is a highly polymorphic gene located on human chromosome 9 (9q31.1) containing 50 exons [27]. According to the NCBI genetic variation database (https://www.ncbi.nlm.nih.gov/SNP), over 5000 polymorphisms have been reported in or near this gene. Several of these variants (intronic, missense and located in the promoter region) have important effects on the expression and function of the ABCA1 protein [28,29].

Both rare and common genetic variations in *ABCA1* contribute to circulating levels of HDL-cholesterol in population-based studies. Genome-wide association studies (GWAS) have consistently identified *ABCA1* as a locus associated with HDL-C levels in various ethnic groups [30,31]. Three nonsynonymous *ABCA1* polymorphisms have been extensively studied in terms of their associations with plasma lipid levels and CHD risk over the past two decades: rs2230806 (R219K) [32–37], rs2066714 (I883M) [33,35,38–40] and rs2230808 (R1587K) [33,38]. A recent meta-analysis confirmed the association of these three variants with plasma lipid levels [27]. Notably, a functional *ABCA1* missense variant (rs9282541; R230C) that was found to be private to the Americas was strongly associated with low HDL-C levels in Mexican mestizos and Native American populations [41,42]. This variant is of particular interest because it decreases cholesterol efflux capacity of the protein, is relatively frequent in the Mexican mestizo population (minor allele frequency is approximately 10%) and the sole presence of the risk allele explains almost 4% of plasma HDL-C variation.

Few studies have reported interactions between *ABCA1* gene variants and dietary macronutrient proportions affecting plasma lipid levels. In the Mexican population, two independent studies observed that the inverse correlation between carbohydrate intake and HDL-C concentrations was of higher magnitude in premenopausal women bearing the *ABCA1*/R230C variant [43,44]. Jacobo-Albavera et al. also reported that premenopausal women carrying the *ABCA1*/R230C risk allele, and consuming lower fat and higher carbohydrate dietary proportions, showed an overall unfavorable metabolic pattern including lower HDL-C levels. This suggests that gene-diet interactions play a role in inter-individual lipid level variations and may provide information useful to design diet intervention studies. In this regard, a study in Mexican individuals with hyperlipidemia reported that those bearing the *ABCA1*/R230C variant showed lower HDL concentrations and were better responders to a dietary portfolio treatment designed to increase plasma HDL-C concentrations [45]. Altogether, these studies demonstrate the relevance of the *ABCA1*/R230C variant on the regulation of HDL-C levels in the Mexican population.

3.3. ABCA1, miRNAs and HDL-C Levels

MicroRNAs regulate the expression of most genes associated with HDL metabolism, including *ABCA1*, *ABCG1* and the scavenger receptor *SRB1*. This implies that miRNAs regulate HDL biogenesis, cellular cholesterol efflux and HDL-C hepatic uptake, thereby controlling all steps involved in RCT [46].

Several miRNAs targeting *ABCA1* and regulating HDL-C plasma levels have been identified. miR-33a and miR-33b are embedded in intronic regions of the *SREBF2* and *SREBF1* genes which encode the SREBP2 and SREBP1 transcription factors that control the expression of genes involved in cholesterol and fatty acid synthesis [47,48]. Both miR-33a and miR-33b are coregulated with their host genes and repress gene programs that oppose SREBP functions like cholesterol efflux and fatty acid oxidation. The physiological relevance of miR-33 targeting of *ABCA1* was initially demonstrated using miR-33 inhibitors, which caused a two-fold increase of cholesterol efflux from hepatocytes to ApoA-I in vitro [47] and a 30% increase of plasma HDL-C levels in mice [48]. Moreover, targeted deletion of miR-33 caused a 25% increase in plasma HDL-C in male, and a 40% increase in female miR-33 null mice [49].

ABCA1 has a long 3′ UTR (>3.3 kb), making it especially susceptible to miRNA post-transcriptional control. miR-758 [50], miR-26 [51] and miR-106b [52] have also been found to repress ABCA1 and cholesterol efflux in vitro. In addition, two independent research groups stated that miR-144, an intergenic miRNA present in the miR-451 bicistronic cluster, also targets liver ABCA1 and modulates HDL-cholesterol plasma levels [53,54]. Moreover, in vivo activation of the farnesoid X nuclear receptor (FXR) increased hepatic miR-144 levels, which, in turn, decreased hepatic ABCA1 and plasma HDL-C levels. In vitro miR-144 overexpression decreased both cellular ABCA1 protein and cholesterol efflux to lipid-poor ApoA-I, while in vivo overexpression reduced hepatic ABCA1 protein and plasma HDL-cholesterol. Conversely, hepatic ABCA1 protein and HDL-cholesterol were increased by silencing miR-144 in mice. In addition, studies in tissue-specific FXR deficient mice showed that hepatic but not intestinal FXR is essential for induction of miR-144 and FXR-dependent hypolipidemia. Interestingly, miR-144 was found to have sex-specific silencing effects [55]. Finally, miR-148a was found to control in vivo hepatic ABCA1 expression and circulating HDL-C levels, revealing a role for miR-148a as a key regulator of hepatic LDL-C clearance through direct modulation of LDLR expression, and showing the therapeutic potential of miR-148a inhibition to improve the elevated LDL-C/HDL-C ratio, a significant risk factor for cardiovascular disease [56].

Overall, these findings suggest that deregulated miRNAs can impact ABCA1 and RCT gene networks. These observations have generated singular interest in identifying novel targets for epigenetic regulation that may lead to novel strategies to raise functional HDL, promote RCT and help prevent atherosclerosis and CHD, which remains an essential challenge.

3.4. ABCA1, LDL-C and Triglyceride (TG) Serum Levels

While the effect of ABCA1 on HDL-C plasma levels is clear, the effect of ABCA1 loss of function on other lipid traits is less evident. Several authors report that TD patients have significantly elevated plasma TG levels and reduced LDL-C concentrations than normal controls, although plasma TG levels vary in these patients [57–60]. Clee et al. reported overall higher TG levels in subjects heterozygous for *ABCA1* mutations than in controls, although TG levels were variable and not elevated in all mutation carriers [61]. Using an extreme phenotype approach, Frikke-Schmidt et al. described nine patients heterozygous for *ABCA1* mutations with very low HDL-C levels, six of which had elevated TG levels (>2.2 mmol/L) [12]. In contrast, heterozygous carriers of *ABCA1* mutations have no significant change in LDL-C levels [62–64].

Most GWAS have not reported *ABCA1* as a locus associated with TG and LDL-C levels [65–68]. However, recent multiethnic GWAS, including hundreds of thousands of cases and controls, have identified different *ABCA1* variants associated with TG (rs2575876, rs1799777, rs1883025 and rs1800978) and LDL-C levels (rs7873387, rs2575876, rs2740488, rs11789603, rs2066714) with genome-wide significance, although with small effect sizes [69–72]. In one of these studies, associations with TG levels were observed in current drinkers and/or regular drinkers [69]. In addition, several candidate gene studies also reported *ABCA1* SNPs associated with LDL-C and TG levels, with inconsistent results. While various studies failed to find associations of *ABCA1* gene variation with TG and/or LDL-C levels [40,73–75], *ABCA1* polymorphisms were associated with TG levels but not with LDL-C levels in Brazilians [76], Chinese [77], Turkish [78], Iranians [79] and Mexican school-aged children [80]. The functional *ABCA1*/R230C variant was associated with lower triglyceride levels only in Pimas and Mayans, but not in Mexican mestizos [41]. Moreover, *ABCA1* gene variants have been associated with LDL-C but not TG levels in a cohort of Greek nurses [81], and in male individuals with hypercholesterolemia [82]. In a large multiethnic cohort studying over 150 common variants, *ABCA1* was associated with both TG and LDL-C levels [83].

Postprandial hypertriglyceridemia is an important factor in developing atherosclerotic plaque and is closely related to the occurrence of cardiovascular events [84,85]. Although TG levels are usually estimated in a fasting state, several epidemiological studies have demonstrated that non-fasting hyperlipidemia is more harmful [86–88]. The high interindividual variability of TG levels observed in TD may be due to the inherent heterogeneity in individual triglyceride levels in different postprandial dietary lipid absorption states [89,90]. While several studies have documented single candidate SNPs associated with postprandial TG metabolism modulation [91,92], studies analyzing the effect of *ABCA1* gene variants on postprandial lipid metabolism are scarce. A recent study identified that most of the interindividual variability in the postprandial chylomicron TG response to dietary fat in healthy male adults could be explained by a combination of 42 SNPs in 23 genes, including *ABCA1* [91]. Moreover, Delgado-Lista et al. showed that major allele homozygotes for rs2575875 and rs4149272 had lower postprandial increases in TG and large-triglyceride rich lipoproteins, suggesting these variants may regulate the clearance of postprandial triglycerides [93].

It is evident that altered ABCA1 function and gene variation do not always affect TG levels. This may have to do with ethnicity, sex-specific effects and with interactions with other gene variants and environmental factors. In this regard, a small number of studies have reported interactions between *ABCA1* gene variants and dietary macronutrient proportions affecting plasma TG levels. In the Mexican population, premenopausal women carrying the *ABCA1*/R230C risk allele and consuming higher carbohydrate/lower fat diets showed an unfavorable metabolic pattern including higher TG levels, with a statistically significant interaction [43]. An independent study in the Inuit population also reported an interaction between the *ABCA1/R219K* variant with saturated fat intake affecting plasma TG levels [94]. These facts indicate that gene-diet interactions may help better predict inter-individual variations in plasma lipid levels and may provide information useful to design diet intervention studies.

3.5. ABCA1 Gene Variation and Coronary Heart Disease

Because low HDL-C levels are a well-established independent risk factor for CHD, genetic variants known to increase HDL-C levels would be expected to decrease CHD risk, and variants associated with lower HDL-C levels would increase CHD risk. However, high HDL-C levels are not always protective of CHD, and Mendelian randomization studies suggest that the inverse relationship between HDL-C levels and CHD risk is not causal [95,96]. Possible explanations are differences in the functionality of HDL-C particles, and pleiotropic effects of ABCA1 [97]. Interestingly the R230C variant was found to be associated with both lower HDL-C levels and lower risk of premature coronary artery disease in the Mexican population [98]. While the possible effects of this and other variants on HDL-C functionality require further study, it is possible that the paradoxical effect of this variant could be due to a pleiotropic effect on platelet, endothelial and leukocyte-derived microparticle formation, all involved in atherosclerosis and CHD pathogenesis. This is the matter of ongoing research by our group.

4. ABCA1, Glucose Metabolism and Type 2 Diabetes

β-cell failure and insulin resistance in muscle and liver represent the core pathophysiologic defects in type 2 diabetes [99]. Although ABCA1 and cholesterol homeostasis are critical in β-cell function and play a role in insulin resistance, global loss of ABCA1 function is not enough to cause type 2 diabetes (T2D). Diabetes is not a characteristic feature of Tangier disease and was not a feature reported in global $Abca1^{-/-}$ mice [100], although some consider diabetes as a complication of Tangier [101]. Moreover, while several patients suffering simultaneously from both diseases have been reported in the medical literature, particularly in the Japanese population [60,102–104], there are no reports on whether the prevalence of T2D is higher in Tangier patients. Still, several lines of evidence including tissue-specific *Abca1* KO models, human gene variation and ABCA1 expression studies point to a strong role of cholesterol homeostasis and ABCA1 in β-cell organization, function, and survival. Additionally, studies in muscle cell, hepatocyte and adipocyte-specific KO models, and some studies in humans, have shown ABCA1 also plays a role in peripheral insulin resistance. Altogether, this suggests that ABCA1 function is one of many factors which, acting in concert, contribute to the etiology of T2D.

4.1. ABCA1, Cholesterol and β-Cell Function

In addition to free fatty acid and triglyceride-mediated lipotoxicity, cholesterol toxicity is known to affect β-cell function and survival. β-cells are remarkably influenced by both intracellular cholesterol content and cholesterol distribution in the plasma membrane. Several transgenic and KO models have shown that increased cholesterol levels in β-cells reduce islet function, islet mass, and reduce insulin secretion by interfering with normal insulin secretory pathways [105–109]. In addition, cholesterol is important for maintaining the cholesterol-rich lipid rafts in the β-cell plasma membrane. By mediating the action of voltage-gated calcium channels and SNARE proteins, these lipid rafts mediate secretory stimuli and granule exocytosis/insulin secretion [110–113]. Being a cholesterol transporter affecting intracellular cholesterol concentrations and cholesterol membrane distribution, ABCA1 is thus expected to play a critical role in islet cholesterol homeostasis, β-cell function, insulin resistance and T2D.

Pancreatic β-cell specific *Abca1* KO mice (*Abca1*$^{-P/-P}$) showed age-related and gene-dose-dependent accumulation of cholesterol in β-cells. In addition, these mice showed significantly decreased insulin secretion in response to an acute glucose challenge in vivo, along with progressive glucose tolerance impairment, which was not related to islet development or β-cell mass. The lack of ABCA1 in β-cells was later found to disrupt insulin granule exocytosis [109]. After loading $Abca1^{-/-}$ β-cells with cholesterol, Ca^{2+} influx in response to glucose stimulation decreased. These cells had a defective depolarization of the membrane and KCl-induced exocytosis. Interestingly, cholesterol depletion rescued the exocytotic defect in β-cells lacking ABCA1, supporting the notion that cholesterol accumulation plays an important role in the dysfunction of insulin secretion [109].

It is noteworthy that mice lacking *Abca1* specifically in β-cells have a more severe impairment in β-cell function compared with mice lacking *Abca1* globally. Because $Abca1^{-P/-P}$ mice have higher levels of total plasma cholesterol than global *Abca1* KO mice, the degree of β-cell dysfunction caused by *Abca1* deficiency may be related to the level of plasma cholesterol to which the islets are exposed [105]. Thus, beneficial reductions in plasma lipids may limit the extent of β-cell damage [114].

4.2. ABCA1 and Insulin Sensitivity

Although global $Abca1^{-/-}$ KO mice did not show alterations in insulin sensitivity, multiple lines of evidence suggest that ABCA1 is involved in this trait. The interaction of HDL particles and ApoA-I results in the phosphorylation and activation of AMP-activated protein kinase (AMPK), a key metabolic enzyme that increases glucose uptake in murine endothelial cells, monocytes and skeletal muscle cells [115,116]. Similarly, in primary skeletal muscle cell cultures from T2D patients, HDL/ApoA-I bound to muscle cell surface receptors (including ABCA1), inducing intracellular Ca^{2+} mobilization, AMPK activation and glucose uptake. Antibody-mediated ABCA1 blockade inhibited HDL/ApoA-I glucose uptake and Ca^{2+} release in vitro, suggesting that HDL/ApoA-I modulates skeletal muscle glucose uptake in an ABCA1-dependent manner [117]. More recently, lipid-free ApoA-I was found to increase insulin-dependent and insulin-independent glucose uptake in primary human skeletal muscle cells, which were regulated by both ABCA1 and SR-B1, and this regulation seemed to be independent of ApoA-I acting as an acceptor of cellular cholesterol [118].

Moreover, observations in adipocyte-specific $Abca1^{-ad/-ad}$ KO mice suggest a critical role for adipocyte intracellular cholesterol and ABCA1 in whole-body glucose homeostasis. These mice showed impaired glucose tolerance and lower muscle insulin sensitivity, along with significant changes in the adipose tissue expression of genes involved in cholesterol and glucose homeostasis, including *ldlr*, *abcg1*, *glut-4*, visfatin, adiponectin, and leptin. They also showed lower glucose-stimulated insulin secretion from β-cells ex vivo. Notably, reduced muscle-tissue insulin sensitivity and glucose tolerance were observed in *Abca1*-deficient mice fed a high fat, high cholesterol diet, suggesting that adipocyte ABCA1 is crucial for proper adipose tissue function in response to dietary fat and cholesterol [119]. Moreover, hepatocyte-specific *Abca1* KO mice (HSKO) produced a form of selective insulin resistance, suppressing lipogenesis but with normal glucose metabolism [120]. HSKO mice had reduced hepatic insulin-stimulated Akt phosphorylation, decreased SREBP-1c activation and reduced expression of lipogenic genes, but normal glucose and insulin tolerance.

4.3. ABCA1 Gene Variation and T2D

There are few studies analyzing β-cell function and insulin sensitivity in human heterozygotes for loss-of-function *ABCA1* mutations, most likely because these mutations are extremely scarce. In consistency with the mouse model, a small study (15 individuals with loss-of function *ABCA1* mutations vs 14 family controls) reported that heterozygosity for these mutations was associated with impaired insulin secretion, mild hyperglycemia and reduced first-phase insulin response to hyperglycemia. However, hyperglycemic clamp studies showed that mutation carriers had normal insulin secretion in response to an oral glucose challenge and had normal insulin sensitivity [64]. Notably, none of the *ABCA1* mutation carriers had diabetes, suggesting that heterozygosity alone confers a relatively mild susceptibility for diabetes. In contrast, a large study including 94 *ABCA1* heterozygotes from the Copenhagen City Heart and the Copenhagen General Population Studies did not find an association with increased T2D risk [121].

Similarly, associations of *ABCA1* polymorphisms with T2D are not always consistent. ABCA1 has not been reported to be significantly associated with T2D in genome-wide association studies [122–124]. However, candidate-gene studies have reported associations of *ABCA1* polymorphisms with T2D mostly in Asian and Latin American populations. Notably, several of these studies are small, including only hundreds of cases and controls. Daimon et al. (2005) were the first to report an association of *ABCA1* gene polymorphisms (a 34-SNP haplotype of the promoter region) with T2D in a small sample of the Japanese population [125]. A few years later, a functional variant (R230C), which decreases ABCA1 cholesterol efflux capability, was associated with early-onset T2D in two independent small cohorts of the Mexican population [126]. Interestingly, R230C was only marginally associated with T2D in Pimas [41], but significantly associated with T2D in Mayan individuals [127], and was not found to be associated with T2D in a case-control study of the Colombian population [128]. Moreover, the missense rs2230806 (R219K), frequently associated with higher HDL-C levels, was found to be associated with decreased T2D risk in a recent meta-analysis including Korean, Chinese and Indian individuals [129]. Several small studies have sought to associate rs1800997, a 5′UTR variant known as the C69T polymorphism, with T2D, with inconsistent results. The minor *ABCA1*/C69T allele was associated with decreased T2D risk in Turkish [130], Saudi [131], and Chinese Han [132] individuals, while the intronic rs4149313 variant was associated with increased T2D risk in a study including 8842 Koreans [133].

In addition to small sample sizes, which may limit statistical power, other factors could explain inconsistencies in studies seeking associations of *ABCA1* gene variation with T2D. According to observations in global and β-cell specific *Abca1* KO models, differences in serum lipid levels may be a determinant factor. Dyslipidemia is highly prevalent in the Mexicans [134], which is consistent with the association of the *ABCA1*/R230C variant with T2D in this population. Likewise, the association of this variant with lower total cholesterol and TG levels found in Pimas could be a factor explaining why it was only marginally associated with T2D in this group [41]. In addition, *Abca1* adipocyte and hepatocyte-specific KO models have shown that a high fat high cholesterol diet may influence the effect of ABCA1 impairment on certain traits [119,120]. In this regard, dietary macronutrient proportions have been found to modulate the effect of the *ABCA1*/R230C not only on lipid levels, but on other metabolic parameters such as homeostasis assessment model for insulin resistance (HOMA-IR), serum adiponectin levels and visceral to subcutaneous abdominal fat ratio [43]. In this study, lower proportions of dietary carbohydrate and higher proportions of dietary fat were associated with a more favorable metabolic profile in premenopausal women bearing the R230C variant. Because these gene-diet interactions were observed only in premenopausal women, gender effects on the associations with T2D are also likely.

5. ABCA1 and Liver Disease

The ABCA1 transporter is ubiquitous, is expressed in a wide variety of tissues and contributes importantly to the plasma HDL-C pool. Hepatic ABCA1 expression promotes cellular free cholesterol flow and improves RCT, transferring excess cholesterol from peripheral tissues to HDL and finally to the liver for the synthesis and excretion of bile acids [135,136].

Although *Abca1* gene inactivation in mice may increase lipid storage in hepatocytes and leads to the accumulation of sterols in some tissues [137–139], rare or common *ABCA1* gene variation seems not to be associated with nonalcoholic fatty liver disease (NAFLD). However, increased lipid and liver cholesterol deposition are known to play a role in the progression of steatosis to nonalcoholic steatohepatitis (NASH) [140–142]. Likewise, in patients with morbid obesity, Vega-Badillo et al. reported that miR-33a/144 hepatic expression and their target ABCA1 are associated with NASH [143]. Additional research is needed to conclude the role of ABCA1 in liver disease including its association with NAFLD/NASH.

6. ABCA1 in Neurological Disease

Cholesterol homeostasis is essential for the central nervous system (CNS). Approximately 23% of total body cholesterol is found in the CNS. Brain cholesterol is mainly synthesized in situ, as essentially no cholesterol enters the brain from the peripheral circulation [144]. Moreover, CNS growth and differentiation requires cholesterol produced by de novo synthesis [144,145]. The capability of neurons to biosynthesize cholesterol decreases in adulthood and depends mainly on glial cells [146]. ABCA1 is expressed in neurons and astrocytes, where it promotes the efflux of phospholipids and unesterified cholesterol to glia-derived apolipoprotein E (apoE) [147]. ApoE is the main apolipoprotein found and synthesized in the brain and is found in the interstitial and cerebrospinal fluid in the form of lipid-rich ApoE particles. The density and size of these particles are similar to those of plasma HDL [148]. ABCA1 contributes to cholesterol homeostasis and participates in the pathophysiology of neurological diseases involving the accumulation of proteins in brain cells, such as traumatic brain injury, stroke sequelae, Parkinson's disease, and Alzheimer's disease (AD) [149–157].

AD is a neurodegenerative disorder clinically characterized by progressive memory loss, disorientation and cognitive decline [158]. At the histopathological level, characteristic amyloid plaques and neurofibrillary tangles are found in the brain tissue [159–161]. Amyloid plaques develop from the accumulation of amyloid β peptide (Aβ) [161]. ApoE plays a crucial role in the proteolytic degradation of soluble forms of Aβ, and this effect is dependent of apoE lipidation by ABCA1-mediated cholesterol and phospholipid transfer [162]. The ABCA1 protein participates in this process by regulating apoE levels and function in the CNS [163–167].

In murine models, ABCA1 deficiency ($Abca1^{-/-}$) was found to reduce apoE protein levels in the brain, to decrease lipidation of astrocyte-secreted apoE and to favor rapid apoE degradation [167,168]. *Abca1* deficiency may also increase amyloid burden in certain AD mouse models. Specifically, in a transgenic AD mouse model (APP23), targeted *Abca1* disruption (APP23/$Abca1^{-/-}$) increased amyloid deposition, increased the level of cerebral amyloid angiopathy, exacerbated cerebral amyloid angiopathy-related microhemorrhage, and caused a sharp decrease of soluble, but not of insoluble brain apoE levels [167]. Conversely, selective ABCA1 overexpression in AD mouse models led to increased CNS apoE lipidation and sharply decreased amyloid deposition [168], while ABCA1 upregulation by miRNA-33 inhibition was found to increase apoE lipidation and to decrease Aβ levels in the brain [169]. Notably, Fitz et al. reported that while *Abca1* deletion in transgenic APP mice caused cognitive deficit at a stage of early amyloid pathology, these characteristics were not observed in $Abca1^{-/-}$/wildtype mice. However, intra-hippocampal infusion of scrambled A oligomers affected cognitive performance of *Abca1* KO mice, which also showed altered neurite architecture in the hippocampus, suggesting that mice lacking ABCA1 have basal cognitive deficits that prevent them from coping with additional stressors [170].

Neuroinflammation and glucose metabolism are also important pathophysiological features in AD. Aβ deposits induce infiltration of immune cells such as T-helper 17 to the brain parenchyma and the secretion of proinflammatory cytokines such as interleukin 17A (IL-17A), which contribute to AD progression [171,172]. Interestingly, Yang et al. demonstrated that intracranial IL-17A overexpression increased ABCA1 protein levels in the hippocampus protein but not in cortex, decreased soluble Aβ levels in the hippocampus and cerebrospinal fluid, and improved glucose metabolism, suggesting that IL-17A may play a protective role in the pathogenesis of AD [173]. Moreover, hyperglycemic states are associated with greater severity of AD [174,175]. In this context, Lee et al. reported that in Zucker diabetic fatty rats (fa^-/fa^-) and in human neuroblastoma cells, exposure to high glucose levels increased Aβ deposition in the brain and decreased ABCA1 expression through JNK-reduced LXRα expression and binding to the *abca1* gene promoter [176].

Genetic studies support a role of ABCA1 in AD. Firstly, loss-of-function *ABCA1* mutations (N1800H) have been associated with low plasma apoE and increased AD risk in humans [177,178]. Moreover, although GWAS have consistently shown the crucial relevance of the *APOE4* variant in increasing AD risk across populations, *ABCA1* gene variation (rs3905000, rs27772082, rs2740488) has also been found to contribute to AD susceptibility in some GWAS [179–185]. Candidate gene studies analyzing the R219K polymorphism (rs2230806) and AD risk have reported conflicting results. This variant was associated with an increased risk of AD in Caucasian [186–189] and Chinese [51] populations, found to be a protective variant to AD in Chinese-Han and Hungarian individuals [190,191], and found not to be associated with AD risk in the German population [192]. However, two meta-analyses failed to find significant associations between *ABCA1* polymorphisms and AD [193,194].

It has been suggested that upregulation of ABCA1 expression or function may be a therapeutic target for AD and other diseases where Aβ plays a pathophysiological role. Interestingly, ABCA1 mediates the effect of some drugs proposed for AD treatment, such as bexarotene [195] and the liver X receptor agonist GW-3965 [164]. In addition, cyclodextrin [196], ondansetron [197], prostaglandin A1 [198], the purinergic receptor antagonist P2X7 [199], and the CS-6253 peptide [200] increase *ABCA1* gene expression in brain cells, although not all of these drugs improved cognitive function in vivo. Furthermore, Sarlak Z et al. reported that aerobic exercise significantly increases *Abca1* mRNA expression and decreases soluble Aβ1-42 in the hippocampus of rats with and without AD diagnosis. Aerobic training also improved cognitive function (learning and memory) [201]. ABCA1 and ApoE are currently the matter of intensive research for AD treatment [202].

7. ABCA1 and Microparticles

Microparticle (MP) release is a means for cell communication and cell-cell interaction, in addition to direct interaction and release of signaling molecules. MPs are small vesicles released from activated and/or apoptotic cells with substantial heterogeneity in size (50–250 nm). MPs include intracellular components involved in cell signaling and have membrane proteins characteristic of the original parent cell. It has been stablished that MPs are both biomarkers and cell signaling effectors that contribute to maintain and/or initiate cell dysfunction [203]. In a wide variety of thrombotic disorders, platelet and endothelial-derived MP levels are increased, with an interesting association between MP levels and pathophysiology, activity or progression of the disease [204]. MPs have procoagulant activity in several diseases including myocardial infarction [205,206], and may play a role in mediating inflammation-induced vascular calcification [207].

ABCA1 has a main role in facilitating outward bending or bulging of the plasma membrane [208]. It is currently known that the C-terminal of ABCA1 separately regulates its cholesterol floppase activity and cholesterol efflux activity [209]. Membrane dynamics are a prerequisite for HDL biogenesis and may also be required to release MPs to the medium [210]. ABCA1 and ApoA-I contribute to MP formation, mediating the production of MPs containing cholesterol. The addition of ApoA-I to human monocyte-derived macrophages markedly increased MP release, while ABCA1 inhibition with probucol and methyl-β cyclodextrin-induced membrane cholesterol depletion markedly reduced MP release and nascent HDL formation. MPs do not contain ApoA-I, but contain the plasma membrane marker flotilin-2, and CD63, an exosome marker. ABCA1 promotes cholesterol efflux, reduces cellular cholesterol accumulation and regulates anti-inflammatory activities in an ApoAI or annexin A1 (ANXA1)-dependent manner. ABCA1 anti-inflammatory activity seems to occur by mediating the efflux of ANXA1, which plays a critical role in anti-inflammatory effects, cholesterol transport, exosome and microparticle secretion and apoptotic cell clearance [211].

Although many studies have shown the importance of *ABCA1* gene variation in serum HDL-C levels, very few studies have reported the effect of gene variants on MP formation and their possible clinical consequences. It is known that ABCA1 participates in infectious and/or thrombotic disorders involving vesiculation [212], and in vitro studies and animal models indicate that ABCA1 also plays an important role in MP formation [21,208]. In Hamster kidney cells and mouse macrophages, ABCA1 was found not only to promote cholesterol efflux towards ApoA-I forming nascent HDL, but it also promoted the formation of ApoA-I-free MPs. This study also demonstrated that the *ABCA1* A937V mutation altered the formation of HDLs and concurrently reduced the release of MPs [208]. Moreover, in an experimental mouse model of cerebral malaria, Combes et al. evaluated the pathogenic implications of MP using *Abca1* deficient mice. Upon infection by *Plasmodium berghei ANKA*, these mice showed complete resistance to cerebral malaria, and MPs purified from infected animals were able to reduce normal plasma clotting time and to significantly enhance tumor necrosis factor release from naive macrophages [213,214]. *ABCA1* promoter variants associated with increased atherosclerotic burden [73] were found to be associated with decreased MP levels and were more prevalent in patients with uncomplicated malaria, suggesting that these polymorphisms have a protective effect against severe malaria in humans [215].

Calcium-dependent cytoskeleton proteolysis causes an eventual transient phospholipid density imbalance between the two plasma membrane leaflets driven by swift phosphatidylserine (PS) egress and lower reverse transport of phosphatidylcholine and sphingomyelin. This imbalance causes local instability of the plasma membrane and MP release upon raft clustering. The calcium-dependent channel TMEM16F plays a crucial role in calcium-induced phospholipid scrambling in the release of MPs exposing PS. TMEM16F mutations cause Scott Syndrome, a rare bleeding disorder characterized by defective platelet PS membrane exposure and MP shedding [216–218]. Because ABCA1 is known to have a role in exofacial PS translocation, Albrecht et al. analyzed the role of this protein in the pathophysiology of a Scott Syndrome patient who carried an *ABCA1* mutation (R1925Q). In vitro expression studies revealed that the 1925Q variant showed impaired trafficking to the plasma membrane, while wild-type ABCA1 overexpression in Scott Syndrome lymphocytes complemented the calcium-dependent PS exposure at the cell surface. Thus, this *ABCA1* mutation contributed to the defective PS translocation phenotype [219].

Abca1-deficient mice show alterations in PS exposure and significant reductions in circulating levels of MPs [212,220]. Moreover, silencing of *ABCA1* in human umbilical cord endothelial cell (HUVEC) cultures significantly reduced the release of MPs when subjected to frictional forces [221]. In this study, atheroprone shear stress conditions stimulated the formation and release of endothelial-derived MPs and hemodynamic forces were identified as an important determinant of MP plasma levels in healthy subjects. Sustained exposure to atheroprone low shear stress conditions increased both endothelial apoptosis and the release of MPs in the medium, when compared with physiological high shear stress conditions. Moreover, downregulation of ABCA1 expression by endogenously released nitric oxide (NO) contributed to limit the release of endothelial-derived MPs in HUVECs exposed to high shear stress [221].

8. ABCA1 in Infectious Diseases

The ABCA1 protein plays an important role in the development of some infectious diseases because of its role in cholesterol metabolism [222]. ABCA1 expression can be altered by some viruses, parasites and bacteria including components of the intestinal microbiota, [1,155,212,223–226], and some authors have proposed ABCA1 as a possible therapeutic target for these infections [212,227,228]. The entry and exit sites of some viral agents such as the human immunodeficiency virus, hepatitis C virus and cytomegalovirus occur in cholesterol, phospholipid and transporter enriched microdomains called lipid rafts [227,229–231]. The interaction of ABCA1 with these viruses alters lipid metabolism and intracellular signaling pathways [225,231,232].

8.1. Human Immunodeficiency Virus (HIV)

HIV is a retrovirus that infects and depletes CD4 T lymphocytes, causing slowly progressive immunodeficiency [225]. Despite antiretroviral therapy, people infected with HIV continue to develop comorbidities such as dyslipidemia, atherosclerosis and diabetes [228].

The role of the viral negative factor (Nef) protein and its association with cardiometabolic comorbidities has become of great interest in recent years (Figure 2). Nef is a multifunctional viral protein that alters the expression of different macromolecules on the surface of the host cell [233]. Nef decreases *ABCA1* gene expression, increases ABCA1 protein degradation in lysosomes and proteasomes by displacing it from the lipid rafts, and alters its maturation and folding in the endoplasmic reticulum by blocking its interaction with calnexin [234–238]. These events induce the accumulation of intracellular cholesterol in the host cell and increase the number of nonfunctional lipid rafts allowing virus survival and increasing virion production [228,239–242]. In addition, recent studies have shown that Nef can be released from infected cells through extracellular vesicles altering cholesterol metabolism in uninfected recipient cells [237,243–245].

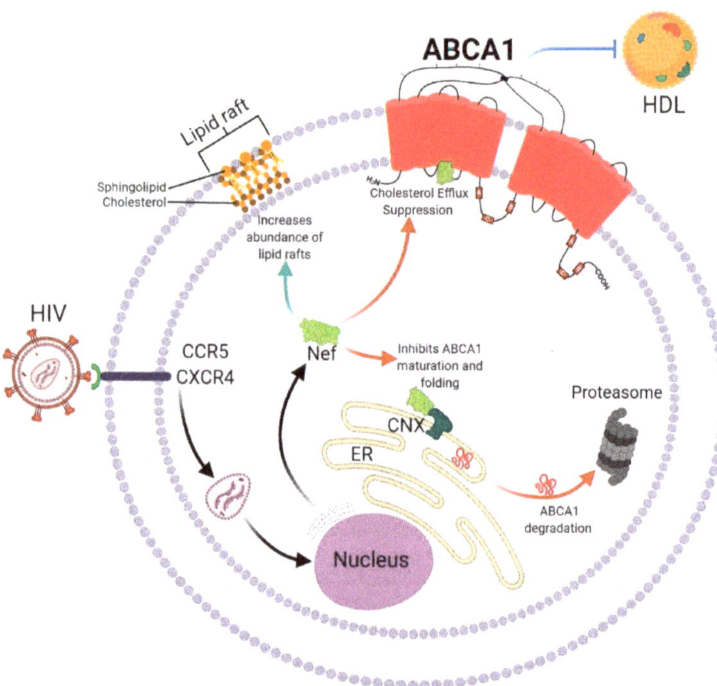

Figure 2. Effects of the viral negative factor (Nef) protein in human immunodeficiency virus (HIV)-infected cells. HIV enters cells by binding to the chemokine receptor 5 (CCR5) and chemokine receptor type 4 (CXCR4) and uses the host cell machinery to synthesize viral proteins such as Nef. Nef increases cholesterol biosynthesis and induces lipid raft formation, required to produce new virions. Nef also inhibits cholesterol efflux by suppressing ABCA1 activity, inducing structural and functional modifications of high-density lipoproteins (HDL). In addition, Nef blocks the interaction of the endoplasmic reticulum (ER) chaperon calnexin (CNX) with ABCA1, altering its folding and maturation. Nonfunctional and misfolded ABCA1 is retained in the ER and degraded in the proteasome, resulting in further accumulation of intracellular cholesterol, creating a favorable microenvironment for viral replication and release.

It is well known that HIV patients develop dyslipidemia, and their HDL-C plasma concentrations can be as low as those of TD patients [60,246,247]. The Nef protein causes dyslipidemia, as it affects cholesterol efflux by reducing the expression of ABCA1 in in vitro and in vivo models [237,248–250]. Similarly, the accumulation of cholesterol in pancreatic β-cells alters their function, decreasing insulin release, predisposing HIV patients to diabetes [105,109,251]. In this context, some studies have shown that antiretroviral therapy not only reduces the viral load [249] but also increases ABCA1 expression, restoring cholesterol efflux and increasing HDL-C plasma concentrations [249,252]. A recent prospective study reported that new antiretroviral therapies mitigate the cardiometabolic effects of HIV, at least in the short term [246]. However, not all studies report cardiometabolic improvement [253,254]. Moreover, long-term antiretroviral therapy is associated with dyslipidemia, although it does not occur in all patients [255]. A study assessing the impact of 192 SNPs in HIV patients receiving antiretroviral therapy identified that the *ABCA1* rs4149313 was associated with decreased TG and increased HDL-C circulating levels [256], while an independent study reported that *ABCA1* rs2066714 was associated with a greater risk of dyslipidemia in patients under antiretroviral treatment [257]. In addition, because of the role of ABCA1 in viral replication, ongoing studies are also investigating whether functional ABCA1 gene variants affect HIV progression or severity.

8.2. Hepatitis C Virus (HCV)

HCV belongs to the *Flaviviridae* family, has marked tropism for liver parenchymal cells, and chronic HCV infection leads to liver cirrhosis and hepatocellular carcinoma [227,232,257]. ABCA1 was found to have low expression levels in hepatocellular carcinoma samples from patients with a history of HCV infection [257]. Consistently, HCV infection was found to increase miR-27a expression in vitro. This miRNA binds to the 3′UTR sequence of *ABCA1* mRNA decreasing ABCA1 protein levels [258]. These events increase cellular cholesterol content and promote virus replication [232,259]. Furthermore, pharmacologically-induced *ABCA1* gene expression caused lipid raft reorganization in human hepatocytes, inhibiting HCV infection [227].

8.3. Human Cytomegalovirus (HCMV) and Other Viruses

HCMV is an opportunistic pathogen associated with an increased risk of atherothrombosis. In vitro models demonstrated that HCMV infection and the viral US28 protein decrease ABCA1 expression in the host cell, altering the distribution of lipid-rich microdomains in the plasma membrane [229,260]. Finally, although cholesterol metabolism has been found to be altered in other viral infections such as dengue [261,262] and chikungunya [263], the direct participation of ABCA1 in these diseases has not been demonstrated [261,263,264].

8.4. Malaria

Malaria is a parasitic disease caused by *Plasmodium* infection, transmitted to humans through the bite of infected female *Anopheles* mosquitoes [265]. Cerebral malaria is a severe complication, occurring in approximately 1% of those infected, and is the main cause of death [266]. The role of ABCA1 in microvesicle formation seems to be relevant in the pathogenesis of cerebral malaria. *Abca1* KO mice showed lower plasma concentrations of microvesicles, and when infected with *Plasmodium berghei ANKA*, these mice show complete resistance to cerebral malaria. In addition, plasma tumor necrosis factor alpha concentrations were reduced, decreasing the proinflammatory and prothrombotic state [212]. Furthermore, *ABCA1* promoter variants were associated with increased microvesicle production and a higher risk of developing severe malaria in humans, suggesting that *ABCA1* genetic variation may confer susceptibility to the development of malaria and its complications [215]. Because miR-27a was found to inhibit *ABCA1* expression in vitro, to abolish microvesicle production and inhibit apoptotic mechanisms, this miRNA has been proposed as protective against cerebral malaria [267,268]. Interestingly, in a murine model of malaria infection during pregnancy, *Abca1* expression was increased in the endothelial cells of the yolk sac. This event may be the result of a compensatory mechanism to maintain cholesterol homeostasis and favor the development and survival of the fetus [269]. Thus, ABCA1 may have a dual role, sometimes favoring infection and sometimes conferring protection.

Further research is required to fully elucidate how ABCA1 and cholesterol homeostasis are involved in infections, and to establish whether ABCA1 can in fact be a therapeutic target.

9. ABCA1, Age-Related Macular Disease and Glaucoma

Age-related macular degeneration (AMD) is a leading cause of visual impairment and severe vision loss in individuals above 50 years of age. AMD is a multifactorial and complex disorder, where immunological factors, inflammation, lipid and cholesterol metabolism, angiogenesis and extracellular matrix are involved in the disease pathogenesis [270]. Early disease is characterized by the presence of cholesterol-rich extracellular deposits similar to atherosclerotic plaques underneath the retinal pigment epithelium (RPE) or in the subretinal space, called drusen or drusenoid deposits [271–276], which may lead to atrophic neurodegeneration or pathologic angiogenesis. Drusen contain polar lipids such as free cholesterol and phosphatidylcholine, as well as neutral lipids such as cholesteryl esters and apolipoproteins [277]; while drusenoid deposits seem to contain only free cholesterol and apolipoproteins [278].

Several lines of evidence suggest that cholesterol metabolism and ABCA1 are involved in AMD pathogenesis. Systemic disturbances in cholesterol metabolism causing altered lipoprotein subtype levels have been associated with AMD [279]. Moreover, while retinal abnormalities have not been reported in Tangier disease, GWAS and candidate gene studies have shown that *ABCA1* gene variation contributes to AMD susceptibility, although to a lesser degree than the complement factor H (*CFH*) Y402H and age-related macular susceptibility-2 (*ARMS2*) A69S polymorphisms, which are well established risk factors for AMD [280–285]. In addition, human RPE cells express *ABCA1* and other genes involved in lipid metabolism such as *SRBI*, and glyburide-mediated inhibition of ABCA1 and SRBI activity was found to abolish HDL-stimulated basal efflux of photoreceptor-derived lipids in cultured human RPE cells, supporting a role of RCT regulation in the pathogenesis of AMD [286]. Finally, murine KO models also support the role of ABCA1 and cholesterol metabolism in AMD pathogenesis. Targeted deletion of macrophage *Abca1* and *Abcg1* in mice led to age-associated extracellular cholesterol-rich deposits underneath the neurosensory retina similar to the drusenoid deposits observed in early stages of human AMD, and the mice developed impaired dark adaptation and rod photoreceptor dysfunction [287,288].

Glaucoma is the world's leading cause of irreversible blindness [289]. It is a degenerative optic neuropathy characterized by the progressive degeneration of retinal ganglion cells (RGC) and the retinal nerve fiber layer (RNFL), leading to visual impairment and eventually to blindness. Elevated intraocular pressure (IOP) is a major risk factor for most types of glaucoma. Primary open-angle glaucoma (POAG) is characterized by increased resistance to aqueous fluid outflow through the trabecular meshwork and is the most common form of glaucoma worldwide. Primary angle-closure glaucoma is caused by blocked access to the outflow tracks; and secondary exfoliation glaucoma is a sequela of exfoliation syndrome characterized by accumulation of a characteristic fibrillar material on the ocular lens and trabecular meshwork [290]. Mendelian forms of glaucoma are caused by mutations in *MYOC*, *OPTN* and *TBK1* genes [291]. However most cases of glaucoma are multifactorial, and various biological processes including lipid metabolism, cytokine signaling, membrane biology, extracellular matrix, fucose and mannose metabolism, cell and ocular development are involved in the pathophysiology of the disease.

Several lines of evidence including GWAS, animal models and in vitro studies suggest ABCA1 plays an important role in the pathophysiology of glaucoma, mainly POAG. Although glaucoma is not a characteristic of Tangier disease, GWAS for IOP and POAG have identified common variants in or near *ABCA1* (rs2472493 and rs2487032) among more than 50 loci in Asian and European Caucasian populations [292–295]. However, while ABCA1 and other genes involved in lipid metabolism were found to be associated with IOP and POAG, a Mendelian randomization study did not find any evidence for a causal association between plasma lipid levels and POAG risk [296].

ABCA1 is highly expressed in retinal ganglion cells and its expression is significantly higher in individuals with glaucoma and upregulated in high-IOP glaucoma murine models. This suggests that ABCA1 is involved in the normal biological functions and cell death of ganglion cells. A recent study reported evidence of a novel role for ABCA1 in IOP modulation via the regulation of the Cav1/eNOS/NO signaling, which is likely to be an important mechanism of pathogenesis in patients with POAG. Based on their findings, the authors suggest that enhancing the ABCA1 signaling pathway could be of therapeutic value in the treatment of glaucoma and ocular hypertension [297].

10. ABCA1 in Cancer

Cellular cholesterol homeostasis is highly regulated to maintain cell membrane integrity and to promote membrane-anchored signaling pathways, and this homeostasis is altered during cancer cell proliferation. Epidemiologic studies have associated high serum total cholesterol concentrations with decreased risk of cancer [298,299]. Furthermore, tumor cells have been found to show high levels of cholesterol, suggesting that cholesterol metabolism is increased in proliferating cancer tissues [300,301]. ABCA1-mediated cholesterol efflux is one of the major regulation pathways of cholesterol. Moreover, as cancer is a highly cooperative process of oncogenic mutations that causes multiple metabolic changes including changes in gene expression patterns, *ABCA1* was identified as one of the cooperation response genes, nonmutant genes synergistically downregulated by multiple cancer gene mutations in the processes of malignant cell transformation [302]. Thus, numerous studies have investigated the role of ABCA1 in cancer development.

There may be a dual role of ABCA1 in cancer, as several studies suggest that ABCA1 function has anticancer properties, although there is also epidemiological and experimental evidence suggesting it may be involved in progression of certain types of cancer. On one hand, diminished *ABCA1* expression in neoplastic breast and prostate tissue was associated with an increased rate of cancer cell proliferation [303,304]. Likewise, ABCA1 downregulation caused by *ABCA1* promoter hypermethylation, miR-183 degradation or loss of function mutations, led to elevated cholesterol levels in cancer cells, enhanced cell proliferation and inhibited apoptosis [305–309]. On the other hand, *ABCA1* has been classified as a member of a lipid metabolism gene expression signature (ColoLipidGene) related to poor prognosis in patients with colorectal cancer (CRC). This signature includes four overexpressed lipid metabolism-related genes [310]. ABCA1 has been proposed as a specific marker of triple-negative breast cancer (TNBC) since its expression was higher in TNBC tissues compared with noncancerous mammary tissues [311]. Additionally, high-level expression of ABCA1 in primary tumors of serous ovarian cancer was associated with reduced survival of the patients and enhanced tumor cell growth and migration [312].

Several groups have provided insights into the molecular mechanisms of ABCA1 in cancer biology to help understand its pathophysiology and to identify potential therapeutic targets. Among the mechanisms proposed for ABCA1 anticancer activity are the following. (1) Deficient ABCA1-mediated cholesterol efflux increases intracellular and mitochondrial cholesterol levels, which decreases mitochondrial membrane fluidity and inhibits mitochondrial permeability transition. This avoids the release of cell death-promoting molecules such as cytochrome c and the apoptosis-inducing factor [308,313]. (2) ABCA1 activity has been linked to lipid raft disruption, by redistributing cholesterol and sphingomyelin from raft to nonraft domains. This results in reduced Akt signaling activation, which is sensitive to raft integrity. Akt upregulation has been associated with prostate cancer progression [314,315]. In other words, ABCA1 downregulation causes Akt upregulation, which in turn promotes cancer cell growth. (3) ABCA1 is known to suppress hematopoietic cell proliferation. Somatic *ABCA1* mutations found in chronic myelomonocytic leukemia patients were found to impair cholesterol efflux and increase cell proliferation by enhancing the cholesterol-dependent IL3-receptor β pathway, which activates and protein-tyrosine kinase Janus kinase 2 (JAK2) and mitogen-activated protein kinase (MAPK) signaling [309] (Figure 3).

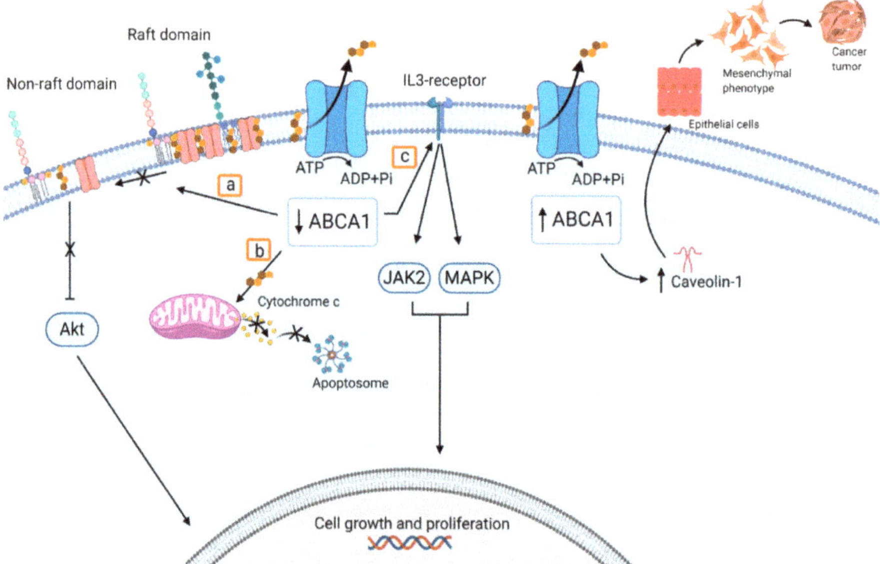

Figure 3. Proposed pathways of ABCA1 involvement in cancer. Downregulation of ABCA1 promotes cell growth and proliferation. a: ABCA1 downregulation avoids raft domain disruption and Akt pathway activation; b: ABCA1 downregulation causes cholesterol accumulation in the mitochondrial membrane inhibiting cytochrome c release and apoptosome formation; and c: ABCA1 downregulation activates the IL3-receptor, activating Janus kinase 2 (JAK2) and mitogen-activated protein kinase (MAPK) pathways. Upregulation of ABCA1 stabilizes caveolin-1 promoting epithelial-mesenchymal transition, and thus cell migration an invasion.

In contrast, other studies proposed mechanisms associated with ABCA1 activity in favor of cell cancer proliferation. For example, in colorectal cancer cell lines, ABCA1 overexpression led to an epithelial-to-mesenchymal transition and stabilized caveolin-1, known to promote cell migration, invasion, and has been proposed to be involved in tumor cell metastasis [316]. In addition, downregulated ABCA1 expression was found to prevent melanoma and bladder tumor growth in a syngeneic murine melanoma tumor model with a myeloid-specific *Abca1* deletion. Lack of Abca1 inhibited tumor bed accumulation of myeloid derived suppressor cells, known to promote tumor angiogenesis, metastasis and immune evasion, resulting in tumor growth inhibition [317].

Summarizing, although there is no conclusive evidence that ABCA1 is involved in the carcinogenesis process, unlike other members of the ABC transporter family (reviewed in [318]), it seems to play an important role in proliferation and survival of cancer cells. Moreover, while most studies suggest that ABCA1 activity is protective of cancer progression, there is also evidence of ABCA1 facilitating cell proliferation and tumor growth. Thus, the consequences of ABCA1 down or upregulation should be thoroughly investigated in different types and stages of cancer. Since intracellular cholesterol accumulation plays a key role in cancer progression, ABCA1 has been proposed as a potential therapeutic target; nevertheless, this subject needs further investigation.

11. Concluding Remarks

By regulating cholesterol homeostasis and plasma membrane dynamics, ABCA1 is involved in many physiological and pathological processes. ABCA1 protects cells from cholesterol toxicity by promoting cholesterol efflux. In addition, by regulating plasma membrane dynamics, ABCA1 plays a role in cell signaling and microparticle formation. Through these mechanisms, ABCA1 is involved in the pathogenesis of a broad array of diseases including dyslipidemia, atherosclerosis, coronary heart disease, type 2 diabetes, thrombosis, neurological disorders, age-related macular degeneration, glaucoma, viral infection, and cancer progression. However, most of these diseases have a complex etiology, where ABCA1 function is one of several factors playing a role in the pathophysiological process. Finally, although ABCA1 has been proposed as a therapeutic target for Alzheimer's disease, age-related macular degeneration, viral infections and other diseases, because the level of pleiotropy of this protein is high, tissue specific ABCA1 targeting may be important to achieve the desired therapeutic effect. A great deal of research is needed to further understand its physiological and pathological role, and the possibilities of targeting ABCA1 for therapy.

Funding: This research received no external funding.

Acknowledgments: All figures and graphical abstract were created with BioRender.com.

Conflicts of Interest: The authors declare no conflict of interest.

References

1. Simons, K.; Toomre, D. Lipid rafts and signal transduction. *Nat. Rev. Mol. Cell Biol.* **2000**, *1*, 31–39. [CrossRef]
2. Simons, K.; Ikonen, E. How cells handle cholesterol. *Science* **2000**, *290*, 1721–1726. [CrossRef]
3. Yeagle, P.L. Modulation of membrane function by cholesterol. *Biochimie* **1991**, *73*, 1303–1310. [CrossRef]
4. Tabas, I. Consequences of cellular cholesterol accumulation: Basic concepts and physiological implications. *J. Clin. Investig.* **2002**, *110*, 905–911. [CrossRef] [PubMed]
5. Oram, J.F.; Vaughan, A.M. ATP-Binding cassette cholesterol transporters and cardiovascular disease. *Circ. Res.* **2006**, *99*, 1031–1043. [CrossRef] [PubMed]
6. Marques, L.R.; Diniz, T.A.; Antunes, B.M.; Rossi, F.E.; Caperuto, E.C.; Lira, F.S.; Goncalves, D.C. Reverse Cholesterol Transport: Molecular Mechanisms and the Non-medical Approach to Enhance HDL Cholesterol. *Front. Physiol.* **2018**, *9*, 1–11. [CrossRef] [PubMed]
7. Burnett, J.R.; Hooper, A.J.; McCormick, S.P.A.; Hegele, R.A. Tangier Disease. In *GeneReviews((R))*; Adam, M.P., Ardinger, H.H., Pagon, R.A., Wallace, S.E., Bean, L.J.H., Stephens, K., Amemiya, A., Eds.; University of Washington: Seattle, WA, USA, 1993.
8. Hooper, A.J.; Hegele, R.A.; Burnett, J.R. Tangier disease: Update for 2020. *Curr. Opin. Lipidol.* **2020**, *31*, 80–84. [CrossRef]
9. Mercan, M.; Yayla, V.; Altinay, S.; Seyhan, S. Peripheral neuropathy in Tangier disease: A literature review and assessment. *J. Peripher. Nerv. Syst.* **2018**, *23*, 88–98. [CrossRef]
10. Schaefer, E.J.; Brousseau, M.E.; Diffenderfer, M.R.; Cohn, J.S.; Welty, F.K.; O'Connor, J., Jr.; Dolnikowski, G.G.; Wang, J.; Hegele, R.A.; Jones, P.J. Cholesterol and apolipoprotein B metabolism in Tangier disease. *Atherosclerosis* **2001**, *159*, 231–236. [CrossRef]
11. Puntoni, M.; Sbrana, F.; Bigazzi, F.; Sampietro, T. Tangier disease: Epidemiology, pathophysiology, and management. *Am. J. Cardiovasc. Drugs* **2012**, *12*, 303–311. [CrossRef]
12. Frikke-Schmidt, R.; Nordestgaard, B.G.; Jensen, G.B.; Tybjærg-Hansen, A. Genetic variation in ABC transporter A1 contributes to HDL cholesterol in the general population. *J. Clin. Investig.* **2004**, *114*, 1343–1353. [CrossRef] [PubMed]
13. Minuz, P.; Meneguzzi, A.; Femia, E.A.; Fava, C.; Calabria, S.; Scavone, M.; Benati, D.; Poli, G.; Zancanaro, C.; Calandra, S.; et al. Reduced platelet count, but no major platelet function abnormalities, are associated with loss-of-function ATP-binding cassette-1 gene mutations. *Clin. Sci.* **2017**, *131*, 2095–2107. [CrossRef] [PubMed]
14. Oram, J.F. Tangier disease and ABCA1. *Biochim Biophys Acta* **2000**, *1529*, 321–330. [CrossRef]
15. Muratsu, J.; Koseki, M.; Masuda, D.; Yasuga, Y.; Tomoyama, S.; Ataka, K.; Yagi, Y.; Nakagawa, A.; Hamada, H.; Fujita, S.; et al. Accelerated Atherogenicity in Tangier Disease. *J. Atheroscler. Thromb.* **2018**, *25*, 1076–1085. [CrossRef]
16. Fasano, T.; Zanoni, P.; Rabacchi, C.; Pisciotta, L.; Favari, E.; Adorni, M.P.; Deegan, P.B.; Park, A.; Hlaing, T.; Feher, M.D.; et al. Novel mutations of ABCA1 transporter in patients with Tangier disease and familial HDL deficiency. *Mol. Genet. Metab.* **2012**, *107*, 534–541. [CrossRef]
17. Oram, J.F.; Vaughan, A.M. ABCA1-mediated transport of cellular cholesterol and phospholipids to HDL apolipoproteins. *Curr. Opin. Lipidol.* **2000**, *11*, 253–260. [CrossRef]

18. Timmins, J.M.; Lee, J.Y.; Boudyguina, E.; Kluckman, K.D.; Brunham, L.R.; Mulya, A.; Gebre, A.K.; Coutinho, J.M.; Colvin, P.L.; Smith, T.L.; et al. Targeted inactivation of hepatic Abca1 causes profound hypoalphalipoproteinemia and kidney hypercatabolism of apoA-I. *J. Clin. Investig.* **2005**, *115*, 1333–1342. [CrossRef]

19. Brunham, L.R.; Kruit, J.K.; Iqbal, J.; Fievet, C.; Timmins, J.M.; Pape, T.D.; Coburn, B.A.; Bissada, N.; Staels, B.; Groen, A.K.; et al. Intestinal ABCA1 directly contributes to HDL biogenesis in vivo. *J. Clin. Investig.* **2006**, *116*, 1052–1062. [CrossRef]

20. Nagao, K.; Tomioka, M.; Ueda, K. Function and regulation of ABCA1–membrane meso-domain organization and reorganization. *FEBS J.* **2011**, *278*, 3190–3203. [CrossRef]

21. Duong, P.T.; Collins, H.L.; Nickel, M.; Lund-Katz, S.; Rothblat, G.H.; Phillips, M.C. Characterization of nascent HDL particles and microparticles formed by ABCA1-mediated efflux of cellular lipids to apoA-I. *J. Lipid Res.* **2006**, *47*, 832–843. [CrossRef]

22. Sorci-Thomas, M.G.; Owen, J.S.; Fulp, B.; Bhat, S.; Zhu, X.; Parks, J.S.; Shah, D.; Jerome, W.G.; Gerelus, M.; Zabalawi, M.; et al. Nascent high density lipoproteins formed by ABCA1 resemble lipid rafts and are structurally organized by three apoA-I monomers. *J. Lipid Res.* **2012**, *53*, 1890–1909. [CrossRef]

23. Hassan, H.H.; Denis, M.; Lee, D.Y.; Iatan, I.; Nyholt, D.; Ruel, I.; Krimbou, L.; Genest, J. Identification of an ABCA1-dependent phospholipid-rich plasma membrane apolipoprotein A-I binding site for nascent HDL formation: Implications for current models of HDL biogenesis. *J. Lipid Res.* **2007**, *48*, 2428–2442. [CrossRef]

24. Vedhachalam, C.; Ghering, A.B.; Davidson, W.S.; Lund-Katz, S.; Rothblat, G.H.; Phillips, M.C. ABCA1-induced cell surface binding sites for ApoA-I. *Arterioscler. Thromb. Vasc. Biol.* **2007**, *27*, 1603–1609. [CrossRef]

25. Gillotte, K.L.; Zaiou, M.; Lund-Katz, S.; Anantharamaiah, G.M.; Holvoet, P.; Dhoest, A.; Palgunachari, M.N.; Segrest, J.P.; Weisgraber, K.H.; Rothblat, G.H.; et al. Apolipoprotein-mediated plasma membrane microsolubilization. Role of lipid affinity and membrane penetration in the efflux of cellular cholesterol and phospholipid. *J. Biol. Chem.* **1999**, *274*, 2021–2028. [CrossRef]

26. Ishigami, M.; Ogasawara, F.; Nagao, K.; Hashimoto, H.; Kimura, Y.; Kioka, N.; Ueda, K. Temporary sequestration of cholesterol and phosphatidylcholine within extracellular domains of ABCA1 during nascent HDL generation. *Sci. Rep.* **2018**, *8*, 1–10. [CrossRef] [PubMed]

27. Lu, Z.; Luo, Z.; Jia, A.; Yu, L.; Muhammad, I.; Zeng, W.; Song, Y. Associations of the ABCA1 gene polymorphisms with plasma lipid levels: A meta-analysis. *Medicine* **2018**, *97*, 1–14. [CrossRef]

28. Qi, L.P.; Chen, L.F.; Dang, A.M.; Li, L.Y.; Fang, Q.; Yan, X.W. Association between the ABCA1-565C/T gene promoter polymorphism and coronary heart disease severity and cholesterol efflux in the Chinese Han population. *Genet. Test. Mol. Biomarkers* **2015**, *19*, 347–352. [CrossRef]

29. Slatter, T.L.; Jones, G.T.; Williams, M.J.; van Rij, A.M.; McCormick, S.P. Novel rare mutations and promoter haplotypes in ABCA1 contribute to low-HDL-C levels. *Clin. Genet.* **2008**, *73*, 179–184. [CrossRef] [PubMed]

30. Teslovich, T.M.; Musunuru, K.; Smith, A.V.; Edmondson, A.C.; Stylianou, I.M.; Koseki, M.; Pirruccello, J.P.; Ripatti, S.; Chasman, D.I.; Willer, C.J.; et al. Biological, clinical and population relevance of 95 loci for blood lipids. *Nature* **2010**, *466*, 707–713. [CrossRef]

31. Willer, C.J.; Schmidt, E.M.; Sengupta, S.; Peloso, G.M.; Gustafsson, S.; Kanoni, S.; Ganna, A.; Chen, J.; Buchkovich, M.L.; Mora, S.; et al. Discovery and refinement of loci associated with lipid levels. *Nat. Genet.* **2013**, *45*, 1274–1283. [CrossRef] [PubMed]

32. Andrikovics, H.; Pongracz, E.; Kalina, E.; Szilvasi, A.; Aslanidis, C.; Schmitz, G.; Tordai, A. Decreased frequencies of ABCA1 polymorphisms R219K and V771M in Hungarian patients with cerebrovascular and cardiovascular diseases. *Cerebrovasc. Dis.* **2006**, *21*, 254–259. [CrossRef]

33. Clee, S.M.; Zwinderman, A.H.; Engert, J.C.; Zwarts, K.Y.; Molhuizen, H.O.; Roomp, K.; Jukema, J.W.; van Wijland, M.; van Dam, M.; Hudson, T.J.; et al. Common genetic variation in ABCA1 is associated with altered lipoprotein levels and a modified risk for coronary artery disease. *Circulation* **2001**, *103*, 1198–1205. [CrossRef]

34. Evans, D.; Beil, F.U. The association of the R219K polymorphism in the ATP-binding cassette transporter 1 (ABCA1) gene with coronary heart disease and hyperlipidaemia. *J. Mol. Med.* **2003**, *81*, 264–270. [CrossRef] [PubMed]

35. Harada, T.; Imai, Y.; Nojiri, T.; Morita, H.; Hayashi, D.; Maemura, K.; Fukino, K.; Kawanami, D.; Nishimura, G.; Tsushima, K.; et al. A common Ile 823 Met variant of ATP-binding cassette transporter A1 gene (ABCA1) alters high density lipoprotein cholesterol level in Japanese population. *Atherosclerosis* **2003**, *169*, 105–112. [CrossRef]

36. Peloso, G.M.; Demissie, S.; Collins, D.; Mirel, D.B.; Gabriel, S.B.; Cupples, L.A.; Robins, S.J.; Schaefer, E.J.; Brousseau, M.E. Common genetic variation in multiple metabolic pathways influences susceptibility to low HDL-cholesterol and coronary heart disease. *J. Lipid Res.* **2010**, *51*, 3524–3532. [CrossRef]

37. Tregouet, D.A.; Ricard, S.; Nicaud, V.; Arnould, I.; Soubigou, S.; Rosier, M.; Duverger, N.; Poirier, O.; Mace, S.; Kee, F.; et al. In-depth haplotype analysis of ABCA1 gene polymorphisms in relation to plasma ApoA1 levels and myocardial infarction. *Arterioscler. Thromb. Vasc. Biol.* **2004**, *24*, 775–781. [CrossRef]

38. Jensen, M.K.; Pai, J.K.; Mukamal, K.J.; Overvad, K.; Rimm, E.B. Common genetic variation in the ATP-binding cassette transporter A1, plasma lipids, and risk of coronary heart disease. *Atherosclerosis* **2007**, *195*, e172–e180. [CrossRef]

39. Song, C.; Pedersen, N.L.; Reynolds, C.A.; Sabater-Lleal, M.; Kanoni, S.; Willenborg, C.; Consortium, C.A.D.; Syvanen, A.C.; Watkins, H.; Hamsten, A.; et al. Genetic variants from lipid-related pathways and risk for incident myocardial infarction. *PLoS ONE* **2013**, *8*, e60454. [CrossRef] [PubMed]

40. Tan, J.H.; Low, P.S.; Tan, Y.S.; Tong, M.C.; Saha, N.; Yang, H.; Heng, C.K. ABCA1 gene polymorphisms and their associations with coronary artery disease and plasma lipids in males from three ethnic populations in Singapore. *Hum. Genet.* **2003**, *113*, 106–117. [CrossRef] [PubMed]

41. Acuna-Alonzo, V.; Flores-Dorantes, T.; Kruit, J.K.; Villarreal-Molina, T.; Arellano-Campos, O.; Hunemeier, T.; Moreno-Estrada, A.; Ortiz-Lopez, M.G.; Villamil-Ramirez, H.; Leon-Mimila, P.; et al. A functional ABCA1 gene variant is associated with low HDL-cholesterol levels and shows evidence of positive selection in Native Americans. *Hum. Mol. Genet.* **2010**, *19*, 2877–2885. [CrossRef]

42. Villarreal-Molina, M.T.; Aguilar-Salinas, C.A.; Rodriguez-Cruz, M.; Riano, D.; Villalobos-Comparan, M.; Coral-Vazquez, R.; Menjivar, M.; Yescas-Gomez, P.; Konigsoerg-Fainstein, M.; Romero-Hidalgo, S.; et al. The ATP-binding cassette transporter A1 R230C variant affects HDL cholesterol levels and BMI in the Mexican population: Association with obesity and obesity-related comorbidities. *Diabetes* **2007**, *56*, 1881–1887. [CrossRef]

43. Jacobo-Albavera, L.; Posadas-Romero, C.; Vargas-Alarcon, G.; Romero-Hidalgo, S.; Posadas-Sanchez, R.; Gonzalez-Salazar Mdel, C.; Carnevale, A.; Canizales-Quinteros, S.; Medina-Urrutia, A.; Antunez-Arguelles, E.; et al. Dietary fat and carbohydrate modulate the effect of the ATP-binding cassette A1 (ABCA1) R230C variant on metabolic risk parameters in premenopausal women from the Genetics of Atherosclerotic Disease (GEA) Study. *Nutr. Metab.* **2015**, *12*, 1–11. [CrossRef]

44. Romero-Hidalgo, S.; Villarreal-Molina, T.; Gonzalez-Barrios, J.A.; Canizales-Quinteros, S.; Rodriguez-Arellano, M.E.; Yanez-Velazco, L.B.; Bernal-Alcantara, D.A.; Villa, A.R.; Antuna-Puente, B.; Acuna-Alonzo, V.; et al. Carbohydrate intake modulates the effect of the ABCA1-R230C variant on HDL cholesterol concentrations in premenopausal women. *J. Nutr.* **2012**, *142*, 278–283. [CrossRef]

45. Guevara-Cruz, M.; Tovar, A.R.; Larrieta, E.; Canizales-Quinteros, S.; Torres, N. Increase in HDL-C concentration by a dietary portfolio with soy protein and soluble fiber is associated with the presence of the ABCA1R230C variant in hyperlipidemic Mexican subjects. *Mol. Genet. Metab.* **2010**, *101*, 268–272. [CrossRef] [PubMed]

46. Canfran-Duque, A.; Ramirez, C.M.; Goedeke, L.; Lin, C.S.; Fernandez-Hernando, C. microRNAs and HDL life cycle. *Cardiovasc. Res.* **2014**, *103*, 414–422. [CrossRef] [PubMed]

47. Najafi-Shoushtari, S.H.; Kristo, F.; Li, Y.; Shioda, T.; Cohen, D.E.; Gerszten, R.E.; Naar, A.M. MicroRNA-33 and the SREBP host genes cooperate to control cholesterol homeostasis. *Science* **2010**, *328*, 1566–1569. [CrossRef] [PubMed]

48. Rayner, K.J.; Suarez, Y.; Davalos, A.; Parathath, S.; Fitzgerald, M.L.; Tamehiro, N.; Fisher, E.A.; Moore, K.J.; Fernandez-Hernando, C. MiR-33 contributes to the regulation of cholesterol homeostasis. *Science* **2010**, *328*, 1570–1573. [CrossRef]

49. Horie, T.; Ono, K.; Horiguchi, M.; Nishi, H.; Nakamura, T.; Nagao, K.; Kinoshita, M.; Kuwabara, Y.; Marusawa, H.; Iwanaga, Y.; et al. MicroRNA-33 encoded by an intron of sterol regulatory element-binding protein 2 (Srebp2) regulates HDL in vivo. *Proc. Natl. Acad. Sci. USA* **2010**, *107*, 17321–17326. [CrossRef]

50. Ramirez, C.M.; Davalos, A.; Goedeke, L.; Salerno, A.G.; Warrier, N.; Cirera-Salinas, D.; Suarez, Y.; Fernandez-Hernando, C. MicroRNA-758 regulates cholesterol efflux through posttranscriptional repression of ATP-binding cassette transporter A1. *Arterioscler. Thromb. Vasc. Biol.* **2011**, *31*, 2707–2714. [CrossRef]

51. Sun, Y.M.; Li, H.L.; Guo, Q.H.; Wu, P.; Hong, Z.; Lu, C.Z.; Wu, Z.Y. The polymorphism of the ATP-binding cassette transporter 1 gene modulates Alzheimer disease risk in Chinese Han ethnic population. *Am. J. Geriatr. Psychiatry* **2012**, *20*, 603–611. [CrossRef]

52. Kim, J.; Yoon, H.; Ramirez, C.M.; Lee, S.M.; Hoe, H.S.; Fernandez-Hernando, C.; Kim, J. MiR-106b impairs cholesterol efflux and increases Aβ levels by repressing ABCA1 expression. *Exp. Neurol.* **2012**, *235*, 476–483. [CrossRef]

53. de Aguiar Vallim, T.Q.; Tarling, E.J.; Kim, T.; Civelek, M.; Baldan, A.; Esau, C.; Edwards, P.A. MicroRNA-144 regulates hepatic ATP binding cassette transporter A1 and plasma high-density lipoprotein after activation of the nuclear receptor farnesoid X receptor. *Circ. Res.* **2013**, *112*, 1602–1612. [CrossRef] [PubMed]

54. Ramirez, C.M.; Rotllan, N.; Vlassov, A.V.; Davalos, A.; Li, M.; Goedeke, L.; Aranda, J.F.; Cirera-Salinas, D.; Araldi, E.; Salerno, A.; et al. Control of cholesterol metabolism and plasma high-density lipoprotein levels by microRNA-144. *Circ. Res.* **2013**, *112*, 1592–1601. [CrossRef] [PubMed]

55. Cheng, J.; Cheng, A.; Clifford, B.L.; Wu, X.; Hedin, U.; Maegdefessel, L.; Pamir, N.; Sallam, T.; Tarling, E.J.; de Aguiar Vallim, T.Q. MicroRNA-144 Silencing Protects Against Atherosclerosis in Male, but Not Female Mice. *Arterioscler. Thromb. Vasc. Biol.* **2020**, *40*, 412–425. [CrossRef] [PubMed]

56. Goedeke, L.; Rotllan, N.; Canfran-Duque, A.; Aranda, J.F.; Ramirez, C.M.; Araldi, E.; Lin, C.S.; Anderson, N.N.; Wagschal, A.; de Cabo, R.; et al. MicroRNA-148a regulates LDL receptor and ABCA1 expression to control circulating lipoprotein levels. *Nat. Med.* **2015**, *21*, 1280–1289. [CrossRef]

57. Bodzioch, M.; Orso, E.; Klucken, J.; Langmann, T.; Bottcher, A.; Diederich, W.; Drobnik, W.; Barlage, S.; Buchler, C.; Porsch-Ozcurumez, M.; et al. The gene encoding ATP-binding cassette transporter 1 is mutated in Tangier disease. *Nat. Genet.* **1999**, *22*, 347–351. [CrossRef]

58. Brooks-Wilson, A.; Marcil, M.; Clee, S.M.; Zhang, L.H.; Roomp, K.; van Dam, M.; Yu, L.; Brewer, C.; Collins, J.A.; Molhuizen, H.O.; et al. Mutations in ABC1 in Tangier disease and familial high-density lipoprotein deficiency. *Nat. Genet.* **1999**, *22*, 336–345. [CrossRef]

59. Liu, M.; Chung, S.; Shelness, G.S.; Parks, J.S. Hepatic ABCA1 deficiency is associated with delayed apolipoprotein B secretory trafficking and augmented VLDL triglyceride secretion. *Biochim. Biophys. Acta Mol. Cell Biol. Lipids* **2017**, *1862*, 1035–1043. [CrossRef]

60. Serfaty-Lacrosniere, C.; Civeira, F.; Lanzberg, A.; Isaia, P.; Berg, J.; Janus, E.D.; Smith, M.P., Jr.; Pritchard, P.H.; Frohlich, J.; Lees, R.S.; et al. Homozygous Tangier disease and cardiovascular disease. *Atherosclerosis* **1994**, *107*, 85–98. [CrossRef]

61. Clee, S.M.; Kastelein, J.J.; van Dam, M.; Marcil, M.; Roomp, K.; Zwarts, K.Y.; Collins, J.A.; Roelants, R.; Tamasawa, N.; Stulc, T.; et al. Age and residual cholesterol efflux affect HDL cholesterol levels and coronary artery disease in ABCA1 heterozygotes. *J. Clin. Investig.* **2000**, *106*, 1263–1270. [CrossRef]

62. Bochem, A.E.; van Wijk, D.F.; Holleboom, A.G.; Duivenvoorden, R.; Motazacker, M.M.; Dallinga-Thie, G.M.; de Groot, E.; Kastelein, J.J.; Nederveen, A.J.; Hovingh, G.K.; et al. ABCA1 mutation carriers with low high-density lipoprotein cholesterol are characterized by a larger atherosclerotic burden. *Eur. Heart J.* **2013**, *34*, 286–291. [CrossRef]

63. Kuivenhoven, J.A.; Hovingh, G.K.; van Tol, A.; Jauhiainen, M.; Ehnholm, C.; Fruchart, J.C.; Brinton, E.A.; Otvos, J.D.; Smelt, A.H.; Brownlee, A.; et al. Heterozygosity for ABCA1 gene mutations: Effects on enzymes, apolipoproteins and lipoprotein particle size. *Atherosclerosis* **2003**, *171*, 311–319. [CrossRef] [PubMed]

64. Vergeer, M.; Brunham, L.R.; Koetsveld, J.; Kruit, J.K.; Verchere, C.B.; Kastelein, J.J.; Hayden, M.R.; Stroes, E.S. Carriers of loss-of-function mutations in ABCA1 display pancreatic β-cell dysfunction. *Diabetes Care* **2010**, *33*, 869–874. [CrossRef] [PubMed]

65. Below, J.E.; Parra, E.J.; Gamazon, E.R.; Torres, J.; Krithika, S.; Candille, S.; Lu, Y.; Manichakul, A.; Peralta-Romero, J.; Duan, Q.; et al. Meta-analysis of lipid-traits in Hispanics identifies novel loci, population-specific effects, and tissue-specific enrichment of eQTLs. *Sci. Rep.* **2016**, *6*, 1–13. [CrossRef]

66. Kathiresan, S.; Melander, O.; Guiducci, C.; Surti, A.; Burtt, N.P.; Rieder, M.J.; Cooper, G.M.; Roos, C.; Voight, B.F.; Havulinna, A.S.; et al. Six new loci associated with blood low-density lipoprotein cholesterol, high-density lipoprotein cholesterol or triglycerides in humans. *Nat. Genet.* **2008**, *40*, 189–197. [CrossRef]

67. Proust, C.; Empana, J.P.; Boutouyrie, P.; Alivon, M.; Challande, P.; Danchin, N.; Escriou, G.; Esslinger, U.; Laurent, S.; Li, Z.; et al. Contribution of Rare and Common Genetic Variants to Plasma Lipid Levels and Carotid Stiffness and Geometry: A Substudy of the Paris Prospective Study 3. *Circ. Cardiovasc. Genet.* **2015**, *8*, 628–636. [CrossRef] [PubMed]

68. Weissglas-Volkov, D.; Aguilar-Salinas, C.A.; Nikkola, E.; Deere, K.A.; Cruz-Bautista, I.; Arellano-Campos, O.; Munoz-Hernandez, L.L.; Gomez-Munguia, L.; Ordonez-Sanchez, M.L.; Reddy, P.M.; et al. Genomic study in Mexicans identifies a new locus for triglycerides and refines European lipid loci. *J. Med. Genet.* **2013**, *50*, 298–308. [CrossRef]

69. de Vries, P.S.; Brown, M.R.; Bentley, A.R.; Sung, Y.J.; Winkler, T.W.; Ntalla, I.; Schwander, K.; Kraja, A.T.; Guo, X.; Franceschini, N.; et al. Multiancestry Genome-Wide Association Study of Lipid Levels Incorporating Gene-Alcohol Interactions. *Am. J. Epidemiol.* **2019**, *188*, 1033–1054. [CrossRef]

70. Klarin, D.; Damrauer, S.M.; Cho, K.; Sun, Y.V.; Teslovich, T.M.; Honerlaw, J.; Gagnon, D.R.; DuVall, S.L.; Li, J.; Peloso, G.M.; et al. Genetics of blood lipids among ~300,000 multi-ethnic participants of the Million Veteran Program. *Nat. Genet.* **2018**, *50*, 1514–1523. [CrossRef]

71. Richardson, T.G.; Sanderson, E.; Palmer, T.M.; Ala-Korpela, M.; Ference, B.A.; Davey Smith, G.; Holmes, M.V. Evaluating the relationship between circulating lipoprotein lipids and apolipoproteins with risk of coronary heart disease: A multivariable Mendelian randomisation analysis. *PLoS Med.* **2020**, *17*, e1003062. [CrossRef]

72. Ripatti, P.; Ramo, J.T.; Mars, N.J.; Fu, Y.; Lin, J.; Soderlund, S.; Benner, C.; Surakka, I.; Kiiskinen, T.; Havulinna, A.S.; et al. Polygenic Hyperlipidemias and Coronary Artery Disease Risk. *Circ. Genom. Precis Med.* **2020**, *13*, 59–65. [CrossRef] [PubMed]

73. Lutucuta, S.; Ballantyne, C.M.; Elghannam, H.; Gotto, A.M., Jr.; Marian, A.J. Novel polymorphisms in promoter region of atp binding cassette transporter gene and plasma lipids, severity, progression, and regression of coronary atherosclerosis and response to therapy. *Circ. Res.* **2001**, *88*, 969–973. [CrossRef] [PubMed]

74. Souverein, O.W.; Jukema, J.W.; Boekholdt, S.M.; Zwinderman, A.H.; Tanck, M.W. Polymorphisms in APOA1 and LPL genes are statistically independently associated with fasting TG in men with CAD. *Eur. J. Hum. Genet.* **2005**, *13*, 445–451. [CrossRef] [PubMed]

75. Tao, F.; Weinstock, J.; Venners, S.A.; Cheng, J.; Hsu, Y.H.; Zou, Y.; Pan, F.; Jiang, S.; Zha, X.; Xu, X. Associations of the ABCA1 and LPL Gene Polymorphisms With Lipid Levels in a Hyperlipidemic Population. *Clin. Appl. Thromb. Hemost.* **2018**, *24*, 771–779. [CrossRef]

76. Genvigir, F.D.; Soares, S.A.; Hirata, M.H.; Willrich, M.A.; Arazi, S.S.; Rebecchi, I.M.; Oliveira, R.; Bernik, M.M.; Dorea, E.L.; Bertolami, M.C.; et al. Effects of ABCA1 SNPs, including the C-105T novel variant, on serum lipids of Brazilian individuals. *Clin. Chim Acta* **2008**, *389*, 79–86. [CrossRef]

77. Zhang, Z.; Tao, L.; Chen, Z.; Zhou, D.; Kan, M.; Zhang, D.; Li, C.; He, L.; Liu, Y. Association of genetic loci with blood lipids in the Chinese population. *PLoS ONE* **2011**, *6*, e27305. [CrossRef]

78. Coban, N.; Onat, A.; Komurcu Bayrak, E.; Gulec, C.; Can, G.; Erginel Unaltuna, N. Gender specific association of ABCA1 gene R219K variant in coronary disease risk through interactions with serum triglyceride elevation in Turkish adults. *Anadolu Kardiyol. Derg.* **2014**, *14*, 18–25. [CrossRef]

79. Babashamsi, M.M.; Halalkhor, S.; Moradi Firouzjah, H.; Parsian, H.; Jalali, S.F.; Babashamsi, M. Association of ATP-Binding Cassette Transporter A1 (ABCA1)-565 C/T Gene Polymorphism with Hypoalphalipoproteinemia and Serum Lipids, IL-6 and CRP Levels. *Avicenna J. Med. Biotechnol.* **2017**, *9*, 38–43.

80. Gamboa-Melendez, M.A.; Galindo-Gomez, C.; Juarez-Martinez, L.; Gomez, F.E.; Diaz-Diaz, E.; Avila-Arcos, M.A.; Avila-Curiel, A. Novel association of the R230C variant of the ABCA1 gene with high triglyceride levels and low high-density lipoprotein cholesterol levels in Mexican school-age children with high prevalence of obesity. *Arch. Med. Res.* **2015**, *46*, 495–501. [CrossRef]

81. Kolovou, V.; Marvaki, A.; Karakosta, A.; Vasilopoulos, G.; Kalogiani, A.; Mavrogeni, S.; Degiannis, D.; Marvaki, C.; Kolovou, G. Association of gender, ABCA1 gene polymorphisms and lipid profile in Greek young nurses. *Lipids Health Dis.* **2012**, *11*, 1–7. [CrossRef]

82. Cenarro, A.; Artieda, M.; Castillo, S.; Mozas, P.; Reyes, G.; Tejedor, D.; Alonso, R.; Mata, P.; Pocovi, M.; Civeira, F.; et al. A common variant in the ABCA1 gene is associated with a lower risk for premature coronary heart disease in familial hypercholesterolaemia. *J. Med. Genet.* **2003**, *40*, 163–168. [CrossRef] [PubMed]

83. Do, R.; Willer, C.J.; Schmidt, E.M.; Sengupta, S.; Gao, C.; Peloso, G.M.; Gustafsson, S.; Kanoni, S.; Ganna, A.; Chen, J.; et al. Common variants associated with plasma triglycerides and risk for coronary artery disease. *Nat. Genet.* **2013**, *45*, 1345–1352. [CrossRef] [PubMed]

84. Patsch, J.R.; Miesenbock, G.; Hopferwieser, T.; Muhlberger, V.; Knapp, E.; Dunn, J.K.; Gotto, A.M., Jr.; Patsch, W. Relation of triglyceride metabolism and coronary artery disease. Studies in the postprandial state. *Arterioscler. Thromb.* **1992**, *12*, 1336–1345. [CrossRef] [PubMed]

85. Tanaka, A. Postprandial hyperlipidemia and atherosclerosis. *J. Atheroscler. Thromb.* **2004**, *11*, 322–329. [CrossRef] [PubMed]

86. Langsted, A.; Freiberg, J.J.; Nordestgaard, B.G. Fasting and nonfasting lipid levels: Influence of normal food intake on lipids, lipoproteins, apolipoproteins, and cardiovascular risk prediction. *Circulation* **2008**, *118*, 2047–2056. [CrossRef] [PubMed]

87. Mora, S.; Rifai, N.; Buring, J.E.; Ridker, P.M. Fasting compared with nonfasting lipids and apolipoproteins for predicting incident cardiovascular events. *Circulation* **2008**, *118*, 993–1001. [CrossRef]

88. Nordestgaard, B.G.; Benn, M.; Schnohr, P.; Tybjaerg-Hansen, A. Nonfasting triglycerides and risk of myocardial infarction, ischemic heart disease, and death in men and women. *JAMA* **2007**, *298*, 299–308. [CrossRef] [PubMed]

89. Kolovou, G.; Daskalova, D.; Anagnostopoulou, K.; Hoursalas, I.; Voudris, V.; Mikhailidis, D.P.; Cokkinos, D.V. Postprandial hypertriglyceridaemia in patients with Tangier disease. *J. Clin. Pathol.* **2003**, *56*, 937–941. [CrossRef] [PubMed]

90. Lopez-Miranda, J.; Williams, C.; Lairon, D. Dietary, physiological, genetic and pathological influences on postprandial lipid metabolism. *Br. J. Nutr.* **2007**, *98*, 458–473. [CrossRef] [PubMed]

91. Desmarchelier, C.; Martin, J.C.; Planells, R.; Gastaldi, M.; Nowicki, M.; Goncalves, A.; Valero, R.; Lairon, D.; Borel, P. The postprandial chylomicron triacylglycerol response to dietary fat in healthy male adults is significantly explained by a combination of single nucleotide polymorphisms in genes involved in triacylglycerol metabolism. *J. Clin. Endocrinol. Metab.* **2014**, *99*, E484–E488. [CrossRef]

92. Perez-Martinez, P.; Lopez-Miranda, J.; Perez-Jimenez, F.; Ordovas, J.M. Influence of genetic factors in the modulation of postprandial lipemia. *Atheroscler. Suppl.* **2008**, *9*, 49–55. [CrossRef]

93. Delgado-Lista, J.; Perez-Martinez, P.; Perez-Jimenez, F.; Garcia-Rios, A.; Fuentes, F.; Marin, C.; Gomez-Luna, P.; Camargo, A.; Parnell, L.D.; Ordovas, J.M.; et al. ABCA1 gene variants regulate postprandial lipid metabolism in healthy men. *Arterioscler. Thromb. Vasc. Biol.* **2010**, *30*, 1051–1057. [CrossRef]

94. Rudkowska, I.; Dewailly, E.; Hegele, R.A.; Boiteau, V.; Dube-Linteau, A.; Abdous, B.; Giguere, Y.; Chateau-Degat, M.L.; Vohl, M.C. Gene-diet interactions on plasma lipid levels in the Inuit population. *Br. J. Nutr.* **2013**, *109*, 953–961. [CrossRef]

95. Dumitrescu, L.; Carty, C.L.; Taylor, K.; Schumacher, F.R.; Hindorff, L.A.; Ambite, J.L.; Anderson, G.; Best, L.G.; Brown-Gentry, K.; Buzkova, P.; et al. Genetic determinants of lipid traits in diverse populations from the population architecture using genomics and epidemiology (PAGE) study. *PLoS Genet.* **2011**, *7*, e1002138. [CrossRef] [PubMed]

96. Voight, B.F.; Peloso, G.M.; Orho-Melander, M.; Frikke-Schmidt, R.; Barbalic, M.; Jensen, M.K.; Hindy, G.; Holm, H.; Ding, E.L.; Johnson, T.; et al. Plasma HDL cholesterol and risk of myocardial infarction: A mendelian randomisation study. *Lancet* **2012**, *380*, 572–580. [CrossRef]

97. Pappa, E.; Elisaf, M.S.; Kostara, C.; Bairaktari, E.; Tsimihodimos, V.K. Cardioprotective Properties of HDL: Structural and Functional Considerations. *Curr. Med. Chem.* **2020**, *27*, 2964–2978. [CrossRef] [PubMed]

98. Villarreal-Molina, T.; Posadas-Romero, C.; Romero-Hidalgo, S.; Antunez-Arguelles, E.; Bautista-Grande, A.; Vargas-Alarcon, G.; Kimura-Hayama, E.; Canizales-Quinteros, S.; Juarez-Rojas, J.G.; Posadas-Sanchez, R.; et al. The ABCA1 gene R230C variant is associated with decreased risk of premature coronary artery disease: The genetics of atherosclerotic disease (GEA) study. *PLoS ONE* **2012**, *7*, e49285. [CrossRef]

99. Defronzo, R.A. Banting Lecture. From the triumvirate to the ominous octet: A new paradigm for the treatment of type 2 diabetes mellitus. *Diabetes* **2009**, *58*, 773–795. [CrossRef]

100. Aiello, R.J.; Brees, D.; Francone, O.L. ABCA1-deficient mice: Insights into the role of monocyte lipid efflux in HDL formation and inflammation. *Arterioscler. Thromb. Vasc. Biol.* **2003**, *23*, 972–980. [CrossRef]

101. Alshaikhli, A.; Bordoni, B. Tangier Disease. In *StatPearls*; StatPearls Publishing: Treasure Island, FL, USA, 2020.

102. Koseki, M.; Matsuyama, A.; Nakatani, K.; Inagaki, M.; Nakaoka, H.; Kawase, R.; Yuasa-Kawase, M.; Tsubakio-Yamamoto, K.; Masuda, D.; Sandoval, J.C.; et al. Impaired insulin secretion in four Tangier disease patients with ABCA1 mutations. *J. Atheroscler. Thromb.* **2009**, *16*, 292–296. [CrossRef]

103. Saito, M.; Eto, M.; Nisimatsu, S.; Kume, Y.; Kawasaki, H.; Yoneda, M.; Matsuda, M.; Matsuki, M.; Kaku, K. Case of familial hypoalphalipoproteinemia, type 2 diabetes mellitus and markedly advanced atherosclerosis with ABCAlexon 4 minus transcript in macrophages. *Nihon Naika Gakkai Zasshi* **2002**, *91*, 2762–2764. [CrossRef] [PubMed]

104. Guo, Z.; Inazu, A.; Yu, W.; Suzumura, T.; Okamoto, M.; Nohara, A.; Higashikata, T.; Sano, R.; Wakasugi, K.; Hayakawa, T.; et al. Double deletions and missense mutations in the first nucleotide-binding fold of the ATP-binding cassette transporter A1 (ABCA1) gene in Japanese patients with Tangier disease. *J. Hum. Genet.* **2002**, *47*, 325–329. [CrossRef]

105. Brunham, L.R.; Kruit, J.K.; Pape, T.D.; Timmins, J.M.; Reuwer, A.Q.; Vasanji, Z.; Marsh, B.J.; Rodrigues, B.; Johnson, J.D.; Parks, J.S.; et al. β-cell ABCA1 influences insulin secretion, glucose homeostasis and response to thiazolidinedione treatment. *Nat. Med.* **2007**, *13*, 340–347. [CrossRef]

106. Fryirs, M.; Barter, P.J.; Rye, K.A. Cholesterol metabolism and pancreatic β-cell function. *Curr. Opin. Lipidol.* **2009**, *20*, 159–164. [CrossRef]

107. Hao, M.; Head, W.S.; Gunawardana, S.C.; Hasty, A.H.; Piston, D.W. Direct effect of cholesterol on insulin secretion: A novel mechanism for pancreatic β-cell dysfunction. *Diabetes* **2007**, *56*, 2328–2338. [CrossRef]

108. Ishikawa, M.; Iwasaki, Y.; Yatoh, S.; Kato, T.; Kumadaki, S.; Inoue, N.; Yamamoto, T.; Matsuzaka, T.; Nakagawa, Y.; Yahagi, N.; et al. Cholesterol accumulation and diabetes in pancreatic β-cell-specific SREBP-2 transgenic mice: A new model for lipotoxicity. *J. Lipid Res.* **2008**, *49*, 2524–2534. [CrossRef]

109. Kruit, J.K.; Wijesekara, N.; Fox, J.E.; Dai, X.Q.; Brunham, L.R.; Searle, G.J.; Morgan, G.P.; Costin, A.J.; Tang, R.; Bhattacharjee, A.; et al. Islet cholesterol accumulation due to loss of ABCA1 leads to impaired exocytosis of insulin granules. *Diabetes* **2011**, *60*, 3186–3196. [CrossRef] [PubMed]

110. Larsson, S.; Wierup, N.; Sundler, F.; Eliasson, L.; Holm, C. Lack of cholesterol mobilization in islets of hormone-sensitive lipase deficient mice impairs insulin secretion. *Biochem. Biophys. Res. Commun.* **2008**, *376*, 558–562. [CrossRef] [PubMed]

111. Vikman, J.; Jimenez-Feltstrom, J.; Nyman, P.; Thelin, J.; Eliasson, L. Insulin secretion is highly sensitive to desorption of plasma membrane cholesterol. *FASEB J.* **2009**, *23*, 58–67. [CrossRef] [PubMed]

112. Wiser, O.; Trus, M.; Hernandez, A.; Renstrom, E.; Barg, S.; Rorsman, P.; Atlas, D. The voltage sensitive Lc-type Ca2+ channel is functionally coupled to the exocytotic machinery. *Proc. Natl. Acad. Sci. USA* **1999**, *96*, 248–253. [CrossRef]

113. Xia, F.; Gao, X.; Kwan, E.; Lam, P.P.; Chan, L.; Sy, K.; Sheu, L.; Wheeler, M.B.; Gaisano, H.Y.; Tsushima, R.G. Disruption of pancreatic β-cell lipid rafts modifies Kv2.1 channel gating and insulin exocytosis. *J. Biol. Chem.* **2004**, *279*, 24685–24691. [CrossRef]

114. Brunham, L.R.; Kruit, J.K.; Verchere, C.B.; Hayden, M.R. Cholesterol in islet dysfunction and type 2 diabetes. *J. Clin. Investig.* **2008**, *118*, 403–408. [CrossRef]

115. Drew, B.G.; Fidge, N.H.; Gallon-Beaumier, G.; Kemp, B.E.; Kingwell, B.A. High-density lipoprotein and apolipoprotein AI increase endothelial NO synthase activity by protein association and multisite phosphorylation. *Proc. Natl. Acad. Sci. USA* **2004**, *101*, 6999–7004. [CrossRef]

116. Han, R.; Lai, R.; Ding, Q.; Wang, Z.; Luo, X.; Zhang, Y.; Cui, G.; He, J.; Liu, W.; Chen, Y. Apolipoprotein A-I stimulates AMP-activated protein kinase and improves glucose metabolism. *Diabetologia* **2007**, *50*, 1960–1968. [CrossRef]

117. Drew, B.G.; Duffy, S.J.; Formosa, M.F.; Natoli, A.K.; Henstridge, D.C.; Penfold, S.A.; Thomas, W.G.; Mukhamedova, N.; de Courten, B.; Forbes, J.M.; et al. High-density lipoprotein modulates glucose metabolism in patients with type 2 diabetes mellitus. *Circulation* **2009**, *119*, 2103–2111. [CrossRef]

118. Tang, S.; Tabet, F.; Cochran, B.J.; Cuesta Torres, L.F.; Wu, B.J.; Barter, P.J.; Rye, K.A. Apolipoprotein A-I enhances insulin-dependent and insulin-independent glucose uptake by skeletal muscle. *Sci. Rep.* **2019**, *9*, 1–9. [CrossRef]

119. de Haan, W.; Bhattacharjee, A.; Ruddle, P.; Kang, M.H.; Hayden, M.R. ABCA1 in adipocytes regulates adipose tissue lipid content, glucose tolerance, and insulin sensitivity. *J. Lipid Res.* **2014**, *55*, 516–523. [CrossRef] [PubMed]

120. Key, C.C.; Liu, M.; Kurtz, C.L.; Chung, S.; Boudyguina, E.; Dinh, T.A.; Bashore, A.; Phelan, P.E.; Freedman, B.I.; Osborne, T.F.; et al. Hepatocyte ABCA1 Deletion Impairs Liver Insulin Signaling and Lipogenesis. *Cell Rep.* **2017**, *19*, 2116–2129. [CrossRef] [PubMed]

121. Schou, J.; Tybjaerg-Hansen, A.; Moller, H.J.; Nordestgaard, B.G.; Frikke-Schmidt, R. ABC transporter genes and risk of type 2 diabetes: A study of 40,000 individuals from the general population. *Diabetes Care* **2012**, *35*, 2600–2606. [CrossRef]

122. Diabetes Genetics Initiative of Broad Institute of Harvard and MIT; Lund University; Novartis Institutes of BioMedical Research; Saxena, R.; Voight, B.F.; Lyssenko, V.; Burtt, N.P.; de Bakker, P.I.; Chen, H.; Roix, J.J.; et al. Genome-wide association analysis identifies loci for type 2 diabetes and triglyceride levels. *Science* **2007**, *316*, 1331–1336. [CrossRef] [PubMed]

123. Sladek, R.; Rocheleau, G.; Rung, J.; Dina, C.; Shen, L.; Serre, D.; Boutin, P.; Vincent, D.; Belisle, A.; Hadjadj, S.; et al. A genome-wide association study identifies novel risk loci for type 2 diabetes. *Nature* **2007**, *445*, 881–885. [CrossRef]

124. Steinthorsdottir, V.; Thorleifsson, G.; Reynisdottir, I.; Benediktsson, R.; Jonsdottir, T.; Walters, G.B.; Styrkarsdottir, U.; Gretarsdottir, S.; Emilsson, V.; Ghosh, S.; et al. A variant in CDKAL1 influences insulin response and risk of type 2 diabetes. *Nat. Genet.* **2007**, *39*, 770–775. [CrossRef]

125. Daimon, M.; Kido, T.; Baba, M.; Oizumi, T.; Jimbu, Y.; Kameda, W.; Yamaguchi, H.; Ohnuma, H.; Tominaga, M.; Muramatsu, M.; et al. Association of the ABCA1 gene polymorphisms with type 2 DM in a Japanese population. *Biochem. Biophys. Res. Commun.* **2005**, *329*, 205–210. [CrossRef]

126. Villarreal-Molina, M.T.; Flores-Dorantes, M.T.; Arellano-Campos, O.; Villalobos-Comparan, M.; Rodriguez-Cruz, M.; Miliar-Garcia, A.; Huertas-Vazquez, A.; Menjivar, M.; Romero-Hidalgo, S.; Wacher, N.H.; et al. Association of the ATP-binding cassette transporter A1 R230C variant with early-onset type 2 diabetes in a Mexican population. *Diabetes* **2008**, *57*, 509–513. [CrossRef]

127. Lara-Riegos, J.C.; Ortiz-Lopez, M.G.; Pena-Espinoza, B.I.; Montufar-Robles, I.; Pena-Rico, M.A.; Sanchez-Pozos, K.; Granados-Silvestre, M.A.; Menjivar, M. Diabetes susceptibility in Mayas: Evidence for the involvement of polymorphisms in HHEX, HNF4α, KCNJ11, PPARγ, CDKN2A/2B, SLC30A8, CDC123/CAMK1D, TCF7L2, ABCA1 and SLC16A11 genes. *Gene* **2015**, *565*, 68–75. [CrossRef] [PubMed]

128. Campbell, D.D.; Parra, M.V.; Duque, C.; Gallego, N.; Franco, L.; Tandon, A.; Hunemeier, T.; Bortolini, C.; Villegas, A.; Bedoya, G.; et al. Amerind ancestry, socioeconomic status and the genetics of type 2 diabetes in a Colombian population. *PLoS ONE* **2012**, *7*, e33570. [CrossRef]

129. Jung, D.; Cao, S.; Liu, M.; Park, S. A Meta-Analysis of the Associations Between the ATP-Binding Cassette Transporter ABCA1 R219K (rs2230806) Polymorphism and the Risk of Type 2 Diabetes in Asians. *Horm. Metab. Res.* **2018**, *50*, 308–316. [CrossRef] [PubMed]

130. Ergen, H.A.; Zeybek, U.; Gok, O.; Karaali, Z.E. Investigation of ABCA1 C69T polymorphism in patients with type 2 diabetes mellitus. *Biochem. Med.* **2012**, *22*, 114–120. [CrossRef] [PubMed]

131. Alharbi, K.K.; Khan, I.A.; Al-Daghri, N.M.; Munshi, A.; Sharma, V.; Mohammed, A.K.; Wani, K.A.; Al-Sheikh, Y.A.; Al-Nbaheen, M.S.; Ansari, M.G.; et al. ABCA1 C69T gene polymorphism and risk of type 2 diabetes mellitus in a Saudi population. *J. Biosci.* **2013**, *38*, 893–897. [CrossRef]

132. Li, C.; Fan, D. Association between the ABCA1 rs1800977 polymorphism and susceptibility to type 2 diabetes mellitus in a Chinese Han population. *Biosci. Rep.* **2018**, *38*, 1–5. [CrossRef]

133. Kim, D.S.; Kim, B.C.; Daily, J.W.; Park, S. High genetic risk scores for impaired insulin secretory capacity doubles the risk for type 2 diabetes in Asians and is exacerbated by Western-type diets. *Diabetes Metab. Res. Rev.* **2018**, *34*, 1–9. [CrossRef]

134. ENSANUT | Encuesta Nacional de Salud y Nutrición 2016. Available online: http://fmdiabetes.org/wp-content/uploads/2017/04/ENSANUT2016-mc.pdf (accessed on 30 December 2020).

135. Parolini, C.; Caligari, S.; Gilio, D.; Manzini, S.; Busnelli, M.; Montagnani, M.; Locatelli, M.; Diani, E.; Giavarini, F.; Caruso, D.; et al. Reduced biliary sterol output with no change in total faecal excretion in mice expressing a human apolipoprotein A-I variant. *Liver Int.* **2012**, *32*, 1363–1371. [CrossRef]

136. Wang, B.; Tontonoz, P. Liver X receptors in lipid signalling and membrane homeostasis. *Nat. Rev. Endocrinol.* **2018**, *14*, 452–463. [CrossRef]

137. Christiansen-Weber, T.A.; Voland, J.R.; Wu, Y.; Ngo, K.; Roland, B.L.; Nguyen, S.; Peterson, P.A.; Fung-Leung, W.P. Functional loss of ABCA1 in mice causes severe placental malformation, aberrant lipid distribution, and kidney glomerulonephritis as well as high-density lipoprotein cholesterol deficiency. *Am. J. Pathol.* **2000**, *157*, 1017–1029. [CrossRef]

138. Orso, E.; Broccardo, C.; Kaminski, W.E.; Bottcher, A.; Liebisch, G.; Drobnik, W.; Gotz, A.; Chambenoit, O.; Diederich, W.; Langmann, T.; et al. Transport of lipids from golgi to plasma membrane is defective in tangier disease patients and Abc1-deficient mice. *Nat. Genet.* **2000**, *24*, 192–196. [CrossRef]

139. McNeish, J.; Aiello, R.J.; Guyot, D.; Turi, T.; Gabel, C.; Aldinger, C.; Hoppe, K.L.; Roach, M.L.; Royer, L.J.; de Wet, J.; et al. High density lipoprotein deficiency and foam cell accumulation in mice with targeted disruption of ATP-binding cassette transporter-1. *Proc. Natl. Acad. Sci. USA* **2000**, *97*, 4245–4250. [CrossRef]

140. Bellentani, S. The epidemiology of non-alcoholic fatty liver disease. *Liver Int.* **2017**, *37* (Supp. S1), 81–84. [CrossRef] [PubMed]

141. Chalasani, N.; Younossi, Z.; Lavine, J.E.; Charlton, M.; Cusi, K.; Rinella, M.; Harrison, S.A.; Brunt, E.M.; Sanyal, A.J. The diagnosis and management of nonalcoholic fatty liver disease: Practice guidance from the American Association for the Study of Liver Diseases. *Hepatology* **2018**, *67*, 328–357. [CrossRef] [PubMed]

142. Dominguez-Perez, M.; Simoni-Nieves, A.; Rosales, P.; Nuno-Lambarri, N.; Rosas-Lemus, M.; Souza, V.; Miranda, R.U.; Bucio, L.; Uribe Carvajal, S.; Marquardt, J.U.; et al. Cholesterol burden in the liver induces mitochondrial dynamic changes and resistance to apoptosis. *J. Cell Physiol.* **2019**, *234*, 7213–7223. [CrossRef]

143. Vega-Badillo, J.; Gutierrez-Vidal, R.; Hernandez-Perez, H.A.; Villamil-Ramirez, H.; Leon-Mimila, P.; Sanchez-Munoz, F.; Moran-Ramos, S.; Larrieta-Carrasco, E.; Fernandez-Silva, I.; Mendez-Sanchez, N.; et al. Hepatic miR-33a/miR-144 and their target gene ABCA1 are associated with steatohepatitis in morbidly obese subjects. *Liver Int.* **2016**, *36*, 1383–1391. [CrossRef] [PubMed]

144. Mahley, R.W. Central Nervous System Lipoproteins: ApoE and Regulation of Cholesterol Metabolism. *Arterioscler. Thromb. Vasc. Biol.* **2016**, *36*, 1305–1315. [CrossRef] [PubMed]

145. Dietschy, J.M.; Turley, S.D. Thematic review series: Brain Lipids. Cholesterol metabolism in the central nervous system during early development and in the mature animal. *J. Lipid Res.* **2004**, *45*, 1375–1397. [CrossRef] [PubMed]

146. Funfschilling, U.; Saher, G.; Xiao, L.; Mobius, W.; Nave, K.A. Survival of adult neurons lacking cholesterol synthesis in vivo. *BMC Neurosci.* **2007**, *8*, 1. [CrossRef] [PubMed]

147. Vitali, C.; Wellington, C.L.; Calabresi, L. HDL and cholesterol handling in the brain. *Cardiovasc Res.* **2014**, *103*, 405–413. [CrossRef] [PubMed]

148. Chang, T.Y.; Yamauchi, Y.; Hasan, M.T.; Chang, C. Cellular cholesterol homeostasis and Alzheimer's disease. *J. Lipid Res.* **2017**, *58*, 2239–2254. [CrossRef]

149. Koldamova, R.P.; Lefterov, I.M.; Ikonomovic, M.D.; Skoko, J.; Lefterov, P.I.; Isanski, B.A.; DeKosky, S.T.; Lazo, J.S. 22R-hydroxycholesterol and 9-cis-retinoic acid induce ATP-binding cassette transporter A1 expression and cholesterol efflux in brain cells and decrease amyloid β secretion. *J. Biol. Chem.* **2003**, *278*, 13244–13256. [CrossRef] [PubMed]

150. Castranio, E.L.; Wolfe, C.M.; Nam, K.N.; Letronne, F.; Fitz, N.F.; Lefterov, I.; Koldamova, R. ABCA1 haplodeficiency affects the brain transcriptome following traumatic brain injury in mice expressing human APOE isoforms. *Acta Neuropathol. Commun.* **2018**, *6*, 1–13. [CrossRef]
151. Cui, X.; Chopp, M.; Zacharek, A.; Karasinska, J.M.; Cui, Y.; Ning, R.; Zhang, Y.; Wang, Y.; Chen, J. Deficiency of brain ATP-binding cassette transporter A-1 exacerbates blood-brain barrier and white matter damage after stroke. *Stroke* **2015**, *46*, 827–834. [CrossRef]
152. Lefterov, I.; Fitz, N.F.; Cronican, A.; Lefterov, P.; Staufenbiel, M.; Koldamova, R. Memory deficits in APP23/Abca1+/− mice correlate with the level of Aβ oligomers. *ASN Neuro* **2009**, *1*, 65–76. [CrossRef]
153. Li, B.; Xia, Y.; Hu, B. Infection and atherosclerosis: TLR-dependent pathways. *Cell Mol. Life Sci.* **2020**, *77*, 2751–2769. [CrossRef]
154. Loane, D.J.; Washington, P.M.; Vardanian, L.; Pocivavsek, A.; Hoe, H.S.; Duff, K.E.; Cernak, I.; Rebeck, G.W.; Faden, A.I.; Burns, M.P. Modulation of ABCA1 by an LXR agonist reduces beta-amyloid levels and improves outcome after traumatic brain injury. *J. Neurotrauma* **2011**, *28*, 225–236. [CrossRef]
155. Pals, P.; Lincoln, S.; Manning, J.; Heckman, M.; Skipper, L.; Hulihan, M.; Van den Broeck, M.; De Pooter, T.; Cras, P.; Crook, J.; et al. α-Synuclein promoter confers susceptibility to Parkinson's disease. *Ann. Neurol.* **2004**, *56*, 591–595. [CrossRef]
156. Wang, X.; Li, R.; Zacharek, A.; Landschoot-Ward, J.; Wang, F.; Wu, K.H.; Chopp, M.; Chen, J.; Cui, X. Administration of Downstream ApoE Attenuates the Adverse Effect of Brain ABCA1 Deficiency on Stroke. *Int. J. Mol. Sci.* **2018**, *19*, 3368. [CrossRef]
157. Ya, L.; Lu, Z. Differences in ABCA1 R219K Polymorphisms and Serum Indexes in Alzheimer and Parkinson Diseases in Northern China. *Med. Sci. Monit.* **2017**, *23*, 4591–4600. [CrossRef]
158. Braak, H.; Braak, E. Diagnostic criteria for neuropathologic assessment of Alzheimer's disease. *Neurobiol. Aging* **1997**, *18*, S85–S88. [CrossRef]
159. Bancher, C.; Brunner, C.; Lassmann, H.; Budka, H.; Jellinger, K.; Wiche, G.; Seitelberger, F.; Grundke-Iqbal, I.; Iqbal, K.; Wisniewski, H.M. Accumulation of abnormally phosphorylated tau precedes the formation of neurofibrillary tangles in Alzheimer's disease. *Brain Res.* **1989**, *477*, 90–99. [CrossRef]
160. Wildsmith, K.R.; Holley, M.; Savage, J.C.; Skerrett, R.; Landreth, G.E. Evidence for impaired amyloid β clearance in Alzheimer's disease. *Alzheimers Res. Ther.* **2013**, *5*, 1–6. [CrossRef]
161. Braak, H.; Alafuzoff, I.; Arzberger, T.; Kretzschmar, H.; Del Tredici, K. Staging of Alzheimer disease-associated neurofibrillary pathology using paraffin sections and immunocytochemistry. *Acta Neuropathol.* **2006**, *112*, 389–404. [CrossRef]
162. Jiang, Q.; Lee, C.Y.; Mandrekar, S.; Wilkinson, B.; Cramer, P.; Zelcer, N.; Mann, K.; Lamb, B.; Willson, T.M.; Collins, J.L.; et al. ApoE promotes the proteolytic degradation of Abeta. *Neuron* **2008**, *58*, 681–693. [CrossRef]
163. Hirsch-Reinshagen, V.; Zhou, S.; Burgess, B.L.; Bernier, L.; McIsaac, S.A.; Chan, J.Y.; Tansley, G.H.; Cohn, J.S.; Hayden, M.R.; Wellington, C.L. Deficiency of ABCA1 impairs apolipoprotein E metabolism in brain. *J. Biol. Chem.* **2004**, *279*, 41197–41207. [CrossRef]
164. Donkin, J.J.; Stukas, S.; Hirsch-Reinshagen, V.; Namjoshi, D.; Wilkinson, A.; May, S.; Chan, J.; Fan, J.; Collins, J.; Wellington, C.L. ATP-binding cassette transporter A1 mediates the beneficial effects of the liver X receptor agonist GW3965 on object recognition memory and amyloid burden in amyloid precursor protein/presenilin 1 mice. *J. Biol. Chem.* **2010**, *285*, 34144–34154. [CrossRef] [PubMed]
165. Hirsch-Reinshagen, V.; Maia, L.F.; Burgess, B.L.; Blain, J.F.; Naus, K.E.; McIsaac, S.A.; Parkinson, P.F.; Chan, J.Y.; Tansley, G.H.; Hayden, M.R.; et al. The absence of ABCA1 decreases soluble ApoE levels but does not diminish amyloid deposition in two murine models of Alzheimer disease. *J. Biol. Chem.* **2005**, *280*, 43243–43256. [CrossRef] [PubMed]
166. Wahrle, S.E.; Jiang, H.; Parsadanian, M.; Kim, J.; Li, A.; Knoten, A.; Jain, S.; Hirsch-Reinshagen, V.; Wellington, C.L.; Bales, K.R.; et al. Overexpression of ABCA1 reduces amyloid deposition in the PDAPP mouse model of Alzheimer disease. *J. Clin. Investig.* **2008**, *118*, 671–682. [CrossRef] [PubMed]
167. Koldamova, R.P.; Lefterov, I.M.; Staufenbiel, M.; Wolfe, D.; Huang, S.; Glorioso, J.C.; Walter, M.; Roth, M.G.; Lazo, J.S. The liver X receptor ligand T0901317 decreases amyloid β production in vitro and in a mouse model of Alzheimer's disease. *J. Biol. Chem.* **2005**, *280*, 4079–4088. [CrossRef]
168. Wahrle, S.E.; Jiang, H.; Parsadanian, M.; Hartman, R.E.; Bales, K.R.; Paul, S.M.; Holtzman, D.M. Deletion of Abca1 increases Aβ deposition in the PDAPP transgenic mouse model of Alzheimer disease. *J. Biol. Chem.* **2005**, *280*, 43236–43242. [CrossRef] [PubMed]
169. Kim, J.; Yoon, H.; Horie, T.; Burchett, J.M.; Restivo, J.L.; Rotllan, N.; Ramirez, C.M.; Verghese, P.B.; Ihara, M.; Hoe, H.S.; et al. microRNA-33 Regulates ApoE Lipidation and Amyloid-β Metabolism in the Brain. *J. Neurosci.* **2015**, *35*, 14717–14726. [CrossRef] [PubMed]
170. Fitz, N.F.; Carter, A.Y.; Tapias, V.; Castranio, E.L.; Kodali, R.; Lefterov, I.; Koldamova, R. ABCA1 Deficiency Affects Basal Cognitive Deficits and Dendritic Density in Mice. *J. Alzheimers Dis.* **2017**, *56*, 1075–1085. [CrossRef]
171. McManus, R.M.; Higgins, S.C.; Mills, K.H.; Lynch, M.A. Respiratory infection promotes T cell infiltration and amyloid-β deposition in APP/PS1 mice. *Neurobiol. Aging* **2014**, *35*, 109–121. [CrossRef] [PubMed]
172. Zhang, J.; Ke, K.F.; Liu, Z.; Qiu, Y.H.; Peng, Y.P. Th17 cell-mediated neuroinflammation is involved in neurodegeneration of aβ1-42-induced Alzheimer's disease model rats. *PLoS ONE* **2013**, *8*, e75786. [CrossRef]
173. Yang, J.; Kou, J.; Lalonde, R.; Fukuchi, K.I. Intracranial IL-17A overexpression decreases cerebral amyloid angiopathy by upregulation of ABCA1 in an animal model of Alzheimer's disease. *Brain Behav. Immun.* **2017**, *65*, 262–273. [CrossRef]

174. Yang, Y.; Wu, Y.; Zhang, S.; Song, W. High glucose promotes Aβ production by inhibiting APP degradation. *PLoS ONE* **2013**, *8*, e69824. [CrossRef]

175. Macauley, S.L.; Stanley, M.; Caesar, E.E.; Yamada, S.A.; Raichle, M.E.; Perez, R.; Mahan, T.E.; Sutphen, C.L.; Holtzman, D.M. Hyperglycemia modulates extracellular amyloid-β concentrations and neuronal activity in vivo. *J. Clin. Investig.* **2015**, *125*, 2463–2467. [CrossRef] [PubMed]

176. Lee, H.J.; Ryu, J.M.; Jung, Y.H.; Lee, S.J.; Kim, J.Y.; Lee, S.H.; Hwang, I.K.; Seong, J.K.; Han, H.J. High glucose upregulates BACE1-mediated Aβ production through ROS-dependent HIF-1α and LXRα/ABCA1-regulated lipid raft reorganization in SK-N-MC cells. *Sci. Rep.* **2016**, *6*, 1–15. [CrossRef] [PubMed]

177. Nordestgaard, L.T.; Tybjaerg-Hansen, A.; Nordestgaard, B.G.; Frikke-Schmidt, R. Loss-of-function mutation in ABCA1 and risk of Alzheimer's disease and cerebrovascular disease. *Alzheimers Dement.* **2015**, *11*, 1430–1438. [CrossRef]

178. Lupton, M.K.; Proitsi, P.; Lin, K.; Hamilton, G.; Daniilidou, M.; Tsolaki, M.; Powell, J.F. The role of ABCA1 gene sequence variants on risk of Alzheimer's disease. *J. Alzheimers Dis.* **2014**, *38*, 897–906. [CrossRef]

179. Furney, S.J.; Simmons, A.; Breen, G.; Pedroso, I.; Lunnon, K.; Proitsi, P.; Hodges, A.; Powell, J.; Wahlund, L.O.; Kloszewska, I.; et al. Genome-wide association with MRI atrophy measures as a quantitative trait locus for Alzheimer's disease. *Mol. Psychiatry* **2011**, *16*, 1130–1138. [CrossRef]

180. Gusareva, E.S.; Carrasquillo, M.M.; Bellenguez, C.; Cuyvers, E.; Colon, S.; Graff-Radford, N.R.; Petersen, R.C.; Dickson, D.W.; Mahachie John, J.M.; Bessonov, K.; et al. Genome-wide association interaction analysis for Alzheimer's disease. *Neurobiol. Aging* **2014**, *35*, 2436–2443. [CrossRef]

181. Lambert, J.C.; Ibrahim-Verbaas, C.A.; Harold, D.; Naj, A.C.; Sims, R.; Bellenguez, C.; DeStafano, A.L.; Bis, J.C.; Beecham, G.W.; Grenier-Boley, B.; et al. Meta-analysis of 74,046 individuals identifies 11 new susceptibility loci for Alzheimer's disease. *Nat. Genet.* **2013**, *45*, 1452–1458. [CrossRef]

182. Schott, J.M.; Crutch, S.J.; Carrasquillo, M.M.; Uphill, J.; Shakespeare, T.J.; Ryan, N.S.; Yong, K.X.; Lehmann, C.M.; Ertekin-Taner, N.; Graff-Radford, N.R.; et al. Genetic risk factors for the posterior cortical atrophy variant of Alzheimer's disease. *Alzheimers Dement.* **2016**, *12*, 862–871. [CrossRef]

183. Kamboh, M.I.; Demirci, F.Y.; Wang, X.; Minster, R.L.; Carrasquillo, M.M.; Pankratz, V.S.; Younkin, S.G.; Saykin, A.J.; Jun, G.; Baldwin, C.; et al. Genome-wide association study of Alzheimer's disease. *Transl. Psychiatry* **2012**, *2*, e117. [CrossRef]

184. Jiang, S.; Zhang, C.Y.; Tang, L.; Zhao, L.X.; Chen, H.Z.; Qiu, Y. Integrated Genomic Analysis Revealed Associated Genes for Alzheimer's Disease in APOE4 Non-Carriers. *Curr. Alzheimer Res.* **2019**, *16*, 753–763. [CrossRef]

185. Zhu, Z.; Lin, Y.; Li, X.; Driver, J.A.; Liang, L. Shared genetic architecture between metabolic traits and Alzheimer's disease: A large-scale genome-wide cross-trait analysis. *Hum. Genet.* **2019**, *138*, 271–285. [CrossRef] [PubMed]

186. Katzov, H.; Chalmers, K.; Palmgren, J.; Andreasen, N.; Johansson, B.; Cairns, N.J.; Gatz, M.; Wilcock, G.K.; Love, S.; Pedersen, N.L.; et al. Genetic Variants of ABCA1 Modify Alzheimer Disease Risk and Quantitative Traits Related to β-Amyloid Metabolism. *Hum. Mutat.* **2004**, *23*, 358–367. [CrossRef] [PubMed]

187. Rodriguez-Rodriguez, E.; Mateo, I.; Llorca, J.; Sanchez-Quintana, C.; Infante, J.; Garcia-Gorostiaga, I.; Sanchez-Juan, P.; Berciano, J.; Combarros, O. Association of genetic variants of ABCA1 with Alzheimer's disease risk. *Am. J. Med. Genet. B Neuropsychiatr Genet.* **2007**, *144B*, 964–968. [CrossRef] [PubMed]

188. Sundar, P.D.; Feingold, E.; Minster, R.L.; DeKosky, S.T.; Kamboh, M.I. Gender-specific association of ATP-binding cassette transporter 1 (ABCA1) polymorphisms with the risk of late-onset Alzheimer's disease. *Neurobiol. Aging* **2007**, *28*, 856–862. [CrossRef] [PubMed]

189. Wavrant-De Vrieze, F.; Compton, D.; Womick, M.; Arepalli, S.; Adighibe, O.; Li, L.; Perez-Tur, J.; Hardy, J. ABCA1 polymorphisms and Alzheimer's disease. *Neurosci. Lett.* **2007**, *416*, 180–183. [CrossRef] [PubMed]

190. Feher, A.; Giricz, Z.; Juhasz, A.; Pakaski, M.; Janka, Z.; Kalman, J. ABCA1 rs2230805 and rs2230806 common gene variants are associated with Alzheimer's disease. *Neurosci. Lett.* **2018**, *664*, 79–83. [CrossRef] [PubMed]

191. Xiao, Z.; Wang, J.; Chen, W.; Wang, P.; Zeng, H.; Chen, W. Association studies of several cholesterol-related genes (ABCA1, CETP and LIPC) with serum lipids and risk of Alzheimer's disease. *Lipids Health Dis.* **2012**, *11*, 1–14. [CrossRef]

192. Kolsch, H.; Lutjohann, D.; Jessen, F.; Von Bergmann, K.; Schmitz, S.; Urbach, H.; Maier, W.; Heun, R. Polymorphism in ABCA1 influences CSF 24S-hydroxycholesterol levels but is not a major risk factor of Alzheimer's disease. *Int. J. Mol. Med.* **2006**, *17*, 791–794. [CrossRef]

193. Jiang, M.; Lv, L.; Wang, H.; Yang, X.; Ji, H.; Zhou, F.; Zhu, W.; Cai, L.; Gu, X.; Sun, J.; et al. Meta-analysis on association between the ATP-binding cassette transporter A1 gene (ABCA1) and Alzheimer's disease. *Gene* **2012**, *510*, 147–153. [CrossRef]

194. Wang, X.F.; Cao, Y.W.; Feng, Z.Z.; Fu, D.; Ma, Y.S.; Zhang, F.; Jiang, X.X.; Shao, Y.C. Quantitative assessment of the effect of ABCA1 gene polymorphism on the risk of Alzheimer's disease. *Mol. Biol. Rep.* **2013**, *40*, 779–785. [CrossRef]

195. Corona, A.W.; Kodoma, N.; Casali, B.T.; Landreth, G.E. ABCA1 is Necessary for Bexarotene-Mediated Clearance of Soluble Amyloid Beta from the Hippocampus of APP/PS1 Mice. *J. Neuroimmune Pharmacol.* **2016**, *11*, 61–72. [CrossRef] [PubMed]

196. Yao, J.; Ho, D.; Calingasan, N.Y.; Pipalia, N.H.; Lin, M.T.; Beal, M.F. Neuroprotection by cyclodextrin in cell and mouse models of Alzheimer disease. *J. Exp. Med.* **2012**, *209*, 2501–2513. [CrossRef] [PubMed]

197. Shinohara, M.; Shinohara, M.; Zhao, J.; Fu, Y.; Liu, C.C.; Kanekiyo, T.; Bu, G. 5-HT3 Antagonist Ondansetron Increases apoE Secretion by Modulating the LXR-ABCA1 Pathway. *Int. J. Mol. Sci.* **2019**, *20*, 1488. [CrossRef] [PubMed]

198. Xu, G.B.; Yang, L.Q.; Guan, P.P.; Wang, Z.Y.; Wang, P. Prostaglandin A1 Inhibits the Cognitive Decline of APP/PS1 Transgenic Mice via PPARγ/ABCA1-dependent Cholesterol Efflux Mechanisms. *Neurotherapeutics* **2019**, *16*, 505–522. [CrossRef]
199. Fan, J.; Zhao, R.Q.; Parro, C.; Zhao, W.; Chou, H.Y.; Robert, J.; Deeb, T.Z.; Raynoschek, C.; Barichievy, S.; Engkvist, O.; et al. Small molecule inducers of ABCA1 and apoE that act through indirect activation of the LXR pathway. *J. Lipid Res.* **2018**, *59*, 830–842. [CrossRef]
200. Boehm-Cagan, A.; Bar, R.; Liraz, O.; Bielicki, J.K.; Johansson, J.O.; Michaelson, D.M. ABCA1 Agonist Reverses the ApoE4-Driven Cognitive and Brain Pathologies. *J. Alzheimers Dis.* **2016**, *54*, 1219–1233. [CrossRef]
201. Sarlak, Z.; Moazzami, M.; Attarzadeh Hosseini, M.; Gharakhanlou, R. The effects of aerobic training before and after the induction of Alzheimer's disease on ABCA1 and APOE mRNA expression and the level of soluble Abeta1-42 in the hippocampus of male Wistar rats. *Iran. J. Basic Med. Sci.* **2019**, *22*, 399–406. [CrossRef]
202. Suidan, G.L.; Ramaswamy, G. Targeting Apolipoprotein E for Alzheimer's Disease: An Industry Perspective. *Int. J. Mol. Sci.* **2019**, *20*, 2161. [CrossRef] [PubMed]
203. Martinez, M.C.; Tual-Chalot, S.; Leonetti, D.; Andriantsitohaina, R. Microparticles: Targets and tools in cardiovascular disease. *Trends Pharmacol. Sci.* **2011**, *32*, 659–665. [CrossRef]
204. Santilli, F.; Marchisio, M.; Lanuti, P.; Boccatonda, A.; Miscia, S.; Davi, G. Microparticles as new markers of cardiovascular risk in diabetes and beyond. *Thromb. Haemost.* **2016**, *116*, 220–234. [CrossRef] [PubMed]
205. Piccin, A.; Murphy, W.G.; Smith, O.P. Circulating microparticles: Pathophysiology and clinical implications. *Blood Rev.* **2007**, *21*, 157–171. [CrossRef]
206. Marco, A.; Brocal, C.; Marco, P. Measurement of procoagulant activity of microparticles in plasma: Feasibility of new functional assays. *Thromb. Res.* **2014**, *134*, 1363–1364. [CrossRef]
207. Buendia, P.; Montes de Oca, A.; Madueno, J.A.; Merino, A.; Martin-Malo, A.; Aljama, P.; Ramirez, R.; Rodriguez, M.; Carracedo, J. Endothelial microparticles mediate inflammation-induced vascular calcification. *FASEB J.* **2015**, *29*, 173–181. [CrossRef] [PubMed]
208. Nandi, S.; Ma, L.; Denis, M.; Karwatsky, J.; Li, Z.; Jiang, X.C.; Zha, X. ABCA1-mediated cholesterol efflux generates microparticles in addition to HDL through processes governed by membrane rigidity. *J. Lipid Res.* **2009**, *50*, 456–466. [CrossRef]
209. Okamoto, Y.; Tomioka, M.; Ogasawara, F.; Nagaiwa, K.; Kimura, Y.; Kioka, N.; Ueda, K. C-terminal of ABCA1 separately regulates cholesterol floppase activity and cholesterol efflux activity. *Biosci. Biotechnol. Biochem.* **2020**, *84*, 764–773. [CrossRef] [PubMed]
210. Hafiane, A.; Genest, J. ATP binding cassette A1 (ABCA1) mediates microparticle formation during high-density lipoprotein (HDL) biogenesis. *Atherosclerosis* **2017**, *257*, 90–99. [CrossRef]
211. Shen, X.; Zhang, S.; Guo, Z.; Xing, D.; Chen, W. The crosstalk of ABCA1 and ANXA1: A potential mechanism for protection against atherosclerosis. *Mol. Med.* **2020**, *26*, 1–8. [CrossRef]
212. Combes, V.; Coltel, N.; Alibert, M.; van Eck, M.; Raymond, C.; Juhan-Vague, I.; Grau, G.E.; Chimini, G. ABCA1 gene deletion protects against cerebral malaria: Potential pathogenic role of microparticles in neuropathology. *Am. J. Pathol.* **2005**, *166*, 295–302. [CrossRef]
213. Grau, G.E.; Chimini, G. Immunopathological consequences of the loss of engulfment genes: The case of ABCA1. *J. Soc. Biol.* **2005**, *199*, 199–206. [CrossRef]
214. Combes, V.; Coltel, N.; Faille, D.; Wassmer, S.C.; Grau, G.E. Cerebral malaria: Role of microparticles and platelets in alterations of the blood-brain barrier. *Int. J. Parasitol.* **2006**, *36*, 541–546. [CrossRef] [PubMed]
215. Sahu, U.; Mohapatra, B.N.; Kar, S.K.; Ranjit, M. Promoter polymorphisms in the ATP binding cassette transporter gene influence production of cell-derived microparticles and are highly associated with susceptibility to severe malaria in humans. *Infect. Immun.* **2013**, *81*, 1287–1294. [CrossRef]
216. Bevers, E.M.; Williamson, P.L. Getting to the Outer Leaflet: Physiology of Phosphatidylserine Exposure at the Plasma Membrane. *Physiol. Rev.* **2016**, *96*, 605–645. [CrossRef]
217. Ridger, V.C.; Boulanger, C.M.; Angelillo-Scherrer, A.; Badimon, L.; Blanc-Brude, O.; Bochaton-Piallat, M.L.; Boilard, E.; Buzas, E.I.; Caporali, A.; Dignat-George, F.; et al. Microvesicles in vascular homeostasis and diseases. Position Paper of the European Society of Cardiology (ESC) Working Group on Atherosclerosis and Vascular Biology. *Thromb. Haemost.* **2017**, *117*, 1296–1316. [CrossRef]
218. Burger, D.; Schock, S.; Thompson, C.S.; Montezano, A.C.; Hakim, A.M.; Touyz, R.M. Microparticles: Biomarkers and beyond. *Clin. Sci.* **2013**, *124*, 423–441. [CrossRef]
219. Albrecht, C.; McVey, J.H.; Elliott, J.I.; Sardini, A.; Kasza, I.; Mumford, A.D.; Naoumova, R.P.; Tuddenham, E.G.; Szabo, K.; Higgins, C.F. A novel missense mutation in ABCA1 results in altered protein trafficking and reduced phosphatidylserine translocation in a patient with Scott syndrome. *Blood* **2005**, *106*, 542–549. [CrossRef] [PubMed]
220. Leroyer, A.S.; Ebrahimian, T.G.; Cochain, C.; Recalde, A.; Blanc-Brude, O.; Mees, B.; Vilar, J.; Tedgui, A.; Levy, B.I.; Chimini, G.; et al. Microparticles from ischemic muscle promotes postnatal vasculogenesis. *Circulation* **2009**, *119*, 2808–2817. [CrossRef] [PubMed]
221. Vion, A.C.; Ramkhelawon, B.; Loyer, X.; Chironi, G.; Devue, C.; Loirand, G.; Tedgui, A.; Lehoux, S.; Boulanger, C.M. Shear stress regulates endothelial microparticle release. *Circ. Res.* **2013**, *112*, 1323–1333. [CrossRef]
222. Sviridov, D.; Bukrinsky, M. Interaction of pathogens with host cholesterol metabolism. *Curr. Opin. Lipidol.* **2014**, *25*, 333–338. [CrossRef] [PubMed]
223. Maitra, U.; Li, L. Molecular mechanisms responsible for the reduced expression of cholesterol transporters from macrophages by low-dose endotoxin. *Arterioscler. Thromb. Vasc. Biol.* **2013**, *33*, 24–33. [CrossRef] [PubMed]

224. Soto-Acosta, R.; Mosso, C.; Cervantes-Salazar, M.; Puerta-Guardo, H.; Medina, F.; Favari, L.; Ludert, J.E.; del Angel, R.M. The increase in cholesterol levels at early stages after dengue virus infection correlates with an augment in LDL particle uptake and HMG-CoA reductase activity. *Virology* **2013**, *442*, 132–147. [CrossRef] [PubMed]

225. Pirillo, A.; Catapano, A.L.; Norata, G.D. HDL in infectious diseases and sepsis. *Handb. Exp. Pharmacol.* **2015**, *224*, 483–508. [CrossRef] [PubMed]

226. Leopold Wager, C.M.; Arnett, E.; Schlesinger, L.S. Macrophage nuclear receptors: Emerging key players in infectious diseases. *PLoS Pathog.* **2019**, *15*, e1007585. [CrossRef]

227. Bocchetta, S.; Maillard, P.; Yamamoto, M.; Gondeau, C.; Douam, F.; Lebreton, S.; Lagaye, S.; Pol, S.; Helle, F.; Plengpanich, W.; et al. Up-regulation of the ATP-binding cassette transporter A1 inhibits hepatitis C virus infection. *PLoS ONE* **2014**, *9*, e92140. [CrossRef] [PubMed]

228. Sviridov, D.; Mukhamedova, N.; Makarov, A.A.; Adzhubei, A.; Bukrinsky, M. Comorbidities of HIV infection: Role of Nef-induced impairment of cholesterol metabolism and lipid raft functionality. *AIDS* **2020**, *34*, 1–13. [CrossRef] [PubMed]

229. Low, H.; Mukhamedova, N.; Cui, H.L.; McSharry, B.P.; Avdic, S.; Hoang, A.; Ditiatkovski, M.; Liu, Y.; Fu, Y.; Meikle, P.J.; et al. Cytomegalovirus Restructures Lipid Rafts via a US28/CDC42-Mediated Pathway, Enhancing Cholesterol Efflux from Host Cells. *Cell Rep.* **2016**, *16*, 186–200. [CrossRef]

230. Thangavel, S.; Mulet, C.T.; Atluri, V.S.R.; Agudelo, M.; Rosenberg, R.; Devieux, J.G.; Nair, M.P.N. Oxidative Stress in HIV Infection and Alcohol Use: Role of Redox Signals in Modulation of Lipid Rafts and ATP-Binding Cassette Transporters. *Antioxid. Redox. Signal.* **2018**, *28*, 324–337. [CrossRef]

231. Saulle, I.; Ibba, S.V.; Vittori, C.; Fenizia, C.; Mercurio, V.; Vichi, F.; Caputo, S.L.; Trabattoni, D.; Clerici, M.; Biasin, M. Sterol metabolism modulates susceptibility to HIV-1 Infection. *AIDS* **2020**, *34*, 1593–1602. [CrossRef]

232. Aizawa, Y.; Seki, N.; Nagano, T.; Abe, H. Chronic hepatitis C virus infection and lipoprotein metabolism. *World J. Gastroenterol.* **2015**, *21*, 10299–10313. [CrossRef]

233. Jacob, D.; Hunegnaw, R.; Sabyrzyanova, T.A.; Pushkarsky, T.; Chekhov, V.O.; Adzhubei, A.A.; Kalebina, T.S.; Bukrinsky, M. The ABCA1 domain responsible for interaction with HIV-1 Nef is conformational and not linear. *Biochem. Biophys. Res. Commun.* **2014**, *444*, 19–23. [CrossRef]

234. van 't Wout, A.B.; Swain, J.V.; Schindler, M.; Rao, U.; Pathmajeyan, M.S.; Mullins, J.I.; Kirchhoff, F. Nef induces multiple genes involved in cholesterol synthesis and uptake in human immunodeficiency virus type 1-infected T cells. *J. Virol.* **2005**, *79*, 10053–10058. [CrossRef]

235. Mujawar, Z.; Rose, H.; Morrow, M.P.; Pushkarsky, T.; Dubrovsky, L.; Mukhamedova, N.; Fu, Y.; Dart, A.; Orenstein, J.M.; Bobryshev, Y.V.; et al. Human immunodeficiency virus impairs reverse cholesterol transport from macrophages. *PLoS Biol.* **2006**, *4*, e40365. [CrossRef]

236. Mujawar, Z.; Tamehiro, N.; Grant, A.; Sviridov, D.; Bukrinsky, M.; Fitzgerald, M.L. Mutation of the ATP cassette binding transporter A1 (ABCA1) C-terminus disrupts HIV-1 Nef binding but does not block the Nef enhancement of ABCA1 protein degradation. *Biochemistry* **2010**, *49*, 8338–8349. [CrossRef] [PubMed]

237. Cui, H.L.; Grant, A.; Mukhamedova, N.; Pushkarsky, T.; Jennelle, L.; Dubrovsky, L.; Gaus, K.; Fitzgerald, M.L.; Sviridov, D.; Bukrinsky, M. HIV-1 Nef mobilizes lipid rafts in macrophages through a pathway that competes with ABCA1-dependent cholesterol efflux. *J. Lipid Res.* **2012**, *53*, 696–708. [CrossRef]

238. McCaffrey, K.; Braakman, I. Protein quality control at the endoplasmic reticulum. *Essays Biochem.* **2016**, *60*, 227–235. [CrossRef] [PubMed]

239. Adzhubei, A.A.; Anashkina, A.A.; Tkachev, Y.V.; Kravatsky, Y.V.; Pushkarsky, T.; Kulkarni, A.; Makarov, A.A.; Bukrinsky, M.I. Modelling interaction between HIV-1 Nef and calnexin. *AIDS* **2018**, *32*, 2103–2111. [CrossRef] [PubMed]

240. Hunegnaw, R.; Vassylyeva, M.; Dubrovsky, L.; Pushkarsky, T.; Sviridov, D.; Anashkina, A.A.; Uren, A.; Brichacek, B.; Vassylyev, D.G.; Adzhubei, A.A.; et al. Interaction Between HIV-1 Nef and Calnexin: From Modeling to Small Molecule Inhibitors Reversing HIV-Induced Lipid Accumulation. *Arterioscler. Thromb. Vasc. Biol.* **2016**, *36*, 1758–1771. [CrossRef]

241. Jennelle, L.; Hunegnaw, R.; Dubrovsky, L.; Pushkarsky, T.; Fitzgerald, M.L.; Sviridov, D.; Popratiloff, A.; Brichacek, B.; Bukrinsky, M. HIV-1 protein Nef inhibits activity of ATP-binding cassette transporter A1 by targeting endoplasmic reticulum chaperone calnexin. *J. Biol. Chem.* **2014**, *289*, 28870–28884. [CrossRef] [PubMed]

242. Wilflingseder, D.; Stoiber, H. Float on: Lipid rafts in the lifecycle of HIV. *Front. Biosci.* **2007**, *12*, 2124–2135. [CrossRef]

243. Olivetta, E.; Arenaccio, C.; Manfredi, F.; Anticoli, S.; Federico, M. The Contribution of Extracellular Nef to HIV-Induced Pathogenesis. *Curr. Drug Targets* **2016**, *17*, 46–53. [CrossRef]

244. Ditiatkovski, M.; Mukhamedova, N.; Dragoljevic, D.; Hoang, A.; Low, H.; Pushkarsky, T.; Fu, Y.; Carmichael, I.; Hill, A.F.; Murphy, A.J.; et al. Modification of lipid rafts by extracellular vesicles carrying HIV-1 protein Nef induces redistribution of amyloid precursor protein and Tau, causing neuronal dysfunction. *J. Biol. Chem.* **2020**, *295*, 13377–13392. [CrossRef]

245. McNamara, R.P.; Costantini, L.M.; Myers, T.A.; Schouest, B.; Maness, N.J.; Griffith, J.D.; Damania, B.A.; MacLean, A.G.; Dittmer, D.P. Nef Secretion into Extracellular Vesicles or Exosomes Is Conserved across Human and Simian Immunodeficiency Viruses. *mBio* **2018**, *9*, 1–20. [CrossRef]

246. Low, H.; Hoang, A.; Pushkarsky, T.; Dubrovsky, L.; Dewar, E.; Di Yacovo, M.S.; Mukhamedova, N.; Cheng, L.; Downs, C.; Simon, G.; et al. HIV disease, metabolic dysfunction and atherosclerosis: A three year prospective study. *PLoS ONE* **2019**, *14*, e215620. [CrossRef] [PubMed]

247. Siegel, M.O.; Borkowska, A.G.; Dubrovsky, L.; Roth, M.; Welti, R.; Roberts, A.D.; Parenti, D.M.; Simon, G.L.; Sviridov, D.; Simmens, S.; et al. HIV infection induces structural and functional changes in high density lipoproteins. *Atherosclerosis* **2015**, *243*, 19–29. [CrossRef] [PubMed]

248. Asztalos, B.F.; Mujawar, Z.; Morrow, M.P.; Grant, A.; Pushkarsky, T.; Wanke, C.; Shannon, R.; Geyer, M.; Kirchhoff, F.; Sviridov, D.; et al. Circulating Nef induces dyslipidemia in simian immunodeficiency virus-infected macaques by suppressing cholesterol efflux. *J. Infect. Dis.* **2010**, *202*, 614–623. [CrossRef]

249. Lo, J.; Rosenberg, E.S.; Fitzgerald, M.L.; Bazner, S.B.; Ihenachor, E.J.; Hawxhurst, V.; Borkowska, A.H.; Wei, J.; Zimmerman, C.O.; Burdo, T.H.; et al. High-density lipoprotein-mediated cholesterol efflux capacity is improved by treatment with antiretroviral therapy in acute human immunodeficiency virus infection. *Open Forum Infect. Dis.* **2014**, *1*, 1–9. [CrossRef]

250. Mukhamedova, N.; Hoang, A.; Dragoljevic, D.; Dubrovsky, L.; Pushkarsky, T.; Low, H.; Ditiatkovski, M.; Fu, Y.; Ohkawa, R.; Meikle, P.J.; et al. Exosomes containing HIV protein Nef reorganize lipid rafts potentiating inflammatory response in bystander cells. *PLoS Pathog.* **2019**, *15*, e1007907. [CrossRef]

251. Bogan, J.S.; Xu, Y.; Hao, M. Cholesterol accumulation increases insulin granule size and impairs membrane trafficking. *Traffic* **2012**, *13*, 1466–1480. [CrossRef]

252. Pou, J.; Rebollo, A.; Roglans, N.; Sanchez, R.M.; Vazquez-Carrera, M.; Laguna, J.C.; Pedro-Botet, J.; Alegret, M. Ritonavir increases CD36, ABCA1 and CYP27 expression in THP-1 macrophages. *Exp. Biol. Med.* **2008**, *233*, 1572–1582. [CrossRef]

253. Piconi, S.; Parisotto, S.; Rizzardini, G.; Passerini, S.; Meraviglia, P.; Schiavini, M.; Niero, F.; Biasin, M.; Bonfanti, P.; Ricci, E.D.; et al. Atherosclerosis is associated with multiple pathogenic mechanisms in HIV-infected antiretroviral-naive or treated individuals. *AIDS* **2013**, *27*, 381–389. [CrossRef] [PubMed]

254. Rose, H.; Hoy, J.; Woolley, I.; Tchoua, U.; Bukrinsky, M.; Dart, A.; Sviridov, D. HIV infection and high density lipoprotein metabolism. *Atherosclerosis* **2008**, *199*, 79–86. [CrossRef]

255. Obirikorang, C.; Acheampong, E.; Quaye, L.; Yorke, J.; Amos-Abanyie, E.K.; Akyaw, P.A.; Anto, E.O.; Bani, S.B.; Asamoah, E.A.; Batu, E.N. Association of single nucleotide polymorphisms with dyslipidemia in antiretroviral exposed HIV patients in a Ghanaian population: A case-control study. *PLoS ONE* **2020**, *15*, e0227779. [CrossRef] [PubMed]

256. Egana-Gorrono, L.; Martinez, E.; Cormand, B.; Escriba, T.; Gatell, J.; Arnedo, M. Impact of genetic factors on dyslipidemia in HIV-infected patients starting antiretroviral therapy. *AIDS* **2013**, *27*, 529–538. [CrossRef] [PubMed]

257. Selitsky, S.R.; Dinh, T.A.; Toth, C.L.; Kurtz, C.L.; Honda, M.; Struck, B.R.; Kaneko, S.; Vickers, K.C.; Lemon, S.M.; Sethupathy, P. Transcriptomic Analysis of Chronic Hepatitis B and C and Liver Cancer Reveals MicroRNA-Mediated Control of Cholesterol Synthesis Programs. *mBio* **2015**, *6*, e01500–e01515. [CrossRef] [PubMed]

258. Shirasaki, T.; Honda, M.; Shimakami, T.; Horii, R.; Yamashita, T.; Sakai, Y.; Sakai, A.; Okada, H.; Watanabe, R.; Murakami, S.; et al. MicroRNA-27a regulates lipid metabolism and inhibits hepatitis C virus replication in human hepatoma cells. *J. Virol.* **2013**, *87*, 5270–5286. [CrossRef]

259. Heaton, N.S.; Randall, G. Multifaceted roles for lipids in viral infection. *Trends Microbiol.* **2011**, *19*, 368–375. [CrossRef] [PubMed]

260. Sanchez, V.; Dong, J.J. Alteration of lipid metabolism in cells infected with human cytomegalovirus. *Virology* **2010**, *404*, 71–77. [CrossRef]

261. Marin-Palma, D.; Sirois, C.M.; Urcuqui-Inchima, S.; Hernandez, J.C. Inflammatory status and severity of disease in dengue patients are associated with lipoprotein alterations. *PLoS ONE* **2019**, *14*, e0214245. [CrossRef]

262. Tree, M.O.; Londono-Renteria, B.; Troupin, A.; Clark, K.M.; Colpitts, T.M.; Conway, M.J. Dengue virus reduces expression of low-density lipoprotein receptor-related protein 1 to facilitate replication in Aedes aegypti. *Sci. Rep.* **2019**, *9*, 1–14. [CrossRef]

263. Hwang, J.; Wang, Y.; Fikrig, E. Inhibition of Chikungunya Virus Replication in Primary Human Fibroblasts by Liver X Receptor Agonist. *Antimicrob. Agents Chemother.* **2019**, *63*, 1–5. [CrossRef]

264. Hapugaswatta, H.; Amarasena, P.; Premaratna, R.; Seneviratne, K.N.; Jayathilaka, N. Differential expression of microRNA, miR-150 and enhancer of zeste homolog 2 (EZH2) in peripheral blood cells as early prognostic markers of severe forms of dengue. *J. Biomed. Sci.* **2020**, *27*, 1–9. [CrossRef] [PubMed]

265. Aly, A.S.; Vaughan, A.M.; Kappe, S.H. Malaria parasite development in the mosquito and infection of the mammalian host. *Annu. Rev. Microbiol.* **2009**, *63*, 195–221. [CrossRef] [PubMed]

266. Babatunde, K.A.; Yesodha Subramanian, B.; Ahouidi, A.D.; Martinez Murillo, P.; Walch, M.; Mantel, P.Y. Role of Extracellular Vesicles in Cellular Cross Talk in Malaria. *Front. Immunol.* **2020**, *11*, 1–13. [CrossRef] [PubMed]

267. Wah, S.T.; Hananantachai, H.; Patarapotikul, J.; Ohashi, J.; Naka, I.; Nuchnoi, P. microRNA-27a and microRNA-146a SNP in cerebral malaria. *Mol. Genet. Genomic Med.* **2019**, *7*, 1–11. [CrossRef]

268. Zhang, M.; Wu, J.F.; Chen, W.J.; Tang, S.L.; Mo, Z.C.; Tang, Y.Y.; Li, Y.; Wang, J.L.; Liu, X.Y.; Peng, J.; et al. MicroRNA-27a/b regulates cellular cholesterol efflux, influx and esterification/hydrolysis in THP-1 macrophages. *Atherosclerosis* **2014**, *234*, 54–64. [CrossRef]

269. Martinelli, L.M.; Fontes, K.N.; Reginatto, M.W.; Andrade, C.B.V.; Monteiro, V.R.S.; Gomes, H.R.; Silva-Filho, J.L.; Pinheiro, A.A.S.; Vago, A.R.; Almeida, F.; et al. Malaria in pregnancy regulates P-glycoprotein (P-gp/Abcb1a) and ABCA1 efflux transporters in the Mouse Visceral Yolk Sac. *J. Cell Mol. Med.* **2020**, *24*, 10636–10647. [CrossRef] [PubMed]

270. Mitchell, P.; Liew, G.; Gopinath, B.; Wong, T.Y. Age-related macular degeneration. *Lancet* **2018**, *392*, 1147–1159. [CrossRef]

271. Curcio, C.A.; Messinger, J.D.; Sloan, K.R.; McGwin, G.; Medeiros, N.E.; Spaide, R.F. Subretinal drusenoid deposits in non-neovascular age-related macular degeneration: Morphology, prevalence, topography, and biogenesis model. *Retina* **2013**, *33*, 265–276. [CrossRef]

272. Greferath, U.; Guymer, R.H.; Vessey, K.A.; Brassington, K.; Fletcher, E.L. Correlation of Histologic Features with In Vivo Imaging of Reticular Pseudodrusen. *Ophthalmology* **2016**, *123*, 1320–1331. [CrossRef]

273. Khan, K.N.; Mahroo, O.A.; Khan, R.S.; Mohamed, M.D.; McKibbin, M.; Bird, A.; Michaelides, M.; Tufail, A.; Moore, A.T. Differentiating drusen: Drusen and drusen-like appearances associated with ageing, age-related macular degeneration, inherited eye disease and other pathological processes. *Prog. Retin. Eye Res.* **2016**, *53*, 70–106. [CrossRef]

274. Sene, A.; Apte, R.S. Eyeballing cholesterol efflux and macrophage function in disease pathogenesis. *Trends Endocrinol. Metab.* **2014**, *25*, 107–114. [CrossRef]

275. Sene, A.; Chin-Yee, D.; Apte, R.S. Seeing through VEGF: Innate and adaptive immunity in pathological angiogenesis in the eye. *Trends Mol. Med.* **2015**, *21*, 43–51. [CrossRef]

276. Zweifel, S.A.; Spaide, R.F.; Curcio, C.A.; Malek, G.; Imamura, Y. Reticular pseudodrusen are subretinal drusenoid deposits. *Ophthalmology* **2010**, *117*, 303–312e1. [CrossRef] [PubMed]

277. Curcio, C.A.; Johnson, M.; Rudolf, M.; Huang, J.D. The oil spill in ageing Bruch membrane. *Br. J. Ophthalmol.* **2011**, *95*, 1638–1645. [CrossRef]

278. Spaide, R.F.; Ooto, S.; Curcio, C.A. Subretinal drusenoid deposits AKA pseudodrusen. *Surv. Ophthalmol.* **2018**, *63*, 782–815. [CrossRef] [PubMed]

279. Cheung, C.M.G.; Gan, A.; Fan, Q.; Chee, M.L.; Apte, R.S.; Khor, C.C.; Yeo, I.; Mathur, R.; Cheng, C.Y.; Wong, T.Y.; et al. Plasma lipoprotein subfraction concentrations are associated with lipid metabolism and age-related macular degeneration. *J. Lipid Res.* **2017**, *58*, 1785–1796. [CrossRef] [PubMed]

280. Chen, W.; Stambolian, D.; Edwards, A.O.; Branham, K.E.; Othman, M.; Jakobsdottir, J.; Tosakulwong, N.; Pericak-Vance, M.A.; Campochiaro, P.A.; Klein, M.L.; et al. Genetic variants near TIMP3 and high-density lipoprotein-associated loci influence susceptibility to age-related macular degeneration. *Proc. Natl. Acad. Sci. USA* **2010**, *107*, 7401–7406. [CrossRef]

281. Dietzel, M.; Pauleikhoff, D.; Arning, A.; Heimes, B.; Lommatzsch, A.; Stoll, M.; Hense, H.W. The contribution of genetic factors to phenotype and progression of drusen in early age-related macular degeneration. *Graefes Arch. Clin. Exp. Ophthalmol.* **2014**, *252*, 1273–1281. [CrossRef]

282. Merle, B.M.; Maubaret, C.; Korobelnik, J.F.; Delyfer, M.N.; Rougier, M.B.; Lambert, J.C.; Amouyel, P.; Malet, F.; Le Goff, M.; Dartigues, J.F.; et al. Association of HDL-related loci with age-related macular degeneration and plasma lutein and zeaxanthin: The Alienor study. *PLoS ONE* **2013**, *8*, e79848. [CrossRef]

283. Neale, B.M.; Fagerness, J.; Reynolds, R.; Sobrin, L.; Parker, M.; Raychaudhuri, S.; Tan, P.L.; Oh, E.C.; Merriam, J.E.; Souied, E.; et al. Genome-wide association study of advanced age-related macular degeneration identifies a role of the hepatic lipase gene (LIPC). *Proc. Natl. Acad. Sci. USA* **2010**, *107*, 7395–7400. [CrossRef]

284. Wang, Y.; Wang, M.; Han, Y.; Zhang, R.; Ma, L. ABCA1 rs1883025 polymorphism and risk of age-related macular degeneration. *Graefes Arch. Clin. Exp. Ophthalmol.* **2016**, *254*, 323–332. [CrossRef] [PubMed]

285. Yu, Y.; Reynolds, R.; Fagerness, J.; Rosner, B.; Daly, M.J.; Seddon, J.M. Association of variants in the LIPC and ABCA1 genes with intermediate and large drusen and advanced age-related macular degeneration. *Invest. Ophthalmol. Vis. Sci.* **2011**, *52*, 4663–4670. [CrossRef]

286. Duncan, K.G.; Hosseini, K.; Bailey, K.R.; Yang, H.; Lowe, R.J.; Matthes, M.T.; Kane, J.P.; LaVail, M.M.; Schwartz, D.M.; Duncan, J.L. Expression of reverse cholesterol transport proteins ATP-binding cassette A1 (ABCA1) and scavenger receptor BI (SR-BI) in the retina and retinal pigment epithelium. *Br. J. Ophthalmol.* **2009**, *93*, 1116–1120. [CrossRef]

287. Ban, N.; Lee, T.J.; Sene, A.; Choudhary, M.; Lekwuwa, M.; Dong, Z.; Santeford, A.; Lin, J.B.; Malek, G.; Ory, D.S.; et al. Impaired monocyte cholesterol clearance initiates age-related retinal degeneration and vision loss. *JCI Insight* **2018**, *3*. [CrossRef] [PubMed]

288. Storti, F.; Klee, K.; Todorova, V.; Steiner, R.; Othman, A.; van der Velde-Visser, S.; Samardzija, M.; Meneau, I.; Barben, M.; Karademir, D.; et al. Impaired ABCA1/ABCG1-mediated lipid efflux in the mouse retinal pigment epithelium (RPE) leads to retinal degeneration. *Elife* **2019**, *8*. [CrossRef]

289. Quigley, H.A.; Broman, A.T. The number of people with glaucoma worldwide in 2010 and 2020. *Br. J. Ophthalmol.* **2006**, *90*, 262–267. [CrossRef] [PubMed]

290. Dietze, J.; Blair, K.; Havens, S.J. Glaucoma. In *StatPearls*; StatPearls Publishing: Treasure Island, FL, USA, 2020.

291. Sears, N.C.; Boese, E.A.; Miller, M.A.; Fingert, J.H. Mendelian genes in primary open angle glaucoma. *Exp. Eye Res.* **2019**, *186*, 1–10. [CrossRef] [PubMed]

292. Eliseeva, N.V.; Churnosov, M.I. [Genome-wide studies of primary open-angle glaucoma]. *Vestn. Oftalmol.* **2020**, *136*, 129–135. [CrossRef]

293. Zukerman, R.; Harris, A.; Vercellin, A.V.; Siesky, B.; Pasquale, L.R.; Ciulla, T.A. Molecular Genetics of Glaucoma: Subtype and Ethnicity Considerations. *Genes* **2020**, *12*, 55. [CrossRef]

294. Hysi, P.G.; Cheng, C.Y.; Springelkamp, H.; Macgregor, S.; Bailey, J.N.C.; Wojciechowski, R.; Vitart, V.; Nag, A.; Hewitt, A.W.; Hohn, R.; et al. Genome-wide analysis of multi-ancestry cohorts identifies new loci influencing intraocular pressure and susceptibility to glaucoma. *Nat. Genet.* **2014**, *46*, 1126–1130. [CrossRef]

295. Shiga, Y.; Akiyama, M.; Nishiguchi, K.M.; Sato, K.; Shimozawa, N.; Takahashi, A.; Momozawa, Y.; Hirata, M.; Matsuda, K.; Yamaji, T.; et al. Genome-wide association study identifies seven novel susceptibility loci for primary open-angle glaucoma. *Hum. Mol. Genet.* **2018**, *27*, 1486–1496. [CrossRef] [PubMed]

296. Xu, M.; Li, S.; Zhu, J.; Luo, D.; Song, W.; Zhou, M. Plasma lipid levels and risk of primary open angle glaucoma: A genetic study using Mendelian randomization. *BMC Ophthalmol.* **2020**, *20*, 390. [CrossRef] [PubMed]

297. Hu, C.; Niu, L.; Li, L.; Song, M.; Zhang, Y.; Lei, Y.; Chen, Y.; Sun, X. ABCA1 Regulates IOP by Modulating Cav1/eNOS/NO Signaling Pathway. *Invest. Ophthalmol. Vis. Sci.* **2020**, *61*, 1–9. [CrossRef]

298. Ahn, J.; Lim, U.; Weinstein, S.J.; Schatzkin, A.; Hayes, R.B.; Virtamo, J.; Albanes, D. Prediagnostic total and high-density lipoprotein cholesterol and risk of cancer. *Cancer Epidemiol. Biomarkers Prev.* **2009**, *18*, 2814–2821. [CrossRef]

299. Jafri, H.; Alsheikh-Ali, A.A.; Karas, R.H. Baseline and on-treatment high-density lipoprotein cholesterol and the risk of cancer in randomized controlled trials of lipid-altering therapy. *J. Am. Coll. Cardiol.* **2010**, *55*, 2846–2854. [CrossRef] [PubMed]

300. Dessi, S.; Batetta, B.; Pulisci, D.; Spano, O.; Anchisi, C.; Tessitore, L.; Costelli, P.; Baccino, F.M.; Aroasio, E.; Pani, P. Cholesterol content in tumor tissues is inversely associated with high-density lipoprotein cholesterol in serum in patients with gastrointestinal cancer. *Cancer* **1994**, *73*, 253–258. [CrossRef]

301. Kolanjiappan, K.; Ramachandran, C.R.; Manoharan, S. Biochemical changes in tumor tissues of oral cancer patients. *Clin. Biochem.* **2003**, *36*, 61–65. [CrossRef]

302. McMurray, H.R.; Sampson, E.R.; Compitello, G.; Kinsey, C.; Newman, L.; Smith, B.; Chen, S.R.; Klebanov, L.; Salzman, P.; Yakovlev, A.; et al. Synergistic response to oncogenic mutations defines gene class critical to cancer phenotype. *Nature* **2008**, *453*, 1112–1116. [CrossRef]

303. Fukuchi, J.; Hiipakka, R.A.; Kokontis, J.M.; Hsu, S.; Ko, A.L.; Fitzgerald, M.L.; Liao, S. Androgenic suppression of ATP-binding cassette transporter A1 expression in LNCaP human prostate cancer cells. *Cancer Res.* **2004**, *64*, 7682–7685. [CrossRef]

304. Schimanski, S.; Wild, P.J.; Treeck, O.; Horn, F.; Sigruener, A.; Rudolph, C.; Blaszyk, H.; Klinkhammer-Schalke, M.; Ortmann, O.; Hartmann, A.; et al. Expression of the lipid transporters ABCA3 and ABCA1 is diminished in human breast cancer tissue. *Horm. Metab. Res.* **2010**, *42*, 102–109. [CrossRef]

305. Bi, D.P.; Yin, C.H.; Zhang, X.Y.; Yang, N.N.; Xu, J.Y. MiR-183 functions as an oncogene by targeting ABCA1 in colon cancer. *Oncol. Rep.* **2016**, *35*, 2873–2879. [CrossRef] [PubMed]

306. Chou, J.L.; Huang, R.L.; Shay, J.; Chen, L.Y.; Lin, S.J.; Yan, P.S.; Chao, W.T.; Lai, Y.H.; Lai, Y.L.; Chao, T.K.; et al. Hypermethylation of the TGF-β target, ABCA1 is associated with poor prognosis in ovarian cancer patients. *Clin. Epigenetics* **2015**, *7*, 1–11. [CrossRef] [PubMed]

307. Lee, B.H.; Taylor, M.G.; Robinet, P.; Smith, J.D.; Schweitzer, J.; Sehayek, E.; Falzarano, S.M.; Magi-Galluzzi, C.; Klein, E.A.; Ting, A.H. Dysregulation of cholesterol homeostasis in human prostate cancer through loss of ABCA1. *Cancer Res.* **2013**, *73*, 1211–1218. [CrossRef] [PubMed]

308. Smith, B.; Land, H. Anticancer activity of the cholesterol exporter ABCA1 gene. *Cell Rep.* **2012**, *2*, 580–590. [CrossRef]

309. Viaud, M.; Abdel-Wahab, O.; Gall, J.; Ivanov, S.; Guinamard, R.; Sore, S.; Merlin, J.; Ayrault, M.; Guilbaud, E.; Jacquel, A.; et al. ABCA1 Exerts Tumor-Suppressor Function in Myeloproliferative Neoplasms. *Cell Rep.* **2020**, *30*, 3397–3410. [CrossRef]

310. Vargas, T.; Moreno-Rubio, J.; Herranz, J.; Cejas, P.; Molina, S.; Gonzalez-Vallinas, M.; Mendiola, M.; Burgos, E.; Aguayo, C.; Custodio, A.B.; et al. ColoLipidGene: Signature of lipid metabolism-related genes to predict prognosis in stage-II colon cancer patients. *Oncotarget* **2015**, *6*, 7348–7363. [CrossRef]

311. Pan, H.; Zheng, Y.; Pan, Q.; Chen, H.; Chen, F.; Wu, J.; Di, D. Expression of LXRβ, ABCA1 and ABCG1 in human triplenegative breast cancer tissues. *Oncol. Rep.* **2019**, *42*, 1869–1877. [CrossRef]

312. Hedditch, E.L.; Gao, B.; Russell, A.J.; Lu, Y.; Emmanuel, C.; Beesley, J.; Johnatty, S.E.; Chen, X.; Harnett, P.; George, J.; et al. ABCA transporter gene expression and poor outcome in epithelial ovarian cancer. *J. Natl. Cancer Inst.* **2014**, *106*, 1–11. [CrossRef]

313. Crompton, M. Mitochondrial intermembrane junctional complexes and their role in cell death. *J. Physiol.* **2000**, *529 Pt 1*, 11–21. [CrossRef]

314. Graff, J.R.; Konicek, B.W.; McNulty, A.M.; Wang, Z.; Houck, K.; Allen, S.; Paul, J.D.; Hbaiu, A.; Goode, R.G.; Sandusky, G.E.; et al. Increased AKT activity contributes to prostate cancer progression by dramatically accelerating prostate tumor growth and diminishing p27Kip1 expression. *J. Biol. Chem.* **2000**, *275*, 24500–24505. [CrossRef] [PubMed]

315. Landry, Y.D.; Denis, M.; Nandi, S.; Bell, S.; Vaughan, A.M.; Zha, X. ATP-binding cassette transporter A1 expression disrupts raft membrane microdomains through its ATPase-related functions. *J. Biol. Chem.* **2006**, *281*, 36091–36101. [CrossRef] [PubMed]

316. Aguirre-Portoles, C.; Feliu, J.; Reglero, G.; Ramirez de Molina, A. ABCA1 overexpression worsens colorectal cancer prognosis by facilitating tumour growth and caveolin-1-dependent invasiveness, and these effects can be ameliorated using the BET inhibitor apabetalone. *Mol. Oncol.* **2018**, *12*, 1735–1752. [CrossRef] [PubMed]

317. Zamanian-Daryoush, M.; Lindner, D.J.; DiDonato, J.A.; Wagner, M.; Buffa, J.; Rayman, P.; Parks, J.S.; Westerterp, M.; Tall, A.R.; Hazen, S.L. Myeloid-specific genetic ablation of ATP-binding cassette transporter ABCA1 is protective against cancer. *Oncotarget* **2017**, *8*, 71965–71980. [CrossRef]

318. Nobili, S.; Lapucci, A.; Landini, I.; Coronnello, M.; Roviello, G.; Mini, E. Role of ATP-binding cassette transporters in cancer initiation and progression. *Semin. Cancer Biol.* **2020**, *60*, 72–95. [CrossRef] [PubMed]

International Journal of
Molecular Sciences

Review

Sitosterolemia: Twenty Years of Discovery of the Function of *ABCG5 ABCG8*

Kori Williams [1], Allison Segard [1] and Gregory A. Graf [1,2,3,*]

[1] Department of Pharmaceutical Sciences, College of Pharmacy, University of Kentucky, Lexington, KY 40536, USA; kswilliams3@uky.edu (K.W.); allison.segard@uky.edu (A.S.)
[2] Saha Cardiovascular Research Center, Lexington, KY 40536, USA
[3] Barnstable Brown Diabetes and Obesity Center, Lexington, KY 40536, USA
* Correspondence: Gregory.Graf@uky.edu; Tel.: +1-859-797-2463

Abstract: Sitosterolemia is a lipid disorder characterized by the accumulation of dietary xenosterols in plasma and tissues caused by mutations in either *ABCG5* or *ABCG8*. *ABCG5 ABCG8* encodes a pair of ABC half transporters that form a heterodimer (G5G8), which then traffics to the surface of hepatocytes and enterocytes and promotes the secretion of cholesterol and xenosterols into the bile and the intestinal lumen. We review the literature from the initial description of the disease, the discovery of its genetic basis, current therapy, and what has been learned from animal, cellular, and molecular investigations of the transporter in the twenty years since its discovery. The genomic era has revealed that there are far more carriers of loss of function mutations and likely pathogenic variants of *ABCG5 ABCG8* than previously thought. The impact of these variants on G5G8 structure and activity are largely unknown. We propose a classification system for *ABCG5 ABCG8* mutants based on previously published systems for diseases caused by defects in ABC transporters. This system establishes a framework for the comprehensive analysis of disease-associated variants and their impact on G5G8 structure–function.

Keywords: phytosterol; xenosterol; cholesterol; atherosclerosis; gall stone; ABC; transporter

Citation: Williams, K.; Segard, A.; Graf, G.A. Sitosterolemia: Twenty Years of Discovery of the Function of *ABCG5 ABCG8. Int. J. Mol. Sci.* **2021**, *22*, 2641. https://doi.org/10.3390/ijms22052641

Academic Editor: Thomas Falguières

Received: 7 January 2021
Accepted: 26 February 2021
Published: 5 March 2021

1. Discovery to Therapy

In 1974, Bhattacharyya and Conner described a new lipid storage disorder in two sisters who presented with tendon and tuberous xanthomas and elevated plasma levels of phytosterols, sitosterol, campesterol, and stigmasterol [1]. Absorption of radiolabeled β-sitosterol was reported to be thirty-five times greater than that of normal subjects. They named their new lipid disorder β-sitosterolemia (hereon referred to as sitosterolemia), but it would be another 26 years before the discovery of *ABCG5 ABCG8* as the causative gene defect. Subsequent case reports established recessive genetics of the disease and greatly expanded its potential clinical presentation, which may include elevated low density lipoprotein (LDL) cholesterol, premature coronary artery disease and death, hemolytic anemia, macrothrombocytopenia, splenomegaly, adrenal dysfunction, elevated liver function tests, and cirrhosis [2–13]. Clinical studies in individuals with sitosterolemia revealed reductions in cholesterol synthesis, biliary cholesterol secretion, plasma clearance, and fecal elimination of neutral sterols [8,9,14,15]. Despite the absorptive phenotype and metabolism of phytosterols to bile acids, the clinical management of these patients with low sterol diets and bile acid binding resins resulted in modest and inconsistent reductions in plasma phytosterols [5,16].

Following the elimination of key genes in the esterification, absorbance, biosynthesis, and regulation of cholesterol metabolism, the sitosterolemia locus was mapped to a 0.5 centimorgan region on chromosome 2p21 [17,18]. The breakthrough would come two years later when investigators were studying agonists for liver X receptors (LXR-α NR1H3, LXR-β NR1H2), sensors of excess cholesterol that promote cholesterol mobilization from macrophages, metabolism to primary bile acids, and the excretion of neutral and acidic sterols [19,20]. A microarray screen of liver and intestinal transcripts from mice treated with an LXR agonist (T0901317) revealed a modest induction (2.5-fold) of the mouse brown gene, so named as it is the ortholog of the ABC transporter that determines brown eye color in *Drosophila* [21,22]. The human ortholog to mouse brown mapped to 2p21, *STSL* (OMIM: *STSL1*: 210250/*STSL2*:618666). At virtually the same time, independent groups identified individuals in multiple kindreds harboring nonsense mutations in either *ABCG5* or *ABCG8* with sitosterolemia, but not their unaffected family members [22,23]. *STSL1* and *STSL2* encode a pair of ATP binding cassette (ABC) half transporters in the G-subfamily. They reside on opposite strands of the DNA with initiation codons separated by a mere 374 base pairs. In these early reports, transcripts were restricted to the liver and intestine, the abundance of which increased in response to dietary cholesterol, suggesting the function of the transporters was to oppose intestinal absorption and promote biliary secretion of neutral sterols.

The discovery of ezetimibe as the inhibitor of Neiman–Pick C-1-Like 1 (NPC1L1), and cholesterol absorption was a breakthrough in the clinical management of sitosterolemia [24,25]. Ezetimibe (10 mg/day) was tested as a cholesterol-lowering agent in healthy subjects with moderate hypercholesterolemia. A detailed analysis of plasma sterols revealed that in addition to cholesterol, plasma phytosterols were also reduced, indicating that phytosterols and cholesterol shared a common, ezetimibe-sensitive pathway for absorption, thus suggesting the drug might be effective in the treatment of sitosterolemia (Figure 1 (1)(2)). Ezetimibe was subsequently shown to reduce plasma phytosterols in sitosterolemic subjects by >20% after eight weeks [26]. In a two-year follow-up study, plasma sitosterol and campesterol levels were reduced by 44% and 51%, respectively [27]. It should be noted that phytosterol levels in these subjects remained well above normal. However, the reductions observed with ezetimibe as a single agent or as an adjunctive therapy resolves many of the clinical manifestations of sitosterolemia (reviewed in [28]).

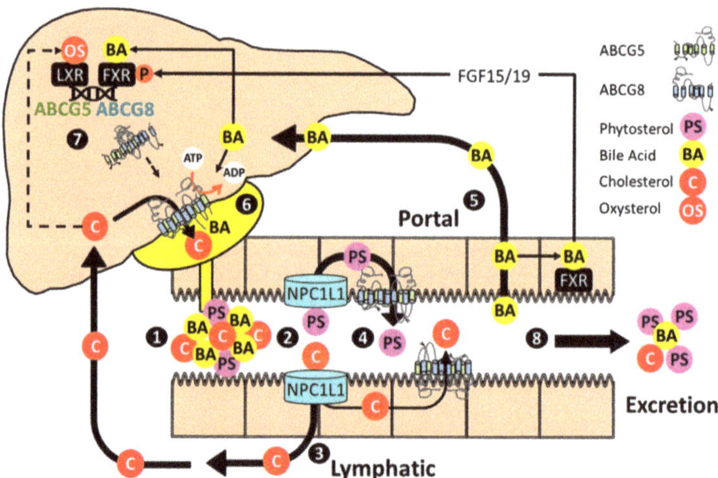

Figure 1. Enterohepatic sterol flux and regulation of *ABCG5 ABCG8*. (1) Bile acid micelles facilitate the solubilization of dietary and endogenous sterols in the proximal small intestine. Phospholipids not depicted. (2) NPC1L1 facilitates uptake of cholesterol and phytosterols into intestinal enterocytes (3) Cholesterol is incorporated into chylomicrons, delivered to the plasma compartment through the lymphatic system, and cleared by the liver. ABCG5 ABCG8 also promotes cholesterol secretion into the intestinal lumen. (4) Phytosterols are poorly absorbed and largely returned to the intestinal lumen by ABCG5 ABCG8. (5) Bile acids are reabsorbed in the distal small intestine, stimulate FXR-

dependent expression of FGF15/19, and are returned to the liver through the portal system. (6) In the liver, bile acids stimulate ABCG5 ABCG8 catalytic activity and promote the formation of bile acid micelles that serve as acceptors for ABCG5 ABCG8 mediated biliary cholesterol secretion. (7) Cholesterol metabolites (oxysterols), through LXR, and bile acids, through FXR and in cooperation with FGF15/19, activate ABCG5 ABCG8. The half transporters heterodimerize, traffic to the canalicular surface, and promote biliary phytosterol and cholesterol secretion. (8) Excess cholesterol, phytosterols and bile acids that are not absorbed/reabsorbed are eliminated from the body.

2. Animal Models of Sitosterolemia

Three independent mouse models of sitosterolemia have been developed which lack functional *Abcg5*, *Abcg8*, or both half transporters [29–31]. In each model, the G5G8 heterodimer is absent. These models largely phenocopy one another and share many phenotypes with sitosterolemia in humans. These include elevated plasma and tissue levels of phytosterols, reduced biliary cholesterol, and repression of the cholesterol biosynthetic pathway. Since the discovery of *ABCG5 ABCG8*, phenotypes in spontaneous rodent models have been attributed to defects in *Abcg5 Abcg8* or the accumulation of phytosterols. The Spontaneous hypertensive rat (SHR) harbors a glycine 583 cysteine mutation in *Abcg5*, which segregates with elevated phytosterol levels in plasma, but not hypertension [32]. Similarly, a premature stop codon in *Abcg5* is present in the thrombocytopenia and cardiomyopathy (trac) mouse [33]. Plasma phytosterols and platelet counts were rescued by crossing this strain with mice harboring a human *ABCG5 ABCG8* transgene. Mice with a targeted disruption in either *Abcg5* or *Abcg8* also display platelet dysfunction, effects that are reversed with ezetimibe or low phytosterol-containing diets, respectively [34,35]. Other phenotypes in mice lacking functional G5G8 are reversed by ezetimibe treatment, genetic inactivation of its pharmacological target, Neiman–Pick C1-Like-1 (NPC1L1), or being fed a diet that lacks phytosterols [36–38]. Conversely, feeding diets enriched in phytosterols exacerbate phenotypes and results in sudden death [36,39].

Collectively, the available data indicates that the presentation of sitosterolemia in both humans and rodent models is a function of the type and abundance of xenosterols present in the diet that ultimately accumulates in plasma and tissues. This complicates interpretations of *ABCG5 ABCG8* physiology with respect to cholesterol metabolism, as phytosterols are known to produce a myriad of biological effects, including disruptions of sterol sensing by sterol receptor element binding protein 2 (SREBP-2), LXR, and the bile acid receptor (farnesoid X receptor, FXR NR1H4) [40–43].

While sitosterolemia may present with normal or only modestly elevated plasma cholesterol, cholesterol levels are generally lower in mice lacking functional G5G8 than their wild-type counterparts. The precise mechanism accounting for this difference has not been investigated but may be due to the paucity of ApoB containing lipoproteins in plasma in mice compared to humans. An alternative explanation is a species difference in the role of G5G8 in cholesterol absorption. Given its abundant expression in the intestinal epithelium, its role in biliary cholesterol secretion, and the common pathway for cholesterol and phytosterol absorption (NPC1L1), it stands to reason that G5G8 would oppose cholesterol absorption. However, studies in both humans and mice have produced inconsistent results. In both humans and mice, phytosterol absorption is clearly elevated [30,31,44,45]. Sitosterolemics appear on the higher end of the range for cholesterol absorption in humans, a trait that maps to the *ABCG5 ABCG8* locus [18,23]. However, the magnitude of the increase is relatively modest compared to the increase in the absorption of phytosterols. Mice lacking G5G8 do not show substantial increases in cholesterol absorption as assessed by the dual isotope method [29,31]. When monitoring the appearance of radiolabeled sterol in lymph, the absence of G5G8 either reduced or increased cholesterol absorption across the intestinal epithelium [46–48]. Expression of a human transgene under the control of its own promoter reduced fractional cholesterol absorption by 50% [49]. Consequently, the role of G5G8 in cholesterol absorption remains unclear, and like other phenotypes, may depend on dietary phytosterols.

Deletion of *Abcg5 Abcg8* reduces biliary cholesterol secretion by 70–90%. Agents that stimulate biliary cholesterol secretion are generally G5G8-dependent, including LXR and FXR agonists, thyroid hormone, and choleretic agents (diosgenin, tauroursodeoxycholate) [29,48,50–54]. Heterologous expression of Niemann–Pick C2 (NPC2) increases biliary secretion of the protein and promotes G5G8-dependent biliary cholesterol secretion, suggesting a role for this sterol-trafficking protein as a mediator of G5G8 activity or perhaps as a source of, or acceptor for G5G8 substrates [55]. G5G8-independent biliary cholesterol secretion is observed under some experimental conditions, including depletion and over-expression of class B, type-1 scavenger receptor (SR-BI), infusion or feeding high levels of cholate, *Atp8b1* deficiency, and in lactating rats [29,56–60]. Some fraction of residual biliary cholesterol secretion in the absence of G5G8 is likely mediated by detergent extraction. The extent to which other enzymes contribute to a G5G8-independent pathway, and if such a pathway might be targeted to increase biliary cholesterol secretion remains unknown.

3. Heterologous Expression of G5G8

The first *ABCG5 ABCG8* transgenic strain contained an estimated 14 copies of the human gene, increased biliary cholesterol concentrations five to six-fold, and reduced susceptibility to experimental hypercholesterolemia and atherosclerosis [49,61]. A liver-specific transgenic failed to protect mice from atherosclerosis unless combined with the cholesterol absorption inhibitor ezetimibe [62,63]. The protective effects of G5G8 are presumably due to its role as the mediator of the final step of reverse cholesterol transport (RCT). Pharmacological stimuli of RCT, including ezetimibe and LXR and FXR agonists, require G5G8 to promote fecal neutral sterol loss and macrophage to feces RCT [50,64,65]. However, acute adenoviral-mediated overexpression of hepatic G5G8 fails to stimulate macrophage to feces RCT [56]. Further, adenoviral expression of G5G8 paradoxically increases plasma cholesterol, an effect blocked by ezetimibe [66]. This indicates that a substantial amount of biliary cholesterol is reabsorbed in the small intestine and illustrates the cooperative nature of hepatic and intestinal G5G8 in order to oppose hypercholesterolemia and promote RCT. This, however, is not the case for preventing phytosterolemia as tissue-selective deletion of intestinal or hepatic G5G8 results in only modest elevations in plasma phytosterols [67].

4. Beyond Phytosterols

Given the roles of G5G8 in opposing dietary sterol accumulation and biliary cholesterol secretion, it is unsurprising that both rare and common variants have been associated with plasma cholesterol, non-cholesterol sterols, low-density lipoprotein cholesterol (LDL-C), and atherosclerotic and gallbladder disease across a large number of studies and populations ([68–80], incomplete list). Other associations are intriguing and include insulin resistance (G8:D19H) and type 2 diabetes (G5:G604E, G8:Y54C) and its renal complications (G8:T400K) [81–83]. Mouse models of metabolic syndrome which lack leptin or its receptor have diminished hepatic G5G8 and reduced biliary cholesterol secretion [66,84]. Rescue of G5G8 in these models with chemical chaperones, adenoviral expression of the molecular chaperone GRP78/binding immunoglobulin protein (BiP), or adenoviral expression of G5G8 itself accelerates biliary cholesterol secretion, restores glycemic control, and reduces plasma triglycerides [66,84,85]. Conversely, mice lacking G5G8 are more susceptible to diet-induced obesity, insulin resistance, and hepatic steatosis when maintained on phytosterol-free diets [86]. Collectively, these studies suggest an unappreciated relationship between biliary cholesterol secretion, triglyceride metabolism, and insulin signaling. There also appears to be a role for intestinal G5G8 in the absorption of triglycerides and chylomicron assembly, which may influence metabolic phenotypes, but the underlying mechanism for this phenotype remains unclear [46,47].

Independent of phytosterol accumulation, genome-wide association studies (GWAS) and animal model data support an anti-atherosclerotic role for G5G8. This is presumably due to the combined effects of limiting dietary cholesterol accumulation and promoting

RCT. Classically, the final steps of RCT are hepatobiliary secretion of neutral and acidic sterols. However, disruptions in biliary cholesterol secretion do not result in concomitant reductions in cholesterol excretion, indicating the presence of a compensatory, non-biliary pathway which has been labeled transintestinal cholesterol elimination (TICE) [87]. Tissue-specific deletion of G5G8 in the intestine reduced excretion of radiolabeled sterol from the plasma compartment to feces, indicating a role for intestinal G5G8 in cholesterol excretion [67]. The mediators of TICE and the relative contribution of G5G8 to intestinal cholesterol excretion under a variety of pharmacological conditions remain to be fully elucidated. These studies and the potential for therapeutic development have recently been reviewed and are beyond the scope of this discussion [88,89].

5. Transcriptional Regulation of G5G8

ABCG5 ABCG8 is effectively a single gene with a common promoter that regulates expression of both transcripts encoding each half of the transporter. To the best of our knowledge, there are no reports of differential regulation of *Abcg5* and *Abcg8* transcripts. Transcriptional regulation of *Abcg5 Abcg8* by a small molecule LXR agonist (T0901317) precipitated its discovery as the defective gene in sitosterolemia [22]. Expression of both mRNA and protein increases in liver and intestine in response to small molecule LXR agonists and dietary cholesterol, which promotes the accumulation of endogenous LXR agonists, oxysterols (Figure 1 (7)) [22,29–31,49,90]. Liver receptor homolog 1 (LRH1), GATA binding factor 4 (GATA-4), and hepatocyte nuclear factor 4α (HNF4α) binding sites map to the 374 base pair intergenic promoter that separates the initiation codons for each protein, the latter of which synergize with LXREs located in distal regions of the gene, to activate the promoter and increase expression of both transcripts [91,92]

Hepatic expression of *Abcg5 Abcg8* is also induced by FXR agonists and bile acids, and where examined, in an FXR-dependent fashion (Figure 1 (7)) [54,93–95]. However, regulation by bile acid FXR agonists is far more complicated. FXR-mediated activation of *Abcg5 Abcg8* requires fibroblast growth factor 15/19 (FGF15/19), which is itself an FXR target gene that is secreted from the ileum in response to bile acids and promotes Src-mediated phosphorylation of hepatic FXR, and FXR binding to the *Abcg5 Abcg8* promoter [96]. *FNDC5*/Irisin is an FXR target gene that increases *Abcg5 Abcg8* mRNA in both the livers and intestines of transgenic mice, but the extent to which the endogenous gene plays a role in *Abcg5 Abcg8* regulation and sterol homeostasis mice or humans is not known [97]. Whereas cholesterol and its metabolites tend to increase expression of *Abcg5 Abcg8* mRNA, agonists for the constitutive androstane receptor (CAR) repress expression of the transporter under conditions of elevated exogenous or endogenously-derived FXR agonists [98,99].

Transcriptional regulation of *Abcg5 Abcg8* by LXR and FXR fits well with its central role in opposing the accumulation of excess cholesterol. Expansion of whole body neutral and/or acid sterol pools increases expression [31,54,84,93,94,100,101]. Conversely, blocking absorption of cholesterol or bile acids reduces expression in either liver or intestine [37,101–103]. Less clear is the physiological benefit, if any, for alterations in *Abcg5 Abcg8* expression by regulators of metabolism. Thyroid hormone increases *Abcg5 Abcg8* mRNA, biliary cholesterol secretion, and fecal sterol excretion in both intact and hypophysectomized rats [104]. Hepatic *Abcg5 Abcg8* mRNA and protein are upregulated in the absence of insulin signaling in mice, an effect attributed to disinhibition of Forkhead box protein O1 (FOXO-1) [105,106]. The opposite was observed in a type 1 diabetic model in rats [107]. The insulin-sensitizing drug, metformin, increased *Abcg5 Abcg8* mRNA and protein, an effect attributed to reduced period 2 occupancy of the *ABCG5 ABCG8* promoter and disinhibition of gene expression [108]. Indeed, *Abcg5 Abcg8* mRNA exhibits a robust circadian rhythm at the transcriptional level (not observed for protein level, unpublished observation) and hepatic *Abcg5 Abcg8* mRNA and biliary cholesterol are reduced in *Bmal1*-deficient mice [108,109]

Alterations in *Abcg5 Abcg8* expression have been reported by a variety of nutritional cues, including upregulation in response elevated n-3 polyunsaturated fatty acids [110–112]. While upregulation of *ABCG5 ABCG8* is generally observed in high fat diets containing cholesterol and cholesterol-free, high fat diets, a single oral gavage of triacylglycerols robustly repressed intestinal *Abcg5 Abcg8* mRNA in mice [113]. In an independent study, suppression of *Abcg5 Abcg8* mRNA following high fat, high sucrose feeding was not observed, but the diet used in this study contained both added cholesterol and cholate, and thus hepatic cholesterol was increased five-fold compared to mice fed the control diet [114]. Diets containing high levels of sucrose robustly repressed expression of hepatic, but not intestinal *Abcg5 Abcg8* in rats [115]. Collectively, the data suggests that dietary repression of *ABCG5 ABCG8* may reflect reductions in the abundance of LXR and FXR agonists rather than active repression by sucrose or triglyceride. Alternatively, differences across studies may be associated with species and strain differences, both of which have been reported for various ligands [116,117]. Diets supplemented with soy protein increased hepatic *Abcg5 Abcg8* mRNA in rats [118]. The mechanisms for such an effect is not known but may include modulation of the intestinal microbiota. Germ free mice exhibited elevations in both intestinal and hepatic *Abcg5 Abcg8* mRNA relative to specific pathogen free mice in the absence and presence of ezetimibe [119]. Depletion of dietary iron upregulated hepatic *Abcg5 Abcg8* mRNA and promotes increased biliary cholesterol secretion [120]. Dietary calcium supplementation was shown to increased intestinal *Abcg5 Abcg8* mRNA and fecal neutral sterol excretion in a hamster model of menopause [121].

Female biological sex has long been associated with increased biliary cholesterol. Female mice had modest, but significant increases in biliary cholesterol and *ABCG5 ABCG8* mRNA in the human transgenic strain [49]. Ovariectomy was subsequently shown to reduce, and estrogen replacement to increase *Abcg5 Abcg8* mRNA across the intestine in independent strains of mice [116,122]. Diosgenin is a choleretic compound with estrogenic properties that increases biliary cholesterol secretion. Its ability to increase biliary cholesterol is largely G5G8-dependent, but reports on its impact on *Abcg5 Abcg8* expression are conflicting, showing no change in mice but an increase in both liver and intestine in rats [54,123]

6. Post-Transcriptional Regulation

Less is known about the post-transcriptional regulation of the G5G8. The half transporters are retained within the endoplasmic reticulum (ER) unless co-expressed [124,125]. Formation of the complex appears to be relatively inefficient in cultured cells, is dependent upon the presence of N-linked glycans that reside in the third extracellular loop of each protein, and can be enhanced by the expression of the lectin chaperones, Calnexin and Calreticulin [124–127]. Using chimeric approaches, the ER-retention motif was localized to the N-terminal, cytosolic domain, but has yet to be defined [128]. Failure to form complexes within the ER results in rapid degradation of each half transporter [127,129]. At the cell surface, the mature G5G8 complex resides within apical membranes of both hepatocytes and enterocytes [124]. There is also evidence of an intracellular, recruitable pool of G5G8 that translocates to the canalicular surface in response to cAMP and in response to diets containing cholate and cholesterol [114,130]. However, the stimuli and signaling pathways involved in intracellular trafficking of G5G8 have yet to be elucidated.

A number of approaches have been utilized to investigate the activity of G5G8. Heterologous expression in HEK293 and dog gall bladder epithelial cells demonstrated G5G8-dependent cholesterol efflux to bile acid micelles, but not HDL or apolipoprotein A1 [131,132]. Native mouse and recombinant human and mouse G5G8 have been purified to varying degrees from liver, rat hepatocytes, Sf9 insect cells, and *Pichia pastoris* [133–136]. These studies demonstrated ATP- and magnesium-dependent, vanadate-sensitive ATPase, and sterol transport activity. Various bile acids stimulate ATP hydrolysis. Among the species tested, G5G8 activity was most sensitive to cholate [133]. Perhaps surprisingly, neither cholesterol nor phytosterols stimulated ATPase or sterol transport activity in preparations from Sf9 cells [135,136]. Using inside-out vesicles in this same system, Wang et al. showed that other nucleotides could support sterol transfer, albeit less efficiently [135]

The nucleotide binding sites of G5G8 were proposed, and later confirmed by crystallography, to be comprised of Walker A and B domains of one partner and the signature motif of the other [137,138]. The Walker A and B domains of G8 juxtaposed to the signature motif of G5 were designated nucleotide binding site (NBS) 1. While both NBSs bind 8-Azido ATP, mutations in highly conserved residues within the Walker A and B domain of G5 (NBS2), but not G8 (NBS1), abolished ATP binding and hydrolysis. These findings were confirmed for G5G8-mediated biliary cholesterol secretion by expressing the mutants in G5G8-defecient mice. Domain swapping experiments between G5 and G8 confirmed that ATP hydrolysis in NBS2 is indispensable for activity [139]. G5G8 was crystallized as a heterodimer in lipid bilayers (bicelles) in the presence of cholesterol in the nucleotide free state to a resolution of four angstroms [138]. G5G8 was designated as a Type II Exporter. Key molecular interactions inferred from this structure were validated as essential for cholesterol transport in vivo by expressing recombinant mutants in G5G8-deficient mice. Naturally occurring missense variants and mutants can provide mechanistic insight to protein structure–function. The potential impact of mutations and polymorphisms on G5G8 structure function inferred from the available crystal structure were recently reviewed [140]. However, formal investigations into the impact of missense variants of any type on G5G8 trafficking, stability, and activity have been limited to only a few.

7. Sitosterolemia and/or Familial Hypercholesterolemia

Challenges in the proper diagnosis of sitosterolemia include heterogeneity of the clinical presentation of the disease, the lack of genotype–phenotype correlations, and the inability of clinical laboratory assays to distinguish phytosterols from cholesterol [28,141,142]. Studies among hypercholesterolemic subjects suggest sitosterolemia is significantly underdiagnosed [143,144]. Genome and exome sequence analysis of large populations indicates that carriers of loss of function mutations are far more common than previously thought and are at an elevated risk of coronary artery disease [72,80]. Exome sequence analysis of over 60,000 individuals across multiple populations reveals 57 and 58 predicted loss of function alleles for *ABCG5* and *ABCG8*, respectively (https://gnomad.broadinstitute.org/ (accessed on 3 January 2021) v2.1.1) [145]. This analysis did not include missense variants, 37 of which have been described for sitosterolemia [141]. At the time of the preparation of this review, 619 and 1307 missense variants have been catalogued in dbSNP for *ABCG5* and *ABCG8*, respectively. Most are predicted to be benign, but virtually none of those likely to be pathogenic or of uncertain significance have been experimentally or clinically validated. The number of *ABCG5 ABCG8* variants will undoubtedly grow as additional genomes and exomes are sequenced, as will the need for better tools to predict which variants are pathogenic.

An analysis of selected missense mutants of *ABCG5* and *ABCG8* suggests that the majority of dysfunctional alleles are due to the inability of *ABCG5 ABCG8* to heterodimerize and traffic beyond the endoplasmic reticulum [127]. The development of correctors and potentiators of the cystic fibrosis transmembrane conductance regulator (CFTR, *ABCC7*, OMIM: 602421) and the rescue of mutants of *ABCB4*, defective in progressive familial intrahepatic cholestasis type 3 (PFIC3, OMIM: 602347), suggest these or perhaps other small molecule chaperones may facilitate maturation and rescue function of G5G8 [146,147]. However, no systematic approach to classify the types of mutations that cause sitosterolemia has been made. We propose a draft of such a classification system for sitosterolemia (Table 1). We based our initial draft on the system established for PFIC3 due to the fact that both are biliary lipid transporters [148]. Our system may need future revision as additional mutations are identified and the impact of these variants/mutants on G5G8 formation, trafficking, and activity are determined. Nonetheless, this classification system provides a framework for characterizing *ABCG5 ABCG8* mutants that cause sitosterolemia and a basis for the systematic investigation of compounds that may potentially rescue G5G8 function.

Table 1. Classification system for experimentally verified sitosterolemia mutations.

Typical	No Mutation	ABCG5	ABCG8
Class I	Nonsense, Frameshift, Deletions	57 known or predicted	58 known or predicted
Class II	Maturation	R389H, R419P, R419H, N437K	R189H, P231T, R263Q, L501P, G574E, L596R
Class III	Activity		
Class IV	Stability		
Class V	No Detectible Defect		
Unclassified	Inconclusive Results	E146Q	R543S, G574R

https://gnomad.broadinstitute.org/ (accessed on 3 January 2021) and Reference [127].

8. Conclusions and Future Directions

The last twenty years have revealed a great deal about the role of *ABCG5 ABCG8* in cholesterol metabolism and in the defense against the accumulation of dietary xenosterols. The master regulators of neutral and acidic sterol metabolism modulate G5G8 abundance and activity as a component of the integrated machinery that maintains sterol homeostasis. Much, however, remains unknown about the hormonal and intracellular signals that promote G5G8 translocation to the biliary surface and G5G8-mediated cholesterol secretion. Beyond G5G8, the hepatocyte orchestrates the clearance of excess cellular cholesterol by metabolism to bile acid or incorporation into very low density and high-density lipoproteins, either on the surface of the particles or in the hydrophobic core following esterification. What regulates and under what conditions does the intracellular flux of cholesterol favor G5G8-dependent biliary secretion? Investigations of intestinal G5G8 regulation and activity have been more limited than in the liver. In what ways does the regulation of G5G8 in the enterocyte differ, if any, from the hepatocyte? Targeting G5G8 to promote TICE is conceptually attractive to promote RCT, but studies have yet to reveal a route to a potential therapeutic and the molecular mechanisms that mediate TICE remain elusive.

Sequencing of large numbers of genomes and exomes reveals that disease-causing mutants in *ABCG5 ABCG8* are significantly more common that previously appreciated. The combination of the required instrumentation, expertise, and cost for routine clinical laboratory analysis of plasma xenosterols is presently impractical. Conversely, genetic screening has become substantially less costly and increasingly common across healthcare systems. Genetic testing may offer a more practical means to identify and diagnose sitosterolemics. Such an approach requires the cataloguing and inclusion of *ABCG5 ABCG8* mutants among the various genetic testing platforms in use. A major limitation for such an approach is the lack of validation of suspected and likely pathogenic variants that may disrupt G5G8 function. Frameshift mutations in exon 13 of both *ABCG5* and *ABCG8* cause sitosterolemia, indicating little tolerance for truncation of the protein. Thus, deletions or frameshifts are expected to be Class I mutants. A significant number of splice donor and acceptor variants have also been identified and are likely pathogenic, but have yet to be formally analyzed. Of the 37 known missense variants, only 13 have been analyzed for maturation. Although most failed to form mature complexes and have been designated as Class II, three retained at least some degree of G5G8 maturation (G5:E146Q, G8:R543S, G8:G574R). Future investigations of the structural and functional impact of these and other missense mutants and variants will advance our understanding of sterol transport, the molecular dynamics of the transporter, and potential therapeutics for sitosterolemia as well as gall bladder and cardiovascular risk reduction.

Author Contributions: A.S. and K.W. crafted the initial draft. G.A.G. revised and edited the final draft. All authors have read and agreed to the published version of the manuscript.

Funding: This research was funded by the National Institutes of Health, grant numbers 1R01DK113625, 1P20GM130456, 5P30GM127211.

Institutional Review Board Statement: Not applicable.

Informed Consent Statement: Not applicable.

Conflicts of Interest: The authors declare no conflict of interest.

References

1. Bhattacharyya, A.K.; Connor, W.E. Beta-sitosterolemia and xanthomatosis. A newly described lipid storage disease in two sisters. *J. Clin. Investig.* **1974**, *53*, 1033–1043. [CrossRef] [PubMed]
2. Miettinen, T.A. Phytosterolaemia, xanthomatosis and premature atherosclerotic arterial disease: A case with high plant sterol absorption, impaired sterol elimination and low cholesterol synthesis. *Eur. J. Clin. Investig.* **1980**, *10*, 27–35. [CrossRef]
3. Kwiterovich, P.O., Jr.; Bachorik, P.S.; Smith, H.H.; McKusick, V.A.; Connor, W.E.; Teng, B.; Sniderman, A.D. Hyperapobetalipoproteinaemia in two families with xanthomas and phytosterolaemia. *Lancet* **1981**, *1*, 466–469. [CrossRef]
4. Shulman, R.S.; Bhattacharyya, A.K.; Connor, W.E.; Fredrickson, D.S. Beta-sitosterolemia and xanthomatosis. *N. Engl. J. Med.* **1976**, *294*, 482–483. [CrossRef] [PubMed]
5. Lin, H.J.; Wang, C.; Salen, G.; Lam, K.C.; Chan, T.K. Sitosterol and cholesterol metabolism in a patient with coexisting phytosterolemia and cholestanolemia. *Metab. Clin. Exp.* **1983**, *32*, 126–133. [CrossRef]
6. Beaty, T.H.; Kwiterovich, P.O., Jr.; Khoury, M.J.; White, S.; Bachorik, P.S.; Smith, H.H.; Teng, B.; Sniderman, A. Genetic analysis of plasma sitosterol, apoprotein B, and lipoproteins in a large Amish pedigree with sitosterolemia. *Am. J. Hum. Genet.* **1986**, *38*, 492–504.
7. Salen, G.; Horak, I.; Rothkopf, M.; Cohen, J.L.; Speck, J.; Tint, G.S.; Shore, V.; Dayal, B.; Chen, T.; Shefer, S. Lethal atherosclerosis associated with abnormal plasma and tissue sterol composition in sitosterolemia with xanthomatosis. *J. Lipid Res.* **1985**, *26*, 1126–1133. [CrossRef]
8. Nguyen, L.B.; Salen, G.; Shefer, S.; Tint, G.S.; Shore, V.; Ness, G.C. Decreased cholesterol biosynthesis in sitosterolemia with xanthomatosis: Diminished mononuclear leukocyte 3-hydroxy-3-methylglutaryl coenzyme A reductase activity and enzyme protein associated with increased low-density lipoprotein receptor function. *Metab. Clin. Exp.* **1990**, *39*, 436–443. [CrossRef]
9. Bhattacharyya, A.K.; Connor, W.E.; Lin, D.S.; McMurry, M.M.; Shulman, R.S. Sluggish sitosterol turnover and hepatic failure to excrete sitosterol into bile cause expansion of body pool of sitosterol in patients with sitosterolemia and xanthomatosis. *Arterioscler. Thromb. A J. Vasc. Biol./Am. Heart Assoc.* **1991**, *11*, 1287–1294. [CrossRef] [PubMed]
10. Rees, D.C.; Iolascon, A.; Carella, M.; O'Marcaigh, A.S.; Kendra, J.R.; Jowitt, S.N.; Wales, J.K.; Vora, A.; Makris, M.; Manning, N.; et al. Stomatocytic haemolysis and macrothrombocytopenia (Mediterranean stomatocytosis/macrothrombocytopenia) is the haematological presentation of phytosterolaemia. *Br. J. Haematol.* **2005**, *130*, 297–309. [CrossRef]
11. Mushtaq, T.; Wales, J.K.; Wright, N.P. Adrenal insufficiency in phytosterolaemia. *Eur. J. Endocrinol.* **2007**, *157* (Suppl. 1), S61–S65. [CrossRef]
12. Wang, Z.; Cao, L.; Su, Y.; Wang, G.; Wang, R.; Yu, Z.; Bai, X.; Ruan, C. Specific macrothrombocytopenia/hemolytic anemia associated with sitosterolemia. *Am. J. Hematol.* **2014**, *89*, 320–324. [CrossRef] [PubMed]
13. Bazerbachi, F.; Conboy, E.E.; Mounajjed, T.; Watt, K.D.; Babovic-Vuksanovic, D.; Patel, S.B.; Kamath, P.S. Cryptogenic Cirrhosis and Sitosterolemia: A Treatable Disease If Identified but Fatal If Missed. *Ann. Hepatol.* **2017**, *16*, 970–978. [CrossRef]
14. Salen, G.; Shore, V.; Tint, G.S.; Forte, T.; Shefer, S.; Horak, I.; Horak, E.; Dayal, B.; Nguyen, L.; Batta, A.K.; et al. Increased sitosterol absorption, decreased removal, and expanded body pools compensate for reduced cholesterol synthesis in sitosterolemia with xanthomatosis. *J. Lipid Res.* **1989**, *30*, 1319–1330. [CrossRef]
15. Cobb, M.M.; Salen, G.; Tint, G.S. Comparative effect of dietary sitosterol on plasma sterols and cholesterol and bile acid synthesis in a sitosterolemic homozygote and heterozygote subject. *J. Am. Coll. Nutr.* **1997**, *16*, 605–613.
16. Nguyen, L.B.; Cobb, M.; Shefer, S.; Salen, G.; Ness, G.C.; Tint, G.S. Regulation of cholesterol biosynthesis in sitosterolemia: Effects of lovastatin, cholestyramine, and dietary sterol restriction. *J. Lipid Res.* **1991**, *32*, 1941–1948. [CrossRef]
17. Patel, S.B.; Honda, A.; Salen, G. Sitosterolemia: Exclusion of genes involved in reduced cholesterol biosynthesis. *J. Lipid Res.* **1998**, *39*, 1055–1061. [CrossRef]
18. Patel, S.B.; Salen, G.; Hidaka, H.; Kwiterovich, P.O.; Stalenhoef, A.F.; Miettinen, T.A.; Grundy, S.M.; Lee, M.H.; Rubenstein, J.S.; Polymeropoulos, M.H.; et al. Mapping a gene involved in regulating dietary cholesterol absorption. The sitosterolemia locus is found at chromosome 2p21. *J. Clin. Investig.* **1998**, *102*, 1041–1044. [CrossRef] [PubMed]
19. Peet, D.J.; Turley, S.D.; Ma, W.; Janowski, B.A.; Lobaccaro, J.M.; Hammer, R.E.; Mangelsdorf, D.J. Cholesterol and bile acid metabolism are impaired in mice lacking the nuclear oxysterol receptor LXR alpha. *Cell* **1998**, *93*, 693–704. [CrossRef]
20. Repa, J.J.; Turley, S.D.; Lobaccaro, J.A.; Medina, J.; Li, L.; Lustig, K.; Shan, B.; Heyman, R.A.; Dietschy, J.M.; Mangelsdorf, D.J. Regulation of absorption and ABC1-mediated efflux of cholesterol by RXR heterodimers. *Science* **2000**, *289*, 1524–1529. [CrossRef]
21. Ewart, G.D.; Cannell, D.; Cox, G.B.; Howells, A.J. Mutational analysis of the traffic ATPase (ABC) transporters involved in uptake of eye pigment precursors in Drosophila melanogaster. Implications for structure-function relationships. *J. Biol. Chem.* **1994**, *269*, 10370–10377. [CrossRef]
22. Berge, K.E.; Tian, H.; Graf, G.A.; Yu, L.; Grishin, N.V.; Schultz, J.; Kwiterovich, P.; Shan, B.; Barnes, R.; Hobbs, H.H. Accumulation of dietary cholesterol in sitosterolemia caused by mutations in adjacent ABC transporters. *Science* **2000**, *290*, 1771–1775. [CrossRef]

23. Lee, M.H.; Lu, K.; Hazard, S.; Yu, H.; Shulenin, S.; Hidaka, H.; Kojima, H.; Allikmets, R.; Sakuma, N.; Pegoraro, R.; et al. Identification of a gene, ABCG5, important in the regulation of dietary cholesterol absorption. *Nat. Genet.* **2001**, *27*, 79–83. [CrossRef]

24. Van Heek, M.; France, C.F.; Compton, D.S.; McLeod, R.L.; Yumibe, N.P.; Alton, K.B.; Sybertz, E.J.; Davis, H.R., Jr. In vivo metabolism-based discovery of a potent cholesterol absorption inhibitor, SCH58235, in the rat and rhesus monkey through the identification of the active metabolites of SCH48461. *J. Pharm. Exp. Ther.* **1997**, *283*, 157–163.

25. Davis, H.R., Jr.; Zhu, L.J.; Hoos, L.M.; Tetzloff, G.; Maguire, M.; Liu, J.; Yao, X.; Iyer, S.P.; Lam, M.H.; Lund, E.G.; et al. Niemann-Pick C1 Like 1 (NPC1L1) is the intestinal phytosterol and cholesterol transporter and a key modulator of whole-body cholesterol homeostasis. *J. Biol. Chem.* **2004**, *279*, 33586–33592. [CrossRef] [PubMed]

26. Salen, G.; von Bergmann, K.; Lutjohann, D.; Kwiterovich, P.; Kane, J.; Patel, S.B.; Musliner, T.; Stein, P.; Musser, B.; Multicenter Sitosterolemia Study Group. Ezetimibe effectively reduces plasma plant sterols in patients with sitosterolemia. *Circulation* **2004**, *109*, 966–971. [CrossRef]

27. Lutjohann, D.; von Bergmann, K.; Sirah, W.; Macdonell, G.; Johnson-Levonas, A.O.; Shah, A.; Lin, J.; Sapre, A.; Musliner, T. Long-term efficacy and safety of ezetimibe 10 mg in patients with homozygous sitosterolemia: A 2-year, open-label extension study. *Int J. Clin. Pract.* **2008**, *62*, 1499–1510. [CrossRef] [PubMed]

28. Escola-Gil, J.C.; Quesada, H.; Julve, J.; Martin-Campos, J.M.; Cedo, L.; Blanco-Vaca, F. Sitosterolemia: Diagnosis, investigation, and management. *Curr. Atheroscler. Rep.* **2014**, *16*, 424. [CrossRef]

29. Plosch, T.; Bloks, V.W.; Terasawa, Y.; Berdy, S.; Siegler, K.; Van Der Sluijs, F.; Kema, I.P.; Groen, A.K.; Shan, B.; Kuipers, F.; et al. Sitosterolemia in ABC-transporter G5-deficient mice is aggravated on activation of the liver-X receptor. *Gastroenterology* **2004**, *126*, 290–300. [CrossRef]

30. Klett, E.L.; Lu, K.; Kosters, A.; Vink, E.; Lee, M.H.; Altenburg, M.; Shefer, S.; Batta, A.K.; Yu, H.; Chen, J.; et al. A mouse model of sitosterolemia: Absence of Abcg8/sterolin-2 results in failure to secrete biliary cholesterol. *BMC Med.* **2004**, *2*, 5. [CrossRef]

31. Yu, L.; Hammer, R.E.; Li-Hawkins, J.; Von Bergmann, K.; Lutjohann, D.; Cohen, J.C.; Hobbs, H.H. Disruption of Abcg5 and Abcg8 in mice reveals their crucial role in biliary cholesterol secretion. *Proc. Natl. Acad. Sci. USA* **2002**, *99*, 16237–16242. [CrossRef]

32. Chen, J.; Batta, A.; Zheng, S.; Fitzgibbon, W.R.; Ullian, M.E.; Yu, H.; Tso, P.; Salen, G.; Patel, S.B. The missense mutation in Abcg5 gene in spontaneously hypertensive rats (SHR) segregates with phytosterolemia but not hypertension. *BMC Genet.* **2005**, *6*, 40.

33. Chase, T.H.; Lyons, B.L.; Bronson, R.T.; Foreman, O.; Donahue, L.R.; Burzenski, L.M.; Gott, B.; Lane, P.; Harris, B.; Ceglarek, U.; et al. The mouse mutation "thrombocytopenia and cardiomyopathy" (trac) disrupts Abcg5: A spontaneous single gene model for human hereditary phytosterolemia/sitosterolemia. *Blood* **2010**, *115*, 1267–1276. [CrossRef] [PubMed]

34. Kruit, J.K.; Drayer, A.L.; Bloks, V.W.; Blom, N.; Olthof, S.G.; Sauer, P.J.; de Haan, G.; Kema, I.P.; Vellenga, E.; Kuipers, F. Plant sterols cause macrothrombocytopenia in a mouse model of sitosterolemia. *J. Biol. Chem.* **2008**, *283*, 6281–6287. [CrossRef]

35. Kanaji, T.; Kanaji, S.; Montgomery, R.R.; Patel, S.B.; Newman, P.J. Platelet hyperreactivity explains the bleeding abnormality and macrothrombocytopenia in a murine model of sitosterolemia. *Blood* **2013**, *122*, 2732–2742. [CrossRef] [PubMed]

36. Solca, C.; Tint, G.S.; Patel, S.B. Dietary xenosterols lead to infertility and loss of abdominal adipose tissue in sterolin-deficient mice. *J. Lipid Res.* **2013**, *54*, 397–409. [CrossRef] [PubMed]

37. Yu, L.; von Bergmann, K.; Lutjohann, D.; Hobbs, H.H.; Cohen, J.C. Ezetimibe normalizes metabolic defects in mice lacking ABCG5 and ABCG8. *J. Lipid Res.* **2005**, *46*, 1739–1744. [CrossRef]

38. Tang, W.; Ma, Y.; Jia, L.; Ioannou, Y.A.; Davies, J.P.; Yu, L. Genetic inactivation of NPC1L1 protects against sitosterolemia in mice lacking ABCG5/ABCG8. *J. Lipid Res.* **2009**, *50*, 293–300. [CrossRef] [PubMed]

39. McDaniel, A.L.; Alger, H.M.; Sawyer, J.K.; Kelley, K.L.; Kock, N.D.; Brown, J.M.; Temel, R.E.; Rudel, L.L. Phytosterol feeding causes toxicity in ABCG5/G8 knockout mice. *Am. J. Pathol.* **2013**, *182*, 1131–1138. [CrossRef]

40. Yang, C.; McDonald, J.G.; Patel, A.; Zhang, Y.; Umetani, M.; Xu, F.; Westover, E.J.; Covey, D.F.; Mangelsdorf, D.J.; Cohen, J.C.; et al. Sterol intermediates from cholesterol biosynthetic pathway as liver X receptor ligands. *J. Biol. Chem.* **2006**, *281*, 27816–27826. [CrossRef] [PubMed]

41. Carter, B.A.; Taylor, O.A.; Prendergast, D.R.; Zimmerman, T.L.; Von Furstenberg, R.; Moore, D.D.; Karpen, S.J. Stigmasterol, a soy lipid-derived phytosterol, is an antagonist of the bile acid nuclear receptor FXR. *Pediatr. Res.* **2007**, *62*, 301–306. [CrossRef]

42. Sabeva, N.S.; McPhaul, C.M.; Li, X.; Cory, T.J.; Feola, D.J.; Graf, G.A. Phytosterols differentially influence ABC transporter expression, cholesterol efflux and inflammatory cytokine secretion in macrophage foam cells. *J. Nutr. Biochem.* **2011**, *22*, 777–783. [CrossRef] [PubMed]

43. Plat, J.; Nichols, J.A.; Mensink, R.P. Plant sterols and stanols: Effects on mixed micellar composition and LXR (target gene) activation. *J. Lipid Res.* **2005**, *46*, 2468–2476. [CrossRef]

44. Salen, G.; Tint, G.S.; Shefer, S.; Shore, V.; Nguyen, L. Increased sitosterol absorption is offset by rapid elimination to prevent accumulation in heterozygotes with sitosterolemia. *Arterioscler. Thromb. A J. Vasc. Biol. Am. Heart Assoc.* **1992**, *12*, 563–568. [CrossRef] [PubMed]

45. Lütjohann, D.; Björkhem, I.; Ose, L. Phytosterolaemia in a Norwegian family: Diagnosis and characterization of the first Scandinavian case. *Scand. J. Clin. Lab. Investig.* **1996**, *56*, 229–240. [CrossRef]

46. Zhang, L.S.; Xu, M.; Yang, Q.; Lou, D.; Howles, P.N.; Tso, P. ABCG5/G8 deficiency in mice reduces dietary triacylglycerol and cholesterol transport into the lymph. *Lipids* **2015**, *50*, 371–379. [CrossRef]

47. Nguyen, T.M.; Sawyer, J.K.; Kelley, K.L.; Davis, M.A.; Kent, C.R.; Rudel, L.L. ACAT2 and ABCG5/G8 are both required for efficient cholesterol absorption in mice: Evidence from thoracic lymph duct cannulation. *J. Lipid Res.* **2012**, *53*, 1598–1609. [CrossRef] [PubMed]

48. Wang, H.H.; Patel, S.B.; Carey, M.C.; Wang, D.Q. Quantifying anomalous intestinal sterol uptake, lymphatic transport, and biliary secretion in Abcg8(-/-) mice. *Hepatology* **2007**, *45*, 998–1006. [CrossRef] [PubMed]

49. Yu, L.; Li-Hawkins, J.; Hammer, R.E.; Berge, K.E.; Horton, J.D.; Cohen, J.C.; Hobbs, H.H. Overexpression of ABCG5 and ABCG8 promotes biliary cholesterol secretion and reduces fractional absorption of dietary cholesterol. *J. Clin. Investig.* **2002**, *110*, 671–680. [CrossRef]

50. Jakulj, L.; Vissers, M.N.; van Roomen, C.P.; van der Veen, J.N.; Vrins, C.L.J.; Kunne, C.; Stellaard, F.; Kastelein, J.J.P.; Groen, A.K. Ezetimibe stimulates faecal neutral sterol excretion depending on abcg8 function in mice. *FEBS Lett.* **2010**, *584*, 3625–3628. [CrossRef]

51. de Boer, J.F.; Schonewille, M.; Boesjes, M.; Wolters, H.; Bloks, V.W.; Bos, T.; van Dijk, T.H.; Jurdzinski, A.; Boverhof, R.; Wolters, J.C.; et al. Intestinal Farnesoid X Receptor Controls Transintestinal Cholesterol Excretion in Mice. *Gastroenterology* **2017**, *152*, 1126–1138.e6. [CrossRef] [PubMed]

52. Bonde, Y.; Plosch, T.; Kuipers, F.; Angelin, B.; Rudling, M. Stimulation of murine biliary cholesterol secretion by thyroid hormone is dependent on a functional ABCG5/G8 complex. *Hepatology* **2012**, *56*, 1828–1837. [CrossRef]

53. Kosters, A.; Frijters, R.J.; Kunne, C.; Vink, E.; Schneiders, M.S.; Schaap, F.G.; Nibbering, C.P.; Patel, S.B.; Groen, A.K. Diosgenin-induced biliary cholesterol secretion in mice requires Abcg8. *Hepatology* **2005**, *41*, 141–150. [CrossRef] [PubMed]

54. Yu, L.; Gupta, S.; Xu, F.; Liverman, A.D.; Moschetta, A.; Mangelsdorf, D.J.; Repa, J.J.; Hobbs, H.H.; Cohen, J.C. Expression of ABCG5 and ABCG8 is required for regulation of biliary cholesterol secretion. *J. Biol. Chem.* **2005**, *280*, 8742–8747. [CrossRef]

55. Yamanashi, Y.; Takada, T.; Yoshikado, T.; Shoda, J.; Suzuki, H. NPC2 regulates biliary cholesterol secretion via stimulation of ABCG5/G8-mediated cholesterol transport. *Gastroenterology* **2011**, *140*, 1664–1674. [CrossRef]

56. Dikkers, A.; de Boer, J.F.; Groen, A.K.; Tietge, U.J. Hepatic ABCG5/G8 overexpression substantially increases biliary cholesterol secretion but does not impact in vivo macrophage-to-feces RCT. *Atherosclerosis* **2015**, *243*, 402–406. [CrossRef]

57. Coy, D.J.; Wooton-Kee, C.R.; Yan, B.; Sabeva, N.; Su, K.; Graf, G.; Vore, M. ABCG5/ABCG8-independent biliary cholesterol excretion in lactating rats. *Am. J. Physiol. Gastrointest. Liver Physiol.* **2010**, *299*, G228–G235. [CrossRef] [PubMed]

58. Groen, A.; Kunne, C.; Jongsma, G.; van den Oever, K.; Mok, K.S.; Petruzzelli, M.; Vrins, C.L.; Bull, L.; Paulusma, C.C.; Oude Elferink, R.P. Abcg5/8 independent biliary cholesterol excretion in Atp8b1-deficient mice. *Gastroenterology* **2008**, *134*, 2091–2100. [CrossRef]

59. Wiersma, H.; Gatti, A.; Nijstad, N.; Oude Elferink, R.P.; Kuipers, F.; Tietge, U.J. Scavenger receptor class B type I mediates biliary cholesterol secretion independent of ATP-binding cassette transporter g5/g8 in mice. *Hepatology* **2009**, *50*, 1263–1272. [CrossRef]

60. Wang, H.H.; Li, X.; Patel, S.B.; Wang, D.Q. Evidence that the adenosine triphosphate-binding cassette G5/G8-independent pathway plays a determinant role in cholesterol gallstone formation in mice. *Hepatology* **2016**, *64*, 853–864. [CrossRef]

61. Wilund, K.R.; Yu, L.; Xu, F.; Hobbs, H.H.; Cohen, J.C. High-level expression of ABCG5 and ABCG8 attenuates diet-induced hypercholesterolemia and atherosclerosis in Ldlr-/-mice. *J. Lipid Res.* **2004**, *45*, 1429–1436. [CrossRef]

62. Wu, J.E.; Basso, F.; Shamburek, R.D.; Amar, M.J.; Vaisman, B.; Szakacs, G.; Joyce, C.; Tansey, T.; Freeman, L.; Paigen, B.J.; et al. Hepatic ABCG5 and ABCG8 overexpression increases hepatobiliary sterol transport but does not alter aortic atherosclerosis in transgenic mice. *J. Biol. Chem.* **2004**, *279*, 22913–22925. [CrossRef] [PubMed]

63. Basso, F.; Freeman, L.A.; Ko, C.; Joyce, C.; Amar, M.J.; Shamburek, R.D.; Tansey, T.; Thomas, F.; Wu, J.; Paigen, B.; et al. Hepatic ABCG5/G8 overexpression reduces apoB-lipoproteins and atherosclerosis when cholesterol absorption is inhibited. *J. Lipid Res.* **2007**, *48*, 114–126. [CrossRef] [PubMed]

64. Calpe-Berdiel, L.; Rotllan, N.; Fievet, C.; Roig, R.; Blanco-Vaca, F.; Escola-Gil, J.C. Liver X receptor-mediated activation of reverse cholesterol transport from macrophages to feces in vivo requires ABCG5/G8. *J. Lipid Res.* **2008**, *49*, 1904–1911. [CrossRef]

65. Altemus, J.B.; Patel, S.B.; Sehayek, E. Liver-specific induction of Abcg5 and Abcg8 stimulates reverse cholesterol transport in response to ezetimibe treatment. *Metab. Clin. Exp.* **2014**, *63*, 1334–1341. [CrossRef]

66. Su, K.; Sabeva, N.S.; Wang, Y.; Liu, X.; Lester, J.D.; Liu, J.; Liang, S.; Graf, G.A. Acceleration of biliary cholesterol secretion restores glycemic control and alleviates hypertriglyceridemia in obese db/db mice. *Arter. Thromb. Vasc. Biol.* **2014**, *34*, 26–33. [CrossRef]

67. Wang, J.; Mitsche, M.A.; Lutjohann, D.; Cohen, J.C.; Xie, X.S.; Hobbs, H.H. Relative roles of ABCG5/ABCG8 in liver and intestine. *J. Lipid Res.* **2015**, *56*, 319–330. [CrossRef] [PubMed]

68. Berge, K.E.; von Bergmann, K.; Lutjohann, D.; Guerra, R.; Grundy, S.M.; Hobbs, H.H.; Cohen, J.C. Heritability of plasma noncholesterol sterols and relationship to DNA sequence polymorphism in ABCG5 and ABCG8. *J. Lipid Res.* **2002**, *43*, 486–494. [CrossRef]

69. Buch, S.; Schafmayer, C.; Volzke, H.; Becker, C.; Franke, A.; von Eller-Eberstein, H.; Kluck, C.; Bassmann, I.; Brosch, M.; Lammert, F.; et al. A genome-wide association scan identifies the hepatic cholesterol transporter ABCG8 as a susceptibility factor for human gallstone disease. *Nat. Genet.* **2007**, *39*, 995–999. [CrossRef]

70. Grunhage, F.; Acalovschi, M.; Tirziu, S.; Walier, M.; Wienker, T.F.; Ciocan, A.; Mosteanu, O.; Sauerbruch, T.; Lammert, F. Increased gallstone risk in humans conferred by common variant of hepatic ATP-binding cassette transporter for cholesterol. *Hepatology* **2007**, *46*, 793–801. [CrossRef]

71. Stender, S.; Frikke-Schmidt, R.; Nordestgaard, B.G.; Tybjaerg-Hansen, A. The ABCG5/8 cholesterol transporter and myocardial infarction versus gallstone disease. *J. Am. Coll. Cardiol.* **2014**, *63*, 2121–2128. [CrossRef]

72. Helgadottir, A.; Thorleifsson, G.; Alexandersson, K.F.; Tragante, V.; Thorsteinsdottir, M.; Eiriksson, F.F.; Gretarsdottir, S.; Bjornsson, E.; Magnusson, O.; Sveinbjornsson, G.; et al. Genetic variability in the absorption of dietary sterols affects the risk of coronary artery disease. *Eur. Heart J.* **2020**, *41*, 2618–2628. [CrossRef]

73. Jiang, Z.Y.; Cai, Q.; Chen, E.Z. Association of three common single nucleotide polymorphisms of ATP binding cassette G8 gene with gallstone disease: A meta-analysis. *PLoS ONE* **2014**, *9*, e87200.

74. Viturro, E.; de Oya, M.; Lasuncion, M.A.; Gorgojo, L.; Moreno, J.M.; Benavente, M.; Cano, B.; Garces, C. Cholesterol and saturated fat intake determine the effect of polymorphisms at ABCG5/ABCG8 genes on lipid levels in children. *Genet. Med.* **2006**, *8*, 594–599. [CrossRef]

75. Jakulj, L.; Vissers, M.N.; Tanck, M.W.; Hutten, B.A.; Stellaard, F.; Kastelein, J.J.; Dallinga-Thie, G.M. ABCG5/G8 polymorphisms and markers of cholesterol metabolism: Systematic review and meta-analysis. *J. Lipid Res.* **2010**, *51*, 3016–3023. [CrossRef] [PubMed]

76. von Kampen, O.; Buch, S.; Nothnagel, M.; Azocar, L.; Molina, H.; Brosch, M.; Erhart, W.; von Schonfels, W.; Egberts, J.; Seeger, M.; et al. Genetic and functional identification of the likely causative variant for cholesterol gallstone disease at the ABCG5/8 lithogenic locus. *Hepatology* **2013**, *57*, 2407–2417. [CrossRef]

77. Ma, L.; Yang, J.; Runesha, H.B.; Tanaka, T.; Ferrucci, L.; Bandinelli, S.; Da, Y. Genome-wide association analysis of total cholesterol and high-density lipoprotein cholesterol levels using the Framingham heart study data. *BMC Med. Genet.* **2010**, *11*, 55. [CrossRef]

78. Li, Q.; Yin, R.X.; Wei, X.L.; Yan, T.T.; Aung, L.H.; Wu, D.F.; Wu, J.Z.; Lin, W.X.; Liu, C.W.; Pan, S.L. ATP-binding cassette transporter G5 and G8 polymorphisms and several environmental factors with serum lipid levels. *PLoS ONE* **2012**, *7*, e37972. [CrossRef]

79. Wu, G.; Li, G.B.; Yao, M.; Zhang, D.Q.; Dai, B.; Ju, C.J.; Han, M. ABCG5/8 variants are associated with susceptibility to coronary heart disease. *Mol. Med. Rep.* **2014**, *9*, 2512–2520. [CrossRef] [PubMed]

80. Nomura, A.; Emdin, C.A.; Won, H.H.; Peloso, G.M.; Natarajan, P.; Ardissino, D.; Danesh, J.; Schunkert, H.; Correa, A.; Bown, M.J.; et al. Heterozygous ATP-binding Cassette Transporter G5 Gene Deficiency and Risk of Coronary Artery Disease. *Circ. Genom. Precis. Med.* **2020**, *13*, 417–423. [CrossRef]

81. Chen, Z.C.; Shin, S.J.; Kuo, K.K.; Lin, K.D.; Yu, M.L.; Hsiao, P.J. Significant association of ABCG8:D19H gene polymorphism with hypercholesterolemia and insulin resistance. *J. Hum. Genet.* **2008**, *53*, 757–763. [CrossRef]

82. Gok, O.; Karaali, Z.E.; Acar, L.; Kilic, U.; Ergen, A. ABCG5 and ABCG8 gene polymorphisms in type 2 diabetes mellitus in the Turkish population. *Can. J. Diabetes* **2015**, *39*, 405–410. [CrossRef] [PubMed]

83. Nicolas, A.; Fatima, S.; Lamri, A.; Bellili-Munoz, N.; Halimi, J.M.; Saulnier, P.J.; Hadjadj, S.; Velho, G.; Marre, M.; Roussel, R.; et al. ABCG8 polymorphisms and renal disease in type 2 diabetic patients. *Metab. Clin. Exp.* **2015**, *64*, 713–719. [CrossRef]

84. Sabeva, N.S.; Rouse, E.J.; Graf, G.A. Defects in the leptin axis reduce abundance of the ABCG5-ABCG8 sterol transporter in liver. *J. Biol. Chem.* **2007**, *282*, 22397–22405. [CrossRef] [PubMed]

85. Wang, Y.; Su, K.; Sabeva, N.S.; Ji, A.; van der Westhuyzen, D.R.; Foufelle, F.; Gao, X.; Graf, G.A. GRP78 rescues the ABCG5 ABCG8 sterol transporter in db/db mice. *Metab. Clin. Exp.* **2015**, *64*, 1435–1443. [CrossRef] [PubMed]

86. Su, K.; Sabeva, N.S.; Liu, J.; Wang, Y.; Bhatnagar, S.; van der Westhuyzen, D.R.; Graf, G.A. The ABCG5 ABCG8 sterol transporter opposes the development of fatty liver disease and loss of glycemic control independently of phytosterol accumulation. *J. Biol. Chem.* **2012**, *287*, 28564–28575. [CrossRef]

87. Van der Velde, A.E.; Vrins, C.L.J.; van den Oever, K.; Kunne, C.; Oude Elferink, R.P.J.; Kuipers, F.; Groen, A.K. Direct Intestinal Cholesterol Secretion Contributes Significantly to Total Fecal Neutral Sterol Excretion in Mice. *Gastroenterology* **2007**, *133*, 967–975. [CrossRef] [PubMed]

88. Grefhorst, A.; Verkade, H.J.; Groen, A.K. The TICE Pathway: Mechanisms and Lipid-Lowering Therapies. *Methodist Debakey Cardiovasc. J.* **2019**, *15*, 70–76.

89. Nakano, T.; Inoue, I.; Murakoshi, T. A Newly Integrated Model for Intestinal Cholesterol Absorption and Efflux Reappraises How Plant Sterol Intake Reduces Circulating Cholesterol Levels. *Nutrients* **2019**, *11*, 310. [CrossRef] [PubMed]

90. Repa, J.J.; Berge, K.E.; Pomajzl, C.; Richardson, J.A.; Hobbs, H.; Mangelsdorf, D.J. Regulation of ATP-binding cassette sterol transporters ABCG5 and ABCG8 by the liver X receptors alpha and beta. *J. Biol. Chem.* **2002**, *277*, 18793–18800. [CrossRef]

91. Freeman, L.A.; Kennedy, A.; Wu, J.; Bark, S.; Remaley, A.T.; Santamarina-Fojo, S.; Brewer, H.B., Jr. The orphan nuclear receptor LRH-1 activates the ABCG5/ABCG8 intergenic promoter. *J. Lipid Res.* **2004**, *45*, 1197–1206. [CrossRef] [PubMed]

92. Sumi, K.; Tanaka, T.; Uchida, A.; Magoori, K.; Urashima, Y.; Ohashi, R.; Ohguchi, H.; Okamura, M.; Kudo, H.; Daigo, K.; et al. Cooperative interaction between hepatocyte nuclear factor 4 alpha and GATA transcription factors regulates ATP-binding cassette sterol transporters ABCG5 and ABCG8. *Mol. Cell Biol.* **2007**, *27*, 4248–4260. [CrossRef] [PubMed]

93. Li, T.; Matozel, M.; Boehme, S.; Kong, B.; Nilsson, L.M.; Guo, G.; Ellis, E.; Chiang, J.Y. Overexpression of cholesterol 7alpha-hydroxylase promotes hepatic bile acid synthesis and secretion and maintains cholesterol homeostasis. *Hepatology* **2011**, *53*, 996–1006. [CrossRef]

94. Wang, J.; Einarsson, C.; Murphy, C.; Parini, P.; Bjorkhem, I.; Gafvels, M.; Eggertsen, G. Studies on LXR- and FXR-mediated effects on cholesterol homeostasis in normal and cholic acid-depleted mice. *J. Lipid Res.* **2006**, *47*, 421–430. [CrossRef]

95. Liu, J.; Lu, H.; Lu, Y.F.; Lei, X.; Cui, J.Y.; Ellis, E.; Strom, S.C.; Klaassen, C.D. Potency of individual bile acids to regulate bile acid synthesis and transport genes in primary human hepatocyte cultures. *Toxicol. Sci.* **2014**, *141*, 538–546. [CrossRef]

96. Byun, S.; Jung, H.; Chen, J.; Kim, Y.C.; Kim, D.H.; Kong, B.; Guo, G.; Kemper, B.; Kemper, J.K. Phosphorylation of hepatic farnesoid X receptor by FGF19 signaling-activated Src maintains cholesterol levels and protects from atherosclerosis. *J. Biol. Chem.* **2019**, *294*, 8732–8744. [CrossRef] [PubMed]

97. Li, H.; Shen, J.; Wu, T.; Kuang, J.; Liu, Q.; Cheng, S.; Pu, S.; Chen, L.; Li, R.; Li, Y.; et al. Irisin Is Controlled by Farnesoid X Receptor and Regulates Cholesterol Homeostasis. *Front. Pharm.* **2019**, *10*, 548. [CrossRef]

98. Cheng, S.; Zou, M.; Liu, Q.; Kuang, J.; Shen, J.; Pu, S.; Chen, L.; Li, H.; Wu, T.; Li, R.; et al. Activation of Constitutive Androstane Receptor Prevents Cholesterol Gallstone Formation. *Am. J. Pathol.* **2017**, *187*, 808–818. [CrossRef]

99. Sberna, A.L.; Assem, M.; Gautier, T.; Grober, J.; Guiu, B.; Jeannin, A.; Pais de Barros, J.P.; Athias, A.; Lagrost, L.; Masson, D. Constitutive androstane receptor activation stimulates faecal bile acid excretion and reverse cholesterol transport in mice. *J. Hepatol.* **2011**, *55*, 154–161. [CrossRef]

100. Gooijert, K.E.; Havinga, R.; Wolters, H.; Wang, R.; Ling, V.; Tazuma, S.; Verkade, H.J. The mechanism of increased biliary lipid secretion in mice with genetic inactivation of bile salt export pump. *Am. J. Physiol. Gastrointest. Liver Physiol.* **2015**, *308*, G450–G457. [CrossRef] [PubMed]

101. Wang, Y.; Liu, X.; Pijut, S.S.; Li, J.; Horn, J.; Bradford, E.M.; Leggas, M.; Barrett, T.A.; Graf, G.A. The combination of ezetimibe and ursodiol promotes fecal sterol excretion and reveals a G5G8-independent pathway for cholesterol elimination. *J. Lipid Res.* **2015**, *56*, 810–820. [CrossRef]

102. Wang, H.H.; Portincasa, P.; Mendez-Sanchez, N.; Uribe, M.; Wang, D.Q. Effect of ezetimibe on the prevention and dissolution of cholesterol gallstones. *Gastroenterology* **2008**, *134*, 2101–2110. [CrossRef] [PubMed]

103. Kamisako, T.; Ogawa, H.; Yamamoto, K. Effect of cholesterol, cholic acid and cholestyramine administration on the intestinal mRNA expressions related to cholesterol and bile acid metabolism in the rat. *J. Gastroenterol. Hepatol.* **2007**, *22*, 1832–1837. [CrossRef] [PubMed]

104. Galman, C.; Bonde, Y.; Matasconi, M.; Angelin, B.; Rudling, M. Dramatically increased intestinal absorption of cholesterol following hypophysectomy is normalized by thyroid hormone. *Gastroenterology* **2008**, *134*, 1127–1136. [CrossRef] [PubMed]

105. Aleksunes, L.M.; Xu, J.; Lin, E.; Wen, X.; Goedken, M.J.; Slitt, A.L. Pregnancy represses induction of efflux transporters in livers of type I diabetic mice. *Pharm. Res.* **2013**, *30*, 2209–2220. [CrossRef]

106. Biddinger, S.B.; Haas, J.T.; Yu, B.B.; Bezy, O.; Jing, E.; Zhang, W.; Unterman, T.G.; Carey, M.C.; Kahn, C.R. Hepatic insulin resistance directly promotes formation of cholesterol gallstones. *Nat. Med.* **2008**, *14*, 778–782. [CrossRef]

107. Li, J.; Wang, X.; Liu, H.; Guo, H.; Zhang, M.; Mei, D.; Liu, C.; He, L.; Liu, L.; Liu, X. Impaired hepatic and intestinal ATP-binding cassette transporter G5/8 was associated with high exposure of beta-sitosterol and the potential risks to blood-brain barrier integrity in diabetic rats. *J. Pharm. Pharm.* **2014**, *66*, 428–436. [CrossRef]

108. Molusky, M.M.; Hsieh, J.; Lee, S.X.; Ramakrishnan, R.; Tascau, L.; Haeusler, R.A.; Accili, D.; Tall, A.R. Metformin and AMP Kinase Activation Increase Expression of the Sterol Transporters ABCG5/8 (ATP-Binding Cassette Transporter G5/G8) With Potential Antiatherogenic Consequences. *Arter. Thromb. Vasc. Biol.* **2018**, *38*, 1493–1503. [CrossRef]

109. Pan, X.; Bradfield, C.A.; Hussain, M.M. Global and hepatocyte-specific ablation of Bmal1 induces hyperlipidaemia and enhances atherosclerosis. *Nat. Commun.* **2016**, *7*, 13011. [CrossRef]

110. Nishimoto, T.; Pellizzon, M.A.; Aihara, M.; Stylianou, I.M.; Billheimer, J.T.; Rothblat, G.; Rader, D.J. Fish oil promotes macrophage reverse cholesterol transport in mice. *Arterioscler. Thromb. Vasc. Biol.* **2009**, *29*, 1502–1508. [CrossRef]

111. Kamisako, T.; Tanaka, Y.; Ikeda, T.; Yamamoto, K.; Ogawa, H. Dietary fish oil regulates gene expression of cholesterol and bile acid transporters in mice. *Hepatol. Res. Off. J. Jpn. Soc. Hepatol.* **2012**, *42*, 321–326. [CrossRef]

112. Kim, E.H.; Bae, J.S.; Hahm, K.B.; Cha, J.Y. Endogenously synthesized n-3 polyunsaturated fatty acids in fat-1 mice ameliorate high-fat diet-induced non-alcoholic fatty liver disease. *Biochem. Pharm.* **2012**, *84*, 1359–1365. [CrossRef]

113. De Vogel-van den Bosch, H.M.; de Wit, N.J.; Hooiveld, G.J.; Vermeulen, H.; van der Veen, J.N.; Houten, S.M.; Kuipers, F.; Muller, M.; van der Meer, R. A cholesterol-free, high-fat diet suppresses gene expression of cholesterol transporters in murine small intestine. *Am. J. Physiol. Gastrointest. Liver Physiol.* **2008**, *294*, G1171–G1180. [CrossRef] [PubMed]

114. Yamazaki, Y.; Hashizume, T.; Morioka, H.; Sadamitsu, S.; Ikari, A.; Miwa, M.; Sugatani, J. Diet-induced lipid accumulation in liver enhances ATP-binding cassette transporter g5/g8 expression in bile canaliculi. *Drug Metab. Pharm.* **2011**, *26*, 442–450. [CrossRef]

115. Apro, J.; Beckman, L.; Angelin, B.; Rudling, M. Influence of dietary sugar on cholesterol and bile acid metabolism in the rat: Marked reduction of hepatic Abcg5/8 expression following sucrose ingestion. *Biochem. Biophys. Res. Commun.* **2015**, *461*, 592–597. [CrossRef]

116. Duan, L.P.; Wang, H.H.; Ohashi, A.; Wang, D.Q. Role of intestinal sterol transporters Abcg5, Abcg8, and Npc1l1 in cholesterol absorption in mice: Gender and age effects. *Am. J. Physiol. Gastrointest. Liver Physiol.* **2006**, *290*, G269–G276. [CrossRef] [PubMed]

117. Dieter, M.Z.; Maher, J.M.; Cheng, X.; Klaassen, C.D. Expression and regulation of the sterol half-transporter genes ABCG5 and ABCG8 in rats. *Comp. Biochem. Physiol. C Toxicol. Pharm.* **2004**, *139*, 209–218. [CrossRef] [PubMed]

118. Ikeda, I.; Kudo, M.; Hamada, T.; Nagao, K.; Oshiro, Y.; Kato, M.; Sugawara, T.; Yamahira, T.; Ito, H.; Tamaru, S.; et al. Dietary soy protein isolate and its undigested high molecular fraction upregulate hepatic ATP-binding cassette transporter G5 and ATP-binding cassette transporter G8 mRNA and increase biliary secretion of cholesterol in rats. *J. Nutr. Sci. Vitaminol.* **2009**, *55*, 252–256. [CrossRef] [PubMed]

119. Zhong, C.Y.; Sun, W.W.; Ma, Y.; Zhu, H.; Yang, P.; Wei, H.; Zeng, B.H.; Zhang, Q.; Liu, Y.; Li, W.X.; et al. Microbiota prevents cholesterol loss from the body by regulating host gene expression in mice. *Sci. Rep.* **2015**, *5*, 10512. [CrossRef]

International Journal of
Molecular Sciences

Article

Transmembrane Polar Relay Drives the Allosteric Regulation for ABCG5/G8 Sterol Transporter

Bala M. Xavier [1,†,‡], Aiman A. Zein [1,†,‡], Angelica Venes [1,2], Junmei Wang [3,*] and Jyh-Yeuan Lee [1,*]

[1] Department of Biochemistry, Microbiology and Immunology, Faculty of Medicine, University of Ottawa, Ottawa, ON K1H 8M5, Canada; bxavier@uottawa.ca (B.M.X.); azein035@uottawa.ca (A.A.Z.); avene063@uottawa.ca (A.V.)

[2] Biomedical Sciences Program, Department of Biology, Faculty of Science, University of Ottawa, Ottawa, ON K1N 6N5, Canada

[3] Department of Pharmaceutical Sciences, School of Pharmacy, University of Pittsburgh, Pittsburgh, PA 15206, USA

* Correspondence: junmei.wang@pitt.edu (J.W.); Jyh-Yeuan.Lee@uOttawa.ca (J.-Y.L.); Tel.: +1-412-383-3268 (J.W.); +1-613-562-5800 (ext. 8308) (J.-Y.L.)

† These authors contributed equally to this work.

‡ Current address: Department of Cellular and Molecular Medicine, Faculty of Medicine, University of Ottawa, Ottawa, ON K1H 8M5, Canada.

Received: 7 October 2020; Accepted: 17 November 2020; Published: 19 November 2020

Abstract: The heterodimeric ATP-binding cassette (ABC) sterol transporter, ABCG5/G8, is responsible for the biliary and transintestinal secretion of cholesterol and dietary plant sterols. Missense mutations of ABCG5/G8 can cause sitosterolemia, a loss-of-function disorder characterized by plant sterol accumulation and premature atherosclerosis. A new molecular framework was recently established by a crystal structure of human ABCG5/G8 and reveals a network of polar and charged amino acids in the core of the transmembrane domains, namely, a polar relay. In this study, we utilize genetic variants to dissect the mechanistic role of this transmembrane polar relay in controlling ABCG5/G8 function. We demonstrated a sterol-coupled ATPase activity of ABCG5/G8 by cholesteryl hemisuccinate (CHS), a relatively water-soluble cholesterol memetic, and characterized CHS-coupled ATPase activity of three loss-of-function missense variants, R543S, E146Q, and A540F, which are respectively within, in contact with, and distant from the polar relay. The results established an in vitro phenotype of the loss-of-function and missense mutations of ABCG5/G8, showing significantly impaired ATPase activity and loss of energy sufficient to weaken the signal transmission from the transmembrane domains. Our data provide a biochemical evidence underlying the importance of the polar relay and its network in regulating the catalytic activity of ABCG5/G8 sterol transporter.

Keywords: ABCG5; ABCG8; ATP-binding cassette transporter; cholesterol; polar relay; sitosterolemia

1. Introduction

All living cells depend on the ability to translocate nutrients, metabolites, and other molecules across their membranes. One major way to achieve this is through membrane-anchored transporter proteins. The evolutionarily conserved ATP-binding cassette (ABC) transporter superfamily, for example, carries out ATP-dependent and active transport of a wide range of substances across cellular membranes, including both hydrophilic and hydrophobic molecules such as sugars, peptides, antibiotics, or cholesterol [1–4]. As a key component of cellular membranes, cholesterol constitutes ~50% of cellular lipid content; it is also the precursor of steroid hormones that modulate gene regulation and bile acids that enable nutrient absorption. Translocation of cholesterol molecules on biological membranes plays an essential role in maintaining cellular and whole-body cholesterol homeostasis. Thus, excess cholesterol needs to be eliminated from cells and tissues through either sterol acceptors in the circulation or direct excretion into the bile or the gut [5,6]. A large body of evidence indicates that ABC sterol transporters regulate cholesterol metabolism, and their defects are associated with dysregulation of whole-body cholesterol homeostasis, a major risk factor for cardiovascular diseases [7,8]. Yet, we have almost no understanding of how these transporters actually translocate cholesterol molecules and how the sterol-transport process is controlled by ATP catalysis. Given the dysregulation of cholesterol metabolism as a major risk factor for cardiovascular disease, there is a pressing need to elucidate of mechanism of these transporters in moving molecules across the cell membranes.

Recent progress in solving a heterodimeric crystal structure of human ABCG5 and ABCG8 established a new molecular framework toward such a mechanistic understanding of ABC sterol transporters. ABCG5 and ABCG8 are half-sized ABC sterol transporters and co-expressed on the apical surface of the hepatocytes along the bile ducts and the enterocytes from the intestinal brush-border membranes [9,10]. ABCG5 and ABCG8 function as obligate heterodimers (ABCG5/G8) and serve as the primary and indispensable sterol-efflux pump that effectively exports excess cholesterol, non-cholesterol sterols, and dietary plant sterols into the bile and the intestinal lumen. In mammals, most cholesterol is eliminated via its metabolism into bile acids or via biliary secretion as free cholesterol. The latter is considered as the last step of reverse cholesterol transport (RCT), where ABCG5/G8 accounts for more than 75% biliary cholesterol secretion [11–14]. Recent studies have shown that, in human subjects and animal models, ABCG5/G8 is also responsible for eliminating neutral sterols via the transintestinal cholesterol efflux (TICE), a cholesterol-lowering process independent of RCT [15]. Thus, physiologically, ABCG5/G8 plays an essential role in controlling cholesterol homeostasis in our bodies.

In general, the smallest functional unit of an ABC transporter consists of two transmembrane domains (TMD1 and TMD2) and two nucleotide-binding domains (NBD1 and NBD2), and both NBDs concertedly bind and hydrolyze ATP to provide the energy and drive substrate transport. The TMDs, on the other hand, have been shown to share low sequence similarity in the amino-acid sequences and three-dimensional structural folds, suggesting substrate-specific mechanisms for individual transporters [16]. Mechanistic analyses of ABC cholesterol transporters have largely centered on sequence requirement at the canonical ATP-binding sites [17–20], whereas little is known about the sterol–protein interaction and its relationship with ATP catalysis. Recent progress solving a crystal structure of human ABCG5/G8 revealed a unique TMD fold and several structural motifs [21]. In particular, for each subunit, a network of polar and charged amino acids is present in the core of the TMD, namely, a polar relay, whose role remains to be characterized. A triple-helical bundle is located at the transmission interface between the NBD and the TMD and consists of an elbow connecting helix, a hotspot helix (also known as an E-helix), and an intracellular loop-1 (ICL1) coupling helix. However, on the triple-helical bundle or the transmembrane polar relay, several residues have been shown to bear disease-causing missense mutations from sitosterolemia or other metabolic disorders with lipid phenotypes (Figure 1A). Notably, several disease-causing mutations are clustered in the membrane-spanning region or at the NBD–TMD interface [8,22]. This suggests the unique roles of these structural motifs in regulating the ABCG5/G8 function, yet no prior knowledge was available to explain the role of these structural motifs in the sterol-transport function.

Loss-of-function (LOF) mutations in *ABCG5* or *ABCG8* are linked to sitosterolemia, a rare autosomal recessive disease, while several other missense mutations are also associated with other lipid disorders, such as gallstone formation or elevated low-density lipoprotein (LDL) cholesterol [23–28]. At the cellular level, many of the missense mutations lead to defects in post-translational trafficking of ABCG5/G8 from the endoplasmic reticulum (ER), an abnormality commonly observed in other ABC transporters with missense mutations, e.g., ΔF508 mutation in the cystic fibrosis transmembrane conductance regulator (CFTR or ABCC7) [29,30]. However, specific missense mutants of ABCG5/G8 heterodimers have shown no defect in protein maturation [29], suggesting alternative disease-causing mechanisms.

Therefore, studies of these mutants will not only show how they alter the transporter activity, but also provide mechanistic insights into the function of wild-type (WT) ABCG5/G8 sterol transporter.

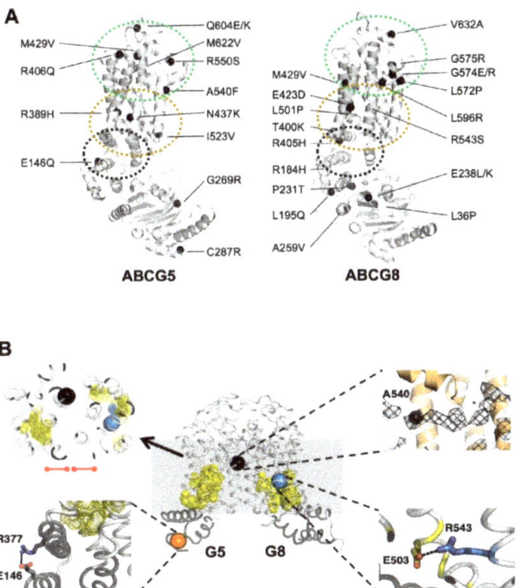

Figure 1. Disease-causing mutations and single-nucleotide polymorphisms (SNPs) in ATP-binding cassette (ABC) sterol transporters (ABCG5/G8). (**A**) Localization of ABCG5/G8 residues carrying missense mutations. The positions of disorder-related polymorphisms or mutations are highlighted in black spheres on the structures of ABCG5 (Protein Data Bank (PDB) identifier (ID): 5D07, chain C) and ABCG8 (PDB ID 5D07, chain D). Structural motifs are indicated in dashed ovals: triple-helical bundle (black), transmembrane domain (TMD) polar relay (yellow), and extracellular domain with re-entry helices (green). (**B**) Microenvironment of G5-E146, G5-A540, and G8-R543. (*Middle*) The transmembrane domains (white) and the triple-helical bundle (gray) are plotted in tube-styled cartoon presentation, showing the α-carbons (spheres) of G5-E146 (orange), G8-R543 (blue), and G5-A540 (black). The polar relays are plotted in dotted yellow spheres. (*Top left*) Slapped top view shows G5-A540 situated more than 10 Å away from the polar relay of either subunit (red dot-ended lines). (*Top right*) Near G5-A540 shows a cholesterol-shaped electron density (mesh) in the crystal structure of ABCG5/G8. The Fo−Fc difference electron density map was contoured at 3.0 σ. (*Bottom left*) At the triple helical bundle of ABCG5, E146 interacts with R377 through their side-chain termini in a distance of hydrogen bonding, 3.5 Å (black dashed line). (*Bottom right*) In the ABCG8 polar relay, R543 interacts E503 through their side-chain termini in a distance of hydrogen bonding, 3.1 Å (black dashed line).

Disease mutations are instrumental in studying the mechanisms of affected proteins in vitro, e.g., familial hypercholesterolemia mutations for proteins involved in low-density lipoprotein metabolism [31]. Guided by the structural framework of ABCG5/G8, we can now investigate its mechanisms using enzymological approaches with purified proteins. For this, we first need to establish at least one robust and consistent in vitro functional assay. Using ATPase activity as the functional benchmark in this study, we optimized an in vitro colorimetric ATPase assay that allows high-throughput activity assessment of detergent-purified ABCG5/G8. Using a soluble cholesterol memetic, cholesteryl hemisuccinate (CHS), we report here the CHS-stimulated ATP hydrolysis by ABCG5/G8 proteo-micelles, consisting of phospholipids, cholate, and dodecyl-maltoside (DDM), and we present an enzymatic analysis for the sterol-coupled ATPase activity on ABCG5/G8 sterol transporter. Using ATPase activity as functional readout of ABCG5/G8, we show differentially inhibition of the CHS-stimulated ATPase activity by three LOF missense mutants, two sitosterolemia mutations, and one sterol-binding mutation, where residues bearing the two disease mutations are located along the polar relay. Our data hereby demonstrate the mechanistic basis on regulating ABCG5/G8 function by the transmembrane polar relay (Figure 1B).

2. Results

2.1. CHS Stimulates ATP Hydrolysis by Wild-Type (WT) ABCG5/G8

Despite the known physiological role of ABCG5/G8 in biliary and intestinal cholesterol secretion, only indirect evidence of sterol-coupled transporter activity was detected by using steroid mimetics, such as androstane or bile acids [32,33]. In this study, we investigated a direct sterol-coupled ATPase activity by using CHS, a cholesterol mimetic that is more soluble in aqueous solution. First, to overcome low sensitivity of detecting the ABCG5/G8 ATPase activity using previous protocols, we optimized the ATPase assay for ABCG5/G8 by adopting a previous assay [34] and a colorimetric bismuth citrate-based detection approach [35]. As described and explained in Section 4, this optimized assay significantly reduces the background noise due to cloudiness by phospholipid/cholate/DDM mixtures, which improves the detecting sensitivity of liberated inorganic phosphate within the first few minutes and allows us to calculate more accurate rates of ATP hydrolysis. We show here that CHS can significantly stimulate ABCG5/G8-mediated ATP hydrolysis when co-incubated with sodium cholate (a bile acid) and *Escherichia coli* polar lipids (Figure 2). Using 5 mM ATP, the basal activity of ABCG5/G8 was calculated as 160 ± 15 nmol/min/mg ($n = 4$), similar to reported values, whereas, in the presence of CHS, the specific ATPase activity of ABCG5/G8 reached 565 ± 30 nmol/min/mg ($n = 8$), 3–4-fold higher than that in the absence of CHS (Figure 3A,B). Absence of cholate was unable to activate the ATP hydrolysis, consistent with the previous studies (data not shown) [33]. In addition, the activity was inhibited either by orthovanadate, an ATPase inhibitor [36] (Figure 3C), or by a catalytically deficient mutant ABCG5$_{WT}$/G8$_{G216D}$ (G8-G216D) [18], which displayed no ATP hydrolysis (Figure 3A,B). The specific activity of ATP hydrolysis by ABCG5/G8 is by far the highest in comparison with the previously reported values [33].

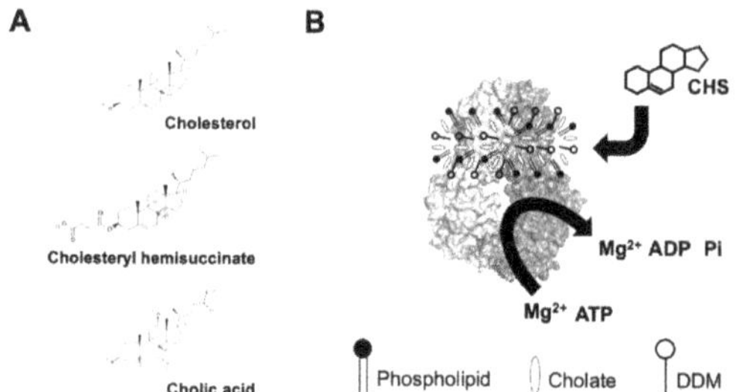

Figure 2. (**A**) Chemical structures of cholesterol, cholesteryl hemisuccinate (CHS), and cholic acid (cholate). Source: PubChem. (**B**) Schematic illustration of sterol-coupled ATPase activity of ABCG5/G8. Dodecyl-maltoside (DDM)-purified ABCG5/G8 (light/dark-gray surface) is preincubated with phospholipids and cholate. Addition of CHS (four-ringed steroid structure) stimulates hydrolysis of ATP to ADP and inorganic phosphate (Pi) in the presence of the divalent magnesium ions (Mg^{2+}). Using the colorimetric and bismuth citrate-based assay, the liberated Pi is then captured by ammonium molybdate in the presence ascorbic acid. The color is developed upon mixing with bismuth citrate and sodium citrate, and the absorbance was measured at 695 nm. See details in Section 4.

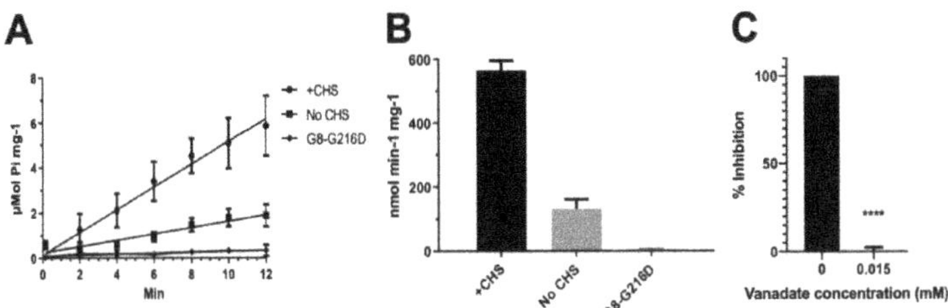

Figure 3. ATPase activity of ABCG5/G8. The ATP hydrolysis was used as a measure of ABCG5/G8 ATPase activity at 37 °C in conditions with 5 mM ATP and 4.1 mM CHS. The protocol is entailed in Section 4. (**A**) Data points are presented as the means ± standard deviations from 4–8 independent experiments using 2–4 independently purified proteins; where not visible, the error bars are covered by the plot symbols. A linear regression, plotted from the first 12 min, is used to calculate the specific activities. (**B**) Bar graphs show the specific activities of ATP hydrolysis by wild type (WT) in the presence and absence of CHS and the catalytically deficient mutant G8-G216D in the presence of CHS. The specific activity of WT in the absence of CHS is regarded as the basal ABCG5/G8 ATPase activity. (**C**) Bar graphs represent the percentage inhibition of ABCG5/G8 ATPase activity by 0.015 mM orthovanadate, where a *p*-value of <0.0001 (marked as ****) was obtained using ordinary one-way ANOVA (Prism 8).

2.2. The Lipid Environments Fine-Tune ABCG5/G8 ATPase Activity

ABC transporters need to function in a phospholipid-embedded environment. However, it is unknown whether the ABCG5/G8 function is controlled by phospholipids of specific headgroups or in specific lipid compositions. Because a high concentration of bile acids is required to activate ABCG5/G8 ATPase activity, attempts to use reconstituted proteoliposomes failed due to the immediate solubilization of the reconstituted proteins. To facilitate the assessment of mutant functions, we evaluated the lipid environments to obtain the most optimal assay conditions. To study the effect of lipid conditions and phospholipid species on the ABCG5/G8 function, we analyzed the CHS-coupled ATPase activity in the presence of two polar lipid extracts under conditions of fixed concentrations of sodium cholate and CHS (see Section 4). Using *E. coli* polar lipids, we carried out an ATP concentration-dependent ATPase assay to determine the Michaelis–Menten kinetic parameters of CHS-stimulated ATP hydrolysis. We observed the maximal ATP hydrolysis by ABCG5/G8 at concentrations slightly over 2.5 mM of ATP with a V_{max} of 677.1 ± 25.6 nmol/min/mg, a K_M(ATP) of 0.60 mM, and a k_{cat} of 1.69 s^{-1}. When using bovine liver polar lipids, we observed ~3.5-fold lower catalytic rate of ATP hydrolysis and ~50% higher K_M(ATP) (Figure 4A and Table 1). In the current study, polar lipids, cholate (bile acid), and CHS were all present in the reaction, indicating that the presence of *E. coli* polar lipids results in higher ATP association and, consequently, better stimulates ABCG5/G8 ATPase activity. When comparing the calculated values of k_{cat} and k_{cat}/K_M, we indeed observed an overall fivefold higher turnover rate in the presence of *E. coli* polar lipids than liver polar lipids (Table 1).

To determine the dependence of phospholipid headgroups, we tested the three most abundant phospholipids in either lipid extract on the ATP hydrolysis by ABCG5/G8, i.e., phosphatidylethanolamine (PE), phosphatidylcholine (PC), and phosphatidylglycerol (PG) (see Section 4). Preincubation with egg PE resulted in the highest specific activity, while the use of soy PC or egg PG only led to slightly higher ATP hydrolysis than the basal activity (Figure 4B). Interestingly, PE, the phospholipid found in both *E. coli* and liver lipids, is sufficient to stimulate ATP hydrolysis in ABCG5/G8 to almost the highest specific activity, as reported here. In the meantime, using PC or PG alone, the specific activity of ABCG5/G8 was also higher than that obtained with the liver polar lipid mixture. These results suggest phospholipid headgroups in regulating the ABCG5/G8 ATPase activity. Further investigations are necessary to pinpoint the effects of individual types of phospholipids on the sterol transporter function.

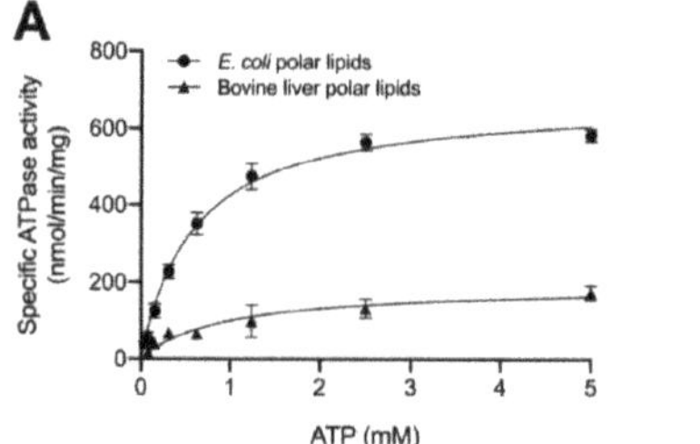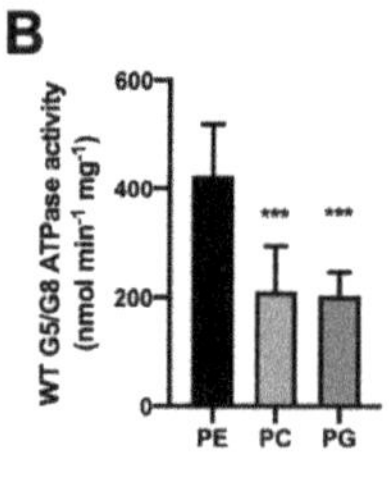

Figure 4. Lipid dependence of ABCG5/G8 ATPase activity. (**A**) Purified ABCG5/G8 was assayed in the presence of either *Escherichia coli* or bovine liver polar lipids, and the specific activities of ATP hydrolysis were obtained by the ATP concentration-dependent experiments (0–5 mM ATP). Both curves are fitted to the Michaelis–Menten equation (Prism 8), and, using two independently purified proteins, the means of at least three independent experiments along with standard deviations are plotted here. The kinetic parameters are listed in Table 1. (**B**) In conditions of 5 mM ATP and 4.1 mM CHS, ATP hydrolysis of purified ABCG5/G8 was assayed in the presence of egg phosphatidylethanolamine (PE), soy phosphatidylcholine (PC), or egg phosphatidylglycerol (PG), where *p*-values of 0.0006 and 0.0003 (marked as ***), respectively, were obtained using ordinary one-way ANOVA (Prism 8).

Table 1. Dependence of ABCG5/G8 ATPase activity on ATP.

	V_{max} [a] (nmol/min/mg)	K_M (ATP) (mM)	k_{cat} [b] (s^{-1})	k_{cat}/K_M $(M^{-1}·s^{-1})$	$\Delta\Delta G_{MUT}$ [c] (kJ/mol)	n [e]
WT (liver polar lipids)	192.8 ± 17.9	0.93 ± 0.25	0.48 ± 0.04	0.52×10^3	-	4
WT (*E. coli* polar lipids)	677.1 ± 25.6	0.60 ± 0.07	1.69 ± 0.06	2.8×10^3	-	6
G5-E146Q [d]	167.1 ± 0.05	0.51 ± 0.05	0.41 ± 0.00	0.82×10^3	11.7	5
G8-R543S [d]	150.7 ± 3.7	0.42 ± 0.04	0.38 ± 0.01	0.90×10^3	12.3	3
G5-A540F [d]	101.2 ± 4.2	0.58 ± 0.08	0.25 ± 0.01	0.43×10^3	15.8	5

[a] Standard errors were calculated from the fits shown in Figures 3A and 5 using Prism 8 (GraphPad Software, San Diego, CA, USA). [b] Turnover rates, k_{cat}, were calculated using the following formula: $V_{max} = k_{cat} \times [E]$, where [E] is the protein concentration of ABCG5/G8 (363.1 nM). [c] Differential Gibbs free energy was calculated according to the following formula: $\Delta\Delta G_{MUT} = -RT\ln(k_{MUT}/k_{WT})$, where k_{MUT} is the k_{cat} of mutants, k_{WT} is the k_{cat} of WT, $R = 8.314$ $J·mol^{-1}·K^{-1}$ (R: gas constant), and $T = 310.15$ K (37 °C). [d] Mutants were all assayed in the presence of *E. coli* polar lipids. [e] Number of independent experiments.

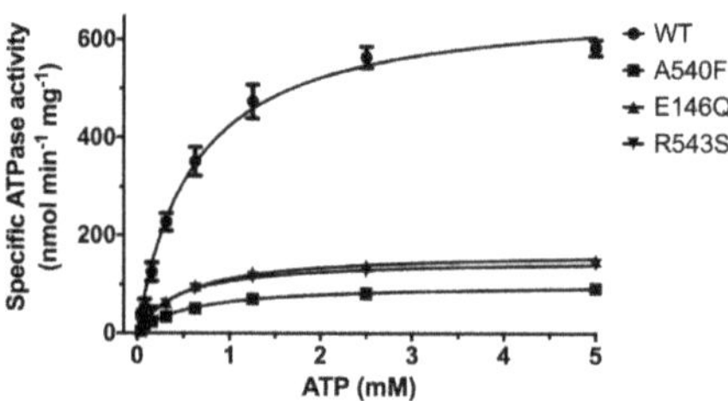

Figure 5. ATP dependence of ABCG5/G8 ATPase activity. Purified proteins were assayed in the presence of *E. coli* polar lipids, and the specific activities of ATP hydrolysis were obtained from the ATP concentration-dependent experiments (0–5 mM ATP). The curves are fitted to the Michaelis–Menten equation (Prism 8), and, using two-to-four independently purified proteins, the means of at least three independent experiments along with standard deviations are plotted here. The kinetic parameters are listed in Table 1.

2.3. Missense Mutants Impair CHS-Coupled ATPase Activity of ABCG5/G8

Using the CHS-coupled ATPase activity as the functional readout, we initiated studies in the catalytic mechanism of ABCG5/G8 by exploiting the transporter's missense mutations that undergo proper trafficking to post-ER cell membranes (ER-escaped mutants). In this study, we used *Pichia pastoris* yeast and expressed recombinant proteins of G8-G216D, a catalytically deficient mutant [18], ABCG5$_{E146Q}$/G8$_{WT}$ (G5-E146Q) and ABCG5$_{WT}$/G8$_{R543S}$ (G8-R543S), two loss-of-function/sitosterolemia missense mutants [22,37], and ABCG5$_{A540F}$/G8$_{WT}$ (G5-A540F), a loss-of-function mutant with putative sterol-binding defect [21] (Figure 1B and Figure S1). The purified mutants were preincubated with *E. coli* polar lipids and sodium cholate as described above. As shown in Figure 5, when compared with WT, the sitosterolemia missense mutants, G5-E146Q and G8-R543S, showed a ~80% reduction of the specific activity in CHS-coupled ATP hydrolysis (160 ± 15 nmol/min/mg and 150 ± 5 nmol/min/mg, respectively). The sterol-binding mutant G5-A540F, when compared to WT, showed a ~90% reduction of the specific activity in CHS-coupled ATP hydrolysis (90 ± 10 nmol/min/mg). Similar levels of activity reduction were also observed for non-CHS-coupled ATP hydrolysis (Figure S2). We then performed ATP concentration-dependent experiments and analyzed the Michaelis–Menten kinetics for these three mutants. For all mutants, K_M(ATP) remained nearly the same as compared to WT, but the mutants displayed a 40–60% reduction in the catalytic rate (Table 1). This result suggests that the mutants do not alter their ability of the nucleotide association, and other molecular events contribute to the reduction of the specific ATPase activity.

The effects of CHS on ABCG5/G8 WT and mutants were further investigated by measuring the ATP hydrolysis in the CHS concentration-dependent manner at a saturated ATP concentration (5 mM here). Purified proteins were preincubated with *E. coli* polar lipids, sodium cholate, and a wide range of CHS concentrations (0.064 mM to 4.1 mM). For WT, we obtained a V_{max} of 702.9 ± 50.7 nmol/min/mg, a K_M(CHS) of 0.79 mM, and a k_{cat} of 1.74 s^{-1} (Figure 6 and Table 2). In the presence of *E. coli* polar lipids, the catalytic rates were similar between the CHS and ATP-dependent ATPase activities, with a V_{max} of ~700 nmol/min/mg, which is about four times higher than that in the presence of liver polar lipids (Tables 1 and 2) and more than twofold higher than the previously reported value, ~290 nmol/min/mg [33]. The catalytic rates of the mutants decreased by 70–90%, except for G5-A540F, whereas both G5-E146Q and G8-R543S displayed significantly larger K_M(CHS), up to a twofold increase. This suggests a more profound impact of sitosterolemia mutations on the ABCG5/G8 ATPase activity through sterol–protein interaction or structural changes.

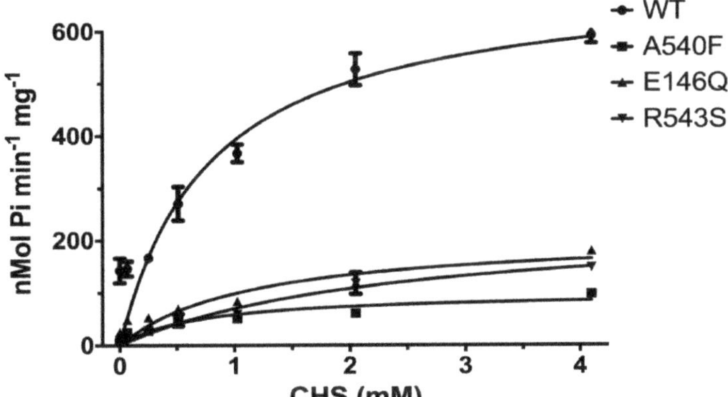

Figure 6. CHS dependence of ABCG5/G8 ATPase activity. Purified proteins were assayed in the presence of *E. coli* polar lipids, and the specific activities of ATP hydrolysis were obtained by the CHS concentration-dependent experiments (0–4.1 mM CHS). The curves are fitted to the Michaelis–Menten equation (Prism 8), and, using two independently purified proteins, the means of at least two independent experiments along with standard deviations are plotted here. The kinetic parameters are listed in Table 2.

Table 2. Dependence of ABCG5/G8 ATPase activity on cholesteryl hemisuccinate.

	V_{max} [a] (nmol/min/mg)	K_M (CHS) (mM)	k_{cat} [b] (s^{-1})	k_{cat}/K_M ($M^{-1} \cdot s^{-1}$)	$\Delta\Delta G_{MUT}$ [c] (kJ/mol)	n [e]
WT [d]	702.9 ± 50.7	0.79 ± 0.17	1.74 ± 0.13	2.2×10^3	-	6
G5-E146Q [d]	210.0 ± 33.2	1.13 ± 0.45	0.52 ± 0.08	0.46×10^3	10.0	2
G8-R543S [d]	237.1 ± 33.4	2.38 ± 0.67	0.59 ± 0.08	0.25×10^3	9.0	2
G5-A540F [d]	99.8 ± 11.4	0.70 ± 0.24	0.25 ± 0.03	0.36×10^3	16.1	4

[a] Standard errors were calculated from the fits shown in Figure 6 using GraphPad Prism 8. [b] Turnover rates, k_{cat}, were calculated using the following formula: $V_{max} = k_{cat} \times [E]$, where [E] is the protein concentration of ABCG5/G8 (363.1 nM). [c] Differential Gibbs free energy was calculated according to the following formula: $\Delta\Delta G_{MUT} = -RT\ln(k_{MUT}/k_{WT})$, where k_{MUT} is the k_{cat} of mutants, k_{WT} is the k_{cat} of WT, $R = 8.314$ $J \cdot mol^{-1} \cdot K^{-1}$ (R: gas constant), and $T = 310.15$ K (37 °C). [d] Both WT and mutants were assayed in the presence of *E. coli* polar lipids. [e] Number of independent experiments.

2.4. Missense Mutations Cause Conformational Changes at the ATP-Binding Site

To examine the relationship between structural changes of missense mutations and their impact on the ATPase activity, we performed molecular dynamics (MD) simulations for the WT and three mutants in this study. We then analyzed the MD structures to understand how the mutations could lead to different conformation around the hypothetical surrounding residues at the nucleotide-binding sites (NBS). These residues were obtained through a structural comparison between the crystal structure of ABCG5/G8 (Protein Data Bank (PDB) identifier (ID): 5DO7) and a cryo-EM structure of ABCG2 (PDB ID: 6HBU) for which two ATPs were bound in the homodimer [21,38].

To identify which residues are important for the ATP binding, we conducted MD simulations for the ABCG2 system. We calculated the ligand–residue MM-GBSA (Molecular Mechanics-Generalized Born Surface Area) free energies ($\Delta G_{lig-res}$) for the 32 surrounding residues and identified eight hotspot residues which have $\Delta G_{lig-res}$ better than −7.0 kcal/mol (Table S1). Although those hotspots were identified for ABCG2, it is reasonable to assume they are also hotspots for ABCG5/G8 given the apparent structural and sequence similarity (only one hotspot has different amino-acid types). The root-mean-square deviation (RMSD) for the main-chain atoms was 2.60 Å, and the corresponding amino acid types of both proteins are listed in Table S1. The detailed interactions between ATP and ABCG2 revealed by a representative MD structure are shown in Figure S3. In this study, we focused on the active nucleotide-binding site (known as NBS2) in ABCG5/G8 [21] and analyzed residues 88–103, 246–251 of ABCG5, and 210–220 and 237–245 of ABCG8. Those residues were recognized as the surrounding residues of the NBS2 in ABCG5/G8.

As shown in Figure 7, the mutations at the three sites could lead to global changes in the overall ABCG5/G8 structure, with RMSD values larger than 2.0 Å. The difference between the RMSDs of the secondary structures was smaller, probably because more obvious changes needed a longer simulation time to manifest. We were especially interested in the mutational effect on the ATP-binding site and generated RMSD vs. simulation time curves for those hypothetic surrounding residues (Figure S4). We observed that the RMSDs with and without least-square (LS) fitting were very stable for the WT, whereas, for G5-E146Q and G5-A540F, both the LS fitting and no-fitting RMSD were significantly larger. However, the G8-R543S mutation did not lead to significantly larger RMSD. This is because the distance between the mutation site and ATP binding site is greater and much longer MD simulations are required. Indeed, the RMSD had an increasing trend along the MD simulation time for G8-R543S (Figure S4D). We then conducted correlation analysis using an internal program to identify possible interaction pathways between the two sites. As shown in Figure S5, the shortest path contained R543, E474, N155, V205, and L213. L213 is linked to four key residues for ATP binding. It is understandable that a perturbation at R543 needs a long simulation time to reach the ATP binding site, given that the shortest interaction path contains six residues including two ends. Overall, we observed a significant perturbation on the conformations of the putative surrounding residues due to the mutations at G5-E146Q and G5-A540F. We anticipated that the G8-R543S mutation could lead to a significant conformational change at NBS2 in much longer MD simulations.

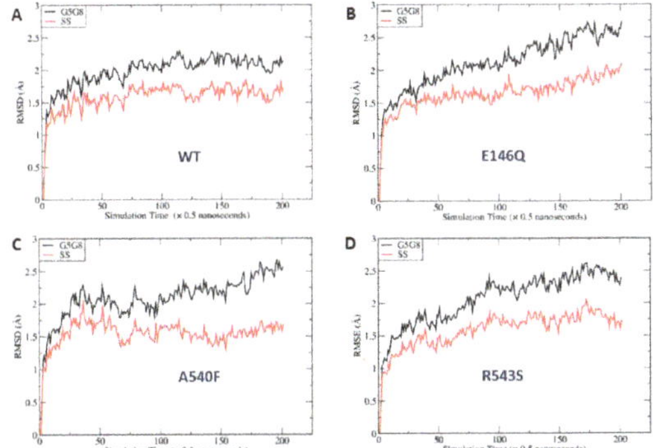

Figure 7. Fluctuation of root-mean-square deviations (RMSDs) along molecular dynamics (MD) simulation time course. RMSDs were calculated using the main-chain atoms of all residues (black lines) or secondary structures only (red lines): (**A**) wild type; (**B**) E146Q mutant in ABCG5; (**C**) A540F mutant in ABCG5; (**D**) R543S mutant in ABCG8. G5G8: ABCG5/G8; SS: secondary structure.

Next, we identified representative MD conformations for all four ABCG5/G8 protein systems for comparison (Figure 8). It was observed that the hotspot residues were overlaid very well between the crystal and MD structures for the WT (Figure 8E) and R543S mutant (Figure 8H), except for R211, while, for the other two mutants, the RMSDs were significantly larger (Figure 8F,G). This observation is expected, and the reason was explained above. Interestingly, the side-chain of R211 underwent dramatic changes for all four protein systems during MD simulations. If R211 was omitted, the main-chain RMSDs became much smaller. In summary, the conformational changes from our molecular modeling could qualitatively explain why the three mutations can lead to impaired ATPase activity. Of particular note, G5-K92, the hotspot residue that has the strongest interaction with ATP, is a part of the Walker A motif at the active nucleotide-binding site and required for ABCG5/G8 functions [18,33].

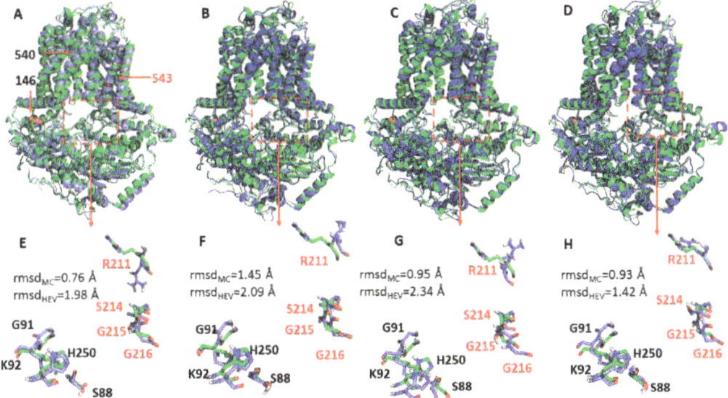

Figure 8. Representative structures of the WT ABCG5/G8 and its three missense mutants. The representative structures (shown as blue cartoons and bluish sticks) were aligned to the crystal structure (green cartoons, and greenish lines). The three mutation residues, E146Q, A540F, and R543S, are shown as spheres. The hypothetical surrounding residues of ATP are shown as dashed rectangles. (**A,E**) Wild type; (**B,F**) E146Q; (**C,G**) A540F; (**D,H**) R543S. G5: ABCG5; G8: ABCG8. Residues in G5 and G8 are separately colored in black and red. Root-mean-square deviations (RMSDs) for the main-chain atoms (rmsd$_{MC}$) and all heavy atoms (rmsd$_{HEV}$) are shown in the lower panels. If R211 is omitted from RMSD calculations, RMSDs of the main-chain atoms are 0.69, 1.30, 0.88, and 0.78 Å for WT, E146Q, A540, and R543S, respectively; the corresponding RMSDs of heavy atoms are 0.85, 1.42, 1.13, and 0.96 Å.

3. Discussion

In this study, we show that CHS stimulates the ATPase activity of the human ABCG5/G8 sterol transporter to a much higher specific activity, as compared to previously reported data (Tables 1 and 2). The much increased CHS-coupled ATPase activity indicates that ABCG5/G8 may need such a high ATP catalytic rate to achieve the sterol-transport function across the cellular membranes. CHS is a relatively water-soluble cholesterol analogue and is used to mimic cholesterol in membrane protein crystallization [21,39]. Our results showing CHS-stimulated ATPase activity suggest that the sterol molecules may have played a role in promoting an active conformation for the ATPase and/or enhancing the stability of ABCG5/G8. This idea of protein stability is supported by recent findings showing that CHS stabilizes a variety of human membrane proteins toward active conformations [40]. In the crystallographic study, >2% cholesterol was necessary to produce crystals capable of diffracting X-ray to better than 4 Å, and several sterol-like electron densities were suspected on the crystal structure of ABCG5/G8 [21]. Building upon previous work using bile acids [33] and androstane [32], our enzymatic results should come with no surprise that the WT ABCG5/G8 functionality and its active conformation are directly coupled with cholesterol analogues.

For ABCG5/G8-mediated ATP catalysis, we observed similar catalytic rates from the CHS and ATP concentration-dependent experiments, with a V_{max} of ~700 nmol/min/mg, whereas the K_M values were very similar to each other, $K_M(ATP) = 0.60$ mM and $K_M(CHS) = 0.79$ mM (Tables 1 and 2). $K_M(ATP)$ and $K_M(CHS)$ can be used to implicate ATP and sterol association to the transporters during the ATP catalytic process, respectively. We, therefore, speculate that one ATP usage is required for sterol–protein association for one CHS (or cholesterol) molecule. Because ABCG5/G8 is believed to contain only one active NBS [18], such 1:1 stoichiometry of ATP and cholesterol for ABCG5/G8 may reflect the sterol transport rate by the single active site on this ABC transporter. An in vitro sterol-binding or transport assay, in need of development, will be necessary to directly address such a relationship. In addition to sterols, it is intriguing that PE, PC, or PG alone was sufficient to support ATPase activity of ABCG5/G8, with PE-driven activity being the highest (Figure 4). PE is the major phospholipid of the *E. coli* polar lipids, ~60%, and the second most abundant phospholipid in the bile canalicular membranes and the small intestine brush-border membranes, ~25% and ~40% respectively, of total phospholipids [41,42]. It has been shown that PE preferentially fits the headgroup-binding sites on integral membrane proteins [43]; thus, PE may be recruited as better phospholipids to support ABCG5/G8 function in the cell membranes. The approximate ratio of lipids for either *E. coli* or liver polar lipids may contribute to the apparent difference in activity, but it remains unknown how phospholipid composition regulates the transporter function. It is worth noting that specific phospholipids were shown to regulate the ATPase activity of other ABC sterol transporters, such as sphingomyelin, although the mechanistic details are not clear [19]. These individual lipids will be subjected to further examination to define the phospholipid specificity on the ABCG5/G8 ATPase activity and/or sterol-transport function.

By mapping disease-carrying residues on the apo structure of ABCG5/G8, we found that most missense variants occur within or near the structural motifs consisting of several conserved amino acids [22]. Several missense mutations (ER-trapped) prevent protein maturation from the endoplasmic reticulum (ER), but at least five mutations (ER-escaped) have been shown to undergo proper trafficking to post-ER cell membranes [29]. So far, no report has shown the impact of these ER-escaped missense mutants on ABCG5/G8 function using either in vitro or in vivo models. In this study, we used purified proteins from *Pichia pastoris* to investigate the functional activity of ABCG5/G8 in vitro and aimed to establish the mechanistic basis of ABCG5/G8 through analyzing the structure–function relationship of its loss-of-function missense mutations. The sitosterolemia missense mutants G5-E146Q and G8-R543S showed a reduction in CHS-coupled ATP hydrolysis, but retained ~20% activity as compared to WT, while the putative sterol-binding mutant G5-A540F showed further reduction to ~10% of WT ATPase activity (Figures 5 and 6). With such activity reduction, the mutant proteins maintained ATPase activity similar to the basal level, as shown by WT, suggesting a remote and allosteric regulation to keep ATPase active during the reaction.

It is not uncommon that reagents such as CHS may be used as protein stabilizers for disease-causing missense variants. Here, in the absence of CHS-coupled stimulation, the mutants showed a similar level of reduced ATPase activity, arguing for a more profound effect from impaired allosteric regulation on the catalytic activity of the mutants, rather than CHS-driven stability for mutant proteins. As predicted by MD simulation, the ATP-bound homology model underwent global conformational changes upon introducing the mutations (Figure 7). These mutations, albeit relatively far away from the nucleotide-binding site, can cause significant structural rearrangement of the residues within the region that encompasses the active NBS2 (Figure 8). Such conformational changes may alter responses to the sterol–protein interaction necessary for maximal ATPase activity.

In the atomic model of ABCG5/G8 (PDB ID: 5D07), G5-E146 is located on the hotspot helix of the triple-helical bundle and in proximity to ABCG5's polar relay, while G8-R543 is part of ABCG8's polar relay in the core of TMD (Figure 1). Both the triple-helical bundle and the polar relay are believed to form a network of hydrogen bonding and salt bridges and play an important role in interdomain communication during the transporter function [21]. G5-E146 and G8-R543 are found in the proximity of hydrogen-bonding distance with arginine 377 of ABCG5 (G5-R377) and glutamate 503 of ABCG8 (G8-503), respectively (Figure 1B). On the basis of the ATP-dependent experiments (Figure 5 and Table 1), we obtained the changes in Gibbs free energy from WT to each mutant ($\Delta\Delta G_{MUT}$) as $\Delta\Delta G_{E146Q}$ = ~11.7 kJ/mol and $\Delta\Delta G_{R543S}$ = ~12.3 kJ/mol. Such energetic loss is in the range of intramolecular hydrogen-bonding potential observed on transmembrane α-helical bundles [44]. Therefore, the results support the hypothesis that the hotspot helix and the polar relay are responsible for transmitting signals between NBD and TMD. Slightly lower $\Delta\Delta G_{MUT}$ was observed from CHS-dependent experiments (Figure 6 and Table 2), with $\Delta\Delta G_{E146Q}$ = ~10.0 kcal/mol and $\Delta\Delta G_{R543S}$ = ~9.0 kcal/mol. This falls in the range of hydrophobic interaction and argues for weakened sterol-transporter interaction due to these disease mutations. As for the sterol-binding mutant, we obtained higher energetic loss, but similar $\Delta\Delta G_{MUT}$ from ATP- or CHS-dependent analysis, with $\Delta\Delta G_{A540F}$ = ~15.8 or ~16.1 kJ/mol, respectively. This likely indicates a strong hydrophobic interaction between sterols and the transporter, as no obvious hydrogen donors/acceptors can be found at the putative sterol-binding site on the crystal structure. In addition, G5-A540 is distant from the polar relay (>10 Å away); thus, these data suggest a remote contact by sterol molecules to control the sterol-coupled signaling, likely through the polar relay in the transmembrane domains. In the ATP concentration-dependent experiments, the K_M values for ATP remained almost the same (Table 1), suggesting that ATP binding was not affected by these mutants. The K_M values for CHS were significantly increased in the disease mutants, but not the sterol-binding mutant (Table 2), suggesting that CHS interacts with ABCG5/G8 and remotely regulates the turnover of ATP hydrolysis in either a sequential (Mode 1) or a concerted (Mode 2) pathway (Figure 9). Collectively, these results argue that a working network of hotspot helix and polar relay is essential to maintain the communication between ATPase and sterol-binding activities in ABCG5/G8, which are impaired by the loss-of-function missense mutations. As G8-R532S is the only known ER-escaped disease mutant, we will expect more insight in such polar relay-driven allosteric regulation by investigating other polar relay residues with site-directed mutagenesis.

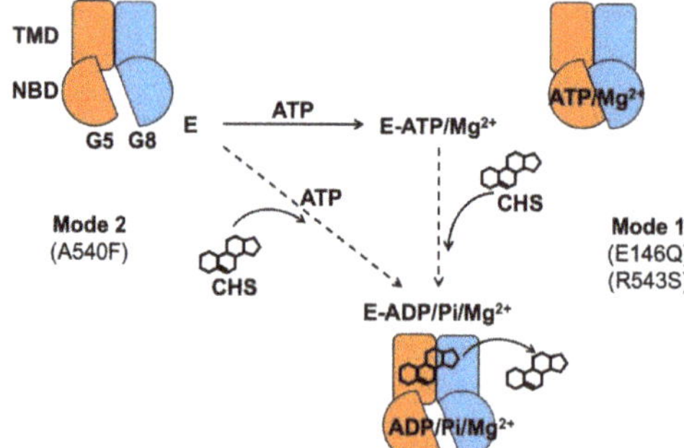

Figure 9. Proposed mechanism of sterol-coupled ATP catalysis by ABCG5/G8. (Mode 1) A sequential pathway is derived from experiments on the disease mutants, G5-E146Q and G8-R543S. ABCG5/G8 first recruits ATP and Mg^{2+} ions, likely causing a conformational change of the nucleotide-binding domain (NBD) for ATP binding. CHS/sterol then binds the transporter and triggers ATP hydrolysis that may result in its dissociation. (Mode 2) A concerted pathway is derived from experiments on the putative sterol-binding mutant, G5-A540F. ABCG5/G8 simultaneously recruits CHS, ATP, and Mg^{2+} ions, induces a transient conformational change of the NBD, and activates ATP hydrolysis and CHS/sterol dissociation from the transporter. G5: ABCG5; G8: ABCG8; E: ABCG5/G8 heterodimer; Pi: inorganic phosphate.

In conclusion, these studies show that CHS stimulates ABCG5/G8 ATPase activity and may promote an active conformation for ABCG5/G8-mediated sterol transport. The enzymatic characterization of three loss-of-function missense variants provides a mechanistic basis for how the polar relay contributes to the interdomain communication for the sterol-coupled ATPase activity in ABCG5/G8 and may be directly involved in such ligand–protein interactions. Further studies will reveal more insight into these molecular events and enable sterol-lowering therapeutics to treat sitosterolemia and hypercholesterolemia.

4. Materials and Methods

4.1. Materials

E. coli polar lipids (Cat. #: 100600C) and bovine liver polar lipids (Cat. #: 181108C) were from Avanti Polar Lipids, Inc. (Alabaster, AL, USA). Cholesterol, cholesteryl hemisuccinate (CHS), and *n*-dodecyl β-D-maltopyranoside (DDM) were from Anatrace (Maumee, OH, USA). The nickel–nitrilotriacetic acid (Ni–NTA) agarose resin was from Qiagen Toronto (Toronto, ON, Canada), while the calmodulin (CBP) affinity resin, zeocin, and ampicillin were from Agilent (Santa Clara, CA, USA). Imidazole, ε-aminocaproic acid, sucrose, yeast extract, tryptone, peptone, yeast nitrogen base (YNB), and ammonium sulfate were obtained from Wisent Bioproducts (St-Bruno, QC, Canada). ATP disodium trihydrate, Tris-(2-carboxyethyl)-phosphine (TCEP), sodium chloride, glycerol, ethylenediaminetetraacetic acid (EDTA), ethylene glycol-bis(β-aminoethyl ether)-*N,N,N′,N′*-tetraacetic acid (EGTA), sodium dodecyl sulfate (SDS), Ponceau S solution, sodium azide, Bradford reagents, Tween-20, magnesium chloride, calcium chloride, and all protease inhibitors were obtained from Bioshop Canada (Burlington, ON, Canada). Biotin, sodium cholate hydrate, L-ascorbic acid, ammonium molybdate, bismuth citrate, sodium citrate, methanol, ammonium hydroxide, hydrochloric acid, and acetic acid were obtained from MilliporeSigma (Oakville, ON, Canada). Dithiothreitol (DTT), Tris base, and Tris acetate were obtained from ThermoFisher (Ottawa, ON, Canada). Clarity Western enhanced chemiluminescence (ECL) substrates, 30% acrylamide, agarose, and ammonium persulfate were obtained from Bio-Rad (Hercules, CA, USA). Restriction enzymes were obtained from New England Biolabs (NEB) (Ipswitch, MA, USA), Promega (Madison, WI, USA), and ThermoFisher

(Ottawa, ON, Canada). The following media were used: yeast extract peptone dextrose (YPD), yeast extract peptone dextrose sorbitol (YPDS), minimal glycerol yeast nitrogen base (MGY), and minimal protease inhibitor buffer for yeast cell lysis (mPIB), consisting of 0.33 M sucrose, 0.3 M Tris-HCl (pH 7.5), 1 mM EDTA, 1 mM EGTA, 100 mM ε-aminocaproic acid, and ddH$_2$O to a final volume of 1 L and stored at 4 °C.

4.2. Cloning of ABCG5/G8 Missense Mutants

The expression vectors (pLIC and pSGP18), carrying human ABCG5 (NCBI accession number NM_022436) and human ABCG8 (NCBI accession number NM_022437), were derived from pPICZB (Invitrogen) as described [33,45], pLIC-ABCG5, and pSGP18-ABCG8, respectively. A tandem array of six histidines separated by glycine (His$_6$GlyHis$_6$) was added to the C terminus of ABCG5, and a tag encoding a rhinovirus 3C protease site followed by a calmodulin binding peptide (CBP) was added to the C terminus of ABCG8. To generate the missense mutants in this study, we performed site-directed mutagenesis by using WT ABCG5 or ABCG8 as the templates and the following codon-optimized oligonucleotide primers (Eurofins Genomics Canada). G5-A540F: CCATTTTTGGGGTGCTTGTTGGATCTGGATTCCTCAG (forward) and GCACCCC AAAAATGGACAGCAGAGCCACTACAC (reverse); G5-E146Q: GCGCCAAACGCTGCACTACACC GCGCTGC (forward) and CAGCGTTTGGCGCACGGTGAGGCTGCTCAG (reverse); G8-R543S: GTT GCTCTATTATGGCCCTGGCCGCCGC (forward) and GCCATAATAGAGCAACAGAAGACCACCAG CCAC (reverse); G8-G216D: ACGAGCGCAGGAGAGTCAGCATTGGGGTGCAG (forward) and CTCTCCTGCGCTCGTCCCCCGACAACCCC (reverse). The polymerase chain reaction (PCR) included 1× Phusion High-Fidelity DNA Polymerase (New England Biolabs), 1× Phusion buffer, 200 mM dNTP, 2% (*v/v*) DMSO, 100 ng DNA templates, and 0.4 mM forward and reverse primers. Each mutant-containing plasmid was amplified by the following PCR settings: initial DNA denaturation (98 °C, 2 min), followed by 30 cycles of denaturation (98 °C, 15 s)/primer annealing (55 °C, 30 s)/DNA extension (72 °C, 3 min), and then final extension (72 °C, 20 min). Next, 5 μL of the PCR products was run on a 1% agarose gel to confirm the amplification, and 1 μL of *Dnp*I restriction enzymes (20 units) was used to digest the WT templates overnight at 37 °C. The modified plasmids were cleaned up using the ethanol acetate precipitation technique. Then, 5 μL of 3 M sodium acetate was added to each 50 μL PCR product. Next, 200 μL of 100% ethanol was added to each tube, vortexed, and left at room temperature for 10 min. At max speed in a table centrifuge for 10 min, the plasmids were pelleted; then, the supernatant was removed and washed with 75% ethanol. Residual ethanol was dried by a Speed-Vac at the maximal speed for 20 min at room temperature. The pellet was resuspended in ddH$_2$O. Mutants plasmids were cloned into XL1-Blue competent *E. coli* cells by the heat-shock approach as described in the supplier's manual (Novagen/Agilent, Santa Clara, CA, USA) and by antibiotic selection using Zeocin (Invitrogen/ThermoFisher, Ottawa, ON, Canada). Using PureYield Plasmid Midiprep kit (Promega, Madison, WI, USA), DNA preparations of selected clones were subjected to sequencing at Eurofins Genomics Canada.

4.3. Expression of ABCG5/G8 Missense Mutants in Pichia pastoris Yeast

Both WT and mutant plasmids (20 mg each plasmid) were linearized using *Pme*I and co-transformed into the *Pichia* strain KM71H by electroporation. Immediately, the cells were resuspended with 1 mL of ice-cold 1 M sorbitol and incubated at 30 °C for 1 h. Then, 5 mL of fresh YPD was added and incubated for 6 h at 250 rpm and 30 °C. The cells were then centrifuged at 3000× *g* for 10 min and resuspended with 200 μL of YPD. Next, 100 μL of transformants were plated on YPDS plates containing 100 (low), 500 (medium), or 1000 (high) μg/mL Zeocin to screen for successful transformation. Seven colonies were picked and grown in 10 mL of MGY medium for 24 h in sterile 50 mL tubes at 250 rpm and 30 °C. The cells were centrifuged for 10 min at 3000× *g* and then resuspended with 10 mL of minimal methanol (MM) medium. Then, 50 μL of methanol was added to the medium and once again after 12 h. The cells were harvested after 24 h incubation at

250 rpm and 30 °C, resuspended in 600 µL mPIB buffer, and transferred into a 1.5 mL Eppendorf tube. After adding 500 µL of glass beads, protease inhibitors, and 10 mM DTT, the cells were lysed using a mini-bead beater (Biospec), with 1.5 min beating and 1.5 min rest on ice for three cycles. The unbroken cells and beads were pelleted by centrifugation at 5000× g for 5 min at 4 °C, followed by 21,130× g for 5 min at 4 °C. The supernatant was collected, and the concentration of the total proteins was estimated by Bradford assay. Next, 1 µL of cell lysate and 0–10 µg BSA standards were separately added to 200 µL of Bradford reagent on a 96-well plate. Absorbance at 595 nm was used to measure the protein concentrations using a Synergy H1 Hybrid reader (BioTek/Agilent, Santa Clara, CA, USA). The cell lysates (20 or 30 µg of total proteins) were resolved by SDS–PAGE, and protein expression was analyzed by immunoblotting using monoclonal anti-RGS-His antibodies (Qiagen Toronto, Toronto, ON, Canada) to detect ABCG5 and polyclonal anti-hABCG8 antibodies (Novus Biologicals, Centennial, CO, USA) to detect ABCG8. The clones expressing the highest level for both subunits were selected and stored in 20% glycerol at −75 °C.

4.4. Cell Culture and Microsomal Membrane Preparation

The conditions for cell growth and WT protein induction were as previously described [21]. Briefly, cells were initially grown at 30 °C to accumulate cell mass in an Innova R43 shaker (Eppendorf) at 250 rpm for 24–48 h with the pH maintained at pH 5–6. To induce protein expression, cells were left fasting for 6–12 h, and then incubated with 0.1% (v/v) methanol for 6–12 h at 20 or 28 °C. The methanol concentration was increased to 0.5% (v/v) by adding methanol every 12 h for 48–60 h. Cell pellets were collected and resuspended in mPIB and stored at −75 °C. Approximately 45 ± 10 g of cell mass was typically obtained from 1 L of cultured cells. The frozen cells were thawed and lysed using a C3-Emulsifier (Avestin, Ottawa, ON, Canada) in mPIB in the presence of 10 mM DTT and protease inhibitors (1 µg/mL leupeptin, 1 µg/mL pepstatin A, 1 µg/mL aprotinin, and 2 mM PMSF(phenylmethylsulfonyl fluoride). The microsomal membranes were then prepared as previously described [21].

4.5. Purification of ABCG5/G8 and Its Mutants

Both WT and mutants were purified following a protocol described previously [21], with minor modification. Briefly, DDM-solubilized membranes were subjected to a tandem affinity column chromatography, first using Ni–NTA and then CBP. The *N*-linked glycans and the CBP tag remained on the purified heterodimers, and the CBP eluates were further purified by gel-filtration chromatography using a Superdex 200 Increase 10/300 GL column (Cytiva) on an KTA Pure purification system (Cytiva, formerly GE Healthcare Life Sciences). The proteins in the peak fractions were collected and concentrated to 1–3 mg/mL for storage at −75 °C. Noticeably, the final yield for mutants was lower than WT, in a range of 400–800 µg per 6 L of cells. The expression level of the mutant proteins in the microsomes and their solubility were slightly lower than for WT. Some proteins were also lost during Ni–NTA binding and imidazole wash. The profile of the gel-filtration chromatography often showed a higher peak at the void volume than dimeric proteins. These factors collectively suggest that the mutant proteins were more prone to aggregation, thus explaining the lower yields.

4.6. ATPase Assay

We consistently observed a strong cloudiness in the assay solution when using previous protocols, consequently resulting in low sensitivity when detecting the ABCG5/G8 ATPase activity. Because a high concentration of bile acids is required, we reasoned that the high content of detergents, both in the assay solution and in the protein preparations, may have caused either high background upon quenching the reaction in the Malachite Green-based assay [33] or poor organic–aqueous phase separation [21]. The measurement of ATPase activity, thus, becomes inconsistent from one protein preparation to another. To overcome this issue, we first optimized the ATPase assay by adopting a colorimetric and bismuth citrate-based approach [35], which also allows high-throughput detection of the liberated inorganic phosphate by a microplate reader. The ATPase assay was performed in a 65 µL final reaction volume containing 2 mg/mL *E. coli* or liver polar lipids or designated phospholipids, 1.5% sodium cholate, 0.2% (4.11 mM) CHS, and 2 mM DTT in Buffer A (50 mM Tris/Cl pH 7.5, 100 mM NaCl, 10% glycerol, 0.1% DDM). The lipid/CHS/DTT mixture was thoroughly sonicated and preincubated with ABCG5/G8 proteins (0.3 to 1.5 µg) for 5 min at room temperature. The catalytically deficient G8-G216D was used as the negative control.

The enzymatic activity of ABCG5/G8 was initiated upon the addition of the 10× ATP cocktail (6.5 µL) and incubated at 37 °C. Aliquots (8.5 µL) were removed every 2 min and added to the prechilled quencher wells to stop the reaction. The quencher solution was made of 5% SDS in 5 mM HCl, which, together with the smaller reaction volume, contributed to a significant reduction in cloudiness for inorganic phosphate detection. Lipid mixtures were prepared at 30 mg/mL (~20 mM) in Buffer A containing 7% sodium cholate. CHS stock solution (1%, *w/v*) was prepared in a Buffer A and 4.5% sodium cholate, whereas 10× Mg/ATP cocktail contained 50 mM ATP, 75 mM $MgCl_2$, and 100 mM NaN_3 in a buffer containing 50 mM Tris/Cl pH 7.5. To detect the liberated inorganic phosphate, 50 µL of freshly made Solution II (142 mM ascorbic acid, 0.42 M HCl, 4.2% Solution I (10% ammonium molybdate)) was added to plate wells and left on ice for 10 min. Then, 75 µL of Solution III (88 mM bismuth citrate, 120 mM sodium citrate, 1 M HCl) was added to plate wells and placed at 37 °C for 10 min. The absorbance was measured at 695 nm using a Synergy H1 Hybrid reader (BioTek/Agilent, Santa Clara, CA, USA). For the phosphate standards, 1 M monobasic or dibasic sodium or potassium phosphate in 50 mM Tris/Cl pH 7.5 was prepared, and six standard inorganic phosphate solutions (0 µM, 12.5 µM, 25 µM, 50 µM, 100 µM, or 200 µM) were used in every experiment. The linear range of each reaction was used to calculate the initial rate of ATP hydrolysis. GraphPad Prism 8 was used to perform nonlinear regression and ordinary one-way ANOVA, with a *p*-value of ≤0.05 considered significant from at least three independent experiments. The kinetic parameters were calculated by nonlinear Michaelis–Menten curve fitting using GraphPad Prism 8.

4.7. Computational Methods

We studied four ABCG5/G8 protein systems including the WT and the E146Q, A540F, and R543S mutants. Each MD system consisted of one copy of ABCG5/G8 heterodimer, 320 1,2-dimyristoyl-*sn*-glycero-3-phosphocholine (DMPC) lipids, 16 cholesterols, 43,621 TIP3P [46] water molecules, and 103 Cl^- and 83 Na^+ to neutralize the MD systems. AMBER ff14SB [47], Lipid14 [48], and GAFF [49] force fields were used to model proteins, DMPC lipids, and cholesterols, respectively. The residue topology of cholesterol was prepared using the Antechamber module [48]. MD simulation was performed to produce isothermal–isobaric ensembles using the pmemd.cuda program in AMBER 18 [50]. The particle mesh Ewald (PME) method [51] was used to accurately calculate the electrostatic energies with the long-ranged correction taken into account. All bonds were constrained using the SHAKE algorithm [52] in both the minimization and MD simulation stages following a computational protocol described in our previous publication [21]. Briefly, there were three stages in a series of constant-pressure and -temperature MD simulations, including the relaxation phase, the equilibrium phase, and the sampling phase. In the relaxation phase, the simulation system was heated progressively from 50 K to 250 K at steps of 50 K, and a 1 ns MD simulation was run at each temperature. In the next equilibrium phase, the system was equilibrated at 298 K, 1 bar for 10 ns. Finally, a 100 ns MD simulation was performed at 298 K, 1 bar to produce isothermal–isobaric ensemble ensembles. In total, 1000 snapshots were recorded from the last phase simulation for post-analysis using the "cpptray" module implemented in the AMBER software package. Binding free energy decomposition and correlation analysis were performed using an internal program and the detailed elsewhere [53,54].

Supplementary Materials: Supplementary Materials can be found at http://www.mdpi.com/1422-0067/21/22/8747/s1.

Int. J. Mol. Sci. **2020**, 21, 8747

Author Contributions: B.M.X. optimized the CHS-stimulated ATPase assay for ABCG5/G8; A.A.Z. generated and validated the mutant constructs; B.M.X., A.A.Z. and A.V. purified the proteins and carried out the ATPase assays and data analysis; J.W. performed the molecular dynamics simulation; J.-Y.L. oversaw the project; J.W. and J.-Y.L. wrote the manuscript. All authors have read and agreed to the published version of the manuscript.

Funding: This research was funded by a startup grant from the University of Ottawa, a Discovery Grant from the Natural Sciences and Engineering Research Council (RGPIN 2018-04070), and a National New Investigator Award from the Heart and Stroke Foundation of Canada to J.-Y.L., as well as the National Science Foundation (NSF 1955260) and National Institutes of Health (R01-GM079383) grants to J.W. B.M.X. is a recipient of the Travel Awards from the Canadian Society of Molecular Biosciences (2018) and Biophysical Society of Canada (2019).

Acknowledgments: We thank William Jennings, Gloria Ihirwe, Midhet Hajira, and Chloé van de Panne for technical assistance. We also thank Donna Clary and Hui Li for their technical help in utilizing the common core facilities. We are indebted to critical feedback on reviewing and editing the manuscript from Vicky Brandt, and John Baenziger, Jean-François Couture, Gregory Graf, and Xiaohui Zha. This work is partially based on the theses that were submitted to fulfill in part the requirement for the degrees of Master of Science (A.A.Z.) and Honors Bachelor of Science (A.V.).

Conflicts of Interest: The authors declare no conflict of interest. The funders had no role in the design of the study; in the collection, analyses, or interpretation of data; in the writing of the manuscript, or in the decision to publish the results.

Abbreviations

ABC	ATP-binding cassette
ABCC7	ATP-binding cassette sub-family C member 7
ABCG5	ATP-binding cassette sub-family G member 5
ABCG8	ATP-binding cassette sub-family G member 8
ATP	Adenosine triphosphate
CBP	Calmodulin-binding peptide
CFTR	Cystic fibrosis transmembrane conductance regulator
CHS	Cholesteryl hemisuccinate
DDM	Dodecyl maltoside or *n*-dodecyl β-D-maltopyranoside
DMPC	1,2-Dimyristoyl-*sn*-glycero-3-phosphocholine
DNA	Deoxyribonucleic acid
DTT	Dithiothreitol
EDTA	Ethylenediaminetetraacetic acid
EGTA	Ethylene glycol-bis(β-aminoethyl ether)-*N,N,N′,N′*-tetraacetic acid
ER	Endoplasmic reticulum
ICL	Intracellular loop
LDL	Low-density lipoprotein
LOF	Loss of function
LS	Least square
MD	Molecular dynamics
MGY	Minimal glycerol yeast nitrogen base
MM	Minimal methanol
mPIB	Minimum protease inhibitor buffer
NBD	Nucleotide-binding domain
NBS	Nucleotide-binding site
Ni-NTA	Nickel–nitrilotriacetic acid
PC	Phosphatidylcholine
PCR	Polymerase chain reaction
PDB	Protein Data Bank
PE	Phosphatidylethanolamine
PG	Phosphatidylglycerol
RCT	Reverse cholesterol transport
RMSD	Root-mean-square deviation
SDS	Sodium dodecyl sulfate
TCEP	Tris-(2-carboxyethyl)-phosphine
TICE	Transintestinal cholesterol efflux
TMD	Transmembrane domain
WT	Wild type
YNB	Yeast nitrogen base
YPD	Yeast extract peptone dextrose
YPDS	Yeast extract peptone dextrose sorbitol

References

1. Dean, M.; Allikmets, R. Evolution of ATP-binding cassette transporter genes. *Curr. Opin. Genet. Dev.* **1995**, *5*, 779–785. [CrossRef]
2. Dean, M.; Hamon, Y.; Chimini, G. The human ATP-binding cassette (ABC) transporter superfamily. *J. Lipid Res.* **2001**, *42*, 1007–1017. [CrossRef] [PubMed]
3. Linton, K.J.; Higgins, C.F. The Escherichia coli ATP-binding cassette (ABC) proteins. *Mol. Microbiol.* **1998**, *28*, 5–13. [CrossRef] [PubMed]
4. Hwang, J.U.; Song, W.Y.; Hong, D.; Ko, D.; Yamaoka, Y.; Jang, S.; Yim, S.; Lee, E.; Khare, D.; Kim, K.; et al. Plant ABC Transporters Enable Many Unique Aspects of a Terrestrial Plant's Lifestyle. *Mol. Plant* **2016**, *9*, 338–355. [CrossRef]

5. Li, G.; Gu, H.M.; Zhang, D.W. ATP-binding cassette transporters and cholesterol translocation. *IUBMB Life* **2013**, *65*, 505–512. [CrossRef] [PubMed]

6. Patel, S.B.; Graf, G.A.; Temel, R.E. ABCG5 and ABCG8: More than a defense against xenosterols. *J. Lipid Res.* **2018**, *59*, 1103–1113. [CrossRef]

7. Borst, P.; Zelcer, N.; Van Helvoort, A. ABC transporters in lipid transport. *Biochim. Biophys. Acta Mol. Cell Biol. Lipids* **2000**, *1486*, 128–144. [CrossRef]

8. Xavier, B.M.; Jennings, W.J.; Zein, A.A.; Wang, J.; Lee, J.Y. Structural snapshot of the cholesterol-transport ATP-binding cassette proteins. *Biochem. Cell Biol.* **2019**, *97*, 224–233. [CrossRef]

9. Graf, G.A.; Li, W.-P.; Gerard, R.D.; Gelissen, I.; White, A.; Cohen, J.C.; Hobbs, H.H. Coexpression of ATP-binding cassette proteins ABCG5 and ABCG8 permits their transport to the apical surface. *J. Clin. Investig.* **2002**, *110*, 659–669. [CrossRef]

10. Graf, G.A.; Yu, L.; Li, W.P.; Gerard, R.; Tuma, P.L.; Cohen, J.C.; Hobbs, H.H. ABCG5 and ABCG8 Are Obligate Heterodimers for Protein Trafficking and Biliary Cholesterol Excretion. *J. Biol. Chem.* **2003**, *278*, 48275–48282. [CrossRef]

11. Bhattacharyya, A.K.; Connor, W.E.; Lin, D.S.; McMurry, M.M.; Shulman, R.S. Sluggish sitosterol turnover and hepatic failure to excrete sitosterol into bile cause expansion of body pool of sitosterol in patients with sitosterolemia and xanthomatosis. *Arterioscler. Thromb. J. Vasc. Biol.* **1991**, *11*, 1287–1294. [CrossRef] [PubMed]

12. Yu, L.; Hammer, R.E.; Li-Hawkins, J.; Von Bergmann, K.; Lutjohann, D.; Cohen, J.C.; Hobbs, H.H. Disruption of Abcg5 and Abcg8 in mice reveals their crucial role in biliary cholesterol secretion. *Proc. Natl. Acad. Sci. USA* **2002**, *99*, 16237–16242. [CrossRef] [PubMed]

13. Klett, E.L.; Lu, K.; Kosters, A.; Vink, E.; Lee, M.-H.H.; Altenburg, M.; Shefer, S.; Batta, A.K.; Yu, H.; Chen, J.; et al. A mouse model of sitosterolemia: Absence of Abcg8/sterolin-2 results in failure to secrete biliary cholesterol. *BMC Med.* **2004**, *2*, 5. [CrossRef]

14. Plösch, T.; Bloks, V.W.; Terasawa, Y.; Berdy, S.; Siegler, K.; Van Der Sluijs, F.; Kema, I.P.; Groen, A.K.; Shan, B.; Kuipers, F.; et al. Sitosterolemia in ABC-Transporter G5-deficient mice is aggravated on activation of the liver-X receptor. *Gastroenterology* **2004**, *126*, 290–300. [CrossRef] [PubMed]

15. Jakulj, L.; van Dijk, T.H.; de Boer, J.F.; Kootte, R.S.; Schonewille, M.; Paalvast, Y.; Boer, T.; Bloks, V.W.; Boverhof, R.; Nieuwdorp, M.; et al. Transintestinal Cholesterol Transport Is Active in Mice and Humans and Controls Ezetimibe-Induced Fecal Neutral Sterol Excretion. *Cell Metab.* **2016**, *24*, 783–794. [CrossRef]

16. Ford, R.C.; Beis, K. Learning the ABCs one at a time: Structure and mechanism of ABC transporters. *Biochem. Soc. Trans.* **2019**, *47*, 23–36. [CrossRef]

17. Takahashi, K.; Kimura, Y.; Kioka, N.; Matsuo, M.; Ueda, K. Purification and ATPase activity of human ABCA1. *J. Biol. Chem.* **2006**, *281*, 10760–10768. [CrossRef]

18. Zhang, D.W.; Graf, G.A.; Gerard, R.D.; Cohen, J.C.; Hobbs, H.H. Functional asymmetry of nucleotide-binding domains in ABCG5 and ABCG8. *J. Biol. Chem.* **2006**, *281*, 4507–4516. [CrossRef]

19. Hirayama, H.; Kimura, Y.; Kioka, N.; Matsuo, M.; Ueda, K. ATPase activity of human ABCG1 is stimulated by cholesterol and sphingomyelin. *J. Lipid Res.* **2013**, *54*, 496–502. [CrossRef]

20. Wang, J.; Grishin, N.; Kinch, L.; Cohen, J.C.; Hobbs, H.H.; Xie, X.S. Sequences in the nonconsensus nucleotide-binding domain of ABCG5/ABCG8 required for sterol transport. *J. Biol. Chem.* **2011**, *286*, 7308–7314. [CrossRef]

21. Lee, J.-Y.Y.; Kinch, L.N.; Borek, D.M.; Wang, J.J.; Wang, J.J.; Urbatsch, I.L.; Xie, X.-S.S.; Grishin, N.V.; Cohen, J.C.; Otwinowski, Z.; et al. Crystal structure of the human sterol transporter ABCG5/ABCG8. *Nature* **2016**, *533*, 561–564. [CrossRef] [PubMed]

22. Zein, A.A.; Kaur, R.; Hussein, T.O.K.; Graf, G.A.; Lee, J.Y. ABCG5/G8: A structural view to pathophysiology of the hepatobiliary cholesterol secretion. *Biochem. Soc. Trans.* **2019**, *47*, 1259–1268. [CrossRef] [PubMed]

23. Berge, K.E.; Tian, H.; Graf, G.A.; Yu, L.; Grishin, N.V.; Schultz, J.; Kwiterovich, P.; Shan, B.; Barnes, R.; Hobbs, H.H. Accumulation of dietary cholesterol in sitosterolemia caused by mutations in adjacent ABC transporters. *Science* **2000**, *290*, 1771–1775. [CrossRef] [PubMed]

24. Lu, K.; Lee, M.H.; Hazard, S.; Brooks-Wilson, A.; Hidaka, H.; Kojima, H.; Ose, L.; Stalenhoef, A.F.H.; Mietinnen, T.; Bjorkhem, I.; et al. Two genes that map to the STSL locus cause sitosterolemia: Genomic structure and spectrum of mutations involving sterolin-1 and sterolin-2, encoded by ABCG5 and ABCG8, respectively. *Am. J. Hum. Genet.* **2001**, *69*, 278–290. [CrossRef] [PubMed]

25. Buch, S.; Schafmayer, C.; Völzke, H.; Becker, C.; Franke, A.; Von Eller-Eberstein, H.; Kluck, C.; Bässmann, I.; Brosch, M.; Lammert, F.; et al. A genome-wide association scan identifies the hepatic cholesterol transporter ABCG8 as a susceptibility factor for human gallstone disease. *Nat. Genet.* **2007**, *39*, 995–999. [CrossRef]

26. Kuo, K.-K.; Shin, S.-J.; Chen, Z.-C.; Yang, Y.-H.; Yang, J.F.; Hsiao, P.J. Significant association of ABCG5 604Q and ABCG8 D19H polymorphisms with gallstone disease. *Br. J. Surg.* **2008**, *95*, 1005–1011. [CrossRef]

27. Chen, Z.C.; Shin, S.J.; Kuo, K.K.; Lin, K.D.; Yu, M.L.; Hsiao, P.J. Significant association of ABCG8:D19H gene polymorphism with hypercholesterolemia and insulin resistance. *J. Hum. Genet.* **2008**, *53*, 757–763. [CrossRef]

28. Kajinami, K.; Brousseau, M.E.; Ordovas, J.M.; Schaefer, E.J. Interactions between common genetic polymorphisms in ABCG5/G8 and CYP7A1 on LDL cholesterol-lowering response to atorvastatin. *Atherosclerosis* **2004**, *175*, 287–293. [CrossRef]

29. Graf, G.A.; Cohen, J.C.; Hobbs, H.H. Missense mutations in ABCG5 and ABCG8 disrupt heterodimerization and trafficking. *J. Biol. Chem.* **2004**, *279*, 24881–24888. [CrossRef]

30. Lukacs, G.L.; Mohamed, A.; Kartner, N.; Chang, X.B.; Riordan, J.R.; Grinstein, S. Conformational maturation of CFTR but not its mutant counterpart (ΔF508) occurs in the endoplasmic reticulum and requires ATP. *EMBO J.* **1994**, *13*, 6076–6086. [CrossRef]

31. Kizhakkedath, P.; John, A.; Al-Sawafi, B.K.; Al-Gazali, L.; Ali, B.R. Endoplasmic reticulum quality control of LDLR variants associated with familial hypercholesterolemia. *FEBS Open Bio* **2019**, *9*, 1994–2005. [CrossRef] [PubMed]

32. Müller, M.; Klein, I.; Kopácsi, S.; Remaley, A.T.; Rajnavölgyi, E.; Sarkadi, B.; Váradi, A. Co-expression of human ABCG5 and ABCG8 in insect cells generates an androstan stimulated membrane ATPase activity. *FEBS Lett.* **2006**, *580*, 6139–6144. [CrossRef] [PubMed]

33. Johnson, B.J.H.; Lee, J.Y.; Pickert, A.; Urbatsch, I.L. Bile acids stimulate atp hydrolysis in the purified cholesterol transporter ABCG5/G8. *Biochemistry* **2010**, *49*, 3403–3411. [CrossRef] [PubMed]

34. Sarkadi, B.; Price, E.M.; Boucher, R.C.; Germann, U.A.; Scarborough, G.A. Expression of the human multidrug resistance cDNA in insect cells generates a high activity drug-stimulated membrane ATPase. *J. Biol. Chem.* **1992**, *267*, 4854–4858.

35. Cariani, L.; Thomas, L.; Brito, J.; Del Castillo, J.R. Bismuth citrate in the quantification of inorganic phosphate and its utility in the determination of membrane-bound phosphatases. *Anal. Biochem.* **2004**, *324*, 79–83. [CrossRef]

36. Davies, D.R.; Hol, W.G.J. The power of vanadate in crystallographic investigations of phosphoryl transfer enzymes. *FEBS Lett.* **2004**, *577*, 315–321. [CrossRef]

37. Khunweeraphong, N.; Mitchell-White, J.; Szöllősi, D.; Hussein, T.; Kuchler, K.; Kerr, I.D.; Stockner, T.; Lee, J. Picky ABCG5/G8 and promiscuous ABCG2—A tale of fatty diets and drug toxicity. *FEBS Lett.* **2020**. [CrossRef]

38. Manolaridis, I.; Jackson, S.M.; Taylor, N.M.I.; Kowal, J.; Stahlberg, H.; Locher, K.P. Cryo-EM structures of a human ABCG2 mutant trapped in ATP-bound and substrate-bound states. *Nature* **2018**, *563*, 426–430. [CrossRef]

39. Hanson, M.A.; Cherezov, V.; Griffith, M.T.; Roth, C.B.; Jaakola, V.P.; Chien, E.Y.T.; Velasquez, J.; Kuhn, P.; Stevens, R.C. A Specific Cholesterol Binding Site Is Established by the 2.8 Å Structure of the Human β2-Adrenergic Receptor. *Structure* **2008**, *16*, 897–905. [CrossRef]

40. Neumann, J.; Rose-Sperling, D.; Hellmich, U.A. Diverse relations between ABC transporters and lipids: An overview. *Biochim. Biophys. Acta Biomembr.* **2017**, *1859*, 605–618. [CrossRef]

41. Kurumi, Y.; Adachi, Y.; Itoh, T.; Kobayashi, H.; Nanno, T.; Yamamoto, T. Novel high-performance liquid chromatography for determination of membrane phospholipid composition of rat hepatocytes. *Gastroenterol. Jpn.* **1991**, *26*, 628–632. [CrossRef] [PubMed]

42. Hauser, H.; Howell, K.; Dawson, R.M.C.; Bowyer, D.E. Rabbit small intestinal brush border membrane. Preparation and lipid composition. *Biochim. Biophys. Acta Biomembr.* **1980**, *602*, 567–577. [CrossRef]

43. Lee, A.G. How lipids affect the activities of integral membrane proteins. *Biochim. Biophys. Acta Biomembr.* **2004**, *1666*, 62–87. [CrossRef] [PubMed]

44. Feldblum, E.S.; Arkin, I.T. Strength of a bifurcated H bond. *Proc. Natl. Acad. Sci. USA* **2014**, *111*, 4085–4090. [CrossRef]

45. Wang, Z.; Stalcup, L.D.; Harvey, B.J.; Weber, J.; Chloupkova, M.; Dumont, M.E.; Dean, M.; Urbatsch, I.L. Purification and ATP hydrolysis of the putative cholesterol transporters ABCG5 and ABCG8. *Biochemistry* **2006**, *45*, 9929–9939. [CrossRef]

46. Jorgensen, W.L.; Chandrasekhar, J.; Madura, J.D.; Impey, R.W.; Klein, M.L. Comparison of simple potential functions for simulating liquid water. *J. Chem. Phys.* **1983**, *79*, 926–935. [CrossRef]

47. Maier, J.A.; Martinez, C.; Kasavajhala, K.; Wickstrom, L.; Hauser, K.E.; Simmerling, C. ff14SB: Improving the Accuracy of Protein Side Chain and Backbone Parameters from ff99SB. *J. Chem. Theory Comput.* **2015**, *11*, 3696–3713. [CrossRef]

48. Dickson, C.J.; Madej, B.D.; Skjevik, Å.A.; Betz, R.M.; Teigen, K.; Gould, I.R.; Walker, R.C. Lipid14: The amber lipid force field. *J. Chem. Theory Comput.* **2014**, *10*, 865–879. [CrossRef]

49. Wang, J.; Wolf, R.M.; Caldwell, J.W.; Kollman, P.A.; Case, D.A. Development and testing of a general Amber force field. *J. Comput. Chem.* **2004**, *25*, 1157–1174. [CrossRef]

50. Case, D.A.; Ben-Shalom, I.Y.; Brozell, S.R.; Cerutti, D.S.; Cheatham, I.T.E.; Cruzeiro, V.W.D.; Darden, T.A.; Duke, R.E.; Ghoreishi, D.; Gilson, M.K.; et al. *AMBER 2018*; University of California: San Francisco, CA, USA, 2018.

51. Darden, T.; Perera, L.; Li, L.; Lee, P.; Pedersen, L. New tricks for modelers from the crystallography toolkit: The particle mesh Ewald algorithm and its use in nucleic acid simulations. *Structure* **1999**, *7*, R55–R60. [CrossRef]

52. Miyamoto, S.; Kollman, P.A. Settle: An analytical version of the SHAKE and RATTLE algorithm for rigid water models. *J. Comput. Chem.* **1992**, *13*, 952–962. [CrossRef]

53. Kong, Y.; Karplus, M. Signaling pathways of PDZ2 domain: A molecular dynamics interaction correlation analysis. *Proteins* **2009**, *74*, 145–154. [CrossRef] [PubMed]

54. Kim, P.; Li, H.; Wang, J.; Zhao, Z. Landscape of drug-resistance mutations in kinase regulatory hotspots. *Brief. Bioinform.* **2020**. [CrossRef] [PubMed]

Publisher's Note: MDPI stays neutral with regard to jurisdictional claims in published maps and institutional affiliations.

International Journal of
Molecular Sciences

Review

The Bile Salt Export Pump: Molecular Structure, Study Models and Small-Molecule Drugs for the Treatment of Inherited BSEP Deficiencies

Muhammad Imran Sohail [1,†], Yaprak Dönmez-Cakil [2,†], Dániel Szöllősi [3], Thomas Stockner [3,*] and Peter Chiba [4,*]

[1] Department of Zoology, Government College University, Lahore 54000, Pakistan; imransohail@gcu.edu.pk
[2] Department of Histology and Embryology, Faculty of Medicine, Maltepe University, Maltepe, 34857 Istanbul, Turkey; yaprak.cakil@maltepe.edu.tr
[3] Institute of Pharmacology, Center for Physiology and Pharmacology, Medical University of Vienna, Waehringerstrasse, 13A, 1090 Vienna, Austria; daniel.szoelloesi@meduniwien.ac.at
[4] Institute of Medical Chemistry, Center for Pathobiochemistry and Genetics, Medical University of Vienna, Waehringerstrasse, 10, 1090 Vienna, Austria
* Correspondence: thomas.stockner@meduniwien.ac.at (T.S.); peter.chiba@meduniwien.ac.at (P.C.); Tel.: +43-1-40160-31215 (T.S.); +43-1-40160-38005 (P.C.)
† Equally contributing first authors.

Abstract: The bile salt export pump (BSEP/ABCB11) is responsible for the transport of bile salts from hepatocytes into bile canaliculi. Malfunction of this transporter results in progressive familial intrahepatic cholestasis type 2 (PFIC2), benign recurrent intrahepatic cholestasis type 2 (BRIC2) and intrahepatic cholestasis of pregnancy (ICP). Over the past few years, several small molecular weight compounds have been identified, which hold the potential to treat these genetic diseases (chaperones and potentiators). As the treatment response is mutation-specific, genetic analysis of the patients and their families is required. Furthermore, some of the mutations are refractory to therapy, with the only remaining treatment option being liver transplantation. In this review, we will focus on the molecular structure of ABCB11, reported mutations involved in cholestasis and current treatment options for inherited BSEP deficiencies.

Keywords: BSEP; ABCB11; bile salts; intrahepatic cholestasis; chaperones; PFIC2; BRIC

Citation: Sohail, M.I.; Dönmez-Cakil, Y.; Szöllősi, D.; Stockner, T.; Chiba, P. The Bile Salt Export Pump: Molecular Structure, Study Models and Small-Molecule Drugs for the Treatment of Inherited BSEP Deficiencies. *Int. J. Mol. Sci.* **2021**, *22*, 784. https://doi.org/10.3390/ijms22020784

Received: 29 December 2020
Accepted: 12 January 2021
Published: 14 January 2021

Publisher's Note: MDPI stays neutral with regard to jurisdictional claims in published maps and institutional affiliations.

1. Introduction

The ATP-binding cassette (ABC) proteins constitute one of the largest families of membrane proteins. They are universally present in all kingdoms of life. In humans, 48 functional genes encode for ABC proteins, which on the basis of structural and sequence similarity are categorized into seven subfamilies, designated as ABCA through G [1]. Most of these proteins transport substrates across cellular membranes. A functional ABC transporter comprises at least four domains: two transmembrane domains (TMDs) and two nucleotide-binding domains (NBDs), as shown in Figure 1. In the ABCB subfamily, each of the TMDs consists of six membrane-spanning helices. Five of these extend into the cytoplasm to form an expansive intracellular domain. The two TMDs are responsible for substrate binding and translocation. The NBDs form two composite nucleotide-binding sites (NBSs) at their interface, which bind and hydrolyze ATP, and thereby provide the energy for substrate transport. These NBSs are formed by the Walker A and Walker B motifs, as well as the A-, Q- and H-loops of one NBD and the signature motif and D-loop of the other NBD [2,3].

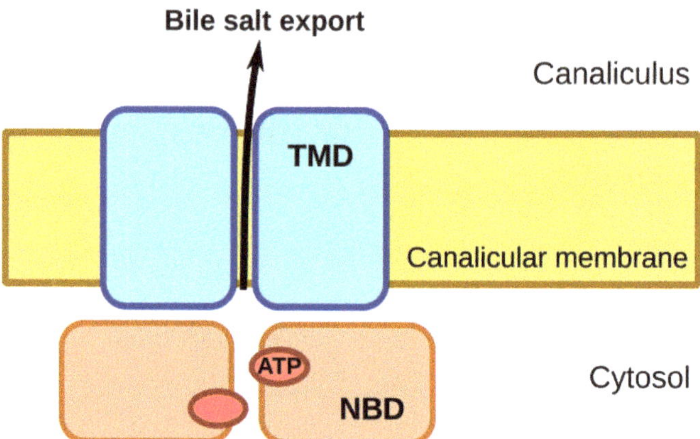

Figure 1. Schematic of the four-domain architecture of an ATP-binding cassette (ABC) transporter, as illustrated for the bile salt export pump BSEP. Hydrolysis of ATP by the nucleotide binding domains (NBD) energizes transport of bile salts from the hepatocyte cytoplasm to the bile canaliculus. Transmembrane domains (TMD) form the path for bile salt transport across the canalicular membrane.

The ABCB subfamily is one of the most diverse groups of ABC proteins, as it includes dimeric half transporters, monomers of which are each composed of one TMD and one NBD, but also full-length transporters, in which all four domains are fused into a single polypeptide chain. The former group comprises the homodimeric transporters ABCB6, ABCB7, ABCB8, ABCB9 and ABCB10, and the heterodimeric transporter ABCB2/ABCB3. The four full-length transporters of the ABCB subfamily are ABCB1, ABCB4, ABCB5 and ABCB11. The heterodimeric ABCB2/ABCB3 and full-length ABCB11 differ from the other members as they contain only one canonical NBS rather than two [4].

2. BSEP/ABCB11: Physiological Role

The bile salt export pump (BSEP) is expressed in hepatocytes. While high levels of BSEP mRNA were detected in testes and lower levels were reported in other extrahepatic tissues, including trachea, prostate, lungs, thymus, kidney and colon [5,6], plasma membrane expression of functional protein was found in liver cells only [5–8]. Adjacent hepatocytes form tight junctions to enclose functional structures called bile canaliculi, to which BSEP is targeted. Bile salts undergo an enterohepatic circulation, which depends on active transport systems in the liver and intestine. In the course of this process, newly synthesized and recycled bile salts are secreted from hepatocytes into bile canaliculi by BSEP and via bile ducts reach the duodenum. In the ileum, these bile salts are reabsorbed by the apical sodium-dependent bile salt transporter in intestinal epithelial cells (ASBT/SLC10A2). From the intestine, bile salts return to the liver via the superior mesenteric and portal veins, which carry the blood that feeds liver sinusoids. Uptake into hepatocytes is mediated by the sodium taurocholate co-transporting polypeptide (NTCP/SLC10A1) and organic anion transporters (OATPs). Bile salt transport by BSEP constitutes the rate-limiting step in bile formation and provides the major driving force for enterohepatic circulation [9].

The bile salt pool is recycled from the intestine to the liver six to eight times a day [10], resulting in daily bile salt excretion of about 20–40 g [11]. Impairment of BSEP results in the failure to maintain physiological bile flow, resulting in a clinical condition called intrahepatic cholestasis. BSEP has narrow specificity for its substrate bile salts, but the drugs pravastatin, vinblastine and fexofenadine are reported to be non-physiological substrates [12–14].

3. Transcriptional Regulation

BSEP expression is highly regulated by transcriptional mechanisms, and a wide inter-individual variability has been described at the mRNA and protein levels [15].

Expression of BSEP is regulated by a major ligand-activated transcription factor, farnesoid X receptor (FXR, NR1H4), which forms a signaling-competent nuclear receptor heterodimer with the retinoid X receptor (RXR) (Figure 2). Bile acids, such as chenodeoxycholic acid (CDCA), deoxycholic acid (DCA) and cholic acid (CA), are endogenous ligands of FXR with varying potential for activation [16–19]. Upon ligand binding, the FXR/RXR heterodimer binds to an FXR response element (FXRE) in the promoter region of BSEP, thereby inducing the expression of the transporter [20]. Additionally, components of the activating signal cointegrator-2-containing complex (ASCOM) interact with FXR to enhance BSEP expression. Ananthanarayanan and co-workers [21] showed that the recruitment of ASCOM to the BSEP promoter was disrupted in cholestasis, which was induced by common bile duct ligation. Furthermore, co-activator-associated arginine methyltransferase 1 (CARM1) also regulates FXR/RXR-dependent BSEP transcription [22]. Similarly, steroid receptor co-activator 2 (SRC2) knockout mice showed reduced expression of BSEP [23], indicating its involvement in transcriptional regulation of the transporter.

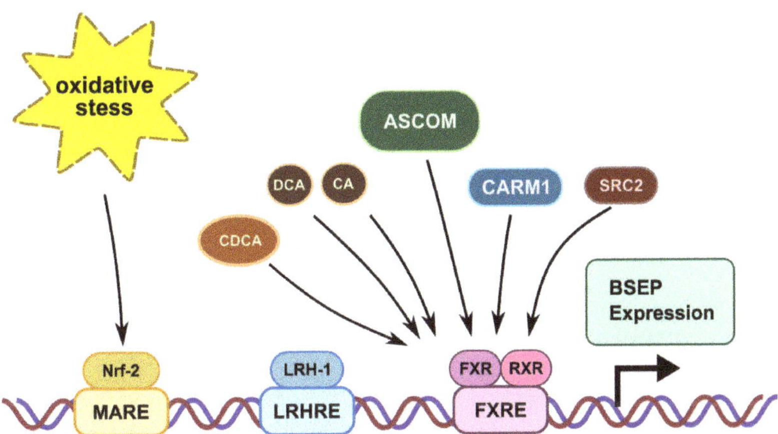

Figure 2. Transcriptional regulation of BSEP via recognition sequences in the promoter region of the BSEP gene: MARE: Maf recognition element; LRHRE: liver receptor homolog-1 responsive element; FXRE: farnesoid X receptor responsive element. The farnesoid receptor FXR binds bile salts after heterodimerization with the retinoid X receptor. The ligand with the highest affinity for FXR is chenodeoxycholic acid (CDCA): however deoxycholic (DCA) and cholic acid (CA) also increase BSEP expression. The co-activators 'activating signal cointegrator-2-containing complex' (ASCOM), 'co-activator-associated arginine methyltransferase 1' (CARM1) and 'steroid receptor co-activator SRC2' (SRC2) increase BSEP expression via FXR. Liver receptor homolog 1 (LHR-1) and nuclear factor erythroid 2-related factor (Nrf-2), a sensor for oxidative stress, also increase BSEP expression.

Hepatocyte-specific liver receptor homolog-1 (LRH-1, NR5A2) is another transcription factor involved in modulation of BSEP expression. LRH-1 plays a supporting role for FXR [24]. The absence of LRH-1 is associated with reduced BSEP expression and an altered BA composition, with disappearance of CA and taurocholic acid (TCA) [25]. BSEP promoter activity is also stimulated by nuclear factor erythroid 2-related factor 2 (Nrf2), a positive transcriptional regulator, which acts as a sensor for oxidative stress. Nrf2 regulates the expression of BSEP, but also that of a number of hepatic phase I and II enzymes and other hepatic efflux transporters such as MRP3 (ABCC3) and MRP4 (ABCC4) [26].

4. Processing and Trafficking of BSEP

Membrane insertion and folding occur at the level of the endoplasmic reticulum (ER) [27]. Insertion into the ER membrane is facilitated by the protein transport protein SEC61, which assists transmembrane portions of nascent proteins to adopt helicity prior to domain folding. Correct positioning of domains or subdomains relative to each other typically occurs late in the folding trajectory of a multidomain membrane protein. Of all ABC proteins, the folding trajectory of cystic fibrosis transmembrane conductance regulator (CFTR, ABCC7) has been studied the most [28,29]. The intracellular loops (ICLs) play a critical role in transporter folding by contributing to the formation of the functionally important TMD/NBD coupling interface [30]. Furthermore, the involvement of molecular chaperons is required, as they sense the presence of hydrophobic helices in the cytosol, and thus contribute to obtaining the folding endpoint [31].

In the ER, newly synthesized and correctly folded BSEP undergoes N-linked core glycosylation. The sugar moieties are added at four conserved asparagine residues in extracellular loop 1 (ECL1), namely Asn109, 116, 122 and 125, and then are subject to subsequent modifications while traveling through the Golgi stacks. N-linked core glycosylation in the ER lumen plays a pivotal role in ER protein folding by mediating interactions with the lectin chaperones calnexin and calreticulin and by increasing the folding efficiency [32]. Only correctly folded proteins are trafficked to the Golgi apparatus in clathrin-coated COPII vesicles. Aberrantly folded proteins are identified by the endoplasmic-reticulum-associated degradation (ERAD) machinery and retro-translocated to the cytoplasm for degradation in the 26S proteasome following ubiquitination [33]. A number of BSEP mutants, including G238V, D482G, G982R, R1153C and R1286Q, are predominantly degraded by ERAD, thus leading to a PFIC2 phenotype [27,34,35]. Different ERAD E3 ubiquitin ligases are thought to recognize and ubiquitinate different mutants of BSEP [27]. Before trafficking to the canalicular membrane, BSEP is fully glycosylated in the Golgi apparatus through trimming to the core structure and extension from the core [34]. Glycosylation directly impacts protein stability and at least two of the four glycans are required for BSEP trafficking to the canalicular membrane [35]. Using enhanced green fluorescent protein (EGFP)-tagged mouse BSEP, it was shown that the partial glycosylation of the PFIC2-related mutant D482G causes an unstable BSEP protein and reduces levels of the mature protein at the canalicular membrane [36].

The majority of integral plasma membrane proteins of polarized hepatic cells are distributed from the basolateral membrane to the appropriate apical cell surface location via transcytosis. In contrast, ABC transporters targeted to the canalicular membrane use the non-transcytotic direct route from the Golgi apparatus via Rab11a-positive apical endosomes [37,38]. Under physiological conditions, the apical pool of BSEP is strictly regulated by the demand for biliary excretion of bile salts. The intracellular endosomal pool is thought to exceed that at the canalicular membrane [39] by at least 6-fold. Internalization of BSEP is mediated by clathrin-coated vesicles and is dependent on the highly conserved endocytic cargo motif (Trp-Lys-Leu-Val) [40]. This trafficking motif is recognized by adaptor protein 2 (AP-2), which modulates the internalization process and expression of cell-surface-resident BSEP through direct interaction [41]. Moreover, trafficking of BSEP through the endosomal system to the canalicular membrane is a microtubule-dependent process and requires the myosin light chain [42], myosin Vb [43] and Rab11a. The latter two components were shown to also be associated with canalicular biogenesis by maintaining proper trafficking of Rab11a–myosin Vb-containing membranes to the canalicular membrane in polarized WIF-B9 cells [43].

Continuous cycling of BSEP between the apical and intracellular pools is disrupted in most human cholestatic liver diseases. Shifting the balance towards endocytic internalization results in impaired bile salt secretion [44]. A causative role of enhanced retrieval into the subapical endosomal compartment was demonstrated for estradiol 17 β-D-glucuronide (E17G)-induced cholestasis, an experimental model for pregnancy-related cholestasis [45,46]. In this model, BSEP was found to co-localize with clathrin, AP-2 and Rab5 as evidence for clathrin-mediated endocytosis [46]. Classical (Ca^{2+}-dependent) protein kinase C (cPKC)–p38, mitogen-activated protein kinase (MAPK) and phosphoinositide 3-kinase (PI3K)–ERK1/2 signaling pathways are thought to be involved in (E17G)-induced cholestasis [47–49].

TCA, the major bile acid in mammals, as well as cyclic adenosine monophosphate (cAMP) are known to increase the apical pool of BSEP within minutes by promoting its cellular relocation [37]. Moreover, TCA was demonstrated to induce the formation of bile canaliculi in mice via the liver kinase B1 (LKB1)–AMP-activated protein kinase (AMPK) pathway [50]. A subsequent publication showed that knocking out LKB1, the upstream serine–threonine kinase, which is implicated in regulation of cellular energy metabolism, impairs both canalicular biogenesis and intracellular trafficking of BSEP. On the other hand, cAMP induces BSEP trafficking through a PKA-mediated pathway, which does not involve AMPK activation [51]. Unlike TCA-mediated trafficking, this process is PI3K-independent [52]. Similar to TCA, the conjugated bile salt tauroursodeoxycholate (TUDCA) also promotes the relocation of BSEP to the canalicular membrane through activation of the p38 MAPK [53,54]. Similar to other conjugated bile salts, TUDCA stimulates the ATPase activity of BSEP [55].

Ubiquitination is another modification, which changes the expression of cell-surface-resident BSEP. The half-life of BSEP in the canaliculi is shortened by modification with two to three ubiquitin molecules. This induces the removal of the protein from the cell surface, whereby the rates are governed by the degree of ubiquitination. While the PFIC2-related mutations E297G and D482G cause short-chain ubiquitination, thereby shortening the half-life of cell-surface-resident BSEP, the chemical chaperone 4-phenylbutyrate (4-PB) reduces its degradation rate [56]. In a later study, ubiquitination of canalicular BSEP was shown to act as a signal for internalization by promoting clathrin-mediated endocytosis. After internalization, BSEP is either recycled back to the canalicular membrane in a Rab11-dependent manner or degraded through a ubiquitination-independent pathway [57]. Degradation was suggested to be lysosome-mediated and dependent on a sorting signal from within the endosomal compartment [58].

5. Structural Models of BSEP

Before publication of the cryo-electron microscopy (cryo-EM) structure of BSEP [59], several homology models of the transporter were presented. Kubitz and colleagues [60,61] generated an outward-facing homology model of ABCB11 by using the Sav1866 structure (PDB: 2HYD [62]) as a template. This model was used to show a putative antibody binding site at the long ECL1, as well as to indicate positions of disease-causing mutations. The model compares well at the individual domain level with the cryo-EM structure that has recently become available. Giovannoni et al. [63] created a model based on the corrected mouse ABCB1 structure (PDBID: 4M1M [64]). Here, loops that could not be matched to the template were not modeled (ABCB11 residues 102–120 and 659–728). The model was used to localize positions of disease-causing mutations. Again, this model agrees well with the cryo-EM structure within the resolution at which cryo-EM data were presented. Notably, an overall structural alignment of mouse ABCB1 (PDBID: 4M1M [64]) and the cryo-EM structure of ABCB11 results in an RMSD of 0.36 nm, with a better fit of the TMDs. Dröge et al. [65] also used the mouse ABCB1 structure as a template. This model was used to locate the positions of the most commonly occurring PFIC2 missense mutations. Moreover, in the process of structure evaluation, different web services were used to predict the influence of missense mutations on protein function. The authors provided a list of possible effects of newly identified mutations included in their study. However, the accuracy of this prediction was not evaluated. In a different study, Jain and co-workers [66] generated an ABCB11 homology model also using the inward-facing mouse ABCB1 structure as the template (PDB: 4M1M [64]). Extensive docking studies with 405 inhibitor compounds and 807 non-inhibitors were performed in order to explore the interaction with small molecules. Prediction accuracy results of 81% in the training set and 73% in two external test sets were obtained. In addition to standard scoring functions, the homology model used for docking was validated by molecular dynamics (MD) simulation. The use of MD simulations in protein stability checks is a well-established procedure, however it is computationally expensive. Short simulations were performed for structure validation of membrane-inserted ABCB11 homology models. We also previously [67] generated a homology model based on the X-ray structure of Sav1866 as a template (PDB ID: 2ONJ [62]) to elucidate the NBD–NBD interdomain communication of the transporter. The model allowed us to infer the potential roles of conserved motifs of the nucleotide-binding domains in ATP hydrolysis and the transmission of conformational changes from the NBDs to the TMDs.

In 2020, Wang et al. [59] determined the human BSEP structure using cryo-EM (PDBID: 6LR0). This structure has an average resolution of 0.35 nm, with the TMDs being resolved at 0.33 nm. The protein shows an inward open state in the absence of nucleotides or other small molecules. According to this structure, ABCB11 closely resembles ABCB1, with the most similar structure found in the Protein Data Bank [68] being the apo inward-open ABCB1 structure (PDBID: 6GDI, RMSD: 0.206 nm [69]. Thus, it shares the typical type I exporter fold with the other members of the ABCB subfamily. The domain-swapped transmembrane helices are connected by coupling helices 2 and 4, which are embedded in grooves formed between the core and the helical domain of the NBDs, supporting their crucial role in interdomain communication. Interestingly, inside the central cavity, contiguous electron density was found, into which the N-terminus of the protein could be fitted. Currently, no biochemical data are available with respect to any putative auto-inhibition of BSEP by this N-terminus. The poor resolution of ECL1 (Q101-I134) reflects its highly dynamic nature. It contains the four known glycosylation sites (N109, 116, 122 and 125) [70].

BSEP harbors two ATP-binding sites, one of which is canonical and capable of ATP hydrolysis. When compared to ABCB1, only four amino acids differ in NBS1. These are E502, M584, R1221 and E1223 in BSEP corresponding to S474, E556, G1178 and Q1180 in the ABCB1 protein sequence [71]. The amino acid changes result in a catalytically inactive ATP binding site (NBS1), which is also variably called "degenerate" NBS. It has been suggested that the degenerate site imparts extended functionality to the transporter. The mechanistic details are currently missing, although the role of ATP hydrolysis in each of the two NBDs has been elucidated in greater detail [67].

6. Experimental Model Systems

In vitro and in vivo models have been developed for the study of BSEP function, folding and trafficking, as well as the actions of drug candidates with the potential to treat the malfunction or incorrect cellular routing of the transporter. These model systems are discussed below with respect to their potentials and limitations.

6.1. In Vitro Models

6.1.1. Membrane Vesicles

For the study of substrate transport and inhibition by drugs and metabolites, membrane vesicles represent the most commonly used model system. Vesicles are either prepared from BSEP-transfected insect cell lines (Sf9 and Sf21), which give higher protein yields, or mammalian cell lines (including CHO, HeLa, MDCK, LLC-PK1 and HEK cells). Despite having lower protein expression, mammalian cells are often preferred for functional studies, as insect cells show a different lipid membrane composition and only core glycosylated protein is produced. In order to overcome lower expression levels in mammalian cells, the Bac/Mam gene transfer system has been advocated [72]. For experimental details on the preparation of membrane vesicles, readers are referred to [13,69,73]. As a mixture of inside-out and right-side-out vesicles is obtained, a protocol for increasing the yield of inside-out vesicles has been published [74]. In addition, a protocol for preparation of membrane vesicles from canalicular membranes of rat hepatocytes has been reported [73]. These vesicles were used for the identification and characterization of BSEP substrates and inhibitors [75,76].

High-quality membrane vesicles represent an ideal experimental system for transport and transport inhibition studies, but cannot be used to address aspects of trafficking or cellular metabolism [77]. As in inside-out vesicles, the BSEP NBDs are exposed towards the medium, ATP and substrates can be added into the transport medium and accumulation of substrates in vesicles can be monitored by rapid filtration. We previously used membrane vesicles from plasmid-transfected HEK cells to study the domain interaction and the roles of the canonical and non-canonical NBS in supporting BSEP substrate transport [67].

6.1.2. Polarized Cell Lines Expressing BSEP

Polarized MDCK and LLC-PK1 cells have been used to study BSEP function in intact cells. The experimental system requires double transfection with BSEP and the hepatocyte bile salt uptake transporter NTCP, as BSEP-mediated efflux can only be monitored after bile salt substrates have been taken-up into cells. Physiologically, NTCP enables the reentry of bile salts from the circulation into hepatocytes in the context of the enterohepatic circulation of these compounds between the intestine and the liver. Polarized cells are grown on a permeable membrane in a hang-in assembly. BSEP is localized on the apical surface. The function of BSEP is determined from the ratio of basal-to-apical as compared to apical-to-basal transport of substrates [78,79].

6.1.3. Primary Hepatocyte Cultures

When cultured in the appropriate medium, hepatocytes form tight junctions to generate sealed tube-like structures, which resemble bile canaliculi [80,81]. Therefore, this system provides the possibility of assessing the excretion of drugs and bile components into bile canaliculi [82–84]. Hepatocytes in suspension also provide a suitable option for studying drug metabolism and drug transport [85,86]. One major advantage of hepatocyte suspension cultures is their easy, quick and high-yield preparation. Furthermore, this assay does not require radiolabeled substrates [87]. Therefore, this system is often used for large-scale screening of drugs for assessment of drug-induced liver injury (DILI). It has to be kept in mind, however, that the presence of multiple transporters in primary hepatocytes limits their use for assessing BSEP-specific substrates [88].

6.2. BSEP Knockout Animals

6.2.1. Rodents

In order to investigate the mechanisms involved in innate and acquired intrahepatic cholestasis, BSEP knockout animal models (mice and rat) have been established [89,90]. Recently, the CRISPR/cas9 technology has been employed to knock out the BSEP gene in adult mice [91,92]. In all models, expression of BSEP was strongly reduced, thus providing an alternative experimental model for studying intrahepatic cholestasis and putative therapeutic intervention in rodents. Interestingly, the mouse models do not show signs of severe cholestasis as seen in humans, because these mice produce a large amount of poly-hydroxylated bile acids, which are excreted renally [93]. Wang et al. used this model system to suggest that P-glycoprotein (ABCB1) can act as a compensatory bile salt transporter, which alleviates the severity of cholestasis in BSEP knockout mice [94].

6.2.2. Zebrafish

Recently, Ellis et al. generated an abcb11b knockout zebrafish by using the CRISPR/Cas-9 gene editing technology [95]. Abcb11b is the orthologue of the human BSEP gene in zebrafish. The histological and ultrastructural analysis showed a morphological hepatocyte injury pattern similar to that seen in patients with PFIC2. Similar to the situation in humans, BSEP deficiency induced autophagy in zebrafish hepatocytes. Treatment with rapamycin restored bile acid excretion, attenuated hepatocyte damage and extended the life span of abcb11b mutant zebrafish. These effects were paralleled by a recovery of the correct canalicular localization of multidrug resistance protein 1.

Due to the transparency of these fish, the system allows monitoring of the bile flow in the intact animal with fluorescently labeled bile salt analogs. Furthermore, P-glycoprotein, which is reported to play a compensatory role (by transporting bile acids, and thereby protecting the hepatocytes from cholestasis induced injury) in BSEP knockout mice [96], is mislocalized to the hepatocyte cytoplasm in mutant zebrafish.

Animal models play an important role in studying cholestatic liver disease [93]. In contrast, identification of drug candidates for the treatment of folding and functionally deficient BSEP-mutants usually relies on in vitro model systems. For such studies, the patient-specific mutations are generated and expressed in cell lines. The impact of small molecules on the structure and function of BSEP is evaluated. Once the drugs have proven a potential in cell models, they are translated to a clinical setting [97,98].

Hydrodynamic tail vein injection in combination with the CRISPR/Cas9 technology has been used to specifically delete BSEP in mice and to study the consequences of the loss of an enzyme of the urea cycle (argininosuccinate lyase) [91]. In a similar way, such a BSEP knockout model system may be used for the study of mice expressing mutant forms of human BSEP in the liver and to monitor the effects of drug candidates on the folding, trafficking and function of the transporter.

7. Treatment Options for BSEP-Related Diseases

Impairment in the expression or function of BSEP leads to one of three human disease phenotypes of differing severity: PFIC2, BRIC2 and intrahepatic cholestasis of pregnancy (ICP). Several drugs with the potential to enhance the expression and function of the transporter have been reported. Disease-causing mutations and potential correctors are listed in Table 1 and depicted in the ABCB11 cryo-EM structure in Figure 3.

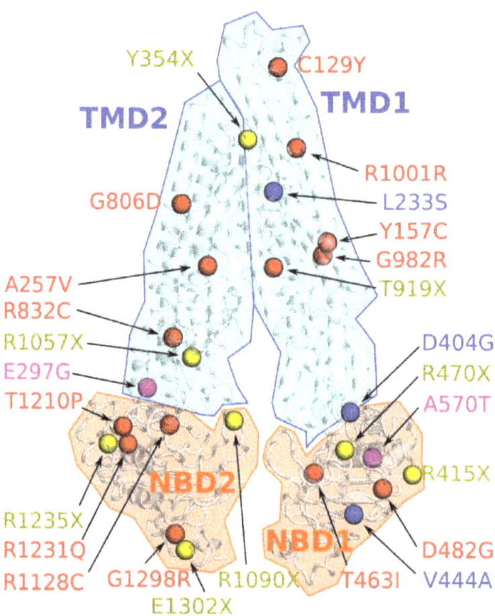

Figure 3. The positions of disease-causing mutations contained in Table 1 are shown in the ABCB11 cryo-EM structure. C-alpha atoms are shown as spheres and colored according to the resulting disease phenotype (PFIC2: red; BRIC2: blue; either PFIC2 or BRIC2: magenta; nonsense mutations: yellow). Outlines and filled colors indicate the domain organization of the transporter in accordance with Figure 1.

Table 1. Synopsis of a subset of disease-causing mutations, for which potential therapeutic interventions have been proposed. The defects, model systems and associated disease phenotypes are indicated. With the exception of the C129Y and G806D variants, the listed mutants represent a subset of a total of 192 disease-associated missense or nonsense mutations that have been identified to date [99]. PFIC2: progressive familial intrahepatic cholestasis type 2, BRIC2: benign recurrent intrahepatic cholestasis type 2,4-PB: 4-phenylbutyrate, CA: cholic acid, CDCA: chenodeoxycholic acid, DCA: deoxycholic acid, UDCA: ursodeoxycholic acid, FXR: farnesoid X receptor.

Nucleotide Change	Type of Mutation	Amino Acid	Defect	Potential Corrective Therapy	Cell Line/Organism	Disease	References
c.386GA	Missense	C129Y [1]	Impaired membrane trafficking, reduced level of mature protein	4-PB	HEK293T	PFIC2	[100]
c.470AG c.3892GA	Missense	Y157C G1298R	Reduced/absent BSEP activity	4-PB in combination with oxcarbazepine and maralixibat	Patient with 2 heterozygous missense mutations	PFIC2	[101]
c.698TC	Missense	L233S	-	Methylprednisolone	Patient with heterozygosity in ABCB11, as well as in CFTR, NPHP4 and A1ATD	BRIC2	[102]
c.890AG	Missense	E297G [2]	Protein instability, ubiquitin-dependent degradation [103], impaired membrane trafficking, reduced level of mature protein	4-PB	Madin-Darby canine kidney (MDCK) II cells and Sprague–Dawley rats	BRIC2, PFIC2	[104]
				Glycerol, glycerol at 28 °C	CHO-K1 cells		[105]
				CA, CDCA, DCA, UDCA, GW4064 (FXR agonist)	MDCK II cells		[106,107]
				Butyrate and octanoic acid	MDCK II cells		[108]
c.1211AG	Missense	D404G	Reduced level of mature protein, ER-like distribution	4-PB	HEK293T cells	BRIC2	[109]
c.1211AG c.1331TC	Missense	D404G V444A	Reduced level of mature protein	4-PB	Patient compound heterozygous for D404G and homozygous for V444A mutations	BRIC2	[109]
c.1388CT	Missense	T463I	Impaired ATP-binding, BSEP dysfunction	Ivacaftor	MDCK II cells	PFIC2	[110]

Table 1. *Cont.*

Nucleotide Change	Type of Mutation	Amino Acid	Defect	Potential Corrective Therapy	Cell Line/Organism	Disease	References
c.1445AG	Missense	D482G [2]	Protein instability, ubiquitin-dependent [103], impaired membrane trafficking, reduced level of mature protein, severe differential splicing [105]	4-PB	MDCK II cells and Sprague–Dawley rats	PFIC2	[104]
				Sodium butyrate and 4-PB	HEK293T cells		[103]
				Butyrate and octanoic acid	MDCK II cells		[108]
c.1708GA	Missense	A570T	Reduced level of mature protein, reduced BSEP activity [105]	UDCA	MDCK II cells	BRIC2 PFIC2	[111]
				Glycerol at 28 °C	CHO-K1 cells		[105]
c.2417GA	Missense	G806D	Reduced level of mature protein, aberrant splicing	4-PB	BSEP-deficient hepatocyte-like cells	PFIC2	[112]
c.-24CA	5′-UTR (five prime untranslated region)						
c.2494CT	Missense	R832C	Differential splice products [105]	Steroid	Patient with compound heterozygosity	PFIC2	[113]
c.150+3AC	Splice-site mutation		Partial exon skipping [114]				
c.2756_2758delCCA	Deletion	T919del	Reduced BSEP activity [115]	Steroid	Patient with compound heterozygosity	PFIC2	[113]
c.3703CT	Nonsense	R1235X	Truncated, non-functional transporter [115]				
c.2944GA	Missense	G982R	Retention in ER, reduced level of mature protein	UDCA, 4-PB single agents or in combination	Can 10 cells	PFIC2	[97]
c.2944GA	Missense	G982R	Retention in ER, reduced level of mature protein	4-PB	Patient with compound heterozygosity	PFIC2	[97]
c.770CT	Missense	A257V	Normal canalicular expression of BSEP				
c.2944GA	Missense	G982R	Retention in ER, reduced level of mature protein	4-PB	Patient with compound heterozygosity	PFIC2	[97]
c.3003AG	Silent	R1001R	Abnormal splicing [116]				

Table 1. *Cont.*

Nucleotide Change	Type of Mutation	Amino Acid	Defect	Potential Corrective Therapy	Cell Line/Organism	Disease	References
c.3382CT	Missense	R1128C	Retention in ER, reduced level of mature protein, Mild exon skipping [105]	UDCA, 4-PB single agents or in combination	Can 10 cells	PFIC2	[97]
				4-PB	Patient homozygous for R1128C		
c.3628AC	Missense	T1210P	Retention in ER, reduced level of mature protein	UDCA, 4-PB single agents or in combination	Can 10 cells	PFIC2	[97]
				4-PB	Can 10 cells		[98]
				4-PB	Patient with homozygous mutation		[97,98]
c.3692GA	Missense	R1231Q	Retention in ER [117], no splicing, immature protein [105]	4-PB	HEK293T cells, McA-RH7777 cells, patient with homozygous mutation	PFIC2	[117]
c.1062TA	Nonsense	Y354X	Premature termination codon	G418, gentamicin	NIH3T3 cells (increased readthrough)	PFIC2	[118]
c.1243CT	Nonsense	R415X	Premature termination codon	G418, gentamicin	NIH3T3 cells (increased readthrough)	PFIC2	[118]
				Gentamicin	HEK293 cells (production of a full-length BSEP protein)		
c.1408CT	Nonsense	R470X	Premature termination codon	G418, gentamicin	NIH3T3 cells (increased readthrough)	PFIC2	[118]
				Gentamicin	HEK293 cells (production of a full-length BSEP protein)		
c.3169CT	Nonsense	R1057X	Premature termination codon	G418, gentamicin	NIH3T3 cells (increased readthrough)	PFIC2	[118]
				Gentamicin	HEK293 cells (production of a full-length BSEP protein)		

Table 1. *Cont.*

Nucleotide Change	Type of Mutation	Amino Acid	Defect	Potential Corrective Therapy	Cell Line/Organism	Disease	References
c.3268CT	Nonsense	R1090X	Premature termination codon	G418, gentamicin	NIH3T3 cells (increased readthrough)	PFIC2	[118]
				Gentamicin	HEK293, Can10 and HepG2 cells (production of a full-length BSEP protein and localization at the PM of HEK293 and at the CM of Can 10 and HepG2 cells)		
				Gentamicin treatment with UDCA, 4-PB and UDCA + 4-PB, gentamicin at 27 °C	Can10 cells (increased canalicular expression)		
				Gentamicin, gentamicin with 4-PB, gentamicin at 27 °C	NTCP expressing MDCK cells (significantly increased transport of [3H]TC)		
c.3904GT	Nonsense	E1302X	Premature termination codon	G418, gentamicin, PTC124	NIH3T3 cells (increased readthrough)	PFIC2	[118]

Note: [1] Most frequently reported in Japan. [2] E297G and D482G mutations account for 58% of PFIC2 cases in the Western population [119].

7.1. Transcriptional Modulators

FXR is the major ligand-activated transcription factor controlling BSEP expression, which makes it a possible target for therapeutic intervention. Furthermore, 6α-ethyl-CDCA (obeticholic acid, OCA), a derivative of the primary human bile acid CDCA and an FXR agonist, was approved by the FDA for the treatment of primary biliary cholangitis (PBC) either in monotherapy or in combination with UDCA in adults, depending on UDCA responsivity and tolerability [120]. OCA's 100-fold higher FXR-activating potential (as compared to the natural ligand CDCA) formed the basis for advocating it as a novel therapeutic treatment strategy for PBC [121]. Its long-term efficacy and safety profile were reported recently [122]. OCA was also introduced for the treatment of non-alcoholic steatohepatitis (NASH), whereby the interim analysis from a phase 3 trial demonstrated clinical improvement and partial reversal of histopathological features [123]. Several other steroidal and non-steroidal FXR agonists, such as EDP-305 and tropifexor (LJN4524), are currently being investigated in a clinical setting for the treatment of NASH [124].

Garzel and co-workers evaluated the effects of 30 BSEP inhibitors on BSEP expression and FXR activation in human primary hepatocytes to understand the underlying mechanisms of drug-induced liver injury (DILI) [125]. Among five potent transcriptional repressors, lopinavir and troglitazone were shown to mediate their effects by reducing the FXR activity. The latter drug was previously withdrawn from the market because of DILI [125]. A number of natural FXR agonists or antagonists were reported to modulate FXR activity in a variety of model systems, as reviewed in detail by Hiebl et al. [126]. A natural product, geniposide, was reported to modulate the expression of BSEP via the FXR, as well as via the Nrf2 signaling pathways [127]. On the other hand, 9-cis retinoic acid (9cRA) is an RXR agonist, which when co-administered with CDCA, represses FXR/RXR-mediated expression of BSEP, thus exerting an opposite effects on BSEP transcription [128].

7.2. Ursodeoxycholic Acid (UDCA)

UDCA is one of the most commonly used agents for the treatment of cholestatic disorders. It showed promising results in animal models and in patients by alleviating disease symptoms. Although the exact mechanism of action of UDCA is not known, it was reported that the compound may act by correcting a potential trafficking deficiency of BSEP mutants, as well as by reducing the internalization of the transporter [112,129,130]. Furthermore, UDCA also reduces the overall hydrophobicity of the bile acid pool, thereby protecting hepatocytes from damage [131]. However, in some patients UDCA failed alleviate disease symptoms. Likely this finding reflects the genetic diversity of the underlying disease [129].

A UDCA derivative, norUDCA, is also being used for treatment of BSEP-related diseases. Because of its capacity for cholehepatic shunting (i.e., bypassing the normal enterohepatic circulation), norUDCA counteracts bile duct damage via bicarbonate-rich choleresis. Furthermore, norUDCA has antifibrotic, antiproliferative and anti-inflammatory properties and propagates bile acid detoxification through elimination via the urine [131–133].

7.3. Chemical Correction with 4-PB

A large number of mutations have been reported to interfere with either BSEP folding or its correct trafficking to the canalicular membrane. Among these are the E297G and D482G mutations, which account for approximately 60% of PFIC2 cases in the European population [119]. The underlying hypothesis behind the concept of chemical correction is that these mutants, when rescued to the canalicular membrane, would function normally, and thus the disease phenotype would be alleviated. Furthermore, 4-PB, an FDA-approved drug for the treatment of urea cycle disorders, functions as a chemical chaperone for folding-deficient BSEP variants [56]. Indeed, in vitro studies in HEK293 and MDCKII cell lines indicated that upon treatment with 4-PB, these mutants would show enhanced surface expression, as well as increased TCA transport activity [103,104]. Furthermore, in support of the validity of this concept, 4-PB treatment also increased the biliary excretion of TCA in animal models [104].

Recently, use of 4-PB showed promising results in a clinical setting. Gonzales and co-workers [97,98] showed that treating PFIC2 patients carrying at least one mutation (out of p.G982R, p.R1128C and p.T1210P) with 4-PB led to an improvement in serum liver parameters, including the serum bile acid concentration, and a reduction in the pruritus score. Similarly, a preterm infant diagnosed with BSEP-related cholestasis was treated with 4-PB and showed an improvement of the disease symptoms [134]. In addition, BRIC2 patients have also been treated successfully with 4-PB [109]. The therapeutic doses ranged from 150 to 500 mg/kg/day. In some patients, a 4-PB dose of $\leq$ 350 mg/kg/day had no beneficial effect, while a high dose regimen (500 mg/kg/day) improved disease symptoms [109,117]. In a recent study, two PFIC II patients were given a combination of 4-PB, oxcarbazepine (a peripheral nerve stabilizer reducing pruritus) and maralixibat (an apical sodium-dependent bile acid transporter inhibitor), which had a beneficial effect on disease markers [101]. Most of these studies did not show apparent side effects, even in the high-dose regimen. However, psychological disorders (bipolar and related disorders) have been connected to the use of 4-PB in a clinical setting [135].

7.4. Potentiation with Ivacaftor

UDCA and 4-PB have been shown to correct the misfolding of trafficking-deficient variants. However, several disease-associated mutations (especially those in the NBDs) do not interfere with folding, trafficking and canalicular localization, but rather lead to impairment of transporter function [65,119]. Ivacaftor (VX-770) a potentiator has been approved by the FDA for treatment of some class III (gating deficient) CFTR mutants [136,137]. In these variants, an improvement of respiratory function could be demonstrated. Similar to clinical results in CFTR patients, ivacaftor was shown to rescue the function of missense mutations in the NBDs of ABCB4/MDR3 [138]. Recently, Mareux and co-workers showed that ivacaftor also rescued the function of an NBD missense mutation (T463I) of BSEP [110]. The mechanism of action of ivacaftor presently remains unclear.

7.5. Readthrough Therapy with Gentamicin

Nonsense mutations result in an in-frame premature termination codon and the absence of functional protein. This results in severe phenotypes of the disease and an increased risk for the development of hepatocellular carcinoma [139]. The aminoglycoside antibiotic gentamicin binds to ribosomes and induces a translational readthrough at the premature termination codon, thereby leading to fractional restoration of synthesis of the full-length protein [140,141]. In a recent study, Amzal et al. [118] evaluated the impact of gentamicin on six BSEP nonsense mutations (Y354X, R415X, R470X, R1057X, R1090X and E1302X) in vitro. Readthrough results were significantly increased for all mutations. The strongest responses were seen for the R1090X mutation, with partial restoration and correct localization at the plasma membrane of HepG2 and Can 10 cells. The rescued protein was shown to mediate transcellular transport of [^{3}H]TC in MDCK cells. Expression of the R1090X mutant was shown to be further enhanced by simultaneous treatment with 4-PB [118].

8. Summary and Conclusions

Despite the relative infrequency of the inherited forms of progressive intrahepatic cholestasis, the symptoms are severe, and about half of the patients progress to a stage of the disease that makes them candidates for liver transplantation. Therefore, the quest for identification of causal therapies that go beyond the purely symptomatic treatment of pruritus is an important objective. Attempts to alleviate disease symptoms by transcriptional upregulation are directed towards missense mutations with preserved functionality. Similarly, those mutants that are folding- and consequently trafficking-deficient have successfully been treated with folding correctors. Recent evidence points to yet another therapeutic path, in which the function of impaired BSEP mutants is potentiated by drugs that were initially developed to treat different disease entities, which are also associated with a malfunction of human ABC proteins. However, BSEP missense or deletion mutations and mutants with compromised functionality will only be amenable to therapy using gene editing. Current advances in gene editing technologies have not been considered in this review, as they are subject to another article in this Special Issue. The availability of the recently published cryo-EM structure of BSEP can be considered an important basis for structure-based drug design. Moreover, structural data, MD-simulations and site-directed mutagenesis studies continuously expand our understanding of the functional biology of BSEP.

We also briefly summarized available in vitro model systems for the functional characterization of BSEP, animal models and case reports discussing emerging clinical therapies. The perspective that treatment regimens combining small molecules with different mechanisms of action will ultimately lead to an improvement in the quality of life and life span of affected individuals appears promising.

Author Contributions: Conceptualization, P.C. and T.S.; writing—original draft preparation, M.I.S., Y.D.-C. and D.S.; writing—review and editing, M.I.S., Y.D.-C., D.S., T.S. and P.C.; visualization, D.S., Y.D.-C., T.S.; project administration, T.S. and P.C.; funding acquisition, P.C. and T.S. All authors have read and agreed to the published version of the manuscript.

Funding: Open Access Funding by the Austrian Science Fund (FWF), grant numbers SFB3509 (to P.C.), SFB3524 (to T.S.) and (P32017 to T.S.).

Institutional Review Board Statement: Not applicable.

Informed Consent Statement: Not applicable.

Data Availability Statement: Not available.

Conflicts of Interest: The authors declare no conflict of interest. The funders had no role in the design of the study; in the collection, analyses, or interpretation of data; in the writing of the manuscript, or in the decision to publish the results.

Abbreviations

ABC	ATP-binding cassette
AMPK	AMP-activated protein kinase
AP-2	adaptor protein 2
ASCOM	activating signal cointegrator-2-containing complex
BSEP (ABCB11)	bile salt export pump
BRIC2	benign recurrent intrahepatic cholestasis type 2
CA	cholic acid
cAMP	cyclic adenosine monophosphate
CARM1	co-activator-associated arginine methyltransferase 1
CFTR (ABCC7)	cystic fibrosis transmembrane conductance regulator
CDCA	chenodeoxycholic acid
CM	canalicular membrane
cPKC	classical (Ca2+-dependent) protein kinase C
cryo-EM	cryo electron microscopy
DCA	deoxycholic acid
DILI	drug-induced liver injury
ECL	extracellular loop
Epac	exchange protein directly activated by cAMP
ICL	intracellular loop
ICP	intrahepatic cholestasis of pregnancy
ERAD	Endoplasmic-reticulum-associated degradation
E17G	estradiol 17 β-D-glucuronide
FXR (NR1H4)	farnesoid X receptor
FXRE	FXR response element

LKB1	liver kinase B1
LRH-1 (NR5A2)	liver receptor homolog-1
MAPK	Mitogen-activated protein kinase
MRP3 (ABCC3)	Multidrug-resistance-associated protein 3
MRP4 (ABCC4)	Multidrug-resistance-associated protein 4
NASH	non-alcoholic steatohepatitis
NBD	nucleotide-binding domain
NBS	nucleotide-binding site
Nrf2	nuclear factor erythroid 2-related factor 2
NTCP (SLC10A1)	sodium taurocholate co-transporting polypeptide
OCA	6α-ethyl-CDCA (obeticholic acid)
4-PB	4-phenylbutyrate
PBC	primary biliary cholangitis
PFIC2	progressive familial intrahepatic cholestasis type 2
PI3K	phosphoinositide 3-kinase
PM	plasma membrane
RXR	retinoid X receptor
SRC2	steroid receptor co-activator 2
TCA	taurocholic acid
TMD	transmembrane domain
TUDCA	tauroursodeoxycholic acid
UDCA	ursodeoxycholic acid

References

1. Dean, M.; Rzhetsky, A.; Allikmets, R. The human ATP-binding cassette (ABC) transporter superfamily. *Genome Res.* **2001**, *11*, 1156–1166. [CrossRef] [PubMed]
2. Schmitt, L. Structure and mechanism of ABC transporters. *Curr. Opin. Struct. Biol.* **2002**, *12*, 754–760. [CrossRef]
3. Shintre, C.A.; Pike, A.C.W.; Li, Q.; Kim, J.-I.; Barr, A.J.; Goubin, S.; Shrestha, L.; Yang, J.; Berridge, G.; Ross, J.; et al. Structures of ABCB10, a human ATP-binding cassette transporter in apo- and nucleotide-bound states. *Proc. Natl. Acad. Sci. USA* **2013**, *110*, 9710–9715. [CrossRef] [PubMed]
4. Procko, E.; O'Mara, M.L.; Bennett, W.F.D.; Tieleman, D.P.; Gaudet, R. The mechanism of ABC transporters: General lessons from structural and functional studies of an antigenic peptide transporter. *FASEB J.* **2009**, *23*, 1287–1302. [CrossRef]
5. Langmann, T.; Mauerer, R.; Zahn, A.; Moehle, C.; Probst, M.; Stremmel, W.; Schmitz, G. Real-Time Reverse Transcription-PCR Expression Profiling of the Complete Human ATP-Binding Cassette Transporter Superfamily in Various Tissues. *Clin. Chem.* **2003**, *49*, 230–238. [CrossRef]
6. Stieger, B. The Role of the Sodium-Taurocholate Cotransporting Polypeptide (NTCP) and of the Bile Salt Export Pump (BSEP) in Physiology and Pathophysiology of Bile Formation. In *Drug Transporters*; Springer: Berlin/Heidelberg, Germany, 2011; pp. 205–259.
7. Lagana, S.M.; Salomao, M.; Remotti, H.E.; Knisely, A.S.; Moreira, R.K. Bile salt export pump: A sensitive and specific immunohistochemical marker of hepatocellular carcinoma. *Histopathology* **2015**, *66*, 598–602. [CrossRef]
8. Hilgendorf, C.; Ahlin, G.; Seithel, A.; Artursson, P.; Ungell, A.-L.; Karlsson, J. Expression of Thirty-six Drug Transporter Genes in Human Intestine, Liver, Kidney, and Organotypic Cell Lines. *Drug Metab. Dispos.* **2007**, *35*, 1333–1340. [CrossRef]
9. Stieger, B.; Beuers, U. The Canalicular Bile Salt Export Pump BSEP (ABCB11) as a Potential Therapeutic Target. *Curr. Drug Targets* **2011**, *12*, 661–670. [CrossRef]
10. Trauner, M.; Fuchs, C.D.; Halilbasic, E.; Paumgartner, G. New therapeutic concepts in bile acid transport and signaling for management of cholestasis. *Hepatology* **2017**, *65*, 1393–1404. [CrossRef]
11. Trauner, M.; Boyer, J.L. Bile Salt Transporters: Molecular Characterization, Function, and Regulation. *Physiol. Rev.* **2003**, *83*, 633–671. [CrossRef]
12. Lecureur, V.; Sun, D.; Hargrove, P.; Schuetz, E.G.; Kim, R.B.; Lan, L.B.; Schuetz, J.D. Cloning and expression of murine sister of P-glycoprotein reveals a more discriminating transporter than MDR1/P-glycoprotein. *Mol. Pharmacol.* **2000**, *57*, 24–35. [PubMed]
13. Hirano, M.; Maeda, K.; Hayashi, H.; Kusuhara, H.; Sugiyama, Y. Bile Salt Export Pump (BSEP/ABCB11) Can Transport a Nonbile Acid Substrate, Pravastatin. *J. Pharmacol. Exp. Ther.* **2005**, *314*, 876–882. [CrossRef] [PubMed]
14. Matsushima, S.; Maeda, K.; Hayashi, H.; Debori, Y.; Schinkel, A.H.; Schuetz, J.D.; Kusuhara, H.; Sugiyama, Y. Involvement of Multiple Efflux Transporters in Hepatic Disposition of Fexofenadine. *Mol. Pharmacol.* **2008**, *73*, 1474–1483. [CrossRef] [PubMed]
15. Ho, R.H.; Leake, B.F.; Kilkenny, D.M.; Meyer zu Schwabedissen, H.E.; Glaeser, H.; Kroetz, D.L.; Kim, R.B. Polymorphic variants in the human bile salt export pump (BSEP; ABCB11): Functional characterization and interindividual variability. *Pharmacogenet. Genom.* **2010**, *20*, 45–57. [CrossRef] [PubMed]
16. Makishima, M. Identification of a Nuclear Receptor for Bile Acids. *Science* **1999**, *284*, 1362–1365. [CrossRef] [PubMed]

17. Wang, H.; Chen, J.; Hollister, K.; Sowers, L.C.; Forman, B.M. Endogenous Bile Acids Are Ligands for the Nuclear Receptor FXR/BAR. *Mol. Cell* **1999**, *3*, 543–553. [CrossRef]
18. Parks, D.J. Bile Acids: Natural Ligands for an Orphan Nuclear Receptor. *Science* **1999**, *284*, 1365–1368. [CrossRef]
19. Baghdasaryan, A.; Chiba, P.; Trauner, M. Clinical application of transcriptional activators of bile salt transporters. *Mol. Aspects Med.* **2014**, *37*, 57–76. [CrossRef]
20. Plass, J.R.M.; Mol, O.; Heegsma, J.; Geuken, M.; Faber, K.N.; Jansen, P.L.M.; Müller, M. Farnesoid X receptor and bile salts are involved in transcriptional regulation of the gene encoding the human bile salt export pump. *Hepatology* **2002**, *35*, 589–596. [CrossRef]
21. Ananthanarayanan, M.; Li, Y.; Surapureddi, S.; Balasubramaniyan, N.; Ahn, J.; Goldstein, J.A.; Suchy, F.J. Histone H3K4 trimethylation by MLL3 as part of ASCOM complex is critical for NR activation of bile acid transporter genes and is downregulated in cholestasis. *Am. J. Physiol. Liver Physiol.* **2011**, *300*, G771–G781. [CrossRef]
22. Ananthanarayanan, M.; Li, S.; Balasubramaniyan, N.; Suchy, F.J.; Walsh, M.J. Ligand-dependent Activation of the Farnesoid X-receptor Directs Arginine Methylation of Histone H3 by CARM1. *J. Biol. Chem.* **2004**, *279*, 54348–54357. [CrossRef] [PubMed]
23. Chopra, A.R.; Kommagani, R.; Saha, P.; Louet, J.-F.; Salazar, C.; Song, J.; Jeong, J.; Finegold, M.; Viollet, B.; DeMayo, F.; et al. Cellular Energy Depletion Resets Whole-Body Energy by Promoting Coactivator-Mediated Dietary Fuel Absorption. *Cell Metab.* **2011**, *13*, 35–43. [CrossRef] [PubMed]
24. Song, X.; Kaimal, R.; Yan, B.; Deng, R. Liver receptor homolog 1 transcriptionally regulates human bile salt export pump expression. *J. Lipid Res.* **2008**, *49*, 973–984. [CrossRef] [PubMed]
25. Mataki, C.; Magnier, B.C.; Houten, S.M.; Annicotte, J.-S.; Argmann, C.; Thomas, C.; Overmars, H.; Kulik, W.; Metzger, D.; Auwerx, J.; et al. Compromised Intestinal Lipid Absorption in Mice with a Liver-Specific Deficiency of Liver Receptor Homolog 1. *Mol. Cell. Biol.* **2007**, *27*, 8330–8339. [CrossRef] [PubMed]
26. Weerachayaphorn, J.; Cai, S.-Y.; Soroka, C.J.; Boyer, J.L. Nuclear factor erythroid 2-related factor 2 is a positive regulator of human bile salt export pump expression. *Hepatology* **2009**, *50*, 1588–1596. [CrossRef] [PubMed]
27. Wang, L.; Dong, H.; Soroka, C.J.; Wei, N.; Boyer, J.L.; Hochstrasser, M. Degradation of the bile salt export pump at endoplasmic reticulum in progressive familial intrahepatic cholestasis type II. *Hepatology* **2008**, *48*, 1558–1569. [CrossRef]
28. Kleizen, B.; van Vlijmen, T.; de Jonge, H.R.; Braakman, I. Folding of CFTR Is Predominantly Cotranslational. *Mol. Cell* **2005**, *20*, 277–287. [CrossRef]
29. Rudashevskaya, E.L.; Stockner, T.; Trauner, M.; Freissmuth, M.; Chiba, P. Pharmacological correction of misfolding of ABC proteins. *Drug Discov. Today Technol.* **2014**, *12*, e87–e94. [CrossRef]
30. Chiba, P.; Freissmuth, M.; Stockner, T. Defining the blanks—Pharmacochaperoning of SLC6 transporters and ABC transporters. *Pharmacol. Res.* **2014**, *83*, 63–73. [CrossRef]
31. Guna, A.; Hegde, R.S. Transmembrane Domain Recognition during Membrane Protein Biogenesis and Quality Control. *Curr. Biol.* **2018**, *28*, R498–R511. [CrossRef]
32. Printsev, I.; Curiel, D.; Carraway, K.L. Membrane Protein Quantity Control at the Endoplasmic Reticulum. *J. Membr. Biol.* **2017**, *250*, 379–392. [CrossRef] [PubMed]
33. Needham, P.G.; Guerriero, C.J.; Brodsky, J.L. Chaperoning Endoplasmic Reticulum–Associated Degradation (ERAD) and Protein Conformational Diseases. *Cold Spring Harb. Perspect. Biol.* **2019**, *11*, a033928. [CrossRef] [PubMed]
34. Clarke, J.D.; Novak, P.; Lake, A.D.; Hardwick, R.N.; Cherrington, N.J. Impaired N-linked glycosylation of uptake and efflux transporters in human non-alcoholic fatty liver disease. *Liver Int.* **2017**, *37*, 1074–1081. [CrossRef] [PubMed]
35. Mochizuki, K.; Kagawa, T.; Numari, A.; Harris, M.J.; Itoh, J.; Watanabe, N.; Mine, T.; Arias, I.M. Two N-linked glycans are required to maintain the transport activity of the bile salt export pump (ABCB11) in MDCK II cells. *Am. J. Physiol. Liver Physiol.* **2007**, *292*, G818–G828. [CrossRef]
36. Plass, J.R.; Mol, O.; Heegsma, J.; Geuken, M.; de Bruin, J.; Elling, G.; Müller, M.; Faber, K.N.; Jansen, P.L. A progressive familial intrahepatic cholestasis type 2 mutation causes an unstable, temperature-sensitive bile salt export pump. *J. Hepatol.* **2004**, *40*, 24–30. [CrossRef]
37. Kipp, H.; Pichetshote, N.; Arias, I.M. Transporters on Demand. *J. Biol. Chem.* **2001**, *276*, 7218–7224. [CrossRef]
38. Wakabayashi, Y.; Lippincott-Schwartz, J.; Arias, I.M. Intracellular Trafficking of Bile Salt Export Pump (ABCB11) in Polarized Hepatic Cells: Constitutive Cycling between the Canalicular Membrane and rab11-positive Endosomes. *Mol. Biol. Cell* **2004**, *15*, 3485–3496. [CrossRef] [PubMed]
39. Lam, P.; Soroka, C.; Boyer, J. The Bile Salt Export Pump: Clinical and Experimental Aspects of Genetic and Acquired Cholestatic Liver Disease. *Semin. Liver Dis.* **2010**, *30*, 125–133. [CrossRef]
40. Lam, P.; Xu, S.; Soroka, C.J.; Boyer, J.L. A C-terminal tyrosine-based motif in the bile salt export pump directs clathrin-dependent endocytosis. *Hepatology* **2012**, *55*, 1901–1911. [CrossRef]
41. Hayashi, H.; Inamura, K.; Aida, K.; Naoi, S.; Horikawa, R.; Nagasaka, H.; Takatani, T.; Fukushima, T.; Hattori, A.; Yabuki, T.; et al. AP2 adaptor complex mediates bile salt export pump internalization and modulates its hepatocanalicular expression and transport function. *Hepatology* **2012**, *55*, 1889–1900. [CrossRef]
42. Chan, W.; Calderon, G.; Swift, A.L.; Moseley, J.; Li, S.; Hosoya, H.; Arias, I.M.; Ortiz, D.F. Myosin II Regulatory Light Chain Is Required for Trafficking of Bile Salt Export Protein to the Apical Membrane in Madin-Darby Canine Kidney Cells. *J. Biol. Chem.* **2005**, *280*, 23741–23747. [CrossRef] [PubMed]

43. Wakabayashi, Y.; Dutt, P.; Lippincott-Schwartz, J.; Arias, I.M. Rab11a and myosin Vb are required for bile canalicular formation in WIF-B9 cells. *Proc. Natl. Acad. Sci. USA* **2005**, *102*, 15087–15092. [CrossRef] [PubMed]
44. Crocenzi, F.A. Localization status of hepatocellular transporters in cholestasis. *Front. Biosci.* **2012**, *17*, 1201. [CrossRef] [PubMed]
45. Crocenzi, F.A.; Mottino, A.D.; Cao, J.; Veggi, L.M.; Pozzi, E.J.S.; Vore, M.; Coleman, R.; Roma, M.G. Estradiol-17β-D-glucuronide induces endocytic internalization of Bsep in rats. *Am. J. Physiol. Liver Physiol.* **2003**, *285*, G449–G459. [CrossRef]
46. Miszczuk, G.S.; Barosso, I.R.; Larocca, M.C.; Marrone, J.; Marinelli, R.A.; Boaglio, A.C.; Sánchez Pozzi, E.J.; Roma, M.G.; Crocenzi, F.A. Mechanisms of canalicular transporter endocytosis in the cholestatic rat liver. *Biochim. Biophys. Acta Mol. Basis Dis.* **2018**, *1864*, 1072–1085. [CrossRef]
47. Crocenzi, F.A.; Sánchez Pozzi, E.J.; Ruiz, M.L.; Zucchetti, A.E.; Roma, M.G.; Mottino, A.D.; Vore, M. Ca^{2+}-dependent protein kinase C isoforms are critical to estradiol 17β-D-glucuronide-induced cholestasis in the rat. *Hepatology* **2008**, *48*, 1885–1895. [CrossRef]
48. Boaglio, A.C.; Zucchetti, A.E.; Sánchez Pozzi, E.J.; Pellegrino, J.M.; Ochoa, J.E.; Mottino, A.D.; Vore, M.; Crocenzi, F.A.; Roma, M.G. Phosphoinositide 3-kinase/protein kinase B signaling pathway is involved in estradiol 17β-d-glucuronide-induced cholestasis: Complementarity with classical protein kinase c. *Hepatology* **2010**, *52*, 1465–1476. [CrossRef]
49. Boaglio, A.C.; Zucchetti, A.E.; Toledo, F.D.; Barosso, I.R.; Sánchez Pozzi, E.J.; Crocenzi, F.A.; Roma, M.G. ERK1/2 and p38 MAPKs Are Complementarily Involved in Estradiol 17ß-d-Glucuronide-Induced Cholestasis: Crosstalk with cPKC and PI3K. *PLoS ONE* **2012**, *7*, e49255. [CrossRef]
50. Fu, D.; Wakabayashi, Y.; Lippincott-Schwartz, J.; Arias, I.M. Bile acid stimulates hepatocyte polarization through a cAMP-Epac-MEK-LKB1-AMPK pathway. *Proc. Natl. Acad. Sci. USA* **2011**, *108*, 1403–1408. [CrossRef]
51. Homolya, L.; Fu, D.; Sengupta, P.; Jarnik, M.; Gillet, J.-P.; Vitale-Cross, L.; Gutkind, J.S.; Lippincott-Schwartz, J.; Arias, I.M. LKB1/AMPK and PKA Control ABCB11 Trafficking and Polarization in Hepatocytes. *PLoS ONE* **2014**, *9*, e91921. [CrossRef]
52. Misra, S.; Varticovski, L.; Arias, I.M. Mechanisms by which cAMP increases bile acid secretion in rat liver and canalicular membrane vesicles. *Am. J. Physiol. Liver Physiol.* **2003**, *285*, G316–G324. [CrossRef] [PubMed]
53. Kurz, A.K.; Graf, D.; Schmitt, M.; Vom Dahl, S.; Häussinger, D. Tauroursodesoxycholate-induced choleresis involves p38MAPK activation and translocation of the bile salt export pump in rats. *Gastroenterology* **2001**, *121*, 407–419. [CrossRef] [PubMed]
54. Kubitz, R.; Sütfels, G.; Kühlkamp, T.; Kölling, R.; Häussinger, D. Trafficking of the bile salt export pump from the Golgi to the canalicular membrane is regulated by the p38 MAP kinase. *Gastroenterology* **2004**, *126*, 541–553. [CrossRef] [PubMed]
55. Noe, J. Characterization of the mouse bile salt export pump overexpressed in the baculovirus system. *Hepatology* **2001**, *33*, 1223–1231. [CrossRef]
56. Hayashi, H.; Sugiyama, Y. Short-Chain Ubiquitination Is Associated with the Degradation Rate of a Cell-Surface-Resident Bile Salt Export Pump (BSEP/ABCB11). *Mol. Pharmacol.* **2009**, *75*, 143–150. [CrossRef]
57. Aida, K.; Hayashi, H.; Inamura, K.; Mizuno, T.; Sugiyama, Y. Differential Roles of Ubiquitination in the Degradation Mechanism of Cell Surface–Resident Bile Salt Export Pump and Multidrug Resistance–Associated Protein 2. *Mol. Pharmacol.* **2014**, *85*, 482–491. [CrossRef]
58. Czuba, L.C.; Hillgren, K.M.; Swaan, P.W. Post-translational modifications of transporters. *Pharmacol. Ther.* **2018**, *192*, 88–99. [CrossRef]
59. Wang, L.; Hou, W.-T.; Chen, L.; Jiang, Y.-L.; Xu, D.; Sun, L.; Zhou, C.-Z.; Chen, Y. Cryo-EM structure of human bile salts exporter ABCB11. *Cell Res.* **2020**, *30*, 623–625. [CrossRef]
60. Keitel, V.; Burdelski, M.; Vojnisek, Z.; Schmitt, L.; Häussinger, D.; Kubitz, R. De novo bile salt transporter antibodies as a possible cause of recurrent graft failure after liver transplantation: A novel mechanism of cholestasis. *Hepatology* **2009**, *50*, 510–517. [CrossRef]
61. Kubitz, R.; Dröge, C.; Stindt, J.; Weissenberger, K.; Häussinger, D. The bile salt export pump (BSEP) in health and disease. *Clin. Res. Hepatol. Gastroenterol.* **2012**, *36*, 536–553. [CrossRef]
62. Dawson, R.J.P.; Locher, K.P. Structure of the multidrug ABC transporter Sav1866 from Staphylococcus aureus in complex with AMP-PNP. *FEBS Lett.* **2007**, *581*, 935–938. [CrossRef] [PubMed]
63. Giovannoni, I.; Callea, F.; Bellacchio, E.; Torre, G.; De Ville De Goyet, J.; Francalanci, P. Genetics and Molecular Modeling of New Mutations of Familial Intrahepatic Cholestasis in a Single Italian Center. *PLoS ONE* **2015**, *10*, e0145021. [CrossRef] [PubMed]
64. Li, J.; Jaimes, K.F.; Aller, S.G. Refined structures of mouse P-glycoprotein. *Protein Sci.* **2014**, *23*, 34–46. [CrossRef] [PubMed]
65. Dröge, C.; Bonus, M.; Baumann, U.; Klindt, C.; Lainka, E.; Kathemann, S.; Brinkert, F.; Grabhorn, E.; Pfister, E.-D.; Wenning, D.; et al. Sequencing of FIC1, BSEP and MDR3 in a large cohort of patients with cholestasis revealed a high number of different genetic variants. *J. Hepatol.* **2017**, *67*, 1253–1264. [CrossRef] [PubMed]
66. Jain, S.; Grandits, M.; Richter, L.; Ecker, G.F. Structure based classification for bile salt export pump (BSEP) inhibitors using comparative structural modeling of human BSEP. *J. Comput. Aided. Mol. Des.* **2017**, *31*, 507–521. [CrossRef] [PubMed]
67. Sohail, M.I.; Schmid, D.; Wlcek, K.; Spork, M.; Szakács, G.; Trauner, M.; Stockner, T.; Chiba, P. Molecular Mechanism of Taurocholate Transport by the Bile Salt Export Pump, an ABC Transporter Associated with Intrahepatic Cholestasis. *Mol. Pharmacol.* **2017**, *92*, 401–413. [CrossRef]
68. Berman, H.M. The Protein Data Bank. *Nucleic Acids Res.* **2000**, *28*, 235–242. [CrossRef]
69. Thonghin, N.; Collins, R.F.; Barbieri, A.; Shafi, T.; Siebert, A.; Ford, R.C. Novel features in the structure of P-glycoprotein (ABCB1) in the post-hydrolytic state as determined at 7.9 Å resolution. *BMC Struct. Biol.* **2018**, *18*, 17. [CrossRef]

70. Kagawa, T.; Watanabe, N.; Mochizuki, K.; Numari, A.; Ikeno, Y.; Itoh, J.; Tanaka, H.; Arias, I.M.; Mine, T. Phenotypic differences in PFIC2 and BRIC2 correlate with protein stability of mutant Bsep and impaired taurocholate secretion in MDCK II cells. *Am. J. Physiol. Liver Physiol.* **2008**, *294*, G58–G67. [CrossRef]

71. Goda, K.; Dönmez-Cakil, Y.; Tarapcsák, S.; Szalóki, G.; Szöllősi, D.; Parveen, Z.; Türk, D.; Szakács, G.; Chiba, P.; Stockner, T. Human ABCB1 with an ABCB11-like degenerate nucleotide binding site maintains transport activity by avoiding nucleotide occlusion. *PLoS Genet.* **2020**, *16*, e1009016. [CrossRef]

72. Shukla, S.; Schwartz, C.; Kapoor, K.; Kouanda, A.; Ambudkar, S.V. Use of Baculovirus BacMam Vectors for Expression of ABC Drug Transporters in Mammalian Cells. *Drug Metab. Dispos.* **2012**, *40*, 304–312. [CrossRef] [PubMed]

73. Stieger, B.; Fattinger, K.; Madon, J.; Kullak-Ublick, G.A.; Meier, P.J. Drug- and estrogen-induced cholestasis through inhibition of the hepatocellular bile salt export pump (Bsep) of rat liver. *Gastroenterology* **2000**, *118*, 422–430. [CrossRef]

74. Kondo, T.; Dale, G.L.; Beutler, E. Simple and rapid purification of inside-out vesicles from human erythrocytes. *Biochim. Biophys. Acta Biomembr.* **1980**, *602*, 127–130. [CrossRef]

75. Guyot, C.; Stieger, B. Interaction of bile salts with rat canalicular membrane vesicles: Evidence for bile salt resistant microdomains. *J. Hepatol.* **2011**, *55*, 1368–1376. [CrossRef] [PubMed]

76. Horikawa, M.; Kato, Y.; Tyson, C.A.; Sugiyama, Y. Potential Cholestatic Activity of Various Therapeutic Agents Assessed by Bile Canalicular Membrane Vesicles Isolated from Rats and Humans. *Drug Metab. Pharmacokinet.* **2003**, *18*, 16–22. [CrossRef] [PubMed]

77. Stieger, B.; Mahdi, Z.M. Model Systems for Studying the Role of Canalicular Efflux Transporters in Drug-Induced Cholestatic Liver Disease. *J. Pharm. Sci.* **2017**, *106*, 2295–2301. [CrossRef] [PubMed]

78. Mita, S.; Suzuki, H.; Akita, H.; Hayashi, H.; Onuki, R.; Hofmann, A.F.; Sugiyama, Y. Vectorial transport of unconjugated and conjugated bile salts by monolayers of LLC-PK1 cells doubly transfected with human NTCP and BSEP or with rat Ntcp and Bsep. *Am. J. Physiol. Liver Physiol.* **2006**, *290*, G550–G556. [CrossRef] [PubMed]

79. Montanari, F.; Pinto, M.; Khunweeraphong, N.; Wlcek, K.; Sohail, M.I.; Noeske, T.; Boyer, S.; Chiba, P.; Stieger, B.; Kuchler, K.; et al. Flagging Drugs That Inhibit the Bile Salt Export Pump. *Mol. Pharm.* **2016**, *13*, 163–171. [CrossRef]

80. Kenna, J.G.; Taskar, K.S.; Battista, C.; Bourdet, D.L.; Brouwer, K.L.R.; Brouwer, K.R.; Dai, D.; Funk, C.; Hafey, M.J.; Lai, Y.; et al. Can Bile Salt Export Pump Inhibition Testing in Drug Discovery and Development Reduce Liver Injury Risk? An International Transporter Consortium Perspective. *Clin. Pharmacol. Ther.* **2018**, *104*, 916–932. [CrossRef]

81. Brouwer, K.L.R.; Keppler, D.; Hoffmaster, K.A.; Bow, D.A.J.; Cheng, Y.; Lai, Y.; Palm, J.E.; Stieger, B.; Evers, R. In Vitro Methods to Support Transporter Evaluation in Drug Discovery and Development. *Clin. Pharmacol. Ther.* **2013**, *94*, 95–112. [CrossRef]

82. Yang, K.; Guo, C.; Woodhead, J.L.; St. Claire, R.L.; Watkins, P.B.; Siler, S.Q.; Howell, B.A.; Brouwer, K.L.R. Sandwich-Cultured Hepatocytes as a Tool to Study Drug Disposition and Drug-Induced Liver Injury. *J. Pharm. Sci.* **2016**, *105*, 443–459. [CrossRef] [PubMed]

83. De Bruyn, T.; Chatterjee, S.; Fattah, S.; Keemink, J.; Nicolaï, J.; Augustijns, P.; Annaert, P. Sandwich-cultured hepatocytes: Utility for in vitro exploration of hepatobiliary drug disposition and drug-induced hepatotoxicity. *Expert Opin. Drug Metab. Toxicol.* **2013**, *9*, 589–616. [CrossRef] [PubMed]

84. Swift*, B.; Pfeifer*, N.D.; Brouwer, K.L.R. Sandwich-cultured hepatocytes: An in vitro model to evaluate hepatobiliary transporter-based drug interactions and hepatotoxicity. *Drug Metab. Rev.* **2010**, *42*, 446–471. [CrossRef]

85. Li, A.P.; Gorycki, P.D.; Hengstler, J.G.; Kedderis, G.L.; Koebe, H.G.; Rahmani, R.; de Sousas, G.; Silva, J.M.; Skett, P. Present status of the application of cryopreserved hepatocytes in the evaluation of xenobiotics: Consensus of an international expert panel. *Chem. Biol. Interact.* **1999**, *121*, 117–123. [CrossRef]

86. Lundquist, P.; Englund, G.; Skogastierna, C.; Lööf, J.; Johansson, J.; Hoogstraate, J.; Afzelius, L.; Andersson, T.B. Functional ATP-Binding Cassette Drug Efflux Transporters in Isolated Human and Rat Hepatocytes Significantly Affect Assessment of Drug Disposition. *Drug Metab. Dispos.* **2014**, *42*, 448–458. [CrossRef] [PubMed]

87. Yucha, R.W.; He, K.; Shi, Q.; Cai, L.; Nakashita, Y.; Xia, C.Q.; Liao, M. In Vitro Drug-Induced Liver Injury Prediction: Criteria Optimization of Efflux Transporter IC50 and Physicochemical Properties. *Toxicol. Sci.* **2017**, *157*, 487–499. [CrossRef]

88. Cheng, Y.; Woolf, T.F.; Gan, J.; He, K. In vitro model systems to investigate bile salt export pump (BSEP) activity and drug interactions: A review. *Chem. Biol. Interact.* **2016**, *255*, 23–30. [CrossRef]

89. Cheng, Y.; Freeden, C.; Zhang, Y.; Abraham, P.; Shen, H.; Wescott, D.; Humphreys, W.G.; Gan, J.; Lai, Y. Biliary excretion of pravastatin and taurocholate in rats with bile salt export pump (Bsep) impairment. *Biopharm. Drug Dispos.* **2016**, *37*, 276–286. [CrossRef]

90. Wang, R. Targeted inactivation of sister of P-glycoprotein gene (spgp) in mice results in nonprogressive but persistent intrahepatic cholestasis. *Proc. Natl. Acad. Sci. USA* **2001**, *98*, 2011–2016. [CrossRef]

91. Pankowicz, F.P.; Barzi, M.; Kim, K.H.; Legras, X.; Martins, C.S.; Wooton-Kee, C.R.; Lagor, W.R.; Marini, J.C.; Elsea, S.H.; Bissig-Choisat, B.; et al. Rapid Disruption of Genes Specifically in Livers of Mice Using Multiplex CRISPR/Cas9 Editing. *Gastroenterology* **2018**, *155*, 1967–1970.e6. [CrossRef]

92. Alves-Bezerra, M.; Furey, N.; Johnson, C.G.; Bissig, K.-D. Using CRISPR/Cas9 to model human liver disease. *JHEP Rep.* **2019**, *1*, 392–402. [CrossRef] [PubMed]

93. Fuchs, C.D.; Paumgartner, G.; Wahlström, A.; Schwabl, P.; Reiberger, T.; Leditznig, N.; Stojakovic, T.; Rohr-Udilova, N.; Chiba, P.; Marschall, H.-U.; et al. Metabolic preconditioning protects BSEP/ABCB11−/− mice against cholestatic liver injury. *J. Hepatol.* **2017**, *66*, 95–101. [CrossRef] [PubMed]
94. Wang, R.; Chen, H.-L.; Liu, L.; Sheps, J.A.; Phillips, M.J.; Ling, V. Compensatory role of P-glycoproteins in knockout mice lacking the bile salt export pump. *Hepatology* **2009**, *50*, 948–956. [CrossRef] [PubMed]
95. Ellis, J.L.; Bove, K.E.; Schuetz, E.G.; Leino, D.; Valencia, C.A.; Schuetz, J.D.; Miethke, A.; Yin, C. Zebrafish abcb11b mutant reveals strategies to restore bile excretion impaired by bile salt export pump deficiency. *Hepatology* **2018**, *67*, 1531–1545. [CrossRef] [PubMed]
96. Lam, P.; Wang, R.; Ling, V. Bile Acid Transport in Sister of P-Glycoprotein (ABCB11) Knockout Mice. *Biochemistry* **2005**, *44*, 12598–12605. [CrossRef] [PubMed]
97. Gonzales, E.; Grosse, B.; Schuller, B.; Davit-Spraul, A.; Conti, F.; Guettier, C.; Cassio, D.; Jacquemin, E. Targeted pharmacotherapy in progressive familial intrahepatic cholestasis type 2: Evidence for improvement of cholestasis with 4-phenylbutyrate. *Hepatology* **2015**, *62*, 558–566. [CrossRef]
98. Gonzales, E.; Grosse, B.; Cassio, D.; Davit-Spraul, A.; Fabre, M.; Jacquemin, E. Successful mutation-specific chaperone therapy with 4-phenylbutyrate in a child with progressive familial intrahepatic cholestasis type 2. *J. Hepatol.* **2012**, *57*, 695–698. [CrossRef]
99. The Human Gene Mutation Database. Available online: http://www.hgmd.cf.ac.uk/ac/gene.php?gene=ABCB11 (accessed on 12 November 2020).
100. Imagawa, K.; Hayashi, H.; Sabu, Y.; Tanikawa, K.; Fujishiro, J.; Kajikawa, D.; Wada, H.; Kudo, T.; Kage, M.; Kusuhara, H.; et al. Clinical phenotype and molecular analysis of a homozygous ABCB11 mutation responsible for progressive infantile cholestasis. *J. Hum. Genet.* **2018**, *63*, 569–577. [CrossRef]
101. Malatack, J.J.; Doyle, D. A Drug Regimen for Progressive Familial Cholestasis Type 2. *Pediatrics* **2018**, *141*, e20163877. [CrossRef]
102. Arthur Lorio, E.; Valadez, D.; Alkhouri, N.; Loo, N. Cholestasis in Benign Recurrent Intrahepatic Cholestasis 2. *ACG Case Rep. J.* **2020**, *7*, e00412. [CrossRef]
103. Lam, P.; Pearson, C.L.; Soroka, C.J.; Xu, S.; Mennone, A.; Boyer, J.L. Levels of plasma membrane expression in progressive and benign mutations of the bile salt export pump (Bsep/Abcb11) correlate with severity of cholestatic diseases. *Am. J. Physiol. Physiol.* **2007**, *293*, C1709–C1716. [CrossRef]
104. Hayashi, H.; Sugiyama, Y. 4-phenylbutyrate enhances the cell surface expression and the transport capacity of wild-type and mutated bile salt export pumps. *Hepatology* **2007**, *45*, 1506–1516. [CrossRef] [PubMed]
105. Byrne, J.A.; Strautnieks, S.S.; Ihrke, G.; Pagani, F.; Knisely, A.S.; Linton, K.J.; Mieli-Vergani, G.; Thompson, R.J. Missense mutations and single nucleotide polymorphisms in ABCB11 impair bile salt export pump processing and function or disrupt pre-messenger RNA splicing. *Hepatology* **2009**, *49*, 553–567. [CrossRef] [PubMed]
106. Misawa, T.; Hayashi, H.; Sugiyama, Y.; Hashimoto, Y. Discovery and structural development of small molecules that enhance transport activity of bile salt export pump mutant associated with progressive familial intrahepatic cholestasis type 2. *Bioorg. Med. Chem.* **2012**, *20*, 2940–2949. [CrossRef] [PubMed]
107. Misawa, T.; Hayashi, H.; Makishima, M.; Sugiyama, Y.; Hashimoto, Y. E297G mutated bile salt export pump (BSEP) function enhancers derived from GW4064: Structural development study and separation from farnesoid X receptor-agonistic activity. *Bioorg. Med. Chem. Lett.* **2012**, *22*, 3962–3966. [CrossRef] [PubMed]
108. Kato, T.; Hayashi, H.; Sugiyama, Y. Short- and medium-chain fatty acids enhance the cell surface expression and transport capacity of the bile salt export pump (BSEP/ABCB11). *Biochim. Biophys. Acta Mol. Cell Biol. Lipids* **2010**, *1801*, 1005–1012. [CrossRef] [PubMed]
109. Hayashi, H.; Naoi, S.; Hirose, Y.; Matsuzaka, Y.; Tanikawa, K.; Igarashi, K.; Nagasaka, H.; Kage, M.; Inui, A.; Kusuhara, H. Successful treatment with 4-phenylbutyrate in a patient with benign recurrent intrahepatic cholestasis type 2 refractory to biliary drainage and bilirubin absorption. *Hepatol. Res.* **2016**, *46*, 192–200. [CrossRef]
110. Mareux, E.; Lapalus, M.; Amzal, R.; Almes, M.; Aït-Slimane, T.; Delaunay, J.; Adnot, P.; Collado-Hilly, M.; Davit-Spraul, A.; Falguières, T.; et al. Functional rescue of an ABCB11 mutant by ivacaftor: A new targeted pharmacotherapy approach in bile salt export pump deficiency. *Liver Int.* **2020**, *40*, 1917–1925. [CrossRef]
111. Kagawa, T.; Orii, R.; Hirose, S.; Arase, Y.; Shiraishi, K.; Mizutani, A.; Tsukamoto, H.; Mine, T. Ursodeoxycholic acid stabilizes the bile salt export pump in the apical membrane in MDCK II cells. *J. Gastroenterol.* **2014**, *49*, 890–899. [CrossRef]
112. Imagawa, K.; Takayama, K.; Isoyama, S.; Tanikawa, K.; Shinkai, M.; Harada, K.; Tachibana, M.; Sakurai, F.; Noguchi, E.; Hirata, K.; et al. Generation of a bile salt export pump deficiency model using patient-specific induced pluripotent stem cell-derived hepatocyte-like cells. *Sci. Rep.* **2017**, *7*, 41806. [CrossRef]
113. Engelmann, G.; Wenning, D.; Herebian, D.; Sander, O.; Droge, C.; Kluge, S.; Kubitz, R. Two Case Reports of Successful Treatment of Cholestasis with Steroids in Patients with PFIC-2. *Pediatrics* **2015**, *135*, e1326–e1332. [CrossRef]
114. Dröge, C.; Schaal, H.; Engelmann, G.; Wenning, D.; Häussinger, D.; Kubitz, R. Exon-skipping and mRNA decay in human liver tissue: Molecular consequences of pathogenic bile salt export pump mutations. *Sci. Rep.* **2016**, *6*, 24827. [CrossRef] [PubMed]
115. Ellinger, P.; Stindt, J.; Dröge, C.; Sattler, K.; Stross, C.; Kluge, S.; Herebian, D.; Smits, S.H.J.; Burdelski, M.; Schulz-Jürgensen, S.; et al. Partial external biliary diversion in bile salt export pump deficiency: Association between outcome and mutation. *World J. Gastroenterol.* **2017**, *23*, 5295. [CrossRef] [PubMed]

116. Davit-Spraul, A.; Oliveira, C.; Gonzales, E.; Gaignard, P.; Thérond, P.; Jacquemin, E. Liver transcript analysis reveals aberrant splicing due to silent and intronic variations in the ABCB11 gene. *Mol. Genet. Metab.* **2014**, *113*, 225–229. [CrossRef] [PubMed]
117. Naoi, S.; Hayashi, H.; Inoue, T.; Tanikawa, K.; Igarashi, K.; Nagasaka, H.; Kage, M.; Takikawa, H.; Sugiyama, Y.; Inui, A.; et al. Improved Liver Function and Relieved Pruritus after 4-Phenylbutyrate Therapy in a Patient with Progressive Familial Intrahepatic Cholestasis Type 2. *J. Pediatr.* **2014**, *164*, 1219–1227.e3. [CrossRef] [PubMed]
118. Amzal, R.; Thébaut, A.; Lapalus, M.; Almes, M.; Grosse, B.; Mareux, E.; Collado-Hilly, M.; Davit-Spraul, A.; Bidou, L.; Namy, O.; et al. Pharmacological premature termination codon readthrough of ABCB11 in bile salt export pump deficiency: An in vitro study. *Hepatology* **2020**, hep.31476. [CrossRef]
119. Strautnieks, S.S.; Byrne, J.A.; Pawlikowska, L.; Cebecauerová, D.; Rayner, A.; Dutton, L.; Meier, Y.; Antoniou, A.; Stieger, B.; Arnell, H.; et al. Severe Bile Salt Export Pump Deficiency: 82 Different ABCB11 Mutations in 109 Families. *Gastroenterology* **2008**, *134*, 1203–1214.e8. [CrossRef]
120. Highlights of Prescribing Information. Available online: https://www.accessdata.fda.gov/drugsatfda_docs/label/2018/207999s003lbl.pdf (accessed on 6 November 2020).
121. Fiorucci, S.; Antonelli, E.; Rizzo, G.; Renga, B.; Mencarelli, A.; Riccardi, L.; Orlandi, S.; Pellicciari, R.; Morelli, A. The nuclear receptor SHP mediates inhibition of hepatic stellate cells by FXR and protects against liver fibrosis. *Gastroenterology* **2004**, *127*, 1497–1512. [CrossRef]
122. Trauner, M.; Nevens, F.; Shiffman, M.L.; Drenth, J.P.H.; Bowlus, C.L.; Vargas, V.; Andreone, P.; Hirschfield, G.M.; Pencek, R.; Malecha, E.S.; et al. Long-term efficacy and safety of obeticholic acid for patients with primary biliary cholangitis: 3-year results of an international open-label extension study. *Lancet Gastroenterol. Hepatol.* **2019**, *4*, 445–453. [CrossRef]
123. Younossi, Z.M.; Ratziu, V.; Loomba, R.; Rinella, M.; Anstee, Q.M.; Goodman, Z.; Bedossa, P.; Geier, A.; Beckebaum, S.; Newsome, P.N.; et al. Obeticholic acid for the treatment of non-alcoholic steatohepatitis: Interim analysis from a multicentre, randomised, placebo-controlled phase 3 trial. *Lancet* **2019**, *394*, 2184–2196. [CrossRef]
124. Fiorucci, S.; Biagioli, M.; Sepe, V.; Zampella, A.; Distrutti, E. Bile acid modulators for the treatment of nonalcoholic steatohepatitis (NASH). *Expert Opin. Investig. Drugs* **2020**, *29*, 623–632. [CrossRef]
125. Garzel, B.; Yang, H.; Zhang, L.; Huang, S.-M.; Polli, J.E.; Wang, H. The Role of Bile Salt Export Pump Gene Repression in Drug-Induced Cholestatic Liver Toxicity. *Drug Metab. Dispos.* **2014**, *42*, 318–322. [CrossRef]
126. Hiebl, V.; Ladurner, A.; Latkolik, S.; Dirsch, V.M. Natural products as modulators of the nuclear receptors and metabolic sensors LXR, FXR and RXR. *Biotechnol. Adv.* **2018**, *36*, 1657–1698. [CrossRef]
127. Wu, G.; Wen, M.; Sun, L.; Li, H.; Liu, Y.; Li, R.; Wu, F.; Yang, R.; Lin, Y. Mechanistic insights into geniposide regulation of bile salt export pump (BSEP) expression. *RSC Adv.* **2018**, *8*, 37117–37128. [CrossRef]
128. Hoeke, M.O.; Plass, J.R.M.; Heegsma, J.; Geuken, M.; van Rijsbergen, D.; Baller, J.F.W.; Kuipers, F.; Moshage, H.; Jansen, P.L.M.; Faber, K.N. Low retinol levels differentially modulate bile salt-induced expression of human and mouse hepatic bile salt transporters. *Hepatology* **2009**, *49*, 151–159. [CrossRef]
129. Telbisz, Á.; Homolya, L. Recent advances in the exploration of the bile salt export pump (BSEP/ABCB11) function. *Expert Opin. Ther. Targets* **2016**, *20*, 501–514. [CrossRef]
130. Halilbasic, E.; Steinacher, D.; Trauner, M. Nor-Ursodeoxycholic Acid as a Novel Therapeutic Approach for Cholestatic and Metabolic Liver Diseases. *Dig. Dis.* **2017**, *35*, 288–292. [CrossRef]
131. Kim, D.J.; Yoon, S.; Ji, S.C.; Yang, J.; Kim, Y.-K.; Lee, S.; Yu, K.-S.; Jang, I.-J.; Chung, J.-Y.; Cho, J.-Y. Ursodeoxycholic acid improves liver function via phenylalanine/tyrosine pathway and microbiome remodelling in patients with liver dysfunction. *Sci. Rep.* **2018**, *8*, 11874. [CrossRef]
132. Halilbasic, E.; Fiorotto, R.; Fickert, P.; Marschall, H.-U.; Moustafa, T.; Spirli, C.; Fuchsbichler, A.; Gumhold, J.; Silbert, D.; Zatloukal, K.; et al. Side chain structure determines unique physiologic and therapeutic properties of norursodeoxycholic acid in Mdr2−/− mice. *Hepatology* **2009**, *49*, 1972–1981. [CrossRef]
133. Moustafa, T.; Fickert, P.; Magnes, C.; Guelly, C.; Thueringer, A.; Frank, S.; Kratky, D.; Sattler, W.; Reicher, H.; Sinner, F.; et al. Alterations in Lipid Metabolism Mediate Inflammation, Fibrosis, and Proliferation in a Mouse Model of Chronic Cholestatic Liver Injury. *Gastroenterology* **2012**, *142*, 140–151.e12. [CrossRef]
134. Ito, S.; Hayashi, H.; Sugiura, T.; Ito, K.; Ueda, H.; Togawa, T.; Endo, T.; Tanikawa, K.; Kage, M.; Kusuhara, H.; et al. Effects of 4-phenylbutyrate therapy in a preterm infant with cholestasis and liver fibrosis. *Pediatr. Int.* **2016**, *58*, 506–509. [CrossRef]
135. Vitale, G.; Simonetti, G.; Pirillo, M.; Taruschio, G.; Pietro, A. Bipolar and Related Disorders Induced by Sodium 4-Phenylbutyrate in a Male Adolescent with Bile Salt Export Pump Deficiency Disease. *Psychiatry Investig.* **2016**, *13*, 580. [CrossRef]
136. Van Goor, F.; Hadida, S.; Grootenhuis, P.D.J.; Burton, B.; Cao, D.; Neuberger, T.; Turnbull, A.; Singh, A.; Joubran, J.; Hazlewood, A.; et al. Rescue of CF airway epithelial cell function in vitro by a CFTR potentiator, VX-770. *Proc. Natl. Acad. Sci. USA* **2009**, *106*, 18825–18830. [CrossRef]
137. De Boeck, K.; Munck, A.; Walker, S.; Faro, A.; Hiatt, P.; Gilmartin, G.; Higgins, M. Efficacy and safety of ivacaftor in patients with cystic fibrosis and a non-G551D gating mutation. *J. Cyst. Fibros.* **2014**, *13*, 674–680. [CrossRef]
138. Delaunay, J.; Bruneau, A.; Hoffmann, B.; Durand-Schneider, A.-M.; Barbu, V.; Jacquemin, E.; Maurice, M.; Housset, C.; Callebaut, I.; Aït-Slimane, T. Functional defect of variants in the adenosine triphosphate–binding sites of ABCB4 and their rescue by the cystic fibrosis transmembrane conductance regulator potentiator, ivacaftor (VX-770). *Hepatology* **2017**, *65*, 560–570. [CrossRef]

139. Van Wessel, D.B.E.; Thompson, R.J.; Gonzales, E.; Jankowska, I.; Sokal, E.; Grammatikopoulos, T.; Kadaristiana, A.; Jacquemin, E.; Spraul, A.; Lipiński, P.; et al. Genotype correlates with the natural history of severe bile salt export pump deficiency. *J. Hepatol.* **2020**, *73*, 84–93. [CrossRef]
140. Dabrowski, M.; Bukowy-Bieryllo, Z.; Zietkiewicz, E. Advances in therapeutic use of a drug-stimulated translational readthrough of premature termination codons. *Mol. Med.* **2018**, *24*, 25. [CrossRef]
141. Cuyx, S.; De Boeck, K. Treating the Underlying Cystic Fibrosis Transmembrane Conductance Regulator Defect in Patients with Cystic Fibrosis. *Semin. Respir. Crit. Care Med.* **2019**, *40*, 762–774. [CrossRef]

International Journal of
Molecular Sciences

MDPI

Review

Molecular Regulation of Canalicular ABC Transporters

Amel Ben Saad [1,2,†], Alix Bruneau [2,3,†], Elodie Mareux [1,†], Martine Lapalus [1], Jean-Louis Delaunay [2], Emmanuel Gonzales [1,4], Emmanuel Jacquemin [1,4], Tounsia Aït-Slimane [2,‡] and Thomas Falguières [1,*,‡]

[1] Physiopathogénèse et Traitement des Maladies du foie, Inserm, Université Paris-Saclay, UMR_S 1193, Hepatinov, 91400 Orsay, France; amel.ben-saad@inserm.fr (A.B.S.); elodie.mareux@universite-paris-saclay.fr (E.M.); martine.lapalus@universite-paris-saclay.fr (M.L.); emmanuel.gonzales@aphp.fr (E.G.); emmanuel.jacquemin@aphp.fr (E.J.)

[2] Centre de Recherche Saint-Antoine (CRSA), Inserm, Sorbonne Université, UMR_S 938, Institute of Cardiometabolism and Nutrition (ICAN), 75012 Paris, France; alix.bruneau@charite.de (A.B.); jean-louis.delaunay@sorbonne-universite.fr (J.-L.D.); tounsia.ait_slimane@sorbonne-universite.fr (T.A.-S.)

[3] Department of Hepatology and Gastroenterology, Charité Universitäts Medizin Berlin, 13353 Berlin, Germany

[4] Paediatric Hepatology and Paediatric Liver Transplant Department, Reference Center for Rare Paediatric Liver Diseases, FILFOIE, ERN RARE LIVER, Assistance Publique-Hôpitaux de Paris, Faculté de Médecine Paris-Saclay, CHU Bicêtre, 94270 Le Kremlin-Bicêtre, France

* Correspondence: thomas.falguieres@inserm.fr; Tel.: +33-(0)1-69-15-62-94

† These authors share first co-authorship.

‡ These authors share last co-authorship.

Abstract: The ATP-binding cassette (ABC) transporters expressed at the canalicular membrane of hepatocytes mediate the secretion of several compounds into the bile canaliculi and therefore play a key role in bile secretion. Among these transporters, ABCB11 secretes bile acids, ABCB4 translocates phosphatidylcholine and ABCG5/G8 is responsible for cholesterol secretion, while ABCB1 and ABCC2 transport a variety of drugs and other compounds. The dysfunction of these transporters leads to severe, rare, evolutionary biliary diseases. The development of new therapies for patients with these diseases requires a deep understanding of the biology of these transporters. In this review, we report the current knowledge regarding the regulation of canalicular ABC transporters' folding, trafficking, membrane stability and function, and we highlight the role of molecular partners in these regulating mechanisms.

Keywords: bile secretion; ABCB1; ABCB4; ABCB11; ABCC2; ABCG5/G8; molecular partners

Citation: Ben Saad, A.; Bruneau, A.; Mareux, E.; Lapalus, M.; Delaunay, J.-L.; Gonzales, E.; Jacquemin, E.; Aït-Slimane, T.; Falguières, T. Molecular Regulation of Canalicular ABC Transporters. *Int. J. Mol. Sci.* **2021**, *22*, 2113. https://doi.org/10.3390/ijms22042113

Academic Editor: Jose J. G. Marin

Received: 1 February 2021
Accepted: 18 February 2021
Published: 20 February 2021

Publisher's Note: MDPI stays neutral with regard to jurisdictional claims in published maps and institutional affiliations.

1. Introduction

One of the liver's main functions is bile production and secretion. In addition to its digestive function, bile plays an important role in detoxification. Bile secretion is mediated by several ATP-binding cassette (ABC) transporters, which are expressed at the canalicular membrane of hepatocytes. The main canalicular ABC transporters are the bile salt export pump (BSEP, ABCB11), which transports bile acids (BAs), ABCB4 also known as multidrug resistance protein 3 (MDR3) translocating phosphatidylcholine (PC) and ABCG5/G8 excreting cholesterol [1]. BA, PC and cholesterol form mixed micelles in the aqueous environment of bile. In addition to these compounds, bile contains a wide variety of drugs and organic anions, which are secreted by ABCB1, also known as multidrug resistance protein 1 (MDR-1, or P-glycoprotein) and ABCC2 (multidrug resistance-associated protein 2, MRP2), respectively [2,3]. ABC transporters share a common basic architecture and similar ATP-driven functions. They are organized in two repeats, each containing a membrane-spanning domain (MSD) with six transmembrane (TM) helices and a cytoplasmic nucleotide-binding domain (NBD), those two moieties being connected by an intracellular linker. The MSDs ensure substrate recognition and translocation, whereas NBDs, which are highly conserved among all ABC transporters, provide the energy for this process [3]. In contrast to other canalicular ABC transporters, ABCG5 and ABCG8 are half transporters that require heterodimerization to ensure their function, and ABCC2 has a third MSD at its N-terminus [4]. The key role of canalicular ABC transporters in bile secretion is highlighted by their implication in a wide range of diseases such as cholestasis (ABCB4, ABCB11), sitosterolemia (ABCG5/G8), Dubin–Johnson syndrome (ABCC2) and cancer (ABCB1) (Figure 1) [3]. To develop new therapies for patients with diseases related to deficient canalicular ABC transporters, it is crucial to better understand the molecular mechanisms regulating the traffic and function of these transporters. Several studies have reported that the biosynthesis, trafficking and activity of ABC transporters are regulated by numerous molecular partners, most of which have been identified by two-hybrid screens using liver banks [3]. Targeting these interactors represents a potential therapeutic option for patients. This review focuses on molecular regulators of canalicular ABC transporters involved in bile formation.

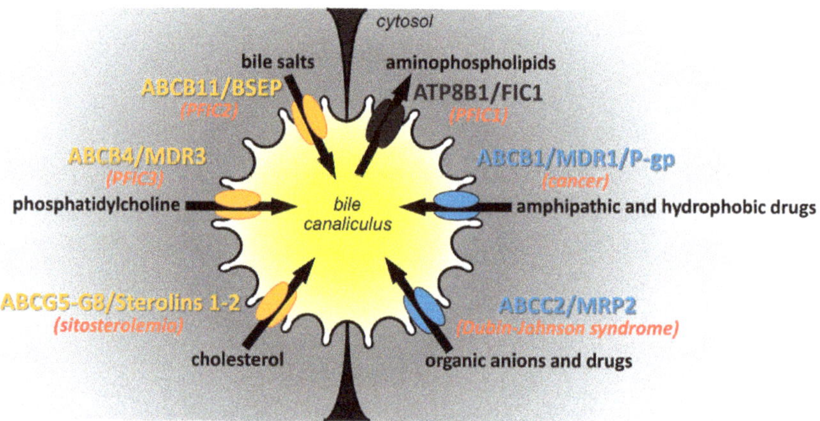

Figure 1. ATP-binding cassette (ABC) transporters at the canalicular membrane of hepatocytes. The bile canaliculus is formed by the canalicular membrane of hepatocytes. The main canalicular ABC transporters are indicated according to the nature of their substrates: in yellow for hydrophobic substrates and in blue for drugs. Note that ATP8B1 is not an ABC transporter but a P-type ATPase. However, its function is tightly related to the other canalicular ABC transporters. The substrates of these transporters are shown in black, and their flows are indicated by black arrows. The main diseases associated with functional defects of these transporters are indicated in red. PFIC: progressive familial intrahepatic cholestasis.

2. Folding and Glycosylation of Canalicular ABC Transporters

Protein folding is a highly regulated process that is mediated by numerous factors, including folding proteins and molecular chaperones [5]. Some of these proteins have been shown as interactors of canalicular ABC transporters, controlling their biosynthesis and folding. As most transmembrane proteins, ABC transporter biosynthesis starts with their cotranslational translocation and insertion into the endoplasmic reticulum (ER) membrane through the Sec61 translocon complex [6]. Numerous accessory factors were described to facilitate the translocation process, including the translocating chain-associated membrane protein (TRAM) and the translocon-associated protein (TRAP) [7]. Interestingly, TRAM and three subunits of the TRAP complex (SSR1, SSR3 and SSR4) were found to interact and coprecipitate specifically with ABCB11 [8].

Once in the ER lumen, nascent ABC transporters are N-glycosylated. Glycans are added to their extracellular asparagine residues [9] and play a critical role in protein folding, stability and interaction with some chaperones [10]. Two main chaperone families exist in the ER: the heat shock protein (HSP) family, which promotes the folding of a wide variety of proteins, and the lectin chaperones, which recognize and fold specifically glycosylated proteins [11]. Calnexin (CNX) and its soluble homolog calreticulin are lectin chaperones that were shown to bind both ABCG5 and ABCG8 and stimulate their folding and assembly. Okiyoneda and colleagues showed that the silencing of either CNX or calreticulin decreases the expression of ABCG5/G8 [12]. It has also been reported that CNX and the heat shock cognate 71 kDa protein (Hsc70) interact with ABCB1 [13,14].

In addition to their folding function, chaperones are central players in the quality control process. They evaluate the folding state of proteins and regulate their ER retention or ER exit [15]. They allow only properly folded proteins to exit the ER, and contrariwise, they retain abnormally folded proteins longer before their targeting to the ER-associated degradation (ERAD) pathway. Some variations in canalicular ABC transporter-encoding genes were described to affect the folding of these transporters, thereby leading to their retention in the ER. Indeed, a prolonged association between misfolded ABCB1 variants and CNX has been observed [13]. Interestingly, we and others have demonstrated that several small molecules known as pharmacological and chemical chaperones can facilitate the folding and exit of defective ABCB4 and ABCB11 variants from the ER [16–18]. Another key component of the quality control system is the B-lymphocyte receptor-associated protein (BAP), which was shown to control the folding state and sorting of many proteins in the ER [19,20]. The BAP29 and BAP31 isoforms were described to interact with the N-terminal domain of ABCB1 and ABCB11, respectively [8,21]. More interestingly, it has been shown that some mutations in the *BAP31* gene are associated with liver dysfunction and cholestasis [22].

Using an immunoprecipitation assay combined with mass spectrometry analysis and yeast two-hybrid screens, Przybylla and colleagues identified several novel potential ER-resident partners for ABCB11, including the receptor expression-enhancing proteins (REEPs) involved in ER shaping, the immediate early response 3 interacting protein 1 (IER3IP1), the transmembrane proteins 205 (TMEM205) and 14A (TMEM14A) and the bile Acyl-CoA synthetase (BAC) [8]. However, their role in the regulation of the folding and/or the trafficking of ABCB11 has not been studied yet.

3. From the Endoplasmic Reticulum to the Plasma Membrane

Correctly folded canalicular ABC transporters leave the ER to reach the Golgi apparatus, where they undergo further post-translational modifications. However, little is known about the molecular players regulating their trafficking from the ER to the Golgi. Involvement of the coat protein complex II (COP II) machinery in ABCC7/cystic fibrosis transmembrane conductance regulator (CFTR) and ABCA1 exit from the ER has been documented [23,24]. Given the homology between these proteins, the same pathway may be involved in the sorting and traffic of canalicular ABC transporters. Once in the Golgi apparatus, canalicular ABC transporters undergo more complex glycosylation [25,26]; then, they are sorted and packaged into secretory vesicles and further delivered to the canalicular membrane [25,26].

Unlike other apical proteins in liver cells, canalicular ABC transporters do not undergo transcytosis after their sorting from the trans-Golgi network (TGN), but they are directly targeted to the canalicular membrane or subapical compartments (SACs) [27,28]. The labeling of newly synthesized ABC transporters has shown that ABCB1 is directly delivered to the canalicular membrane, whereas ABCB11 is targeted to the SAC before reaching the canalicular membrane [29]. Kipp and coworkers also described the involvement of many intracellular components, such as cyclic adenosine monophosphate (cAMP), taurocholate and Ca^{2+} in the vesicular trafficking of canalicular ABC transporters. Indeed, they showed that the administration of these components into the perfused liver or directly in cells increases the amount of ABC transporters present at the canalicular membrane as well as bile secretion [28].

In addition to these components, many interacting proteins, including specific GTPases, kinases, molecular motors and other factors, have been shown to associate with canalicular ABC transporters and promote their exocytosis and/or endocytosis. Indeed, CFTR-associated ligand (CAL), a Golgi-associated protein, has been found to interact with

ABCC2 and regulate its plasma membrane targeting [30]. Some members of the Ras-related in brain (RAB) GTPase family have also been identified as ABCB1-interacting proteins. The overexpression of RAB4, RAB5 or their constitutively active forms increases the presence of ABCB1 at the cell surface [31,32]. Moreover, the motor protein myosin II regulatory light chain (MLC2) was reported as a prominent regulator of canalicular ABC transporters. Using a yeast two-hybrid screen of a rat liver cDNA library, MLC2 was found to interact with the linker domains of ABCB1, ABCB4 and ABCB11 [33]. Based on immunofluorescence and biochemical experiments, Chan and colleagues showed that the inhibition of MLC2 or the expression of its dominant negative form leads to a decrease in ABCB11 levels at the apical membrane [33].

Furthermore, other studies have revealed that many kinases are important for the exocytosis of ABC transporters. These include the p38 mitogen-activated protein kinase (MAPK) [34,35], protein kinase A (PKA [36], protein kinase C (PKC) [37], proto-oncogene serine/threonine-protein kinase (Pim-1) [38] and phosphoinositide 3-kinase (PI3K) [39]. Misra and coworkers showed that the administration of wortmannin, a specific inhibitor of PI3K, resulted in a decrease in the amounts of ABCB11 and ABCC2 present at the canalicular membrane [39,40]. Another kinase, the liver kinase B1 (LKB1), was shown as a key regulator of ABCB11 trafficking. In LKB1 knockout (KO) mice, an altered distribution of ABCB11, as well as an impaired bile formation, was observed [41,42]. Ursodeoxycholic acid (UDCA), used as a treatment for patients with cholestasis, was also shown to stimulate the targeting of ABCC2 and ABCB11 transporters to the plasma membrane [43,44].

4. Membrane Stability of Canalicular ABC Transporters

Membrane protein turnover through uninterrupted synthesis and degradation is essential to provide a functional set of proteins and ensure cell function. Tight regulation of protein stability at the plasma membrane is fundamental for cell homeostasis and relies on environmental signals and/or post-translational modifications such as phosphorylation/dephosphorylation and ubiquitination/deubiquitination cycles. Indeed, on the one hand, the accumulation of some proteins at the plasma membrane can be deleterious for cells and result, for instance, in a multidrug resistance (MDR) phenotype, a true obstacle in cancer treatment, caused by the development of chemoresistance [45]. On the other hand, defects in the expression level or stability of ABC transporters can also contribute to the development of human diseases, including cystic fibrosis [46], neuropathies [47] or cholestasis [48]. The regulation of the stability/turnover of proteins such as canalicular ABC transporters remains poorly understood, mostly due to technical limitations. ABC transporter stability is yet essential to regulate the spatiotemporal availability of a given protein at the bile canaliculi (e.g., between meals, the need for bile is reduced, and the amount of ABC transporters at the plasma membrane must be regulated accordingly). On the contrary, a decrease in the stability of numerous transporters at the plasma membrane, such as ABCB1, ABCC1 (MRP1) or ABCG2 (BCRP), would be necessary to improve the efficiency of cancer treatments facing ABC transporter-mediated MDR.

Several kinases are involved in the regulation of ABC transporter stability. The atypical Pim-1 kinase coimmunoprecipitates with and phosphorylates ABCB1. Pim-1 downregulation by siRNA diminishes ABCB1 maturation and favors its degradation through the ubiquitin–proteasome system, indicating that Pim-1 may stabilize ABCB1 at the plasma membrane [38]. A yeast two-hybrid screen using the linker domain of ABCB4 allowed the identification of receptor for activated C-kinase 1 (RACK1) as an interacting partner of this transporter. Moreover, RACK1 has been reported to activate two isoforms of PKC and be involved in the regulation of membrane stability for many proteins, thus playing a determinant role in fundamental cellular activities [49]. Following RACK1 knockdown, ABCB4 is no longer localized at the plasma membrane but mainly relocalized in cytosolic compartments [50].

PDZ (postsynaptic density protein (*PSD95*), Drosophila disc large tumor suppressor (*Dlg1*) and zonula occludens-1 protein (*ZO-1*))-domain-containing proteins are well known for their function in protein stabilization at membranes. They act as scaffolds by linking transmembrane proteins to the cytoskeleton and thus regulate their subcellular localization and stability at the plasma membrane [51]. The PDZ-domain-containing protein ezrin–radixin–moesin (ERM)-binding phosphoprotein 50 (EBP50), also known as sodium–hydrogen exchanger regulatory factor-1 (NHERF1), interacts with both ABCC2 and ABCB4 through their C-terminal PDZ-binding motif [52,53]. In the absence of EBP50, ABCB4 and ABCC2 are both targeted to the plasma membrane, but their presence is drastically reduced, therefore demonstrating that EBP50 plays a crucial role in the regulation of membrane stability for both ABCC2 and ABCB4 [52,53]. PDZK1 (NHERF3), another PDZ domain-containing protein, interacts with ABCC2 and increases its plasma membrane stability. Indeed, the expression of a dominant negative form of PDZK1 leads to a decrease in ABCC2 membrane expression and its accumulation in intracellular compartments [54,55].

Radixin is part of the ERM protein family, which is involved in actin cytoskeleton remodeling, e.g., to organize submembranous cortical actin or microvillosities [56]. Radixin KO mice develop a phenotype comparable to Dubin–Johnson syndrome. Indeed, these mice show a severe reduction in ABCC2 protein expression at bile the canaliculi without any change at the mRNA level. Importantly, this effect is specific to ABCC2 as no effect was observed for other canalicular ABC transporters such as ABCB1, ABCB11 or ABCB4 [57]. Moreover, a direct interaction between ABCC2 and radixin has been confirmed by GST-pulldown [57].

5. Endocytosis and Membrane Recycling of Canalicular ABC Transporters

Small GTPases, protein kinases, class V myosins and adaptor proteins have been identified as molecular players in the regulation of canalicular ABC transporter endocytosis and recycling [3]. The existence of ABCB11 intracytoplasmic reservoirs is well known [28,29], but the nature of those compartments long remained uncharacterized until a YFP-tagged ABCB11 was detected in the SAC, a RAB11-positive compartment. Indeed, RAB11 and myosin VB (MYO5B) are established regulators of the recycling of several proteins from the SAC to the plasma membrane [58]. ABCB11 continuously cycles between the canalicular membrane of hepatocytes and the SAC [59]. This constant exchange allows tight regulation of ABCB11 availability at the bile canaliculi. The perturbation of actin cytoskeleton or microtubules inhibits this traffic [59]. These results were corroborated as ABCB11 apical targeting is considerably slowed down in WIF-B9 cells expressing RAB11- or MYO5B-dominant negative constructs [60]. In the presence of a mutated or truncated MYO5B, ABCC2 displays an intracellular localization in RAB8- and RAB11-positive compartments, suggesting defects in canalicular transporter recycling [61]. Recently, mutations in the *MYO5B* gene, identified in patients, have been associated with a progressive familial intrahepatic cholestasis (PFIC)-like phenotype, further proposed as PFIC6 [61,62].

The ERM protein family has also been involved in the endocytic process of several ABC transporters. Coimmunoprecipitation performed with human liver lysates highlighted the interaction between ABCC2 and ezrin, and additional experiments showed that ezrin phosphorylation on its Thr567 regulates this interaction, thus further controlling the amounts of ABCC2 present at the plasma membrane [63].

Several isoforms of PKC, as well as PKA, PI3K, Pim-1 or Fyn kinases, play a role in the regulation of ABCB1, ABCC2 and ABCB11 membrane targeting or endocytosis [33,39,41,42,64–66]. Cantore and colleagues have reported that the Src family kinase Fyn induces ABCC2 and ABCB11 retrieval from the canalicular membrane by increasing cortactin phosphorylation [66]. Schonhoff and colleagues observed that taurolithocholate-activated PKCε phosphorylates and activates myristoylated alanine-rich C-kinase substrate (MARCKS), a membrane-bound F-actin crosslinking protein [67]. MARCKS is a crucial regulator of molecular interactions and cytoskeletal reorganization. In a nonphosphorylated state, MARCKS is associated with the cytosolic leaflet of the plasma membrane and can serve as a stabilizer for transmembrane proteins, whereas after phosphorylation, MARCKS is released in the cytosol, where it can interact with other proteins [68].. MARCKS has been shown to regulate the endocytosis of ABCC2 and ABCB1 [67,69]. Indeed, in colon carcinoma cells, MARCKS expression has been associated with the reduced export function of ABCB1 [69].

The hematopoietic cell-specific Lyn substrate 1 associated protein X-1 (HAX-1) is a small protein abundantly expressed in the liver, regulating cortical actin organization. This protein has been identified as an interactor of the linker domain of ABCB1, ABCB4 and ABCB11 [70]. Through this interaction, HAX-1 has been proposed to stabilize ABCB11 at the plasma membrane [70]. However, the role of HAX-1 in other canalicular ABC transporter endocytosis has not been further investigated.

A tyrosine motif has been identified in the ABCB11 cytoplasmic tail [71] along with one of its partners, the clathrin adaptor protein complex 2 (AP2) [72]. AP2 is localized at the plasma membrane and binds tyrosine-based internalization motifs of proteins, including ABCB11, thus allowing its internalization from the canalicular membrane through clathrin-dependent endocytosis [72]. The ubiquitination of ABCB11 and ABCC2 has also been shown to be essential for clathrin-mediated endocytosis and degradation of these transporters [73].

Hormones and intracellular signaling molecules also play a role in canalicular ABC transporter internalization. In a model of estradiol-induced cholestasis, authors showed that following treatment with estradiol-17β-d-glucuronide (E217G), ABCB11 and ABCC2 are relocalized from canalicular membranes to intracytoplasmic compartments. The same group demonstrated later that ABCB11 and ABCC2 endocytosis is mediated by PKC, which is activated by E217G, and that PKC inhibitors prevent the internalization of both transporters after treatment with estradiol [74,75]. In the same model, Zuchetti and colleagues established that glucagon and an adrenaline analog mediate cAMP activation, thus preventing ABCB11 and ABCC2 membrane retrieval [76]. Moreover, they showed that this E217G-induced endocytosis is AP2- and clathrin-dependent [77]. It has also been suggested that lipopolysaccharides act as signals for ABCC2 and ABCB11 endocytosis as their canalicular expression is reduced, with no mRNA decrease in an in vitro cholestatic model [78].

6. Regulation of the Transport Activity of Canalicular ABC Transporters

At the canalicular membrane, ABCB1, ABCB4, ABCB11, ABCC2 and ABCG5 have been proposed to mostly reside within glycosphingolipids-, cholesterol- and caveolin-1 (Cav-1)-enriched raft microdomains [27,79–81]. These domains could provide a favorable environment for the regulation of the activity of canalicular ABC transporters. Indeed, the shift of ABCB1 and ABCB11 from cholesterol-enriched microdomains to low-cholesterol domains lowered their transport activity [82–85]. Moreover, accumulating evidence indicates that phospholipids and cholesterol are required for the proper function of ABCB1, ABCB4, ABCB11 and ABCC2 [86–93]. In purified membrane vesicles, delipidation due to detergent action inactivates ABCB1, whereas phospholipid addition fully restores the ATPase activity of the transporter [94,95]. Phosphoinositides, lipid products from PI3K-mediated activities, are required for ABCB11 and ABCC2 activation because their addition reverses the negative effect of PI3K inhibitors on the activity of these transporters [40,96].

The importance of membrane cholesterol content has been highlighted in Atp8b1-deficient mice [92]. Indeed, in these mice, the normal phospholipid asymmetry of the canalicular membrane is lost, thereby enhancing sensitivity to cholesterol extraction by hydrophobic BA and subsequent loss of ABCB11 and ABCC2 activity [92]. How exactly membrane cholesterol influences the transport activity of these transporters is not known, but this may involve allosteric modulations and/or indirect means such as changes in membrane fluidity. Several studies have proposed that cholesterol directly interacts with the ABCB1 substrate binding site and thereby facilitates the recognition of small drugs (<500 Da) [89,97,98]. Cyclodextrin treatment or Cav-1 overexpression leads to cholesterol depletion from the plasma membrane, which inhibits ABCB1 transport activity by increasing membrane fluidity and loosening lipid packing density [99]. However, Moreno and colleagues showed that Cav-1 overexpression in mice increases both bile flow and the biliary secretion of phospholipids, BA and cholesterol, suggesting a positive role of Cav-1 in ABCB11 transport activity [100]. Considering the role of Cav-1 in intracellular cholesterol trafficking [101,102], some of the Cav-1 effects may be indirect and mediated through cholesterol homeostasis. Alternatively, Cav-1 may also directly bind ABCB1 and inhibit its transport activity [99,103–105]. It has been reported that the binding capacity of Cav-1 to ABCB1 is negatively modulated by Src kinase-mediated Cav-1 phosphorylation, a process facilitated by RACK1, which interacts with both Src and ABCB1 [105–107].

Several studies suggest a role for phosphorylation in the regulation of canalicular ABC transporters. Phosphorylation sites in the linker domain of ABCB1 have been well documented at Ser661, Ser667, Ser671, Ser675, and Ser683 [108–112]. Likewise, six potential phosphorylation sites have been found in the linker region of ABCC2 at Ser904, Ser912, Ser916, Ser917, Ser922 and Ser926 [113] and in the N-terminal domain of ABCB4 at Thr34, Thr44 and Ser49 [114]. An analysis of ABCB11 amino acid sequence also predicted multiple potential serine/threonine phosphorylation sites [115]. Overwhelming evidence indicates that PKC is a major player in ABC transporter phosphorylation and activity regulation. Phosphorylation within the linker domain of ABCB1 is specific for PKCα in purified vesicles from Sf9 cells [116]. The coexpression of PKCα and ABCB1 increases the ATPase activity of the transporter in insect and ovarian cells [116,117], while PKC inhibitor treatment did not alter ATPase activity in MCF-7 cells [118]. It has been shown that PKCα also mediates ABCB11 phosphorylation [115], as well as ABCC2 phosphorylation, resulting in stimulation of the intrinsic transport activity of ABCC2 [113]. PKC-dependent phosphorylation has also been shown to regulate ABCB4-mediated PC secretion [114]. The variation-induced impairment of ABCB4 N-terminal phosphorylation may also be related to a decrease in ABCB4-mediated PC secretion [114]. The substitution of all conserved serines in the linker domain of ABCC2 by non-phosphorylatable alanines significantly reduces the basal transport activity of ABCC2, while substitution into aspartates (mimicking constitutive phosphorylation of the residues) increases it [113]. Conversely, the role of ABCB1 phosphorylation in the regulation of its transport activity is less obvious because the substitution of potentially phosphorylatable residues by aspartates or non-phosphorylatable residues has no effect [119,120].

7. Ubiquitination and Degradation of Canalicular ABC Transporters

To target proteins for degradation, cells mostly use the endolysosomal pathway vs. proteasomal degradation, related to the monoubiquitination or polyubiquitination of their substrates, respectively [121,122].

The lysosomal pathway is the main way by which cells turn over plasma membrane proteins. Indeed, ABCB1 colocalizes with lysosomal-associated membrane protein 1 (LAMP1) in human colorectal cancer HTC15 cells [123]. In addition, the half-life of ABCB1 and ABCC2 is extended in cells treated with lysosomal inhibitors alone but not proteasomal inhibitors alone, suggesting the involvement of the lysosomal pathway in the degradation of these transporters [123,124]. However, ABCB11 expression is unaffected by treatment with lysosomal inhibitors, indicating that this transporter may use another degradation pathway [125,126]. Indeed, it has been shown that the inhibition of proteasomal degradation stabilizes wild-type (WT) and mutated ABCB11 in MDCK and HEK cells, suggesting that ABCB11 degradation involves the proteasome [125,127].

Several E3 ubiquitin ligases (E3 Ubl) may be involved in canalicular ABC transporter degradation. Ring finger protein 2 (RNF2) has E3 Ubl activity and may mediate the ubiquitination of ABCB1 [128]. E3 Ubl FBXO21 is involved in the proteasome-mediated degradation of ABCB1 [129]. Additionally, the E2-conjugating enzyme UBE2R1 (also named CDC34 or UBC3) and the E3 complex Skp1–Cullin–FBOX15 (SCFFbx15) are both implicated in ABCB1 ubiquitination [130]. Coprecipitation assays revealed that FBX015/Fbx15 (a member of the SCFFbx15 E3 complex) and UBE2R1 both interact with ABCB1, and their knockdown is associated with a decrease in ubiquitination and subsequent degradation of ABCB1. By contrast, FBX015 expression enhances ABCB1 ubiquitination and degradation [130].

ABCB1 ubiquitination may be modulated by the MAPK pathway [131,132]. Indeed, the inhibition of MEK or the downregulation of its downstream effectors, such as ERK and p90 ribosomal S6 kinases (RSKs), lower ABCB1 protein expression in HTC15 cells [131,133]. Pulse-chase labeling experiments revealed that MEK inhibitor-mediated downregulation of ABCB1 is caused by the increase of its degradation [131]. The same team has shown that RSK1 induces self-ubiquitination of UBE2R1, followed by its proteasomal degradation in a phosphorylation-dependent manner, thus resulting in the protection of ABCB1 against degradation [132].

Some variations in ABC transporter genes are responsible for the production of an unstable protein which is retained in the ER and subsequently degraded in the cytosol by the ERAD system [134]. Some misfolded ABCB11 variants appear to be more ubiquitinated than the WT transporter [126]. The RING finger proteins Rma1, TEB4 and HRD1 are all E3 Ubl involved in the ubiquitination of ABCB11-WT and its variants but with a folding sensitivity as the knockdown of each E3 Ubl stabilizes different ABCB11 variants [126]. HRD1 targets proteins with defects in the ER lumen side, while TEB4 and Rma1 target proteins with defects in their moieties facing the cytosol. Likewise, E3 Ubl seems to exhibit sensitivity towards ABC transporters. GP78, rather than TEB4 and HDR1, plays an important role in the ubiquitination of ABCC2, as shown in patients with obstructive cholestasis and in rifampicin-treated HepG2 cells [63,135]. Proteins can escape ubiquitination through small ubiquitin-like modifier (SUMO) modification as both processes compete on the same residues. Using a protein–protein interaction assay, a number of SUMO-related proteins (including SUMO-1 and ubiquitin carrier protein 9/Ubc9) were pulled down using the linker region of the rat ABCC2 [136]. Moreover, the knockdown of SUMO-related enzymes in hepatoma cells reduces ABCC2 protein expression but not its mRNA expression or canalicular localization [136]. Proteins can escape degradation subsequent to their ubiquitination by reversing ubiquitination thanks to deubiquitinating enzymes (DUBs). As an example, the DUB ubiquitin-specific protease 19 (USP19), through TEB4 stabilization, negatively regulates the expression of a defective ABCB11 variant [137].

Manipulation of the ER quality control system might be combined with chemical or pharmacological chaperones to stabilize variants and restore the cell surface expression of ABC transporters. Indeed, cell surface biotinylation assays revealed that the most frequent ABCB11 variants found in patients with PFIC2, E297G and D482G are highly ubiquitinated [138], and this induces their internalization [73]. Additionally, the half-life of ABCC2 is extended in cells overexpressing a dominant negative form of ubiquitin due to the inhibition of ABCC2 degradation [73,124]. Therefore, by reducing susceptibility to ubiquitination, the chemical chaperone 4-phenylbutyrate (4-PB) extends the half-life of both ABCB11 and ABCC2 expressed at the cell surface [124,138,139]. However, since 4-PB has no effect on ABCB1 [139], the 4-PB mechanism of action would involve interaction with specific E3 Ubl or adaptor protein(s) for both ABCB11 and ABCC2. For instance, 4-PB downregulates Hsc70 (Hsp73) which plays a role in the lysosomal degradation of intracellular proteins and was shown to be required for the ubiquitin-dependent degradation of several proteins ([140]. and references therein).

8. Conclusions

Over the last decade, proteomic studies have become an important means for understanding the biology and pathophysiology of many proteins. The characterization and identification of key players prompted the understanding of the molecular basis of pathologies and helped the development of improved therapeutic approaches for patients. We described here the molecular partners that interact either directly or indirectly with the five canalicular ABC transporters (ABCB11, ABCB4, ABCG5/G8, ABCB1 and ABCC2) and regulate their folding, trafficking, stability and function (Table 1). Nowadays, an important amount of information regarding the genetics of ABC transporters is gathered.

We expect that proteomic approaches merged with genomic studies will be a powerful tool in the development of personalized treatment for patients with biliary diseases related to canalicular ABC transporter defects.

Table 1. Molecular partners of canalicular ABC transporters.

Proteins [1]	Interacting ABC Transporters	Subcellular Localization	Functions	References
AP2	ABCB11	Plasma membrane	Clathrin-dependent endocytosis	[71–73,77]
BACs	ABCB11	ER	Conjugation of bile acids	[8]
BAP29	ABCB1	ER	Controls protein sorting from the ER	[21]
BAP31	ABCB11	ER	Controls protein sorting from the ER	[8]
CAL	ABCC2	Golgi	Golgi sorting	[30]
Calnexin	ABCG5/G8 ABCB1	ER	Assists glycoprotein folding	[12,13,16]
Calreticulin	ABCG5/G8	ER	Assists glycoprotein folding	[12]
Cav-1	ABCB1 ABCB4 ABCB11 ABCC2 ABCG5/G8	Plasma membrane	Scaffold protein	[99,100,103–105,107]
CD44	ABCB1	Plasma membrane	Inhibitor of FBX021	[129]
EBP50	ABCB4 ABCC2	Plasma membrane	Scaffold protein	[52,53]
Ezrin	ABCC2	Plasma membrane Associated with the cytoskeleton	Endocytosis	[63]
FBXO21	ABCB1	Cytosol	E3 ubiquitin ligase	[129]
Fyn	ABCC2 ABCB11	Plasma membrane	Endocytosis	[66]
GP78	ABCC2	ER	SUMO-related proteins	[63,135]
HAX-1	ABCB1 ABCB4 ABCB11	Cytosol Associated with cortical actin	Clathrin-dependent endocytosis	[70]
Hsc70	ABCB1	ER	Chaperone Assists protein folding	[14,16]
IER3IP1	ABCB11	ER	Implicated in apoptosis and protein transport from the ER to the Golgi	[8]
LKB1	ABCB11	Cytoplasm	Intracellular traffic	[41,42]
MARCKS	ABCC2 ABCB1	Cytosol Plasma membrane	Endocytosis	[67]
Myosin Vb	ABCB11	Cytosol Recycling endosomes Plasma membrane	Recycling to the plasma membrane	[60]
MLC2	ABCB1 ABCB4 ABCB11	Cytosol	Motor protein	[33]
PDZK1	ABCC2	Cytosol	Promotes membrane stability	[54,55]
Pim-1	ABCB1	Cytosol	Promotes membrane stability	[38]

Table 1. *Cont.*

Proteins [1]	Interacting ABC Transporters	Subcellular Localization	Functions	References
PI3K	ABCB4 ABCB11 ABCC2	Plasma membrane Cytosol	Protein kinase	[39]
PKA and PKC	ABCB1 ABCB11 ABCC2	Plasma membrane	Protein kinase	[113–118]
RAB4	ABCB1	Endosomes Plasma membrane Cytosol	Vesicular trafficking	[31]
RAB5	ABCB1	Endosomes Plasma membrane Cytosol	Vesicular trafficking	[32]
RAB8	ABCC2	Endosomes Plasma membrane Cytosol	Vesicular trafficking	[61]
RAB11	ABCB11	Endosomes Plasma membrane Cytosol	Vesicular trafficking	[60]
RACK1	ABCB1 ABCB4	Plasma membrane	Scaffold protein	[107]
Radixin	ABCC2	Cytosol Plasma membrane	Promotes membrane stability	[57]
REEP	ABCB11	ER	ER shaping and remodeling	[8]
Rma1, TEB4 and HRD1	ABCB11	ER	E3 ubiquitin ligases	[126]
RNF2	ABCB1	Cytosol	E3 ubiquitin ligase	[128]
RSK1	ABCB1	Cytosol	Kinase	[131,133]
SCFFbx15	ABCB1	Cytosol	E3 ubiquitin ligase	[130]
Src kinase	ABCB1	Plasma membrane	Protein kinase	[105–107]
TMEM14A	ABCB11	ER	Implicated in apoptosis	[8]
TMEM205	ABCB11	ER	Drug resistance	[8]
TRAM/TRAP	ABCB11	ER	Accessory protein in the Sec61 translocon complex	[8]
UBC9	ABCC2	Cytosol	SUMO-related protein	[136]
UBE2R1	ABCB1	Cytosol	E2 ubiquitin-conjugating enzyme	[130,132]
USP19	ABCB11	ER	Deubiquitinating enzyme	[137]

[1] See the main text for full names of the proteins.

Author Contributions: A.B.S., A.B. and E.M. designed and wrote the manuscript. T.A.-S. and T.F. supervised this work. M.L., J.-L.D., E.G. and E.J. provided significant intellectual contributions. All authors have read and agreed to the published version of the manuscript.

Funding: A.B.S. and E.M. were supported by the Ministère de l'Enseignement Supérieur, de la Recherche et de l'Innovation. A.B was supported by the German Research Foundation, Grant Number DFG Project-ID 403224013, SFB 1382, and CRC296. T.F. was supported by grants from the Agence Nationale de la Recherche (ANR-15-CE14-0008-01) and the French Association for the Study of the Liver (AFEF). T.A.-S. was supported by grants from Fondation pour la Recherche Médicale (FRM-EQU-2020-03010517) and the Association Mucoviscidose-ABCF2.

Institutional Review Board Statement: Not applicable.

Informed Consent Statement: Not applicable.

Data Availability Statement: Not applicable.

Acknowledgments: We thank Association Maladie Foie Enfants (AMFE) (Malakoff, France), Monaco Liver Disorder (MLD) (Monaco), Association "Pour Louis 1000 Foie Merci" (Fournet-Luisans, France), Association "Il était un foie" (Plouescat, France), Fondation Rumsey-Cartier (Genève, Switzerland) and FILFOIE for their support.

Conflicts of Interest: The authors declare no conflict of interest.

Abbreviations

4-PB	4-phenylbutyrate
ABC	ATP-binding cassette
AP2	Adaptor protein complex 2
BA	Bile acids
Cav-1	Caveolin-1
CNX	Calnexin
E217G	Estradiol-17β-d-glucuronide
E3 Ubl	E3 ubiquitin ligase
ER	Endoplasmic reticulum
ERM	Ezrin–radixin–moesin
ERAD	ER-associated degradation
MARCKS	Myristoylated alanine-rich C-kinase substrate
MLC2	Myosin regulatory light chain 2
PC	Phosphatidylcholine
PDZ	Postsynaptic density protein (PSD95), Drosophila disc large tumor suppressor (Dlg1) and zonula occludens-1 protein (ZO-1)
PFIC	Progressive familial intrahepatic cholestasis
PI3K	Phosphoinositide 3-kinase
PKA/C	Protein kinase A/C
RAB	Ras-related in brain
RACK1	Receptor for activated C-kinase 1
SAC	Subapical compartment
WT	Wild type

References

1. Boyer, J.L. Bile formation and secretion. *Compr. Physiol.* **2013**, *3*, 1035–1078.
2. Small, D.M. Role of ABC transporters in secretion of cholesterol from liver into bile. *Proc. Natl. Acad. Sci. USA* **2003**, *100*, 4–6. [CrossRef]
3. Kroll, T.; Prescher, M.; Smits, S.H.J.; Schmitt, L. Structure and Function of Hepatobiliary ATP Binding Cassette Transporters. *Chem. Rev.* **2020**, in press. [CrossRef] [PubMed]
4. Fernandez, S.B.; Hollo, Z.; Kern, A.; Bakos, E.; Fischer, P.A.; Borst, P.; Evers, R. Role of the N-terminal transmembrane region of the multidrug resistance protein MRP2 in routing to the apical membrane in MDCKII cells. *J. Biol. Chem.* **2002**, *277*, 31048–31055. [CrossRef] [PubMed]
5. Gething, M.J.; Sambrook, J. Protein folding in the cell. *Nature* **1992**, *355*, 33–45. [CrossRef]
6. Shao, S.; Hegde, R.S. Membrane protein insertion at the endoplasmic reticulum. *Annu. Rev. Cell Dev. Biol.* **2011**, *27*, 25–56. [CrossRef]
7. Gemmer, M.; Förster, F. A clearer picture of the ER translocon complex. *J. Cell Sci.* **2020**, *133*, 3. [CrossRef]
8. Przybylla, S.; Stindt, J.; Kleinschrodt, D.; Schulte Am Esch, J.; Häussinger, D.; Keitel, V.; Smits, S.H.; Schmitt, L. Analysis of the Bile Salt Export Pump (ABCB11) Interactome Employing Complementary Approaches. *PLoS ONE* **2016**, *11*, e0159778. [CrossRef] [PubMed]
9. Czuba, L.C.; Hillgren, K.M.; Swaan, P.W. Post-translational modifications of transporters. *Pharmacol. Ther.* **2018**, *192*, 88–99. [CrossRef]
10. Jayaprakash, N.G.; Surolia, A. Role of glycosylation in nucleating protein folding and stability. *Biochem. J.* **2017**, *474*, 2333–2347. [CrossRef] [PubMed]
11. Ellis, R.J.; van der Vies, S.M. Molecular chaperones. *Annu. Rev. Biochem.* **1991**, *60*, 321–347. [CrossRef]

12. Okiyoneda, T.; Kono, T.; Niibori, A.; Harada, K.; Kusuhara, H.; Takada, T.; Shuto, T.; Suico, M.A.; Sugiyama, Y.; Kai, H. Calreticulin facilitates the cell surface expression of ABCG5/G8. *Biochem. Biophys. Res. Commun.* **2006**, *347*, 67–75. [CrossRef] [PubMed]

13. Loo, T.W.; Clarke, D.M. Prolonged association of temperature-sensitive mutants of human P-glycoprotein with calnexin during biogenesis. *J. Biol. Chem.* **1994**, *269*, 28683–28689. [CrossRef]

14. Loo, T.W.; Clarke, D.M. P-glycoprotein. Associations between domains and between domains and molecular chaperones. *J. Biol. Chem.* **1995**, *270*, 21839–21844. [CrossRef]

15. Adams, B.M.; Oster, M.E.; Hebert, D.N. Protein Quality Control in the Endoplasmic Reticulum. *Protein J.* **2019**, *38*, 317–329. [CrossRef] [PubMed]

16. Gautherot, J.; Durand-Schneider, A.M.; Delautier, D.; Delaunay, J.L.; Rada, A.; Gabillet, J.; Housset, C.; Maurice, M.; Aït-Slimane, T. Effects of cellular, chemical, and pharmacological chaperones on the rescue of a trafficking-defective mutant of the ATP-binding cassette transporter proteins ABCB1/ABCB4. *J. Biol. Chem.* **2012**, *287*, 5070–5078. [CrossRef] [PubMed]

17. Gonzales, E.; Grosse, B.; Schuller, B.; Davit-Spraul, A.; Conti, F.; Guettier, C.; Cassio, D.; Jacquemin, E. Targeted pharmacotherapy in progressive familial intrahepatic cholestasis type 2: Evidence for improvement of cholestasis with 4-phenylbutyrate. *Hepatology* **2015**, *62*, 558–566. [CrossRef]

18. Vauthier, V.; Ben Saad, A.; Elie, J.; Oumata, N.; Durand-Schneider, A.M.; Bruneau, A.; Delaunay, J.L.; Housset, C.; Aït-Slimane, T.; Meijer, L.; et al. Structural analogues of roscovitine rescue the intracellular traffic and the function of ER-retained ABCB4 variants in cell models. *Sci. Rep.* **2019**, *9*, 6653. [CrossRef] [PubMed]

19. Wakana, Y.; Takai, S.; Nakajima, K.; Tani, K.; Yamamoto, A.; Watson, P.; Stephens, D.J.; Hauri, H.P.; Tagaya, M. Bap31 is an itinerant protein that moves between the peripheral endoplasmic reticulum (ER) and a juxtanuclear compartment related to ER-associated Degradation. *Mol. Biol. Cell* **2008**, *19*, 1825–1836. [CrossRef] [PubMed]

20. Ladasky, J.J.; Boyle, S.; Seth, M.; Li, H.; Pentcheva, T.; Abe, F.; Steinberg, S.J.; Edidin, M. Bap31 enhances the endoplasmic reticulum export and quality control of human class I MHC molecules. *J. Immunol.* **2006**, *177*, 6172–6181. [CrossRef] [PubMed]

21. Rao, P.S.; Bickel, U.; Srivenugopal, K.S.; Rao, U.S. Bap29varP, a variant of Bap29, influences the cell surface expression of the human P-glycoprotein. *Int. J. Oncol.* **2008**, *32*, 135–144. [CrossRef] [PubMed]

22. Van de Kamp, J.M.; Errami, A.; Howidi, M.; Anselm, I.; Winter, S.; Phalin-Roque, J.; Osaka, H.; van Dooren, S.J.; Mancini, G.M.; Steinberg, S.J.; et al. Genotype-phenotype correlation of contiguous gene deletions of SLC6A8, BCAP31 and ABCD1. *Clin. Genet.* **2015**, *87*, 141–147. [CrossRef] [PubMed]

23. Wang, X.; Matteson, J.; An, Y.; Moyer, B.; Yoo, J.S.; Bannykh, S.; Wilson, I.A.; Riordan, J.R.; Balch, W.E. COPII-dependent export of cystic fibrosis transmembrane conductance regulator from the ER uses a di-acidic exit code. *J. Cell Biol.* **2004**, *167*, 65–74. [CrossRef]

24. Tanaka, A.R.; Kano, F.; Ueda, K.; Murata, M. The ABCA1 Q597R mutant undergoes trafficking from the ER upon ER stress. *Biochem. Biophys. Res. Commun.* **2008**, *369*, 1174–1178. [CrossRef]

25. Draheim, V.; Reichel, A.; Weitschies, W.; Moenning, U. N-glycosylation of ABC transporters is associated with functional activity in sandwich-cultured rat hepatocytes. *Eur. J. Pharm. Sci. Off. J. Eur. Fed. Pharm. Sci.* **2010**, *41*, 201–209. [CrossRef] [PubMed]

26. Stanley, P. Golgi glycosylation. *Cold Spring Harb. Perspect. Biol.* **2011**, *3*, a005199. [CrossRef]

27. Slimane, T.A.; Trugnan, G.; Van, I.S.C.; Hoekstra, D. Raft-mediated trafficking of apical resident proteins occurs in both direct and transcytotic pathways in polarized hepatic cells: Role of distinct lipid microdomains. *Mol. Biol. Cell* **2003**, *14*, 611–624. [CrossRef]

28. Kipp, H.; Arias, I.M. Newly synthesized canalicular ABC transporters are directly targeted from the Golgi to the hepatocyte apical domain in rat liver. *J. Biol. Chem.* **2000**, *275*, 15917–15925. [CrossRef]

29. Kipp, H.; Arias, I.M. Intracellular trafficking and regulation of canalicular ATP-binding cassette transporters. *Semin. Liver Dis.* **2000**, *20*, 339–351. [CrossRef]

30. Li, M.; Soroka, C.J.; Harry, K.; Boyer, J.L. CFTR-associated ligand is a negative regulator of Mrp2 expression. *Am. J. Physiol. Cell Physiol.* **2017**, *312*, C40–C46. [CrossRef] [PubMed]

31. Ferrándiz-Huertas, C.; Fernández-Carvajal, A.; Ferrer-Montiel, A. Rab4 interacts with the human P-glycoprotein and modulates its surface expression in multidrug resistant K562 cells. *Int. J. Cancer* **2011**, *128*, 192–205. [CrossRef]

32. Fu, D.; van Dam, E.M.; Brymora, A.; Duggin, I.G.; Robinson, P.J.; Roufogalis, B.D. The small GTPases Rab5 and RalA regulate intracellular traffic of P-glycoprotein. *Biochim. Biophys. Acta* **2007**, *1773*, 1062–1072. [CrossRef]

33. Chan, W.; Calderon, G.; Swift, A.L.; Moseley, J.; Li, S.; Hosoya, H.; Arias, I.M.; Ortiz, D.F. Myosin II regulatory light chain is required for trafficking of bile salt export protein to the apical membrane in Madin-Darby canine kidney cells. *J. Biol. Chem.* **2005**, *280*, 23741–23747. [CrossRef] [PubMed]

34. Schonhoff, C.M.; Park, S.W.; Webster, C.R.; Anwer, M.S. MAPK α and β isoforms differentially regulate plasma membrane localization of MRP2. *Am. J. Physiol. Gastrointest. Liver Physiol.* **2016**, *310*, G999–G1005. [CrossRef]

35. Kubitz, R.; Sütfels, G.; Kühlkamp, T.; Kölling, R.; Häussinger, D. Trafficking of the bile salt export pump from the Golgi to the canalicular membrane is regulated by the p38 MAP kinase. *Gastroenterology* **2004**, *126*, 541–553. [CrossRef] [PubMed]

36. Wojtal, K.A.; de Vries, E.; Hoekstra, D.; van Ijzendoorn, S.C. Efficient trafficking of MDR1/P-glycoprotein to apical canalicular plasma membranes in HepG2 cells requires PKA-RIIalpha anchoring and glucosylceramide. *Mol. Biol. Cell* **2006**, *17*, 3638–3650. [CrossRef]

37. Anwer, M.S. Role of protein kinase C isoforms in bile formation and cholestasis. *Hepatology* **2014**, *60*, 1090–1097. [CrossRef] [PubMed]

38. Xie, Y.; Burcu, M.; Linn, D.E.; Qiu, Y.; Baer, M.R. Pim-1 kinase protects P-glycoprotein from degradation and enables its glycosylation and cell surface expression. *Mol. Pharmacol.* **2010**, *78*, 310–318. [CrossRef] [PubMed]

39. Misra, S.; Ujházy, P.; Gatmaitan, Z.; Varticovski, L.; Arias, I.M. The role of phosphoinositide 3-kinase in taurocholate-induced trafficking of ATP-dependent canalicular transporters in rat liver. *J. Biol. Chem.* **1998**, *273*, 26638–26644. [CrossRef] [PubMed]

40. Misra, S.; Ujhazy, P.; Varticovski, L.; Arias, I.M. Phosphoinositide 3-kinase lipid products regulate ATP-dependent transport by sister of P-glycoprotein and multidrug resistance associated protein 2 in bile canalicular membrane vesicles. *Proc. Natl. Acad. Sci. USA* **1999**, *96*, 5814–5819. [CrossRef]

41. Homolya, L.; Fu, D.; Sengupta, P.; Jarnik, M.; Gillet, J.P.; Vitale-Cross, L.; Gutkind, J.S.; Lippincott-Schwartz, J.; Arias, I.M. LKB1/AMPK and PKA control ABCB11 trafficking and polarization in hepatocytes. *PLoS ONE* **2014**, *9*, e91921. [CrossRef]

42. Woods, A.; Heslegrave, A.J.; Muckett, P.J.; Levene, A.P.; Clements, M.; Mobberley, M.; Ryder, T.A.; Abu-Hayyeh, S.; Williamson, C.; Goldin, R.D.; et al. LKB1 is required for hepatic bile acid transport and canalicular membrane integrity in mice. *Biochem. J.* **2011**, *434*, 49–60. [CrossRef]

43. Dumont, M.; Jacquemin, E.; Erlinger, S. Effect of Ursodeoxycholic Acid on the Expression of the Hepatocellular Bile Acid Transporters (Ntcp and bsep) in Rats with Estrogen-Induced Cholestasis. *J. Pediatric Gastroenterol. Nutr.* **2002**, *35*, 185–191.

44. Glasova, H.; Berghaus, T.M.; Kullak-Ublick, G.A.; Paumgartner, G.; Beuers, U. Tauroursodeoxycholic acid mobilizes alpha-PKC after uptake in human HepG2 hepatoma cells. *Eur. J. Clin. Investig.* **2002**, *32*, 437–442. [CrossRef]

45. Gottesman, M.M.; Fojo, T.; Bates, S.E. Multidrug resistance in cancer: Role of ATP-dependent transporters. *Nat. Rev. Cancer* **2002**, *2*, 48–58. [CrossRef]

46. Veit, G.; Avramescu, R.G.; Chiang, A.N.; Houck, S.A.; Cai, Z.; Peters, K.W.; Hong, J.S.; Pollard, H.B.; Guggino, W.B.; Balch, W.E.; et al. From CFTR biology toward combinatorial pharmacotherapy: Expanded classification of cystic fibrosis mutations. *Mol. Biol. Cell* **2016**, *27*, 424–433. [CrossRef] [PubMed]

47. Ruskamo, S.; Nieminen, T.; Kristiansen, C.K.; Vatne, G.H.; Baumann, A.; Hallin, E.I.; Raasakka, A.; Joensuu, P.; Bergmann, U.; Vattulainen, I.; et al. Molecular mechanisms of Charcot-Marie-Tooth neuropathy linked to mutations in human myelin protein P2. *Sci. Rep.* **2017**, *7*, 6510. [CrossRef]

48. Delaunay, J.L.; Durand-Schneider, A.M.; Dossier, C.; Falguières, T.; Gautherot, J.; Davit-Spraul, A.; Aït-Slimane, T.; Housset, C.; Jacquemin, E.; Maurice, M. A functional classification of ABCB4 variations causing progressive familial intrahepatic cholestasis type 3. *Hepatology* **2016**, *63*, 1620–1631. [CrossRef] [PubMed]

49. Adams, D.R.; Ron, D.; Kiely, P.A. RACK1, A multifaceted scaffolding protein: Structure and function. *Cell Commun. Signal.* **2011**, *9*, 22. [CrossRef] [PubMed]

50. Ikebuchi, Y.; Takada, T.; Ito, K.; Yoshikado, T.; Anzai, N.; Kanai, Y.; Suzuki, H. Receptor for activated C-kinase 1 regulates the cellular localization and function of ABCB4. *Hepatol. Res.: Off. J. Jpn. Soc. Hepatol.* **2009**, *39*, 1091–1107. [CrossRef] [PubMed]

51. Bezprozvanny, I.; Maximov, A. PDZ domains: More than just a glue. *Proc. Natl. Acad. Sci. USA* **2001**, *98*, 787–789. [CrossRef] [PubMed]

52. Li, M.; Wang, W.; Soroka, C.J.; Mennone, A.; Harry, K.; Weinman, E.J.; Boyer, J.L. NHERF-1 binds to Mrp2 and regulates hepatic Mrp2 expression and function. *J. Biol. Chem.* **2010**, *285*, 19299–19307. [CrossRef] [PubMed]

53. Venot, Q.; Delaunay, J.L.; Fouassier, L.; Delautier, D.; Falguières, T.; Housset, C.; Maurice, M.; Aït-Slimane, T. A PDZ-Like Motif in the Biliary Transporter ABCB4 Interacts with the Scaffold Protein EBP50 and Regulates ABCB4 Cell Surface Expression. *PLoS ONE* **2016**, *11*, e0146962. [CrossRef] [PubMed]

54. Emi, Y.; Nomura, S.; Yokota, H.; Sakaguchi, M. ATP-binding cassette transporter isoform C2 localizes to the apical plasma membrane via interactions with scaffolding protein. *J. Biochem.* **2011**, *149*, 177–189. [CrossRef] [PubMed]

55. Kocher, O.; Comella, N.; Gilchrist, A.; Pal, R.; Tognazzi, K.; Brown, L.F.; Knoll, J.H. PDZK1, a novel PDZ domain-containing protein up-regulated in carcinomas and mapped to chromosome 1q21, interacts with cMOAT (MRP2), the multidrug resistance-associated protein. *Lab. Investig. A J. Tech. Methods Pathol.* **1999**, *79*, 1161–1170.

56. Bretscher, A.; Reczek, D.; Berryman, M. Ezrin: A protein requiring conformational activation to link microfilaments to the plasma membrane in the assembly of cell surface structures. *J. Cell Sci.* **1997**, *110 Pt 24*, 3011–3018.

57. Kikuchi, S.; Hata, M.; Fukumoto, K.; Yamane, Y.; Matsui, T.; Tamura, A.; Yonemura, S.; Yamagishi, H.; Keppler, D.; Tsukita, S.; et al. Radixin deficiency causes conjugated hyperbilirubinemia with loss of Mrp2 from bile canalicular membranes. *Nat. Genet.* **2002**, *31*, 320–325. [CrossRef] [PubMed]

58. Lapierre, L.A.; Kumar, R.; Hales, C.M.; Navarre, J.; Bhartur, S.G.; Burnette, J.O.; Provance, D.W., Jr.; Mercer, J.A.; Bähler, M.; Goldenring, J.R. Myosin vb is associated with plasma membrane recycling systems. *Mol. Biol. Cell* **2001**, *12*, 1843–1857. [CrossRef]

59. Wakabayashi, Y.; Lippincott-Schwartz, J.; Arias, I.M. Intracellular trafficking of bile salt export pump (ABCB11) in polarized hepatic cells: Constitutive cycling between the canalicular membrane and rab11-positive endosomes. *Mol. Biol. Cell* **2004**, *15*, 3485–3496. [CrossRef] [PubMed]

60. Wakabayashi, Y.; Dutt, P.; Lippincott-Schwartz, J.; Arias, I.M. Rab11a and myosin Vb are required for bile canalicular formation in WIF-B9 cells. *Proc. Natl. Acad. Sci. USA* **2005**, *102*, 15087–15092. [CrossRef] [PubMed]

61. Overeem, A.W.; Li, Q.; Qiu, Y.L.; Cartón-García, F.; Leng, C.; Klappe, K.; Dronkers, J.; Hsiao, N.H.; Wang, J.S.; Arango, D.; et al. A Molecular Mechanism Underlying Genotype-Specific Intrahepatic Cholestasis Resulting From MYO5B Mutations. *Hepatology* **2020**, *72*, 213–229. [CrossRef]

62. Gonzales, E.; Taylor, S.A.; Davit-Spraul, A.; Thébaut, A.; Thomassin, N.; Guettier, C.; Whitington, P.F.; Jacquemin, E. MYO5B mutations cause cholestasis with normal serum gamma-glutamyl transferase activity in children without microvillous inclusion disease. *Hepatology* **2017**, *65*, 164–173. [CrossRef] [PubMed]

63. Chai, J.; Cai, S.Y.; Liu, X.; Lian, W.; Chen, S.; Zhang, L.; Feng, X.; Cheng, Y.; He, X.; He, Y.; et al. Canalicular membrane MRP2/ABCC2 internalization is determined by Ezrin Thr567 phosphorylation in human obstructive cholestasis. *J. Hepatol.* **2015**, *63*, 1440–1448. [CrossRef]

64. Kubitz, R.; Huth, C.; Schmitt, M.; Horbach, A.; Kullak-Ublick, G.; Häussinger, D. Protein kinase C-dependent distribution of the multidrug resistance protein 2 from the canalicular to the basolateral membrane in human HepG2 cells. *Hepatology* **2001**, *34*, 340–350. [CrossRef] [PubMed]

65. Kurz, A.K.; Graf, D.; Schmitt, M.; Vom Dahl, S.; Häussinger, D. Tauroursodesoxycholate-induced choleresis involves p38(MAPK) activation and translocation of the bile salt export pump in rats. *Gastroenterology* **2001**, *121*, 407–419. [CrossRef] [PubMed]

66. Cantore, M.; Reinehr, R.; Sommerfeld, A.; Becker, M.; Häussinger, D. The Src family kinase Fyn mediates hyperosmolarity-induced Mrp2 and Bsep retrieval from canalicular membrane. *J. Biol. Chem.* **2011**, *286*, 45014–45029. [CrossRef] [PubMed]

67. Schonhoff, C.M.; Webster, C.R.; Anwer, M.S. Taurolithocholate-induced MRP2 retrieval involves MARCKS phosphorylation by protein kinase Cε in HUH-NTCP Cells. *Hepatology* **2013**, *58*, 284–292. [CrossRef]

68. El Amri, M.; Fitzgerald, U.; Schlosser, G. MARCKS and MARCKS-like proteins in development and regeneration. *J. Biomed. Sci.* **2018**, *25*, 43. [CrossRef]

69. Wenzel, T.; Büch, T.; Urban, N.; Weirauch, U.; Schierle, K.; Aigner, A.; Schaefer, M.; Kalwa, H. Restoration of MARCKS enhances chemosensitivity in cancer. *J. Cancer Res. Clin. Oncol.* **2020**, *146*, 843–858. [CrossRef]

70. Ortiz, D.F.; Moseley, J.; Calderon, G.; Swift, A.L.; Li, S.; Arias, I.M. Identification of HAX-1 as a protein that binds bile salt export protein and regulates its abundance in the apical membrane of Madin-Darby canine kidney cells. *J. Biol. Chem.* **2004**, *279*, 32761–32770. [CrossRef]

71. Lam, P.; Xu, S.; Soroka, C.J.; Boyer, J.L. A C-terminal tyrosine-based motif in the bile salt export pump directs clathrin-dependent endocytosis. *Hepatology* **2012**, *55*, 1901–1911. [CrossRef] [PubMed]

72. Hayashi, H.; Inamura, K.; Aida, K.; Naoi, S.; Horikawa, R.; Nagasaka, H.; Takatani, T.; Fukushima, T.; Hattori, A.; Yabuki, T.; et al. AP2 adaptor complex mediates bile salt export pump internalization and modulates its hepatocanalicular expression and transport function. *Hepatology* **2012**, *55*, 1889–1900. [CrossRef] [PubMed]

73. Aida, K.; Hayashi, H.; Inamura, K.; Mizuno, T.; Sugiyama, Y. Differential roles of ubiquitination in the degradation mechanism of cell surface-resident bile salt export pump and multidrug resistance-associated protein 2. *Mol. Pharmacol.* **2014**, *85*, 482–491. [CrossRef]

74. Crocenzi, F.A.; Mottino, A.D.; Cao, J.; Veggi, L.M.; Pozzi, E.J.; Vore, M.; Coleman, R.; Roma, M.G. Estradiol-17beta-D-glucuronide induces endocytic internalization of Bsep in rats. *Am. J. Physiol. Gastrointest. Liver Physiol.* **2003**, *285*, G449–G459. [CrossRef] [PubMed]

75. Crocenzi, F.A.; Sánchez Pozzi, E.J.; Ruiz, M.L.; Zucchetti, A.E.; Roma, M.G.; Mottino, A.D.; Vore, M. Ca(2+)-dependent protein kinase C isoforms are critical to estradiol 17beta-D-glucuronide-induced cholestasis in the rat. *Hepatology* **2008**, *48*, 1885–1895. [CrossRef]

76. Zucchetti, A.E.; Barosso, I.R.; Boaglio, A.; Pellegrino, J.M.; Ochoa, E.J.; Roma, M.G.; Crocenzi, F.A.; Sánchez Pozzi, E.J. Prevention of estradiol 17beta-D-glucuronide-induced canalicular transporter internalization by hormonal modulation of cAMP in rat hepatocytes. *Mol. Biol. Cell* **2011**, *22*, 3902–3915. [CrossRef]

77. Miszczuk, G.S.; Barosso, I.R.; Larocca, M.C.; Marrone, J.; Marinelli, R.A.; Boaglio, A.C.; Sánchez Pozzi, E.J.; Roma, M.G.; Crocenzi, F.A. Mechanisms of canalicular transporter endocytosis in the cholestatic rat liver. *Biochim. Biophys. Acta. Mol. Basis Dis.* **2018**, *1864*, 1072–1085. [CrossRef]

78. Elferink, M.G.; Olinga, P.; Draaisma, A.L.; Merema, M.T.; Faber, K.N.; Slooff, M.J.; Meijer, D.K.; Groothuis, G.M. LPS-induced downregulation of MRP2 and BSEP in human liver is due to a posttranscriptional process. *Am. J. Physiol. Gastrointest. Liver Physiol.* **2004**, *287*, G1008–G1016. [CrossRef]

79. Nyasae, L.K.; Hubbard, A.L.; Tuma, P.L. Transcytotic efflux from early endosomes is dependent on cholesterol and glycosphingolipids in polarized hepatic cells. *Mol. Biol. Cell* **2003**, *14*, 2689–2705. [CrossRef]

80. Ismair, M.G.; Hausler, S.; Stuermer, C.A.; Guyot, C.; Meier, P.J.; Roth, J.; Stieger, B. ABC-transporters are localized in caveolin-1-positive and reggie-1-negative and reggie-2-negative microdomains of the canalicular membrane in rat hepatocytes. *Hepatology* **2009**, *49*, 1673–1682. [CrossRef] [PubMed]

81. Guyot, C.; Stieger, B. Interaction of bile salts with rat canalicular membrane vesicles: Evidence for bile salt resistant microdomains. *J. Hepatol.* **2011**, *55*, 1368–1376. [CrossRef]

82. Luker, G.D.; Pica, C.M.; Kumar, A.S.; Covey, D.F.; Piwnica-Worms, D. Effects of cholesterol and enantiomeric cholesterol on P-glycoprotein localization and function in low-density membrane domains. *Biochemistry* **2000**, *39*, 7651–7661. [CrossRef] [PubMed]

83. Troost, J.; Lindenmaier, H.; Haefeli, W.E.; Weiss, J. Modulation of cellular cholesterol alters P-glycoprotein activity in multidrug-resistant cells. *Mol. Pharmacol.* **2004**, *66*, 1332–1339. [CrossRef] [PubMed]

84. Ghetie, M.A.; Marches, R.; Kufert, S.; Vitetta, E.S. An anti-CD19 antibody inhibits the interaction between P-glycoprotein (P-gp) and CD19, causes P-gp to translocate out of lipid rafts, and chemosensitizes a multidrug-resistant (MDR) lymphoma cell line. *Blood* **2004**, *104*, 178–183. [CrossRef] [PubMed]
85. Marrone, J.; Soria, L.R.; Danielli, M.; Lehmann, G.L.; Larocca, M.C.; Marinelli, R.A. Hepatic gene transfer of human aquaporin-1 improves bile salt secretory failure in rats with estrogen-induced cholestasis. *Hepatology* **2016**, *64*, 535–548. [CrossRef]
86. Urbatsch, I.L.; Senior, A.E. Effects of lipids on ATPase activity of purified Chinese hamster P-glycoprotein. *Arch. Biochem. Biophys.* **1995**, *316*, 135–140. [CrossRef]
87. Rothnie, A.; Theron, D.; Soceneantu, L.; Martin, C.; Traikia, M.; Berridge, G.; Higgins, C.F.; Devaux, P.F.; Callaghan, R. The importance of cholesterol in maintenance of P-glycoprotein activity and its membrane perturbing influence. *Eur. Biophys. J.* **2001**, *30*, 430–442. [CrossRef] [PubMed]
88. Gayet, L.; Dayan, G.; Barakat, S.; Labialle, S.; Michaud, M.; Cogne, S.; Mazane, A.; Coleman, A.W.; Rigal, D.; Baggetto, L.G. Control of P-glycoprotein activity by membrane cholesterol amounts and their relation to multidrug resistance in human CEM leukemia cells. *Biochemistry* **2005**, *44*, 4499–4509. [CrossRef] [PubMed]
89. Kimura, Y.; Kioka, N.; Kato, H.; Matsuo, M.; Ueda, K. Modulation of drug-stimulated ATPase activity of human MDR1/P-glycoprotein by cholesterol. *Biochem. J.* **2007**, *401*, 597–605. [CrossRef]
90. Ito, K.; Hoekstra, D.; van Ijzendoorn, S.C. Cholesterol but not association with detergent resistant membranes is necessary for the transport function of MRP2/ABCC2. *Febs Lett.* **2008**, *582*, 4153–4157. [CrossRef]
91. Kis, E.; Ioja, E.; Nagy, T.; Szente, L.; Heredi-Szabo, K.; Krajcsi, P. Effect of membrane cholesterol on BSEP/Bsep activity: Species specificity studies for substrates and inhibitors. *Drug Metab. Dispos.* **2009**, *37*, 1878–1886. [CrossRef] [PubMed]
92. Paulusma, C.C.; de Waart, D.R.; Kunne, C.; Mok, K.S.; Elferink, R.P. Activity of the bile salt export pump (ABCB11) is critically dependent on canalicular membrane cholesterol content. *J. Biol. Chem.* **2009**, *284*, 9947–9954. [CrossRef] [PubMed]
93. Guyot, C.; Hofstetter, L.; Stieger, B. Differential effects of membrane cholesterol content on the transport activity of multidrug resistance-associated protein 2 (ABCC2) and of the bile salt export pump (ABCB11). *Mol. Pharmacol.* **2014**, *85*, 909–920. [CrossRef] [PubMed]
94. Doige, C.A.; Yu, X.; Sharom, F.J. The effects of lipids and detergents on ATPase-active P-glycoprotein. *Biochim. Biophys. Acta* **1993**, *1146*, 65–72. [CrossRef]
95. Sharom, F.J. The P-glycoprotein multidrug transporter: Interactions with membrane lipids, and their modulation of activity. *Biochem. Soc. Trans.* **1997**, *25*, 1088–1096. [CrossRef]
96. Misra, S.; Varticovski, L.; Arias, I.M. Mechanisms by which cAMP increases bile acid secretion in rat liver and canalicular membrane vesicles. *Am. J. Physiol. Gastrointest. Liver Physiol.* **2003**, *285*, G316–G324. [CrossRef]
97. Clay, A.T.; Lu, P.; Sharom, F.J. Interaction of the P-Glycoprotein Multidrug Transporter with Sterols. *Biochemistry* **2015**, *54*, 6586–6597. [CrossRef]
98. Domicevica, L.; Koldso, H.; Biggin, P.C. Multiscale molecular dynamics simulations of lipid interactions with P-glycoprotein in a complex membrane. *J. Mol. Graph. Model.* **2018**, *80*, 147–156. [CrossRef]
99. Cai, C.; Zhu, H.; Chen, J. Overexpression of caveolin-1 increases plasma membrane fluidity and reduces P-glycoprotein function in Hs578T/Dox. *Biochem. Biophys. Res. Commun.* **2004**, *320*, 868–874. [CrossRef]
100. Moreno, M.; Molina, H.; Amigo, L.; Zanlungo, S.; Arrese, M.; Rigotti, A.; Miquel, J.F. Hepatic overexpression of caveolins increases bile salt secretion in mice. *Hepatology* **2003**, *38*, 1477–1488. [CrossRef]
101. Smart, E.J.; Ying, Y.; Donzell, W.C.; Anderson, R.G. A role for caveolin in transport of cholesterol from endoplasmic reticulum to plasma membrane. *J. Biol. Chem.* **1996**, *271*, 29427–29435. [CrossRef]
102. Roy, S.; Luetterforst, R.; Harding, A.; Apolloni, A.; Etheridge, M.; Stang, E.; Rolls, B.; Hancock, J.F.; Parton, R.G. Dominant-negative caveolin inhibits H-Ras function by disrupting cholesterol-rich plasma membrane domains. *Nat. Cell Biol.* **1999**, *1*, 98–105. [CrossRef] [PubMed]
103. Jodoin, J.; Demeule, M.; Fenart, L.; Cecchelli, R.; Farmer, S.; Linton, K.J.; Higgins, C.F.; Beliveau, R. P-glycoprotein in blood-brain barrier endothelial cells: Interaction and oligomerization with caveolins. *J. Neurochem.* **2003**, *87*, 1010–1023. [CrossRef] [PubMed]
104. Ronaldson, P.T.; Bendayan, M.; Gingras, D.; Piquette-Miller, M.; Bendayan, R. Cellular localization and functional expression of P-glycoprotein in rat astrocyte cultures. *J. Neurochem.* **2004**, *89*, 788–800. [CrossRef] [PubMed]
105. Barakat, S.; Demeule, M.; Pilorget, A.; Regina, A.; Gingras, D.; Baggetto, L.G.; Beliveau, R. Modulation of p-glycoprotein function by caveolin-1 phosphorylation. *J. Neurochem.* **2007**, *101*, 1–8. [CrossRef] [PubMed]
106. Labrecque, L.; Nyalendo, C.; Langlois, S.; Durocher, Y.; Roghi, C.; Murphy, G.; Gingras, D.; Beliveau, R. Src-mediated tyrosine phosphorylation of caveolin-1 induces its association with membrane type 1 matrix metalloproteinase. *J. Biol. Chem.* **2004**, *279*, 52132–52140. [CrossRef] [PubMed]
107. Fan, Y.; Si, W.; Ji, W.; Wang, Z.; Gao, Z.; Tian, R.; Song, W.; Zhang, H.; Niu, R.; Zhang, F. Rack1 mediates Src binding to drug transporter P-glycoprotein and modulates its activity through regulating Caveolin-1 phosphorylation in breast cancer cells. *Cell Death Dis.* **2019**, *10*, 394. [CrossRef]
108. Chambers, T.C.; Pohl, J.; Raynor, R.L.; Kuo, J.F. Identification of specific sites in human P-glycoprotein phosphorylated by protein kinase C. *J. Biol. Chem.* **1993**, *268*, 4592–4595. [CrossRef]
109. Chambers, T.C.; Pohl, J.; Glass, D.B.; Kuo, J.F. Phosphorylation by protein kinase C and cyclic AMP-dependent protein kinase of synthetic peptides derived from the linker region of human P-glycoprotein. *Biochem. J.* **1994**, *299*, 309–315. [CrossRef]

110. Chambers, T.C. Identification of phosphorylation sites in human MDR1 P-glycoprotein. *Methods Enzym.* **1998**, *292*, 328–342.
111. Orr, G.A.; Han, E.K.; Browne, P.C.; Nieves, E.; O'Connor, B.M.; Yang, C.P.; Horwitz, S.B. Identification of the major phosphorylation domain of murine mdr1b P-glycoprotein. Analysis of the protein kinase A and protein kinase C phosphorylation sites. *J. Biol. Chem.* **1993**, *268*, 25054–25062. [CrossRef]
112. Sachs, C.W.; Chambers, T.C.; Fine, R.L. Differential phosphorylation of sites in the linker region of P-glycoprotein by protein kinase C isozymes alpha, betaI, betaII, gamma, delta, epsilon, eta, and zeta. *Biochem. Pharm.* **1999**, *58*, 1587–1592. [CrossRef]
113. Ito, K.; Wakabayashi, T.; Horie, T. Mrp2/Abcc2 transport activity is stimulated by protein kinase Calpha in a baculo virus co-expression system. *Life Sci.* **2005**, *77*, 539–550. [CrossRef] [PubMed]
114. Gautherot, J.; Delautier, D.; Maubert, M.A.; Ait-Slimane, T.; Bolbach, G.; Delaunay, J.L.; Durand-Schneider, A.M.; Firrincieli, D.; Barbu, V.; Chignard, N.; et al. Phosphorylation of ABCB4 impacts its function: Insights from disease-causing mutations. *Hepatology* **2014**, *60*, 610–621. [CrossRef] [PubMed]
115. Noe, J.; Hagenbuch, B.; Meier, P.J.; St-Pierre, M.V. Characterization of the mouse bile salt export pump overexpressed in the baculovirus system. *Hepatology* **2001**, *33*, 1223–1231. [CrossRef]
116. Idriss, H.; Urquidi, V.; Basavappa, S. Selective modulation of P-glycoprotein's ATPase and anion efflux regulation activities with PKC alpha and PKC epsilon in Sf9 cells. *Cancer Chemother. Pharm.* **2000**, *46*, 287–292. [CrossRef]
117. Ahmad, S.; Safa, A.R.; Glazer, R.I. Modulation of P-glycoprotein by protein kinase C alpha in a baculovirus expression system. *Biochemistry* **1994**, *33*, 10313–10318. [CrossRef]
118. Sachs, C.W.; Ballas, L.M.; Mascarella, S.W.; Safa, A.R.; Lewin, A.H.; Loomis, C.; Carroll, F.I.; Bell, R.M.; Fine, R.L. Effects of sphingosine stereoisomers on P-glycoprotein phosphorylation and vinblastine accumulation in multidrug-resistant MCF-7 cells. *Biochem. Pharm.* **1996**, *52*, 603–612. [CrossRef]
119. Goodfellow, H.R.; Sardini, A.; Ruetz, S.; Callaghan, R.; Gros, P.; McNaughton, P.A.; Higgins, C.F. Protein kinase C-mediated phosphorylation does not regulate drug transport by the human multidrug resistance P-glycoprotein. *J. Biol. Chem.* **1996**, *271*, 13668–13674. [CrossRef] [PubMed]
120. Germann, U.A.; Chambers, T.C.; Ambudkar, S.V.; Licht, T.; Cardarelli, C.O.; Pastan, I.; Gottesman, M.M. Characterization of phosphorylation-defective mutants of human P-glycoprotein expressed in mammalian cells. *J. Biol. Chem.* **1996**, *271*, 1708–1716. [CrossRef] [PubMed]
121. Pickart, C.M. Ubiquitin enters the new millennium. *Mol. Cell* **2001**, *8*, 499–504. [CrossRef]
122. Clague, M.J.; Urbe, S. Ubiquitin: Same molecule, different degradation pathways. *Cell* **2010**, *143*, 682–685. [CrossRef]
123. Katayama, K.; Kapoor, K.; Ohnuma, S.; Patel, A.; Swaim, W.; Ambudkar, I.S.; Ambudkar, S.V. Revealing the fate of cell surface human P-glycoprotein (ABCB1): The lysosomal degradation pathway. *Biochim. Biophys. Acta* **2015**, *1853*, 2361–2370. [CrossRef] [PubMed]
124. Hayashi, H.; Mizuno, T.; Horikawa, R.; Nagasaka, H.; Yabuki, T.; Takikawa, H.; Sugiyama, Y. 4-Phenylbutyrate modulates ubiquitination of hepatocanalicular MRP2 and reduces serum total bilirubin concentration. *J. Hepatol.* **2012**, *56*, 1136–1144. [CrossRef]
125. Kagawa, T.; Watanabe, N.; Mochizuki, K.; Numari, A.; Ikeno, Y.; Itoh, J.; Tanaka, H.; Arias, I.M.; Mine, T. Phenotypic differences in PFIC2 and BRIC2 correlate with protein stability of mutant Bsep and impaired taurocholate secretion in MDCK II cells. *Am. J. Physiol. Gastrointest. Liver Physiol.* **2008**, *294*, G58–G67. [CrossRef]
126. Wang, L.; Dong, H.; Soroka, C.J.; Wei, N.; Boyer, J.L.; Hochstrasser, M. Degradation of the bile salt export pump at endoplasmic reticulum in progressive familial intrahepatic cholestasis type II. *Hepatology* **2008**, *48*, 1558–1569. [CrossRef] [PubMed]
127. Hayashi, H.; Takada, T.; Suzuki, H.; Akita, H.; Sugiyama, Y. Two common PFIC2 mutations are associated with the impaired membrane trafficking of BSEP/ABCB11. *Hepatology* **2005**, *41*, 916–924. [CrossRef]
128. Rao, P.S.; Mallya, K.B.; Srivenugopal, K.S.; Balaji, K.C.; Rao, U.S. RNF2 interacts with the linker region of the human P-glycoprotein. *Int. J. Oncol* **2006**, *29*, 1413–1419. [CrossRef] [PubMed]
129. Ravindranath, A.K.; Kaur, S.; Wernyj, R.P.; Kumaran, M.N.; Miletti-Gonzalez, K.E.; Chan, R.; Lim, E.; Madura, K.; Rodriguez-Rodriguez, L. CD44 promotes multi-drug resistance by protecting P-glycoprotein from FBXO21-mediated ubiquitination. *Oncotarget* **2015**, *6*, 26308–26321. [CrossRef] [PubMed]
130. Katayama, K.; Noguchi, K.; Sugimoto, Y. FBXO15 regulates P-glycoprotein/ABCB1 expression through the ubiquitin–proteasome pathway in cancer cells. *Cancer Sci.* **2013**, *104*, 694–702. [CrossRef]
131. Katayama, K.; Yoshioka, S.; Tsukahara, S.; Mitsuhashi, J.; Sugimoto, Y. Inhibition of the mitogen-activated protein kinase pathway results in the down-regulation of P-glycoprotein. *Mol. Cancer* **2007**, *6*, 2092–2102. [CrossRef]
132. Katayama, K.; Fujiwara, C.; Noguchi, K.; Sugimoto, Y. RSK1 protects P-glycoprotein/ABCB1 against ubiquitin-proteasomal degradation by downregulating the ubiquitin-conjugating enzyme E2 R1. *Sci. Rep.* **2016**, *6*, 36134. [CrossRef]
133. Romeo, Y.; Zhang, X.; Roux, P.P. Regulation and function of the RSK family of protein kinases. *Biochem. J.* **2012**, *441*, 553–569. [CrossRef] [PubMed]
134. Smith, M.H.; Ploegh, H.L.; Weissman, J.S. Road to ruin: Targeting proteins for degradation in the endoplasmic reticulum. *Science* **2011**, *334*, 1086–1090. [CrossRef]
135. Xu, B.Y.; Tang, X.D.; Chen, J.; Wu, H.B.; Chen, W.S.; Chen, L. Rifampicin induces clathrin-dependent endocytosis and ubiquitin-proteasome degradation of MRP2 via oxidative stress-activated PKC-ERK/JNK/p38 and PI3K signaling pathways in HepG2 cells. *Acta Pharm. Sin.* **2020**, *41*, 56–64. [CrossRef]

136. Minami, S.; Ito, K.; Honma, M.; Ikebuchi, Y.; Anzai, N.; Kanai, Y.; Nishida, T.; Tsukita, S.; Sekine, S.; Horie, T.; et al. Posttranslational regulation of Abcc2 expression by SUMOylation system. *Am. J. Physiol. Gastrointest. Liver Physiol.* **2009**, *296*, G406–G413. [CrossRef]
137. Nakamura, N.; Harada, K.; Kato, M.; Hirose, S. Ubiquitin-specific protease 19 regulates the stability of the E3 ubiquitin ligase MARCH6. *Exp. Cell Res.* **2014**, *328*, 207–216. [CrossRef]
138. Hayashi, H.; Sugiyama, Y. Short-chain ubiquitination is associated with the degradation rate of a cell-surface-resident bile salt export pump (BSEP/ABCB11). *Mol. Pharmacol.* **2009**, *75*, 143–150. [CrossRef]
139. Hayashi, H.; Sugiyama, Y. 4-phenylbutyrate enhances the cell surface expression and the transport capacity of wild-type and mutated bile salt export pumps. *Hepatology* **2007**, *45*, 1506–1516. [CrossRef] [PubMed]
140. Rubenstein, R.C.; Zeitlin, P.L. Sodium 4-phenylbutyrate downregulates Hsc70: Implications for intracellular trafficking of DeltaF508-CFTR. *Am. J. Physiol. Cell Physiol.* **2000**, *278*, C259–C267. [CrossRef] [PubMed]

International Journal of
Molecular Sciences

Article

RAB10 Interacts with ABCB4 and Regulates Its Intracellular Traffic

Amel Ben Saad [1,2], Virginie Vauthier [2,3], Martine Lapalus [1], Elodie Mareux [1], Evangéline Bennana [4], Anne-Marie Durand-Schneider [2], Alix Bruneau [2,5], Jean-Louis Delaunay [2], Emmanuel Gonzales [1,6], Chantal Housset [2,7], Tounsia Aït-Slimane [2], François Guillonneau [4], Emmanuel Jacquemin [1,6] and Thomas Falguières [1,*]

[1] Inserm, Université Paris-Saclay, Physiopathogénèse et traitement des maladies du foie, UMR_S 1193, Hepatinov, 91400 Orsay, France; amel.ben-saad@inserm.fr (A.B.S.); martine.lapalus@universite-paris-saclay.fr (M.L.); elodie.mareux@universite-paris-saclay.fr (E.M.); emmanuel.gonzales@aphp.fr (E.G.); emmanuel.jacquemin@aphp.fr (E.J.)

[2] Inserm, Sorbonne Université, Centre de Recherche Saint-Antoine (CRSA), UMR_S 938, Institute of Cardiometabolism and Nutrition (ICAN), 75012 Paris, France; vauthier.v@gmail.com (V.V.); anne-marie.durand-schneider@inserm.fr (A.-M.D.-S.); alix.bruneau@charite.de (A.B.); jean-louis.delaunay@sorbonne-universite.fr (J.-L.D.); chantal.housset@inserm.fr (C.H.); tounsia.ait_slimane@sorbonne-universite.fr (T.A.-S.)

[3] Université de Paris, Institut Cochin, Inserm U1016, CNRS UMR 8104, 75014 Paris, France

[4] 3P5-Proteom'IC platform, Université de Paris, Institut Cochin, Inserm U1016, CNRS UMR8104, 75014 Paris, France; at.evangeline@gmail.com (E.B.); francois.guillonneau@u-paris.fr (F.G.)

[5] Department of Hepatology & Gastroenterology, Charité Universitätsmedizin Berlin, 13353 Berlin, Germany

[6] Assistance Publique—Hôpitaux de Paris, Paediatric Hepatology & Paediatric Liver Transplant Department, Reference Center for Rare Paediatric Liver Diseases, FILFOIE, ERN Rare-Liver, Faculté de Médecine Paris-Saclay, CHU Bicêtre, 94270 Le Kremlin-Bicêtre, France

[7] Assistance Publique—Hôpitaux de Paris, Hôpital Saint-Antoine, Reference Center for Inflammatory Biliary Diseases and Autoimmune Hepatitis, FILFOIE, ERN Rare-Liver, 75012 Paris, France

* Correspondence: thomas.falguieres@inserm.fr; Tel.: +33-(0)1-69-15-62-94

Citation: Ben Saad, A.; Vauthier, V.; Lapalus, M.; Mareux, E.; Bennana, E.; Durand-Schneider, A.-M.; Bruneau, A.; Delaunay, J.-L.; Gonzales, E.; Housset, C.; et al. RAB10 Interacts with ABCB4 and Regulates Its Intracellular Traffic. *Int. J. Mol. Sci.* **2021**, *22*, 7087. https://doi.org/10.3390/ijms22137087

Academic Editor: Jose J.G. Marin

Received: 27 May 2021
Accepted: 29 June 2021
Published: 30 June 2021

Publisher's Note: MDPI stays neutral with regard to jurisdictional claims in published maps and institutional affiliations.

Abstract: ABCB4 (ATP-binding cassette subfamily B member 4) is an ABC transporter expressed at the canalicular membrane of hepatocytes where it ensures phosphatidylcholine secretion into bile. Genetic variations of ABCB4 are associated with several rare cholestatic diseases. The available treatments are not efficient for a significant proportion of patients with ABCB4-related diseases and liver transplantation is often required. The development of novel therapies requires a deep understanding of the molecular mechanisms regulating ABCB4 expression, intracellular traffic, and function. Using an immunoprecipitation approach combined with mass spectrometry analyses, we have identified the small GTPase RAB10 as a novel molecular partner of ABCB4. Our results indicate that the overexpression of wild type RAB10 or its dominant-active mutant significantly increases the amount of ABCB4 at the plasma membrane expression and its phosphatidylcholine floppase function. Contrariwise, RAB10 silencing induces the intracellular retention of ABCB4 and then indirectly diminishes its secretory function. Taken together, our findings suggest that RAB10 regulates the plasma membrane targeting of ABCB4 and consequently its capacity to mediate phosphatidylcholine secretion.

Keywords: bile secretion; intracellular traffic; MDR3; phosphatidylcholine; RAB GTPase

1. Introduction

ABCB4 (ATP-binding cassette subfamily B member 4), also named MDR3 (MultiDrug Resistance 3), is one of the main biliary transporters [1]. It is expressed at the canalicular membrane of hepatocytes where it mediates phosphatidylcholine (PC) secretion into bile [2]. In the aqueous environment of bile, PC plays a critical role in cholesterol solubilization as well as bile acid neutralization [3]. Variations in the *ABCB4* gene are associated with several rare cholestatic diseases, including Progressive Familial Intrahepatic Cholestasis type 3 (PFIC3), Low Phospholipid-Associated Cholelithiasis (LPAC) syndrome, and Intrahepatic Cholestasis of Pregnancy (ICP) [4,5]. While treatment with ursodeoxycholic acid remains efficient for the majority of patients with milder forms of ABCB4-related diseases, this is not the case for patients with PFIC3, the most severe form of these diseases, who most often require liver transplantation [6,7], stressing the unmet need for new therapeutic options. In the frame of personalized medicine, new targeted pharmacotherapies for ABCB4-related diseases have been proposed from in vitro studies and include the potentiator Ivacaftor/VX-770 as well as structural analogues of roscovitine [8,9]. In addition, AAV8- or mRNA-mediated gene therapy constitute an interesting alternative to rescue ABCB4 deficiency, as recently described in mouse models by several research groups [10–13]. In order to better characterize and understand the mode of action of new therapies, a better understanding of the molecular mechanisms regulating ABCB4 traffic and function is required.

Over the last decade, proteomic studies have become a powerful tool to decipher molecular mechanisms at the subcellular level and to provide new insights into protein biology. In the last years, a limited number of ABCB4 interacting partners were identified by yeast two-hybrid screens, including HS1-Associated protein X-1 (HAX1), the motor protein Myosin II regulatory Light Chain 2 (MLC2), Receptor for Activated C-kinase 1 (RACK1), and ERM-Binding Phosphoprotein of 50 kDa (EBP50) [14–18]. Aiming at identifying novel molecular partners of ABCB4, we used an immunoprecipitation approach combined with HPLC-coupled tandem mass spectrometry analyses (usually defined as "Affinity Purification-Mass Spectrometry" or AP-MS). This allowed us to identify the small GTPase RAB10 as a potential ABCB4 interactor. RAB10 belongs to the Ras-related in brain (Rab) family of proteins, which are well known as master regulators of intracellular traffic and sorting processes [19]. RAB proteins are implicated in the majority of vesicular transport steps, including vesicle formation, motility, and tethering, as well as their fusion with target membranes [19]. To date, more than 60 RAB proteins have been identified in mammals. These soluble proteins are ubiquitously expressed and can be membrane-associated with many subcellular compartments thanks to their post-translational geranyl-geranylation [20]. RAB10 was first cloned from Madin-Darby Canine Kidney cells [21]. It is mainly involved in protein trafficking from the Golgi apparatus to the plasma membrane [22]. Its implication in GLUT4 traffic as well as TLR4 exocytosis is well documented [23,24]. RAB10 also plays a key role in ciliogenesis, neuronal development, and basolateral recycling [22].

In the present study, we used biochemical and morphological approaches in cell models, namely HEK and HeLa cells. These cells allow high transfection rate and high expression level of transgenes, and they are validated models for investigating the cell biology of ABC transporters [25,26]. We found that the overexpression of RAB10-wild type (WT) or its constitutively active mutant increases ABCB4 membrane expression and function, whereas RAB10 silencing attenuates ABCB4 cell surface expression, as well as its PC secretion function. Taken together, our results indicate that RAB10 is an important regulator of ABCB4 traffic, by promoting its transport from the Golgi apparatus to the plasma membrane.

2. Results

2.1. Identification of RAB10 as a New Molecular Partner of ABCB4

Despite the important role played by ABCB4 in bile secretion, little is known about the molecular mechanisms regulating its expression, intracellular traffic, and function. To clarify some of these mechanisms, the aim of this study was to identify key players implicated in ABCB4 regulation. For this task, an AP-MS approach was used (see details in the Section 4). Among several ABCB4 candidate interactors (Supplementary Table S1), we found the small GTPases RAB10 and RAB13 of particular interest, as RAB11 has been involved in the traffic of the canalicular bile salt export pump ABCB11 [27]. Our preliminary results indicated less encouraging results for RAB13 (data not shown), so we decided to further investigate the role of RAB10 in the intracellular traffic and function of ABCB4. To confirm the interaction between ABCB4 and RAB10, co-immunoprecipitation experiments were performed in primary human hepatocytes. RAB10 was detected after ABCB4-specific immunoprecipitation but not in the control condition with unspecific antibodies (Figure 1A). Conversely, ABCB4 was specifically detected from RAB10-immunoprecipitated complexes (Figure 1B). The same results were observed in HEK cells co-transfected with ABCB4 and RAB10-GFP (Supplementary Figure S1). These results confirm the interaction between ABCB4 and RAB10.

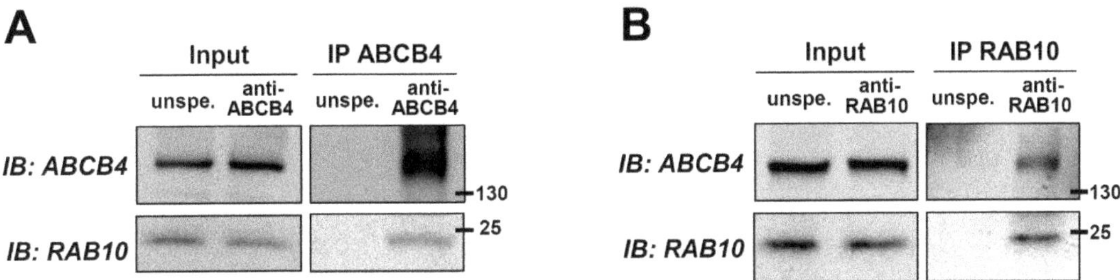

Figure 1. Co-immunoprecipitation of ABCB4 and RAB10 in primary human hepatocytes. (**A,B**) ABCB4 (A) or RAB10 (B) were immunoprecipitated from primary human hepatocyte lysates using specific antibodies. Controls were performed using unspecific antibodies (unspe.). After SDS-PAGE, the presence of ABCB4 and RAB10 in the lysates (Input) and the immunoprecipitates (IP) was detected by immunoblot (IB) as indicated. Molecular weight markers (in kDa) are indicated. These panels are representative of at least three independent experiments per condition. The presented data were cropped from full immunoblots shown in Supplementary Figure S2.

2.2. Overexpression of RAB10-WT or Its Dominant-Active Mutant Increases ABCB4 Plasma Membrane Expression and Function

To investigate the functional role of RAB10 in ABCB4 regulation, we first studied the effect of RAB10 overexpression on ABCB4 expression. HEK cells, co-expressing ABCB4 and different forms of RAB10-GFP (WT or its constitutively active Q68L and inactive T23N forms), or GFP alone (as control), were used. RAB10 (endogenous and GFP-tagged) and ABCB4 expression was detected by immunoblot (Figure 2A). The densitometry analyses indicated that the transient expression of the different forms of RAB10 had no effect on total ABCB4 expression (Figure 2B). However, when we studied ABCB4 function, we observed an important increase of ABCB4-mediated PC secretion in HEK cells expressing RAB10-WT or the constitutively active RAB10-Q68L, but not in cells expressing the inactive RAB10-T23N form (Figure 2C). In the absence of ABCB4-WT expression, no significant increase of basal PC efflux was observed when RAB10-WT alone was expressed (Supplementary Figure S3).

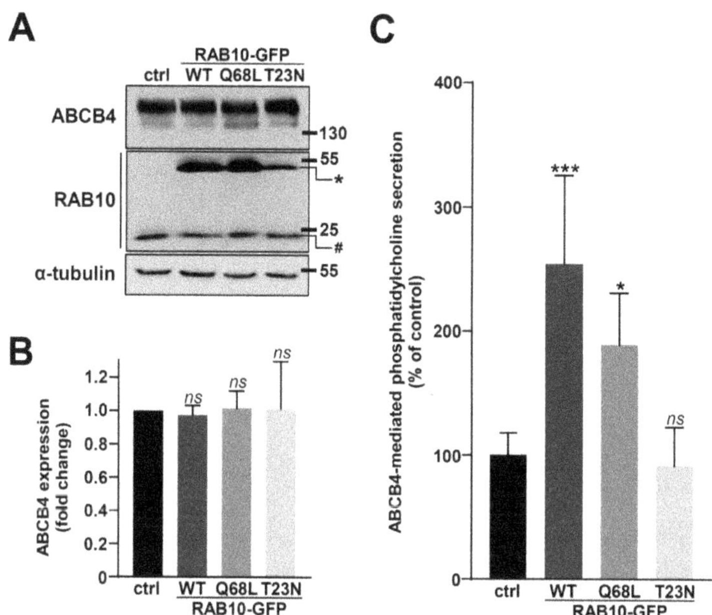

Figure 2. Analysis of ABCB4 expression and function after RAB10 overexpression in HEK cells. (**A**) ABCB4 and the indicated forms of RAB10-GFP were transiently expressed in HEK cells. Empty EGFP-C1 plasmid was used as control (ctrl). Forty-eight hours post-transfection, cell lysates were prepared and analyzed by immunoblot using the indicated antibodies. Molecular weight markers (in kDa) are indicated, as well as endogenous (#) and GFP-tagged (*) RAB10. This panel is representative of three independent experiments. The presented data were cropped from full immunoblots shown in Supplementary Figure S4. (**B**) Densitometry analysis of (A). The amount of ABCB4 was quantified, normalized to the amount of tubulin, and then expressed as fold change compared to the control condition (ctrl). Means (±SD) of three independent experiments are shown; ns: not significant. (**C**) HEK cells expressing both ABCB4 and the indicated forms of RAB10-GFP were used to measure ABCB4-mediated PC secretion. PC secretion was represented as a percentage of the activity for control cells (ctrl, expressing GFP alone) after background subtraction. Means (±SD) of three independent experiments performed in triplicate for each tested condition are shown; * $p < 0.05$; *** $p < 0.005$; ns: not significant.

Considering the key role played by RAB proteins in intracellular protein trafficking, we explored the hypothesis that the increased ABCB4-mediated PC secretion in cells overexpressing RAB10-WT or -Q68L might be correlated with an increase of ABCB4 expression at the cell surface. To verify this hypothesis, in a first attempt, we used polarized HepG2 cells to quantify amounts of ABCB4 present at bile canaliculi (Supplementary Figure S5). However, this did not allow an appropriate quantification of fluorescence intensities at the canalicular membrane, due to the important clustering of ABCB4-associated fluorescence in the very small area of bile canaliculus. Alternatively, we studied ABCB4 immunolocalization in HeLa cells co-expressing a modified version of ABCB4 with a FLAG tag in its first extracellular loop (Figure 3A) and the different forms of RAB10-GFP as above. We used HeLa cells for this approach since they are much more spread than HEK cells and thus allow better quantification analyses. In the absence of cell permeabilization and using anti-FLAG antibodies, this strategy allows for exclusively detecting cell surface ABCB4 (Figure 3A,B, red), while the use of anti-ABCB4 antibodies after cell permeabilization reveals the ABCB4 total population (Figure 3A,B, blue). The specificity of the plasma membrane staining of ABCB4 using anti-FLAG antibodies is supported by the observation that the intracellularly retained ABCB4-I541F variant [28] with the same FLAG tag (ABCB4-I541F-FLAG) does not display this specific signal, while this is partially rescued upon treatment with cyclosporin A (Supplementary Figure S6), known to rescue the plasma membrane targeting of this variant [29]. In line with our hypothesis and in agreement with results regarding ABCB4 function (Figure 2B), we observed that the amount of ABCB4 at the cell surface was markedly increased in cells expressing RAB10-WT or -Q68L in comparison to the control condition with GFP alone (Figure 3B). To confirm these observations, we quantified the fluorescence ratio of cell surface/total ABCB4 in the different experimental conditions and we observed a significant increase of this fluorescence ratio when RAB10-WT or its dominant-active form were expressed (Figure 3C). It is important to note that the reference for these quantifications of fluorescence intensities was "total ABCB4", a parameter that is not influenced by the subcellular localization of ABCB4. Altogether, these results suggest that RAB10 is involved in the plasma membrane targeting of ABCB4, an obligatory process allowing its function of PC secretion towards the extracellular environment.

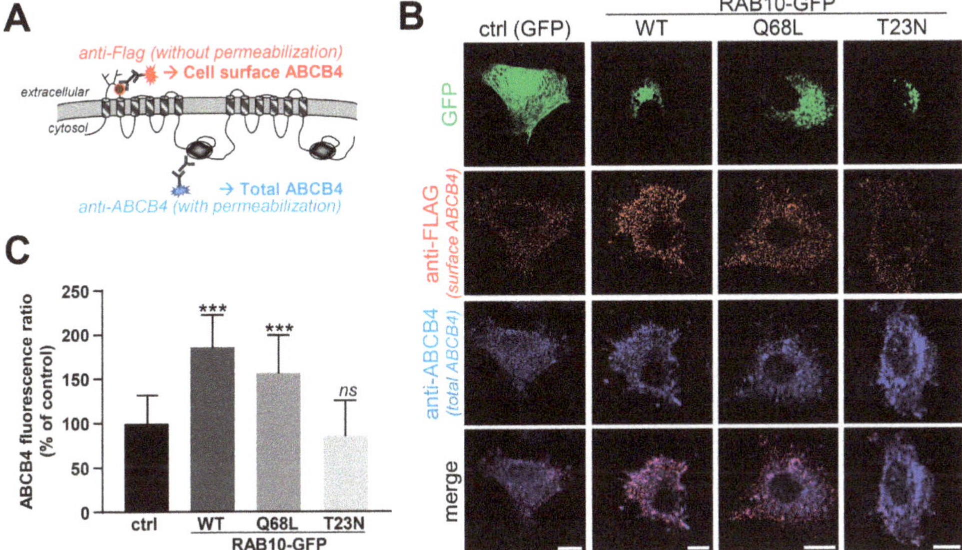

Figure 3. Analysis of ABCB4 localization at the plasma membrane after RAB10 overexpression in HeLa cells. (**A**) Schematic representation of the indirect immunofluorescence approach allowing the discrimination of cell surface (red) vs. total (blue) ABCB4 using anti-FLAG antibodies before permeabilization and anti-ABCB4 antibodies after permeabilization, respectively. (**B**) ABCB4-FLAG and the indicated RAB10 forms were expressed in HeLa cells. After 48 h of expression, the cell surface and total ABCB4 were labeled as schematized in (A). After immunolabeling, cell surface ABCB4 (red), total ABCB4 (blue) and RAB10-GFP (green; GFP alone for control) were visualized by confocal microscopy. This panel is representative of three independent experiments. Bars: 10 μm. (**C**) Quantification of (B). Fluorescence ratio of cell surface ABCB4/total ABCB4 was determined and represented as a percentage of the control condition (GFP alone-expressing cells). For each condition, at least 90 independent cells from three independent experiments were analyzed. Means (±SD) are represented; *** $p < 0.005$; ns: not significant.

2.3. *RAB10 Silencing Reduces ABCB4 Plasma Membrane Expression and Decreases Its Function*

Then, we investigated the effect of RAB10 silencing on ABCB4 cell surface expression using a siRNA approach. The transfection of specific anti-RAB10 siRNAs in HeLa cells induced an important knockdown of endogenous RAB10 (Figure 4A). The quantification of these experiments indicated that RAB10 expression was decreased by more than 90% in comparison with cells transfected with control siRNAs (Figure 4B). In this RAB10 knockdown condition, using the specific plasma membrane labeling described above, we observed that less ABCB4 was localized at the plasma membrane than in the control condition (Figure 4C), which was confirmed by the quantification of these experiments (Figure 4D).

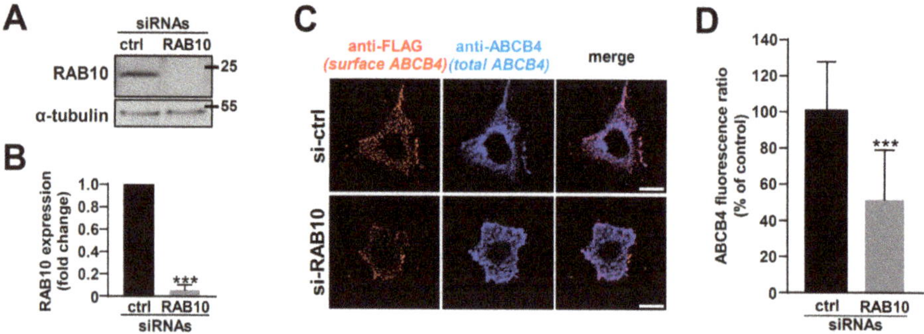

Figure 4. Effect of RAB10 knockdown on ABCB4 localization in HeLa cells. (**A**) Forty-eight hours after siRNA transfection in HeLa cells (Control, ctrl or anti-RAB10), RAB10 expression was analyzed by immunoblot, as in Figure 2A. This panel is representative of three independent experiments. The presented data were cropped from full immunoblots shown in Supplementary Figure S7. (**B**) Densitometry analysis of (A), as performed in Figure 2B. Means ($\pm$SD) of at least three independent experiments are shown; *** $p < 0.005$. (**C**) HeLa cells treated as in (A) were used to analyze ABCB4-FLAG localization at the plasma membrane, as performed in Figure 3B. This panel is representative of three independent experiments. Bars: 10 µm. (**D**) Quantification of (C), as performed in Figure 3C. Means ($\pm$SD) are represented; *** $p < 0.005$.

To confirm the effect of RAB10 silencing on ABCB4 cell surface expression, we used RAB10-KO HEK cells generated by the CRISPR/Cas9 approach. This strategy was complementary to siRNA transfection and less aggressive for the cells in the frame of assays aiming at measuring ABCB4-mediated PC secretion. Immunoblot analyses of RAB10 expression indicated an important knockdown of protein expression in HEK cells (Figure 5A) to more than 90% compared to control cells (Figure 5B). Then, ABCB4 cell surface expression was determined as aforementioned. In line with our previous results in HeLa cells, an important decrease in ABCB4 cell surface expression was detected in RAB10-KO HEK cells, in comparison to control cells (Figure 5C; quantification in Figure 5D). We also examined the effect of RAB10 silencing by CRISPR/Cas9 on ABCB4 function in HEK cells: ABCB4-mediated PC secretion was strongly impaired in RAB10-KO cells in comparison to control cells (Figure 5E). Altogether, these results indicate that RAB10 knockdown diminishes plasma membrane expression levels of ABCB4, and thus indirectly its PC secretion function towards the extracellular environment.

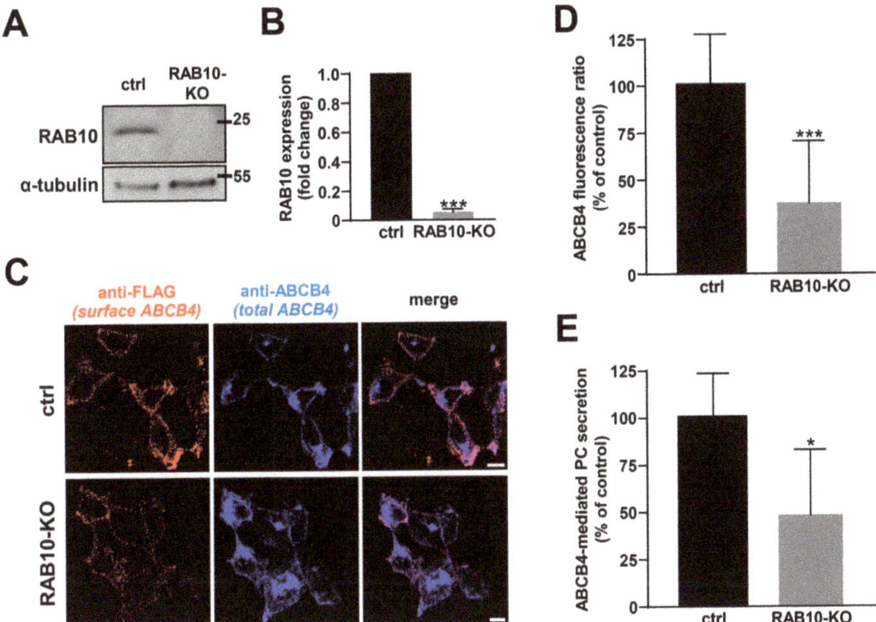

Figure 5. Effect of RAB10 knockdown on ABCB4 localization and function in HEK cells. (**A,B**) RAB10 expression in control (ctrl) and RAB10-KO HEK cells was analyzed (A) and quantified (B) as in Figure 4A,B. (A) is representative of three independent experiments and (B) represents the means (±SD) of three independent experiments; *** $p < 0.005$. The presented data were cropped from full immunoblots shown in Supplementary Figure S7. (**C**) HEK cells treated as in (A) were used to analyze ABCB4-FLAG localization at the plasma membrane, as performed in Figure 3B. This panel is representative of three independent experiments. Bars: 10 μm. (**D**) Quantification of (C), as performed in Figure 3C. Means (±SD) are represented; *** $p < 0.005$. (**E**) Control (ctrl) and RAB10-KO HEK cells were used to measure ABCB4-mediated PC secretion, as performed in Figure 2C. Means (±SD) of three independent experiments are shown; * $p < 0.05$.

2.4. RAB10 Promotes ABCB4 Trafficking from the Golgi Apparatus to the Plasma Membrane

Since we observed a considerable reduction of ABCB4 expression at the cell surface after RAB10 knockdown, both in HeLa and HEK cells (see Figures 4 and 5) without significant modification of total ABCB4 expression (Supplementary Figure S8), we inferred that RAB10 knockdown caused an intracellular retention of ABCB4. As previously reported [22], we found that RAB10 was preferentially associated with the Golgi apparatus, colocalizing with giantin (Supplementary Figure S9). We thus hypothesized that the plasma membrane targeting of ABCB4 might be impaired at post-Golgi steps when RAB10 is knocked down. The total ABCB4 distribution was examined by confocal microscopy in both HeLa and HEK cells. Under control conditions, ABCB4 displayed a significant plasma membrane staining (Figure 6A,B, upper panels). However, after RAB10 knockdown, ABCB4 was less present at the plasma membrane with an increased colocalization with the Golgi marker giantin, both in HeLa cells (Figure 6A, lower panels) and HEK cells (Figure 6B, lower panels). These results provide evidence that RAB10 is involved in ABCB4 trafficking from the Golgi apparatus to the plasma membrane.

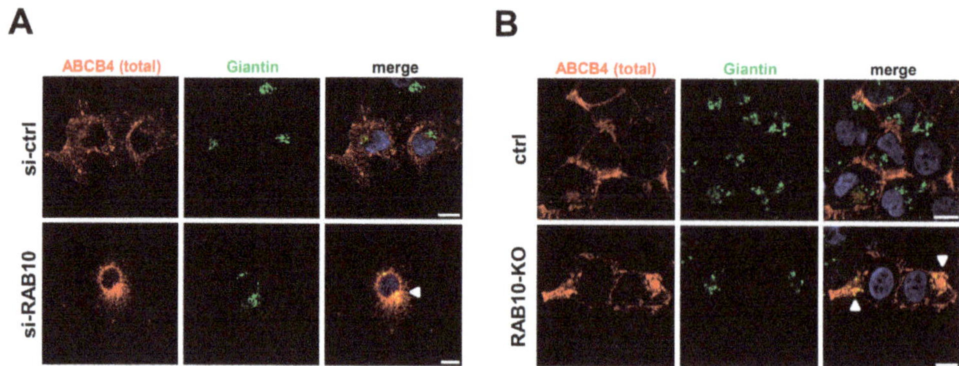

Figure 6. Effect of RAB10 depletion on total ABCB4 subcellular distribution. (**A**,**B**) HeLa cells transfected with control (si-ctrl) or anti-RAB10 (si-RAB10) (A), and control (ctrl) vs. RAB10-KO HEK cells (B) were used to analyze ABCB4 localization by indirect immunofluorescence and confocal microscopy. ABCB4 (red) and giantin (green) were labeled using specific antibodies while nuclei were stained with Hoechst 33,342 (blue). Each panel is representative of at least three independent experiments per condition. White arrows indicate colocalization of ABCB4 with giantin at the Golgi apparatus. Bars: 10 μm.

3. Discussion

ABCB4 plays an important role in bile secretion [1,5]. However, little is known regarding its molecular regulation. In the present study, to further understand ABCB4 biology, we explored ABCB4 interactome. Using an AP-MS screen, we identified the small GTPase RAB10 as a novel ABCB4 binding partner. However, using this method, we did not identify the few other known partners of ABCB4 [18]. This could be explained by: (i) a weak sensitivity of the technique; (ii) the stringent conditions of the immunoprecipitation; or (iii) the poor overlap of the different techniques and studies, as already reported for the analysis of ABCB11 interactome [30]. RAB10 belongs to the RAB protein family, known as key players in intracellular traffic processes [19]. Interestingly, it has already been reported that several RAB proteins interact with and regulate the intracellular traffic of several ABC transporters. Indeed, RAB11 colocalizes with both ABCB11 and ABCC2 and regulates their recycling to the canalicular membrane [27,31]. RAB4 and RAB5 were also reported as ABCB1 regulators [32,33]; and RAB5a, RAB7, RAB4, RAB11a, and RAB27a are implicated in the regulation of ABCC7/CFTR traffic and function [34]. More recently, RAB10 has also been shown to be implicated in CFTR targeting to the plasma membrane [35].

In the present study, we show that the transient overexpression of RAB10-WT or its dominant-active mutant (RAB10-Q68L) significantly increases ABCB4 expression at the cell surface and consequently its ability to secrete PC outside the cells. In contrast, the overexpression of the dominant-inactive form RAB10-T23N had no significant effect on ABCB4 localization or function, except for a tendency to decrease ABCB4 expression at the cell surface and its function. These results might be explained by the fact that this constitutively inactive form is less expressed than the other RAB10 forms (see Figure 2A), which might be due to its reduced stability, as previously reported [36]. This hypothesis is strengthened by the fact that RAB10 depletion, by both siRNA or CRISPR/Cas9 approaches, reduces ABCB4 cell surface expression and function as well as it induces its intracellular accumulation in the Golgi apparatus. Since RAB10 has been implicated in post-Golgi trafficking of GLUT4 and TLR4 [23,24], we can speculate that this RAB protein is also necessary for vesicular trafficking of ABCB4 from the Golgi apparatus to the plasma membrane. The fact that we only observed a partial decrease of ABCB4 cell surface expression and function in RAB10-knocked down and RAB10-KO cells (see Figures 4 and 5) might be due to the redundancy of RAB10 function with other RAB proteins. Indeed, functional similarity and redundancy between RAB8, RAB10, and RAB13 have been reported [22]. Further investigation will be required for a deeper analysis of the role of RAB protein redundancy on ABCB4 intracellular traffic and function. It is also interesting to note that ABCB4 staining at the plasma membrane appears as punctuated (ABCB4-FLAG staining in Figures 3B and 4C), suggesting its partition in specialized microdomains at the plasma membrane. Whether the transporter is localized in "raft-like" structures or not, the latest having already been suggested [37], would require further investigation.

ABCB4 defects are associated with several cholestatic liver diseases [2,4]. Interestingly, during the last years, many cholestatic like-phenotypes were associated with mutations in genes encoding traffic regulators, such as the motor protein MYO5B, the vacuole protein sorting-associated protein VPS33B, and its interacting protein VIPAS39 [38–41]. This highlights and supports the importance of correct ABCB4 trafficking to ensure its function and thereby a normal bile flow [41]. Thus, it seems important to more deeply explore molecular mechanisms regulating ABCB4 intracellular traffic. Moreover, based on the present study, it is tempting to speculate that *RAB10* mutations or malfunction could lead to diseases mimicking ABCB4 deficiency, arguing for the research of *RAB10* mutations in patients with unexplained cholestatic diseases. In this respect, it is of interest that RAB10 dysregulation was reported in some cases of hepatocellular carcinoma, which may support its key role in protein trafficking in hepatocytes [42].

In conclusion, using cell models, we report here that RAB10 is a novel ABCB4 molecular partner and is a key regulator of its intracellular traffic from the Golgi apparatus to the plasma membrane. Further work will be necessary to determine the physiological relevance of this interaction and its role in polarized hepatocytes. Finally, we expect that an improved understanding of ABCB4 regulation will help the development of new therapeutic options for patients with cholestatic diseases related to ABCB4 defects.

4. Materials and Methods

4.1. Plasmids, Cell Culture and Transfection

Two different constructs encoding human ABCB4 were used: pcDNA3-ABCB4, which was previously described [28], and pcDNA3-ABCB4-FLAG, a modified version of ABCB4 with a FLAG tag (DYKDDDDK) within its first extracellular loop (between Ser 99 and Leu 100), prepared by Genscript (Piscataway, NJ, USA). Constructs encoding human RAB10-WT (pEGFP-RAB10-WT) and its constitutively active (pEGFP-RAB10-Q68L) and inactive (pEGFP-RAB10-T23N) forms, all with N-terminal GFP, were a kind gift from Mark McNiven (Department of Biochemistry and Molecular Biology, Mayo Clinic, Rochester, MN, USA), which were prepared as described [43].

Primary human hepatocytes were prepared as described [44] and kindly provided by the Human HepCell platform (ICAN, Paris, France). Human embryonic kidney cells (HEK-293, herein referred to as HEK; ATCC®-CRL-1573TM) and HeLa cells (ATTC®-CCL-2TM) were grown at 37 °C with 5% CO_2 as described [45]. HEK and HeLa cells were transfected using Turbofect (Thermo Fisher Scientific, Villebon-sur-Yvette, France) at a ratio of reagent:DNA of 2:1 according to the manufacturer's instructions. JetPrime (PolyPlus Transfection, Illkirch, France) was used for small interfering RNA (siRNA) transfection according to the manufacturer's instructions.

4.2. RNA Interference and CRISPR/Cas9 System

For RAB10 silencing experiments, anti-RAB10 siRNA (ON TARGETplus human RAB10 siRNA SMARTpool) and scramble control siRNA (ON TARGETplus non-targeting control pool) from Horizon Discovery (Cambridge, UK) were used. RAB10-knockout (KO) HEK cells were established using the CRISPR/Cas9 technology as previously described [46]. Briefly, sgRNA guide sequences targeting human RAB10 were designed using the CRISPR design tool from Horizon Discovery (sense, 5′–GCGTACGTCTTCTTCGCCATT–3′ and antisense, 5′–AATGGCGAAGAAGACGTACGC–3′) and cloned into pSpCas9(BB)-2A-Puro plasmid (PX459, Addgene). After sequence verification, the resulting plasmid was transfected into HEK cells, and 24 h later, 2 µg/mL puromycin (Gibco) was added to the culture medium for the selection of transfected cells.

4.3. Immunoprecipitation and Mass Spectrometry Analyses (AP-MS)

For this approach, ABCB4 was immunoprecipitated from primary human hepatocytes and ABCB4-expressing HEK cells, then potential interactors were identified by tandem mass spectrometry. Primary human hepatocytes (2×10^7 cells) and HEK cells expressing ABCB4 (10^7 cells) were lysed in lysis buffer (20 mM Tris-HCl pH 8.0, 137 mM NaCl, 2 mM EDTA, 1% Triton X-100) containing protease inhibitors (cOmplete™ Protease Inhibitor Cocktail, Sigma, Saint-Quentin-Fallavier, France). The lysates were centrifuged at 14,000 rpm, 10 min at 4 °C, then supernatants were collected. After supernatant preclearing 1 h at 4 °C with 40 µL Protein-A Sepharose alone (VWR, Fontenay-sous-Bois, France), samples were incubated for 1 h at 4 °C with agitation in the presence of 0.5 µg of anti-ABCB4 (P3II-26 – Enzo Life Sciences, Villeurbanne, France) or unspecific IgG2b (MPC-11, BioLegend) antibodies, and then overnight at 4 °C with 40 µL Protein A-Sepharose. As a negative control allowing the elimination of unspecific contaminating proteins, we also included a condition with non-ABCB4-expressing HEK cells. Immunoprecipitates were recovered by centrifugation (1500 rpm, 5 min, 4 °C), washed three times with lysis buffer, and twice with ice-cold PBS. Bound proteins were eluted with 50 mM Tris-HCl pH 8.5 containing 2% SDS. As a control of immunoprecipitation efficacy, 30% of the eluted fraction was subjected to SDS-PAGE and analyzed by immunoblot with appropriate antibodies. The rest of the eluted samples was processed using the filtered-aided sample preparation (FASP) method to collect peptides as previously described [47]. Stage-tip-desalted peptides were analyzed by LC-MS-MS using an Ultimate 3000 Rapid Separation liquid chromatographic system coupled to an Orbitrap Fusion mass spectrometer (both from Thermo Fisher Scientific) as follows: peptides were loaded on a C18 reverse phase pre-column (3 µm particle size, 100 Å pore size, 75 µm inner diameter, 2 cm length; Thermo Fischer Scientific) using loading solvent (1% Acetonitrile and 0.1% trifluoroacetic acid in milliQ water) for 3 min at 5 µL.min^{-1}, then separated on an analytical C18 reverse phase column (2 µm particle size, 100 Å pore size, 75 µm internal diameter, 25 cm length) with a 45 min effective gradient from 99% A (0.1% formic acid in milliQ water) to 50% B (80% Acetonitrile, 0.085% formic acid in milliQ water) at 400 nL.min^{-1}. The mass spectrometer acquired data throughout the LC elution process and operated in a data-dependent scheme with full MS scans acquired with the orbitrap, followed by HCD fragmentation and ion trap fragment detection (top speed mode in 5 s) on the most abundant ions detected in the MS scan. Mass spectrometer settings were for full scan MS: AGC: 2.0E4, target resolution: 60,000, m/z range was 350–1500, maximum ion injection time: 60 ms; for HCD MS/MS: quadrupole filtering, normalized collision energy: 27. Ion trap rapid detection, intensity threshold: 1.0E4, isolation window: 1.6 m/z, dynamic exclusion time: 30 s, AGC Target: 2.0E4 and maximum injection time: 100 ms. The fragmentation was permitted for precursor with charge state of 2 to 7. Proteome discoverer 1.4 (Thermo Fisher Scientific) was used to generate. mgf peaklists files.

Peptides were identified as follows. Experimental mass lists were used to perform comparisons with theoretical mass lists from the *Homo sapiens* taxon of the Swiss-Prot database (May 2017, 20,204 sequences) using Mascot version 2.5.1 (www.matrixscience.com, accessed on 30 June 2021). The cleavage specificity set was the trypsin with maximum 2 missed cleavages. The precursor mass tolerance was set to 4 ppm and the MS/MS mass tolerance to 0.55 Da. Cystein carbamidomethylation was set as a constant modification while methionine oxidation was set as variable modification. With these settings, peptide identifications were considered as valid whenever their scores reached a minimum of 25, thus meeting the *p*-values criteria less than 0.05. The sample comparison was performed with MyPROMS software [48]. Identified proteins with at least 2 distinct peptides in at least one sample were considered positive.

4.4. Immunoblots, Immunofluorescence and Measurement of ABCB4-Mediated Phosphatidylcholine Secretion

Immunoblots were performed as previously described [9,45], using cells co-transfected with ABCB4- and RAB10-encoding constructs at a 1:1 ratio with the following primary antibodies: anti-ABCB4 (clone P3II-26) from Enzo Life Sciences, anti-α-tubulin (clone 1E4C11) from ProteinTech (Manchester, UK), anti-RAB10 (clone D36C4) from Cell Signaling (Danvers, MA, USA) and anti-GFP (clone AB10145) from Sigma. Immunoblots were quantified in the linear range of detection using ImageJ 1.50i software (U.S. National Institutes of Health, Bethesda, MD, USA).

For indirect immunofluorescence experiments, cells were grown on glass coverslips, fixed with 4% paraformaldehyde (Thermo Fisher Scientific) for 10 min, and permeabilized with 0.05% saponin in PBS containing 0.2% of bovine serum albumin (BSA). The coverslips were then incubated with the following primary antibodies: anti-ABCB4 and anti-RAB10 (see above), anti-calnexin (SPA-865, Enzo Life Sciences), anti-giantin (PRB-114C, Covance, Rueil-Malmaison, France), anti-EEA1 (sc-6415, Santa Cruz Biotechnology, Santa Cruz, CA, USA) and anti-LAMP1 (sc-20011, Santa Cruz Biotechnology); and subsequently incubated for 1 hour with appropriate Alexa Fluor-conjugated secondary antibodies (ThermoFisher Scientific). Nuclei were labeled using Hoechst 33342 (ThermoFisher Scientific). Immunofluorescence images were acquired using a confocal microscope (Eclipse TE-2000-Nikon-C2) equipped with a 60X objective, serial xy optical sections with a z-step of 0.3 μm were acquired using Nikon NIS-Elements software version AR 4.50 with constant settings (laser powers and correction of signal intensities) and treated using Adobe Photoshop version 8.0.1.

Alternatively, for specific cell surface labeling of ABCB4, non-permeabilized HeLa cells expressing ABCB4-FLAG were incubated 1 h with polyclonal anti-FLAG antibodies (F7425, Sigma) prior to fixation. Then, after fixation and permeabilization, total vs. cell surface ABCB4 were revealed using the standard procedure described above. A cell by cell quantification of cell surface vs. total ABCB4 signal intensities was performed as follows: individual ABCB4-expressing cells were randomly selected and delineated in ImageJ software, version 1.50i using the total ABCB4 signal. For each segmented cell, the fluorescence intensities associated with total ABCB4 (shown in blue) or plasma membrane ABCB4 (shown in red) were measured in each channel using the 'Measure' function of ImageJ. Then, after background subtraction (measured in areas with no apparent fluorescence) for each fluorescence intensity, plasma membrane over total ABCB4 fluorescence ratios were expressed as percentages of the mean ratio calculated for the reference condition (control).

The measurement of ABCB4-mediated PC secretion using a fluoro-enzymatic assay was performed as described [49] and results were analyzed as published [45].

4.5. Statistical Analyses

Graphics and one-way ANOVA tests were performed using Prism version 7.00 (GraphPad Software, La Jolla, CA, USA). A p value of less than 0.05 was considered significant with *: $p < 0.05$; **: $p < 0.01$; ***: $p < 0.005$; ns: not significant. Symbols always indicate the comparison between the control and the other tested conditions.

Supplementary Materials: The following are available online at https://www.mdpi.com/article/10.3390/ijms22137087/s1. Figure S1: Co-immunoprecipitation of ABCB4 and RAB10 in HEK cells. HEK cells co-expressing RAB10-GFP and ABCB4 were used. GFP (A) and ABCB4 (B) were immunoprecipitated using specific antibodies. Controls were performed using unspecific antibodies (unspe.). After SDS-PAGE, the presence of ABCB4 and RAB10-GFP in the lysates (Input) and the immunoprecipitates (IP) was detected by immunoblot (IB) as indicated. Molecular weight markers (in kDa) are shown. These panels are representative of three independent experiments per condition. Figure S2: Full immunoblots related to Figure 1A,B. Results shown in Figure 1A,B are delineated by dotted rectangles. MW (in kDa) are indicated. Figure S3: Phosphatidylcholine efflux after RAB10-WT expression. The phosphatidylcholine efflux from HEK cells expressing ABCB4-WT or RAB10-WT-GFP was measured as in Figure 2C. Note that for these experiments, means could not be normalized to ABCB4 expression levels. Means (± SD) of three independent experiments are shown. Figure S4: Full immunoblots related to Figure 2A. Results shown in Figure 2A are delineated by dotted rectangles. MW (in kDa) are indicated. Figure S5: ABCB4 immunolocalization in HepG2 cells. ABCB4 and the indicated forms of RAB10-GFP (GFP alone as control, ctrl) were co-expressed in

HepG2 cells. Total ABCB4 and RAB10-GFP localization was analyzed as in Figure 3B. This figure is representative of three independent experiments. Bars: 10 μm. Figure S6: ABCB4-I541F-FLAG immunolocalization in HeLa cells. ABCB4-WT-FLAG or ABCB4-I541F-FLAG were expressed in HeLa cells. After treatment with vehicle (DMSO) or 10 μM cyclosporin A, cell surface and total ABCB4 were immunolabeled and visualized as in Figure 3B. This figure is representative of three independent experiments. Bars: 10 μm. Figure S7: Full immunoblots related to Figures 4A and 5A. Results shown in Figures 4A and 5A are delineated by dotted rectangles. MW (in kDa) are indicated. Figure S8: ABCB4 expression after RAB10 knockdown. Cell lysates from HeLa cells (A) and HEK cells (B) treated as in Figures 4 and 5, respectively, were analyzed by immunoblot using the indicated antibodies (upper panels). MW (in kDa) are indicated. These panels are representative of three independent experiments per condition. ABCB4 signal intensities were quantified and means (± SD) of three independent experiments are shown (lower panels). Figure S9: Confocal microscopy analysis of RAB10-WT subcellular localization in HeLa cells. HeLa cells were transfected with a RAB10-WT-GFP-encoding construct (green) and the following intracellular compartments were immunolabeled (red): calnexin for endoplasmic reticulum, giantin for the Golgi apparatus, EAA1 for early endosomes, LAMP1 for late endosomes and lysosomes. This figure is representative of three independent experiments. Bars: 10 μm.

Author Contributions: A.B.S., V.V., and T.F. designed the study. A.B.S., V.V., E.B., A.-M.D.-S. and T.F. performed the experiments. A.B.S., V.V., M.L., E.M., A.-M.D.-S., A.B., J. L.D., E.G., C.H., T.A.-S., F.G., E.J. and T.F. analyzed the data and provided significant intellectual contribution. A.B.S. and T.F. wrote the manuscript, which was reviewed and approved by all authors. All authors have read and agreed to the published version of the manuscript.

Funding: A.B.S. and E.M. were supported by the « Ministère de l'Enseignement Supérieur, de la Recherche et de l'Innovation ». T.F. was supported by grants from the Agence Nationale de la Recherche (ANR-15-CE14-0008-01), the French Association for the Study of the Liver (AFEF) and FILFOIE (Filière de santé des maladies rares du foie, Paris, France). T.A.-S. was supported by grants from the Association Mucoviscidose-ABCF2, Fondation pour la Recherche Médicale (EQU2020-03010517) and FILFOIE. The Orbitrap Fusion mass spectrometer was acquired with funds from the FEDER through the «Operational Programme for Competitiveness Factors and employment 2007-2013», and from the «Cancéropôle Ile-de-France».

Institutional Review Board Statement: Not applicable.

Informed Consent Statement: Not applicable.

Data Availability Statement: The datasets generated and analyzed during the current study are available from the corresponding author on reasonable request.

Acknowledgments: We are grateful to FILFOIE, AMFE (Association Maladie Foie Enfants, Malakoff, France), MLD (Monaco Liver Disorder, Monaco), Association "Pour Louis 1000 Foie Merci" (Fournet-Luisans, France), Association "Il était un foie" (Plouescat, France) and Fondation Rumsey-Cartier (Genève, Switzerland) for their financial support. We also thank Mark McNiven (Department of Biochemistry and Molecular Biology, Mayo Clinic, Rochester, USA) for providing RAB10-encoding plasmids, Wadih Ghattas and Charles Skarbek (Institut de Chimie Moléculaire et des Matériaux d'Orsay, Université Paris-Saclay, Orsay, France) for their help with multiplate fluorescence reading, as well as Lynda Aoudjehane (Human HepCell–ICAN, Saint-Antoine Research Center, Paris, France) for providing primary human hepatocytes, Jérémie Gautheron (Saint-Antoine Research Center, Paris, France) for his help with the CRISPR/Cas9 strategy and Isabelle Garcin (UMR_S 1193 Inserm/Université Paris-Saclay, Orsay, France) for her help with imaging studies.

Abbreviations

ABC: ATP-Binding Cassette; AP-MS: Affinity Purification-Mass Spectrometry; HEK: Human Embryonic Kidney; KO: KnockOut; PC: PhosphatidylCholine; RAB: RAs-related in Brain; siRNA: small interfering RNA; WT: Wild Type

References

1. Kroll, T.; Prescher, M.; Smits, S.H.J.; Schmitt, L. Structure and function of hepatobiliary atp binding cassette transporters. *Chem. Rev.* **2020**, *121*, 5240–5288. [CrossRef]
2. Falguières, T.; Aït-Slimane, T.; Housset, C.; Maurice, M. Abcb4: Insights from pathobiology into therapy. *Clin. Res. Hepatol. Gastroenterol.* **2014**, *38*, 557–563. [CrossRef]
3. Boyer, J.L. Bile formation and secretion. *Compr. Physiol.* **2013**, *3*, 1035–1078. [PubMed]
4. Davit-Spraul, A.; Gonzales, E.; Baussan, C.; Jacquemin, E. The spectrum of liver diseases related to abcb4 gene mutations: Pathophysiology and clinical aspects. *Semin. Liver Dis.* **2010**, *30*, 134–146. [CrossRef]
5. Stättermayer, A.F.; Halilbasic, E.; Wrba, F.; Ferenci, P.; Trauner, M. Variants in abcb4 (mdr3) across the spectrum of cholestatic liver diseases in adults. *J. Hepatol.* **2020**, *73*, 651–663. [CrossRef]
6. Jacquemin, E.; Hermans, D.; Myara, A.; Habes, D.; Debray, D.; Hadchouel, M.; Sokal, E.M.; Bernard, O. Ursodeoxycholic acid therapy in pediatric patients with progressive familial intrahepatic cholestasis. *Hepatology* **1997**, *25*, 519–523. [CrossRef] [PubMed]
7. Poupon, R. Ursodeoxycholic acid and bile-acid mimetics as therapeutic agents for cholestatic liver diseases: An overview of their mechanisms of action. *Clin. Res. Hepatol. Gastroenterol.* **2012**, *36* (Suppl. 1), S3–S12. [CrossRef]
8. Delaunay, J.L.; Bruneau, A.; Hoffmann, B.; Durand-Schneider, A.M.; Barbu, V.; Jacquemin, E.; Maurice, M.; Housset, C.; Callebaut, I.; Ait-Slimane, T. Functional defect of variants in the adenosine triphosphate-binding sites of abcb4 and their rescue by the cystic fibrosis transmembrane conductance regulator potentiator, ivacaftor (vx-770). *Hepatology* **2017**, *65*, 560–570. [CrossRef] [PubMed]
9. Vauthier, V.; Ben Saad, A.; Elie, J.; Oumata, N.; Durand-Schneider, A.M.; Bruneau, A.; Delaunay, J.L.; Housset, C.; Aït-Slimane, T.; Meijer, L.; et al. Structural analogues of roscovitine rescue the intracellular traffic and the function of er-retained abcb4 variants in cell models. *Sci. Rep.* **2019**, *9*, 6653. [CrossRef]
10. Aronson, S.J.; Bakker, R.S.; Shi, X.; Duijst, S.; Ten Bloemendaal, L.; de Waart, D.R.; Verheij, J.; Elferink, R.P.O.; Beuers, U.; Paulusma, C.C.; et al. Liver-directed gene therapy results in long term correction of progressive familial intrahepatic cholestasis type 3 in mice. *J. Hepatol.* **2019**, *71*, 153–162. [CrossRef]
11. Siew, S.M.; Cunningham, S.C.; Zhu, E.; Tay, S.S.; Venuti, E.; Bolitho, C.; Alexander, I.E. Prevention of cholestatic liver disease and reduced tumorigenicity in a murine model of pfic type 3 using hybrid aav-piggybac gene therapy. *Hepatology* **2019**, *70*, 2047–2061. [CrossRef]
12. Weber, N.D.; Odriozola, L.; Martínez-García, J.; Ferrer, V.; Douar, A.; Bénichou, B.; González-Aseguinolaza, G.; Smerdou, C. Gene therapy for progressive familial intrahepatic cholestasis type 3 in a clinically relevant mouse model. *Nat. Commun.* **2019**, *10*, 5694. [CrossRef] [PubMed]
13. Wei, G.; Cao, J.; Huang, P.; An, P.; Badlani, D.; Vaid, K.A.; Zhao, S.; Wang, D.Q.; Zhuo, J.; Yin, L.; et al. Synthetic human abcb4 mrna therapy rescues severe liver disease phenotype in a balb/c.Abcb4(-/-) mouse model of pfic3. *J. Hepatol.* **2020**, *74*, 1416–1428. [CrossRef] [PubMed]
14. Ortiz, D.F.; Moseley, J.; Calderon, G.; Swift, A.L.; Li, S.; Arias, I.M. Identification of hax-1 as a protein that binds bile salt export protein and regulates its abundance in the apical membrane of madin-darby canine kidney cells. *J. Biol. Chem.* **2004**, *279*, 32761–32770. [CrossRef] [PubMed]
15. Chan, W.; Calderon, G.; Swift, A.L.; Moseley, J.; Li, S.; Hosoya, H.; Arias, I.M.; Ortiz, D.F. Myosin ii regulatory light chain is required for trafficking of bile salt export protein to the apical membrane in madin-darby canine kidney cells. *J. Biol. Chem.* **2005**, *280*, 23741–23747. [CrossRef]
16. Ikebuchi, Y.; Takada, T.; Ito, K.; Yoshikado, T.; Anzai, N.; Kanai, Y.; Suzuki, H. Receptor for activated c-kinase 1 regulates the cellular localization and function of abcb4. *Hepatol. Res.* **2009**, *39*, 1091–1107. [CrossRef]
17. Venot, Q.; Delaunay, J.L.; Fouassier, L.; Delautier, D.; Falguieres, T.; Housset, C.; Maurice, M.; Ait-Slimane, T. A pdz-like motif in the biliary transporter abcb4 interacts with the scaffold protein ebp50 and regulates abcb4 cell surface expression. *PLoS ONE* **2016**, *11*, e0146962. [CrossRef]
18. Ben Saad, A.; Bruneau, A.; Mareux, E.; Lapalus, M.; Delaunay, J.L.; Gonzales, E.; Jacquemin, E.; Aït-Slimane, T.; Falguières, T. Molecular regulation of canalicular abc transporters. *Int. J. Mol. Sci.* **2021**, *22*, 2113. [CrossRef]
19. Stenmark, H. Rab gtpases as coordinators of vesicle traffic. *Nat. Rev. Mol. Cell Biol.* **2009**, *10*, 513–525. [CrossRef]
20. Homma, Y.; Hiragi, S.; Fukuda, M. Rab family of small gtpases: An updated view on their regulation and functions. *FEBS J.* **2021**, *288*, 36–55. [CrossRef]
21. Chavrier, P.; Vingron, M.; Sander, C.; Simons, K.; Zerial, M. Molecular cloning of ypt1/sec4-related cdnas from an epithelial cell line. *Mol. Cell Biol.* **1990**, *10*, 6578–6585. [CrossRef] [PubMed]
22. Chua, C.E.L.; Tang, B.L. Rab 10-a traffic controller in multiple cellular pathways and locations. *J. Cell Physiol.* **2018**, *233*, 6483–6494. [CrossRef] [PubMed]

23. Wang, D.; Lou, J.; Ouyang, C.; Chen, W.; Liu, Y.; Liu, X.; Cao, X.; Wang, J.; Lu, L. Ras-related protein rab10 facilitates tlr4 signaling by promoting replenishment of tlr4 onto the plasma membrane. *Proc. Natl. Acad. Sci. USA* **2010**, *107*, 13806–13811. [CrossRef] [PubMed]

24. Chen, Y.; Lippincott-Schwartz, J. Rab10 delivers glut4 storage vesicles to the plasma membrane. *Commun. Integr. Biol.* **2013**, *6*, e23779. [CrossRef] [PubMed]

25. Lam, P.; Pearson, C.L.; Soroka, C.J.; Xu, S.; Mennone, A.; Boyer, J.L. Levels of plasma membrane expression in progressive and benign mutations of the bile salt export pump (bsep/abcb11) correlate with severity of cholestatic diseases. *Am. J. Physiol. Cell Physiol.* **2007**, *293*, C1709–C1716. [CrossRef]

26. Szeri, F.; Niaziorimi, F.; Donnelly, S.; Orndorff, J.; van de Wetering, K. Generation of fully functional fluorescent fusion proteins to gain insights into abcc6 biology. *FEBS Lett.* **2020**, *595*, 799–810. [CrossRef]

27. Wakabayashi, Y.; Lippincott-Schwartz, J.; Arias, I.M. Intracellular trafficking of bile salt export pump (abcb11) in polarized hepatic cells: Constitutive cycling between the canalicular membrane and rab11-positive endosomes. *Mol. Biol. Cell* **2004**, *15*, 3485–3496. [CrossRef]

28. Delaunay, J.L.; Durand-Schneider, A.M.; Delautier, D.; Rada, A.; Gautherot, J.; Jacquemin, E.; Ait-Slimane, T.; Maurice, M. A missense mutation in abcb4 gene involved in progressive familial intrahepatic cholestasis type 3 leads to a folding defect that can be rescued by low temperature. *Hepatology* **2009**, *49*, 1218–1227. [CrossRef]

29. Gautherot, J.; Durand-Schneider, A.M.; Delautier, D.; Delaunay, J.L.; Rada, A.; Gabillet, J.; Housset, C.; Maurice, M.; Aït-Slimane, T. Effects of cellular, chemical and pharmacological chaperones on the rescue of a trafficking-defective mutant of the atp-binding cassette transporters abcb1/abcb4. *J. Biol. Chem.* **2012**, *287*, 5070–5078. [CrossRef]

30. Przybylla, S.; Stindt, J.; Kleinschrodt, D.; Schulte Am Esch, J.; Haussinger, D.; Keitel, V.; Smits, S.H.; Schmitt, L. Analysis of the bile salt export pump (abcb11) interactome employing complementary approaches. *PLoS ONE* **2016**, *11*, e0159778. [CrossRef]

31. Park, S.W.; Schonhoff, C.M.; Webster, C.R.; Anwer, M.S. Rab11, but not rab4, facilitates cyclic amp- and tauroursodeoxycholate-induced mrp2 translocation to the plasma membrane. *Am. J. Physiol. Gastrointest. Liver Physiol.* **2014**, *307*, G863–G870. [CrossRef]

32. Fu, D.; van Dam, E.M.; Brymora, A.; Duggin, I.G.; Robinson, P.J.; Roufogalis, B.D. The small gtpases rab5 and rala regulate intracellular traffic of p-glycoprotein. *Biochim. Biophys. Acta* **2007**, *1773*, 1062–1072. [CrossRef] [PubMed]

33. Ferrándiz-Huertas, C.; Fernández-Carvajal, A.; Ferrer-Montiel, A. Rab4 interacts with the human p-glycoprotein and modulates its surface expression in multidrug resistant k562 cells. *Int. J. Cancer* **2011**, *128*, 192–205. [CrossRef] [PubMed]

34. Farinha, C.M.; Matos, P. Rab gtpases regulate the trafficking of channels and transporters—A focus on cystic fibrosis. *Small GTPases* **2018**, *9*, 136–144. [CrossRef]

35. Lucken-Ardjomande Hässler, S.; Vallis, Y.; Pasche, M.; McMahon, H.T. Graf2, wdr44, and mical1 mediate rab8/10/11-dependent export of e-cadherin, mmp14, and cftr δf508. *J. Cell Biol.* **2020**, *219*, e201811014. [CrossRef]

36. Schuck, S.; Gerl, M.J.; Ang, A.; Manninen, A.; Keller, P.; Mellman, I.; Simons, K. Rab10 is involved in basolateral transport in polarized madin-darby canine kidney cells. *Traffic* **2007**, *8*, 47–60. [CrossRef]

37. Morita, S.Y.; Tsuda, T.; Horikami, M.; Teraoka, R.; Kitagawa, S.; Terada, T. Bile salt-stimulated phospholipid efflux mediated by abcb4 localized in nonraft membranes. *J. Lipid Res.* **2013**, *54*, 1221–1230. [CrossRef]

38. Gissen, P.; Johnson, C.A.; Morgan, N.V.; Stapelbroek, J.M.; Forshew, T.; Cooper, W.N.; McKiernan, P.J.; Klomp, L.W.; Morris, A.A.; Wraith, J.E.; et al. Mutations in vps33b, encoding a regulator of snare-dependent membrane fusion, cause arthrogryposis-renal dysfunction-cholestasis (arc) syndrome. *Nat. Genet.* **2004**, *36*, 400–404. [CrossRef]

39. Cullinane, A.R.; Straatman-Iwanowska, A.; Zaucker, A.; Wakabayashi, Y.; Bruce, C.K.; Luo, G.; Rahman, F.; Gürakan, F.; Utine, E.; Ozkan, T.B.; et al. Mutations in vipar cause an arthrogryposis, renal dysfunction and cholestasis syndrome phenotype with defects in epithelial polarization. *Nat. Genet.* **2010**, *42*, 303–312. [CrossRef]

40. Gonzales, E.; Taylor, S.A.; Davit-Spraul, A.; Thébaut, A.; Thomassin, N.; Guettier, C.; Whitington, P.F.; Jacquemin, E. Myo5b mutations cause cholestasis with normal serum gamma-glutamyl transferase activity in children without microvillous inclusion disease. *Hepatology* **2017**, *65*, 164–173. [CrossRef] [PubMed]

41. Li, Q.; Sun, Y.; van Ijzendoorn, S.C. A link between intrahepatic cholestasis and genetic variations in intracellular trafficking regulators. *Biology* **2021**, *10*, 119. [CrossRef] [PubMed]

42. He, H.; Dai, F.; Yu, L.; She, X.; Zhao, Y.; Jiang, J.; Chen, X.; Zhao, S. Identification and characterization of nine novel human small gtpases showing variable expressions in liver cancer tissues. *Gene Expr.* **2002**, *10*, 231–242. [CrossRef]

43. Li, Z.; Schulze, R.J.; Weller, S.G.; Krueger, E.W.; Schott, M.B.; Zhang, X.; Casey, C.A.; Liu, J.; Stöckli, J.; James, D.E.; et al. A novel rab10-ehbp1-ehd2 complex essential for the autophagic engulfment of lipid droplets. *Sci. Adv.* **2016**, *2*, e1601470. [CrossRef]

44. Aoudjehane, L.; Gautheron, J.; Le Goff, W.; Goumard, C.; Gilaizeau, J.; Nget, C.S.; Savier, E.; Atif, M.; Lesnik, P.; Morichon, R.; et al. Novel defatting strategies reduce lipid accumulation in primary human culture models of liver steatosis. *Dis. Models Mech.* **2020**, *13*, dmm042663. [CrossRef]

45. Ben Saad, A.; Vauthier, V.; Tóth, Á.; Janaszkiewicz, A.; Durand-Schneider, A.M.; Bruneau, A.; Delaunay, J.L.; Lapalus, M.; Mareux, E.; Garcin, I.; et al. Effect of cftr correctors on the traffic and the function of intracellularly retained abcb4 variants. *Liver Int.* **2021**, *41*, 1344–1357. [CrossRef]

46. Ran, F.A.; Hsu, P.D.; Wright, J.; Agarwala, V.; Scott, D.A.; Zhang, F. Genome engineering using the crispr-cas9 system. *Nat. Protoc.* **2013**, *8*, 2281–2308. [CrossRef] [PubMed]

47. Wiśniewski, J.R.; Zougman, A.; Nagaraj, N.; Mann, M. Universal sample preparation method for proteome analysis. *Nat. Methods* **2009**, *6*, 359–362. [CrossRef] [PubMed]
48. Poullet, P.; Carpentier, S.; Barillot, E. Myproms, a web server for management and validation of mass spectrometry-based proteomic data. *Proteomics* **2007**, *7*, 2553–2556. [CrossRef] [PubMed]
49. Gautherot, J.; Delautier, D.; Maubert, M.A.; Aït-Slimane, T.; Bolbach, G.; Delaunay, J.L.; Durand-Schneider, A.M.; Firrincieli, D.; Barbu, V.; Chignard, N.; et al. Phosphorylation of abcb4 impacts its function: Insights from disease-causing mutations. *Hepatology* **2014**, *60*, 610–621. [CrossRef]

Review

Gene Therapy for Progressive Familial Intrahepatic Cholestasis: Current Progress and Future Prospects

Piter J. Bosma *, Marius Wits and Ronald P. J. Oude-Elferink

Tytgat Institute for Liver and Intestinal Research and Department of Gastroenterology and Hepatology, AGEM, Amsterdam UMC, University of Amsterdam, 1105 BK Amsterdam, The Netherlands; marius_wits@hotmail.com (M.W.); r.p.oude-elferink@amsterdamumc.nl (R.P.J.O.-E.)
* Correspondence: p.j.bosma@amsterdamumc.nl

Abstract: Progressive Familial Intrahepatic Cholestasis (PFIC) are inherited severe liver disorders presenting early in life, with high serum bile salt and bilirubin levels. Six types have been reported, two of these are caused by deficiency of an ABC transporter; ABCB11 (bile salt export pump) in type 2; ABCB4 (phosphatidylcholine floppase) in type 3. In addition, ABCB11 function is affected in 3 other types of PFIC. A lack of effective treatment makes a liver transplantation necessary in most patients. In view of long-term adverse effects, for instance due to life-long immune suppression needed to prevent organ rejection, gene therapy could be a preferable approach, as supported by proof of concept in animal models for PFIC3. This review discusses the feasibility of gene therapy as an alternative for liver transplantation for all forms of PFIC based on their pathological mechanism. **Conclusion:** Using presently available gene therapy vectors, major hurdles need to be overcome to make gene therapy for all types of PFIC a reality.

Keywords: gene therapy; AAV; PFIC

Citation: Bosma, P.J.; Wits, M.; Oude-Elferink, R.P.J. Gene Therapy for Progressive Familial Intrahepatic Cholestasis: Current Progress and Future Prospects. *Int. J. Mol. Sci.* **2021**, 22, 273. https://doi.org/10.3390/ijms22010273

Received: 26 November 2020
Accepted: 26 December 2020
Published: 29 December 2020

Publisher's Note: MDPI stays neutral with regard to jurisdictional claims in published maps and institutional affiliations.

1. Introduction

1.1. Clinical Challenge

Progressive Familial Intrahepatic Cholestasis (PFIC) is a heterogenic group of recessively inherited severe liver disorders [1,2]. All types present during infancy or childhood with increased serum bile salts and bilirubin, and pruritus, having a major impact on health-related quality of life. All forms of PFIC are rare autosomal recessive diseases and deficiency of the bile acid export pump (BSEP) activity impairing bile salt handling is seen in several forms. Deficiency can be caused by mutations in the *ABCB11* gene encoding BSEP for instance or by loss of BSEP expression due to mutations in the *NR1H4* gene, encoding the Farnesoid X Receptor (FXR) that is essential for BSEP expression [3–6]. A loss of BSEP presence in the canalicular membrane due to mutations in Myosin VB (*MYO5B*), involved in its intracellular transport, or reduced BSEP activity, due to canalicular membrane integrity and composition, impair bile salt export from the hepatocyte [7–9]. In addition to impaired bile salt export, PFIC can be caused by ABCB4 deficiency involved in maintaining phosphatidylcholine presence in bile, needed to moderate the detergent effect of bile salt by forming mixed micelles [10]. Recently, mutations in the Tight Junction Protein 2 encoding gene (*TJP2*) were identified to cause PFIC [11]. Although *TJP2* mutations may affect tight junction integrity, these patients neither suffer from cholestasis nor other cholangiopathies, as would be expected in leakage of bile and has been observed in Claudin deficiency, for instance [1]. All types of PFIC are mono-genetic disorders and most progress to end stage liver disease making a liver transplant inevitable. In view of the adverse effects of this highly invasive treatment, developing novel treatment options, such as liver directed gene therapy, are warranted.

1.2. Presentation and Current Treatment

All forms of PFIC present with jaundice and elevated bile acid levels in serum [1]. The disease onset varies between the different types of PFIC. ABCB11/BSEP (PFIC2), TJP2 (PFIC4) and FXR (PFIC5) deficiency, present in the first months after birth and are rapidly progressing. Presentation of ATP8B1 deficiency (PFIC1) is often months after birth and the progression is moderate. Both other forms, ABCB4 (PFIC3) and MYO5B (PFIC6) deficiency, are diagnosed at a later more variable age with moderate to slow progression [1]. Upon diagnosing cholestasis; serum parameters, liver biopsies and genetic analyses are all used to identify the genetic cause of cholestasis. For instance, high gamma GT is only seen in ABCB4 deficiency (PFIC3) [12] while FXR deficiency results in persistent increased Alpha FetoProtein (AFP) levels in serum [6]. Liver biopsies can be used to verify the absence of proteins involved in physiologic bile formation in the canaliculi or other mechanisms like intracellular protein accumulation that cause hepatocyte and cholangiocyte damage. Identification of the cause allows treatment to begin, albeit all current treatments for PFIC, at best, only slow down disease progression. Ursodeoxycholic acid (UDCA) treatment, reducing the hydrophobicity of the bile salt pool, is a first treatment option for all types of PFIC. The efficacy of this treatment depends on the type of mutation, with patients having missense mutations showing a better response compared to those with complete deficiency [5,13]. In patients with missense mutations in the *ATP8B1* or the *ABCB11* gene, that affect protein folding, 4-phenylbutyrate, restored presence of these transporters in the canalicular membrane, albeit at a low level [14–16]. Symptoms are mitigated by a reduction in the bile salt pool by biliary diversion or by preventing intestinal uptake by binding of bile acids to cholestyramine [1]. The efficacy and safety of pharmacological inhibition bile acid uptake using inhibitors of the apical sodium dependent bile transporter (ASBT) may be another approach to reduce the bile acid pool [17,18]. Diarrhea is an adverse effect of several of these approaches due to increasing amounts of bile acids reaching the colon. In view of this, FXR activation to lower bile acid synthesis in the hepatocytes maybe an option to overcome this hurdle and in combination with ASBT inhibition could be an effective approach [19,20]. Several of these treatments do relieve the symptoms and slow disease progression but none of these is curative. For the more recently identified genetic causes of PFIC, namely TJP2-, FXR- and MYO5B-deficiency no treatment options have been reported [1].

This lack of effective treatment options results in disease progression in all patients suffering from PFIC. Upon reaching end-stage liver disease, a liver transplantation is needed for patient survival.

Progress in the performance of liver transplantations over the past decades has led to an established, albeit highly invasive, procedure called orthotopic liver transplantation (OLT) [21]. Long-term complications consist of graft loss and an increased risk for infections and cancer development due to immunosuppressive treatment. The early severe liver damage in many patients suffering from PFIC renders liver transplantation during childhood necessary. This procedure has become the standard of care to treat children with end-stage liver disease with a 20 year survival of over 80% [22,23]. Partial liver transplantation improves size matching between grafts and recipients and alleviates the shortage of size matched donor livers without compromising survival [24]. In addition, this also allows living family-related donor transplantations.

For all severely affected patients, a liver transplant is currently the only curative therapy which has been successfully performed in children suffering from PFIC [25]. Additional complications of this major procedure were also reported in these patients. Patients with mutations that completely abolish ATP8B1 function developed hepatic steatosis in the donor liver. This can be prevented when combining the liver transplant with total internal biliary diversion, but extrahepatic symptoms like severe diarrhea persist [2,25,26]. Also in patients with FXR deficiency complications occurred upon transplantation, these patients also developed a fatty liver [1]. In patients with ABCB11 (BSEP) deficiency, the cholestasis symptoms re-occurred after OLT due to an immune response inhibiting BSEP function [27].

These additional complications, the lack of sufficient donor organs, the risk of life-long immune suppression to prevent organ rejection, and an increased risk for infection and cancer, warrant the development of novel treatment options for these devastating diseases. Gene therapy seems one of the potential treatments for these monogenic recessive disorders.

2. Gene Therapy

2.1. Clinical Applications

Gene addition therapies driven by Adeno Associated Viral (AAV) vectors for recessive monogenic diseases have been developed and, upon showing safety and efficacy, have received market approval, albeit in a limited number. Presently, three AAV-vector based gene therapeutics are available as Glybera, treating Lipoprotein Lipase (LPL)-deficiency, Luxturna [28], treating Retinal Pigment Epithelial (RPE) 65-related retinal dystrophy, and Zolgensma, treating Spinal Muscular Atrophy (SMA) [28–30]. The promising safety and efficacy data obtained in small clinical AAV gene therapy studies resulted in their application for registration to obtain market approval. For instance this approach as a treatment for Choroidenemia [31], aromatic L-amino acid decarboxylase deficiency [32], Pompe's Disease [33,34], Duchenne Muscular Dystrophy (ongoing: NCT03375164), Becker's Muscular Dystrophy [35], Limb-Girdle Muscular Dystrophy [36], X-linked myotubular myopathy (ongoing: NCT03199469), haemophilia A [37] and haemophilia B [38,39] all show efficacy. However, some safety issues have occurred, especially in clinical trials for diseases requiring systemic high vector doses. Liver toxicity has been reported for AAV doses above 1×10^{-14} vg/kg, indicating that this will be a problem for some applications [40]. Gene therapies for ocular diseases have relatively few hurdles to surmount because the eye is easily accessible, highly compartmentalized, immune-privileged [41], and less invasive as the injection occurs non-systemically. SMA treatment with Zolgensma, targeting the motor neurons, urges the use of intrathecal injections and a high vector doses, causing transient liver inflammation [42,43]. Recently, in a trial targeting X-linked myotubular myopathy, there have been two tragic deaths of pediatric patients in the cohort receiving the highest vector dose [44]. Both patients died due to progressive liver dysfunction and subsequently fatal sepsis upon receiving 3×10^{-14} vg/kg. Patients treated with a lower vector dose, 1×10^{-14} vg/kg, did not experience any severe adverse effects. The liver tropism of AAV vectors, resulting in toxic vector levels in the hepatocytes complicates gene therapy of disorders requiring a high vector dose in the systemic circulation, like for instance all muscular disorders. Within the scope of this review, looking at gene therapy for PFIC, AAV-mediated gene transfer directed to the liver is especially relevant. When treating inherited liver disorders, the AAV tropism for the liver is a major advantage since lower vector doses are needed for therapeutic efficacy. A number of clinical trials to treat Factor VIII or Factor IX deficiency in hemophilia A or B patients have our equitable interest.

The three year follow-up of a dose escalation study to treat haemophilia A, performed by Pasi and colleagues [37], showed long-term efficacy. All 15 adult patients received doses of 6×10^{-12} vg/kg up to 6×10^{-13} vg/kg of AAV5-h*FVIII* intravenously. No hepatotoxicity was observed, although in some cases elevated Aspartate Transaminase (AST) levels occurred and were successfully treated with glucocorticoids. Factor VIII plasma activity was restored leading to strong decreases in bleeding incidences, and recombinant Factor VIII use. Although after three years, factor VIII plasma levels were still therapeutic, a gradual decrease over time was seen. A longer follow up is needed to learn if factor VIII levels will remain therapeutic or if re-treatment will be needed. Dose escalation studies to treat Factor IX deficiency also proved to be safe and effective since levels up to 40–50% of normal Factor IX activity were reached [39]. Long-term follow up of an older study for Factor IX deficiency showed long-term efficacy [38]. In these patients the AAV8 vector was injected over nine years ago and they still have clinically relevant Factor IX production. To conclude, AAV-mediated liver-directed gene therapy as a one-time treatment proved to be safe and to provide long-term correction in adults, indicating that this treatment strategy is feasible for inherited severe liver disorders.

AAV vectors do not integrate actively in the host genome but persist in the nucleus in episomal form [45,46]. An important advantage of this is the lack of genotoxicity which can for instance result in tumour formation, as seen with integrating retroviral vectors [47,48]. During cell division, these episomes are not copied and not distributed to the daughter cells. This renders AAV vectors less suitable for treatment of liver disorders early after birth. The growth of the liver results in loss of the initial efficacy [49,50]. Hepatocyte proliferation is also induced upon liver damage for instance after a partial hepatectomy [51]. This complicates the application of AAV-mediated gene therapy for disorders resulting in hepatocyte damage, such as Fumarylacetoacetate hydrolase (FAH) deficiency and PFIC. Ongoing safety and efficacy studies for AAV gene therapy are performed in adults targeting diseases that do not cause liver damage. Recent pre-clinical studies do indicate that AAV-mediated gene therapy for PFIC may be feasible.

2.2. Pre-clinical AAV-mediated Gene Therapy for PFIC3

ABCB4 deficiency seems the best option to investigate the feasibility of gene therapy as a treatment option for PFIC. This is because of the availability of a suitable animal model and a previous study showing that partial correction by hepatocyte transplantation provides therapeutic correction [52]. The $Mdr2^{-/-}$ ($Abcb4^{-/-}$) mouse has been demonstrated to be a relevant model for PFIC3 [10,53]. The severity of the pathology depends on the mouse strain, with FVB mice displaying a more severe phenotype compared to the C57Bl/6 mice. Adjustments via the diet, such as cholate supplementation, increase the mouse bile hydrophobicity which worsens the disease phenotype [54]. In a recent study we investigated the feasibility of AAV8-h*ABCB4* mediated correction in adult $Abcb4^{-/-}$ mice with a C57Bl/6 background [55]. Upon vector administration, the efficacy was monitored over time until 6 months using dietary cholate administration to mimic the human bile toxicity [54]. This study provided proof of concept by demonstrating long-term correction, as shown by normalization of the liver damage parameters AST, Alanine Transaminase (ALT) and Alkaline Phosphatase (ALP), and the absence of fibrosis. Restoring a sufficient Phosphatidylcholine (PC) content in bile is a pre-requisite for prolonged correction using AAV. In case of insufficient correction, the ongoing hepatocyte proliferation will result in loss of AAV vectors, thereby further reducing PC presence in bile and thus increasing the liver damage. This was demonstrated by Weber et al. with $Abcb4^{-/-}$ FVB mice displaying a more severe phenotype [56]. These mice were treated with an AAV vector, consisting of a codon-optimized h*ABCB4* at week 2 after birth and the effect was monitored for 12 weeks. In males, this treatment appeared effective, but in 50% of the females the correction was lost. To overcome this, a second cohort was injected two times, first at week 2 and three weeks later a second dose was given. This protocol provided prolonged correction and is, with regard to severity of the pathological model, more comparable to PFIC3 patients. Both studies do indicate that long-term correction of a disease causing hepatocyte proliferation using non-integrating AAV vectors is feasible but only if the efficacy is sufficient.

In PFIC3 patients, the disease onset is observed in young children and therefore ideally the therapy should be given at an early stage. This complicates the use of a non-integrating vector like AAV. Several studies have shown that the initial correction is lost over time when treating neonatal or juvenile animals [49,50]. Moreover, the induction of proliferation upon (partial) loss of correction will contribute to episomal transgene loss and will accelerate the decrease in therapeutic efficacy. This hurdle can be overcome by strategies that aim for integration of the therapeutic transgene. Siew et al. tested integration of a therapeutic construct, consisting of a liver specific promoter and a codon-optimized human *ABCB4* transgene, flanked by the piggyBac transposase short terminal repeats [57]. Co-administration of this construct with an AAV2/8 vector containing a piggybac expression cassette to juvenile FVB $Abcb4^{-/-}$ mice resulted in the integration of the therapeutic construct providing lifelong hABCB4 expression and correction of the disease. Integration is non-random, but transposons do cause integration in close proximity of actively transcribed gene regions, transcription start sites, and open chromatin structures [58,59]. None of the analyzed unique integration regions were linked to genes known to play a role in hepatocellular carcinomas [59]. Nevertheless, this cannot be excluded. Targeting the transgene integration to a safe genomic locus could overcome this safety issue.

An obvious target to ensure integration in an active region in hepatocytes is the Albumin locus. A recent study aimed at Homologous Directed Repair (HDR) mediated integration of a promoterless Factor IX coding region in this locus without using a nuclease [60]. As the episomal construct lacks a promoter, and a self-cleaving protein is added to the transgene, its expression is regulated by the expression of albumin. The absence of a promoter and a nuclease increases safety of this "generide" strategy. The basis of this approach is to deliver a Factor IX encoding transgene flanked by two homologous arms to the albumin gene with complimentary sequences coding for the desired integration region. Using the generide strategy Muro et al. demonstrated that this could be used to correct a metabolic disorder, albeit with a low efficacy [61]. Inducing double strand breaks in this locus will stimulate HDR and integration efficacy. In a follow up study, increased correction was established by combining the generide approach with an AAV-*saCas9* guided to the albumin locus [62]. In all studies neonatal mice were used, modelling the use early after birth, when the liver is actively growing and hepatocytes are proliferating. The latter is a major advantage because during cell division the HDR system is active whereas it is inactive in quiescent cells [63]. In this study, life-long therapeutic correction of the pathology was established with a treatment shortly after birth. Further, no off-target integrations were seen in predicted sites which underlines the safety profile of this gene modulation system [62]. This targeted integration strategy seems a feasible option to treat PFIC shortly after birth.

2.3. Prospects of AAV-Mediated Gene Therapy for PFIC1, 2, 4, 5, and 6

The feasibility of in vivo AAV-mediated non-integrating gene therapy has only been reported for PFIC3. In this disease, the detergent activity of bile salts, that causes the pathology, can be neutralized by partial correction of phosphatidylcholine levels in bile [64]. Establishing ABCB4 expression in a sufficient percentage of the hepatocytes will prevent further damage, halt disease progression and stop hepatocyte proliferation, thereby, preventing the loss of episomal AAV vector genomes caused by cell division [55,56]. AAV-mediated in vivo gene therapy using non-integrating strategies seems less feasible for other types of PFIC.

In the case of PFIC1 and 2, their pathophysiology is driven by individual cellular stressors. These stresses are caused by decreased membrane integrity (in PFIC1) [65] and intracellular bile salt accumulation (in PFIC1, and 2) [66]. Therefore, all hepatocytes need to be corrected to stop damage induced hepatocyte proliferation. The high vector dose to ensure transduction of all hepatocytes may cause liver toxicity. The subsequent proliferation to compensate hepatocyte loss will result in loss of non-integrated AAV vector copies causing loss of correction. For both types an integrating gene therapy strategy, preventing the loss of the vector genome upon cell division, seems a pre-requisite for sustained correction. The current state of the art integrating AAV-mediated gene therapy systems, as described above, have a low efficacy but when performed in neonatal mice the percentage of corrected cells can go up to 24% of total hepatocytes [62]. This indicates therapeutic correction in PFIC1, 2 and 3, seems feasible because of the survival benefit of cells with a correctly integrated copy of the relevant gene. In contrast to the episomal AAV vector genomes, these integrated genes are copied during cell division and transferred to both daughter cells. Due to this transfer of the survival benefit to their descendants a limited number of corrected hepatocytes will repopulate the liver as shown for transplanted hepatocytes [52,67].

The pathophysiologic mechanisms causing liver damage in PFIC4, 5 and 6 have been clarified more recently. In PFIC4, the lack of functional TJP2 results in claudin mislocalization leading to paracellular bile leakage [68]. Patients with PFIC5, are deficient for the nuclear FXR, that has a central role in bile acid (BA) synthesis and homeostasis. Since FXR signaling is also needed for the expression of both ABCB4 and ABCB11, its deficiency results in lack of expression of these two transporters resulting in intracellular BA accumulation as seen in PFIC1 and 2 [69]. Recently, PFIC6 has been described to be caused by bi-allelic missense mutations in *MYO5B*. This protein plays a central role in intracellular transport of membrane proteins and these MYO5B mutants lead to ABCB11 mislocalization and, as a consequence, results in intracellular accumulation of bile salts [7,9]. Based on the pathophysiology of these three disease types, only integrating gene therapy strategies seem suited as possible treatment because the survival benefit of corrected hepatocytes may lead to liver repopulation and subsequent correction of the disease.

Animal models are essential to investigate the feasibility by showing proof of concept of gene therapy for these types of PFIC. For PFIC4, a TJP2 deficient model is needed. The whole body knock-out has been generated but appeared to be lethal during the embryonal period [11]. Therefore additional models, such as a hepatocyte specific knock-out by crossing a *TJP2* gene floxed mouse with an Alb-*Cre* mouse or even a conditionally inducible model need to be generated [70]. FXR-deficient mice have been generated, but in addition to a liver phenotype, this model displays a wide array of symptoms in agreement with the central role of this nuclear receptor in many processes. In addition to impaired liver regeneration, increased hepatic tumorigenesis and cholestasis, intestinal pathology, atherosclerosis and neurological malfunction is seen [71–74]. This suggests that gene therapy that only targets the liver may be partly therapeutic. Mouse models for MYO5B-deficiency have been generated and the predominant symptom in this model is Microvillus Inclusion Disease (MID). A recent study shows that the MID mouse, having total body knock-out of MYO5B, is not suitable to model cholestasis. The aberrant protein trafficking to the apical membrane of hepatocytes, resulting in the PFIC6 phenotype, is caused by missense mutations affecting the motor domain but not by complete MYO5B-deficiency [75]. The presence of wildtype MYO5B partially corrects the mislocalization of apical proteins in a hepatoma cell line, suggesting gene addition therapy seems a feasible approach. Proof of concept for this strategy would require a new mouse model expressing one of the specific missense mutations identified in PFIC6 patients.

The genome of wild-type AAV consist of a 4.8 Kb long single stranded DNA chain. Although AAV vectors can package a somewhat longer genome, longer genomes result in less efficient packaging or packaging of partial constructs. The maximal capacity that allows efficient packaging is limited to 5.2 Kb [76]. The coding sequences for the proteins deficient in the different types of PFIC are 3753 bp for hATP8B1 (PFIC1), 3963 bp for hABCB11 (PFIC2), 3858 bp for hABCB4 (PFIC3), 3570 bp for hTJP2 (PFIC4), and 1458 bp for hFXR (PFIC5). In PFIC6, the canonical spliced MYO5B variant consist of 5544 bp, but smaller splice variants consist of 2889 and of 1257 bp. As only specific missense mutations in the motor domain of MYO5B cause cholestasis, gene addition therapy of such a smaller splice variant, if functional, may in theory be an option, but only in combination with deleting the expression of the endogenous mutated MYO5B protein. The MYO5B domain binding Rab11 plays a crucial role in the mislocalization of apical proteins in hepatocytes resulting in PFIC6 [75]. This suggests that using AAV-mediated gene therapy to knock-out the expression of this domain using for instance CRISPR/Cas could be an option. Such a treatment would not require homologous repair nor the delivery of a donor template, that both limit the efficacy of in vivo gene correction therapy. Importantly, besides the transgene, AAV constructs consist of one poly A consensus sequence, a promoter, when using a non-integrating strategy, or two homologous arms, in the case of an integrating approach, and the inverted terminal repeats needed for packaging. These requirements make the development of both integrating and non-integrating therapeutic constructs within the limited packaging capacity of AAV challenging, but feasible for PFIC1 to 5. The absence of intrahepatic cholestasis in patients with complete MYO5B deficiency, suggest that a gene therapy that blocks the expression of MYO5B with motor domain missense mutations may partly correct the mislocalization of apical proteins and in theory could be a feasible approach to treat PFIC6.

2.4. Prospects of Ex Vivo Gene Therapy for PFIC

Applying gene therapy to isolated patient cells has major important advantages over in vivo approaches, like absence of interference of pre-existing immunity, the lack of an immune response towards the vector, no vector toxicity, exclusive targeting of the affected cell type, and the less invasive procedure [28]. Ex vivo gene therapy was the first form to show therapeutic efficacy in correcting inherited diseases of the hematopoietic stem cells (HSCs) [77]. This ex vivo correction benefited largely from the extensive experience with bone-marrow transplantations for tissue culture and pre-conditioning of the patient. Several HSC ex vivo gene therapies have shown long-term therapeutic efficacy and one has been registered, namely Strimvelis [78–80]. Transplantation of donor derived hepatocytes has been applied in patients suffering from different forms of inherited liver disorders [81]. Several studies report a partial correction of the disease indicating functionality of the transplanted hepatocytes. In all studies, this effect appeared to be transient, most likely due to immune responses towards the donor derived hepatocytes. Correcting patient derived hepatocytes would, by overcoming immune mediated loss of transplanted hepatocytes, provide long term correction. However, a small clinical study to treat familial hypercholesterolemia caused by low-density lipoprotein receptor (LDLR) deficiency did not provide therapeutic efficacy [82]. In part, this low efficacy could be explained by the use of a retroviral vector that, in contrast to a lentiviral vector, does not transduce non-dividing cells, such as mature hepatocytes. In addition, this trial also showed the complexity of the procedure including partial resection of the liver to isolate, culture and transduce the amount of hepatocytes needed for therapeutic efficacy. Further, in contrast to bone marrow transplantation, the experience with conditioning of the recipient is limited. All taken together, this approach only renders feasibility if corrected hepatocytes have a survival advantage resulting in a partial repopulation of the liver by the corrected cells. The liver damage observed in all types of PFIC does indicate that corrected hepatocytes may be able to repopulate the liver in these patients. Studies in relevant pre-clinical models for PFIC2 and 3 demonstrate that transplanted hepatocytes indeed repopulate the liver [52,67]. For the other types of PFIC a significant survival advantage of corrected cells has not been demonstrated, hence it is less clear if this approach will be suitable.

Correction of patient hepatocytes using ex vivo gene therapy and subsequent transplantation does require integration of the therapeutic construct in, or correction of, the host genome. The most effective vector for this type of gene therapy are the lentiviral vectors that transduce non-dividing cells such as hepatocytes. These vectors do integrate mostly in active genes which could increase the risk of tumorigenesis. For traditional retroviral vectors this risk indeed resulted in the development of acute leukemia, as shown in two trials [47,83]. In comparison to the vector used in these older studies, the safety of lentiviral vectors has been improved [84]. These third generation lentiviral vectors have been applied in several clinical trials and no tumor formation has been seen [85–87]. Ex vivo gene therapy using mature hepatocytes obtained from the patient liver will be a very challenging procedure especially in view of the large number of cells needed. To render this method applicable, cell proliferation will be essential to provide sufficient numbers for effective gene therapy. In this respect the use of stem cells, either induced pluripotent stem cells (iPSCs) or from the liver, remains necessary. In combination with a survival effect, these ex vivo corrected cells can be effective. A recent study for a severe skin disease resulted in complete recovery upon transplanting corrected skin stem cells [88]. For the liver this will be more complicated, but may be possible in future perspective.

2.5. Clinical Feasibility

To summarize, in vivo gene therapy using AAV vectors shows safety and efficacy in the clinic for several inherited diseases, including disorders affecting the liver. Within these therapies, the non-integrating approach has been extensively explored. In view of the early onset of severe damage and the persistent proliferation in case of insufficient efficacy, for all PFIC, integrating approaches are necessary for improved clinical application. Ex vivo gene therapy in adult hepatocytes followed by transplantation seems a potential option due to the survival benefit of the corrected cells that will promote liver repopulation. In view of the complicated procedure and inefficient grafting, in vivo gene therapy currently looks more promising for all types of PFIC.

The predominant concern for integrating gene therapy is induction of tumorigenesis, fueled by the early trials in severe combined immunodeficiency (SCID) patients [77]. The current adequate focus on tumorigenesis within gene therapy research will accelerate the development of safe, genome integrating gene therapeutics. State of the art integrating in vivo gene therapy approaches, including CRISPR/Cas9 show great promise in pre-clinical research because integration is guided, and therefore controlled, by Cas9 [56]. Currently, a first AAV-directed Cas9 driven gene therapy clinical trial for Leber's congenital amaurosis 10 (LCA10) has been approved and is recruiting (NCT03872479). Particularly safety assessments will determine conclusive outcomes for future in vivo application of integrating gene therapies in other inherited diseases.

Easily accessible organs like the eye, cervix, and liver are good targets for integrating gene therapy and future perspectives likely include new therapeutics for disorders within these target organs [89]. Proliferation advantages of corrected hepatocytes in stressed and diseased livers represent an important argument to choose for integrating gene therapy in certain types of PFIC.

Feasibility of gene therapy as a curative treatment differs between the types of PFIC. Deficiency in ABCB11 (PFIC2) show a complicated clinical expectation as the disease is cell autonomous and patients can develop hepatocellular carcinomas (HCC) or cholangiocarcinoma's (CCC) [90–92]. Although integrating gene therapy results in liver repopulation, presence of non-corrected cells cannot be excluded and maintains a permanent risk. Therefore even upon effective treatment in these patients, the risk of developing HCC and subsequent OLT remains [93]. Current available gene therapy methods seem an effective option to prevent liver failure until a suitable donor liver is available, but cannot replace liver transplantation in PFIC2 patients.

In PFIC5, deficiency of FXR, the potential role of gene therapy is comparable to that in patients with PFIC2. FXR is essential for transcription of the *ABCB11* gene and the expression of BSEP in all hepatocytes. Like PFIC2 patients, PFIC5 patients are at risk to develop HCC at an early age. Even upon effective gene therapy, presence of non-corrected hepatocytes results in a persisting risk for HCC development. In addition, the broad range of processes regulated by FXR results in a complicated disease and therefore correction of the diseased liver will not cure the pathology in all affected tissues. PFIC6 is caused by mutations in specific parts of the MYO5B protein, affecting its translocational capacity of proteins to the apical membrane [75]. Most MYO5B-deficient patients dominantly suffer from Microvillus Inclusion Disease [7]. Only for a sub-group of patients suffering from PFIC6, in which the liver pathology is the main symptom, liver directed gene therapy deleting the expression of the mutated MYO5B seems a feasible option.

In view of currently available methods, gene therapy to treat PFIC1, 3, and 4 seems the most feasible for clinical translation. Deficiencies of ABCB4 is the best studied type where three different pre-clinical studies all showed efficacy and proof of concept for liver directed in vivo gene therapy. The liver pathology caused by ATP8B1 deficiency (PFIC1) is also an attractive target for integrating gene therapy since the survival benefit of the corrected cells will promote liver repopulation. However, this approach will not address the extra-hepatic symptoms and it is not clear if an effective treatment will result in the development of a fatty liver, as seen upon OLT in these PFIC patients. In patients suffering from PFIC4, re-expression of TJP2 needed for the formation of functional tight junction formation in the liver, may also result in a survival benefit of the corrected cells [11]. Since pathology caused by TJP2 deficiency seems restricted to the liver, this appears a good target for in vivo gene therapy. Proof of concept studies require the development of a conditional liver specific TJP2 knock-out model for PFIC4 because the whole body knock-out is not viable.

3. Conclusions

In this review, we discussed the most recent in vivo and ex vivo approaches for using gene therapy in patients with PFIC, and addressed future developments within this field. Further, we elaborated on the clinical feasibility for all types of PFIC. The safety and efficacy results from many clinical trials conducting gene therapy are sufficiently positive to push the application of gene therapy for PFIC patients closer to the clinic. We can conclude that state of the art in vivo non-integrating gene therapy approaches will most likely not provide lifelong correction, due to a loss of AAV vector genomes caused by hepatocyte proliferation. However, the transient correction can slow down the decrease in liver function. Only integrating strategies that provide survival benefit to the corrected cells seem suitable for sustained efficacy in PFIC.

The feasibility to translate state of the art in vivo integrating gene therapies, upon showing proof of concept in pre-clinical models, into the clinic depends on the type of PFIC. Although integrating gene therapy would cure the PFIC2 and 5 phenotype, these patients remain at risk to develop HCC or CCC. Gene therapy for PFIC6 will only be effective for the small group of patients with the mutations in the motor domain of MYO5B that dominantly suffer from liver symptoms. Although MYO5B is too large to be packaged into AAVs, treating PFIC6 by knocking out gene expression of the mutated protein in the liver using AAV-mediated gene therapy or another approach to deliver CRISPR/Cas9 specifically to the hepatocytes, in theory could be a promising option. In most of MYO5B deficient patients, the pathology in the intestine is much more severe and will not be mitigated by AAV-mediated in vivo liver-directed gene therapy. PFIC1, 3 and 4 do seem good candidates for integrating gene therapy strategies targeting the diseased hepatocytes and could provide life-long correction in patients suffering from these severe life-threatening disorders.

Author Contributions: Conceptualization, M.W. and P.J.B.; writing—original draft preparation, M.W. and P.J.B.; writing—review and editing, R.P.J.O.-E. All authors have read and agreed to the published version of the manuscript

Funding: This research received no external funding.

Conflicts of Interest: The authors declare no conflict of interest.

References

1. Bull, L.N.; Thompson, R.J. Progressive Familial Intrahepatic Cholestasis. *Clin. Liver Dis.* **2018**, *22*, 657–669. [CrossRef] [PubMed]
2. Baker, A.; Kerkar, N.; Todorova, L.; Kamath, B.M.; Houwen, R.H.J. Systematic review of progressive familial intrahepatic cholestasis. *Clin. Res. Hepatol. Gastroenterol.* **2019**, *43*, 20–36. [CrossRef] [PubMed]
3. Strautnieks, S.S.; Bull, L.N.; Knisely, A.S.; Kocoshis, S.A.; Dahl, N.; Arnell, H.; Sokal, E.; Dahan, K.; Childs, S.; Ling, V.; et al. A gene encoding a liver-specific ABC transporter is mutated in progressive familial intrahepatic cholestasis. *Nat. Genet.* **1998**, *20*, 233–238. [CrossRef] [PubMed]
4. Pawlikowska, L.; Strautnieks, S.; Jankowska, I.; Czubkowski, P.; Emerick, K.; Antoniou, A.; Wanty, C.; Fischler, B.; Jacquemin, E.; Wali, S.; et al. Differences in presentation and progression between severe FIC1 and BSEP deficiencies. *J. Hepatol.* **2010**, *53*, 170–178. [CrossRef]
5. Davit-Spraul, A.; Fabre, M.; Branchereau, S.; Baussan, C.; Gonzales, E.; Stieger, B.; Bernard, O.; Jacquemin, E. ATP8B1 and ABCB11 Analysis in 62 children with normal gamma-glutamyl transferase Progressive Familial Intrahepatic Cholestasis (PFIC): Phenotypic differences between PFIC1 and PFIC2 and natural history. *Hepatology* **2010**, *51*, 1645–1655. [CrossRef]
6. Gomez-Ospina, N.; Potter, C.J.; Xiao, R.; Manickam, K.; Kim, M.S.; Kim, K.H.; Shneider, B.L.; Picarsic, J.L.; Jacobson, T.A.; Zhang, J.; et al. Mutations in the nuclear bile acid receptor FXR cause progressive familial intrahepatic cholestasis. *Nat. Commun.* **2016**, *7*, 1–8. [CrossRef]

7. Gonzales, E.; Taylor, S.A.; Davit-Spraul, A.; Thébaut, A.; Thomassin, N.; Guettier, C.; Whitington, P.F.; Jacquemin, E. MYO5B mutations cause cholestasis with normal serum gamma-glutamyl transferase activity in children without microvillous inclusion disease. *Hepatology* **2017**, *65*, 164–173. [CrossRef]

8. Paulusma, C.C.; de Waart, D.R.; Kunne, C.; Mok, K.S.; Oude Elferink, R.P.J. Activity of the bile salt export pump (ABCB11) is critically dependent on canalicular membrane cholesterol content. *J. Biol. Chem.* **2009**, *284*, 9947–9954. [CrossRef]

9. Qiu, Y.L.; Gong, J.Y.; Feng, J.Y.; Wang, R.X.; Han, J.; Liu, T.; Lu, Y.; Li, L.T.; Zhang, M.H.; Sheps, J.A.; et al. Defects in myosin VB are associated with a spectrum of previously undiagnosed low γ-glutamyltransferase cholestasis. *Hepatology* **2017**, *65*, 1655–1669. [CrossRef]

10. Smit, J.J.M.; Groen, K.; Mel, C.A.A.M.; Ottenhoff, R.; Roan, M.A.Van; Valk, M.A.Van Der; Diseases, L.; Centre, A.M.; Schinkel, A.H.; Elferink, R.P.J.O.; et al. Homozygous disruption of the murine MDR2 P-glycoprotein gene leads to a complete absence of phospholipid from bile and to liver disease. *Cell* **1993**, *75*, 451–462. [CrossRef]

11. Sambrotta, M.; Strautnieks, S.; Papouli, E.; Rushton, P.; Clark, B.E.; Parry, D.A.; Logan, C.V.; Newbury, L.J.; Kamath, B.M.; Ling, S.; et al. Mutations in TJP2 cause progressive cholestatic liver disease. *Nat. Genet.* **2014**, *46*, 326–328. [CrossRef] [PubMed]

12. De Vree, J.M.L.; Jacquemin, E.; Sturm, E.; Cresteil, D.; Bosma, P.J.; Aten, J.; Deleuze, J.F.; Desrochers, M.; Burdelski, M.; Bernard, O.; et al. Mutations in the MDR3 gene cause progressive familial intrahepatic cholestasis. *Proc. Natl. Acad. Sci. USA* **1998**, *95*, 282–287. [CrossRef] [PubMed]

13. Jacquemin, E.; DeVree, J.M.L.; Cresteil, D.; Sokal, E.M.; Sturm, E.; Dumont, M.; Scheffer, G.L.; Paul, M.; Burdelski, M.; Bosma, P.J.; et al. The wide spectrum of multidrug resistance 3 deficiency: From neonatal cholestasis to cirrhosis of adulthood. *Gastroenterology* **2001**, *120*, 1448–1458. [CrossRef] [PubMed]

14. Van Der Velden, L.M.; Stapelbroek, J.M.; Krieger, E.; Van Den Berghe, P.V.E.; Berger, R.; Verhulst, P.M.; Holthuis, J.C.M.; Houwen, R.II.J.; Klomp, L.W.J.; Van De Graaf, S.F.J. Folding defects in P-type ATP 8B1 associated with hereditary cholestasis are ameliorated by 4-phenylbutyrate. *Hepatology* **2010**, *51*, 286–296. [CrossRef]

15. Gonzales, E.; Grosse, B.; Schuller, B.; Davit-Spraul, A.; Conti, F.; Guettier, C.; Cassio, D.; Jacquemin, E. Targeted pharmacotherapy in progressive familial intrahepatic cholestasis type 2: Evidence for improvement of cholestasis with 4-phenylbutyrate. *Hepatology* **2015**, *62*, 558–566. [CrossRef]

16. Hasegawa, Y.; Hayashi, H.; Naoi, S.; Kondou, H.; Bessho, K.; Igarashi, K.; Hanada, K.; Nakao, K.; Kimura, T.; Konishi, A.; et al. Intractable itch relieved by 4-phenylbutyrate therapy in patients with progressive familial intrahepatic cholestasis type 1. *Orphanet J. Rare Dis.* **2014**, *9*, 1–9. [CrossRef]

17. Hegade, V.S.; Kendrick, S.F.W.; Dobbins, R.L.; Miller, S.R.; Thompson, D.; Richards, D.; Storey, J.; Dukes, G.E.; Corrigan, M.; Oude Elferink, R.P.J.; et al. Effect of ileal bile acid transporter inhibitor GSK2330672 on pruritus in primary biliary cholangitis: A double-blind, randomised, placebo-controlled, crossover, phase 2a study. *Lancet* **2017**, *389*, 1114–1123. [CrossRef]

18. Gonzales, E.; Sturm, E.; Stormon, M.; Sokal, E.; Hardikar, W.; Lacaille, F.; Gliwicz, D.; Hierro, L.; Jaecklin, T.; Gu, J.; et al. PS-193-Phase 2 open-label study with a placebo-controlled drug withdrawal period of the apical sodium-dependent bile acid transporter inhibitor maralixibat in children with Alagille Syndrome: 48-week interim efficacy analysis. *J. Hepatol.* **2019**, *70*, e119–e120. [CrossRef]

19. Roscam Abbing, R.L.P.; de Lannoy, L.M.; van de Graaf, S.F.J. The case for combining treatments for primary sclerosing cholangitis. *Lancet Gastroenterol. Hepatol.* **2018**, *3*, 526–528. [CrossRef]

20. Jansen, P.L.M. New therapies target the toxic consequences of cholestatic liver disease. *Expert Rev. Gastroenterol. Hepatol.* **2018**, *12*, 277–285. [CrossRef]

21. Meirelles Júnior, R.F.; Salvalaggio, P.; Rezende, M.B.D.; Evangelista, A.S.; Guardia, B.D.; Matielo, C.E.; Neves, D.B.; Pandullo, F.L.; Felga, G.E.; Alves, J.A.; et al. Liver transplantation: History, outcomes and perspectives. *Einstein* **2015**, *13*, 149–152. [CrossRef]

22. Pham, Y.H.; Miloh, T. Liver Transplantation in Children. *Clin. Liver Dis.* **2018**, *22*, 807–821. [CrossRef] [PubMed]

23. Kelly, D.; Verkade, H.J.; Rajanayagam, J.; McKiernan, P.; Mazariegos, G.; Hübscher, S. Late graft hepatitis and fibrosis in pediatric liver allograft recipients: Current concepts and future developments. *Liver Transplant.* **2016**, *22*, 1593–1602. [CrossRef] [PubMed]

24. Otte, J.B.; De Ville De Goyet, J.; Sokal, E.; Alberti, D.; Moulin, D.; De Hemptinne, B.; Veyckemans, F.; Van Obbergh, L.; Carlier, M.; Clapuyt, P.; et al. Size reduction of the donor liver is a safe way to alleviate the shortage of size-matched organs in pediatric liver transplantation. *Ann. Surg.* **1990**, *211*, 146–157. [CrossRef] [PubMed]

25. Bull, L.N.; Pawlikowska, L.; Strautnieks, S.; Jankowska, I.; Czubkowski, P.; Dodge, J.L.; Emerick, K.; Wanty, C.; Wali, S.; Blanchard, S.; et al. Outcomes of surgical management of familial intrahepatic cholestasis 1 and bile salt export protein deficiencies. *Hepatol. Commun.* **2018**, *2*, 515–528. [CrossRef] [PubMed]

26. Mali, V.P.; Fukuda, A.; Shigeta, T.; Uchida, H.; Hirata, Y.; Rahayatri, T.H.; Kanazawa, H.; Sasaki, K.; de Ville de Goyet, J.; Kasahara, M. Total internal biliary diversion during liver transplantation for type 1 progressive familial intrahepatic cholestasis: A novel approach. *Pediatr. Transplant.* **2016**, *20*, 981–986. [CrossRef]

27. Jara, P.; Hierro, L.; Martínez-Fernández, P.; Alvarez-Doforno, R.; Yánez, F.; Diaz, M.C.; Camarena, C.; De La Vega, A.; Frauca, E.; Muñoz-Bartolo, G.; et al. Recurrence of bile salt export pump deficiency after liver transplantation. *N. Engl. J. Med.* **2009**, *361*, 1359–1367. [CrossRef]

28. Anguela, X.M.; High, K.A. Entering the Modern Era of Gene Therapy. *Annu. Rev. Med.* **2019**, *70*, 273–288. [CrossRef]

29. Gaudet, D.; Méthot, J.; Déry, S.; Brisson, D.; Essiembre, C.; Tremblay, G.; Tremblay, K.; De Wal, J.; Twisk, J.; Van Den Bulk, N.; et al. Efficacy and long-term safety of alipogene tiparvovec (AAV1-LPL S447X) gene therapy for lipoprotein lipase deficiency: An open-label trial. *Gene Ther.* **2013**, *20*, 361–369. [CrossRef]
30. Russell, S.; Bennett, J.; Wellman, J.A.; Chung, D.C.; Yu, Z.F.; Tillman, A.; Wittes, J.; Pappas, J.; Elci, O.; McCague, S.; et al. Efficacy and safety of voretigene neparvovec (AAV2-hRPE65v2) in patients with RPE65-mediated inherited retinal dystrophy: A randomised, controlled, open-label, phase 3 trial. *Lancet* **2017**, *390*, 849–860. [CrossRef]
31. Lam, B.L.; Davis, J.L.; Gregori, N.Z.; MacLaren, R.E.; Girach, A.; Verriotto, J.D.; Rodriguez, B.; Rosa, P.R.; Zhang, X.; Feuer, W.J. Choroideremia Gene Therapy Phase 2 Clinical Trial: 24-Month Results. *Am. J. Ophthalmol.* **2019**, *197*, 65–73. [CrossRef] [PubMed]
32. Kojima, K.; Nakajima, T.; Taga, N.; Miyauchi, A.; Kato, M.; Matsumoto, A.; Ikeda, T.; Nakamura, K.; Kubota, T.; Mizukami, H.; et al. Gene therapy improves motor and mental function of aromatic l-amino acid decarboxylase deficiency. *Brain* **2019**, *142*, 322–333. [CrossRef] [PubMed]
33. Smith, B.K.; Collins, S.W.; Conlon, T.J.; Mah, C.S.; Lawson, L.A.; Martin, A.D.; Fuller, D.D.; Cleaver, B.D.; Clément, N.; Phillips, D.; et al. Phase I/II trial of adeno-associated virus-mediated alpha-glucosidase gene therapy to the diaphragm for chronic respiratory failure in pompe disease: Initial safety and ventilatory outcomes. *Hum. Gene Ther.* **2013**, *24*, 630–640. [CrossRef] [PubMed]
34. Corti, M.; Liberati, C.; Smith, B.K.; Lawson, L.A.; Tuna, I.S.; Conlon, T.J.; Coleman, K.E.; Islam, S.; Herzog, R.W.; Fuller, D.D.; et al. Safety of Intradiaphragmatic Delivery of Adeno-Associated Virus-Mediated Alpha-Glucosidase (rAAV1-CMV-hGAA) Gene Therapy in Children Affected by Pompe Disease. *Hum. Gene Ther. Clin. Dev.* **2017**, *28*, 208–218. [CrossRef] [PubMed]
35. Mendell, J.R.; Sahenk, Z.; Malik, V.; Gomez, A.M.; Flanigan, K.M.; Lowes, L.P.; Alfano, L.N.; Berry, K.; Meadows, E.; Lewis, S.; et al. A phase 1/2a follistatin gene therapy trial for becker muscular dystrophy. *Mol. Ther.* **2015**, *23*, 192–201. [CrossRef] [PubMed]
36. Mendell, J.R.; Rodino-Klapac, L.R.; Rosales, X.Q.; Coley, B.D.; Galloway, G.; Lewis, S.; Malik, V.; Shilling, C.; Byrne, B.J.; Conlon, T.; et al. Sustained alpha-sarcoglycan gene expression after gene transfer in limb-girdle muscular dystrophy, type 2D. *Ann. Neurol.* **2010**, *68*, 629–638. [CrossRef]
37. Pasi, K.J.; Rangarajan, S.; Mitchell, N.; Lester, W.; Symington, E.; Madan, B.; Laffan, M.; Russell, C.B.; Li, M.; Pierce, G.F.; et al. Multiyear follow-up of aav5-hfviii-sq gene therapy for hemophilia a. *N. Engl. J. Med.* **2020**, *382*, 29–40. [CrossRef]
38. Nathwani, A.C.; Reiss, U.M.; Tuddenham, E.G.D.; Rosales, C.; Chowdary, P.; McIntosh, J.; Della Peruta, M.; Lheriteau, E.; Patel, N.; Raj, D.; et al. Long-term safety and efficacy of factor IX gene therapy in hemophilia B. *N. Engl. J. Med.* **2014**, *371*, 1994–2004. [CrossRef]
39. George, L.A.; Sullivan, S.K.; Giermasz, A.; Rasko, J.E.J.; Samelson-Jones, B.J.; Ducore, J.; Cuker, A.; Sullivan, L.M.; Majumdar, S.; Teitel, J.; et al. Hemophilia B gene therapy with a high-specific-activity factor IX variant. *N. Engl. J. Med.* **2017**, *377*, 2215–2227. [CrossRef]
40. Hinderer, C.; Katz, N.; Buza, E.L.; Dyer, C.; Goode, T.; Bell, P.; Richman, L.K.; Wilson, J.M. Severe Toxicity in Nonhuman Primates and Piglets Following High-Dose Intravenous Administration of an Adeno-Associated Virus Vector Expressing Human SMN. *Hum. Gene Ther.* **2018**, *29*, 285–298. [CrossRef]
41. Liu, M.M.; Tuo, J.; Chan, C.C. Gene therapy for ocular diseases. *Br. J. Ophthalmol.* **2011**, *95*, 604–612. [CrossRef] [PubMed]
42. Chand, D.; Mohr, F.; McMillan, H.; Tukov, F.F.; Montgomery, K.; Kleyn, A.; Sun, R.; Tauscher-Wisniewski, S.; Kaufmann, P.; Kullak-Ublick, G. Hepatotoxicity following administration of onasemnogene abeparvovec (AVXS-101) for the treatment of spinal muscular atrophy. *J. Hepatol.* **2020**. [CrossRef] [PubMed]
43. Harada, Y.; Rao, V.K.; Arya, K.; Kuntz, N.L.; DiDonato, C.J.; Napchan-Pomerantz, G.; Agarwal, A.; Stefans, V.; Katsuno, M.; Veerapandiyan, A. Combination molecular therapies for type 1 spinal muscular atrophy. *Muscle Nerve* **2020**, *62*, 550–554. [CrossRef] [PubMed]
44. Harrison, C. High-dose AAV gene therapy deaths. *Nat. Biotechnol.* **2020**, *38*, 905–908. [CrossRef] [PubMed]
45. Daya, S.; Berns, K.I. Gene therapy using adeno-associated virus vectors. *Clin. Microbiol. Rev.* **2008**, *21*, 583–593. [CrossRef]
46. Monahan, P.E.; Samulski, R.J. Adeno-associated virus vectors for gene therapy: More pros than cons? *Mol. Med. Today* **2000**, *6*, 433–440. [CrossRef]
47. Hacein-Bey-Abina, S.; Garrigue, A.; Wang, G.P.; Soulier, J.; Lim, A.; Morillon, E.; Clappier, E.; Caccavelli, L.; Delabesse, E.; Beldjord, K.; et al. Insertional oncogenesis in 4 patients after retrovirus-mediated gene therapy of SCID-X1 Find the latest version: Insertional oncogenesis in 4 patients after retrovirus-mediated gene therapy of SCID-X1. *J. Clin. Investig.* **2008**, *118*, 3132–3142. [CrossRef]
48. Braun, C.J.; Boztug, K.; Paruzynski, A.; Witzel, M.; Schwarzer, A.; Rothe, M.; Modlich, U.; Beier, R.; Göhring, G.; Steinemann, D.; et al. Gene therapy for Wiskott-Aldrich syndrome-long - Term efficacy and genotoxicity (Science Translational Medicine). *Sci. Transl. Med.* **2014**, *6*, 1–15. [CrossRef]
49. Wang, L.; Bell, P.; Lin, J.; Calcedo, R.; Tarantal, A.F.; Wilson, J.M. AAV8-mediated hepatic gene transfer in infant rhesus monkeys (macaca mulatta). *Mol. Ther.* **2011**, *19*, 2012–2020. [CrossRef]
50. Cunningham, S.C.; Spinoulas, A.; Carpenter, K.H.; Wilcken, B.; Kuchel, P.W.; Alexander, I.E. AAV2/8-mediated correction of OTC deficiency is robust in adult but not neonatal Spfash mice. *Mol. Ther.* **2009**, *17*, 1340–1346. [CrossRef]
51. Michalopoulos, G.K. Liver regeneration after partial hepatectomy: Critical analysis of mechanistic dilemmas. *Am. J. Pathol.* **2010**, *176*, 2–13. [CrossRef] [PubMed]

52. De Vree, J.M.L.; Ottenhoff, R.; Bosma, P.J.; Smith, A.J.; Aten, J.; Oude Elferink, R.P.J. Correction of liver disease by hepatocyte transplantation in a mouse model of progressive familial intrahepatic cholestasis. *Gastroenterology* **2000**, *119*, 1720–1730. [CrossRef] [PubMed]

53. Oude Elferink, R.P.J.; Paulusma, C.C. Function and pathophysiological importance of ABCB4 (MDR3 P-glycoprotein). *Pflugers Arch. Eur. J. Physiol.* **2007**, *453*, 601–610. [CrossRef] [PubMed]

54. Van Nieuwkerk, C.M.J.; Elferink, R.P.J.O.; Groen, A.K.; Ottenhoff, R.; Tytgat, G.N.J.; Dingemans, K.P.; Van den Bergh Weerman, M.A.; Offerhaus, G.J.A. Effects of ursodeoxycholate and cholate feeding on liver disease in FVB mice with a disrupted mdr2 P-glycoprotein gene. *Gastroenterology* **1996**, *111*, 165–171. [CrossRef]

55. Aronson, S.J.; Bakker, R.S.; Shi, X.; Duijst, S.; ten Bloemendaal, L.; de Waart, D.R.; Verheij, J.; Ronzitti, G.; Oude Elferink, R.P.; Beuers, U.; et al. Liver-directed gene therapy results in long-term correction of progressive familial intrahepatic cholestasis type 3 in mice. *J. Hepatol.* **2019**, *71*, 153–162. [CrossRef]

56. Weber, N.D.; Odriozola, L.; Martínez-García, J.; Ferrer, V.; Douar, A.; Bénichou, B.; González-Aseguinolaza, G.; Smerdou, C. Gene therapy for progressive familial intrahepatic cholestasis type 3 in a clinically relevant mouse model. *Nat. Commun.* **2019**, *10*. [CrossRef]

57. Siew, S.M.; Cunningham, S.C.; Zhu, E.; Tay, S.S.; Venuti, E.; Bolitho, C.; Alexander, I.E. Prevention of Cholestatic Liver Disease and Reduced Tumorigenicity in a Murine Model of PFIC Type 3 Using Hybrid AAV-piggyBac Gene Therapy. *Hepatology* **2019**, *70*, 2047–2061. [CrossRef]

58. Meir, Y.J.J.; Weirauch, M.T.; Yang, H.S.; Chung, P.C.; Yu, R.K.; Wu, S.C.Y. Genome-wide target profiling of piggyBac and Tol2 in HEK 293: Pros and cons for gene discovery and gene therapy. *BMC Biotechnol.* **2011**, *11*. [CrossRef]

59. Cunningham, S.C.; Siew, S.M.; Hallwirth, C.V.; Bolitho, C.; Sasaki, N.; Garg, G.; Michael, I.P.; Hetherington, N.A.; Carpenter, K.; de Alencastro, G.; et al. Modeling correction of severe urea cycle defects in the growing murine liver using a hybrid recombinant adeno-associated virus/piggyBac transposase gene delivery system. *Hepatology* **2015**, *62*, 417–428. [CrossRef]

60. Barzel, N.K. Paulk, Y. Shi, Y. Huang, K. Chu, F. Zhang, P.N. Valdmanis, L.P. Spector, M.H. Porteus, K.M. Gaensler, and M.A.K. Promoterless gene targeting without nucleases ameliorates haemophilia B in mice. *Physiol. Behav.* **2015**, *517*, 360–364. [CrossRef]

61. Porro, F.; Bortolussi, G.; Barzel, A.; De Caneva, A.; Iaconcig, A.; Vodret, S.; Zentilin, L.; Kay, M.A.; Muro, A.F. Promoterless gene targeting without nucleases rescues lethality of a Crigler-Najjar syndrome mouse model. *EMBO Mol. Med.* **2017**, *9*, 1346–1355. [CrossRef] [PubMed]

62. De Caneva, A.; Porro, F.; Bortolussi, G.; Sola, R.; Lisjak, M.; Barzel, A.; Giacca, M.; Kay, M.A.; Vlahovicek, K.; Zentilin, L.; et al. Coupling AAV-mediated promoterless gene targeting to SaCas9 nuclease to efficiently correct liver metabolic diseases. *JCI Insight* **2019**, *4*. [CrossRef] [PubMed]

63. Liu, M.; Rehman, S.; Tang, X.; Gu, K.; Fan, Q.; Chen, D.; Ma, W. Methodologies for improving HDR efficiency. *Front. Genet.* **2019**, *10*, 1–9. [CrossRef] [PubMed]

64. Davit-Spraul, A.; Gonzales, E.; Baussan, C.; Jacquemin, E. Progressive familial intrahepatic cholestasis. *Orphanet J. Rare Dis.* **2009**, *4*, 1–12. [CrossRef]

65. Groen, A.; Romero, M.R.; Kunne, C.; Hoosdally, S.J.; Dixon, P.H.; Wooding, C.; Williamson, C.; Seppen, J.; Van Den Oever, K.; Mok, K.S.; et al. Complementary functions of the flippase ATP8B1 and the floppase ABCB4 in maintaining canalicular membrane integrity. *Gastroenterology* **2011**, *141*, 1927–1937. [CrossRef]

66. Telbisz, Á.; Homolya, L. Recent advances in the exploration of the bile salt export pump (BSEP/ABCB11) function. *Expert Opin. Ther. Targets* **2016**, *20*, 501–514. [CrossRef]

67. Chen, H.L.; Chen, H.L.; Yuan, R.H.; Wu, S.H.; Chen, Y.H.; Chien, C.S.; Chou, S.P.; Wang, R.; Ling, V.; Chang, M.H. Hepatocyte transplantation in bile salt export pump-deficient mice: Selective growth advantage of donor hepatocytes under bile acid stress. *J. Cell. Mol. Med.* **2012**, *16*, 2679–2689. [CrossRef]

68. Grosse, B.; Cassio, D.; Yousef, N.; Bernardo, C.; Jacquemin, E.; Gonzales, E. Claudin-1 involved in neonatal ichthyosis sclerosing cholangitis syndrome regulates hepatic paracellular permeability. *Hepatology* **2012**, *55*, 1249–1259. [CrossRef]

69. Cariello, M.; Piccinin, E.; Garcia-Irigoyen, O.; Sabbà, C.; Moschetta, A. Nuclear receptor FXR, bile acids and liver damage: Introducing the progressive familial intrahepatic cholestasis with FXR mutations. *Biochim. Biophys. Acta - Mol. Basis Dis.* **2018**, *1864*, 1308–1318. [CrossRef]

70. Zhang, J.; Zhao, J.; Jiang, W.J.; Shan, X.W.; Yang, X.M.; Gao, J.G. Conditional gene manipulation: Cre-ating a new biological era. *J. Zhejiang Univ. Sci. B* **2012**, *13*, 511–524. [CrossRef]

71. Kim, I.; Ahn, S.H.; Inagaki, T.; Choi, M.; Ito, S.; Guo, G.L.; Kliewer, S.A.; Gonzalez, F.J. Differential regulation of bile acid homeostasis by the farnesoid X receptor in liver and intestine. *J. Lipid Res.* **2007**, *48*, 2664–2672. [CrossRef] [PubMed]

72. Raman, R.; Thomas, R.G.; Weiner, M.W.; Jack, C.R.; Ernstrom, K.; Aisen, P.S.; Tariot, P.N.; Quinn, J.F. Effects of FXR in Foam-cell Formation and Atherosclerosis Development. *Biochim. Biophys. Acta* **2010**, *23*, 333–336. [CrossRef]

73. Fan, M.; Wang, X.; Xu, G.; Yan, Q.; Huang, W. Bile acid signaling and liver regeneration. *Biochim. Biophys. Acta Gene Regul. Mech.* **2015**, *1849*, 196–200. [CrossRef] [PubMed]

74. Huang, F.; Wang, T.; Lan, Y.; Yang, L.; Pan, W.; Zhu, Y.; Lv, B.; Wei, Y.; Shi, H.; Wu, H.; et al. Deletion of mouse FXR gene disturbs multiple neurotransmitter systems and alters neurobehavior. *Front. Behav. Neurosci.* **2015**, *9*, 1–10. [CrossRef] [PubMed]

75. Overeem, A.W.; Li, Q.; Qiu, Y.L.; Cartón-García, F.; Leng, C.; Klappe, K.; Dronkers, J.; Hsiao, N.H.; Wang, J.S.; Arango, D.; et al. A Molecular Mechanism Underlying Genotype-Specific Intrahepatic Cholestasis Resulting From MYO5B Mutations. *Hepatology* **2020**, *72*, 213–229. [CrossRef] [PubMed]

76. Wu, Z.; Yang, H.; Colosi, P. Effect of genome size on AAV vector packaging. *Mol. Ther.* **2010**, *18*, 80–86. [CrossRef] [PubMed]

77. Aiuti, A.; Slavin, S.; Aker, M.; Ficara, F.; Deola, S.; Mortellaro, A.; Morecki, S.; Andolfi, G.; Tabucchi, A.; Carlucci, F.; et al. Correction of ADA-SCID by stem cell gene therapy combined with nonmyeloablative conditioning. *Science* **2002**, *296*, 2410–2413. [CrossRef]

78. Deichmann, A.; Hacein-Bey-Abina, S.; Schmidt, M.; Garrigue, A.; Brugman, M.H.; Hu, J.; Glimm, H.; Gyapay, G.; Prum, B.; Fraser, C.C.; et al. Vector integration is nonrandom and clustered and influences the fate of lymphopoiesis in SCID-X1 gene therapy. *J. Clin. Investig.* **2007**, *117*, 2225–2232. [CrossRef]

79. Alessandro Aiuti, M.D.P.D.; Federica Cattaneo, M.D.; Stefania Galimberti, P.D.; Ulrike Benninghoff, M.D.; Barbara Cassani, Ph.D.; Luciano Callegaro, R.N.; Samantha Scaramuzza, P.D.; Andolfi, G.; Massimiliano Mirolo, B.S.; Immacolata Brigida, Ph.D.; et al. Gene therapy for immunodeficiency due to adenosine deaminase deficiency. *N. Engl. J. Med.* **2009**, *360*, 447–458.

80. Shahryari, A.; Jazi, M.S.; Mohammadi, S.; Nikoo, H.R.; Nazari, Z.; Hosseini, E.S.; Burtscher, I.; Mowla, S.J.; Lickert, H. Development and clinical translation of approved gene therapy products for genetic disorders. *Front. Genet.* **2019**, *10*. [CrossRef]

81. Iansante, V.; Mitry, R.R.; Filippi, C.; Fitzpatrick, E.; Dhawan, A. Human hepatocyte transplantation for liver disease: Current status and future perspectives. *Pediatr. Res.* **2018**, *83*, 232–240. [CrossRef] [PubMed]

82. Al-Allaf, F.A.; Coutelle, C.; Waddington, S.N.; David, A.L.; Harbottle, R.; Themis, M. LDLR-Gene therapy for familial hypercholesterolaemia: Problems, progress, and perspectives. *Int. Arch. Med.* **2010**, *3*, 36. [CrossRef] [PubMed]

83. Hacein-Bey-Abina, S.; Hauer, J.; Lim, A.; Picard, C.; Wang, G.P.; Berry, C.C.; Martinache, C.; Rieux-Laucat, F.; Latour, S.; Belohradsky, B.H.; et al. Efficacy of gene therapy for X-linked severe combined immunodeficiency. *N. Engl. J. Med.* **2010**, *363*, 355–364. [CrossRef] [PubMed]

84. Escors, D.; Breckpot, K. Lentiviral vectors in gene therapy: Their current status and future potential. *Arch. Immunol. Ther. Exp.* **2011**, *58*, 107–119. [CrossRef] [PubMed]

85. Sessa, M.; Lorioli, L.; Fumagalli, F.; Acquati, S.; Redaelli, D.; Baldoli, C.; Canale, S.; Lopez, I.D.; Morena, F.; Calabria, A.; et al. Lentiviral haemopoietic stem-cell gene therapy in early-onset metachromatic leukodystrophy: An ad-hoc analysis of a non-randomised, open-label, phase 1/2 trial. *Lancet* **2016**, *388*, 476–487. [CrossRef]

86. McGarrity, G.J.; Hoyah, G.; Winemiller, A.; Andre, K.; Stein, D.; Blick, G.; Greenberg, R.N.; Kinder, C.; Zolopa, A.; Binder-Scholl, G.; et al. Patient monitoring and follow-up in lentiviral clinical trials. *J. Gene Med.* **2013**, *15*, 78–82. [CrossRef]

87. Cavazzana-Calvo, M.; Payen, E.; Negre, O.; Wang, G.; Hehir, K.; Fusil, F.; Down, J.; Denaro, M.; Brady, T.; Westerman, K.; et al. Transfusion independence and HMGA2 activation after gene therapy of human β-thalassaemia. *Nature* **2010**, *467*, 318–322. [CrossRef]

88. Rouanet, S.; Warrick, E.; Gache, Y.; Scarzello, S.; Avril, M.F.; Bernerd, F.; Magnaldo, T. Genetic correction of stem cells in the treatment of inherited diseases and focus on Xeroderma pigmentosum. *Int. J. Mol. Sci.* **2013**, *14*, 20019–20036. [CrossRef]

89. Hirakawa, M.P.; Krishnakumar, R.; Timlin, J.A.; Carney, J.P.; Butler, K.S. Gene editing and CRISPR in the clinic: Current and future perspectives. *Biosci. Rep.* **2020**, *40*. [CrossRef]

90. Knisely, A.S.; Strautnieks, S.S.; Meier, Y.; Stieger, B.; Byrne, J.A.; Portmann, B.C.; Bull, L.N.; Pawlikowska, L.; Bilezikçi, B.; Özçay, F.; et al. Hepatocellular carcinoma in ten children under five years of age with bile salt export pump deficiency. *Hepatology* **2006**, *44*, 478–486. [CrossRef]

91. Al Salloom, A. Hepatocellular Carcinoma in a Boy with Progressive Familial Intrahepatic Cholestasis Type II: Challenging Identification: Case Report. *Int. J. Health Sci.* **2013**, *7*, 252–255. [CrossRef] [PubMed]

92. Scheimann, A.O.; Strautnieks, S.S.; Knisely, A.S.; Byrne, J.A.; Thompson, R.J.; Finegold, M.J. Mutations in Bile Salt Export Pump (ABCB11) in Two Children with Progressive Familial Intrahepatic Cholestasis and Cholangiocarcinoma. *J. Pediatr.* **2007**. [CrossRef] [PubMed]

93. Santopaolo, F.; Lenci, I.; Milana, M.; Manzia, T.M.; Baiocchi, L. Liver transplantation for hepatocellular carcinoma: Where do we stand? *World J. Gastroenterol.* **2019**, *25*, 2591–2602. [CrossRef] [PubMed]

MDPI

St. Alban-Anlage 66

4052 Basel

Switzerland

Tel. +41 61 683 77 34

Fax +41 61 302 89 18

www.mdpi.com

International Journal of Molecular Sciences Editorial Office

E-mail: ijms@mdpi.com

www.mdpi.com/journal/ijms